FUNDAMENTALS OF SOLID MECHANICS

Fundamentals of
Solid Mechanics

A Treatise on
Strength of Materials

M.L. Gambhir
Formerly, Professor and Head of Civil Engineering Department
and Dean, Planning and Resource Generation
Thapar University, Patiala

PHI Learning Private Limited
Delhi-110092
2025

*In fond memory of **Shri Asoke K. Ghosh** (October 1942 – February 2024), Founder Chairman and Managing Director of PHI Learning, whose vision endlessly inspires.*

The Legacy Continues....

Published by Pushpita Ghosh, PHI Learning Private Limited, Rimjhim House, 111, Patparganj Industrial Estate, Delhi-110092 and Printed by Multi Colour Services, I-45, DLF Industrial Area, Sector-32, Faridabad, Haryana-121003.

₹1795.00

FUNDAMENTALS OF SOLID MECHANICS
M.L. Gambhir

ISBN-978-81-203-3870-8 (Print Book)
ISBN-978-93-5443-361-0 (e-Book)

Contents

6. Torsion 207–262

7. Shearing Force and Bending Moment Diagrams 263–322

8. Bending of Simple Beams 323–366

14. **Deflections** (Direct Integration and Geometrical Methods) 528–594

15. **Deflections** (Energy Methods) 595–657

16. Members Subjected to Compression: Columns and Struts 658–717

17. Unsymmetrical (Skew) Bending 718–745

18. Curved Members Subjected to Bending 746–764

Preface

This text is written for the students in engineering and technology programmes at the undergraduate level studying the core course called Solid Mechanics, Strength of Materials or Mechanics of Materials. It comprehensively covers the topics generally prescribed in engineering curricula of most colleges and universities.

A number of books on the subject are available. However, most of the books are examination-oriented and lack in the fundamental aspects of the subject matter. In this book, an attempt has been made to rectify many of such deficiencies, and present the material in a simple, practical, logical and cogent manner so as to provide the reader with an insight into the subject matter, and at the same time develop some proficiency in solving a variety of problems encountered in engineering practice.

The book is intended to develop the ability of the student to analyze the engineering objects subjected to different types of forces using the basic principles of statics. Since the students are expected to be able to apply the principles of statics, the main techniques of analysis of forces and moments are discussed in Chapter 1. This book is therefore written to achieve the following objectives:

1. To present a text to cater mainly to the classroom or self-study needs of undergraduate students pursuing courses in mechanical, civil, chemical and aeronautical engineering. The book is designed to provide good readability and sound coverage of the principles of solid mechanics.
2. To provide a large number of worked examples in each chapter in a simple, efficient and systematic manner to illustrate the application of the subject matter to practical problems. The

example problems are carefully selected to help the reader master the process of calculations, manipulations in the equations, seek appropriate data and make appropriate inferences.

3. To provide an extensive set of problems at the end of each chapter for classroom or self-study practice to achieve a high level of comprehension of fundamentals covered in the chapters and to rationally assess the situation. The readers are advised to solve, as many as possible, the problems given at the end of chapters.
4. To provide a wide variety of well-labelled diagrams throughout the text to enable the reader to achieve a level of comprehension to handle mechanical, structural and construction problems.
5. To provide a dependable companion for teachers and design engineers. Most of the examples and chapter-end problems have been formulated in parametric form to enable the instructor to formulate as many problems as required for class work. The results are independent of units making such problems useful for practical applications.

In writing this text, I have mainly drawn on the experience accumulated over the years as a result of teaching, research and consultancy. The book is arranged in twenty-one chapters and two appendices. A brief description of each chapter is given below:

The first chapter deals with the introductory concepts of applied mechanics to help readers in brushing up the fundamentals required for the study of ensuing chapters.

Analysis of plane frames and cables is given in Chapter 2. Chapter 3 deals with the stresses and strains in simple systems. The stress–strain relations and the relations among the elastic constants have been derived in the chapter. The composite systems subjected to external loads and temperature variations are discussed in Chapter 4.

The biaxial and triaxial stresses and strains in thin and thick pressure vessels are presented in Chapter 5. Chapter 6 is devoted to the computation of shear stresses due to torsion. The stress analysis of shafts with different types of cross-sections is presented with adequate examples.

The bending moment and shear force diagrams for different types of beams, including rectangular and non-rectangular simple frames, are presented with a large number of examples in Chapter 7. However, the computation of bending and shearing stresses in the beams having a wide variety of cross-sections is covered in Chapters 8 and 9, respectively. The stresses in the springs are explained in Chapter 10. In Chapter 11, the combined stresses developed in chimneys, dams and retaining walls due to gravity and lateral forces are discussed.

Chapters 12 and 13 describe the riveted and bolted, and welded connections, respectively, to transmit the force from one member to another.

Chapter 14 covers the deflection analysis of simple beams and frames subjected to a wide variety of loads. However, more advanced energy methods used for the computation of deflection of beams and frames are described in Chapter 15.

Chapter 16 discusses members subjected to compression, i.e. columns and struts and provides the analysis and design of columns with different end and load conditions.

Chapter 17 is devoted to the analysis of beams subjected to unsymmetrical bending. The concept of shear centre has been explained in detail.

Chapter 18 discusses curved members with different cross-sections subjected to bending. Chapter 19 is devoted to the plastic bending of beams, wherein the concept of plastic hinge, limit torque and residual stresses are explained in detail. The concepts are illustrated by means of example problems.

Chapter 20 deals with the principal stresses and maximum shear stresses, a sound knowledge of which is essential for the safe and economical design of structural and machine components. A large number of examples are presented to illustrate the concepts.

Chapter 21 is devoted to the special topics like dynamic loading, stress concentration, fatigue, etc.

The author thanks all those who have assisted in various ways in preparation of the text. In particular, the author wishes to acknowledge the assistance rendered by Dr. Puneet Gambhir, Er. Mohit Gambhir and Er. Atul Jamwal, CE, MBA. The author expresses his gratitude to Mrs. Neha Jamwal, M.Tech. Struct (NIT), who assisted in the preparation of the manuscript and in solving some of the problems. The author is extremely grateful to his wife Saroj for continued cooperation and encouragement while he was busy completing the job.

The author welcomes suggestions from the readers for improvement in the subject matter and presentation of the book

M.L. Gambhir

Chapter 1

Introductory Concepts

1.1 GENERAL

Mechanics is a branch of applied sciences (based on *physics* and *mathematics*) concerned with the behaviour of solid matter under external actions (e.g., external *forces*, temperature changes, applied displacements, etc). It is a part of a broader *study* known as *continuum mechanics*. Thus mechanics deals with forces and motions. The part of mechanics specifically concerned with the relation between forces acting on rigid bodies that are *in equilibrium* and *at rest*, is termed *statics*. The part of mechanics dealing with rigid bodies in motion is termed *dynamics*. If the inertia forces are taken into account, the bodies in motion can also be considered to be in equilibrium.

A material has a rest shape (the condition existing when the system of forces acting on the body is in a state of balance) and its shape departs away from the rest shape when subjected to stresses. The amount of departure from rest shape is called *deformation*; the proportion of deformation to its original size is called *strain*. If the applied stress is sufficiently low (or the imposed strain is small enough), almost all solid materials behave in such a way that the strain is directly proportional to the *stress*; the coefficient of the proportionality is called the modulus of elasticity. This region of deformation is known as the *linearly elastic region*.

The field of study that addresses the relationship between the external forces acting on a body and the internal effects (including the deformations in the body) produced by these forces is referred to as **strength of materials** or **solid mechanics**.

The solids or bodies discussed in this text are generally defined as *physical objects* or structures primarily created to function as integral units comprising of constituent elements positioned in space to accomplish the desired objectives. The constituent elements are invariably so positioned and interconnected as to enable the overall object to function as an integral unit in channelling loads or forces resulting from its use under service conditions to the supports. Designing a body is the act of determining a configuration of loaded system and formulating interconnections with the objective of imparting the solid certain load-carrying attributes. The desired attributes are the *conditions of equilibrium*, *compatibility* and *force-displacement relations of the materials*.

1.2 CLASSIFICATION OF SOLIDS OR OBJECTS

The classification scheme based on shape implies that complex objects are the result of additive aggregation of elements. However, each addition should be within the desired objective.

1.2.1 Classification Based on Geometry

In terms of basic geometries, the solid forms are generally classified as those composed of linear elements or those composed of surface elements. The linear elements are further distinguished as straight or curved. The surface-forming elements are either planar or curved. The curved-surface elements can be of either single or double curvature.

1.2.2 Classification Based on Stiffness

This classification is based on stiffness characteristics of the elements, i.e. whether the element is rigid or flexible. The *rigid elements* do not undergo appreciable change in shape under the action of loads. They usually bend to a small degree by the action of load. The typical examples are beams, floors, etc.

The *flexible elements* change shape drastically with the changing loads. However, irrespective of the shape they assume, the physical integrity is maintained. The typical example is *cable supported system*. Usually the rigidity or flexibility of an element is associated with the material of construction. Some materials, such as timber, are inherently rigid. On the other hand, materials such as steel can make either flexible or rigid members. For example, a steel cable or chain is a flexible member and a beam is a rigid member.

1.2.3 Classification Based on the Nature of Internal Forces

Based on the nature of predominant internal forces developed within the body due to externally applied loads, the load carrying elements or members are usually classified as,

1. *Axial force member*: A member that is subjected either to axial tensile or axial compressive force is referred to as an axial force member, e.g. a cable and truss member.
2. *Bending and shear resisting members*: These members are subjected predominantly to bending or flexural actions, e.g. a member subjected to loads transverse to its length.
3. *Members subjected to torsion*: These members are subjected predominantly to torsion or twisting actions, e.g. a shaft transmitting motion from one shaft to another.
4. *Members subjected to a combined action*: These members are subjected to any combination of axial force, bending moment, shear force and torsion.

1.2.4 Classification Based on Materials of Construction

In terms of materials used in construction, the objects are commonly classified as timber, steel and concrete objects.

1.3 BASIC RIGID ELEMENTS

The commonly used rigid elements include beams, struts or columns, arches, flat plates, singly curved plates and shells. Flexible elements include cables and membranes (planar, singly curved, and doubly curved). These elements are used to generate other types of objects or structures such as trusses, frames, grids, geodesic domes, cablenets, etc.

Beams and columns: The *beams* are the elements, generally horizontal, which carry loads applied transversely to their lengths and transfer them to the supporting vertical columns or other supports. The beams bend under transverse loads and are said to carry loads by bending. The elements carrying axial compressive forces are termed *struts*, when vertical they are termed *columns*.

Frames: A framed object or structure is made by assembling beam and column elements with rigid joints. The joints rigidity imparts a measure of stability against lateral forces. In a framed object, both beams and columns bend under the loads acting on the object and hence subjected to bending actions.

Trusses: The *truss* is composed of short and straight discrete elements arranged into triangulated patterns. The truss is non-rigid, but it maintains its shape as a result of the exact way the individual line elements are positioned relative to one another. A truss deflects in an overall way under the action of applied loads. The individual truss members, however, are not subjected to bending but are subjected to either compression or tension.

Arches: An *arch* is a curved line-forming structural member spanning between two points and carry the loads to the supports while being subjected predominantly to axial compression. The rigid arches are often characterized by their support conditions as two-hinged, three-hinged, etc.

Walls and Plates: These are rigid *surface elements*. A load-bearing wall can typically carry both vertically and laterally acting loads along its length. Resistance to out of plane forces in masonry walls is marginal.

Cylindrical shells: Cylindrical barrel *shells* are singly curved-plate structures spanning longitudinally such that the curve is perpendicular to the span. When fairly long, a barrel shell behaves much like a beam with a curved cross-section.

Spherical Shells and Domes: These doubly curved *surface objects* include objects that are portions of spheres. These curved surfaces are generated by rotating curved elements.

Cables: These are flexible structural elements which assume a shape depending upon the nature and magnitude of the load. When a cable spanning between two points carries a series of external point loads, it deforms into a shape made up of a series of straight line segments. The self-weight of the cable results in a *catenary* curve. Suspension cables are used to span very large distances particularly in bridges.

1.4 BASIC REQUIREMENTS OF SOLIDS

Physical solid objects subjected to vertical and horizontal forces shall be shaped and sized keeping in mind three concerns that are:

(i) Overall stability of the object, i.e. sliding or overturning of the object as a whole, shall be prevented by providing proper support system.

(ii) Internal or relational stability, i.e. the assembly should be internally stable that the parts of a solid should be properly arranged in space or inter-connected appropriately, to avoid internal collapse involving large relative movements.

(iii) Strength and stiffness of constituent elements, i.e. components breaking apart or deforming excessively due to actual stress developed in the components exceeding the permissible limits should be controlled by limiting the stresses to safe levels.

It is evident from the above discussion that there are two types of failures associated with a physical object namely material failure and configuration failure. In the former, the stresses exceed the permissible values which may result in the formation of cracks. In the latter case, even though the stresses are within permissible range, the object is unable to maintain its designed configuration under the externally applied loads due to large deformations.

1.5 INTERNAL FORCES

Internal forces or stress resultants develop within the objects due to externally applied loads. The most common internal force states are those of tension, compression, bending, shear, torsion and bearing. Associated with each of these force states are related internal stresses and strains.

(i) *Tension forces* tend to pull an element apart. The strength of a tension member is generally independent of its length and tension stresses are uniformly distributed across the cross section of the member, i.e. stress = force/area.

(ii) *Compression forces* tend to crush or buckle the element. Short members tend to crush and have higher strength compared to a tension member. The load-carrying capacity of a long member, however, decreases with the increase in the length. The long compression members may become unstable and may suddenly snap out from beneath the load at certain critical load levels. This phenomenon is called *buckling*. Because of this buckling phenomenon, long compression members are not capable of carrying very high loads.

(iii) *Bending force* is a force state associated with bending of a member. The bending action causes fibres on one face of the member to elongate, and hence are in tension, and fibres on the opposite face to compress. Thus both tensile and compressive stresses are developed at the same cross-section. These stresses act perpendicular to the face of the section. A member subjected to bending can carry only a small amount of load in comparison to a member carrying purely tensile forces. The strength of a member in bending is highly dependent on the amount and distribution of material in a cross-section.

(iv) *Shearing force* is a force state associated with the action of opposing forces that tend to cause one part of the member to slide with respect to the adjacent part. The stresses those act tangential to the sliding surface are called *shear stresses*. Shear stresses are common in beams.

(v) *Torsion* is a twisting action. Both tension and compression stresses are normally developed in the member subjected to torsion.

(vi) *Bearing stresses* exist at the interface between two members when forces are transferred from one member to another. They act perpendicular to the *contact surfaces*. The bearing stresses are also developed at the ends of beams where they rest on the walls.

1.5.1 Stresses and Strains

Each type of internal force cause *internal stresses* within the object. The *stress*, or force intensity per unit area, is a measure of the severity of force distributed over an area. Force acting over a small area produces higher unit stresses than does the same force acting over a larger area. These stresses deform the object and tend to cause the material to fail in tension or compression or through the shearing action. Different materials have different capacities, determined experimentally for carrying different types of stresses. Thus, by knowing the stress level to which a material is likely to be subjected, the member can be adequately sized to the stress safely. Corresponding to these stresses are related *strains* or deformations per unit length of material. The strains contribute to the overall deformed shape or total deflection of object.

1.6 PRINCIPLES OF SOLID MECHANICS

In the following sections, an overview of the basic issues involved in statics and strength of materials are presented.

1.6.1 Forces and Moments

Forces: A *force* is a directed interaction between bodies which may cause changes in shape or motion, or both, of the bodies involved. The force is typically a vector quantity characterized in terms of both magnitude and direction and hence can be represented by a line.

Since most solids are essentially considered rigid bodies that deform only slightly under the application of a force, the point of application of a given force may be transferred to any other point on the line of action without altering the *translatory* and *rotational effects* of the force acting on the body, i.e. *force may be regarded as acting anywhere along its line of action.*

To study the effect of several forces acting on a body, it is essential to determine the net result of the interaction of these forces. This net result of the interaction of forces can be studied in terms of the laws of vector addition which is basically based on the *parallelogram law*. The parallelogram law states that when the lines of action of two forces intersect, there is a single force, or resultant, exactly equivalent to these forces, which can be represented by the diagonal of the parallelogram formed by using the force vectors as sides of parallelogram as illustrated in Figure 1.1. This treatment can be extended for finding the resultant force of several force vectors whose lines of action intersect. In general, *a resultant force is the simplest force system to which a more complex set of forces may be reduced and still produce the same effect on the body.*

Conversely, a single force can be broken up into two or more separate forces that form a force system equivalent to the initial force. This process is usually referred to as resolving a force into components. In solid mechanics, it is often convenient to resolve a force into rectangular (x', y') or Cartesian components (x, y). as shown in Figure 1.1(a). It is to be noted that a right angle is a special form of a parallelogram. In a right angle reference axes, the components can be found by an algebraic method using simple trigonometric functions. For example, consider a force F making an angle θ with X-axis to be resolved into components on X- and Y-axes as shown in Figure 1.1(b). The components are:

$$F_x = F(\cos\theta) \text{ and } F_y = F(\sin\theta) \tag{1.1a}$$

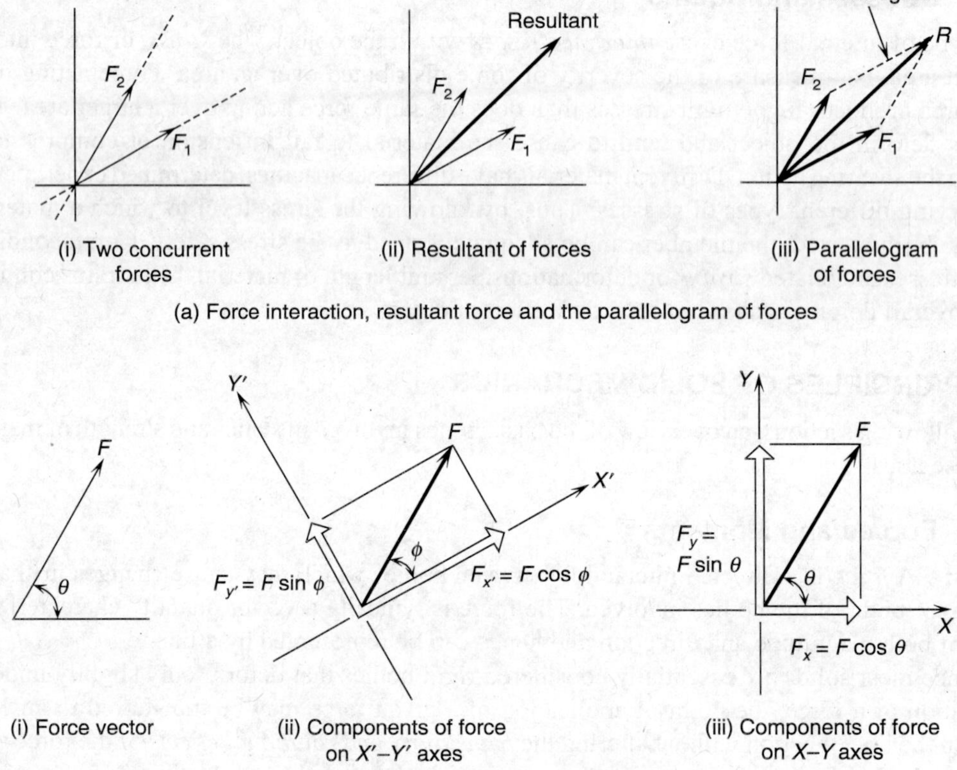

Figure 1.1 Resultant of forces and resolution of a force into components.

Conversely, if F_x and F_y are given, the resultant force F can be determined as

$$F = \sqrt{F_x^2 + F_y^2} \quad \text{and} \quad \theta = \tan^{-1}(F_y / F_x) \tag{1.1b}$$

The process of composition of forces as explained above for two components of a force can be extended to a system of concurrent forces (in which, forces intersect at a common point) which do not tend to produce rotational effects. The algebraic sum of the components, F_x and F_y of the forces in two directions, are: ΣF_x and ΣF_y, respectively. The resultant force F is then given by:

$$F = \sqrt{\left(\sum F_x\right)^2 + \left(\sum F_y\right)^2} \tag{1.2}$$

And its orientation is,

$$\theta_x = \tan^{-1}\left(\frac{\sum F_y}{\sum F_x}\right) \tag{1.3}$$

Thus a system of forces acting on a body can be replaced by another statically equivalent system of forces applied to the same body without causing any net change in the translational or rotational effects on the body. A resultant force, for example, is statically equivalent to the force system from which it was derived.

Moments: Depending upon its direction and point of application, the force acting on the body may tend to cause the body to rotate in addition to translate. The action of force tending to rotate the body is called *moment of the force*. The magnitude of this turning or rotational tendency is equal to the product of the magnitude of the force and the perpendicular distance of the line of action of the force from the axis about which turning tendency is considered as shown in Figure 1.2(a). The moment M of a force F about a point O on the axis of rotation is simply $M_o = F \times r$, where r, called the moment arm of the force, is the perpendicular distance of line of action of F from the point O.

The total rotational effect produced by several forces, e.g. two forces shown in Figure 1.2(b) about the same axis is the algebraic sum of their individual moments about the axis. Thus,

$$M_o = (F_1 \times r_1) + (F_2 \times r_2) + \cdots + (F_n \times r_n) = \sum_{i=1}^{n} (F_i \times r_i) \tag{1.4}$$

In case of a *couple*, i.e. a force system consisting two forces equal in magnitude but opposite in sense and having parallel lines of action that are not on the same straight line, the rotational effect

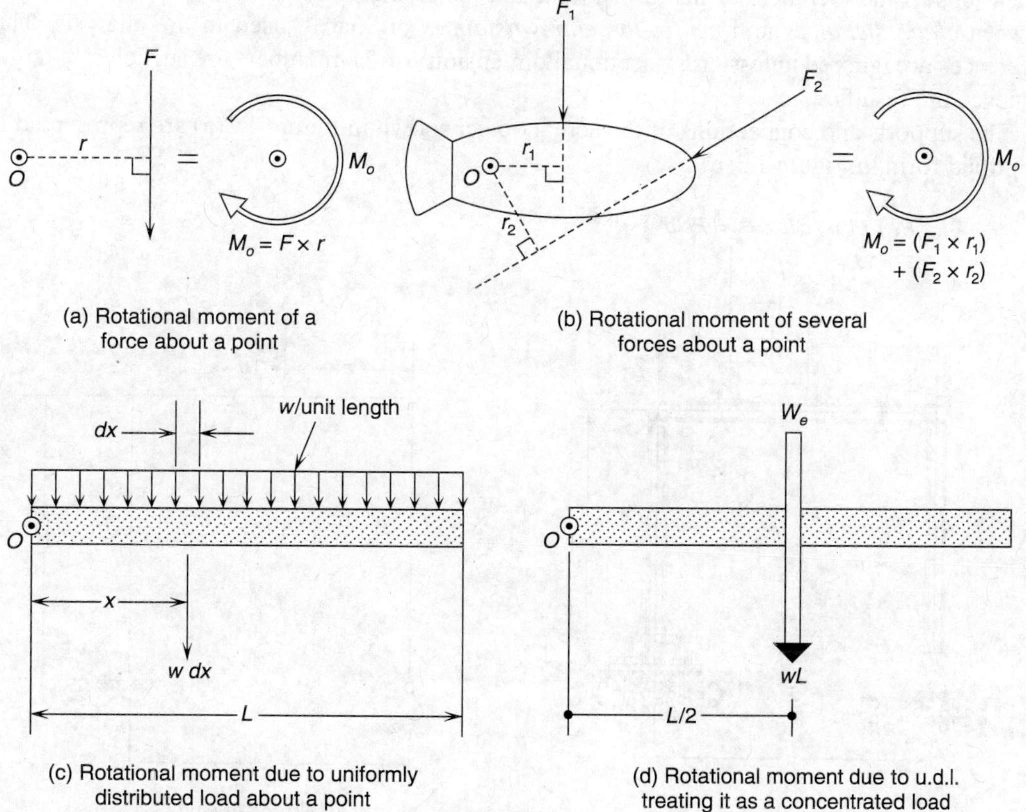

(a) Rotational moment of a force about a point

(b) Rotational moment of several forces about a point

(c) Rotational moment due to uniformly distributed load about a point

(d) Rotational moment due to u.d.l. treating it as a concentrated load

Figure 1.2 Rotational moment of concentrated and distributed forces about a point.

is simply the product of one of the forces times the perpendicular distance between the two forces. The couple does not cause translation. Moreover, the moment of a couple is independent of reference axis and its rotational effect on the body is also independent of its point of application on the body.

In case of *continuous* or *distributed load* of intensity w per unit length acting on a body as shown in Figure 1.2(c), an elemental portion of load $w\,dx$ is considered. This elemental load produces moment of $w\,dx\,(x)$ about point O. The total moment of the entire load about the point O is thus,

$$M_o = \int_0^L wx\,dx = \frac{wL^2}{2} = (wL) \times \frac{L}{2} = \frac{W_e L}{2} \tag{1.5}$$

Thus, *the magnitude of moment is the same as obtained by considering the total continuous load to be acting at the centre of the mass of load* as shown in Figure 1.2(d).

1.7 IDEALIZATION OF SOLID OBJECTS

The primary aim of solid mechanics is to determine the reactions, internal forces and deformation at any point of the given body caused by applied loads and forces. To achieve this objective, it becomes necessary to idealize a body in a simplified form emendable to analysis procedures. The members are normally represented by their *centroidal axes*. This naturally does not consider the dimensions of the members or depth of joints and hence there may be considerable differences between *clear distances* and *centre-to-centre distances* ordinarily used in the analysis. These differences are ignored unless cross-sectional dimensions of the members are sufficiently large to influence the results.

The supports and connections of the actual object shown in Figure 1.3(a) are represented in a simplified form in Figure 1.3(b).

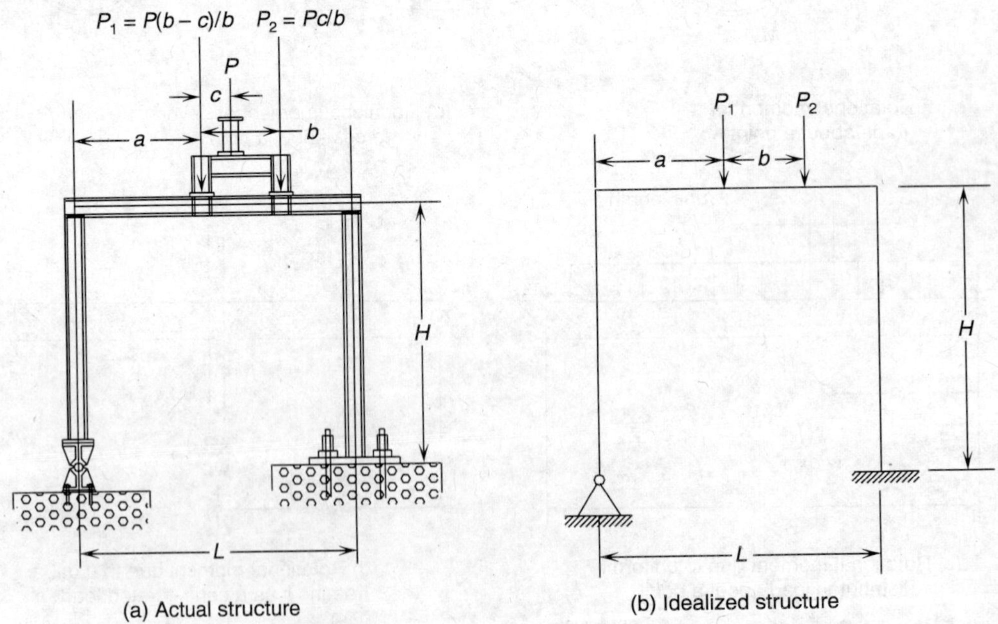

Figure 1.3 Idealization of a structure.

1.8 EQUILIBRIUM OF SOLIDS

The principle of *equilibrium* is the most important of the concepts constituting theoretical base for solid mechanics. It provides the conditions for determining the relationships that must necessarily exist between applied loads and resulting stress resultants i.e. internal forces, moments and external reactions; between stress resultants and stresses; even among internal and external energies. It is assumed that the student has been introduced to this concept in the previous course work (applied mechanics, physics, etc). In this section, only a selective review highlighting key points have been provided.

A body in general is subjected to a set of forces which include external or applied forces, internal forces or reactions that are developed within the body at connection points and gravity forces caused by the mass of the elements. The body must be in the state of static equilibrium with respect to these forces, i.e. the net translational effect of all force components acting on the body must have a zero sum along any axis or in any direction. This fundamental requirement is formally expressed as: $\Sigma F_r = 0$. Likewise, there must be no net rotational effect of forces about a point of support. The *rotational effect* of a force about a point of support, called a *moment* ($M = F \times r$), may act in clockwise or counterclockwise sense. For the body to be in rotational equilibrium, the net sum of these rotational moments must be zero, i.e. $\Sigma M_r = 0$.

For a 3-dimensional situation, the conditions of equilibrium are normally established with reference to a Cartesian coordinate system X, Y and Z, and in the component form these two equations can be written in their more common form:

(i) Translational equilibrium: $\Sigma F_x = 0, \Sigma F_y = 0$ and $\Sigma F_z = 0$ (1.6)

(ii) Rotational equilibrium: $\Sigma M_x = 0, \Sigma M_y = 0$ and $\Sigma M_z = 0$ (1.7)

The summation ΣF_j includes all the forces or components of forces (including reactions) that exist in jth direction, while ΣM_j is the sum of all moments that exist about any general axis j perpendicular to the plane of object about which the sum of the moments of forces must vanish. However, the selection of different moment centres in the plane of object depends upon the system being analysed and preference of the user. The only requirement is that the three equations must be independent.

The condition of equilibrium in X-direction expresses the fact that there is no net unbalanced force to move the body in that direction. Thus for static equilibrium, the algebraic sum of all the forces along the co-ordinate axis X must be zero. Mathematically, it has been expressed as $\Sigma F_x = 0$. Same holds good along co-ordinate axes Y and Z. On the other hand, the three additional rotational equilibrium conditions state that the object or element does not spin or rotate about any of the three axes due to unbalanced moments.

For a 2-dimensional planar body lying in XY plane there is no force acting in Z-direction or any moment about X- and Y-directions (axes). The moment ΣM_z represents moment about an axis perpendicular to XY plane. Thus for a planar structure, the equilibrium conditions are:

$$\Sigma F_x = 0, \Sigma F_y = 0 \text{ and } \Sigma M_z = 0 \tag{1.8}$$

If a concurrent force system is in equilibrium, its resultant must be zero, i.e. $\Sigma F_x = 0$ and $\Sigma F_y = 0$. Thus, the algebraic sum of all the components of the forces applied to a particle in X- and Y-directions must be zero. Here X and Y represent any *orthogonal set of axes* irrespective of their orientation. In more general terms, the conditions $\Sigma F_x = 0$, $\Sigma F_y = 0$ and $\Sigma F_z = 0$ are necessary and sufficient to ensure equilibrium in a concurrent force system. A force system satisfying these conditions will not cause the particle to translate (the rotational effect is already zero as all forces pass through the same point).

1.8.1 Special Equilibrium Cases

When a rigid body is subjected to only two forces, the forces cannot have arbitrary magnitudes and lines of action if the member is to be in equilibrium. It can be seen that for net translation or rotation of the body to be zero, the two forces must be collinear. They must also be equal in magnitude but opposite in sense. This observation is important in the analysis of trusses.

As in the case of element with two forces, three forces acting on a member cannot have random magnitudes and orientation if the member is to be in equilibrium. For the member to be in rotational equilibrium, the lines of action of all the three forces must pass through a common point.

EXAMPLE 1.1

In the straight member AOB of uniform cross-section shown in Figure 1.4(a), the part AO of length $2a$ is made of timber and the part OB is made of aluminium. If the unit weight of aluminium is three times that of timber, determine the length of part OB for the member AOB to be in exact balance about the fulcrum O.

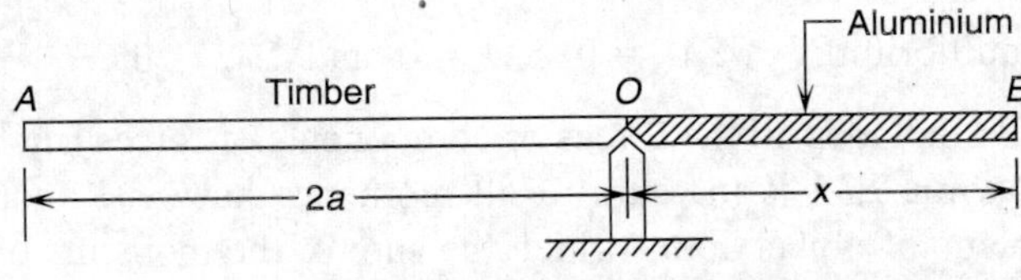

(a) Compound bar of uniform cross-section

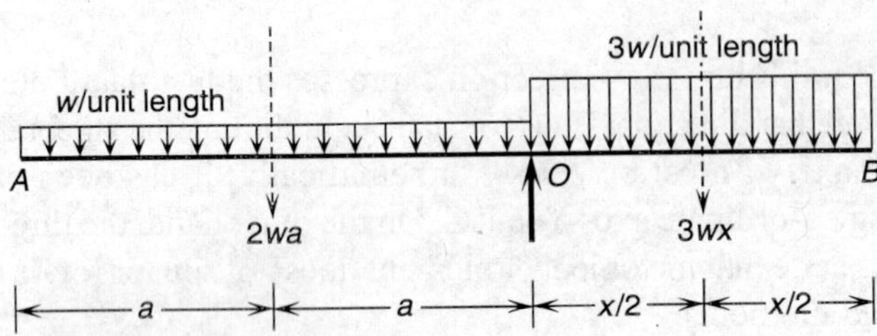

(b) Uniformly distributed loads concentrated at mid-points

Figure 1.4 Balancing of compound bar of Example 1.1.

Solution: For exact balance or rotational equilibrium the rotational moment of each part about the fulcrum must exactly balance each other. For convenience, uniformly distributed loads on two portions may be considered to be concentrated at their mid-points as shown in Figure 1.4(b). Let the length of aluminium part is x, then:

Rotational moment of timber part = Rotational moment of aluminium part

$$(2wa) \times a = (3wx) \times \frac{x}{2} \quad \text{i.e.} \quad x = \frac{2a}{\sqrt{3}}$$

1.8.2 Sign Convention

When working with a general force system, it is convenient to use normal sign convention adopted for a 3-dimensional right-handed system of Cartesian or rectangular coordinate axes OX, OY, and OZ shown in Figure 1.5(a) with origin O on the extreme left of the structure. The forces measured from the origin towards positive directions of axes are always positive as shown in Figure 1.5(b). Similarly, *rotational moments* expressed in vector form pointing towards positive directions of the axes are positive as shown in Figure 1.5(c). Thus, moments that tend to produce counterclockwise rotations are considered positive and those tend to produce clockwise rotations are considered negative. Once the positive directions of any two of the axes, say OX and OY are selected, the positive direction of the third axis OZ will be that in which right-handed screw would advance when turned from X-axis to Y-axis. These sign conventions are for equilibrium calculations involving external forces only. Other sign conventions based on the type of object and involving internal forces will be described later.

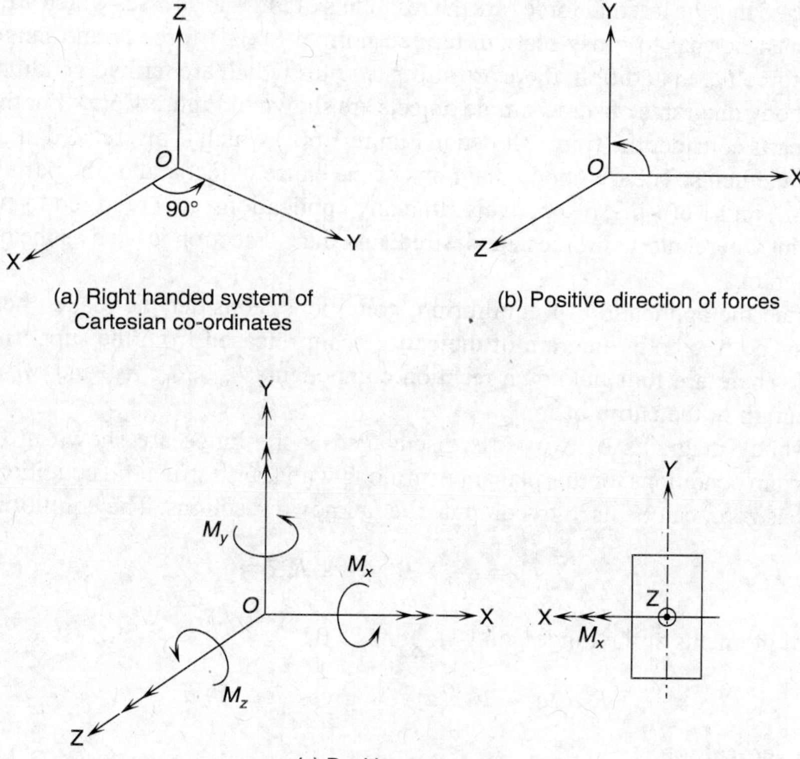

Figure 1.5 Right-handed Cartesian or rectangular co-ordinate axes.

1.9 FREE-BODY DIAGRAMS

For static analysis of bodies subjected to external loads, *analytical diagrams* that illustrate the force systems acting on the objects, called **equilibrium** or **free-body diagrams**, are generally constructed. Using equilibrium concepts, the numerical values of *reactions* that occur at supports and hence *internal forces*, i.e. *axial forces*, *shear forces* and *bending moments* can be determined. The major application of equilibrium analysis is in the evaluation of reactions and internal forces by representing an object by a series of free-body diagrams.

The analysis of all the objects is based on the concept that any part of an object is in equilibrium along with the object as a whole. This concept is used to determine the internal forces in an object by drawing free-body diagrams for the parts of the object. These are obtained by cutting the object hypothetically or disengaging some connections and supports. In constructing a free-body diagram, the correct depiction of all the possible forces in the object at the cuts and disengaged connections by appropriate force vectors is of extreme importance. At this stage, the correct direction of the internal forces is not obvious. Once the values of these quantities are ascertained by statics, the proper direction (sense) of each force component can be established. All the external forces acting on the body in its original state must also be depicted on the diagram. This procedure can be applied to each of the free-body diagrams into which the object has been discretised or broken down. However, in dealing with the forces acting on the free bodies, the internal forces common to two free bodies are double action forces denoted as equal but oppositely directed force vectors. It should be realized that the internal forces are the resultants of internal stresses which are decomposed into components, normal to cross-section, termed normal (axial) force N and tangent to cross-section shear force F_y. In addition, there are stress couples which are termed bending moment M.

The free-body diagrams of some simple objects are shown in Figure 1.6(a). For the free bodies of individual parts consider the frame shown in Figure 1.6(b) which is discretized into a number of free-bodies or elements. The free-body diagrams of the entire object and of the parts are shown in Figures 1.6(b-ii) and 1.6(b-iii), respectively. In many applications, it is required to cut the section only at the point where internal forces are desired, and the cut section is used as the one end of the free-body diagram.

To illustrate the application of equilibrium conditions, consider the loaded beam shown in Figure 1.7(a). The free body diagram of the entire beam released from the supports is shown in Figure 1.7(b). There are four unknown reaction components R_{0x}, R_{0y}, R_{2y} and M_0 acting on the free-body diagram of the entire beam.

The free-body diagrams of two parts disengaged at the hinge are shown in Figure 1.7(c). Three equilibrium conditions for this planar beam along with fourth structural condition that moment at the hinge 1 is zero, can be used to compute the unknown reactions. The equilibrium condition

$$\sum F_x = R_{0x} = 0 \text{ gives } R_{0x} = 0 \tag{1.9}$$

Summation of moments at the hinge point 1, $\sum \bar{M}_1 = 0$:

$$R_{2y}(2a) - 2wa(a) = 0 \text{ gives } R_{2y} = wa \tag{1.10}$$

Summing up vertical forces

$$\sum F_y = R_{0y} + R_{2y} - 2wa = 0$$

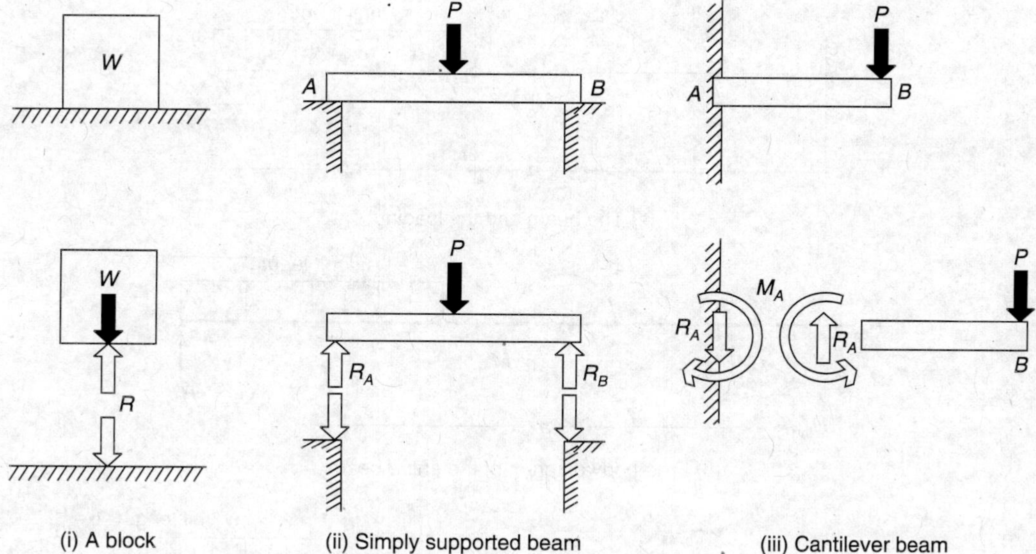

(i) A block (ii) Simply supported beam (iii) Cantilever beam

(a) Free-body diagrams of a block, simply supported beam and cantilever beam

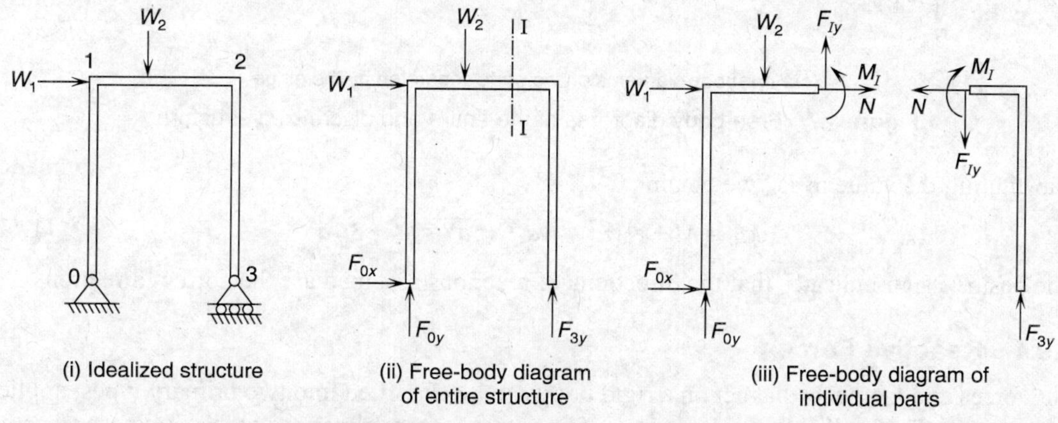

(i) Idealized structure (ii) Free-body diagram of entire structure (iii) Free-body diagram of individual parts

(b) Free-body diagrams of the entire and discritized frame

Figure 1.6 Free-body diagrams of simple and frame structures.

Therefore,

$$R_{0y} + R_{2y} = 2wa \quad \text{or} \quad R_{0y} = wa \tag{1.11}$$

and

$$\sum \bar{M}_0 = M_0 - M - 2wa(2a + a + a) + R_{2y}(5a) = 0$$

or

$$M_0 = M + 8wa^2 - R_{2y}(5a)$$

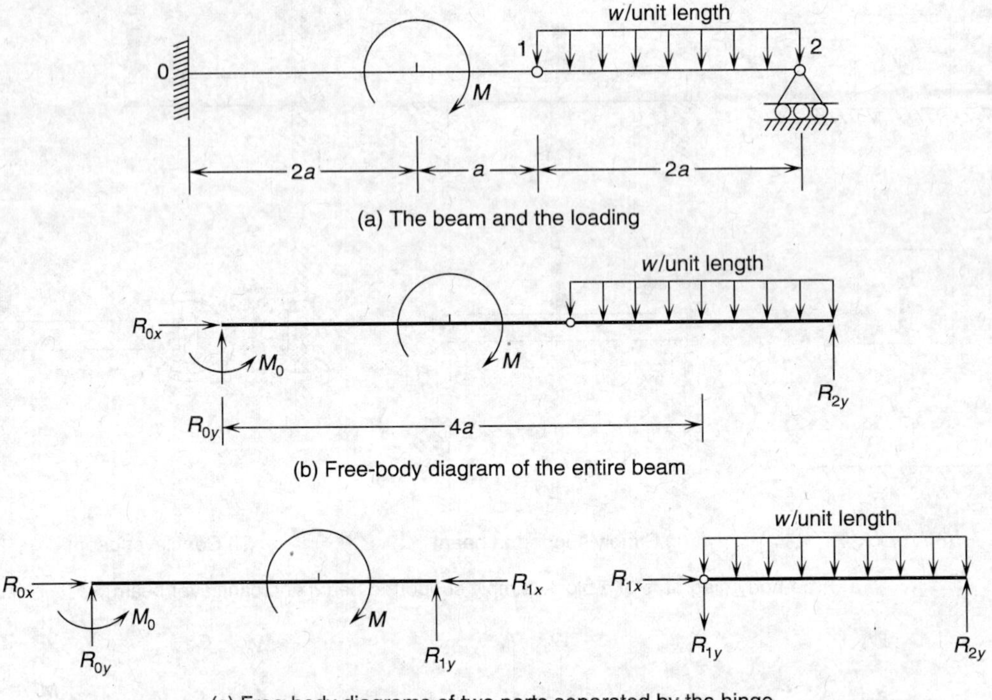

Figure 1.7 Free-body diagrams of the entire and discretized structure.

Substituting the value of R_{2y} we obtain

$$M_0 = M + 8wa^2 - wa \times (5a) = M + 3wa^2 \tag{1.12}$$

The positive sign indicates that the directions of reactions assumed are the correct directions.

1.9.1 Reactive Forces

The forces and moments that act on a rigid body can be classified into two primary types; applied and reactive. The applied forces include the forces that act directly on the object, whereas, *reactive forces* are generated by the action of one body on another and hence typically occur at connections and supports. The reactive forces are based on *Newton's third law* which states that whenever one body exerts a force on another, the second always exerts on the first a force which is equal in magnitude, and opposite in direction, i.e. has the same line of action. For example, consider the case of the beam exerting downward forces on the supports and upward reactive forces are consequently developed. Thus, a pair of action and reaction forces exists at each interface between the beam and its supports. In some cases, the moments also form a part of reactive system as shown in Figures 1.6(b) and (c).

For overall equilibrium of the body, the magnitude and direction of any reactive force developed must be such that the equilibrium is maintained and thus necessarily depend on the characteristics of the applied force system. The whole system of applied and reactive forces acting on the body, as represented by the free-body-diagram, must be in the state of equilibrium. The free-body-diagrams, consequently called *equilibrium diagrams*, are useful in solid mechanics. Internal forces are obtained by cutting the object being analysed at the point where internal forces are desired, and using the cut section as one end of the free-body diagram.

1.9.2 Support Conditions

The nature of reactive forces developed on a loaded body depends on the type of support or connection to other bodies. Figure 1.8 illustrates relations between the type of support condition and type of reactive forces developed. There are several types of support conditions. Of primary importance are pinned or hinged connections, roller supports, and fixed connections. The idealized connections are shown in Figure 1.8.

Pinned connection. In this connection joint allows attached member to rotate freely but does not allow translation in any direction. Consequently, the joint cannot provide moment resistance but can provide force resistance in any direction.

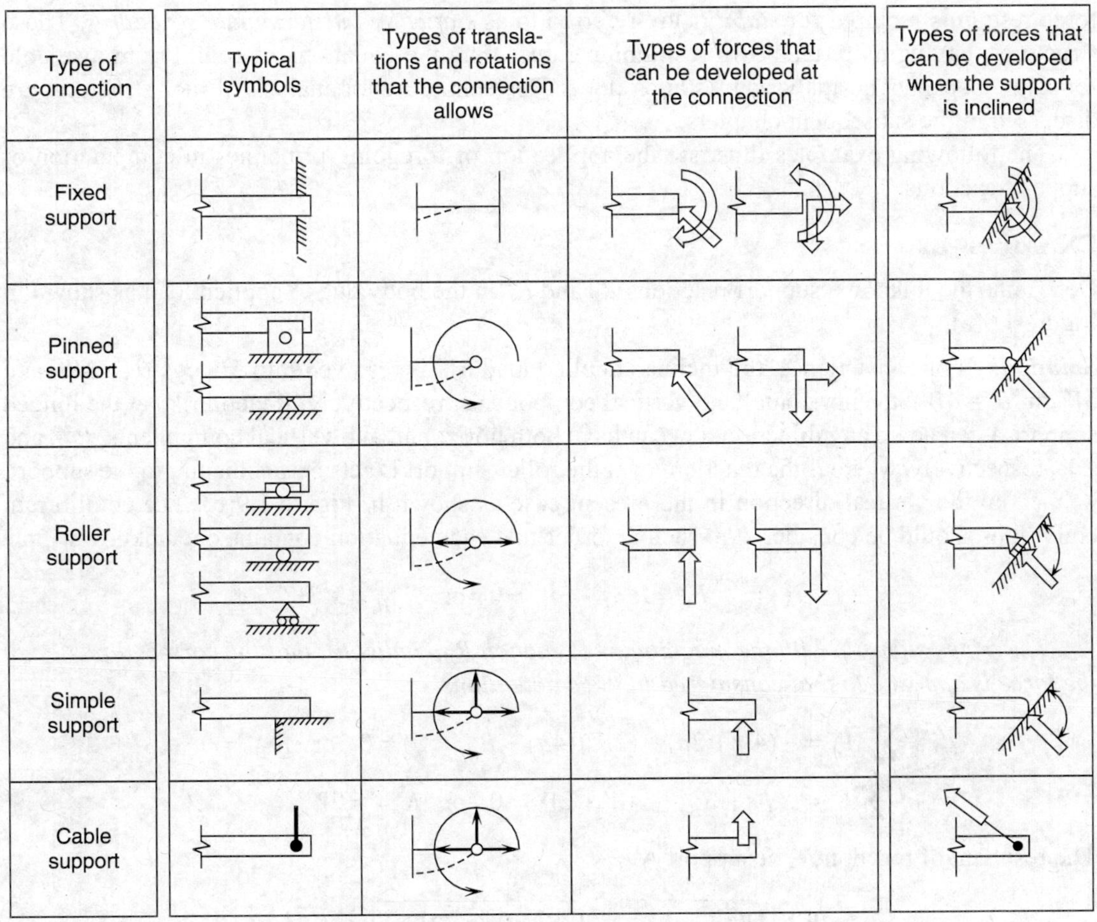

Type of connection	Typical symbols	Types of translations and rotations that the connection allows	Types of forces that can be developed at the connection	Types of forces that can be developed when the support is inclined
Fixed support				
Pinned support				
Roller support				
Simple support				
Cable support				

Figure 1.8 Idealized support conditions.

Roller connection. In addition to rotation, this connection also allows attached member to translate freely parallel to the surface of the support, i.e. does not provide any force resistance parallel to the surface of the support. However, the joint resists translations in the direction perpendicular to the surface of the support.

Fixed connection. This connection completely restrains rotations and translations of the attached members in any direction. Consequently, the joint can provide moment and force resistances in any direction.

For an object to be in stable equilibrium, the supports must be capable of providing specific minimum number of force restraints. For example, for a simple beam subjected to the vertical and horizontal forces, the supports must provide three force restraints for its equilibrium corresponding to three conditions of equilibrium namely $\Sigma F_x = 0$, $\Sigma F_y = 0$ and $\Sigma M_z = 0$ for this type of object. This requirement can be met by providing a fixed connection at one end or alternatively by using a pinned connection on one end and a roller connection on the other.

The objects having connections or supports that provide more than the minimum number of force restraints required for *stability* are referred to as *statically indeterminate structures*. These structures, having more unknown restraining forces than the equations of equilibrium available for solution, cannot be analysed by statics alone. The techniques for analysis of such structures are discussed in the subsequent chapters.

The following examples illustrate the application of foregoing principles in computation of support reactions.

EXAMPLE 1.2

Determine the unknown support reactions R_A and R_B in the body due to applied load as shown in Figure 1.9(a).

Solution: For convenience, the inclined applied load $5W$ is resolved into $5W \cos\theta (= 3W)$ and $5W \sin\theta (= 4W)$, the horizontal and vertical components, respectively. Reaction R_A at the hinged support A can act in any direction, i.e. can have both horizontal and vertical components, R_{AH} and R_{AV}, respectively, whereas the reaction R_B at the roller support B acts perpendicular to the support, i.e. acts in the vertical direction in the present case as shown in Figure 1.9(b). The equilibrium conditions should be considered in such an order that each equation contains one unknown:

$$\rightarrow(+) \sum F_x = R_{AH} + 3W = 0 \quad \text{or} \quad R_{AH} = -3W$$

The minus sign (–), with the magnitude of reaction R_{AH} indicates that the correct direction of the force is opposite to that considered in the calculations.

$$\sum \bar{M}_A = -(4W)(2a) + (2W)(4a) + R_{BV}(5a) = 0 \quad \text{or} \quad R_{BV} = 0$$

$$\uparrow(+) \sum F_y = R_{AV} + R_{BV} - 4W + 2W = 0 \quad \text{or} \quad R_{AV} = 2W$$

The resultant of reaction R_A at support A:

$$R_A = \sqrt{R_{AH}^2 + R_{AV}^2} = \sqrt{[(-3W)^2 + (2W)^2]} = \sqrt{13}\,W$$

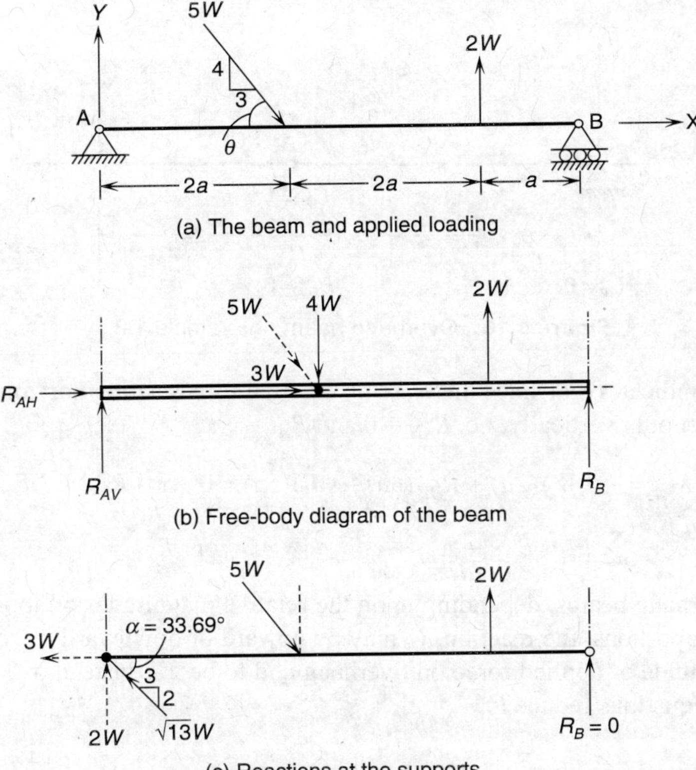

Figure 1.9 Analysis of the beam of Example 1.2.

and $$\alpha = \tan^{-1}\left(\frac{R_{AV}}{R_{AH}}\right) = \tan^{-1}\left(\frac{2W}{3W}\right) = \tan^{-1}\left(\frac{2}{3}\right) = 33.69°$$

It should be noted that in this case, there is no reaction component at the support B. The magnitude of reactions and their directions are shown in Figure 1.9(c).

In the moment calculations, several forces passed through the moment centre and hence their moment arms had zero length resulting in zero rotational effects. Generally, the forces passing through the moment centre are not included in the equilibrium calculations.

EXAMPLE 1.3

Determine the unknown support reactions R_A and R_B in the structure with an overhang due to applied loads as shown in Figure 1.10.

Solution: Reaction R_A at hinged support A has both horizontal and vertical components, R_{AH} and R_{AV}, respectively, whereas the reaction R_B at the roller support B acts in vertical direction as shown in Figure 1.8(c). Consider equilibrium conditions in the following order to obtain equations each containing one unknown:

$$\rightarrow(+) \sum F_x = R_{AH} = 0$$

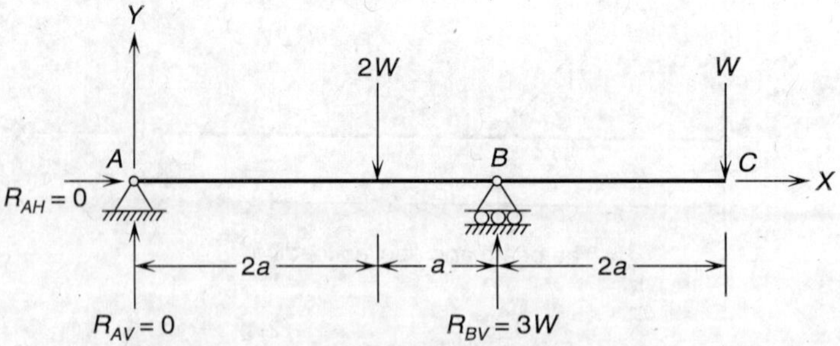

Figure 1.10 Overhang beam of Example 1.3.

This equation indicates that since there are no external forces acting horizontally, reaction R_A must necessarily act only vertically, i.e. $R_{AH} = 0$ and $R_{AV} = R_A$.

$$\sum \bar{M}_A = -(2W)(2a) + R_{BV}(3a) - (W)(5a) = 0 \quad \text{or} \quad R_{BV} = 3W \uparrow$$

$$\uparrow(+) \sum F_y = R_{AV} + R_{BV} - 2W - W = 0 \quad \text{or} \quad R_{AV} = 0$$

In case of overhang beams, depending upon the relative magnitudes of applied forces on the span and overhang portions, the reaction R_A may be upward or downward or zero. For example, consider the magnitude of applied force on overhang end to be $2W$ instead of W, the second and third equilibrium equations reduce to,

$$\sum \bar{M}_A = -(2W)(2a) + R_{BV}(3a) - (2W)(5a) = 0 \quad \text{or} \quad R_{BV} = \frac{14W}{3} \uparrow$$

and

$$\uparrow(+) \sum F_y = R_{AV} + \frac{14W}{3} - 2W - 2W = 0 \quad \text{or} \quad R_{AV} = -\frac{2W}{3} \downarrow$$

The minus sign (–), with the magnitude of reaction R_{AV}, indicates that the actual direction of the force is opposite to that considered in the calculations, i.e. it acts downward.

EXAMPLE 1.4

Determine the unknown support reactions R_A and R_B in the structure with bends as shown in Figure 1.11(a).

Solution: The hinged support A can potentially provide a force resistance in any direction, i.e. can have both horizontal and vertical components, R_{AH} and R_{AV}, respectively, whereas the roller support B transmits reaction R_B in vertical direction only. For convenience, the forces applied at the ends of the bends can be transferred to the centroidal axis of the beam as the equivalent horizontal and vertical components, and moment for the same effect as shown in Figure 1.11(b). The equilibrium equations are:

$$\rightarrow(+) \sum F_x = R_{AH} + W + 2W = 0 \quad \text{or} \quad R_{AH} = -3W$$

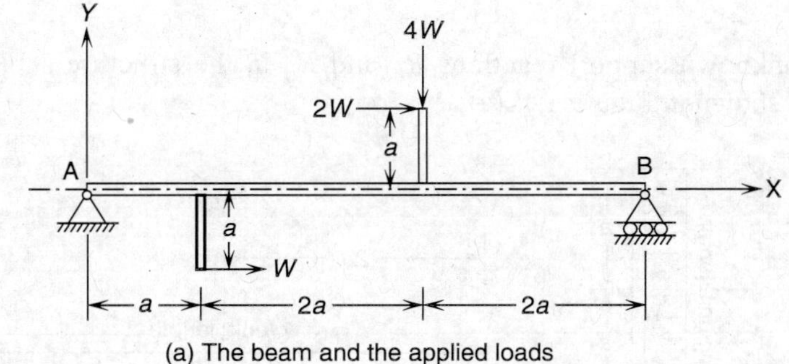

(a) The beam and the applied loads

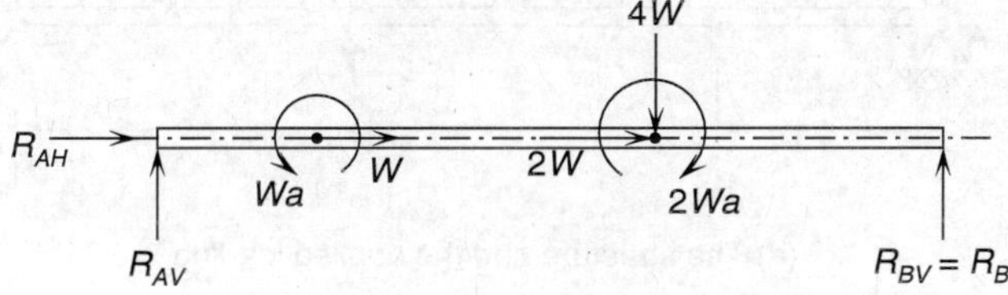

(b) Free-body diagram of the beam

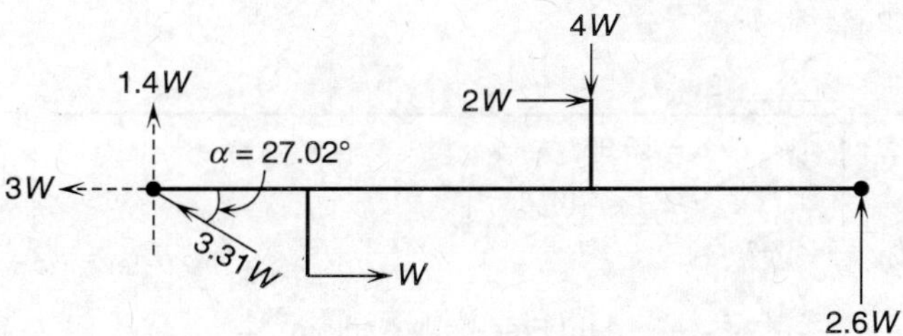

(c) Reactions at the supports

Figure 1.11 Analysis of beam of Example 1.4.

The minus sign (–), with the magnitude of reaction R_{AH}, indicates that the correct direction of the force is opposite to that considered in the calculations.

$$\sum \bar{M}_A = Wa - 2Wa - (4W)(3a) + R_{BV}(5a) = 0 \quad \text{or} \quad R_{BV} = \frac{13W}{5} = 2.6W\uparrow$$

$$\uparrow(+) \sum F_y = R_{AV} - 4W + R_{BV} = 0 \quad \text{or} \quad R_{AV} = 1.4W$$

The resultant of reaction R_A at support A:

$$R_A = \sqrt{R_{Ax}^2 + R_{Ay}^2} = W\sqrt{[3^2 + 1.4^2]} = 3.31W$$

and

$$\alpha = \tan^{-1}\left(\frac{R_{Ay}}{R_{Ax}}\right) = \tan^{-1}\left(\frac{1.4}{3}\right) = 27.02°$$

The magnitude of reactions and their directions are shown in Figure 1.11(c).

EXAMPLE 1.5

Determine the unknown support reactions R_A and R_B in the structure subjected to uniformly distributed loads shown in Figure 1.12(a).

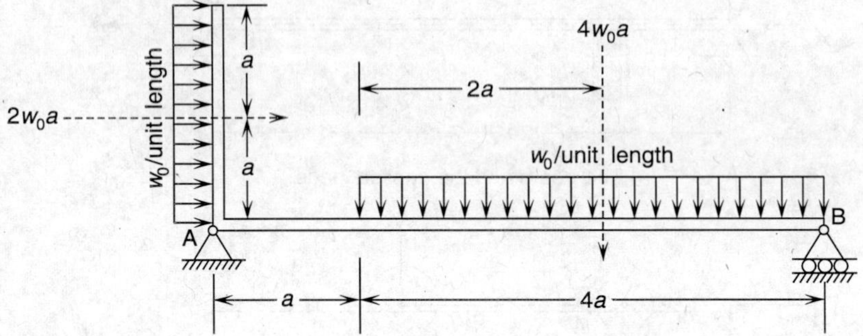

(a) The structure and the applied loading

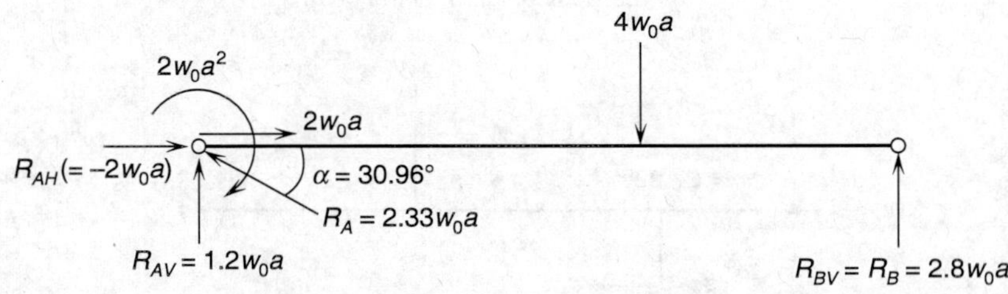

(b) Free-body diagram

Figure 1.12 Analysis of the structure of Example 1.5.

Solution: The total load acting downward due to distributed load w_0 is the load per unit length multiplied by the length over which the load acts. For computations, consider the uniformly distributed loads to be concentrated at the mid-points of loaded lengths as shown in Figure 1.12(b). The equilibrium equations are:

$$\rightarrow(+)\ \sum F_x = R_{AH} + (w_0)(2a) = 0 \quad \text{or} \quad R_{AH} = -2w_0 a$$

The minus sign indicates that the correct direction of the reaction R_{AH} is opposite to that considered in the calculations.

$$\sum \bar{M}_A = -(2w_0 a)(a) - (4w_0 a)(3a) + R_{BV}(5a) = 0 \quad \text{or} \quad R_{BV} = 2.8w_0 a \uparrow$$

$$\uparrow(+)\ \sum F_y = R_{AV} - (4w_0 a) + R_{BV} = 0 \quad \text{or} \quad R_{AV} = 1.2w_0 a$$

The resultant of reaction R_A at support A:

$$R_A = \sqrt{R_{AH}^2 + R_{AV}^2} = w_0 a\sqrt{[(2.0)^2 + (1.2)^2]} = 2.33w_0 a$$

and

$$\alpha = \tan^{-1}\left(\frac{R_{AV}}{R_{AH}}\right) = \tan^{-1}\left(\frac{1.2}{2.0}\right) = 30.96°.$$

EXAMPLE 1.6

Determine the support reactions R_A and R_B due to uniformly distributed load acting on the structure with roller resting on a sloped support as shown in Figure 1.13(a).

Solution: As usual the hinged support A can have both horizontal and vertical components of reaction, R_{AH} and R_{AV}, respectively, whereas the roller support B transmits reaction R_B only perpendicular to the surface on which they roll. For convenience, the reaction R_B can be resolved into $R_B \sin\theta$ and $R_B \cos\theta$, the horizontal and vertical components, respectively.

For computation of reactions consider the uniformly distributed load to be concentrated at the mid-point of loaded length. Thus the equilibrium equations are (see Figure 1.13(b)):

$$\sum \bar{M}_A = -(2w_0a)(a) - (4w_0a)(4a) + R_{BV}(6a) = 0 \quad \text{or} \quad R_{BV} = 3w_0a\uparrow$$

$$\rightarrow(+)\sum F_x = R_{AH} - R_{BH} = 0 \quad \text{or} \quad R_{AH} = R_{BH}$$

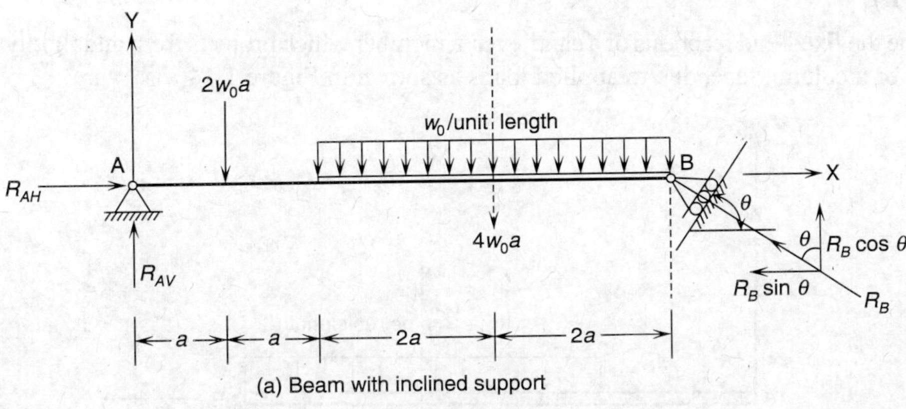

(a) Beam with inclined support

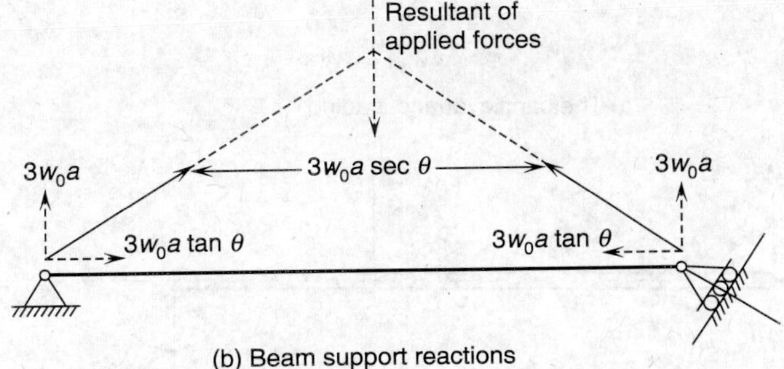

(b) Beam support reactions

Figure 1.13 Analysis of beam with inclined support.

Therefore,

$$R_{AH} = R_{BH} = R_{BV} \tan \theta = 3w_0 a \tan \theta$$

$$\uparrow(+) \sum F_y = R_{AV} - (2w_0 a) - (4w_0 a) + R_{BV} = 0 \quad \text{or} \quad R_{AV} = 3w_0 a \uparrow$$

Thus, the resultant reaction R_A at support A:

$$R_A = \sqrt{R_{AH}^2 + R_{AV}^2} = (3w_0 a)\sqrt{(\tan^2 \theta + 1)} = (3w_0 a) \sec \theta$$

and

$$\alpha = \tan^{-1}\left(\frac{R_{AV}}{R_{AH}}\right) = \tan^{-1}\left(\frac{3w_0 a}{3w_0 a \tan \theta}\right) = \tan^{-1} \cot \theta$$

$$\alpha = 90° - \theta$$

Similarly,

$$R_B = \sqrt{R_{BH}^2 + R_{BV}^2} = 3w_0 a\sqrt{(\tan^2 \theta + 1)} = (3w_0 a) \sec \theta$$

EXAMPLE 1.7

Determine the fixed-end reactions of a cantilever, a member which projects from and rigidly attached to a wall or a column face, due to applied loads as shown in Figure 1.14(a).

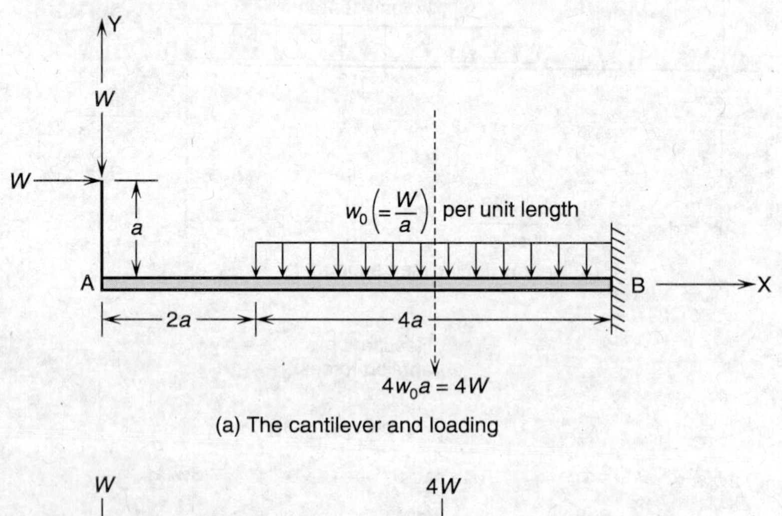

(a) The cantilever and loading

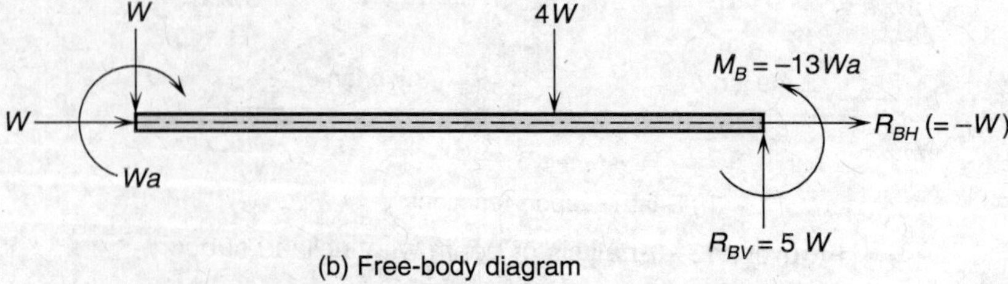

(b) Free-body diagram

Figure 1.14 Analysis of cantilever beam of Example 1.7.

Solution: The fixed end connection completely restrains the end of the member from either translation or rotation. Thus, the end develops horizontal and vertical reactive forces, R_{BH} and R_{BV}, respectively, that prevent the member from translating, and restraining moment M_B that prevents the end of the member from rotating. The restraining moment that balances the externally applied moments at the end is typically called *fixed-end moment*. As the free end does not have any restraint, no reactive force is developed there. The reactions are shown in Figure 1.14(b). The equilibrium equations are:

$$\rightarrow(+)\sum F_x = W + R_{BH} = 0 \quad \text{or} \quad R_{BH} = -W$$

The minus sign indicates that the correct direction of the reaction R_{BH} is opposite to that presumed in the calculations.

$$\uparrow(+)\sum F_y = -W - 4W + R_{BV} = 0 \quad \text{or} \quad R_{AV} = 5W\uparrow$$

$$\sum \bar{M}_B = -Wa + (W)(6a) + (4W)(2a) + M_B = 0 \quad \text{or} \quad M_B = -13Wa$$

The minus sign (–), with the magnitude of reaction M_B, indicates that the actual direction of the moment is opposite to that presumed in the calculations, i.e. it acts clockwise.

1.9.3 Forces and Reactions Due to Overturning

In the situations where the applied forces have significant horizontal components, there is a tendency of *overturning* of the objects. The rigid bodies resist this overturning via the *stabilizing* or *resisting moment* provided by the reactive forces developed at the supports or that associated with the gravity loads and the size of body. The following example illustrates the development of tie-down reactions.

EXAMPLE 1.8

Determine the support reactions R_A and R_B in the body shown in Figure 1.15(a).

Solution: The potential reaction components R_{AH} and R_{AV} at support A; and R_B at support B, respectively, are shown in Figure 1.15(b). The equilibrium conditions are as follows.
Moment equilibrium about an axis through point A,

$$(+)\sum \bar{M}_A = -W \times a - 3W \times 2a + R_B \times 2a = 0 \quad \text{i.e.} \quad R_B = 3.5W$$

Equilibrium in vertical direction,

$$\uparrow(+)\sum F_y = R_{AV} - 3W + R_B = 0 \quad \text{or} \quad R_{AV} = -0.5W\ (\downarrow)$$

Equilibrium in horizontal direction,

$$\rightarrow(+)\sum F_x = R_{AH} + W = 0 \quad \text{i.e.} \quad R_{AH} = -W(\leftarrow)$$

The correct directions of R_{AH} and R_{AV} are opposite to that considered in the calculations. The resultant tie down reaction R_A at support A:

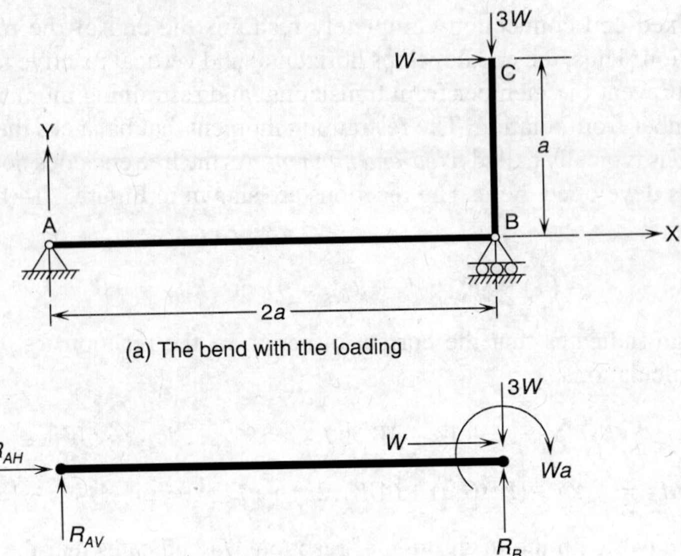

(a) The bend with the loading

(b) Free-body diagram of the bend support reactions

Figure 1.15 Analysis of bend of Example 1.8.

$$R_A = \sqrt{R_{AH}^2 + R_{AV}^2} = W\sqrt{(-1)^2 + (-0.5)^2} = 1.12W$$

$$\alpha = \tan^{-1}\left(\frac{R_{AV}}{R_{AH}}\right) = \tan^{-1}\left(\frac{0.5W}{W}\right) = 26.57°$$

1.9.4 Bodies with Cable Supported End

The bodies supported on one end by cables are commonly encountered. The reactive forces developed within the cables provide equilibrium forces for the supported body. *The forces developed in a cable are in tension and inherently directed along its length.* Thus the direction of reactive force is predetermined. The internal force in the cable is treated as an external reaction for calculating equilibrium of the body. The following examples illustrate the procedure for the static analysis of a simple *cable-stayed body*.

EXAMPLE 1.9

Determine the reactive forces developed in the cable-supported beam shown in Figure 1.16(a).

Solution: For convenience, the reactive force in the cable T at the end B can be resolved into $T(\cos\theta)$ and $T(\sin\theta)$, the horizontal and vertical components, respectively. The cable force is considered to be in tension, as it must be. These components along with the potential reaction components R_{AH} and R_{AV} at supports A are shown in Figure 1.16(b). The equilibrium conditions are as follows.

Moment equilibrium about an axis through the point A,

$$+\sum \bar{M}_A = -W(a) + T(\sin\theta)(2a) = 0 \quad \text{or} \quad T = \left(\frac{W}{2}\right)\text{cosec}\,\theta$$

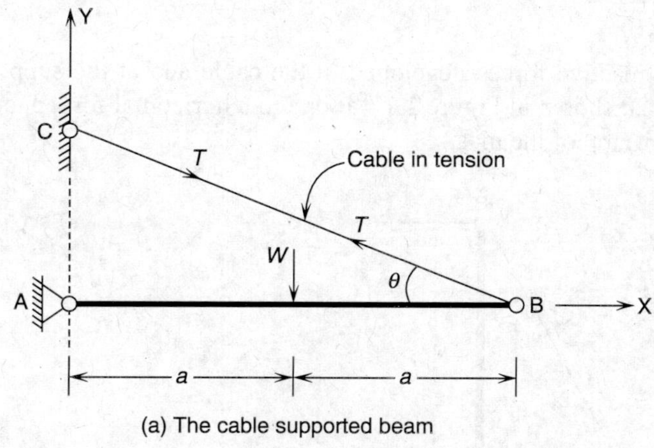

(a) The cable supported beam

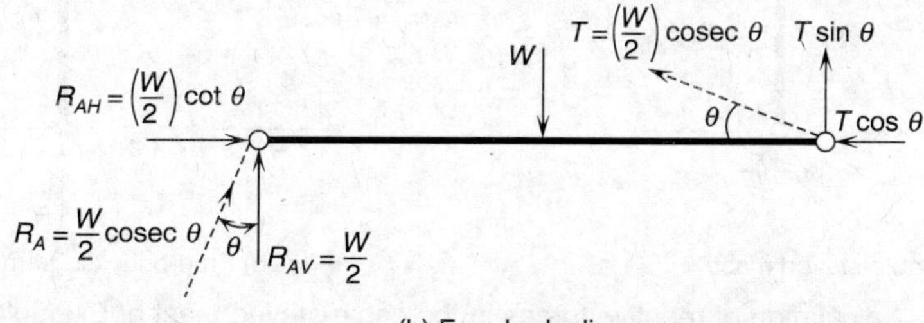

(b) Free-body diagram

Figure 1.16 Analysis of cable supported beam of Example 1.9.

Equilibrium in vertical direction,

$$\uparrow + \sum F_y = R_{AV} - W + T\,(\sin\theta) = 0 \quad \text{i.e.} \quad R_{AV} = \frac{W}{2}$$

Equilibrium in horizontal direction,

$$\rightarrow + \sum F_x = R_{AH} - T\,(\cos\theta) = 0 \quad \text{i.e.} \quad R_{AH} = \left(\frac{W}{2}\right)\cot\theta$$

The resultant reaction R_A at support A:

$$R_A = \sqrt{R_{AH}^2 + R_{AV}^2} = \left(\frac{W}{2}\right)\sqrt{\cot^2\theta + (1)^2} = \left(\frac{W}{2}\right)\text{cosec}\,\theta$$

$$\alpha = \tan^{-1}\left(\frac{R_{Ay}}{R_{Ax}}\right) = \tan^{-1}\left(\frac{(W/2)}{(W/2)\cot\theta}\right) = \theta$$

It should be noted that the cable induces a compressive force in the beam which is equal to the horizontal component of the cable force, and the reactive force at B is equal and opposite of the force in the cable.

EXAMPLE 1.10

Determine the reactive forces developed in the cable and at the support points of cable-stayed mast of height $2a$ shown in Figure 1.17(a) due to a horizontal force F acting at: (i) mid-height of the mast, and (ii) top of the mast.

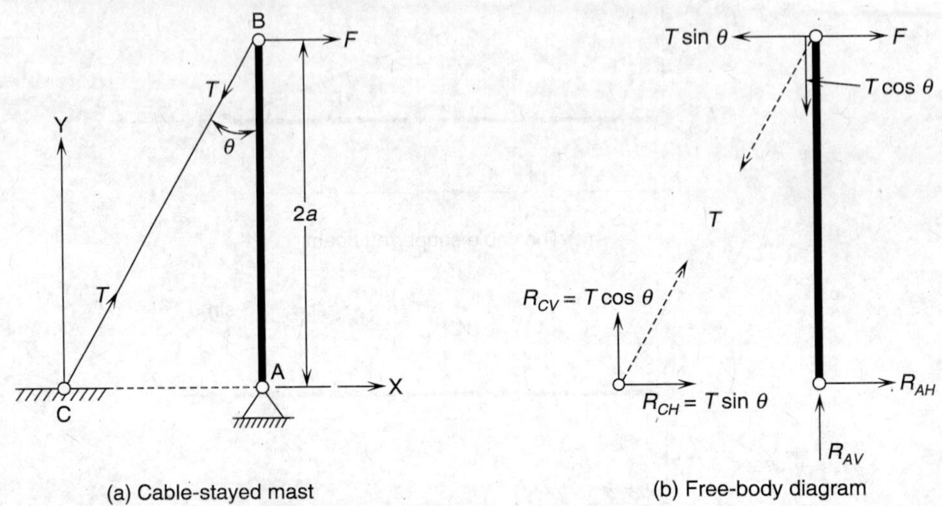

Figure 1.17 Calculation of reactive forces in the cable-stayed mast of Example 1.10.

Solution: Consider inclination of the cable to be θ with the vertical.

Case-I: This case is exactly the same if the beam of Example 1.9 is rotated through 90°. Thus,

$$T = \left(\frac{F}{2}\right) \text{cosec } \theta, \; R_{AV} = \frac{F}{2} \; \text{ and } \; R_{AH} = \left(\frac{F}{2}\right) \cot \theta$$

Therefore,

$$R_A = \left(\frac{F}{2}\right) \text{cosec } \theta \text{ inclined at an angle } \alpha = 0$$

Case-II: In this the horizontal force F acts at the top of the mast. Following the procedure used in Example 1.9, consider equilibrium conditions of the mast. The moment equilibrium of the mast about an axis through the point A (see Figure 1.17(b)):

$$+\sum \bar{M}_A = -F(2a) + T\,(\sin\theta)(2a) = 0 \quad \text{or} \quad T = F \text{ cosec } \theta$$

Equilibrium in vertical direction,

$$\uparrow + \sum F_y = R_{AV} - T\,(\cos\theta) = 0 \quad \text{i.e.} \quad R_{AV} = F \cot \theta$$

Equilibrium in horizontal direction,

$$\rightarrow + \sum F_x = R_{AH} + F - T\,(\sin\theta) = 0 \quad \text{i.e.} \quad R_{AH} = 0$$

The resultant reaction R_A at support A:

$$R_A = \sqrt{R_{AH}^2 + R_{AV}^2} = F \cot\theta$$

$$\alpha = \tan^{-1}\left(\frac{R_{AV}}{R_{AH}}\right) = \tan^{-1}\infty = 90°$$

The reaction at B is numerically equal to the force T developed in the cable.

EXAMPLE 1.11

Determine the reactive forces in the cables and at the support points of cable-stayed mast of height $4a$ and the projecting member or beam of length $2a$ as shown in Figure 1.18(a).

Solution: For the whole body, the support points are at A and C.

For overall equilibrium of the body (see Figure 1.18(b))

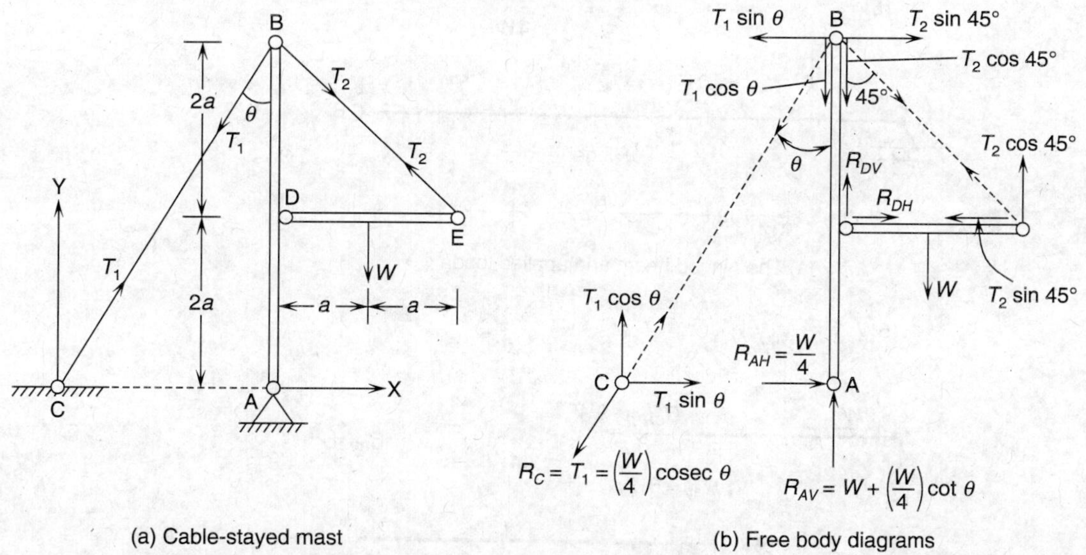

Figure 1.18 Analysis of cable-stayed mast with projecting member of Example 1.11.

$$(+)\sum \bar{M}_A = -W(a) - T_1(\cos\theta)(4a\tan\theta) = 0 \quad \text{or} \quad T_1 = -\left(\frac{W}{4}\right)\text{cosec}\,\theta$$

$$\rightarrow (+)\sum F_x = R_{AH} + T_1(\sin\theta) = 0 \quad \text{i.e.} \quad R_{AH} = \frac{W}{4}$$

$$\uparrow (+)\sum F_y = R_{AV} + T_1(\cos\theta) - W = 0 \quad \text{i.e.} \quad R_{AV} = W + \left(\frac{W}{4}\right)\cot\theta$$

For equilibrium of the beam:

$$(+)\sum \bar{M}_D = -W(a) + T_2\,(\cos 45^\circ)\,(2a) = 0 \quad \text{or} \quad T_2 = \frac{W}{\sqrt{2}}$$

$$\rightarrow(+)\sum F_x = R_{DH} - T_2\,(\sin 45^\circ) = 0 \quad \text{i.e.} \quad R_{DH} = \frac{W}{2}$$

$$\uparrow(+)\sum F_y = R_{DV} + T_2\,(\cos 45^\circ) - W = 0 \quad \text{i.e.} \quad R_{DV} = \frac{W}{2}$$

Therefore, $R_D = \sqrt{2}W$ inclined at an angle 45°

EXAMPLE 1.12

Determine the reactive forces developed in the hinged beam illustrated in Figure 1.19(a).

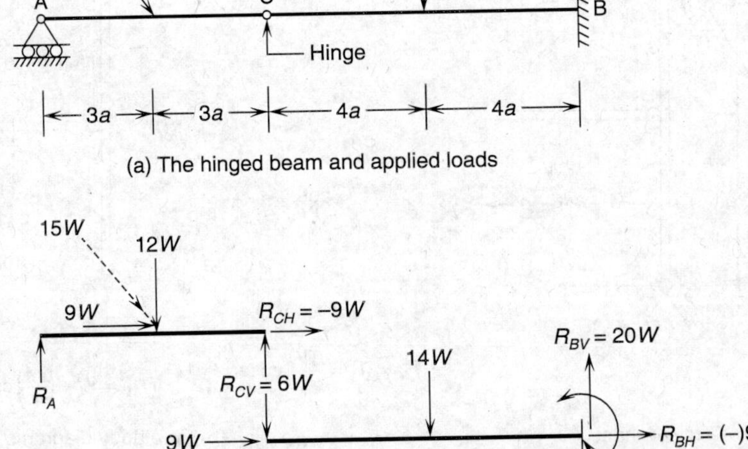

Figure 1.19 Analysis of a beam with a hinge of Example 1.12.

Solution: For convenience, the inclined applied load $15W$ is resolved into $15W \cos\theta\,(=9W)$ and $15W \sin\theta\,(=12W)$, the horizontal and vertical components, respectively. The free-body diagram shown in Figure 1.19(b) represents the interaction of two separate elements of the hinged cantilevered beam. Because of boundary conditions that exist at sections A and C, the element AC can be isolated and treated as simply supported beam resting at the ends A of the body and C of the cantilever CB, respectively. Consider the equilibrium of the segment AC and compute the reactions as follows,

$$\sum \bar{M}_C = 0 = (12W)(3a) - R_{AV}(6a) \quad \text{or} \quad R_{AV} = R_A = 6W\,(\uparrow)$$

$$\uparrow(+)\sum F_y = 0 = R_A + R_{CV} - 12W \quad \text{or} \quad R_{CV} = 12W - 6W = 6W\,(\uparrow)$$

$$\rightarrow(+)\sum F_x = 0 = 9W + R_{CH} \quad \text{or} \quad R_{CH} = -\,9W\,(\leftarrow)$$

The minus sign indicates that the correct direction is opposite to that presumed in the calculations.

Now consider the equilibrium of the cantilever segment CB, the equilibrium equations are:

$$\rightarrow(+)\sum F_x = 9W + R_{BH} = 0 \quad \text{or} \quad R_{BH} = -\,9W$$

The minus sign indicates that the correct direction of the reaction R_{BH} is opposite to that presumed in the calculations.

$$\uparrow(+)\sum F_y = -\,6W - 14W + R_{BV} = 0 \quad \text{or} \quad R_{BV} = 20W\,(\uparrow)$$

$$(+)\sum \bar{M}_B = (6W)(8a) + (14W)(4a) + M_B = 0 \quad \text{or} \quad M_B = -104W_0 a$$

The minus sign (–), with the magnitude of reaction M_B, indicates that the actual direction of the moment is opposite to that presumed in the calculations, i.e. it acts clockwise.

1.9.5 Multi-span Statically Determinate Beams or Cantilevered Beams

The *multi-span statically determinate beam* is a geometrically stable structure consisting of a series of simply supported beams with or without overhangs connected together by means of hinged joints as shown in Figure 1.20. Such beams are also called *multi-span hinged* or *cantilevered beams*. These beams are usually more economical due to substantial reduction in bending moments than a series of independent simply supported beams spanning the same opening. In addition to the simplicity of computation of the determinate structures, these beams have the advantage over conventional continuous beams as they are not influenced by any *settlement of the supports*. These beams are widely used in engineering structures.

A *statically indeterminate continuous beam* can be transformed into a determinate multi-span beam by introducing a number of hinges. Each hinge introduced either at the span or over a support of a continuous beam provides for one additional equilibrium equation. This equation expresses the condition that the sum of moments of all the external forces applied to the beam either to the left or to the right of the hinge about its centre is zero. Thus, the number of hinges introduced must be equal to the *degree of redundancy* of the continuous beam. As in other statically determinate structures, in multi-span cantilevered beam, the unknowns can be obtained with the application of equations of statics alone. However, to ensure stability of the structure, the hinges must be distributed along the beam in such a way that each part of the structure should become statically determinate and remain geometrically stable. Several ways of transforming the continuous beam of Figure 1.20(a) into a statically determinate one are illustrated in Figure 1.20(b). Diagrams schematically representing the interaction of its separate parts are beneficial in investigating the stability of the transformed beam.

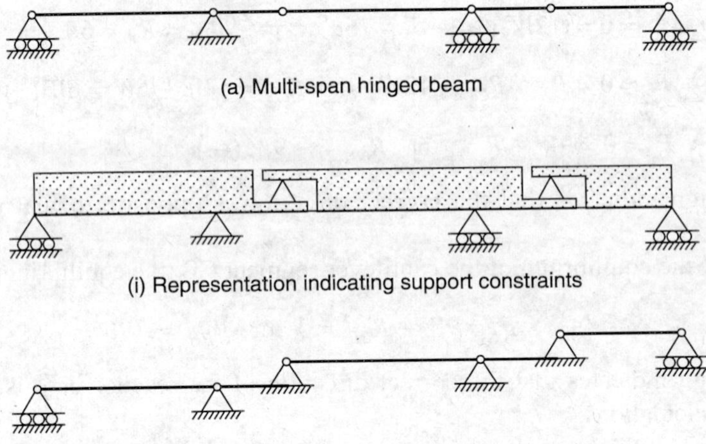

(a) Multi-span hinged beam

(i) Representation indicating support constraints

(ii) Representation indicating reactions at the hinges

(b) Representation of multi-span hinged beams

Figure 1.20 Schematic representation of interaction of separate parts of multispan beam with a hinge in each span for checking its geometric stability.

The most commonly used multi-span hinged beam is presented in Figure 1.21(a). The interaction of its elements is represented in Figure 1.21(b). The beam of Figure 1.21(a) is characterized by alternating double-hinged spans and the spans without any hinges. Thus it consists of a series of beams with two overhangs supporting suspended simple beams. The following rules regarding distribution of hinges in the beams which have no built-in ends may be used as a guide:

1. There may not be more than two hinges in a span;
2. There must be no hinges in the spans adjacent to the one provided with two hinges and
3. Spans containing one hinge only may follow each other with the exception of one of the extreme spans where there should be no hinge at all.

The following example will illustrate the procedure for analysis of such beams.

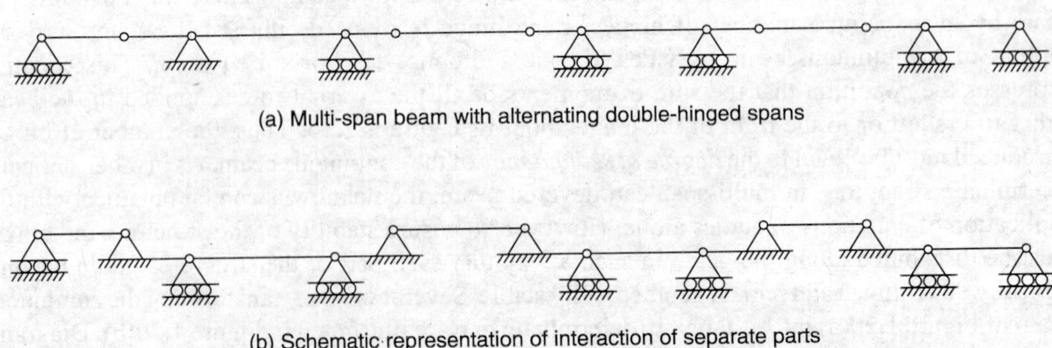

(a) Multi-span beam with alternating double-hinged spans

(b) Schematic representation of interaction of separate parts

Figure 1.21 Schematic representation of beams with two overhangs supporting suspended simple beams (two hinged portion).

EXAMPLE 1.13

Determine all the reactive forces developed in the two-span hinged or cantilevered beams shown in Figure 1.22(a).

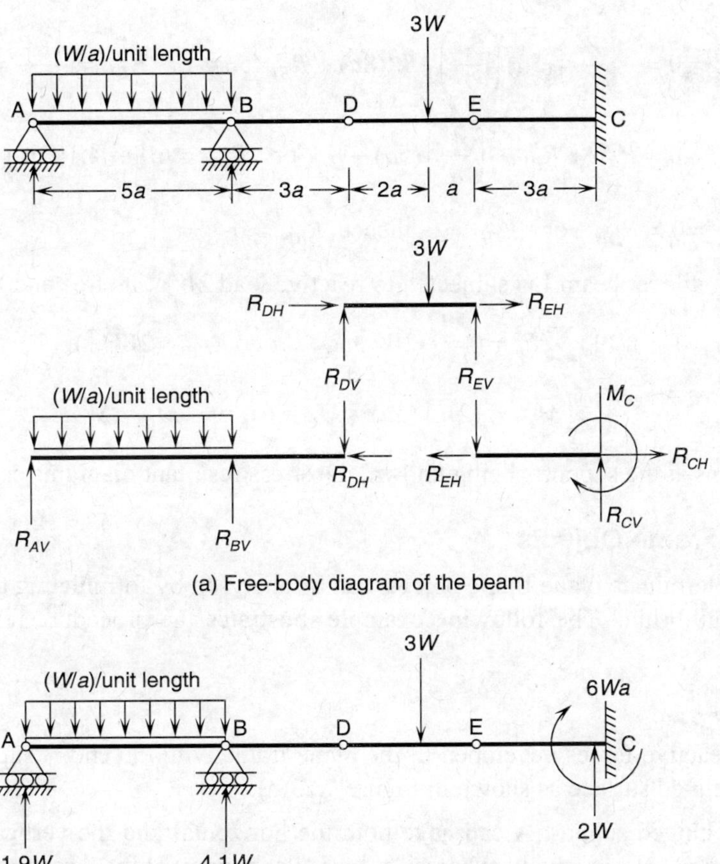

Figure 1.22 Computation of reactive forces in the cantilever beam of Example 1.13.

Solution: The free-body diagram shown in Figure 1.22(b) is a schematic sketch representing the interaction of separate elements of two-span cantilevered beam. Because of symmetry conditions that exist at sections D and E the element DE can be isolated and treated as simply supported beam resting at the ends D and E of overhanged beam ABD and the cantilever EC, respectively. The reactions at the hinges can be computed as,

$$(+)\sum \bar{M}_D = 0 = -3W(2a) + R_{EV}(3a) \quad \text{or} \quad R_{EV} = 2W\,(\uparrow)$$

$$\uparrow(+)\sum F_y = 0 = R_{DV} + R_{EV} - 3W \quad \text{or} \quad R_{DV} = 3W - 2W = W(\uparrow)$$

$$\rightarrow(+)\sum F_x = 0 = R_{DH} + R_{EH} = 0 \quad \text{or} \quad R_{DH} = -R_{EH}\,(\rightarrow)$$

Once reactive forces in the central segment have been determined, consider left segment ABD consisting of simply supported beam with an overhang. This beam is subjected to the applied loads and reaction as indicated in Figure 1.22(b). Consider the moment equilibrium about an axis through section A.

$$(+)\sum \bar{M}_A = 0 = -\left(\frac{W}{a}\right)(5a)\left(\frac{5a}{2}\right) - W(8a) + R_{BV}(5a) \quad \text{or} \quad R_{BV} = \frac{41W}{10} = 4.1W\,(\uparrow)$$

$$\uparrow(+)\sum F_y = 0 = R_{AV} + R_{BV} - \left(\frac{W}{a}\right)(5a) - W \quad \text{or} \quad R_{AV} = 6W - 4.1W = 1.9W\,(\uparrow)$$

$$\rightarrow(+)\sum F_x = 0 = R_{DH} \quad \text{or} \quad R_{DH} = 0, \text{ hence } R_{BH} = 0$$

Now consider cantilever beam EC subjected to reactive load $2W$ at its free end E,

$$\uparrow(+)\sum F_y = 0 = -2W + R_{CV} \quad \text{or} \quad R_{CV} = 2W\,(\uparrow)$$

$$\sum \bar{M}_C = (2W)(3a) - M_C = 0 \quad \text{or} \quad M_C = 6Wa$$

With the reactions at the supports being known, the stress resultant diagrams can be constructed.

1.9.6 Plane Frame Objects

The statically determinate plane frame objects can be analysed by formulating three independent equations of equilibrium. The following example illustrates the procedure for computation of reactive forces.

EXAMPLE 1.14

Determine the reactive forces developed in the plane frame with the end B supported by rollers resting on an inclined surface as shown in Figure 1.23(a).

Solution: The hinged support A can have both the horizontal and the vertical components of reaction, R_{AH} and R_{AV}, respectively, whereas at the roller support B the direction of reaction R_B is perpendicular to the supporting surface. The reaction has a slope of three vertical to four horizontal (3:4) which is perpendicular to the slope of supporting surface of four to three (4:3) as shown in Figure 1.23(b). For convenience, the reaction R_B can be resolved into $R_B \sin\theta\,(= 0.8R_B)$ and $R_B \cos\theta\,(= 0.6R_B)$, the horizontal and vertical components, respectively.

Reactions: Consider $\sum \bar{M}_A = 0$ to compute the reaction at B:

$$\sum \bar{M}_A = -W(a) - (4W)(a) - (6W)(2a) + (0.6R_B)(3a) + (0.8R_B)(2a) = 0 \quad \text{or} \quad R_B = 5W$$

As the computed magnitude of R_B is positive, the assumed direction is correct.

For computation of vertical component of reaction at A consider the condition $\sum F_y = 0$:

$$R_{AV} - 4W - 6W + 3W\,(= 0.6R_B) = 0 \quad \text{or} \quad R_{AV} = 7W$$

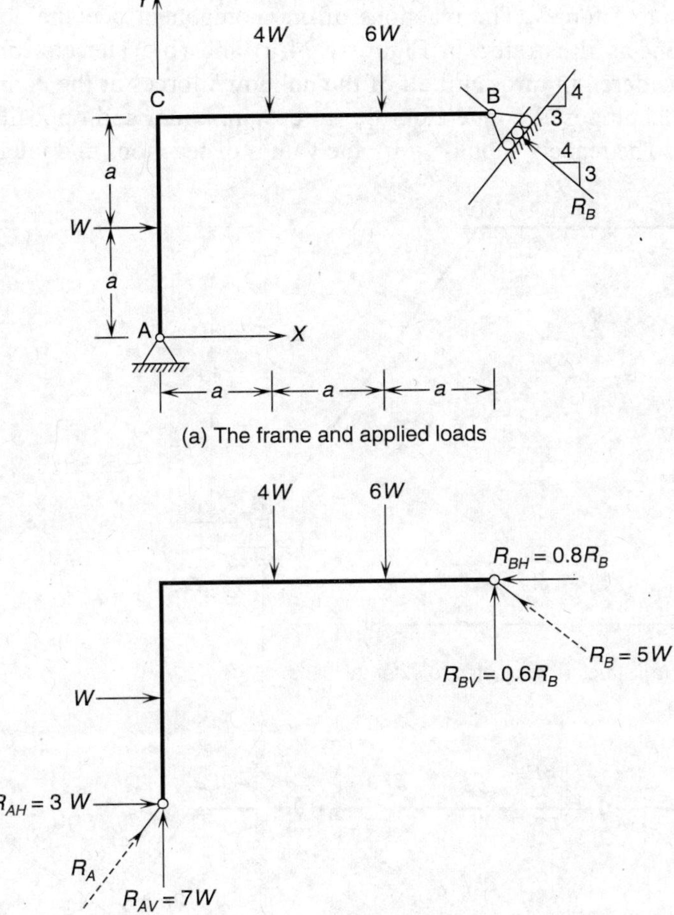

Figure 1.23 Analysis of rigid frame with an end supported by rollers at an inclined surface of Example 1.14.

Due to positive sign of magnitude of R_{AV} the assumed direction is correct. Consider $\sum F_x = 0$ to compute the horizontal component of reaction at A,

$$R_{AH} + W - 4W\,(= 0.8R_B) = 0 \quad \text{or} \quad R_{AH} = 3W$$

1.10 ANALYSIS OF STRUCTURES

In the preceding sections, the computation of reactions of overall object has been discussed. This is the first step towards analysis of the object. Once the reactions are known, the body is decomposed into its basic components (free bodies) for its analysis. Each component is subjected to external and internal or reactive forces. The reactions of one component become applied forces to the adjacent component as illustrated in Figure 1.24(a) and (b). The equilibrium of different components is considered in turn until all of the unknown forces at the connections have been determined. At a decomposed connection, the forces are equal and opposite between the two adjacent members. The readers should verify the values of reactions and internal forces.

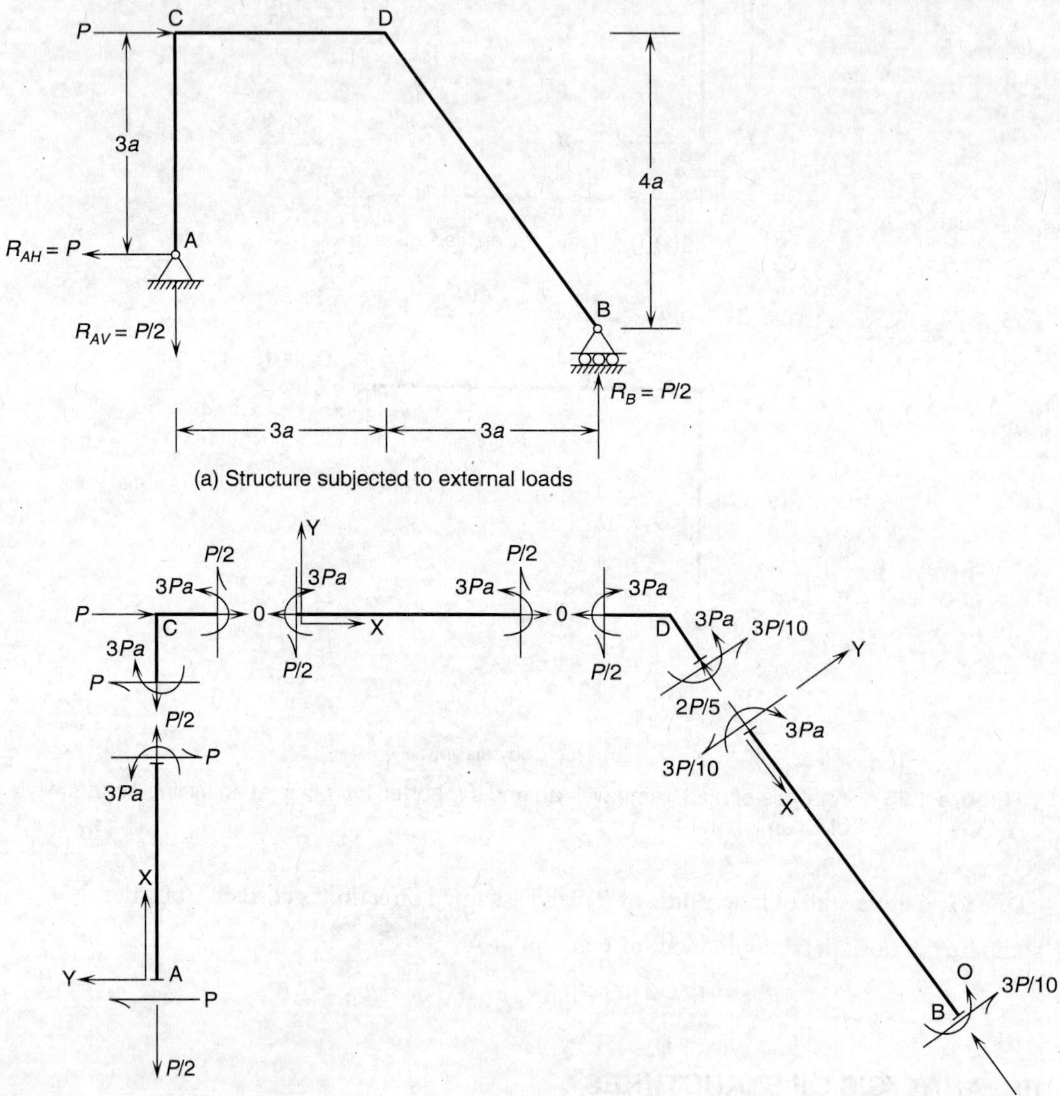

Figure 1.24 Analysis of a structure.

The analysis of axially loaded objects like the cable and the plane trusses is illustrated in Chapter 2.

1.11 PROBLEMS

1.1 Determine the unknown support reactions R_A and R_B in the structure with an overhang, due to applied loads as shown in Figure 1.25.

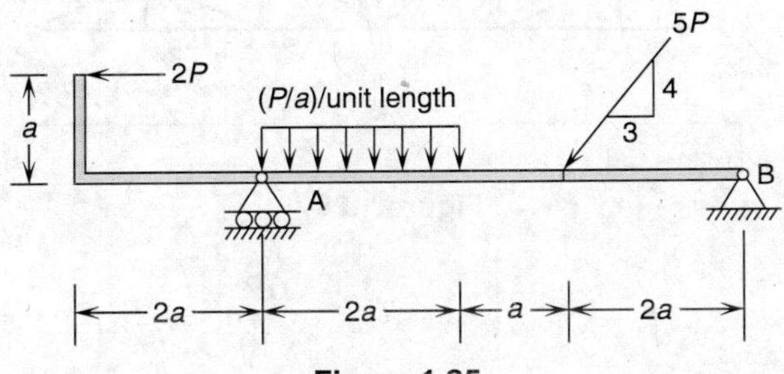

Figure 1.25

[**Ans.** $R_A = 3.6P(\uparrow)$, $R_{BH} = 5P(\rightarrow)$ and $R_{BV} = 2.4P(\uparrow)$]

1.2 Determine the fixed-end reactions of a cantilever beam, due to applied loads as shown in Figure 1.26.

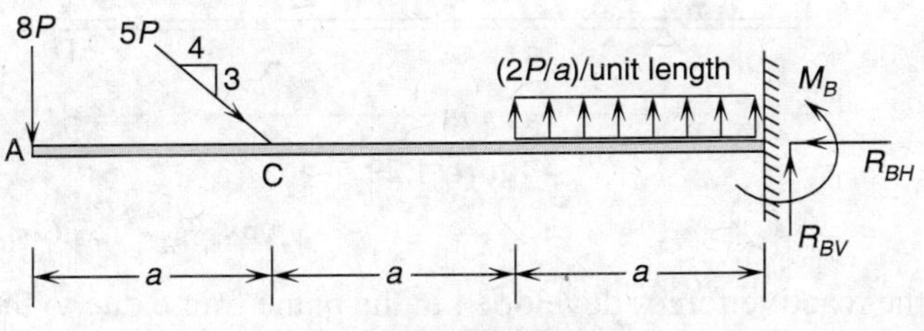

Figure 1.26

[**Ans.** $R_{BV} = 9P$, $R_{BH} = 4P$ and $M_B = -29Pa$]

1.3 Determine the reactive forces developed in the hinged beam illustrated in Figure 1.27.

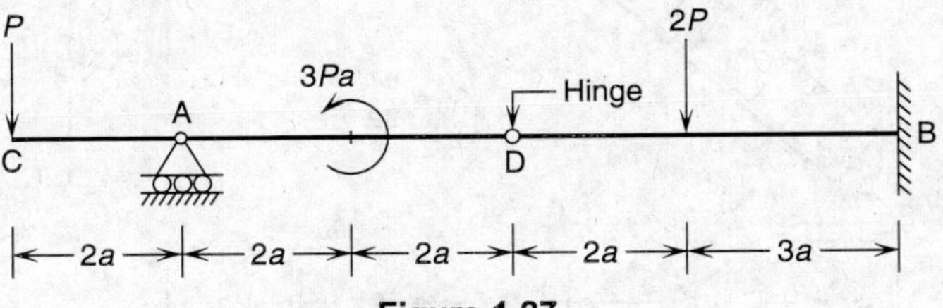

Figure 1.27

$$\left[\textbf{Ans. } R_A = \frac{9Pa}{4},\ R_{BV} = \frac{3Pa}{4},\ M_B = \frac{Pa}{4}(\curvearrowleft)\right]$$

1.4 Determine the support reactions R_A and R_B in the plane truss due to applied loads as shown in Figure 1.28.

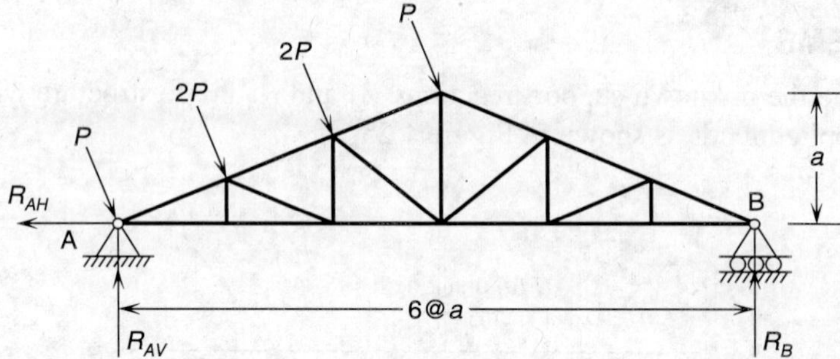

Figure 1.28

$$\left[\textbf{Ans. } R_{AH} = \frac{6P}{\sqrt{10}} (\leftarrow),\ R_{AV} = \frac{13P}{\sqrt{10}} \uparrow \text{ and } R_B = \frac{\sqrt{10}P}{2}\right]$$

1.5 Determine the unknown support reactions R_A and R_B in the structure with overhangs at both the ends due to distributed loads as shown in Figure 1.29.

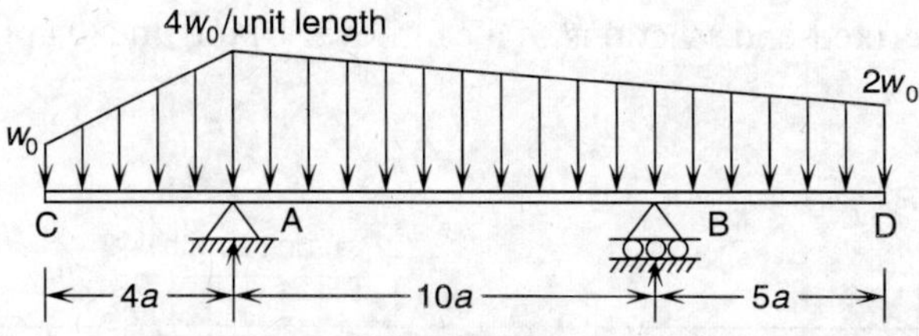

Figure 1.29

[**Ans.** $R_A = 26.6w_0a$ and $R_B = 28.4w_0a$]

1.6 Determine the reactive forces developed in the plane frame due to the load system shown in Figure 1.30.

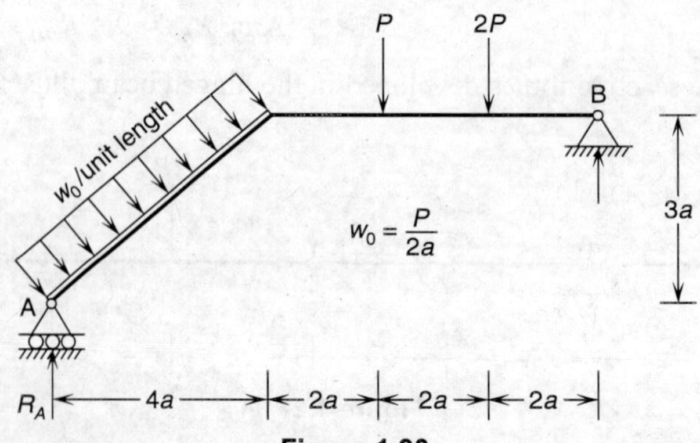

Figure 1.30

$$\left[\textbf{Ans. } R_A = \frac{21P}{8},\ R_{BV} = \frac{19P}{8} \text{ and } R_{BH} = -\frac{3P}{2}\right]$$

1.7 Determine the unknown reactions R_A and R_B developed in the loaded beam with the end B supported by rollers resting on an inclined surface as shown in Figure 1.31.

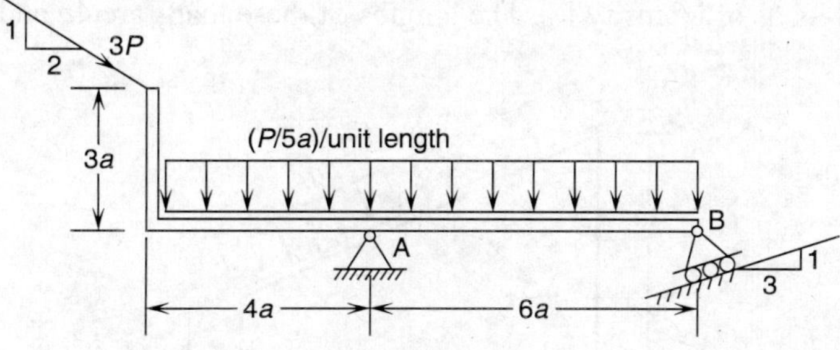

Figure 1.31

$$\left[\textbf{Ans. } R_{AV} = \left(\frac{5\sqrt{5}+6}{3\sqrt{5}}\right)P;\ R_{AH} = \frac{51-\sqrt{5}}{9\sqrt{5}};\quad R_{BV} = \left(\frac{\sqrt{5}+3}{3\sqrt{5}}\right)P \text{ and } R_{BH} = \left(\frac{\sqrt{5}+3}{9\sqrt{5}}\right)P\right]$$

1.8 Determine the reactive forces developed in the plane frame with the end B supported by rollers resting on an inclined surface due to the distributed load shown in Figure 1.32.

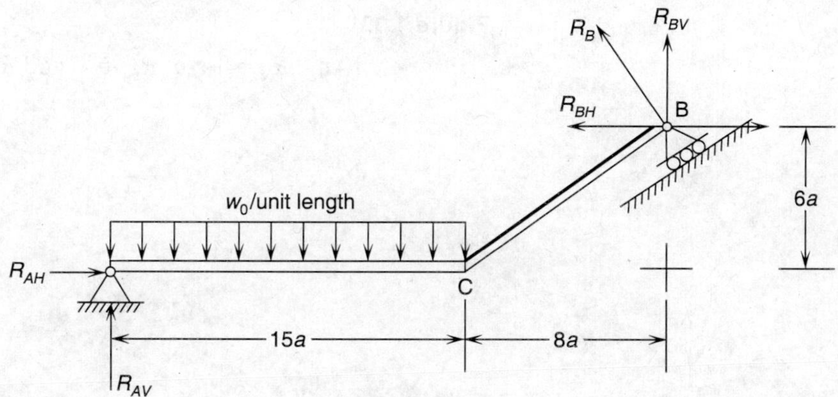

Figure 1.32

$$\left[\textbf{Ans. } R_{AV} = \frac{120w_0a}{11},\ R_{AH} = \frac{135w_0a}{44},\quad R_{BV} = \frac{45w_0a}{11} \text{ and } R_{BH} = -\frac{135w_0a}{44}\right]$$

1.9 Determine the reactive forces developed in the plane frame due to distributed load shown in Figure 1.33.

[**Hint:** Resolve inclined distributed load into horizontally acting and vertically acting distributed loads of intensity w_0. The lengths of these loads are $4a$ and $2a$, respectively.]

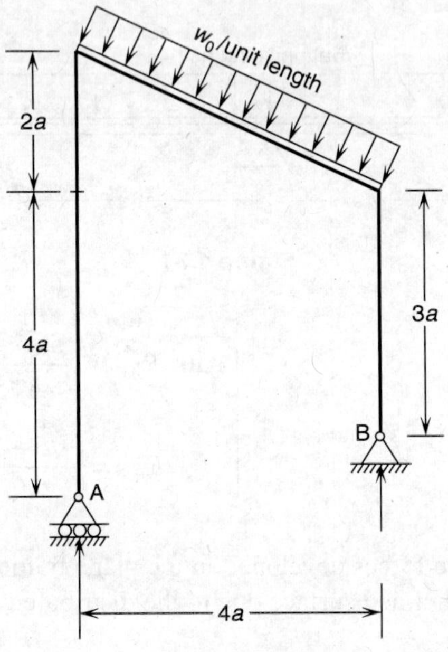

Figure 1.33

[**Ans.** $R_A = 4w_0a$, $R_{BV} = 0$ and $R_{BH} = 2w_0a$]

Chapter 2

Analysis for Axial Forces: The Cable and the Plane Truss

2.1 INTRODUCTION

In the most commonly encountered situations, it is imperative to determine the maximum normal stress. The determination of this stress requires the maximum axial load carried by the member. In practice, there is a class of objects where the axial forces are dominant; the typical examples are the *cables* and the pin-jointed frames or *trusses*. This chapter is exclusively devoted to illustrate general concepts for computation of the axial loads by the *methods of joints* and the *method of sections* for the cables and trusses.

2.2 THE CABLE

Cables provide an efficient means for supporting loads from bridges and roofs and are commonly used for cable car systems at sky slopes around the world, guys for derricks, radio towers, etc. The steel cables manufactured from high-strength steel wire may provide the lowest cost-to-strength ratio of any common structural members. They are easily handled and positioned, even for long spans.

Since a freely deforming cable subjected to the loads takes the shape of *funicular curve* (a term meaning *rope* in Latin), a profile assumed naturally, it develops only tensile forces. The absence of significant undesirable bending in the structure, results in an efficient use of the material. Inverting this structural form yields a new structure called *arch* which is analogous to the cable structure except that the forces developed are in a state of compression rather than pure tension. These two structures are collectively referred to as a group of *funicular structures*.

A cable of constant cross-section supporting only its own dead weight deforms into a *catenary*, while a cable carrying the load that is uniformly distributed along the horizontal projection of the cable, as in the case of *suspension bridge*, deforms into a parabola. Cables carrying concentrated point loads (ignoring the dead weight of the cable itself) will deform into a series of straight-line segments.

The magnitude of the forces developed in a cable or an arch depends on the relative height or depth of funicular shape in relation to its length as well as the magnitude and location of applied loads. The greater the sag of a cable or the rise of an arch, the smaller are the internal forces developed in the object, and vice versa.

2.3 THE ANALYSIS OF A CABLE

2.3.1 Cable with the Supports at the Same Level

For analysis, a cable is generally conceived as a continuous series of discrete elements joined together by pinned connections, i.e. it is analogous to a chain. As discussed earlier, the reactive force at each end has components one each in the vertical and horizontal directions. Thus, a total of four unknown components are present; but only three independent equations of statics are available for their determination. The fourth condition is provided by the fact that the cable does not carry any bending moment, i.e. the sum of all rotational effects produced by external and internal forces at a location on the cable must be zero. *In a cable, the external shear at a section is balanced by an internal resisting shear that is provided by the vertical components of the internal axial force (tension) in the cable*, thus the cable must have a vertical projection in order to support the loads. The greater the vertical projection, the smaller will be the cable tension. The overall external bending moment at the same section is balanced by an internal resisting moment that is provided by the couple formed between the horizontal component of the tension in the cable and the horizontal component of support reaction.

The resultant tension at any point is obtained from:

$$F_T = \sqrt{F_H^2 + F_V^2} \tag{2.1}$$

where F_H and F_V are, respectively, the horizontal and vertical components of tensile force in the cable at that point. It is evident that the tension varies along the length of the cable. In case of vertical loads, only the value of F_H will be constant throughout the cable and the maximum tension will occur at the point where the vertical force is the maximum.

The following example illustrates the determination of cable geometries, end reactions, and internal forces.

EXAMPLE 2.1

Determine the deflected shape, reactions at the supports, internal forces or tensions in the various segments and the total length of the cable structure carrying two concentrated loads at one-third span points as shown in Figure 2.1(a). The maximum depth under the first load is given.

Solution: Support reactions R_A and R_B:

Vertical reactions: Consider the moment equilibrium of the whole structure about A or B

$$+\sum \bar{M}_A = -2W \times 3a - W \times 6a + R_{BV} \times 9a = 0 \text{ i.e. } R_{BV} = (4/3)W \uparrow$$

$$+\sum \bar{M}_B = -R_{AV} \times 9a + 2W \times 6a + W \times 3a = 0 \text{ i.e. } R_{AV} = (5/3)W \uparrow$$

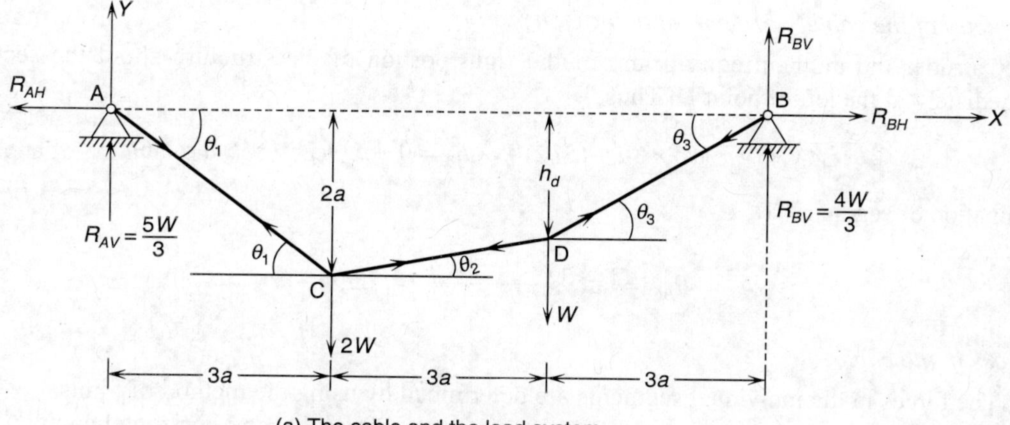

(a) The cable and the load system

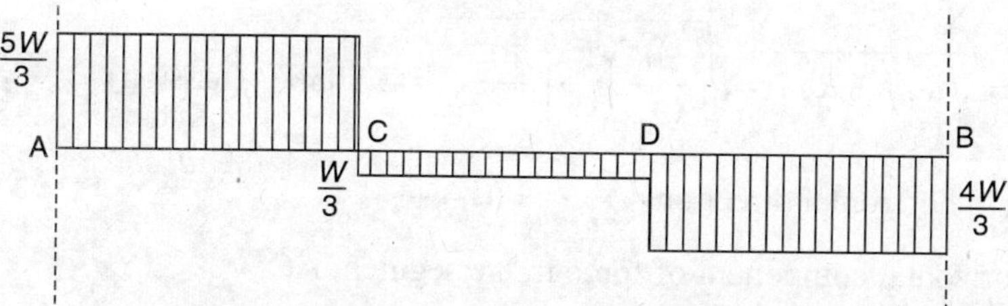

(b) Shear force diagram as a simply supported beam

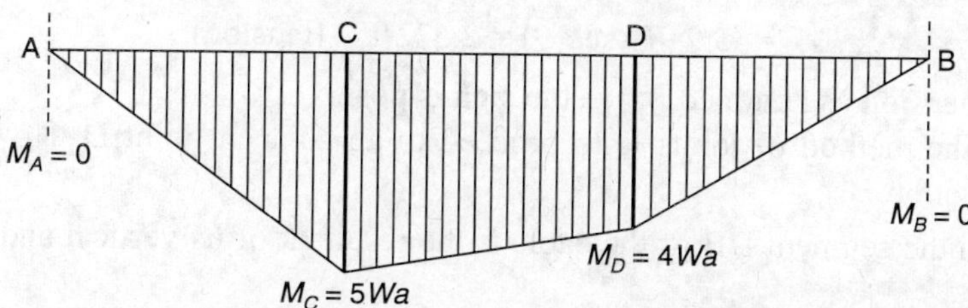

(c) Bending moment diagram as a simply supported beam

Figure 2.1 Analysis of cable supporting concentrated loads of Example 2.1.

Horizontal reactions: Consider the moment equilibrium of the left portion of the structure about the section immediately to the right of the location of the maximum depth, i.e. point C of known depth.

$$+\sum \bar{M}_C = -(5/3)W \times 3a + R_{AH} \times 2a = 0 \text{ i.e. } R_{AH} = (5/2)W \leftarrow$$

Consider the translational equilibrium of the whole structure

$$\rightarrow + \sum F_x = -R_{AH} + R_{BH} = 0 \text{ i.e. } R_{BH} = R_{AH} = (5/2)W \rightarrow$$

Geometry of the cable, i.e. depth at point D, h_d:

Consider the moment equilibrium of the right portion of the structure about the section immediately to the left of point D. Thus,

$$+\sum \bar{M}_D = (4/3)W \times 3a - (5/2)W \times h_d = 0 \text{ i.e. } h_d = (8/5)a = 1.6a$$

Inclination of segment CD,

$$\theta_{BC} = \tan^{-1}\left(\frac{2a - 1.6a}{3a}\right) = 7.59°$$

Forces in the cable:

The forces in the individual segments are determined by using the method of joints.

Joint A: The force in cable segment AC is the resultant of vertical and horizontal components. Thus,

$$F_{AC} = \sqrt{R_{AH}^2 + R_{AV}^2} = \sqrt{\left(\frac{5W}{2}\right)^2 + \left(\frac{5W}{3}\right)^2} = 3.005W \quad \text{(tension)}$$

Joint C: Equilibrium in horizontal direction, $\sum F_x = 0 \rightarrow +$:

– Horizontal component of force in the segment AC

+ Horizontal component of force in the segment CD = 0

$$-(5/2)W + F_{CD}(\cos\theta) = 0$$

or

$$F_{CD} = (5/2)W(\sec\theta) = 2.522W \quad \text{(tension)}$$

The member CD is in tension as assumed, since the sign is positive.

The next step in the method of joints is to proceed to an adjacent joint D. However, it is convenient to consider joint B.

Joint B: The force in the segment DB of the cable is the resultant of its vertical and horizontal components. Thus,

$$F_{DB} = \sqrt{R_{BH}^2 + R_{BV}^2} = \sqrt{\left(\frac{5W}{2}\right)^2 + \left(\frac{4W}{3}\right)^2} = 2.833W \quad \text{(tension)}$$

It should be noted that if the loads are vertical, the horizontal component of the force in the cable is always equal to the horizontal reaction because of translatory equilibrium in the horizontal direction.

Alternatively, the analysis can be carried out on the basis of load transfer mechanism, i.e. the external shear is balanced by the vertical component of the force in the cable. The external bending moment is balanced by the couple formed between the horizontal component of the force in the cable and the horizontal component of support reaction. This method uses the shear force and the bending moment diagrams for the externally applied load system and for the method of sections to calculate the vertical and horizontal components of cable forces.

Horizontal reaction: Consider moment transfer criteria at point C:

Internal resisting couple = External bending moment

or $R_{AH} \times (2a) = 5Wa$ (from moment diagram) i.e. $R_{AH} = (5/2)W$

Depth at point D, h_d: Use moment transfer criteria at point D:

$(5/2)W \times h_d = 4Wa$ (from moment diagram), i.e. $h_c = (8/5)a$

Forces in the cable:

Vertical component of the force in the cable segment = External shear force

Segment AC: $\theta_{AC} = \tan^{-1}\left(\dfrac{2a}{3a}\right) = 33.69°$

$F_{AC}\,(\sin\theta_{AC}) = (5/3)W$ i.e. $F_{AC} = +3.005W$ (tension)

Segment CD: $\theta_{BC} = 7.59°$

$F_{CD}\,(\sin\theta_{CD}) = (1/3)W$ i.e. $F_{CD} = +2.524W$ (tension)

Segment DB: $\theta_{DB} = \tan^{-1}\left(\dfrac{1.6a}{3a}\right) = 28.07°$

$F_{DB}\,(\sin\theta_{DB}) = (4/3)W$ i.e. $F_{DB} = +2.834W$ (tension)

It should be noted that the reactions and the internal forces depend on the sag of the cable.

Total length of chord = Sum of lengths of segments AC, CD and DB:

$$S = 3a \times (\sec\theta_{AC} + \sec\theta_{CD} + \sec\theta_{DB})$$
$$= 3a \times (1.2018 + 1.0088 + 1.1333) = 10.032a$$

Shear force and Bending moment diagrams are shown in Figures 2.1(b) and (c).

EXAMPLE 2.2

Determine the deflected shape, reaction components and internal forces in the symmetrical cable of span L and dip h_{max} at the mid-span carrying uniformly distributed load w per unit run on the horizontal span as shown in Figure 2.2(a).

Solution: Support reactions R_A and R_B:

Vertical reactions: Consider moment equilibrium of the whole structure about A or B:

$$+\sum \bar{M}_A: \quad -wL \times (L/2) + R_{BV} \times L = 0 \quad \text{i.e.} \quad R_{BV} = (wL/2)\uparrow$$

$$\uparrow + \sum F_y: \quad R_{AV} + R_{BV} - wL = 0 \quad \text{i.e.} \quad R_{AV} = (wL/2)\uparrow$$

These values could also have known by inspection.

Horizontal reactions: Horizontal component of support reactions can be found by passing a section through a point where sag is known or specified. Consider moment equilibrium of left portion about mid-point *C*, a point of known sag,

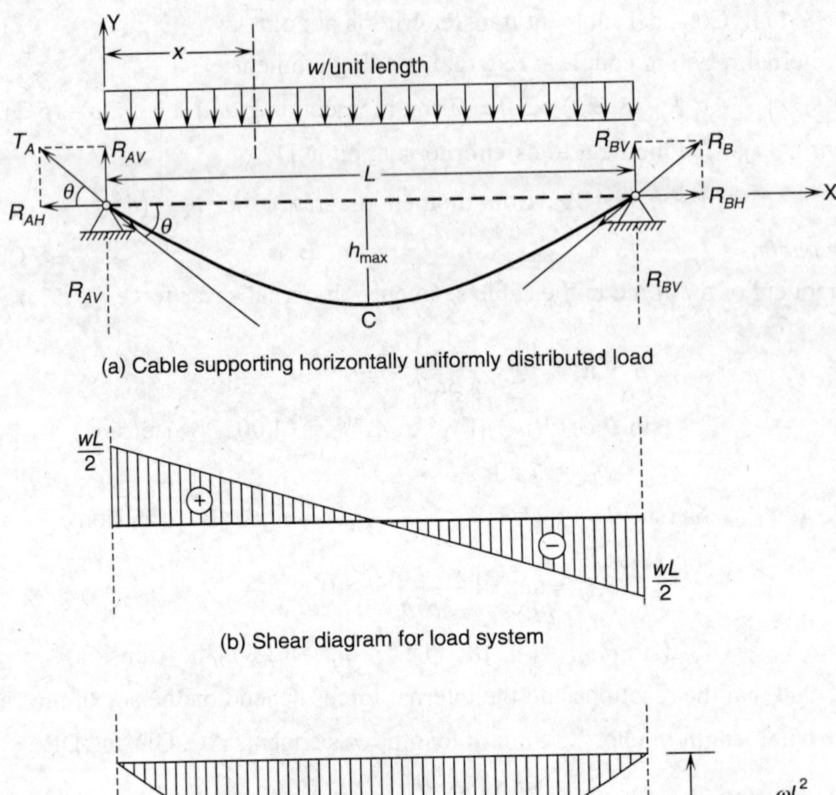

(a) Cable supporting horizontally uniformly distributed load

(b) Shear diagram for load system

(c) Bending moment diagram for the load system

Figure 2.2 The stress resultants in a cable carrying uniformly distributed loads.

$$\sum \bar{M}_C : \ -\left(\frac{wL}{2}\right)\left(\frac{L}{2}\right) + R_{AH} \times h_{\max} + \left(\frac{wL}{2}\right)\left(\frac{L}{4}\right) = 0 \quad \text{or} \quad R_{AH} = \frac{wL^2}{8h_{\max}}$$

Alternatively, from moment transfer mechanism

$$R_{AH} \times h_{\max} = \frac{wL^2}{8} = 0 \quad \text{or} \quad R_{AH} = \frac{wL^2}{8h_{\max}}$$

Geometry of the cable: Consider the equilibrium of a section of the cable at distance x from the left support,

$$\sum \bar{M}_x : \ -\left(\frac{wL}{2}\right)(x) + \left(\frac{wL^2}{8h_{\max}}\right)(y) + (wx)\left(\frac{x}{2}\right) = 0$$

or
$$y = \frac{4h_{max}(Lx - x^2)}{L^2} = \frac{4h_{max}x(L - x)}{L^2}$$

This is an equation of parabola, and the slopes at the ends are given by

$$\left(\frac{dy}{dx}\right)_{x=0} = \tan\theta_A = \frac{4h_{max}}{L} \quad \text{and} \quad \left(\frac{dy}{dx}\right)_{x=L} = \tan\theta_B = -\frac{4h_{max}}{L}$$

or alternatively,
$$\tan\theta_A = \frac{R_{AV}}{R_{AH}} = \frac{(wL/2)}{(wL^2/8h_{max})} = \frac{4h_{max}}{L}$$

With mid-point as origin, the equation of the cable will be, $y = \left(\frac{4h_{max}}{L^2}\right)x^2$

Forces in the cable: The cable force at the either end can be determined by using the method of joints or alternatively, the cable force is given by the resultant of the vertical and horizontal components of the end reactions,

$$F_A = F_{max} = \sqrt{R_{AH}^2 + R_{AV}^2} = \sqrt{\left(\frac{wL^2}{8h_{max}}\right)^2 + \left(\frac{wL}{2}\right)^2}$$

$$= \left(\frac{wL^2}{8h_{max}}\right)\sqrt{1 + \left(\frac{4h_{max}}{L}\right)^2} \quad \text{or} \quad \left(\frac{wL}{2}\right)\sqrt{1 + \left(\frac{L^2}{16h_{max}^2}\right)}$$

Maximum force in a cable

Consider vertical and horizontal equilibrium of the section to the left or the right of mid-point. The vertical component of the cable force is zero and the horizontal component is equal to the horizontal reaction. Thus the cable force at mid-span is equal to the horizontal reaction and less than that at any other section. The cable diameter design is based on maximum cable force at the end.

2.3.2 Cable with Supports at Different Heights

A more general cable which is not symmetrically loaded or have supports at different levels can be analysed in much the same way as the cases discussed above. To determine support reactions of the cable shown in Figure 2.3, consider the total moment equilibrium about point A.

$$\sum \bar{M}_{A,\text{external}} = -[W_1x_1 + W_2x_2 + \cdots] - R_{BH} \times h + R_{BV} \times L = 0$$

$$R_{BV} = \frac{\sum \bar{M}_{A,\text{external}}}{L} + \frac{R_{BH} \times h}{L} \tag{2.2}$$

where
$$\sum \bar{M}_{A,\text{external}} = W_1x_1 + W_2x_2 + \cdots \tag{2.3}$$

To determine R_{BH} consider the moment equilibrium of the right portion about a point on the cable where cable sag is known or specified.

$$\sum \bar{M}_x = -\sum \bar{M}_{x,\text{external}} - R_{BH} \times \left(h - \frac{hx}{L} + h_x\right) + R_{BV} \times (L - x) = 0$$

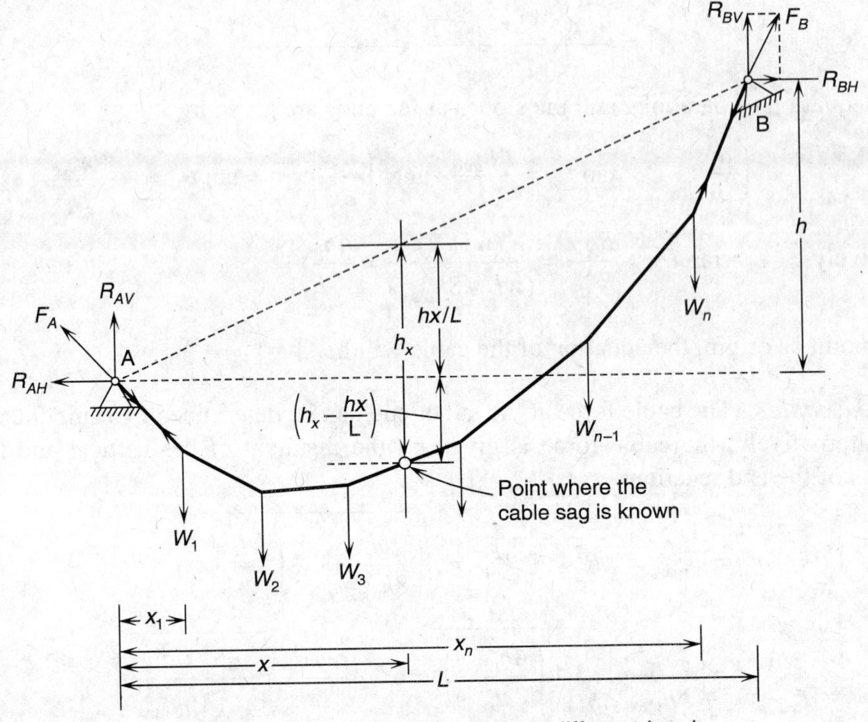

Figure 2.3 Cable with supports at different levels.

Substituting R_{BV} from Eq. (2.2) yields,

$$R_{BH} \times (h_x) = \left(\sum \bar{M}_{A,\text{external}} - \sum \bar{M}_{x,\text{external}}\right) - \sum \bar{M}_{A,\text{external}} \left(\frac{x}{L}\right) \tag{2.4}$$

where h_x is the cable sag measured from the line joining two end points of the cable. If h_x is known at a point R_{BH} can be evaluated. Once cable reactions are known, the unknown cable sags, and cable forces can be found by equilibrium as described previously.

2.3.3 Cable Lengths

The cable length can be evaluated from the expression of the deformed shape of the cable. For example, consider a uniformly loaded symmetrical cable having both supports on the same level with total length, horizontal span and maximum sag of S, L and $h_{\max}$, respectively. The equation of the cable with the lowest point C as its origin is,

$$y = \left(\frac{4h_{\max}}{L^2}\right)x^2 \quad \text{and} \quad \left(\frac{dy}{dx}\right) = \frac{8h_{\max}x}{L^2} \tag{2.5}$$

Consider elemental length ds of curved cable with projections dx and dy along X- and Y-directions. Thus,

$$ds = \sqrt{(dx)^2 + (dy)^2} \quad \text{or} \quad ds = dx\sqrt{1 + \left(\frac{dy}{dx}\right)^2}$$

$$\left(\frac{ds}{dx}\right) = \sqrt{1+\left(\frac{dy}{dx}\right)^2} = \left[1+\frac{64h_{max}^2 x^2}{L^4}\right]^{1/2}$$

$$= (1+\theta)^{1/2} = \left[(1)^{1/2} + \frac{1}{2}(1)^{-1/2}\theta + \frac{1}{2}\left(-\frac{1}{2}\right)(1)^{-3/2}\theta^2 + \cdots\right]$$

$$\approx 1 + \frac{1}{2}\left(\frac{64h_{max}^2 x^2}{L^4}\right) - \frac{1}{4}\left(\frac{64h_{max}^2 x^2}{L^4}\right)^2 + \cdots$$

Therefore,

$$ds = \left[1 + \frac{32h_{max}^2 x^2}{L^4} - \frac{1024h_{max}^4 x^4}{L^8}\right]dx \tag{2.6}$$

The terms of the expansion which are very small have been neglected.
The total length is given by,

$$S = 2\int_0^{L/2}\left[1 + \frac{32h_{max}^2}{L^4}x^2 - \frac{1024h_{max}^4}{L^8}x^4\right]dx$$

$$= 2\left[\frac{L}{2} + \frac{32h_{max}^2}{L^4}\left(\frac{1}{3}\right)\left(\frac{L^3}{8}\right) - \frac{1024h_{max}^4}{L^8}\left(\frac{1}{5}\right)\left(\frac{L^5}{32}\right)\right]$$

$$= L\left[1 + \frac{8}{3}\left(\frac{h_{max}}{L}\right)^2 - \frac{32}{5}\left(\frac{h_{max}}{L}\right)^4\right] \tag{2.7}$$

Generally, the first two terms of the expression are adequate as the third term is very small. Thus,

$$S = L\left[1 + \frac{8}{3}\left(\frac{h_{max}}{L}\right)^2\right] = L + \frac{8}{3}\left(\frac{h_{max}^2}{L}\right) \tag{2.8}$$

The sag-to-span ratio, h_{max}/L, is an important parameter which defines the geometry of cable and critically influences the forces in the cables and hence their sizes. The forces in the supporting masts or *tie-back cables* or *guy cables* (which transfer the force to the ground) and hence their sizes are controlled by this ratio. As discussed above, the horizontal component of the force in a uniformly loaded level ended cable is given by $F_H = wL^2/8h_{max}$, where h_{max} is the maximum sag of the cable. It is evident that the cable force increases with the decrease in h_{max}, and vice versa.

2.4 TEMPERATURE STRESSES IN THE CABLE

An increase or decrease in temperature causes change in the length and hence in the sag or the sag-to-span ratio h_{max}/L of the cable. As discussed above, the sag-to-span ratio is a critical parameter which controls the forces in the cables and hence their sizes. In general, the tension in the cable increases with the decrease in sag h_{max}, and vice versa. The relation between change in length dS and the corresponding change in sag dh_{max} can be obtained from Eq. (2.8) as:

$$dS = \frac{16}{3}\left(\frac{h_{max}}{L}\right)dh_{max} \quad \text{or} \quad dh_{max} = \frac{3}{16}\left(\frac{L}{h_{max}}\right)dS \tag{2.9}$$

Consider the case when temperature rises by t°C, the length of the cable increases by:

$$dS = \alpha t S = \alpha t \left[L + \frac{8}{3}\left(\frac{h_{max}^2}{L}\right)\right] = \alpha t L + \frac{8}{3}\left(\frac{h_{max}^2}{L}\right)\alpha t \tag{2.10}$$

Since the second term is very small, it can be ignored without much error, i.e. $dS \approx \alpha tL$. Therefore,

$$dh_{max} = \frac{3}{16}\left(\frac{L}{h_{max}}\right)\alpha t L = \frac{3\alpha t}{16}\left(\frac{L^2}{h_{max}}\right) \tag{2.11}$$

For a cable carrying uniformly distributed load w per unit horizontal run over the whole span, $F_H = \frac{wL^2}{8h_{max}}$. Differentiating this relation,

$$dF_H = -\left(\frac{wL^2}{8h_{max}^2}\right)dh_{max} \quad \text{or} \quad \frac{dF_H}{F_H} = -\frac{dh_{max}}{h_{max}} \tag{2.12}$$

When the sag is small, $F_{max} = F_H$ and the stress in the cable can be expressed as

$$\sigma = \frac{F_{max}}{A} = \frac{F_H}{A} \tag{2.13}$$

From Eqns. (2.11), (2.12) and (2.13), the fraction change in the stress in the cable is given by:

$$\frac{d\sigma}{\sigma} = \frac{dF_H}{F_H} = -\frac{dh_{max}}{h_{max}} = -\frac{3\alpha t}{16}\left(\frac{L^2}{h_{max}^2}\right) \tag{2.14}$$

EXAMPLE 2.3

A cable ACB of span L has its supports A and B at heights h_1 and h_2 above the lowest point C of the cable as shown in Figure 2.4. It carries uniformly distributed load of intensity w per unit run over the whole span. Determine the horizontal and vertical reactions at each support.

Solution: Location of the lowest point C:

Consider the lowest point C to be at a distance L_1 from the support A. The equation of parabola with origin at C is:

$$y = \left(\frac{4h}{L^2}\right)x^2 \quad \text{i.e.} \quad \frac{x^2}{y} = \text{Constant} \quad \text{or} \quad \frac{x}{\sqrt{y}} = \text{Constant}$$

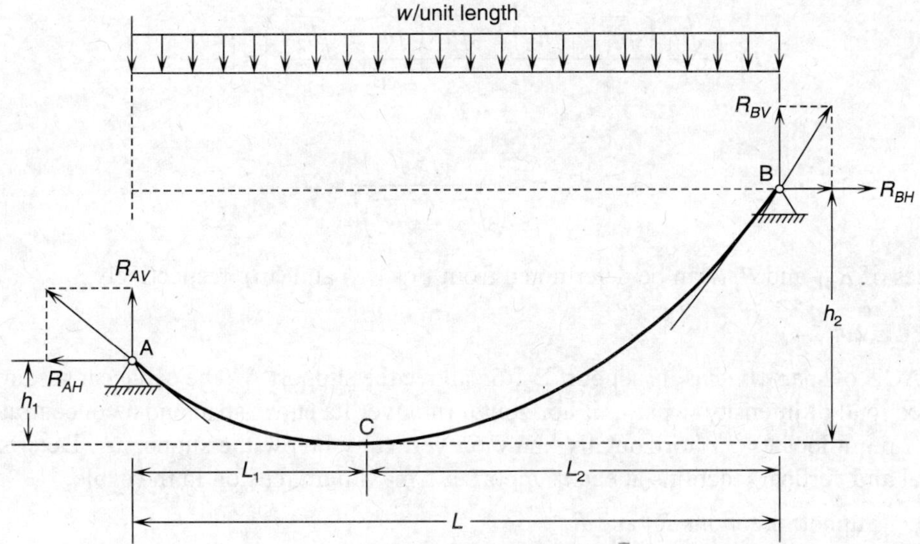

Figure 2.4 Cable with supports at different heights for Example 2.3.

At supports,
$$\frac{L_1}{\sqrt{h_1}} = \frac{(L - L_1)}{\sqrt{h_2}} = \frac{L_1 + (L - L_1)}{\sqrt{h_1} + \sqrt{h_2}} = \frac{L}{\sqrt{h_1} + \sqrt{h_2}}$$

Therefore,

$$L_1 = \frac{L\sqrt{h_1}}{\sqrt{h_1} + \sqrt{h_2}} \quad \text{and} \quad (L - L_1) = \frac{L\sqrt{h_2}}{\sqrt{h_1} + \sqrt{h_2}}$$

Support reactions R_A and R_C:

Consider the equilibrium of whole structure,

$$\uparrow + \sum F_y: \ R_{AV} + R_{BV} - wL = 0 \quad \text{or} \quad R_{AV} + R_{BV} = wL \tag{i}$$

Consider the moment equilibrium about C of the forces on the left of C,

$$R_{AV}L_1 = \frac{wL_1^2}{2} + F_H h_1 \quad \text{or} \quad R_{AV} = \frac{wL_1}{2} + F_H\left(\frac{h_1}{L_1}\right) \tag{ii}$$

Consider the moment equilibrium about C of the forces on the right of C,

$$R_{BV}(L - L_1) = \frac{w(L - L_1)^2}{2} + F_H h_2 \quad \text{or} \quad R_{BV} = \frac{w(L - L_1)}{2} + F_H\left(\frac{h_2}{L - L_1}\right) \tag{iii}$$

Substitute from Eqs. (ii) and (iii) in Eq. (i):

$$R_{AV} + R_{BV} = \left(\frac{wL}{2}\right) + F_H\left(\frac{h_1}{L_1} + \frac{h_2}{(L - L_1)}\right) = wL \quad \text{or} \quad F_H\left(\frac{h_1}{L_1} + \frac{h_2}{(L - L_1)}\right) = \left(\frac{wL}{2}\right)$$

or
$$F_H\left(\frac{\sqrt{h_1}\left(\sqrt{h_1}+\sqrt{h_2}\right)}{L}+\frac{\sqrt{h_2}\left(\sqrt{h_1}+\sqrt{h_2}\right)}{L}\right)=\left(\frac{wL}{2}\right)$$

Therefore,

$$F_H=\frac{wL^2}{2\left(\sqrt{h_1}+\sqrt{h_2}\right)^2}$$

The values of R_{AV} and R_{BV} can be determined from Eqs. (ii) and (iii), respectively.

EXAMPLE 2.4

A cable ACB of span $10a$ has its support B $1.6a$ above the support A. The cable carries uniformly distributed load of intensity w per unit horizontal run over its entire span and a concentrated load $24wa$ at a point located $3a$ horizontally and $0.6a$ vertically below the support A. Determine the horizontal and vertical reactions at each support and maximum tension in the cable.

Solution: Support reactions R_A and R_B,
Consider the equilibrium of whole structure,

$$\uparrow+\sum F_y:\ R_{AV}+R_{BV}-w\times(10a)-(24wa)=0 \quad \text{or} \quad R_{AV}+R_{BV}=34wa \tag{i}$$

Consider the moment equilibrium about C of the forces on left of C,

$$-R_{AV}\times 3a+\frac{w(3a)^2}{2}+F_H\times(0.6a)=0 \quad \text{or} \quad R_{AV}=\frac{3wa}{2}+F_H\left(\frac{1}{5}\right) \tag{ii}$$

Consider the moment equilibrium about C of the forces on the right of C,

$$R_{BV}\times(7a)-\frac{w(7a)^2}{2}-F_H(0.6a+1.6a)=0 \quad \text{or} \quad R_{BV}=\frac{7wa}{2}+F_H\left(\frac{2.2}{7}\right) \tag{iii}$$

Substitute the values from Eqs. (ii) and (iii) in Eq. (i):

$$R_{AV}+R_{BV}=\frac{10wa}{2}+\left(\frac{18}{35}\right)F_H=34wa \quad \text{or} \quad F_H=56.39wa$$

From Eq. (ii):
$$R_{AV}=\frac{3wa}{2}+F_H\left(\frac{1}{5}\right)=12.778wa$$

From Eq. (iii):
$$R_{BV}=\frac{7wa}{2}+F_H\left(\frac{2.2}{7}\right)=21.2226wa$$

Maximum tension occurs at the higher support B,

$$F_B=F_{\max}=\sqrt{F_H^2+R_{BV}^2}=\sqrt{(56.39wa)^2+(21.2226wa)^2}=60.25wa$$

EXAMPLE 2.5

A steel cable hangs between two level supports which are 100 m apart with central dip of one-tenth of its span. The cable carries uniformly distributed load of 24 kN/m over its entire horizontal span. Determine the change in tension due to a rise of temperature of 12.5°C, when coefficient of thermal expansion of steel is $\alpha = 12 \times 10^{-6}$ per °C.

Solution: The horizontal reaction, $F_H = \dfrac{wL^2}{8h_{max}} = \dfrac{24 \times 100^2}{8 \times 10} = 3000$ kN

If dF_H is the change in the horizontal force due to the rise in temperature, then from Eq. (2.14),

$$\frac{dF_H}{F_H} = -\frac{3\alpha t}{16}\left(\frac{L}{h_{max}}\right)^2 = -\frac{3 \times (12 \times 10^{-6}) \times 12.5 \times (10)^2}{16} = -2.8125 \times 10^{-3}$$

The negative sign indicates a decrease in the horizontal reaction. The change in the horizontal reaction is: $dF_H = 2.8125 \times 10^{-3} \times 3000 = 8.4375$ kN. This is also practically the change in the tension in the cable.

EXAMPLE 2.6

A flexible steel cable hangs between two level supports with central dip or sag of one-tenth of its span. Determine the maximum possible span that can be provided if the permissible stress in the cable is 150 MPa. The unit weight of the cable 78 kN/m³.

Solution: Consider the maximum span to be L metre and cross-sectional area of cable be A mm². The length of the cable is given by,

$$S = L\left[1 + \frac{8}{3}\left(\frac{h_{max}}{L}\right)^2 - \frac{32}{5}\left(\frac{h_{max}}{L}\right)^4\right] = L\left[1 + \frac{8}{3}\left(\frac{1}{10}\right)^2 - \frac{32}{5}\left(\frac{1}{10}\right)^4\right] = 1.026L \text{ m}$$

Total weight of the cable,

$$W = 78 \times (A \times 10^{-6}) \times (1.026L) = 80.028 \times 10^{-6} \times (AL) \text{ kN}$$

Vertical components of reactions, $R_{AV} = R_{BV} = (W/2)$

Horizontal component of reactions,

$$R_{AH} \times h_{max} = \frac{wL^2}{8} = 0 \quad \text{or} \quad R_{AH} = F_H = \frac{wL}{8}\left(\frac{L}{h_{max}}\right) = \frac{W}{8}\left(\frac{L}{h_{max}}\right) = \frac{5W}{4}$$

Maximum tension occurs at the supports,

$$F_A = F_{max} = \sqrt{F_H^2 + R_{AV}^2} = \sqrt{\left(\frac{5W}{4}\right)^2 + \left(\frac{W}{2}\right)^2} = 1.3463W \text{ kN}$$

Maximum stress, $f_{max} = \dfrac{F_{max}}{A} = \dfrac{1.3463W \times 10^3}{A} = \dfrac{1.3463 \times 80.028 \times 10^{-3} \times AL}{A}$

Therefore,

$$1.3463 \times 80.028 \times 10^{-3} \times L \times 1 = 150 \quad \text{or} \quad L = 1392.218 \text{ m}$$

2.5 CABLE SUPPORTS

The forces transmitted to the supporting piers depend on the type of support mechanism provided at the ends of the cable. At the ends cables are generally passed over *guide pulleys* or clamped to the rollers based *saddles* mounted on the top of piers. The cable or the saddle is anchored on the other side of the piers as shown in Figure 2.5.

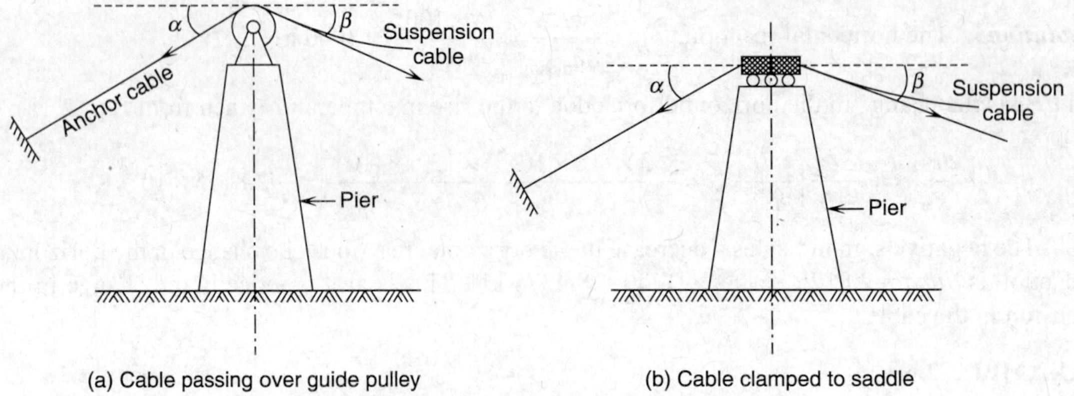

Figure 2.5 Cable support arrangement provided at the top of supporting piers.

2.5.1 Cable Passing over Guide Pulleys

In this case, the cable at the end is passed over guide pulley provided at the top of the supporting *pier* and is anchored on the other side of the pier (see Figure 2.5(a)). The *anchor cable* is treated just as the continuation of the suspension cable itself and the tension F_T in both the cables will be the same. If α and β are the inclinations of anchor and suspension cables, respectively, with the horizontal, then

Vertical load transmitted to the pier $= F_T \sin\alpha + F_T \sin\beta = F_T(\sin\alpha + \sin\beta)$ (2.15)

Net horizontal force transferred to the top of the pier $= F_T(\cos\alpha - \cos\beta)$ (2.16)

where $F_T = \sqrt{F_H^2 + R_{AV}^2}$ and $\tan\beta = \dfrac{R_{AV}}{F_H}$ (2.17)

The maximum bending moment in the pier = Net horizontal force at the top of the pier × Height of the pier

2.5.2 Cable Clamped to Saddle on Smooth Rollers

In this case, the suspension cable and anchor cable are different with tensions F_T and F_{TA}, respectively, where $F_T = \sqrt{F_H^2 + R_{AV}^2}$ and $\tan\beta = R_{AV}/F_H$ as in the previous case (see Figure 2.5(b)).

For the horizontal equilibrium of the saddle: $F_{TA}\cos\alpha = F_T\cos\beta$ (2.18)

Thus, for the known inclination of the anchor cable, the tension F_{TA} in the anchor cable can be determined. Since the horizontal components of F_{TA} and F_T balance each other, there will be no bending moment in the pier.

Total vertical load transmitted to the pier $= F_{TA}\sin\alpha + F_T\sin\beta$ (2.19)

EXAMPLE 2.7

A cable of span $15a$ and dip $1.5a$ carries a uniformly distributed load of intensity w per unit horizontal run over its entire span. Determine the maximum tension in the cable and its inclination at the support. Also calculate horizontal and vertical forces transmitted to the supporting pier of height $2a$ when: (i) the cable passes over smooth rollers provided on the top of the pier, and (ii) the cable is clamped to a saddle with smooth rollers mounted on the top of the pier. The inclination of the anchor cable is 30° to the horizontal in both the cases. What will be the bending moment developed in the pier in Case (I)?

Solution: Support reactions R_A and R_B,

Vertical reactions: Consider the suspension cable between the supports

$$R_{AV} = R_{BV} = \frac{wL}{2} = \frac{15aw}{2} \uparrow$$

Horizontal reactions: Horizontal component of support reaction at each end

$$R_{AH} = F_H = \frac{wL^2}{8h} = \frac{w \times (15a)^2}{8 \times (1.5a)} = \frac{150wa}{8}$$

Maximum tension in the suspension cable which occurs at the supports,

$$F_T = F_{\max} = \sqrt{F_{HA}^2 + R_{AV}^2} = \sqrt{\left(\frac{150wa}{8}\right)^2 + \left(\frac{15wa}{2}\right)^2} = 20.1944wa$$

The inclinations β of suspension cables with the horizontal

$$\tan\beta = \frac{R_{AV}}{F_H} = \frac{(15/2)\,wa}{(150/8)\,wa} = \frac{2}{5} \quad \text{or} \quad \beta = 21.8°$$

Case-I: Cable passed over the smooth rollers

The tension in the anchor cable is the same as in the suspension cable.

Total vertical load carried by the pier

$$= F_{\max}\sin\alpha + R_{AV} = 20.1944wa \sin 30° + 7.5wa = 17.597wa$$

Net horizontal force carried by the pier at its top

$$= F_{\max}(\cos\alpha - \cos\beta) = 20.1944wa\,(\cos 30° - \cos 21.8°)$$

$$= -1.2614wa \text{ (acting inward)}$$

The maximum bending moment in the pier = $1.2614wa \times 2a = 2.5227wa^2$

Case-II: Cable is clamped to a saddle

For the horizontal equilibrium of the saddle:

$$F_{TA}\cos\alpha = F_H \quad \text{or} \quad F_{TA} = F_H \sec\alpha$$

$$F_{TA} = 18.75wa \sec 30° = 21.651wa$$

Total vertical load carried by the pier

$$= F_{TA}\sin\alpha + R_{AV} = 21.651wa \sin 30° + 7.5wa = 18.3255wa$$

There is no horizontal force carried by the pier as horizontal components of F_{TA} and F_T balance each other.

2.6 THE TRUSSES

The *trusses* are used for the same purpose as *beams* and girders, except that the spans trusses cover are usually much larger. The *beams are generally* 1-*dimensional linear elements* whereas the *trusses are* 2-*dimensional articulated structural forms*. This 2-dimensional structural form is an assemblage of individual linear elements arranged in triangle or combination of triangles to form a stable framework that cannot be deformed by the application of external forces without straining one or more of its members. The individual members are assumed to be joined at their intersections with pinned connections. The members are generally so arranged that the loads and reactions occur only at these joints. This arrangement results in a state of either pure tension or compression in the individual members, the bending is absent as long as external loads act at the joints.

The trusses in which all the members are situated in one and the same plane are called *two-dimensional* or *plane trusses*. On the other hand, the framed structures in which the members are situated in several planes are called *three-dimensional* or *space trusses*.

2.6.1 Components of a Truss

The *span* of a truss is the distance between the supports. The lower and upper longitudinal members form lower and upper chords of the truss while the members which connect the two chords are called web members. The latter may be subdivided into verticals and diagonals or into *struts* and *ties*, the *struts being always in compression* and *the ties in tension*. A *counterbrace* is a member designed to resist both tensile and compressive forces. The distance between two adjacent joints measured along the horizontal is called a *panel*, the joints themselves being frequently referred to as *panel points*.

2.6.2 Member Forces

As mentioned above, with loads acting at the joints, all the members of a truss are subjected either to direct extension or compression which ensures a far better utilization of materials, the force diagram for each of these members being practically rectangular. Therefore, the trusses are always much lighter than solid web beams of the same span and height.

For common types of trusses carrying vertical loads, compressive forces develop in the upper chord members and tensile stresses in the lower chord members as is the case with top and bottom fibers of a beam. In an interstitial member, either type of force may develop. However, usually there exists an alternating pattern of tensile and compressive forces. In some simple truss configurations, the basic sense (tension, compression and neutral or zero) of the forces can be determined through intuitive approaches, but more complex trusses require application of the equilibrium equations.

Techniques like visualizing the probable deformed shape of structure it would attain if the member under consideration is imagined to be removed; and the nature of the force along member's direction required to prevent the deformation will provide the answer. Another way of visualizing the nature of force developed in different members of a truss is to use a cable or an arch analogy. The set of members imagined to be a cable will be in tension and those conceived to form a simple linear arch in compression. Other members serve the roles related to maintaining the equilibrium of the basic configuration providing load transfer or reactive functions.

2.6.3 Classification of Trusses

The classification of trusses is based on the following criteria:

(a) The shape of the upper and lower chords, e.g. trusses with parallel and non-parallel chords;
(b) The type of the web, e.g. truss with web members forming a letter K, called K-truss;
(c) The condition of the support, e.g. the cantilever trusses cantilevering over one or both supports;
(d) The application of the structure, e.g. roof trusses, bridge trusses and miscellaneous trusses used in crane construction, etc;
(e) The level of floor, e.g. through-bridge trusses with load carried directly by the bottom chord joints and the deck-bridge trusses where the load is carried directly by the upper chord or their joints.

The labelling scheme for the joints of a bridge truss is illustrated in Figure 2.6.

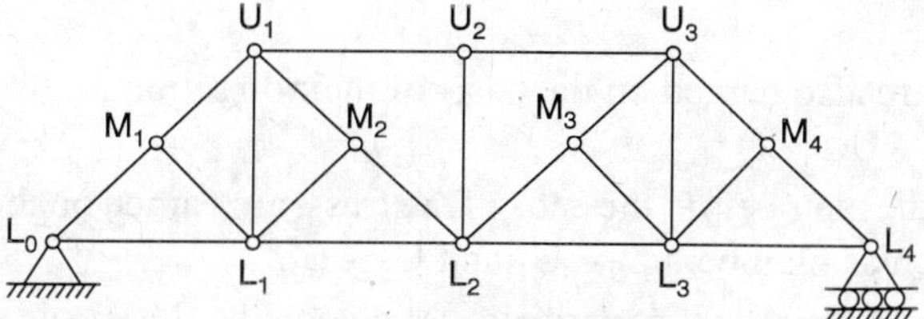

Figure 2.6 Labelling scheme for the joints of a truss.

2.7 ANALYSIS OF SIMPLE PLANAR TRUSSES

The analysis of two-dimensional or planar trusses of stable configuration, formed by a group of members arranged in the shape of one or more triangles connected by frictionless hinges, which are statically determinate will be discussed in this chapter. The hinge connected simple triangle is the only stable basic shape which is used as a repetitive element for building planar trusses. Such a truss is usually called *simple truss*.

2.7.1 Assumptions

The following simplifying assumptions are made in the analysis of the trusses:

1. The truss members are connected together with frictionless hinges so that they are free to rotate.
2. The truss members are straight between the joints.
3. The truss members are so arranged that the loads and reactions occur only at the truss joints.

4. The deformation of a truss under applied loads caused by change in length of the individual members is small enough to cause appreciable change in the overall shape and dimensions of the truss.

The assumptions 1 to 3 ensure that the externally applied loads result in the axial forces only, i.e. the individual members are in a state of either pure tension or compression, the bending is absent. These simplifications in analysis are quite satisfactory despite the fact that in actual practice welded and bolted connections are used.

2.7.2 Notation

The joints are numbered from left to right. Those on the bottom or lower chord are levelled L. The joints on the top or upper chord are levelled U. The joints between the lower and upper chords are levelled M, for middle. Thus, various members of a truss are often referred to by the following names as shown in Figure 2.6.

1. The chord members forming the outline of the truss are named in the direction of truss, such as members $U_1 - U_2$ and $L_2 - L_3$.
2. The verticals are named on the basis of their direction in the truss, such as members $U_1 - L_1$ and $U_2 - L_2$.
3. The diagonals are also named on the basis of their direction in the truss, such as members $U_1 - L_2$ and $L_2 - U_3$.
4. The end posts, the members at the ends of the truss, are named on the basis of their direction in the truss, such as members $L_0 - U_1$ and $U_3 - L_4$.
5. The web members include the verticals and diagonals of a truss.

2.7.3 Geometric Stability

In general, any truss composed of an aggregation of basic triangular shapes will be a *stable structure*. In case of any deviation in a bar or member pattern, the structure should be looked into carefully. To ascertain the stability of configurations, expressions are developed which relate the number of joints (j) in a truss to the number of members (m) necessary for stability. As discussed earlier, a basic planar triangle has three members ($m = 3$) and three joints ($j = 3$). Adding a single new connection or joint, beyond the original three, requires the addition of two new members. Hence the total number of members (m) in a truss having a number of connections (j) is given by: $m = 3 + 2(j - 3) = 2j - 3$. This gives the minimum number of members required for the geometric or configuration stability, fewer members will result in an *unstable structure*, more may indicate a structure with *redundant members*. The expression, however, is not foolproof and should not be used as a replacement for careful visual inspection. The expression serves as an indicator whether or not the internal forces in a structure can be calculated by the equations of statics.

For a truss connected directly to the foundation, above relationship is modified to $m = 2j$ where m is the total number of members and j the total number of joints excluding those at the foundation.

2.7.4 Statical Determinacy

Since all the j joints of the structure are in equilibrium and there are two equilibrium equations involving the summation of forces in X- and Y-directions, namely, $\Sigma F_x = 0$ and $\Sigma F_y = 0$ at each joint, 2j *equations of static equilibrium are available for the entire structure to compute the support reactions and internal forces in all the members.* If the number of unknown member forces and support reactions in a truss are m and r, respectively, then for the adequacy of the number of the available equilibrium equations to compute the support reactions and internal forces in all the members, $2j = (m + r)$. Such a structure is called *statically determinate* and unknown can be obtained from 2j equations. This equation is more commonly written as

$$m = 2j - r \tag{2.20}$$

If $2j < m + r$, there are more unknowns than the number of equilibrium equations, the structure is *statically indeterminate*. The *degree of indeterminacy* is $n = (m + r) - 2j$. The structure requires n additional relationships such as *compatibility conditions* for its analysis. The truss may be *redundant internally* or *externally* or *both*.

If $2j > m + r$, there are more equilibrium equations available than the number of unknowns, such a structure is a *mechanism* and always *unstable*. The structure does not have unique solution. *Existence of more than one solution indicates instability*.

EXAMPLE 2.8

Determine whether the trusses shown in Figure 2.7 are stable. If stable, then find whether they are statically determinate or indeterminate.

Solution: The truss of Figure 2.7(a) is stable as $r > 3$ and $m > (2j - r)$. It is statically indeterminate externally because $r > 3$, but statically-determinate internally because $m = (2j - 3)$. In the relation $m = 2j - r$, $r = 3$ has been used since that is the minimum number of reaction components required for the stability and statically determinacy. The degree of indeterminacy is

$$n = (m + r) - 2j = (7 + 4) - 2 \times 5 = 1$$

The truss shown in Figure 2.7(b) is stable as $r > 3$ and $m > (2j - r)$. It is statically-indeterminate externally as well as internally because $r > 3$, and also $m > (2j - r)$. The degree of indeterminacy is

$$n = (m + r) - 2j = (18 + 4) - 2 \times 10 = 2$$

The truss of Figure 2.7(c) is stable and statically determinate because $m = (2j - r)$ and $r = 3$. However, an inspection of the structure indicates that it is geometrically unstable due to improper arrangement of members. The third panel consists of a hinged rectangle which does not provide any shearing resistance and may take shape of parallelogram. If one of the diagonal members from the first panel of truss is shifted to the third panel, the structure will acquire configuration stability. It should be noted that it is not mere the number of members but also their arrangement that defines the stability of the structure.

The truss shown in Figure 2.7(d) is stable as $r > 3$ and $m > 2j - r$. It is statically indeterminate externally because $r > 3$, and also statically-indeterminate internally because $m + r > 2j$. The degree of indeterminacy is

$$n = (m + r) - 2j = (21 + 6) - 2 \times 13 = 1$$

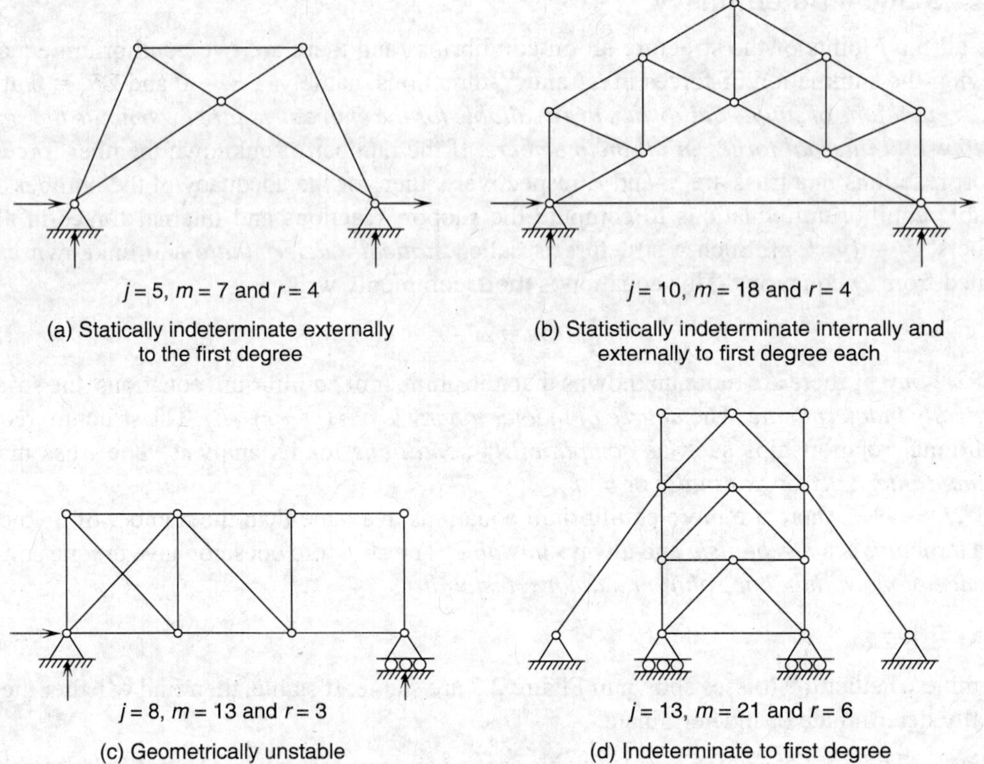

Figure 2.7 Typical examples of plane trusses.

2.8 THE PRINCIPLE OF ANALYSIS

The principle underlying the analytical methods is that any structure, or any portion of the structure must be in the state of equilibrium, i.e. any set of forces acting on the structure, or its portion, including both the externally applied and internally developed, must form a system whose resultant or net result is zero. The extent of the portion of the structure selected for analysis is not restricted. A whole segment or *sub-assembly* consisting of several members and joints could be considered or attention could be limited to a single joint or member. Each decomposed portion is considered to be a *free- body* in a state of equilibrium under the action of force system present on that portion.

Thus in truss analysis, a part of the truss is usually considered to be isolated with an imaginary section as illustrated in Figure 2.8(a). This isolated free body acted upon by a set of forces is studied independently. The set of forces applied to the free body includes the axial forces in the members that have been cut by the section and any load or reaction that is applied externally as shown in Figures 2.8(b),(c) and (d). The application of equations of static equilibrium to isolated free-body enables the determination of the forces in the individual cut members. The free bodies are carefully selected so that the sections do not cut more than three members since there are only three equations of statics. Thus the total number independent equilibrium equations are $2j - 3$. It is usual to start the analysis with the determination of the support reactions for which three equilibrium equations are written for the truss as a whole.

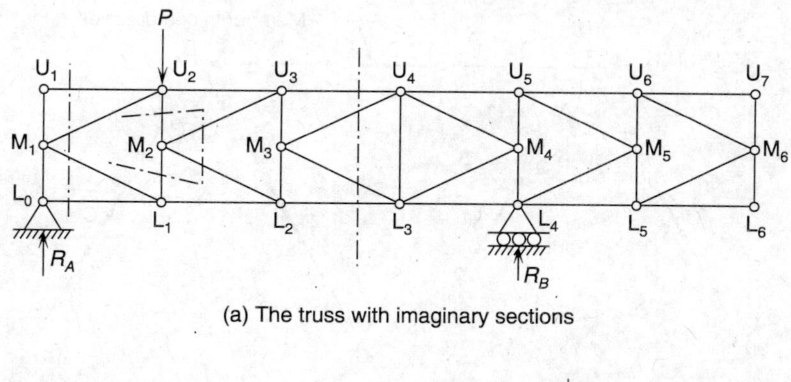

(a) The truss with imaginary sections

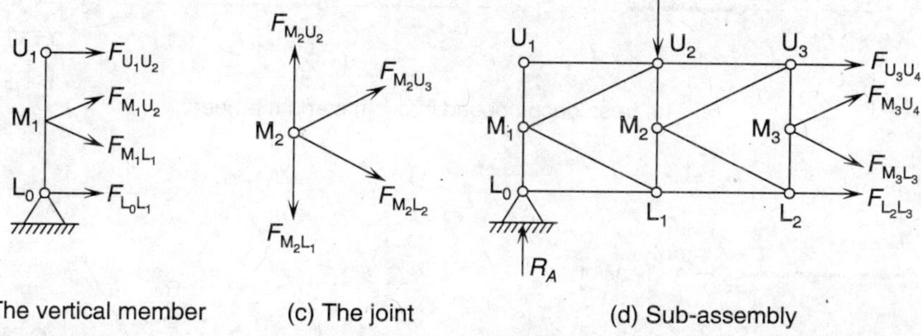

(b) The vertical member (c) The joint (d) Sub-assembly

Figure 2.8 Free-body diagrams of a typical truss member, truss joint and a truss sub-assembly.

2.8.1 Methods for Analysis and Sign Conventions

As discussed above the truss analysis consists in dividing the truss into parts, constructing free-body diagram for each part, then evaluating the member forces using free-body diagrams. The forces being vector quantities involve nature or directions of forces. To avoid errors, it is desirable to adopt a sign convention for forces in the members of truss.

2.8.2 Member Force Notation

In this chapter, all unknown forces in a truss are considered to be *tensile forces*. Further, tensile and compressive forces are accompanied with plus and minus signs (+) and (–), respectively. Arrows are generally used to represent the character of forces. On the isolated members diagrams, the arrows indicate the action of the joints on the members as shown in Figure 2.9(a). However, the arrows marked on the members in a frame as illustrated in Figure 2.9(c) indicate the action of the member on the joints, e.g. *a member with arrows pointing towards the end joints indicates that a state of compression exists in the member* and vice-versa, i.e. the arrow convention of Figure 2.9(b-i) is used for member in tension and of Figure 2.9(b-ii) for compression member in a truss. It is the joint action on the member which determines the nature of force in the member. The action of joints on the member and member's reaction on the joints are illustrated in Figure 2.9(a). Thus, *on an isolated joint, the arrows indicate members' action on the joint.*

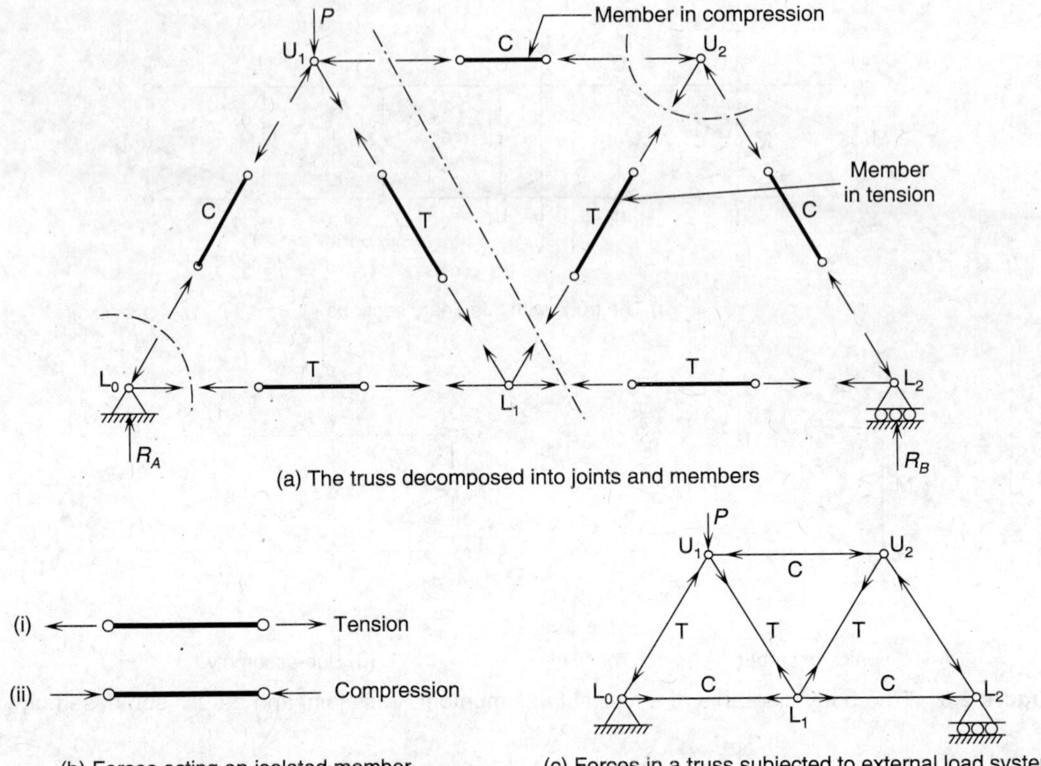

(b) Forces acting on isolated member (arrows indicate joints action on the member)

(c) Forces in a truss subjected to external load system (arrows indicate members action on its joints)

Figure 2.9 Free-body diagrams for solution of member forces by method of joints.

The following methods are generally used for the analysis of trusses:

1. The method of joint equilibrium
2. The method of sections or moments and shears.

Frequently, both the methods are used simultaneously in the analysis of trusses. Depending upon the geometry of the truss, some member forces are more easily calculated with the method of sections than others that are more easily calculated using the method of joints.

2.9 METHOD OF JOINTS EQUILIBRIUM

In this method, the equilibrium of each joint is considered separately, the joint being isolated completely from the remainder of the truss by the imaginary section passing completely around it. The isolated joint reduces to a free body that is in equilibrium under the action of externally applied joint loads including the reaction, if any, and the axial forces in the members that have been cut by the section. The equilibrium equations $\Sigma F_x = 0$ and $\Sigma F_y = 0$ are applied at the joint to determine the unknown forces in the members meeting there. It should be noted that no more than two unknowns can be determined at a joint with these two equations.

Normally analysis of forces in a truss is started at a support with two members meeting there and where reactions have already been determined by considering the rigid body equilibrium of the whole structure. Once all the forces acting on the initial joint have been found, it is possible to proceed to adjacent joint. The equations are then applied in turn to each joint. In most of the cases, it would be possible to determine all member forces without solving more than two simultaneous equations at a time. The procedure is illustrated in Figure 2.10(a).

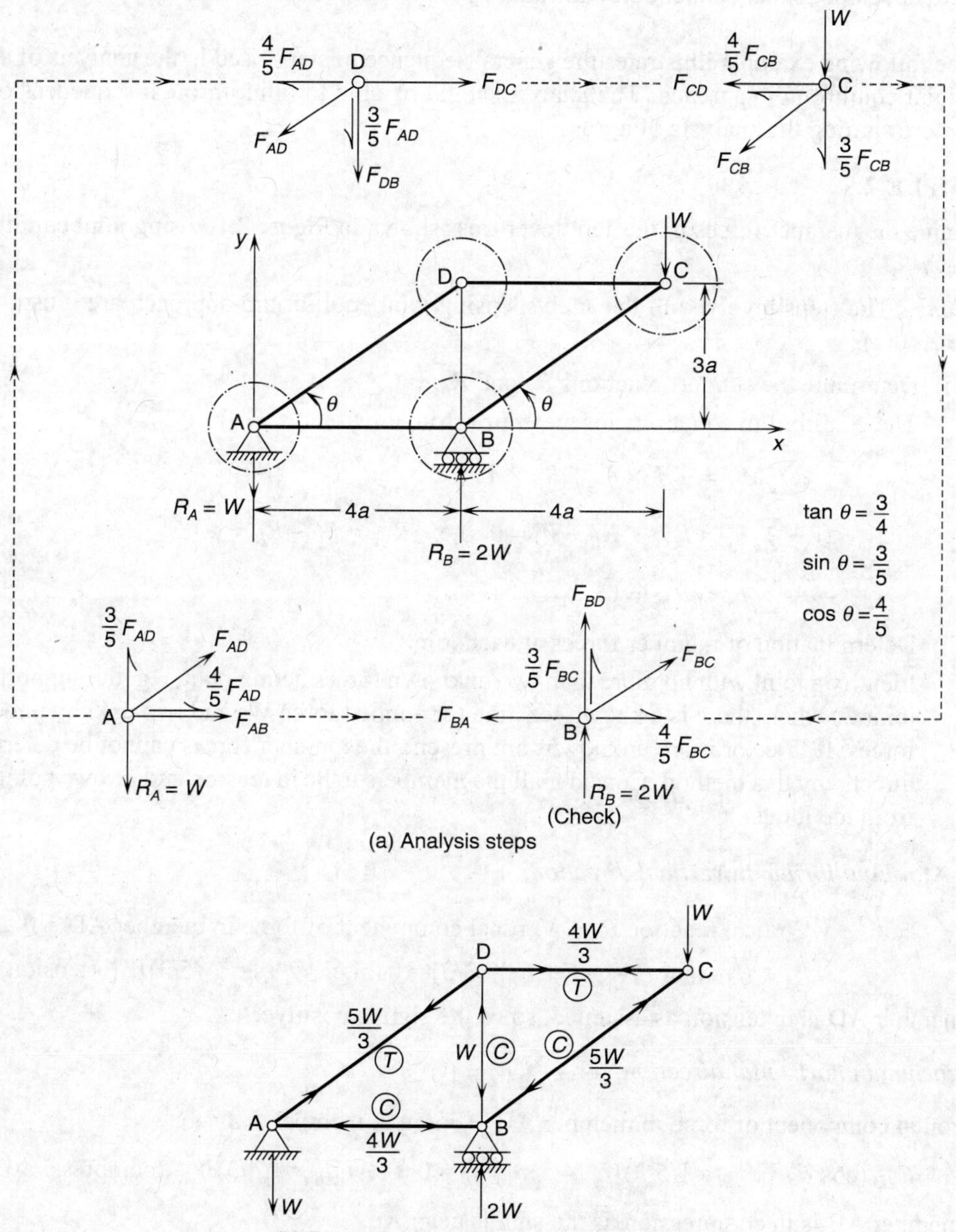

Figure 2.10 Graphical representations of the steps for the analysis of truss by joint equilibrium method.

In drawing the free-body diagrams and writing equilibrium equations, the nature of an unknown member force is arbitrarily considered to be the tension. If the computed value is positive, it indicates that the member is in tension and negative result indicates that the member is in compression. However, the numerical value is correct regardless of the sign. The nature and direction of forces developed in an element are indicated by arrows at the ends of the member as discussed earlier.

The following example illustrates the general sequence of steps used in the analysis of a truss using joint equilibrium approach. The analyst should be able to think through sequences of this type before starting the analysis of a truss.

EXAMPLE 2.9

Determine the member forces in the cantilever truss shown in Figure 2.10 using joint equilibrium approach.

Solution: The steps involved in the analysis using joint equilibrium approach are illustrated in Figure 2.10(a).

(i) Determine the support reactions R_A and R_B:
The equilibrium equations for the entire structure are:

$$+\sum M_A = -W \times 8a + R_B \times 4a = 0 \quad \text{i.e.} \quad R_B = 2W \uparrow$$

$$\uparrow + \sum F_y = R_{Ay} + R_B - W = 0 \quad \text{i.e.} \quad R_{Ay} = W - R_B = -W \downarrow$$

$$\rightarrow + \sum F_x = R_{Ax} = 0$$

(ii) Determination of member forces at each joint:
Identify a joint with no more than two unknown forces acting on it, e.g. the support A (or joint C) where there is a known reaction (or applied force W) and two unknown member forces. If three or more unknowns are present, the member forces cannot be determined directly by this method. Consider all the members to be in tension and arrows point away from the joints.

Joint A: *Equilibrium in vertical direction,* $\uparrow + \sum F_y = 0$:

–Vertical reaction R_A + Vertical component of force in member AD = 0

$$-R_A + F_{AD}(\sin\theta) = -W + F_{AD}(3/5) = 0 \quad \text{i.e.} \quad F_{AD} = +(5/3)W \quad \text{(tension)}$$

The member AD is in tension as assumed, since the sign is positive.

Equilibrium in horizontal direction, $\rightarrow + \sum F_x = 0$:

Horizontal component of force in member AD + Force in member AB = 0

$$+F_{AD}(\cos\theta) + F_{AB} = [(5/3)W](4/5) + F_{AB} = 0 \quad \text{i.e.} \quad F_{AB} = -(4/3)W \quad \text{(compression)}$$

The member AB is in compression as the sign is negative.

The next step is to proceed to an adjacent joint. Both joints B and D involve two unknown forces. Hence, any of the two can be selected for next joint analysis.

Joint B: *Equilibrium in horizontal direction,* $\rightarrow + \sum F_x = 0$:

– Force in member AB + Horizontal component of force in member BC = 0

$-F_{AB} + F_{BC}$ (cos θ) = (4/3)W + F_{BC} (4/5) = 0 i.e. F_{BC} = –(5/3)W (compression)

The member BC is in compression since the sign is negative.

Equilibrium in vertical direction, $\uparrow + \sum F_y = 0$:

Vertical reaction R_B + Vertical component of force in member BC + Force in member BD = 0

$$R_B + F_{BC} (\sin \theta) + F_{BD} = 2W - [(5/3)W] (3/5) + F_{BD} = 0$$

i.e. $F_{BD} = -W$ (compression)

Joint C: *Equilibrium in horizontal direction,* $\rightarrow + \sum F_x = 0$:

– Force in member CD – horizontal component of force in member BC = 0

$-F_{CD} - F_{BC}$ (cos θ) = $-F_{CD}$ + [(5/3)W] (4/5) = 0 i.e. F_{BC} = +(4/3)W (tension)

All the forces are now known. The application of equilibrium equations to the joint D can serve as a check on the analysis. Alternatively, the sequence of application of equilibrium equations to the joints may be A – D – C – B as illustrate in Figure 2.10(a).

It should be noted that the nature of forces could have been predetermined by inspection.

A close look at this cantilever type structure suggests that the top chord members AD and DC are in tension and those in the bottom chord, i.e. AB and BC in compression. The diagonal member BD has to be in compression to balance the vertical components of tension members meeting at joint D.

In many trusses, the analysis may involve solving simultaneous equations as the individual equations cannot be solved directly. Forces in the members and their nature are shown in Figure 2.10(b).

EXAMPLE 2.10

Determine the magnitude and nature of forces developed in the member of the truss shown in Figure 2.11 using method of joints equilibrium.

Solution: (i) Support reactions R_A and R_B:

$$R_A = \left(\frac{3}{4}\right)W \text{ and } R_B = \left(\frac{1}{4}\right)W$$

(ii) Member forces at the joints:

All the members are considered to be in tension as illustrated in Figure 2.11(a).

Joint L_0:

$$\sum F_y = 0; \quad \frac{3W}{4} + F_{L_0-U_1} \sin 45° = 0 \text{ i.e. } F_{L_0-U_1} = -\frac{3\sqrt{2}W}{4} \quad \text{(compression)}$$

$$\sum F_x = 0: \quad F_{L_0-L_1} + F_{L_0-U_1} \cos 45° = 0 \text{ i.e. } F_{L_0-L_1} = \frac{3W}{4} \quad \text{(tension)}$$

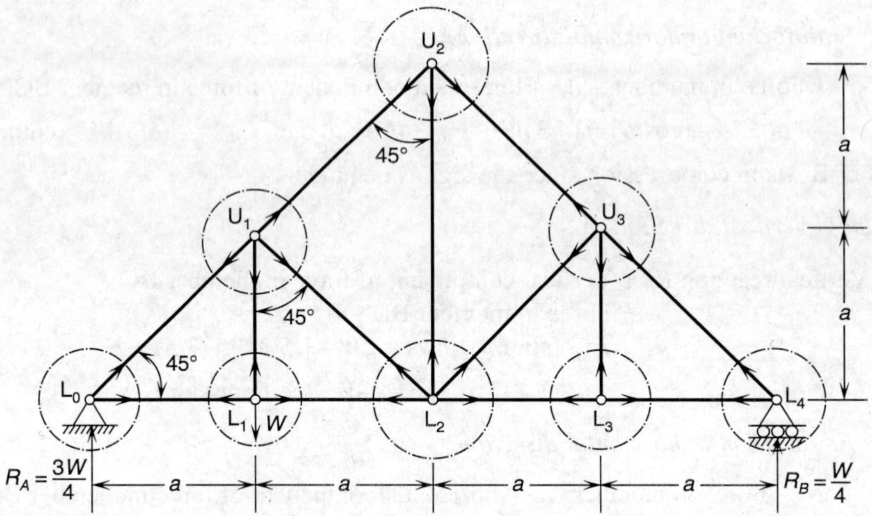

(a) The truss, its joints and assumed direction of forces in the members (tension)

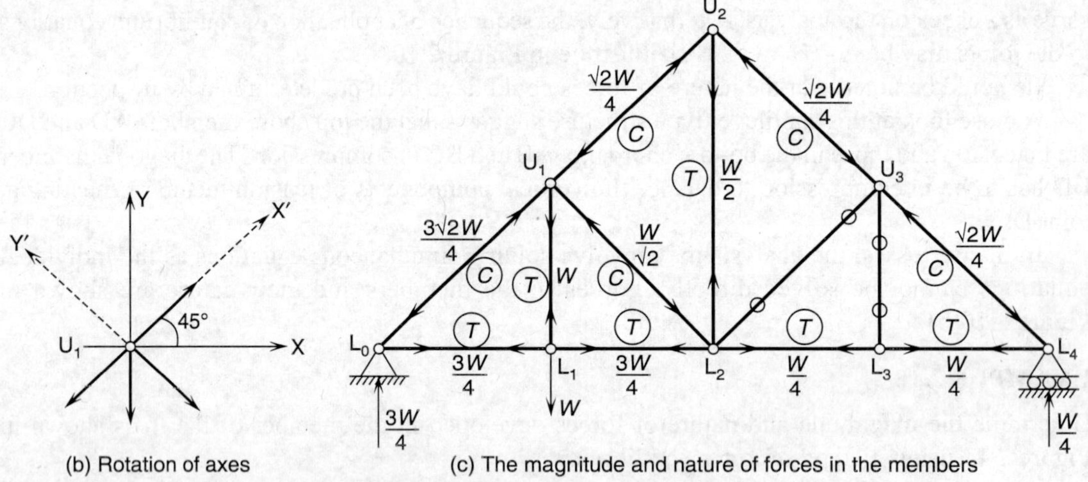

(c) The magnitude and nature of forces in the members

Figure 2.11 Analysis of truss of Example 2.10.

Joint L$_1$:

$$\sum F_y = 0: \; -W + F_{L_1-U_1} = 0 \text{ i.e. } F_{L_1-U_1} = W \qquad \text{(tension)}$$

$$\sum F_x = 0: \; -F_{L_0-L_1} + F_{L_1-L_2} = 0 \text{ i.e. } F_{L_1-L_2} = \frac{3W}{4} \qquad \text{(tension)}$$

Joint U$_1$:

$$\sum F_y = 0: \; -F_{L_0-U_1} \sin 45° - F_{L_1-U_1} - F_{U_1-L_2} \sin 45° + F_{U_1-U_2} \sin 45° = 0$$

$$- F_{L_0-U_1} - F_{L_1-U_1} \text{ cosec } 45° - F_{U_1-L_2} + F_{U_1-U_2} = 0$$

$$\frac{3\sqrt{2}W}{4} - W\sqrt{2} - F_{U_1-L_2} + F_{U_1-U_2} = 0 \qquad \text{(a)}$$

$$\sum F_x = 0: \quad -F_{L_0-U_1} \cos 45° + F_{U_1-L_2} \cos 45° + F_{U_1-U_2} \cos 45° = 0$$

$$\frac{3\sqrt{2}\,W}{4} + F_{U_1-L_2} + F_{U_1-U_2} = 0 \qquad \text{(b)}$$

Solve Eqs. (a) and (b) simultaneously as follows,

Operate (a) – (b): $\Rightarrow \quad F_{U_1-L_2} = -\frac{W}{\sqrt{2}}$ (compression)

Operate (a) + (b): $\Rightarrow \quad F_{U_1-U_2} = \frac{\sqrt{2}\,W}{4}$ (tension)

The computation of member forces can be simplified by using the rotated reference axis system (X', Y') instead of usual vertical and horizontal (X, Y) system as shown in Figure 2.11(b). Thus, using the rotated reference system:

$$\sum F_{y'} = 0: \quad -F_{L_1-U_1} \cos 45° - F_{U_1-L_2} = -\frac{W}{\sqrt{2}} - F_{U_1-L_2} = 0$$

or $\quad F_{U_1-L_2} = -\frac{W}{\sqrt{2}}$ (compression)

$$\sum F_{x'} = 0: \quad -F_{L_0-U_1} - F_{U_1-L_1} \cos 45° + F_{U_1-U_2} = 0$$

$$\frac{3\sqrt{2}\,W}{4} - \frac{W}{\sqrt{2}} + F_{U_1-U_2} = 0 \text{ or } F_{U_1-U_2} = -\frac{\sqrt{2}\,W}{4} \text{ (compression)}$$

It should be noted that the use of rotated reference system has obviated the need of solving the equations simultaneously.

Joint U$_2$:

$$\sum F_x = 0: \quad F_{U_2-U_3} \cos 45° - F_{U_1-U_2} \cos 45° = 0 \text{ or } F_{U_2-U_3} - F_{U_1-U_2} = 0$$

Therefore,

$$F_{U_2-U_3} = F_{U_1-U_2} = -\left(\frac{\sqrt{2}}{4}\right) W \text{ (compression)}$$

$$\sum F_y = 0: \quad -F_{U_1-U_2} \sin 45° - F_{U_2-L_2} - F_{U_2-U_3} \sin 45° = \frac{W}{2} - F_{U_2-U_3} = 0$$

or $\quad F_{U_2-U_3} = \frac{W}{2}$ (tension)

Joint L_2:

$$\sum F_y = 0: \qquad F_{U_1-L_2} \sin 45° + F_{U_2-L_2} + F_{L_2-U_3} \sin 45°$$

$$= -\frac{W}{2} + \frac{W}{2} - \frac{1}{\sqrt{2}} F_{L_2-U_3} = 0 \quad \text{or} \quad F_{L_2-U_3} = 0$$

$$\sum F_x = 0: \qquad -F_{L_1-L_2} - F_{U_1-L_2} \cos 45° + F_{L_2-U_3} \cos 45° + F_{L_2-L_3}$$

$$= -\frac{3W}{4} + \frac{W}{2} + 0 + F_{L_2-L_3} = 0 \quad \text{or} \quad F_{L_2-L_3} = \frac{W}{4} \quad \text{(tension)}$$

Joint L_3:

$$\sum F_y = 0: \qquad F_{L_3-U_3} = 0 \quad \text{i.e.} \quad F_{L_3-U_3} = 0$$

$$\sum F_x = 0: \qquad -F_{L_2-L_3} + F_{L_3-L_4} = 0 \quad \text{i.e.} \quad F_{L_3-L_4} = \frac{W}{4} \quad \text{(tension)}$$

Joint L_4:

$$\sum F_y = 0: \qquad \frac{W}{4} + F_{U_3-L_4} \sin 45° = 0 \quad \text{i.e.} \quad F_{U_3-L_4} = -\frac{\sqrt{2}W}{4} \quad \text{(compression)}$$

$$\sum F_x = 0: \qquad -F_{L_3-L_4} - F_{U_3-L_4} \cos 45° = 0 \quad \text{i.e.} \quad F_{L_3-L_4} = \frac{W}{4} \quad \text{(tension)}$$

An inspection of the structure suggests that the top chord and the bottom chord members may be considered in compression and tension, respectively. The inclined members are alternatively in compression and tension when moving from left to right. The magnitude and the nature of forces are shown in Figure 2.11(c).

2.9.1 Simplifying Conditions

Sometimes special conditions that frequently arise enable to pick out the members and more evident joints for quick determination of member forces by qualitative inspection. The selected joints are isolated and inspected for equilibrium. Thinking in terms of rotated axes system, can also provide useful information. Normally the zero force members can be identified by mentally observing the equilibrium of joints. In addition to facilitating calculations, these techniques help in the development of a more intuitive understanding of force distribution in a truss. Consider the truss shown in Figure 2.12 and refer to the circled joints which have been picked out for inspection.

Joint L_1: inspection of the vertical equilibrium indicates, $F_{L_1-U_2} = W$ (tension)

Joint L_3: observing the vertical equilibrium indicates $F_{L_3-U_4} = 0$, i.e. $L_3 - U_4$ is a zero-force member.

Joint U_3: considering vertical equilibrium indicates $F_{L_2-U_3} = 0$, i.e. $L_2 - U_3$ is a zero-force member.

Joint U_1: observing equilibrium in both the horizontal and vertical directions points out that both $L_0 - U_1$ and $U_1 - U_2$ are zero-force members.

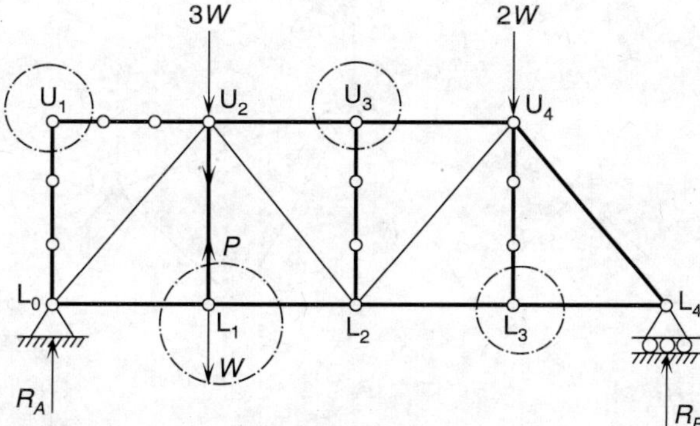

Figure 2.12 Identifiable zero-force member in a truss.

Thus, if at a joint a member has a component in a direction in which no other member or force has the component; the force in the member will be zero.

2.10 METHOD OF SECTIONS OR MOMENTS

This method is a very powerful tool for the analysis and design of many structures. The method is an extension of the method of joints where the elemental portions of the truss considered for equilibrium were the joint themselves, whereas in the method of sections, the elemental portion is a complete subassembly consisting of several joints and members. In this method, an imaginary section line is passed through the truss, dividing it in two segments. The method of joints could be thought of in terms of passing a section line around each joint. Both can be interchangeably used in analyzing trusses. The method of joints is generally preferred for use when all the member forces in a truss are to be determined. The method of sections is particularly useful when only a limited number of forces are to be determined.

In the method of section, as mentioned earlier, an imaginary section is passed through the truss to divide it into two free-bodies (subassemblies), each of which is in equilibrium under the action of externally applied forces (including reactions) and the internal forces in cut members as illustrated in Figure 2.13(b). However, the section is so placed that it cuts the members in which forces are to be calculated and as few other members as possible.

As in the method of joints free-body diagrams of the subassemblies under consideration are drawn as shown in Figures 2.14(b) and (c) for sections I-I and II-II, respectively, and analyzed for translational and rotational equilibriums under the action of complete force system acting on them. Both translational and rotational equilibriums must be considered, since the force system acting on the truss is coplanar, but nonconcurrent and nonparallel. Thus, there are three independent basic equilibrium conditions $\Sigma F_x = 0$; $\Sigma F_y = 0$ and $\Sigma M = 0$, which are available for the analysis. Hence the subassemblies must be isolated from the structure such that no more than three unknowns are involved.

For moment equilibrium $\Sigma M = 0$, the moment of all forces applied to the free-body under consideration about any point in the plane of truss must be zero. The point is usually so selected that it is located on the line of action of one or more of the member forces. Sometimes only one unknown force appears in the equation and the value of force can be obtained directly.

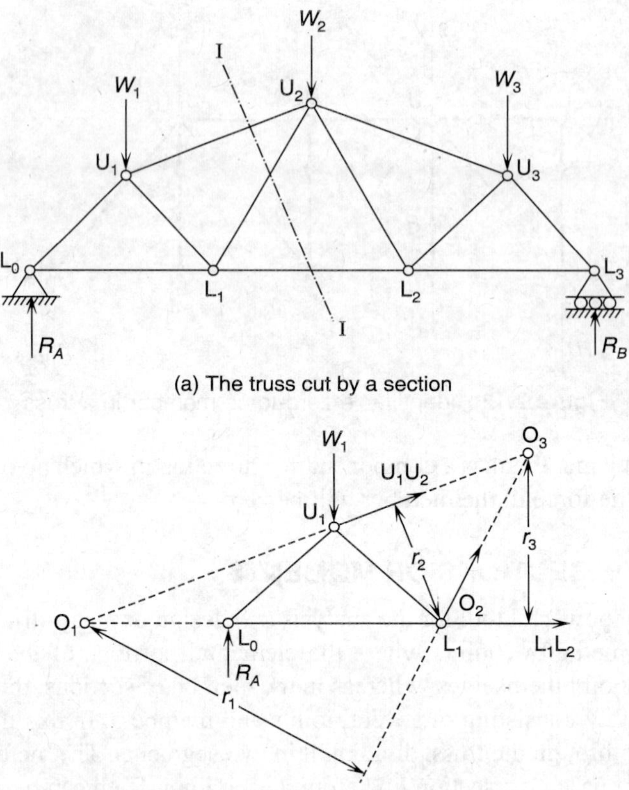

Figure 2.13 The points of intersection of two members or moment centres.

For illustration, consider the truss shown in Figure 2.13(a) where it is desired to compute the forces in the members $L_1 - L_2$, $U_2 - L_1$ and $U_1 - U_2$. To achieve the objective, a section I-I is passed through the members dividing the truss into two parts or sub-assemblies. Consider the moment equilibrium of left hand part of the truss. For computing the force in the member $L_1 - U_2$, the point O_1 is selected as a moment centre for convenience, as the unknown forces $F_{L_1-L_2}$ and $F_{U_1-U_2}$ pass through the point and hence do not enter into moment equation. The force in the member $F_{L_1-U_2}$ is given by,

$$F_{L_1-U_2} = \pm \frac{M_{O_1,\text{external}}}{r_1} \tag{2.21}$$

where $M_{O_1,\text{external}}$ is the moment of the forces acting on the part of the truss under consideration about the moment centre O_1 and r_1 is the lever arm of the member force about the moment centre. Similarly, for computation of force in thc member $F_{U_1-U_2}$, O_2 or L_1 and r_2 are the moment centre and lever arm of the force, respectively as shown in Figure 2.13(b).

Thus, the main advantage of this method lies in the fact that it enables the determination of force in a particular member directly in most cases without first calculating forces in other members.

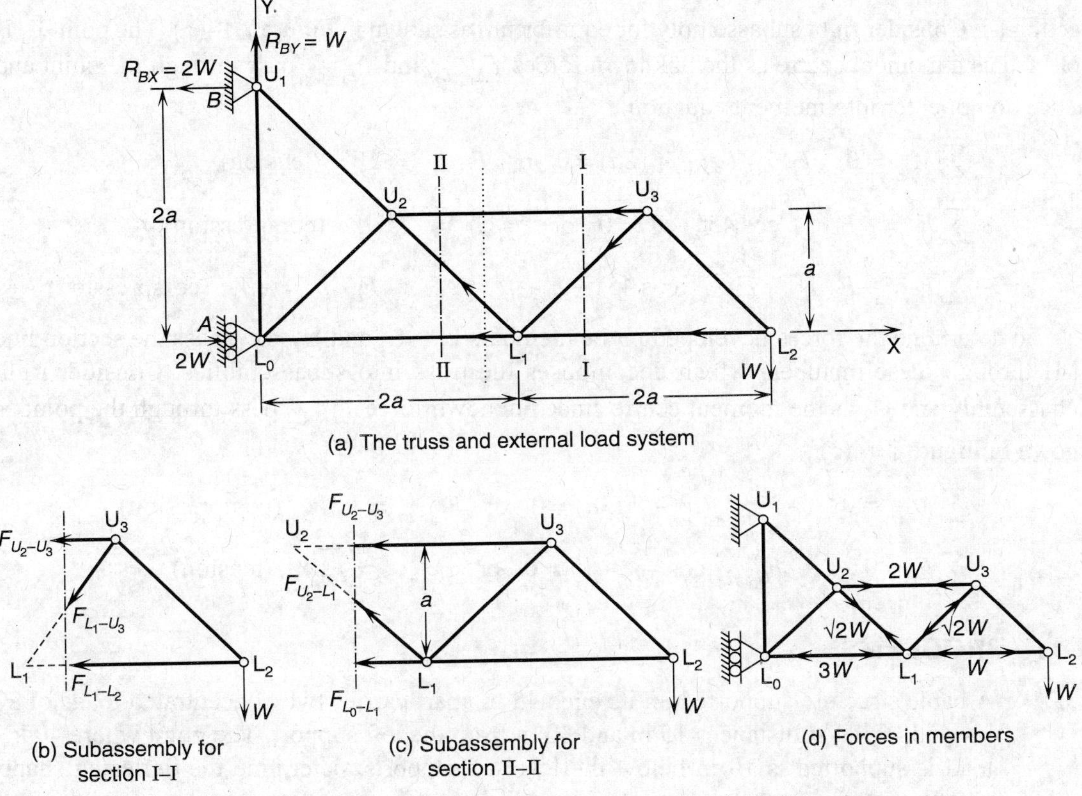

Figure 2.14 Analysis of truss of example by method of sections.

As with the method of joints, the unknown member forces are considered to be acting in tension. If the computed values are positive, they indicate that the members are in tension and negative results indicate that the members are in compression. However, the numerical values are correct regardless of the sign. Since predetermination of the nature of member forces may be useful, an experienced analyst may attempt to determine the nature of member forces by qualitative inspection before starting the analysis.

The application of the method is illustrated in the following example of determination of member forces by the method of sections.

EXAMPLE 2.11

Determine the magnitude and nature forces developed in the members $L_0 - L_1$, $L_1 - L_2$, $U_2 - L_1$, $L_1 - U_3$ and $U_2 - U_3$ of the truss shown in Figure 2.14(a) using method of sections.

Solution: (i) Support reactions R_A and R_B:

$$R_A = 2W, R_{BX} = -2W \quad \text{and} \quad R_{BY} = W$$

The unknown member forces are considered to be acting in tension. For the determination of the forces in the members $L_1 - U_3$, $L_1 - L_2$ and $U_2 - U_3$, the truss is decomposed through the section I-I. Consider right subassembly for equilibrium as shown in Figure 2.14(b). The point L_1 is selected as a moment centre as the unknown forces $F_{L_1-L_2}$ and $F_{L_1-U_3}$ pass through the point and hence do not enter into moment equation.

$$+\sum M_{L_1} = 0: \; F_{U_2-U_3}(a) - W(2a) = 0 \quad \text{or} \quad F_{U_2-U_3} = 2W \quad \text{(tension)}$$

$$+\sum F_y = -F_{L_1-U_3}\cos 45° - W = 0 \quad \text{or} \quad F_{L_1-U_3} = -\sqrt{2}W \quad \text{(compression)}$$

$$+\sum F_x = -F_{L_1-L_2} - F_{L_1-U_3}\cos 45° - F_{U_2-U_3} = 0 \quad \text{or} \quad F_{L_1-L_2} = -W \quad \text{(compression)}$$

To determine the forces developed in the members $L_0 - L_1$ and $U_2 - L_1$, pass the section line II-II through these members which decomposes the truss into subassemblies. Consider right subassembly and U_2 as the moment centre since unknown force $F_{L_1-U_2}$ pass through the point as shown in Figure 2.14(c).

$$+\sum M_{U_2} = 0: \quad -F_{L_0-L_1}(a) - W(3a) = 0 \quad \text{or} \quad F_{L_0-L_1} = -3W \quad \text{(compression)}$$

$$+\sum F_y = 0: \quad F_{U_2-L_1}\cos 45° - W = 0 \quad \text{or} \quad F_{U_2-L_1} = \sqrt{2}W \quad \text{(tension)}$$

2.11 PROBLEMS

2.1 A cable structure supported on its ends 45 m apart carries two concentrated loads of 20 kN and 40 kN at distances 15 m and 30 m from the left support. The point where 40 kN load is supported is 10 m below the level of supports, determine the deflected shape, reactions at the supports, internal forces or tensions in the various segments and the total length of the cable.

[**Ans:** Depth at 20 kN load = 8 m; F_H = 50 kN; tensions in three segments starting from left end are: 56.66, 50.44 and 60.10 kN; and the length of cable = 50.16 m]

2.2 A cable of 60 m span has its left and right supports at heights 9 and 4 m, respectively, above the lowest point of the cable. If the cable carries uniformly distributed load of intensity 5 kN per metre run over the whole span, determine the horizontal and vertical reactions at each supports.

[**Ans:** At the left support, R_V = 180 kN; F_H = 360 kN]

2.3 A cable with span of 50 m has its left support 8 m above the right support. The cable carries uniformly distributed load of intensity 1.0 kN/m over its entire span and a concentrated load 120 kN at a point located 35 m horizontally and 11 m vertically below the left support. Determine the horizontal and vertical reactions at each support and maximum tension in the cable.

[**Ans:** F_H = 281.95 kN; R_{AV} = 106.11 kN; R_{BV} = 63.89 kN and F_{max} = 301.25 kN]

2.4 A steel cable hangs between two level supports which are 120 m apart with central dip of 10 m. The cable carries uniformly distributed load of 20 kN/m over its entire horizontal span. Determine the change in tension due to a rise of temperature of 15°C, when coefficient of thermal expansion of steel is $\alpha = 12 \times 10^{-6}$ per °C.

[**Ans:** Change in horizontal tension dF_H = 17.496 kN]

2.5 A flexible steel cable hangs between two level supports with central dip or sag of one-tenth of its span. Determine the maximum possible span that can be provided if the permissible stress in the cable is 145 MPa. The unit weight of the cable 75 kN/m^3.

[**Ans:** Length = 1399.64 m]

2.6 A cable of span 120 m with the span-by-dip ratio of 10 carries a uniformly distributed load of intensity 5 kN per unit horizontal run over its entire span. Determine the maximum tension in the cable and its inclination at the support; horizontal and vertical forces transmitted to the supporting pier of height 16, when supporting arrangement on the top of the pier is: (i) the smooth pulley, and (ii) the saddle with smooth rollers. The inclination of the anchor cable is 30° to the horizontal in both the cases. Also determine the bending moment developed in the pier in the case (i).

[**Ans:** F_H = 750 kN; (i) vertical force = 703.88 kN, horizontal force = –50.456 kN and maximum bending moment = 807.264 kN.m; and (ii) vertical force = 733.02 kN, horizontal force = 0, and maximum bending moment = 0].

In Problems 2.7 through 2.18, check the stability of the structures and if stable, determine whether they are statically determinate or indeterminate. If indeterminate, calculate internal and external redundancies. Reduce the structures to the stable and statically determinate ones.

2.7

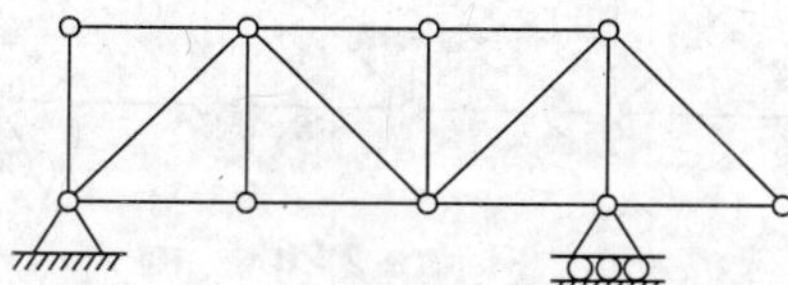

Figure 2.15

[**Ans:** Statically determinate externally and internally]

2.8

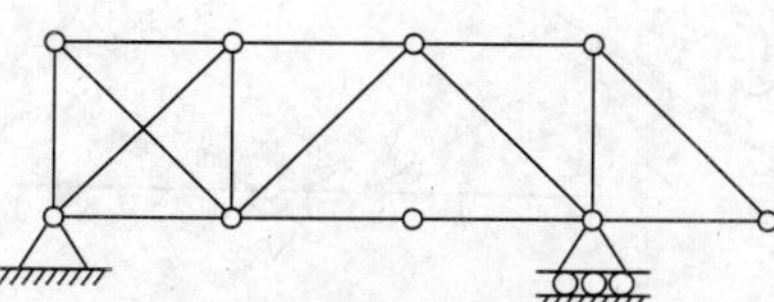

Figure 2.16

[**Ans:** Unstable]

2.9

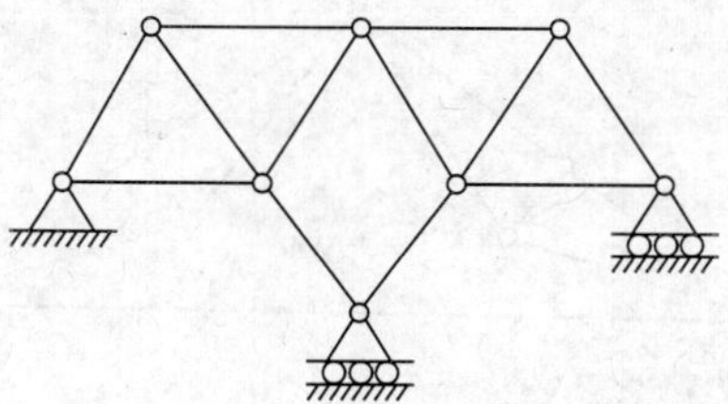

Figure 2.17

[**Ans:** Geometrically stable and statically determinate]

2.10

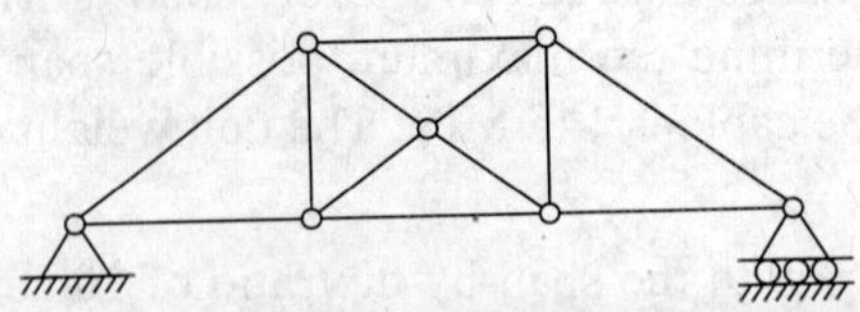

Figure 2.18

[**Ans:** Statically indeterminate internally to the first degree]

2.11

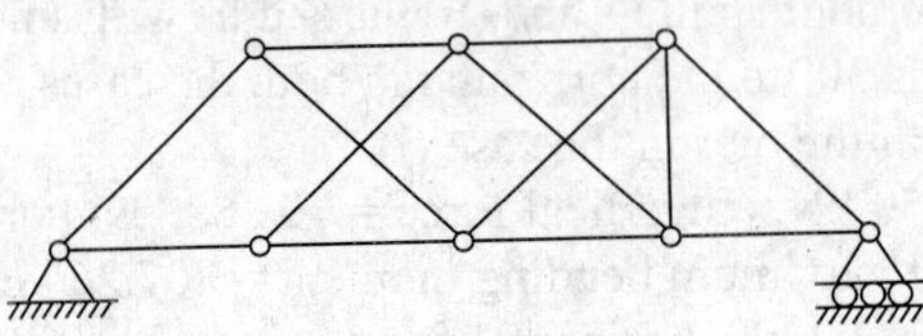

Figure 2.19

[**Ans:** Statically determinate externally and internally]

2.12

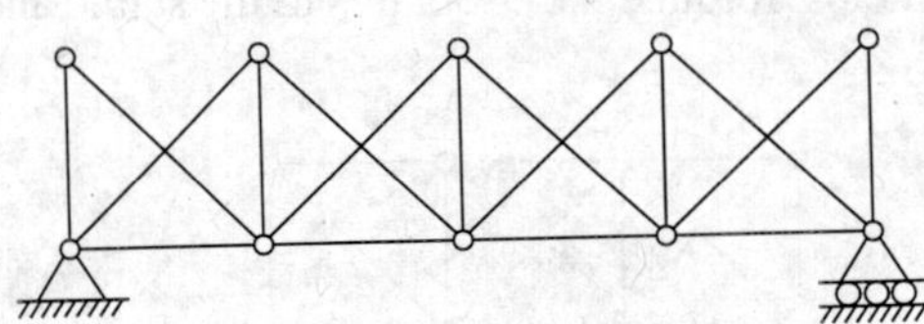

Figure 2.20

[**Ans:** Statically determinate externally and internally]

2.13

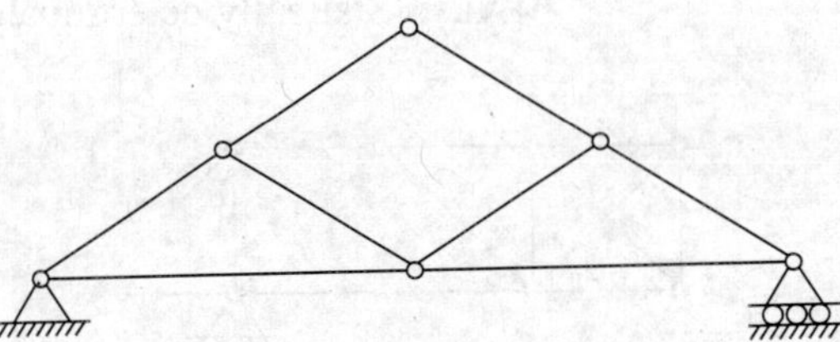

Figure 2.21

[**Ans:** Unstable internally]

2.14

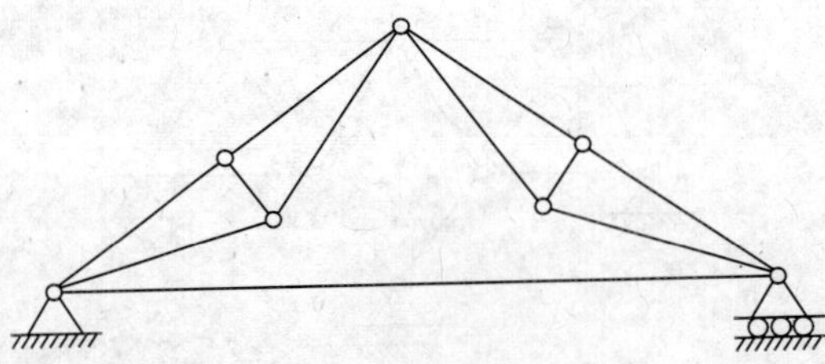

Figure 2.22

[**Ans:** Stable determinate externally and internally]

2.15

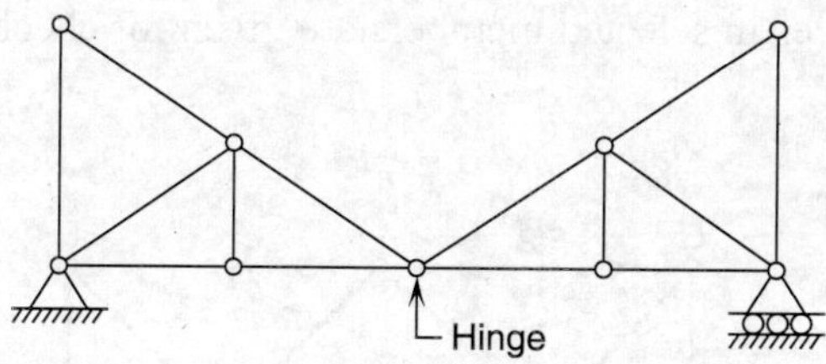

Figure 2.23

[**Ans:** Unstable externally]

2.16

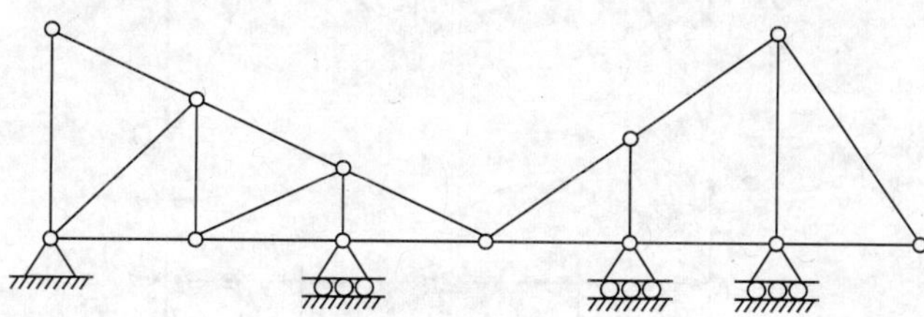

Figure 2.24

[**Ans:** Statically determinate externally and internally]

2.17

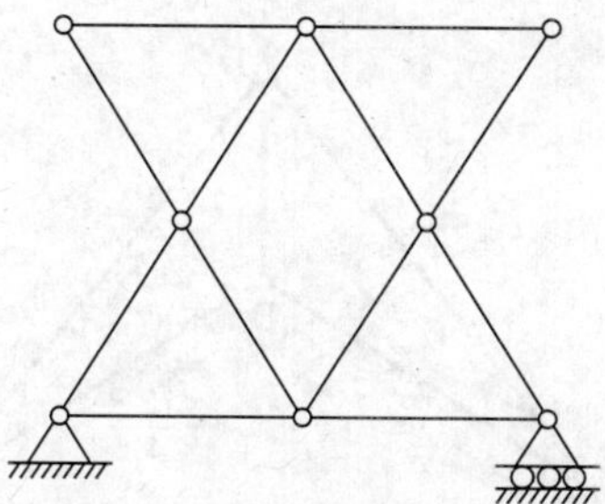

Figure 2.25

[**Ans:** Unstable internally]

2.18

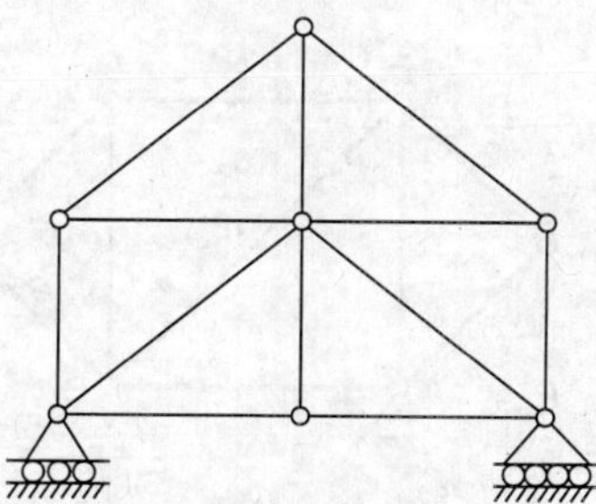

Figure 2.26

[**Ans:** Unstable externally but statically determinate internally]

In Problems 2.19 to 2.22, calculate the forces in all the members of the trusses using the method of joints. [Answers to the forces in selected members are given for checking purposes].

2.19

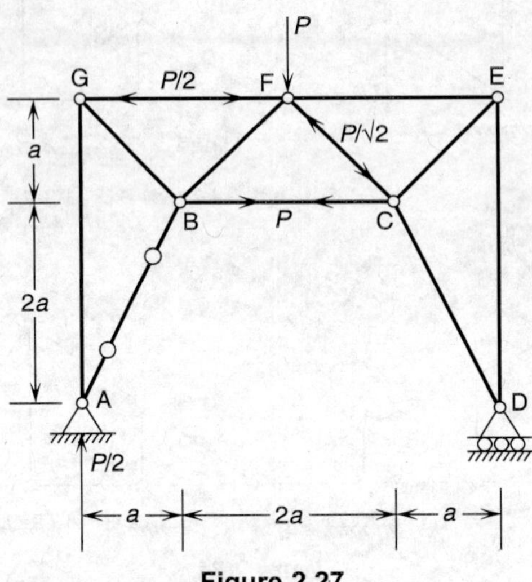

Figure 2.27

2.20

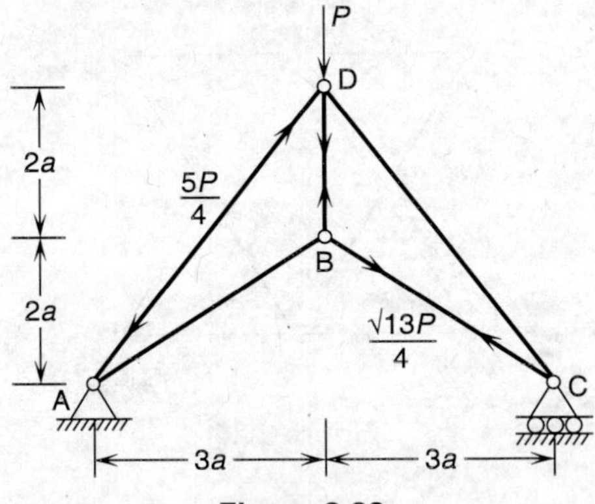

Figure 2.28

2.21

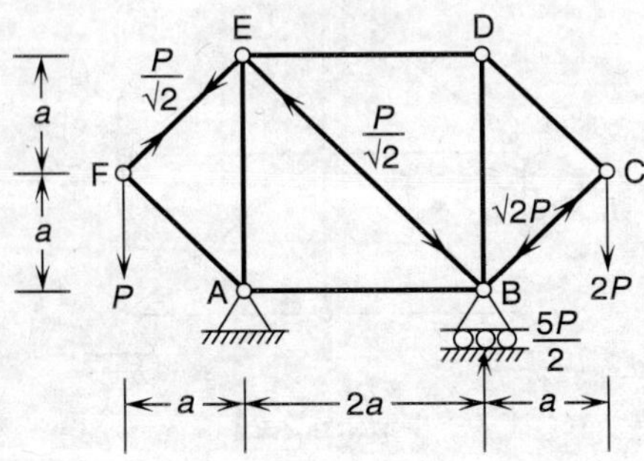

Figure 2.29

2.22

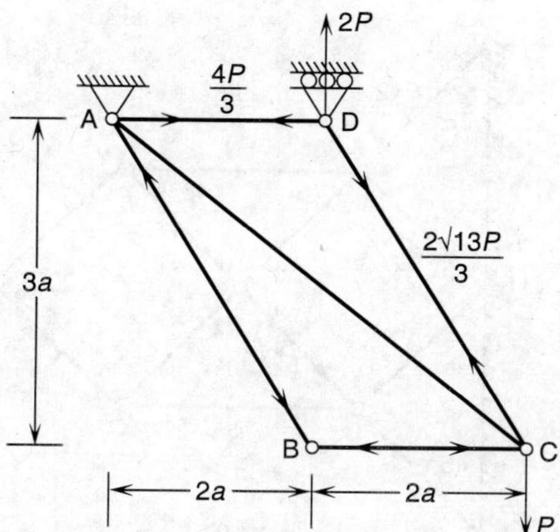

Figure 2.30

In Problems 2.23 to 2.30, calculate the forces in all the members of these trusses using the method of sections or method of joints, as appropriate. [Answers to the forces in selected members are given for checking purposes].

2.23

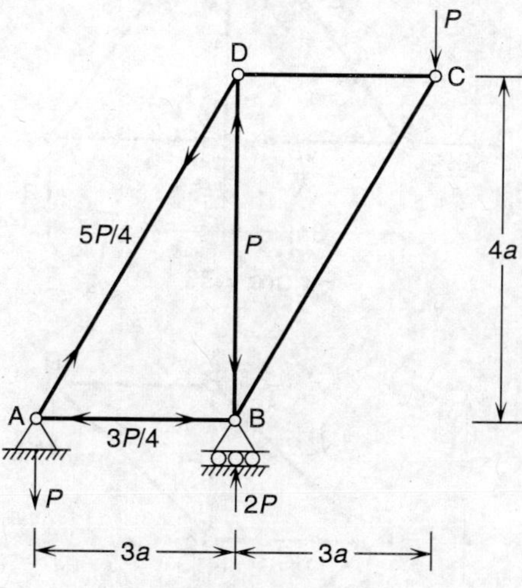

Figure 2.31

2.24

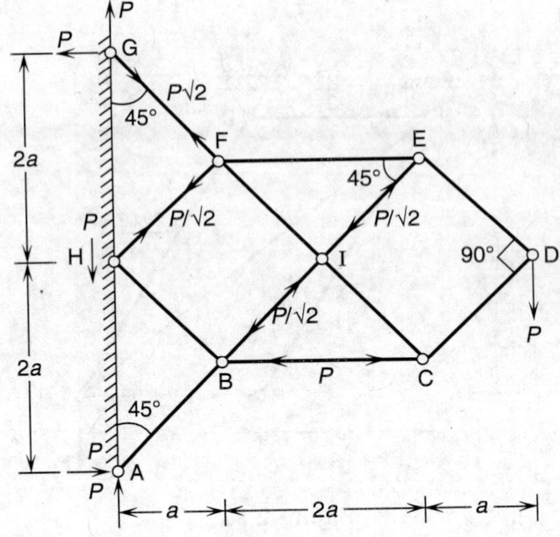

Figure 2.32

2.25

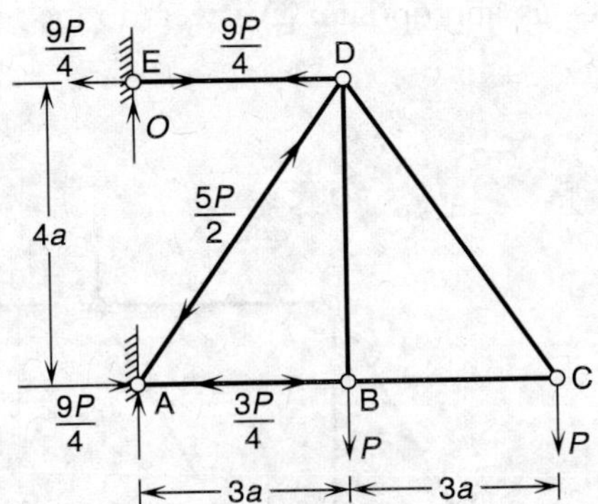

Figure 2.33

2.26

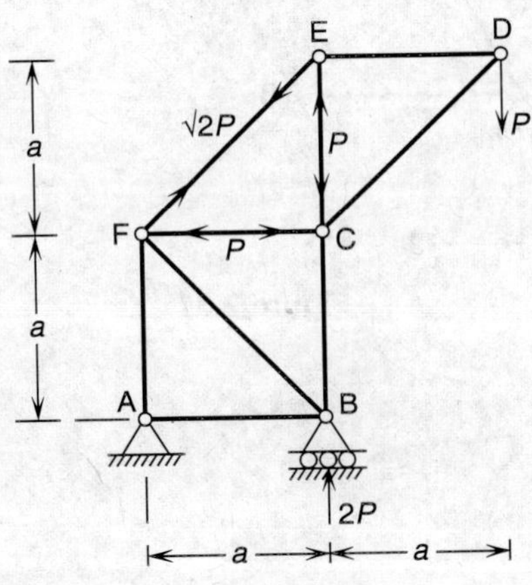

Figure 2.34

2.27

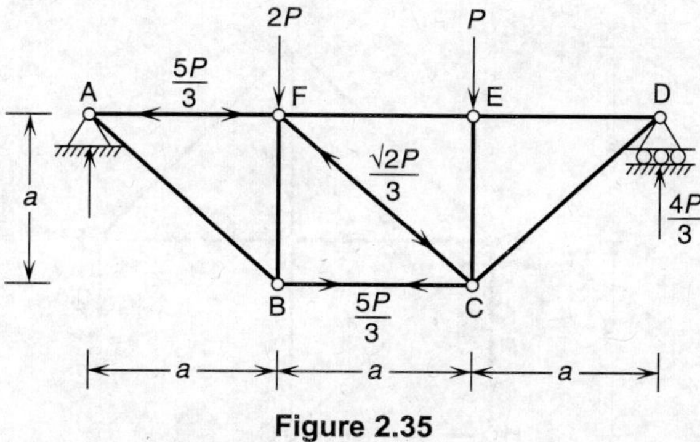

Figure 2.35

2.28

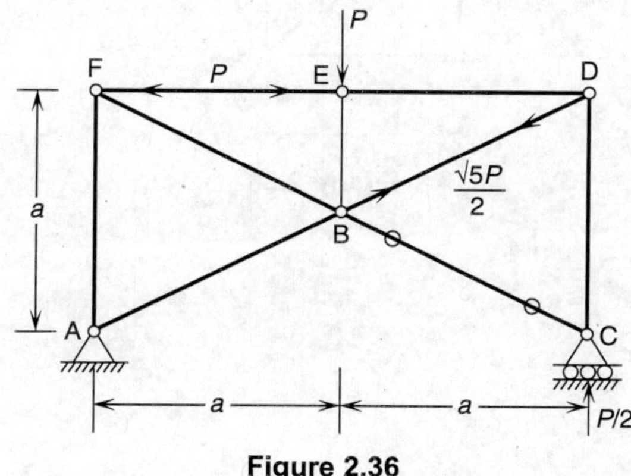

Figure 2.36

2.29

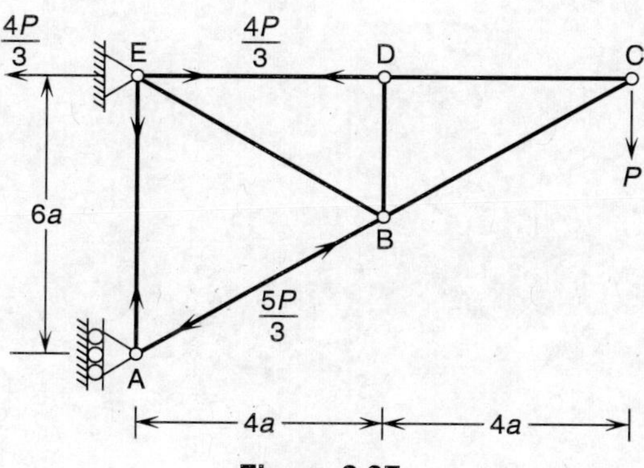

Figure 2.37

2.30

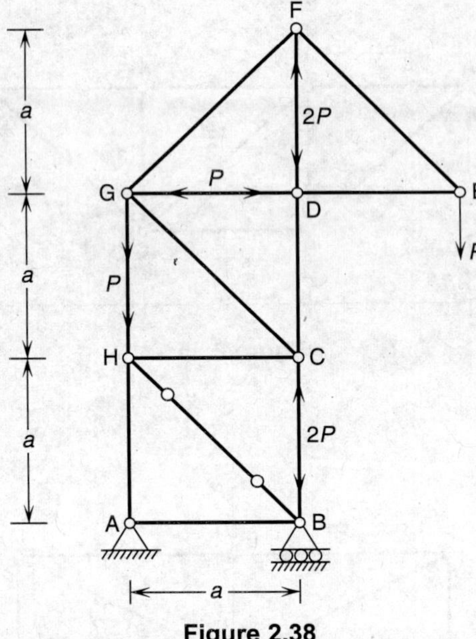

Figure 2.38

Chapter 3

Analysis for Deformation of Bars: Statically Determinate Systems (Simple Stresses and Strains—Elastic Constants)

3.1 INTRODUCTION

The application of a load to a member invariably produces dimensional changes in the member which include size or shape or both. The dimensional changes in the member are generally of two types namely, *elastic* and *plastic*, these occur sequentially with increasing loads. At smaller loads, the deformations are in the elastic range of the materials. In the elastic range, the member generally returns to its original dimensions if the applied loads are removed, i.e. it behaves like a spring. Deformations in the elastic range of the materials depend directly on the magnitude of the stress present in the member. As the loads increase, the deformations in the material move into plastic range with the stress in the material reaching a sufficiently high level to *cause a permanent change in the internal structure of the material*. At this stage, the original state cannot be exactly regained, even if the load is removed completely, i.e. the material has undergone irreversible dimensional changes and *permanent set* shall exists. In the plastic range, the deformations are very large relative to those in the elastic range and vary nonlinearly with the load or stress level present in the material.

It should be noted that not all materials demonstrate both elastic and plastic behaviour under increasing loads, typically steel demonstrate both elastic and plastic behaviours and concrete does not. From structural design point of view the *strength* and *load-deformation* properties of the material are of prime interest. The strength is dependent on load-deformation properties of the material.

3.2 STRESSES

The material of a body while undergoing a deformation due to an external load system offers a resistance which tends to prevent the deformation. This resistance is offered by the material by virtue of its strength. In the elastic range, the resistance offered by the material is proportional to the deformation induced in the material by the external load system. In a loaded member, the resistance offered by the material against deformation and the applied load are in equilibrium. When the member is incapable of offering the necessary resistance against external forces, the deformation continues leading to the *failure* of the member.

The each type of resisting force offered by a body against the deformation namely tension, compression, shear, bending moment, etc. cause internal stresses within the material of the body. *The intensity of the resistance per unit area of the section is called the stress or unit stress*. These stresses deform structure and tend to cause the material to fail in tension or compression or through the shearing action. The capacities of different materials for carrying different types of stresses are determined experimentally. Thus, by knowing the stress level to which a material is likely to be subjected, the size of the member can be determined to carry the stress safely.

3.2.1 Tensile and Compressive Stresses

When the resistance offered by a section of a member is against an increase in length, the section is said to offer a *tensile stress*. On the other hand, if the resistance developed is against a decrease in length, the section is said to offer a *compressive stress* as illustrated in Figure 3.1(a).

A member carrying external tensile or compressive force along its axis and acting at the centroid or the point of symmetry of the cross-section is subjected to stresses which are uniform across the cross-section of the member and is said to have *stresses of uniform intensity*. The internal resisting force F is the resultant of all internal stresses on the section. These members are called *axially loaded members*. Consider section I-I of the loaded member shown in Figure 3.1(b-i). Since the resistance offered by the material F against the deformation and the applied force P are in equilibrium, the resultant of internal stresses developed would have the same line of action as the externally applied force as shown in Figure 3.1(b-iii). The magnitude of uniformly distributed stresses is given by:

$$\text{Intensity of stress} = \frac{\text{Resisting force}}{\text{Area of cross-section}} \quad \text{or} \quad \sigma_o = \frac{F}{A} = \frac{P}{A} \tag{3.1}$$

where σ_o is the stress, the force intensity per unit area as shown in Figure 3.1(b-iv); P is the applied axial force, and A is the area of cross-section. The stresses of this type are generally called axial or normal stresses. Note that this *stress in the member depends only on the applied force and cross-sectional area and is independent of the material of the member*. More complex non-uniformly distributed stress states associated with shear and bending will be discussed later in the chapter.

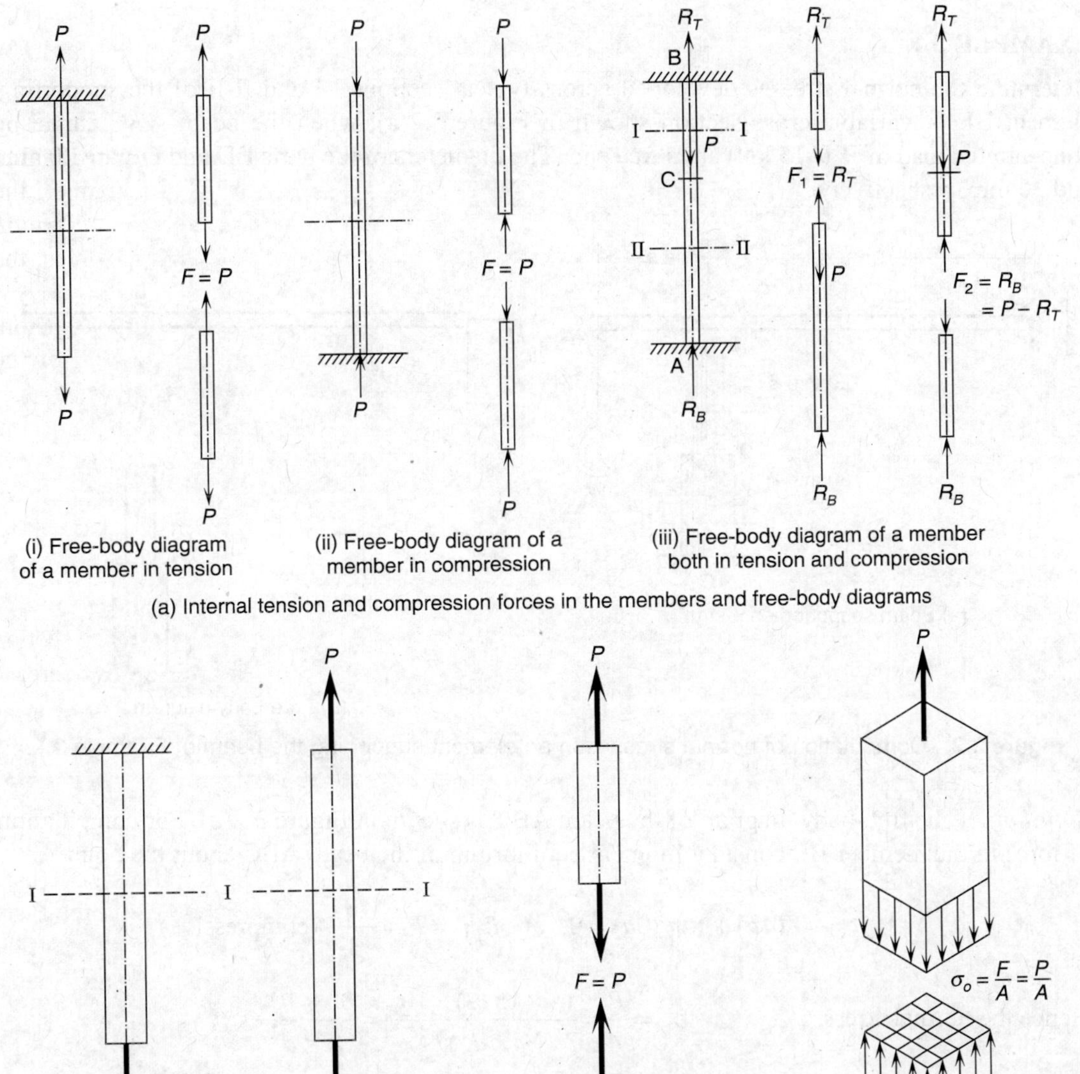

(a) Internal tension and compression forces in the members and free-body diagrams

(b) Computation of internal stresses developed in an axially loaded member at a section I–I

Figure 3.1 Free-body diagrams and internal stresses in axially loaded members.

EXAMPLE 3.1

Determine the normal stresses developed across typical sections I-I and II-II of the supporting element BE of variable cross-section shown in Figure 3.2(a), when the beam ABC carries a concentrated load of W (=15 kN) at its free end. The diameters of the parts BD and DE are 20 mm and 30 mm, respectively.

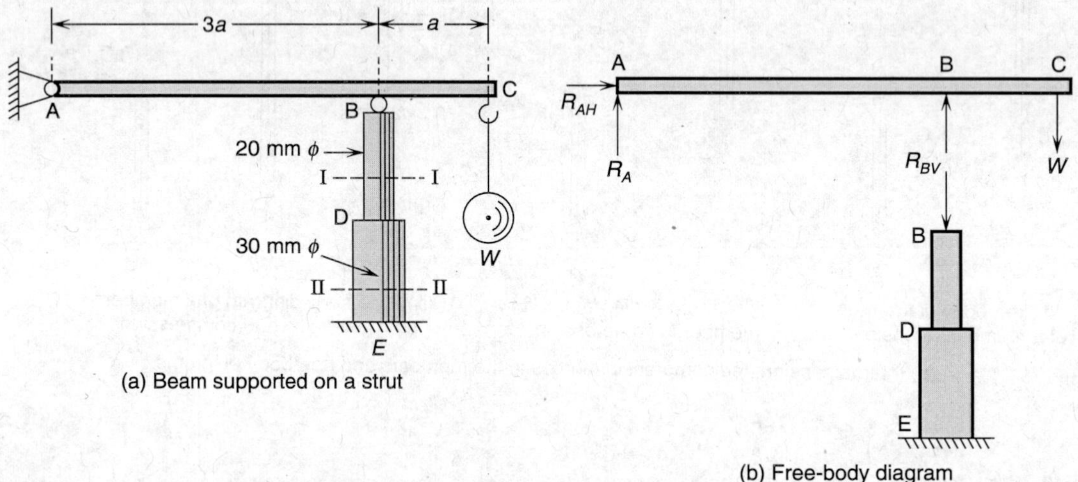

Figure 3.2 Computation of normal stresses in an element supporting the beam of Example 3.1.

Solution: The free-body diagram of the beam ABC is shown in Figure 3.2(b). For computation of force in the member BE consider moment equilibrium of the beam ABC about the point A,

$$\sum M_A = -W(4a) + R_{BV}(3a) = 0 \quad \text{or} \quad R_{BV} = P = \frac{4W}{3} \quad \text{(compressive)}$$

Hence the normal stress,

$$\sigma_o = \frac{P}{A} = \frac{(4 \times 15/3) \times 10^3}{(\pi \times d^2)/4} = \frac{80 \times 10^3}{\pi \times d^2}$$

At the section I-I:

$$\sigma_{o,I} = -\frac{80 \times 10^3}{\pi \times 20^2} = -63.66 \text{ N/mm}^2 = -63.66 \text{ MPa}$$

At the section II-II:

$$\sigma_{o,II} = -\frac{80 \times 10^3}{\pi \times 30^2} = -28.29 \text{ N/mm}^2 = -28.29 \text{ MPa}$$

EXAMPLE 3.2

A variable cross-section concrete bridge pier with top and bottom widths of 600 mm and 1800 mm, respectively, carries a uniformly distributed load of 25 kN/m^2 at its top as shown in Figures 3.3(a) and (b). The unit weight of concrete is 25 kN/m^3. Determine the normal stress developed across the section at a distance 1.0 m from the top, i.e. at mid-height of the pier.

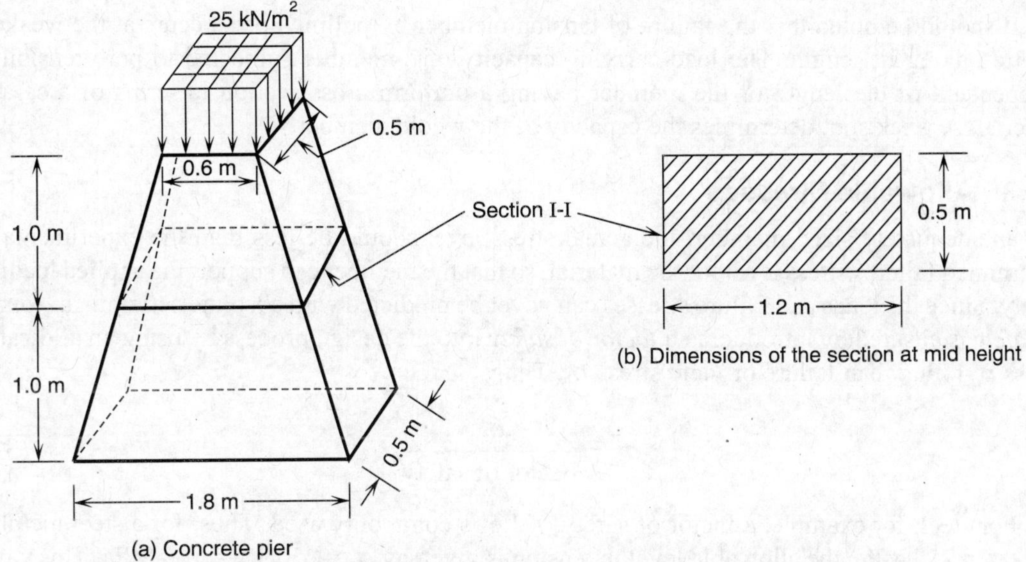

Figure 3.3 Normal stress at a section of concrete bridge pier of Example 3.2.

Solution: The total force acting at the section I-I is:

$F_{I\text{–}I}$ = Uniformly distributed load (u.d.l.) + Weight of concrete above the section I–I

= (Area of pier at the top × Intensity of u.d.l.) + (Volume of concrete × Unit weight)

$$= (0.6 \times 0.5) \times 25 + \left[\left\{\frac{1}{2} \times (0.6 + 1.2) \times 1.0\right\} \times 0.5\right] \times 25 = 18.75 \text{ kN (compression)}$$

Area of section at I-I, $A = 1.2 \times 0.5 = 0.6 \text{ m}^2$

Considering the total force acting at the section I-I to be uniformly distributed, the normal stress is:

$$\sigma_o = \frac{P}{A} = -\frac{F}{A} = -\frac{18.75 \times 10^3}{0.6 \times 10^6} = -0.03125 \text{ MPa (compressive)}$$

The corresponding free body diagram is shown in Figure 3.2(b).

3.2.2 Strength of Tension Member

In an axially loaded member, if the material of the member can sustain the actual stress developed which is given by $\sigma_o = P/A$, the member will carry the load. With the increasing loads, the stress levels increase and there comes a stage when the stress intensity exceeds the capacity of the material to withstand the pulling apart. This is the *failure stress level* of the material which depends upon its properties. The failure stress level of a material is determined experimentally. For mild steel, the commonly used material for structural steel sections, the apparent tensile stress level associated with the beginning of the material pulling apart or yielding is f_y = 250 MPa.

It should be noted that the failure of tension member by pulling apart occurs at the weakest location along its length. The load-carrying capacity of a member subjected to pure tension is independent of the length of the member having a uniform cross-section in terms of area and material. A weak spot determines the capacity of the whole member.

3.2.3 Allowable Stresses

For an adequately sized member, the actual stress level should be less than the experimentally determined failures stress level for the material, so that the member can support the applied loading safely. Since the loads and failure stresses can never be predicted with absolute certainty, a *margin of safety* is ensured by introducing a *factor of safety* into the design process by using an allowable stress σ_a rather than failure or yield stress σ_y. Thus,

$$\sigma_a = \frac{\sigma_y}{\text{Factor of safety}} \tag{3.2}$$

For steel, for example, a factor of safety of 1.5 is commonly used. Thus, for a steel member with $\sigma_y = 250$ MPa, the allowable level in tension is given by $\sigma_a = 250/1.5 = 166$ MPa. This value can be used in computing the required cross-sectional area of a steel member in tension, thus,

$$A_{\text{required}} = \frac{\text{Tensile force}}{\text{Allowable stress}} = \frac{P}{\sigma_a} \tag{3.3}$$

EXAMPLE 3.3

Determine the diameter of steel rod required to safely support an axial tension load of 60 kN when stresses are limited to 165 MPa.

Solution: The area required:

$$A = \frac{P}{\sigma_a} = \frac{60 \times 10^3}{165} = 363.64 \text{ mm}^2$$

The diameter of the rod is,

$$d = \sqrt{\frac{4A}{\pi}} = \sqrt{\frac{4 \times 363.64}{\pi}} = 21.52 \text{ mm}$$

3.3 NORMAL STRAINS

As illustrated in Figure 3.4(a), a member subjected to an axial force undergoes elastic changes in the direction of applied load and in its lateral dimensions. The lateral dimensions of a member decrease when the member is subjected to a *tensile force* and increase when the member is subjected to a *compressive force*. The actual deformations in an element of the stressed body shown in Figure 3.4(b) can be represented in a simplified form as in Figure 3.4(c). This simplification has been followed in the text.

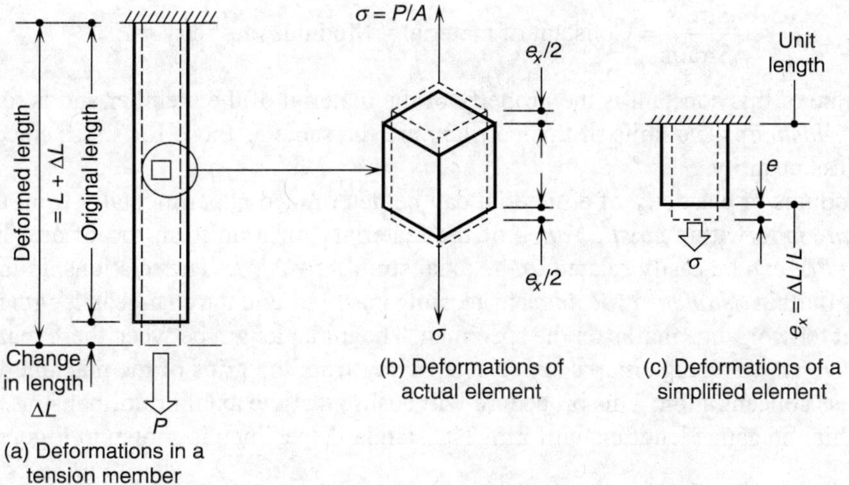

Figure 3.4 Deformations of a simple tension member.

The behaviour of a body subjected to a load is generally described in terms of *strain* ε which is defined as the *ratio of the change in size or shape of an element, subjected to stress, to the original size or shape.* In the present case, since the strain is associated with the normal stress, it is usually called *normal strain*. *Being a ratio, it is dimensionless quantity*, i.e. *it has no physical dimensions*. However, it is useful to consider strain as the amount of deformation per unit of length, thus its dimensions would be millimetres/millimetres or mm/mm, m/m. Sometimes the strain is given as a percentage of the original length.

If a prism of uniform material is subjected to a uniform normal stress, it will elongate by an amount e as shown in Figure 3.4(b). The normal strain ε is defined as the extension (or compression) per unit length, i.e. $\varepsilon = e/L$.

3.4 ELASTIC LIMIT

A material is said to be elastic when it undergoes a deformation under the action of external loads such that the deformation disappears on the removal of the loads. With the removal of the loads, the resisting force and hence the stress will vanish along with the deformation. However, this is true when the deformation caused by the loads is within a certain limit. For every material, the property of regaining its previous shape and size is exhibited on the removal loads, when the stress intensity is within a certain limit called *elastic limit*. As the loads increase, the intensity of stress exceeds the elastic limit and the member loses to some extent the property of elasticity. At this stage, the original state cannot be exactly regained, even if the load is removed completely, i.e. the material has undergone irreversible dimensional changes and a *residual strain* or *permanent set* remains.

3.4.1 Modulus of Elasticity

A general relationship between stress and strain for elastic material was first postulated by Robert Hooke and is known as *Hooke's law*. Hooke's law states that for an elastic body, the ratio of the stress present in an element to the strain produced is a constant. Thus,

$$\frac{\text{Stress}}{\text{Strain}} = \text{Constant of material} = \text{Modulus elasticity} = E \tag{3.4}$$

The value of this constant is the property of the material of the member and is referred to as *modulus of elasticity*. The units of this constant are the same as those for stress since strain is a dimensionless quantity.

The modulus of elasticity of a material can be determined experimentally from the slope of *stress–strain curve* within *elastic range* of the material. For a uniform rod of area A, the axial stress, $\sigma_o = P/A$ can be easily related to the axial strain, $\varepsilon = \Delta L/L$. The relationship in practice is obtained by pulling a *standard specimen* in a testing machine, and the change in length is measured with respect to two gauge marks on the specimen. The initial length between these marks is called the *gauge length*. The gauge marks are selected away from the grips of the machine where there may be stress concentration. This procedure will ensure that the axial or normal stress across any section within the gauge length is uniform. The standard specimen is shown in Figure 3.5(a).

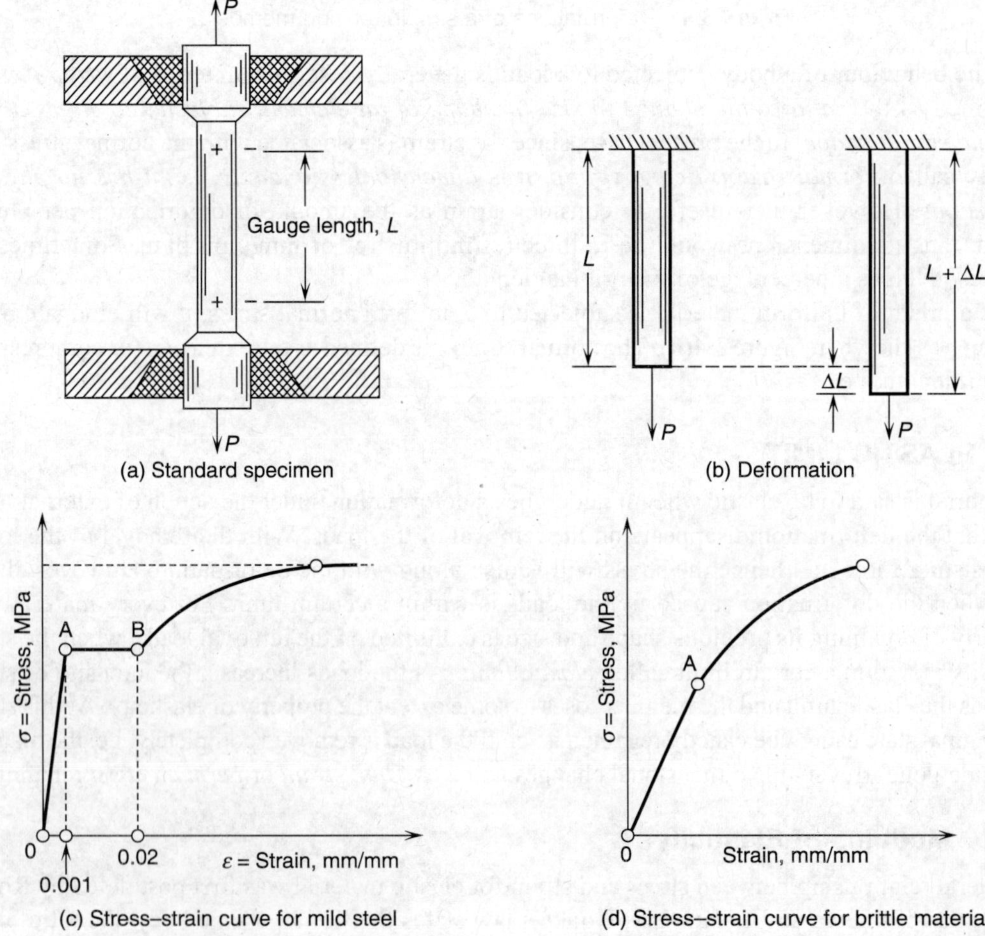

Figure 3.5 Determination of modulus of elasticity of materials.

As the load P is gradually increased, the length also changes gradually (see Figure 3.5(b)). Although, P can be plotted against ΔL but for generality the stress, σ is plotted against strain ε. The typical stress–strain curve shown in Figure 3.5(c) represents stress–strain curve for mild steel, a ductile material; whereas the curve shown in Figure 3.5(d) is for a brittle material like cast iron, concrete, etc.

It should be noted that as the stress level present in a member continues to increase upto the point A, the curve is a straight line, indicating that stress is linearly proportional to strain. This point is termed as the *proportionality limit of the material.* The region OA is known as *elastic zone*. Beyond the point A, the strains developed are no longer linearly dependent on the stress and the concept of constant modulus of elasticity is not valid. It should be noted that in the stress–strain curve for steel shown in Figure 3.5(c), the region AB is approximately horizontal, i.e. the strain ε increases without much change in the magnitude of the stress σ_o. This is known as *yield zone* or *plastic range*. In the plastic range, the deformations are very large as compared to those in elastic range. The aluminium do not have pronounced proportional limit, while materials like cast iron, concrete, etc do not exhibit any plastic deformation.

As discussed above for a limited range from the origin, the experimental values of stress versus strain lie essentially on a straight line. For all practical purposes, the relationship between stress and strain is taken linear for all structural materials. The Moduli of elasticity E of some of the more commonly used materials are:

TABLE 3.1 Moduli of elasticity E of common materials

Material	*Modulus of elasticity E*, MPa	(GPa)
Steel	2.0 to 2.1 × 10^5	(200 to 210)
Copper	1.2 × 10^5	(120)
Brass	1.0 to 1.2 × 10^5	(100 to 120)
Aluminium	0.7 to 0.8 × 10^5	(70 to 80)
Gray cast iron	0.7 × 10^5	(70)
Concrete	0.21 × 10^5	(21)
Wood (parallel to grains)	0.08 to 0.12 × 10^5	(8 to 12)

3.4.2 Deformations in the Materials

Once the value of E for a material is known, deformations in the materials under different stress conditions can be predicted. For computation of displacement, consider an infinitesimal differential element of length dz of the axially loaded member i–j having gradually varying cross-sectional area A_z and original length L_{i-j} subjected to a force system shown in Figure 3.6. The axial deformation of the element is given by:

$$du = \varepsilon_z \, dz$$

where ε_z is the normal strain in the Z-direction due to applied forces, u is the absolute displacement of a point on the member from an initial fixed position, and du is the axial deformation of the element. Consider the origin O at the point i and integrating

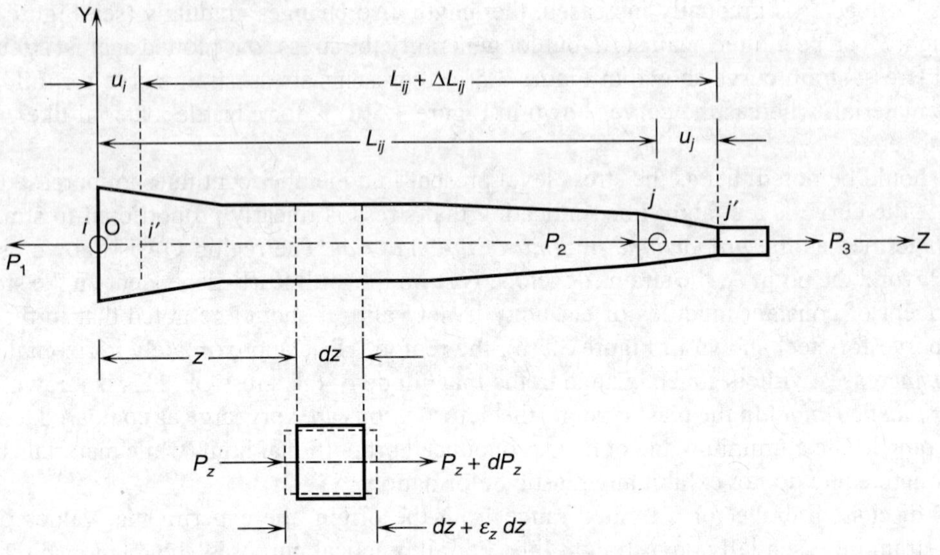

Figure 3.6 Notations for an axially loaded member.

$$\int_0^{L_{i-j}} du = u_j - u_i = \int_0^{L_{i-j}} \varepsilon_z \, dz$$

Therefore,
$$u_j = u_i + \int_0^{L_{i-j}} \varepsilon_z \, dz \tag{3.5}$$

where u_j and u_i are the absolute displacements of points j and i respectively. The quantity $u_j - u_i$ represents the change in length ΔL between the points j and i. For linearly elastic materials, the strain is given by Hooke's law, $\varepsilon_z = \sigma_z/E_z$, where $\sigma_z = F_z/A_z$. Therefore, change in length of the member is given by

$$\Delta L_{i-j} = \int_0^{L_{i-j}} \left(\frac{F_z}{A_z E_z} \right) dz$$

Thus,
$$u_j = u_i + \Delta L_{i-j} = u_i + \int_0^{L_{i-j}} \left(\frac{F_z}{A_z E_z} \right) dz \tag{3.6}$$

where the resisting force $F_z = F(z)$, the cross-sectional area $A_z = A(z)$ and the elastic modulus $E_z = E(z)$ can vary along the length of the member.

For a homogeneous member of constant cross-sectional area A, subjected to a force P, applied at the end, the strain is given by

$$\text{Strain, } \varepsilon = \frac{\text{Increase in length}}{\text{Original length}} = \frac{\Delta L}{L} \tag{3.7}$$

Considering the stress to be constant over the cross-section, the *nominal* or *engineering stress* is given by dividing the applied force by the original cross-sectional area, i.e. $\sigma_o = \dfrac{F(=P)}{A}$. The modulus of elasticity can therefore be expressed as:

$$E = \frac{\sigma_o}{\varepsilon} = \frac{(P/A)}{(\Delta L/L)} = \frac{PL}{A\Delta L}$$

or
$$\Delta L = \frac{PL}{AE} \tag{3.8}$$

This equation indicates that deflection of an axially loaded member is directly proportional to the applied force and length, and is inversely proportional to A and E. The equation can also be expressed as

$$P = \left(\frac{AE}{L}\right)\Delta L \quad \text{or} \quad \frac{P}{\Delta L} = \left(\frac{AE}{L}\right)$$

The quantity $P/\Delta L$ which represents the force required to produce a unit deflection is called axial *stiffness* k of an axially loaded member. This is analogous to spring constant of a linear spring, defined as

$$k = \frac{P}{\Delta L} = \left(\frac{AE}{L}\right) \text{ [N/m]} \tag{3.9}$$

The reciprocal of k is defined as *flexibility* f i.e.

$$f = \frac{1}{k} = \frac{\Delta}{P} = \left(\frac{L}{AE}\right) \text{ [m/N]} \tag{3.10}$$

These concepts are widely used in structural analysis.

3.5 ANALYSIS OF SIMPLE SYSTEMS

The systems composed of components of the same material in which forces can be determined by satisfying the equilibrium requirements are called *simple systems*. These systems do not require the application of compatibility conditions or geometrical requirements for the analysis for forces. These systems are generally called statically-determinate systems.

The following simple examples will illustrate computation of stress and displacements for members subjected to axial tension and compression.

EXAMPLE 3.4

Draw diagrams indicating variation of normal forces, stresses and displacements along the length of a stepped member of Figure 3.7(a) carrying a force P at its end. Also determine the maximum stress and displacement, when the steel member with cross-section A equal to 200 mm^2 carries an axial tension load of 60 kN over a length of 2.4 m. The modulus of elasticity of the material is $E_s = 2 \times 10^5$ MPa.

Solution: It is evident from the free-body diagram shown in Figure 3.7(b) that the normal resisting force F at any cross-section of the stepped beam is numerically equal to the external force P. The variation of normal force is shown in Figure 3.7(c). The normal stress at any cross-section is,

$$\sigma_o = \frac{P}{A}$$

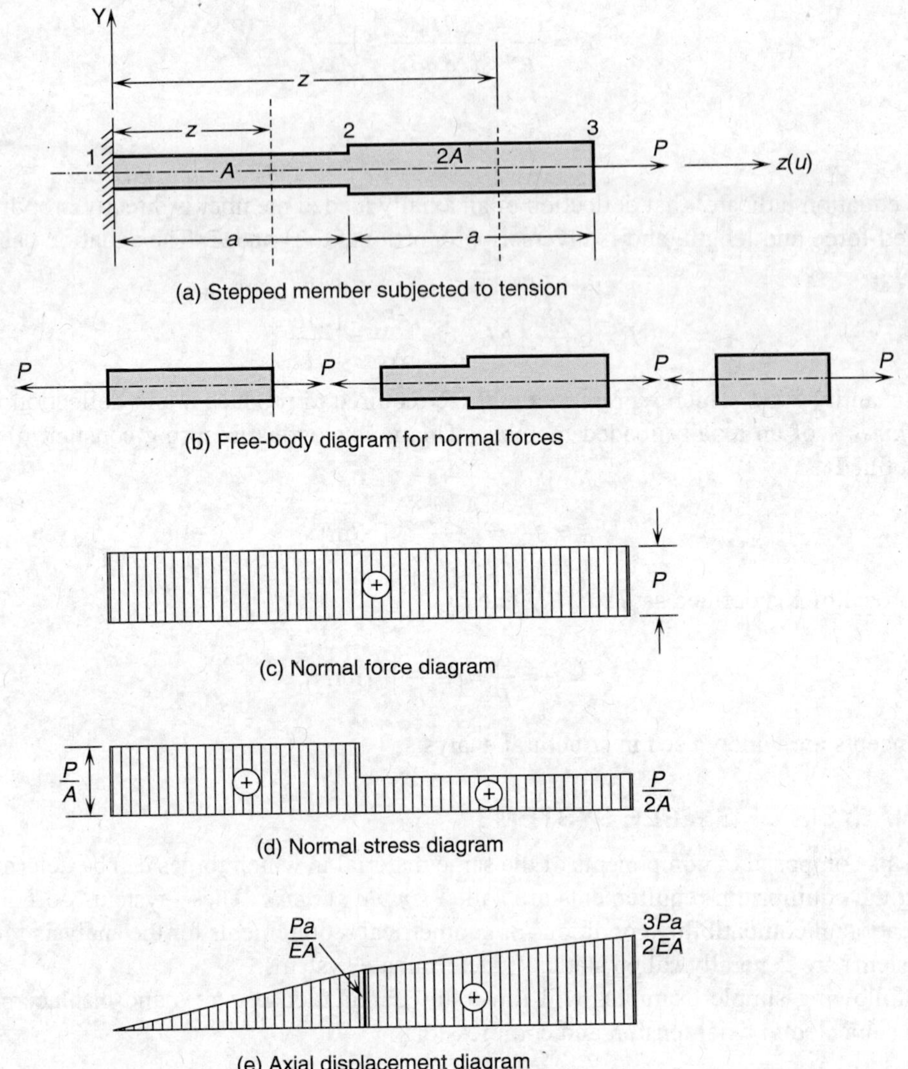

Figure 3.7 Variation of normal forces, stresses and displacements along the length of a stepped member subjected to a force P at its end.

i.e. stress at any cross-section is inversely proportional to the magnitude of A. The maximum value of stress $\sigma_{o,\max}$ is

$$\sigma_{o,\max} = \frac{P}{A} = \frac{60 \times 10^3}{200} = 300 \text{ MPa}$$

The displacement or movement at a section at distance z from point 1 is equal to the elongation of the segment of length z, thus, $u = \dfrac{Pz}{AE}$. Therefore,

In the segments:

$$0 \le z \le a \qquad u = \frac{Pz}{AE}$$

$$a \le z \le 2a \qquad u = \frac{Pa}{AE} + \frac{P(z-a)}{(2A)E}$$

The variation of displacement is shown in Figure 3.7(e). The maximum displacement occurs at the end of the member, i.e. at $z = 2a$,

$$u_{\max} = \frac{Pa}{AE} + \frac{P(a)}{(2A)E} = \frac{3Pa}{2AE}$$

$$= \frac{3 \times (60 \times 10^3) \times 1200}{2 \times 200 \times (2 \times 10^5)} = 2.7 \text{ mm}$$

Axial displacement diagram is given in Figure 3.7(e).

EXAMPLE 3.5

A steel bar shown in Figure 3.8(a) consists of two parts AB and BC 1.5 m long each having cross-sectional areas of 480 mm^2 and 600 mm^2, respectively. The bar is rigidly fixed at its upper end A. Before application of load there is clearance of 1.0 mm between its lower end C and a rigid horizontal support. Determine the reactions at the supports and the stresses in the two parts due to an axial load of 125 kN applied vertically downward at B. The modulus of elasticity of the material is $E_s = 2 \times 10^5$ MPa.

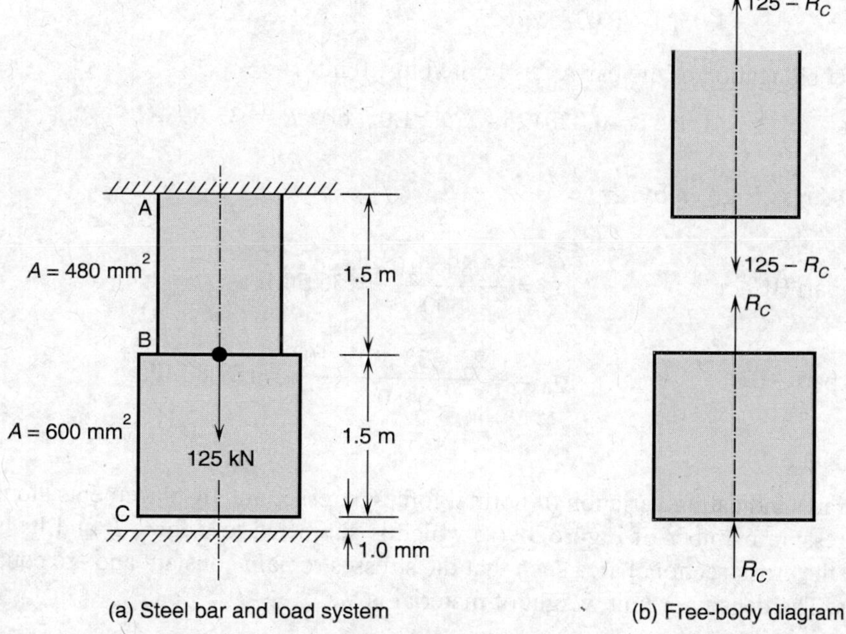

(a) Steel bar and load system (b) Free-body diagram

Figure 3.8 The steel bar of variable cross-section subjected to an axial load at mid-height.

Solution: Free elongation of the bar in the absence of support near the end C:

$$\Delta L = (\Delta L)_{AB} + (\Delta L)_{BC} = \frac{P_{AB}L}{A_{AB}E} + \frac{P_{BC}L}{A_{BC}E} = \frac{(125\times10^3)\times(1500)}{480\times(2\times10^5)} + 0 = 1.9531 \text{ mm}$$

However, due to the presence of horizontal support near the end C, the total elongation of the bar AC is limited to 1.0 mm. Since the bar is not allowed to elongate beyond this limit, the bar is subjected to an upward reaction R_c. The deformation produced by R_c when acting alone on the freely expanded bar should be equal to the excess deformation of the bar, i.e. (1.9531–1.0) mm. Thus,

$$\Delta L = \frac{(R_c\times10^3)\times(1500)}{480\times(2\times10^5)} + \frac{(R_c\times10^3)\times(1500)}{600\times(2\times10^5)} = \frac{(R_c\times10^3)\times(1500)}{(2\times10^5)}\times\left(\frac{1}{480}+\frac{1}{600}\right)$$

$$= 1.9531 - 1.0 \quad \text{or} \quad 0.028125 R_c = 0.9531$$

Therefore, $R_c = 33.89$ kN.

Alternatively, consider the combined effect of the externally applied force and the reaction. The part BC of the bar is subjected to an upward compressive force R_c and the part AB of the bar will be subjected to a tensile force of $125 - R_c$ as shown in Figure 3.8(b). Thus, the net deformation

$$\Delta L = \frac{\{(125-R_c)\times10^3\}\times(1500)}{480\times(2\times10^5)} - \frac{(R_c\times10^3)\times(1500)}{600\times(2\times10^5)}$$

$$= \frac{125\times10^3 0\times(1500)}{480\times(2\times10^5)} - \frac{(R_c\times10^3)\times(1500)}{(2\times10^5)}\times\left(\frac{1}{480}+\frac{1}{600}\right)$$

$$= 1.9531 - 0.028125\times R_c$$

Since the net elongation of the bar AC is limited to 1.0 mm,

$$1.9531 - 0.028125\times R_c = 1.0 \quad \text{or} \quad R_c = 33.89 \text{ kN}$$

The normal stress is given by, $\sigma_o = \dfrac{P}{A}$

In the part BC, $$\sigma_o = \frac{33.89\times10^3}{600} = 56.5 \text{ MPa}$$

In the part AB, $$\sigma_o = \frac{(125-33.89)\times10^3}{480} = 189.8 \text{ MPa}$$

EXAMPLE 3.6

Draw diagrams indicating variation of normal forces, stresses and displacements along the length of a compression member of Figure 3.9(a) which is subjected to a force P and its own weight. Determine the cross-sectional area such that the stresses remain constant and are equal to P/A_o at all sections. The density or unit weight of material is γ_o.

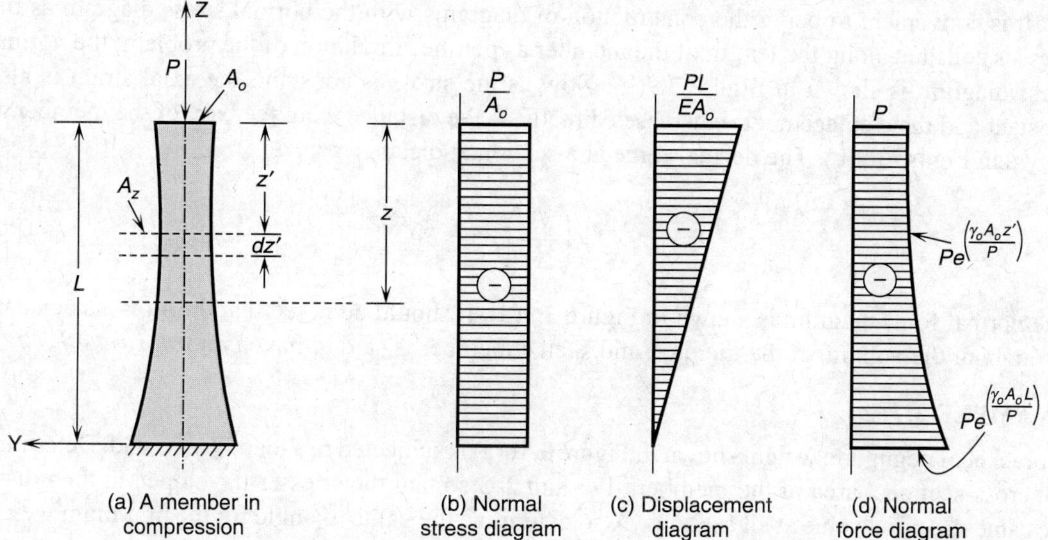

Figure 3.9 Computation of variation of stress and displacement in compression member of Example 3.6.

Solution: The normal compressive force F at a distance z from the end is

$$F_z = P + \gamma_o \int_0^z A_z \, dz'$$

Therefore, normal stress

$$\sigma_z = \frac{F_z}{A_z} = \frac{P + \gamma_o \int_0^z A_z \, dz'}{A_z} = \frac{P}{A_o} = \text{constant}$$

or
$$P + \gamma_o \int_0^z A_z \, dz' = \frac{PA_z}{A_o}$$

Differentiate both the sides with respect to z,

$$\gamma_o A_z = \frac{P}{A_o}\frac{dA_z}{dz} \quad \text{or} \quad dz = \frac{P}{\gamma_o A_o}\frac{dA_z}{A_z}$$

On integrating with respect to z: $z = \dfrac{P}{\gamma_o A_o}\log_e A_z + C$

At $z = 0$, $A_z = A_o$, hence $C = -\dfrac{P}{\gamma_o A_o}\log_e A_o$ and $z = \dfrac{P}{\gamma_o A_o}(\log_e A_z - \log_e A_o)$

Therefore,

$$\log_e \frac{A_z}{A_o} = \frac{\gamma_o A_o z}{P} \quad \text{or} \quad A_z = A_o e^{\frac{\gamma_o A_o z}{P}}$$

It is convenient to begin the construction of diagrams with the normal stress diagram as the stress is constant along the length of the member as per the stipulation of the problem; the normal stress diagram is shown in Figure 3.9(b). Now as the stress is constant, the axial strain is also constant and the displacement *u* increases directly as the distance from the base of the member as shown in Figure 3.9(c). The normal force at a section at distance *z* is,

$$F = \sigma_z A_z = \left(\frac{P}{A_o}\right) A_z = Pe^{\frac{\gamma_o A_o z}{P}}$$

The normal force diagram is shown in Figure 3.9(d). It should be noted that the stress is constant throughout the volume of the member and such a member is said to have *uniform strength.*

EXAMPLE 3.7

A bracket of negligible weight shown in Figure 3.10(a) is subjected to a load *W* at its end. Determine the cross-sectional area of the members 1–3 and 2–3 so that the stresses developed in them have the same specified value at all sections. Also determine the value of angle for the minimum weight of structure.

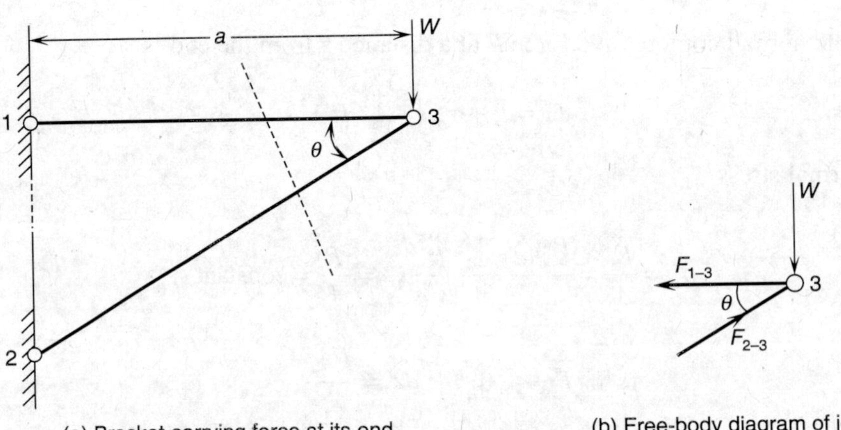

(a) Bracket carrying force at its end (b) Free-body diagram of joint C

Figure 3.10 Bracket problem of Example 3.10.

Solution: From the free-body diagram of the joint 3 shown in Figure 3.10(b), the normal forces in the members by method of joints are,

$$F_{1\text{-}3} = W \cot \theta \quad \text{and} \quad F_{2\text{-}3} = W \operatorname{cosec} \theta$$

The cross-sectional area of the members for the specified stress level σ_o at all sections,

$$A_{1\text{-}3} = \frac{F_{1\text{-}3}}{\sigma_o} = \frac{W \cot \theta}{\sigma_o} \quad \text{and} \quad A_{2\text{-}3} = \frac{F_{2\text{-}3}}{\sigma_o} = \frac{W \operatorname{cosec} \theta}{\sigma_o}$$

The weight of structure is proportional to its volume,

$$\text{Volume } V = A_{1-3}(a) + A_{2-3}\,(a\sec\theta) = \left(\frac{W\cot\theta}{\sigma_o}\right)(a) + \left(\frac{W\operatorname{cosec}\theta}{\sigma_o}\right)(a\sec\theta)$$

$$= \left(\frac{Wa}{\sigma_o}\right)\left(\cot\theta + \frac{1}{\sin\theta\cos\theta}\right)$$

For the volume to be the minimum,

$$\frac{dV}{d\theta} = 0 \quad \text{or} \quad \cos^2\theta = \frac{1}{3} \quad \text{i.e.} \quad \theta = 55°$$

EXAMPLE 3.8

A straight homogeneous member of negligible weight is rigidly fixed at both the ends as shown in Figure 3.11(a) and subjected to an axial force P applied at a distance of one-third of the length from the top end. Determine the maximum stresses developed in the member.

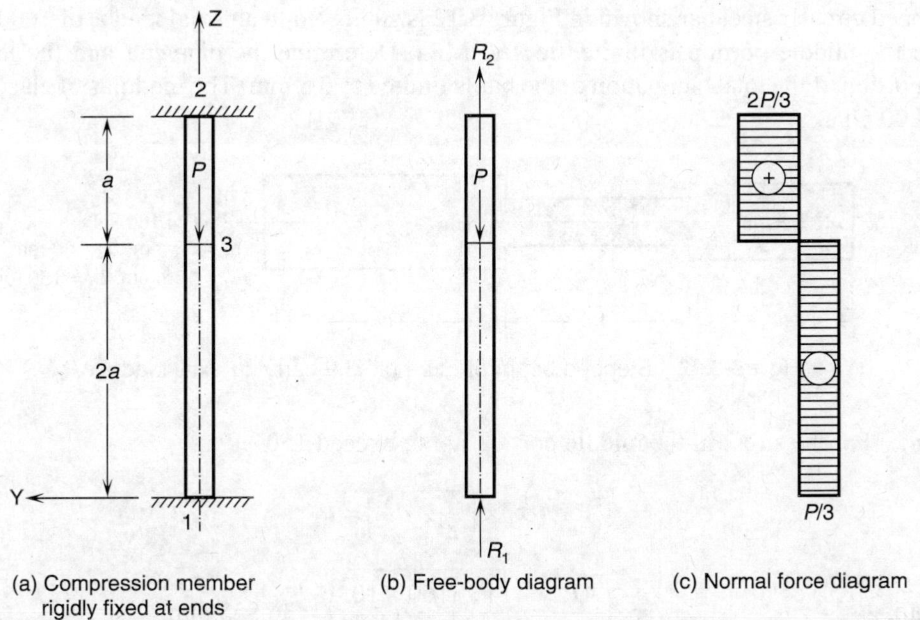

(a) Compression member rigidly fixed at ends (b) Free-body diagram (c) Normal force diagram

Figure 3.11 The fixed ended member of Example 3.11, carrying axial force at an intermediate point.

Solution: The structure is statical indeterminate to the first degree since two collinear reactions R_1 and R_2 of supports cannot be determined from one equilibrium condition,

$$\sum F_z = R_1 + R_2 - P = 0 \quad \text{or} \quad R_1 + R_2 = P$$

However, the condition that the overall length of the member remains unchanged, i.e. the lower portion shortens by the amount the upper portion lengthens or the net change in length is zero.

Using the normal forces in the segments obtained from the free-body diagram shown in Figure 3.11(b), the total elongation of the member is given by:

$$\Delta L = \sum \frac{FL}{AE} = \frac{R_2 a}{AE} + \frac{(R_2 - P)(2a)}{AE} = 0 \quad \text{or} \quad R_2 + 2(R_2 - P) = 0$$

or $$3R_2 = 2P \quad \text{i.e.} \quad R_2 = \frac{2P}{3}$$

Therefore, $$R_1 = P - R_2 = P - \frac{2P}{3} = \frac{P}{3}$$

The normal force diagram is shown in Figure 3.11(c). The maximum value of stress $\sigma_{o,\max}$ is

$$\sigma_{o,\max} = \frac{2P}{3A}$$

EXAMPLE 3.9

The stepped circular steel bar shown in Figure 3.12 is subjected to an axial load P of 160 kN. The stress in the middle portion is limited to 150 MPa. Determine the diameter and the length of middle portion, if the total elongation of the bar is limited to 0.3 mm. The modulus of elasticity for steel is 200 GPa.

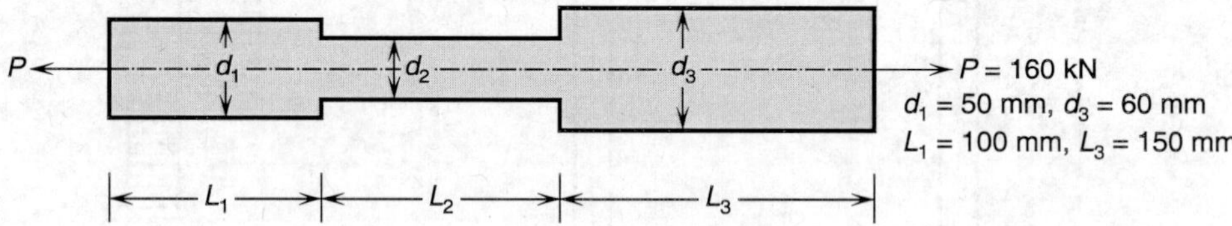

Figure 3.12 Stepped beam of Example 3.9 carrying axial load.

Solution: For the stress in the middle portion not to exceed 150 MPa,

$$\sigma_a A_s = P \quad \text{or} \quad \sigma_a \times \left(\frac{\pi}{4} \times d^2 \right) = P$$

Therefore, $$d = \sqrt{\frac{4P}{\pi \sigma_a}} = \sqrt{\frac{4 \times (160 \times 10^3)}{\pi \times 150}} = 36.853 \text{ mm}$$

Since the normal force F at any cross-section of the stepped beam is numerically equal to the external force P the stresses in the end portions are:

$$\sigma_1 = \frac{4P}{\pi d_1^2} = \frac{4 \times (160 \times 10^3)}{\pi \times (50)^2} = 81.487 \text{ MPa}$$

$$\sigma_3 = \frac{4P}{\pi d_3^2} = \frac{4 \times (160 \times 10^3)}{\pi \times (60)^2} = 56.588 \text{ MPa}$$

Total elongation of the bar,

$$\Delta = \sum_{i=1}^{3}\left(\frac{\sigma_i L_i}{E}\right) \quad \text{or} \quad \sum_{i=1}^{3} \sigma_i L_i = E\,\Delta$$

Therefore,

$$81.487 \times 100 + 150 \times L_2 + 56.588 \times 150 = (200 \times 10^3) \times 0.3$$

or $$150L_2 = 60000 - 8148.7 - 8488.2 \quad \text{or} \quad L_2 = 289.087 \text{ mm}$$

The diameter and the length of middle portion are 36.9 mm and 289.0 mm, respectively.

EXAMPLE 3.10

A rigid bar AB hinged at A is supported by a steel tie rod CD having a diameter of 20 mm as shown in Figure 3.13(a). Determine the maximum load W the bar can support at its free end B, if the modulus of elasticity and allowable stress for the material of the tie rod are 200 GPa and 150 MPa, respectively. Also compute the deflection at the free end B.

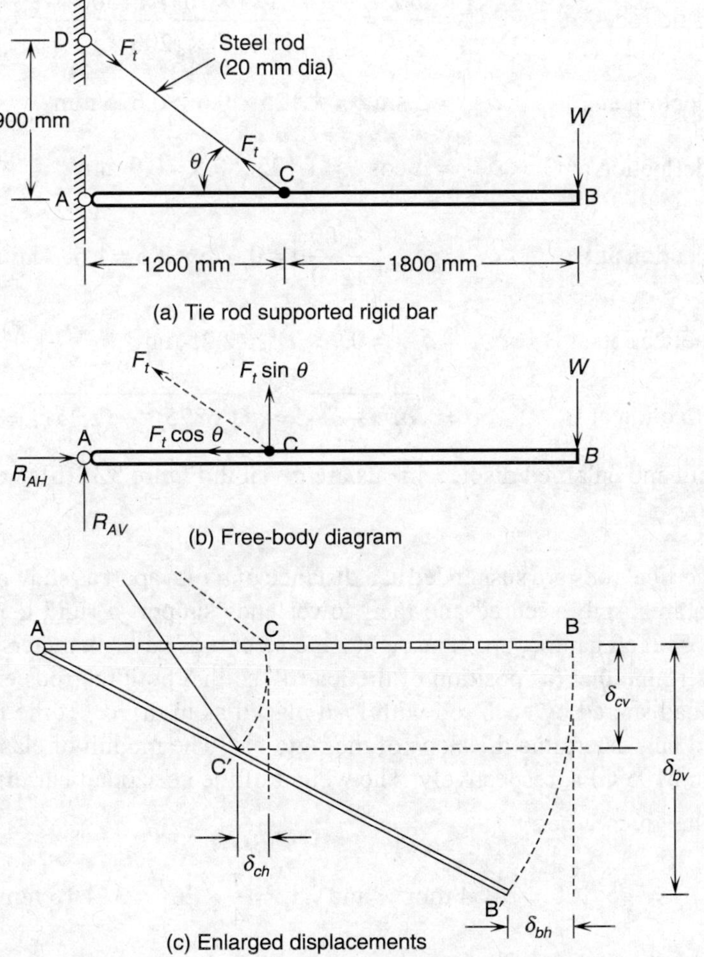

Figure 3.13 Analysis of cable supported rigid beam of Example 3.10.

Solution: Length of tie rod, $L = \sqrt{1200^2 + 900^2} = 1500$ mm

Inclination of tie rod, $\tan\theta = \dfrac{900}{1200} = 0.75$ and $\sin\theta = 0.6$; $\cos\theta = 0.8$

The permissible load in the tie rod,

$$F_t = \sigma_a A_s = 150 \times \left(\frac{\pi}{4} \times 20^2\right) = 47123.89\ \text{N} = 47.124\ \text{kN}$$

The vertical component of force in the tie rod, $F_{t,v} = 0.6 \times 47.124 = 28.274$ kN
Take moment about point A

$$\sum M_A = F_{t,v}(1200) - W(3000) = 0 \quad \text{or} \quad W = 0.4F_{t,v}$$

Therefore, $W = 0.4 \times 28.274 = 11.3096$ kN

Extension of tie rod, $$\Delta = \frac{F_t L}{AE} = \frac{(47.124 \times 10^3) \times 1500}{(\pi \times 20^2/4) \times (200 \times 10^3)} = 1.125\ \text{mm}$$

Vertical deflection at C, $\delta_{c,v} = \Delta \sin\theta = 1.125 \times 0.6 = 0.675$ mm

Horizontal deflection at C, $\delta_{c,h} = \Delta \cos\theta = 1.125 \times 0.8 = 0.9$ mm

Vertical deflection at B, $$\delta_{b,v} = \delta_{c,v}\left(\frac{3000}{1200}\right) = 0.675 \times 2.5 = 1.6875\ \text{mm}$$

Horizontal deflection at B, $\delta_{b,h} = \delta_{c,h} = 0.9 \times 2.5 = 2.25$ mm

Resultant deflection at B, $$\delta = \sqrt{\delta_{b,v}^2 + \delta_{b,h}^2} = \sqrt{(1.6875)^2 + (2.25)^2} = 2.8125\ \text{mm}$$

Free-body diagram and enlarged displacements are given in Figure 3.13(b) and (c).

EXAMPLE 3.11

Two 5.0 m long vertical rods are suspended at a distance of *a* mm apart as shown in Figure 3.14(a). Their upper ends are firmly secured and their lower ends support a rigid horizontal bar which carries a load *W*. The left hand copper and right hand steel rods have diameters of 10 and 20 mm, respectively. Determine the: (a) position of the load *W* so that both the rods extend by the same amount and (b) load shared by each rod with load placed as in (a). (c) If the load is hung at the centre of the rigid bar, determine the slope of the rigid bar. The moduli of elasticity for steel and copper are 210 and 105 GPa, respectively. The weight of the horizontal bar may be neglected.

Solution: For the given wires,

$$A_c = \frac{\pi}{4} \times 10^2 = 78.54\ \text{mm}^2 \quad \text{and} \quad A_s = \frac{\pi}{4} \times 20^2 = 314.16\ \text{mm}^2$$

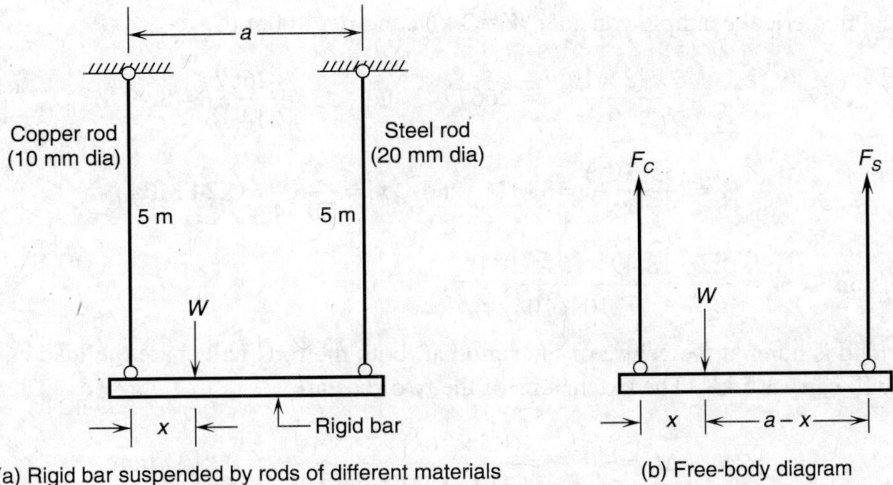

(a) Rigid bar suspended by rods of different materials

(b) Free-body diagram

Figure 3.14 Rigid bar of Example 3.11 suspended by two rods of different materials.

(a) Consider the load W to be placed at distance x from the copper rod, and F_c and F_s kN be the tensions in the copper and steel rods, respectively. The free-body diagram is shown in Figure 3.14(b). Consider moment about copper wire,

$$Wx = F_s \times a \quad \text{or} \quad F_s = \frac{Wx}{a}$$

and

$$F_c = W - F_s = W - \frac{Wx}{a} = \frac{W(a-x)}{a}$$

Therefore,

$$\frac{F_s}{F_c} = \frac{x}{a-x}$$

Since the rods are of same length and extend by the same amount, i.e. the rods to have same strain, the load is shared in the ratio of their (EA) values

$$\frac{x}{a-x} = \frac{F_s}{F_c} = \frac{E_s A_s}{E_c A_c} \quad \text{or} \quad \frac{x}{a-x} = \frac{(210 \times 10^3) \times 314.16}{(105 \times 10^3) \times 78.54} = 8.0$$

Therefore, the position of the load W is given by, $x = \dfrac{8}{9}a$

(b) The loads shared by two rods,

$$F_s = \frac{Wx}{a} = \frac{W(8a/9)}{a} = \frac{8W}{9}$$

$$F_c = \frac{W(a-x)}{a} = \frac{W(a/9)}{a} = \frac{W}{9}$$

As a typical numerical example consider $W = 3$ kN, and $a = 400$ mm

$$F_s = \frac{8\times(3\times10^3)}{9} = 2667\ \text{N} \quad \text{and} \quad \sigma_s = \frac{2667}{314.16} = 8.49\ \text{MPa}$$

$$F_c = \frac{(3\times10^3)}{9} = 333\ \text{N} \quad \text{and} \quad \sigma_c = \frac{333}{78.54} = 4.24\ \text{MPa}$$

$$\text{Extension,} \quad \Delta l = \frac{\sigma_s L}{E_s} = \frac{8.49\times(5\times10^3)}{(210\times10^3)} = 0.202\ \text{mm}$$

(c) If the load is hung at the centre of the rigid bar, both the rods will share the load equally, i.e. each rod will carry 1.5 kN. The extensions of the two rods are,

$$\Delta l_s = \frac{F_s L}{A_s E_s} = \frac{(1.5\times10^3)\times(5\times10^3)}{314.16\times(210\times10^3)} = 0.1137\ \text{mm}$$

$$\Delta l_c = \frac{F_c L}{A_c E_c} = \frac{(1.5\times10^3)\times(5\times10^3)}{78.54\times(105\times10^3)} = 0.9095\ \text{mm}$$

Slope of the rigid bar,

$$\tan\theta = \frac{\Delta l_c - \Delta l_s}{a} = \frac{0.9095 - 0.1137}{400} = 1.9895\times10^{-3} \quad \text{or} \quad \theta = 0.114°$$

3.6 LATERAL STRAIN AND POISSON'S RATIO

As discussed earlier, when a body is subjected to a force system, its dimensions change. The change in a dimension per unit dimension is termed the *strain*. The strain may be tensile or compressive depending upon whether the dimension under consideration increases or decreases. Thus the strain is a dimensionless quantity.

3.6.1 Lateral Strain

A longitudinal stress in a member subjected to axial load not only produces a strain in its own direction but also produces a strain in the lateral direction called *lateral strain* as shown in Figure 3.15. The lateral dimensions of the member decrease when the member is subjected to a tensile force and increase when the member is subjected to a compressive force.

For illustration, consider a cylindrical rod subjected to an axial tensile load P. Under the action of applied load, the length of the rod will increase, but the diameter of the rod will decrease. In case rectangular rod of width b, depth d and length l subjected to an axial tensile load, the member will deform such that the length of the member will increase, the lateral dimensions namely the width and the depth will decrease. Let δl, $-\delta b$ and $-\delta d$ be the increase in length, decrease in width and decrease depth, respectively. The ratio $\delta l/l$ is called *longitudinal strain* or linear strain ε_x whereas $-\left(\frac{\delta d}{d}\right)$ and $-\left(\frac{\delta b}{b}\right)$ are called the *lateral strains* ε_y *and* ε_z.

3.6.2 Poisson's Ratio

A constant relation exists between these lateral changes and those in the direction of applied load. When a member deforms within *elastic limit*, the ratio lateral strain to the longitudinal strain is a constant for the material of the member. This constant is usually called *Poisson's ratio*. It is denoted by ν (nu) and is defined as follows:

$$\nu = \text{Poisson's ratio} = -\frac{\text{Lateral strain}}{\text{Axial or longitudinal strain}} = -\frac{\varepsilon_y}{\varepsilon_x} \tag{3.11}$$

The minus indicates that the lateral and axial strains are always of opposite sign for *uniaxial stress*. Hence,

$$\text{Lateral strain} = (-)\,\nu \times \text{Longitudinal strain}$$

Thus the Poisson's ratio gives relationship between the longitudinal and lateral strains. For most of the materials, its value lies between 0.25 and 0.33.

For illustration consider a rod of uniform rectangular cross-section of size $b \times d$ subjected to an axial tensile load P. If the length of rod is L, the strains and hence the changes in dimensions can be determined as follows.

Longitudinal and lateral strains,

$$\varepsilon_1 = \frac{\sigma_o}{E} = \frac{P}{AE} \quad \text{and} \quad \varepsilon_b = \varepsilon_d = -\nu\varepsilon_l = -\frac{\nu P}{AE}$$

Increase in the length of the rod, $\delta l = \varepsilon_l \times l = \dfrac{Pl}{AE}$

If, in addition to a force producing the stress σ_x, there is a force on the perpendicular plane producing a stress σ_y, then the strains ε_x and ε_y due to combined stresses are given by

$$\varepsilon_x = \text{Longitudinal strain due to } \sigma_x + \text{Lateral strain due to } \sigma_y = \frac{1}{E}(\sigma_x - \nu\sigma_y)$$

Similarly,

$$\varepsilon_y = \frac{1}{E}(\sigma_y - \nu\sigma_x) \tag{3.12}$$

Percentage elongation: It is the change in the length per unit original length of the test specimen expressed as a percentage, i.e.

$$\text{Percentage elongation} = \frac{\delta l}{l} \times 100 \tag{3.13}$$

Percentage reduction of area: It is the change in the area per unit original area expressed as a percentage, i.e.

$$\text{Percentage reduction of area} = \frac{A_o - A}{A_o} \times 100 \tag{3.14}$$

3.7 SHEAR STRAIN AND MODULUS OF RIGIDITY

3.7.1 Shear Strain

It is the strain produced under the action of *shearing stresses*. Consider a small rectangular element of size $a \times b \times c$ subjected to a pure shear stress τ_1 at the top face as shown in Figure 3.15(a). For convenience in understanding, rectangular element is shown as a plane element ABCD in

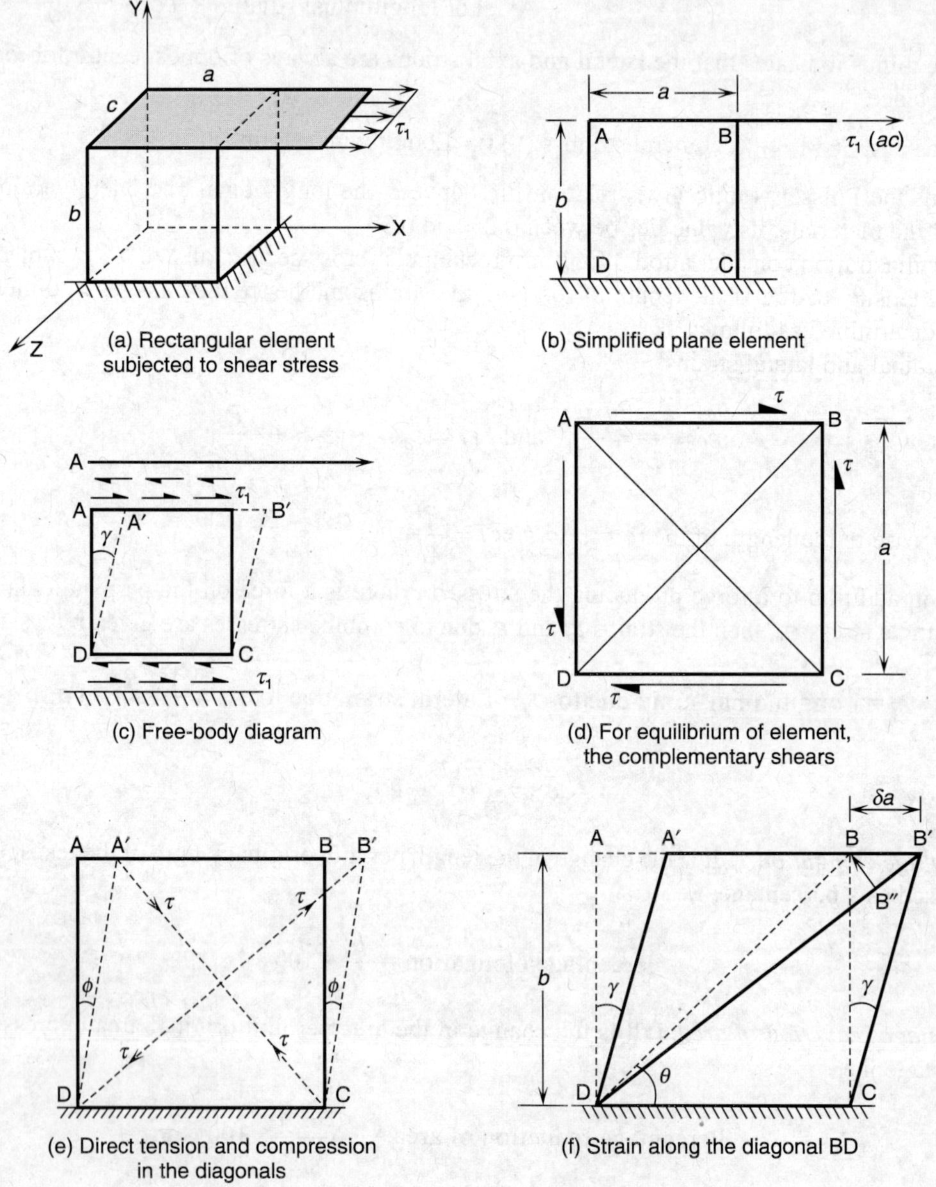

Figure 3.15 Shear stresses and strains in an element.

Figure 3.15(b), subjected to pure shear. The plane element has a thickness c perpendicular to the plane of the element. It should be noted from the free-body diagram illustrated in Figure 3.15(c) that a reactive force develops on the bottom face, the two shear stresses having no resultant force form a couple of magnitude $(\tau_1\, ac) \times b = \tau_1\, abc$. For equilibrium a counter-couple with no resultant force is required, i.e. two more shear stresses τ_2 develop, one on right face and the other on the left face as shown in Figure 3.15(d). The shears developed on these vertical faces form a couple of magnitude $(\tau_2\, bc) \times a = \tau_2\, abc$. For equilibrium $\tau_1\, abc = \tau_2\, abc$, i.e. $\tau_1 = \tau_2 = \tau$ and thus the element is subjected to equal shear τ on all four faces as shown in Figure 3.15(d). The shear stresses developed on the vertical faces are called *complementary shear stresses*.

The stresses acting on the rectangular element cause the element to deform to a parallelogram shown by dotted lines with B moving to B′ in Figure 3.15(e). The angle between two adjacent sides has changed from 90° to $(90 \pm \phi)°$. In engineering applications, it is convenient to consider a reference axis parallel to one of the deformed sides, DC is taken as an axis in this case as shown in Figure 3.15(e). If δa (= BB′) represents the shift of the face AB under the action of shear stress τ, then

$$\text{Shear strain, } \gamma = \frac{\delta a}{b} = \tan \phi$$

For small strain,

$$\tan \phi \approx \phi, \quad \text{thus,} \quad \phi = \frac{\delta a}{b} = \gamma \tag{3.15}$$

Thus, the *shear strain* γ is measured by the change in the angle between two adjacent sides and like normal strain it is dimensionless quantity.

3.7.2 Modulus of Rigidity

For a member deformed within elastic limit, the ratio of *shear stress* to the *shear strain* is a constant for the material of the member. This ratio is termed *modulus of rigidity* and is denoted by G. Therefore,

$$\frac{\text{Shearing stress}}{\text{Shearing strain}} = \frac{\tau}{\gamma} = G = \text{Modulus of rigidity} \tag{3.16}$$

The modulus of rigidity is generally related to the modulus of elasticity and Possion's ratio. Consider a cube of side a, the shear stresses τ acting on each face produce a direct tension and compression of magnitude τ on two diagonals as illustrated in Figure 3.15(e).
The strain on the diagonal BD (see Figure 3.15(f)) is given by

$$\varepsilon_{bd} = \frac{1}{E}(\tau + \nu\tau) = \frac{\tau}{E}(1 + \nu)$$

Shear strain in the element, $\gamma = \tau / G$

The original length of diagonal BD is $\sqrt{2}a$, and increase in its length is

$$\delta l_{bd} = l_{b'd} - l_{b''d} = l_{b'b''} = l_{bb'} \cos 45° = \frac{\delta a}{\sqrt{2}}$$

Therefore, $$\delta l_{bd} = \frac{(a\phi)}{\sqrt{2}} = \frac{a\gamma}{\sqrt{2}} = \frac{a\tau}{\sqrt{2}G}$$

and corresponding strain is

$$\varepsilon_{bd} = \frac{\delta l_{bd}}{l_{bd}} = \frac{\delta l_{bd}}{\sqrt{2}a} = \frac{a\tau}{\sqrt{2}G} \times \frac{1}{\sqrt{2}a} = \frac{\tau}{2G}$$

Equating two values of ε_{bd}

$$\frac{\tau}{E}(1+\nu) = \frac{\tau}{2G}$$

Therefore, $$G = \frac{E}{2(1+\nu)} \tag{3.17}$$

3.8 VOLUMETRIC STRAIN AND BULK MODULUS

3.8.1 Volumetric Strain

When a body is subjected to a force system, its dimensions change and hence its volume also changes. The ratio of the change in the volume to the original volume is termed *volumetric strain.* This is usually denoted by ε_v i.e.

$$\frac{\text{Change in volume}}{\text{Original volume}} = \frac{\delta V}{V} = \varepsilon_v = \text{Volumetric strain} \tag{3.18}$$

(a) *Rectangular rod:* Consider a rod of length l, width b and depth d subjected to a load system which undergoes small changes of δl, δb and δd, respectively, in length, width and depth. Then

Original volume, $$V = l\,b\,d$$

Final volume, $$V + \delta V = (l + \delta l)(b + \delta b)(d + \delta d)$$

$$= l\,b\,d + b\,d\,\delta l + l\,d\,\delta b + l\,b\,\delta d$$

In the above expression, the products of small quantities have been ignored. Therefore,

$$\varepsilon_v = \frac{\delta V}{V} = \frac{b\,d\,\delta l + l\,d\,\delta b + l\,b\,\delta d}{l\,b\,d} = \frac{\delta l}{l} + \frac{\delta b}{b} + \frac{\delta d}{d} = \varepsilon_l + \varepsilon_b + \varepsilon_d \tag{3.19}$$

Therefore, the volumetric strain is the sum of longitudinal strain, and the lateral strains along the width and the depth of the member.

For illustration, consider a rod of uniform rectangular cross-section of area A subjected to an axial tensile force P, the change in volume can be determined as follows.

Longitudinal and lateral strains,

$$\varepsilon_l = \frac{\sigma_o}{E} = \frac{P}{AE} \quad \text{and} \quad \varepsilon_b = \varepsilon_d = -\nu\varepsilon_l = -\frac{\nu P}{AE}$$

Volumetric strain, $\varepsilon_v = \varepsilon_l + \varepsilon_b + \varepsilon_d = \dfrac{P}{AE} - \dfrac{\nu P}{AE} - \dfrac{\nu P}{AE} = \dfrac{P}{AE}(1 - 2\nu)$

Change volume, $\delta V = \varepsilon_v V = \dfrac{P}{AE}(1 - 2\nu) \times (AL) = \dfrac{PL}{E}(1 - 2\nu)$ (3.20)

(b) *Cylindrical rod*: Consider a cylindrical rod of length l and diameter d which undergoes small changes δl and δd in the length and the diameter, respectively. Then

$$\text{Original volume, } V = \frac{\pi}{4}(d^2 l)$$

$$\text{Final volume, } V + \delta V = \frac{\pi}{4}(d + \delta d)^2 \, (l + \delta l)$$

$$= \frac{\pi}{4}(d^2 l + d^2 \delta l + 2ld\,\delta d)$$

In the above expression, the products of small quantities have been ignored. Therefore,

$$\varepsilon_v = \frac{\delta V}{V} = \frac{(d^2 l + d^2 \delta l + 2ld\,\delta d) - d^2 l}{d^2 l} = \frac{\delta l}{l} + 2\frac{\delta d}{d} = \varepsilon_l + 2\varepsilon_d \tag{3.21}$$

Therefore, the volumetric strain in this case is equal to the sum of longitudinal strain and twice the diametrical strain.

(c) *Solid sphere*: Consider a sphere of diameter d which undergoes small change δd in the diameter. Then

Original volume, $V = \dfrac{4\pi}{3}(r^3) = \dfrac{\pi}{6}(d^3)$

Final volume, $V + \delta V = \dfrac{\pi}{6}(d + \delta d)^3 = \dfrac{\pi}{6}(d^3 + 3d^2\ \delta d)$

In the above expression, higher powers of δd have been ignored. Therefore,

$$\varepsilon_v = \frac{\delta V}{V} = \frac{(d^3 + 3d^2\ \delta d) - d^3}{d^3} = 3 \times \frac{\delta d}{d} = 3\varepsilon_d$$

Alternatively, consider the diameter of the sphere to change by δd so that the volume changes by,

$$\delta V = \frac{\pi}{6} \times (3d^2)\,\delta d$$

Therefore, volumetric strain, $\varepsilon_v = \dfrac{\delta V}{V} = 3\dfrac{\delta d}{d} = 3\varepsilon_d$ (3.22)

Therefore, the volumetric strain of a sphere is equal to thrice the diametrical strain.

(d) *Rectangular block subjected to direct* or *normal stresses on all its faces*: Consider a rectangular block with the dimensions $a \times b \times c$ measured along X, Y and Z directions subjected to direct stresses σ_x, σ_y and σ_z acting on the planes normal to X-, Y- and Z-axes, respectively, as shown in Figure 3.16.

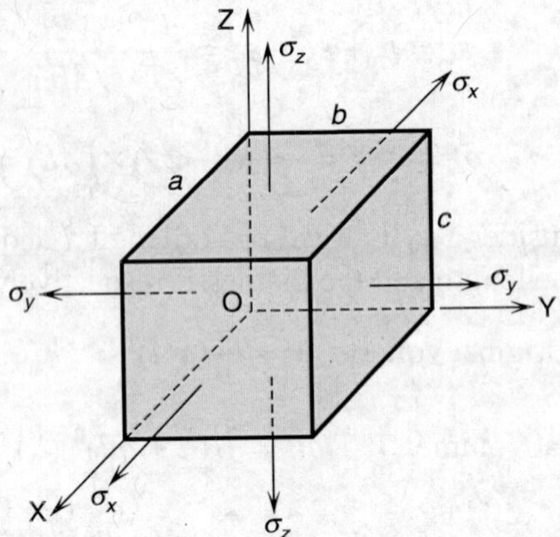

Figure 3.16 Rectangular block subjected to direct stresses on all its faces.

The strains along each dimension can be calculated as the algebraic sum of the strains produced by the stresses σ_x, σ_y and σ_z. Therefore, strain along X-axis

$$\varepsilon_x = \text{Longitudinal strain due to } \sigma_x + \text{Lateral strains due to } \sigma_y \text{ and } \sigma_z$$

Therefore,
$$\varepsilon_x = \frac{\sigma_x}{E} - \frac{\nu\sigma_y}{E} - \frac{\nu\sigma_z}{E} = \frac{1}{E}[\sigma_x - \nu(\sigma_y + \sigma_z)]$$

Similarly, the strains along X- and Y-axes are

$$\varepsilon_y = \frac{1}{E}[\sigma_y - \nu(\sigma_z + \sigma_x)] \quad \text{and} \quad \varepsilon_z = \frac{1}{E}[\sigma_z - \nu(\sigma_x + \sigma_y)]$$

The changes in the dimensions are

$$\delta a = \varepsilon_x \times a, \quad \delta b = \varepsilon_y \times b \quad \text{and} \quad \delta c = \varepsilon_z \times c$$

The volumetric strain,

$$\varepsilon_v = \varepsilon_x + \varepsilon_y + \varepsilon_z = \frac{1}{E}(\sigma_x + \sigma_y + \sigma_z) - \frac{2\nu}{E}(\sigma_x + \sigma_y + \sigma_z)$$

Therefore,
$$\varepsilon_v = \frac{(1 - 2\nu)}{E}(\sigma_x + \sigma_y + \sigma_z) \tag{3.23}$$

For the material to be incompressible, $\varepsilon_v = 0$. Since for the metals, the value of ν lies between 0.25 and 0.33, the term $(1 - 2\nu)$ is always positive and is never equal to zero. Therefore, for $\varepsilon_v = 0$, either $(\sigma_x + \sigma_y + \sigma_z) = 0$ or $(1 - 2\nu) = 0$. If $(\sigma_x + \sigma_y + \sigma_z) \neq 0$, then

$$(1 - 2\nu) = 0 \quad \text{or} \quad \nu = 0.5$$

Change in the volume of block, $\delta V = \varepsilon_v V$

3.8.2 Bulk Modulus

Consider an elastic body subjected to equal and alike direct stresses $\sigma_x = \sigma_y = \sigma_z = \sigma_o$ acting along three mutually perpendicular directions, say along X-, Y- and Z-axes. When the deformations of the body are within the elastic limit, the ratio of the direct stress to the corresponding volumetric strain is constant for a given material. This ratio is termed *bulk modulus of rigidity* and is usually denoted by K. Therefore,

$$K = \frac{\text{Direct stress}}{\text{Volumetric strain}} = \frac{\sigma_o}{\varepsilon_v} \tag{3.24}$$

As a typical example, consider the elastic body to be a rectangular block having each side of length a, i.e. a cube with each face subjected to direct stress of intensity σ.

The strains along each dimension can be calculated as the algebraic sum of the strains produced by the direct stresses of intensity σ acting in three directions. Therefore, strain along X-axis

$$\varepsilon_x = \frac{\sigma}{E} - \frac{\nu\sigma}{E} - \frac{\nu\sigma}{E} = \frac{\sigma}{E}(1-2\nu)$$

Similarly strains along Y- and Z-axes are

$$\varepsilon_y = \frac{\sigma}{E}(1-2\nu) \quad \text{and} \quad \varepsilon_z = \frac{\sigma}{E}(1-2\nu)$$

The volumetric strain, $\varepsilon_v = \varepsilon_x + \varepsilon_y + \varepsilon_z = \dfrac{3\sigma}{E}(1-2\nu)$

Therefore, the bulk modulus

$$K = \frac{\sigma}{\varepsilon_v} = \frac{E}{3(1-2\nu)} \tag{3.25}$$

3.9 RELATIONS BETWEEN ELASTIC CONSTANTS

From Eqs. (3.17) and (3.25)

$$G = \frac{E}{2(1+\nu)} \quad \text{or} \quad \nu = \frac{E}{2G} - 1 = \frac{E-2G}{2G} \tag{i}$$

$$K = \frac{E}{3(1-2\nu)} \quad \text{or} \quad \nu = \frac{1}{2} \times \left(1 - \frac{E}{3K}\right) = \frac{3K-E}{6K} \tag{ii}$$

Substitute Eq. (i) in Eq. (ii) to eliminate ν:

$$2\left(\frac{E}{2G} - 1\right) = \left(1 - \frac{E}{3K}\right)$$

$$E\left(\frac{1}{G} + \frac{1}{3K}\right) = 3 \quad \text{or} \quad E = \frac{9GK}{G+3K} \tag{iii}$$

Eliminate E between Eqs. (i) and (ii):

$$2(1+\nu)G = 3(1-2\nu)K \quad \text{or} \quad \nu(2G+6K) = 3K-2G$$

Therefore,
$$\nu = \frac{3K-2G}{6K+2G} \tag{iv}$$

EXAMPLE 3.12

A 20 mm diameter steel bar of length 6 m is subjected to an axial tensile load 30 kN. Determine the change in length, diameter and the volume of the bar. The modulus of elasticity and Poisson's ratio for steel are 200 GPa and $\nu = 0.25$, respectively.

Solution: Area of the bar, $A = \dfrac{\pi \times 20^2}{4} = 314.16 \text{ mm}^2$

Direct tensile stress, $\sigma_o = \dfrac{30 \times 10^3}{314.16} = 95.49$ MPa

Longitudinal and lateral strains,

$$\varepsilon_l = \frac{\sigma_o}{E} = \frac{95.49}{200 \times 10^3} = (+)0.000477$$

and
$$\varepsilon_d = -\nu\varepsilon_l = (-)0.000477 \times 0.25 = (-)0.000119$$

Increase in the length, $\delta l = \varepsilon_l l = 0.000477 \times (6 \times 10^3) = (+)\ 2.862$ mm

Decrease in the diameter, $\delta d = \varepsilon_d d = 0.000119 \times 20 = (-)\ 0.00238$ mm

Volumetric strain, $\varepsilon_v = \varepsilon_l + 2\varepsilon_d = 0.000477 - 2 \times 0.000119 = (+)\ 0.000239$

Increase in the volume, $\delta V = \varepsilon_v V = 0.000239 \times (314.16 \times 6000) = (+)\ 450.5 \text{ mm}^3$

EXAMPLE 3.13

A 75 mm square steel bar of length 500 mm is subjected to an axial compressive load of 450 kN. Determine the change in length and volume if all lateral strain in one direction is prevented by application of suitable compressive stress.

The modulus of elasticity and Poisson's ratio for steel are 200 GPa and $\nu = 0.25$, respectively.

Solution: Direct longitudinal stress, $\sigma_x = \dfrac{450 \times 10^3}{75 \times 75} = 80$ MPa

Consider compressive stress to prevent lateral strain in one direction (say y) to be f_y, then

$$\varepsilon_y = \frac{1}{E}(-\sigma_y + \nu\sigma_x) = 0 \quad \text{or} \quad \sigma_y = \nu\sigma_x = 0.25 \times 80 = 20 \text{ MPa}$$

Longitudinal strain, $\varepsilon_x = \dfrac{1}{E}(-\sigma_x + \nu\sigma_y) = \dfrac{1}{E}(-80 + 0.25 \times 20) = -\dfrac{75}{E}$

Therefore, decrease in length of the bar, $\delta l = \varepsilon_x l = \dfrac{75 \times 500}{(200 \times 10^3)} = 0.1875$ mm

Lateral strain in Z-direction,

$$\varepsilon_z = \frac{1}{E}[\sigma_z + \nu(\sigma_x + \sigma_y)]$$

$$= \frac{1}{E}[0 - 0.25 \times (-80 - 20)] = \frac{25}{E}$$

Volumetric strain, $$\varepsilon_v = \frac{\delta V}{V} = \varepsilon_x + \varepsilon_y + \varepsilon_z$$

$$= -\frac{75}{E} + 0 + \frac{25}{E} = -\frac{50}{E}$$

Therefore, change in volume, $$\delta V = \frac{50 \times 500 \times 75 \times 75}{200 \times 10^3} = 703.125 \text{ mm}^3$$

EXAMPLE 3.14

A 50 mm square steel bar of length 150 mm is subjected to a tensile load of 315 kN along the longitudinal axis in X-direction and tensile loads of 785 kN and 625 kN along Y- and Z-directions, respectively, on the lateral faces. Determine: (a) the change in the dimensions and volume of the bar, and (b) the magnitude of longitudinal tensile load which acting alone can produce the same longitudinal strain as in (a). The modulus of elasticity and Poisson's ratio for steel are 200 GPa and $\nu = 0.3$, respectively.

Solution:

(a) Direct longitudinal stress, $\sigma_x = \dfrac{315 \times 10^3}{50 \times 50} = 126$ MPa

$$\sigma_y = \frac{785 \times 10^3}{50 \times 150} = 104.67 \text{ MPa} \quad \text{and} \quad \sigma_x = \frac{625 \times 10^3}{50 \times 150} = 83.33 \text{ MPa}$$

Therefore, longitudinal strain, i.e. along X-axis

$$\varepsilon_x = \frac{1}{E}[\sigma_x - \nu(\sigma_y + \sigma_z)] = \frac{1}{(200 \times 10^3)}[126 - 0.3 \times (104.67 + 83.33)]$$

$$= 3.48 \times 10^{-4}$$

Similarly strains along Y- and Z-axes are

$$\varepsilon_y = \frac{1}{E}[\sigma_y - \nu(\sigma_z + \sigma_x)] = \frac{1}{(200 \times 10^3)}[104.67 - 0.3 \times (83.33 + 126)]$$

$$= 2.0936 \times 10^{-4}$$

$$\varepsilon_z = \frac{1}{E}[\sigma_z - \nu(\sigma_x + \sigma_y)] = \frac{1}{(200 \times 10^3)}[83.33 - 0.3 \times (126 + 104.67)]$$

$$= 7.0745 \times 10^{-5}$$

The changes in the dimensions are:

Increase in length of the bar, $\delta l = \varepsilon_x l = (3.48 \times 10^{-4}) \times 150 = 0.0522$ mm

Increase in width of the bar, $\delta b = \varepsilon_y b = (2.0936 \times 10^{-4}) \times 50 = 0.01047$ mm

Increase in depth of the bar, $\delta d = \varepsilon_z d = (7.0745 \times 10^{-5}) \times 50 = 0.003537$ mm

The volumetric strain,

$$\varepsilon_v = \frac{(1-2\nu)}{E}(\sigma_x + \sigma_y + \sigma_z) = \frac{(1 - 2 \times 0.3)}{200 \times 10^3} \times (126 + 104.67 + 83.33)$$

$$= 6.28 \times 10^{-4}$$

Increase in the volume of the bar,

$$\delta V = \varepsilon_v V = (6.28 \times 10^{-4}) \times (50 \times 50 \times 150) = 235.5 \text{ mm}^3$$

(b) If σ is the magnitude of longitudinal tensile stress which would produce the same longitudinal strain as in (a). Then

$$\frac{\sigma}{E} = \varepsilon_1 \quad \text{or} \quad \sigma = \varepsilon_l E = (3.48 \times 10^{-4}) \times (200 \times 10^3) = 69.6 \text{ MPa}$$

Therefore, longitudinal tensile load,

$$P = \sigma \times A = 69.6 \times (50 \times 50) = 174000 \text{ N} = 174 \text{ kN}$$

EXAMPLE 3.15

A rectangular block is subjected to a direct longitudinal compressive stress σ_1. External pressures of intensity σ_2 and σ_3 are applied in the lateral directions perpendicular to the longitudinal direction as shown in Figure 3.17, so that the net lateral strains in the directions along σ_2 and σ_3 are reduced to one-half and one-third, respectively, of the lateral strains that would have occurred if σ_1 alone was acting. Determine: (a) σ_2 and σ_3 in terms of σ_1 and the strain in its own direction, (b) modified value of modulus of elasticity, if the modulus of elasticity and Poisson's ratio for the material of the block are 200 GPa and $\nu = 0.25$, respectively, and (c) show that a material with Possion's ratio 0.5 is practically incompressible.

Solution:

(a) Lateral strain without lateral pressure $= \dfrac{\nu\sigma_1}{E}$

Lateral strain with lateral pressure,

$$\varepsilon_2 = \frac{1}{E}[-\sigma_2 + \nu(\sigma_1 + \sigma_3)] = \frac{1}{2} \times \frac{\nu\sigma_1}{E}$$

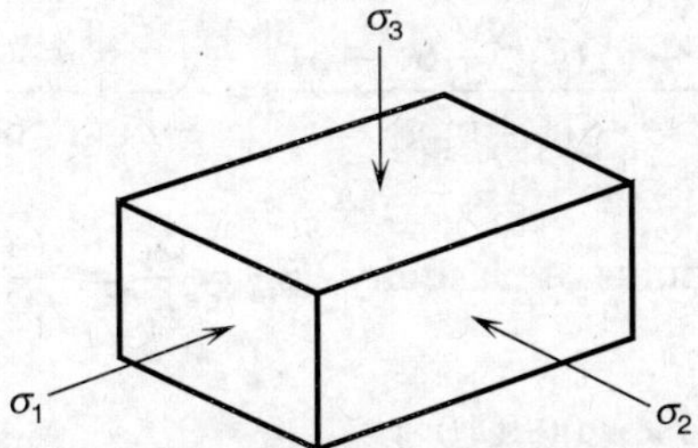

Figure 3.17 Rectangular block (of Example 3.15) subjected to compressive stresses on all the faces.

or $$-\sigma_2 + \frac{1}{2}\nu\sigma_1 + \nu\sigma_3 = 0 \qquad \text{(i)}$$

$$\varepsilon_3 = \frac{1}{E}[-\sigma_3 + \nu(\sigma_1 + \sigma_2)] = \frac{1}{3} \times \frac{\nu\sigma_1}{E}$$

or $$-\sigma_3 + \frac{2}{3}\nu\sigma_1 + \nu\sigma_2 = 0 \qquad \text{(ii)}$$

Operate on Eqs. (i) and (ii), as (i) + ν(ii)

$$\left(-\sigma_2 + \frac{1}{2}\nu\sigma_1 + \nu\sigma_3\right) + \nu\left(-\sigma_3 + \frac{2}{3}\nu\sigma_1 + \nu\sigma_2\right) = -6(1-\nu^2)\sigma_2 + \nu(3+4\nu)\sigma_1 = 0$$

Therefore, $$\sigma_2 = \frac{\nu(3+4\nu)}{6(1-\nu^2)}\sigma_1 \qquad \text{(iii)}$$

From Eqs. (ii) and (iii),

$$\sigma_3 = \frac{2}{3}\nu\sigma_1 + \nu\sigma_2 \quad \text{or} \quad \sigma_3 = \frac{2}{3}\nu\sigma_1 + \frac{\nu^2(3+4\nu)}{6(1-\nu^2)}\sigma_1$$

Thus, $$\sigma_3 = \frac{4\nu(1-\nu^2) + \nu^2(3+4\nu)}{6(1-\nu^2)}\sigma_1$$

$$= \frac{\nu(4+3\nu)}{6(1-\nu^2)}\sigma_1 \qquad \text{(iv)}$$

The strain in longitudinal direction with lateral pressures,

$$\varepsilon_1 = \frac{1}{E}[-\sigma_1 + \nu(\sigma_2 + \sigma_3)] = \frac{1}{E}\left[-\sigma_1 + \nu\left\{\frac{\nu(3+4\nu)}{6(1-\nu^2)}\sigma_1 + \frac{\nu(4+3\nu)}{6(1-\nu^2)}\sigma_1\right\}\right]$$

$$= \frac{\sigma_1}{E}\left[\frac{-6(1-\nu^2) + \nu^2(3+4\nu) + \nu^2(4+3\nu)}{6(1-\nu^2)}\right] = \frac{\sigma_1}{E}\left[\frac{7\nu^3 + 13\nu^2 - 6}{6(1-\nu^2)}\right]$$

$$\varepsilon_1 = \frac{\sigma_1}{E}\left[\frac{(1+\nu)(7\nu^2+6\nu-6)}{6(1-\nu^2)}\right] = -\frac{\sigma_1}{E}\left[\frac{6-6\nu-7\nu^2}{6(1-\nu)}\right] \quad \text{(v)}$$

(b) Therefore, the modified modulus of elasticity, $E_m = \frac{\sigma_1}{\varepsilon_1} = \frac{6(1-\nu)E}{7\nu^2+6\nu-6}$

$$E_m = \frac{6\times(1-0.25)\times(200\times10^3)}{7\times(0.25)^2+6\times0.25-6} = -221538.46 \text{ MPa} = -221.54 \text{ GPa}$$

Negative sign indicates that the elastic modulus is in compression.

(c) The volumetric strain, $\varepsilon_v = \varepsilon_1 + 2\varepsilon_2 = \frac{\sigma_1}{E} - 2\nu\frac{\sigma_1}{E} = \frac{\sigma_1}{E}(1-2\nu)$

For the material to be incompressible, $\varepsilon_v = 0$ i.e. $(1-2\nu) = 0$ or $\nu = 0.5$.

EXAMPLE 3.16

A 30 mm square bar of 160 mm length is subjected to an axial compressive load of 150 kN. Determine the change in length, if all lateral strains are prevented by the application of uniform lateral external pressure of suitable intensity. The modulus of elasticity and Poisson's ratio for the material are 110 GPa and $\nu = 0.25$, respectively.

Solution: Direct longitudinal stress, $\sigma_x = \frac{150\times10^3}{30\times30} = 166.67$ MPa

The strains under longitudinal force and lateral pressures are,

$$\varepsilon_x = \frac{1}{E}[-\sigma_x + \nu(\sigma_y + \sigma_z)]$$

$$\varepsilon_y = \frac{1}{E}[-\sigma_y + \nu(\sigma_z + \sigma_x)]$$

$$\varepsilon_z = \frac{1}{E}[-\sigma_z + \nu(\sigma_x + \sigma_y)]$$

Since the strains in the Y- and Z-directions are prevented, i.e. ε_y and ε_z are equal, and each of them vanishes, thus

$$-\sigma_y + \nu(\sigma_z + \sigma_x) = -\sigma_z + \nu(\sigma_x + \sigma_y) \quad \text{or} \quad -\sigma_y + \nu\sigma_z = -\sigma_z + \nu\sigma_y$$

Therefore, $(1+\nu)\sigma_z = (1+\nu)\sigma_y$ or $\sigma_z = \sigma_y$

$$\varepsilon_y = \frac{1}{E}[-\sigma_y + \nu(\sigma_y + \sigma_x)] = 0 \quad \text{or} \quad -(1-\nu)\sigma_y + \nu\sigma_x = 0$$

or $\sigma_y = \frac{\nu}{(1-\nu)}\sigma_x$

Longitudinal strain, $\varepsilon_x = \dfrac{1}{E}[-\sigma_x + 2\nu\sigma_y] = \dfrac{1}{E}\left[-\sigma_x + \dfrac{2\nu^2}{(1-\nu)}\sigma_x\right] = -\dfrac{\sigma_x}{E}\left[\dfrac{1-\nu-2\nu^2}{(1-\nu)}\right]$

Therefore, $\varepsilon_x = -\dfrac{166.67}{(110\times10^3)} \times \left[\dfrac{1-0.25-2\times(0.25)^2}{(1-0.25)}\right] = -1.263\times10^{-3}$

Contraction in length, $\delta l = \varepsilon_x l = (1.263 \times 10^{-3}) \times 160 = 0.2021$ mm.

EXAMPLE 3.17

A steel bar of cross-section 100 × 60 mm in the unstrained condition is fitted in a 100 × 60 mm hole in a rigid block, with a clearance of 0.0125 mm on all sides. Determine the axial force which must be applied to reduce the clearance to zero, (a) in 100 mm direction, and (b) in both directions. In case of (b) also determine the pressure between the surfaces that are first to come into contact. The modulus of elasticity and Poisson's ratio for steel are 200 GPa and $\nu = 0.25$, respectively.

Solution:

(a) To reduce the clearance in 100 mm direction to zero, consider σ_x to be the axial compressive stress applied to the bar, then

Axial and lateral strains, $\varepsilon_x = -\dfrac{\sigma_x}{E}$ and $\varepsilon_y = \dfrac{\nu\sigma_x}{E}$

Increase in the 100 mm dimension is equal to the total clearance, Δ,

$$\delta d = (2 \times 0.0125) = 0.025 \text{ mm}$$

$$= \varepsilon_y d = \left(\frac{\nu\sigma_x}{E}\right)d$$

Therefore, $\sigma_x = \dfrac{\delta d\, E}{\nu d} = \dfrac{0.025\times(200\times10^3)}{0.25\times100} = 200$ MPa

Axial force applied, $P = \sigma_x A = (200 \times 60 \times 100) \times 10^{-3} = 1200$ kN

(b) To reduce the clearance in both the directions to zero, let σ_x to be the applied axial compressive stress and σ_y the lateral compressive stress, then

$$\text{Lateral strain, } \varepsilon_l = \frac{1}{E}(-\sigma_x + \nu\sigma_y)$$

The changes in the lateral dimensions δd and δb are equal to the clearance, Δ i.e.,

$$\delta d = \varepsilon_d \times d = \frac{1}{E}(-\sigma_y + \nu\sigma_x)\times d = \Delta \quad \text{or} \quad (-\sigma_y + \nu\sigma_x) = \frac{E\Delta}{d}$$

and, $\delta b = \varepsilon_b \times b = \dfrac{\nu}{E}(\sigma_x + \sigma_y)\times b = \Delta \quad \text{or} \quad (\sigma_x + \sigma_y) = \dfrac{E\Delta}{\nu b}$

Therefore, $-\sigma_y + 0.25\sigma_x = \dfrac{(200\times10^3)\times0.025}{100} = 50$ (i)

$$\sigma_x + \sigma_y = \frac{(200 \times 10^3) \times 0.025}{0.25 \times 60} = \frac{1000}{3} \quad \text{(ii)}$$

Operate Eqs. (i) and (ii) as, (i) + (ii)

$$1.25\sigma_x = 50 + \frac{1000}{3} \quad \text{or} \quad \sigma_x = 306.67 \text{ MPa}$$

And from Eq. (ii) $\quad \sigma_y = \frac{1000}{3} - \sigma_x \quad \text{or} \quad \sigma_y = \frac{1000}{3} - 306.67 = 26.66 \text{ MPa}$

Axial force applied, $\quad P = (306.67 \times 60 \times 100) \times 10^{-3} = 1840 \text{ kN}$

Pressure between the surfaces, $\sigma_y = 22.66$ MPa

EXAMPLE 3.18

A metal rectangular block of size $a \times b \times c$ is fitted in a rigid channel of width a, and is subjected to a pressure p on the top face $a \times b$ as shown in Figure 3.18(a) and (b). Determine the: (a) stresses on the three faces in contact with the rigid channel and (b) changes in the dimensions of the block.

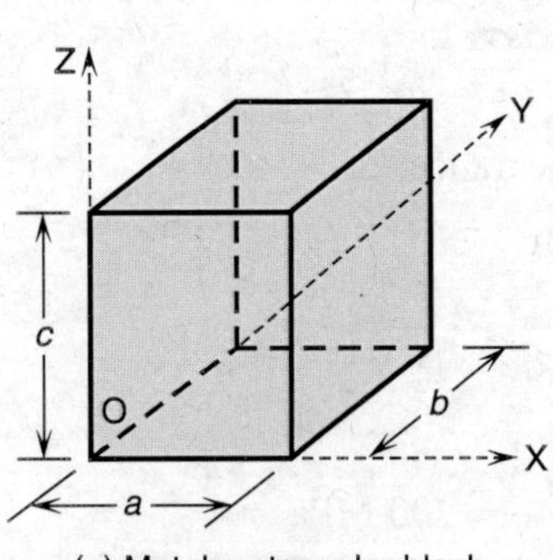

(a) Metal rectangular block

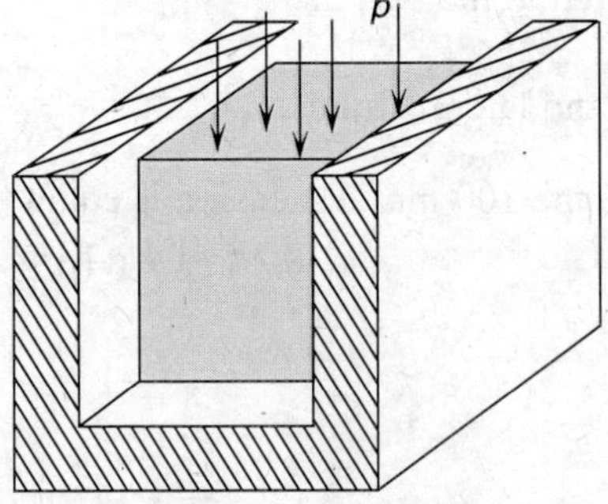

(b) Block fitted in rigid channel

Figure 3.18 Rectangular block fitted in a rigid channel of Example 3.18.

Solution: Consider the stresses in the X-, Y- and Z-directions to be tensile. The corresponding strains are,

$$\varepsilon_x = \frac{1}{E}[\sigma_x - \nu(\sigma_y + \sigma_z)];\ \varepsilon_y = \frac{1}{E}[\sigma_y - \nu(\sigma_z + \sigma_x)] \text{ and } \varepsilon_z = \frac{1}{E}[\sigma_z - \nu(\sigma_x + \sigma_y)]$$

(a) Due to the presence of rigid channel, the strain ε_x in the X-direction is prevented, i.e. the dimension a does not undergo any change. Whereas expansion or contraction can occur in the Y- and Z-directions. Since there is no restraint in the Y-direction, $\sigma_y = 0$ and the pressures $\sigma_z = -p$. Therefore,

$$\varepsilon_x = \frac{1}{E}[\sigma_x - \nu(0 - p)] = 0 \quad \text{or} \quad \sigma_x = -\nu p \quad \text{(compressive)}$$

$$\varepsilon_y = \frac{1}{E}[0 - \nu(-p - \nu p)] = \frac{p\nu(1+\nu)}{E} \quad \text{(tensile)}$$

$$\varepsilon_z = \frac{1}{E}[-p - \nu(-\nu p + 0)] = -\frac{p}{E}(1 - \nu^2) \quad \text{(compressive)}$$

The corresponding deformations are

$$\delta a = 0; \delta b = \frac{p\nu(1+\nu)b}{E} \text{ (expansion) and } \delta c = -\frac{p(1-\nu^2)c}{E} \text{ (contraction)}$$

EXAMPLE 3.19

Compare the stresses in a metal prism subjected to a pressure p on the top (say Z-direction) in two cases: (a) the prism is compressed freely and (b) the prism is compressed in a rigid enclosure which prevents it from expanding laterally.

Solution:

(a) Since there is no restraint in the lateral directions, $\sigma_x = \sigma_y = 0$ and the pressures $\sigma_z = -p$.

(b) Due to the presence of rigid enclosure, the strains ε_x and ε_y in the lateral X- and Y-directions are prevented, and thus $\sigma_x = \sigma_y \neq 0$. However, since there is no restraint in Z-direction, the pressure $\sigma_z = -p$. Therefore,

$$\varepsilon_x = \frac{1}{E}[\sigma_x - \nu(\sigma_y + \sigma_z)] = \frac{1}{E}[\sigma_x - \nu(\sigma_x - p)] = 0 \quad \text{(compressive)}$$

Hence lateral stress is

$$\sigma_x = -\left(\frac{\nu}{1-\nu}\right)p = \sigma_y \quad \text{(compressive)}$$

EXAMPLE 3.20

A bar specimen of 36 mm diameter was subjected to a pull of 90 kN during a tension test. The extension on a gauge length of 200 mm was measured to be 0.089 mm and the change in diameter of 0.0046 mm. Determine the Poisson's ratio and the values of three moduli of elasticity.

Solution: For the given specimen,

Gauge length, $l = 200$ mm, $A = \frac{\pi}{4} \times 36^2 = 1017.88$ mm^2, $\delta l = 0.089$ mm and $\delta d = 0.0046$ mm

Direct tensile stress, $\sigma_o = \frac{P}{A} = \frac{90 \times 10^3}{1017.88} = 88.42$ MPa

Longitudinal strain, $\varepsilon_l = \frac{\delta l}{l} = \frac{0.089}{200} = 0.000445$

Lateral strain, $\varepsilon_d = \frac{\delta d}{d} = \frac{0.0046}{36} = 0.0001278$

Poisson's ratio, $\nu = \frac{\varepsilon_d}{\varepsilon_l} = \frac{0.0001278}{0.000445} = 0.287$

Modulus of elasticity,

$$E = \frac{\sigma_o}{\varepsilon_l} = \frac{88.42}{0.000445} = 198696.63 \text{ MPa}$$

$$= 198696.63 \times (10^{-3}) = 198.697 \text{ GPa}$$

Modulus of rigidity, $G = \dfrac{E}{2(1+\nu)} = \dfrac{198.697}{2 \times (1+0.287)} = 77.194 \text{ GPa}$

Bulk modulus, $K = \dfrac{E}{3(1-2\nu)} = \dfrac{198.697}{3 \times (1 - 2 \times 0.287)} = 155.47 \text{ GPa}$

EXAMPLE 3.21

At a certain point in a steel structure, the directions of principal stresses σ_1 and σ_2 are known. The strain gauges mounted on the structure indicate a tensile strain of 8.5×10^{-4} in the direction of σ_1 and a compressive strain of 6.0×10^{-4} in the direction of σ_2. Determine the magnitudes of σ_1 and σ_2, and the value of the maximum shear stress. The modulus of elasticity and Poisson's ratio for steel are 200 GPa and $\nu = 0.25$, respectively.

Solution: At the given point on the structure, $\varepsilon_1 = 0.00085$ and $\varepsilon_2 = -0.0006$

The strains are related to the principal stresses as,

$$\varepsilon_1 = \frac{1}{E}(\sigma_1 - \nu\sigma_2) \quad \text{or} \quad (\sigma_1 - \nu\sigma_2) = E\varepsilon_1 \tag{i}$$

$$\varepsilon_2 = \frac{1}{E}(\sigma_2 - \nu\sigma_1) \quad \text{or} \quad (\sigma_2 - \nu\sigma_1) = E\varepsilon_2 \tag{ii}$$

Eliminate σ_2 by operating on Eqs. (i) and (ii) as, (i) + ν (ii):

$$(\sigma_1 - \nu\sigma_2) + \nu(\sigma_2 - \nu\sigma_1) = E(\varepsilon_1 + \nu\varepsilon_2) \quad \text{or} \quad \sigma_1 = \frac{E}{1-\nu^2}(\varepsilon_1 + \nu\varepsilon_2)$$

Operate on Eqs. (i) and (ii) as, ν (i) + (ii):

$$\nu(\sigma_1 - \nu\sigma_2) + (\sigma_2 - \nu\sigma_1) = E(\nu\varepsilon_1 + \varepsilon_2) \quad \text{or} \quad \sigma_2 = \frac{E}{1-\nu^2}(\nu\varepsilon_1 + \varepsilon_2)$$

Therefore, the principal stresses are

$$\sigma_1 = \frac{(210 \times 10^3)}{(1 - 0.25^2)}(8.5 - 0.25 \times 6.0) \times 10^{-4} = 156.8 \text{ MPa (tension)}$$

and $$\sigma_2 = \frac{(210 \times 10^3)}{(1 - 0.25^2)}(0.25 \times 8.5 - 6.0) \times 10^{-4} = -86.8 \text{ MPa (compressive)}$$

The maximum shear stress,

$$\tau_{\max} = \frac{\sigma_1 - \sigma_2}{2} = \frac{156.8 + 86.8}{2} = 121.8 \text{ MPa}$$

EXAMPLE 3.22

In order to determine the principal stresses, two strain gauges are mounted at a certain point in a structural member at 30° to the known directions of the principal stresses. The strain gauges indicate a tensile strain of 3.5×10^{-4} and a compressive strain of -5.0×10^{-5} in the two directions, respectively. Determine the magnitude of the principal stresses σ_1 and σ_2, and value of the maximum shear stress. The modulus of elasticity and Poisson's ratio for material are 200 GPa and $\nu = 0.25$, respectively.

Solution: At the given point on the structure,

$$\varepsilon_{\theta=30^\circ} = 3.5 \times 10^{-4} \quad \text{and} \quad \varepsilon_{\theta=120^\circ} = -5.0 \times 10^{-5}$$

The measured strain is related to the principal strains as

$$\varepsilon_\theta = \varepsilon_1 \cos^2\theta + \varepsilon_2 \sin^2\theta$$

Therefore,

$$\varepsilon_1 \cos^2 30^\circ + \varepsilon_2 \sin^2 30^\circ = 0.75\varepsilon_1 + 0.25\varepsilon_2 = 3.50 \times 10^{-4} \tag{i}$$

$$\varepsilon_1 \cos^2 120^\circ + \varepsilon_2 \sin^2 120^\circ = 0.25\varepsilon_1 + 0.75\varepsilon_2 = -5.0 \times 10^{-5} \tag{ii}$$

Eliminate ε_2 by operating on Eqs. (i) and (ii) as, 3(i) – (ii):

$$(3 \times 0.75 - 0.25)\,\varepsilon_1 = (3 \times 350 + 50) \times 10^{-6} \quad \text{or} \quad \varepsilon_1 = 550 \times 10^{-6}$$

Similarly, to eliminate ε_1 operate on Eqs. (i) and (ii) as, (i) – 3(ii):

$$(0.25 - 3 \times 0.75)\,\varepsilon_1 = (350 + 3 \times 50) \times 10^{-6} \quad \text{or} \quad \varepsilon_2 = -250 \times 10^{-6}$$

The principal strains are related to the principal stresses as

$$\varepsilon_1 = \frac{1}{E}(\sigma_1 - \nu\sigma_2) \quad \text{or} \quad (\sigma_1 - \nu\sigma_2) = E\varepsilon_1 \tag{iii}$$

$$\varepsilon_2 = \frac{1}{E}(\sigma_2 - \nu\sigma_1) \quad \text{or} \quad (\sigma_2 - \nu\sigma_1) = E\varepsilon_2 \tag{iv}$$

Eliminate σ_2 by operating on Eqs. (iii) and (iv) as, (iii) + ν(iv):

$$(\sigma_1 - \nu\sigma_2) + \nu(\sigma_2 - \nu\sigma_1) = E(\varepsilon_1 + \nu\varepsilon_2) \quad \text{or} \quad \sigma_1 = \frac{E}{1-\nu^2}(\varepsilon_1 + \nu\varepsilon_2)$$

Similarly,

$$\sigma_2 = \frac{E}{1-\nu^2}(\nu\varepsilon_1 + \varepsilon_2)$$

Therefore, the principal stresses are

$$\sigma_1 = \frac{(200 \times 10^3)}{(1 - 0.25^2)}(550 - 0.25 \times 250) \times 10^{-6} = 104 \text{ MPa (tension)}$$

and

$$\sigma_2 = \frac{(200 \times 10^3)}{(1 - 0.25^2)}(0.25 \times 550 - 250) \times 10^{-6} = -24 \text{ MPa (compressive)}$$

The maximum shear stress, $\tau_{max} = \dfrac{\sigma_1 - \sigma_2}{2} = \dfrac{104 + 24}{2} = 64$ MPa

3.10 PROBLEMS

3.1 Draw diagrams of normal forces, stresses and axial displacements for a freely suspended elastic bar of uniform cross-section caused by its own weight, i.e. the gravity forces. The length, cross-sectional area and unit weight of the material of the bar are L, A and γ_o, respectively.

$$\left[\textbf{Ans. } F_{max} = \gamma_o AL, \sigma_{max} = \gamma_o L, u_x = \frac{\gamma_o}{2E}(L^2 - x^2) \text{ and } u_{max} = \frac{\gamma_o L^2}{2E}\right]$$

3.2 An elastic member of constant cross-section, fixed at one end is axially loaded at several points as shown in Figure 3.19. Draw normal force and axial displacement diagrams.

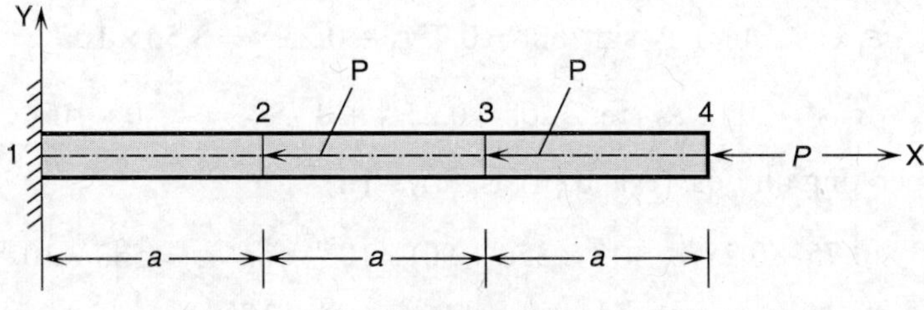

Figure 3.19

$$\left[\textbf{Ans. } F_{max} = 3P \text{ and } u_{max} = -\frac{6Pa}{AE}\right]$$

3.3 A straight homogeneous elastic member of variable cross-sectional area, rigidly fixed at both the ends, is loaded as shown in Figure 3.20. Determine the reactions and draw the normal force and axial displacement diagrams.

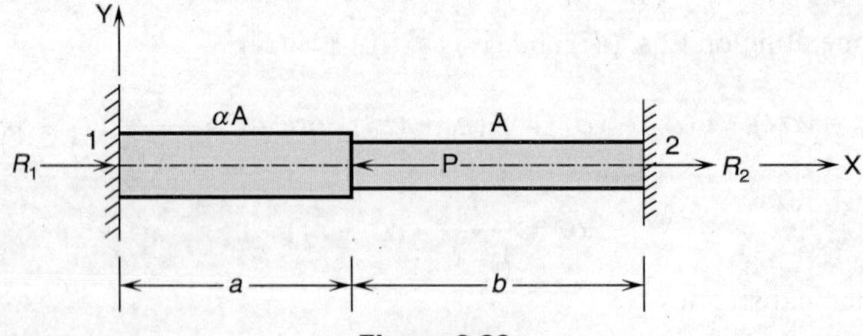

Figure 3.20

$$\left[\textbf{Ans. } R_1 = \frac{P}{1 + [a/(\alpha b)]} \text{ and } R_2 = \frac{P}{1 + [b/(\alpha b)]}\right]$$

3.4 An elastic stepped member, rigidly fixed at both the ends, is axially loaded at two points as shown in Figure 3.21. Determine the reactions, and draw the normal force and axial displacement diagrams.

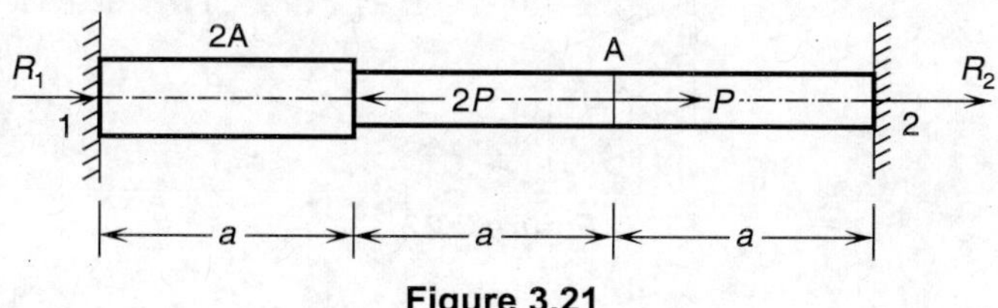

Figure 3.21

$$\left[\textbf{Ans. } R_1 = \frac{6P}{5} \text{ and } R_2 = \frac{P}{5}\right]$$

3.5 The rectangular cross-section of a straight steel bar of length 4.0 m varies uniformly from 100 × 10 at one end to 40 × 10 at the other. Determine the change in the length of the bar when it is subjected to an axial tensile load of 30 kN. The modulus of elasticity for steel is 210 GPa.

[**Hint:** Consider an elementary strip *dx* at a section at distance *x* mm from the smaller end, the width $b_x = 40 + \dfrac{(100-40)x}{4000} = 40 + 0.015x$; extension of elementary strip,

$$\delta x = \frac{(30 \times 10^3)\,dx}{[(40 + 0.015x) \times 10]\,E}$$

Total extension

$$= \int_0^{4000} \frac{(30 \times 10^3)}{[(40 + 0.015x) \times 10]E}\,dx = \frac{3000}{E} \times \frac{1}{0.015} \left| \log_e (40 + 0.015x) \right|_0^{4000}$$

$= 0.873$ mm]

3.6 A rectangular block of elastic material is subjected to a direct longitudinal compressive stress of α_1. Compressive stresses of intensity σ_2 and σ_3 are applied in the lateral directions perpendicular to the longitudinal direction, so that the net lateral strain in the directions along σ_2 and σ_3 is reduced to one-half and one-fourth of the lateral strains that would have occurred if σ_1 alone was acting. Determine: (a) σ_2 and σ_3 in terms of σ_1 and (b) net strain in the longitudinal direction. The Poisson's ratio for material of the block is $\nu = 0.25$.

$$\left[\textbf{Ans. } \sigma_2 = \left(\frac{11}{60}\right)\sigma_1;\ \sigma_3 = \left(\frac{7}{30}\right)\sigma_1 \text{ and } \varepsilon_1 = -\left(\frac{43}{48}\right)\frac{\sigma_1}{E}\right]$$

3.7 A block of elastic material shown in Figure 3.22 is subjected to a direct longitudinal compressive of σ_1. Suitable lateral compressive stresses of intensity σ_2 are applied along the other two the lateral directions so that the net strain in each of the lateral directions is one-half that would have occurred if σ_1 alone was acting. Determine the intensity of σ_2 and the net strain in the longitudinal direction.

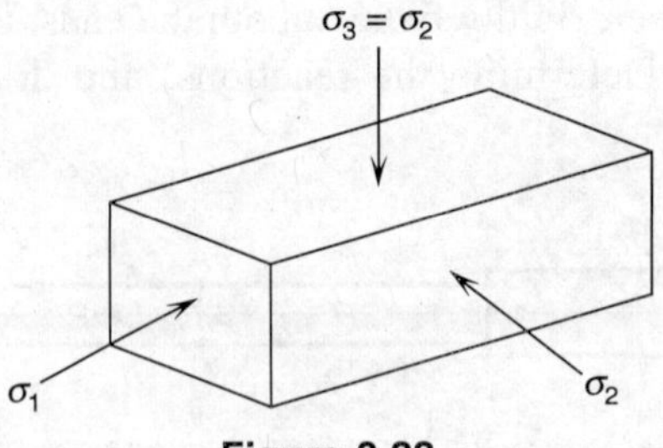

Figure 3.22

$$\left[\textbf{Ans. } \sigma_2 = \frac{\nu}{2(1-\nu)}\sigma_1; \text{ and } \varepsilon_1 = -\left(\frac{1-\nu-\nu^2}{1-\nu}\right)\frac{\sigma_1}{E}\right]$$

3.8 A 150 mm long steel bar of 50 × 50 cross-section is subjected to an axial compressive load of 250 kN. The lateral strains are prevented by application of uniform external pressure. If the modulus of elasticity and the Poisson's ratio for steel are E_s = 200 GPa and ν = 0.3, respectively, determine: (a) change in the length of the bar, and (b) change in the length of the bar when only one-half the lateral strain is prevented.

[**Ans.** (a) Decrease in length, δl = 0.0557 mm and (b) δl = 0.06536 mm]

3.9 A 300 mm long steel block of 60 × 60 mm cross-section is subjected to a tensile stress of 60 MPa in the direction of its length and lateral compressive stresses of 30 MPa on one set of lateral faces and a lateral tensile stress of 15 MPa on the other set of lateral faces. Determine the change in the dimensions and the volume of the block. The modulus of elasticity and the Poisson's ratio for steel are E_s = 200 GPa and ν = 0.40, respectively.

[**Ans.** δl = 1.32 mm; δb = 0.012 mm; δd = (–)0.24 mm and change in the volume, δV = 648 mm^3]

3.10 A 300 mm long steel bar of 120 × 60 mm cross-section is subjected to tensile load of 500 kN in the direction of its length, a compressive load of 5000 kN on its 120 × 300 mm faces and a tensile load of 2500 kN on its 60 × 300 mm faces. Determine: (a) change in the volume of the bar, and (b) the change required in the load of 5000 kN when there is no change in the volume of the bar, i.e. volumetric strain is prevented. The modulus of elasticity and the Poisson's ratio for steel are E_s = 200 GPa and ν = 0.25, respectively.
[**Hint:** For no change in volume $(\sigma_x + \sigma_y + \sigma_z) = 0$]

[**Ans.** (a) Change in volume, δV = 375 mm^3 and (b) δF_z = 2500 MPa (Comp)]

3.11 A rod of 5 × 5 mm cross-section is subjected to an axial tensile load of 3.5 kN. It is observed that the cross-section has changed to 4.99925 × 4.99925 mm. Determine the Poisson's ratio and the modulus of elasticity, if the modulus of rigidity for the material of the rod is G = 80 GPa. [**Ans.** ν = 0.3158 and E = 210 GPa]

3.12 A 300 mm long bar of 40 × 40 mm cross-section is subjected to an axial pull of 160 kN. The extension of the bar has been observed to be 0.150 mm, while the decrease in each of the lateral dimensions is observed to be 0.005 mm, determine the Young's modulus, Poisson's ratio, modulus of rigidity and bulk modulus for the material of the bar.

[**Ans.** E = 200 GPa; ν = 0.25; G = 80 GPa and K = 133.3 GPa]

3.13 A 40 mm square bar of 160 mm length is subjected to an axial compressive load of 20 kN. If the lateral strains are prevented by application of a uniform external pressure of suitable intensity on its lateral faces, compute: (a) the intensity of pressure and (b) the change in the volume. The modulus of elasticity and Poisson's ratio for the material are 200 GPa and $\nu = 0.25$, respectively.

[**Ans.** $\sigma_y = \sigma_z = -4.167$ MPa and $\delta V = -24$ mm^3]

3.14 A metal bar of square section is subjected to an axial compressive stress. The modulus of elasticity and Poisson's ratio for the material are E and ν, respectively. If the lateral strains are prevented by application of a uniform external pressure of suitable intensity on the lateral faces of the bar show that under these conditions the axial stress and strain in the bar give a modified elastic modulus of $E_m = \left(\dfrac{1-\nu}{1-\nu-2\nu^2}\right)E$.

3.15 (a) Determine the percentage change in the volume of a 125 mm square steel bar of 3.5 m length when subjected to an axial compressive load of 35 kN.

(b) What change in volume of 125 mm cube of steel would occur when it is placed at a depth of 6 km in the sea water? The unit weight of sea water may be taken as 9.807 kN/m^3. The moduli of elasticity and rigidity for steel are 200 GPa and 80 GPa, respectively.

[**Hint:** Compute ν and K; $\nu = 0.25$ and $K = 133.33$ GPa, Percent change in volume $= \varepsilon_v \times 100$ and Pressure, $\sigma = (9.807 \times 10^3) \times (6000) \times 10^{-6} = 58.842$ MPa]

[**Ans.** (a) δV (%) $= 0.56 \times 10^{-3}$ and (b) $\delta V = 861.965$ mm^3]

3.16 In order to determine the principal stresses at a point on a pressure vessel, the electrical gauges recorded the following strains: in the X-direction, $\varepsilon_x = 400 \times 10^{-6}$; in the Y-direction, $\varepsilon_y = 80 \times 10^{-6}$ and in the direction at 45° with the X-direction, $\varepsilon_\theta = -160 \times 10^{-6}$. Determine at the point under consideration the magnitude of: (a) the principal strains, (b) the shearing strain and (c) principal stresses. The modulus of elasticity and Poisson's ratio for material of the pressure vessel are 210 GPa and $\nu = 0.3$, respectively.

[**Ans.** (a) $\varepsilon_1 = 560 \times 10^{-6}$; $\varepsilon_2 = -240 \times 10^{-6}$, (b) $\varepsilon_s = 640 \times 10^{-6}$;
(c) $\sigma_1 = 112.62$ MPa and $\sigma_2 = -16.62$ MPa]

Chapter 4

Analysis for Axial Deformation of Bars:

Statically Indeterminate Systems

(Composite Systems and Thermal Stresses)

4.1 INTRODUCTION

In this chapter, statically indeterminate composite systems are discussed. These systems are composed of more than one component of the same or different material rigidly connected in such a way that when subjected to loads or temperature variations, each individual component undergoes change of same magnitude. These systems are called *composite systems*.

The forces in and deformations of a structure must satisfy two conditions:

1. Equilibrium, and
2. Compatibility

The first condition ensures that the forces satisfy the *equilibrium* requirements; the second that the deformations satisfy the geometrical requirements.

In the analysis of *statically-determinate* structures no consideration is given to the *compatibility* because the components of such structures always fit together without any straining, and a slight change in the dimension only produces small alterations in the geometry, with the corresponding changes in the forces. However, in the analysis of *statically-indeterminate* structures, the compatibility is important since even a slight change in the dimensions will produce significant alterations in the geometry, with large changes in the forces. Thus while analyzing the statically-indeterminate structures both the conditions have to be taken into account.

EXAMPLE 4.1

The planar system of three elastic members shown in Figure 4.1(a) is subjected to a force W at the joint 1. Determine the vertical deflection at joint 1 and forces in the members developed due to load W. The cross-sectional area A and the elastic modulus E of each member is same.

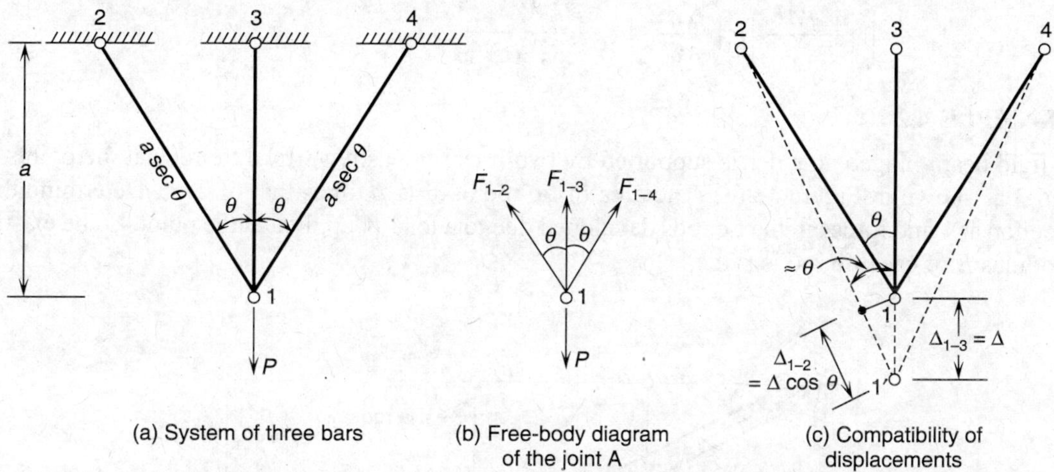

Figure 4.1 Analysis of three bar system of Example 4.1.

Solution: The lengths of three members are, $L_{1-2} = L_{1-4} = a \sec \theta$, and $L_{1-3} = a$.

The structure is statically indeterminate to the first degree since three unknown concurrent member forces (meeting at a point) cannot be determined from two equilibrium conditions, $\Sigma F_V = 0$ and $\Sigma F_H = 0$. However, the displacement compatibility condition provides the third equation for solution.

The relations between normal forces obtained from equilibrium conditions using the free-body diagram shown in Figure 4.1(b) are:

$$F_{1-2} = F_{1-4} \tag{i}$$

and

$$2F_{1-2} \cos \theta + F_{1-3} = W \tag{ii}$$

In terms of vertical displacement Δ of the joint 1, the axial deformations in the three members (see Figure 4.1(c)) can be expressed for displacement compatibility as follows,

$$\Delta_{1-2} = \Delta_{1-4} = \Delta \cos \theta \text{ and } \Delta_{1-3} = \Delta$$

From Eq. (ii),

$$2\left(\frac{\Delta_{1-2}\, AE}{L_{1-2}}\right) \cos \theta + \left(\frac{\Delta_{1-3}\, AE}{L_{1-3}}\right) = 2\left(\frac{\Delta \cos \theta\, AE}{a \sec \theta}\right) \cos \theta + \left(\frac{\Delta\, AE}{a}\right) = W$$

Therefore,

$$\Delta = \frac{W}{2 \cos^3 \theta + 1}\left(\frac{a}{AE}\right)$$

The normal forces in the members are;

$$F_{1-4} = F_{1-2} = \left(\frac{\Delta_{1-2}\, AE}{L_{1-2}}\right) = \left(\frac{\Delta \cos\theta\, AE}{a \sec\theta}\right) = \Delta \cos^2\theta \left(\frac{AE}{a}\right) = \frac{W \cos^2\theta}{2\cos^3\theta + 1}$$

and $$F_{1-3} = \left(\frac{\Delta_{1-3}\, AE}{L_{1-3}}\right) = \left(\frac{\Delta\, AE}{a}\right) = \frac{W}{2\cos^3\theta + 1}$$

EXAMPLE 4.2

A rigid beam hinged at end 1 is supported by two linearly elastic vertical tie rods at the points 2 and 3 as shown in Figure 4.2(a). The area of tie rod 3–3′ is β times that of 2–2′. Determine the reaction at 1 and forces in the tie rods developed due to a load W applied at the point 4. The elastic modulus E of each tie rod is same.

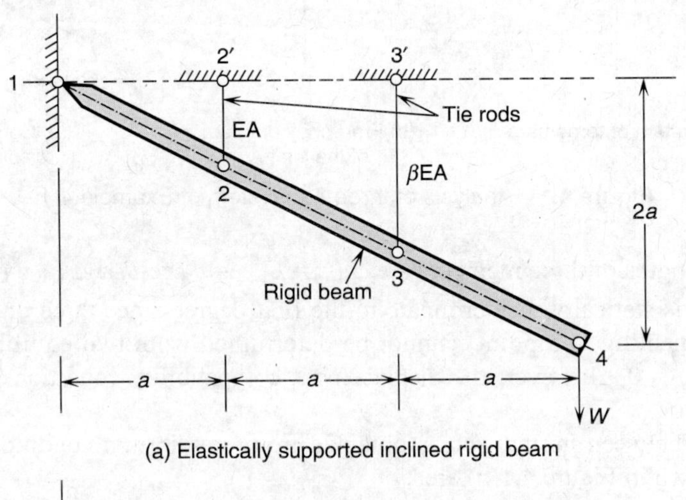

(a) Elastically supported inclined rigid beam

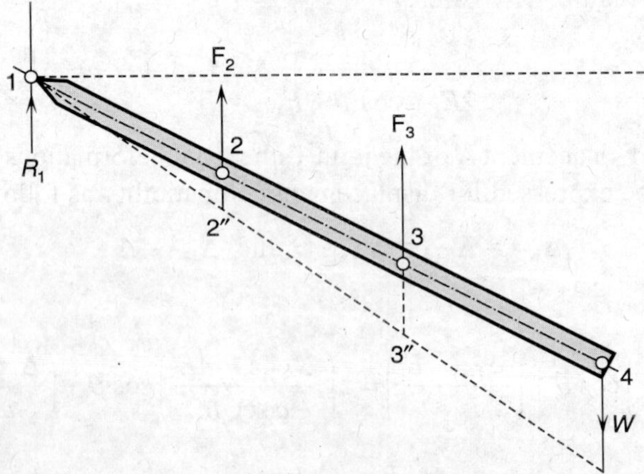

(b) Free-body diagram of the inclined rigid beam

Figure 4.2 Rigid beam supported by a pin and elastic tie rods.

Solution: The lengths of tie rods are, $L_{2-2'} = 2a/3$, and $L_{3-3'} = 4a/3$.

The structure is statically indeterminate to the first degree since three unknown parallel forces cannot be determined from two equilibrium conditions, $\Sigma F_V = 0$ and $\Sigma M_1 = 0$. However, the displacement compatibility condition provides the third equation for solution. From equilibrium conditions using the free-body diagram shown in Figure 4.2(b) are:

$$\sum F_V = R_1 + F_2 + F_3 - W = 0 \quad \text{or} \quad R_1 + F_2 + F_3 = W \tag{i}$$

$$\sum M_1 = F_2(a) + F_3(2a) - W(3a) = 0 \quad \text{or} \quad F_2 + 2F_3 = 3W \tag{ii}$$

For displacement compatibility (see Figure 4.1(c)) the axial deformations in tie rod 3–3′ is twice of that in 2–2′, thus,

$$\Delta_{3-3'} = 2 \times \Delta_{2-2'} \quad \text{or} \quad \frac{F_3(4a/3)}{\beta AE} = 2 \times \frac{F_2(2a/3)}{AE} \quad \text{or} \quad F_3 = \beta F_2 \tag{iii}$$

From Eqs. (ii) and (iii):

$$F_2 + 2(\beta F_2) = 3W \quad \text{or} \quad F_2 = \frac{3W}{1+2\beta}$$

Therefore,

$$F_3 = \frac{3\beta W}{1+2\beta}$$

From Eq. (i):

$$R_1 = W - F_2 - F_3 = W - \frac{3W}{1+2\beta} - \frac{3\beta W}{1+2\beta} = -\frac{(2+\beta)W}{1+2\beta}$$

4.2 SYSTEMS CONSISTING OF ELEMENTS OF EQUAL LENGTHS

Consider a bar system comprising of n bars of equal length with areas of cross-section, moduli of elasticity and loads carried by bars of (A_1, E_1, F_1), ..., (A_i, E_i, F_i), ..., (A_n, E_n, F_n) subjected to a load W as shown in Figure 4.3(a).

For equilibrium,

$$F_1 + \cdots + F_i + \cdots + F_n = W \quad \text{or} \quad \sum_{i=1}^{n} F_i = W \tag{4.1}$$

Since all the bars are rigidly connected together, they all undergo equal change in the length ΔL and since length of each bar is same (see Figure 4.3(b)), therefore, strains in the bars $\left(\frac{\Delta L}{L}\right)$ are same.

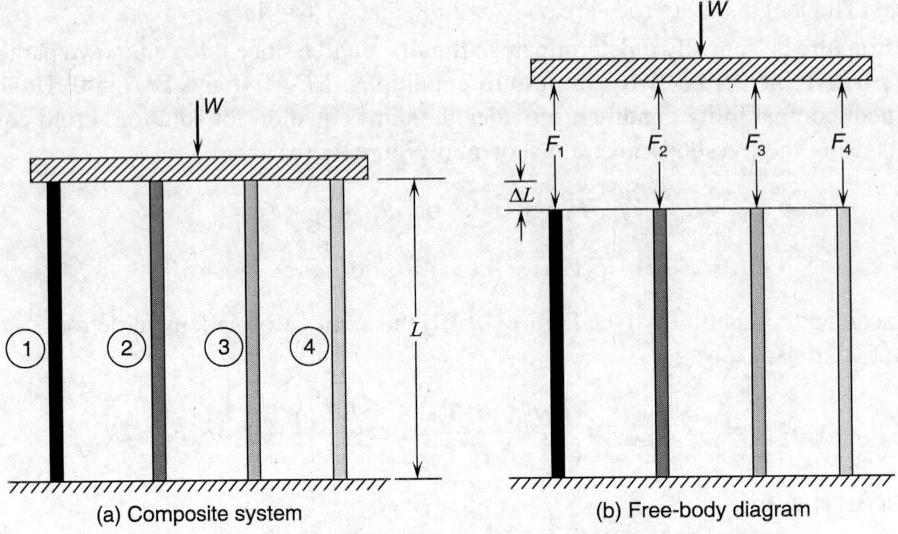

Figure 4.3 Composite system composed of elements of equal length.

Load carried by ith bar, F_i = Stress × Area = (Strain × E) × Area

$$= \left(\frac{\Delta L}{L}\right) \times E_i A_i$$

Therefore,
$$\sum_{i=1}^{n} F_i = \left(\frac{\Delta L}{L}\right)\left(\sum_{i=1}^{n} E_i A_i\right) = W \quad \text{or} \quad \Delta L = \frac{WL}{\left(\sum_{i=1}^{n} E_i A_i\right)} \tag{4.2}$$

and
$$F_i = \frac{E_i A_i}{\left(\sum_{i=1}^{n} E_i A_i\right)} \times W \tag{4.3}$$

Thus, the load is shared in the ratio of their (EA) values.

$$\text{Stress in the } i\text{th bar, } \sigma_i = \frac{W}{\dfrac{1}{E_i}\left(\sum_{i=1}^{n} E_i A_i\right)} \tag{4.4}$$

It is usually convenient to consider the above composite system as a single element of an imaginary material with an equivalent or *composite modulus* E_e.

Load carrying capacity of the equivalent element

$$W = F_1 + F_2 + \cdots + F_n = \sum_{i=1}^{n} F_i$$

If σ_e is the stress in the equivalent single element, then

$$\sigma_e (A_1 + A_2 + \cdots + A_n) = \sigma_1 A_1 + \sigma_2 A_2 + \cdots + \sigma_n A_n$$

Dividing throughout by the common strain ε,

$$\frac{\sigma_e}{\varepsilon}(A_1 + A_2 + \cdots + A_n) = \left(\frac{\sigma_1}{\varepsilon}\right) A_1 + \left(\frac{\sigma_2}{\varepsilon}\right) A_2 + \cdots + \left(\frac{\sigma_n}{\varepsilon}\right) A_n$$

or $$E_e (A_1 + A_2 + \cdots + A_n) = E_1 A_1 + E_2 A_2 + \cdots + E_n A_n \quad \text{or} \quad E_e \sum_{i=1}^{n} A_i = \sum_{i=1}^{n} E_i A_i$$

Therefore, $$E_e = \sum_{i=1}^{n} E_i A_i \Big/ \sum_{i=1}^{n} A_i \tag{4.5}$$

Where, total area of cross-section,

$$A = A_1 + A_2 + \cdots + A_n = \sum_{i=1}^{n} A_i$$

Stress in the equivalent composite,

$$\sigma_e = \frac{W}{A} = \frac{W}{\sum_{i=1}^{n} A_i} \tag{4.6}$$

Strain in the equivalent composite,

$$\varepsilon_e = \frac{W}{E_e \sum_{i=1}^{n} A_i} = \frac{\Delta L}{L} \tag{4.7}$$

Common extension,

$$\Delta L = \frac{WL}{E_e \sum_{i=1}^{n} A_i} \tag{4.8}$$

4.3 SYSTEMS CONSISTING OF ELEMENTS OF UNEQUAL LENGTHS

Consider a system comprised of a solid cylinder placed inside a tube. The two elements are of different materials having cross-sectional areas A_1 and A_2, and unequal lengths. The system is subjected to a load W as shown in Figure 4.4(a). Since the cylinder is shorter than the outer tube by an amount ΔL, therefore, upon application of load W, the outer tube supports the entire load until it is shortened by ΔL. Thereafter, the analysis is similar to that for a composite system with equal lengths as shown in Figure 4.4(b) and the corresponding free-body diagram is shown in Figure 4.4(c). Thus, the compressive load carried by the tube during shortening by ΔL is:

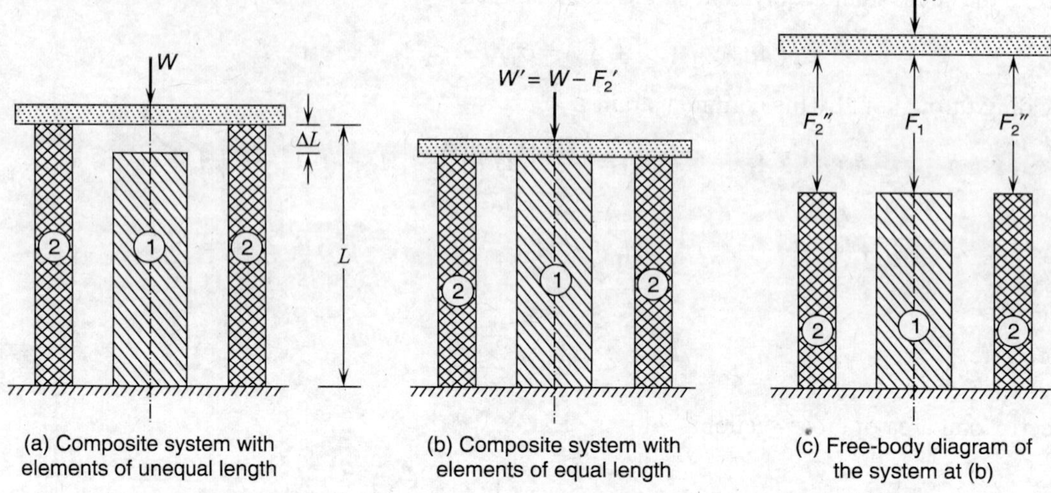

Figure 4.4 Composite system composed of elements of unequal lengths.

$$F_2' = \left(\frac{\Delta L}{L}\right) \times E_2 A_2$$

The balance load to be shared between both the components is,

$$W' = W - F_2' = W - \left(\frac{\Delta L}{L}\right) \times E_2 A_2$$

The load W' is shared in the ratio of (EA) values of the components,

$$F_1 = \left(\frac{E_1 A_1}{E_1 A_1 + E_2 A_2}\right) \times W' = \left(\frac{E_1 A_1}{E_1 A_1 + E_2 A_2}\right) \times \left[W - \left(\frac{\Delta L}{L}\right) \times E_2 A_2\right] \tag{4.9}$$

and

$$F_2'' = \left(\frac{E_2 A_2}{E_1 A_1 + E_2 A_2}\right) \times W' = \left(\frac{E_2 A_2}{E_1 A_1 + E_2 A_2}\right) \times \left[W - \left(\frac{\Delta L}{L}\right) \times E_2 A_2\right]$$

Therefore,

$$F_2 = F_2' + F_2'' = \left(\frac{\Delta L}{L}\right) \times E_2 A_2 + \left(\frac{E_2 A_2}{E_1 A_1 + E_2 A_2}\right) \times \left[W - \left(\frac{\Delta L}{L}\right) \times E_2 A_2\right] \tag{4.10}$$

EXAMPLE 4.3

A 100 mm long steel bolt of 10 mm diameter having a thread pitch of 1.6 mm passes concentrically through a copper spacing tube of internal and external diameters of 12 and 18 mm, respectively, as shown in Figure 4.5(a). If the nut is initially tightened up by hand so that it does not cause any stress in the spacing tube. Determine the stresses induced in the tube and in the bolt if a spanner is then used to turn the nut through 90°. The moduli of elasticity for steel and copper are 209 and 100 GPa, respectively.

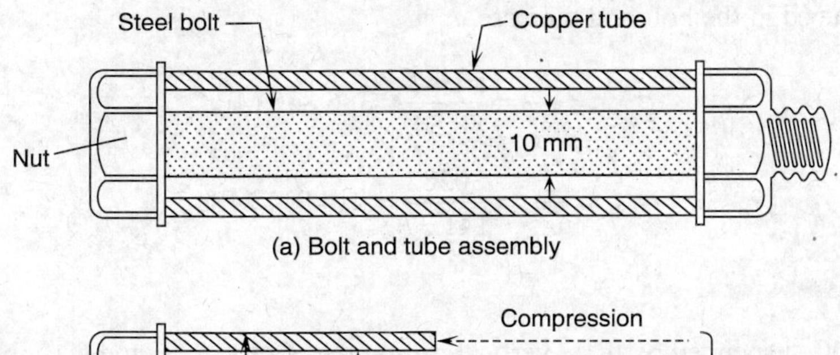

(a) Bolt and tube assembly

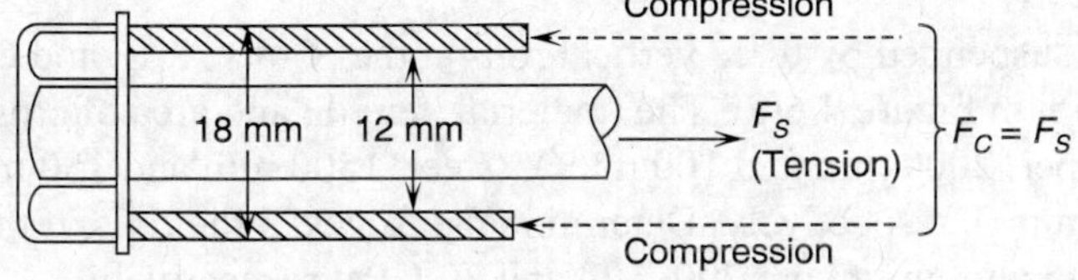

(b) Self-balancing of the forces

Figure 4.5 Bolt and tube assembly compressed by tightened by the nut (Example 4.3).

Solution: For the given bolt and tube assembly,

$$A_s = \frac{\pi}{4} \times 10^2 = 78.54 \text{ mm}^2 \quad \text{and} \quad A_c = \frac{\pi}{4} \times (18^2 - 12^2) = 141.37 \text{ mm}^2$$

In this case, the assembly is not subjected to any external force; the forces acting on the elements balance each other as shown in Figure 4.5(b), i.e. it is a self-restrained system. Here, when the nut is tightened up by the spanner, the spacing tube is compressed and the bolt is elongated. Moreover, the decrease in the length of tube plus extension of the bolt should be equal to the distance moved by the nut. Therefore,

For internal equilibrium:

Tensile force in the steel bolt = Compressive force in the copper tube, i.e., $F_s = F_c = F$

For displacement compatibility:

Increase in the length of bolt + Decrease in the length of tube = Axial displacement of the nut

$$\delta_s + \delta_c = \frac{F_s l}{A_s E_s} + \frac{F_c l}{A_c E_c} = Fl\left(\frac{1}{A_s E_s} + \frac{1}{A_c E_c}\right) = 1.6 \times \left(\frac{90}{360}\right)$$

$$= 0.4 \text{ mm}$$

Therefore,

$$F \times 100 \times \left(\frac{1}{78.54 \times 209 \times 10^3} + \frac{1}{141.37 \times 100 \times 10^3}\right) = 0.4$$

or

$$F = 30382 \text{ N} = 30.382 \text{ kN}$$

The stresses induced in the bolt and tube are,

$$\sigma_{\text{bolt}} = \frac{F}{A_s} = \frac{30382}{78.54} = 386.835 \text{ MPa}$$

$$\sigma_{\text{tube}} = \frac{F}{A_c} = \frac{30382}{141.37} = 214.911 \text{ MPa}$$

EXAMPLE 4.4

A rigid bar ABDC suspended by three vertical rods at the points A, B, and C supports a load W at the point D as shown in Figure 4.6(a). The (material, length, and area of cross-section) of the rods 1, 2 and 3 are (copper, 2000 mm and 100 mm^2), (steel, 1500 mm and 150 mm^2) and (aluminium, 2000 mm and 200 mm^2), respectively. Determine the forces in the rods, if the moduli of elasticity for steel, copper and aluminium are 200, 100 and 70 GPa, respectively.

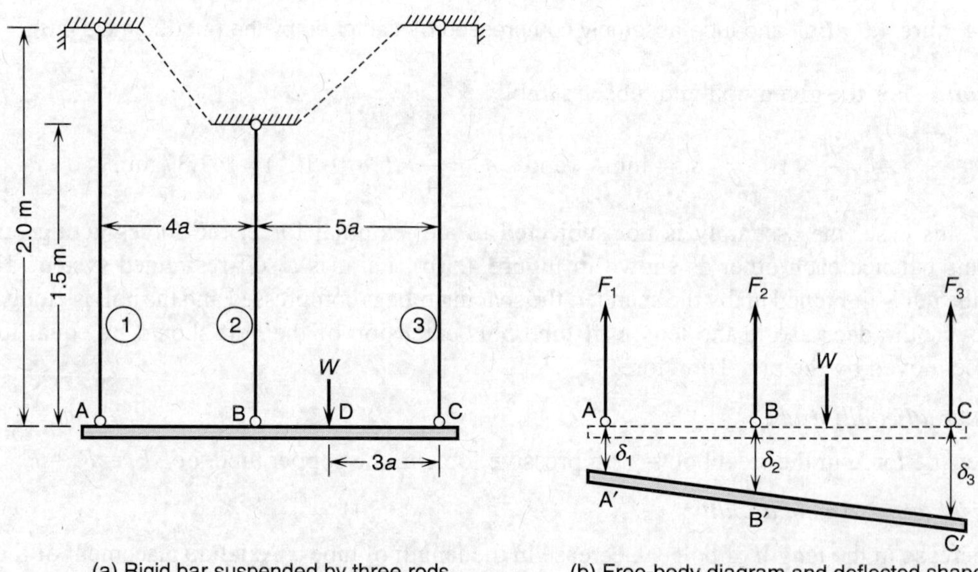

(a) Rigid bar suspended by three rods (b) Free-body diagram and deflected shape

Figure 4.6 Analysis of rigid bar suspended by three rods of Example 4.4.

Solution: Consider tensions in the rods 1, 2 and 3 to be F_1, F_2 and F_3 kN with corresponding extensions of δ_1, δ_2 and δ_3 mm, respectively, as shown in Figure 4.6(b).

(i) Equilibrium: $\sum F_v = F_1 + F_2 + F_3 - W = 0$ or $F_1 + F_2 + F_3 = W$ (i)

$$\sum M_3 = F_1(9a) + F_2(5a) - W(3a) = 0 \quad \text{or} \quad 9F_1 + 5F_2 = 3W \qquad \text{(ii)}$$

(ii) Compatibility: $\dfrac{\delta_2 - \delta_1}{4a} = \dfrac{\delta_3 - \delta_2}{5a}$ or $5\delta_1 - 9\delta_2 + 4\delta_3 = 0$

where

$$\delta_1 = \frac{F_1 L_1}{A_1 E_1} = \frac{(F_1 \times 10^3) \times 2000}{100 \times (100 \times 10^3)} = \frac{F_1}{5}$$

$$\delta_2 = \frac{(F_2 \times 10^3) \times 1500}{150 \times (200 \times 10^3)} = \frac{F_1}{20} \text{ and } \delta_3 = \frac{(F_3 \times 10^3) \times 2000}{200 \times (70 \times 10^3)} = \frac{F_1}{7}$$

Therefore,

$$5 \times \left(\frac{F_1}{5}\right) - 9 \times \left(\frac{F_2}{20}\right) + 4 \times \left(\frac{F_3}{7}\right) = 0 \text{ or } 7F_1 - 3.15F_2 + 4F_3 = 0 \qquad \text{(iii)}$$

From Eqs. (i) and (ii)

$$F_2 = 0.6W - 1.8F_1 \text{ and } F_3 = 0.4W + 0.8F_1$$

From Eq. (iii)

$$7F_1 - 3.15(0.6W - 1.8F_1) + 4(0.4W + 0.8F_1) = 0 \text{ or } F_1 = 0.01827W$$

Therefore,

$$F_2 = 0.56711W \text{ and } F_3 = 0.41462W$$

EXAMPLE 4.5

A wire strand consists of a steel wire 3.0 mm diameter covered by six, 2.5 mm diameter bronze wires. If the permissible stress for bronze wire is 65 MPa, determine the load carrying capacity of the strand and the equivalent tensile modulus of elasticity for the strand. The moduli of elasticity for steel and bronze are 200 and 80 GPa, respectively.

Solution: For the given strand,

$$A_s = \frac{\pi}{4} \times (3.0)^2 = 7.069 \text{ mm}^2 \text{ and } A_b = 6 \times \frac{\pi}{4} \times (2.5^2) = 29.452 \text{ mm}^2$$

For the integrity of the strand, the strains in the steel and bronze wires should be equal, i.e.

$$\frac{\sigma_{\text{steel}}}{E_s} = \frac{\sigma_{\text{bronze}}}{E_b} \text{ or } \sigma_{\text{steel}} = \sigma_{\text{bronze}} \left(\frac{E_s}{E_b}\right) = 65 \times \left(\frac{200 \times 10^3}{80 \times 10^3}\right) = 162.5 \text{ MPa}$$

Load carrying capacity of the strand

$$P = \sigma_{\text{steel}} A_s + \sigma_{\text{bronze}} A_b = 162.5 \times 7.069 + 65 \times 29.452 = 3063.0925 \text{ N}$$

$$= 3.063 \text{ kN}$$

Total area of the cross-section of the strand,

$$A = A_s + A_b = 7.069 + 29.452 = 36.521 \text{ mm}^2$$

Equivalent stress, $\sigma_e = \frac{P}{A} = \frac{3.063 \times 10^3}{36.521} = 83.87 \text{ MPa}$

Strain in the strand,

$$\varepsilon_e = \frac{\sigma_{\text{steel}}}{E_s} = \frac{\sigma_{\text{bronze}}}{E_b} = \left(\frac{65}{80 \times 10^3}\right) = 8.125 \times 10^{-4}$$

Therefore, equivalent tensile modulus of elasticity of the strand,

$$E_e = \frac{\sigma_e}{\varepsilon_e} = \left(\frac{83.87}{8.125 \times 10^{-4}}\right) = 1.03225 \times 10^5 \text{ MPa} = 103.225 \text{ GPa}$$

EXAMPLE 4.6

An aluminium solid cylinder of 75 mm diameter is placed concentrically inside a 2500 mm long steel tube of external and internal diameters of 100 and 80 mm, respectively. The steel tube is 0.2 mm longer than the aluminium cylinder as illustrated in Figure 4.7. Determine the safe load which can be supported on the rigid flat plate placed on the top of the steel tube. The allowable stresses in the steel and aluminium are 95 and 65 MPa, respectively. The moduli of elasticity for steel and aluminium are 210 and 70 GPa, respectively.

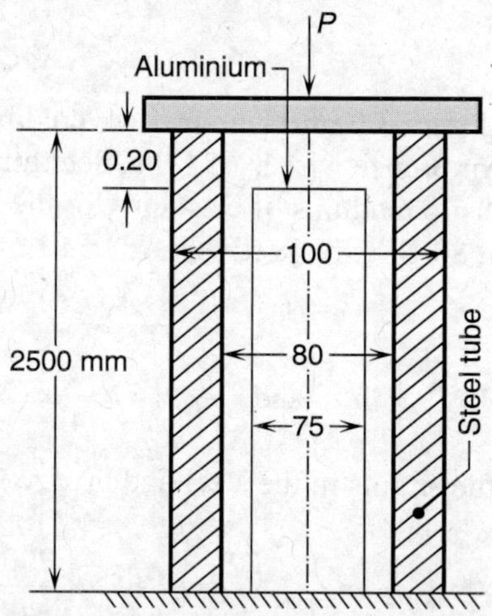

Figure 4.7 Aluminium cylinder and steel tube assembly of Example 4.6.

Solution: For the given cylinder and tube assembly,

$$A_a = \frac{\pi}{4} \times 75^2 = 4418 \text{ mm}^2 \text{ and } A_s = \frac{\pi}{4} \times (100^2 - 80^2) = 2827 \text{ mm}^2$$

The stiffness ratio,

$$\frac{E_s A_s}{E_a A_a} = \frac{(210 \times 10^3) \times 2827 \times 10^{-3}}{(70 \times 10^3) \times 4418 \times 10^{-3}} = 1.92$$

The safe load for the steel tube,

$$W_s = 95 \times 2827 = 268565 \text{ N} = 268.565 \text{ kN}$$

The safe load for the aluminium cylinder,

$$W_a = 65 \times 4418 = 287170 \text{ N} = 287.17 \text{ kN}$$

The compressive load carried by the steel tube during elimination of the initial gap of $\Delta l = 0.2$ mm:

$$W_s' = \left(\frac{0.2}{2500}\right) \times (210 \times 10^3) \times 2827 \times 10^{-3} = 47.494 \text{ kN}$$

The balance load to be shared between the two components of composite system,

$$W' = W - W_s' = (W - 47.494) \text{ kN}$$

The load W' is shared in the ratio of (EA) values of the components,

Aluminium cylinder:

$$W_a = \left(\frac{1}{1+1.92}\right) \times (W - 47.494) = 287.17$$

or
$$W = (287.17) \times (1 + 1.92) + 47.494$$
$$= 886.03 \text{ kN}$$

Steel tube:

$$W_s = W_s' + W_s'' = 47.494 + \left(\frac{1.92}{1+1.92}\right) \times (W - 47.494) = 268.565$$

or
$$W = (268.565 - 47.494) \times \left(\frac{1+1.92}{1.92}\right) + 47.494$$
$$= 383.71 \text{ kN}$$

Therefore, the safe load is $W = 383.71$ kN.

EXAMPLE 4.7

A rigid bar ABCD hinged at A is suspended by three vertical steel rods of 1000 mm length each at the points B, C and D as shown in Figure 4.8(a). The rods are of identical cross-section with an area of 2500 mm^2 each. The middle rod is fabricated shorter by $\Delta l = 0.8$ mm than the outer rods. Determine the stresses in the rods after assembly of the structure, if the modulus of elasticity for steel is 200 GPa.

Solution: Consider the forces induced in the rods 1, 2 and 3 after the structure is assembled to be F_1, F_2 and F_3 kN with corresponding deformations of $-\delta_1$, δ_2 and $-\delta_3$ mm, respectively, as shown in Figure 4.8(b). The corresponding upward displacements are δ_1, $(\Delta l - \delta_2)$ and δ_3 mm, respectively. The forces F_1 and F_3 are compressive and the force F_2 is tensile in nature.

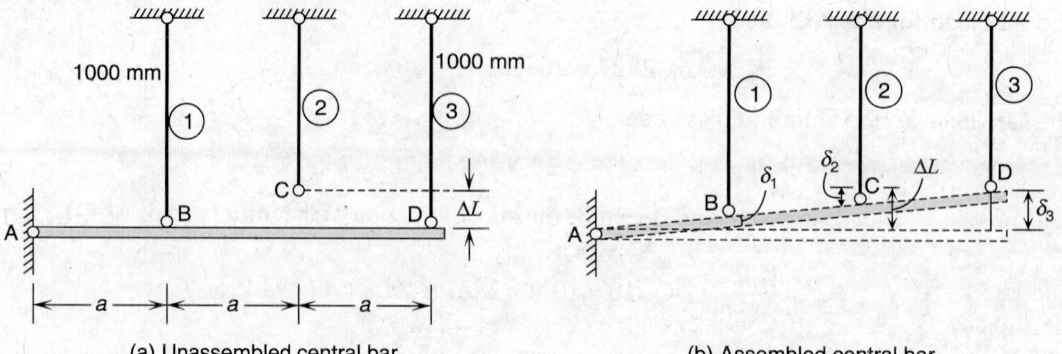

Figure 4.8 Rigid bar with lack of fit in suspenders of Example 4.7.

(a) Equilibrium: $$\sum M_A = F_1(a) - F_2(2a) + F_3(3a) = 0 \quad \text{or} \quad F_1 + 3F_3 = 2F_2 \qquad \text{(i)}$$

(b) Compatibility: (1) $$\frac{\delta_3}{3a} = \frac{\delta_1}{a} \quad \text{or} \quad \frac{\delta_3}{\delta_1} = 3$$

Therefore,

$$\frac{\delta_3}{\delta_1} = \frac{F_3L_3 / AE}{F_1L_1 / AE} = \frac{F_3}{F_1} = 3 \quad \text{or} \quad F_3 = 3F_1 \qquad \text{(ii)}$$

(2) $$\frac{\Delta L - \delta_2}{2a} = \frac{\delta_1}{a} \quad \text{or} \quad \frac{\Delta L - \delta_2}{\delta_1} = 2$$

Therefore,

$$\frac{\Delta l - \delta_2}{\delta_1} = \frac{(\Delta l - F_2L_2 / AE)}{F_1L_1 / AE} = \frac{[(AE\Delta l / L) - F_2]}{F_1} = 2$$

or $$\frac{2500 \times (200 \times 10^3) \times 0.8}{1000} - F_2 = 2F_1 \quad \text{or} \quad F_2 + 2F_1 = 4 \times 10^5 \qquad \text{(iii)}$$

From Eqs. (i) and (ii)

$$5F_1 = F_2$$

From Eq. (iii)

$$5F_1 + 2F_1 = 4 \times 10^5 \quad \text{or} \quad F_1 = \frac{4}{7} \times 10^5 \text{ N}$$

Therefore,

$$F_2 = \frac{20}{7} \times 10^5 \text{ N} \quad \text{and} \quad F_3 = \frac{12}{7} \times 10^5 \text{ N}$$

Stresses: $\sigma_1 = \frac{4}{7} \times \frac{10^5}{2500} = 22.857$ MPa and $\sigma_2 = \frac{20}{7} \times \frac{10^5}{2500} = 114.286$ MPa

$$\sigma_3 = \frac{12}{7} \times \frac{10^5}{2500} = 68.571 \text{ MPa}$$

4.4 THERMAL DEFORMATIONS AND STRAINS

In addition to usual loads, the engineering objects are invariably subjected to temperature variations. The dimensions of objects change with the change in temperature. A body expands with the increase in the temperature and contracts with decrease. Ordinarily, over a limited range of temperature change, this expansion or contraction is linearly related to the temperature increase or decrease that occurs. Consider a bar of length l at temperature t_o subjected to a temperature change of t measured in degree Celsius (°C). If the *coefficient of linear thermal expansion* for the material of the bar is α expressed as strain per degree of temperature (/°C), the change in the length of the bar,

$$\Delta l = \pm \alpha l t \tag{4.11}$$

Final length, $l + \Delta l = l \pm \alpha l t = l(1 \pm \alpha t)$

Strain, $\varepsilon_t = \frac{\pm \alpha l t}{l(1 \pm \alpha t)} = \frac{\pm \alpha t}{1 \pm \alpha t} = \pm \alpha t \times (1 \pm \alpha t)^{-1}$

$$= \pm \alpha t \times (1 \mp \alpha t) \approx \pm \alpha t$$

Since Δl is small as compared to l. Minus sign is used for decrease in temperature.

In many practical problems, *thermal strains* are required to be considered along with those due to applied loading. In such cases, the method of superposition can be used and the strain ε is taken as the sum of the strains that result from loading and pure thermal action,

$$\varepsilon = \frac{\sigma}{E} \pm \alpha t \tag{4.12}$$

For a homogeneous member loaded at its end and heated uniformly the total elongation of a member is given by

$$\Delta L = \frac{PL}{EA} \pm \alpha t L \tag{4.13}$$

Thus the strains due to loading and temperature changes are considered independent, because the modulus of elasticity E of a material varies only slightly with temperature under moderate heating, just as the quantity α is practically independent of stress σ_o.

If the free changes in the dimensions or strains of a body due to a temperature variation are either fully or partially restrained or prevented, the stresses that are induced in the body are called the *temperature stresses*.

For the bar considered, the change in the length or strain given by Eq. (4.11) can be fully prevented, if the supports exert a force F which can produce a deformation equal to Δl, i.e. equal and opposite to the temperature deformation, when acting on the unrestrained expanded member. If Δl is small compared to l, Hooke's gives

$$\varepsilon = \frac{\sigma_{th}}{E} = \mp \alpha t \quad \text{or} \quad \sigma_{th} = \mp \alpha t E$$

Thus, the stress induced due to prevention of entire strain, called thermal stress, is given by,

$$\sigma_{th} = \mp E\alpha t \tag{4.14}$$

When the temperature rises and the body is prevented from expanding, compressive stress is induced in the material which is represented by negative sign. On the other hand, when the temperature decreases and the body is prevented from contracting, tensile stress is developed.

If the deformations are partially restrained due to yielding of supports by an amount δ, actual change in dimension to be restrained after the supports yield by δ,

$$(\Delta l)' = \alpha lt - \delta$$

Actual temperature strain to be restrained, $\varepsilon = \dfrac{\alpha tl - \delta}{l}$

Net force exerted on the supports,

$$F' = AE\varepsilon = AE\left(\frac{\alpha tl - \delta}{l}\right) \tag{4.15}$$

4.5 SYSTEMS SUBJECTED TO TEMPERATURE CHANGES

When a composite system is subjected to a change in temperature, then different components change in length by different amounts due to the difference in their coefficient of linear expansions. A component with material having a higher coefficient of thermal expansion will undergo larger increase in the dimensions than the one with lower coefficient of thermal expansion. If these differential changes in the dimensions are not permitted, the material having higher coefficient of thermal expansion will be in the state of compression and other material with lower coefficient of thermal expansion will be in the state of tension. If no external force is applied, then for internal equilibrium the compressive force F in component having higher coefficient of thermal expansion will be equal to the tensile force F in component with lower coefficient of thermal expansion.

4.5.1 System Consisting of Components of Equal Lengths

Consider a composite system consisting of two elements of equal length l, a cylinder and a tube, as shown in Figure 4.9(a). The areas of cross-section, and modulus of elasticity of the cylinder and the tube are (A_1, E_1) and (A_2, E_2), respectively. If the composite system is subjected to a temperature rise of t°C, the free increase in the lengths of two elements will be $(\Delta l)_1 = \alpha_1 lt$ and $(\Delta l)_2 = \alpha_2 lt$, respectively. With the presumption $\alpha_1 > \alpha_2$, the difference in the increase of lengths is $(\alpha_1 - \alpha_2)tl$.

Since the elements are rigidly connected together, they undergo equal change in the length Δl (see Figure 4.9(b)) such that $(\Delta l)_1 > \Delta l > (\Delta l)_2$. Thus the element 1 is not able undergo its free expansion, i.e. it is compressed by an amount $(\Delta l)_1 - \Delta l$ and is in compression. Whereas the element 2 undergoes an expansion more than its free expansion, i.e. it is pulled by an amount $\Delta l - (\Delta l)_2$ and is thus in tension. Consider F_1 and F_2 to be the forces developed in the two elements as shown in Figure 4.9(c).

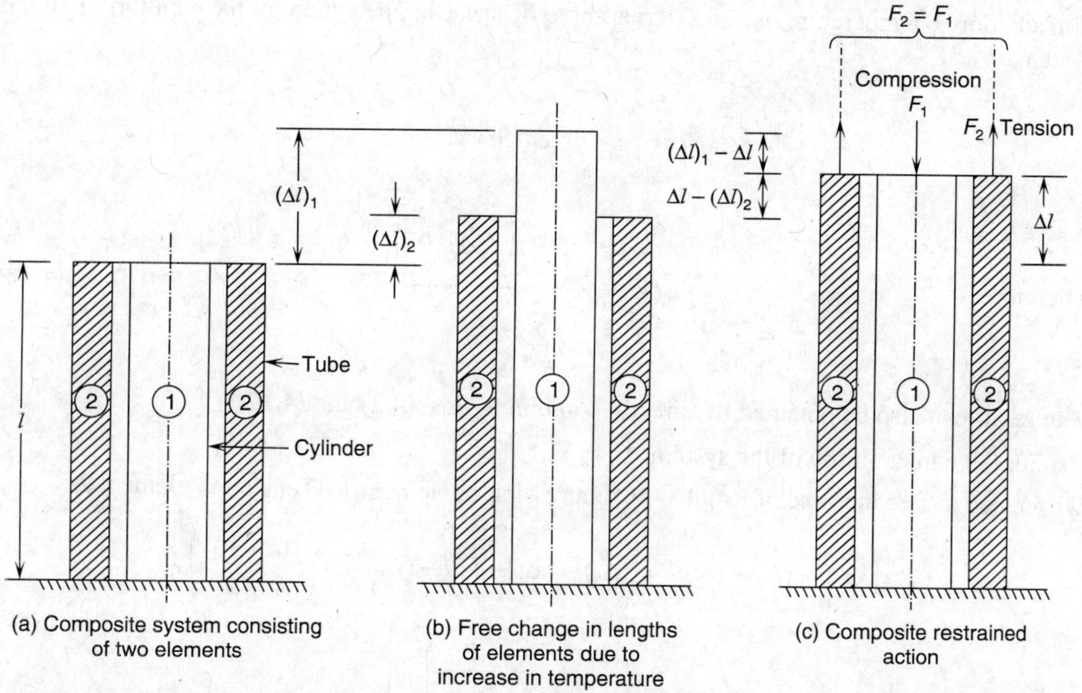

(a) Composite system consisting of two elements

(b) Free change in lengths of elements due to increase in temperature

(c) Composite restrained action

Figure 4.9 Composite system with elements of equal length subjected to variation of temperature.

(i) For the equilibrium of the system, $F_1 = F_2 = F$ i.e.,

$$\left[\frac{\Delta l - (\Delta l)_2}{l}\right] A_2 E_2 = \left[\frac{(\Delta l)_1 - \Delta l}{l}\right] A_1 E_1$$

or
$$\frac{\Delta l}{l}(A_2 E_2 + A_1 E_1) = \frac{(\Delta l)_1}{l} A_1 E_1 + \frac{(\Delta l)_2}{l} A_2 E_2$$

$$= (\alpha_1 t) A_1 E_1 + (\alpha_2 t) A_2 E_2 = t(\alpha_1 A_1 E_1 + \alpha_2 A_2 E_2)$$

i.e.
$$\Delta l = \frac{tl(\alpha_1 A_1 E_1 + \alpha_2 A_2 E_2)}{A_2 E_2 + A_1 E_1} \tag{4.16}$$

The expression can be generalized for the case of n elements of equal length

$$\left(\frac{\Delta l}{l}\right) \sum_{i=1}^{n} A_i E_i = t \sum_{i=1}^{n} \alpha_i A_i E_i$$

or
$$\Delta l = \frac{tl \sum_{i=1}^{n} \alpha_i A_i E_i}{\sum_{i=1}^{n} A_i E_i} \tag{4.17}$$

If in addition to temperature rise an external force W is also applied, then for the equilibrium of the system

$$\left(\frac{\Delta l}{l}\right)\sum_{i=1}^{n} A_i E_i = W + t\sum_{i=1}^{n} \alpha_i A_i E_i$$

Therefore,

$$\Delta l = \frac{l\left[W + t\sum_{i=1}^{n} \alpha_i A_i E_i\right]}{\sum_{i=1}^{n} A_i E_i} \tag{4.18}$$

The result can also be obtained by directly combining Eqs. (4.2) and (4.17).

(ii) For the compatibility of the system,

Difference in free increase of lengths = Compression in element 1+ Tension in element 2

$$(\alpha_1 - \alpha_2)tl = [(\Delta l)_1 - \Delta l] + [\Delta l - (\Delta l)_2] = \left(\frac{\sigma_1}{E_1} + \frac{\sigma_2}{E_2}\right) l$$

$$= Fl\left(\frac{1}{A_1 E_1} + \frac{1}{A_2 E_2}\right)$$

Therefore,

$$\left(\frac{\sigma_1}{E_1} + \frac{\sigma_2}{E_2}\right) = (\alpha_1 - \alpha_2)t \tag{4.19}$$

or

$$F = \frac{(\alpha_1 - \alpha_2)t}{\left(\frac{1}{A_1 E_1} + \frac{1}{A_2 E_2}\right)} \tag{4.20}$$

The stresses induced in the two elements are:

$$\sigma_1 = \frac{F}{A_1} = \frac{(\alpha_1 - \alpha_2)t}{\left(\frac{1}{E_1} + \frac{A_1}{A_2} \times \frac{1}{E_2}\right)} \quad \text{and} \quad \sigma_1 = \frac{F}{A_2} = \frac{(\alpha_1 - \alpha_2)t}{\left(\frac{A_2}{A_1} \times \frac{1}{E_1} + \frac{1}{E_2}\right)} \tag{4.21}$$

Alternatively, the compatibility condition given by Eq. (4.19) can be obtained as follows:

Actual increase in the length of element 1 = Actual increase in the length of element 2

$$\delta_1 = \delta_2 \quad \text{or} \quad \alpha_1 tl - \frac{\sigma_1 l}{E_1} = \alpha_2 tl + \frac{\sigma_2 l}{E_2}$$

or

$$\frac{\sigma_1}{E_1} + \frac{\sigma_2}{E_2} = (\alpha_1 - \alpha_2)t$$

EXAMPLE 4.8

Two parallel but out-of-plumb walls 6 m apart are stayed together by a steel rod of 25 mm diameter at a temperature of 78°C passing through steel plates, washers and nuts on outer side of the wall at each end. Determine the pull exerted by the rod when it is cooled to 24°C. (a) if the walls do not yield, and (b) if the walls yield at the two ends by 2 mm. The modulus of elasticity and coefficient of thermal expansion for steel are 200 GPa, and $\alpha = 11 \times 10^{-6}$ per °C, respectively.

Solution: (a) When the walls do not yield:

Fall in temperature, $t = 78 - 24 = 54°\text{C}$

Temperature strain, $\varepsilon = \alpha t$

Stress developed, $\sigma = E\alpha t$

Pull exerted on walls, $F = AE\alpha t$

$$= \frac{\pi}{4} \times (25)^2 \times (200 \times 10^3) \times (11 \times 10^{-6}) \times 54$$

$$= 58315.814 \text{ N} = 58.316 \text{ kN}$$

(b) When the walls yield at the ends of the rod:

Free decrease in length of rod when temperature falls to 24°C = $\alpha l t$

Actual decrease in length after walls yield by δ, $\Delta l = \alpha l t - \delta$

Actual temperature strain, $\varepsilon = \dfrac{\alpha t l - \delta}{l}$

Pull exerted on the walls, $F = AE\varepsilon = AE\left(\dfrac{\alpha t l - \delta}{l}\right)$

Therefore, $$F = \frac{\pi}{4} \times (25)^2 \times (200 \times 10^3) \times \frac{(11 \times 10^{-6}) \times 54 \times 6000 - 2.00}{6000}$$

$$= 25590.89 \text{ N} = 25.59 \text{ kN}$$

EXAMPLE 4.9

It is required to shrink a thin steel tyre on to a rigid wheel of 2.5 m diameter. If the maximum stress in the tyre is limited to 100 MPa, determine the internal diameter of the tyre. Also determine the least temperature to which the tyre is to be heated so that it fits over the wheel. The modulus of elasticity and coefficient of thermal expansion for steel are 200 GPa, and $\alpha = 10 \times 10^{-6}$ per °C, respectively.

Solution: The heated steel tyre of diameter d when slipped on to the wheel of diameter D and cooled, it is prevented from taking its original diameter by the wheel and in the process it grips the wheel. The tensile stress induced circumferentially along the tyre is called the hoop stress.

Temperature strain, $\varepsilon = \dfrac{\text{Contraction prevented}}{\text{Original length}} = \dfrac{\pi D - \pi d}{\pi d} = \dfrac{D - d}{d}$

Hoop stress, $$\sigma = \varepsilon E = \left(\frac{D-d}{d}\right)E$$

As the maximum hoop stress in the tyre is limited to 100 MPa,

$$\sigma = \left(\frac{D-d}{d}\right)E = \sigma_a \quad \text{i.e.} \quad \left(\frac{2500-d}{d}\right)\times(200\times10^3) = 100$$

$$\frac{2500}{d} = \left(1+\frac{100}{200\times10^3}\right) \quad \text{or} \quad \frac{d}{2500} = \left(1+\frac{100}{200\times10^3}\right)^{-1} \approx \left(1-\frac{100}{200\times10^3}\right)$$

Therefore, $$d = 2500\times\left(1-\frac{100}{200\times10^3}\right) = 2498.75 \text{ mm} = 2.4987 \text{ m}$$

Consider t°C to be the rise of temperature to which the tyre is to be heated such that the inner diameter of the tyre is equal to the diameter of the wheel,

$$\pi D = \pi d\,(1+\alpha t) \quad \text{or} \quad D = d\,(1+\alpha t)$$

or $$\alpha t = \left(\frac{D-d}{d}\right) = \frac{\sigma}{E} \quad \text{or} \quad t = \frac{\sigma}{\alpha E}$$

Therefore, $$t = \frac{100}{(10\times10^{-6})\times(200\times10^3)} = 50°\text{C}$$

EXAMPLE 4.10

Two steel rods of nominal diameters of 75 and 90 mm are connected end to end by means of a turn-buckle as shown in Figure 4.10. The other end of each rod is connected to a rigid support. If the turn-buckle is initially tightened snug (just tight), i.e. it does not cause any tension in the rods. If the effective length of each rod is 5 m, determine the tension induced in the rods if the turn-buckle is given one-third additional turn. The smaller and bigger rods have 5 and 4 threads per 20 mm, respectively. Also, determine the least rise in temperature which would eliminate the tension in the rods. Neglect the extension of turn-buckle. The modulus of elasticity and coefficient of thermal expansion for steel are 200 GPa, and $\alpha = 12 \times 10^{-6}$ per °C, respectively.

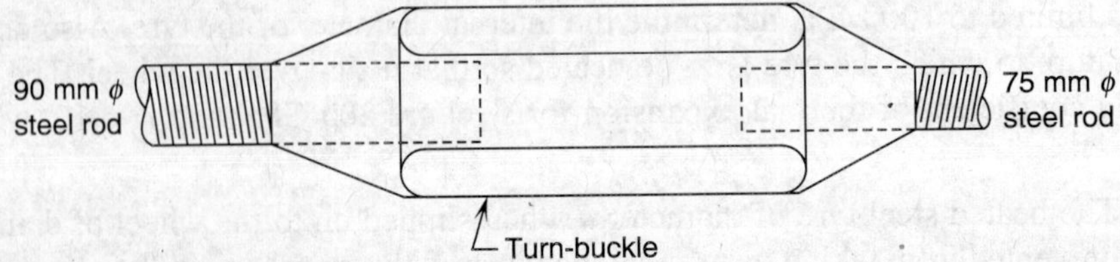

Figure 4.10 Turn-buckle for tightening two different diameter rods of Example 4.10.

Solution: For the given rods,

$$A = \frac{\pi}{4}\times(75)^2 = 4417.9 \text{ mm}^2 \text{ and } A_2 = \frac{\pi}{4}\times(90)^2 = 6361.7 \text{ mm}^2$$

Extension of rods due to turning the turn-buckle by one-third of a turn

$$\delta_1 = \frac{1}{3} \times \frac{20}{5} = \frac{4}{3} \quad \text{and} \quad \delta_2 = \frac{1}{3} \times \frac{20}{4} = \frac{5}{3}$$

Total extension of two rods,
$$\delta = \delta_1 + \delta_2 = \frac{4}{3} + \frac{5}{3} = 3$$

Consider the tension in each rod to be F, then total extension in the two rods,

$$\delta = \frac{FL}{A_1 E} + \frac{FL}{A_2 E} = \frac{FL}{E}\left(\frac{1}{A_1} + \frac{1}{A_2}\right) = 3$$

Therefore,
$$\frac{F \times 5000}{(200 \times 10^3)} \times \left(\frac{1}{4417.9} + \frac{1}{6361.7}\right) = 3$$

Thus,
$$F = 312872.7 \text{ N} = 312.873 \text{ kN}$$

To eliminate the tension in the rods, the rise in temperature should cause total expansion $\delta = 3$ mm,

Therefore,
$$\alpha l t = \delta \quad \text{or} \quad (12 \times 10^{-6}) \times (2 \times 5000)t = 3$$

or
$$t = \frac{3}{(12 \times 10^{-6}) \times (2 \times 5000)} = 25°\text{C}$$

EXAMPLE 4.11

A 1.25 m high concrete member is separated from the 1.0 m deep steel loading assembly by 1.0 mm as shown in Figure 4.11. Determine the force developed in the concrete due to a temperature rise of 100°C in the total system. The area of cross-section, modulus of elasticity and coefficient of thermal expansion for steel frame and concrete member are,

Steel: $A_s = 1.5 \times 10^4$ mm^2 ; $E_s = 210$ GPa and $\alpha_s = 12 \times 10^{-6}$ per °C

Concrete: $A_c = 3.0 \times 10^4$ mm^2 ; $E_c = 21$ GPa and $\alpha_c = 4 \times 10^{-6}$ per °C

Solution: Free expansions of steel assembly and concrete member due to temperature rise are,

$$\Delta l_s = \alpha_s l_s t = (12 \times 10^{-6}) \times (1.0 \times 10^3) \times 100 = 1.2 \text{ mm}$$

$$\Delta l_c = \alpha_c l_c t = (4 \times 10^{-6}) \times (1.25 \times 10^3) \times 100 = 0.5 \text{ mm}$$

Since the total expansion of the steel assembly and concrete member is more than 1.0 mm, both the components will be under compression. Consider the compressive force developed to be F, then total contraction in the components,

$$(\Delta l) = (\Delta l)_s + (\Delta l)_c = -F\left(\frac{l_s}{A_s E_s} + \frac{l_c}{A_c E_c}\right)$$

However, Δl should not exceed $\delta = 1.0 - 1.2 - 0.5 = -0.7$.

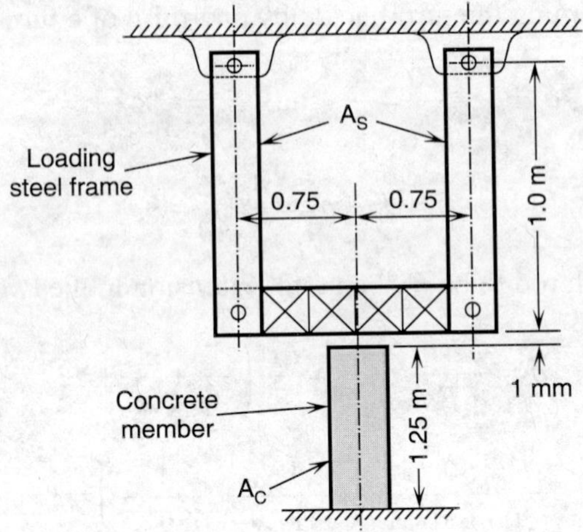

Figure 4.11 The steel loading assembly and concrete member of Example 4.11.

Therefore,

$$-F\left[\frac{1\times10^3}{(2\times1.5\times10^4)\times(210\times10^3)}+\frac{1.25\times10^3}{(3.0\times10^4)\times(21\times10^3)}\right]=-0.7$$

$$-F\left[\frac{1}{10}+\frac{1.25}{1.0}\right]=-0.7\times(3.0\times10^4)\times21$$

Therefore, $F = 326667\text{ N} = 326.67\text{ kN}$

EXAMPLE 4.12

A composite bar is fabricated by connecting end to end three bars one each of steel, aluminium and copper of lengths 300, 600 and 300 mm, respectively, with corresponding nominal diameters of 60, 120 and 60 mm. The composite bar is rigidly attached to the end supports. The temperature of the composite bar is raised by 90°C. Determine the stresses developed in three components of the bar when: (a) the supports are rigid and (b) the supports yield by 0.6 mm. The modulus of elasticity and coefficient of thermal expansion for the materials are,

Steel: $E_s = 200$ GPa, $\alpha_s = 12 \times 10^{-6}$ per °C

Copper: $E_c = 100$ GPa, $\alpha_c = 16 \times 10^{-6}$ per °C

Aluminium: $E_a = 70$ GPa, $\alpha_a = 20 \times 10^{-6}$ per °C

Solution: For the given bars,

$$A_s = A_c = \frac{\pi}{4}\times(60)^2 \quad \text{and} \quad A_a = \frac{\pi}{4}\times(120)^2$$

Total free expansion of the composite bar

$$\Delta l = \Delta l_s + \Delta l_a + \Delta l_c = \alpha_s l_s t + \alpha_a l_a t + \alpha_c l_c t = (\alpha_s l_s + \alpha_a l_a + \alpha_c l_c)t$$

Therefore,

$$\Delta l = (12 \times 300 + 20 \times 600 + 16 \times 300) \times 10^{-6} \times 90 = 1.836 \text{ mm}$$

Since this free expansion is prevented by rigid supports, compressive stresses are induced in the composite bar. Consider compressive force in composite bar to be F, then total contraction in bar,

$$(\Delta l)_1 = (\Delta l)_s + (\Delta l)_a + (\Delta l)_c = \left(\frac{Fl_s}{A_s E_s} + \frac{Fl_a}{A_a E_a} + \frac{Fl_c}{A_c E_c}\right)$$

or
$$(\Delta l)_1 = F\left(\frac{l_s}{A_s E_s} + \frac{l_a}{A_a E_a} + \frac{l_c}{A_c E_c}\right)$$

(a) *When the supports are rigid*:

$$\Delta l = (\Delta l)_1 = F\left(\frac{300}{60^2 \times 200} + \frac{600}{120^2 \times 70} + \frac{300}{60^2 \times 100}\right) \times \frac{4}{\pi \times 10^3} = 1.836$$

Therefore,

$$F\left(\frac{5}{12} + \frac{25}{42} + \frac{5}{6}\right) \times \frac{4}{\pi \times 10^6} = F \times \frac{155}{84} \times \frac{4}{\pi \times 10^6} = 1.836$$

or
$$F = \frac{1.836 \times 84 \times (\pi \times 10^6)}{155 \times 4} = 781466.1 \text{ N} = 781.466 \text{ kN}$$

Stresses,
$$\sigma_s = \sigma_c = \frac{F}{A_s} = \frac{4 \times 781466.1}{\pi \times 60^2} = 276.38 \text{ MPa}$$

$$\sigma_a = \frac{F}{A_a} = \frac{4 \times 781466.1}{\pi \times 120^2} = 69.1 \text{ MPa}$$

(b) *When the supports yield by 0.6 mm*:

$$(\Delta l)_2 = (\Delta l)_1 - 0.6 = 1.836 - 0.6 = 1.236 \text{ mm}$$

$$F = \frac{1.236 \times 84 \times (\pi \times 10^6)}{155 \times 4} = 526085.0 \text{ N} = 526.085 \text{ kN}$$

$$\sigma_s = \sigma_c = \frac{F}{A_s} = \frac{4 \times 526085.0}{\pi \times 60^2} = 186.06 \text{ MPa}$$

$$\sigma_s = \frac{F}{A_a} = \frac{4 \times 526085.0}{\pi \times 120^2} = 46.52 \text{ MPa}$$

EXAMPLE 4.13

An 18 mm diameter steel rod screwed at the ends passes centrally through a copper tube of external and internal diameters of 60 and 48 mm, respectively. The tube is closed at each end by rigid plates of negligible thickness and the nuts on the rod are then tightened lightly on the projecting parts of the rod at 30°C. Determine the stresses induced in the rod and tube when the temperature of the composite assembly is raised to 90°C. The modulus of elasticity and coefficient of thermal expansion for steel and copper are:

$$E_s = 210 \text{ GPa}; \alpha_s = 12 \times 10^{-6} \text{ per °C and } E_c = 105 \text{ GPa}; \alpha_c = 17.5 \times 10^{-6} \text{ per °C}$$

Solution: Though the free expansion/contraction of copper tube is greater than that for the steel rod under rising/falling temperature, but due to restraint provided by the nuts at the ends, each member will expand by the same amount under raised temperature. Hence the copper tube will be subjected to compressive stress while steel rod will be subjected to tensile stress.

For the given rod and tube assembly,

$$A_s = \frac{\pi}{4} \times 18^2 = 254.47 \text{ mm}^2 \text{ and } A_c = \frac{\pi}{4} \times (60^2 - 48^2) = 1017.88 \text{ mm}^2$$

Fall in temperature = 90 – 30 = 60°C

(i) For the equilibrium of the system, $F_1 = F_2$ i.e.

$$\sigma_s A_s = \sigma_c A_c \text{ or } \sigma_s = \sigma_c \left(\frac{A_c}{A_s}\right) = \sigma_c \left(\frac{1017.88}{254.47}\right) = 4\sigma_c$$

(ii) The compatibility condition,

$$\frac{\sigma_s}{E_s} + \frac{\sigma_c}{E_c} = (\alpha_1 - \alpha_2)t \text{ or } \frac{\sigma_s}{(210 \times 10^3)} + \frac{\sigma_c}{(105 \times 10^3)} = [(17.5 - 12) \times 10^{-6}] \times 60$$

Therefore,

$$\sigma_s + 2\sigma_c = [(17.5 - 12) \times 10^{-6}] \times 60 \times (210 \times 10^3) = 69.3$$

$$4\sigma_c + 2\sigma_c = 69.3 \text{ or } \sigma_c = 11.55 \text{ MPa (compression)}$$

$$\sigma_s = 4\sigma_c = 4 \times 11.55 = 46.2 \text{ MPa (tension)}$$

EXAMPLE 4.14

A steel bar of 36 mm diameter is placed concentrically inside a brass tube of external and internal diameters of 50 and 38 mm, respectively. The bar and tube are 450 mm long and are held between two rigid stops 450 mm apart. If the temperature of the composite assembly is raised by 65°C, determine: (a) the stresses induced in the bar and tube if the distance between the stops (i) remains constant, (ii) increases by 0.3 mm, and (b) the increase in the distance between the stops if the force exerted between them is 100 kN. The modulus of elasticity and coefficient of thermal expansion for steel and brass are:

$$E_s = 200 \text{ GPa} \qquad \alpha_s = 12 \times 10^{-6} \text{ per °C}$$

$$E_b = 90 \text{ GPa} \qquad \alpha_b = 21 \times 10^{-6} \text{ per °C}$$

Solution: (a) For the given bar and tube assembly,

$$A_s = \frac{\pi}{4}\times 36^2 = 1017.88 \text{ mm}^2 \text{ and } A_b = \frac{\pi}{4}\times(50^2 - 38^2) = 829.38 \text{ mm}^2$$

(i) The distance between the stops remains constant,

Stress in steel bar, $\sigma_s = \alpha_s t E_s = (12 \times 10^{-6}) \times 65 \times (200 \times 10^3) = 156$ MPa (comp.)

Stress in brass tube, $\sigma_b = \alpha_b t E_b = (21 \times 10^{-6}) \times 65 \times (90 \times 10^3) = 122.85$ MPa (comp.)

(ii) The distance between the stops increases by 0.3 mm,

$$\text{Stress in steel bar, } \sigma_s = \varepsilon_s E_s = \left(\frac{\alpha_s tl - \delta}{l}\right) E_s$$

$$= \left(\frac{(12\times10^{-6})\times 65\times 450 - 0.3}{450}\right)\times(200\times10^3)$$

$$= 22.67 \text{ MPa}$$

$$\text{Stress in brass tube, } \sigma_b = \varepsilon_b E_b = \left(\frac{\alpha_b tl - \delta}{l}\right) E_b$$

$$= \left(\frac{(21\times10^{-6})\times 65\times 450 - 0.3}{450}\right)\times(90\times10^3)$$

$$= 62.85 \text{ MPa}$$

(b) *Yield of stops when the force exerted between them is* 100 kN:

Consider the expansion of the assembly to be δ

$$\text{Stress in the steel bar, } \sigma_s = \varepsilon_s E_s = \left(\frac{\sigma_s tl - \delta}{l}\right) E_s = \left(\alpha_s t - \frac{\delta}{l}\right) E_s$$

$$\text{Stress in the brass tube, } \sigma_b = \varepsilon_b E_b = \left(\frac{\alpha_b tl - \delta}{l}\right) E_b = \left(\alpha_b t - \frac{\delta}{l}\right) E_b$$

For equilibrium,

Force carried by steel bar + Force carried by brass tube = Total force between the stops, F

$$\sigma_s A_s + \sigma_b A_s = F$$

$$\left(\alpha_s t - \frac{\delta}{l}\right) E_s A_s + \left(\alpha_b t - \frac{\delta}{l}\right) E_b A_b = F$$

Therefore,

$$\delta = \frac{lt(\alpha_s E_s A_s + \alpha_b E_b A_b) - Fl}{(E_s A_s + E_b A_b)}$$

where, for steel $E_s A_s = (200 \times 10^3) \times 1017.88 = 2.03576 \times 10^8$ N

and $\alpha_s E_s A_s = (12 \times 10^{-6}) \times (2.03576 \times 10^8) = 0.2442912$ N per °C

For brass tube $E_b A_b = (90 \times 10^3) \times 829.38 = 0.746442 \times 10^8$ N

and $\alpha_b E_b A_b = (21 \times 10^{-6}) \times (0.746442 \times 10^8) = 0.15675282$ N per °C

Therefore, $$\delta = 450 \times \left[\frac{65 \times (0.2442912 + 0.15675282) - (100 \times 10^3)}{(2.03576 + 0.746442) \times 10^8}\right] = -0.1617 \text{ mm}$$

EXAMPLE 4.15

A composite bar of size 26 × 28 mm deep is fabricated by placing a flat bar of aluminium alloy 26 mm wide and 8 mm thick between two steel bars each 26 mm wide and 10 mm thick. The three bars are fastened together at their ends when the temperature is 12°C. (a) Determine the stress in each of the materials when temperature of the composite bar is raised to 52°C. (b) At this stage, a compressive load of 20 kN is applied to the composite bar, compute the final stresses in steel and aluminium. The modulus of elasticity and coefficient of thermal expansion for steel and aluminium are:

$$E_s = 200 \text{ GPa} \qquad \alpha_s = 12 \times 10^{-6} \text{ per °C}$$

$$E_a = 65 \text{ GPa} \qquad \alpha_a = 22 \times 10^{-6} \text{ per °C}$$

Solution: For the given composite bar,

$$A_a = 26 \times 8 = 208 \text{ mm}^2 \text{ and } A_s = 2 \times (26 \times 10) = 520 \text{ mm}^2$$

Rise in the temperature of composite bar, $t = 52 - 12 = 40$°C.

For internal equilibrium:

Total tensile force in the steel bars = Compression in the aluminium bar i.e.

$$\sigma_s A_s = \sigma_a A_a \quad \text{i.e.} \quad \sigma_s \times 520 = \sigma_a \times 208 \quad \text{or} \quad \sigma_s = 0.4\sigma_a \qquad \text{(i)}$$

For displacement compatibility:

Increase in the length of steel bars = Increase in the length of aluminium bar

$$\delta_s = \delta_a \quad \text{or} \quad \alpha_s tl + \frac{\sigma_s l}{E_s} = \alpha_a tl - \frac{\sigma_a l}{E_a}$$

or $$\frac{\sigma_s}{E_s} + \frac{\sigma_a}{E_a} = (\alpha_a - \alpha_s)t$$

Therefore, $$\frac{\sigma_s}{200 \times 10^3} + \frac{\sigma_a}{65 \times 10^3} = [(22 - 12) \times 10^{-6}] \times 40$$

or $$\sigma_s + 3.0769\sigma_a = 80 \qquad \text{(ii)}$$

From Eqs. (i) and (ii):

$$0.4\sigma_a + 3.0769\sigma_a = 80 \quad \text{or} \quad \sigma_a = 23.01 \text{ MPa (comp.)}$$

Therefore, $\sigma_s = 0.4\sigma_a = 0.4 \times 23.01 = 9.204$ MPa (tension)

(b) Stresses due to the external compressive load applied to the composite bar

Consider the stresses in steel and aluminium to be σ_s and σ_a. For strain compatibility, the strain in the steel and aluminium bars should be equal, i.e.

$$\frac{\sigma_s}{E_s} = \frac{\sigma_a}{E_a} \quad \text{or} \quad \sigma_s = \left(\frac{E_s}{E_a}\right)\sigma_a = \left(\frac{200 \times 10^3}{65 \times 10^3}\right)\sigma_a = 3.0769\,\sigma_a \qquad \text{(iii)}$$

For force equilibrium:

Force in the steel bars + Force in the aluminium bar = P

$$\sigma_s A_s + \sigma_a A_a = P \quad \text{i.e.} \quad \sigma_s \times 520 + \sigma_a \times 208 = 20 \times 10^3$$

or

$$\sigma_s + 0.4\sigma_a = 38.46 \qquad \text{(iv)}$$

From Eqs. (iii) and (iv)

$$3.0769\sigma_a + 0.4\sigma_a = 38.46 \quad \text{or} \quad \sigma_a = 11.06 \text{ MPa (comp.)}$$

Therefore, $\sigma_s = 3.0769\sigma_a = 3.0769 \times 11.06 = 34.04$ MPa (comp.)

Final stresses:

$$\sigma_a = 11.06 + 23.01 = 34.17 \text{ MPa (compression)}$$

$$\sigma_s = 34.04 - 9.204 = 24.836 \text{ MPa (compression)}$$

EXAMPLE 4.16

Two flanged copper pipes of 50 mm bore are fastened by six 12.5 mm diameter steel bolts, having six threads per 10 mm, passing through holes in the flanges. The diameter of the 8 mm thick flange is 125 mm. If the nut is initially tightened snug (just tight), determine the increase in stress in the bolt if the nut is turned an additional 10°.

What rise in temperature would produce the same increase in stress? The modulus of elasticity and coefficient of thermal expansion for steel and copper are:

$$E_s = 200 \text{ GPa} \qquad \alpha_s = 11 \times 10^{-6} \text{ per °C}$$

$$E_c = 120 \text{ GPa} \qquad \alpha_c = 17 \times 10^{-6} \text{ per °C}$$

Solution: For the given bolts and pipe flange assembly,

$$A_s = 6 \times \left(\frac{\pi}{4} \times 12.5^2\right) = 736.31 \text{ mm}^2 \quad \text{and} \quad A_c = \frac{\pi}{4} \times (125^2 - 50^2) = 10308.35 \text{ mm}^2$$

(a) Forces developed due to tightening of the nuts

$$\text{Axial displacement of the nut, } \delta = \frac{10}{6} \times \left(\frac{10}{360}\right) = 0.0463 \text{ mm}$$

For displacement compatibility:

Increase in the length of bolt + Decrease in the thickness of flange plates
= Axial displacement of the nut

$$\delta_s + \delta_c = \frac{\sigma_s l}{E_s} + \frac{\sigma_c l}{E_c} = Fl\left(\frac{1}{A_s E_s} + \frac{1}{A_c E_c}\right)$$

$$= 0.0463 \text{ mm}$$

$$F \times (2 \times 8) \times \left[\frac{1}{736.31 \times (200 \times 10^3)} + \frac{1}{10308.35 \times (120 \times 10^3)}\right] = 0.0463 \text{ mm}$$

$$F\left[\frac{1}{736.31 \times (5)} + \frac{1}{10308.35 \times (3)}\right] = \frac{0.0463 \times 10^4}{4}$$

or $$F = 380805.47 \text{ N} = 380.81 \text{ kN}$$

The stresses induced in the bolt and the flange are:

$$\sigma_s = \frac{F}{A_s} = \frac{380805.47}{736.31} = 517.18 \text{ MPa} \quad \text{and} \quad \sigma_c = \frac{F}{A_c} = \frac{380805.47}{10308.35} = 36.94 \text{ MPa}$$

Low stress in the flange indicates that with its comparatively very large area, it is almost incompressible and may be considered to be rigid. Thus, the stress induced in the bolt,

$$\sigma_s = \varepsilon_s E_s = \left(\frac{0.0463}{16}\right) \times (200 \times 10^3) = 578.75 \text{ MPa}$$

Thus, the procedure which ignores deformations in the flanges provides fairly good estimate of stress in the bolts.

(b) Temperature rise to produce the same increase in stress

For force equilibrium:

Tensile force in the steel bolts = Compressive force in copper flange,

or $$F_s = F_c$$

or $$\sigma_s A_s = \sigma_c A_c \quad \text{or} \quad \sigma_c = \left(\frac{A_s}{A_c}\right)\sigma_s = \left(\frac{736.31}{10308.35}\right)\sigma_s = 0.07143\sigma_s \qquad \text{(i)}$$

For displacement compatibility:

Increase in the length of steel bolts = Increase in the thickness of copper flanges

$$\delta_s = \delta_c \quad \text{or} \quad \alpha_s t l + \frac{\sigma_s l}{E_s} = \alpha_c t l - \frac{\sigma_c l}{E_c}$$

or $$\frac{\sigma_s}{E_s} + \frac{\sigma_c}{E_c} = (\alpha_c - \alpha_s)t$$

Therefore, $$\frac{\sigma_s}{200 \times 10^3} + \frac{\sigma_c}{120 \times 10^3} = [(17 - 11) \times 10^{-6}] \times t$$

or $$\sigma_s + 1.667\sigma_c = 1.2t \qquad \text{(ii)}$$

From Eqs. (i) and (ii)

$$\sigma_s + 1.667 \times 0.07143\sigma_s = 1.2t \quad \text{or} \quad t = \frac{1.1191\sigma_s}{1.2}$$

For $\sigma_s = 517.18$ MPa, $$t = \frac{1.1191 \times 517.18}{1.2} = 482.3°\text{C}$$

Alternatively, Eqs. (i) and (ii) can be combined as:

$$t = \sigma_s \left(\frac{1}{E_s} + \frac{A_s}{A_c E_c}\right)\left(\frac{1}{\alpha_c - \sigma_s}\right)$$

EXAMPLE 4.17

A 30 mm diameter threaded steel rod of 300 mm length passes centrally through a copper tube of external and internal diameters of 48 and 36 mm, respectively. The tube is closed at each end by rigid plates of negligible thickness. The rod has four threads per 10 mm and the nut is initially just tight. Determine: (a) the angle through which the nut on the rod is turned through to induce a tensile stress of 100 MPa in the rod, (b) the stress induced in the tube, and (c) the change of stress in the rod and tube when the temperature of the assembly is raised by 40°C. The modulus of elasticity and coefficient of thermal expansion for steel and copper are

$E_s = 200$ GPa $\qquad \alpha_s = 12 \times 10^{-6}$ per °C

$E_c = 100$ GPa $\qquad \alpha_c = 17 \times 10^{-6}$ per °C

Solution: For the given rod and tube assembly,

$$A_s = \frac{\pi}{4} \times 30^2 = 225\pi \text{ mm}^2 \quad \text{and} \quad A_c = \frac{\pi}{4} \times (48^2 - 36^2) = 252\pi \text{ mm}^2$$

(a) Forces developed due to tightening of the nuts, $\sigma_s = 100$ MPa

For force equilibrium,

$$\sigma_s A_s = \sigma_c A_c \quad \text{or} \quad \sigma_c = \sigma_s \left(\frac{A_s}{A_c}\right) = 100 \times \frac{225\pi}{252\pi} = 89.286 \text{ MPa}$$

Axial displacement of the nut, $$\delta = \frac{10}{4} \times \left(\frac{\theta}{360}\right) = \frac{\theta}{144} \text{ mm}$$

For displacement compatibility,

Increase in the length of rod + Decrease in the length of tube
= Distance travelled by the nut

$$\frac{\sigma_s l}{E_s} + \frac{\sigma_c l}{E_c} = \frac{\theta}{144}$$

$$300 \times \left[\frac{100}{(200 \times 10^3)} + \frac{89.286}{(100 \times 10^3)}\right] = \frac{\theta}{144}$$

Therefore, $$\theta = \frac{300 \times 144}{10^5} \times \left[\frac{100}{2} + 89.286\right] = 60.17°$$

(b) The stress induced in the tube, $\sigma_c = 89.286$ MPa

(c) The change of stress in the rod and tube when the temperature is raised by 40°C.

For force equilibrium:

Tensile force in the steel rod = Compressive force in copper tube

$$\sigma_s A_s = \sigma_c A_c \quad \text{or} \quad \sigma_c = \left(\frac{225\pi}{252\pi}\right)\sigma_s = 0.8929\sigma_s \tag{i}$$

For displacement compatibility,

Increase in the length of steel rod = Increase in the length of copper tube

$$\alpha_s tl + \frac{\sigma_s l}{E_s} = \alpha_c tl - \frac{\sigma_c l}{E_c} \quad \text{or} \quad \frac{\sigma_s l}{E_s} + \frac{\sigma_c l}{E_c} = (\alpha_c - \alpha_s)tl$$

Therefore,
$$\frac{\sigma_s}{200 \times 10^3} + \frac{\sigma_c}{100 \times 10^3} = [(17 - 12) \times 10^{-6}] \times 40$$

or
$$\sigma_s + 2\sigma_c = [(17 - 12) \times 10^{-6}] \times 40 \times 200 \times 10^3 = 40 \tag{ii}$$

From Eqs. (i) and (ii)

$$\sigma_s + 2 \times 0.8929\sigma_s = 40 \quad \text{or} \quad \sigma_s = \frac{40}{2.7858} = 14.36 \text{ MPa}$$

Thus
$$\sigma_c = 0.8929\sigma_s = 0.8929 \times 14.36 = 12.82 \text{ MPa}$$

4.6 PROBLEMS

4.1 A rigid weightless beam hinged at the end 1 is supported by two identical linearly elastic vertical tie rods at points 2 and 3 as shown in Figure 4.12. Determine the normal forces induced in the tie rods when they are heated by t°C. The coefficient of thermal expansion for the material of tie rods is α/°C.
[**Hint:** For displacement compatibility extension of tie rod 3–3′ is twice that of 2–2′.]

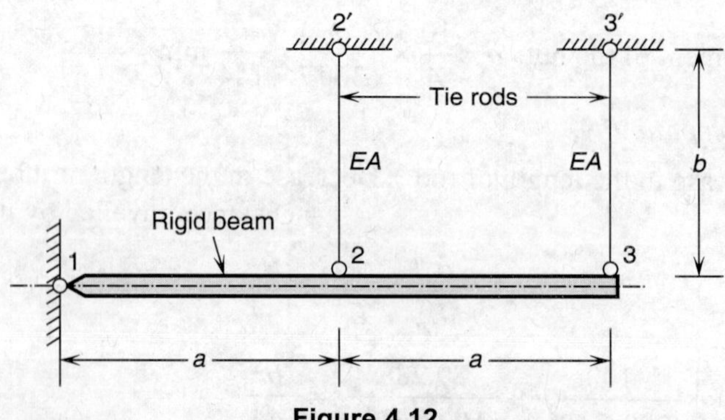

Figure 4.12

$$\left[\textbf{Ans. } F_2 = -\frac{2}{5}(E\,A\,\alpha t) \text{ and } F_3 = \frac{1}{5}(E\,A\,\alpha t)\right]$$

4.2 A rigid beam hinged at one end is supported by two inclined linearly elastic tie rods at points 2 and 3 as illustrated in Figure 4.13. Determine the normal forces in the tie rods induced by applied load W.

[**Hint:** For displacement compatibility, the extensions in the tie rods 2–4 and 3–4 are in the ratio of $2:\sqrt{6}$].

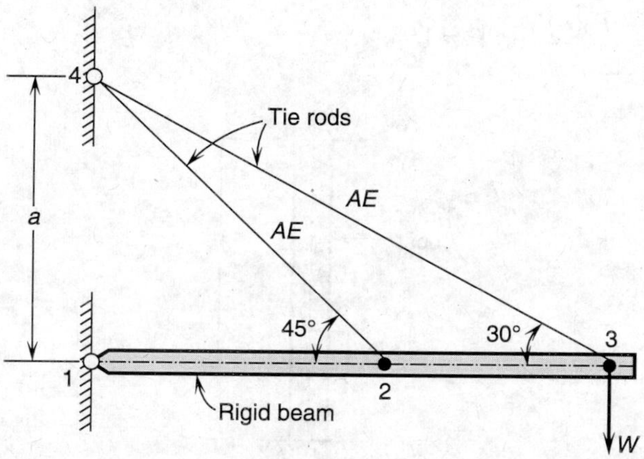

Figure 4.13

4.3 While fabricating a planar system of five elastic members, it was found that the member 3–4 was short in length by Δ as shown in Figure 4.14. The members were put into place by bringing the hinges 4 and 5 to one point. Determine forces in the members developed after assembly of the system. The cross-sectional area A and elastic modulus E of each member is same.

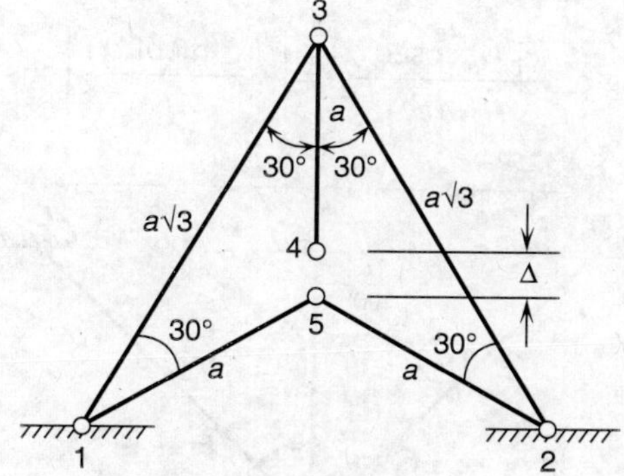

Figure 4.14

$$\left[\textbf{Ans. } F_{1-5} = F_{2-5} = F_{3-4} = \frac{\sqrt{3}}{2+3\sqrt{3}}\left(\frac{\Delta EA}{a}\right) \text{ (tension);} \quad F_{1-3} = F_{2-3} = -\frac{1}{2+3\sqrt{3}}\left(\frac{\Delta EA}{a}\right) \text{(compression)}\right]$$

4.4 A 200 mm square timber post is held in position between two fixed supports A and B, $4a$ apart as shown in Figure 4.15. The post carries a collar C at height a from the lower support B. Determine the stresses induced in the upper and lower portions of the post, if the collar carries a load of 60 kN.

[**Hint:** If P is tensile force in the upper part AC, then $60 - P$ is the compressive force in the lower part CB; for compatibility, the extension of AC must be equal to compression in CB, i.e. $\dfrac{P(3a)}{AE} = \dfrac{(60-P)(a)}{AE}$ or $P = 15$ kN.]

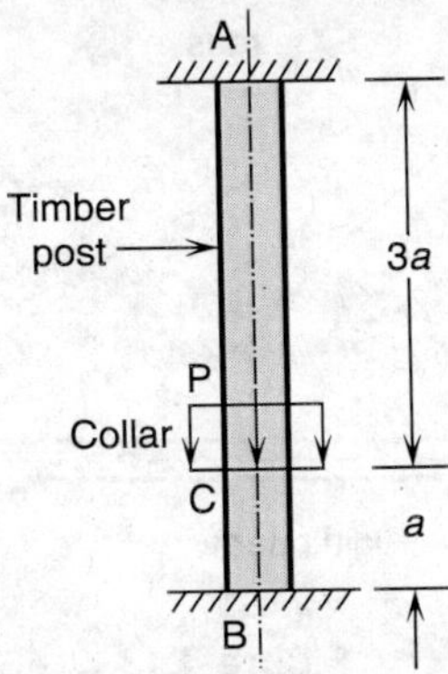

Figure 4.15 Timber post fixed at supports.

[**Ans.** $\sigma_{AC} = 0.375$ MPa and $\sigma_{CB} = 1.125$ MPa]

4.5 Three steel cables AO, BO and CO of same cross-section area support a load W at O as shown in Figure 4.16. Determine the load carried by each cable.

$$\left[\textbf{Hint}: \text{(i) Equilibrium: } \sum F_{o,v} = 2F_1 \cos\alpha + F_2 - W = 0; \text{ and for (ii) Compatibility:}\right.$$

$$\left.\delta_1 = \delta_2 \cos\alpha \text{ or } \frac{F_1\,(a \operatorname{cosec}\alpha)}{AE} = \left[\frac{F_2\,(a\cot\alpha)}{AE}\right]\cos\alpha \text{ i.e. } \frac{F_1}{F_2} = \cos^2\alpha\right]$$

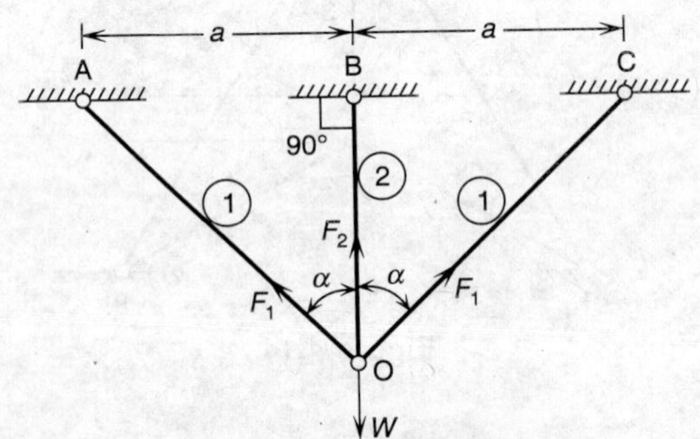

Figure 4.16 Three cables.

$$\left[\textbf{Ans. } F_1 = \frac{W\cos^2\alpha}{1 + 2\cos^3\alpha} \text{ and } F_2 = \frac{W}{1 + 2\cos^3\alpha}\right]$$

4.6 A 350 mm long steel bolt having a nominal diameter of 20 mm and a thread pitch of 2.4 mm is used to connect two plates of 10 mm thickness each. An aluminium spacing tube of internal and external diameters of 22 and 44 mm, respectively, separates the plates as shown in Figure 4.17. If the nut is initially tightened snug (just tight), i.e. it does not cause any stress in the spacing tube. Determine the stresses induced in the tube and in the bolt if the nut is given one-third additional turn. Neglect the deformation of the plates. The moduli of elasticity for steel and aluminium are 207 and 67.5 GPa, respectively.

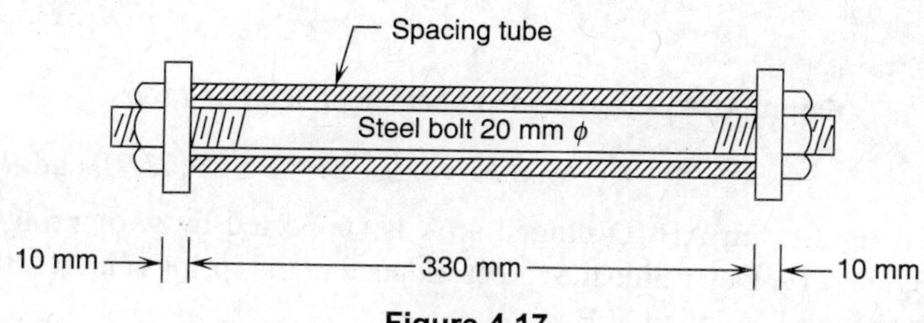

Figure 4.17

[**Ans:** Movement of the nut = 0.8 mm; σ_b = 263.4 MPa and σ_t = 72.56 MPa]

4.7 A 250 mm square short concrete column strengthened by four steel bars in the corners is subjected to an axial load of 300 kN. The total cross-sectional area of steel bars is A_s = 6000 mm^3. Determine the stresses induced in the steel and concrete, if the modulus of elasticity for steel is 15 times that for concrete.

[**Ans:** σ_c = 2.048 MPa and σ_s = 30.716 MPa]

4.8 A horizontal rigid beam ABC, 6*a* long and hinged at A is supported by two vertical props each of length 3 m at the points B and C as shown in Figure 4.18. The cross-sectional area of steel prop BD, is 0.8 times that of copper prop CE. The beam supports a load *W* at the middle point of BC. Determine the forces transmitted to the props, if the moduli of elasticity for steel and copper are 200 and 100 GPa, respectively.

[**Hint:** (i) Equilibrium:

$$\sum F_v = R_A + F_b + F_c - W = 0; \sum M_A = F_b(4a) + F_c(6a) - W(5a) = 0;$$

(ii) Compatibility:

$$\frac{\delta_b}{\delta_c} = \frac{4a}{6a} \text{ or } \frac{(F_b L_b / A_b E_b)}{(F_c L_c / A_c E_c)} = \frac{F_b}{F_c} \times \frac{(A_c E_c / L_c)}{(A_b E_b / L_b)} = \frac{F_b}{F_c} \times \beta = \frac{2}{3} \text{ or } F_b = \frac{2F_c}{3\beta};$$

Therefore,

$$F_b = \left(\frac{5}{4+9\beta}\right) W; F_c = \left(\frac{15\beta}{8+18\beta}\right) W \text{ and } R_A = \left(\frac{3\beta - 2}{8+18\beta}\right) W$$

$$\beta = \frac{(A_c E_c / L_c)}{(A_b E_b / L_b)} = \frac{A_c E_c}{A_b E_b} = \frac{A_c \times 100}{(0.8 A_c) \times 200} = \frac{5}{8}\Bigg]$$

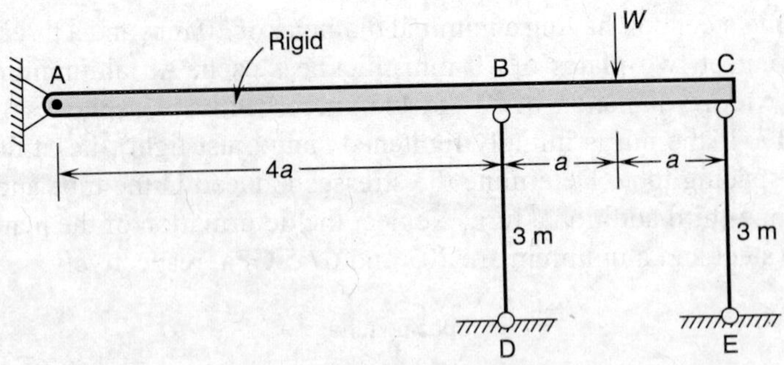

Figure 4.18 Rigid hinged beam supported on props.

[**Ans.** $F_b = 0.5195W$; $F_c = 0.487W$ and $R_A = -0.00649W$]

4.9 A horizontal rigid beam ABCD hinged at A is supported by two springs at C and D as shown in Figure 4.19. Determine the forces induced in the springs and reaction at the hinged support due to load W applied at B.

[**Hint:** (i) Equilibrium:

$$\sum F_v = R_A - W + F_c + F_d = 0; \sum M_A = -W(a) + F_c(2a) + F_d(3a) = 0;$$

(ii) Compatibility: $\dfrac{\delta_c}{\delta_d} = \dfrac{2a}{3a}$ or $\dfrac{(F_c/k_c)}{(F_d/k_d)} = \dfrac{F_c}{F_d} \times \dfrac{k_d}{k_c} = \dfrac{F_c}{F_d} \times \beta = \dfrac{2}{3}$ or $F_c = \dfrac{2F_d}{3\beta}$]

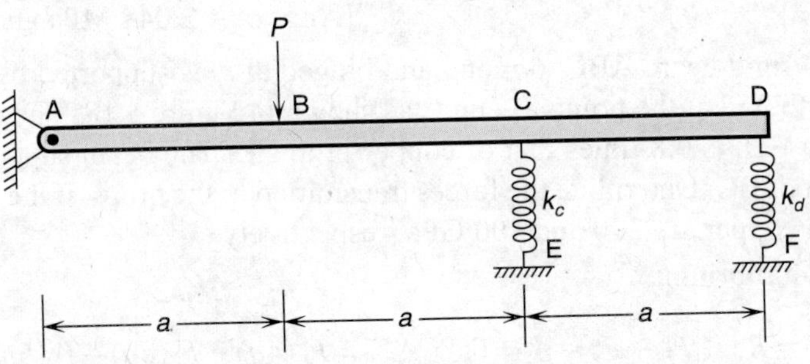

Figure 4.19 Rigid hinged beam supported on the springs.

$$\left[\textbf{Ans. } F_c = \left(\frac{2}{4+9\beta}\right)W;\ F_d = \left(\frac{3\beta}{4+9\beta}\right)W \text{ and } R_A = \left(\frac{2+6\beta}{4+9\beta}\right)W\right]$$

4.10 Two 5.0 m long vertical wires suspended at a distance of 300 mm apart. Their upper ends are firmly secured and their lower ends support a rigid horizontal bar which carries a load 3 kN. The left hand copper and right hand steel wires have diameters of 5 and 15 mm, respectively. Determine the: (a) position of the load so that both the wires extend by the same amount and (b) load, stress and extension of each wire. The moduli of elasticity for steel and copper are 210 and 125 GPa, respectively. The weight of the horizontal bar may be neglected.

[**Ans.** $x = 281.4$ mm; $\sigma_s = 15.92$ MPa; $\sigma_c = 9.47$ MPa and $\Delta l = 0.3788$ mm]

4.11 Two rigid castings are fastened 200 mm by 10 mm nominal diameter steel bolts as shown in Figure 4.20. Determine the force P necessary for drawing the casting fixtures apart so that a 200.3 mm long copper rod having diameter of 28 mm can be freely inserted between them. Determine the distance between the castings after removal of the force P. The moduli of elasticity for steel and copper are 200 and 100 GPa, respectively.

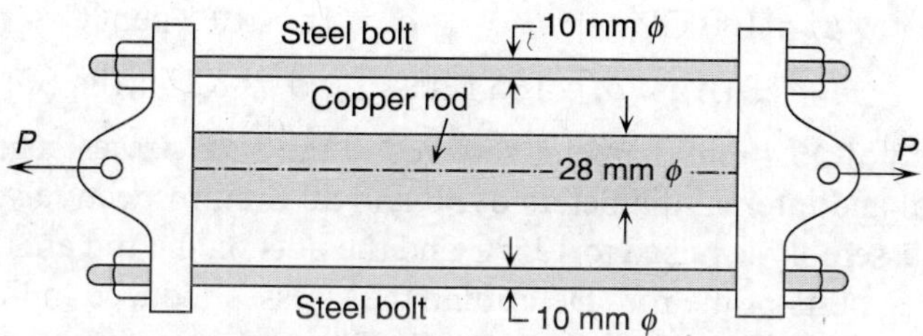

Figure 4.20 Fixing of rigid castings of Problem 4.11.

[**Ans:** $P = 47.124$ kN; the common compression of steel and copper rods, $\Delta l = 0.10135$ mm and the final distance between castings, $l_f = 200.19865$ mm]

4.12 A rigid wheel of 2.1 m diameter is to be provided with a thin steel tyre shrunk on it. If the maximum stress in the tyre is limited to 140 MPa, determine the internal diameter of the tyre. Also determine the least temperature to which the tyre is to be heated so that it fits over the wheel. The modulus of elasticity and coefficient of thermal expansion for steel are 210 GPa, and $\alpha = 11 \times 10^{-6}$ per °C, respectively.

[**Ans:** $d = 2.09859$ m and $t = 60.64$°C]

4.13 A weight of 270 kN is supported by three short pillars, each having cross-section area of 600 mm². The central pillar is of steel and outer ones are of copper. The pillars are so adjusted that at the temperature of 12°C they carry equal loads. The temperature is then raised to 112°C, determine the stress in each pillar at the temperatures of 12°C and 112°C. The modulus of elasticity and coefficient of thermal expansion for steel and copper are:

$$E_s = 200 \text{ GPa}; \alpha_s = 12 \times 10^{-6} \text{ per °C and } E_c = 80 \text{ GPa}; \alpha_c = 18.5 \times 10^{-6} \text{ per °C}$$

[**Ans:** At 12°C, $\sigma_s = \sigma_c = 150$ MPa (C) and at 112°C, $\sigma_c = 178.89$ MPa (C); $\sigma_s = 92.22$ MPa (C)]

4.14 A copper bar of 40 mm diameter is placed centrally inside a steel tube of external and internal diameters of 60 and 52 mm, respectively. The rod and tube are 400 mm long and are rigidly connected together at the ends at a temperature of 25°C. If the temperature of the composite assembly is raised to 175°C, determine: (a) the stresses induced in the bar and tube and (b) the increase in the length of the assembly. The modulus of elasticity and coefficient of thermal expansion for steel and copper are:

$$E_s = 210 \text{ GPa}; \alpha_s = 10.8 \times 10^{-6} \text{ per °C and } E_c = 110 \text{ GPa}; \alpha_c = 17 \times 10^{-6} \text{ per °C}$$

[**Ans.** $\sigma_s = 94.37$ MPa (T); $\sigma_c = 52.87$ MPa (C), and (b) $\Delta l = 0.836$ mm]

4.15 A composite bar is fabricated by placing a flat steel bar between two flat copper bars each having the same area and length as the steel bar at 12°C. At this stage, three bars are rigidly connected at both the ends. When the temperature is raised to 212°C, the length of composite bar increases by 1.0 mm. Determine the original length and stress developed in each of the materials. The modulus of elasticity and coefficient of thermal expansion for steel and copper are:

$E_s = 200$ GPa $\quad \alpha_s = 12.5 \times 10^{-6}$ per °C

$E_c = 100$ GPa $\quad \alpha_c = 18 \times 10^{-6}$ per °C

[**Ans.** $\sigma_s = 165$ MPa (T), $\sigma_c = 82.5$ MPa (C) and $l = 225.99$ mm]

4.16 A gun metal rod of 26.4 mm diameter screwed at the ends passes centrally through a steel tube of external and internal diameters of 36 and 30 mm, respectively. The temperature of the composite assembly is raised to 125°C and the nuts on the rod are then tightened lightly on the projecting parts of the rod. Determine the stresses induced in the rod and tube when the temperature of the assembly falls to 25°C. The modulus of elasticity and coefficient of thermal expansion for steel and gun metal are:

$E_s = 210$ GPa $\quad \alpha_s = 11.5 \times 10^{-6}$ per °C

$E_g = 94$ GPa $\quad \alpha_g = 19.5 \times 10^{-6}$ per °C

[**Ans.** $\sigma_g = 42.06$ MPa (T) and $\sigma_s = 74.03$ MPa (C)]

4.17 Two flanged copper pipes of 60 mm bore are fastened by six 15 mm diameter steel bolts, having six threads per 10 mm, passing through holes in the flanges. The diameter of the flange is 150 mm. Assuming that the compressive stress in the flange is uniformly distributed, determine the increase in stress in the bolts for a rise of temperature of 80°C. The modulus of elasticity and coefficient of thermal expansion for steel and copper are:

$E_s = 200$ GPa; $\alpha_s = 12.5 \times 10^{-6}$ per °C and $E_c = 120$ GPa; $\alpha_c = 17.5 \times 10^{-6}$ per °C

[**Ans.** $\sigma_s = 71.49$ MPa (C)]

4.18 Two copper plates of 48 mm thickness each are fastened together by a steel bolt, having six threads per 10 mm. (a) If the nut is initially tightened snug (just tight), determine the increase in stress in the bolt if the nut is turned an additional 18°. The modulus of elasticity for steel is 200 GPa. (b)What rise in temperature would produce the same increase in stress? The coefficients of thermal expansion for steel and copper are:

$\alpha_s = 12 \times 10^{-6}$ per °C and $\alpha_c = 17.5 \times 10^{-6}$ per °C

[**Ans.** $\sigma_s = 173.61$ MPa (T) and $t = 157.83$°C]

4.19 Determine the temperature rise necessary to induce buckling in a 1.25 m long aluminium bar of 50 mm nominal diameter hinged at its ends. Consider the bar to be uniformly heated. The coefficients of thermal expansion for aluminium is $\alpha_a = 20 \times 10^{-6}$ per °C.

$$\left[\textbf{Hint}: P_{cr} = \frac{\pi^2 EI}{L^2} = (AE)\frac{\Delta L}{L} \text{ or } \Delta L = \frac{\pi^2 I}{AL} = \alpha L t \text{ i.e. } t = \frac{\pi^2 I}{\alpha A L^2}\right]$$

[**Ans.** $t = 49.35$°C]

4.20 A rod of length L and cross-sectional area A fits snugly between a rigid support to its left and a linear spring of stiffness k to its right as shown in Figure 4.21. Determine the stress developed in the rod due to an increase of temperature of t.

$$\left[\textbf{Hint}: \frac{\Delta L}{L} = \frac{\sigma}{E} + \alpha t \text{ where } \Delta L = \frac{F}{k} \text{ and } F = \sigma A\right]$$

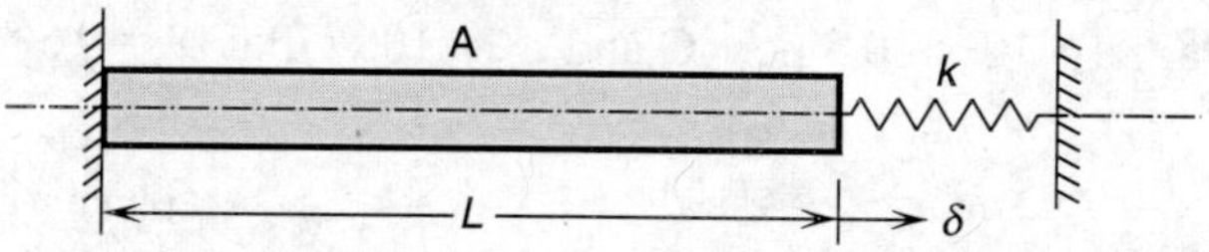

Figure 4.21 Rod of problem snugly fixed between rigid support and linear spring.

$$\left[\textbf{Ans. } \sigma = (-E\alpha t)\left(1 + \frac{EA}{kL}\right)^{-1}\right]$$

4.21 A 36 mm diameter copper rod passes centrally through a 4.6 mm thick steel tube of external diameter of 54 mm. The rod and tube are rigidly connected together at the ends at a temperature of 25°C. The length of the finished composite assembly is 350 mm. If the temperature of the assembly is raised to 175°C, determine: (a) the stresses induced in the rod and the tube and (b) the change in the length of the assembly. The modulus of elasticity and coefficient of thermal expansion for steel and copper are:

$E_s = 210$ GPa $\qquad \alpha_s = 10.8 \times 10^{-6}$ per °C

$E_c = 110$ GPa $\qquad \alpha_c = 17 \times 10^{-6}$ per °C

[**Ans.** (a) $\sigma_s = 94.39$ MPa (T); $\sigma_c = 52.86$ MPa (C), and $\Delta l = 0.724$ mm]

4.22 A 30 mm diameter steel rod passes centrally through a brass tube of external and internal diameters of 36 and 32 mm, respectively. The tube is closed at each end by rigid plates of negligible thickness as shown in Figure 4.22. The nuts on the rod are tightened until a stress of 12 MPa is developed in the rod. Determine the stresses induced in the rod and tube when the temperature of the assembly is raised to 60°C. The modulus of elasticity and coefficient of thermal expansion for steel and brass are:

$E_s = 210$ GPa $\qquad \alpha_s = 11.5 \times 10^{-6}$ per °C

$E_b = 80$ GPa $\qquad \alpha_b = 19 \times 10^{-6}$ per °C

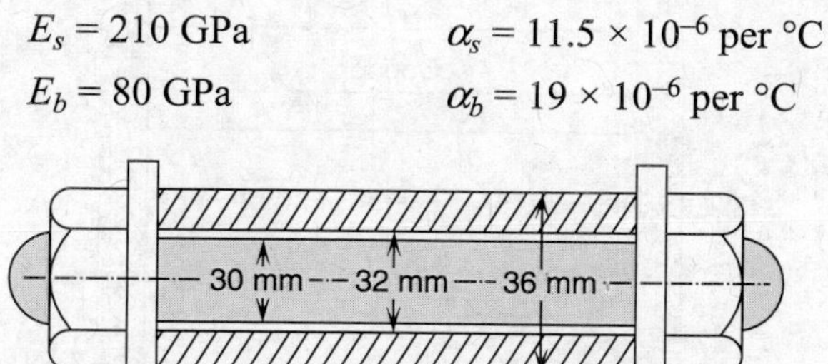

Figure 4.22 Assembly of steel rod and brass tube.

$$\left[\textbf{Ans. } \frac{\Delta l}{l} = 7.5459 \times 10^{-4} \text{ and } \sigma_s = 25.64 \text{ MPa (T)}; \ \sigma_b = 58.11 \text{ MPa (C)}\right]$$

4.23 A 60 mm diameter copper bar is enclosed completely in a steel tube of external and internal diameters of 96 and 60 mm, respectively. A pin of 30 mm diameter is fitted transversely to the axis of the assembly near each end to secure the copper rod to the steel tube as shown in Figure 4.23. Determine the shear stress induced in the pin when the temperature of the assembly is raised by 75°C. The modulus of elasticity and coefficient of thermal expansion for steel and copper are:

$$E_s = 210 \text{ GPa}; \alpha_s = 12 \times 10^{-6} \text{ per °C and } E_c = 100 \text{ GPa}; \alpha_c = 16 \times 10^{-6} \text{ per °C}$$

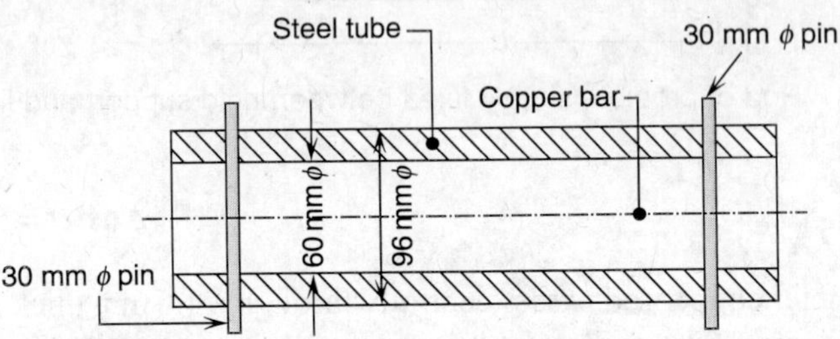

Figure 4.23 Steel tube and copper bar assembly of Problem 4.23.

[**Ans.** $F = 64986$ N and $\sigma_s = 45.97$ MPa]

4.24 A composite bar is fabricated by connecting a copper member and a steel member rigidly fixed at ends as shown in Figure 4.24. The cross-sectional area of the copper member is A mm² whilst that of the steel member is $2A$ mm² for half the length and is A mm² for the other half of the length. Determine the stresses induced in the members due to a temperature rise of T°C. The moduli of elasticity for steel and copper are E and $0.5E$, respectively, while the coefficient of thermal expansion are α and 1.3α, respectively.

$$\left[\textbf{Hint}: P = \frac{(\alpha_c - \alpha_s)tl}{\left[\dfrac{l}{A_cE_c} + \dfrac{l}{A_{es}E_s}\right]} = \frac{(\alpha_c - \alpha_s)tl}{\left[\dfrac{1}{A_cE_c} + \left(\dfrac{0.5l}{A_1E_s} + \dfrac{0.5l}{A_2E_s}\right)\right]} = \frac{(1.3\alpha - \alpha)T}{\left[\dfrac{1}{A(0.5E)} + \left(\dfrac{0.5}{2AE} + \dfrac{0.5}{AE}\right)\right]}\right]$$

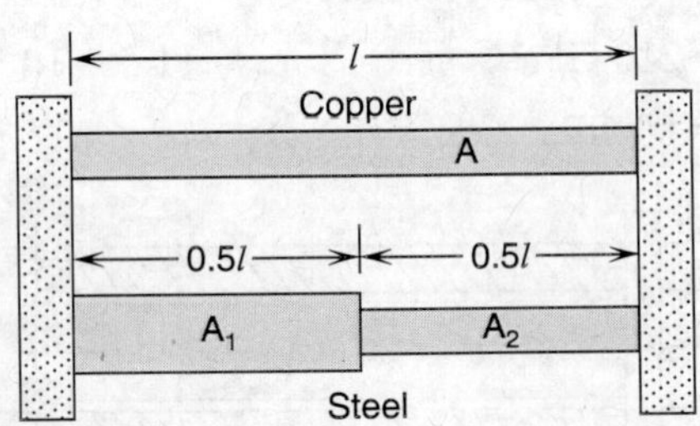

Figure 4.24 Composite bar of Problem 4.24.

[**Ans.** $\sigma_s = \sigma_c = (6E\alpha T)/55$]

Chapter 5

Biaxial and Triaxial Stresses and Strains: Pressure Vessels

5.1 INTRODUCTION

In the preceding chapters, the members subjected to uniaxial stress states due to independent actions of axial tension and compression have been discussed. In this chapter, the typical situations where *biaxial and triaxial states of stress* of pure tension or compression develop are considered. A biaxial state of stress arises when a plate or sheet of material, for example, is subjected to tension or compression in two mutually perpendicular directions. In practice commonly used *cylindrical and spherical thin pressure vessels* are important examples of biaxial state of stress of pure tension or compression. When the wall thickness t, of these *shell structures* is small as compared to the diameter d, the wall is called thin and it has very little resistance to bending and acts as a *membrane*. The membrane stresses are considered to act tangentially to the middle surface and uniformly distributed across the thickness. The stresses are shown in Figures 5.1(b) and (c).

On the other hand, in a thick or compound spherical or cylindrical shells subjected to an internal or external pressure, there is another stress state present which is normal to the plane of the biaxial state of stress developed in thin pressure vessels, thus reducing the biaxial state of stress to a triaxial stress state.

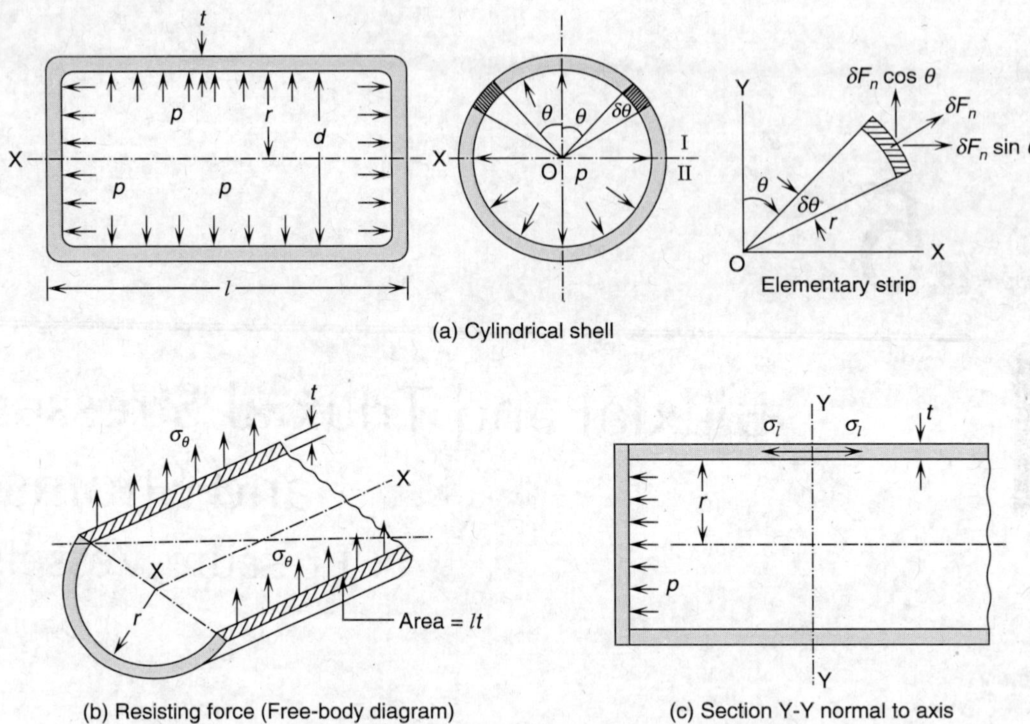

Figure 5.1 Stresses in cylindrical shell due to internal pressure.

5.2 THIN WALLED PRESSURE VESSELS

5.2.1 Thin Cylindrical Pressure Vessels

(a) Stresses: A long cylindrical shell of internal diameter d, wall thickness t, and length l which is closed at the ends and subjected to internal pressure p is shown in Figure 5.1(a). Consider a longitudinal section cut by diametrical horizontal plane X-X through the axis, dividing the shell into two halves I and II. Consider two elementary strips subtending an angle $\delta\theta$ at the centre at an angle θ on either side of the vertical through the centre. Normal force on each strip shown in Figure 5.1(a) is given by,

$$\delta F_n = p(r\,\delta\theta)\,l$$

where r (= d/2) is the radius of the shell. The resultant of the normal forces on two elemental strips acting vertically, i.e. normal to X-X is:

$$\delta F = 2 \times [p\,(r\,\delta\theta)l] \cos\theta$$

Total vertical force on one half on one side of X-X,

$$F = (2prl) \int_0^{\pi/2} \cos\theta \, d\theta = 2prl = p(dl)$$

$$= \text{Intensity of radial pressure} \times \text{Projected area} \qquad (5.1)$$

From the free body diagram, it is evident that the resultant force in the vertical direction due to internal pressure, i.e. the total bursting or splitting force is equal to the resisting force induced in shell thickness across the section X-X. If the intensity of tensile stress induced in the metal over the cross-sectional area of the shell is σ_θ, then the resisting force F_r offered by the section X-X shown in Figure 5.1(b) is:

$$F_r = 2 \times [\sigma_\theta (lt)] \tag{5.2}$$

For equilibrium,

$$F_r = F \quad \text{i.e.} \quad 2 \times \sigma_\theta (lt) = p(dl) \quad \text{or} \quad \sigma_\theta = \frac{pd}{2t} \tag{5.3}$$

As the direction of tensile stress σ_θ, is along the circumference of the shell it is called the *circumferential* or *hoop stress*.

Consider a section Y-Y shown in Figure 5.1(c) normal to the axis of the shell at a distance far away from the end sections, i.e. the sections not affected by end constraints and where uniform stress distribution can be assumed. This stress in the axial direction, or *longitudinal stress*, is designated by σ_l.

From the free body diagram, it can be seen that the bursting or splitting force in the longitudinal direction is equal to resultant force acting on the end cover plate. Thus, total longitudinal or axial force on one side of Y-Y is,

$$F = p\left(\frac{\pi d^2}{4}\right) \tag{5.4}$$

This tensile force is distributed over the circular cross-sectional area of the shell, which is πdt (since t is small). Therefore, the resisting force is

$$F_r = \sigma_l (\pi dt) \tag{5.5}$$

For equilibrium,

$$F_r = F \quad \text{i.e.} \quad \sigma_l (\pi dt) = p\left(\frac{\pi d^2}{4}\right) \quad \text{or} \quad \sigma_l = \frac{pd}{4t} \tag{5.6}$$

Thus, at any point in the material of the shell there are two principal tensile stresses, namely the hoop stress σ_θ, acting circumferentially and the longitudinal stress σ_l acting in the axial direction. The maximum shear stress developed is given by,

$$\tau_{\max} = \frac{1}{2}(\sigma_\theta - \sigma_l) = \frac{1}{2} \times \left(\frac{pd}{2t} - \frac{pd}{4t}\right) = \frac{pd}{8t} \tag{5.7}$$

If the maximum tensile stress in the material is limited to σ_a, the major principal stress σ_θ shall not exceed this value, i.e.

$$\sigma_\theta \le \sigma_a \quad \text{or} \quad t \ge \frac{pd}{2\sigma_a} \tag{5.8}$$

The above relation can be used to compute the thickness of the shell under given radial pressure p.

(b) Strains:

Circumferential strain,

$$\varepsilon_\theta = \frac{\text{Change in circumference}}{\text{Original circumference}} = \frac{\pi\,\delta d}{\pi d} = \frac{\delta d}{d} \tag{5.9}$$

Therefore, change in diameter $\delta d = \varepsilon_\theta d$

In terms of stresses,

$$\varepsilon_\theta = \frac{1}{E}(\sigma_\theta - \nu\sigma_l) = \frac{1}{E}\left[\left(\frac{pd}{2t}\right) - \nu\left(\frac{pd}{4t}\right)\right] = \frac{pd}{4tE}(2-\nu) \tag{5.10}$$

Longitudinal strain,

$$\varepsilon_l = \frac{\text{Change in length}}{\text{Original length}} = \frac{\delta l}{l} \tag{5.11}$$

Therefore, change in length $\delta l = l\varepsilon_l$

In terms of stresses,

$$\varepsilon_l = \frac{1}{E}(\sigma_l - \nu\sigma_\theta) = \frac{1}{E}\left[\left(\frac{pd}{4t}\right) - \nu\left(\frac{pd}{2t}\right)\right] = \frac{pd}{4tE}(1-2\nu) \tag{5.12}$$

where E are ν modulus of elasticity and Poisson's ratio, respectively.

Change in the capacity of the shell,

$$\text{Capacity of the shell, } V = \frac{\pi}{4}d^2 l \tag{5.13}$$

Therefore, change in the capacity of the shell,

$$\delta V = \frac{\pi}{4}(2d\,\delta d)\,l + \frac{\pi}{4}d^2\,(\delta l) \tag{5.14}$$

From Eqs. (5.13) and (5.14):

$$\frac{\delta V}{V} = 2\frac{\delta d}{d} + \frac{\delta l}{l} \quad \text{or} \quad \varepsilon_v = 2\varepsilon_\theta + \varepsilon_l \tag{5.15}$$

$$\frac{\delta V}{V} = 2 \times \frac{pd}{4tE}(2-\nu) + \frac{pd}{4tE}(1-2\nu) = \frac{pd}{4tE}(5-4\nu) \tag{5.16}$$

The expression can be used to compute the change in the capacity of the shell under radial pressure p.

EXAMPLE 5.1

A thin cylindrical shell of 1.0 m internal diameter and 10 mm thickness is subjected to an internal pressure of 4 MPa. If the length of the shell is 3.25 m, determine (a) hoop and longitudinal stresses,

(b) maximum shear stress and (c) the changes in the dimensions of the shell. The modulus of elasticity and Poisson's ratio for the shell material are 200 GPa and 0.3, respectively.

Solution: For the given problem,

$$p = 4 \text{ MPa};\ l = 3250 \text{ mm};\ d = 1000 \text{ mm and } t = 10 \text{ mm}$$

Hoop stress,
$$\sigma_\theta = \frac{pd}{2t} = \frac{4 \times 1000}{2 \times 10} = 200 \text{ MPa (tension)}$$

Longitudinal stress,
$$\sigma_l = \frac{\sigma_\theta}{2} = \frac{200}{2} = 100 \text{ MPa (tension)}$$

$$\tau_{\max} = \frac{1}{2}(\sigma_\theta - \sigma_l) = \frac{1}{2} \times (200 - 100) = 50 \text{ MPa}$$

Strains:

$$\varepsilon_\theta = \frac{1}{E}(\sigma_\theta - \nu\sigma_l) = \frac{(200 - 0.3 \times 100)}{E} = \frac{170}{E}$$

$$\varepsilon_l = \frac{1}{E}(\sigma_l - \nu\sigma_\theta) = \frac{(100 - 0.3 \times 200)}{E} = \frac{40}{E}$$

$$\varepsilon_v = 2\varepsilon_\theta + \varepsilon_l = \left(2 \times \frac{170}{E} + \frac{40}{E}\right) = \frac{380}{E}$$

$$\text{Change in diameter } = \varepsilon_\theta d = \frac{170 \times 1000}{200 \times 10^3} = 0.85 \text{ mm}$$

$$\text{Change in length } = \varepsilon_l l = \frac{40 \times 3250}{200 \times 10^3} = 0.65 \text{ mm}$$

$$\text{Change in capacity } = \varepsilon_v V = \left(\frac{380}{200 \times 10^3}\right) \times \left(\frac{\pi}{4} \times 1000^2 \times 3250\right)$$

$$= 4849834 \text{ mm}^4$$

EXAMPLE 5.2

A steel tube of 18 mm internal diameter and 3 mm thickness is closed at one end and screwed at the other end into a pressure vessel. The projected length of tube is 300 mm. While ignoring any constraints due to end connections, determine the safe internal pressure for the tube, if the maximum tensile stress in the material is limited to 150 MPa. Calculate the increase in its volume under this pressure. The modulus of elasticity and Poisson's ratio for the tube material are 200 GPa and 0.3, respectively.

Solution: $\sigma_\theta = \dfrac{pd}{2t} = \sigma_a = 150$ MPa (tension), thus $\sigma_l = \dfrac{pd}{4t} = 75$ MPa (tension)

Therefore, the safe pressure for the tube, $p = \dfrac{2t\sigma_a}{d} = \dfrac{2 \times 3 \times 150}{18} = 50$ MPa

Strains

$$\varepsilon_\theta = \frac{1}{E}(\sigma_\theta - \nu\sigma_l) = \frac{(150 - 0.3\times 75)}{(200\times 10^3)} = 637.5\times 10^{-6}$$

$$\varepsilon_l = \frac{1}{E}(\sigma_l - \nu\sigma_\theta) = \frac{(75 - 0.3\times 150)}{(200\times 10^3)} = 150\times 10^{-6}$$

$$\varepsilon_v = 2\varepsilon_\theta + \varepsilon_l = (2\times 637.5 + 150)\times 10^{-6} = 1425\times 10^{-6}$$

Increase in volume,

$$\delta V = \varepsilon_v V = (1425\times 10^{-6})\times\left(\frac{\pi}{4}\times 18^2\times 300\right) = 108.78\ \text{mm}^3$$

EXAMPLE 5.3

A thin cylindrical vessel of 150 mm internal diameter and 2.0 mm wall thickness closed at ends by rigid plates is filled with water under pressure. When an external axial pull of 16.5 kN is applied to the ends, the water pressure is observed to fall by 50 kPa. Determine the Poisson's ratio for the material of the vessel when E = 150 GPa and K for water = 2.2 GPa.

Solution: Due to fall in water pressure, the reduction in hoop stress,

$$\sigma_\theta = \frac{pd}{2t} = \frac{(50\times 10^{-3})\times 150}{2\times 2} = 1.875\ \text{MPa}\ (-)$$

Increase in longitudinal stress can be obtained from,

$$(\pi\times 150)\times 2\times \sigma_l = 16.5\times 10^3 - \frac{\pi}{4}\times(150)^2\times(50\times 10^{-3})$$

or $942.48\sigma_l = 16500 - 883.573$ i.e. $\sigma_l = 16.57$ MPa

Change in hoop strain and longitudinal strain are given by

$$\varepsilon_\theta = \frac{1}{E}(\sigma_\theta - \nu\sigma_l) = \frac{1}{E}(-1.875 - 16.57\nu)$$

$$\varepsilon_l = \frac{1}{E}(\sigma_l - \nu\sigma_\theta) = \frac{1}{E}(16.57 + 1.875\nu)$$

Change in volumetric strain of the vessel

$$\varepsilon_v = 2\varepsilon_\theta + \varepsilon_l = \frac{2}{E}(-1.875 - 16.57\nu) + \frac{1}{E}(16.57 + 1.875\nu)$$

$$= \frac{1}{E}(12.82 - 31.265\nu)$$

Change in volumetric strain of the water

$$\varepsilon_{v,\text{water}} = \frac{p}{K} = \frac{50\times 10^{-3}}{(2.2\times 10^3)}$$

Equate the changes in volumetric strain of the vessel and water

$$\frac{1}{(150\times 10^3)}\times(12.82-31.265\nu)=\frac{50\times 10^{-3}}{(2.2\times 10^3)} \quad \text{or} \quad \nu = 0.301$$

The Poisson's ratio for the material of the vessel is 0.301.

EXAMPLE 5.4

A 3 m long cylindrical tank of 1.8 m internal diameter is fabricated from 12 mm thick steel plate. The tank is closed at ends by rigid flat plates and these end plates are joined by nine equally spaced tie bars, each 40 mm diameter. The tie bars are initially stressed to 50 MPa and the tank filled with water at atmospheric pressure.

When the internal pressure is raised to 1.5 MPa, determine: (a) The final stresses in the tie bars and (b) The increase in capacity of the tank.

The modulus of elasticity and Poisson's ratio are 210 GPa and 0.28, respectively.

Solution: For the problem,

$$d = 1.8 \text{ m};\ t = 12 \text{ mm};\ d_t = 40 \text{ mm};\ \sigma_t' = 50 \text{ MPa and } p_i = 1.5 \text{ MPa}$$

When the internal pressure is atmospheric and tie bars are stressed

$\sigma_\theta' = 0$ and longitudinal stress σ_l' in the tank wall is given by,

$$(\pi d t)\sigma_l' = 9\times\left(\frac{\pi}{4}d_t^2\right)\times\sigma_t' \quad \text{or} \quad \sigma_l' = \frac{9}{4}\left(\frac{d_t^2\sigma_t'}{dt}\right)$$

Therefore,

$$\sigma_l' = \frac{9}{4}\times\left(\frac{40^2\times 50}{1800\times 12}\right) = 8.333 \text{ MPa (compression)}$$

When the internal pressure is raised

Consider the final stress in the tie bars and longitudinal stress in the tank wall to be σ_t and σ_l, respectively. For equilibrium in longitudinal direction $F_r = F_p$ i.e.

$$(\pi d t)\sigma_l + 9\times\left(\frac{\pi}{4}d_t^2\right)\times\sigma_t = \left(\frac{\pi}{4}d^2 - 9\times\frac{\pi}{4}d_t^2\right)p_i$$

or

$$\frac{4}{9}\times\left(\frac{dt}{d_t^2}\right)\sigma_l + \sigma_t = \left[\frac{1}{9}\left(\frac{d}{d_t}\right)^2 - 1\right]p_i$$

Therefore,

$$\frac{4}{9}\times\left(\frac{1800\times 12}{40^2}\right)\sigma_l + \sigma_t = \left[\frac{1}{9}\left(\frac{1800}{40}\right)^2 - 1\right]\times 1.5$$

or

$$6\sigma_l + \sigma_t = 336 \tag{i}$$

Circumferential stress,

$$\sigma_\theta = \frac{pd}{2t} = \frac{1.5\times 1800}{2\times 12} = 112.5 \text{ MPa}$$

For compatibility, *the change* in longitudinal strain must be same for both the tank and tie bars,

$$\varepsilon_l = \frac{1}{E}[\sigma_l' + (\sigma_l - \nu\sigma_\theta)] = \frac{1}{E}(\sigma_t - 50) \text{ i.e. } \sigma_l - \sigma_t = -\sigma_l' - 50 + \nu\sigma_\theta$$

Therefore, $\sigma_l - \sigma_t = -8.333 - 50 + 0.28 \times 112.5 = -26.83$ (ii)

Operate on Eqs. (i) and (ii) as: (i) + (ii)

$$7\sigma_l = 336 - 26.83 = 309.17 \quad \text{or} \quad \sigma_l = 44.17 \text{ MPa}$$

$$\sigma_t = \sigma_l + 26.83 = 44.17 + 26.83 = 71 \text{ MPa}$$

Change in circumferential strain and longitudinal strain are given by

$$\varepsilon_\theta = \frac{1}{E}[\sigma_\theta - \nu(\sigma_l + \sigma_l')]$$

$$\varepsilon_l = \frac{1}{E}(\sigma_t - 50)$$

Change in volumetric strain of the vessel

$$\varepsilon_v = 2\varepsilon_\theta + \varepsilon_l = \frac{1}{E}[2\sigma_\theta - 2\nu(\sigma_l + \sigma_l') + (\sigma - 50)]$$

$$= \frac{1}{(210 \times 10^3)} \times [2 \times 112.5 - 2 \times 0.28 \times (44.17 + 8.33) + (71 - 50)]$$

$$= 1.03143 \times 10^{-3}$$

Increase in volume,

$$\delta V = \varepsilon_v V = (1.03143 \times 10^{-3}) \times \left(\frac{\pi}{4} \times 1.8^2 \times 3.0\right) = 7.874 \times 10^{-3} \text{ m}^3$$

Thus the capacity of the tank increases by 7.874×10^{-3} m^3.

EXAMPLE 5.5

A thin cylindrical vessel of 1.25 m mean diameter is fabricated with a 20 mm thick plate. The joints are formed by welding along the helical seams that make an angle of 50° with the axis of cylinder. Determine tensile and shear stresses developed on the welded joint when the vessel is subjected to an internal pressure of 2.0 MPa.

Solution:

The circumferential and longitudinal stresses developed in the material of vessel are,

$$\sigma_\theta = \frac{pd}{2t} = \frac{2 \times 1250}{2 \times 20} = 62.5 \text{ MPa (tension)}$$

$$\sigma_l = \frac{pd}{4t} = \frac{2 \times 1250}{4 \times 20} = 31.25 \text{ MPa (tension)}$$

Stresses developed on the welded joint

Tensile stress:

$$\sigma_t = \sigma_\theta \cos^2 \theta + \sigma_l \sin^2 \theta$$
$$= 62.5 \cos^2 50° + 31.25 \sin^2 50°$$
$$= 18.338 + 25.823 = 44.161 \text{ MPa (tension)}$$

Shear stress:

$$\tau = \left(\frac{\sigma_\theta - \sigma_l}{2}\right) \sin 2\theta = \left(\frac{62.5 - 31.25}{2}\right) \sin 100°$$
$$= 15.38 \text{ MPa}$$

5.2.2 Riveted Cylindrical Shells

Pressure vessels of desired capacity are generally fabricated by bending plates to the required diameter and connecting them usually by a butt joint as shown in Figure 5.2(b). The desired length of the shell can be obtained by connecting individually fabricated smaller length shells usually by lap joint as shown in Figure 5.2(a). Thus both longitudinal and circumferential joints may be present in the shells. Figure 5.2(c) shows shell with hemispherical ends.

In the case of riveted shells, the actual longitudinal and circumferential stresses are greater than that were given in the preceding sections. This *reduction in the strength is due to the weakening*

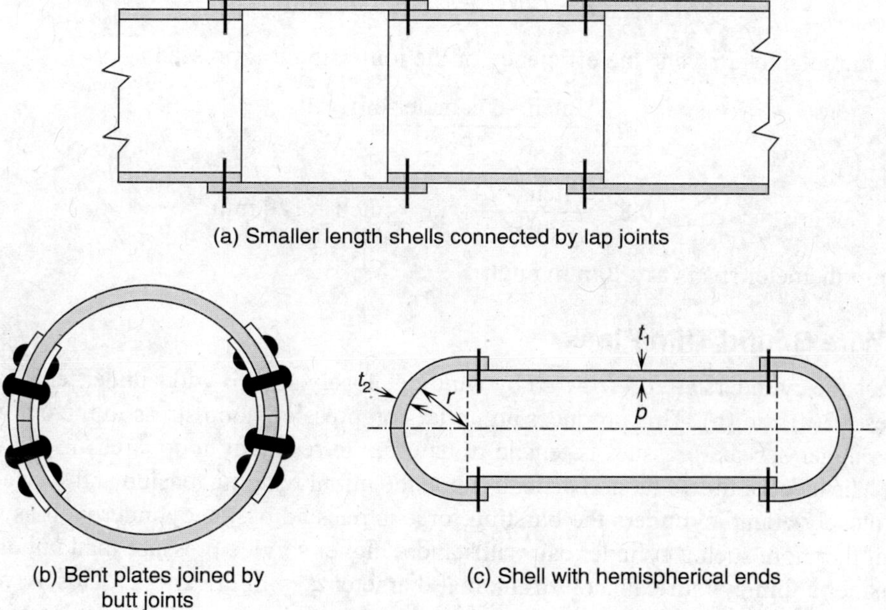

(a) Smaller length shells connected by lap joints

(b) Bent plates joined by butt joints

(c) Shell with hemispherical ends

Figure 5.2 Riveted cylindrical shells.

of the plates due to rivet holes. Consider η_1 and η_2 to be the efficiencies of longitudinal and circumferential joints, respectively. Then the circumferential and longitudinal stresses in the material of the vessel on the sections through rivet holes are, $\dfrac{pd}{2t\eta_1}$ and $\dfrac{pd}{4t\eta_2}$, respectively.

EXAMPLE 5.6

A 3.0 m long thin riveted boiler of 2.25 m internal diameter has to sustain an internal pressure of 0.64 MPa. The efficiencies of longitudinal and circumferential joints are 80 per cent and 85 per cent, respectively. Determine the thickness of the shell and the necessary pitch of rivets for the single riveted butt joints. The allowable tensile stress for the tank material is 60 MPa. Consider the diameter of rivet $d = 6\sqrt{t}$ where t is the thickness of the plate in mm.

Solution: From the considerations of hoop tension,

$$\sigma_\theta = \sigma_a = \frac{pd}{2t\eta_l} \quad \text{or} \quad t = \frac{pd}{2\sigma_a\eta_l} = \frac{0.64 \times 2250}{2 \times 60 \times 0.8} = 15.0 \text{ mm}$$

From the considerations of longitudinal tension

$$\sigma_l = \sigma_a = \frac{pd}{4t\eta_\theta} \quad \text{or} \quad t = \frac{pd}{4\sigma_a\eta_\theta} = \frac{0.64 \times 2250}{4 \times 60 \times 0.85} = 7.06 \text{ mm}$$

Thickness of the shell required is 15 mm. Therefore, diameter of the rivets is

$$d = 6\sqrt{t} = 6\sqrt{15} = 23.2 \text{ mm}$$

Using 24 mm diameter rivets, the efficiency of the joint can be expressed as

$$\eta = \frac{\text{Pitch} - \text{Diameter of rivet}}{\text{Pitch}}$$

or
$$0.8 = \frac{\text{Pitch} - 24}{\text{Pitch}} \quad \text{i.e. pitch} = 120 \text{ mm}$$

Use 24 mm diameter rivets at 120 mm pitch.

5.2.3 Wire Bound Thin Pipes

Sometimes the cylinders are *prestressed* by winding closely spaced wires under tension as shown in Figures 5.3(a) and (b). This produces an initial compressive hoop stress in the cylinder. Since the stress under service pressure is tensile in nature, the resultant hoop stress in the cylinder is reduced. The level of this reduction depends upon the initial winding tension and service pressure. Thus in closely bound cylinders the bursting force is resisted by the cylinder wall as well as the wires and therefore such a cylinder can withstand a higher service pressure than unbound one.

Consider a thin cylindrical shell of internal diameter d, wall thickness t, closely wound by a single layer of steel wire of diameter d_w under constant tensile stress T and subjected to internal pressure p. Consider the modulus of elasticity for the cylinder and wire materials to be E_c and E_w, respectively, and the Poisson's ratio for the cylinder material ν_c.

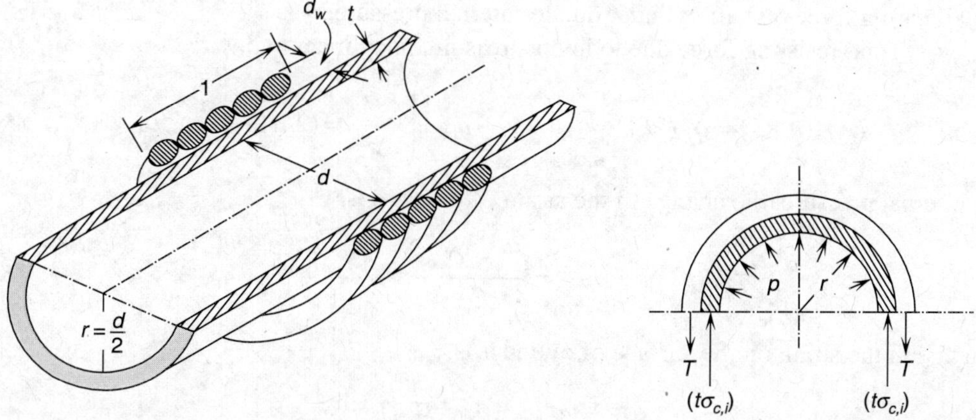

(a) A cylinder bound by closely spaced wires

(b) Initial stresses in wires and cylinder

Figure 5.3 Wire bound thin pipes.

(i) **Winding stresses.** In a unit length of the cylinder, the number of sections of the wound wire,

$$n = \frac{1}{d_w}$$

Thus, due to initial stress in the wire, the tensile force in the wires over the unit length of the cylinder is,

$$F_w = \text{Cross-sectional area of wires} \times \text{Initial stress in wires}$$

$$= n\left(2 \times \frac{\pi}{4} d_w^2\right) \times \sigma_{w,i} = \frac{1}{d_w} \times \left(2 \times \frac{\pi}{4} d_w^2\right) \times \sigma_{w,i} = \frac{\pi d_w \sigma_{w,i}}{2} \tag{5.17}$$

Due to tensile stress in the wires, the compressive hoop force or crushing force induced in the cylinder is given by:

$$F_c = \text{Cross-sectional area of cylinder} \times \text{Crushing stress in walls}$$

$$= (2t \times 1) \times \sigma_{c,i} \tag{5.18}$$

Therefore, for equilibrium

$$F_c + F_w = 0 \text{ i.e. } (2t \times 1) \times \sigma_{c,i} = -\frac{\pi d_w \sigma_{w,i}}{2} \text{ or } \sigma_{c,i} = -\frac{\pi d_w \sigma_{w,i}}{4t} \tag{5.19}$$

(ii) **Final stresses.** Due to an internal pressure the material at a point in the closed cylinder is subjected to 2-dimensional principal stresses σ_1 and σ_2 i.e. hoop σ_θ and longitudinal σ_l stresses, respectively, whereas the wire is subjected to direct stress only. Consider the final stresses in the cylinder and the wire under internal pressure p to be σ_c and σ_w, respectively. Then from the first principles,

Total tearing force on half cylinder due to internal pressure
= Total resisting force due to hoop stress in the cylinder and wire

$$p \times (d \times 1) = \sigma_c (2t) + \left[\sigma_w \left(2 \times \frac{\pi}{4} d_w^2 \right) \right] \times \frac{1}{d_w} = (2t\sigma_c) + \frac{\pi}{2} d_w \sigma_w \quad (5.20)$$

Due to internal pressure, the change in the strain of the wire is,

$$\varepsilon_w = \frac{\sigma_w - \sigma_{w,i}}{E_w} \quad (5.21)$$

The change in the strain on the surface of cylinder is,

$$\varepsilon_\theta = \frac{\sigma_1 - \nu\sigma_2}{E_c} = \frac{(\sigma_c + \sigma_{c,i}) - \nu\sigma_l}{E_c} \quad (5.22)$$

For the circumferential strain compatibility at the interface of wire and tube

$$\frac{\sigma_w - \sigma_{w,i}}{E_w} = \frac{(\sigma_c + \sigma_{c,i}) - \nu\sigma_l}{E_c} \quad (5.23)$$

Equations (5.19), (5.20) and (5.23) can be used to compute the values σ_w, $\sigma_{c,i}$ and σ_c.

EXAMPLE 5.7

A 5 mm diameter steel wire under a tensile stress of 60 MPa is wound around a 12 mm thick cast iron pipe with 200 mm internal diameter to increase the strength of the pipe against internal pressure. Determine the stresses developed in the pipe and steel wire when water under a pressure of 3.5 MPa is admitted in the pipe. Consider the modulus of elasticity for the cast iron and steel to be 105 GPa and 210 GPa, respectively. Poisson's ratio for the cast iron is 0.3.

Solution: *Before water is admitted in the pipe*:

Initial stress in the wire, $\sigma_{w,i} = 60$ MPa

Compressive stress induced in the cylinder, from Eq. (5.19), is,

$$\sigma_{c,i} = -\frac{\pi d_w \sigma_{w,i}}{4t} = -\frac{\pi \times 5 \times 60}{4 \times 12} = -19.635 \text{ MPa (compressive)}$$

After water is admitted in the pipe:

From Eq. (5.20),

$$(2 \times 12 \times \sigma_c) + \frac{\pi}{2} \times 5 \times \sigma_w = 3.5 \times 200 \quad \text{or} \quad 24\sigma_c + 7.854\sigma_w = 700$$

or

$$3.056\sigma_c + \sigma_w = 89.127$$

and

$$\sigma_l = \frac{pd}{4t} = \frac{3.5 \times 200}{4 \times 12} = 14.583 \text{ MPa}$$

From Eq. (5.23):

$$\frac{\sigma_w - 60}{(210 \times 10^3)} = \frac{(\sigma_c + 19.635) - 0.3 \times 14.583}{(105 \times 10^3)}$$

i.e. $\sigma_w - 60 = 2 \times [(\sigma_c + 19.635) - 0.3 \times 14.583]$ or $\sigma_w = 2\sigma_c + 90.510$

Therefore, $\sigma_w = 89.963$ MPa (tension); $\sigma_c = 0.274$ MPa (compression)

5.3 THIN SPHERICAL SHELLS

(i) Stresses: Consider a thin spherical seamless shell of internal diameter d, wall thickness t, and subjected to internal pressure p cut by diametrical horizontal plane X-X through its centre, dividing the shell into two halves I and II as shown in Figure 5.4(a).

Consider the forces acting on an elementary annular ring of the sphere subtending an angle $\delta\theta$ at the centre at an angle θ from the horizontal through the centre.

Normal or radial force on each ring ABCD,

$$\delta F_n = \text{Surface area of the ring ABCD} \times p$$
$$= (\text{Circumference of the ring} \times \text{Width of the ring BC}) \times p$$
$$= [(2\pi r \cos\theta)(r\,\delta\theta)]p = (2\pi r^2 p \cos\theta)\,\delta\theta$$

Vertical component of radial force $\delta F_n = \delta F_n \times \sin\theta$

$$\delta F = [(2\pi r^2 p \cos\theta)\sin\theta]\,\delta\theta = \pi r^2 p \sin 2\theta\, d\theta$$

Total vertical bursting force or splitting on one side of X-X,

$$F = \int_0^{\pi/2} (\pi r^2 p \sin 2\theta)\, d\theta = (\pi r^2 p)\int_0^{\pi/2} \sin 2\theta\, d\theta$$

$$= (\pi r^2 p)\left| -\frac{\cos 2\theta}{2} \right|_0^{\pi/2} = \frac{1}{2}(\pi r^2 p)(1+1) = p(\pi r^2)$$

$$= \text{Intensity of radial pressure} \times \text{Projected area} = p \times \left(\frac{\pi}{4} d^2\right) \quad (5.24)$$

where $r\ (= d/2)$ is the radius of the shell.

From the free-body diagram shown in Figure 5.4(b), it can be seen that the resultant force in the vertical direction due to internal pressure, i.e. the total bursting force is equal to the resisting force induced in shell thickness across the section X-X. This tensile force is distributed over the circular cross-sectional area of the shell, which is πdt (since t is small). If the intensity of tensile stress induced in the metal over the cross-sectional area of the shell is σ_θ, then the resisting force F_r offered by the section X-X:

$$F_r = \sigma_\theta(\pi dt) \quad (5.25)$$

For equilibrium,

$$F_r = F \quad \text{i.e.} \quad \sigma_\theta(\pi dt) = p\left(\frac{\pi d^2}{4}\right) \quad \text{or} \quad \sigma_\theta = \frac{pd}{4t} \quad (5.26)$$

As the direction of tensile stress, σ_θ, is along the circumference of the shell it is called the *circumferential* or *hoop stress*. The same result is obtained for any other section through the centre of the shell. Thus for a thin spherical shell subjected to internal pressure, the principle stresses σ_1 and σ_2 at any point are equal and alike, i.e. every plane through a point is a principal plane with an intensity of stress equal to $\sigma_1 = \sigma_2 = \dfrac{pd}{4t}$. Thus, *no shear stress exists anywhere in the spherical shell.*

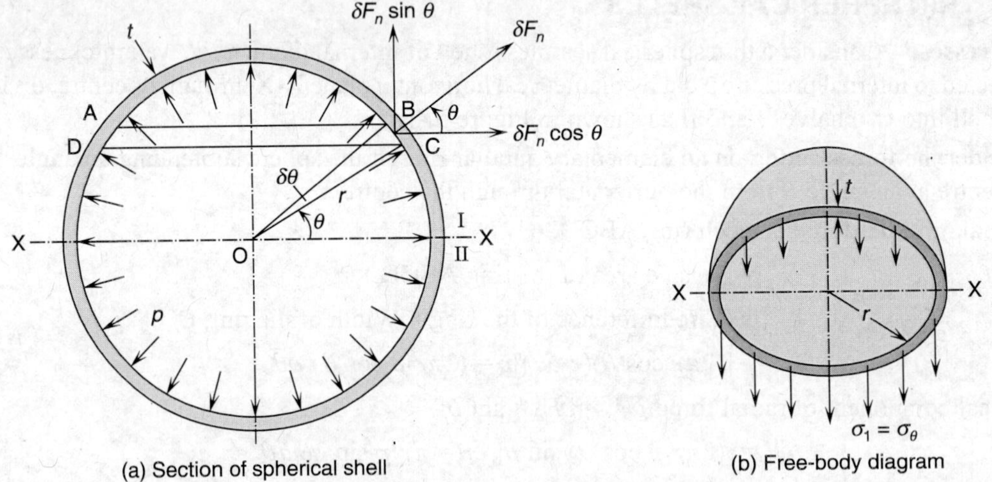

Figure 5.4 Thin spherical shell.

(ii) Strains

Circumferential strain:

$$\varepsilon_\theta = \frac{\text{Change in circumference}}{\text{Original circumference}} = \frac{\pi\,\delta d}{\pi d} = \frac{\delta d}{d}$$

Therefore, change in diameter $\delta d = \varepsilon_\theta d$

In terms of stresses,

$$\varepsilon_\theta = \frac{1}{E}(\sigma_1 - \nu\sigma_2) = \frac{1}{E}\left[\left(\frac{pd}{4t}\right) - \nu\left(\frac{pd}{4t}\right)\right] = \frac{pd}{4tE}(1-\nu) \qquad (5.27)$$

where E and ν are modulus of elasticity and Poisson's ratio, respectively.

Change in the capacity of the shell:

$$\text{Capacity of the shell, } V = \frac{4}{3}\pi r^3 = \frac{\pi d^3}{6} \qquad (5.28)$$

Therefore, change in the capacity of the shell,

$$\delta V = \frac{\pi}{6}(3d^2\,\delta d) = \frac{\pi d^2}{2}\,\delta d \qquad (5.29)$$

From Eqs. (5.28) and (5.29):

$$\frac{\delta V}{V} = 3\frac{\delta d}{d} \quad \text{or} \quad \varepsilon_v = 3\varepsilon_\theta \tag{5.30}$$

Therefore,

$$\frac{\delta V}{V} = \frac{3pd}{4tE}(1-\nu)$$

or

$$\delta V = \frac{\pi d^3}{6} \times \left[\frac{3pd}{4tE}(1-\nu)\right] = \frac{p\pi d^4}{8tE}(1-\nu) \tag{5.31}$$

The expression can be used to compute the change in the capacity of the shell under radial pressure p.

5.4 VESSELS SUBJECTED TO FLUID PRESSURE

When the pressure is increased in a vessel initially filled with a fluid at atmospheric pressure, the fluid itself will change in volume which must be taken into account while calculating the additional quantity v of fluid at atmospheric pressure which has to be pumped into vessel to raise the pressure by a specified amount.

The bulk modulus for a fluid is given by,

$$K = \frac{\text{Volumetric stress}}{\text{Volumetric strain}} = \frac{p}{(\delta V_t / V_t)} = \left(\frac{pV_t}{\delta V_t}\right) \tag{5.32}$$

where $V_t = V + v$. Therefore decrease in the volume of fluid under pressure is:

$$\delta V_2 = \frac{pV_t}{K} \tag{5.33}$$

Hence, the additional fluid to be pumped into vessel to raise the vessel pressure by p:

$$\delta V = \delta V_1 + \delta V_2 \tag{5.34}$$

where, δV_1 and δV_2 are the increase in the volume of boiler due to internal pressure and the decrease in the volume of water due to pressure, respectively.

(i) For a cylindrical shell

$$\delta V_c = \frac{pd}{4tE}(5-4\nu)V + \left(\frac{p}{K}\right)V_t \tag{5.35}$$

(ii) For a spherical shell

$$\delta V_s = \frac{3pd}{4tE}(1-\nu)V + \left(\frac{p}{K}\right)V_t \tag{5.36}$$

EXAMPLE 5.8

A 10 m long cylindrical boiler of 2.5 m internal diameter is fabricated from 20 mm thick steel plates. (a) Determine the allowable pressure, if the maximum tensile stress is limited to 120 MPa. (b) For a hydraulic test, the boiler is first filled with water at atmospheric pressure and then the internal pressure is raised to twice the allowable, pressure by pumping in more water. Determine the additional quantity of water at atmospheric pressure which has to be pumped into the boiler to raise the pressure by a specified amount.

The modulus of elasticity and Poisson's ratio for the material of the boiler are 210 GPa and 0.3, respectively and the bulk modulus for water, $K = 1.84$ GPa.

Solution:

(a) Allowable pressure to develop the maximum tensile stress of 120 MPa,

$$\sigma_\theta = \sigma_a = \frac{pd}{2t} \quad \text{or} \quad p = \frac{\sigma_a \times 2t}{d}$$

Therefore,
$$p = \frac{120 \times 2 \times 20}{2500} = 1.92 \text{ MPa}$$

(b) Increase in the volume of boiler due to internal pressure

$$\delta V_1 = \frac{1}{E}\left(\frac{pd}{4t}\right)(5 - 4\nu)V$$

$$= \frac{1}{(210 \times 10^3)} \times \left(\frac{2 \times 1.92 \times 2500}{4 \times 20}\right) \times (5 - 4 \times 0.3)\left(\frac{\pi}{4} \times 2500^2 \times 10000\right)$$

$$= 106590000 \text{ mm}^3 = 0.10659 \text{ m}^3$$

(c) Decrease in the volume of water due to pressure

Consider $V_t \approx V$, therefore,

$$\delta V_2 = \left(\frac{p}{K}\right)V = \left(\frac{2 \times 1.92}{1.84 \times 10^3}\right) \times \left(\frac{\pi}{4} \times 2500^2 \times 10000\right)$$

$$= 102{,}440{,}000 \text{ mm}^3 = 0.10244 \text{ m}^3$$

Additional quantity of water to be pumped for increasing the internal pressure to twice the allowable pressure,

$$\delta V = \delta V_1 + \delta V_2 = 0.10659 + 0.10244 = 0.20903 \text{ m}^3$$

EXAMPLE 5.9

A thin spherical shell of 500 mm internal diameter fabricated from 3 mm thick steel plates is first filled with water at atmospheric pressure, i.e. 0.1 MPa. Determine the internal pressure developed when 30,000 mm^3 of additional water at atmospheric pressure is pumped into the shell. Determine: (i) the circumferential stress developed in the shell and (ii) the change in the volume of shell, if the modulus of elasticity and Poisson's ratio for the steel are 210 GPa and 0.33, respectively and the bulk modulus for water, $K = 2.361$ GPa.

Solution: The increase in the pressure due to pumping of additional quantity of fluid at atmospheric pressure is given by Eq. (5.36):

$$\delta V_s = \frac{3pd}{4tE}(1-\nu)V + \left(\frac{p}{K}\right)V_t$$

$$= \left[\frac{3\times p\times 500}{4\times 3\times(210\times 10^3)}(1-0.33)\right]\times\left(\frac{\pi}{6}\times 500^3\right)$$

$$+\left(\frac{p}{2.361\times 10^3}\right)\times\left(\frac{\pi}{6}\times 500^3+30000\right)=30000$$

$$= p[(0.39881\times 10^{-3})\times 65449.8\times 10^3]$$

$$+ p[(0.42355\times 10^{-3})\times(65449.8\times 10^3+30000) = 30000$$

or $\quad 53836.00p = 30000 \quad$ i.e. $\quad p = 0.5572$ MPa

The circumferential stress developed in the shell

$$\sigma_\theta = \frac{pd}{4t} = \frac{0.5572\times 500}{4\times 3} = 23.22 \text{ MPa}$$

The change in the volume of shell

$$\delta V = \left[\frac{3pd}{4tE}(1-\nu)\right]V$$

$$= 0.5572\times[(0.39881)\times 10^{-3}]\times 65449.8\times 10^3$$

$$= 14544.05 \text{ mm}^3$$

5.5 CYLINDRICAL SHELL WITH HEMISPHERICAL ENDS

Cylindrical shell with hemispherical ends is advantageous over that with flat ends as the bending stresses are reduced. Consider a thin cylindrical shell of internal diameter d having hemispherical ends with wall thicknesses of t_c and t_s for cylindrical and hemispherical portions, respectively, as shown in Figure 5.5. For the shell subjected to an internal pressure p, the stresses in various parts are:

Cylindrical part of the shell

The circumferential and longitudinal stresses $\sigma_{\theta,c}$ and $\sigma_{l,c}$ developed in the cylindrical part of vessel are $\frac{pd}{2t_c}$ and $\frac{pd}{4t_c}$ respectively. Therefore, the circumferential strain in the cylindrical part is

$$\varepsilon_{\theta,c} = \frac{1}{E}(\sigma_{\theta,c} - \nu\sigma_{l,c}) = \frac{1}{E}\left[\left(\frac{pd}{2t_c}\right) - \nu\left(\frac{pd}{4t_c}\right)\right] = \frac{pd}{4t_cE}(2-\nu) \tag{5.37}$$

where E and ν are modulus of elasticity and Poisson's ratio, respectively.

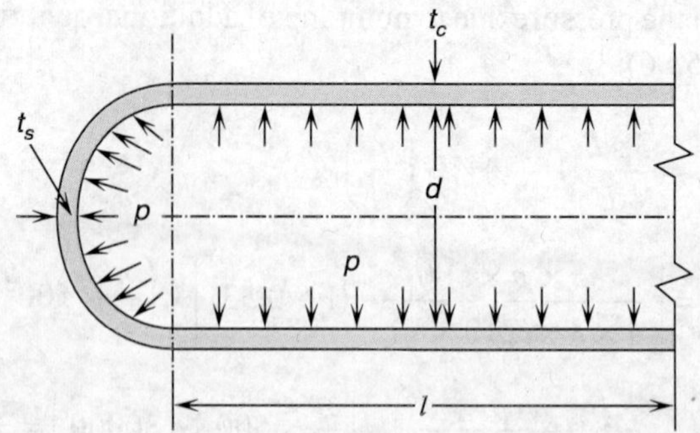

Figure 5.5 Cylindrical shell with hemispherical ends.

Spherical part of the shell

The circumferential stress $\sigma_{\theta,s}$ developed in the spherical part of vessel is $\dfrac{pd}{4t_s}$.

Therefore, the circumferential strain in the hemispherical part is:

$$\varepsilon_{\theta,s} = \frac{1}{E}(\sigma_{\theta,s} - \nu\sigma_{\theta,s}) = \frac{pd}{4t_s E}(1-\nu) \tag{5.38}$$

For the strain compatibility at the junction of cylindrical and hemispherical parts,

$$\varepsilon_{\theta,c} = \varepsilon_{\theta,s} \quad \text{or} \quad \frac{pd}{4t_c E}(2-\nu) = \frac{pd}{4t_s E}(1-\nu)$$

Thus,

$$\frac{t_s}{t_c} = \frac{1-\nu}{2-\nu} \tag{5.39}$$

The condition of same maximum stress at the junction of cylindrical and hemispherical parts,

$$\sigma_{\theta,c} = \sigma_{\theta,s} \quad \text{or} \quad \frac{pd}{2t_c E} = \frac{pd}{4t_s E} \tag{5.40}$$

Therefore,

$$\frac{t_s}{t_c} = \frac{1}{2}$$

EXAMPLE 5.10

A cylindrical vessel of internal diameter 1.2 m and having hemispherical ends is fabricated from thin plates. If the vessel is subjected to an internal pressure of 2 MPa, determine: (i) the thickness of cylindrical and hemispherical parts if the stress in the material is limited to 100 MPa and (ii) the thickness of hemispherical part when $t_c = 10$ mm and the condition of no distortion at the junction is to be satisfied. The Poisson's ratio for the material is 0.3.

Solution: (a) For the permissible stress of 100 MPa, the thickness can be computed as,

Cylindrical part of the shell

$$\sigma_a = \frac{pd}{2t_c} \quad \text{or} \quad t_c = \frac{2 \times 1200}{2 \times 100} = 12 \text{ mm}$$

Spherical part of the shell

$$\sigma_a = \frac{pd}{4t_s} \quad \text{or} \quad t_s = \frac{2 \times 1200}{4 \times 100} = 6 \text{ mm}$$

(b) For the condition of no distortion at the junction

$$\frac{t_s}{t_c} = \frac{1-\nu}{2-\nu} = \frac{1-0.33}{2-0.33} = 0.4$$

Therefore,
$$t_s = 0.4t_c = 0.4 \times 10 = 4 \text{ mm}$$

EXAMPLE 5.11

A 600 mm long (overall) cylindrical steel vessel with hemispherical ends is fabricated from 5 mm thick plates. If the internal diameter of the vessel is 90 mm, determine the change in the volume of vessel when it is subjected to an internal pressure of 15 MPa. The modulus of elasticity and the Poisson's ratio for steel are 200 GPa and 0.28, respectively.

Solution: The length of cylindrical part of the shell is

$$l = 600 - 100 = 500 \text{ mm}$$

Consider δV_c and δV_s to be the increases in the volume of cylindrical and hemispherical parts of the vessel, respectively, due to internal pressure.

(i) Increase in volume for the cylindrical shell from Eq. (5.16):

$$\delta V_c = \frac{pd}{4tE}(5 - 4\nu)V$$

$$= \frac{15 \times 90}{4 \times 5 \times (200 \times 10^3)} \times (5 - 4 \times 0.28) \times \left(\frac{\pi}{4} \times 90^2 \times 500\right)$$

$$= 4165.34 \text{ mm}^3$$

(ii) Increase in volume for the spherical shell from Eq. (5.31):

$$\delta V_s = \frac{3pd}{4tE}(1 - \nu)V$$

$$= \frac{3 \times 15 \times 90}{4 \times 5 \times (200 \times 10^3)} \times (1 - 0.28) \times \left[2 \times \left(\frac{\pi}{6} \times 90^3 \times 0.5\right)\right]$$

$$= 278.26 \text{ mm}^3$$

Total increase in volume

$$\delta V = \delta V_c + \delta V_s = 4165.34 + 278.26 = 4443.60 \text{ mm}^3$$

5.6 THICK PRESSURE VESSEL

In case of thin pressure vessels the stresses are considered to be uniform across the thickness of the vessel. However, in practice the hoop stress is not uniform across the thickness and it varies from a maximum value at the inner circumference to a minimum value at the outer circumference. The stress distribution in the cylinder is based on the following assumptions:

(i) The material of the cylinder is linearly elastic, homogeneous and isotropic.

(ii) The plane transverse sections remain plane under pressure. This assumption is only nearly true at a considerable distance from the ends. As a consequence of this assumption, the longitudinal strain is constant at all points in thickness of the wall, i.e. it is independent of radius.

The longitudinal strain,
$$\varepsilon_l = \frac{1}{E}[\sigma_l - \nu(\sigma_\theta + \sigma_r)] \tag{5.41}$$

Since ε_l, σ_l, E and ν are all constant, $(\sigma_\theta + \sigma_r)$ is also constant.

A thick cylindrical vessel with external radius, inner radius and length of r_e, r_i and l, respectively, subjected to an internal and external pressures of p_i and p_e is shown in Figures 5.6(a) and (b). Consider a concentric elementary ring of radius r and thickness δr with radial pressure intensities of p_r and $(p_r + \delta p_r)$ at its inner and outer circumference, respectively. The total vertical bursting force on one side of longitudinal section I-I is given by,

$$F = \text{Intensity of radial pressure} \times \text{Projected area}$$
$$= p_r \times (2rl) - (p_r + \delta p_r) \times [2(r + \delta r)l]$$
$$= -2[p_r\delta r + r\,\delta p_r + \delta p_r\, dr]l = -2[p_r\delta r + r\,\delta p_r]l$$

(the product of small quantities $\delta p_r \delta r$ is neglected).

If the intensity of tensile hoop stress induced in the elementary ring over its cross-sectional area is $\sigma_{\theta,r}$, the resisting force F_r offered by the section X-X (see Figure 5.6(c)):

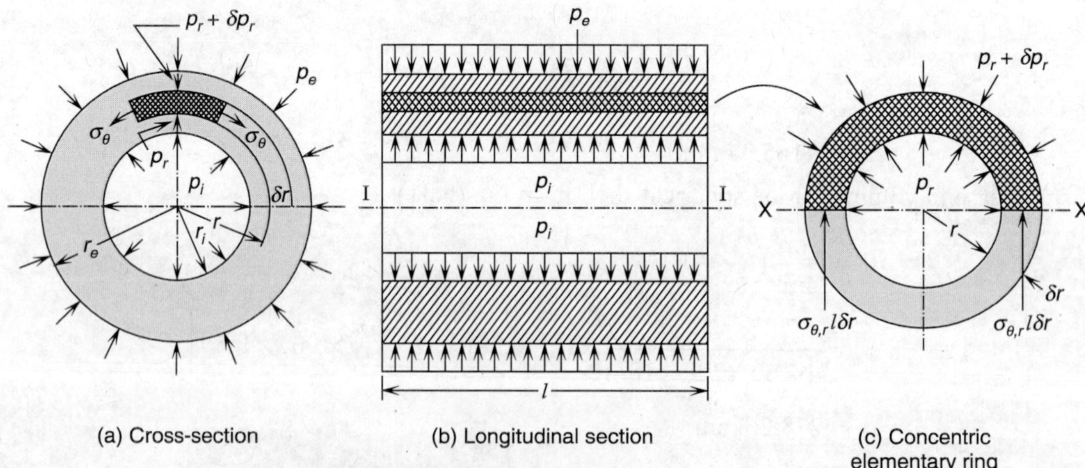

Figure 5.6 Thick cylindrical vessel subjected to internal and external pressures.

$$F_r = \sigma_{\theta,r}\,(2\delta r l) \tag{5.42}$$

For equilibrium, total bursting force must be equal to the resisting force, i.e.

$$F_r = F \quad \text{or} \quad \sigma_{\theta,r}\,(2\delta r l) = -\,2[p_r \delta r + r\delta p_r]l$$

Therefore,

$$\sigma_{\theta,r} = -p_r - r\frac{\delta p_r}{\delta r} \tag{5.43}$$

Thus the radial pressure varies with the radial distance from the centre to the point under consideration. Whereas, the longitudinal strain at any point is assumed to be constant. If the longitudinal stress at any point is $\sigma_{l,r}$ then

$$\varepsilon_{l,r} = \frac{1}{E}[\sigma_{l,r} - \nu(\sigma_{\theta,r} - p_r)] = \text{Constant} \tag{5.44}$$

where $\varepsilon_{l,r}$, $\sigma_{l,r}$ along with E and ν, the modulus of elasticity and Poisson's ratio of the shell material, respectively, are constant. Thus, $(\sigma_{\theta,r} - p_r)$ is also constant

$$\sigma_{\theta,r} - p_r = \text{Constant} = 2a$$

Therefore,

$$\sigma_{\theta,r} = 2a + p_r \tag{5.45}$$

From Eqs. (5.43) and (5.45):

$$-p_r - r\frac{\delta p_r}{\delta r} = 2a + p_r \quad \text{or} \quad -r\frac{\delta p_r}{\delta r} = 2(a + p_r)$$

Hence,

$$\frac{dp_r}{p_r + a} = -\frac{2dr}{r} \tag{5.46}$$

The integration gives,

$$\log_e (p_r + a) = -\,2\log_e r + \log_e b$$

where $\log_e b$ is a constant. Therefore,

$$\log_e (p_r + a) = \log_e \frac{b}{r^2} \quad \text{or} \quad p_r + a = \frac{b}{r^2}$$

or

$$p_r = \frac{b}{r^2} - a \tag{5.47}$$

From Eqs. (5.45) and (5.47):

or

$$\sigma_{\theta,r} = \frac{b}{r^2} + a \tag{5.48}$$

The constants a and b may be determined from the known radial pressure conditions. Equations (5.47) and (5.48) are called *Lame's relations* and can be used to determine radial pressure and hoop stress at any radius.

For illustration, consider a thick cylinder subjected to an internal pressure $p_i = p_i$ and $p_e = 0$. From the known radial pressure conditions:

$$\text{At } r = r_e,\ p_r = 0 \qquad \text{i.e.} \qquad 0 = \frac{b}{r_e^2} - a \tag{5.49}$$

$$\text{At } r = r_i, p_r = p_i \qquad \text{i.e.} \qquad p_i = \frac{b}{r_i^2} - a \tag{5.50}$$

The constants a and b can be evaluated from the above equations,

$$\text{Operate: Eq. (5.50) – Eq. (5.49)} \quad p_i = \left(\frac{1}{r_i^2} - \frac{1}{r_e^2}\right) b \quad \text{or} \quad b = p_i \left(\frac{r_e^2 r_i^2}{r_e^2 - r_i^2}\right) \tag{5.51}$$

$$\text{From Eq. (5.49)} \qquad a = \frac{b}{r_e^2} = p_i \left(\frac{r_i^2}{r_e^2 - r_i^2}\right) \tag{5.52}$$

Therefore, the hoop stress developed in the vessel at any radius r,

$$\sigma_{\theta,r} = \frac{b}{r^2} + a = p_i \left(\frac{r_i^2}{r_e^2 - r_i^2}\right)\left(\frac{r_e^2}{r^2} + 1\right) = p_i \left(\frac{r_i}{r}\right)^2 \left(\frac{r_e^2 + r^2}{r_e^2 - r_i^2}\right) \tag{5.53}$$

The *maximum circumferential stress* is developed in the shell at $r = r_i$,

$$\sigma_{\theta,r_i} = p_i \left(\frac{r_e^2 + r_i^2}{r_e^2 - r_i^2}\right) = p_i \left(\frac{k^2 + 1}{k^2 - 1}\right) \quad \text{where} \quad k = \frac{r_e}{r_i} \tag{5.54}$$

The *minimum circumferential stress* occurs at $r = r_e$,

$$\sigma_{\theta,r_e} = p_i \left(\frac{r_i}{r_e}\right)^2 \left(\frac{r_e^2 + r_e^2}{r_e^2 - r_i^2}\right) = p_i \left(\frac{2r_i^2}{r_e^2 - r_i^2}\right) = p_i \left(\frac{2}{k^2 - 1}\right) \tag{5.55}$$

The radial pressure and hoop stress variations across the section of the cylinder due to internal and external pressures are shown in Figures 5.7(a) and 5.7(b), respectively.

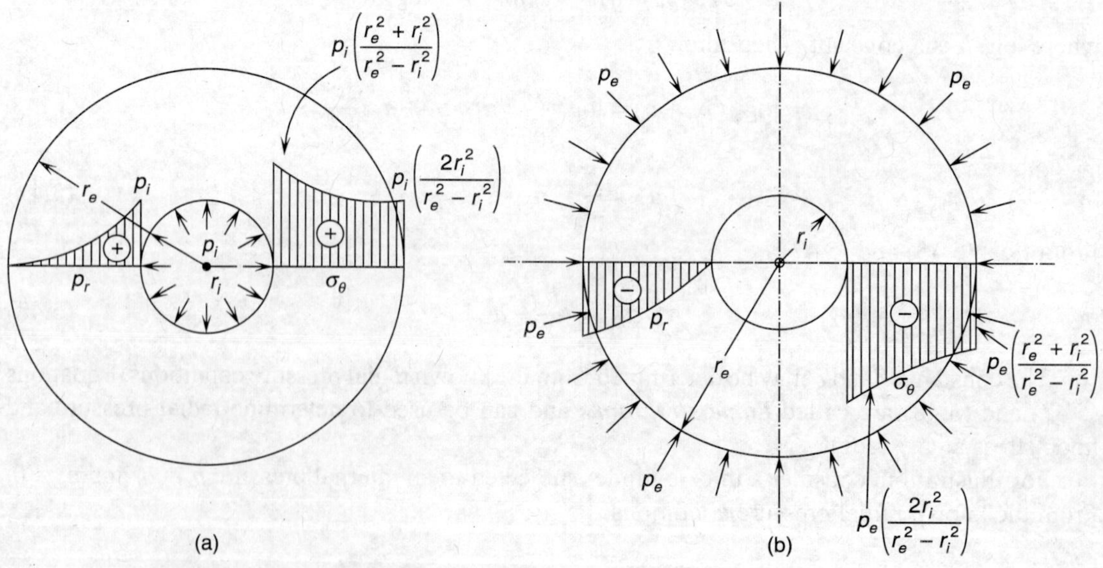

Figure 5.7 Distribution of stresses in a thick cylinder due to a pressure.

EXAMPLE 5.12

A pipe with external and internal diameters of 600 and 400 mm, respectively, contains a fluid at a pressure of 10 MPa. Determine the maximum and minimum hoop stresses across the section. Draw the radial pressure and hoop stress variations across the section.

Solution: For the given problem $r_e = 300$ mm and $r_i = 200$ mm:

Using Lame's relations: $p_r = \dfrac{b}{r^2} - a$

At $r = 300$, $p_r = 0$ i.e. $0 = \dfrac{b}{90000} - a$ (i)

At $r = 200$, $p_r = 10$ i.e. $10 = \dfrac{b}{40000} - a$ (ii)

Operate: Eq. (ii) – Eq. (i)

$$10 = \left(\frac{1}{40000} - \frac{1}{90000}\right)b \quad \text{or} \quad b = 720000$$

From (i) $a = \dfrac{b}{90000} = 8$

Therefore, $a = 8$ and $b = 720000$

Therefore, the hoop stress developed in the vessel at any radius r,

$$\sigma_{\theta,r} = \frac{b}{r^2} + a = \frac{720000}{r^2} + 8$$

The maximum hoop stress, $\sigma_{\theta,200} = \dfrac{720000}{40000} + 8 = 26$ MPa

The minimum hoop stress, $\sigma_{\theta,300} = \dfrac{720000}{90000} + 8 = 16$ MPa

Alternatively, from Eq. (5.53):

$$\sigma_{\theta,r} = p\left(\frac{r_i}{r}\right)^2 \left(\frac{r_e^2 + r^2}{r_e^2 - r_i^2}\right)$$

$$\sigma_{\theta,200} = 10 \times \left(\frac{200}{200}\right)^2 \left(\frac{300^2 + 200^2}{300^2 - 200^2}\right) = 26 \text{ MPa}$$

$$\sigma_{\theta,300} = 10 \times \left(\frac{200}{300}\right)^2 \left(\frac{300^2 + 300^2}{300^2 - 200^2}\right) = 16 \text{ MPa}$$

EXAMPLE 5.13

The thick cylindrical shell of a hydraulic press has to withstand an internal fluid pressure of 15 MPa. If the internal diameter of the cylinder is 150 mm, determine the thickness so that the maximum tensile stress in a section does not exceed 45 MPa.

Solution: Consider the thickness of shell to be t, then $r_i = 75$ mm and $r_e = 75 + t$

The maximum circumferential stress developed in the shell is,

$$\sigma_\theta = \sigma_a = p\left(\frac{k^2+1}{k^2-1}\right) \quad \text{where} \quad k = \frac{r_e}{r_i} = \frac{r_i + t}{r_i}$$

or
$$45 = 15 \times \left(\frac{k^2+1}{k^2-1}\right) \quad \text{i.e.} \quad 3k^2 - 3 = k^2 + 1$$

Therefore,
$$k = \sqrt{2} = 1.414$$

Thus,
$$t = (k-1)r_i = (\sqrt{2}-1) \times 75 = 31.066 \text{ mm}$$

The required thickness is 31.066 mm.

Alternatively using the fundamental relations for radial pressure and hoop stress, i.e. Lame's relations, are:

$$p_r = \frac{b}{r^2} - a \quad \text{and} \quad \sigma_\theta = \frac{b}{r^2} + a$$

For the internal fluid pressure and permissible stress conditions:

$$\text{At } r = 75 \text{ mm}, p_r = 15 \text{ MPa} \quad \text{i.e.} \quad 15 = \frac{b}{75^2} - a \qquad \text{(i)}$$

$$\text{At } r = 75 \text{ mm}, \sigma_\theta = 45 \text{ MPa} \quad \text{i.e.} \quad 45 = \frac{b}{75^2} + a \qquad \text{(ii)}$$

Operate: Eq. (i) + Eq. (ii)

$$\frac{2b}{75^2} = 60 \quad \text{or} \quad b = 30 \times 75^2 = 168750$$

Eq. (ii) – Eq. (i)

$$2a = 30 \quad \text{or} \quad a = 15$$

Thus,
$$a = 15 \quad \text{and} \quad b = 168750$$

For the condition,

$$\text{At } r = r_e, p_r = 0 \quad \text{i.e.} \quad 0 = \frac{b}{r_e^2} - a$$

Therefore,
$$r_e^2 = \frac{b}{a} = \frac{168750}{15} \quad \text{or} \quad r_e = 106.066 \text{ mm}$$

Thus,
$$t = t_e - r_i = 106.066 - 75 = 31.066 \text{ mm}$$

The required the thickness is 31.066 mm.

EXAMPLE 5.14

A cylinder of internal diameters d, wall thickness t is subjected to an internal pressure. If the cylinder is considered to be a thin cylinder, determine the maximum value for the ratio t/d such that the error in the estimated maximum hoop stress does not exceed: (i) 7% and (ii) 5%.

Solution: The hoop stress as a thin cylinder,

$$\sigma_\theta = \frac{pd}{2t} = \frac{pr_i}{t}$$

Whereas, the hoop stress as a thick cylinder is given by:

$$\sigma_{\theta,r} = p\left(\frac{r_i}{r}\right)^2\left(\frac{r_e^2 + r^2}{r_e^2 - r_i^2}\right)$$

The maximum stress occurs at $r = r_i$

$$\sigma_{\theta,\max} = p\left(\frac{r_i}{r_i}\right)^2\left(\frac{r_e^2 + r_i^2}{r_e^2 - r_i^2}\right) = p\left(\frac{r_e^2 + r_i^2}{r_e^2 - r_i^2}\right)$$

The percentage error in treating the cylinder a thin cylinder,

$$\lambda = \frac{\sigma_{\theta,\max} - \sigma_\theta}{\sigma_{\theta,\max}} = 1 - \frac{\sigma_\theta}{\sigma_{\theta,\max}} \quad \text{or} \quad \frac{\sigma_\theta}{\sigma_{\theta,\max}} = 1 - \lambda$$

Therefore, $$1 - \lambda = \left(\frac{pr_i}{t}\right)\left[\frac{r_e^2 - r_i^2}{p(r_e^2 + r_i^2)}\right] = \left(\frac{r_i}{t}\right)\frac{(r_i + t)^2 - r_i^2}{(r_i + t)^2 + r_i^2}$$

or $$1 - \lambda = \frac{r_i}{t} \times \frac{2tr_i + t^2}{2r_i^2 + 2tr_i + t^2} = \frac{2r_i^2 + tr_i}{2r_i^2 + 2tr_i + t^2}$$

$$= \frac{2 + x}{2 + 2x + x^2} \quad \text{where} \quad x = \frac{t}{r_i}$$

Substitute $1 - \lambda = \lambda_1$, then

$$\lambda_1 = \frac{2 + x}{2 + 2x + x^2} \quad \text{or} \quad x^2 + \left(2 - \frac{1}{\lambda_1}\right)x + 2\left(1 - \frac{1}{\lambda_1}\right) = 0$$

(i) For $\lambda = 0.07$, $\lambda_1 = 1 - 0.07 = 0.93$

Therefore, $$x^2 + 0.925x - 0.151 = 0$$

$$x = -\frac{0.925}{2} + \sqrt{\left(\frac{0.925}{2}\right)^2 + 0.151} = 0.14157$$

Thus, $$\frac{t}{d} = \frac{1}{2}\left(\frac{t}{r_i}\right) = \frac{x}{2} = 0.07079$$

(ii) For $\lambda = 0.05$, $\lambda_1 = 1 - 0.05 = 0.95$, and $x^2 + 0.947x - 0.105 = 0$

Therefore, $x = 0.10026$ and $\frac{t}{d} = \frac{x}{2} = 0.0501$

EXAMPLE 5.15

A thick hollow cylinder with internal and external diameters of 80 and 120 mm, respectively, is subjected to an external pressure of 40 MPa when the internal pressure is 120 MPa. Determine: (i) the circumferential stress at the external and internal surfaces and (ii) the radial and circumferential stresses at the mean radius. Draw the radial pressure and hoop stress variations across the section.

Solution: For the given problem $r_e = 60$ mm and $r_i = 40$ mm:

Using Lame's relations: $p_r = \dfrac{b}{r^2} - a$

$$\text{At } r = 60,\ p_r = 40 \quad \text{i.e.} \quad 40 = \frac{b}{3600} - a \qquad \text{(i)}$$

$$\text{At } r = 40,\ p_r = 120 \quad \text{i.e.} \quad 120 = \frac{b}{1600} - a \qquad \text{(ii)}$$

Operate: Eq. (ii) – Eq. (i)

$$80 = \left(\frac{1}{1600} - \frac{1}{3600}\right)b \quad \text{or} \quad b = 230400$$

From Eq. (i)

$$a = \frac{b}{3600} - 40 = \frac{230400}{3600} - 40 = 24$$

Therefore, the hoop stress developed in the vessel at any radius r,

$$\sigma_{\theta,r} = \frac{b}{r^2} + a = \frac{230400}{r^2} + 24$$

(i) The circumferential stress

$$\text{At } r = 40 \qquad \sigma_{\theta,40} = \frac{230400}{1600} + 24 = 168 \text{ MPa}$$

$$\text{At } r = 60 \qquad \sigma_{\theta,60} = \frac{230400}{3600} + 24 = 88 \text{ MPa}$$

(ii) The circumferential and radial stresses at the mean radius

$$\text{At } r = 50 \qquad \sigma_{\theta,50} = \frac{230400}{2500} + 24 = 116.16 \text{ MPa}$$

$$p_{r,50} = \frac{230400}{r^2} - 24 = \frac{230400}{2500} - 24 = 68.16 \text{ MPa}$$

The distribution of stresses can be drawn as shown in Figures 5.7(a) and (b).

EXAMPLE 5.16

On the outer surface of a closed thick cylinder of external-to-internal diameters ratio of 2.5, were fixed strain gauges to measure the circumferential and longitudinal strains. At an internal pressure of 230 MPa, the longitudinal and circumferential strains were recorded as 9.18 × 10^{-6} and 369 × 10^{-6}, respectively. Determine the modulus of rigidity and Poisson's ratio.

Solution: For the given problem,

$$\frac{r_e}{r_i} = 2.5;\ p_i = 230\ \text{MPa};\ \varepsilon_l = 9.18 \times 10^{-6}\ \text{ and }\ \varepsilon_\theta = 369 \times 10^{-6}$$

On the outer surface of the cylinder under internal pressure, $p_r = 0$ and

$$\sigma_\theta = p_i\left(\frac{r_i}{r_e}\right)^2\left(\frac{r_e^2 + r_e^2}{r_e^2 - r_i^2}\right) = \frac{2pr_i^2}{(r_e^2 - r_i^2)} = \frac{2 \times 230}{(2.5^2 - 1)}$$

$$= 87.62\ \text{MPa}$$

$$\sigma_l = \frac{p_i r_i^2}{(r_e^2 - r_i^2)} = \frac{230}{(2.5^2 - 1)} = 43.81\ \text{MPa}$$

The strains in the two directions are,

$$\varepsilon_\theta = \frac{1}{E}(\sigma_\theta - \nu\sigma_l)\ \text{ and }\ \varepsilon_l = \frac{1}{E}(\sigma_l - \nu\sigma_\theta)$$

Therefore,

$$\frac{\varepsilon_\theta}{\varepsilon_l} = \left(\frac{\sigma_\theta - \nu\sigma_l}{\sigma_l - \nu\sigma_\theta}\right) = \frac{2 - \nu}{1 - 2\nu}\quad \text{since}\quad \sigma_\theta = 2\sigma_l$$

or

$$\frac{369 \times 10^{-6}}{9.18 \times 10^{-6}}(= 40.196) = \frac{2 - \nu}{1 - 2\nu}\quad \text{i.e.}\quad 79.392\nu = 38.196$$

Therefore, $\nu = 0.481$

$$\varepsilon_\theta = \frac{1}{E}(\sigma_\theta - \nu\sigma_l) = \frac{\sigma_l}{E}(2 - \nu)\quad \text{or}\quad E = \frac{\sigma_l}{\varepsilon_\theta}(2 - \nu)$$

Thus,

$$E = \frac{43.81}{369 \times 10^{-6}}(2 - 0.481) = 180.333 \times 10^3\ \text{MPa} = 180.333\ \text{GPa}$$

$$G = \frac{E}{2(1 + \nu)} = \frac{180.333}{2 \times (1 + 0.481)} = 60.886\ \text{GPa}$$

The modulus of rigidity and Poisson's ratio are 180.333 and 60.886 GPa, respectively.

5.7 SHRINK-FITTED THIN CYLINDERS

In a thick cylinder subjected to an internal pressure, the maximum hoop stress which occurs on the inner surface determines the intensity of internal pressure so as the permissible tensile stress for the material is not exceeded. In order to reduce stresses inside a thick cylinder subjected to a high internal pressure, as in the case of guns, compound cylinders are commonly used.

The compound cylinder is formed by shrinking a cylindrical jacket, also called tube or tyre, onto another cylinder/tube which initially introduces hoop compressive stresses in the inner tube and hoop tensile stresses in the outer tube. In the case of two cylinders, the outside cylinder has an internal diameter slightly less than the external diameter of the inside cylinder, so that when the cylinder is shrunk on, the required radial pressure p_s is developed at the interface. The outside cylinder is expanded by heating and the two cylinders are assembled together. When the compound cylinder is subjected to an internal pressure, both the inner and outer tubes are subjected to hoop tensile stress due to internal pressure alone. The final hoop stresses in both the tubes are determined by adding the stresses induced during shrink-fit and the stresses developed due to internal pressure alone. Thus this arrangement can be used to make the hoop stresses more or less uniform throughout the metal.

Consider a compound cylinder with outer, inner and interface radii of r_e, r_i and r_s, respectively, formed by shrink-fitting a cylindrical jacket onto another cylinder which induces shrinkage pressure at the interface of intensity p_s. The Lame's relations developed for computing radial pressure and hoop stress induced in a homogeneous vessel at any radius are also applicable to a compound cylinder,

$$p_r = \frac{b_1}{r^2} - a_1 \quad \text{and} \quad \sigma_\theta = \frac{b_1}{r^2} + a_1 \tag{5.56}$$

For the outer tube:

$$\text{At } r = r_e,\ p_r = 0 \qquad \text{i.e.} \qquad 0 = \frac{b_1}{r_e^2} - a_1 \tag{5.57}$$

$$\text{At } r = r_s,\ p_r = p_s \qquad \text{i.e.} \qquad p_s = \frac{b_1}{r_s^2} - a_1 \tag{5.58}$$

The constants a_1 and b_1 as evaluated from the above equation are

$$a_1 = p_s\left(\frac{r_s^2}{r_e^2 - r_s^2}\right) \quad \text{and} \quad b_1 = p_s\left(\frac{r_e^2 r_s^2}{r_e^2 - r_s^2}\right) \tag{5.59}$$

Therefore, the hoop stress developed in the jacket of the vessel at any radius r,

$$\sigma_\theta' = \frac{b_1}{r^2} + a_1 = p_s\left(\frac{r_s^2}{r_e^2 - r_s^2}\right)\left(\frac{r_e^2}{r^2} + 1\right) = p_s\left(\frac{r_s}{r}\right)^2\left(\frac{r_e^2 + r^2}{r_e^2 - r_s^2}\right) \tag{5.60}$$

For the inner tube: The Lame's relations are

$$p_r = \frac{b_2}{r^2} - a_2 \qquad \text{and} \qquad \sigma_\theta = \frac{b_2}{r^2} + a_2$$

At $r = r_i$, $p_r = 0$ i.e. $0 = \dfrac{b_2}{r_i^2} - a_2$

At $r = r_s$, $p_r = -p_s$ i.e. $-p_s = \dfrac{b_2}{r_s^2} - a_2$

Thus, the constants a_2 and b_2 are

$$a_2 = -p_s\left(\frac{r_s^2}{r_i^2 - r_s^2}\right) \quad \text{and} \quad b_2 = -p_s\left(\frac{r_i^2\, r_s^2}{r_i^2 - r_s^2}\right) \tag{5.61}$$

Therefore, the hoop stress in the inner part of the vessel at any radius r,

$$\sigma_\theta' = \frac{b_2}{r^2} + a_2 = -p_s\left(\frac{r_s^2}{r_s^2 - r_i^2}\right)\left(\frac{r_i^2}{r^2} + 1\right) = -p_s\left(\frac{r_s}{r}\right)^2\left(\frac{r_i^2 + r^2}{r_s^2 - r_i^2}\right) \tag{5.62}$$

For, the case when the compound cylinder is subjected to an internal pressure p alone the vessel can be treated as a thick cylinder, and the stresses can be determined using the Lame's relations:

$$p_r = \frac{b}{r^2} - a \quad \text{and} \quad \sigma_\theta = \frac{b}{r^2} + a$$

At $r = r_e$, $p_r = 0$ i.e. $0 = \dfrac{b}{r_e^2} - a$

At $r = r_i$, $p_r = p$ i.e. $p = \dfrac{b}{r_i^2} - a$

The constants a and b are

$$a = p\left(\frac{r_i^2}{r_e^2 - r_i^2}\right) \quad \text{and} \quad b = p\left(\frac{r_e^2\, r_i^2}{r_e^2 - r_i^2}\right) \tag{5.63}$$

Therefore, the hoop stress in the vessel at a radius r,

$$\sigma_\theta'' = \frac{b}{r^2} + a = p\left(\frac{r_i^2}{r_e^2 - r_i^2}\right)\left(\frac{r_e^2}{r^2} + 1\right) = p\left(\frac{r_i}{r}\right)^2\left(\frac{r_e^2 + r^2}{r_e^2 - r_i^2}\right) \tag{5.64}$$

The final hoop stresses are determined by adding the stresses induced due to shrink-fit and internal pressure. The maximum circumferential stress in the outer tube is

$$\sigma_{\theta,s} = \sigma'_{\theta,s} + \sigma''_{\theta,s} \tag{5.65}$$

The maximum circumferential stress in the inner tube

$$\sigma_{\theta,i} = \sigma'_{\theta,i} + \sigma''_{\theta,i} \tag{5.66}$$

The distribution of stresses in a compound cylinder is shown in Figure 5.8(b). Unlike the stress distribution in a thick cylinder shown in Figure 5.8(a), the stress in a compound cylinder is discontinuous at the interface.

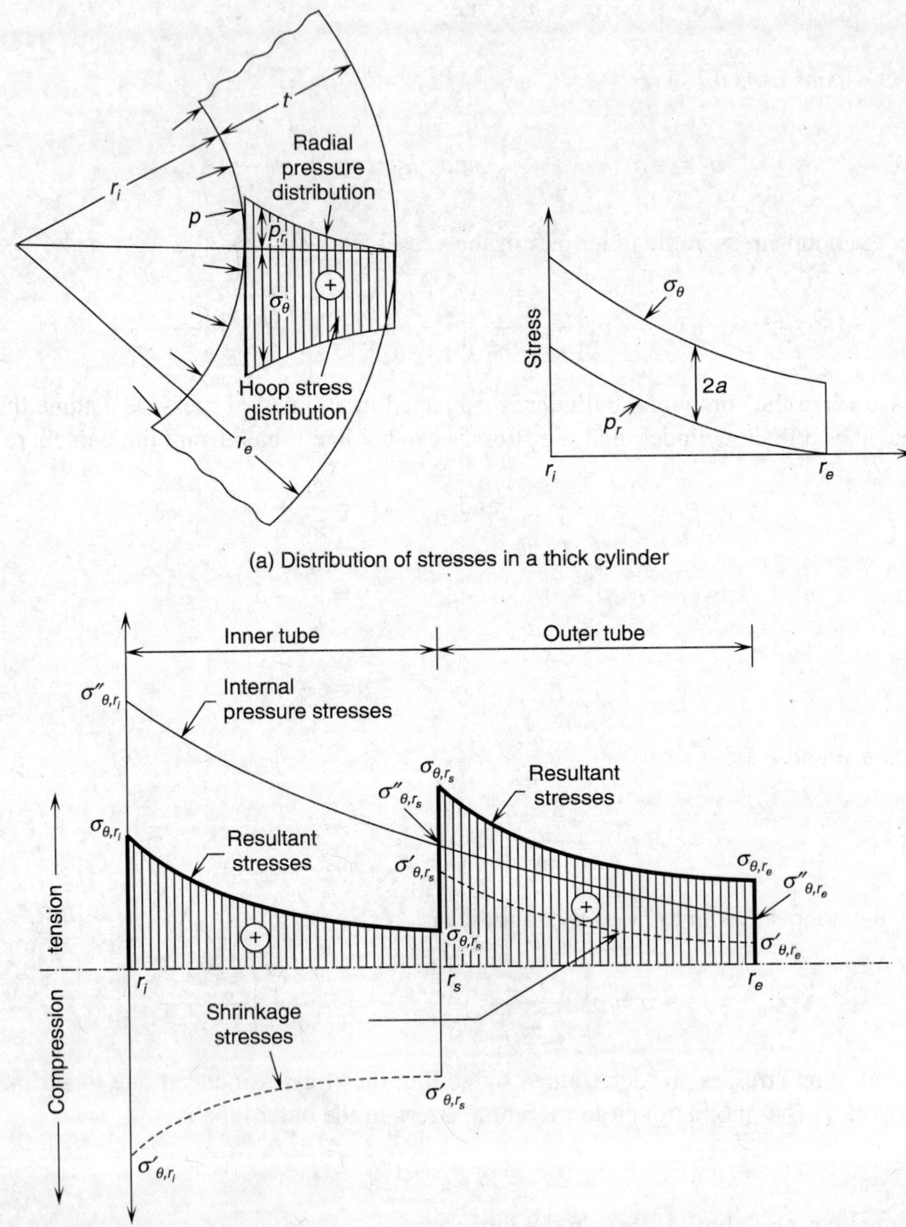

(a) Distribution of stresses in a thick cylinder

(b) Distribution of hoop stress in a compound cylinder

Figure 5.8 Distribution of stresses in cylinders.

5.7.1 Initial Difference in Radii of Two Tubes at the Interface

In order to obtain the pre-stressing of the inner tube by shrinking on the outer tube or jacket over the inner tube, it is necessary that the diameter of the inner face of the outer tube should be slightly less than the diameter of the outer face of the inner tube, so that after the shrink-fit of the tubes the required radial pressure at the interface of the tubes is attained. The original difference of radii of cylinders at the interface is generally called *shrinkage allowance*. Thus for a compound cylinder with interface radius r_s, the shrinkage allowance is the sum of differences of interface radius, and original outer and inner radii of inner and external cylinders, respectively.

$$\delta r_s = \delta r_{s,e} - (-\delta r_{s,i}) = \delta r_{s,e} + \delta r_{s,i}$$

or

$$\frac{\delta r_s}{r_s} = \frac{\delta r_{s,e}}{r_s} + \frac{\delta r_{s,i}}{r_s} \tag{5.67}$$

Shrinkage allowance

Pressures and strains in the inner and outer tubes at the interface

For the inner cylinder, the tangential or circumferential strain at the interface radius r_s, is given by

$$\frac{\delta r_{s,i}}{r_s} = \frac{1}{E_i}(\sigma'_{\theta,s} - \nu_i p_s)$$

where

$$\sigma'_{\theta,s} = -p_s\left(\frac{r_s}{r_s}\right)^2\left(\frac{r_i^2 + r_s^2}{r_s^2 - r_i^2}\right) = -p_s\left(\frac{r_i^2 + r_s^2}{r_s^2 - r_i^2}\right) \text{ (compressive)}$$

Therefore, the tangential strain at interface in the inner tube

$$\frac{\delta r_{s,i}}{r_s} = -\frac{p_s}{E_i}\left[\left(\frac{r_i^2 + r_s^2}{r_s^2 - r_i^2}\right) + \nu_i\right] \tag{5.68}$$

Similarly, for outer cylinder or jacket, the circumferential strain at the interface radius r_s, is given by

$$\frac{\delta r_{s,e}}{r_s} = \frac{p_s}{E_e}\left[\left(\frac{r_e^2 + r_s^2}{r_e^2 - r_s^2}\right) - \nu_e\right] \text{ (tensile)} \tag{5.69}$$

Therefore, the shrink allowance or the resultant strain at the interface is

$$\frac{\delta r_s}{r_s} = \frac{\delta r_{s,e}}{r_s} + \frac{\delta r_{s,i}}{r_s}$$

$$= p_s\left[\frac{1}{E_i}\left\{\left(\frac{r_i^2 + r_s^2}{r_s^2 - r_i^2}\right) + \nu_i\right\} + \frac{1}{E_e}\left\{\left(\frac{r_e^2 + r_s^2}{r_e^2 - r_s^2}\right) - \nu_e\right\}\right] \tag{5.70}$$

When the cylinders are made of same materials, i.e. $E_i = E_e = E$ and $\nu_i = \nu_e = \nu$

$$\frac{\delta r_s}{r_s} = \frac{p_s}{E}\left[\left(\frac{r_i^2 + r_s^2}{r_s^2 - r_i^2}\right) + \left(\frac{r_e^2 + r_s^2}{r_e^2 - r_s^2}\right)\right] \tag{5.71}$$

Therefore, $$\alpha\,\delta T = \frac{\delta r_s}{r_s} \quad \text{or} \quad \delta T = \frac{\delta r_s}{\alpha r_s} \tag{5.72}$$

where α and δT are the coefficient linear thermal expansion and rise in temperature, respectively.

EXAMPLE 5.17

A compound tube is formed by shrinking a 25 mm thick jacket with internal diameter of 250 mm onto a tube which has external and internal diameters of 250 mm and 200 mm, respectively. The radial pressure at the interface is 8 MPa. Determine the variation of circumferential stress over the section of compound tube when it is subjected to an internal pressure of 84.5 MPa.

Solution: For the given problem,

$$r_i = 100 \text{ mm},\ r_s = 125 \text{ mm},\ r_e = 150 \text{ mm},\ p_s = 8 \text{ MPa and } p = 84.5 \text{ MPa}$$

Circumferential stresses due to shrinking the jacket:

Inner tube:

$$\sigma_\theta' = -p_s\left(\frac{r_s}{r}\right)^2\left(\frac{r_i^2 + r^2}{r_s^2 - r_i^2}\right)$$

At $r = r_i$ $$\sigma'_{\theta,i} = -8 \times \left(\frac{125}{100}\right)^2 \times \left(\frac{100^2 + 100^2}{125^2 - 100^2}\right) = -44.44 \text{ MPa (compression)}$$

At $r = r_s$ $$\sigma'_{\theta,s} = -8 \times \left(\frac{125}{125}\right)^2 \times \left(\frac{100^2 + 125^2}{125^2 - 100^2}\right) = -36.44 \text{ MPa (compression)}$$

Outer tube:

$$\sigma_\theta' = p_s\left(\frac{r_s}{r}\right)^2\left(\frac{r_e^2 + r^2}{r_e^2 - r_s^2}\right)$$

At $r = r_s$ $$\sigma'_{\theta,s} = 8 \times \left(\frac{125}{125}\right)^2 \times \left(\frac{150^2 + 125^2}{150^2 - 125^2}\right) = 44.36 \text{ MPa}_s \text{ (tension)}$$

At $r = r_e$ $$\sigma'_{\theta,e} = 8 \times \left(\frac{125}{150}\right)^2 \times \left(\frac{150^2 + 150^2}{150^2 - 125^2}\right) = 36.36 \text{ MPa}_s \text{ (tension)}$$

Stresses due to internal pressure:

$$\sigma_\theta'' = p\left(\frac{r_i}{r}\right)^2\left(\frac{r_e^2 + r^2}{r_e^2 - r_i^2}\right)$$

At $r = r_i$ $$\sigma''_{\theta,i} = 84.5 \times \left(\frac{150^2 + 100^2}{150^2 - 100^2}\right) = 219.70 \text{ MPa (tension)}$$

At $r = r_s$ $$\sigma''_{\theta,s} = 84.5 \times \left(\frac{100}{125}\right)^2 \times \left(\frac{150^2 + 125^2}{150^2 - 100^2}\right) = 164.94 \text{ MPa (tension)}$$

At $r = r_e$ $$\sigma''_{\theta,e} = 84.5 \times \left(\frac{100}{150}\right)^2 \times \left(\frac{150^2 + 150^2}{150^2 - 100^2}\right) = 135.2 \text{ MPa (tension)}$$

The combined stresses due to shrinking the outer tube onto inner tube and internal pressure:

$$\sigma_\theta = \sigma_\theta' + \sigma_\theta''$$

Inner tube:

$$\sigma_{\theta,i} = -44.44 + 219.70 = 175.26 \text{ MPa (tension)}$$

$$\sigma_{\theta,s} = -36.44 + 164.94 = 128.50 \text{ MPa (tension)}$$

Outer tube:

$$\sigma_{\theta,s} = 44.36 + 164.94 = 209.30 \text{ MPa (tension)}$$

$$\sigma_{\theta,e} = 36.36 + 135.20 = 171.36 \text{ MPa (tension)}$$

EXAMPLE 5.18

A steel shaft of 75 mm diameter has an aluminium disc of 300 mm outside diameter shrunk on it. The shrink allowance is 1 part per 1000. Determine tangential and radial stresses at the interface due to shrink fit. The modulus of elasticity for steel and aluminium are 210 GPa and 71 GPa, respectively. The Poisson's ratio for both the metals is 0.3.

Solution: Consider shrinkage pressure at the interface to be p_s, then the pressures and strains in the shaft and disc at the interface

Shaft:

$$\sigma_\theta = \sigma_r = -p_s$$

$$\varepsilon_{\theta,s} = -\frac{p_s}{E_s}(1 - \nu_s) \text{ (compressive)}$$

Disc:

The radial and tangential stresses in the disc,

$$\sigma_r = p_s \text{ at } r = 37.5 \text{ mm}$$

$$\sigma_\theta = p_s\left(\frac{k^2+1}{k^2-1}\right) \quad \text{where} \quad k = \frac{r_e}{r_s} = \frac{150}{37.5} = 4$$

$$\sigma_\theta = \left(\frac{17}{15}\right) p_s \text{ (tension)}$$

The tangential strain in the disc

$$\varepsilon_{\theta,d} = \frac{1}{E_a}\left(\frac{17}{15}p_s + \nu_a p_s\right) = \frac{p_s}{E_a}\left(\frac{17}{15} + \nu_a\right) \text{ (tensile)}$$

Resultant strain at the interface

$$\varepsilon_\theta = \frac{p_s}{E_a}\left(\frac{17}{15} + \nu_a\right) - \left[-\frac{p_s}{E_s}(1-\nu_s)\right]$$

where,

$$\nu_a = \nu_s = 0.3 \quad \text{and} \quad \varepsilon_\theta = \frac{1}{1000}$$

Therefore, $$\frac{1}{1000} = \frac{p_s}{(71\times10^9)} \times \left(\frac{17}{15} + 0.3\right) + \frac{p_s}{(210\times10^9)} \times (1-0.3)$$

or $$p_s = 42.515 \text{ MPa}$$

The tangential and radial stresses at the interface due to shrink fit are 42.515 MPa.

EXAMPLE 5.19

A tyre is shrunk on a wheel of 12 m diameter. Considering the wheel to be rigid, determine the internal diameter of the tyre, if after shrinking the hoop stress in the tyre is 120 MPa. The modulus of elasticity and the coefficient of thermal expansion for steel are 200 GPa and $\alpha = 11.7 \times 10^{-6}$ per °C, respectively. Also determine the least temperature to which the tyre must be heated so that it could be fitted on the wheel.

Solution: For the given problem,

$$r_s = 6 \text{ m}, (\sigma_\theta)_{\text{tyre}} = 120 \text{ MPa}, E = 200 \text{ GPa and } \alpha = 11.7 \times 10^{-6} \text{ per °C}$$

For the tyre

$$\varepsilon_\theta = \frac{\delta r_s}{r_s} = \frac{1}{E}(\sigma_\theta - \nu\sigma_r) \quad \text{where } \sigma_r = 0 \text{ for a thin tyre}$$

Therefore, $$\frac{\delta r_s}{r_s} = \frac{\sigma_\theta}{E} = \frac{120}{(200\times10^3)} = 0.6\times10^{-3}$$

Thus, $$\delta r_s = r_s\left(\frac{\sigma_\theta}{E}\right) = 6\times(0.6\times10^{-3}) = 3.6\times10^{-3} \text{ m}$$

or $$\delta d_s = 2\delta r_s = 7.2 \times 10^{-3} \text{ m}$$

Therefore, internal diameter of the tyre

$$d_s = d_w - \delta d_s = 12 - 7.2 \times 10^{-3} = 11.9928 \text{ m}$$

The temperature δT to which the tyre must be heated is given by:

$$\alpha\,\delta T = \frac{\delta r_s}{r_s} = 0.6 \times 10^{-3} \quad \text{or} \quad \delta T = \frac{0.6 \times 10^{-3}}{11.7 \times 10^{-6}} = 51.282\ ^\circ\text{C}$$

EXAMPLE 5.20

A compound cylinder is formed by shrinking a steel jacket (tube) with external diameter of 200 mm onto a steel tube which has external and internal diameters of 150 mm and 100 mm, respectively. When the compound cylinder is subjected to an internal pressure of 35 MPa, the maximum circumferential stress in both the tubes is same. If the modulus of elasticity of steel is 200 GPa, determine the maximum circumferential stress developed at the inner surface of the jacket.

Solution: For the given problem,

$$r_i = 50 \text{ mm},\ r_s = 75 \text{ mm},\ r_e = 100 \text{ mm},\ p = 35 \text{ MPa and } E = 200 \text{ GPa}$$

Consider shrinkage pressure at the interface to be p_s.
Stresses due to shrinking the jacket:

Inner cylinder:

At $r = r_i$

$$\sigma'_{\theta,i} = -2p_s\left(\frac{r_s^2}{r_s^2 - r_i^2}\right)$$

$$= -2 \times p_s\left(\frac{1}{1-(50/75)^2}\right) = -3.6p_s \text{ (compression)}$$

Outer cylinder:

At $r = r_s$

$$\sigma'_{\theta,s} = p_s\left(\frac{r_e^2 + r_s^2}{r_e^2 - r_s^2}\right) = p_s\left(\frac{1+(75/100)^2}{1-(75/100)^2}\right)$$

$$= 3.57p_s \text{ (tension)}$$

Stresses due to internal pressure:

At $r = r_i$

$$\sigma''_{\theta,i} = p\left(\frac{r_e^2 + r_i^2}{r_e^2 - r_i^2}\right) = 35 \times \left(\frac{1+(50/100)^2}{1-(50/100)^2}\right) = 35 \times \left(\frac{5}{3}\right)$$

$$= 58.33 \text{ MPa (tension)}$$

At $r = r_s$

$$\sigma''_{\theta,s} = p\left(\frac{r_i}{r_s}\right)^2\left(\frac{r_e^2 + r_s^2}{r_e^2 - r_i^2}\right) = 35 \times \left(\frac{50}{75}\right)^2 \times \left(\frac{1+(75/100)^2}{1-(50/100)^2}\right)$$

$$= 32.41 \text{ MPa (tension)}$$

The maximum circumferential stress in the outer tube

$$\sigma_{\theta,s} = \sigma'_{\theta,s} + \sigma''_{\theta,s} = 3.57p_s + 32.41$$

The maximum circumferential stress in the inner tube

$$\sigma_{\theta,i} = \sigma'_{\theta,i} + \sigma''_{\theta,i} = -3.6p_s + 58.33$$

Since the maximum circumferential stress in both the tubes is same

$$3.57p_s + 32.41 = -3.6p_s + 58.33 \quad \text{or} \quad 7.17p_s = 25.92$$

Therefore, $p_s = 3.615$ MPa

The maximum circumferential stress at the internal face of outer tube,

$$(\sigma_{\theta,i})_{\max} = 3.57 \times 3.615 + 32.41 = 45.316 \text{ MPa}$$

EXAMPLE 5.21

A steel tyre of 20 mm thickness is shrunk on a cast iron rim of 480 mm outside diameter and 60 mm thickness. Determine the internal diameter of the tyre, if after shrinking on, the tyre exerts a radial pressure of 50 MPa on the cast iron rim. The modulus of elasticity for the steel and cast iron are 210 GPa and 100 GPa, respectively. The Poisson's ratio for both the materials is 0.25.

Solution: For the given problem,

$r_i = 180$ mm, $r_s = 240$ mm, $r_e = 260$ mm, $p_s = 50$ MPa, $E_s = 210$ GPa, $E_c = 100$ GPa and $\nu = 0.25$

From Eq. (5.70), the shrink allowance or the resultant strain at the interface is

$$\frac{\delta r_s}{r_s} = 50 \times \left[\frac{1}{(100 \times 10^3)} \times \left\{ \left(\frac{180^2 + 240^2}{240^2 - 180^2} \right) - 0.25 \right\} + \frac{1}{(210 \times 10^3)} \left\{ \left(\frac{260^2 + 240^2}{260^2 - 240^2} \right) + 0.25 \right\} \right]$$

$$= \frac{50}{100 \times 10^3} \times \left[\left\{ \left(\frac{180^2 + 240^2}{240^2 - 180^2} \right) - 0.25 \right\} + \frac{1}{2.1} \left\{ \left(\frac{260^2 + 240^2}{260^2 - 240^2} \right) + 0.25 \right\} \right]$$

$$= \frac{50}{100 \times 10^3} \times \left[\{3.571429 - 0.25\} + \frac{1}{2.1} \{12.52 - 0.25\} \right] = 4.7012 \times 10^{-3}$$

Therefore,

$$\frac{\delta r_s}{r_s} = 4.7012 \times 10^{-3} \quad \text{or} \quad \delta r_s = 240 \times 4.7012 \times 10^{-3} = 1.1283 \text{ mm}$$

Therefore, internal diameter of the steel tyre

$$= 2 \times (240 - 1.1283) = 477.74 \text{ mm}$$

5.8 PROPORTIONING OF COMPOUND CYLINDERS

For an economical design, the proportions of a compound cylinder are so fixed that the allowable tensile stresses for the materials of two cylinders reach simultaneously under the internal pressure. Considering the *maximum principal stress theory* expressed as:

$$\tau_{\max} = \sigma_1 - \sigma_3 \quad \text{where } \sigma_1 = \sigma_\theta \text{ and } \sigma_3 = p_r$$

The condition of equal strength of the cylinders can be expressed as,

$$\sigma_{\max,i} = \sigma_{\max,e}$$

The maximum circumferential stress in the outer tube occurs at $r = r_s$

$$\sigma_{\theta,s} = \sigma'_{\theta,s} + \sigma''_{\theta,s} = p_s\left(\frac{r_e^2 + r_s^2}{r_e^2 - r_s^2}\right) + p\left(\frac{r_i}{r_s}\right)^2\left(\frac{r_e^2 + r_s^2}{r_e^2 - r_i^2}\right) \tag{5.73}$$

The maximum circumferential stress in the inner tube occurs at $r = r_i$

$$\sigma_{\theta,i} = \sigma'_{\theta,i} + \sigma''_{\theta,i} = -\,p_s\left(\frac{r_s}{r_i}\right)^2\left(\frac{r_i^2 + r_i^2}{r_s^2 - r_i^2}\right) + p\left(\frac{r_e^2 + r_i^2}{r_e^2 - r_i^2}\right) \tag{5.74}$$

Therefore, the maximum principal stresses are

$$\sigma_{\max,e} = p_s\left(\frac{r_e^2 + r_s^2}{r_e^2 - r_s^2}\right) + p\left(\frac{r_i}{r_s}\right)^2\left(\frac{r_e^2 + r_s^2}{r_e^2 - r_i^2}\right) - (-p_s) \tag{5.75}$$

$$\sigma_{\max,i} = -p_s\left(\frac{r_s}{r_i}\right)^2\left(\frac{r_i^2 + r_i^2}{r_s^2 - r_i^2}\right) + p\left(\frac{r_e^2 + r_i^2}{r_e^2 - r_i^2}\right) - (-p) \tag{5.76}$$

Consider both the cylinders to be of same material, and hence the maximum principal stresses in both the cylinders must be the same. Thus, equate the above expressions

$$p_s\left[\left(\frac{r_e^2 + r_s^2}{r_e^2 - r_s^2}\right) + \left(\frac{r_s}{r_i}\right)^2\left(\frac{r_i^2 + r_i^2}{r_s^2 - r_i^2}\right) + 1\right] = p\left[-\left(\frac{r_i}{r_s}\right)^2\left(\frac{r_e^2 + r_s^2}{r_e^2 - r_i^2}\right) + \left(\frac{r_e^2 + r_i^2}{r_e^2 - r_i^2}\right) + 1\right]$$

$$\text{L.H.S.} = 2p_s\left[\frac{r_s^2\,(r_e^2 - r_s^2) + r_e^2\,(r_s^2 - r_i^2)}{(r_s^2 - r_i^2)(r_e^2 - r_s^2)}\right] = 2p_s\left[\frac{r_s^2}{(r_s^2 - r_i^2)} + \frac{r_e^2}{(r_e^2 - r_s^2)}\right]$$

$$\text{R.H.S.} = 2p\left[\frac{r_e^2\,(r_s^2 - r_i^2)}{r_s^2\,(r_e^2 - r_i^2)}\right]$$

Therefore,

$$2p_s\left[\frac{r_s^2}{(r_s^2 - r_i^2)} + \frac{r_e^2}{(r_e^2 - r_s^2)}\right] = 2p\left[\frac{r_e^2\,(r_s^2 - r_i^2)}{r_s^2\,(r_e^2 - r_i^2)}\right]$$

or
$$p_s = p\left[\frac{r_e^2\,(r_s^2 - r_i^2)}{r_s^2\,(r_e^2 - r_i^2)}\right]\left[\frac{r_s^2}{(r_s^2 - r_i^2)} + \frac{r_e^2}{(r_e^2 - r_s^2)}\right]^{-1}$$

Therefore,

$$\sigma_{\max,i} = -p_s\left(\frac{r_s}{r_i}\right)^2\left(\frac{r_i^2 + r_i^2}{r_s^2 - r_i^2}\right) + p\left(\frac{r_e^2 + r_i^2}{r_e^2 - r_i^2}\right) + p$$

$$= -p_s\left(\frac{r_s}{r_i}\right)^2\left(\frac{r_i^2 + r_i^2}{r_s^2 - r_i^2}\right) + p\left(\frac{2r_e^2}{r_e^2 - r_i^2}\right)$$

$$= -p\left[\frac{2r_e^2}{(r_e^2 - r_i^2)}\right]\left[\frac{r_s^2}{(r_s^2 - r_i^2)} + \frac{r_e^2}{(r_e^2 - r_s^2)}\right]^{-1} + p\left(\frac{2r_e^2}{r_e^2 - r_i^2}\right)$$

Thus
$$\sigma_{\max,i} = p\left(\frac{2r_e^2}{r_e^2 - r_i^2}\right)\left[1 - \left(\frac{r_e^2}{r_e^2 - r_s^2} + \frac{r_s^2}{r_s^2 - r_i^2}\right)^{-1}\right] \tag{5.77}$$

For this quantity to be minimum

$$r_s = \sqrt{r_e r_i} \tag{5.78}$$

The corresponding lowest value of maximum principal stress,

$$(\sigma_{\max,e})_{\min} = p\left(\frac{r_e}{r_e - r_i}\right) \tag{5.79}$$

For design,
$$\sigma_a = (\sigma_{\max,e})_{\min} = p\left(\frac{r_e}{r_e - r_i}\right) \tag{5.80}$$

For the cylinders made of same materials, the required radial negative or shrinkage allowance is

$$\frac{\delta r_s}{r_s} = \frac{p_s}{E}\left[\left(\frac{r_i^2 + r_s^2}{r_s^2 - r_i^2}\right) + \left(\frac{r_e^2 + r_s^2}{r_e^2 - r_s^2}\right)\right]$$

$$\frac{\delta r_s}{r_s} = \frac{2p}{E}\left[\frac{r_e^2\,(r_s^2 - r_i^2)}{r_e^2\,(r_s^2 - r_i^2) + r_s^2\,(r_e^2 - r_s^2)}\right]$$

$$= \frac{2p}{E}\left[\frac{r_e^2\,(r_e r_i - r_i^2)}{r_e^2\,(r_e r_i - r_i^2) + r_e r_i\,(r_e^2 - r_e r_i)}\right] = \frac{2p}{E}\left[\frac{r_e^2\,(r_e r_i - r_i^2)}{2r_e^2\,(r_e r_i - r_i^2)}\right] = \frac{p}{E}$$

Therefore,
$$\frac{\delta r_s}{r_s} = \frac{p}{E} \quad \text{or} \quad \delta r_s = \frac{p}{E}\sqrt{r_e r_i} \tag{5.81}$$

EXAMPLE 5.22

A two layer steel gun barrel of inner diameter of 100 mm is to be subjected to an internal pressure of 200 MPa at the moment of firing. Determine the external and internal diameters of the outer tube and the shrinkage or negative allowance for the assembly. The modulus of elasticity, yield stress and factor of safety for the steel are 200 GPa, 600 MPa and 2, respectively.

Solution: For the given problem,

$$r_i = 50 \text{ mm}, p = 200 \text{ MPa}, E_s = 200 \text{ GPa}, \sigma_y = 600 \text{ MPa and F.O.S.} = 2$$

Therefore,

$$\sigma_a = \frac{\sigma_y}{\text{F.O.S.}} = \frac{600}{2} = 300 \text{ MPa}$$

From Eq. (5.80),

$$300 = 200 \times \left(\frac{r_e}{r_e - 50}\right) \quad \text{or} \quad r_e = 150 \text{ mm}$$

The interface radius is given by

$$r_s = \sqrt{r_e r_i} = \sqrt{150 \times 50} = 86.603 \text{ mm}$$

Thus the internal diameter of the outer tube is 173.21 mm.
The radial shrinkage allowance is

$$\frac{\delta r_s}{r_s} = \frac{p}{E_s}$$

or

$$\delta r_s = \frac{p}{E_s}\sqrt{r_e r_i} = \frac{200}{200 \times 10^3} \times \sqrt{150 \times 50}$$

$$= 0.0865 \text{ mm}$$

5.9 THICK SPHERICAL SHELLS

A thick spherical vessel with external and inner radii of r_e and r_i, respectively, and subjected to an internal pressure of p is shown in Figure 5.9(a). Consider a concentric elemental spherical shell of radius r and thickness δr with radial pressure intensities of p_r and $(p_r + \delta p_r)$ at the inner and outer surfaces of elementary shell, respectively. As in the case of a thin spherical shell, the total vertical bursting force on one side of the diametrical plane

$$F = \text{Intensity of radial pressure} \times \text{Projected area}$$

$$= [p_r \times \pi r^2] - (p_r + \delta p_r) \times [\pi(r + \delta r)^2] = -\pi r\,(r\,\delta p_r + 2\delta r\, p_r)$$

The squares and products of very small quantities have been ignored.

The bursting force is resisted by the circumferential force induced in elemental shell thickness across the diametrical plane. This resisting tensile force is distributed over the circular cross-sectional area of the elemental shell, which is $2\pi r\ \delta r$ (since δr is small). Therefore, the resisting force F_r is:

$$F_r = \sigma_\theta (2\pi r\ \delta r)$$

Equate the bursting force to the resisting tensile force

$$F_r = F \quad \text{i.e.} \quad \sigma_\theta (2\pi r\ \delta r) = -\pi r\,(r\ \delta p_r + 2\delta r\ p_r)$$

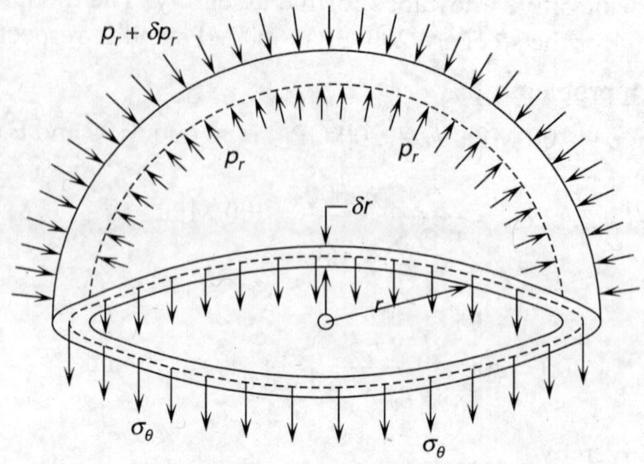

(a) Concentric elementary spherical thin shell

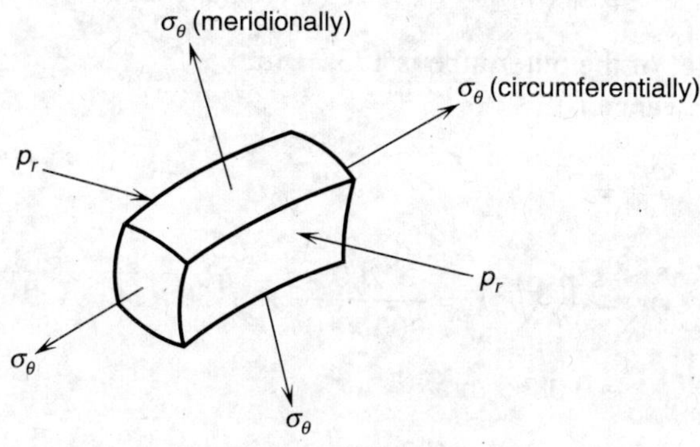

(b) Stress state at a point

Figure 5.9 Analysis of thick spherical shell.

Therefore,

$$\sigma_\theta = -\frac{r}{2}\left(\frac{\delta p_r}{\delta r}\right) - p_r = \frac{r}{2}\left(\frac{dp_r}{dr}\right) - p_r \tag{5.82}$$

(i) **Stresses**

The state of stress at any point at a radius r can be expressed by the three principal stresses as shown in Figure 5.9(b).

(i) The compressive radial pressure, p_r

(ii) The tensile hoop stress, σ_θ

(iii) The tensile hoop stress, σ_r (= σ_θ) on a plane right angle to that in (ii).

(ii) **Strains**

Consider an elementary disc of radius r subtending an angle $\delta\theta$ at the centre cut from the elemental shell. This disc has a thickness δr with radial pressure intensities of p_r and $(p_r + \delta p_r)$ at the inner and outer surfaces of the element, respectively.

Consequent to the internal pressure the radius of disc increases from r to $r + u$, and the corresponding circumferential and radial strains are given by:

Circumferential strain, $$\varepsilon_\theta = \frac{(r+u)\,\delta\theta}{r\,\delta\theta} = \frac{u}{r} \quad \text{or} \quad u = r\varepsilon_\theta \tag{5.83}$$

Radial strain, $$\varepsilon_r = \frac{\delta(r+u) - \delta r}{\delta r} = \frac{\delta u}{\delta r}$$

Therefore, $$\varepsilon_r = \frac{\delta u}{\delta r} = \frac{\delta}{\delta r}(r\,\varepsilon_\theta) = \varepsilon_\theta + r\frac{\delta \varepsilon_\theta}{\delta r} \tag{5.84}$$

In terms of stresses the strains are:

Radial strain, $$\varepsilon_r = -\left(\frac{p_r}{E} + \nu\frac{\sigma_\theta}{E} + \nu\frac{\sigma_r}{E}\right) = -\frac{1}{E}(p_r + 2\nu\sigma_\theta) \text{ (compressive)} \tag{5.85}$$

Circumferential strain,

$$\varepsilon_\theta = \frac{\sigma_r}{E} - \nu\frac{\sigma_\theta}{E} + \nu\frac{p_r}{E} = \frac{1}{E}[\nu p_r + (1-\nu)\sigma_\theta] \text{ (tensile)} \tag{5.86}$$

Substitute for ε_r and ε_θ from Eqs. (5.85) and (5.86) in Eq. (5.84):

$$-\frac{1}{E}(p_r + 2\sigma_\theta) = \frac{1}{E}[\nu p_r + (1-\nu)\sigma_\theta] + \frac{r}{E}\frac{\delta}{\delta r}[\nu p_r + (1-\nu)\sigma_\theta]$$

$$-(p_r + 2\sigma_\theta) = [\nu p_r + (1-\nu)\sigma_\theta] + r\left[\nu\frac{\delta p_r}{\delta r} + (1-\nu)\frac{\delta \sigma_\theta}{\delta r}\right]$$

$$(\nu+1)[p_r + \sigma_\theta] + \nu r\frac{\delta p_r}{\delta r} + (1-\nu)r\frac{\delta \sigma_\theta}{\delta r} = 0$$

Substitute $p_r + \sigma_\theta = -\frac{r}{2}\left(\frac{\delta p_r}{\delta r}\right)$ from Eq. (5.82),

$$(\nu+1)\left[-\frac{r}{2}\left(\frac{\delta p_r}{\delta r}\right)\right] + \nu r\frac{\delta p_r}{\delta r} + (1-\nu)r\frac{\delta \sigma_\theta}{\delta r} = 0$$

or $$-\frac{1}{2}\frac{\delta p_r}{\delta r} + \frac{\delta \sigma_\theta}{\delta r} = 0 \quad \text{i.e.} \quad \frac{d\sigma_\theta}{dr} = \frac{1}{2}\frac{dp_r}{dr} \tag{5.87}$$

Differentiate Eq. (5.82),

$$\frac{d\sigma_\theta}{dr} = -\frac{1}{2}\left[\left(\frac{dp_r}{dr} + r\frac{d^2 p_r}{dr^2}\right)\right] - \frac{dp_r}{dr} \tag{5.88}$$

From Eqs. (5.87) and (5.88),

$$r\frac{d^2 p_r}{dr^2} + 4\frac{dp_r}{dr} = 0 \tag{5.89}$$

This differential equation can be solved by substituting $\frac{dp_r}{dr} = \chi$,

Therefore, $$r\frac{d\chi}{dr} + 4\chi = 0 \quad \text{or} \quad \frac{d\chi}{\chi} = -4\frac{dr}{r}$$

Integration gives, $$\log_e \chi = -4 \log_e r + \log_e C_1 \quad \text{or} \quad \log_e \chi = \log_e \frac{C_1}{r^4}$$

where $\log_e C_1$ is a constant of integration. Thus,

$$\chi = \frac{C_1}{r^4} \quad \text{or} \quad \frac{dp_r}{dr} = \chi = \frac{C_1}{r^4} \quad \text{or} \quad dp_r = C_1\frac{dr}{r^4}$$

Integration gives,

$$p_r = -\frac{C_1}{3r^3} + C_2$$

where C_2 is a constant of integration. However,

$$\sigma_\theta = -\frac{r}{2}\left(\frac{dp_r}{dr}\right) - p_r = -\frac{r}{2}\frac{d}{dr}\left(-\frac{C_1}{3r^3} + C_2\right) + \frac{C_1}{3r^3} - C_2$$

$$= -\frac{C_1}{2r^3} + \frac{C_1}{3r^3} - C_2 = -\frac{C_1}{6r^3} - C_2$$

The substitution of $C_1 = -6b$ and $C_1 = -a$ gives

$$p_r = \frac{2b}{r^3} - a \quad \text{and} \tag{5.90}$$

$$\sigma_\theta = \frac{b}{r^3} + a \tag{5.91}$$

Equations (5.90) and (5.91) are called *initial conditions* and can be used to determine *radial pressure* and *hoop stress* at any radius. The constants a may b be determined from the known radial pressure or boundary conditions.

For illustration, consider a thick spherical shell subjected to an internal pressure p. From the known radial pressure conditions:

At $r = r_e, p_r = 0$ i.e. $0 = \frac{2b}{r_e^3} - a$

At $r = r_i, p_r = p$ i.e. $p = \frac{2b}{r_i^3} - a$

The constants a and b as evaluated from the above equation are,

$$a = \frac{pr_i^3}{r_e^3 - r_i^3} \quad \text{and} \quad b = \frac{pr_e^3 r_i^3}{2(r_e^3 - r_i^3)} \tag{5.92}$$

Therefore,

$$\sigma_\theta = \frac{b}{r^3} + a = \frac{pr_e^3 r_i^3}{2r^3(r_e^3 - r_i^3)} + \frac{pr_i^3}{r_e^3 - r_i^3} = \frac{pr_i^3}{(r_e^3 - r_i^3)}\left[1 + \frac{r_e^3}{2r^3}\right]$$

$$\sigma_{\theta,r_i} = \frac{pr_i^3}{(r_e^3 - r_i^3)}\left[1 + \frac{r_e^3}{2r_i^3}\right] = \frac{p}{2}\left[\frac{2r_i^3 + r_e^3}{r_e^3 - r_i^3}\right] = \frac{p}{2}\left[\frac{2 + k^3}{k^3 - 1}\right] \tag{5.93}$$

EXAMPLE 5.23

The thick spherical shell has to withstand an internal fluid pressure of 40 MPa. If the internal diameter of the shell is 150 mm, determine the thickness so that the maximum tensile stress in a section does not exceed 80 MPa.

Solution: Consider the thickness of shell to be t, then $r_i = 75$ mm and $r_e = 75 + t$

The fundamental relations for radial pressure and hoop stress, i.e. initial conditions, are:

$$p_r = \frac{2b}{r^3} - a \quad \text{and} \quad \sigma_\theta = \frac{b}{r^3} + a$$

For the internal fluid pressure and permissible stress conditions:

At $r = 75$ mm, $p_r = 40$ MPa i.e. $40 = \dfrac{2b}{75^3} - a$ (i)

At $r = 75$ mm, $\sigma_\theta = 80$ MPa i.e. $80 = \dfrac{b}{75^3} + a$ (ii)

Operate: Eq. (i) + Eq. (ii)

$$\frac{3b}{75^3} = 120 \quad \text{or} \quad b = 40 \times 75^3 = 16875000$$

2 × Eq. (ii) – Eq. (i)

$$3a = 120 \quad \text{or} \quad a = 40$$

Thus, $a = 40$ and $b = 16875000$

For the condition,

At $r = r_e$, $p_r = 0$ i.e. $0 = \dfrac{2b}{r_e^3} - a$

Therefore, $r_e^3 = \dfrac{2b}{a} = \dfrac{2 \times 16875000}{40}$ or $r_e = 94.49$ mm

Thus, $t = r_e - r_i = 94.49 - 75 = 19.49$ mm

The required the thickness is 19.49 mm.

EXAMPLE 5.24

The thick spherical shell having internal and external radii of 250 mm and 350 mm, respectively, has to withstand an internal pressure of 18 MPa. Determine the variation of hoop stress and radial pressure in the shell. Also calculate the increase in the internal and external diameters. The modulus of elasticity and Poisson's ratio for the material of the shell are $E = 200$ GPa and $\nu = 0.3$, respectively.

Solution: Consider the fundamental relations for hoop and radial stresses in a thick spherical shell,

$$p_r = \frac{2b}{r^3} - a \quad \text{and} \quad \sigma_\theta = \frac{b}{r^3} + a$$

For the internal fluid pressure conditions:

At $r = 125$ mm, $p_r = 18$ MPa i.e. $$18 = \frac{2b}{125^3} - a \quad \text{(i)}$$

At $r = 175$ mm, $p_r = 0$ i.e. $$0 = \frac{2b}{175^3} - a \quad \text{(ii)}$$

From Eq. (ii):

$$a = \frac{2b}{175^3} \text{ and thus from Eq. (i): } 18 = 2b\left[\frac{1}{125^3} - \frac{1}{175^3}\right]$$

$$a = 10.32 \quad \text{and} \quad b = 27657325$$

Therefore,

$$p_r = \frac{55314650}{r^3} - 10.32 \quad \text{and} \quad \sigma_\theta = \frac{27657325}{r^3} + 10.32$$

The hoop stress and radial pressure at different radii are:

$$p_{125} = -18;\ p_{140} = -9.838;\ p_{160} = -3.1846 \text{ and } p_{175} = 0.000 \text{ (MPa)}$$

$$\sigma_{125} = 24.481;\ \sigma_{140} = 20.399;\ \sigma_{160} = 17.072 \text{ and } \sigma_{175} = 15.481 \text{ (MPa)}$$

Increase in the internal and external diameters:

The circumferential strain,

$$\varepsilon_\theta = \frac{\delta d}{d} \quad \text{or} \quad \delta d = \varepsilon_\theta d$$

Therefore,

$$\delta d = \frac{1}{E}[\sigma_\theta - \nu(\sigma_r + \sigma_\theta)]d$$

$$\delta d_i = \frac{1}{200 \times 10^3} \times [24.481 - 0.3 \times (-18.000 + 24.481)] \times 250$$

$$= 0.002817 \text{ mm}$$

$$\delta d_e = \frac{1}{200 \times 10^3} \times [15.481 - 0.3 \times (0 + 15.481)] \times 350$$

$$= 0.001896 \text{ mm}$$

5.10 PROBLEMS

5.1 A 1.0 m long thin cylindrical pressure vessel of 400 mm internal diameter and 3.0 mm wall thickness which has flat ends is subjected to an internal pressure of 3.0 MPa. Determine: (a) the circumferential and longitudinal stresses developed in the material and (b) the increase in the volume of vessel. The modulus of elasticity and Poisson's ratio for the material of the vessel are $E = 200$ GPa and $\nu = 0.3$, respectively.

[**Ans.** $\sigma_l = 100$ MPa; $\sigma_\theta = 200$ MPa and $\delta V = (+)\ 2.388 \times 10^{-4}$ m^3]

5.2 A thin cylindrical shell of 420 mm internal diameter and 5 mm thickness is subjected to an internal pressure which causes a strain of 4.8×10^{-4} in the diameter. Determine the internal pressure and the consequent hoop and longitudinal stresses. The modulus of elasticity and Poisson's ratio for the shell material are 200 GPa and 0.3, respectively.

$$\left[\textbf{Hint}: \varepsilon_\theta = \frac{pd}{4tE}(2 - \nu) = \frac{\delta d}{d}\right]$$

[**Ans.** $p = 2.689$ MPa; $\sigma_1 = 112.94$ MPa and $\sigma_2 = 56.47$ MPa]

5.3 A thin cylindrical vessel of 1.0 m internal diameter and 10 mm thickness is subjected to an internal pressure of 2.5 MPa. If the length of the shell is 3.0, determine (a) hoop and longitudinal stresses, (b) maximum shear stress and (c) the changes in the dimensions of the shell. The modulus of elasticity and Poisson's ratio for the vessel material are 200 GPa and 0.25, respectively.

[**Ans.** $\sigma_\theta = 125$ MPa; $\sigma_l = 62.5$ MPa; $\tau_{max} = 31.25$ MPa;
$\delta d = 0.4375$ mm; $\delta V = 2945244$ mm^3]

5.4 A seamless spherical shell of 900 mm internal diameter and 10 mm wall thickness is filled with a fluid under pressure such that its volume increases by 1.5×10^5 mm^3. Determine the pressure exerted by the fluid on the shell. The modulus of elasticity and Poisson's ratio for the material of the shell are $E = 200$ GPa and $\nu = 0.3$, respectively.

[**Ans.** $p = 1.6634$ MPa]

5.5 A 6 mm thick cylindrical vessel is filled with an incompressible fluid at a pressure of 5 MPa. The vessel has an internal diameter of 250 mm and is 750 mm long. Find the additional volume of fluid pumped into the vessel. The modulus of elasticity and Poisson's ratio for the metal of the vessel are $E = 1.05 \times 10^5$ MPa and $\nu = 0.3$, respectively.

[**Ans.** $\sigma_\theta = 104.167$ MPa; $\sigma_l = 52.08$ MPa and $\delta V = 69395$ mm^3]

5.6 A 2.4 m long thin cylindrical tank of 1.5 m internal diameter and 12.5 mm wall thickness has a circumferential joint which is 60 per cent efficient and a longitudinal joint which is

85 per cent efficient. Determine the safe internal pressure for the tank if allowable tensile stress for the tank material is 66 MPa.

$$\left[\textbf{Hint and Ans:}\, \sigma_\theta = \sigma_a = \frac{pd}{2t\eta_l}; \sigma_l = \sigma_a = \frac{pd}{4t\eta_\theta} \text{ and } p = 0.935 \text{ MPa}\right]$$

5.7 A thin spherical shell of 375 mm internal diameter with a wall thickness of 10 mm is first filled with fluid at atmospheric pressure. Determine: (i) the internal pressure developed when 6000 mm^3 of additional water at atmospheric pressure is pumped into the shell and (ii) the circumferential stress induced in the shell. The modulus of elasticity and Poisson's ratio for the shell material are 200 GPa and 0.3, respectively.

[**Ans.** $p = 4.31$ MPa and $\sigma_\theta = 40.4$ MPa]

5.8 A thin cylindrical vessel with hemispherical ends is subjected to an internal pressure. Determine the ratio of thicknesses of cylindrical and hemispherical portions so that the junction section remains free from unequal deformation.

$$\left[\textbf{Ans:}\, \frac{t_1}{t_2} = \frac{2-\nu}{1-\nu}\right]$$

5.9 The ends of a thin cylindrical tube of 100 mm internal diameter and 2.0 mm wall thickness are closed by rigid plates. The vessel is then filled with a liquid. When an external axial compressive load of 20 kN is applied to the ends, the water pressure rises by 50 kPa. Determine the bulk modulus of the liquid. The modulus of elasticity and Poisson's ratio for the material of the tube are $E = 200$ GPa and $\nu = 0.28$, respectively.

[**Ans.** Resultant longitudinal stress = $\sigma_l = 31.205$ MPa (compressive); $K = 0.864$ GPa]

5.10 The thick cylindrical shell of internal diameter of 200 mm has to withstand an internal fluid pressure of 30 MPa. Determine the thickness of the metal so that the maximum stress in a section does not exceed tensile 60 MPa.

[**Ans.** $t = 41.4$ mm]

5.11 The cylinder of a hydraulic press having an internal diameter of 300 mm is to be designed to withstand an internal pressure of 10 MPa without the material being stressed over 20 MPa. Determine the thickness of the material and the stress on the outside of the cylinder.

Sketch a diagram showing the variation of radial and hoop stresses across the thickness of the wall of the cylinder.

[**Ans.** $t = 10.98$ mm]

5.12 A thick tube is subjected to an internal pressure of intensity equal to 5/8 of the permissible tangential stress. (a) Determine the ratio of the thickness to internal diameter of the tube. (b) If the internal diameter of such a tube is 100 mm and is subjected to an internal pressure of 100 MPa, calculate the increase in the internal diameter. The modulus of elasticity and Poisson's ratio for the tube material are $E = 200$ GPa and $\nu = 0.286$, respectively.

[**Ans.** (a) $\dfrac{\sigma_\theta}{p} = \dfrac{8}{5}$; $k = 2.082$ and $\dfrac{t}{d_i} = 0.5408$;

(b) $\sigma_\theta = 160$ MPa and $\delta d = 0.0943$ mm]

5.13 A cylindrical vessel with external and internal radii of 100 and 75 mm, respectively, is subjected to an internal fluid pressure. In addition to internal fluid pressure there is a compressive load of 200 kN applied at the ends of the vessel. Determine the maximum fluid pressure the vessel can sustain without exceeding the maximum stress in the material of 42 MPa.

[**Hint:** The maximum fluid pressure is the one that does not allow hoop stress to exceed the maximum stress in the material and resultant longitudinal stress to be compressive]

[**Ans.** $p = 11.765$ MPa]

5.14 A collar having internal and external diameters of d_i and d_e is shrunk on a solid shaft of same material. If the external diameter of the shaft is $d_i + \delta$, show that the radial pressure at the interface is given by

$$p_i = \frac{E\delta}{2}\left(\frac{d_e^2 - d_i^2}{d_e^2\, d_i}\right)$$

where E is the modulus of elasticity for the shaft material.

5.15 A steel hub 200 mm outside diameter and 200 mm long is shrunk on a steel shaft of 100 mm diameter with an interference of 0.1 mm. Determine the magnitude of the maximum shearing stress in the assembly due to shrink fit. The modulus of elasticity and the Poisson's ratio for steel are 210 GPa and 0.3, respectively.

$$\left[\textbf{Hint}: (\sigma_\theta)_h = \left(\frac{5}{3}\right)p_s \text{ and } (\varepsilon_\theta)_{\text{interface}} \left(= \frac{0.1}{100}\right) = \left(\frac{8}{3}\right)\frac{p_s}{E_s}\right]$$

[**Ans.** $p_s = 78.75$ MPa; $(\sigma_\theta)_h = 131.25$ MPa and $\tau_{\max} = [p_s + (\sigma_\theta)_h]/2 = 105$ MPa]

5.16 A compound cylinder is formed by shrinking a cylinder with external and internal diameters of 160 mm and 120 mm, respectively, onto another cylinder which has an internal diameter of 80 mm. After shrinking the radial compression at the interface is 30 MPa. Determine the maximum circumferential stress developed at the inner surface, interface and outer surface of the compound cylinder.

[**Ans.** $\sigma_{\theta 1} = -108$ MPa; $\sigma_{\theta,i} = 29.14$ MPa and $\sigma_{\theta,2} = 77.14$ MPa]

5.17 A 100 mm long steel bush, having 80 mm inside diameter and 40 mm wall thickness, is shrunk fit on a steel shaft with diametrical interference of 0.04 mm. The modulus of elasticity and the Poisson's ratio for steel are 210 GPa and 0.3, respectively. The coefficient of friction between the bush and the shaft is 0.25. Determine the axial force required to remove the bush from the assembly.

[**Ans.** $p_s = 37.5$ MPa and $F_a = (2\pi r_1 l)\,\mu p_s = 235.62$ kN]

5.18 A steel tube of 200 mm external diameter is shrunk fit on another steel tube of 60 mm internal diameter. After shrunk fitting the diameter at the junction is 120 mm with diametrical interference of 0.08 mm. Determine tangential and radial stresses at the interface due to shrink fit. The modulus of elasticity and Poisson's ratio for steel are 200 GPa and 0.3, respectively.

[**Ans.** $p_s \approx 70.3$ MPa $(\sigma_{\theta,i})_s = -117.25$ MPa and $(\sigma_{\theta,e})_s = 149.45$ MPa]

5.19 A steel hoop of 200 mm external and 130 mm internal diameters is shrunk on a hollow steel cylinder which has an internal diameter of 80 mm. After shrinking the radial compression at the interface is 20 MPa. If the assembly is subjected to an internal pressure of 70 MPa, determine: (a) the maximum circumferential stress developed in the cylinder, (b) the radial pressure between the cylinder and the hoop and (c) the maximum circumferential stress developed in the hoop. Consider the stresses induced are within the proportional limit.

[**Ans.** (a) $\sigma_{\theta,c} = 32.286$ MPa; (b) $p_s = -18.227$ MPa and (c) $\sigma_{\theta,h} = 94.155$ MPa]

5.20 A gun metal cylinder of 120 mm external and 79.95 mm internal diameters is forced onto a hollow steel cylinder 80 mm external diameter and 40 mm internal diameter. Determine the maximum stresses developed in the steel and the gun metal. The modulus of elasticity for the steel and gun metal are 200 GPa and 100 GPa, respectively. The Poisson's ratio for both the materials is 0.35.

[**Ans.** $p_s = 17.32$ MPa; $\sigma_{\theta,s,\max} = -46.187$ MPa and $\sigma_{\theta,g,\max} = 45.032$ MPa]

5.21 The thick steel cylinder of internal and external diameters of 100 mm and 160 mm, respectively, is subjected to an internal pressure of 250 MPa. To prevent the maximum hoop stress from exceeding 280 MPa a cylindrical jacket of the same material is shrunk on the steel cylinder. Determine the initial difference between the internal diameter of the jacket and the external diameters of cylinder.

[**Ans.** $(\sigma_\theta')_i = -3.281 p_s$; $(\sigma_\theta'')_i = 570.512$ MPa; $p_s = 88.544$ MPa and $\delta d = -0.140$ mm]

5.22 The thick spherical shell has to withstand an internal fluid pressure of 20 MPa. If the internal diameter of the cylinder is 100 mm, determine the thickness so that the maximum stress in a section does not exceed tensile 85 MPa.

[**Ans.** $a = 50$; $b = 4375000$; $r_e = 55.93$ mm and $t = 5.93$ mm]

Chapter 6

Torsion

6.1 INTRODUCTION

In the previous chapters, the procedures of analysis for stresses and deformations in the members subjected to pure tension and compression were presented. Analogous relations for the members subjected to torsion about their longitudinal axes are developed in this chapter. However, discussions are confined to the effect of torsion alone when other types of actions are absent. The frequently encountered cases where the members are subjected simultaneously to torsion and bending are discussed latter in the book.

The torsion is the twisting due to *torque* or *twisting moment* T, generally caused by the off-balanced load or force application. Since in practice, the members transmitting torque such as shafts for motors, torque tubes for power equipment, etc., are predominantly circular or tubular in cross-section, emphasis is laid on members having circular cross-sections, both solid and tubular. Problems involving non-circular cross-sections are also discussed, to make the reader aware of the differences in their analysis from that for circular sections.

In the analysis of members subjected to torque, the basic *method of sections* along with the moment equilibrium equation $\Sigma M_x = 0$ with X-axis directed along the member length are employed. Therefore, for *statically determinate systems*, there can be only one reactive torque. In *statically indeterminate* problems, the reactions are determined before calculating the internal torques. After determining the *reactive torques*, the member segment on either side of a section perpendicular to the axis of the member can be isolated. The applied twisting moment generates stresses in the member. The force resultant of these internal stresses produces a couple or torque that balances the applied twisting moment. In statically determinate systems, the formal calculation of a reaction may be bypassed by isolating a member segment with the unsupported end. However, equilibrium of the whole system must always be assured. For simplicity, the members are considered to be weightless in this chapter. The torque has both magnitude and sense, i.e. it is a vector quantity and can be conveniently represented by the symbol shown in Figure 6.1.

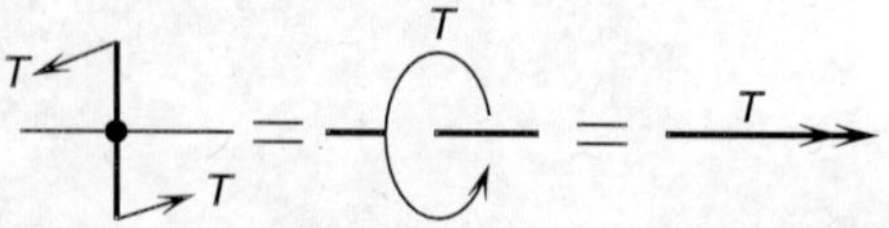

Figure 6.1 Different representations of torque.

6.2 TORSION OF MEMBERS WITH CIRCULAR CROSS-SECTION

The torsional stresses τ developed in a member are dependent on the magnitude of the applied torque or twisting moment, T; the location of the point considered, and properties of the material and cross-section. For a circular section subjected to torsion, the location of the point is defined by the distance r from the centroid, the properties of the cross-section are typically designated by the symbol J, called *polar moment of inertia*.

6.2.1 Basic Assumptions

The relation between the internal torque and the stresses it develops in a member with circular solid or tubular cross-sections is based on the following assumptions:

1. A plane section of material perpendicular to the longitudinal axis of the member remains plane before and after the torque is applied, i.e. there is no warpage or distortion in parallel planes due to the application of torque.
2. In a member with circular solid or tubular cross-section subjected to torque, the shear strain γ varies linearly from zero at the central axis to γ_{max} at the periphery.
3. The material of the member is homogeneous, isotropic and obeys Hooke's law, i.e. shear stress is proportional to shear strain.

6.2.2 The Torsion Formula

Consider an infinitesimal element of length dx isolated from the member AB of circular cross-section by two transverse planes as shown in Figure 6.2(a). The free-body diagram of the isolated element is shown in Figure 6.2(b) and it should be noted that the *torque is constant throughout*. An elementary ring is isolated in turn from the element by two cylindrical surfaces of radii r and $r + dr$ as shown in Figures 6.2(c) and (d). The right end section of the ring rotates, when subjected to torque, with respect to the left end section through an angle $d\theta_x$. The generator AB of the cylinder then rotates through an angle $d\varphi$ and occupies the position AB′. The movement BB′ can be expressed as

$$\text{BB}' = dx\, d\varphi = r\, d\theta_x \quad \text{or} \quad d\varphi = \gamma = r\frac{d\theta_x}{dx} \tag{6.1}$$

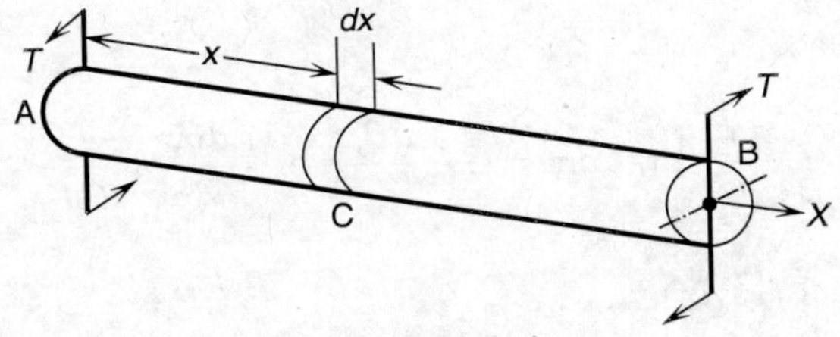

(a) Element of length dx

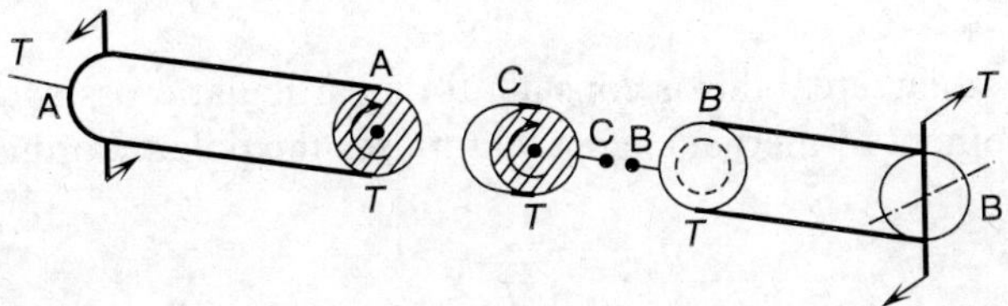

(b) Free-body diagram of the element

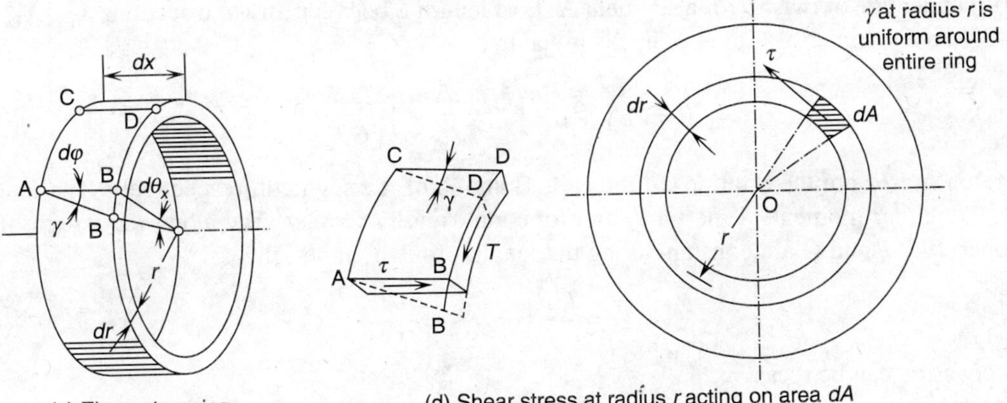

(c) Elementary ring

(d) Shear stress at radius r acting on area dA

Figure 6.2 Torsion of a member of circular cross-section.

The angle $d\varphi$ represents the angle of shear γ of cylindrical surface. The parameter $\dfrac{d\theta_x}{dx}$ is called *angle of twist per unit length*. According to the Hooke's law for shear, the shearing stress on the surface of the member is given by:

$$\tau = G\gamma = Gr\frac{d\theta_x}{dx} \tag{6.2}$$

These stresses act over the cross-sectional area of the member to produce a resisting torque equal and opposite to the applied torque. The internal twisting moment produced by the elementary forces $\tau\, dA$

$$dT_x = (\tau\, dA)r = \left(Gr\frac{d\theta_x}{dx}\, dA \right) r = G(r^2\, dA)\,\frac{d\theta_x}{dx}$$

For whole cross-section area A,

$$T_x = G\frac{d\theta_x}{dx}\int_A (r^2\,dA) = GJ_x\frac{d\theta_x}{dx} \quad \text{or} \quad \frac{d\theta_x}{dx} = \frac{T_x}{GJ_x} \tag{6.3}$$

where

$$J_x = \int_A (r^2\,dA) \tag{6.4}$$

The integral $\int_A (r^2\,dA)$ represents pure geometrical characteristic and is known as *polar moment of inertia* of the cross-sectional area; this is constant for a particular cross-section. It is analogous to the parameter I, the moment of inertia, except that it uses the polar coordinates. Thus,

$$d\theta_x = \frac{T_x}{GJ_x}\,dx \tag{6.5}$$

This gives the relative *angle of twist* of the two adjoining sections at infinitesimal distance dx apart. Total angle of twist θ for a segment A–B of length L between any two sections $X_1 - X_1'$ and $X_2 - X_2'$ on a member of Figure 6.3(a) is given by,

$$\theta = \theta_B - \theta_A = \int_A^B \frac{T_x\,dx}{GJ_x} = \int_L \frac{T_x\,dx}{GJ_x} \tag{6.6}$$

where θ_B and θ_A are the rotations at the ends B and A of the segment, respectively. The internal torque T_x, polar moment of inertia J_x and torsional rigidity G may vary along the length of the member. If T, J and G are constant along the length of the member, then

$$\theta = \frac{TL}{GJ} \quad \text{or} \quad \frac{T}{J} = \frac{G\theta}{L} \tag{6.7}$$

This equation can be recast as,

$$\frac{T}{\theta} = \text{Torque required to cause a rotation of one radian} = \frac{GJ}{L} \tag{6.8}$$

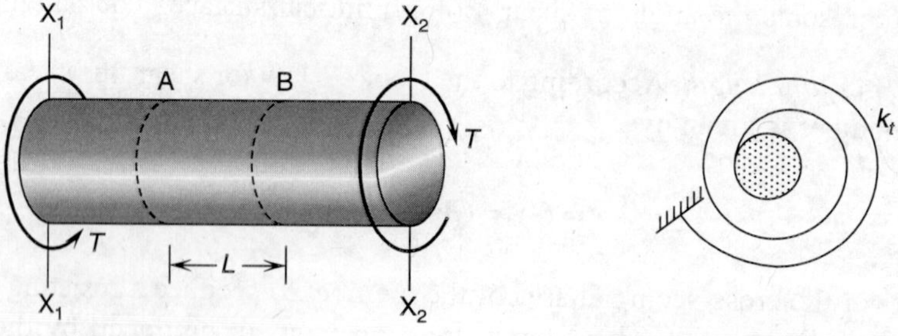

(a) Segment of the member

(b) Representation of spring content of an elastic member

Figure 6.3 Schematic representation of a torsion spring.

This represents the torsion spring constant or torsion stiffness, k_t of the member. The reciprocal of k_t is called the torsion flexibility f_t. The schematic representation of the torsion spring is shown in Figure 6.3(b).

The angle θ_x is measured in radians. Eliminating $\frac{d\theta_x}{dx}$ from Eqs. (6.2) and (6.3):

$$\frac{\tau}{r} = G\frac{d\theta_x}{dx} = \frac{T_x}{J_x} \quad \text{or} \quad \frac{\tau}{r} = \frac{T_x}{J_x} \quad \text{or} \quad \tau = \frac{T_x r}{J_x} \tag{6.9}$$

Thus the torsional stress is distributed linearly over the cross-section. The stresses vary from zero at the centre of circular member to a maximum at the surface of the member as shown in Figure 6.4(a). It should be noted that Eqs. (6.7) and (6.9) are also valid for tubular section as shown in Figures 6.4(b)(i-ii), which follows from the assumptions used in the derivation of torsion formula.

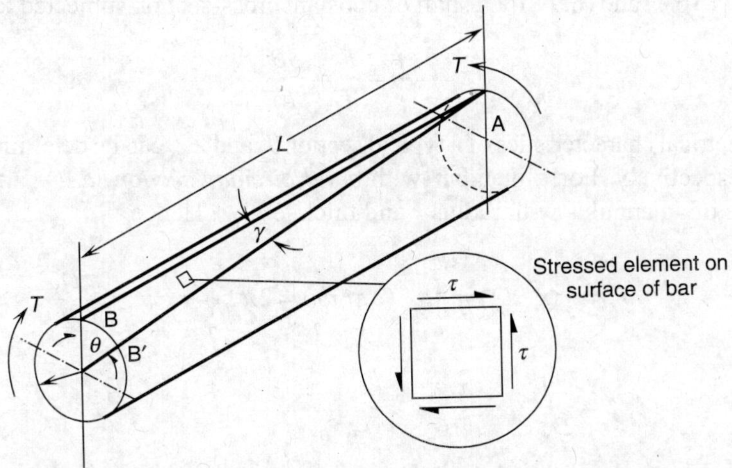

(a) Deformation in the cross-section and stressed element on the surface

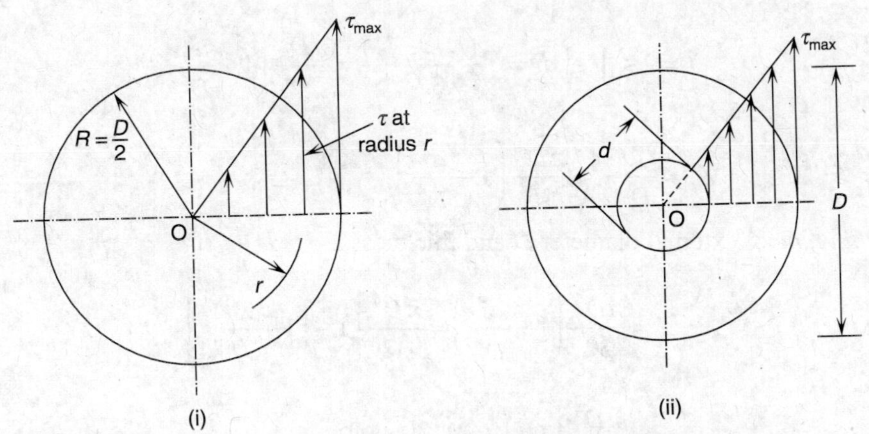

(b) Distribution of shear stress on the solid and tubular cross-section

Figure 6.4 Torsional shear stress and deformation in circular member.

The maximum shearing stress is designated τ_{max}. Then

$$\tau_{max} = \frac{T_x r_{max}}{J_x} = \frac{T_x R_x}{J_x} = \frac{T_x}{J_x / R_x} = \frac{T_x}{Z_{p,x}} \tag{6.10}$$

This expression is analogous to the expression of stresses developed by bending moment. J_x is analogous to the parameter I and is given by $J_x = \int Ar^2\, dA$, except that polar coordinates are used. J_x is called *polar moment of inertia*. The quantity $Z_{p,x}$ is known as *polar section modulus* and is given by

$$Z_{p,x} = \frac{J_x}{R_x} \tag{6.11}$$

Combining Eqs. (6.7) and (6.9), for a shaft of constant cross-section subjected to a constant torque.

$$\frac{\tau}{r} = \frac{T}{J} = \frac{G\theta}{L} \tag{6.12}$$

The geometrical characteristics of a typical section J_x and $Z_{p,x}$ can be determined from Eqs. (6.4) and (6.11), respectively. For a member with *solid circular section*, $dA = 2\pi r\, dr$ where $2\pi r$ is circumference of an annulus with radius r and thickness dr. Hence,

$$J_x = \int_A (r^2\, dA) = \int_0^{R_x} r^2\, (2\pi r\, dr) = 2\pi \int_0^{D_x/2} r^3\, dr = \frac{\pi D_x^4}{32} \tag{6.13}$$

and
$$Z_{p,x} = \frac{J_x}{D_x / 2} = \frac{\pi D_x^3}{16} \tag{6.14}$$

where D_x is the diameter of the section. For a member with uniform *hollow circular* or *tubular section* with central hole of diameter d,

$$J = 2\pi \int_{d/2}^{D/2} r^3\, dr = \frac{\pi (D^4 - d^4)}{32} = \frac{\pi D^4}{32}\left(1 - \frac{d^4}{D^4}\right) \tag{6.15}$$

and
$$Z_p = \frac{J}{D/2} = \frac{\pi D^3}{16}\left(1 - \frac{d^4}{D^4}\right) \tag{6.16}$$

For *very thin tube* of external diameter D and thickness t, $d = D - 2t$

$$J = \frac{\pi D^4}{32}\left(1 - \frac{d^4}{D^4}\right) = \frac{\pi D^4}{32}\left[1 - \left(1 - \frac{2t}{D}\right)^4\right]$$

$$= \frac{\pi D^4}{32}\left[1 - \left(1 - \frac{8t}{D} + \cdots\right)\right]$$

$$= \frac{\pi D^4}{32} \times \frac{8t}{D} = \frac{\pi D^3 t}{4}$$

Therefore,
$$J = \frac{\pi D_{av}^3 t}{4} = 2\pi R_{av}^3 t \tag{6.17}$$

where R_{av} is the centre-line radius of the tube. The torsion carried by the member is

$$T = \frac{\pi D_{av}^3 t}{4} \times \frac{\tau_{max}}{D/2} = \frac{\pi D_{av}^2}{2} \times \frac{D_{av}}{D} \times (t\,\tau_{max})$$

$$\approx \frac{\pi D_{av}^2}{2}(t\,\tau_{max}) = 2A^*(t\,\tau_{max}) \tag{6.18}$$

and total angle of twist for the tube of constant thickness

$$\theta = \frac{TL}{GJ} = \frac{TL}{G(\pi D_{av}^3 t/4)}$$

$$= \frac{TL(\pi D_{av})}{4G(\pi D_{av}^2/4)^2 t} = \frac{TLs}{4G(A^*)^2 t} \tag{6.19}$$

If the thickness t is variable along the arc length of the contour from Eq. (6.19),

$$\theta = \int_L \frac{T_x\,dx}{GJ_x} = \frac{TL}{G(2A^*)^2}\int_s \frac{ds}{t} \tag{6.20}$$

where A^* is the area enclosed by the centre-line of the contour of the cross-section and s is the length of the closed contour. This expression is applicable to all the thin-walled closed cross-sections. The general derivations are given in Section 6.5.3.

6.2.3 Distributed Torque

For the shaft of constant torsional rigidity GJ, subjected to a torque which itself is a continuous function as shown in Figures 6.5(a) and (b), Eq. (6.3) can be recast into a second-order differential equation. The infinitesimal element of Section 6.2.2 is considered to be subjected to end torques T_x and $T_x + dT_x$, and to an applied distributed torque t_x. For the equilibrium of the element

$$t_x\,dx + dT_x = 0 \quad \text{or} \quad \frac{dT_x}{dx} = -t_x$$

On differentiating Eq. (6.3) with respect to x,

$$GJ\frac{d^2\theta_x}{dx^2} = \frac{dT_x}{dx} = -t_x \tag{6.21}$$

The constants of integration appearing in the solution of this equation are determined from the end conditions of the shaft.

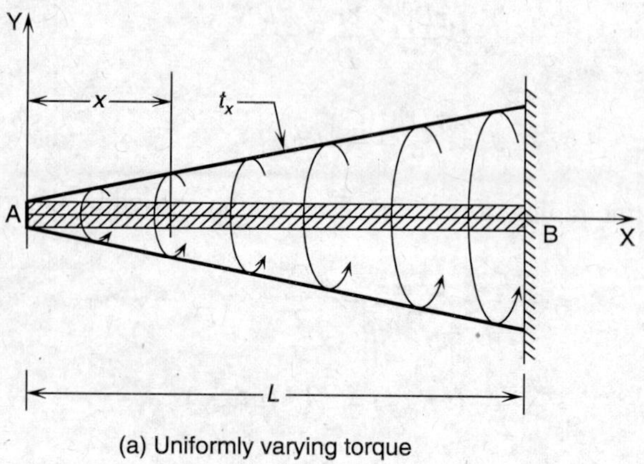

(a) Uniformly varying torque

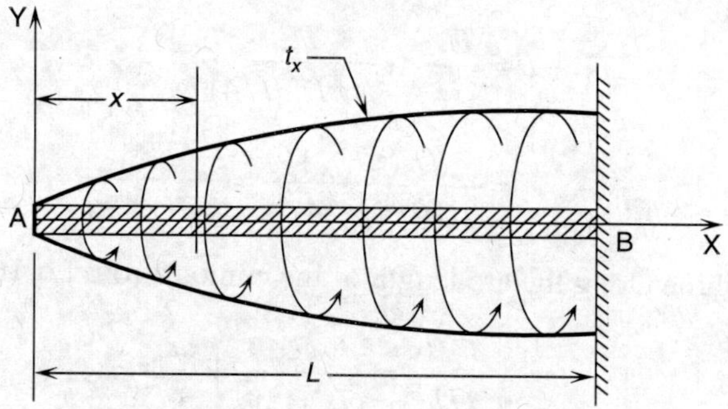

(b) Distributed torque with parabolic variation

Figure 6.5 Typical cases of distributed torque.

For illustration, consider a shaft A–B of length L, free at the end A and built-in at the end B, and subjected to a uniformly varying torque $t_x = kx$, as shown in Figure 6.5(a), where k is a constant. Therefore from Eq. (6.21),

$$GJ\frac{d^2\theta_x}{dx^2} = \frac{dT_x}{dx} = -kx$$

By integrating $$GJ\frac{d\theta_x}{dx} = T_x = -k\frac{x^2}{2} + C_1 \tag{i}$$

At $x = 0$, $$T_A = T(0) = 0 \quad \text{hence } C_1 = 0$$

At $x = L$, $$T_B = T(L) = -\frac{kL^2}{2}$$

The minus sign indicates that the reaction torque is opposite to the applied torque.

By integrating Eq. (i) $$GJ\theta_x = -k\frac{x^3}{6} + C_2 \tag{ii}$$

At $x = L$, $\theta_B = \theta(L) = 0$ hence, $C_2 = \frac{kL^3}{6}$

Therefore, $\theta_x = \frac{k}{6GJ}(L^3 - x^3)$

If both the ends are built-in, by integrating twice

$$GJ\frac{d\theta_x}{dx} = T_x = -k\frac{x^2}{2} + C_1$$

$$GJ\theta_x = -k\frac{x^3}{6} + C_1 x + C_2$$

At $x = 0$, $\theta_A = \theta(0) = 0 \Rightarrow C_2 = 0$

At $x = L$, $\theta_B = \theta(L) = 0 \Rightarrow C_1 = \frac{kL^2}{6}$

Therefore, $\theta_x = \frac{k}{6GJ}(xL^2 - x^3)$

and $T_x = \frac{k}{6}(L^2 - 3x^2)$

At $x = 0$, $T_A = T(0) = \frac{kL^2}{6}$

At $x = L$, $T_B = T(L) = -\frac{kL^2}{3}$

6.3 POWER TRANSMITTED BY A SHAFT

The members subjected to torque are widely used as rotating shafts for transmitting the power. When a shaft transmitting a constant mean torque T_{mean} measured in N.m, undergoes an angular rotation θ, the work done is $T_{\text{mean}}\theta$. The rate at which this work is done is termed the *power* transmitted by the shaft, i.e.

$$\text{Power} = \frac{d}{dt}(T_{\text{mean}}\theta) = T_{\text{mean}}\frac{d\theta}{dt} = T_{\text{mean}}\,\omega$$

where the term $d\theta/dt$ represents the angular velocity ω. For a shaft rotating with frequency f cycles per second (cps) or N revolutions per minute (rpm), the angular velocity is

$$\omega = 2\pi f = \frac{2\pi N}{60} \text{ radian}$$

Therefore, the power P is given by

$$P = \omega T_{\text{mean}} = (2\pi f) T_{\text{mean}} \text{ N.m/s} \tag{6.22}$$

This expression gives the power P in watts (W is the unit symbol for watt in SI units). Thus, the *power P* is defined as the *torque* multiplied by the angle, measured in radians through which shaft rotates per unit of time. Thus

$$P = \frac{2\pi N T_{\text{mean}}}{60} \text{ [N.m/s or W]} = \frac{2\pi N T_{\text{mean}}}{60 \times 10^3} \text{ [kN.m/s or kW]} \tag{6.23}$$

In the case of rotating shafts transmitting power, the conventional unit for power transmission used is *horse power* (hp). The horse power is related to the work done as follows,

$$1 \text{ hp} = 745.7 \text{ N.m/s} = 745.7 \text{ W}$$

Therefore, in terms of conventional unit

$$P = \frac{(2\pi f) T_{\text{mean}}}{745.7} = \frac{(2\pi N) T_{\text{mean}}}{60 \times 645.7} \text{ [hp]} \tag{6.24}$$

It should be noted that *for the given power to be transmitted, higher the frequency of the shaft lower will be the torque*. Thus there is tendency to use *high speed machines* in mechanical equipment, resulting in considerable saving in material of the shaft.

The maximum power transmitted by a shaft is limited by its torque-carrying capacity which is dependent on the allowable shearing stress for the shaft material, and the permitted twist. Thus, for a solid shaft of diameter D mm rotating at N rpm with permissible shearing stress in its material $\tau_{\max}$ MPa, the maximum power transmission capacity P watts is given by

$$P = (T_{\text{mean}}) \frac{2\pi N}{60} = (Z_p \, \tau_{\max}) \frac{2\pi N}{60} = \left(\frac{\pi D^3}{16} \tau_{\max} \right) \frac{2\pi N}{60}$$

or

$$D = \sqrt[3]{\frac{60}{2\pi N} \times \frac{16P}{\pi \tau_{\max}}} = K \times \sqrt[3]{\frac{P}{N}} \tag{6.25}$$

where

$$K = \sqrt[3]{\frac{480}{\pi^2 \tau_{\max}}} \tag{6.26}$$

On the other hand, if the permissible twist is $\theta_{\max}$ for a length L, then the torque-carrying capacity is,

$$T_{\text{mean}} = \left(\frac{GJ}{L} \right) \theta_{\max} = \left(\frac{\pi D^4}{32} \right) \left(\frac{G}{L} \right) \theta_{\max} \tag{6.27}$$

The maximum power transmission capacity is

$$P = T_{\text{mean}} \, \omega = \left(\frac{\pi D^4}{32} \right) \left(\frac{G}{L} \right) \theta_{\max} \left(\frac{2\pi N}{60} \right) \tag{6.28}$$

For illustration, consider the case of the drive shaft for the boat transmitting 80 kW of power while rotating at 525 rpm. The mean torque in the shaft will be,

$$T_{\text{mean}} = \frac{(60 \times 10^3) \times P}{2\pi N} = \frac{(60 \times 10^3) \times 80}{2 \times \pi \times 525} = 1455.13 \text{ N.m}$$

If the allowable shear stress is known, the minimum acceptable diameter of the shaft can be computed. As an example, consider the allowable shear stress to be 40 MPa and the maximum torque to be 20 per cent greater than the mean. Then, from Eq. (6.14):

$$Z_p = \frac{\pi D^3}{16} = \frac{T_{\max}}{\tau_{\max}} \quad \text{or} \quad D = \sqrt[3]{\frac{16 \times (1.2 \times T_{\text{mean}})}{\pi \tau_{\max}}} \tag{6.29}$$

Therefore,

$$D = \sqrt[3]{\frac{16 \times (1.2 \times 1455.13 \times 10^3)}{\pi \times 40}} = 60.58 \text{ mm}$$

A shaft of convenient size say 65 mm may be used.

EXAMPLE 6.1

In a tension test, a metal test specimen of 30 mm diameter, 240 mm gauge length stretched 0.1018 mm under a pull of 60 kN. In a torsion test, an identical specimen of same material twisted 0.0208 radian over a length of 240 mm when a torque of 575 N.m was applied. Determine the elastic constants for the material.

Solution: *Tension test data*, $d = 30$ mm; $L = 240$ mm; $F = 60$ kN and $\Delta L = 0.1018$

From Eq. (3.8):

$$E = \frac{f_n}{\varepsilon} = \frac{(P/A)}{(\Delta L/L)} = \frac{PL}{A\Delta L}$$

Therefore,

$$E = \frac{(60 \times 10^3) \times 240}{(\pi \times 30^2/4) \times 0.1018} = 200116.23 \text{ MPa}$$

$$\approx 2 \times 10^5 \text{ MPa} = 200 \text{ GPa}$$

Torsion test data, $d = 30$ mm; $L = 240$ mm; $T = 575$ N.m and $\theta = 0.0208$ radian

$$G = \frac{TL}{J\theta} = \frac{(575 \times 10^3) \times 240}{(\pi \times 30^4/32) \times 0.0208} = 83431.65 \text{ MPa} \approx 83.4 \text{ GPa}$$

From Eqs. (3.26) and (3.27), the relations between elastic constants are,

$$G = \frac{E}{2(1+\nu)} \quad \text{or} \quad \nu = \frac{E}{2G} - 1 = \frac{200116.23}{2 \times 83431.65} - 1 = 0.1993 \approx 0.2$$

$$K = \frac{E}{3(1-2\nu)} = \frac{200116.23}{3 \times (1 - 2 \times 0.1993)} = 110910.41 \text{ MPa} = 1.11 \times 10^5 \text{ MPa}$$

EXAMPLE 6.2

Two shafts, first one of solid circular section and second of hollow circular cross-section with a ratio of internal and external diameters of 0.9, are considered for the use in an application that has to transmit a maximum torque $T_{max} = 10$ kN.m. Which of these shafts will be economical for equal strength condition? Determine: (a) the per cent saving in the material and (b) the dimensions of the shafts if the shearing stress is limited to 60 MPa.

Solution: The polar section modulus for the solid circular section of diameter D_1 is

$$Z_{p,s} = \frac{\pi D_1^3}{16}$$

For hollow cross-section the polar sectional modulus is given by

$$Z_{p,h} = \left(\frac{\pi D_2^4}{32} - \frac{\pi d^4}{32}\right)\left(\frac{2}{D_2}\right) = \frac{\pi D_2^3}{16}\left[1 - \left(\frac{d}{D_2}\right)^4\right]$$

(a) For equal strength condition, $Z_{p,s} = Z_{p,h}$ i.e.,

$$\left(\frac{D_1}{D_2}\right)^3 = 1 - \left(\frac{d}{D_2}\right)^4 \quad \text{or} \quad \frac{D_1}{D_2} = \sqrt[3]{1 - \left(\frac{d}{D_2}\right)^4}$$

For the ratio, $\dfrac{d}{D_2} = 0.9,\qquad \dfrac{D_1}{D_2} = \sqrt[3]{1 - (0.9)^4} = 0.7006$

The saving in material is governed by the cross-sectional areas:

$$A_s = \frac{\pi D_1^2}{4} \quad \text{and} \quad A_h = \frac{\pi}{4}(D_2^2 - d^2) = \frac{\pi D_2^2}{4}\left[1 - \left(\frac{d}{D_2}\right)^2\right]$$

The saving in material is computed as follows,

$$\left(\frac{A_s - A_h}{A_s}\right) = \left[1 - \left(\frac{D_2}{D_1}\right)^2 \left\{1 - \left(\frac{d}{D_2}\right)^2\right\}\right]$$

Percentage saving in material is,

$$\left(\frac{A_s - A_h}{A_s}\right) \times 100 = \left[1 - \left(\frac{1}{0.7006}\right)^2 \{1 - (0.9)^2\}\right] \times 100 = 61.29 \text{ per cent}$$

(b) Polar section modulus for both the sections,

$$Z_p = \frac{T_{max}}{\tau} = \frac{10 \times 10^6}{60} = 1.67 \times 10^5 \text{ mm}^3$$

For the solid section: $$\frac{\pi D_1^3}{16} = 1.67 \times 10^5 \quad \text{or} \quad D_1 = 94.3 \text{ mm}$$

For the hollow section:

$$\frac{\pi D_2^3}{16}\left[1 - \left(\frac{d}{D_2}\right)^4\right] = \frac{\pi D_2^3}{16}[1 - (0.9)^4] = 1.67 \times 10^5 \quad \text{or} \quad D_2 = 127 \text{ mm}$$

EXAMPLE 6.3

There are two shafts of same weight and material; first one of solid circular section and second of hollow circular cross-section with a ratio of internal and external diameters of 2/3. Compare their torque carrying capacities.

Solution: The geometrical characteristics of the two shafts are,

$$A_h = \left(\frac{\pi D_h^2}{4} - \frac{\pi d^2}{4}\right) = \frac{\pi D_h^2}{4}\left[1 - \left(\frac{d}{D_h}\right)^2\right]$$

$$= \frac{\pi D_h^2}{4}\left[1 - \left(\frac{2}{3}\right)^2\right] = \frac{5\pi D_h^2}{36}$$

and $$A_s = \frac{\pi D_s^2}{4}$$

(i) For the same weight of two shafts of same material

$$A_h = A_s \quad \text{or} \quad \frac{5\pi D_h^2}{36} = \frac{\pi D_s^2}{4}$$

Therefore, $$D_h^2 = \frac{9}{5} D_s^2 \quad \text{or} \quad D_h = 1.3416 D_s$$

The polar moment of inertias,

$$J_h = \frac{\pi}{32}(D_h^4 - d^4) = \frac{\pi D_h^4}{32}\left[1 - \left(\frac{d}{D_h}\right)^4\right]$$

$$= \frac{\pi D_h^4}{32}\left[1 - \left(\frac{2}{3}\right)^4\right] = \frac{\pi D_h^4}{32} \times \frac{65}{81}$$

$$= \frac{\pi}{32} \times \frac{81 D_s^4}{25} \times \frac{65}{81} = \frac{13\pi D_s^4}{160}$$

and $$J_s = \frac{\pi D_s^4}{32}$$

(ii) For the same stress condition in two shafts,

$$\tau_s = \tau_h = \tau \text{ i.e., } \frac{T_{\max,h} D_h}{2J_h} = \frac{T_{\max,s} D_s}{2J_s}$$

$$\frac{T_{\max,h}}{T_{\max,s}} = \frac{D_s}{D_h} \times \frac{J_h}{J_s} = \frac{D_s}{D_h} \times \frac{13\pi D_s^4}{160} \times \frac{32}{\pi D_s^4}$$

$$= \frac{D_s}{D_h} \times \frac{13}{5} = \frac{1}{1.3416} \times \frac{13}{5} = 1.938$$

EXAMPLE 6.4

A circular shaft 50 mm in diameter has a concentric hole drilled for a portion of its length. The shaft carries a torque of 1.67 kN.m. Determine the maximum length and the diameter of the hole when the maximum shearing stress and total angle of twist in a length of 750 mm are limited to 75 MPa and 1.5 degrees, respectively. The elastic modulus of the material is 80 GPa.

Solution: The material properties are:

$$\tau_{\max} = 75 \text{ MPa and } G = 80 \times 10^3 \text{ MPa}$$

The maximum stress develops in the portion with the hole,

$$Z_p = \frac{\pi}{16D}(D^4 - d^4) = \frac{\pi}{16 \times 50}(50^4 - d^4)$$

$$T = \tau_{\max} Z_p \quad \text{or} \quad 1.67 \times 10^6 = 75 \times \frac{\pi}{16 \times 50}(50^4 - d^4)$$

$$(50^4 - d^4) = \frac{(1.67 \times 10^6) \times 16 \times 50}{\pi \times 75}$$

Therefore, $$d^4 = 50^4 - \frac{(1.67 \times 10^6) \times 16 \times 50}{\pi \times 75} \quad \text{i.e. } d = 27.59 \text{ mm}$$

Let L' be the length of the hole, then from relation $\theta = \dfrac{TL}{GJ}$

$$\left(1.5 \times \frac{\pi}{180}\right) = \frac{T}{G}\left(\frac{750 - L'}{(\pi \times 50^4/32)} + \frac{L'}{[\pi \times (50^4 - 27.59^4)/32]}\right)$$

$$\frac{750 - L'}{50^4} + \frac{L'}{(50^4 - 27.59^4)} = \left(1.5 \times \frac{\pi}{180}\right) \times \frac{\pi}{32} \times \frac{80 \times 10^3}{1.67 \times 10^6} = \left(\frac{\pi}{120}\right) \times \frac{\pi}{668}$$

$$(750 - L') \times (50^4 - 27.59^4) + 50^4 \times L' = 750 \times (50^4 - 27.59^4) + 27.59^4 L'$$

$$= \frac{\pi}{120} \times \frac{\pi}{668} \times (50^4 - 27.59^4) \times 50^4$$

$$27.59^4 L' = \left(\frac{\pi}{120} \times \frac{\pi}{668} \times 50^4 - 750\right) \times (50^4 - 27.59^4)$$

Thus, $L' = 191.066$ mm.

EXAMPLE 6.5

A hollow circular marine propeller shaft with internal diameter of 0.6 times the external diameter, propels a vessel at speed 11.5 m/s using 6000 kW shaft power at 100 rpm. If the efficiency of the propeller is 66 per cent, determine (a) the shaft diameter which will enable the shaft to transmit the power when the direct stress due to thrust is limited to 8 MPa and (b) the maximum shearing stress due to torque.

Solution: Let the thrust developed by the propeller be F N. Then the useful work done

$$W = \text{Thrust} \times \text{Distance moved per second}$$

$$= F \times 11.5 \text{ N.m per sec} = 11.5F \text{ watt}$$

$$\text{Efficiency, } \eta = \frac{\text{Useful work done}}{\text{The shaft power}} = \frac{11.5F}{6000 \times 10^3}$$

Therefore,

$$F = \frac{0.66 \times (6000 \times 10^3)}{11.5} = 344347.83 \text{ N}$$

If the external diameter of the shaft is D mm,
The area of the shaft,

$$A = \frac{\pi}{4} D^2 \times (1 - 0.6^2) = 0.5027D^2 \text{ mm}^2$$

The stress due to direct load,

$$f_n = \frac{F}{A} = f_{\max} \quad \text{or} \quad \frac{344347.83}{0.5027D^2} = 8$$

Therefore, $D = 292.62$ mm

and internal diameter

$$d = 0.6D = 0.6 \times 292.62 = 175.57 \text{ mm}$$

$$Z_p = \frac{\pi}{32(D/2)}(D^4 - d^4) = \frac{\pi D^3}{16}(1 - 0.6^4)$$

$$= \frac{\pi \times 292.62^3}{16} \times (1 - 0.6^4) = 4282139.29 \text{ mm}^3$$

The power transmitted is given by

$$\text{Power (watts)} = \frac{2\pi NT}{60} \text{ N.m}$$

Therefore, $$T = \frac{60P}{2\pi N} = \frac{60 \times (6000 \times 10^3)}{2 \times \pi \times 100} = 572957.8 \text{ N.m}$$

$$\tau_{\max} = \frac{T}{Z_p} = \frac{572957.8}{4282139.29} = 133.8 \text{ MPa}$$

EXAMPLE 6.6

A motor drives by means of a set of gears a solid circular shaft A–B–C at 500 r.p.m. as shown in Figure 6.6(a). Powers amounting to 25 kW and 15 kW are delivered at A and C and as indicated in the figure. Determine the maximum shearing stresses and the angle of twist of end A of the shaft relative to end C. The modulus of rigidity of the material of the shaft is 80 GPa.

***Solution*:** The power transmitted is given by

$$\text{Power } P, \text{ (watts)} = \frac{2\pi NT}{60} \text{ N.m}$$

Therefore, $$T = \frac{60P}{2\pi N} \text{ N.m}$$

Torque in shaft A–B

$$T_{A-B} = \frac{60 \times (25 \times 10^3)}{2\pi \times 500} = 477.46 \text{ N.m}$$

Torque in shaft B–C

$$T_{B-C} = \frac{60 \times (15 \times 10^3)}{2\pi \times 500} = 286.48 \text{ N.m}$$

The polar section modulus, $Z_p = \frac{\pi}{16} D^3$

The maximum shear stress is given by $\tau_{\max} = \frac{T}{Z_p} = \frac{16T}{\pi D^3}$

Therefore, $$\tau_{A-B} = \frac{16 \times (477.46 \times 10^3)}{\pi \times 25^3} = 155.63 \text{ MPa}$$

$$\tau_{B-C} = \frac{16 \times (286.48 \times 10^3)}{\pi \times 50^3} = 11.67 \text{ MPa}$$

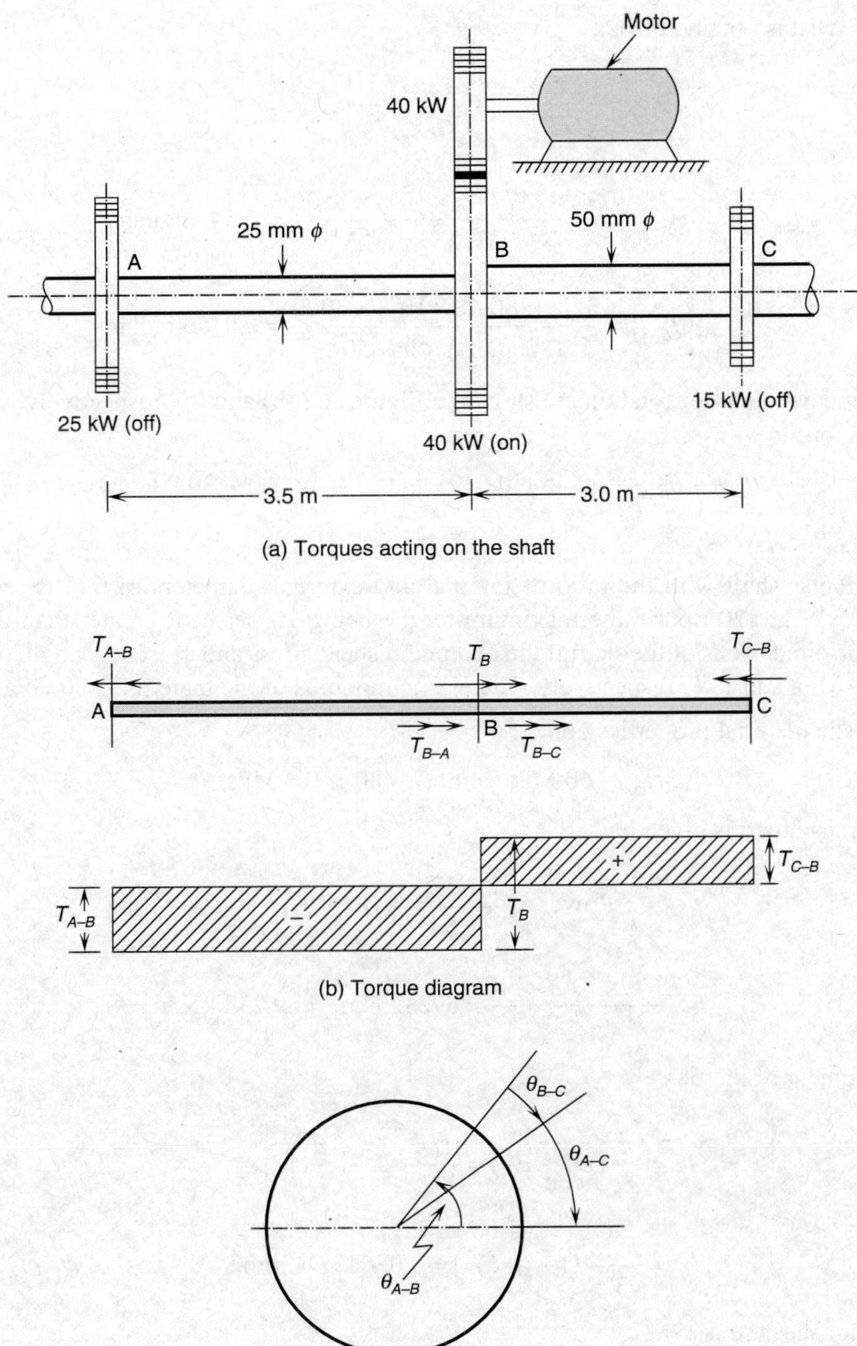

Figure 6.6 Analysis of the shaft of Example 6.6.

The angle of twist is obtained from,

$$\theta_{i-j} = \left(\frac{TL}{GJ}\right)_{i-j} = \left[\frac{32TL}{G(\pi D^4)}\right]_{i-j}$$

$$\theta_{A-B} = \frac{32 \times (477.46 \times 10^3) \times 3500}{(80 \times 10^3) \times \pi \times 25^4} = 0.545 \text{ radians}$$

$$\theta_{B-C} = \frac{32 \times (286.48 \times 10^3) \times 3000}{(80 \times 10^3) \times \pi \times 50^4} = 0.0175 \text{ radians}$$

The torque and geometry of rotation are shown in Figures 6.6(b) and (c), respectively. Relative rotation,

$$\theta_{A-C} = \theta_{A-B} - \theta_{B-C} = 0.545 - 0.0175 = 0.5275 \text{ radian} = 30.22 \text{ degrees.}$$

EXAMPLE 6.7

A hollow circular shaft with the ratio of internal and external diameters of 0.6, is required to transmit 870 kW at 120 r.p.m.; the maximum torque being 20 per cent greater than the mean. Determine the dimensions of the shaft if the maximum shear stress and twist in a length of 3 m are limited to 60 MPa and 1.4°, respectively. The elastic modulus of the material is 80 GPa.

Solution: The material properties are

$$\tau_{max} = 60 \text{ MPa and } G = 80 \times 10^3 \text{ MPa}$$

The power transmitted is given by

$$\text{Power} = \frac{2\pi N T_{mean}}{60} W \text{ [N.m]}$$

Therefore, $$T_{mean} = \frac{(870 \times 10^3) \times 60}{(120 \times 2\pi)} = 69232 \text{ N.m} = 69.232 \text{ kN.m}$$

The polar moment of inertias,

$$J_h = \frac{\pi}{32}(D^4 - d^4) = \frac{\pi D^4}{32}\left[1 - \left(\frac{d}{D}\right)^4\right]$$

$$= \frac{\pi D^4}{32}[1 - (0.6)^4] = 0.08545 D^4 \text{ mm}^4$$

The maximum shear stress,

$$\tau_{max} = \frac{T_{max} D}{2J} \quad \text{or} \quad 60 = \frac{[1.2 \times (69.232 \times 10^6)] \times D}{2 \times 0.08545 D^4}$$

$$D^3 = \frac{1.2 \times (69.232 \times 10^6)}{2 \times 0.08545 \times 60} = 9.7225 \times 10^6$$

Therefore, $D = 213.4$ mm

$$\theta = \frac{TL}{GJ} \quad \text{or} \quad 1.4 \times \frac{\pi}{180} = \frac{(1.2 \times 69.232 \times 10^6) \times (3 \times 10^3)}{(80 \times 10^3) \times 0.08545 D^4}$$

or

$$D^4 = \frac{(1.2 \times 69.232 \times 10^6) \times 3000 \times 180}{(80 \times 10^3) \times 0.08545 \times 1.4 \times \pi} = 1.4921 \times 10^9 \text{ mm}^4$$

Hence $D = 196.54$ mm

A shaft of convenient size say 215 mm may be used.

EXAMPLE 6.8

A circular shaft A–B of length L, free at the end A and built-in at the end B, is subjected to a variable torque $t_x = kx^2/L^2$, as shown in Figure 6.5(b), where x is the distance measured from the free end and k is a constant. Determine the angle of twist of the free end. The torsional rigidity GJ of the shaft is constant.

Solution: From Eq. (6.21),

$$GJ \frac{d^2\theta_x}{dx^2} = \frac{dT_x}{dx} = -k\frac{x^2}{L^2}$$

By integrating

$$GJ \frac{d\theta_x}{dx} = T_x = -k\frac{x^3}{3L^2} + C_1 \tag{i}$$

At $x = 0$, $T_A = T(0) = 0$ hence $C_1 = 0$

At $x = L$, $T_B = T(L) = -\frac{kL}{3}$

The minus sign indicates that the reaction torque is opposite to the applied torque.
By integrating (i)

$$GJ\theta_x = -k\frac{x^4}{12L^2} + C_2 \tag{ii}$$

At $x = L$, $\theta_B = \theta(L) = 0$ hence, $C_2 = \frac{kL^2}{12}$

Therefore,

$$\theta_x = \frac{k}{12L^2 GJ}(L^4 - x^4)$$

The maximum twist,

$$\theta_{max} = \frac{kL^2}{12GJ}$$

6.4 TAPERED CIRCULAR SHAFTS

The expressions derived for the prismatic circular shaft, i.e. the shaft of uniform cross-section, subjected to a constant torque over its entire length, can be applied to the tapered shaft with small taper without much error. However, in case of large taper, the following procedure can be used.

Consider a tapered solid circular shaft of length L with its diameter varying from D_1 at one end to D_2 at the other as shown in Figure 6.7. The shaft is subjected to a torque T which is constant throughout its length. Consider an element of length dx at distance x from the smaller end. The diameter of the elemental slice is,

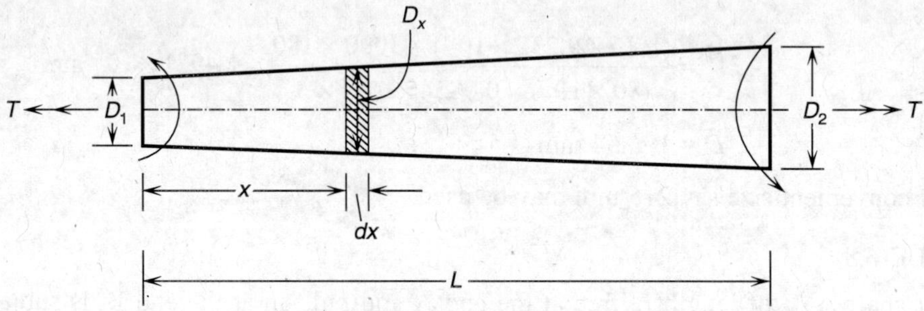

Figure 6.7 Tapered circular shaft.

$$D_x = D_1 + \left(\frac{D_2 - D_1}{L}\right)x = D_1 + kx$$

where $$k = (D_2 - D_1)/L \tag{6.30}$$

Corresponding polar moment of inertia of the shaft at the section under consideration,

$$J_x = \frac{\pi D_x^4}{32} = \frac{\pi \times [D_1 + kx]^4}{32}$$

The angle of twist over the length dx can be obtained from the relation,

$$\frac{T}{J_x} = \frac{G\,d\theta_x}{dx} \quad \text{i.e.} \quad d\theta_x = \frac{T\,dx}{GJ_x}$$

Therefore, $$d\theta_x = \frac{T}{G}\left[\frac{32}{\pi(D_1 + kx)^4}\right]dx$$

The total angle of twist over the entire length is

$$\theta = \frac{32T}{\pi G}\int_0^L \frac{dx}{(D_1 + kx)^4} = \frac{32T}{\pi G}\int_0^L (D_1 + kx)^{-4}\,dx$$

$$= \frac{32T}{\pi G}\left[-\frac{(D_1 + kx)^{-3}}{3k}\right]_0^L = \frac{32T}{3\pi Gk}\left[-\frac{1}{(D_1 + kx)^3}\right]_0^L$$

$$= \frac{32T}{3\pi Gk}\left[-\frac{1}{(D_1+kL)^3}+\frac{1}{D_1^3}\right] = \frac{32T}{3\pi Gk}\left[-\frac{1}{D_2^3}+\frac{1}{D_1^3}\right]$$

$$= \frac{32TL}{3\pi G(D_2-D_1)} \times \frac{(D_2^3-D_1^3)}{D_1^3\,D_2^3}$$

$$= \frac{32TL}{3\pi G} \times \left(\frac{D_2^2+D_1^2+D_1D_2}{D_1^3D_2^3}\right) \tag{6.31}$$

The maximum shear stress $$\tau_{max} = \frac{16T}{\pi D_1^3} \tag{6.32}$$

The following example shall illustrate the procedure.

EXAMPLE 6.9

A tapered solid circular shaft of length L, has cross-section with its diameter varying from D at one end to $2D$ at the other. Determine the angle of rotation when the shaft is subjected to a pair of equal and opposite torques T applied at its ends.

Solution: As the shaft is subjected to a pair of equal and opposite torques T applied at its ends, the torque is constant throughout the length of the shaft. The diameter of the shaft at distance x from the smaller end,

$$D_x = D + \left(\frac{2D-D}{L}\right)x = \frac{D}{L}(L+x)$$

The corresponding polar moment of inertia,

$$J_x = \frac{\pi D_x^4}{32} = \frac{\pi}{32}\left[\frac{D}{L}(L+x)\right]^4 = \frac{\pi D^4}{32L^4}(L+x)^4$$

The angle of twist of the element can be obtained from the relation,

$$\theta = \frac{TL}{GJ} \quad \text{i.e.} \quad d\theta = \frac{T\,dx}{GJ_x}$$

Therefore, $$d\theta = \frac{T}{G}\left[\frac{32L^4}{\pi D^4 (L+x)^4}\right]dx$$

The total angle of twist θ over the entire length is

$$\theta = \frac{32TL^4}{\pi GD^4}\int_0^L \frac{dx}{(L+x)^4} = \frac{32TL^4}{\pi GD^4}\left[-\frac{1}{3(L+x)^3}\right]_0^L$$

$$= \frac{32TL^4}{3\pi GD^4} \times \frac{7}{8L^3} = \frac{28TL}{3\pi GD^4}$$

Alternatively, from Eq. (6.31):

$$\theta = \frac{32TL}{3\pi G} \times \left[\frac{D^2 + (2D)^2 + D(2D)}{D^3(2D)^3} \right]$$

$$= \frac{32TL}{3\pi G} \times \frac{7}{8D^4} = \frac{28TL}{3\pi GD^4}$$

6.5 COMPOUND CIRCULAR SHAFTS

6.5.1 Shafts in Series

These shafts are made of two or more shafts of different sizes and materials connected together in series. At the connecting points for compatibility, the twist in shafts must be equal. The following examples will illustrate the procedure.

EXAMPLE 6.10

A 3 m long steel shaft ABCD consists of three segments of different cross-sections; the segment AB is hollow with internal and external diameters of 60 and 100 mm, respectively (see Figure 6.8); the segments BC and CD are solid having 100 and 85 mm diameters, respectively. The shaft is subjected to equal and opposite torques at the ends and the angle of twist is the same for each segment. Determine: (a) the length of three segments, and (b) the torque that can be transmitted by the shaft and the total angle of twist, if the maximum shearing stress in the hollow segment is limited to 45 MPa. The elastic modulus of rigidity for steel is 80 GPa.

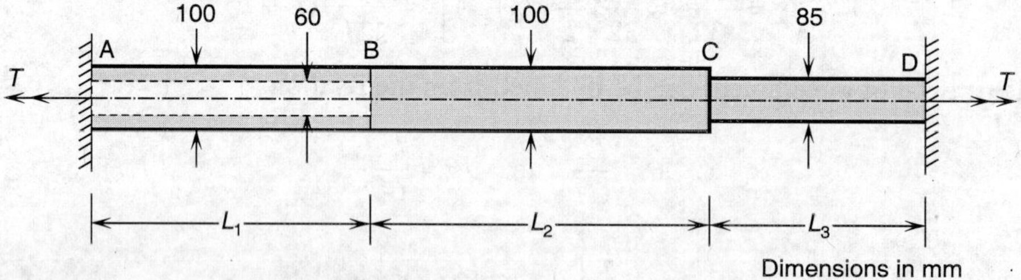

Figure 6.8 Shafts in series of Example 6.10.

Solution: Consider lengths of segments AB, BC and CD to be L_1, L_2 and L_3, respectively. The torque T and elastic modulus G are constant throughout the length of the shaft, and the angle of twist $\theta = TL/GJ$ is same for all the three segments. Thus,

$$\frac{L_1}{J_1} = \frac{L_2}{J_2} = \frac{L_3}{J_3} = \frac{L_1 + L_2 + L_3}{J_1 + J_2 + J_3} = \frac{L}{J_1 + J_2 + J_3}$$

where $$J_1 + J_2 + J_3 = \frac{\pi}{32}[(100^4 - 60^4) + (100)^4 + (85)^4]$$

$$= \frac{\pi}{32}(2.3924 \times 10^8)$$

Therefore,

$$L_1 = \frac{J_1}{J_1 + J_2 + J_3} L = \frac{(100^4 - 60^4) \times 3}{2.3924 \times 10^8} = 1.0915 \text{ m}$$

$$L_2 = \frac{J_2}{J_1 + J_2 + J_3} L = \frac{(100^4) \times 3}{2.3924 \times 10^8} = 1.2540 \text{ m}$$

$$L_3 = \frac{J_3}{J_1 + J_2 + J_3} L = \frac{(85^4) \times 3}{2.3924 \times 10^8} = 0.6546 \text{ m}$$

(b) Torque transmitted by the shaft

$$T = \frac{\tau_{max} J_1}{r_1} = \frac{45 \times [\pi(100^4 - 60^4)/32]}{(100/2)}$$

$$= 7690619 \text{ N.mm} = 7.6906 \text{ kN.m}$$

The angle of twist of a segment

$$\theta_{A-B} = \frac{\tau_{max} L_1}{G r_1} = \frac{45 \times 1091.5}{(80 \times 10^3) \times (100/2)} = 0.01228 \text{ radian}$$

Total angle of twist of the shaft

$$\theta = 3\theta_{A-B} = 3 \times 0.01228 = 0.03684 \text{ radian} = 0.03684 \times \frac{180}{\pi} = 2°6'40''$$

EXAMPLE 6.11

A steel shaft ABCD having a total length of 1530 mm consists of three segments of different lengths and cross-sections; the segment AB is hollow with internal and external diameters of d and 120 mm, respectively; the segments BC and CD are solid having diameters of 120 mm and 105 mm, respectively. The lengths of three segments AB, BC and CD are 360, 450 and 720 mm, respectively. The shaft is subjected to equal and opposite torques at the ends. Determine: (a) the maximum value of d such that the maximum shearing stress in segment AB will not exceed that in the weakest segment CD and (b) the total angle of twist, if the torque transmitted by the shaft is 13 kN.m. The elastic modulus of rigidity for steel is 80 GPa.

Solution: Since the torque T is constant throughout the length of the shaft and the shearing stress is equal in the segments AB and CD, thus

$$\tau_{\max} = \frac{T}{Z_{p,AB}} = \frac{T}{Z_{p,CD}} \quad \text{or} \quad Z_{p,AB} = Z_{p,CD}$$

The polar section moduli for segments AB and CD of the shaft are,

$$Z_{p,AB} = \frac{\pi}{16D}(D^4 - d^4) = \frac{\pi}{16\times 120}(120^4 - d^4)$$

$$Z_{p,CD} = \frac{\pi D^3}{16} = \frac{\pi \times 105^3}{16}$$

$$105^3 = \frac{(120^4 - d^4)}{120} = 120^3 - \frac{d^4}{120} \quad \text{or} \quad d^4 = 120\times(120^3 - 105^3)$$

Therefore, $d = 90.96$ mm.

The total angle of twist in the shaft is

$$\theta = \sum \frac{T_i L_i}{G_i J_i} = \frac{T}{G}\sum \frac{L_i}{J_i} = \frac{T}{G}\left(\frac{L_{AB}}{J_{AB}} + \frac{L_{BC}}{J_{BC}} + \frac{L_{CD}}{J_{CD}}\right)$$

$$= \frac{13\times 10^6}{80\times 10^3} \times \left(\frac{360}{\pi(120^4 - 90.96^4)/32} + \frac{450}{\pi(120^4)/32} + \frac{720}{\pi(105^4)/32}\right)$$

$$= \frac{13\times 10^6 \times 32}{80\times 10^3 \times \pi} \times \left(\frac{360}{120^4 - 90.96^4} + \frac{450}{120^4} + \frac{720}{105^4}\right) = 0.01769 \text{ radian} = 1°0.8'$$

EXAMPLE 6.12

A compound shaft A–B–C consisting of two solid shafts A–B and B–C of aluminium and steel having lengths of 1.6 and 2.2 m, respectively, is fixed to rigid supports 3.8 m apart as shown in Figure 6.9. The diameters of the shafts A–B and B–C are 90 and 60 mm, respectively. Determine the maximum shearing stress developed in the assembly when a torque of 1.5 kN.m is applied at the junction of the two shafts. The elastic moduli of rigidity for aluminium and steel are 28 and 84 GPa, respectively.

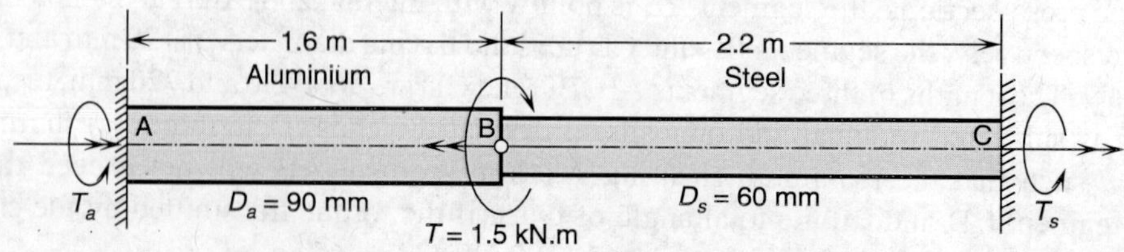

Figure 6.9 Composite shaft of Example 6.12.

Solution: If T_a and T_s are the torques resisted by the shafts, A–B and B–C, respectively, then

$$T_a + T_s = 1.5 \text{ kN.m}$$

For the rotational compatibility at the junction of the two shafts: $\theta_a = \theta_s$ i.e.

$$\frac{T_a L_a}{G_a J_a} = \frac{T_s L_s}{G_s J_s} \quad \text{or} \quad \frac{T_a}{T_s} = \left(\frac{G_a}{G_s}\right)\left(\frac{J_a}{J_s}\right)\left(\frac{L_s}{L_a}\right)$$

Therefore,
$$\frac{T_a}{T_s} = \frac{28 \times 10^3}{84 \times 10^3} \times \frac{(\pi \times 90^4/32)}{(\pi \times 60^4/32)} \times \frac{2200}{1600} = \frac{1}{3} \times \left(\frac{90}{60}\right)^4 \times \frac{11}{8} = \frac{297}{128}$$

or
$$\frac{T_a}{297} = \frac{T_s}{128} = \frac{T_a + T_s}{297 + 128} = \frac{(1.5 \times 10^6)}{425}$$

Thus,

$$T_a = \frac{(1.5 \times 10^6)}{425} \times 297 = 1048235.29 \text{ N.mm} = 1.048 \text{ kN.m}$$

$$T_s = \frac{(1.5 \times 10^6)}{425} \times 128 = 451764.7 \text{ N.mm} = 0.4518 \text{ kN.m}$$

The maximum stress in a shaft is,

$$\tau_{\max} = \frac{T}{Z_p} = \frac{16T}{\pi D^3}$$

$$\tau_{a,\max} = \frac{16T_a}{\pi D_a^3} = \frac{16 \times 1048235.29}{(\pi \times 90^3)} = 7.323 \text{ MPa}$$

$$\tau_{s,\max} = \frac{16T_s}{\pi D_s^3} = \frac{16 \times 451764.7}{(\pi \times 60^3)} = 10.652 \text{ MPa}$$

6.5.2 Compound Sections or Shafts in Parallel

If the cross-section of a member is made of two different materials bonded together as shown in Figure 6.10(a), the *linear strain* assumption used for a solid member still holds good, i.e. at the interface of two materials, the strain is same. If the shear modulus of inner core and outer sleeve or tube are G_i and G_o, respectively, then according to Hooke's law,

$$\left(\frac{\tau}{G}\right)_o = \left(\frac{\tau}{G}\right)_i \quad \text{or} \quad \left(\frac{\theta}{L}\right)_o r_{\text{interface}} = \left(\frac{\theta}{L}\right)_i r_{\text{interface}}$$

Therefore,
$$\frac{\tau_i}{\tau_o} = \frac{G_i}{G_o} \tag{6.33}$$

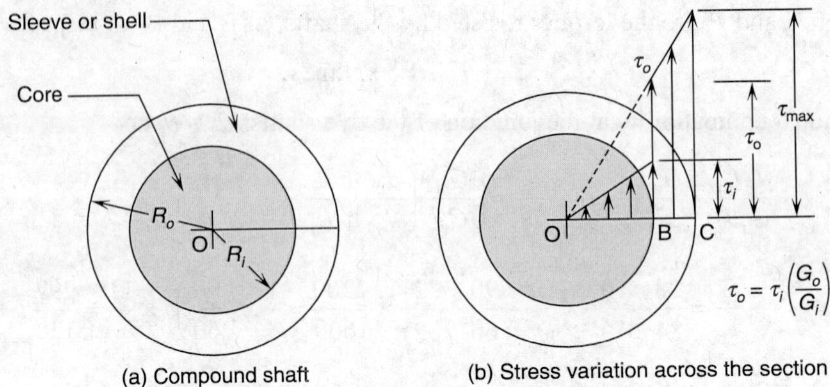

Figure 6.10 Analysis of compound shaft.

Thus, the ratio of the shear stresses in two materials at the interface is G_i/G_o. The shear stress distribution is shown in Figure 6.10(b). If a composite section is subjected to a torque T, then $T = T_o + T_i$. If both the shafts are of same length, then $\theta_i = \theta_o = \theta$

$$T = \frac{G_oJ_o\theta}{L} + \frac{G_iJ_i\theta}{L} = \frac{\theta}{L}(G_oJ_o + G_iJ_i)$$

Therefore,
$$\theta = \frac{TL}{G_oJ_o + G_iJ_i} \tag{6.34}$$

Usually, the inner core is softer. The following examples illustrate the procedure.

EXAMPLE 6.13

A gun metal sleeve fits closely over a steel shaft and compound shaft carries a torque. Determine the ratio of outer diameter of the sleeve to the diameter of the shaft if the torque carried by the sleeve is three times that carried by the shaft. The allowable shear stresses in gun metal and steel are 45 and 80 MPa, respectively. The corresponding elastic moduli for the two materials are 30 and 80 GPa, respectively. Also determine the torque that can be transmitted by the compound shaft when the shaft diameter is 60 mm.

Solution: The polar section modulus for the solid shaft of diameter D_i is

$$Z_{p,s} = \frac{\pi D_i^3}{16}$$

Polar sectional modulus for the sleeve,

$$Z_{p,gm} = \left(\frac{\pi D_o^4}{32} - \frac{\pi D_i^4}{32}\right)\left(\frac{2}{D_o}\right) = \frac{\pi D_o^3}{16}\left[1 - \left(\frac{D_i}{D_o}\right)^4\right]$$

For torque carrying condition,

$$3T_s = T_{gm} \quad \text{i.e.} \quad 3\tau_s Z_{p,s} = \tau_{gm} Z_{p,gm}$$

or $$3\times 80\times\left(\frac{\pi D_i^3}{16}\right)=45\times\frac{\pi D_o^3}{16}\left[1-\left(\frac{D_i}{D_o}\right)^4\right]$$

or $$\left(\frac{D_i}{D_o}\right)^3=\frac{3}{16}\left[1-\left(\frac{D_i}{D_o}\right)^4\right] \text{ or } \left(\frac{D_i}{D_o}\right)^4+\frac{16}{3}\left(\frac{D_i}{D_o}\right)^3-1=0$$

By trial and modification, the lowest real value of the ratio is 0.596. For

$$D_i = 60 \text{ mm}, D_o=\frac{60}{0.596}=100.67 \text{ mm}$$

For compatibility, $$\left(\frac{\theta}{L}\right)_s=\left(\frac{\theta}{L}\right)_{gm} \text{ or } \left(\frac{T}{GJ}\right)_s=\left(\frac{T}{GJ}\right)_{gm}$$

$$\left(\frac{T_s}{(80\times 10^3)}\times\frac{64}{\pi D_i^4}\right)=\left[\frac{T_{gm}}{(30\times 10^3)}\times\frac{64}{\pi(D_o^4-D_i^4)}\right]$$

or $$\frac{T_{gm}}{T_s}=\frac{3}{8}\left(\frac{D_o^4-D_i^4}{D_i^4}\right)=3$$

Therefore, $$\left(\frac{D_o^4}{D_i^4}\right)-1=8 \text{ or } \left(\frac{D_o}{D_i}\right)^4=9 \text{ i.e. } \frac{D_o}{D_i}=1.732$$

For $D_i = 60$ mm, $D_o = 60 \times 1.732 = 103.92$ mm. Adopt larger of the two values.

Hence, $d_o = 103.92$ mm

$$T_s=\tau_s Z_{p,s}=80\times\left(\frac{\pi\times 60^3}{16}\right)=3393\times 10^3 \text{ N.mm} = 3393 \text{ N.m}$$

$$T_{gm}=\tau_{gm}Z_{p,gm}=45\times\frac{\pi}{16}\times(103.92)^3\times\left[1-\left(\frac{1}{\sqrt{3}}\right)^4\right]=8814\times 10^3 \text{ N.mm} = 8814 \text{ N.m}$$

Therefore, total torque, $T = T_s + T_{gm} = 3393 + 8814 = 12207$ N.m

EXAMPLE 6.14

A composite shaft consists of a steel shaft of 100 mm diameter encased in a closely fitting brass sleeve. Determine: (a) the outside diameter of the sleeve if the torque applied to the composite shaft is to be shared equally by the two components and (b) the maximum shearing stress in each material and angle of twist over a length of 3 m, when the composite shaft transmits a torque of 25 kN.m. The elastic moduli for the brass and steel are 40 and 80 MPa, respectively.

Solution: As the torque is shared equally by the steel shaft and brass sleeve

$$T_s = T_b \text{ i.e. } \frac{G_s J_s\theta}{L}=\frac{G_b J_b\theta}{L} \text{ or } \frac{J_b}{J_s}=\frac{G_s}{G_b} \quad \text{(i)}$$

Consider the outside diameter of the brass sleeve to be D, the corresponding polar moments of inertia for the steel shaft and brass sleeve are

$$J_s = \frac{\pi \times 100^4}{32} \quad \text{and} \quad J_b = \frac{\pi}{32}(D^4 - 100^4)$$

Therefore, from Eq. (i):

$$\frac{(D^4 - 100^4)}{100^4} = \frac{80 \times 10^3}{40 \times 10^3} = 2$$

$$D^4 = (2+1) \times 100^4 \quad \text{or} \quad D = 100\sqrt[4]{3} = 131.6 \text{ mm}$$

From torque sharing condition,

$$T_s = T_b = 12.5 \text{ kN.m}$$

$$\tau_s = \frac{T_s(D/2)}{J_s} = \frac{(12.5 \times 10^6) \times 50}{(\pi \times 100^4/32)} = \frac{1.25 \times 32 \times 5}{\pi} = 63.66 \text{ MPa}$$

$$\tau_b = \frac{T_b(D/2)}{J_b} = \frac{(12.5 \times 10^6) \times (131.6/2)}{[\pi \times (131.6^4 - 100^4)/32]} = 41.90 \text{ MPa}$$

(b) Angle of twist,
$$\theta = \frac{T_s L}{G_s J_s} = \frac{(12.5 \times 10^6) \times 3000}{(80 \times 10^3)(\pi \times 100^4/32)}$$

$$= 0.04775 \text{ radian} = 0.04775 \times \frac{180}{\pi} = 2.736° = 2°44'10''$$

6.6 SHAFT COUPLINGS

In practice, frequently situations arise where a long shaft is made up of several pieces joined together by means of so-called *flanged shaft couplings* of the type shown in Figure 6.11. When bolted together, such couplings are termed *rigid*, to differentiate them from the one called *flexible* that provides the misalignment of adjoining shafts. The latter type is mainly used to join the shaft of a motor to the driven or rotating equipment. In this section only flanged shaft couplings are discussed.

In the rigid couplings, the torque at one end of the shaft is transmitted to the other end through the flanges and flange bolts; the bolts are primarily subjected to shear and friction between flanges is neglected. The shear strains in the bolts are assumed to vary linearly with their distance from the axis of the shaft. Therefore, the problem is analogous to that of a circular shaft in torsion, if the bolts are of same material, the elastic shear stresses in the bolts also vary linearly with respective distances from the centre of coupling. However, unlike a continuous cross-section of the circular shaft, a system of discrete points is considered in the coupling. The shearing stress in a bolt is considered to be uniform across its section. *The moment of tangential forces developed in the bolts around the axis of the shaft provide the torque resisting capacity of the coupling.*

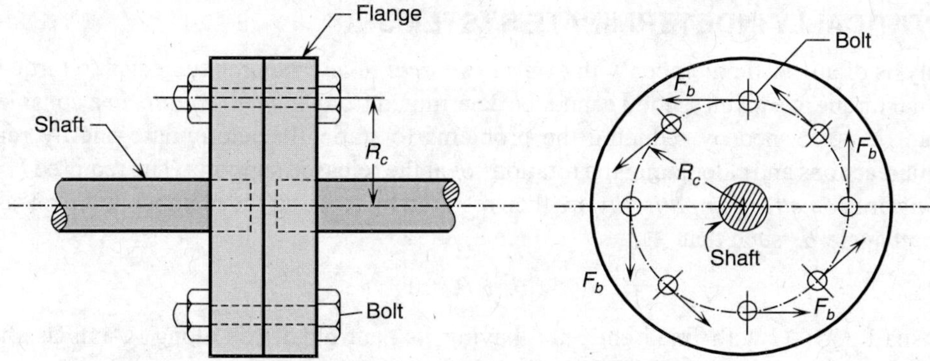

Figure 6.11 Flanged shaft coupling.

The forgoing method of analysis suffers from the fact that the initial tension in the bolts is large enough to cause the entire coupling to act in friction. However, with high tensile strength bolts, the strength of coupling may be greater than it would be if the bolts had to act in shear.

Tangential force in a bolt, $$F_b = A_b \tau_b \tag{6.35}$$

where A_b and τ_b are the area of the bolt and corresponding shear stress in the bolt, respectively. If the coupling has n bolts of diameter of d_b arranged in a *bolt circle* of radius R_c, then the torque resisting capacity of the shaft coupling,

$$T_{\text{coup}} = nF_b R_c = n(A_b \tau_b) R_c = n\left(\frac{\pi}{4} d_b^2\right)(\tau_b) R_c \tag{6.36}$$

EXAMPLE 6.15

A shaft made up of two pieces of 60 mm diameter joined together by means of a rigid flanged coupling transmits a torque such that the allowable shearing stress in the shafts and bolts are limited to 80 MPa and 60 MPa, respectively. Determine the diameter of the bolts if eight bolts are used on the coupling in a bolt circle of 160 mm diameter.

Solution: The torque transmitting capacity of the shaft

$$T_s = \tau_s Z_s = 80 \times \frac{\pi \times 60^3}{16} \text{ N.mm}$$

The torque resisting capacity of the shaft coupling

$$T_{\text{coup}} = 8 \times \left(\frac{\pi}{4} d_b^2\right) \times (60) \times 80$$

Therefore, for $T_{\text{coup}} = T_s$

$$8 \times \left(\frac{\pi}{4} d_b^2\right) \times (60) \times 80 = 80 \times \frac{\pi \times 60^3}{16}$$

or $$d_b^2 = \frac{15^2}{2} \quad \text{i.e.} \quad d_b = \frac{15}{\sqrt{2}} = 10.61 \text{ mm}$$

6.7 STATICALLY INDETERMINATE SYSTEMS

The analysis of an elastic member with *external statical indeterminacy* subjected to torque where the internal torque at a given section cannot be determined from *moment equilibrium* considerations alone can be performed by reducing the problems to statically determinate one by removing redundant reactions and calculating the rotations θ_o at the released supports. The required *boundary conditions* or *geometric compatibility* are then restored by twisting the member at the released end through an angle θ_r, such that

$$\theta_o + \theta_r = 0 \tag{6.37}$$

For the shaft A–B–C with fixed ends and having its centroidal axis along X-axis as shown in Figure 6.12, the moment equilibrium equation for the member can be expressed as:

$$\sum M_x = T_A + T + T_C = 0 \tag{6.38}$$

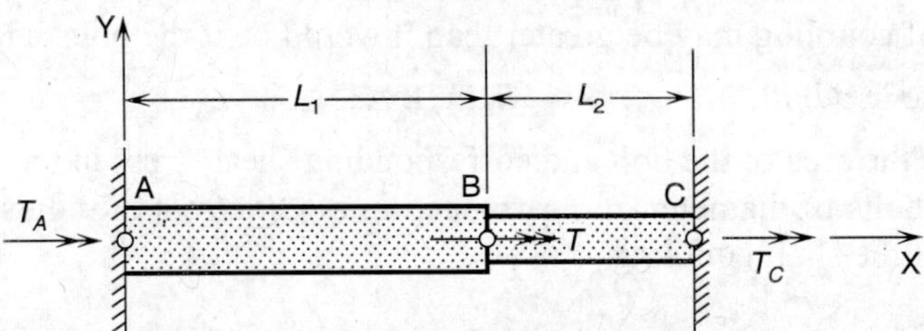

Figure 6.12 Externally statically indeterminate member in torsion.

The geometric compatibility condition requires that the ends A and C do not rotate.

Thus for this shaft, $\theta_{A-C} = \theta_{A-B} + \theta_{B-C} = 0$ (6.39)

The torsion problems with *internal statical indeterminacy* are generally encountered in the composite shafts built up from two or more tubes or materials. In such cases, the angle of twist θ is the same for each constituent part of the member at the common connection. In these problems, the torque T_i for ith constituent part of the shaft is expressed as

$$T_i = \frac{G_i J_i}{L_i}\theta_i = k_i \theta_i \tag{6.40}$$

where is the *rotational stiffness* of the member.

The applied torque T is then the sum of n parts, i.e.

$$T = \sum_{i=1}^{n} T_i = \sum_{i=1}^{n} k_i \theta_i \tag{6.41}$$

EXAMPLE 6.16

Draw the twisting moments, stresses and angles of rotation diagrams for a stepped shaft of solid circular section subjected to a system of torques shown in Figure 6.13(a).

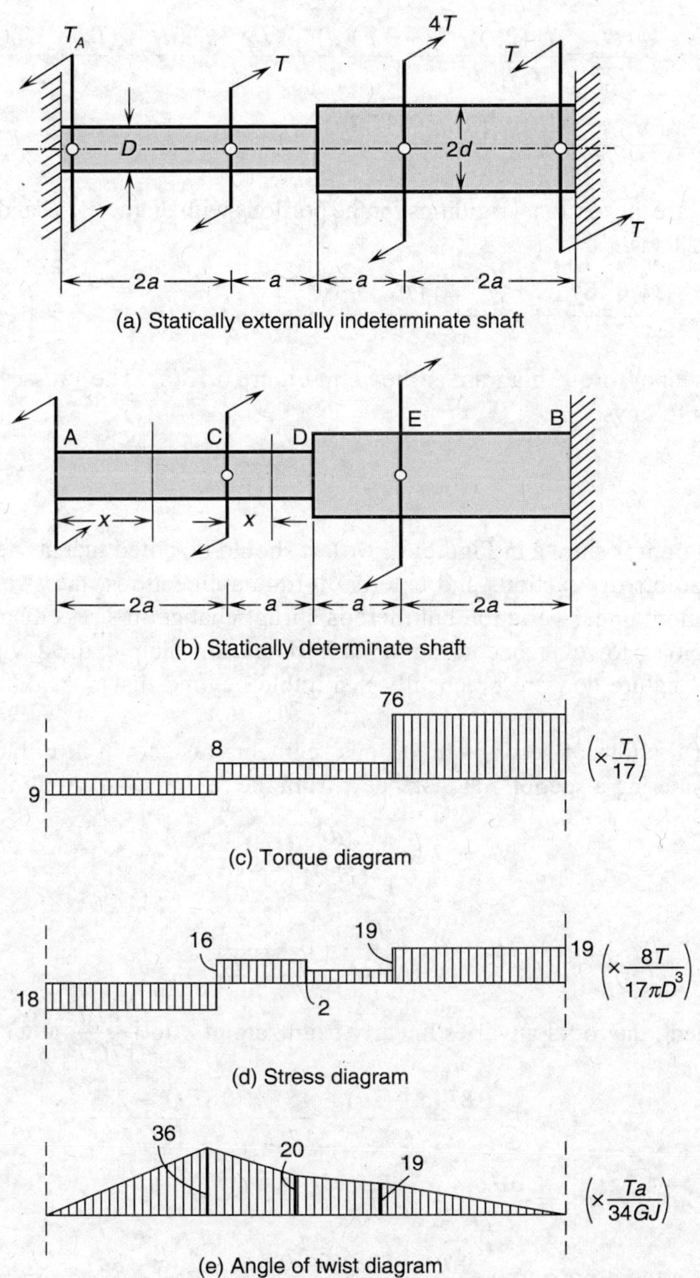

Figure 6.13 Stepped shaft of Example 6.16 subjected to a system of torques.

Solution: The shaft is statically externally indeterminate to the first degree. Consider the fixing moment at the left support T_A to be the unknown reaction. The statically determinate shaft is shown in Figure 6.13(b). The condition that the rotation at the left end A with respect to the right end B is zero can be used to determine unknown reaction T_A. Therefore,

$$\theta_{A-B} = \sum \frac{TL}{GJ} = \frac{T_A(2a)}{GJ_1} + \frac{(T_A - T)(a)}{GJ_1} + \frac{(T_A - T)(a)}{GJ_2} + \frac{(T_A - 5T)(2a)}{GJ_2}$$

$$= \frac{(3T_A - T)(a)}{GJ_1} + \frac{(3T_A - 11T)(a)}{GJ_2} = 0$$

where GJ_1 and GJ_2 are the torsional rigidities for the portions with diameters D and $2D$, respectively, with $GJ_2 = 16GJ_1$. Thus,

$$\theta_{A-B} = \frac{(3T_A - T)}{1} + \frac{(3T_A - 11T)}{16} = \frac{(51T_A - 27T)}{16} = 0 \quad \text{or} \quad T_A = \left(\frac{9}{17}\right)T$$

The twisting moment or torque diagram is shown in Figure 6.13(c). The stresses can be computed from Eq. (6.10) as follows,

$$\tau_{\max} = \frac{T_x}{Z_{p,x}} = \frac{16T_x}{\pi D_x^3}$$

The stress diagram is shown in Figure 6.13(d). It should be noted that at the points of change in cross-section, points of reactions and external torque applications, the stress distribution on sections do not follow linear variation but there is abrupt change in stress distribution, i.e. large local stresses develop. However, according to *Saint-Venant's* principle, the deviations from linear law are of a local nature and are practically negligible beyond distances of the order of shaft diameter.

The angles of rotation of sections in various segments (as shown in Figure 6.13(e)) can be computed by considering a section j at distance x from the reference section i i.e.

$$\theta_j = \theta_i + \int_L \frac{T_x\,dx}{GJ_x}$$

For $0 < x < 2a$ $$\theta_x = \theta_A + \frac{T_A x}{GJ} = 0 + \frac{9Tx}{17GJ}$$

Thus in this segment, the rotation varies linearly from zero at A to $\dfrac{18Ta}{17GJ}$ at C.

For $2a < x < 3a$ $$\theta_x = \theta_c - \frac{8T(x-2a)}{17GJ} = \frac{18Ta}{17GJ} - \frac{8T(x-2a)}{17GJ}$$

Therefore, at $x = 3a$, $$\theta_D = \frac{10Ta}{17GJ}$$

For $3a < x < 4a$ $$\theta_x = \theta_D - \frac{8T(x-3a)}{17(16GJ)} = \frac{10Ta}{17GJ} - \frac{8T(x-3a)}{17(16GJ)}$$

Therefore, at $x = 4a$, $$\theta_E = \frac{19Ta}{34GJ}$$

For $4a < x < 6a$ $$\theta_x = \theta_E - \frac{76T(x-4a)}{17(16GJ)} = \frac{19Ta}{34GJ} - \frac{76T(x-4a)}{17(16GJ)}$$

Therefore, at $x = 6a$, $\theta_B = 0$

The angle of rotation diagram is shown in Figure 6.13(e).

EXAMPLE 6.17

A 1.4 m long composite shaft A–B–C built-in at the ends consists of two segments A–B and B–C of equal length. The segment A–B is comprised of a brass rod of 20 mm diameter placed concentrically in an aluminium tube of external and internal diameters of 60 and 40 mm, respectively, whereas the segment, B–C is a hollow steel tube having external and internal diameters of 40 and 30 mm, respectively, as shown in Figure 6.14. Determine the maximum shearing stress in each material when the composite shaft is subjected to a torque of 7.5 kN.m applied through a thin rigid plate fixed at B. The elastic moduli for the aluminium, brass and steel are 28, 42 and 84 MPa, respectively.

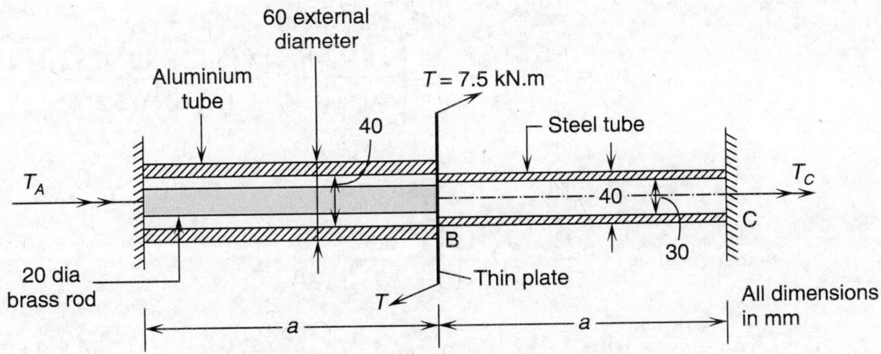

Figure 6.14 Compound shaft of Example 6.17.

Solution: The shaft is statically indeterminate to the first degree. Consider the fixing moment at the left support T_A to be the unknown reaction. From the compatibility condition $\theta_{A-C} = 0$:

$$\theta_{A-C} = \sum \frac{TL}{GJ} = \frac{T_A(a)}{G_1J_1} + \frac{(T_A - T)(a)}{G_2J_2}$$

$$= T_A(a)\left(\frac{1}{G_1J_1} + \frac{1}{G_2J_2}\right) - \frac{T(a)}{G_2J_2} = 0$$

Therefore, $$T_A\left(\frac{1}{G_1J_1} + \frac{1}{G_2J_2}\right) = \frac{T}{G_2J_2} \quad \text{or} \quad T_A\left(\frac{G_2J_2}{G_1J_1} + 1\right) = T \qquad \text{(i)}$$

where G_1J_1 and G_2J_2 are the torsion rigidities for the segments A–B and B–C, respectively.

$$G_1J_1 = G_aJ_a + G_bJ_b = (28 \times 10^3) \times \frac{\pi}{32}(60^4 - 40^4) + (40 \times 10^3) \times \frac{\pi}{32}(20^4)$$

$$= (4 \times 10^3) \times \frac{\pi}{32}[7 \times (60^4 - 40^4) + 10 \times (20^4)]$$

$$G_2J_2 = G_sJ_s = (84 \times 10^3) \times \frac{\pi}{32}(40^4 - 30^4)$$

$$\frac{G_2J_2}{G_1J_1} = \frac{21 \times (40^4 - 30^4)}{[7 \times (60^4 - 40^4) + 10 \times 20^4]} = 0.494$$

From Eq. (i); $T_A(0.494 + 1) = 7.5$ or $T_A = 5.02$ kN.m

Therefore, $T_C = T - T_A = T_s = 7.5 - 5.02 = 2.48$ kN.m

The stresses in the materials

$$\tau_s = \frac{T_s(D_s/2)}{J_s} = \frac{(2.48 \times 10^6) \times 20}{[\pi \times (40^4 - 30^4)/32]} = 288.698 \text{ MPa}$$

For composite shaft, $T_a + T_b = T_A = 5.02$ kN.m

and
$$\frac{T_a}{T_b} = \left(\frac{G_a}{G_b}\right)\left(\frac{J_a}{J_b}\right) = \left(\frac{28 \times 10^3}{40 \times 10^3}\right)\frac{[\pi(60^4 - 40^4)/32]}{[\pi(20^4)/32]}$$

$$= \frac{7}{10} \times (3^4 - 2^4) = \frac{91}{2}$$

Therefore,
$$\frac{T_a}{91} = \frac{T_b}{2} = \frac{T_a + T_b}{91 + 2} = \frac{5.02}{93}$$

$$T_a = \frac{5.02}{93} \times 91 = 4.912 \text{ kN.m} \quad \text{and} \quad T_b = \frac{5.02}{93} \times 2 = 0.10796 \text{ kN.m}$$

$$\tau_a = \frac{T_a(D_a/2)}{J_a} = \frac{(4.912 \times 10^6) \times 30}{[\pi \times (60^4 - 40^4)/32]} = 144.327 \text{ MPa}$$

$$\tau_b = \frac{T_b(D_b/2)}{J_b} = \frac{(0.10796 \times 10^6) \times 10}{[\pi \times (20^4)/32]} = 68.728 \text{ MPa}$$

6.8 TORSION OF A MEMBER OF NON-CIRCULAR CROSS-SECTION

6.8.1 Solid Non-circular Section

The stress analysis of a member of non-circular cross-section subjected to torsion is rather complicated, since the simplifying assumption of plane sections is no more applicable. The sections of the member warp resulting in different stress distribution, and the stresses must be determined as functions of two independent variables (y and z) rather than a single variable r. Moreover, *at the external corners of the cross-section, the shearing stresses vanish*. The *maximum stress* and *angular displacement* can be computed as:

$$\tau_{\max} = \frac{T_x}{Z_e} \quad \text{and} \quad \theta = \frac{T_xL}{GJ_e} \tag{6.42}$$

The quantities, Z_e and J_e are analogous or equivalent to the *polar section modulus* Z_p and *polar moment of inertia* J_p of the circular section, respectively.

Rectangular cross-section. For a member of rectangular cross-section shearing stress diagram based on theory of elasticity is shown in Figure 6.15(c). The stresses at the corners are zero, and the maximum stresses occur at the middle of the long sides, i.e. at the points A

$$\tau_A = \tau_{max} = \frac{T_x}{Z_e} = \frac{T_x}{\alpha a t^2} \tag{6.43}$$

At the point B,

$$\tau_B = \eta \tau_{max} \tag{6.44}$$

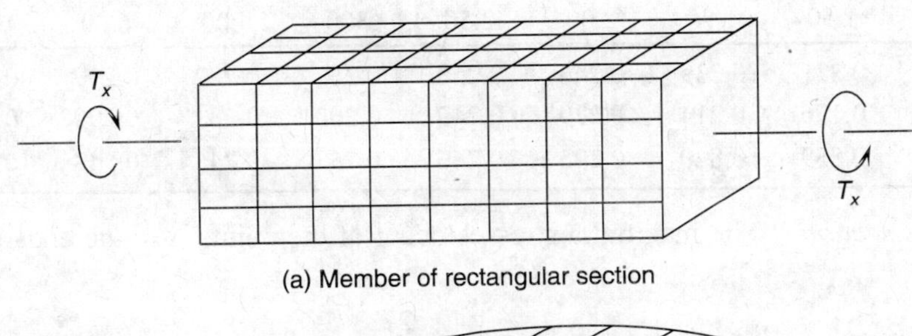

(a) Member of rectangular section

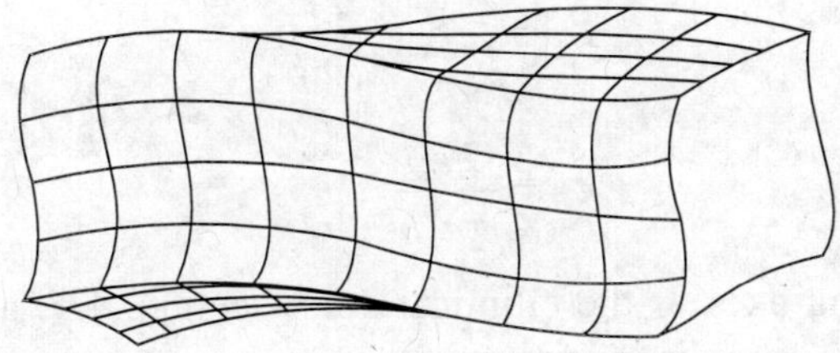

(b) Warping of transverse planes

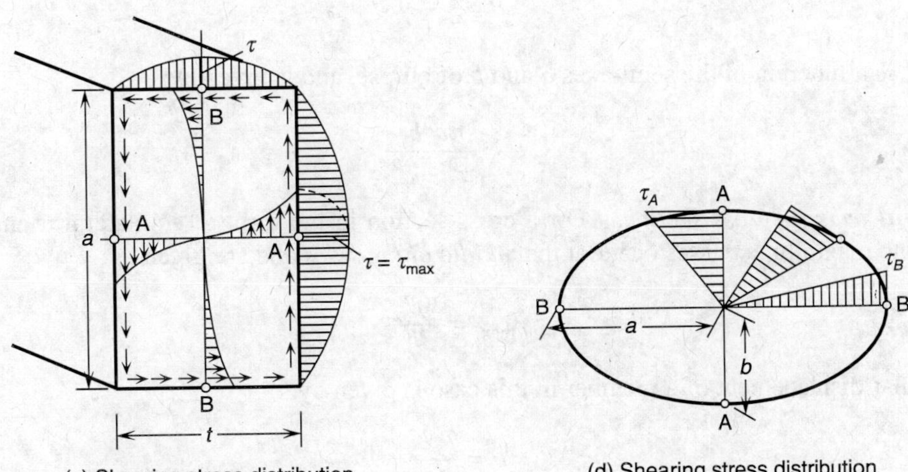

(c) Shearing stress distribution on a rectangular section

(d) Shearing stress distribution on an elliptic section

Figure 6.15 Members with non-circular sections subjected to torsion.

Where a is the longer side and t is the shorten side or thickness of the rectangular cross-section. The factors α and η depend on the ratio of the sides (a/t) and their numerical values are listed in Table 6.1. The angular displacement is given by

$$\theta = \frac{T_x L}{GJ_e} = \frac{T_x L}{G(\beta t^3 a)} \tag{6.45}$$

Here, β is also a function of ratio of the sides (a/t) and its values are given in Table 6.1.

TABLE 6.1 Coefficients for members of rectangular cross-sections

a/t	1.00	1.50	1.75	2.00	2.50	300	4.00	6.00	10.00	∞
α	0.208	0.231	0.239	0.246	0.258	0.267	0.282	0.299	0.313	0.333
β	0.141	0.196	0.214	0.229	0.249	0.263	0.281	0.299	0.313	0.333
η	1.000	0.859	0.820	0.795	0.766	0.753	0.745	0.743	0.742	0.742

Elliptical cross-section. The maximum stresses occur at the points A at the ends of the minor axis.

$$\tau_A = \tau_{\max} = \frac{2T_x}{\pi a b^2} \tag{6.46}$$

The stresses at the points B,

$$\tau_B = \frac{2T_x}{\pi a^2 b} \tag{6.47}$$

where a and b are the semi-axes of the elliptical cross-section. The angular displacement of a member of elliptical cross-section is given by:

$$\theta = \frac{T_x L}{GJ_e} \tag{6.48}$$

where, J_e is a function of the semi-axes a and b of ellipse, and is given by:

$$J_e = \frac{\pi a^3 b^3}{a^2 + b^2} \tag{6.49}$$

Equilateral triangle cross-section. For a cross-section in the form of equilateral triangle with sides a, the maximum stresses occur at the *middle of the sides* and are given by:

$$\gamma_{\max} = \frac{20T_x}{a^3} \tag{6.50}$$

The angular displacement of a member in this case is given by:

$$\theta = \frac{T_x L}{GJ_e}$$

where J_e is given by:

$$J_e = \left(\frac{\sqrt{3}}{80}\right) a^4 \tag{6.51}$$

6.8.2 Geometrical Factors

In Table 6.2 are listed *geometrical factors* for some of the commonly encountered cases of noncircular cross-sections in structural analysis. The *maximum stress* and *angular displacement* can be computed as given below.

TABLE 6.2 Geometrical factors J_e and Z_e for various sections

	Regular sections				*Thin sections*	
	Circular	*Elliptical*	*Rectangular*	*Equilateral triangle*	*Open sections*	*Closed sections*
Section						
J_e	$\frac{\pi D^4}{32}\left(1-\frac{d^4}{D^4}\right)$	$\frac{\pi a^3 b^3}{a^2+b^2}$	$\beta b^3 h$	$\frac{\sqrt{3}a^4}{80}$	$\frac{1}{3}t^3 s$	$\frac{4(A^*)^2}{\int_s \frac{ds}{t}}$
Z_e	$\frac{\pi D^3}{16}\left(1-\frac{d^4}{D^4}\right)$	$\frac{\pi ab^2}{2}$	$\alpha b^2 h$	$\frac{a^3}{20}$	$\frac{1}{3}t^2 s$	$2A^* t_{min}$

$$\tau_{max} = \frac{T_x}{Z_e} \quad \text{and} \quad \theta = \frac{T_x L}{GJ_e} \tag{6.51a}$$

6.8.3 Thin-walled Open Cross Section

The thickness of such members is essentially smaller than the other linear dimensions. This type of members subjected to torsion is commonly used in aircraft construction. Typical shapes of rolled, bent, drawn and extruded sections are shown in Figure 6.16. Thin sections are generally divided into closed and open types as shown in Figures 6.16(a) and 6.16(b), respectively. In case of *closed-section* the region inside the contour is not connected with the outside region and deflects like a closed membrane subjected to internal pressure and stresses are assumed to be uniformly distributed across the thickness of the section.

In case of an *open-section*, if it is assumed that, the stresses in a curvilinear section are nearly the same as in straight one, then from the formula for rectangular cross-section with

$$\frac{a}{t} = \infty \quad \text{i.e.} \quad \alpha = \frac{1}{3}$$

Therefore,

$$\tau_{max} = \frac{T_x}{\alpha a t^2} = \frac{3T_x}{at^2} \quad \text{and} \quad \theta = \frac{3T_x L}{Gt^3 a} \tag{6.52}$$

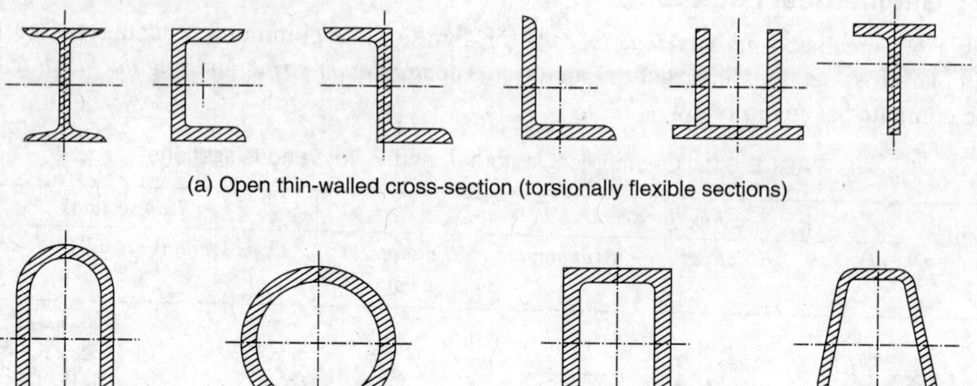

(a) Open thin-walled cross-section (torsionally flexible sections)

(b) Closed thin-walled cross-sections (torsionally stiff sections)

Figure 6.16 Thin-walled cross-sections.

These relations are general whatever be shape of the cross-section provided it can be developed into an *elongated rectangle*. In case when the thin-walled open cross-section is compound and cannot be developed into an elongated rectangle, the torsion can be considered as the sum of the torsion developed in separate portions. Then

$$T = \frac{G\theta}{3L}(t_1^3 a_1 + t_2^3 a_2 + \cdots + t_n^3 a_n) \tag{6.53}$$

and

$$\theta = \frac{3TL}{G(t_1^3 a_1 + t_2^3 a_2 + \cdots + t_n^3 a_n)} \tag{6.54}$$

It should be noted that the portion with the maximum thickness contributes the maximum, i.e. maximum stresses occur in the portion of maximum thickness $t_{\max}$. Equations (6.53) and (6.54) are also valid even if the portions are considered separately, for ith component,

$$\tau_i = \frac{3T_{x,i}}{a_i t_i^2} \quad \text{and} \quad \theta = \frac{3T_{x,i}L}{G a_i t_i^3} \tag{6.55}$$

where $T_{x,i}$ is the fraction of twisting moment corresponding to ith portion, and θ is the angular displacement which is same for all the portions of the cross-section. Eliminate $T_{x,i}$ from these expressions,

$$\tau_i = G\theta\frac{t_i}{L} \quad \text{or} \quad \tau_{\max} = G\theta\frac{t_{\max}}{L} \tag{6.56}$$

Therefore, from Eq. (6.53):

$$\tau_{\max} = \frac{3T_x t_{\max}}{(t_1^3 a_1 + t_2^3 a_2 + \cdots + t_n^3 a_n)} \tag{6.57}$$

It should be noted that this method is approximate since no account is taken of an increase in local stresses at internal corners of broken sections. To minimize local overstresses re-entrant corner in sections are made rounded.

The open, thin sections like the thin plate, the angle, the channel and the Zee shown in Figures 6.17(a), (b), (c) and (e) provide the examples of members having low rigidity or stiffness. They have the same thickness and cross-sectional area, and provide nearly the same low torsion stiffness. If the thin plate of Figure 6.17(a) is formed into a tube-like shape Figure 6.17(d) but with a slit, its stiffness will remain low. However, closing the tube completely by welding or riveting or drawing seamless tube would provide substantially higher torsion stiffness.

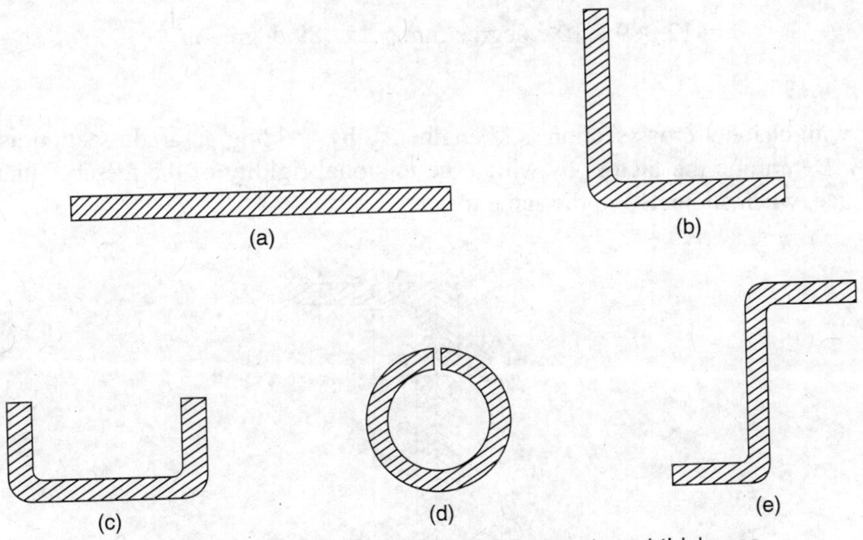

Figure 6.17 Open sections with same length and thickness.

EXAMPLE 6.18

A member of rolled aluminium angle cross-section of size 75 × 50 × 5 mm is subjected to a torque of 30 N.m. Determine the maximum shear stress and the angle of twist per unit length. The elastic modulus for the material of the section is 28 GPa.

Solution: The equivalent polar moment of inertia and section modulus of section are:

$$J_e = \sum_{i=1}^{2} \beta_i a_i t_i^3 = \beta_1 a_1 t_1^3 + \beta_2 a_2 t_2^3 \quad \text{and} \quad Z_e = \sum_{i=1}^{2} \alpha_i a_i t_i^2 = \alpha_1 a_1 t_1^2 + \alpha_2 a_2 t_2^2$$

From Table 6.1, for:

$$\left(\frac{a}{t}\right)_1 = \frac{75}{5} = 15, \quad \alpha = 0.333 \text{ and } \beta = 0.333$$

$$\left(\frac{a}{t}\right)_2 = \frac{50}{5} = 10, \quad \alpha = 0.313 \text{ and } \beta = 0.313$$

Therefore, $J_e = 0.333 \times 75 \times 5^3 + 0.313 \times 50 \times 5^3 = 5078.125 \text{ mm}^4$

$$Z_e = 0.333 \times 75 \times 5^2 + 0.313 \times 50 \times 5^2 = 1015.625 \text{ mm}^3$$

Hence $$\tau_{\max} = \frac{T}{Z_e} = \frac{30 \times 10^3}{1015.625} = 29.538 \text{ MPa}$$

$$\frac{\theta}{L} = \frac{T}{GJ_e} = \frac{30 \times 10^3}{(28 \times 10^3) \times 5078.125} \times \frac{180}{\pi}$$

$$= 12.089 \times 10^{-3} \text{ degree/mm} = 12.089 \text{ degree/m}$$

EXAMPLE 6.19

A member with channel cross-section is strengthened by welding an angle section as shown in Figure 6.18. Determine the factors by which the torsional rigidity of the member increases and stresses reduce, when subjected to the same torque.

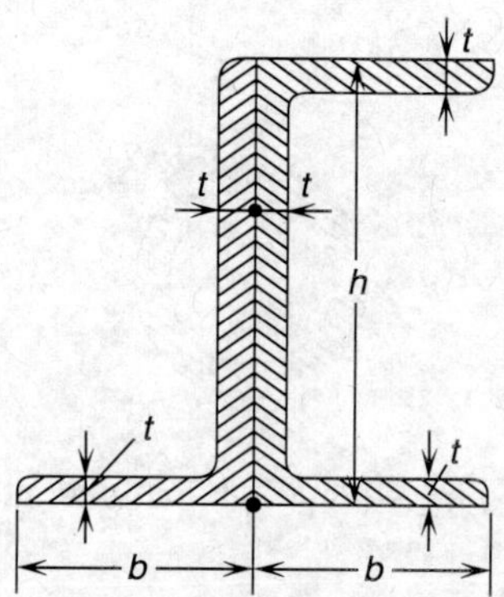

Figure 6.18 Welded member of Example 6.19.

Solution: The torsional rigidity of the member is inversely proportional to the angular displacement or rotation. In case of open thin sections, the angle of rotation or angular displacement is given by Eq. (6.53). Therefore,

(i) For the single channel section, $\theta_c = \dfrac{3TL}{Gt^3\,(2b + h)}$

(ii) For the built-up or compound section, $\theta_b = \dfrac{3TL}{G[h(2t)^3 + 3bt^3]}$

Factor by which the torsional rigidity increases, $\dfrac{\theta_c}{\theta_b} = \dfrac{8h + 3b}{h + 2b}$

The stress in an open thin section is given by Eq. (6.56). Therefore,

$$\tau_{c,\max} = \frac{3Tt}{t^3(2b+h)}$$

$$\tau_{b,\max} = \frac{3T(2t)}{h(2t)^3 + 3bt^3}$$

The factor by which the stresses reduce, $\dfrac{\tau_{c,\max}}{\tau_{b,\max}} = \dfrac{1}{2}\left(\dfrac{8h+3b}{h+2b}\right)$

6.8.4 Members with Thin-walled Closed Cross-sections

Although solid or hollow circular shafts are ideal for transmitting torques, many structural elements in aircraft, buildings and machines are expected to carry twisting moments in addition other types of load. In such applications, generally different cross-sectional profiles having wall thickness which is small as compared to linear dimensions are used. Such structural members are therefore called *thin-walled members*. Such members can be open as shown in Figure 6.16(a) or closed as shown in Figure 6.16(b).

Figure 6.19(a) illustrates a thin walled member of an arbitrary cross-section subjected to a torque T. Consider an elementary prism of length dx of the member with the dimensions of the

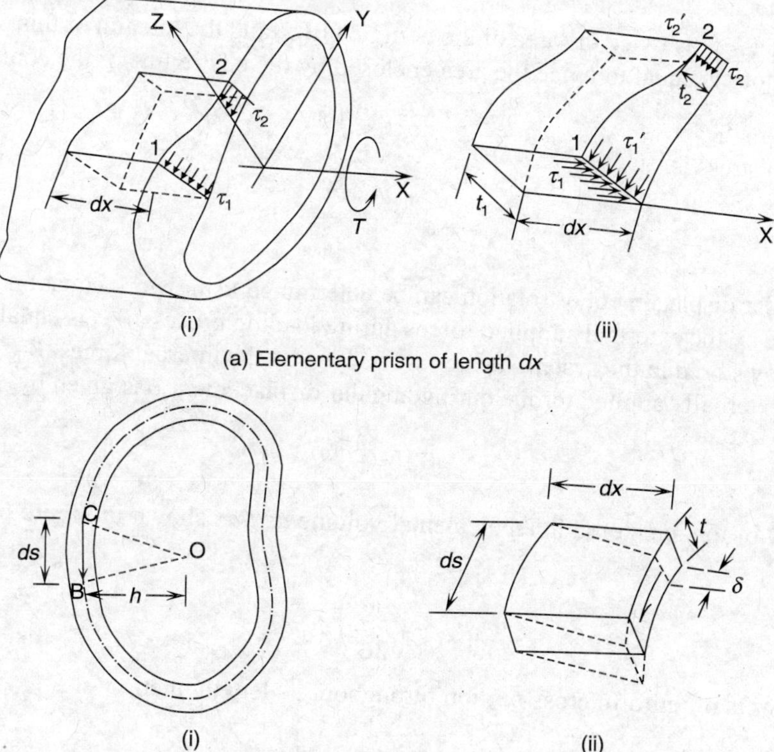

(a) Elementary prism of length dx

(b) Torque of shear force on an elementary length about point O

Figure 6.19 A member with a closed thin-wall section subjected to torsion.

prism, i.e. the distance between points 1and 2, along the arc of contour being arbitrary. The stresses τ_1 and τ_2 developed across the thicknesses t_1 and t_2 of the contour at the points 1 and 2, respectively, are uniformly distributed. Corresponding stresses develop on longitudinal planes. For equilibrium of the isolated element, all forces acting on the element are projected in the direction of the axis of the member. Therefore,

$$\tau_1 t_1\, dx = \tau_2 t_2\, dx \tag{6.58}$$

Since the points 1 and 2 along the arc of contour are taken arbitrarily, τt = constant, i.e. the product τt, called *shear flow* q does not vary along the length of the closed contour. The sections like tubes, pipes, or box sections of thickness t are generally analysed in terms of shear flow $q = \tau t$. The wall thickness is small but need not be uniform, hence the *shear stress* is considered constant across the thickness and tangential to the centre line of the section. It should be noted that thinner the section the greater will be the stress.

For computation of twisting moment, consider an elementary segment of length ds measured along the perimeter as shown in Figure 6.19(b-i). The moment of the force $\tau t\, ds$ representing the contribution of an element to the torque resistance with respect to an arbitrarily taken point O is $(\tau t\, ds)h$.Therefore, the torque resistance of the cross-section is,

$$T = \tau t \oint_s h\, ds \quad \text{(since } \tau t \text{ does not vary along the arc length)} \tag{6.59}$$

The product $h\, ds$ is twice the area of the triangle OBC, and the integral of this product around the closed contour is equal to twice the area enclosed by the centre line of the contour, A^*. Thus,

$$T = \tau t\,(2A^*) \tag{6.60}$$

The maximum stress is,

$$\tau_{\max} = \frac{T}{2A^* t_{\min}} \tag{6.61}$$

The angular displacement or rotation can be determined using work equation $W_{ex} = -W_{in}$ i.e. the work done by the externally applied forces during loading process W_{ex} is equal to the increase in strain energy stored in the system, i.e. the work done by the internal forces, $W_{in} = U_t$. The work done by the externally applied torque during angular displacement θ is given by,

$$W_{ex} = \frac{1}{2} T\theta \tag{6.62}$$

Strain energy of torsion stored in an elemental volume $ds\; t\; dx$ shown in Figure 6.19(b-ii) can be expressed as:

$$dU_t = \left(\frac{\tau^2}{2G}\right) ds\; t\; dx$$

If the member is of uniform cross-section throughout its length, then

$$U_t = \left(\frac{L}{2G}\right) \oint_s \tau^2 t\, ds = \frac{L(\tau t)^2}{2G} \oint_s \frac{ds}{t}$$

The quantity under integral sign represents geometrical property of the section. The parameter τt is defined by Eq. (6.59). Therefore,

$$U_t = \frac{L}{2G}\left(\frac{T}{2A^*}\right)^2 \oint_s \frac{ds}{t} = \frac{T^2 L}{8G(A^*)^2} \oint_s \frac{ds}{t}$$

where, G is modulus of shear or rigidity and T is the twisting moment. From the work equation,

$$\frac{1}{2} T\theta = \frac{T^2 L}{8G(A^*)^2} \oint_s \frac{ds}{t}$$

Therefore,

$$\theta = \frac{TL}{4G(A^*)^2} \oint_s \frac{ds}{t} \tag{6.63}$$

If the thickness t does not vary along the arc length of the contour,

$$\theta = \frac{TLs}{4G(A^*)^2 t} \tag{6.64}$$

where, s is the length of the closed contour or perimeter of the median line.

Closed form sections have better torsional resistance than the open forms such as plates and channel sections.

EXAMPLE 6.20

A thin walled tube rolled up from a sheet as shown in Figure 6.20(a) is subjected to twisting moment. Determine the stresses and the angular displacement when: (a) the edges of sheet are free and (b) the edges of sheet are riveted as shown in Figure 6.20(b).

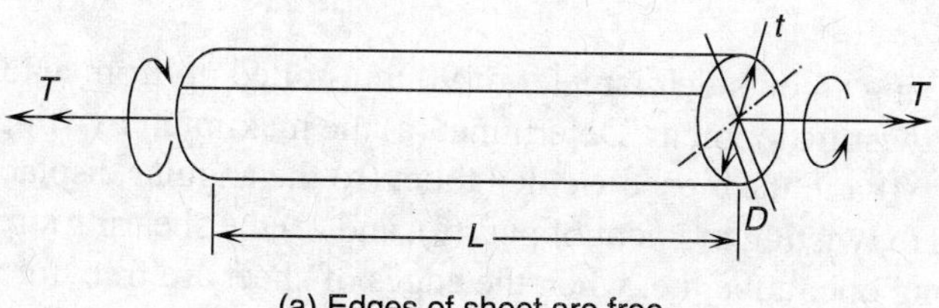

(a) Edges of sheet are free

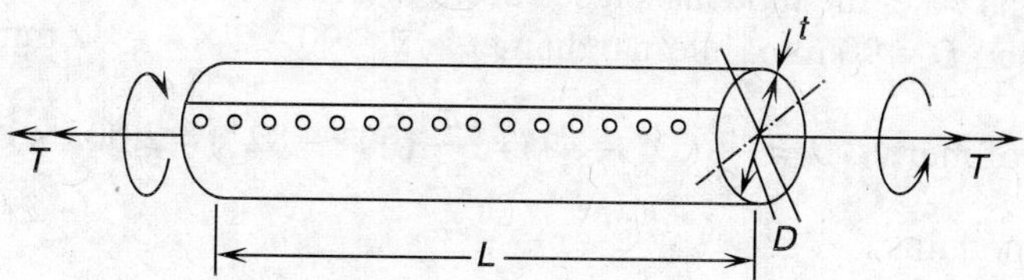

(b) Edges of sheet are rivetted

Figure 6.20 Thin-walled rolled up tube of Example 6.20.

Solution: In the case (a) the section is considered as open. However, the segment of the section in the region of the lap joint is ignored. Therefore, from Eq. (6.51),

$$\tau_{o,\max} = \frac{3T}{at^2} = \frac{3T}{st^2} = \frac{3T}{(\pi D)t^2} \text{ and } \theta_o = \frac{3TL}{Gst^3} = \frac{3TL}{G(\pi D)t^3}$$

where t is the thickness of the section (the shorter side of the developed cross-section) and s is the length of the perimeter of the cross-section (the longer side of the developed cross-section).

(b) In this case, the section is closed. From Eqs. (6.60) and (6.63),

$$\tau_{c,\max} = \frac{T}{2A^* t_{\min}} = \frac{T}{2(\pi D^2/4)t} = \frac{2T}{\pi D^2 t}$$

$$\theta_c = \frac{TLs}{4G(A^*)^2 t} = \frac{TL(\pi D)}{4G(\pi D^2/4)^2 t} = \frac{4TL}{G\pi D^3 t}$$

Factor by which the stresses reduce, $\dfrac{\tau_{c,\max}}{\tau_{o,\max}} = \dfrac{2}{3}\left(\dfrac{t}{D}\right)$

Factor by which the angular displacements reduce, $\dfrac{\theta_c}{\theta_o} = \dfrac{4}{3}\left(\dfrac{t}{D}\right)^2$

Thus the slit tube has high degree of flexibility as compared to a closed tube.

It should be noted that the ratio of the stresses is of the order of (t/D) and the ratio of the angular displacements is of the order of $(t/D)^2$. However, the quantity D is considerably greater than t, thus a closed section is substantially stronger and to an even greater degree more rigid than a similar open section.

EXAMPLE 6.21

A thin-walled tube having an outside diameter of 90 mm rolled up from a 4.0 mm thick flat steel sheet is subjected to a twisting moment. Determine: (a) the maximum twisting moment that would develop a stress of 10 MPa in the closed welded tube; (b) the angular displacement of 1.0 m long welded tube subjected to twisting moment of part (a), and (c) the shearing stress and angle of twist developed in 1.0 m long open tube, i.e. when the edges of sheet are free, for the twisting moment of part (a). The elastic modulus of rigidity of the material is 80 GPa.

Solution: (a) In this case, the tube has closed section. The cross-sectional properties are,

Outside diameter, $D = 90$ mm; internal diameter, $d = 90 - 2 \times 4 = 82$ mm.

The polar moment of inertia, $J = \dfrac{\pi}{32}(D^4 - d^4) = \dfrac{\pi}{32}(90^4 - 82^4) = 2.003 \times 10^6 \text{ mm}^4$

The polar section modulus,

$$Z_p = \frac{J}{D/2} = \frac{2.003 \times 10^6}{45} = 4.45 \times 10^4 \text{ mm}^3$$

Alternatively, $$Z_p = \left(\frac{\pi D^4}{32} - \frac{\pi d^4}{32}\right)\left(\frac{2}{D}\right) = \frac{\pi D^3}{16}\left[1 - \left(\frac{d}{D}\right)^4\right]$$

$$= \frac{\pi \times 90^3}{16}\left[1 - \left(\frac{82}{90}\right)^4\right] = 4.45 \times 10^4 \text{ mm}^3$$

The maximum twisting moment that can be applied,

$$T_{c,\max} = \tau_{\max} Z_p = 10 \times (4.45 \times 10^4) = 4.45 \times 10^5 \text{ N.m}$$

$$\theta_c = \frac{TL}{GJ_e} = \frac{(4.45 \times 10^5) \times 1000}{(80 \times 10^3) \times (2 \times 10^6)} = 0.00278 \text{ radian}$$

$$= \frac{0.00278 \times 180}{\pi} = 0.1594°$$

(c) For the open section, $$\tau_{o,\max} = \frac{3T}{(\pi D)t^2} = \frac{3 \times (4.45 \times 10^5)}{\pi \times 90 \times 4^2} = 295.1 \text{ MPa}$$

and twist $$\theta_o = \frac{3TL}{G(\pi D)t^3} = \frac{3 \times (4.45 \times 10^5) \times 1000}{(80 \times 10^3) \times (\pi \times 90) \times 4^3}$$

$$= 0.922 \text{ radian} = 52.84 \text{ degrees}$$

$$\frac{\tau_{o,\max}}{\tau_{c,\max}} = \frac{295.1}{10} = 29.51 \quad \text{and} \quad \frac{\theta_o}{\theta_c} = \frac{0.922}{0.00278} = 331.65$$

Therefore, the stress is 29.51 times and angle of twist 331.65 times.

6.8.5 Thin-walled Cellular Sections

For discussion, consider a two cell section shown in Figure 6.21. The *shear flow* directions at various locations are shown in the figure. At the tri-junction E, the shear flow q_1 along FE divides into q_2 along ED and q_3 along EB, therefore, for *continuity of shear flow* or *equilibrium of shear stresses*,

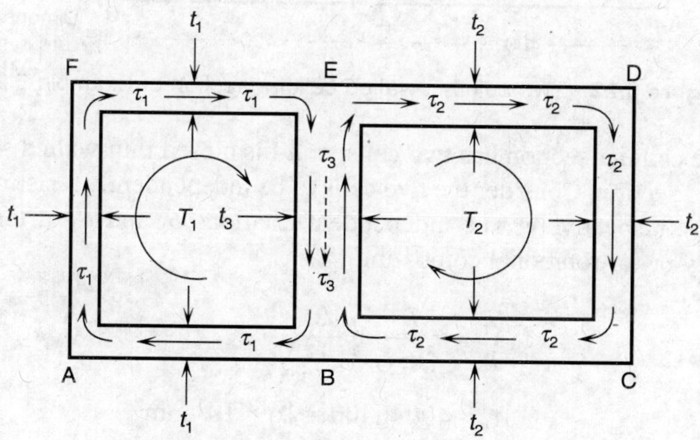

Figure 6.21 Thin-walled cellular section.

$$q_1 = q_2 + q_3 \quad \text{or} \quad \tau_1 t_1 = \tau_2 t_2 + \tau_3 t_3 \tag{6.65}$$

The total torque at the section,

$$T = 2q_1 A_1 + 2q_2 A_2 = 2(\tau_1 t_1 A_1 + \tau_2 t_2 A_2) \tag{6.66}$$

$$\theta = \frac{L}{2G}\left(\frac{\tau_1 s_1 + \tau_3 s_3}{A_1}\right) = \frac{L}{2G}\left(\frac{\tau_2 s_2 - \tau_3 s_3}{A_2}\right) \tag{6.67}$$

where s_1, s_2 and s_3 are the lengths or centroidal perimeters of cell contours AFEB, CBED and EB, respectively. The negative sign appears in the final terms because the shear flow along EB, for CBED cell opposes that in the remainder of perimeter.

$$\frac{2G\theta}{L} = \frac{1}{A_1}(\tau_1 s_1 + \tau_3 s_3) = \frac{1}{A_2}(\tau_2 s_2 - \tau_3 s_3) \tag{6.68}$$

EXAMPLE 6.22

A thin-walled cellular section with thickness of all the walls of 2 mm shown in Figure 6.22 is subjected to a twisting moment of 60 N.m. Determine: (a) the twisting moment shared by two cells, (b) the maximum shearing stress developed in each cell and (c) the angular displacement or twist per unit length. The elastic modulus of rigidity of material is 80 GPa.

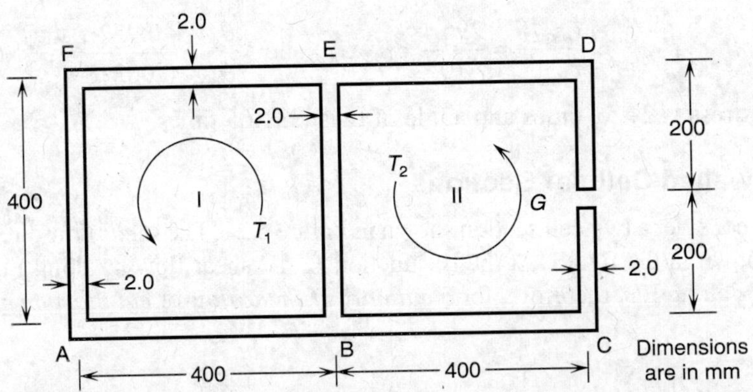

Figure 6.22 Two-cell thin walled cellular section of Example 6.22.

Solution: (a) The cellular section has two cells: cell I is closed thin-walled section and cell II is an open thin-walled section. Consider the two cells to be independently resisting twisting moment of T_1 and T_2 (N.m), respectively. The independent rotations θ_1 and θ_2 of these cells should be equal for continuity or for rotational compatibility.

For cell-I
$$\frac{\theta_1}{L} = \frac{T_1}{4A_1^2 G}\oint \frac{ds}{t}$$

where
$$A_1 = 400 \times 400 = 16 \times 10^4 \text{ mm}^2$$

and for uniform thickness,

$$\oint \frac{ds}{t} = \frac{s}{t} = \frac{4 \times 400}{2} = 800 \text{ mm}$$

For cell-II $$\frac{\theta_2}{L} = \frac{3T_2}{G\sum st^3}$$

$$s = 400 + 400 + 2 \times 200 = 1200 \text{ mm and } t = 2 \text{ mm}$$

For the ratio, $$\frac{s}{t} = \frac{1200}{2} = 600; \qquad \alpha = \beta = 1/3$$

For compatibility,

$$\frac{\theta_1}{L} = \frac{\theta_2}{L}$$

$$\frac{T_1}{4A_1^2 G}\oint \frac{ds}{t} = \frac{3T_2}{G\sum st^3}$$

or $$\frac{T_1}{T_2} = \frac{3 \times 4A_1^2}{\sum st^3 \oint \frac{ds}{t}} = \frac{3 \times 4 \times (16 \times 10^4)^2}{(1200 \times 2^3) \times 800} = 4 \times 10^4$$

$$T_1 = \frac{(60 \times 10^3) \times (4 \times 10^4)}{(4 \times 10^4 + 1)}$$

$$= 5.99985 \times 10^4 \text{ N.mm} = 59.9985 \text{ N.m}$$

$$T_2 = \frac{(60 \times 10^3) \times 1}{(4 \times 10^4 + 1)} = 1.49996 \text{ N.mm}$$

(b) The maximum shearing stress

For the closed cell, $$\tau_{1,\max} = \frac{T_1}{2(A)t} = \frac{5.99985 \times 10^4}{2 \times (16 \times 10^4) \times 2} = 9.375 \times 10^{-2} \text{ MPa}$$

For the open cell, $$\tau_{2,\max} = \frac{3T_2}{(s)t^2} = \frac{3 \times 1.49996}{(1200) \times 2^2} = 9.36875 \times 10^{-4} \text{ MPa}$$

(c) Angular displacement or twist per unit length

$$\frac{\theta_1}{L} = \frac{\theta_2}{L} = \frac{3T_2}{G\sum st^3} = \frac{3 \times (1.49996)}{(80 \times 10^3) \times (1200 \times 2^3)}$$

$$= 5.8592 \times 10^{-9} \text{ rad/mm}$$

It should be noted that the behaviour of noncircular sections subjected to torsion is vastly different from that of circulars sections. The solid sections having the same cross-sectional area are stiffer when their shape more closely approaches a circle. Conversely, a member made up of long, thin section that do not form a closed tube-like shape exhibit the high degree of flexibility in torsion. The commonly used structural shapes like wide-flange beams, standard I-beams, channels, angles, tees, etc. are flexible in torsion, whereas, the pipes, solid circular sections, and structural rectangular tubes have high rigidity or stiffness.

6.9 PROBLEMS

6.1 In a tension test, a specimen of 12 mm diameter, 240 mm gauge length stretched 0.25 mm under a pull of 23.6 kN. In a torsion test, on the same material, a maximum shear stress of 40.71 MPa was measured on a bar of 50 mm diameter, the angle of twist has been observed to be 0°25′12″ over a length of 360 mm. Determine the elastic constants for the material.

[**Ans.** $E = 200.32$ GPa; $G = 79.972$ GPa and $\nu = 0.254$]

6.2 Determine the external diameter of the shaft if the shear strain due to torsion is limited to 0.00086. The modulus of rigidity for steel is 80 GPa.

[**Hint:** $\tau_{max} = G\gamma$; determine T and hence Z_p]

[**Ans.** $D_h = 107.94$ mm]

6.3 Draw the twisting moment, shearing stress and angle of twist diagrams for a stepped shaft of circular section connected rigidly to support at E subjected to a system of torques shown in Figure 6.23.

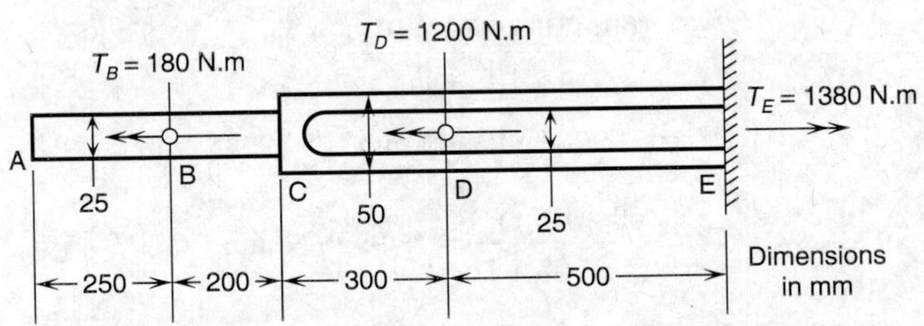

Figure 6.23 Stepped shaft of Problem 6.3.

[**Ans.** $T_A = 0$; $T_B = 180$; $T_D = 180/1380$; $T_A = 1380$ (N.m); $\theta_A = 0.02796$ radian]

6.4 A 50 mm diameter axial hole is bored in a 100 mm diameter solid circular shaft. Determine the percentage reduction in: (a) the torsional strength of the solid shaft and (b) strength–weight ratio.

[**Hint:** $T_s = \tau_{max}\,(\pi D^4/16)$ and $T_h = \tau_{max}\,(\pi D^4/16)\,(0.9375)$]

[**Ans.** (a) 6.25 per cent and (b) 29.69 per cent]

6.5 A tapered solid circular shaft of length L with its diameter varying uniformly from $D - a$ at one end to $D + a$ at the other is subjected to a pair of equal and opposite torques T at its ends. Determine the percentage error in the angle of twist when calculated using the mean radius D of the shaft. Take $a = D/10$.

[**Ans.** 3.34]

6.6 A shaft made up of two lengths of same material of 100 mm diameter joined together by means of a rigid flanged coupling transmits a torque such that the allowable shearing stress in the shafts and bolts are limited to 75 MPa and 60 MPa, respectively. Determine the number of 20 mm diameter bolts equally spaced on a pitch circle of 200 mm diameter required to transmit the torque.

[**Ans.** 8]

6.7 Two shafts, first one of solid circular section of diameter 60 mm and second of hollow cross-section of external diameter 120 mm, are connected end to end by means of a rigid flanged coupling in which there are 12 bolts spaced equally on a pitch circle of 300 mm diameter. Determine: (a) the inner diameter of the hollow shaft and (b) the diameter of bolts, if the maximum shearing stress in the shafts and bolts are limited to 66 MPa and 24 MPa, respectively.

[**Ans.** $d = 106.06$ mm and $d_b = 9.083$ mm]

6.8 Power is transmitted to and taken from a solid circular shaft A–B–C as shown in Figure 6.24. A motor running at 200 r.p.m. transmits a power of 45 kW at A with 15 kW and 30 kW taken off at B and C, respectively. Determine the maximum shearing stress developed in the shaft and the angle of twist (in degree) of the gear A relative to that at C. The modulus of rigidity of the material of the shaft is 85 GPa.

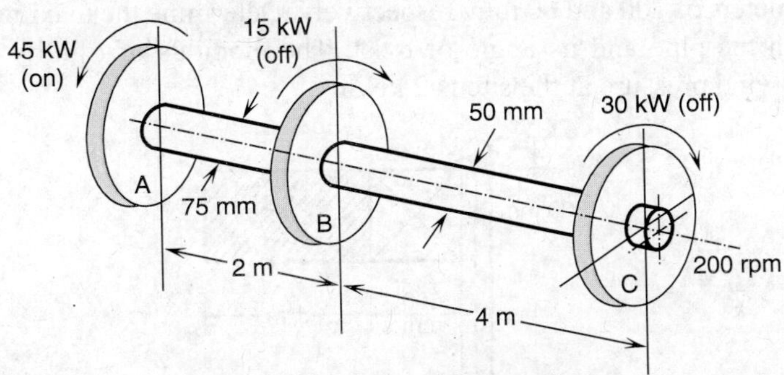

Figure 6.24 Shaft of Problem 6.8.

[**Ans.** Maximum shearing stress occurs in segment A–B, $\tau_{max} = 58.36$ MPa and $\theta_{A\text{–}C} = 7°14'$]

6.9 Two shafts of same material and same length: the first one of solid circular section and the second of hollow circular cross-section are considered for the use in an application that has to transmit a maximum torque T_{max}. If the stress in both the cases is not to exceed the elastic limit, compare the weights of the shafts when the ratio of internal and external diameters of the hollow shaft is: (a) 2/3 and (b) 3/4.

[**Ans.** w_{hollow}/w_{solid} is: (a) 64.35 per cent and (b) 56.3 per cent]

6.10 A solid circular shaft with diameter of 200 mm has the same cross-sectional area as that of a hollow shaft of 150 mm internal diameter. Both the shafts are of same material. Determine the ratio of the power transmitted by the two shafts at the same speed.

[**Hint:** $P_{hollow}/P_{solid} = T_{hollow}/T_{solid} = Z_{hollow}/Z_{solid}$]

[**Ans.** $D_h = 250$ mm and $P_{hollow}/P_{solid} = 1.7$]

6.11 A 20 mm thick hollow circular steel shaft transmits 300 kW of power at 200 r.p.m. Determine the external diameter of the shaft if the shear strain due to torsion is limited to 0.00086. The modulus of rigidity for steel is 80 GPa

[**Hint:** $\tau_{max} = G\gamma$; determine T and hence Z_p]

[**Ans.** $D_h = 107.94$ mm]

6.12 For a shaft in pure torsion, carrying 800 N.m of torque, a design proposal requires the shaft to have a hollow circular cross-section with 60 mm outside diameter. Determine the maximum inside diameter the shaft can have without exceeding the allowable shearing stress of $\tau_a = 32.6$ MPa.

[**Ans.** $d = 48.4$ mm]

6.13 A hollow circular steel shaft of diameters ratio of 3/8 transmits 375 kW of power at 100 r.p.m. The maximum torque is 20 per cent greater than the mean. Determine the external and internal diameters of the shaft when the maximum shearing stress and angle of twist over a length of 4 m are limited to 60 MPa and two degrees, respectively. The modulus of rigidity for steel is 85 GPa.

[**Ans.** $T_{max} = 42.973$ kN.m; $D_h = 156.64$ mm and $d = 58.74$ mm]

6.14 A signboard shown in Figure 6.25 is fixed on the top of a vertical pipe of external and internal diameters of 100 and 60 mm, respectively. Determine the maximum shearing stress developed in the pipe and its angle of twist. The modulus of rigidity of the member is 82 GPa and wind pressure at the site is 2 kN/m^2.

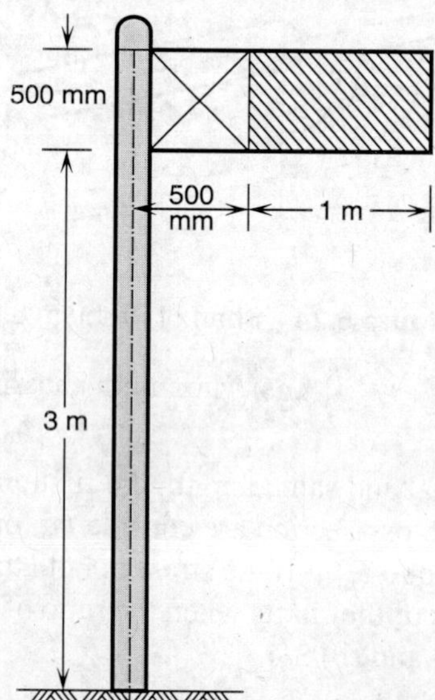

Figure 6.25 Sign board of Problem 6.14.

$$\left[\begin{array}{c}\textbf{Ans. } T_{max} = 1000 \text{ N.m}, M_{max} = 3250 \text{ N.m}, T_e = \sqrt{T_{max}^2 + M_{max}^2} = 3400 \text{ N.m}, \\ J = 8545132 \text{ mm}^4, \gamma_{max} = 19.89 \text{ MPa and } \theta = 0.97^\circ\end{array}\right]$$

6.15 A brass sleeve is securely fixed over a steel shaft of 100 mm diameter. Determine the outside diameter of the sleeve which would ensure that the torque applied to the compound shaft is shared equally by the two components. Determine the maximum shearing stresses in brass and steel, and the angle of twist in a length of 3 m due to a torque of 20 kN.m. The moduli of rigidity for brass and steel are 40 and 80 GPa, respectively.

[**Ans.** $D = 131.6$ mm; and $\theta = 0.0382$ radians = 2.189°]

6.16 A circular stepped shaft of length $2a$ fixed at the ends is subjected to a torque T as shown in Figure 6.26. Determine the ratio of reactions developed at the supports A and B.

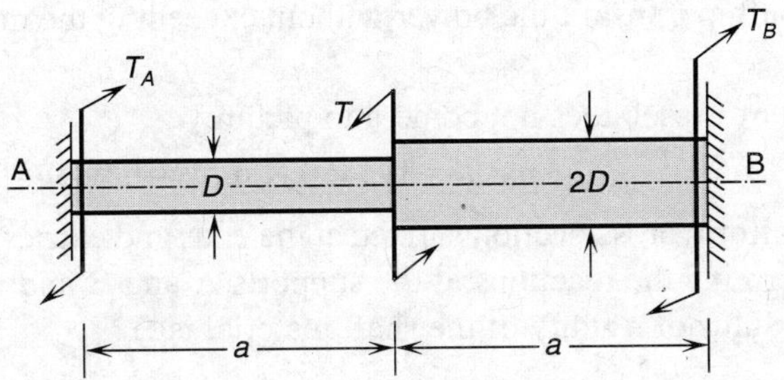

Figure 6.26 Circular stepped shaft of Problem 6.16.

$$\left[\textbf{Ans. } \frac{T_A}{T_B} = \frac{1}{16}\right]$$

6.17 A circular stepped shaft of length $6a$ is fixed at the ends and carries the torques as shown in Figure 6.27. Determine the reactions at the supports A and B, maximum shearing stress induced and rotations at the sections C. The modulus of rigidity of the shaft material is G.

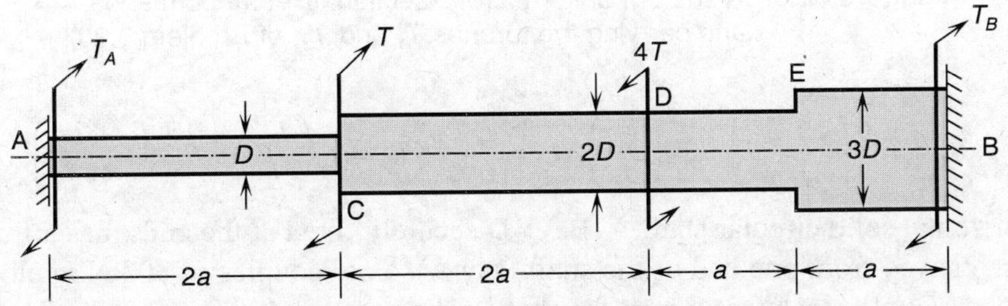

Figure 6.27 Circular stepped shaft of Problem 6.17.

$$\left[\textbf{Ans. } T_A = 0.6818T;\ T_B = -3.6818T;\ \gamma_{\max} = 18.75\left(\frac{T}{D^3}\right) \text{ and } \theta_c = 75\left(\frac{Ta}{GD^4}\right)\right]$$

6.18 A solid alloy shaft 60 mm in diameter is to be coupled in series with a hollow steel shaft of same external diameter. Determine: (a) the internal diameter of the steel shaft if the angle of twist per unit length of the steel shaft is to be 72 per cent of that of the alloy shaft and (b) the speed at which shafts are to be driven to transmit the 22.5 kW power when the maximum shearing stresses in alloy and steel are limited to 50 MPa and 80 MPa, respectively. Take $G_s = 2.3G_a$.

[**Hint:** $\tau_s/\tau_a = (G_s/G_a)(R_s/R_a)(\theta_s/\theta_a) = 2.3 \times 1 \times 0.72 = 1.656$, thus τ_a controls]

[**Ans.** $d = 47.6$ mm; $\tau_a = 48.31$ MPa; $T = 2048.86$ N.m and $N = 104.87$ r.p.m.]

6.19 A hollow circular shaft with internal diameter of 0.6 times the external diameter driving the screw propeller fitted to a vessel moving at speed 10 m/s uses 12,000 shaft horse powers at 150 rpm. If the efficiency of the propeller is 70 per cent, determine the shaft diameter which will enable the shaft to transmit the power without exceeding the maximum shearing stress of 80 MPa.

[**Hint:** The speed of vessel does not come into picture]

[**Ans.** $T = 393450$ N.m, $J = 0.08545D^4$ and $D = 306.44$ mm]

6.20 A shaft A–B of uniform cross-section is fixed at the ends and carries the torques as shown in Figure 6.28. Determine the reactions at the supports A and B and rotations at the sections C and D. The modulus of rigidity of the shaft material is G.

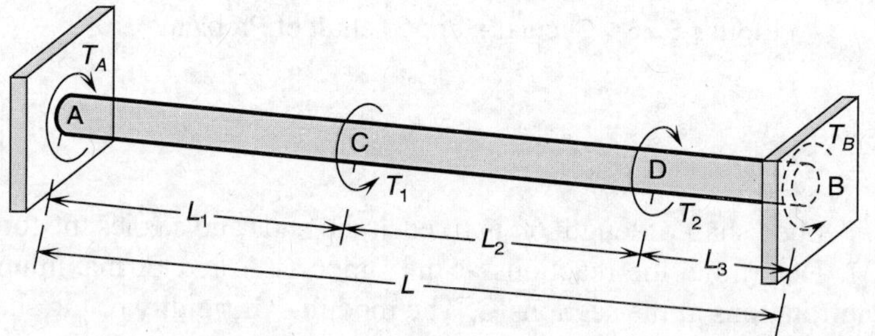

Figure 6.28 A shaft of uniform cross-section, fixed at both ends, and carrying the torques T_1 and T_2 (of Problem 6.20).

$$\left[\textbf{Ans. } T_A = T_1\left(\frac{L_2 + L_3}{L}\right) - T_2\left(\frac{L_3}{L}\right); \theta_C = \frac{L_1 T_A}{GJ}; T_B = T_1\left(\frac{L_1}{L}\right) - T_2\left(\frac{L_1 + L_2}{L}\right); \theta_D = \frac{L_3 T_B}{GJ}\right]$$

6.21 A horizontal solid circular shaft A–B–C–D securely fixed at the ends, has a free length of 10 m. Viewed from one end of the shaft (say A), axial couples of 30 kN.m clockwise and 40 kN.m counterclockwise act on the shaft at distances of 4 m and 7 m, respectively, from the viewed end. Determine: (a) the end fixing couples in magnitude and direction and (b) the diameter of the shaft if maximum shearing stress is limited to 60 MPa.

Also draw a diagram to show how a line on the surface of the shaft, originally parallel to the axis, will appear after the application of the couples, and find the position along the shaft where there is no angular twist.

[**Ans.** $T_A = 6$ kN.m (anticlockwise); $T_A = 16$ kN.m (clockwise); $D = 126.77$ mm; $\theta_B = -24/GJ$; $\theta_C = +48/GJ$ and $x = 5$ m from A]

6.22 A 1.25 m long hollow shaft has an external diameter of 60 mm. It has an internal diameter of 24 mm for a part of its length from one end and 36 mm for the remaining length. The shearing stress in the shaft is limited to 80 MPa. Determine: (a) the maximum power transmitted by the shaft at a speed of 250 rpm and (b) the lengths of two portions if the angle of twist in the two portions of the shaft is equal.

[**Ans.** $T = 2.9532$ kN.m; $P = 77.3146$ kW; $L_1 = 589.77$ mm and $L_2 = 660.23$ mm]

6.23 A solid vertical steel shaft A–B–C is rigidly fixed at its base A as shown in Figure 6.29. The segment A–B is 4 m long and has diameter of 75 mm, while B–C is 2 m long with a diameter of 40 mm. The torques of 2000 N.m and 400 N.m are applied in the same sense at B and C, respectively. Determine: (a) the maximum shearing stress in each segment, (b) the relative twist between the ends of each segment of the shaft, and between the ends of the shaft and (c) the total power transmitted at 100 r.p.m.

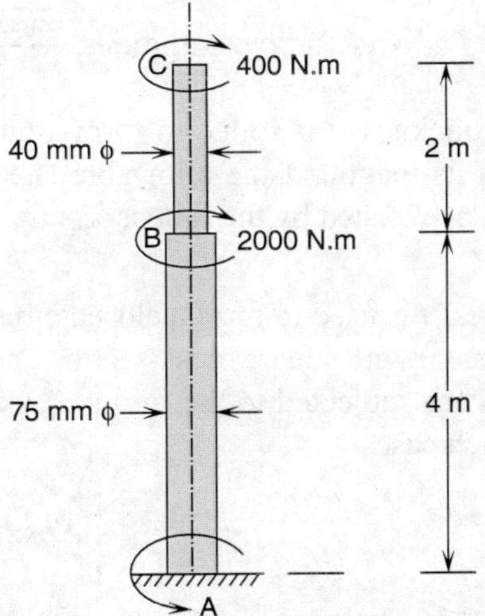

Figure 6.29 Vertical steel shaft of Problem 6.23.

[**Ans.** $\tau_{B\text{–}C} = 31.83$ MPa; $\theta_{B\text{–}C} = 2.309°$; $\tau_{A\text{–}B} = 31.83$ MPa; $\theta_{A\text{–}B} = 2.24°$; $\theta_{C\text{–}A} = 4.55°$ and Power $= 4.189 + 20.943 = 25.13$ kW]

6.24 An 8 m long composite shaft consists of a steel shaft 300 mm in diameter surrounded by a closely fitting 36 mm thick bronze tube. Determine: (a) the maximum power transmitted by the shaft at a speed of 200 r.p.m., if the maximum shearing stress in the steel is limited to 15 MPa. Take $G_s = 2G_a$.

[**Ans.** $T_s/T_b = 1.466$; $T = 133.765$ kN.m and $P = 2801.57$ kW]

6.25 Two similar rolled steel channel sections are to be used for fabricating a torsion member. Evaluate the torsional rigidity of the member when its cross-section is built-in of: (a) channel sections welded back-to-back, (b) channel sections welded face to face at both edges and (c) channel sections welded face to face at one edge as shown in Figure 6.30.

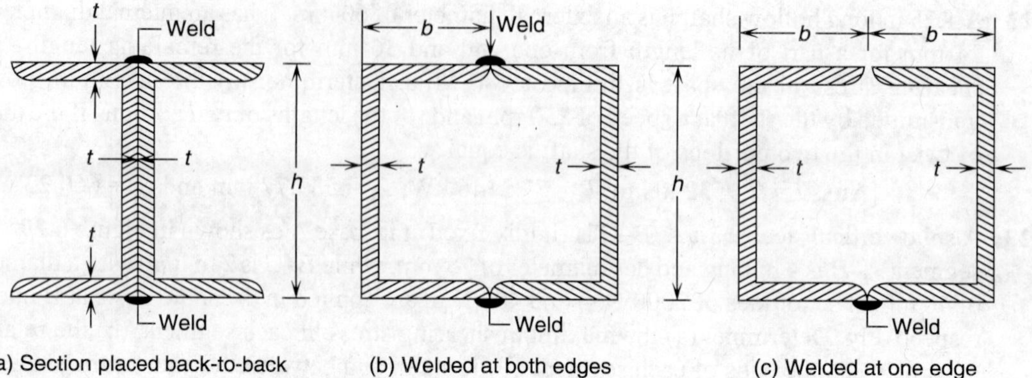

(a) Section placed back-to-back (b) Welded at both edges (c) Welded at one edge

Figure 6.30 Built-in section of Problem 6.25.

$$\left[\textbf{Ans}. \text{ Ratio of rigidities, } \frac{2(2h+b)t^2}{3} : \frac{4h^2b^2}{(2b+h)} : \frac{t^2(2b+h)}{3}\right]$$

6.26 A thin flat steel sheet of thickness t is rolled to form a tube of mean diameter D, and the seam is welded all along its length. If the allowable shearing stress is τ_a, show that the maximum torque that can be resisted by the tube is $T = (\pi D^2 t\tau_a)/2$ and twist per unit length is $4T/(\pi D^3 Gt)$.

6.27 A thin flat sheet of thickness t is used to form a closed circular tube shaft of mean diameter D, and also a box section shaft with square cross-section enclosing the same area as the tube section. If both the shafts are subjected to the torque T, determine the shearing stress and twist per unit length in each case.

$$\left[\textbf{Ans. } \tau_t = \tau_b = \frac{2T}{\pi D^2 t} \text{ and } \frac{\theta_t}{\theta_b} = \frac{\sqrt{\pi}}{2}\right]$$

6.28 A one millimeter thick thin-walled 1.2 m long member has the closed cross-section shown in Figure 6.31. Determine (a) the maximum torque carrying capacity of the member if the angle of twist is limited to 12°, (b) the maximum shear stress developed when the maximum torque computed (a) is applied. The modulus of rigidity of the member is 80 GPa.

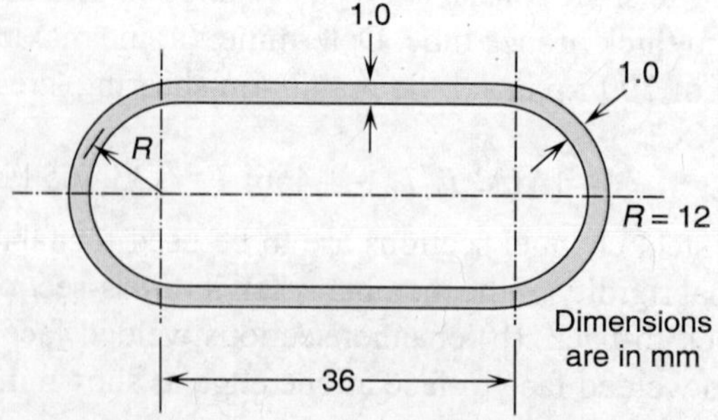

Figure 6.31 Thin-walled closed tube of Problem 6.28.

[**Ans.** $T = 456$ N.m and $\tau_{max} = 173.2$ MPa]

6.29 A thin-walled cellular section shown in Figure 6.32 is subjected to a twisting moment of 360 N.m. Determine: (a) the maximum shearing stress developed in each cell and (b) the angular displacement or twist per unit length. The elastic modulus of material is 30 GPa.

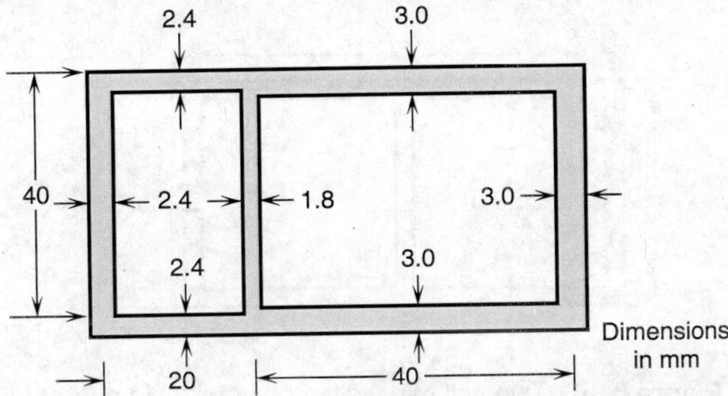

Figure 6.32 Thin-walled cellular cross-section of Problem 6.29.

[**Ans.** $\tau_1 = 31.25$ MPa; $\tau_2 = 41.67$ MPa; $\tau_3 = 0.60$ MPa and $\theta = 3.013$ degree/m]

6.30 Two tubular sections shown in Figure 6.33 have the same wall thickness t and circumference. Neglecting stress concentration, determine the ratio of shear stresses for: the two sections when: (a) subjected to equal twisting moments and (b) the angles of twist is same.

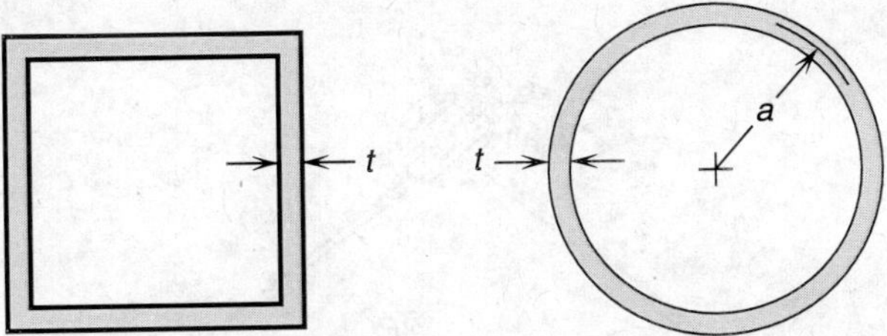

Figure 6.33 Tubular sections of Problem 6.30.

[**Ans.** (a) = (π : 4) and (b) = (4 : π)]

6.31 A thin-walled box section of dimensions $2a \times a \times t$ is to be compared with a solid section of diameter a as shown in Figure 6.34. Determine the wall thickness t, so that the two sections have: (a) the same maximum stress for the same torque and (b) the same stiffness.

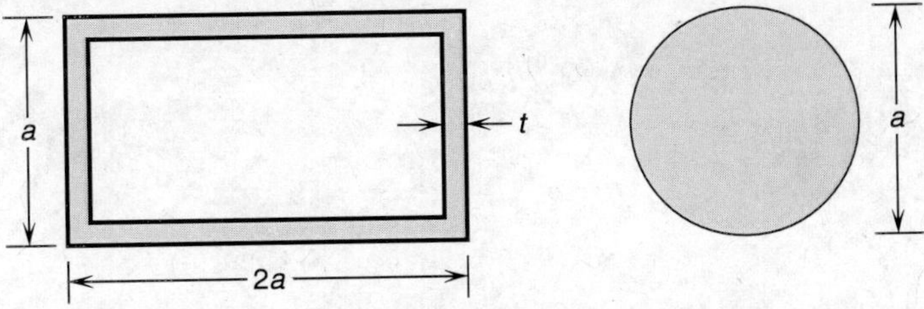

Figure 6.34 Thin-walled box and solid sections of Problem 6.31.

[**Ans.** (a) $t = \pi a/64$ and (b) $t = (3/4)\,(\pi a/64)$]

6.32 A thin-walled box section has two compartments as shown in Figure 6.35. It has constant wall thickness t. Determine (a) the shear stress for a given torque and (b) the stiffness of the section, i.e. the torque required per unit radian of twist.

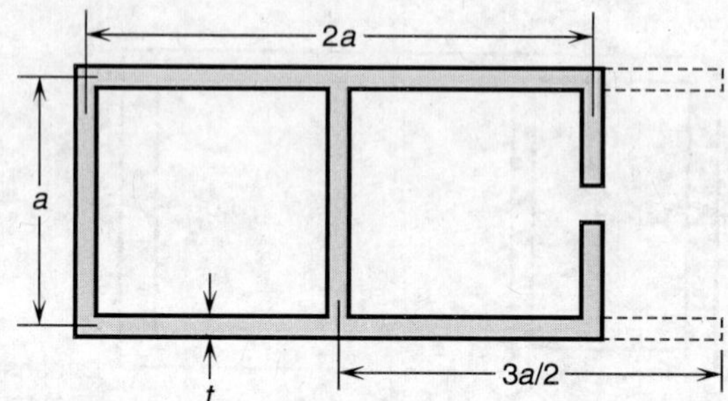

Figure 6.35 Two cell thin-walled box of Problem 6.32.

[**Ans.** For simplicity straighten the sides of the open cell as shown in Figure]

6.33 Torque measuring equipment is modelled as shown in Figure 6.36. Determine the stiffness of the system, i.e. the torque per unit angle of twist of the shaft. Each of the spring bars has length L and the moments of inertia of the bars are I and $2I$, respectively, for bending in the plane of moment.

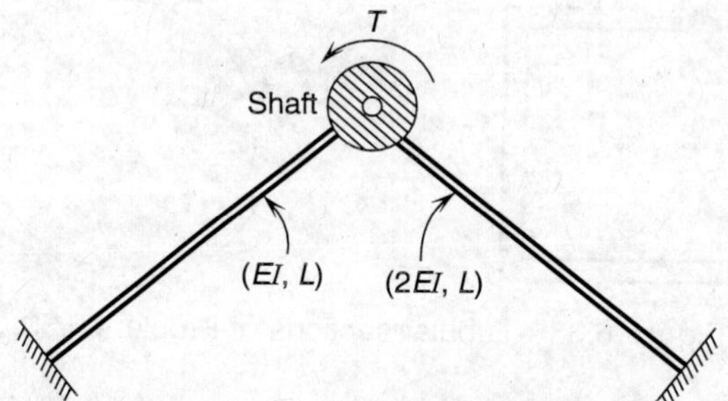

Figure 6.36 Torque measuring arrangement of Problem 6.33.

[**Ans.** $k_t = 12EI/L$]

Chapter 7

Shearing Force and Bending Moment Diagrams

7.1 INTRODUCTION

A member carrying an external force system that acts transversely to its axis is called a *beam*. The beam is a dominant member of modern construction. The beam with all forces acting in the same plane is termed a *planar beam*. In such a system, three internal force components namely *the axial force*, *the shearing force* and *the bending moment*, are developed at a section. *The graphic representation of variation of internal forces along the beam is called stress resultant diagrams*. In this chapter, the discussion is largely confined to beams considered in horizontal position, for convenience. The problems related to the planar frames resisting axial forces, shearing forces and the bending moments are also discussed.

The analysis of all structures begins with determination of *reactions* at the supports. As discussed earlier in Chapter 1, based on the kind of resistance offered by a support, it is classified as pinned or hinged connection, roller support, and fixed connection. It is identified as a *roller* or a *link* if it is capable of resisting a force in only one specific direction coincident with the line of action; as *a hinged* or *pinned support* if it is capable of resisting force acting in any direction of the plane; and as *a fixed support* if it is capable of resisting force acting in any direction and is also capable of resisting a moment. Generally, the reaction at a hinged support may have two components one in the horizontal direction and one in the vertical direction. A system of three forces can exist at a fixed support, two components of force and a moment. The roller and hinged supports, which are not capable of resisting moment, are termed *simple supports*.

7.1.1 Calculation of Beam Reactions

For a beam to be in stable equilibrium, the supports must be capable of providing specific minimum number of force restraints. The techniques for computation of support reactions have already been discussed in Chapter 1. Internal forces developed within a beam due to externally applied loads are those of tension and compression resulting from bending, and shear.

7.1.2 Shearing Forces and Bending Moments

To study the behaviour of a transversely loaded member, consider a cantilever shown in Figure 7.1(a). This member might fail in two ways, in one potential mode of failure two contiguous parts of the member slide relative to each other in a direction parallel to their plane of contact under the action of external force system. This is called a *shear failure* (see Figure 7.1(b)) and the internal force associated with this phenomenon of transverse sliding is called *internal shearing* or *shearing force*. These internal shearing forces balance or resist the net external shearing force tending to cause the sliding. Thus, the internal force V, acting at *right angles* or *transverse* to the axis of a member, is called the *shear* or *shearing force. The shear is numerically equal to the algebraic sum of all the transverse components of the external forces* (including reaction components) *acting on the isolated segment, but opposite in direction* to satisfy equilibrium condition.

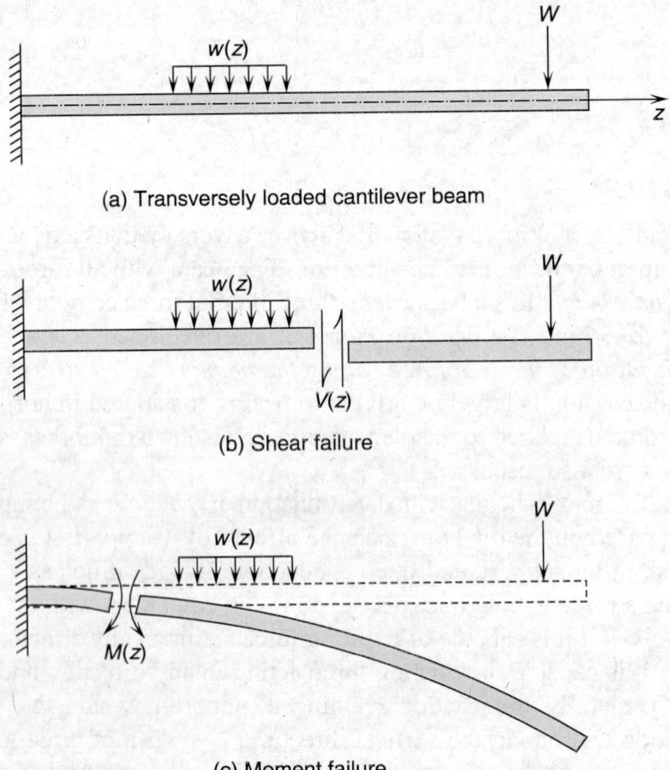

Figure 7.1 Potential modes of failure and types of internal forces generated in a transversely loaded member.

In the second potential mode of failure, the transverse external forces cause part of the structure to bend or rotate. The internal couple or moment associated with this phenomenon of *bending* is called internal moment which is developed to balance or resist the rotational tendency of the external forces (see Figure 7.1(c)). The *internal resisting moment* must act in a direction opposite to the external moment to satisfy the equilibrium equation $\Sigma M_z = 0$. Thus, the magnitude of the internal resisting moment equals the *external moment*. This moment which tends to bend a beam in the plane of the loads is generally referred to as *bending moment*. The magnitude of *the bending moment is* given by *the algebraic sum of the moments of all the forces to the left* or *to the right of the section*, the moments being taken about an axis through the centroid of the cross-section.

If a loaded member is decomposed at a section into two parts, internal shearing force and bending moment develop simultaneously to maintain the translational and rotational equilibrium of each part. These internally developed forces also represent the internal actions and reactions of one part of the member on the other part.

Based on fundamental proposition that any structure, or a part of it must be in a state of equilibrium under the action of the complete force system (including internal as well as external forces and moment), the magnitude of shearing forces and moments can be determined. The external force system is typically known, the part of the system not initially known such as reactions at supports, can be readily determined by the methods discussed earlier in Chapter 1. If the equilibrium of an isolated portion of a structure is considered, the unknown internal shears and moments that must be present at the point of decomposition can be determined. To compute an internal bending moment required to maintain a beam segment in equilibrium, either the left or the right-hand segment of the beam can be considered as a free body. It is often convenient to consider the segment of the beam that has the least number of loads.

7.2 SHEAR FORCE AND BENDING MOMENT DIAGRAMS

The magnitude and sense of shears and moments vary from section to section in a structure. The graphical plots showing the variation of shears and moments in sense and intensity across the structure are called *the shear and bending moment diagrams*, respectively. It is convenient to draw the plots immediately below the free-body diagram of the beam, using the same horizontal scale as for the length of the beam.

The shear and moment diagrams are invaluable aids in the analysis and design of structures since they enable the designer to appreciate the effects of an external force system on any type of structure. With the aid of these diagrams the location of critical sections in the structures and values of stress resultants at the critical sections is immediately known. The magnitudes of shears and moments as represented in the diagrams are used to determine the required size of a member. In the following section, construction of shear and moment diagrams for a simple beam subjected to any loading system is discussed.

For illustration consider a simple beam carrying a concentrated load as shown in Figure 7.2(a). At any section of the beam along its length, there is a net external shearing force present which is the algebraic sum of all upward and downward forces (applied forces, loads and reactions) that exist on the isolated segment. This external shearing force is balanced by an internal *resisting shearing force* $V_z(z)$ that maintains the translational equilibrium of the section. Figure 7.2(b) indicates the directional sense of the internal shearing forces which vary along the length of the beam

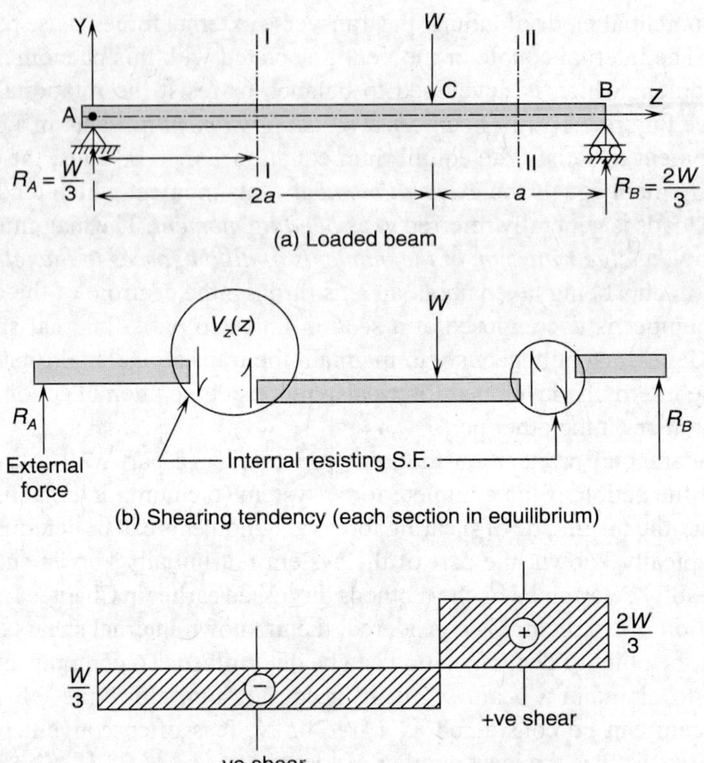

(c) Variation of shear force along the length of the beam, i.e. shearing force diagram

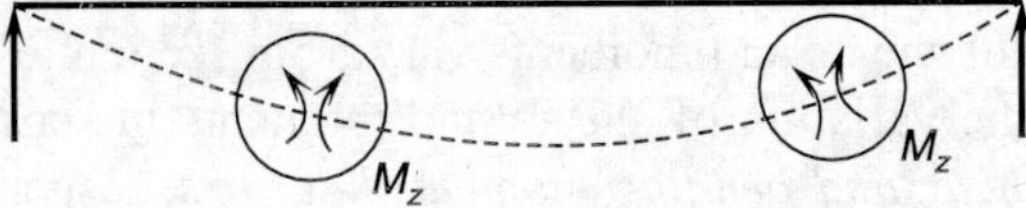

(d) Each beam section is in equilibrium under internal and external moments

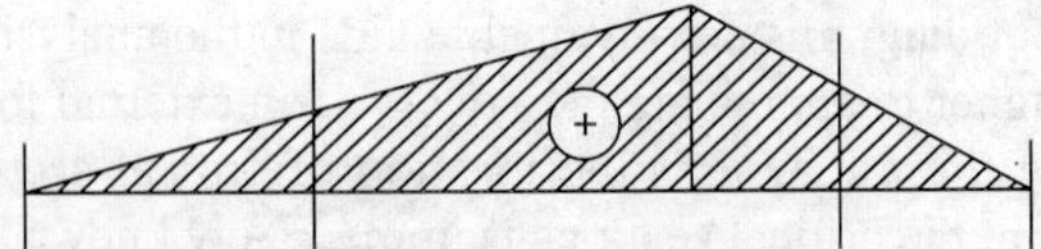

(e) Variation of bending moment along the length of the beam, i.e. bending moment diagram

Figure 7.2 General nature of externally applied and internally resisting shears and moments.

depending upon the location of section under consideration. The shaded diagram shown in Figure 7.2(c) is a graphical depiction of the magnitude of the external shearing force present at each section of the beam along its length. Figure 7.2(d) indicates the tendency of the structure to bend under the action of loads. At any section of the structure along its length there is a net external bending moment that is the algebraic sum of the rotational effects produced by all external forces (applied forces, loads and reactions) to the left (or right) of the section considered . This external bending moment is balanced by an internal resisting bending moment M_z that maintains the rotational equilibrium of each part of the structure. The shaded diagram shown in Figure 7.2(e) is a graphical depiction of the magnitude of the external bending moment present at each section of the beam along its length.

In case of beams, the axial force diagrams are not as commonly used as the shear and the moment diagrams, since the majority of beams carry loads that act perpendicular to their axes.

7.2.1 Sign Convention

The sign convention employed for calculating reactions, where the internal moment or shear is designated, as either positive or negative by its direction only as shown in Figure 7.3(a), is misleading in the present situation because of the fact that with reference to one part of the structure a value would be positive and with reference to the other part the value would be negative. Therefore a different sign convention is adopted, wherein the moment due to external force system is considered positive when it causes *compression on the top face* of the structure and *tension at the bottom*, i.e. the *structure bends concave upwards* as shown in Figure 7.3(b). If the moment due to external

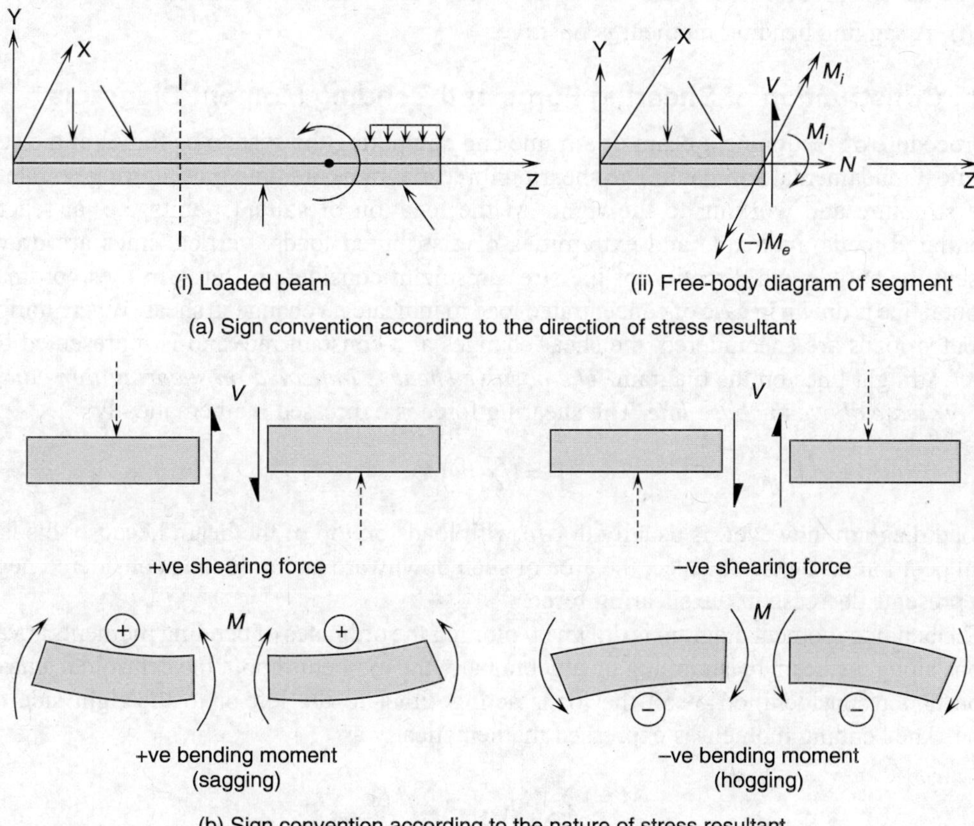

Figure 7.3 Sign conventions for stress resultants.

force system causes reverse bending, i.e. produces convex curvature (concave downwards), it is considered negative. In other words, the upward-acting external forces tend to cause positive moments to develop at a section and the downward-acting external forces develop negative moments. In this text the moments are plotted on the compression face of the structure. If the member under consideration happens to be a vertical member, the bottom side of the member is considered to be the right-hand side with origin at the lower end.

The Z-axis is chosen as longitudinal centroidal axis of the member. If the outward normal to the face of the section is in the positive direction of Z-axis, the positive shearing force vector is in the positive direction of Y-axis and the positive bending moment vector is in positive X-direction as shown in Figure 7.3(a-ii). On the other hand, if the outward normal of the face of the section is in the negative direction of Z-axis, then the positive shearing force vector and the positive bending moment vector are in negative Y- and X-directions, respectively. In case of a horizontal member, if the positive Z-direction is always to right and positive Y-direction is always chosen upward, then the sign convention can be easily interpreted as follows:

(i) When the resultant of the external forces to the left of the section acts downwards, the shearing force is said to be positive. The positive shear is associated with the tendency of the external forces produce down-up relative movement of a section.

(ii) A sagging bending moment is positive.

7.2.2 Construction of Shearing Force and Bending Moment Diagrams

The procedure of sectioning a beam or a frame and computing the system of forces at the section is the most fundamental approach. The shearing force diagrams are drawn by starting with left end of the structure and working to the right. At the location of salient points, i.e. at reactions, concentrated loads, moments and extremities of distributed loads, vertical lines are drawn to represent magnitude and direction of the stress resultant considered. Between these ordinates a horizontal line is drawn in case of concentrated load to indicate no change in shear. Where uniformly distributed loads are encountered, the shear changes at a constant rate and is represented by the inclined straight lines on the diagram. *The positive shear is indicated on shearing force diagram by an ordinate above the base line.* The shearing force is expressed mathematically as

$$V = \left(\sum W\right)_{\text{left}} \tag{7.1}$$

The load diagram, however, is usually drawn with loads on top of the beam because this is their natural position; as a consequence, the area of such downward acting load is considered negative and represents decrease in the shearing force.

The bending moment diagram is drawn by plotting the ordinates of bending moments at various sections along the beam by summing up algebraically the moments about the centroidal axis of the section under consideration of all the loads acting either to the left or to the right side of the section. The bending moment is expressed mathematically as

$$M = \left(\sum M\right)_{\text{left}} = \left(\sum M\right)_{\text{right}} \tag{7.2}$$

The subscript refers to the side of the section on which loads have been considered. It is often convenient to *consider the part of the beam that has the least number of loads.*

For illustration, consider the simply supported beam carrying a concentrated load W as shown in Figure 7.4(a). The free-body diagram of the beam is shown in Figure 7.4(b). Since the beam is

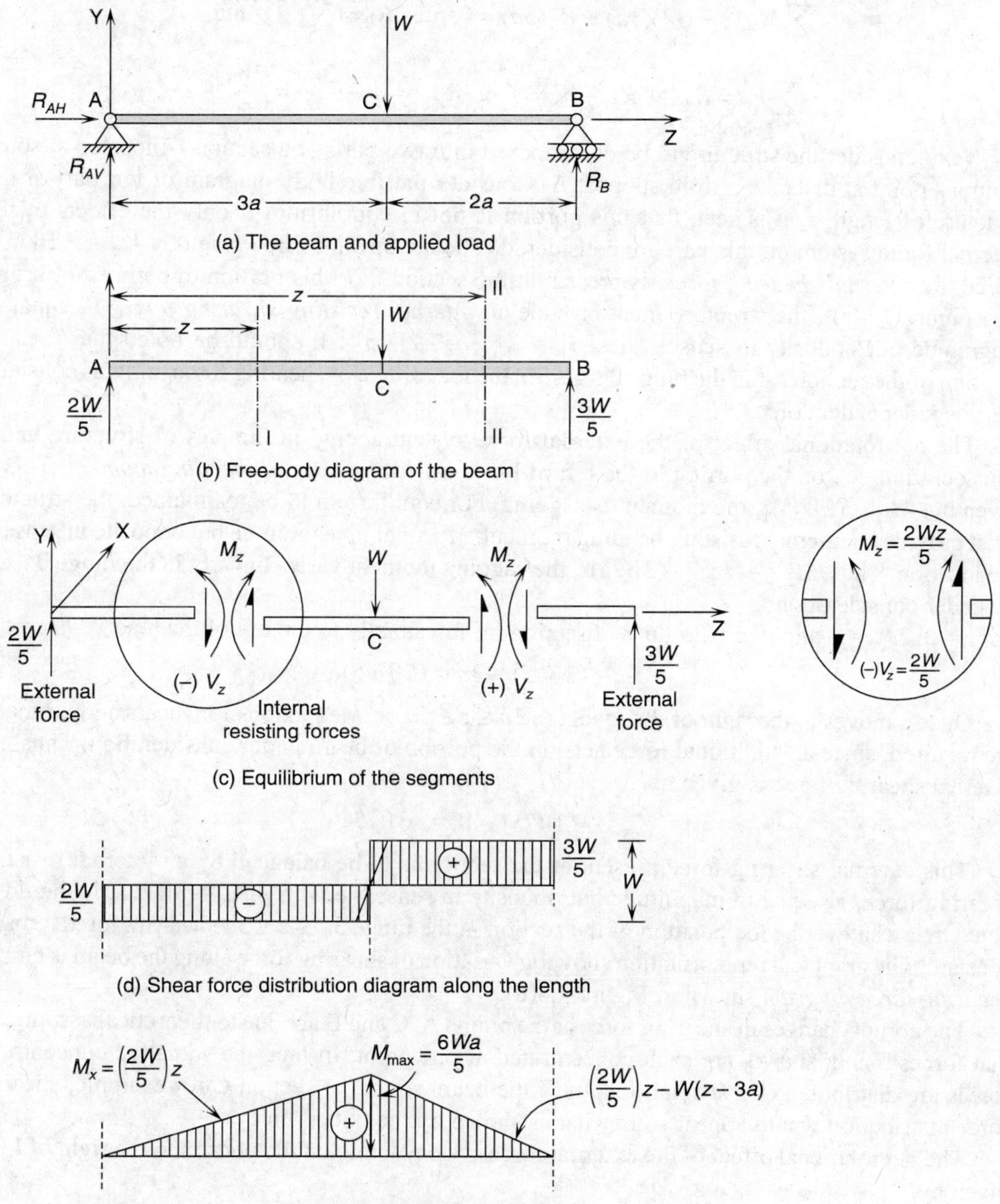

Figure 7.4 General nature of externally applied and internal resisting shear forces and bending moments in structures. Construction of stress resultant diagrams.

in equilibrium, the support reactions can be determined by successive applications of conditions of equilibrium as explained earlier in Chapter 1:

$$\sum M_A = -(W)(3a) + R_B(5a) = 0, \text{ i.e. } R_B = \frac{3W}{5}\uparrow, \text{ and}$$

$$\sum F_y = R_{AV} + R_B - W = 0 \text{ or } R_{AV} = W - \frac{3W}{5} = \frac{2W}{5}\uparrow$$

Next, consider the structure to be decomposed into two parts at a section I-I located at some arbitrary point at distance z from support A. Consider the free-body diagram of left part of the structure of length z, it is seen that this portion is not in equilibrium if only the effects of the external force system on this part are considered. The net upward force given by $V_{ex} = +2W/5\uparrow$, called the *external shearing force,* is present at the section. For this portion to be in equilibrium (see Figure 7.4(c)), the structure must provide an *internal resisting shearing force,* V_z equal in magnitude but opposite in sense. Thus, $V_z = -V_{ex} = -2W/5\downarrow$. It should be noted that for any position of the section I–I in the range $0 \le z \le 3a$, the magnitude of shearing force remains constant, i.e. it is independent of z.

The net rotational effect of the external force system acting on the part of structure under consideration, i.e. on the portion to the left of I–I is called *the external bending moment.* This is given by: $M_{ex} = (2W/5)z$, the moment is sagging. For equilibrium to be maintained, the structure must provide an internal resisting bending moment, M_z equal in magnitude but opposite in sense at the location. Thus, $M_z = -M_{ex} = -(2W/5)z$, the sagging moment varies linearly in the range $0 \le z \le 3a$ under consideration.

For $z = 0$; $M_{ex} = 0$ and if, $z = 3a$ i.e. at a section infinitesimally to the left of the load,

$$M_{ex} = (2W/5) \times 3a = (6Wa/5)$$

Once z moves to the right of the load, i.e. $3a \le z \le 5a$, new equations for shears and moments are required, since an additional force acts on the portion of beam under consideration. Thus, the external shearing force is given by

$$V_{ex} = +(2W/5) - W = -3W/5\downarrow$$

This external shearing force present at the section is to be balanced by an internal resisting shearing force, V_z equal in magnitude but opposite in sense, i.e. $V_z = -V_{ex} = 3W/5\uparrow$. It should be noted that whatever be the position of the section in the range $3a \le z \le 5a$ shearing force remains constant. The graphical representation showing variation of shearing force along the beam is termed *shearing force diagram* shown in Figure 7.4(d).

The abrupt changes in shearing force at the points A, C and B are due to theoretical assumption that forces R_A, W and R_B are each concentrated at one point. In fact, the so called concentrated forces are distributed over a small length of the beam so that at section C for example, shearing force distribution would appear somewhat as shown by the dotted line.

The net rotational effect of the external force system acting on the portion to the right of I–I is given by:

$$M_{I\text{–}I} = +(2W/5)z - W(z - 3a) = -(3W/5)z + 3Wa \quad \text{for } 3a < z < 5a$$

For equilibrium to be maintained, the structure must provide an internal resisting bending moment, M_z equal in magnitude but opposite in sense at the location, i.e.

$$M_z = -M_{ex} = +(3W/5)z - 3Wa$$

Thus, if $z = 3a$ i.e. section is infinitesimally to the right of the load,

$$M_{ex} = -(3W/5)(3a) + 3Wa = 6Wa/5$$

The graphical representation of these bending moments is shown in Figure 7.4(e).

Thus, for the construction of shearing force and bending moment diagrams equations are written for the variation of the external shearing force V_{ex} and the external bending moment M_{ex} as a function of a variable distance z along the length of the member, with the left end of the member as a reference point.

The following examples illustrate the construction of shearing force and bending moment diagrams along the length of the structure. *The shearing force and bending moment diagrams are essentially graphs of the algebraic expressions of external shearing forces and bending moments expressing their variation along the length of the structure.*

EXAMPLE 7.1

Draw shearing force and bending moment diagrams for the structure illustrated in Figure 7.5(a).

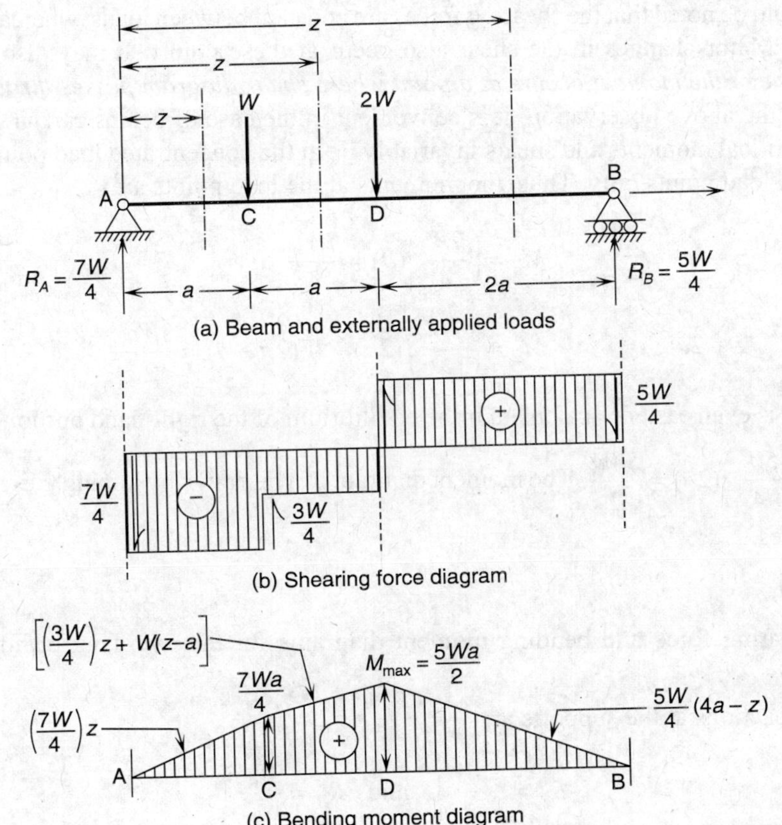

Figure 7.5 Shearing force and bending moment diagrams of beam of Example 7.1.

Solution: Reactions are first calculated as usual:

$$R_A = \frac{7W}{4}\uparrow \text{ and } R_B = \frac{5W}{4}\uparrow$$

To formulate equations for external shearing forces and bending moments, consider the equilibrium of the portion of the structure to the left of a section at distance z from the left support.

For the range $0 \le z \le a$: $\quad V_z = -\frac{7W}{4};\; M_z = \left(\frac{7W}{4}\right)z$

$a \le z \le 2a$: $\quad V_z = -\frac{3W}{4};\; M_z = \left(\frac{7W}{4}\right)z - W(z-a)$

$2a \le z \le 4a$: $\quad V_z = \frac{5W}{4};\; M_z = \left(\frac{7W}{4}\right)z - W(z-a) - 2W(z-2a)$

These equations are plotted to obtain shear and moment diagrams as shown in Figures 7.5(b) and (c). It should be noted that the shearing forces are constant between loads, whereas the moments vary linearly. Major changes in the shear also occur at these points. It may also be noted that *maximum value of the moment occurs at a point where shear diagram passes through zero.*

Based on the above observation, it is convenient in the case of beams carrying concentrated loads where critical moments and shears invariably lie at the concentrated load points, to consider sections at the load points only. Thus, the moments at the load points are:

At $z = a$: $\quad M_z = \left(\frac{7W}{4}\right)(a) = \frac{7Wa}{4}$

$z = 2a$: $\quad M_z = \left(\frac{7W}{4}\right)(2a) - W(2a-a) = \frac{5Wa}{2}$

The latter is even easier to calculate if the equilibrium of the right-hand portion is considered. Thus, $M_D = \left(\frac{5W}{4}\right)(2a) = \frac{5Wa}{2}$. The moment diagram can be drawn by joining these critical points by straight lines.

EXAMPLE 7.2

Construct shearing force and bending moment diagrams for the overhang beam illustrated in Figure 7.6(a).

Solution: Reactions at the supports are:

$$R_A = -\frac{W}{2}\downarrow \text{ and } R_B = \frac{3W}{2}\uparrow$$

Since the external load system consists only of concentrated loads, the shearing force is constant and its diagram is horizontal between the load points (see Figure 7.6(b)), whereas the moments vary linearly. The moment diagram for the beam to the left can be easily drawn by noting that

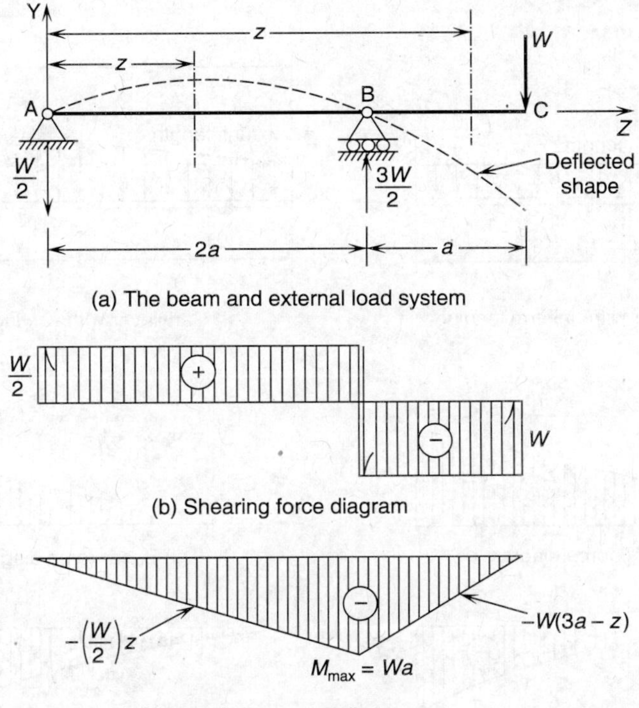

Figure 7.6 Shearing force and bending moment diagrams for the overhang beam of Example 7.2.

bending moments must be zero at member ends and the peak must be at the interior support (see Figure 7.6(c)). Thus the critical value of moment is the product of a single force (either left or right of the support) and the corresponding distance from the support. The value of moment $M_B = \left(\frac{W}{2}\right)(2a)$ [or $(W) \times (a)$] $= Wa$. The moment diagram is obtained by joining this value at critical point by straight lines to the end points. As in previous example, it may also be noted that maximum value of the moment occurs at a point where shear diagram passes through zero.

EXAMPLE 7.3

Develop complete shearing force and bending moment diagrams for a cantilever beam when it carries: (i) uniformly distributed load w_0 over the entire span and (ii) distributed load with intensity varying from w_0 at the free end to $2w_0$ at the fixed end.

Solution: In case of cantilever (as in Figure 7.7(a-i)), the computation of reactions at the support is not a prerequisite for drawing shearing force and bending moment diagrams. The shears and moments at any section defined by a distance z from the free end of the member are determined by considering the equilibrium of the portion of the structure between the section and the free end.

Case-I: As explained earlier the moment due to uniformly distributed load is found by considering total uniformly distributed load to be concentrated at the mid-point of loaded length. Thus,

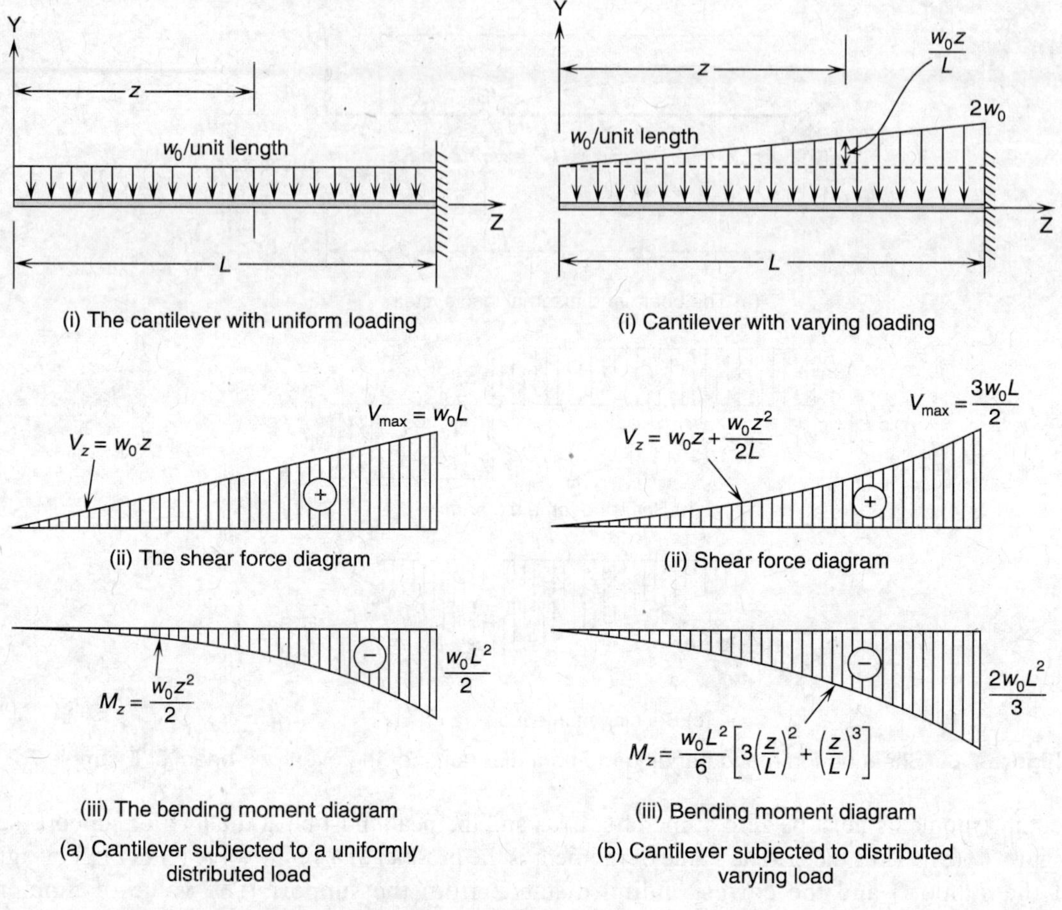

Figure 7.7 Shear and moment diagrams for the cantilever of Example 7.3.

For $0 \le z \le L$:

$$V_z = w_0 z \quad \text{and}$$

$$M_x = -w_0 z\left(\frac{z}{2}\right) = -\frac{w_0 z^2}{2}$$

The maximum value of shear and moment which occur at $x = L$ are,

$$V_{max} = w_0 L \quad \text{and} \quad M_{max} = -\frac{w_0 L^2}{2}$$

It should be noted that shearing forces vary linearly according to the first power of z and that the moments have parabolic variation according to the second power of z. The shearing force and bending moment diagram are illustrated in Figures 7.7(a-ii) and (a-iii).

Case-II: The shear and moment due to distributed varying load can be determined by considering distributed load to consist of uniformly distributed load w_0 over the entire span and triangular load varying from zero at the free end to w_0 at the fixed end as illustrated in Figure 7.7(b-i). Thus,

For $0 \le z \le L$:

$$V_z = w_0 z + \frac{1}{2}(z)\left(\frac{w_0 z}{L}\right) = w_0 z + \frac{w_0 z^2}{2L} \quad \text{and}$$

$$M_z = -(w_0 z)\left(\frac{z}{2}\right) - \left(\frac{w_0 z^2}{2L}\right)\left(\frac{z}{3}\right) = -\left(\frac{w_0 z^2}{2}\right) - \left(\frac{w_0 z^3}{6L}\right)$$

$$= -\frac{w_0 L^2}{6}\left[3\left(\frac{z}{L}\right)^2 + \left(\frac{z}{L}\right)^3\right]$$

The maximum values of shear and moment which occur at $z = L$ are,

$$V_{\max} = \frac{3w_0 L}{2} \quad \text{and} \quad M_{\max} = -\frac{2w_0 L^2}{3}$$

It should be noted that shear has parabolic variation according to the second power of z and the moment has cubic parabolic variation according to the third power of z. The shearing force and bending moment diagrams are illustrated in Figures 7.7(b-ii) and (b-iii).

EXAMPLE 7.4

Draw shearing force and bending moment diagrams for the simply supported beam carrying uniformly distributed load over a part of its length as illustrated in Figure 7.8(a).

Solution: The support reactions are:

$$R_A = \frac{4w_0 a}{3}\uparrow \quad \text{and} \quad R_B = \frac{2w_0 a}{3}\uparrow$$

For equations of shearing forces and bending moments consider the equilibrium of the portion of the structure to the left of the section at distance z from the left support.

For $0 \le z \le 2a$: $\quad V_z = -R_A + w_0 z = -\dfrac{4w_0 a}{3} + wz \quad \text{and} \quad M_z = \dfrac{4w_0 a}{3}z - \dfrac{w_0 z^2}{2}$

For $2a \le z \le 3a$: $\quad V_z = -\left(\dfrac{4w_0 a}{3}\right) + w_0(2a) \quad \text{and} \quad M_z = \left(\dfrac{4w_0 a}{3}\right)z - (2w_0 a)(z - a)$

These expressions are plotted to obtain shear and moment diagrams as shown in Figure 7.8(b) and (c). It should be noted that the shearing forces vary linearly and the moments vary parabolically in the portion of the beam carrying uniformly distributed load. The maximum moment in the structure generally occurs at the point where the shearing force is zero, i.e.

$$V_z = -\frac{4w_0 a}{3} + w_0 z = 0 \quad \text{or} \quad z = \frac{4a}{3}$$

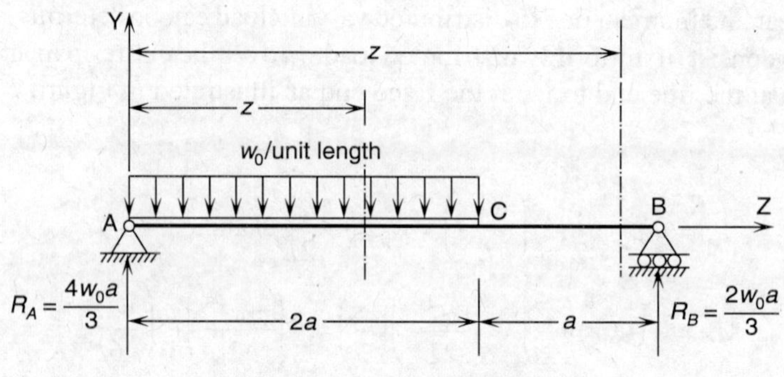

(a) The beam and loading diagram

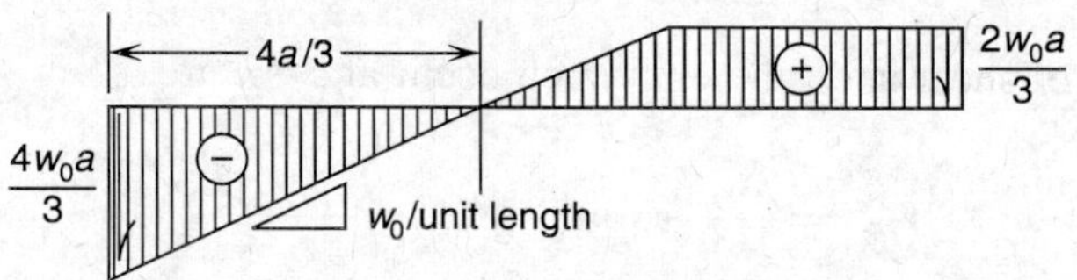

(b) The shearing force diagram

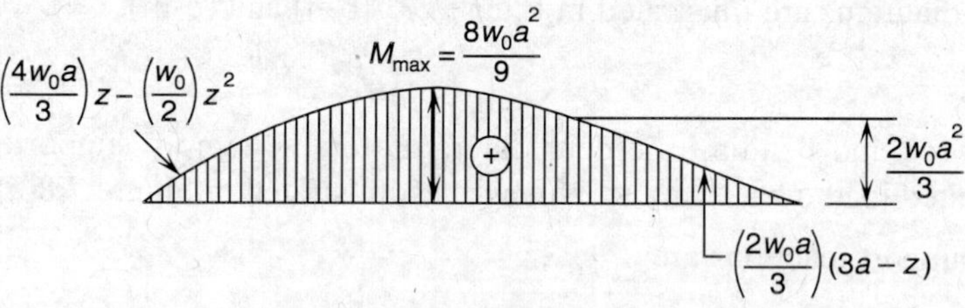

(c) The bending moment diagram

Figure 7.8 Shearing force and bending moment diagram for the partially loaded beam of Example 7.4.

Thus,
$$M_z = \frac{4w_0a}{3}z - \frac{w_0z^2}{2} = \left(\frac{4w_0a}{3}\right)\left(\frac{4a}{3}\right) - \frac{w_0}{2}\left(\frac{4a}{3}\right)^2 = \frac{8w_0a^2}{9}$$

Alternatively, shear and moment diagrams can be constructed by calculating the critical values of shears and moments which occur at $z = 0$, $2a$ and $3a$,

$$V_z = -\frac{4w_0a}{3}, \frac{2w_0a}{3}, \text{ and } \frac{2w_0a}{3}; M_z = 0, \frac{2w_0a^2}{3} \text{ and } 0$$

It should be noted that the effect of uniformly distributed load is to cause the shearing force diagram to decrease at the rate per unit length equal to load per unit length, i.e. the shearing force diagram has a slope equal to the intensity of uniformly distributed load itself.

EXAMPLE 7.5

Develop complete shearing force and bending moment diagrams for the simply supported beam with an overhang which carries distributed load over its entire length as illustrated in Figure 7.9(a).

***Solution*:** The support reactions can be determined as usual:

$$\sum M_B = \left(\frac{1}{2}\right)(w_0)(a)\left(L + \frac{a}{3}\right) - R_{AV}(L) + (w_0)(L)\left(\frac{L}{2}\right) = 0$$

i.e.
$$R_{AV} = \left(\frac{w_0 a}{2} + \frac{w_0 a^2}{6L} + \frac{w_0 L}{2}\right)\uparrow$$

$$\sum F_y = R_{AV} + R_B - \left(\frac{1}{2}\right)(w_0)(a) - (w_0)(L) = 0 \text{ or } R_B = \left(\frac{w_0 L}{2} - \frac{w_0 a^2}{6L}\right)\uparrow$$

To determine the shearing forces and bending moments consider the equilibrium of the portion of the structure to the left of a section at distance z from the left end C.

For $0 \le z \le a$:
$$V_z = \left(\frac{1}{2}\right)(z)\left(\frac{w_0 z}{a}\right) = \frac{w_0 z^2}{2a} \text{ and } M_x = -\left(\frac{w_0 z^2}{2a}\right)\left(\frac{z}{3}\right) = -\frac{w_0 z^3}{6a}$$

Therefore,
$$V_C = 0, V_{A,\text{left}} = \frac{w_0 a}{2}; M_C = 0 \text{ and } M_A = -\frac{w_0 a^2}{6}$$

It will be convenient to consider the right-hand portion of the beam segment AB with origin at B. Consider a section at distance z' from the end B.

For $0 \le z' \le L$:
$$V_z = R_B - w_0 z' = \left(\frac{w_0 L}{2} - \frac{w_0 a^2}{6L}\right) - w_0 z'$$

or for $a \le z \le L + a$:
$$= -\left(\frac{w_0 a^2}{6L} + \frac{w_0 L}{2}\right) + w_0 (z - a)$$

Similarly, for bending moment:

For $0 \le z' \le L$:
$$M_{z'} = \left(\frac{w_0 L}{2} - \frac{w_0 a^2}{6L}\right)(z') - w_0\left(\frac{z'^2}{2}\right)$$

or for $a \le z \le L + a$:
$$= \left(\frac{w_0 L}{2} - \frac{w_0 a^2}{6L}\right)(L + a - z) - w_0 \frac{(L + a - z)^2}{2}$$

The shear and moment diagrams are plotted in Figures 7.9(b) and (c). It should be noted that in the segment AB of the beam carrying uniformly distributed load the variations of shear and moment are linear and parabolic, respectively. Whereas, in the case of segment CA carrying uniformly varying load, the variations of shear and moment are square parabolic and cubic parabolic, respectively. The maximum moment occurs at the point where the shearing force is zero, i.e.

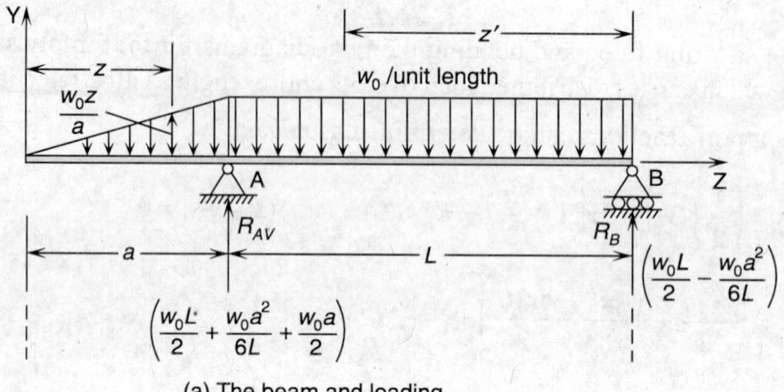

(a) The beam and loading

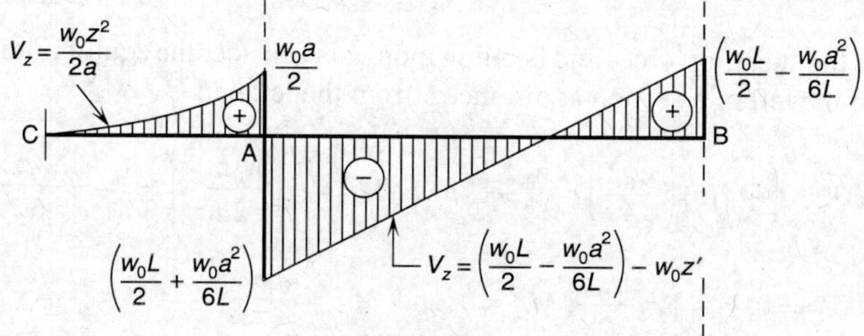

(b) Shear force diagram

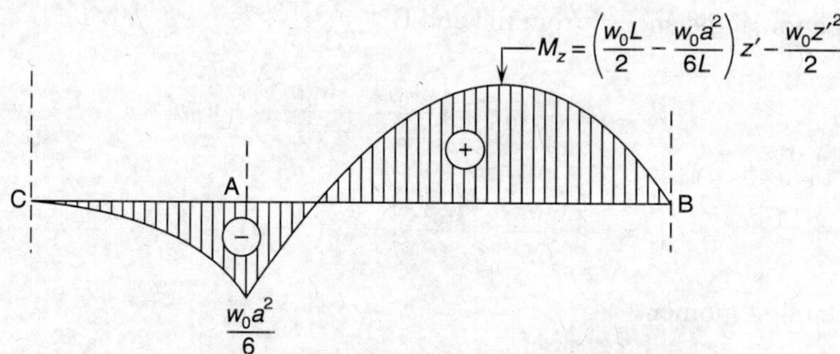

(c) Bending moment diagram

Figure 7.9 Shear force and bending moment diagrams of the beam of Example 7.5.

$$V_{z'} = \left(\frac{w_0L}{2} - \frac{w_0a^2}{6L}\right) - w_0z' = 0 \quad \text{or} \quad z' = \frac{L}{2} - \frac{a^2}{6L}$$

Thus, $$M_{\max} = \left(\frac{w_0L}{2} - \frac{w_0a^2}{6L}\right)(z') - \frac{w_0}{2}(z'^2)$$

$$= \left(\frac{w_0L}{2} - \frac{w_0a^2}{6L}\right)\left(\frac{L}{2} - \frac{a^2}{6L}\right) - \frac{w_0}{2}\left(\frac{L}{2} - \frac{a^2}{6L}\right)^2 = \frac{w_0}{2}\left(\frac{L}{2} - \frac{a^2}{6L}\right)^2$$

EXAMPLE 7.6

A vertical column ACB carries a horizontal bracket connected at the points C and D as shown in Figure 7.10. The bracket carries a concentrated load P at a distance d from the axis of the column. Construct the axial force, shearing force and bending moment diagrams for the column when (i) the column is rigidly fixed at its base and free at the top and (ii) the column has simple supports at the ends.

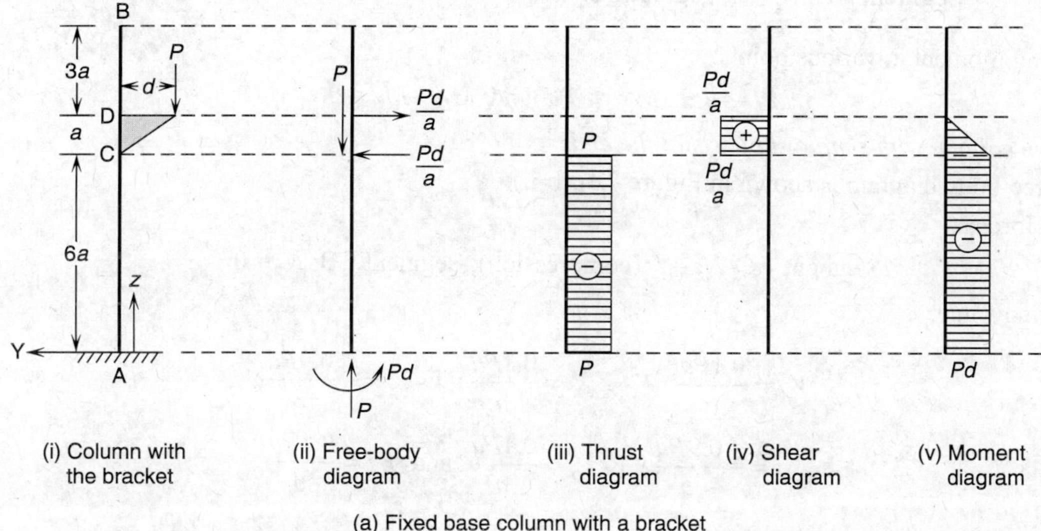

(a) Fixed base column with a bracket

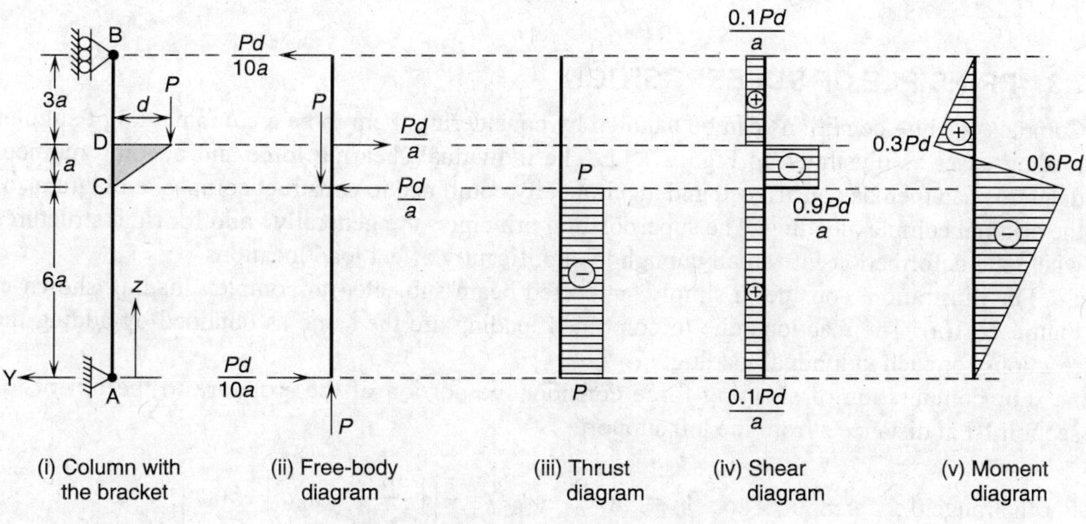

(b) Simply supported column with a bracket

Figure 7.10 Thrust, shearing force and bending moment diagrams for columns of Example 7.6.

Solution: (i) *The column rigidly fixed at its base and free at the top*:
The free-body diagram is shown in Figure 7.10(a-ii).
Axial force:

$$\text{Segment AC, } N_z = -P \text{ (compression); Segment CB, } N_z = 0$$

Shearing force:

$$\text{Segment AC, } V_z = 0\text{; segment CD, } V_z = \frac{Pd}{a} \text{ and segment DB, } V_z = 0$$

Bending moment at various points:

$$M_A = -Pd\,;\ M_C = -Pd \text{ and } M_D = M_B = 0$$

(ii) *The column with simple supports at the ends*:
The free-body diagram is shown in Figure 7.10(b-ii).
Axial force,

$$\text{Segment AC, } N_z = P \text{ (compression); Segment CB: } N_z = 0$$

Shearing force,

$$V_A = \frac{0.1Pd}{a};\ V_{C,\text{down}} = \frac{0.1Pd}{a};\ V_{C,\text{up}} = -\frac{0.9Pd}{a};$$

$$V_{D,\text{down}} = -\frac{0.9Pd}{a};\ V_{D,\text{up}} = \frac{0.1Pd}{a} \text{ and } V_A = \frac{0.1Pd}{a}$$

Bending moments at various points are,

$$M_A = 0;\ M_C = -0.6Pd;\ M_D = 0.3Pd \text{ and } M_B = 0$$

7.3 PRINCIPLE OF SUPERPOSITION

Complex loading conditions can be handled by considering them to be a combination of simpler loading cases as illustrated in Figure 7.11. The individual shearing force and bending moment diagrams can then be constructed and algebraically combined to construct composite diagrams of the original complex loading. The superposition principles are generally valid for rigid structures where the deformations are small enough to significantly affect load locations.

For illustration, consider a simply supported beam subjected to complex loading shown in Figure 7.11(a). The reactions due to combined loading are the same as obtained by adding the reactions for each individual loading.

For computation of shearing force consider the portion of the structure to the left of the section I-I at distance z from the left support.

For the range $0 \le z \le a$:
$$V_z = -(R_A - w_0 z) = -\left[\left(\frac{3W}{4} + 2w_0 a\right) - w_0 z\right]$$

$$= \left[\left(\frac{3W}{4}\right) + (2w_0 a - w_0 z)\right]$$

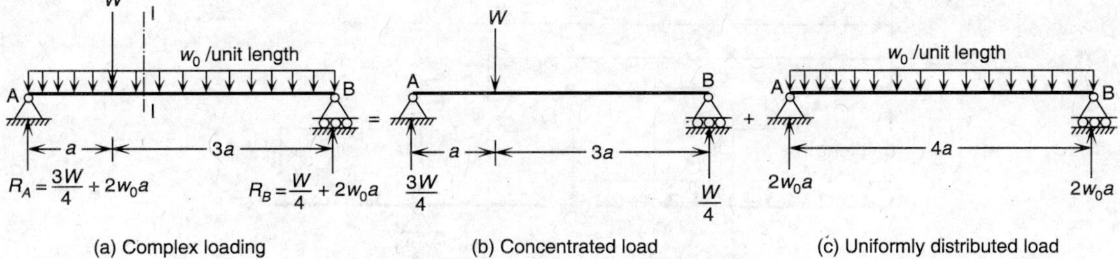

Figure 7.11 Principle of superposition.

For the range $a \le z \le 4a$:
$$V_z = -(R_A - W - w_0 z) = -\left[\left(\frac{3W}{4} + 2w_0 a\right) - W - w_0 z\right]$$

$$= -\left[\left(\frac{3W}{4} - W\right) + (2w_0 a - w_0 z)\right]$$

Thus the shears due to combined loading are the same as obtained by adding the shears due to individual loadings shown in Figures 7.11(b) and (c). Similarly, the moments are:

For the range $0 \le z \le a$:

$$M_z = \left(R_A z - \frac{w_0 z^2}{2}\right) = \left[\left(\frac{3W}{4} + 2w_0 a\right)z - \frac{w_0 z^2}{2}\right]$$

$$= \left[\left(\frac{3W}{4}\right)z + \left((2w_0 a)z - \frac{w_0 z^2}{2}\right)\right]$$

For the range $a \le z \le 4a$:

$$M_z = \left(R_A z - W(z-a) - \frac{w_0 z^2}{2}\right)$$

$$= \left[\left(\frac{3W}{4} + 2w_0 a\right)z - W(z-a) - \frac{w_0 z^2}{2}\right]$$

$$= \left[\left(\frac{3W}{4}z - W(z-a)\right) + (2w_0 a)z - \frac{w_0 z^2}{2}\right]$$

Thus the moments due to combined loading are the same as obtained by adding the moments for the individual loadings shown in Figures 7.11(b) and (c).

EXAMPLE 7.7

Develop shearing force and bending moment diagrams for the structure illustrated in Figure 7.12(a).

Solution: To compute support reactions consider the equilibrium of the forces acting on the structure. Moment equilibrium of the forces about A:

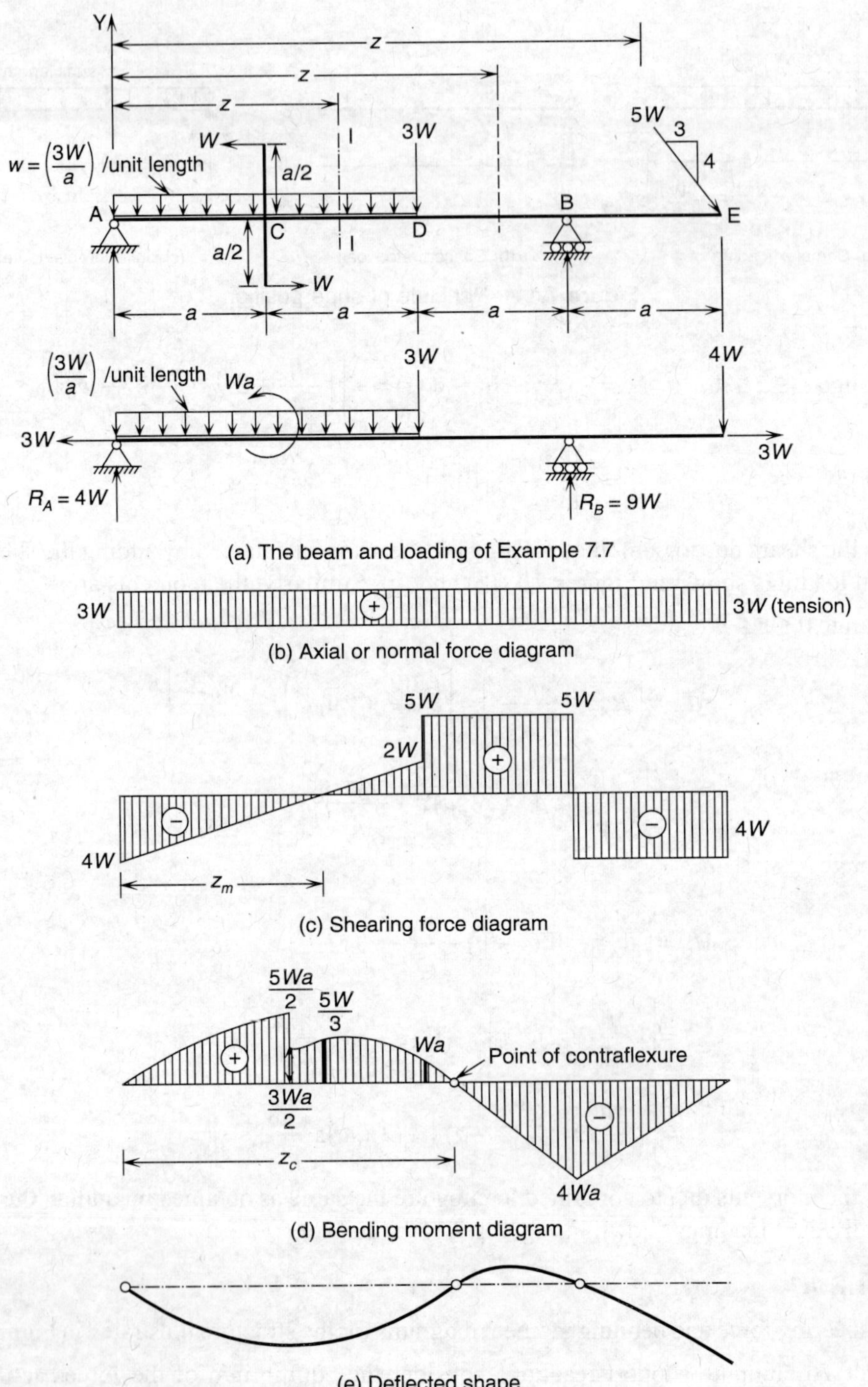

Figure 7.12 Stress resultant diagrams and deflected shape of the beam of Example 7.7.

$$\sum M_A = (2wa)(a) - (W)(a) + (3W)(2a) - R_B(3a) + (4W)(4a) = 0$$

or
$$R_B = \frac{1}{3a}[21Wa + 2wa^2] = \frac{1}{3a}[21Wa + 6Wa] = 9W \uparrow$$

$$\sum F_y = R_{AV} + R_B - \left[\left(\frac{3W}{a}\right)(2a) + (3W) + (5W)\left(\frac{4}{5}\right)\right] = R_{VA} + R_B - 13W = 0$$

Therefore,
$$R_{AV} = 13P - R_B = 4W \uparrow$$

$$\sum F_x = R_{AH} + (5W)\left(\frac{3}{5}\right) = 0 \quad \text{or} \quad R_{AH} = -3W \leftarrow$$

Figure 7.12(b) shows the axial or normal force diagram.

Shearing force: For computation of shearing force consider the portion of the structure to the left of a section I-I at distance z from the left support.

For the range $0 \le z \le 2a$:
$$V_z = -(R_{AV} - wz) = -\left[4W - \left(\frac{3W}{a}\right)z\right]$$

$$V_A = -4W \quad \text{and} \quad V_{D,\text{left}} = 2W$$

For the range $2a \le z \le 3a$:
$$V_z = -(R_{AV} - 6W - 3W) = 5W$$

Therefore,
$$V_{D,\text{right}} = 5W \text{ and } V_{B,\text{left}} = 5W$$

For the range $3a \le z \le 4a$:
$$V_z = -(R_{AV} - 6W - 3W + R_B) = -4W$$

$$V_{B,\text{right}} = -4W \text{ and } V = -4W$$

Bending moment: Bending moment at a section I-I at distance z from the left support is given by:

For the range $0 \le z \le a$:

$$M_z = \left(R_{AV}z - \frac{wz^2}{2}\right) = 4Wz - \left(\frac{3Wz^2}{2a}\right)$$

Hence,
$$M_A = 0 \text{ and } M_{C,\text{left}} = \frac{5Wa}{2}$$

For the range $a \le z \le 2a$:

$$M_z = 4Wz - \left(\frac{3Wz^2}{2a}\right) - Wa$$

$$M_{C,\text{right}} = \frac{3Wa}{2} \text{ and } M_D = Wa$$

For the range $2a \le z \le 3a$:

$$M_z = 4Wz - \left(\frac{3W}{a}\right)(2a)(z-a) - Wa - (3W)(z-2a) \text{ (straight line)}$$

$$M_B = -4Wa$$

The local maximum positive bending moment will occur at the point distant z_m from A where shearing force changes sign, i.e. $z_m = \frac{R_{AV}}{w} = \frac{4W}{(3W/a)} = \frac{4a}{3}$.

The bending moment at this point is given by,

$$M_{\max} = 4W\left(\frac{4a}{3}\right) - \left(\frac{3W}{2a}\right)\left(\frac{4a}{3}\right)^2 - Wa = \frac{5Wa}{3}$$

Alternatively, shear and moment diagrams can be constructed by calculating the critical values of shears and moments which occur at $z = 0$, a, $4a/3$, $2a$, $3a$ and $4a$. The axial force, shearing force and bending moment diagrams are shown in Figures 7.12(b), (c) and (d), respectively. The point of contraflexure, where the bending moment changes sign, occurs in the range $2a \le z \le 3a$.

Thus $$4Wz_c - \left(\frac{3W}{a}\right)(2a)(z_c - a) - Wa - (3W)(z_c - 2a) = 0$$

or $$(4W - 6W - 3W)z_c + (6W - W + 6W)a = 0 \text{ or } z_c = \frac{11a}{5}$$

Figure 7.12(e) shows the deflected shape.

EXAMPLE 7.8

Develop complete shearing force and bending moment diagrams for the composite beam shown in Figure 7.13(a). Find shear and moment at the point G.

Solution: To simplify the analysis the interior hinge is treated as shown in Figure 7.13(b-i). The free-body diagrams of two segments are shown in Figure 7.13(b-ii). To compute support reactions consider the following equilibrium conditions:

The segment AE:

$$\sum M_E = -R_A(12a) + (W)(8a) + 6Wa = 0 \quad \text{or} \quad R_A = \frac{7W}{6}\uparrow$$

$$\sum F_y = R_A + R_E - W = \frac{7W}{6} + R_E - W = 0 \quad \text{or} \quad R_E = -\frac{W}{6}\downarrow$$

The segment EB:

$$\sum F_y = R_E + R_B - 2W = \frac{W}{6} - 2W + R_B = 0 \quad \text{or} \quad R_B = \frac{11W}{6}\uparrow$$

Shearing force: For computation of shearing force, consider the portion of the beam to the left of a section I-I at distance z from the left support A.

For the range $0 \le z \le 4a$: $$V_z = -R_A = -\frac{7W}{6}$$

$$V_A = -\frac{7W}{6}; \text{ and } V_{C,\text{left}} = -\frac{7W}{6}$$

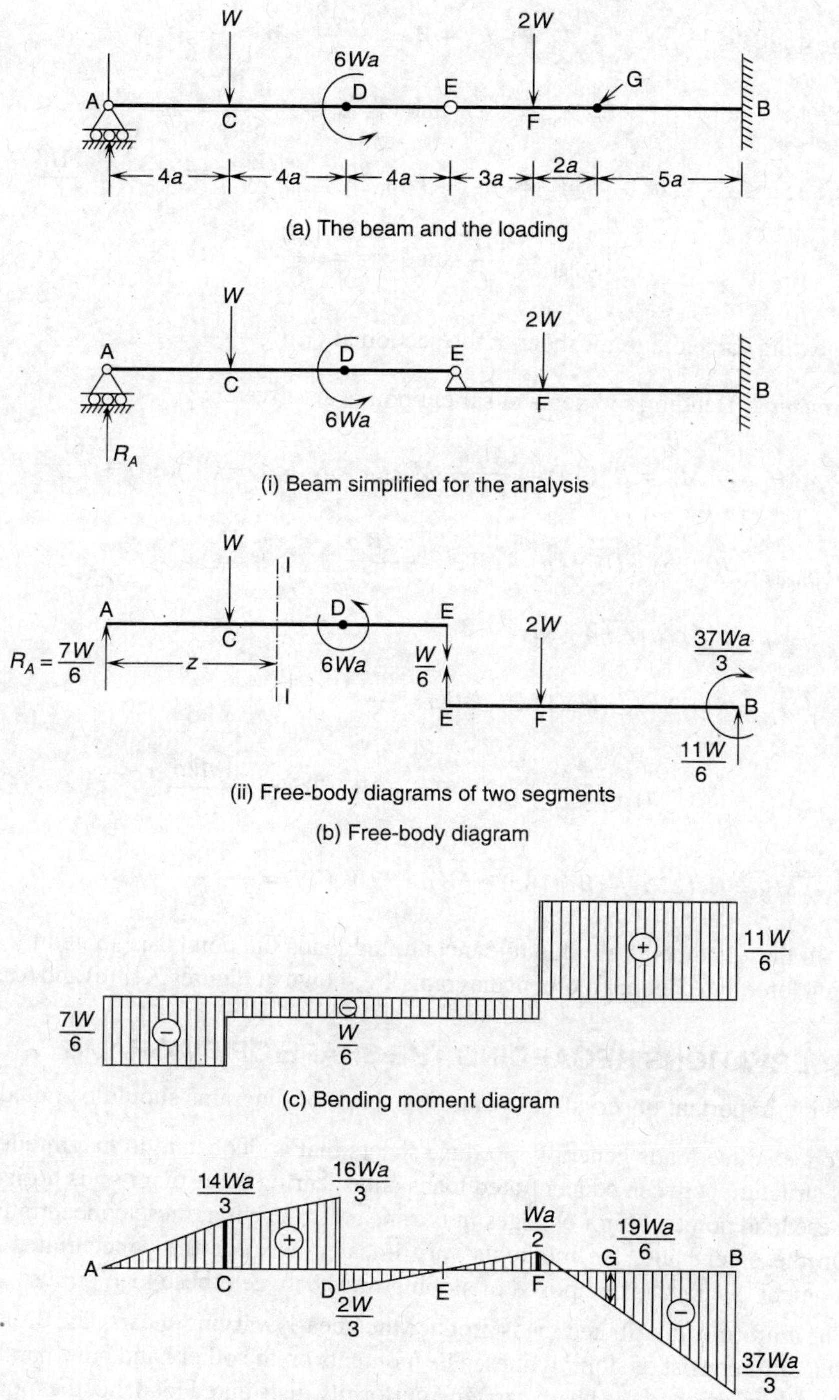

Figure 7.13 Shear force and bending moment diagrams of hinged beam of Example 7.8.

For the range $4a \le z \le 15a$: $$V_z = -R_A + W = -\frac{7W}{6} + W = -\frac{W}{6}$$

Therefore, $$V_{C,\text{right}} = -\frac{W}{6} \text{ and } V_{F,\text{left}} = -\frac{W}{6}$$

For the range $15a \le z \le 22a$: $$V_z = -R_A + W + 2W = -\frac{7W}{6} + W + 2W = \frac{11W}{6}$$

$$V_{F,\text{right}} = \frac{11W}{6} \text{ and } V_B = \frac{11W}{6}$$

From the shearing force diagram, shear at the section at G, $V_G = \dfrac{11W}{6}$.

Bending moment: Bending moments at salient points are given by:

$$M_A = 0, M_C = R_A(4a) = \frac{14Wa}{3}, M_{D,\text{left}} = R_A(8a) - (W)(4a) = \frac{16Wa}{3}$$

$$M_{D,\text{right}} = R_A(8a) - (W)(4a) - 6Wa = -\frac{2Wa}{3}$$

$$M_E = R_A(12a) - (W)(8a) - 6Wa = 0 \quad \text{(check)}$$

$$M_F = R_A(15a) - (W)(11a) - 6Wa = \frac{Wa}{2}$$

$$M_G = R_A(17a) - (W)(13a) - 6Wa - (2W)(2a) = -\frac{19Wa}{6}$$

$$M_B = R_A(22a) - (W)(18a) - 6Wa - (2W)(7a) = -\frac{37Wa}{6}$$

Since all these moments are due to concentrated loads the points are joined by straight lines. The shearing force and bending moment diagrams are shown in Figures 7.13(c) and (d), respectively.

7.4 OBSERVATIONS REGARDING THE SHAPE OF DIAGRAMS

The following important observations regarding shape of diagrams should be noted.

1. Concentrated loads generally produce shears that are constant in magnitude over the part of structure between concentrated loads and shearing force diagram is horizontal between these load points. Major changes in the shear also occur at these concentrated load points. On the other hand, the moments vary linearly between the concentrated loads and the moment diagram is composed of sloping lines between these load points.
2. The uniformly distributed loads produce the linearly varying shears, and the shearing forces diagrams consist of sloping lines. The moments on the other hand have parabolic variation over the portion of the beam carrying uniformly distributed load, i.e. the moment diagrams have curved shapes. The maximum or minimum moments in the structure generally occur at the points where the shearing force diagram passes through zero.

3. The complex loading condition produces combined shape which can be handled by considering it to be a combination of simpler loading cases. The individual shearing force and bending moment diagrams can then be constructed and algebraically combined to construct composite diagrams of the original complex loading.
4. For inclined members, the components of the loads and reactions perpendicular to centroidal axis of the member are used and diagrams are drawn parallel to the member.
5. In some beams forces are applied through the rigid arms or brackets fastened to them. The arrangement induces moment in the structure at the point of attachment. To draw the bending moment diagram at the point of attachment, it is necessary to imagine the moment to be at an infinitesimal distance to the left of the point and then add the moment applied by the arm. The moment acting exactly at the point of attachment is discontinuous and cannot be figured, but moment immediately beyond the point is available.
6. The point where the bending moment is zero is called *point of inflection* or *contra-flexure*.
7. A beam which has one or both supports positioned at the intermediate points along the beam is known as *overhang beam* and in many cases the bending moment changes sign.

7.5 RELATIONSHIPS AMONG LOADS, SHEARS, AND MOMENTS IN STRUCTURES

The differential equation of relationships among loads, shears, and moments in structures can be obtained by using standard equilibrium procedure. Consider the translational and rotational equilibrium of an infinitesimal element of length dz isolated from the structure as shown in Figure 7.14(a). The element carries constant distributed load $w(z)$ within the infinitesimal length of element. Due to this loading force components V_z and M_z will generally vary with z. Thus the element is subjected to shearing force and moment $V_z(z)$ and $M_z(z)$ along with their incremental changes $dV_z(z)$, and $dM_z(z)$ as illustrated in Figure 7.14(b). The equilibrium conditions are:

$$\sum F_y = 0: \; V_z - \{V_z + dV_z\} - w(z)\, dz = 0$$

Dividing by dz,

$$\frac{dV_z}{dz} + w(z) = V_z'(z) + w(z) = 0$$

or

$$V_z' = -w(z) \tag{7.3}$$

In the other words, the rate of change of $V_z(z)$ with respect to z is equal to the negative of loading $w(z)$.

The equilibrium condition $\Sigma M = 0$ about the left end of element gives

$$M_z - \{(V_z + dV_z\}\, dz] - [w(z)\, dz]\left(\frac{dz}{2}\right) - \{M_z + dM_z\} = 0$$

On simplification and dividing by dz we obtain

$$V_z + w(z)\frac{dz}{2} + \frac{dM_z}{dz} = 0$$

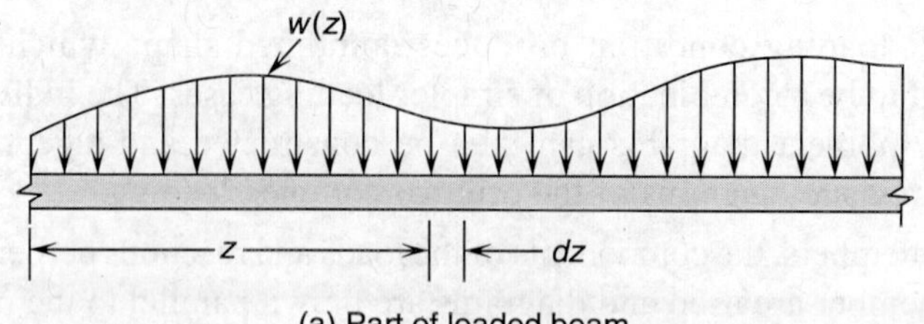

(a) Part of loaded beam

(b) Enlarged free-body diagram of the isolated element of length *dz*

Figure 7.14 Forces acting on an element of a loaded beam.

In the limit dz, the second term vanishes

$$V_z + \frac{dM_z}{dz} = 0 \quad \text{or} \quad \frac{dM_z}{dz} = M_z' = -V_z$$

or
$$\frac{dM_z}{dz} = -V_z \tag{7.4}$$

In other words, the rate of change of M_z with respect to z is equal to the shearing force V_z. For the segment between two points i and j, shown in Figure 7.15(a) from Eq. (7.3):

$$\Delta V = V_j - V_i = \int_i^j w(z)\, dz = (\text{area})_{\text{load},i-j}$$

or
$$V_j = V_i + (\text{area})_{\text{load},i-j} \tag{7.5}$$

i.e. *the change in shearing force between two points on the beam is equal to the area of the loading diagram between the points.*

The relation given by Eq. (7.4) can be expressed as: $dM(z) = V(z)\, dz$ and for the segment between two points i and j,

$$\Delta M = M_j - M_i = \int_i^j V_z(z)\, dz = (\text{area})_{\text{shear},i\text{-}j}$$

or
$$M_j = M_i + (\text{area})_{\text{shear},i\text{-}j} \tag{7.6}$$

i.e. *the change in bending moment between two points on the beam is equal to the area of the shear diagram between the points.*

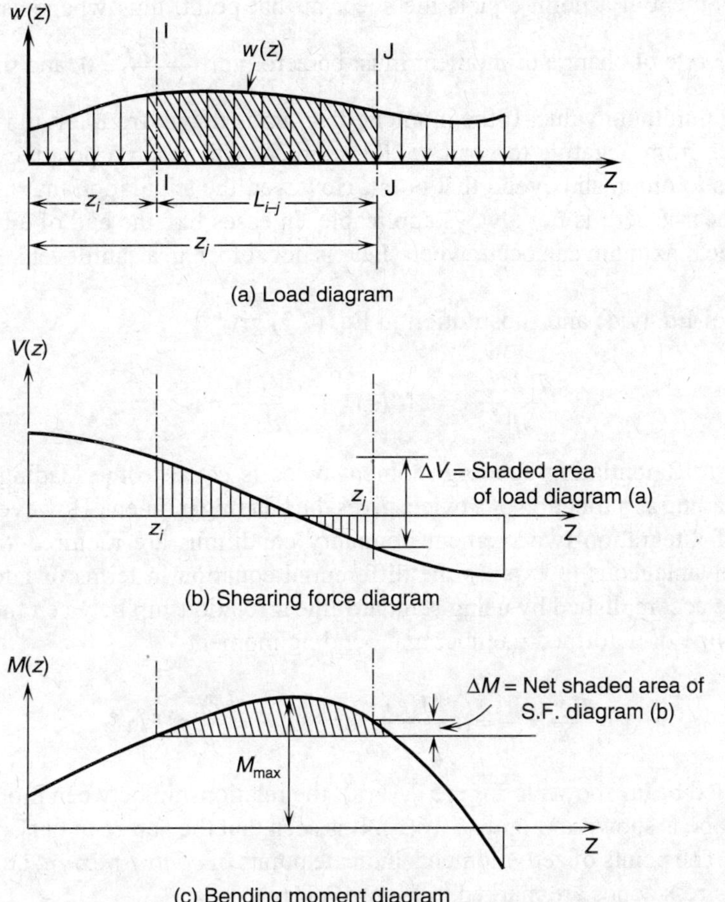

Figure 7.15 Relationships among load, shearing force and bending moment.

For concentrated load, $\int_i^j w(z)\,dz = 0$ as $w(z) = 0$, therefore, $V_j = V_i = V$ i.e. the shearing force has constant value between these points. For the change in bending moment between these two points,

$$M_j - M_i = \int_i^j V(z)\,dz = V\int_i^j dz = V(z_j - z_i)$$

i.e. the bending moment varies linearly between these two points if no load acts between the points.

The foregoing relationship between shear and moment greatly simplifies the construction of the bending moment diagrams. To determine the moment at a particular section, it is only necessary to compute the total area beneath the shear curve either to left or to the right of the section, taking in to account the algebraic signs of various segments of shear diagram. The shear and moment diagrams are self-checking (see Figures 7.15(b) and (c)). Moreover, as mentioned above the rate of change of moment at a point equals the shear at that point, thus whenever the shear passes through zero the rate of change of moment must be zero, i.e. $\frac{dM(z)}{dz} = 0$, and the moment attains its maximum or minimum value. If the moment diagram is drawn from left to right and the shear diagram changes from negative to positive, the moment will reach a positive maximum at that point and begins to diminish beyond that point. However, the stipulation that maximum moment occurs where shear is zero is not always applicable. In cases like the end of a beam, the point of discontinuity, the maximum can occur when shear is not zero, e.g. a cantilever subjected to gravity loads.

Differentiation of Eq. (7.4) and substitution in Eq. (7.3) gives

$$\frac{d^2M(z)}{dz^2} = M''(z) = -\frac{dV(z)}{dz} = w(z) \tag{7.7}$$

This means that differentiating bending moment twice is equal to the loading. In other words, integration of loading as a function of z twice, gives the bending moment. However, for determining the constants of integration two moment boundary conditions are required which may not be obvious. It is advantageous to express the differential equation in terms of lateral displacement $y(z)$. This can be accomplished by using standard linear relationship between the curvature of the curve of centre line of deformed member and bending moment

$$\frac{d^2y}{dz^2} = \frac{1}{R} = -\frac{M(z)}{El} \quad \text{or} \quad M(z) = \frac{El}{R} \equiv El\,y'' \tag{7.8}$$

For the loaded beam shown in Figure 7.16(a), the relationship between moment diagram and the deflected shape is shown in Figure 7.16(b). It is seen that the nature of curvature changes with that of moment. The points of zero moments indicate points of *contra-flexure*, i.e. points of *reverse curvature*. The stress zones are marked in Figure 7.16(c).

The differential equation of elastic beam in the displacement form can be obtained by differentiating Eq. (7.8) and substituting from Eq. (7.7):

$$M''(z) = El\,y'''' = w(z) \quad \text{or} \quad El\,y''''\left(= El\frac{d^4y}{dz^4}\right) = w(z) \tag{7.9}$$

This is an ordinary linear differential equation of fourth-order and therefore requires determination of four constants of integration from the four displacement boundary conditions. It should be noted that a particular value $\frac{dV(z)}{dz}$ or $\frac{dM(z)}{dz}$ is valid only for the portion of the structure at which the function is continuous.

These relations indicate that;

(i) If $w(z) = 0$ shear will remain constant, i.e. if concentrated load is applied, shear on either side of load will be constant. Moreover, the change in the value of shear at that point will equal to the magnitude of the concentrated load.

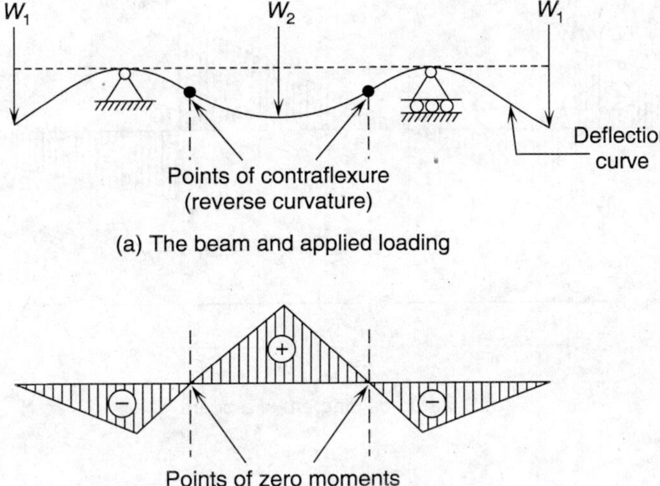

(a) The beam and applied loading

(b) Bending moment diagram

(c) Stress zones

Figure 7.16 Relationship between moment diagram and the deflected shape.

(ii) The value of the distributed load at a point is equal to the slope of the shear diagram at the point.

(iii) If $w(z)$ varies continuously in the segment under consideration, then $V(z)$ will also vary continuously in the segment.

(iv) The value of shear at a point is equal to the slope of the moment diagram at the point.

(v) If $V(z) = 0$ or shear passes through zero at a point in the structure, the moment must be either maximum or minimum since $\frac{dM(z)}{dz} = 0$ at this point.

(vi) The change in moment between any two points on a structure is given by the area under shearing force diagram between these points.

EXAMPLE 7.9

Construct the bending moment and the loading diagrams for a beam from its shearing force shown in Figure 7.17(a).

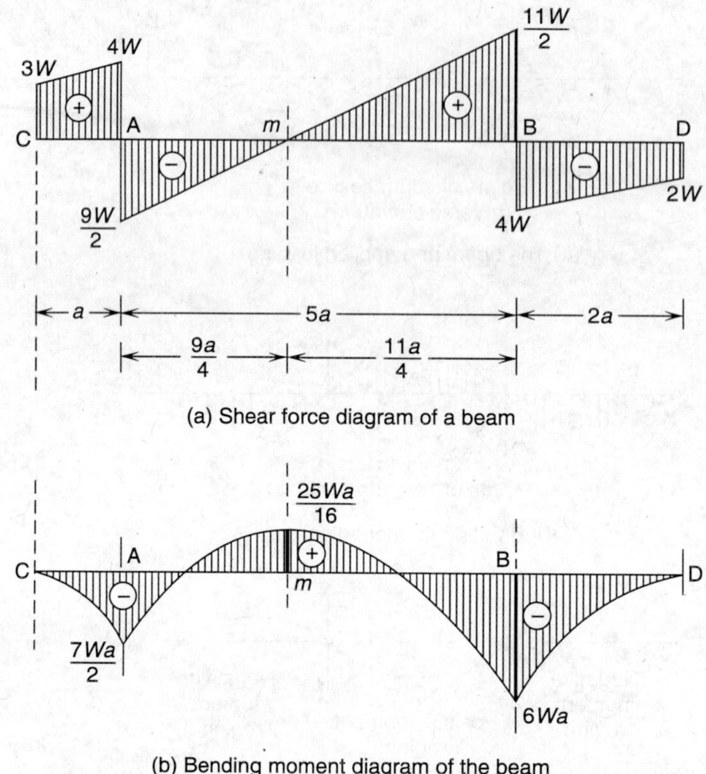

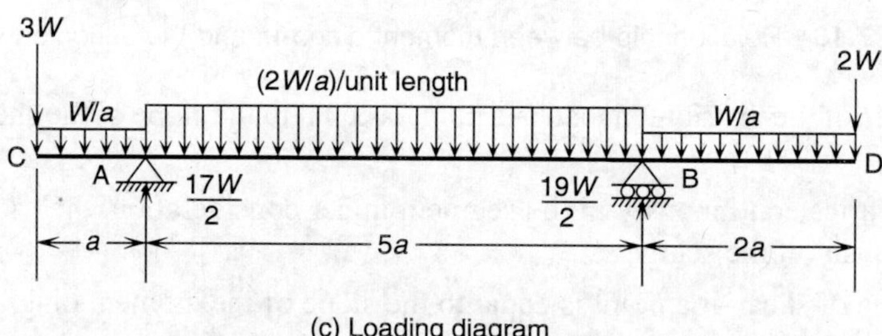

Figure 7.17 Constructing BM and the loading diagrams for a beam from its shearing force diagram (Example 7.9).

Solution: A study of the shearing force diagram indicates that:

(i) There are supports or concentrated loads at the points C, A, B and D. As per the sign convention adopted for shears, the magnitude and nature of loads at these points are, $3W\downarrow$, $-(4 + 4.5)W\uparrow$, $-(5.5 + 4)W\uparrow$ and $2W\downarrow$, respectively. It is evident that there are supports at the points A and B with reactions $8.5W$ and $9.5W$, respectively; and concentrated loads $3W$ and $2W$ act at the overhang points C and D, respectively.

(ii) The sloping lines in the segments CA, AB and BD point out that there are uniformly distributed loads. Based on the rule that *the value of the distributed load at a point is equal to the slope of the shear diagram at the point*, the distributed loads in various segments are:

Segments CA: $$w = \frac{dV}{dx} = \frac{4W - 3W}{a} = (W/a)\downarrow$$

Segments AB: $$w = \frac{5.5W - (-4.5W)}{5a} = \frac{10W}{5a} = (2W/a)\downarrow$$

Segments BD: $$w = \frac{-2W - (-4W)}{2a} = \frac{2W}{2a} = (W/a)\downarrow$$

The loading diagram is shown in Figure 7.17 (c). Now, the bending moment diagram can be constructed from the loading diagram. However, in this problem, it is constructed from the shear diagram, by using the rules:

(i) The maximum/minimum value of moment occurs when the shear diagram passes through zero.

(ii) The change in moment between any two points is given by the area under shearing force diagram between these points, as follows.

Starting from the left end C where bending moment is zero, i.e. $M_C = 0$, the moment at the point A is given by,

$$M_A = M_C + (\text{area})_{\text{shear,C-A}} = 0 - \left(\frac{1}{2}\right)(3W + 4W)(a) = -\frac{7Wa}{2}$$

Similarly, $$M_m = M_A + (\text{area})_{\text{shear,A-m}} = -\frac{7Wa}{2} + \left(\frac{1}{2}\right)\left(\frac{9W}{2}\right)\left(\frac{9a}{4}\right) = \frac{25Wa}{16}$$

$$M_B = M_m + (\text{area})_{\text{shear,m-B}} = \frac{25Wa}{16} - \left(\frac{1}{2}\right)\left(\frac{11W}{2}\right)\left(\frac{11a}{4}\right) = -6Wa$$

The bending moment diagram is shown in Figure 7.17 (b).

In the following problems, the procedures which require the computation of certain quantities before the stress resultant diagrams can be constructed are illustrated.

EXAMPLE 7.10

A beam AB of length $10a$ rests on two supports D and E at distance a and z from the ends A and B respectively. The support E is located in the right half of the beam. The beam carries uniformly distributed load $w(= 2W/a)$ per unit run over its entire length together with a concentrated load $5W$ at the left end A. Determine the value of z, if the mid-point of the beam is a point of contra-flexure. Construct the bending moment diagram for the beam for this arrangement shown in Figure 7.18(a).

Solution: To compute support reactions consider moment equilibrium condition at D:

$$\sum M_D = R_E(10a - a - z) + (5W)(a) - (10a)\left(\frac{2W}{a}\right)(5a - a) = 0 \text{ or } R_E = \frac{75Wa}{9a - z}\uparrow$$

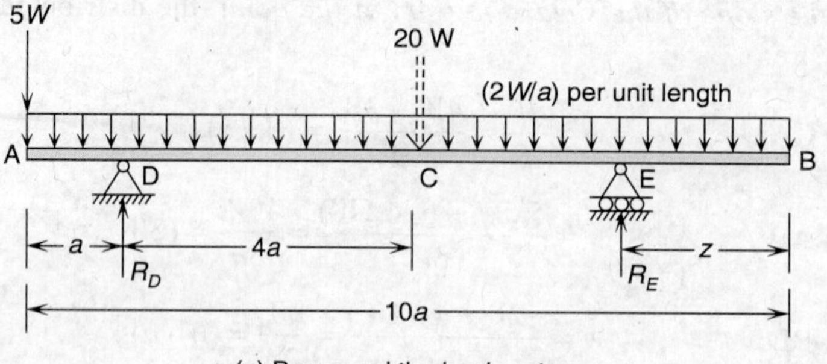

(a) Beam and the load system

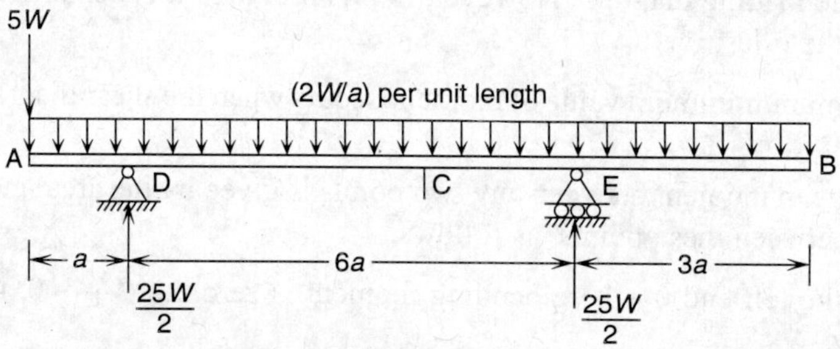

(b) Free-body diagram of the beam

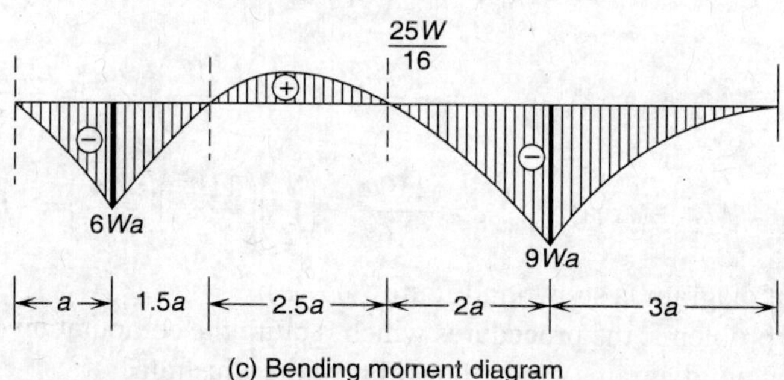

(c) Bending moment diagram

Figure 7.18 Loaded beam of Example 7.10.

The bending moment at the mid-point C,

$$\sum M_{C,\text{right}} = R_E(5a - z) - \left(\frac{1}{2}\right)\left(\frac{2W}{a}\right)(5a)^2 = \left(\frac{75Wa}{9a - z}\right)(5a - z) - 25Wa = 0$$

or
$$3(5a - z) - (9a - z) = 0 \quad \text{or} \quad z = 3a$$

Therefore,
$$R_E = \frac{75Wa}{9a - z} = \frac{25W}{2} \quad \text{and} \quad R_D = 25W - \frac{25W}{2} = \frac{25W}{2}$$

Free-body diagram is shown in Figure 7.18(b).

For constructing the bending moment diagram, the points of contra-flexure, ($M_z = 0$) and location of maximum moment $\left(\dfrac{dM}{dz} = 0\right)$ is required. Consider a point at distance z from A,

$$M_z = (5W)(z) - \left(\frac{25W}{2}\right)(z - a) + \left(\frac{1}{2}\right)\left(\frac{2W}{a}\right)(z^2) = 0$$

Points of contra-flexure:

$$M_z = 5za - \frac{25}{2}a(z - a) + z^2 = 0 \quad \text{or} \quad z^2 - \left(\frac{15a}{2}\right)z + \left(\frac{25}{2}\right)a^2 = 0$$

Therefore,
$$z = \frac{15a}{4} \pm \sqrt{\left(\frac{15a}{4}\right)^2 - \frac{25a^2}{2}} = \frac{15a}{4} \pm \frac{5a}{4} = 5a, 2.5a$$

Maximum bending moment:

$$\frac{dM}{dz} = \frac{d}{dz}\left[z^2 - \left(\frac{15a}{2}\right)z + \left(\frac{25}{2}\right)a^2\right] = 2z - \left(\frac{15a}{2}\right) = 0 \quad \text{or} \quad z = \frac{15a}{4}$$

$$M_{\max} = -(5W)\left(\frac{15a}{4}\right) + \left(\frac{25W}{2}\right)\left(\frac{15a}{4} - a\right) - \left(\frac{1}{2}\right)\left(\frac{2W}{a}\right)\left(\frac{15a}{4}\right)^2 = \frac{25Wa}{16}$$

$$M_D = (5W)(a) + \left(\frac{1}{2}\right)\left(\frac{2W}{a}\right)a^2 = 6Wa$$

$$M_{E,\text{right}} = \left(\frac{1}{2}\right)\left(\frac{2W}{a}\right)(3a)^2 = 9Wa$$

The bending moment diagram is shown in Figure 7.18 (c).

EXAMPLE 7.11

A beam carrying uniformly distributed load rests on two supports b metres apart with equal overhang a metres at each end. Determine the ratio b/a for the maximum bending moment to be as small as possible. Use this result to determine the most economical length for a sleeper if the rail centres are 1.6 m apart (see Figure 7.19(a)). **[I.A.S. 1969]**

Solution: For the maximum bending moment to be as small as possible equate the negative moment at the support to the positive bending moment at the mid-span. The bending moment in general terms is shown in Figure 7.19(b). Thus,

$$\frac{w_0 a^2}{2} = \frac{w_0 b^2}{8} - \frac{w_0 a^2}{2} \quad \text{or} \quad w_0 a^2 = \frac{w_0 b^2}{8}$$

Therefore,
$$\left(\frac{b}{a}\right)^2 = 8 \quad \text{or} \quad \frac{b}{a} = 2\sqrt{2}$$

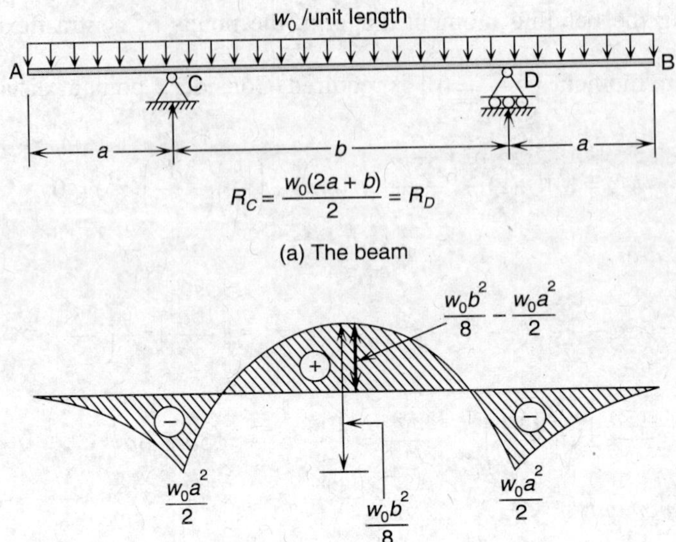

Figure 7.19 The simply supported uniformly loaded beam of Example 7.11.

For the given sleeper, $b = 1.6$ m and thus, $a = \dfrac{1.6}{2\sqrt{2}}$ m.

The total length of the sleeper,

$$L = b + 2a = 1.6 + 2\left(\frac{1.6}{2\sqrt{2}}\right) = 2.73 \text{ m}$$

EXAMPLE 7.12

A beam carrying uniformly distributed load w per unit length rests on two supports $3a$ apart with overhangs of z and a at the left and right ends, respectively (see Figure 7.20(a)). If the reactions at the left and right supports are $5W$ and $4W$, respectively, determine the length z of overhang CA and the intensity of uniformly distributed load w.

Solution: In this particular problem, the reactions are given and it is required to determine the values of z and w to satisfy the equilibrium conditions. Consider:

$$\sum M_D = \frac{1}{2}(w)(4a+z)(4a+z) - (4W)(a) - (5W)(4a) = 0$$

or

$$\frac{w(4a+z)^2}{2} - 24Wa = 0 \qquad \text{(i)}$$

$$\sum F_y = (w)(4a+z) - 9W = 0 \quad \text{or} \quad (w)(4a+z) = 9W \qquad \text{(ii)}$$

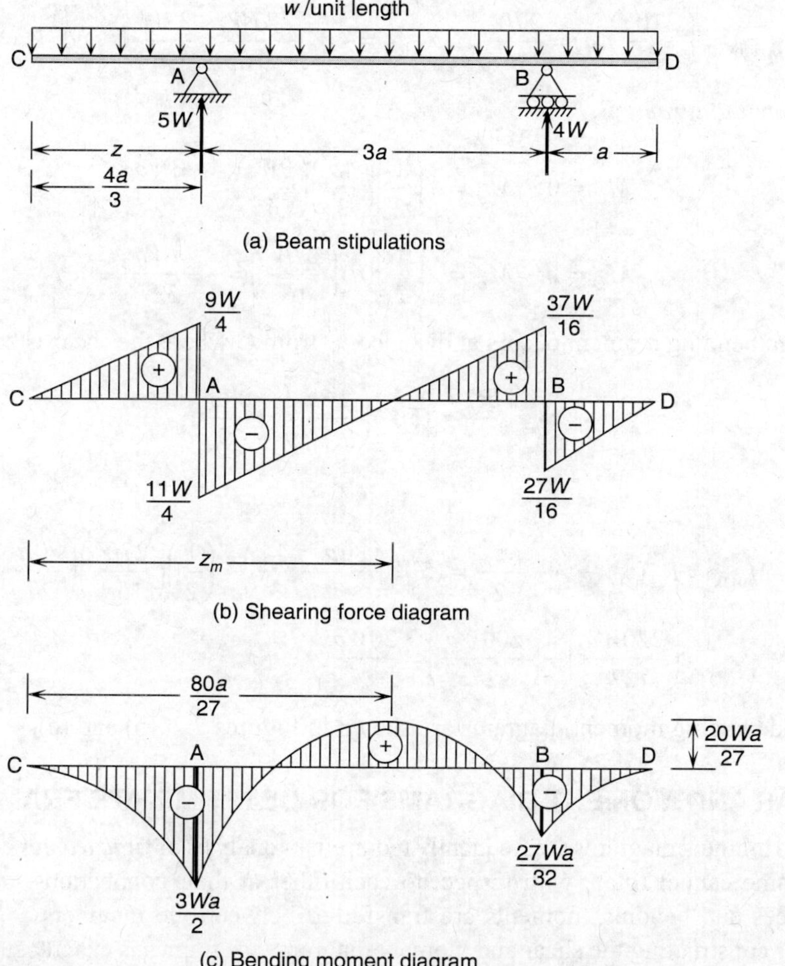

Figure 7.20 Shear force and bending moment diagrams of the beam of Example 7.12.

From Eqs. (i) and (ii):

$$\frac{(4a+z)}{2}(9W) - 24Wa = 0 \quad \text{or} \quad z = \frac{48a}{9} - 4a = \frac{4a}{3}$$

From Eq. (ii):

$$w = \frac{9W}{4a+z} \quad \text{or} \quad w = \frac{9W}{4a+(4a/3)} = \frac{27W}{16a}$$

Shearing force diagram:

$$V_C = 0; V_{A,\text{left}} = \left(\frac{27W}{16a}\right)\left(\frac{4a}{3}\right) = \frac{9W}{4}\downarrow; V_{A,\text{right}} = 5W - \frac{9W}{4} = \frac{11W}{4}\uparrow$$

$$V_{B,\text{right}} = \left(\frac{27W}{16a}\right)(a) = \frac{27W}{16}\downarrow;\ V_{B,\text{left}} = 4W - \frac{27W}{16} = \frac{37W}{16}\uparrow \text{ and } V_D = 0$$

Bending moment diagram:

$$M_C = 0;\ M_A = -\left(\frac{1}{2}\right)\left(\frac{4a}{3}\right)\left(\frac{9W}{4}\right) = -\frac{3Wa}{2}$$

$$M_D = 0;\ M_B = -\left(\frac{1}{2}\right)(a)\left(\frac{27W}{16}\right) = -\frac{27Wa}{32}$$

The maximum bending moment occurs at distance z_m from C where the shear is zero, i.e.

$$z_m = \frac{R_A}{w} = (5W)\left(\frac{16a}{27W}\right) = \frac{80a}{27}$$

Therefore,

$$M_{\max} = R_A(z_m - z) - \frac{wz_m^2}{2} = (5W)\left(\frac{80a}{27} - \frac{4a}{3}\right) - \left(\frac{1}{2}\right)\left(\frac{27W}{16a}\right)\left(\frac{80a}{27}\right)^2$$

$$= \left(\frac{220Wa}{27}\right) - \left(\frac{200Wa}{27}\right) = \frac{20Wa}{27}$$

The shear and bending moment diagrams are shown in Figures 7.20(b) and (c).

7.6 SHEAR AND MOMENT DIAGRAMS FOR DETERMINATE FRAMES

The shear and moment diagrams are frequently required for analysis of *rigid frames*. As the members of a rigid frame cannot rotate with respect to each other at their connections, the axial forces, shearing forces and bending moments are transferred between the members at the joints. The procedure for constructing the shear and moment diagrams for frames is exactly similar to the one for individual members. However, the sign convention used for constructing the shear and moment diagrams need to be examined to be consistent for all frame members. For illustration, consider the bent member shown in Figure 7.21(a). Consider section I-I in the member CB of the frame at distance z from C as shown in free-body diagram of bent in Figure 7.21(b). With the member co-ordinate axes as shown the outward normal to the face of the section is in positive Z-direction, the shear vector V_z is negative. Thus, from equilibrium considerations,

$$V_z = P \sin\alpha \text{ (negative) and } M_z = (P \sin\alpha)(z) \text{ (positive)}$$

Thus, for the member CB, the shear V_z is constant and moment M_z varies linearly from 0 at C to $(P \sin\alpha)b$ at B. At a section II-II in the member BA at distance z from B, the outward normal to the section is in positive Z-direction, the shear vector V_z is in negative and moment M_z vector is in positive X-direction. Thus the shear is negative and moment is positive. From equilibrium,

$$V_z = P \text{ (negative) and } M_z = P(b \sin\alpha + z) \text{ (positive)}$$

The shear V_z therefore is constant and moment M_z varies linearly from $P(b \sin \alpha)$ at B to $P(b \sin \alpha + a)$ at A. The shear force and moment diagrams are shown in Figure 7.21(c).

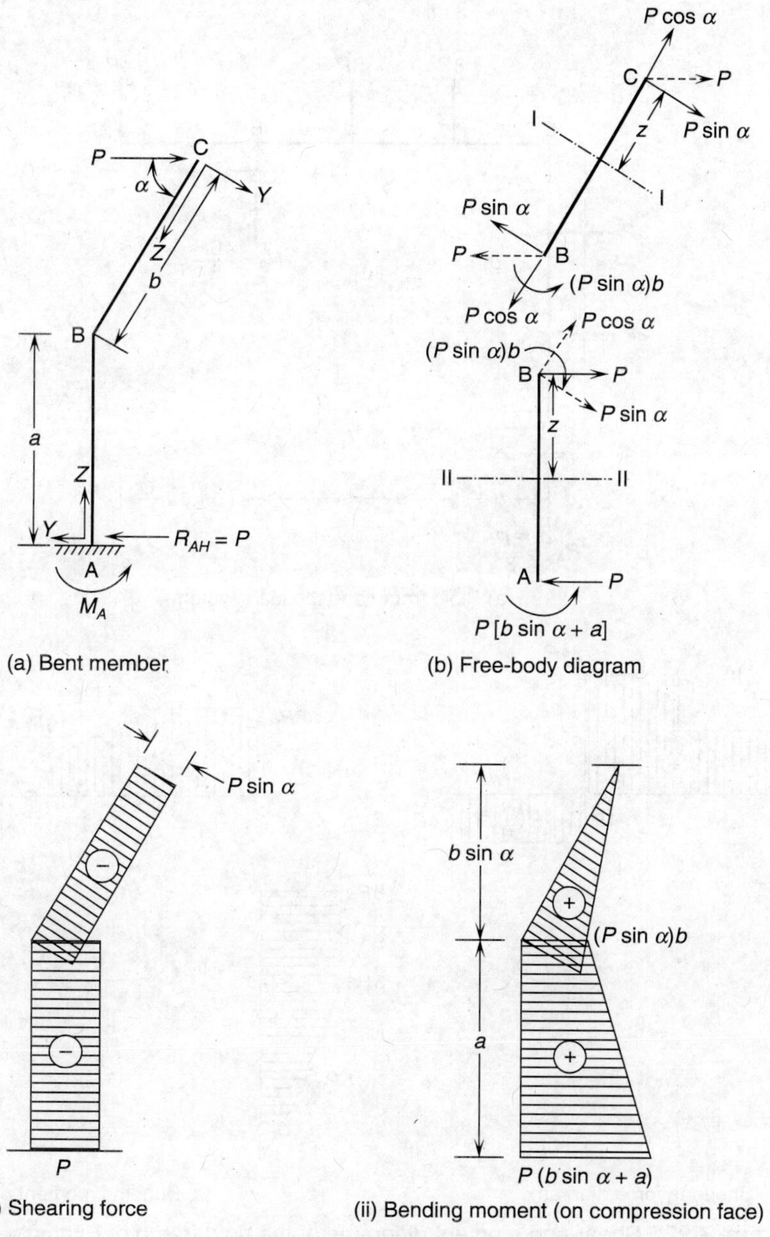

Figure 7.21 Shear force and moment diagrams for a bent member.

EXAMPLE 7.13

Draw shearing force and bending moment diagrams for the determinate rigid frame shown in Figure 7.22(a).

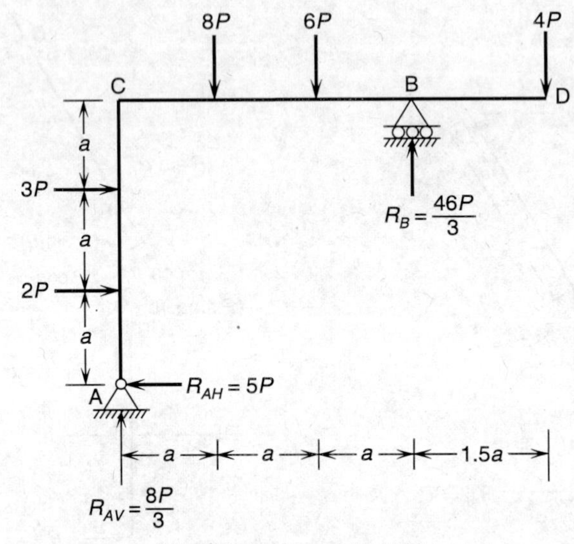

(a) Rigid frame and the load system

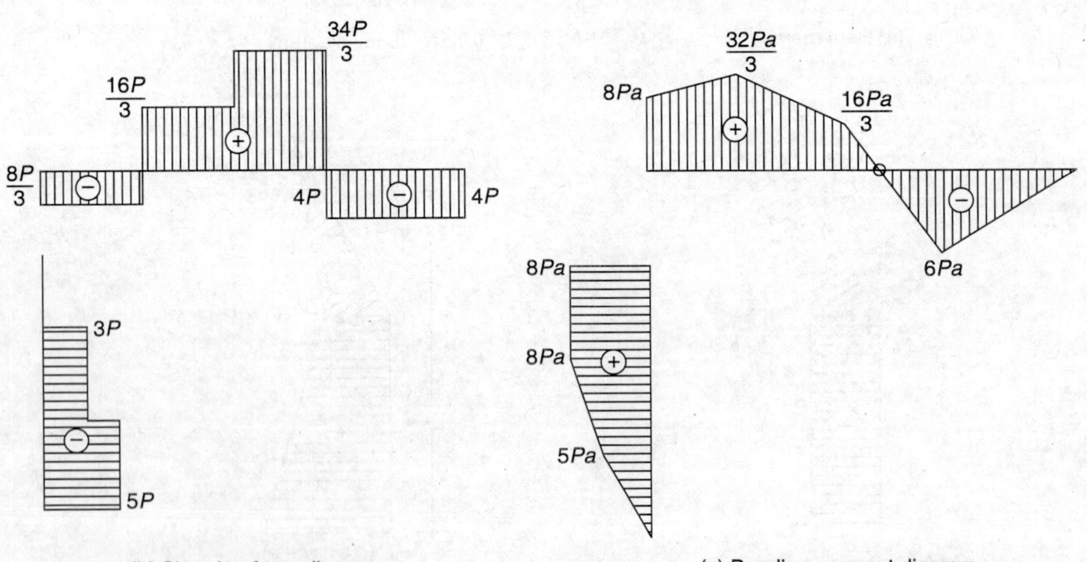

(b) Shearing force diagram (c) Bending moment diagram

Figure 7.22 Shear and moment diagrams of the rigid frame of Example 7.13.

Solution: For support reactions consider moment equilibrium of the entire frame about A:

$$\sum M_A = (2P)(a) + (3P)(2a) + (8P)(a) + (6P)(2a) + (4P)(4.5a) - R_B(3a) = 0$$

or $$R_B = \frac{46Pa}{3a} = \frac{46P}{3a} \uparrow \text{ and } R_A = 18P - \frac{46Pa}{3a} = \frac{8P}{3a} \uparrow$$

$$\sum F_H = R_{AH} + (3P) + (2P) = 0 \text{ or } R_{AH} = -5P\,(\leftarrow)$$

The free-body diagram is shown in Figure 7.22(a).

Shearing force:

For the member AC with origin at A, from A to C the shears at the salient points are:

$$-5P;\ (-5P, -3P);\ (-3P, 0) \text{ and } 0$$

For the member CB with origin at C, from C to B the shears at the salient points are:

$$-\frac{8P}{3};\left(-\frac{8P}{3}, \frac{16P}{3}\right);\left(\frac{16P}{3}, \frac{34P}{3}\right);\left(\frac{34P}{3}, -4P\right) \text{ and } -4P$$

Bending moment:

For the member AC, moments from A to C at salient points are:

$$0;\ -5Pa;\ -8Pa \text{ and } -8Pa$$

For the member CB moments from C to B at salient points are:

$$-8Pa;\ -\frac{32Pa}{3};\ -\frac{16Pa}{3};\ 6Pa, \text{ and } 0$$

The shear and moment diagrams are shown in Figures 7.22(b) and (c).

EXAMPLE 7.14

Draw shearing force and bending moment diagrams for the frame shown in Figure 7.23(a).

Solution: For support reactions, consider the moment equilibrium about A:

$$\sum M_A = -(4P + 2P)(a) - (2P + P)(2a) - \left[\left(\frac{4P}{a}\right)(4a)\right](2a) + R_B(4a) = 0$$

or $$R_B = \frac{44Pa}{4a} = 11P(\uparrow) \text{ and } R_A = \left(\frac{4P}{a}\right)(4a) - 11P = 5P(\uparrow)$$

$$\sum F_x = -R_{AH} + (4P + 2P + 2P + P) = 0 \text{ or } R_{AH} = -9P(\leftarrow)$$

The free-body diagram is shown in Figure 7.23(a).

Shearing forces:

For member AD with origin at A, from A to D the shears at the salient points are:

$$-9P;\ (-9P, -5P) \text{ and } -5P$$

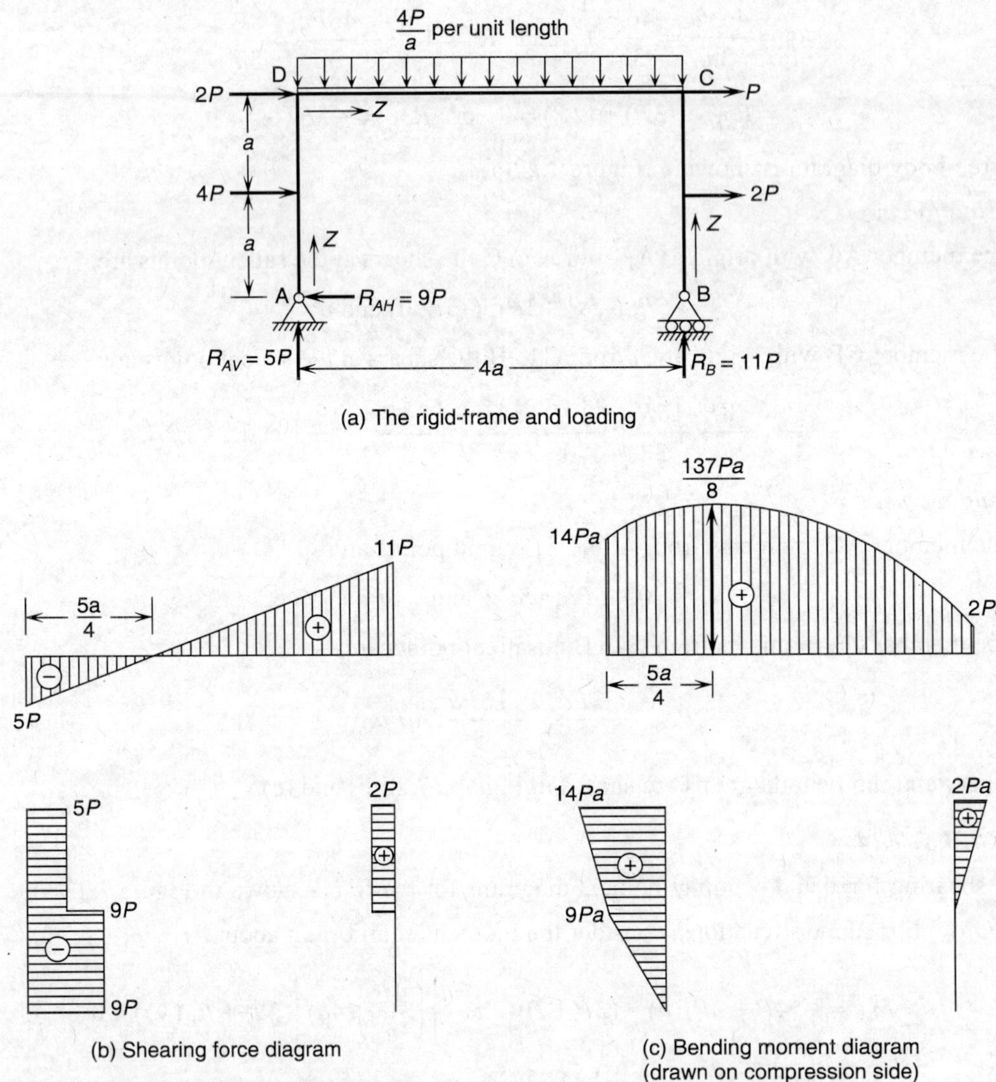

(b) Shearing force diagram

(c) Bending moment diagram (drawn on compression side)

Figure 7.23 Shear and moment diagrams of the frame of Example 7.14.

For member DC with origin at D, from D to C the shears at the salient points are:

$$-5P; 0\left[\text{at } x_m = \frac{5P}{4P/a} = \frac{5a}{4}\right] \text{ and } 11P$$

For member BC with origin at B, from B to C the shears at the salient points are:

$$0; (0, 2P) \text{ and } 2P$$

Bending moments:

For member AD, moments from A to D at salient points are: 0, $9Pa$ and $14Pa$

For member DC moments from D to C at the salient points are:

$$14Pa, \frac{137Pa}{8}\left(\text{at } x_{m} = \frac{5a}{4}\right) \text{ and } 2Pa$$

For member BC with origin at B, from B to C shears at the salient points are:

$$0, 0 \text{ and } 2Pa$$

The shear force and moment diagrams are shown in Figures 7.23(b) and (c).

EXAMPLE 7.15

Draw bending moment diagrams for the vertical post EF and horizontal arm ABC of the crane shown in Figure 7.24(a).

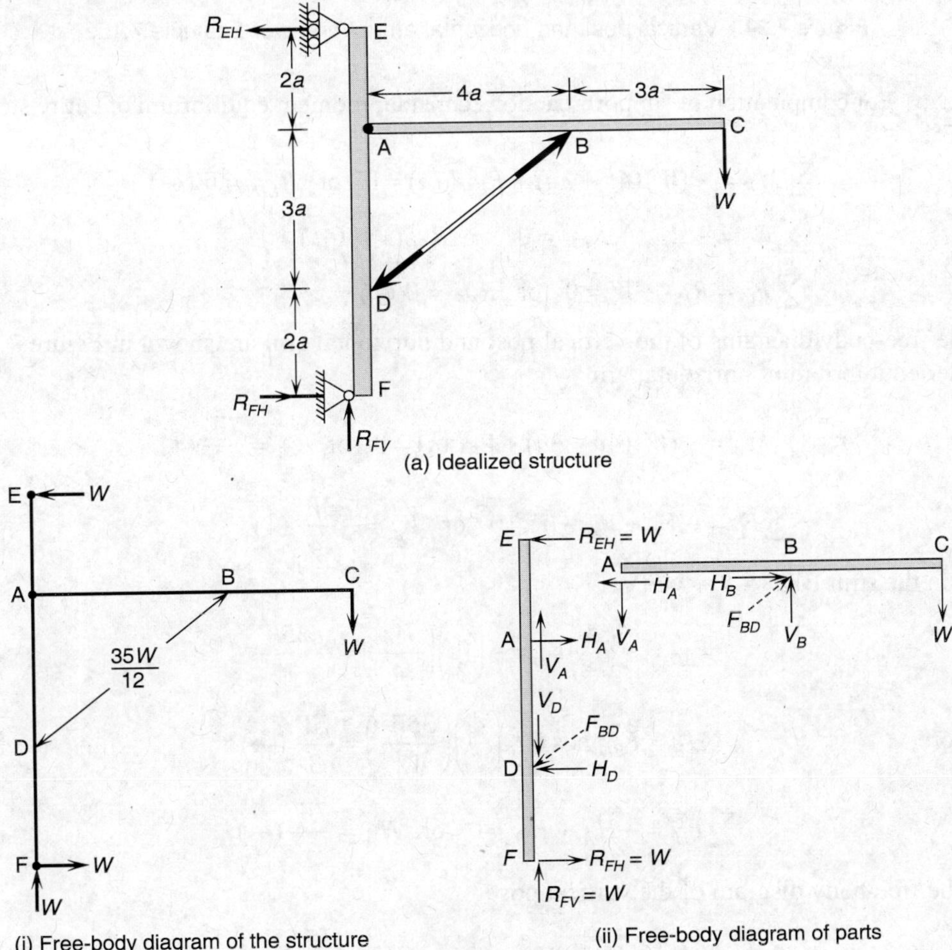

Figure 7.24 (Contd.)

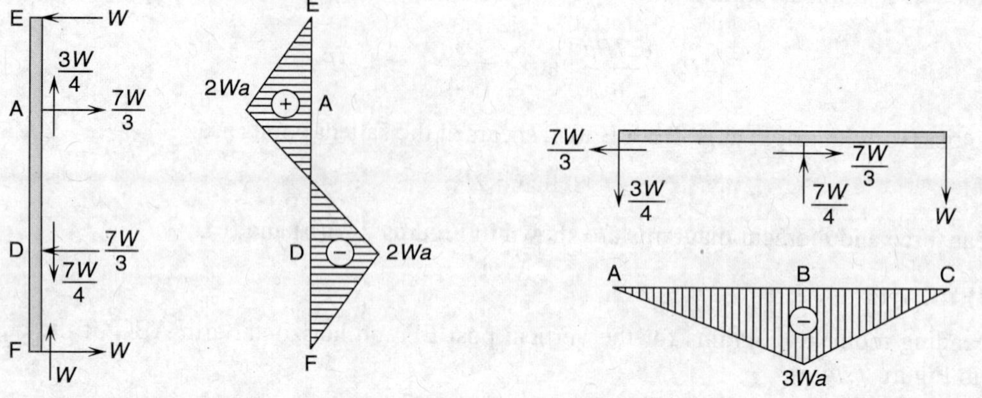

(c) Bending moment diagram of vertical past (d) Bending moment diagram of horizontal arm

Figure 7.24 Vertical post and horizontal arm of crane of Example 7.15.

Solution: For computation of support reactions consider moment equilibrium of entire structure about F:

$$\sum M_F = -(W)(4a+3a) + R_{EH}(7a) = 0 \quad \text{or} \quad R_{EH} = W(\leftarrow)$$

$$\sum F_H = -R_{EH} + R_{FH} = 0 \quad \text{or} \quad R_{FH} = W(\rightarrow)$$

$$\sum F_V = R_{FV} - W = 0 \quad \text{or} \quad R_{FV} = W(\uparrow)$$

The free-body diagrams of the vertical post and horizontal arm are shown in Figure 7.24(b). Consider equilibrium of horizontal arm,

$$\sum M_A = -(W)(4a+3a) + V_B(4a) = 0 \quad \text{or} \quad V_B = \frac{7W}{4}(\uparrow)$$

$$\sum F_V = -V_A + V_B - W = 0 \quad \text{or} \quad V_A = -\frac{3W}{4}(\downarrow)$$

Thrust in the strut BD,

$$F_{BD} = -\left(\frac{5}{3}\right)V_B = -\left(\frac{5}{3}\right)\left(\frac{7W}{4}\right) = -\frac{35W}{12}$$

Therefore,
$$H_B = F_{BD}\cos\theta = \left(\frac{4}{5}\right)\left(\frac{35W}{12}\right) = \frac{7W}{3}(\rightarrow)$$

$$\sum F_H = -H_A + H_B = 0 \quad \text{or} \quad H_A = \frac{7W}{3}(\leftarrow)$$

From the free-body diagram of the vertical post,

$$\sum F_H = H_A - H_D = 0 \quad \text{or} \quad H_D = \frac{7W}{3}(\leftarrow)$$

$$\sum F_V = V_A - V_D + W = 0 \quad \text{or} \quad V_D = -\frac{7W}{4}(\downarrow)$$

The bending moment diagrams for the horizontal arm ABC and vertical post EF are shown in Figures 7.24(c) and (d), respectively.

7.7 PROBLEMS

7.1 State the mathematical relations between the loads, shears and bending moments in a beam and verify the same with respect to a beam shown in Figure 7.25.

[**Hint:** Determine shearing force and bending moment at the sections 1-1 and 2-2 at distances z and $z + dz$, respectively, from A. Ignore the second order quantities].

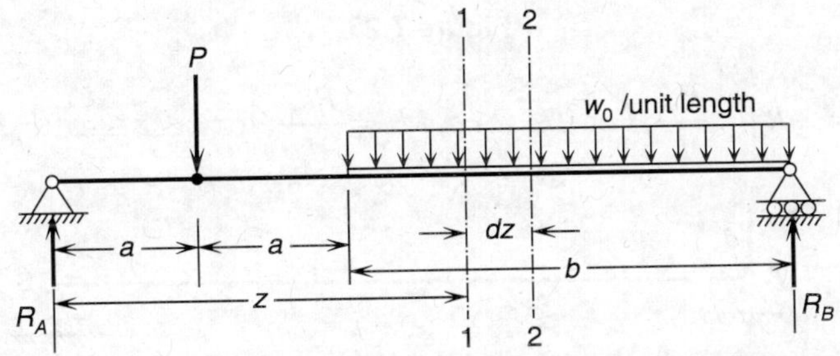

Figure 7.25 Beam with different loads and forces acting on it.

7.2 Develop complete shearing force and bending moment diagrams for the simply supported beams carrying uniformly distributed loads as shown in Figures 7.26 to 7.27.

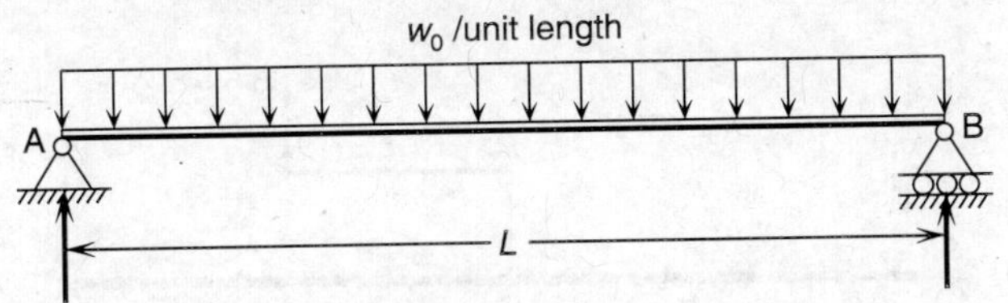

Figure 7.26

$$\left[\textbf{Ans. } R_A = R_A = \frac{w_0 L}{2};\ \frac{w_0 z(L-z)}{2}\right]$$

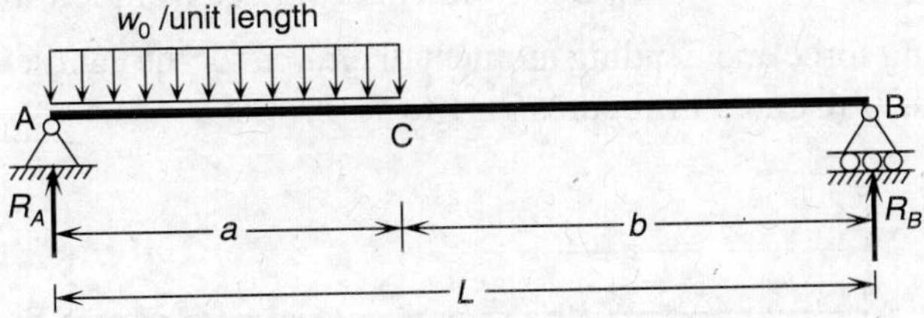

Figure 7.27

$$\left[\begin{array}{l}\textbf{Ans. } \text{For the range } 0 \le z \le a,\ M_x = \dfrac{w_0 z}{2L}[a(L+b) - Lz] \\ \quad \text{for } a \le z \le a+b,\ M_z = \dfrac{w_0 a^2}{2L}(L-z)\end{array}\right]$$

7.3 Construct shearing force and bending moment diagrams for the simply supported beams carrying the load system illustrated in Figures 7.28 to 7.30.

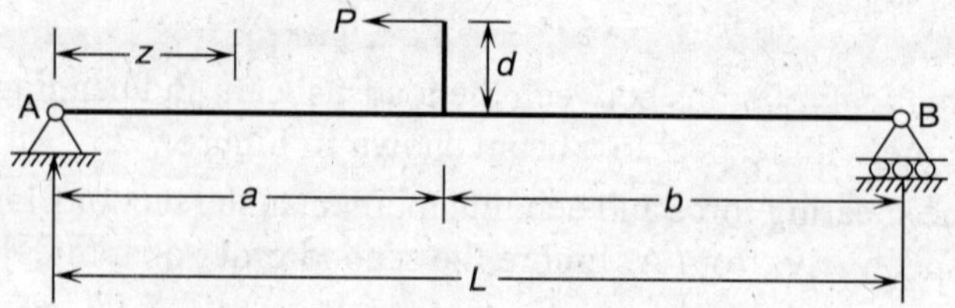

Figure 7.28

$$\left[\textbf{Ans. } R_A = -R_B = \frac{Pd}{L}, \text{ For } 0 \le z \le a, M_z = \frac{Pdz}{L}; \text{ for } a \le z \le a+b, M_z = -\frac{Pd(L-z)}{L}\right]$$

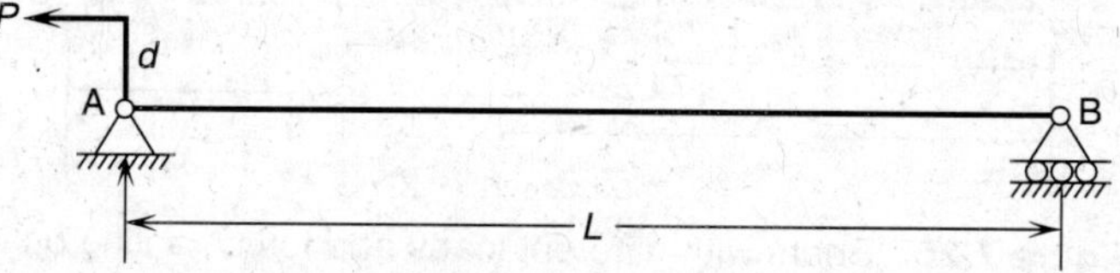

Figure 7.29

$$\left[\textbf{Ans. } M_z = -\frac{Pd}{L}(L-z)\right]$$

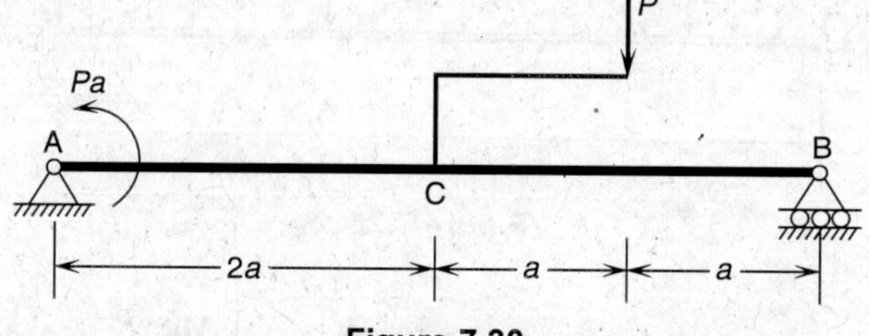

Figure 7.30

[**Ans.** Moment ordinates from left to right $-Pa$, $(0, Pa)$, 0]

7.4 Develop shearing force and bending moment diagrams for the cantilever beams which carry the load systems illustrated in Figures 7.31 to 7.38.

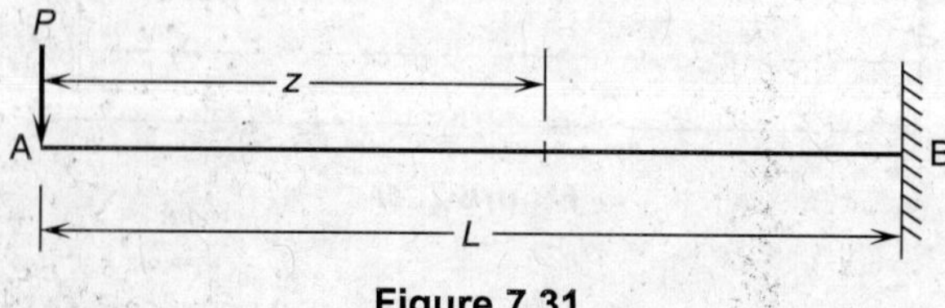

Figure 7.31

[**Ans.** $M_z = -Pz$]

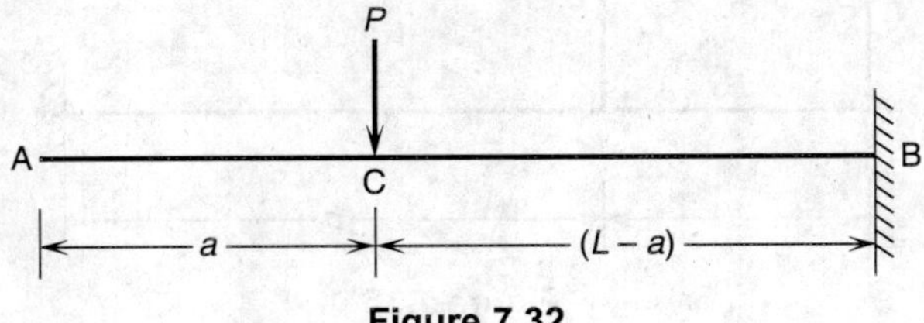

Figure 7.32

[**Ans.** For $0 \leq z \leq a$, $M_z = 0$; for $a \leq z \leq L$, $M_z = -P(z - a)$]

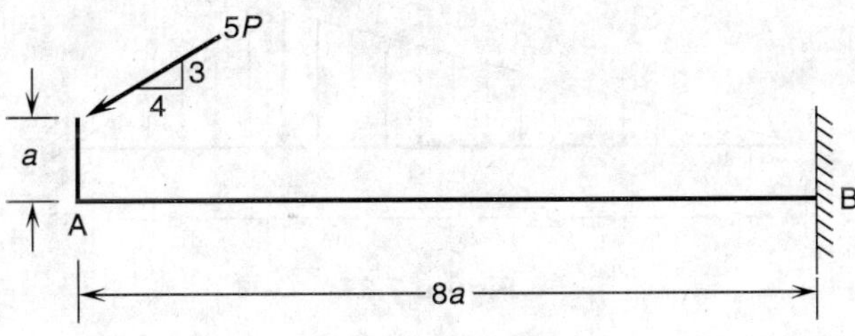

Figure 7.33

[**Ans.** $M_z = -P(3z + 4a)$, $M_{max} = -28Pa$]

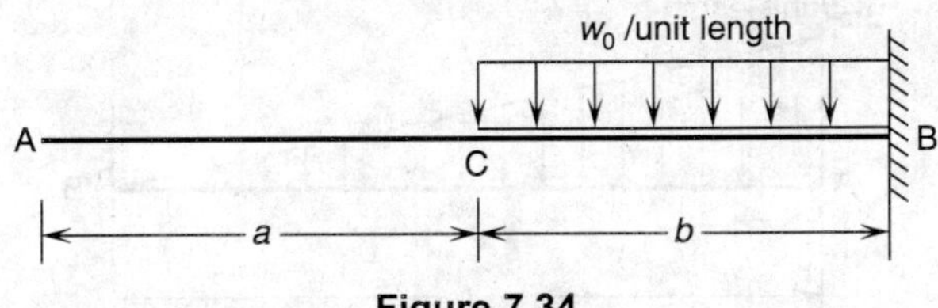

Figure 7.34

$$\left[\textbf{Ans. } 0 \leq z \leq a, M_z = 0; \text{For } a \leq z \leq L, M_z = -\frac{w_0(z-a)^2}{2}\right]$$

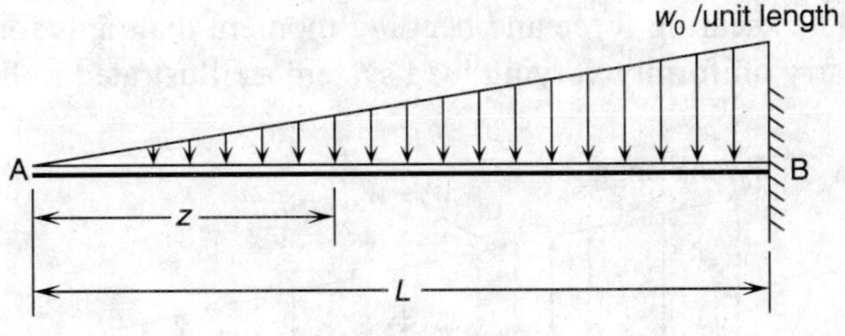

Figure 7.35

$$\left[\textbf{Ans. } M_z = -\frac{w_0 z^3}{6L}, M_{max} = -\frac{w_0 L^2}{6}\right]$$

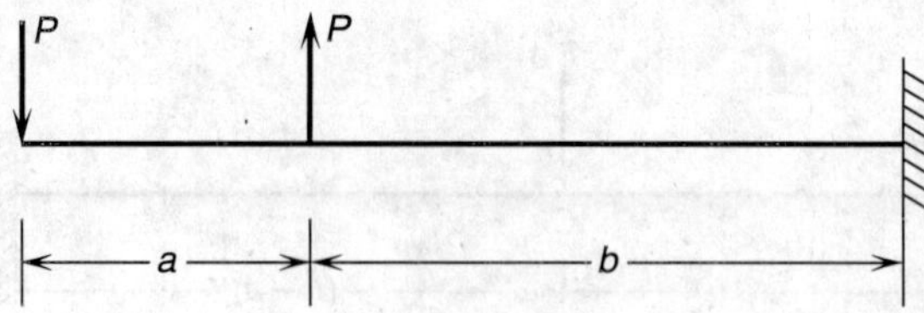

Figure 7.36

[**Ans.** $0 \le z \le a, M_z = -Pz$; $a \le z \le L, M_z = -Pa$]

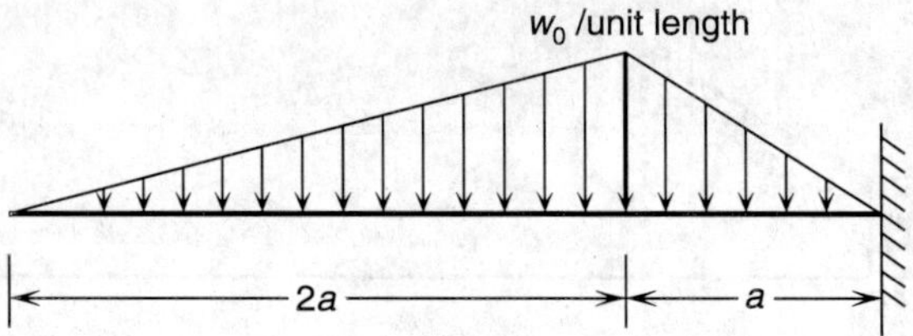

Figure 7.37

$$\left[\textbf{Ans}. \text{ For } 0 \le z \le 2a, M_z = -\frac{w_0 z^3}{12a}, \text{ For } 2a < z \le 3a, M_z = -\frac{w_0}{12a}[z^3 - 3(z-2a)^3\right]$$

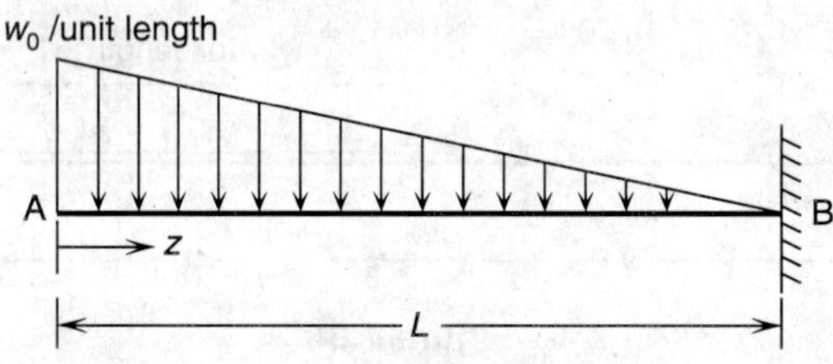

Figure 7.38

$$\left[\textbf{Ans.}\left(\frac{w_0 z^3}{6L} - \frac{w_0 z^2}{2}\right)\right]$$

7.5 Develop complete shearing force and bending moment diagrams for the simply supported beams which carry uniformly varying load system as illustrated in Figures 7.39 to 7.42.

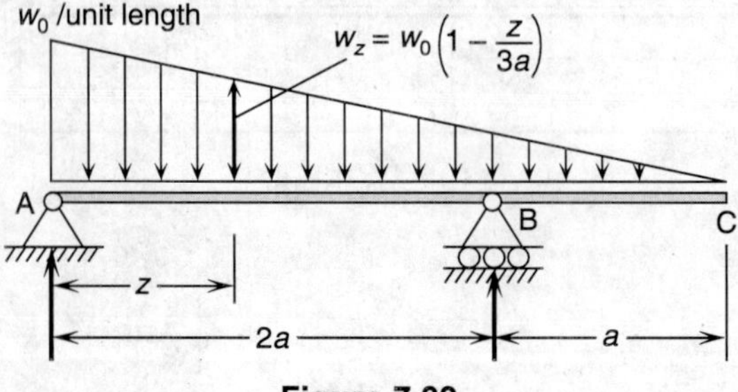

Figure 7.39

$$\left[\textbf{Ans. } R_A = R_B = \frac{3w_0 a}{4}; V_x = -\frac{w_0}{12a}(2x^2 - 12ax + 9a^2) \text{ for } 0 \le x \le 2a,\right.$$

$$V_x = \frac{-w_0}{6a}[x^2 - 6ax + 9a^2] \text{ for } 2a \le x \le 3a;$$

$$M_x = \frac{w_0 x}{36a}(2x^2 - 18ax + 27a^2) \text{ for } 0 \le x \le 2a$$

$$\left.\text{and } M_x = \frac{w_0}{36a}\left[(2x^3 - 18ax^2 + 54a^2x - 54a^3) \text{ for } 2a \le x \le 3a\right]\right]$$

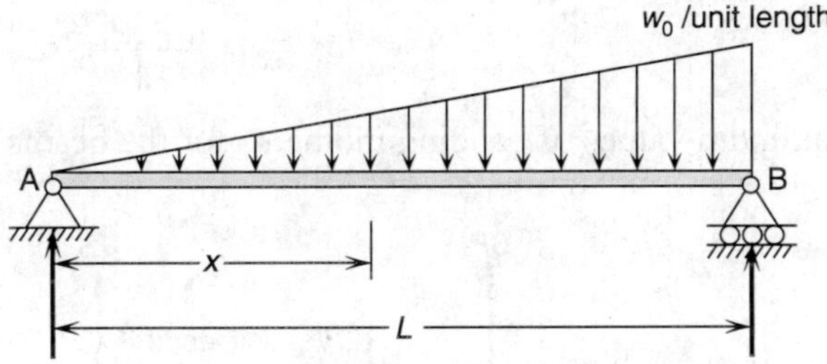

Figure 7.40

$$\left[\textbf{Ans. } R_A = \frac{w_0 L}{6}(\uparrow), R_B = \frac{w_0 L}{3}(\uparrow), V_x = \frac{w_0(3x^2 - L^2)}{6L} \text{ and } M_x = \frac{w_0 x(L^2 - x^2)}{6L},\right.$$

$$\left. x_m = \frac{L}{\sqrt{3}} \text{ and } M_{\max} = \frac{w_0 L^2}{9\sqrt{3}}\right]$$

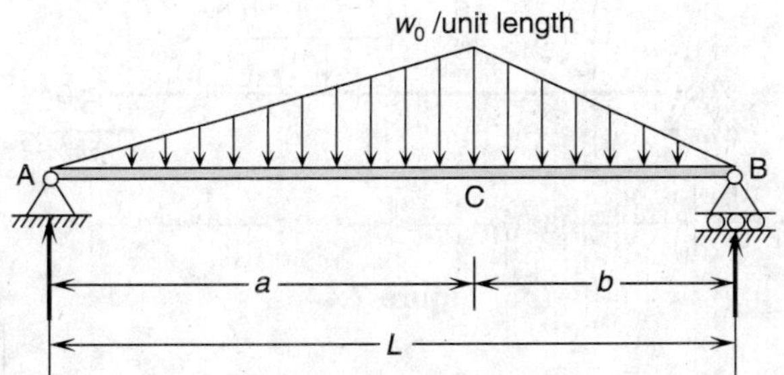

Figure 7.41

$$\left[\textbf{Ans. } R_A = \frac{w_0 L(L+b)}{6}, R_B = \frac{w_0 L(L+a)}{6};\right.$$

$$\text{For } 0 \le x \le a, V_x = \frac{w_0}{2}\left(\frac{x^2}{a} - \frac{L+b}{3}\right), M_x = \frac{w_0 x}{6a}\{(L+b)a - x^2\}$$

$$\left.\text{and with origin at B; For } 0 \le x' \le b, V_{x'} = \frac{w_0}{2}\left(\frac{L+a}{3} - \frac{x'^2}{b}\right)\right]$$

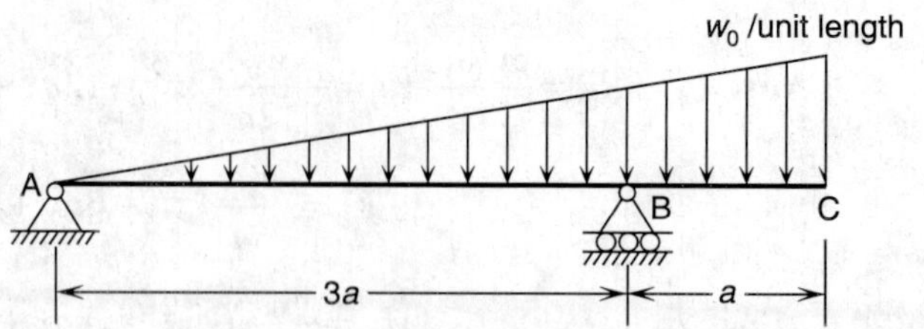

Figure 7.42

$$\left[\textbf{Ans. } R_A = \frac{2w_0 a}{a},\ R_B = \frac{16w_0}{9},\ z_m = \frac{4a}{3},\ M_A = 0,\right.$$

$$\left. M_{\max} = \frac{16w_0 a^2}{81},\ M_B = -\frac{11w_0 a^2}{24},\ M_C = 0\right]$$

7.6 Determine the maximum values of bending moments for the beams shown in Figures 7.43 to 7.45.

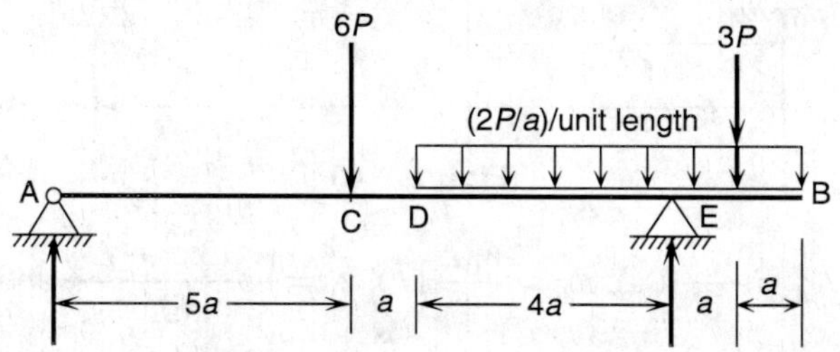

Figure 7.43

[**Ans.** $R_A = 3.9P$; $R_B = 17.1P$; $V_{\max} = 10.1P$ and $M_{\max} = 19.5Pa$]

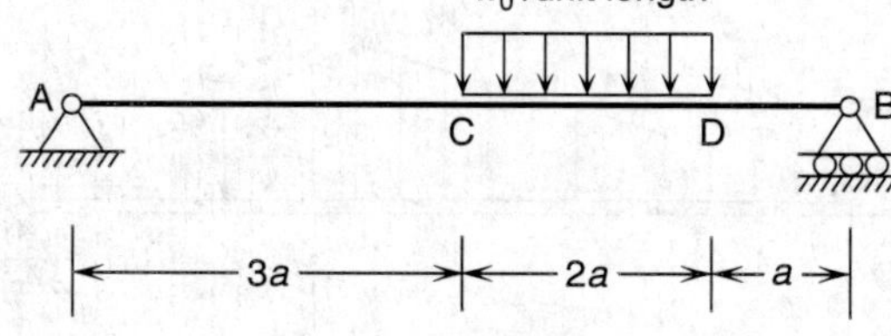

Figure 7.44

$$\left[\textbf{Ans. } M_{\max} = \frac{20w_0 a}{9}\right]$$

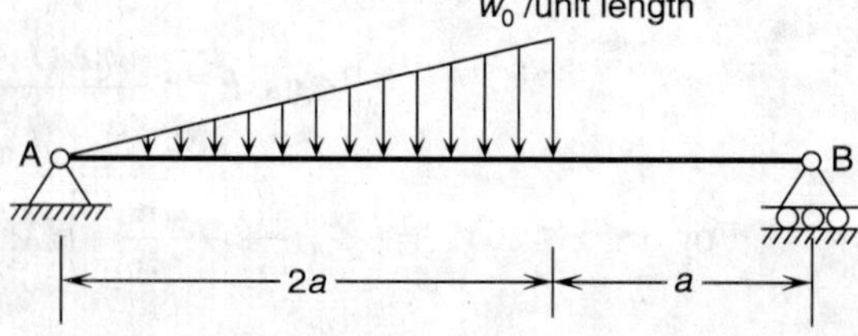

Figure 7.45

$$\left[\textbf{Ans. } M_{\max} = \frac{20\sqrt{5}\, w_0 a^2}{81}\right]$$

7.7 Develop complete shearing force and bending moment diagrams for the simply supported beams with overhang(s) which carry the load system illustrated in Figures 7.46 to 7.49.

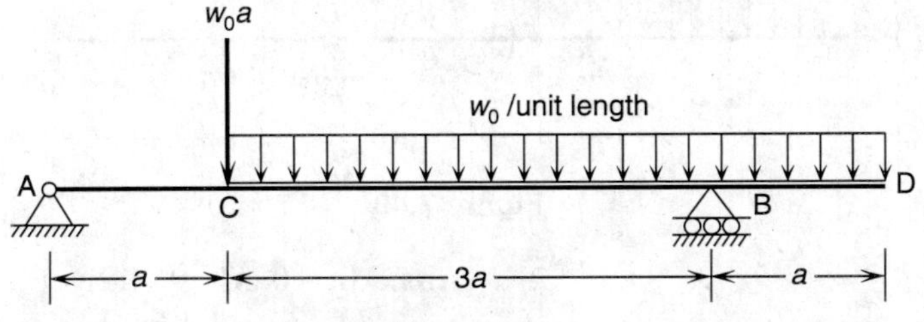

Figure 7.46

$$\left[\textbf{Ans. } V_A = -\frac{7w_0a}{4};\ V_{C,\text{left}} = -\frac{7w_0a}{4},\ V_{C,\text{right}} = -\frac{3w_0a}{4},\ V_{B,\text{left}} = \frac{9w_0a}{4},\right.$$

$$\left.V_{B,\text{right}} = -w_0a;\ M_A = 0,\ M_C = \frac{7w_0a^2}{4},\ M_B = -\frac{w_0a^2}{2} \text{ and } M_D = 0\right]$$

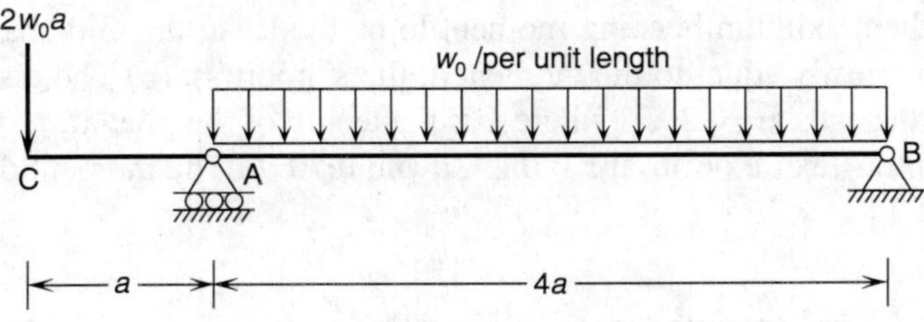

Figure 7.47

$$\left[\textbf{Ans. } V_C = 2w_0a,\ V_{A,\text{left}} = 2w_0a,\ V_{A,\text{right}} = -\frac{5w_0a}{2},\ V_B = \frac{3w_0a}{2};\right.$$

$$\left.M_C = 0,\ M_A = -2w_0a^2,\ M_{\max} = \frac{9w_0a^2}{8}\right]$$

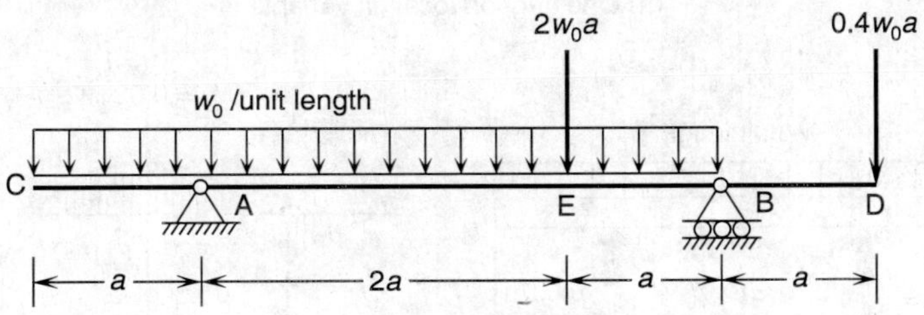

Figure 7.48

$$\left[\textbf{Ans. } V_{C\text{–}A} = 0 - w_0a,\ V_{A\text{–}E} = 2.2w_0a - 0.2w_0a,\ V_{E\text{–}B} = 1.8wa - 2.8w_0a,\right.$$

$$\left.V_{B-D} = 0.4w_0a,\ M_C = 0,\ M_A = -\frac{w_0a^2}{2},\ M_{\max} = +1.9w_0a^2,\ M_B = -2w_0a^2\right]$$

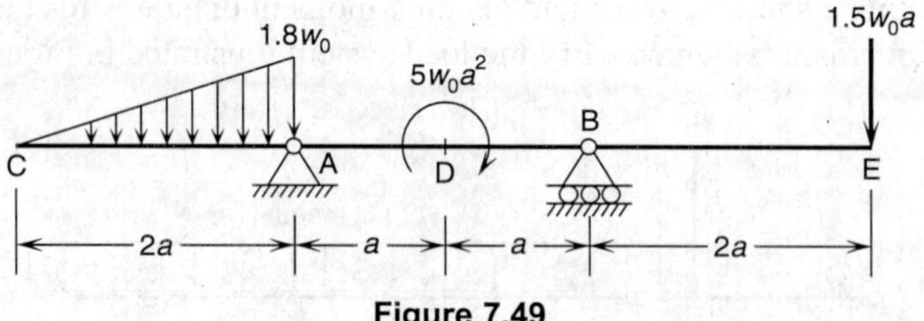

Figure 7.49

[**Ans.** $M_C = 0$, $M_A = -1.2w_0a^2$, $M_{D,\text{left}} = +0.4w_0a^2$, $M_{D,\text{right}} = -4.6w_0a^2$, $M_B = -3w_0a^2$, $M_E = 0$]

7.8 A pre-cast pile of uniform cross-section with self-weight of w_0 per unit length is to be lifted for driving into the ground. Determine the position of the lifting point(s) which will result in the least value of the maximum bending moment when the pile is lifted: (a) at one point located at a distance from the end and (b) horizontally at two points positioned symmetrically. Draw the shearing force and bending moment diagrams for the arrangement, and show the points of contra-flexure.

[**Hint:** For the maximum bending moment to be the least, the mid-span bending moment must be numerically equal to the moment at the support(s). (b) The distance of the lifting points from the ends are $0.207L$ where L is the length of the pile; $M_{\max} = \pm 0.0215 w_0 L^2$; the points of contra-flexure occur at $z = 0.293L$ and $0.707L$. The moment diagrams are shown in Figure 7.50].

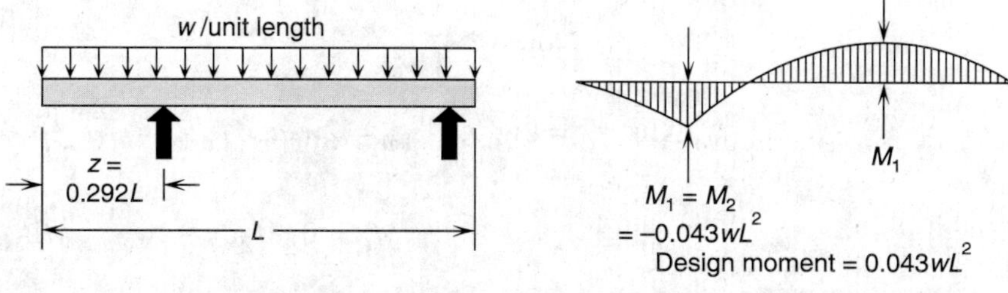

(a) One support location variable

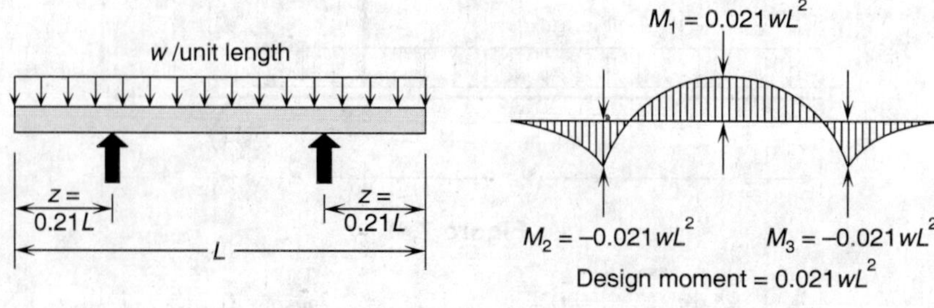

(b) Both support locations variable

Figure 7.50 Optimum support locations to minimize design moments in beams.

7.9 A vertical pole ACB is rigidly fixed at its base A and carries loads shown in Figure 7.51. (i) Draw the thrust, shearing force and bending moment diagrams for the pole, (ii) Determine the magnitude and direction of horizontal force applied at D which would be required for equilibrium if the fixing moment at A is to be zero.

$$\left[\textbf{Hint}: N_{\max} = \frac{(12 + 5\sqrt{3})P}{2} \text{ (comp.)}; V_z = \frac{9P}{2}; M_{\max} = 18Pa \text{ and } P_H = 5P(\leftarrow)\right]$$

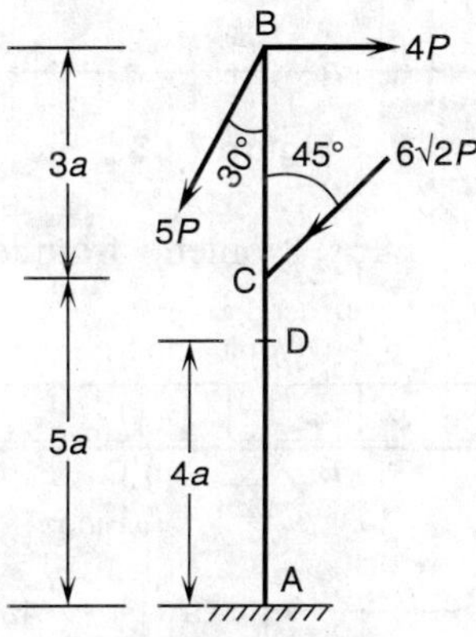

Figure 7.51

7.10 A vertical column ACB of height H between supports carries a horizontal bracket at the point C shown in Figure 7.52. The bracket carries a concentrated load P at a distance d from the axis of the column. Draw the axial force, shearing force and bending moment diagrams for the pole.

$$\left[\textbf{Hint}: N_x = P \text{ (tension in B–C)}; V_x = \left(\frac{Pd}{H}\right) \text{ and } M_{\max} = \left(\frac{Pd}{H}\right) a \text{ (for } a \geq b)\right]$$

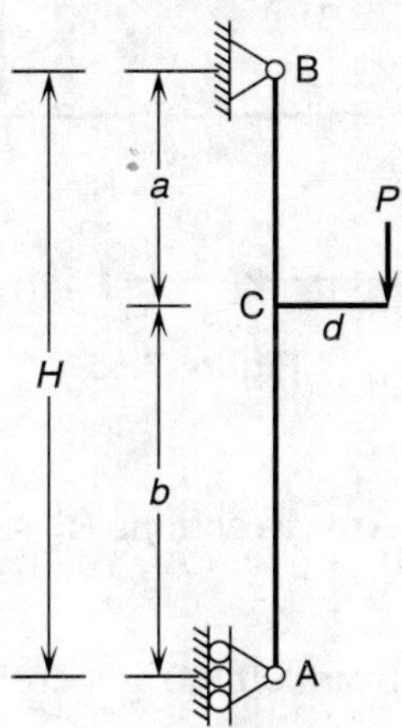

Figure 7.52

7.11 Construct complete shearing force and bending moment diagrams for the beams having interior hinge and carry the load system as shown in Figures 7.53 to 7.56.

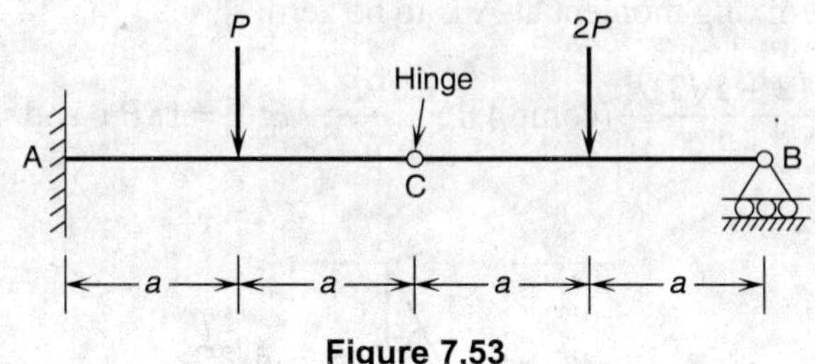

Figure 7.53

[**Ans.** Moments from left to right are: $-3Pa$, $-Pa$, 0, Pa, 0]

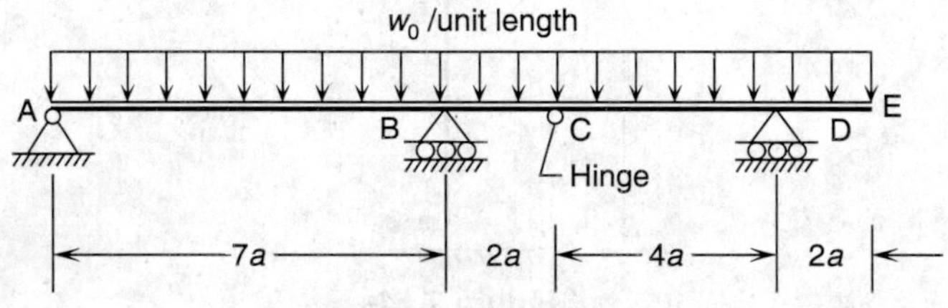

Figure 7.54

$$\left[\textbf{Ans.}\text{ Reactions and moments from left to right are: }\left(\frac{39w_0a}{14}, \frac{54w_0a}{7}, \frac{9w_0a}{2}\right)\right.$$

$$\left. 0, \frac{1521w_0a^2}{392}\text{ (max)}, -5w_0a^2, 0, \frac{9w_0a^2}{8}, -2w_0a^2, 0\right]$$

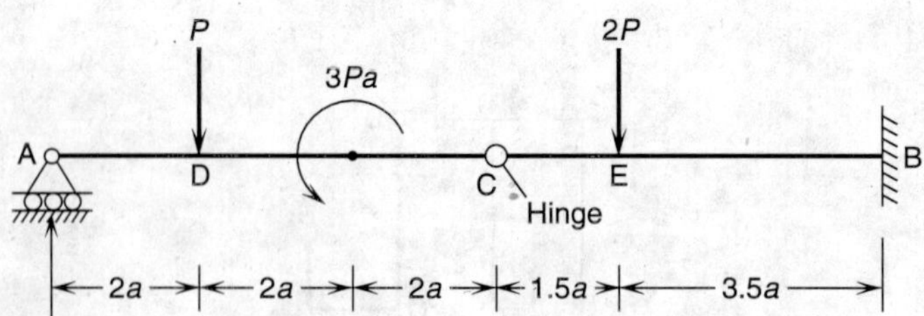

Figure 7.55

$$\left[\textbf{Ans.}\text{ Reactions } R_A = \frac{7P}{6}(\uparrow),\ R_B = \frac{11P}{6}(\uparrow),\ M_A = -\frac{37Pa}{6};\right.$$

$$\left.\text{Moments from left to right are: } 0, \frac{7Pa}{3}, \left(\frac{8Pa}{3}, -\frac{Pa}{3}\right), 0, \frac{Pa}{4}, -\frac{3Pa}{6}\right]$$

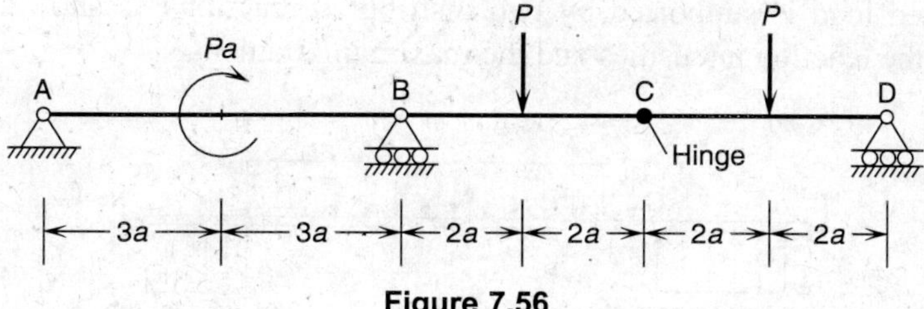

Figure 7.56

$$\left[\textbf{Ans. } R_A = -\frac{5P}{6}\downarrow, R_B = \frac{7P}{3}\uparrow, R_C = \frac{P}{2}, R_D = \frac{P}{2} \text{ and } M_B = 4Pa\right]$$

7.12 Draw the axial force, shearing force and bending moment diagrams for the beam ACDB which is supported by a hinge at A and a roller at B. The beam is subjected to loads through vertical members CE and DF which are rigidly attached to the beam at C and D as shown in Figure 7.57.

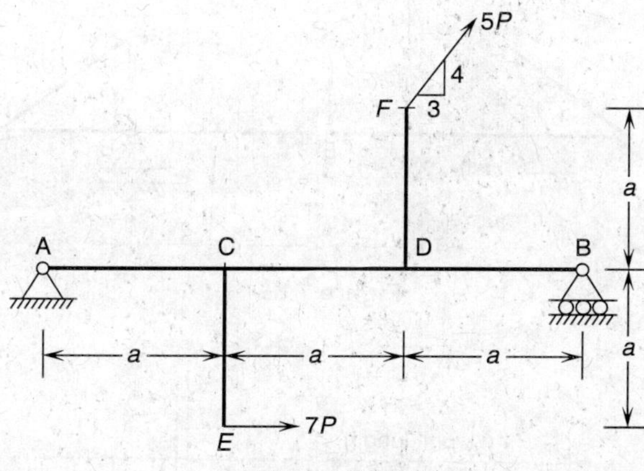

Figure 7.57

[**Ans.** $M_{max} = -7Pa$]

7.13 Distributed load acting on a beam is supported by a distributed reaction as shown in Figure 7.58. Determine the load intensity, the maximum values of shear and moment.

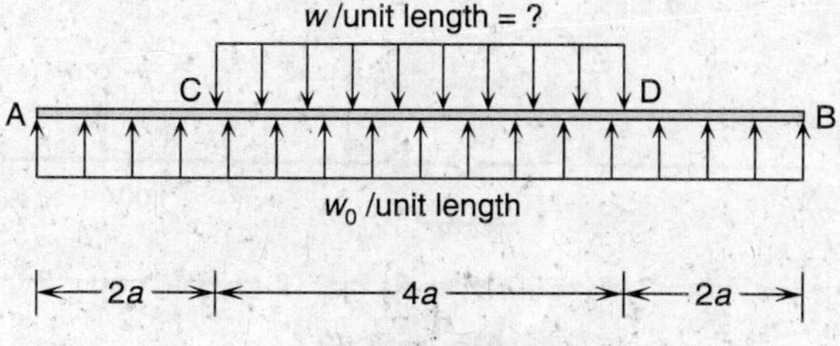

Figure 7.58

[**Ans.** $w = 2w_0$, $V_{max} = 2w_0a$ and $M_{max} = 4w_0a^2$]

7.14 A distributed load is supported by two distributed reactions as shown in Figure 7.59. Determine the reaction intensities and the maximum moment.

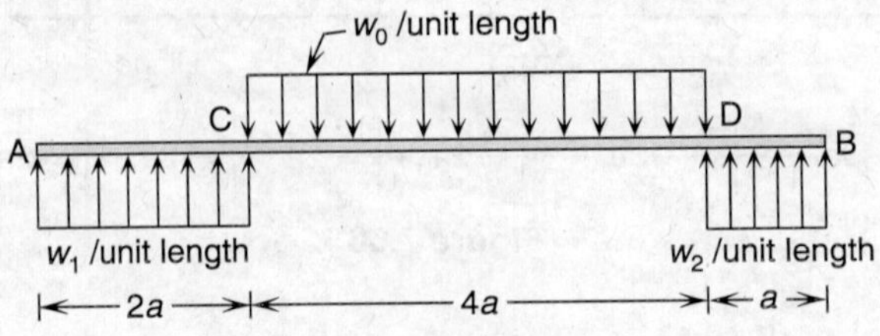

Figure 7.59

$$\left[\textbf{Ans. } w_1 = \frac{10w_0}{11}, w_2 = \frac{24w_0}{11}, M_{max} = \frac{611w_0a^2}{176}\right]$$

7.15 Without writing the shear and moment equations, draw shear force and moment diagrams for the beams shown in Figures 7.60 to 7.62.

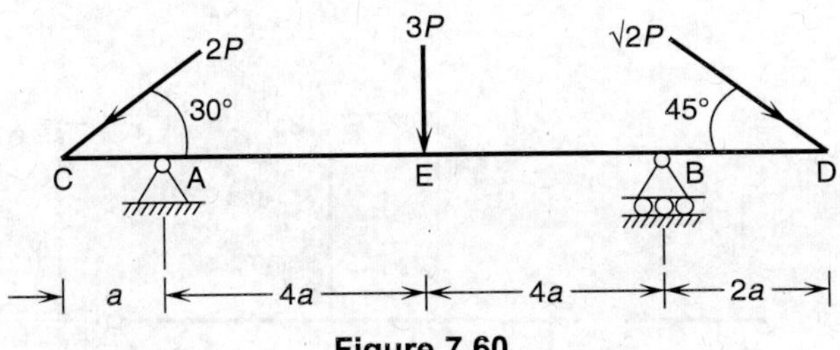

Figure 7.60

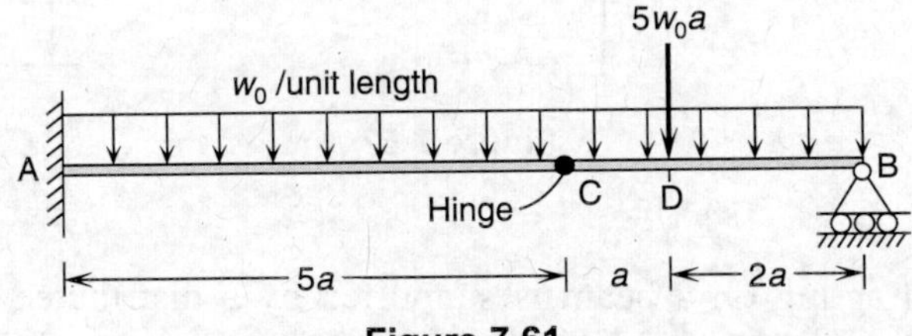

Figure 7.61

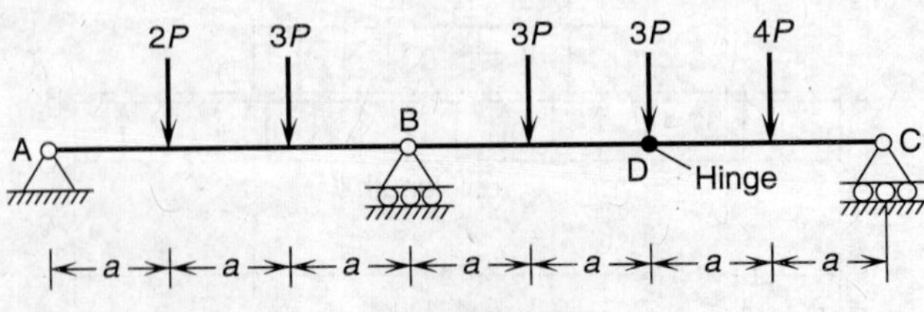

Figure 7.62

[**Ans.** $R_A = -2P$, $R_B = 15P$, $R_C = 2P$, $V_{max} = 4P$ and $M_{max} = -13Pa$]

7.16 Construct the shearing force and the loading diagrams corresponding to the given bending moment diagrams for the beams shown in Figures 7.63 to 7.65. Consider upward forces as reactions.

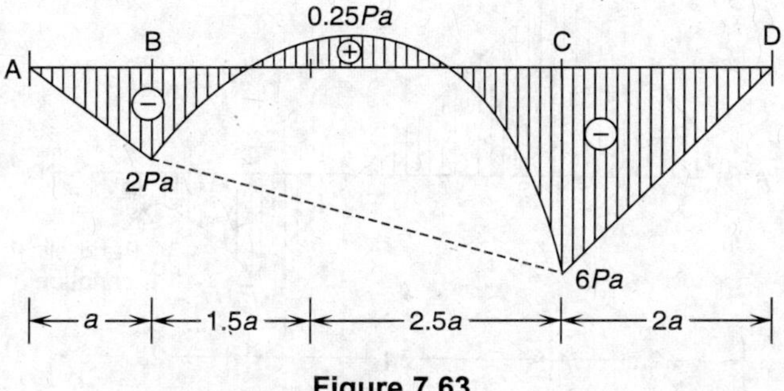

Figure 7.63

[**Ans.** The external actions from left to right are: $P_A = 2P$, $R_B = 5P$, $w_0 = 2P/a$, $R_C = 8P$, $P_D = 3P$]

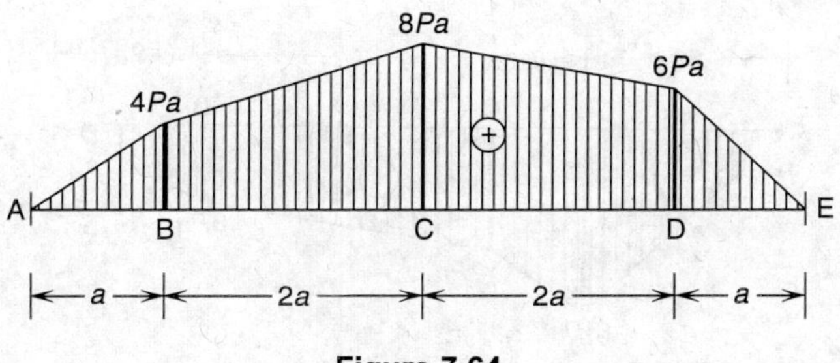

Figure 7.64

[**Ans.** Reactions and loads from left to right are, $4P$, $2P$, $3P$, $5P$ and $6P$]

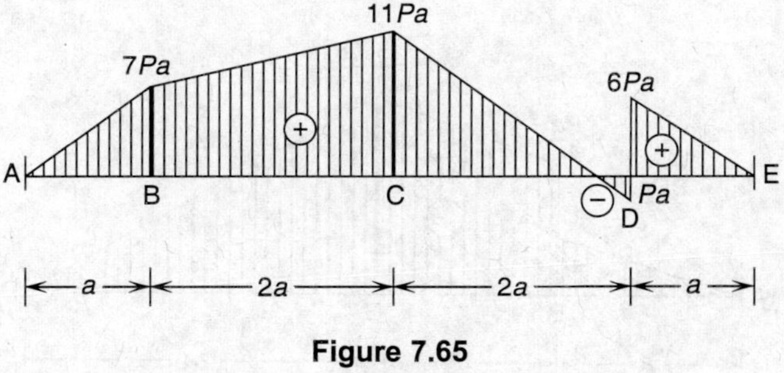

Figure 7.65

[**Ans.** Reactions and loads from left to right are $7P$, $5P$, $8P$, $7P$ and $6P$]

7.17 Construct the shearing force and the loading diagrams corresponding to the given bending moment diagrams for the beams shown in Figures 7.66 to 7.68.

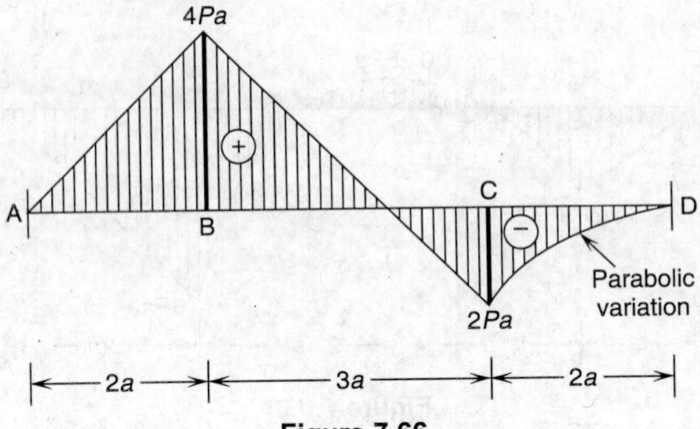

Figure 7.66

[**Ans.** The ordinates of the shear force diagram from left to right are: $-2P$, $(-2P, 2P)$, $(2P, -2P)$, 0]

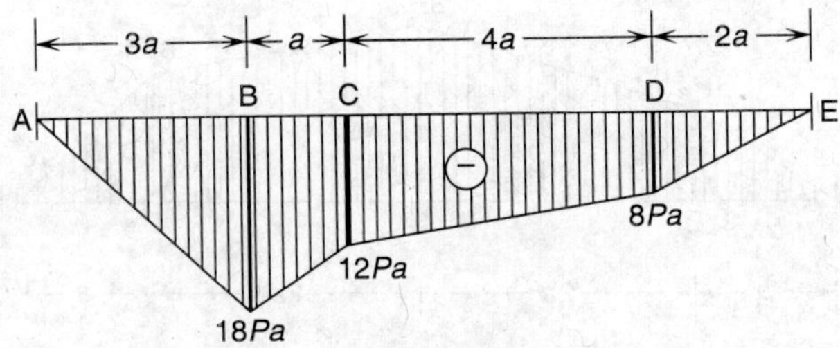

Figure 7.67

[**Ans.** The ordinates of the shear force diagram from left to right are: $6P$, $(6P, -6P)$, $(-6P, -P)$, $(-P, -4P)$ and $-4P$]

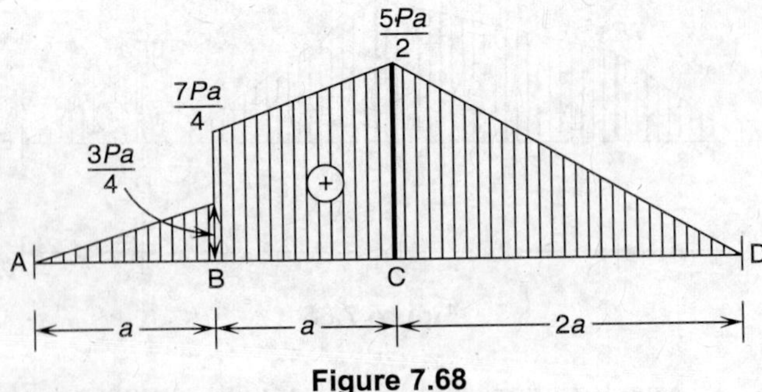

Figure 7.68

[**Ans.** The ordinates of the shear force diagram from left to right are: $-\frac{3P}{4}, \left(-\frac{3P}{4}, \frac{5P}{4}\right), \frac{5P}{4}$]

7.18 Develop the loading and the moment diagrams corresponding to the given shearing force diagram for the beam shown in Figure 7.69.

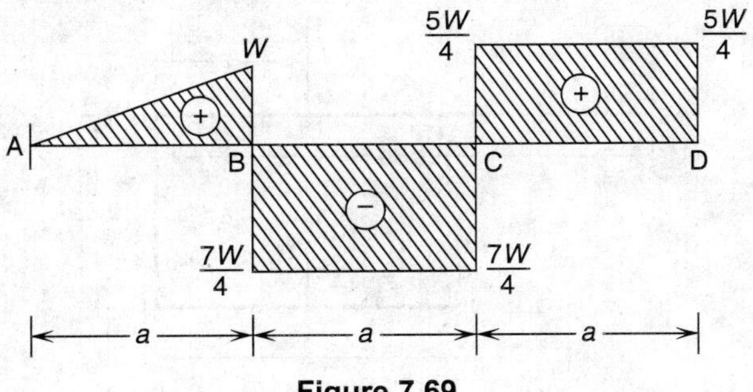

Figure 7.69

[**Ans.** $R_B = \frac{11W}{4}$; $W_C = 3W$; $w = W/a$; $R_D = \frac{5W}{4}$]

7.19 Construct shear and moment diagrams for the continuous beams shown in Figures 7.70 to 7.71 for which reactions are given.

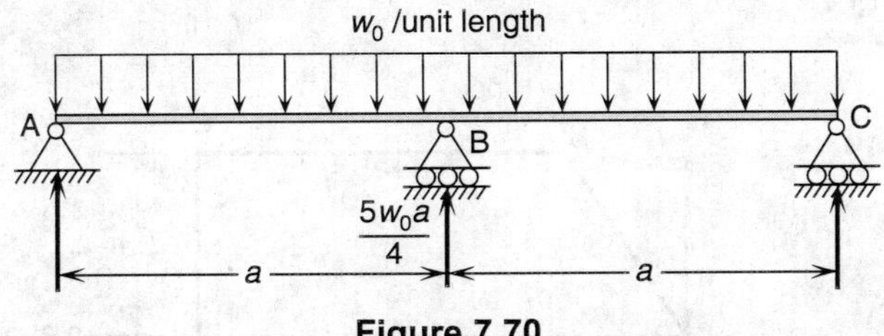

Figure 7.70

[**Ans.** From left to right salient ordinates are:

Shear; $-\frac{3w_0a}{8}, \left(+\frac{5w_0a}{8}, -\frac{5w_0a}{8}\right), \frac{3w_0a}{8}$; Moment; $0, \frac{9w_0a^2}{128}, -\frac{w_0L^2}{8}, \frac{9w_0a^2}{128}, 0$]

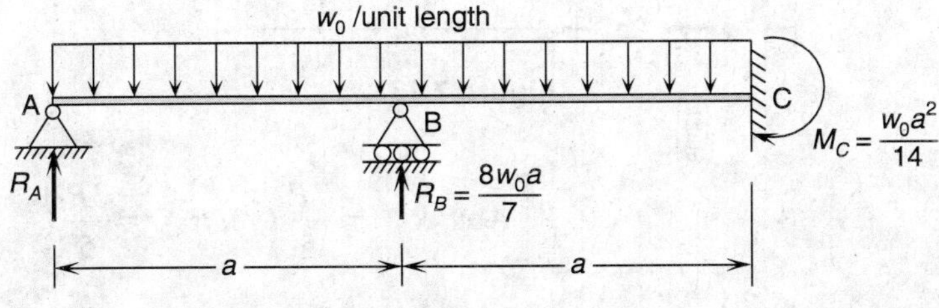

Figure 7.71

[**Ans.** $R_A = \frac{11w_0a}{28}$ (↑), $R_C = \frac{13w_0a}{28}$ (↑), Moments from left to right;

$0, \frac{121w_0a^2}{1568}, -\frac{3w_0a^2}{28}, \frac{57w_0a^2}{1568}, -\frac{w_0a^2}{14}$]

7.20 A cranked bar is fixed horizontally at one end and bent as shown in Figure 7.72. It weighs w_0 per unit length. Draw the bending moment diagram and indicate the maximum values.

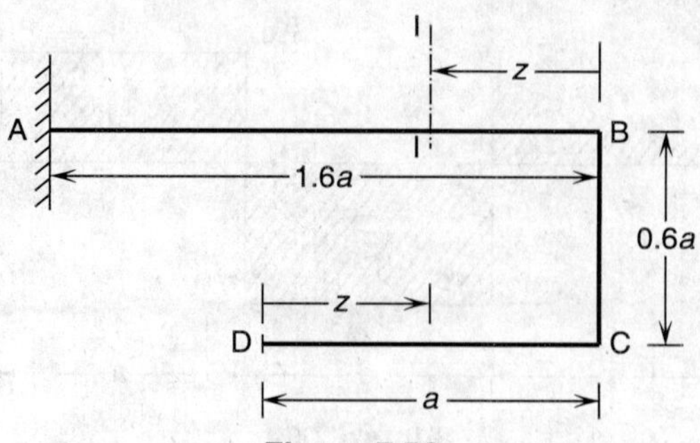

Figure 7.72

$$\left[\textbf{Ans. } M_B = M_C = -\frac{w_0 a^2}{2};\ M_Z = -\frac{w_0 a^2}{10}\left\{5\left(\frac{z}{a}\right)^2 + 8\left(\frac{z}{a}\right) - 5\right\}\right]$$

7.21 Draw shearing force and bending moment diagrams for the determinate rigid frame shown in Figures 7.73 to 7.78.

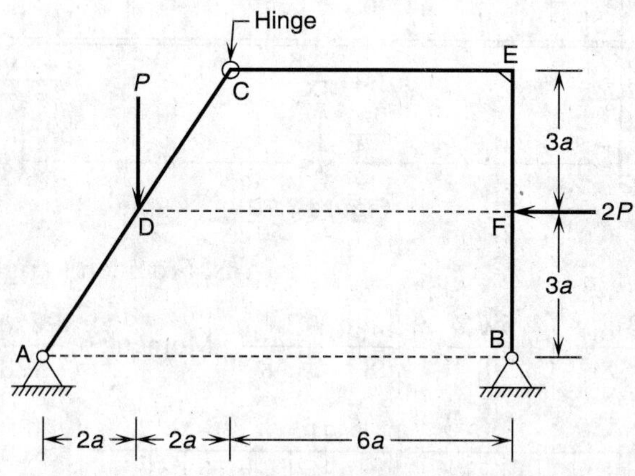

Figure 7.73

$$\left[\textbf{Ans. } R_{AV} = \frac{7P}{5}(\uparrow),\ R_{AH} = \frac{3P}{5}(\rightarrow),\ R_{BV} = \frac{2P}{5}(\downarrow),\right.$$

$$\left. R_{BH} = \frac{7P}{5}(\rightarrow),\ V_{max} = \frac{7P}{5} \text{ and } M_{max} = \frac{21Pa}{5}\right]$$

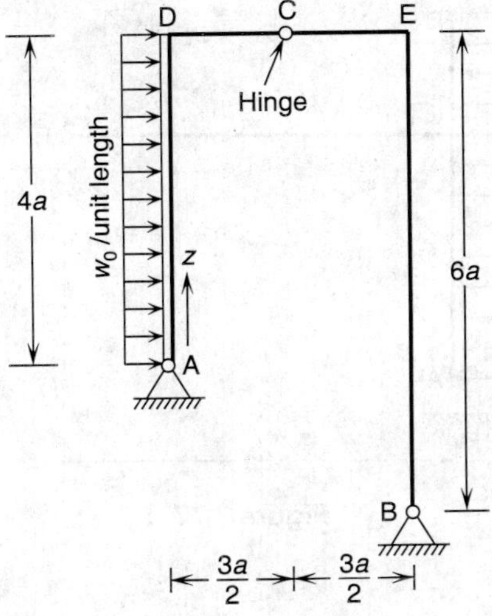

Figure 7.74

[**Ans**. $R_{AV} = 3.2wa(\downarrow)$, $R_{AH} = 3.2wa(\leftarrow)$; $R_{BV} = 3.2wa$, $R_{BH} = 0.8wa$
$M_A = 0$, $M_D = 4.8wa^2$, $M_E = -4.8wa^2$, $z_m = 3.2a$, $M_{\max} = 5.12wa^2$]

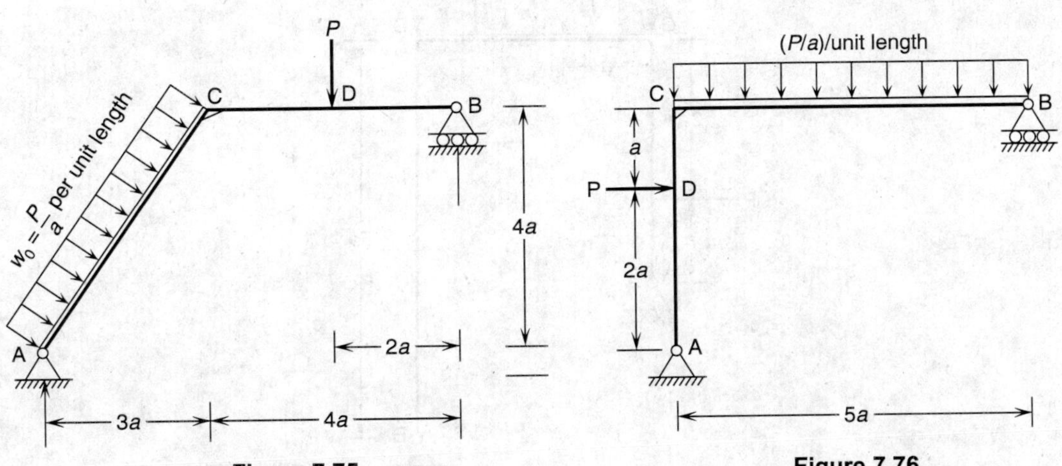

Figure 7.75

Figure 7.76

$\left[\textbf{Ans}.\ R_{AV} = \frac{3P}{2}, R_{AH} = 4P, R_B = \frac{5P}{2}\right]$

$\left[\textbf{Ans}.\ V_{\max} = 2.9P \text{ and } M_{\max} = \frac{841Pa}{200}\right]$

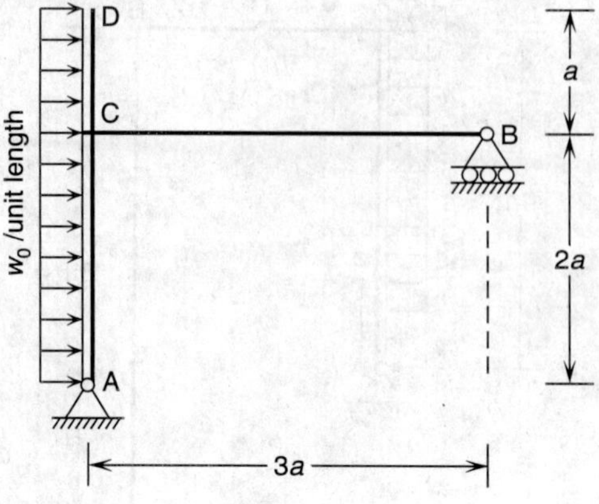

Figure 7.77

$$\left[\textbf{Ans.}\ V_{\max} = 3w_0,\ M_{\max} = \frac{32 w_0 a^2}{3} \right]$$

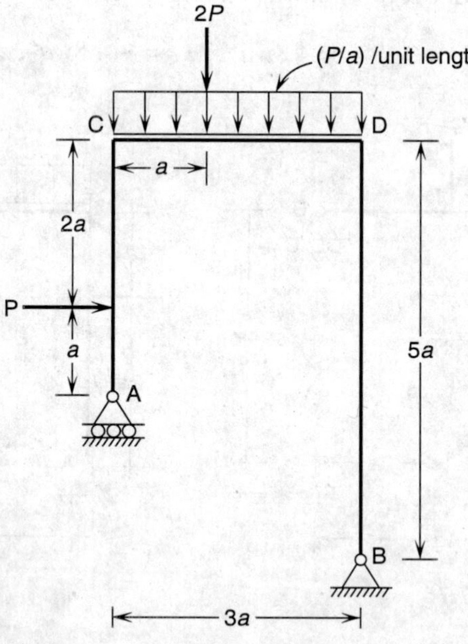

Figure 7.78

Chapter 8

Bending of Simple Beams

8.1 INTRODUCTION

Of the structural elements, the *beam* is the most commonly used element. The more evident post-and-beam assembly, where a horizontal member is simply rested on two vertical supports, serves as basic structural systems. This primary *structural system* is most typically used in a repetitive, pattern-forming style. Thus, constitutes the most widely used construction approach due to the convenience and simplicity of the beam as an element of construction.

In Chapter 1, it is seen that a system of internal forces consisting of an axial force, a shear force, and a bending moment may develop in beams and plane frames. The stresses caused by an axial force have already been discussed in Chapters 2 and 3. In this chapter, the analysis of beams has been considered.

8.2 STRESSES IN BEAMS

The topics which generally comprise the basic issues involved in the analysis of beams made of linearly-elastic materials generally addressed are: (a) bending stresses, (b) shearing stresses, (c) bearing stresses, (d) combined stresses, (e) torsional stresses, (f) principal stresses, (g) shear centre, and (h) deflections. In this chapter, the stresses due to bending of members are discussed. For simplicity, the beam members are generally shown in a horizontal position. When a segment of a beam is in equilibrium under the action of bending moments alone, such a condition is referred to as *pure bending* or *flexure*.

As discussed in the previous chapters, the actions of externally applied forces, which can be described in terms of axial forces, bending moment, twisting moment and shear forces, produce a set of internal forces, related stresses and deformations. Figure 8.1 illustrates the load carrying mechanism in a loaded beam. The net effect of internal stresses developed due to deformation illustrated in Figure 8.1(b) is to produce a rotational moment equal in magnitude but opposite in sense to the external bending, as shown in Figure 8.1(c).

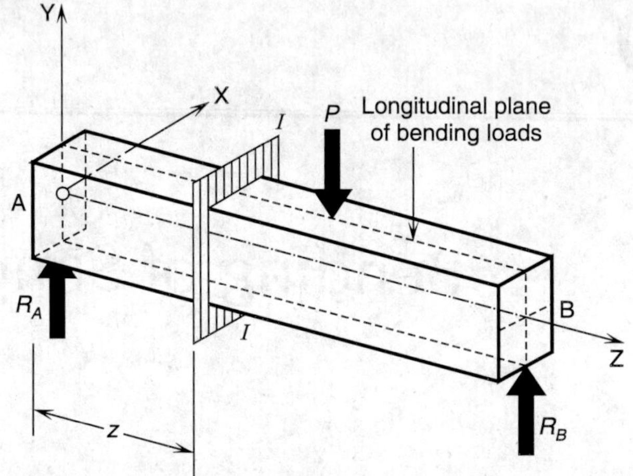

(a) Loaded beam and section I–I under consideration

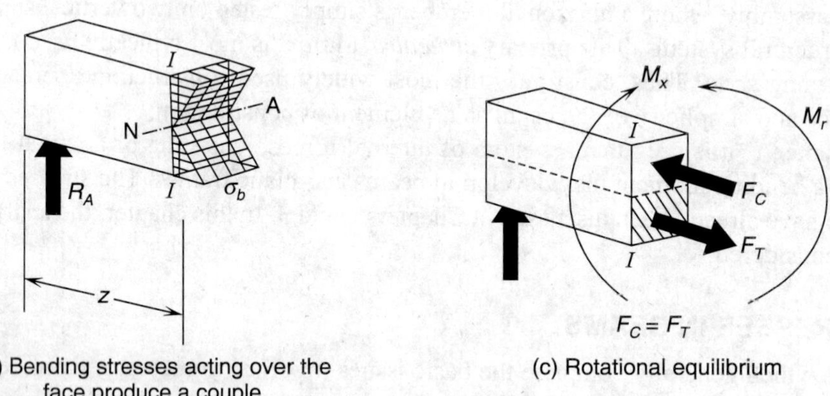

(b) Bending stresses acting over the face produce a couple

(c) Rotational equilibrium

Figure 8.1 Moment carrying mechanism in a beam.

The element of transversely loaded beam of a *linearly-elastic material* shown in Figure 8.2(a), undergoes tensile and compressive strains due to bending action resulting in the stresses which are directly proportional to the deformations, i.e. strains as illustrated in Figures 8.2(b) and (c). Tensile stresses are associated with elongation in the beam fibres, and compressive stresses with contraction. Together these stresses are referred to as *bending stresses*.

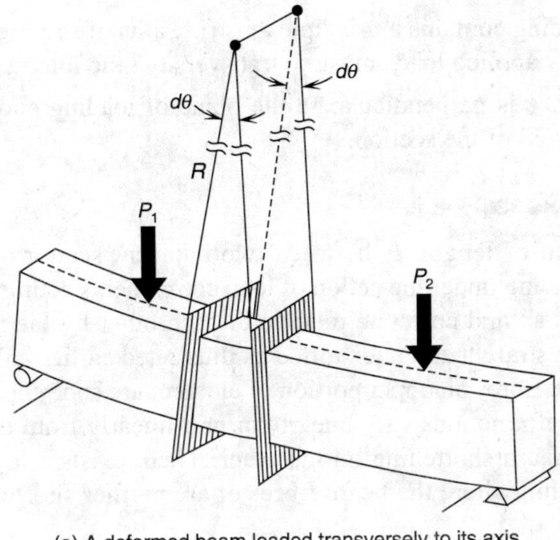

(a) A deformed beam loaded transversely to its axis

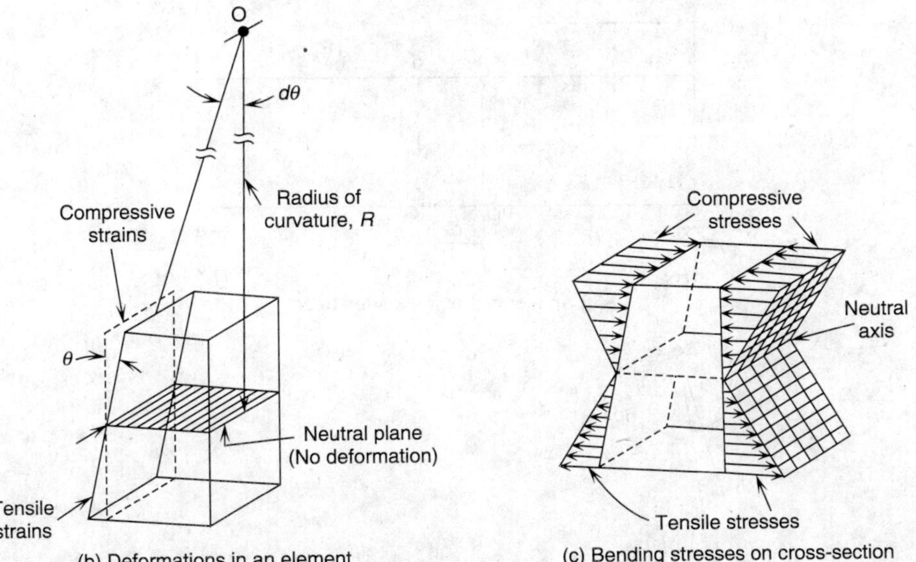

(b) Deformations in an element

(c) Bending stresses on cross-section

Figure 8.2 Bending stresses in a beam.

8.3 FLEXURAL FORMULA

The relation between the bending moment and the bending stresses it causes, is derived using the following assumptions:

1. The plane sections before bending remain plane after bending.
2. The material of the beam is homogeneous, isotropic and obeys Hooke's law.
3. The moduli of elasticity of material for tension and compression are equal.
4. The beam is initially straight and of constant cross-section.

5. The plane of loading contains a principle axis (i.e. axis of symmetry) of the cross-section and the externally applied loads must be transverse to the longitudinal axis of the beam.
6. The neutral surface is perpendicular to the plane of loading and the neutral axis passes through the centroid of the section.

8.3.1 Bending Stresses

Consider a beam element of length dx between two adjacent sections ab and cd as shown in Figure 8.3(a). Due to bending under the action of load components acting transverse to the axis of the element, the sections ab and cd rotate relative to each other by the amount $d\theta$, as shown in Figure 8.3(b), but remain straight and undistorted as illustrated earlier in Figure 8.2(a). The beam element bends, so that the fibres on upper portion of element are shortened and those on the lower portion elongated. The deformations vary linearly or non-linearly from the maximum elongation on one face to the maximum shortening on the other. There exists a layer somewhere near the middle of the beam section where the beam fibres ef are neither shortened nor elongated. The

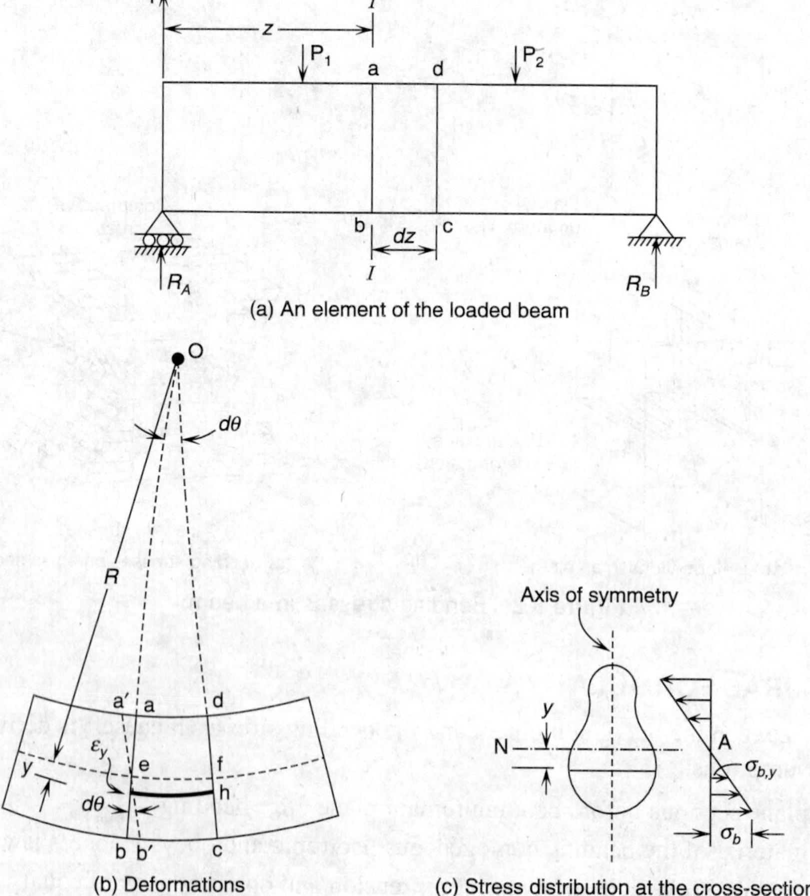

Figure 8.3 Strains and stresses acting on a typical element of a beam.

plane, containing fibres like ef, is called *neutral surface* because such fibres remain unchanged in length and hence carry no stress.

Consider the line a′b′ through *e* parallel to cd indicating that fibres at level ad are shortened by an amount aa′ and are in compression, and those at bc are lengthened by an amount bb′ and are in tension. The deformation (elongation) of a typical fibre gh located at distance y from the neutral surface is given by $\Delta = y\,d\theta$.

The curved length ef which was initially dz is equal to $R\,d\theta$ where R is the radius of curvature of the neutral surface. Thus corresponding strain is

$$\varepsilon_y = \frac{\Delta}{dz} = \frac{y\,d\theta}{R\,d\theta} = \frac{y}{R} \tag{8.1}$$

For the fibre of material, which is homogeneous and elastic the stress is given by

$$\sigma_{b,y} = E\varepsilon_y = \left(\frac{E}{R}\right)y \quad \text{or} \quad \frac{\sigma_{b,y}}{y} = \frac{E}{R} \tag{8.2}$$

Thus the deformations in the beam cause stresses which are compressive on upper face and tensile on the lower. The equation indicates the stress in any fibre varies directly with its distance y from the neutral surface, since it is assumed that the modulus of elasticity E is equal in tension and compression. These stresses caused by the bending moment are known as *bending* or *flexural stresses* and the relation between these stresses and bending moment is expressed by *flexure formula*. The actual stresses developed in a beam depend on the cross-section of the member, i.e. on the amount and distribution of material in the cross-section. The line of intersection between the neutral surface and the transverse section under consideration is called the *neutral axis*, abbreviated NA. Figure 8.3(c) shows the stress distribution in the element.

8.3.2 Equilibrium Conditions

The bending stresses produce an internal resisting couple M_r that balances the externally applied bending moment M_z. The forces of the couple F_T and F_C are produced by tensile and compressive stress fields acting over the portion of the cross-section as shown in Figure 8.1(c). The effect of compressive stresses in the upper part of the area can be described by resultant compressive force, F_C acting at a particular location. The same is true for the tensile stresses. These two resultant forces separated by moment arm provide the *internal resisting moment*.

The bending stresses developed due to applied external forces act perpendicular to the cross-section and consequently do nothing to balance the external shear forces. As discussed in the next chapter, the external *shear force* is balanced by shear stresses which act parallel to the face. The net effect of these stresses acting over the face of the cross-section is to produce a resultant shear force which equilibrates the applied external shear force.

The maximum bending stress on the extreme fibre at a distance $y_{\max}$ from the neutral axis which is represented by $\sigma_{b,\max}$ can be related to the bending stress $\sigma_{b,y}$ at an arbitrary point defined by the distance y from the neutral axis, by Eq. (8.2).

$$\frac{\sigma_{b,y}}{y} = \frac{\sigma_{b,\max}}{y_{\max}} = \frac{E}{R} \quad \text{or} \quad \sigma_{b,y} = \sigma_{b,\max}\left(\frac{y}{y_{\max}}\right) \tag{8.3}$$

At the level defined by y within the cross-section consider an elemental area dA. The force on the element is directly proportional to y, i.e.

$$dF_{b,z} = \sigma_{b,y}\, dA$$

and its moment is

$$dM_x = dF_{b,z} y = (\sigma_{b,y}\, dA) y$$

The total force in the direction of the member axis, i.e. Z-direction, produced by the entire stress field is $F_{b,z} = \int_A \sigma_{b,y}\, dA$. For equilibrium of internal forces in Z-direction, $\sum F_z = 0$. If there are no externally applied horizontal forces acting on the member, the total force produced by the tensile and compressive components (F_T and F_C) must add up to zero. Hence,

$$\int_A \sigma_{b,y}\, dA = \int_A \sigma_{b,\max}\left(\frac{y}{y_{\max}}\right) dA = \left(\frac{\sigma_{b,\max}}{y_{\max}}\right)\int_A y\, dA = 0$$

As $(\sigma_{b,\max}/y_{\max}) \neq 0$, then $\int_A y\, dA = 0$. Since the quantity $y\, dA$ is the moment of the elemental area dA about the neutral axis, the integral $\int_A y\, dA$ is the total moment of area. Hence

$$\int_A y\, dA = A\bar{y} = 0 \tag{8.4}$$

However, since only $\bar{y}$ in this relation can be zero, it is concluded that the distance from the neutral axis (which is the reference axis in this case) to the *centroid* of the cross-sectional area must be zero; that is the neutral axis must contain the centroid of the cross-sectional area.

In other words, the relation $\int_A y\, dA = 0$ means that the neutral axis of the beam corresponds to the centroidal axis of the cross-section of the beam.

For *rotational equilibrium* $\sum M = 0$, the sum of the moments produced by the elemental forces about the neutral axis must be equal to the applied external moment. The moment of an elemental force about the neutral axis

$$dM_r = (\sigma_{b,y}\, dA) y = \left[\sigma_{b,\max}\left(\frac{y}{y_{\max}}\right) dA\right] y = \left(\frac{\sigma_{b,\max}}{y_{\max}}\right) y^2\, dA$$

The sum of moments of all these elemental forces provides internal resisting moment, M_r which balances the external moment, M_x is given by

$$M_r = \left(\frac{\sigma_{\max}}{y_{\max}}\right)\int_A y^2\, dA = \left(\frac{\sigma_{\max}}{y_{\max}}\right) I = M_x \quad \text{or} \quad \sigma_{\max} = \left(\frac{M_x}{I}\right) y_{\max} \tag{8.5}$$

The term $\int_A y^2\, dA$ which represents the total resistance to bending associated with the sum of all elemental areas in the beam cross-section, is commonly called the *second moment* or *moment of inertia I* of an area about a reference axis which in this case is neutral axis. The moment of inertia describes the distribution of material in a cross-section.

The stress at any point located at the distance y from NA is given by

$$\sigma_{b,y} = \left(\frac{M_x}{I}\right) y \quad \text{or} \quad \frac{\sigma_{b,y}}{y} = \frac{M_x}{I} \tag{8.6}$$

Equations (8.2) and (8.6) can be combined and written as follows:

$$\frac{M_x}{I} = \frac{E}{R} = \frac{\sigma_{b,y}}{y} \tag{8.7}$$

The maximum bending stress $\sigma_{b,y} = \sigma_{b,\max}$ in a beam which occurs at $y = y_{\max}$ is given by:

$$\sigma_{b,\max} = \left(\frac{M_x}{I}\right) y_{\max} = \left(\frac{M_x}{Z}\right) \quad \text{where} \quad Z = \left(\frac{I}{y_{\max}}\right) \tag{8.8}$$

The quantity Z is termed *section modulus*. Thus, for the given moment the maximum stress is inversely proportional to the sectional modulus of the cross-section. Since the maximum bending stress at a section is proportional to the bending moment acting on it, hence for the beam of constant cross-section the variation of maximum bending stress along the length of a beam follows the pattern of the bending moment diagram as shown in Figures 8.4(b) and (c). The maximum bending stress occurs at the section of maximum bending moment. Values of section modulus for commonly encountered cross-sections are listed in Table 8.1.

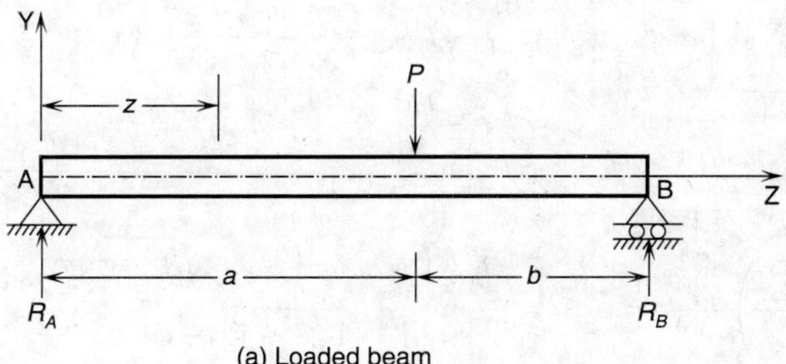

(a) Loaded beam

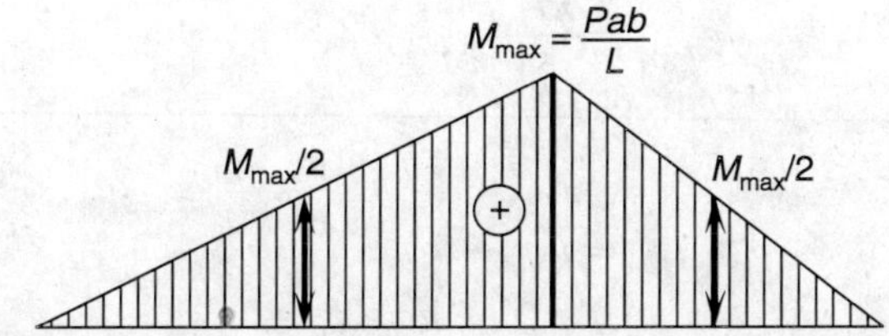

(b) Bending moment diagram (sagging moment is positive)

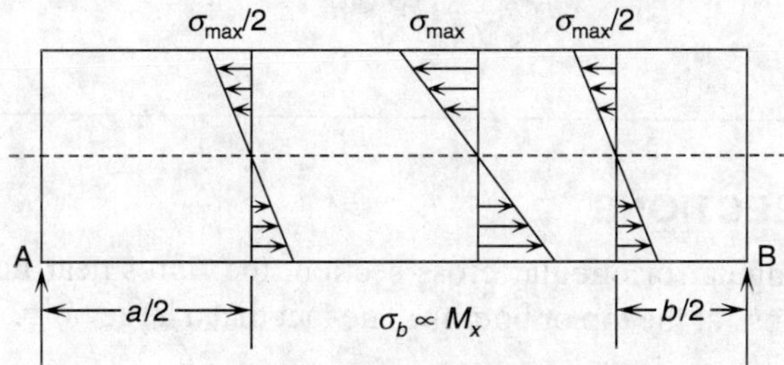

(c) Magnitude of bending stresses along the length of beam

Figure 8.4 Distribution of bending stresses in a beam.

Table 8.1 Moment of inertia and section modulus

Section	Moment of inertia	Section modulus
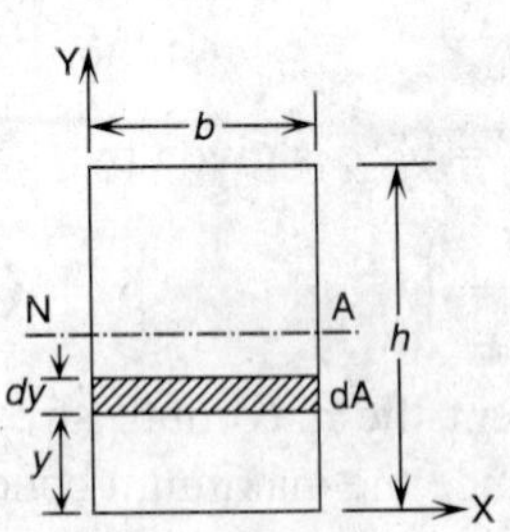	$I_x = \int_A y^2\, dA = \int_0^h (b\,dy) y^2\, dy$ $= \dfrac{bh^3}{3}$ $I_n = I_x - \bar{y}^2 A = \dfrac{bh^3}{3} - \left(\dfrac{h}{2}\right)^2 A$ $= \dfrac{bh^3}{12}$	$Z = \dfrac{I_n}{y_{\max}}$ $= \dfrac{bh^3}{12} \times \dfrac{2}{h}$ $= \dfrac{bh^2}{6}$
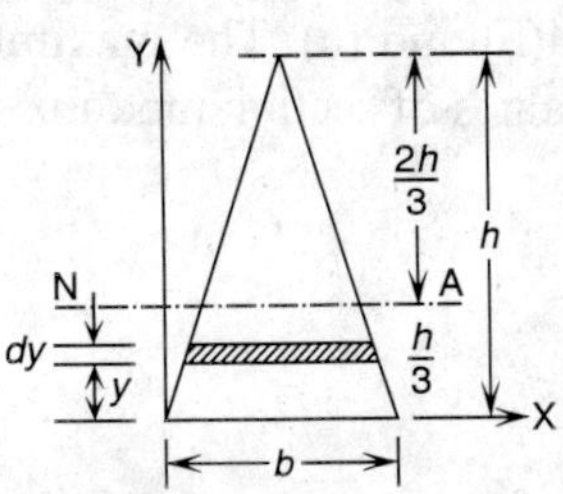	$I_x = \int_A y^2\, dA = \int_0^h \dfrac{b(h-y)}{h} y^2\, dy$ $= \dfrac{bh^3}{12}$ $I_n = I_x - \bar{y}^2 A = \dfrac{bh^3}{12} - \left(\dfrac{h}{3}\right)^2 A$ $= \dfrac{bh^3}{36}$	$Z = \dfrac{bh^3}{36} \times \dfrac{3}{2h}$ $= \dfrac{bh^2}{24}$
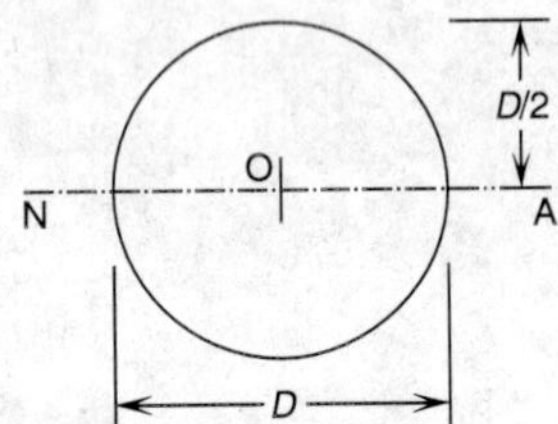	$I_n = \int_A y^2\, dA = \int y^2 \times (2\sqrt{R^2 - y^2})\, dy$ $= 2\int_{-D/2}^{D/2} (\sqrt{R^2 - y^2}) y^2\, dy$ $= \dfrac{\pi D^4}{64}$	$Z = \dfrac{\pi D^4}{64} \times \dfrac{2}{D}$ $= \dfrac{\pi D^3}{32}$
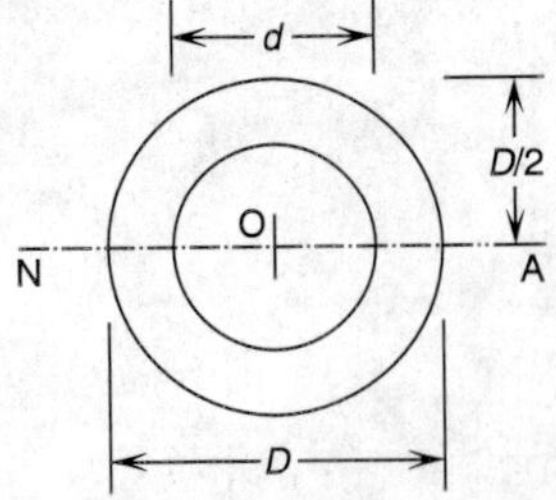	$I_{tube} = I_{outer} - I_{inner}$ $= \dfrac{\pi D^4}{64} - \dfrac{\pi d^4}{64}$ $= \dfrac{\pi}{64}(D^4 - d^4)$	$Z = \dfrac{I_n}{y_{\max}}$ $= \dfrac{\pi(D^4 - d^4)}{64} \cdot \dfrac{2}{D}$ $= \dfrac{\pi(D^4 - d^4)}{32D}$

8.4 ECONOMICAL SECTIONS

In a beam having a rectangular or circular cross-section, the fibres near neutral axis are under-stressed compared with those at the top or bottom. The fact that a large portion of the cross-section is thus under-stressed making it inefficient for resisting flexure. The expression $M_x = \left(\dfrac{\sigma_{b,\max}}{y_{\max}}\right) I$

indicates that if the area of a beam of rectangular cross-section of Figure 8.5(a) could be rearranged so as to maintain the same depth and same over all area but have shape shown in Figure 8.2(b), the moment of inertia would be greatly increased, resulting in greater moment resisting capacity. This increased moment resisting capacity is due to placement of more material at greater distance from the neutral axis, thus the reallocated material carry greater stress with greater moment arm about the neutral axis to resist applied bending moment. Because of its shape, this section is called I-section. The rectangular area joining two extreme rectangular parts called *flanges* is known as *web* which transmits the vertical shear.

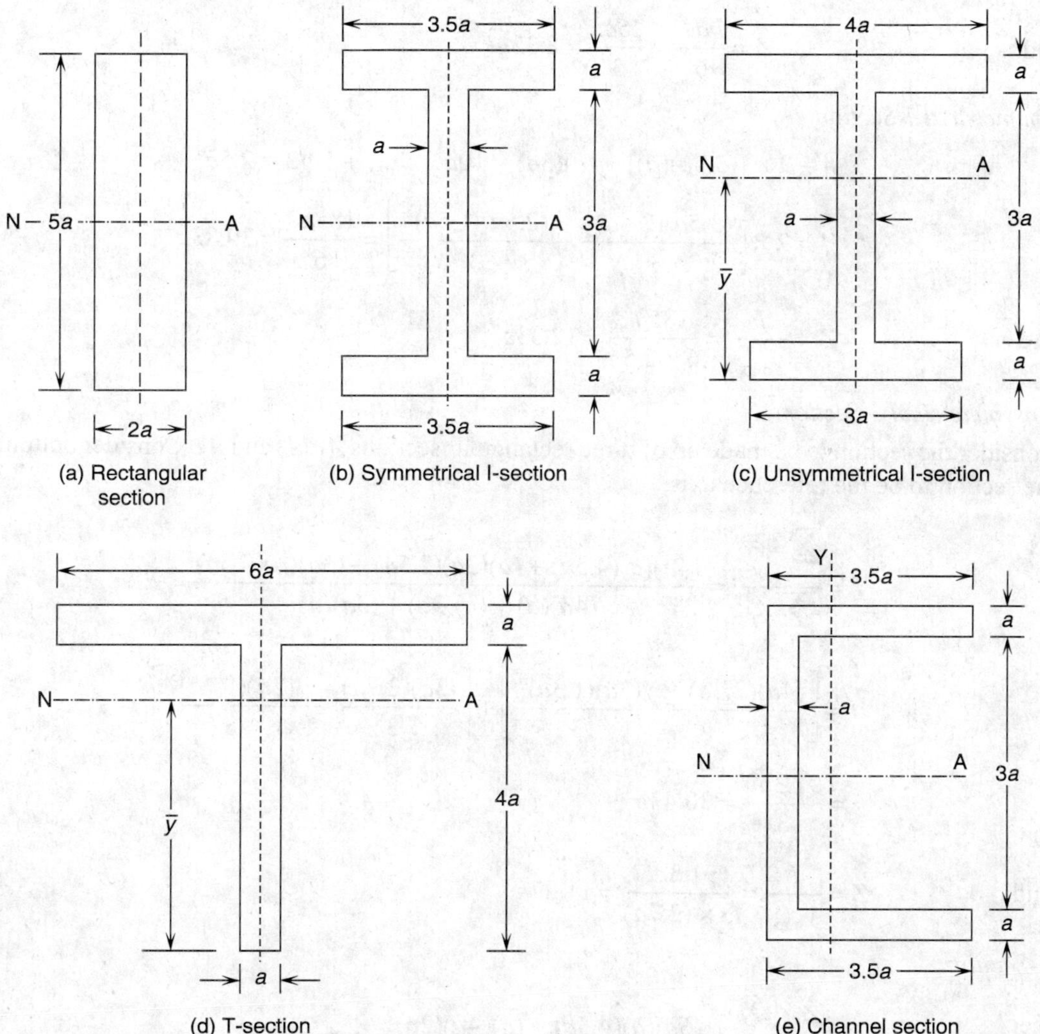

Figure 8.5 Comparison of moment of inertia and section modulus with material disposition in different sections.

To study the effect of reallocation of material on the stiffness and strength of a beam, five cross-sections having same depth shown in Figures 8.5(a), (b), (c), (d) and (e) which employ the same amount of material are considered. Since the stiffness depends on moment of inertia and strength on the section modulus, the values of these parameters have been computed as follows:

Rectangular Section:

$$A = (2a)(5a) = 10a^2,\ y_{max} = \frac{5a}{2},\ I = \frac{(2a)(5a)^3}{12} = \frac{125a^4}{6} = 20.83a^4,$$

and
$$Z = \frac{bd^2}{6} = \frac{25a^3}{3} = 8.33a^3$$

Symmetrical I-Section:

$$A = 2 \times [(3.5a)(a)] + (a)(3a) = 10a^2,\ y_{max} = 5a/2 = 2.5a,$$

$$I = 2 \times \left[\frac{(3.5a)(2.5a)^3}{3} - \frac{(2.5a)(1.5a)^3}{3}\right] = \frac{185a^4}{6} = 30.83a^4$$

and
$$Z = \frac{I}{y_{max}} = \frac{185a^3}{6 \times 2.5a} = 12.33a^3$$

Unsymmetrical I-Section:

Consider the section to be made up of three rectangular sections A_1, A_2 and A_3. Consider bottom of the section to be the reference axis;

$$\bar{y} = \frac{\sum_i A_i \bar{y}_i}{\sum_i A_i} = \frac{(4a)(a)(4.5a) + (a)(3a)(2.5a) + (3a)(a)(0.5a)}{(4a)(a) + (a)(3a) + (3a)(a)} = 2.7a$$

$$I = \left[\frac{(4a)(2.3a)^3}{3} - \frac{(3a)(1.3a)^3}{3}\right] + \left[\frac{(3a)(2.7a)^3}{3} - \frac{(2a)(1.7a)^3}{3}\right]$$

$$= \frac{91.3a^4}{3} = 30.43a^4$$

and
$$Z = \frac{I}{y_{max}} = \frac{91.3a^4}{3 \times (2.7a)} = 11.27a^3$$

T-Section:

$$\bar{y} = \frac{\sum_i A_i \bar{y}_i}{\sum_i A_i} = \frac{(6a)(a)(4.5a) + (a)(4a)(2a)}{(6a)(a) + (a)(4a)} = 3.5a$$

$$I = \frac{(6a)(1.5a)^3}{3} - \frac{(5a)(0.5a)^3}{3} + \frac{(a)(3.5a)^3}{3} = \frac{62.5a^4}{3} = 20.83a^4$$

and $$Z = \frac{I}{y_{max}} = \frac{62.5a^4}{3 \times (3.5a)} = 5.95a^3$$

Channel Section:

$$A = (a)(2.5a + 5a + 2.5a) = 10a^2,\ y_{max} = \frac{5a}{2} = 2.5a$$

$$I = \frac{(3.5a)(5a)^3}{12} - \frac{(2.5a)(3a)^3}{12} = \frac{130a^4}{3} = 30.83a^4$$

and $$Z = \frac{I}{y_{max}} = \frac{30.83a^4}{2.5a} = 12.33a^3$$

It is evident that *symmetrical I-section* and *channel section* have the highest and identical values of moment of inertia and section modulus. Therefore, the stiffness and the strength of these sections are the maximum. However, the single channel section being unsymmetrical about the vertical axis may result in unsymmetrical bending with the result its independent use is restricted. The commercial shapes are shown in Figures 8.6(a), (b) and (c). The I-section called *wide flange section* being the most efficient of above sections is widely used in practice.

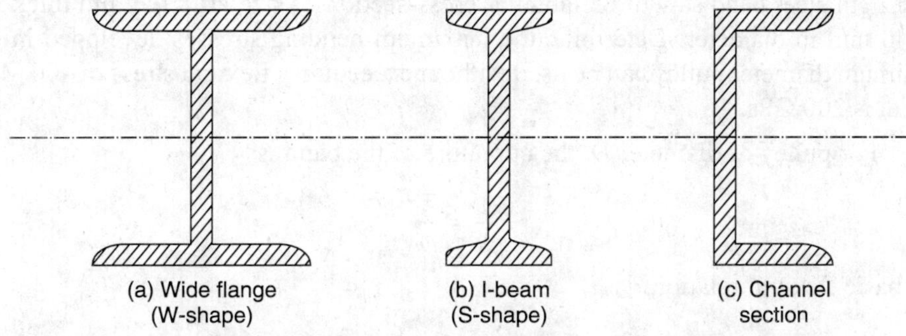

Figure 8.6 Commonly manufactured sections.

A practical range of wide flange beam sections shown in Figure 8.6(b) have been developed which are structurally efficient with minimum weight. This section is specified in terms of depth in millimeters followed by its nominal mass in kilogram per metre. Structural tables list the dimensions and other properties of cross-sectional area, such as moment of inertia (I), section modulus (Z) and radius of gyration (r) for each of the principal axes of the section.

Depending upon the material the allowable stresses in bending $\sigma_{b,a}$ are defined. The maximum bending stress $\sigma_{b,max}$ in a beam occurs at $y = y_{max}$. For the section to be adequate,

$$\sigma_{b,max} \le \sigma_{b,a} \quad \text{i.e.} \quad Z_{required} > \left(\frac{M_{max}}{\sigma_{b,a}}\right) \tag{8.9}$$

Thus, for an application, a structural section for a beam should be so selected that its section modulus is equal to or greater than the ratio of bending moment to allowable stress.

8.5 UNSYMMETRICAL SECTIONS

The discussions so far have been limited to the beams of materials assumed to be equally strong in tension and compression. Thus the sections symmetrical with respect to the neutral (centroidal) axis are preferable. However, for the materials relatively weak in tension and strong in compression, such as cast iron, it is desirable to use beams that are unsymmetrical with respect to the neutral axis. With such cross-section, the stronger fibre can be located at a greater distance from the neutral axis than weaker fibres. *The condition for such materials is to place the centroidal or neutral axis in such a position that the ratio of the distances from it to the fibres in tension and compression is equal to the ratio of allowable stresses in tension and compression.* The allowable stresses thus reach their permitted values simultaneously.

8.6 APPLICATION OF FLEXURE FORMULA

8.6.1 Symmetrical Sections

The following examples illustrate the application of flexural formula to beams having symmetrical cross-sections.

EXAMPLE 8.1

A high strength steel band saw of rectangular cross-section of size 20 × 0.8 mm thick runs over pulleys 640 mm in diameter. Determine the maximum bending stresses developed in the band. What minimum diameter pulley can be used without exceeding a flexural stress of 400 MPa? E for the material is 200 GPa.

Solution: For pulley of diameter D, the curvature of the band is

$$\frac{1}{R} = \frac{2}{D}$$

From the basic flexural relation,

$$\frac{\sigma_b}{y} = \frac{E}{R} \quad \text{or} \quad \sigma_b = \frac{Ey}{R}$$

For the band of thickness d, $y_{\text{max}} = d/2$

Therefore, $$\sigma_{b,\text{max}} = \frac{Ey_{\text{max}}}{R} = E\left(\frac{2}{D}\right)\left(\frac{d}{2}\right) = E\left(\frac{d}{D}\right)$$

$$\sigma_{b,\text{max}} = E\left(\frac{d}{D}\right) = (200 \times 10^3) \times \left(\frac{0.8}{640}\right) = 250 \text{ MPa}$$

For the given size of the band, the maximum bending stress in the band varies inversely with the diameter of the pulley. The diameter of the pulley to develop a maximum flexural stress of 400 MPa in the band is given by,

$$D = E\left(\frac{d}{\sigma_{b,\text{max}}}\right) = (200 \times 10^3) \times \left(\frac{0.8}{400}\right) = 400 \text{ mm}$$

EXAMPLE 8.2

Beams of approximately equal cross-sectional area ($A = 8.64 \times 10^4$ mm^2) but different shapes, i.e. rectangular, circular, or triangular section are considered for the use in an application that has to resist a maximum moment of $M_{\max}$ (= 75 kN.m). Which of these beams will have minimum bending stresses? Consider $h/b = 1.5$.

Solution: The maximum bending stress in a beam is inversely proportional to the sectional modulus of the cross-section and is given by:

$$\sigma_{b,\max} = \left(\frac{M_{\max}}{Z}\right)$$

Rectangular Section:

Sectional Properties: $A = bh;\ y_{\max} = \dfrac{h}{2};\ I = \dfrac{bh^3}{12}$ and $Z = \dfrac{bh^2}{6}$

For $A = 8.64 \times 10^4$ mm^2, $b = 240$ mm and $h = 360$ mm

$$Z = \frac{bh^2}{6} = \frac{240 \times 360^2}{6} = 5184000 \text{ mm}^3$$

$$\sigma_{b,\max} = \frac{M_{\max}}{Z} = \frac{75 \times 10^6}{5184000} = 14.47 \text{ MPa}$$

Triangular Section:

Sectional Properties:

For $A = 8.657 \times 10^4$ mm^2; $b = 340$ mm and $h = 510$ mm

$$Z = \frac{bh^2}{36} = \frac{340 \times 510^2}{36} = 3684750 \text{ mm}^3$$

$$\sigma_{b,\max} = \frac{75 \times 10^6}{3684750} = 20.88 \text{ MPa}$$

Circular Section:

Sectional Properties: $A = \dfrac{\pi d^2}{4};\ y_{\max} = \dfrac{d}{2};\ I = \dfrac{\pi d^4}{64}$ and $Z = \dfrac{\pi d^3}{32}$

For $A = 8.67 \times 10^4$ mm^2 and $d = 332$ mm

$$Z = \frac{\pi d^3}{32} = \frac{\pi \times 332^3}{32} = 3592644 \text{ mm}^3$$

$$\sigma_{b,\max} = \frac{75 \times 10^6}{3592644} = 20.35 \text{ MPa}$$

The variations of bending stresses are shown in Figure 8.7. The rectangular section has the largest sectional modulus, i.e. the smallest stresses are developed and the beam would carry heaviest external loading, followed by triangular and circular sections in the order. It should be noted that though the triangular section has the higher I value than any of the sections but has lower sectional modulus because of its larger y_{max}. On the other hand, the circular section has a lowest Z value, because more material is concentrated near the neutral axis, and it develops the largest stresses.

It should be noted that the *stress depends on the geometrical properties of the cross-section and is independent of the type of material used in the beam*. However, if the value of allowable stress in bending for the beam material is known, the beam can be adequately sized with respect to

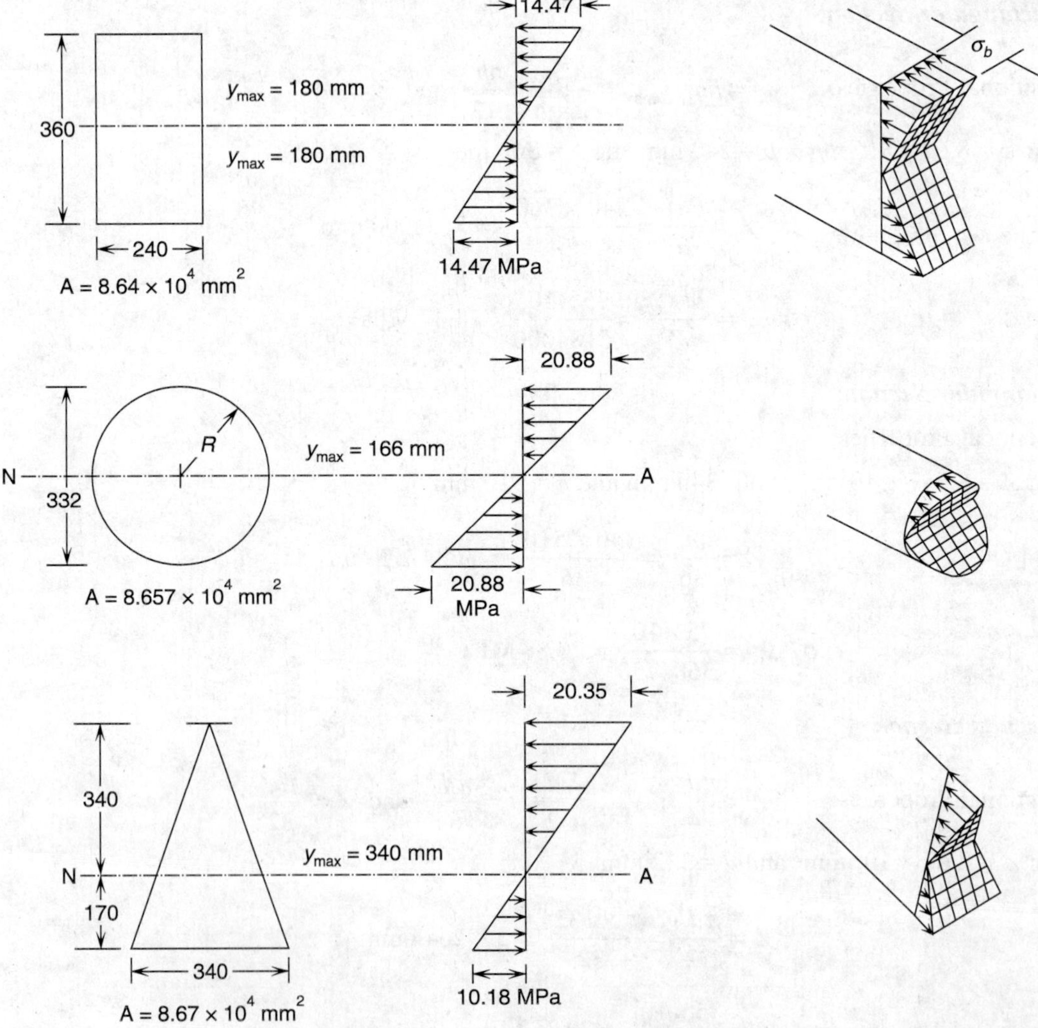

Figure 8.7 Comparison of bending stresses in different beam sections with approximately same areas subjected to the same bending moment.

bending such that bending stress developed is less than this allowable stress. Depending upon the material, the allowable stresses in bending $\sigma_{b,a}$ are defined.

EXAMPLE 8.3

A beam of square cross-section of side a has to resist a maximum moment of M_{max}. The beam can be placed so that the plane of moment (a) is either parallel to the sides of square or (b) coincides with the diagonal. Which of these positions will have minimum bending stress?

Solution: The bending stress in a beam is governed by its section modulus. The section moduli in two cases are:

Case (a):
$$Z = \frac{bh^2}{6} = \frac{a^3}{6}$$

Case (b):
$$y_{max} = \frac{a}{\sqrt{2}};\ I_x = 2\left(\frac{bh^3}{12}\right) = 2 \times \frac{(\sqrt{2}a)(a/\sqrt{2})^3}{12} = \frac{a^4}{12}$$

Thus,
$$Z = \frac{a^3}{6\sqrt{2}}$$

As the section modulus in case (a) is about 40 per cent higher than that in case (b), the beam will develop minimum bending stress in case (a) position.

EXAMPLE 8.4

Two beams, the first one of hollow circular cross-section with a ratio of internal and external diameters of 0.9, and the second of solid circular section are considered for use in an application that has to resist a maximum moment of M_{max}. Which of these beams will be economical for equal strength condition? Determine the per cent saving in the material. If the ratio of internal and external diameters of hollow circular cross-section is modified to 0.8, determine the modified saving in the material.

Solution: The section modulus for the solid circular section of diameter D_1 is

$$Z_s = \frac{\pi D_1^3}{32}$$

For hollow cross-section the sectional modulus is given by

$$Z_h = \left(\frac{\pi D^4}{64} - \frac{\pi d^4}{64}\right)\left(\frac{2}{D}\right) = \frac{\pi D^3}{32}\left[1 - \left(\frac{d}{D}\right)^4\right]$$

For equal strength condition, $Z_s = Z_h$ i.e.,

$$\left(\frac{D_1}{D}\right)^3 = 1 - \left(\frac{d}{D}\right)^4 \quad \text{or} \quad \frac{D_1}{D} = \sqrt[3]{1 - \left(\frac{d}{D}\right)^4}$$

The saving in material is governed by the sectional areas:

$$A_s = \frac{\pi D_1^2}{4} \quad \text{and} \quad A_h = \frac{\pi}{4}(D^2 - d^2) = \frac{\pi D^2}{4}\left[1 - \left(\frac{d}{D}\right)^2\right]$$

Percentage saving in material is computed as follows,

$$\left(\frac{A_s - A_h}{A_s}\right) \times 100 = \left[1 - \left(\frac{D}{D_1}\right)^2 \left\{1 - \left(\frac{d}{D}\right)^2\right\}\right]$$

Case I: For the ratio $\dfrac{d}{D} = 0.9$,

$$\frac{D_1}{D} = \sqrt[3]{1 - (0.9)^4} = 0.7006$$

Percentage saving in material is

$$\left(\frac{A_s - A_h}{A_s}\right) \times 100 = \left[1 - \left(\frac{1}{0.7006}\right)^2 \{1 - (0.9)^2\}\right] \times 100 = 61.29 \text{ per cent}$$

Case II: For the ratio $\dfrac{d}{D} = 0.8$,

$$\frac{D_1}{D} = \sqrt[3]{1 - (0.8)^4} = 0.8389$$

Percentage saving in material

$$\left(\frac{A_s - A_h}{A_s}\right) \times 100 = \left[1 - \left(\frac{1}{0.8389}\right)^2 \{1 - (0.8)^2\}\right] \times 100 = 48.85 \text{ per cent}$$

It should be noted that the saving in material reduces with the decrease in the ratio of internal and external diameters of hollow circular cross-section.

EXAMPLE 8.5

A simply supported timber beam carries a uniform distributed load of 10 kN/m over an effective span of 3.6 m. Determine the required depth of the member if the width is predetermined to be 100 mm. The allowable stress for the type of timber available in bending is 11.0 MPa.

Solution: The maximum bending moment occurs at the mid-span and is given by,

$$M_{\max} = \frac{w_0 L^2}{8} = \frac{(10 \times 3.6^2) \times 10^6}{8} = 16.2 \times 10^6 \text{ N.mm}$$

Required section modulus, $Z_{\text{required}} > \left(\dfrac{M_{\max}}{\sigma_{b,a}}\right) = \dfrac{16.2 \times 10^6}{11.0} = 1.473 \times 10^6$

Therefore,
$$Z_{\text{required}} = \frac{bd^2}{6} = \frac{100 \times d^2}{6}$$
$$= 1.473 \times 10^6 \quad \text{or} \quad d = 297.29 \text{ mm}$$

Provide 100 × 300 mm.

EXAMPLE 8.6

A simply supported timber beam 100 mm wide and 300 mm deep of Figure 8.8(a) carries a uniformly distributed load of w_0 kN/m over its entire span of 6 m and a concentrated load W at distance 4.5 m from the left support. If the allowable stress for timber in bending is 8.0 MPa, determine the value of w_0 and W that will result in zero shear force just to the left of W.

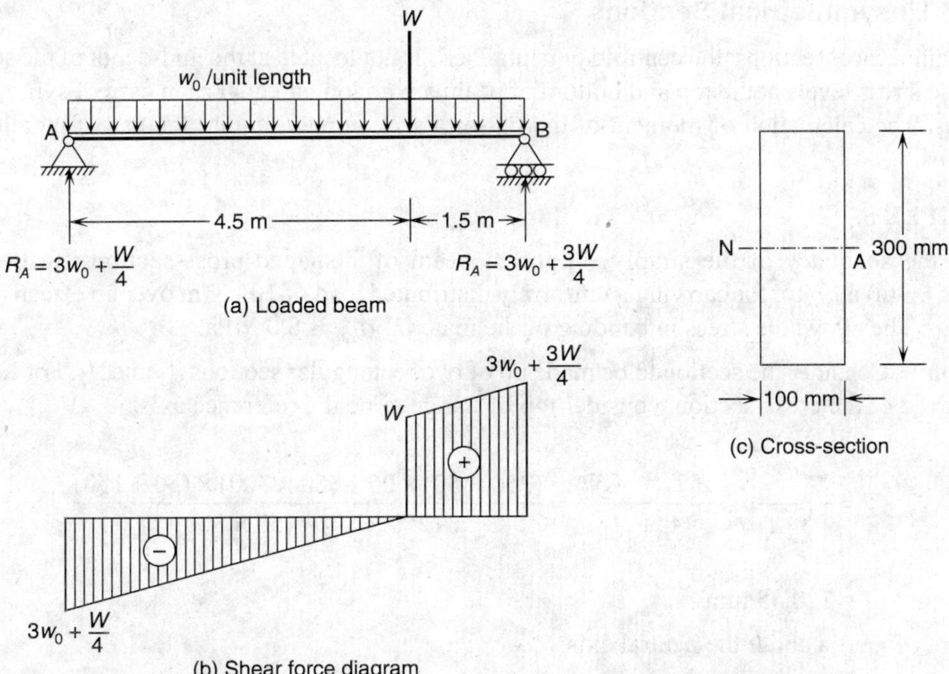

Figure 8.8 The beam and shear diagram of Example 8.6.

Solution: The shear force diagram is shown in Figure 8.8(b) and the cross-section of the member in Figure 8.8(c). For the shear force to be zero, at the location of concentrated load W,

$$-\left(3w_0 + \frac{W}{4}\right) + 4.5w_0 = 0 \quad \text{or} \quad W = 6w_0$$

The maximum bending moment which occurs under load W is given by,

$$M_{\max} = \frac{1}{2} \times \left(3w_0 + \frac{W}{4}\right) \times 4.5 = \frac{81w_0}{8} \text{ kN.m}$$

Moment resisting capacity of the beam,

$$M_r = \sigma_{b,\max}\left(\frac{bd^2}{6}\right) = 8\times\left(\frac{100\times 300^2}{6}\right)\times 10^{-6} = 12 \text{ kN.m}$$

To compute the value of w_0, equate $M_{\max}$ to M_r:

$$\frac{81w_0}{8} = 12 \quad \text{or} \quad w_0 = \frac{32}{27} \text{ kN.m}$$

Therefore, $$W = 6w_0 = 6\times\frac{32}{27} = \frac{64}{9} \text{ kN}$$

8.6.2 Unsymmetrical Sections

In unsymmetrical sections, the centroid or neutral axis is not located at the mid-depth of the section. Thus the stress levels at the top and bottom of beam are no longer equal as in case of symmetrical sections. The calculation of moment of inertia for a section may require the use of parallel axis theorems.

EXAMPLE 8.7

Check the adequacy of the simply supported beam of T-shaped cross-section illustrated in Figures 8.9(a) and (b), for carrying a uniformly distributed load of 10 kN/m over an effective span of 3.6 m. The allowable stress in bending of the material $\sigma_{b,a}$ is 8.0 MPa.

Solution: Consider the section to be made up of two rectangular sections A_1 and A_2. For location of centroid of the cross-section, consider top of flange to be the reference axis;

$$\bar{y} = \frac{\sum_i A_i\bar{y}_i}{\sum_i A_i} = \frac{A_1\bar{y}_1 + A_2\bar{y}_2}{A_1 + A_2} = \frac{(250\times 50)\times 25 + (50\times 300)\times(50+150)}{(250\times 50)+(50\times 300)}$$

$$= 120.45 \text{ mm}$$

Moment of inertia about the neutral axis:

$$I = I_1 + I_2 = \frac{250\times 120.45^3}{3} - \frac{(250-50)\times(120.45-50)^3}{3} + \frac{50\times(350-120.45)^3}{3}$$

$$= 3.7053\times 10^8 \text{ mm}^4$$

The maximum bending moment occurs at the mid-span and is given by

$$M_{\max} = \frac{wl^2}{8} = \frac{(10\times 3.6^2)\times 10^6}{8} = 16.2\times 10^6 \text{ N mm}$$

Bending stresses at the top and bottom faces of the beam are:

$$\sigma_{b,t} = \frac{(16.2\times 10^6)\times 120.45}{3.7053\times 10^8} = 5.27 \text{ MPa}$$

$$\sigma_{b,b} = \frac{(16.2 \times 10^6) \times 229.55}{3.7053 \times 10^8} = 10.04 \text{ MPa}$$

The centroid is located near the top flange of the beam. The stresses are greater at the bottom of the beam than those at the top because of the larger y-distance. The bending moment distribution is shown in Figure 8.9(c).

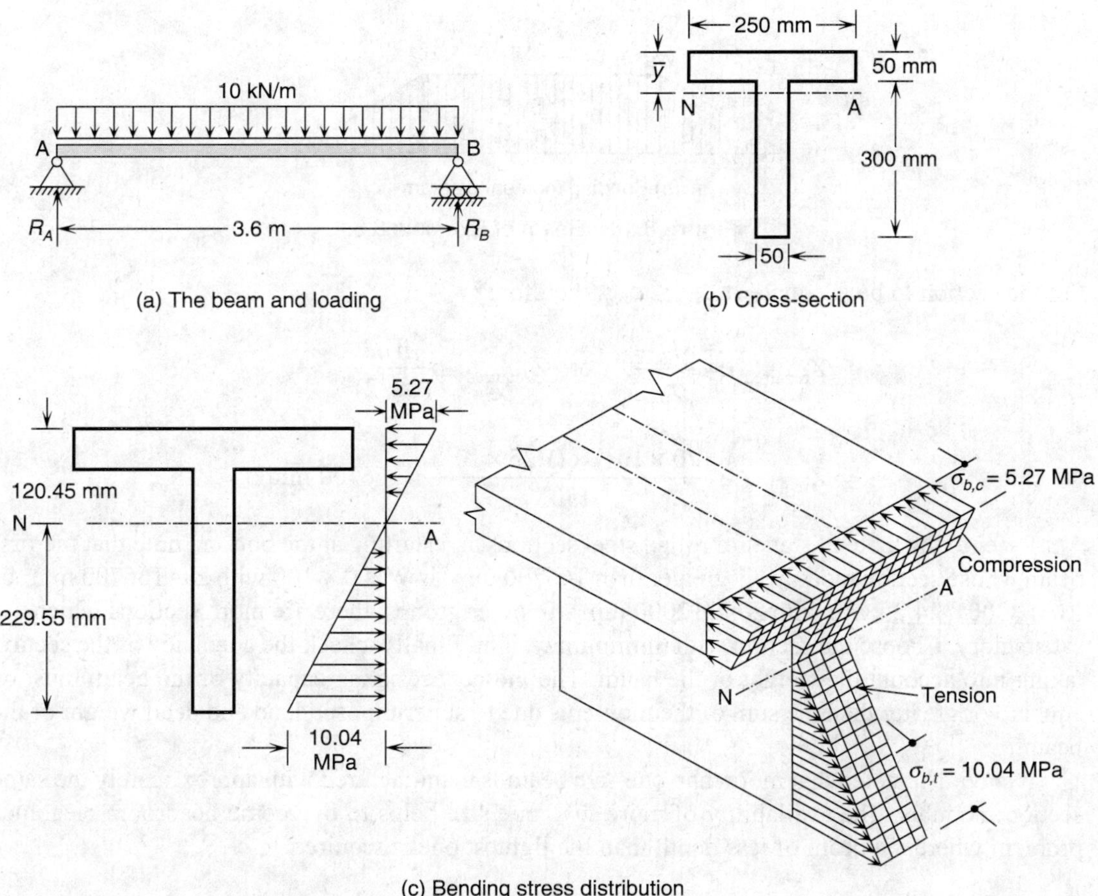

Figure 8.9 Stresses in the T-beam analysed in Example 8.7.

EXAMPLE 8.8

A simply supported rolled steel beam of I-section carries two concentrated loads W at distance a from each support as illustrated in Figure 8.10(a). The effective span of the beam is L. Select a section to provide a factor of safety of 2, when W = 20 kN, L = 6 m, a = 1.25 mm and σ_y = 300 MPa.

Solution: The maximum bending moment of magnitude Wa occurs in the middle portion of pure bending of length 3.5 m (see Figure 8.10(b)).

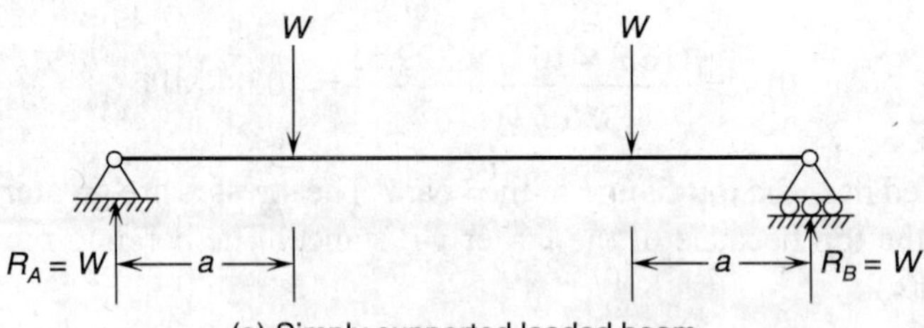

(a) Simply supported loaded beam

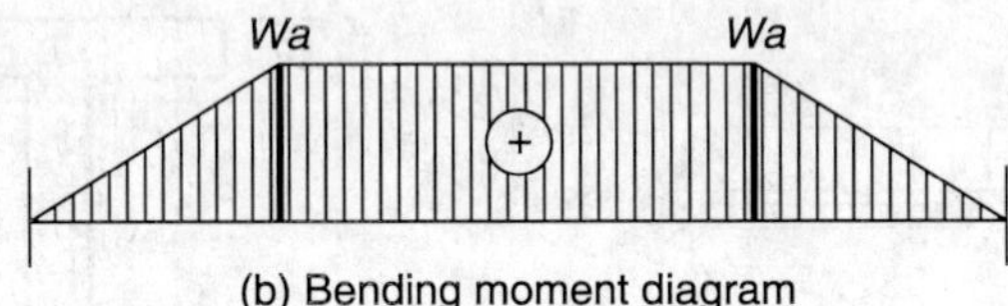

(b) Bending moment diagram

Figure 8.10 Beam of Example 8.8.

For the section to be adequate $\sigma_{b,\max} \le \sigma_{b,a}$, therefore,

$$Z_{\text{required}} \ge \left(\frac{M_{\max}}{\sigma_{b,a}}\right) \quad \text{or} \quad Z_{\text{required}} \ge \frac{Wa}{(\sigma_y/2)}$$

or
$$Z_{\text{required}} \ge \frac{(20 \times 10^3) \times (1.25 \times 10^3)}{150} = 166700 \text{ mm}^3$$

Refer to the table of standard rolled steel sections and starting at the bottom, note that the first beam whose section modulus is greater than 166700 mm^3 is W200 × 100 with Z = 166700 mm^3 or IS MB200@24.2 kg/m with Z = 212000 mm^3. In other groups, there are more sections which are satisfactory. Choose the section with minimum weight. Finally, check the adequacy of the section taking into account the weight of the beam. The moment resisting capacity of the beam must be equal to or greater than the sum of the moments due to superimposed load and dead weight of the beam.

It should be noted that more than one size beam is manufactured with approximately the same section modulus. The availability of more than one size helps to overcome headroom clearance problem wherein a beam of less depth than the lightest one is required.

EXAMPLE 8.9

A simply supported cast-iron beam of inverted T cross-section carries a uniformly distributed load on its entire span as illustrated in Figure 8.11(a). Determine the flange width b of the section so that the allowable stresses, $\sigma_{b,t,a}$ = 40 MPa and $\sigma_{b,c,a}$ = 100 MPa are reached simultaneously. The cross-section of the beam is shown in Figure 8.11(b).

Solution: A simply supported beam bends concave upward with the topmost and bottommost fibres being in compression and tension, respectively. To make allowable stresses in tension and compression to reach their permitted values simultaneously, the ratio of the distances from neutral axis to the bottommost fibres in tension and topmost fibres in compression should be equal to the ratio of allowable stresses. Thus

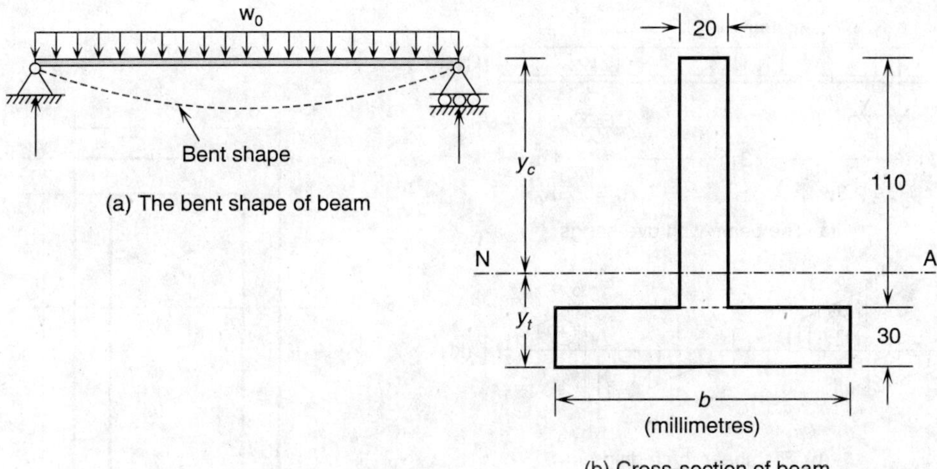

Figure 8.11 The simply supported cast-iron beam and its cross-section of Example 8.9.

$$\frac{y_t}{y_c} = \frac{\sigma_{b,t,a}}{\sigma_{b,c,a}} = \frac{40}{100} = \frac{2}{5}$$

Total depth of the section, $h = y_t + y_c = 110 + 30 = 140$ mm
Therefore,

$$y_t = \frac{2 \times 140}{2+5} = 40 \text{ mm} \quad \text{and} \quad y_c = \frac{5 \times 140}{2+5} = 100 \text{ mm}$$

Since the neutral coincides with the centroidal axis,

$$(A_1 + A_2)\overline{y} = A_1\overline{y}_1 + A_2\overline{y}_2$$

Therefore, with reference to the bottommost fibre,

$$(b \times 30 + 20 \times 110) \times 40 = (b \times 30) \times 15 + (20 \times 110) \times (30 + 55)$$

Hence,
$$b = \frac{20 \times 110 \times 45}{(1200 - 450)} \approx 132 \text{ mm}$$

EXAMPLE 8.10

A cast-iron beam over-hanged at both the ends has inverted U cross-section illustrated in Figure 8.12(d). Determine the maximum uniformly distributed load the beam can support, if the allowable stresses are $\sigma_{b,t,a} = 30$ MPa and $\sigma_{b,c,a} = 90$ MPa.

Solution: To locate centroid of the cross-section, consider the section has been made up of three rectangular sections A_1, A_2 and A_3 (= A_2). Consider top of the flange to be the reference axis:

$$\overline{y} = \frac{\sum_i A_i \overline{y}_i}{\sum_i A_i y_i} = \frac{A_1\overline{y}_1 + 2 \times A_2\overline{y}_2}{A_1 + A_2}$$

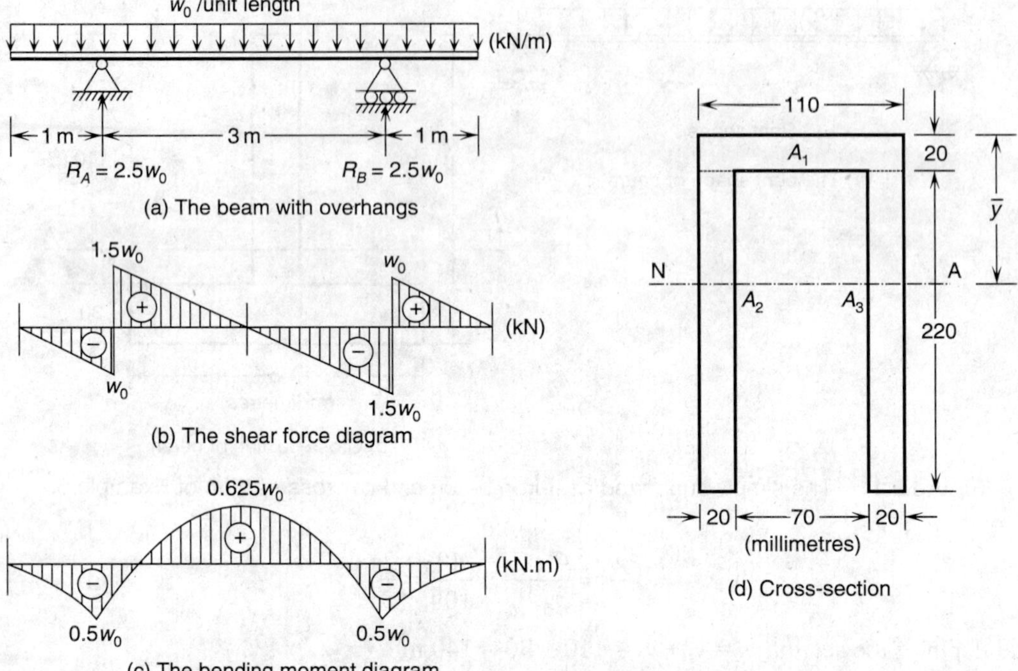

Figure 8.12 The beam and cross-section of Example 8.10.

$$= \frac{(110 \times 20) \times 10 + 2 \times (20 \times 220) \times (20 + 110)}{(110 \times 20) + 2 \times (20 \times 220)} = 106 \text{ mm}$$

Moment of inertia about the neutral axis:

$$I = \frac{110 \times 106^3}{3} - \frac{70 \times 86^3}{3} + 2 \times \frac{20 \times 134^3}{3} = 60.91 \times 10^6 \text{ mm}^4$$

The maximum bending moment: Consider that the maximum uniformly distributed load the beam can support to be w_0. The shear force and bending moment diagrams are shown in Figures 8.12(b) and (c). The maximum positive bending moment at the mid-span and the maximum negative bending moment at the supports are:

$$(-)\, M_{\max} = 0.5 w_0 \quad \text{and} \quad (+)\, M_{\max} = 0.625 w_0$$

The negative bending moment causes tension in upper fibres and the positive bending moment causes tension in bottom fibres.

In the negative bending moment region, the moment resisting capacities of the section in terms of allowable stresses are given by,

$$M_{r,t} = \frac{\sigma_{b,t,a} I}{y_t} = \frac{30 \times (60.91 \times 10^6)}{106} = 17.24 \text{ kN.m}$$

$$M_{r,c} = \frac{\sigma_{b,c,a} I}{y_c} = \frac{90 \times (60.91 \times 10^6)}{134} = 51.72 \text{ kN.m}$$

Safe moment resisting capacity of the section is lower of the two, i.e. $M_r = 17.24$ kN.m. Equate this to the bending moment,

$$M_{max} = M_r \quad \text{or} \quad 0.5w_0 = 17.24 \times 10^3 \quad \text{or} \quad w_0 = 34.48 \text{ kN/m}$$

In the positive bending moment region, the moment resisting capacities of the section are,

$$M_{r,c} = \frac{\sigma_{b,c,a} I}{y_c} = \frac{90 \times (60.91 \times 10^6)}{106} = 40.92 \text{ kN.m}$$

$$M_{r,t} = \frac{\sigma_{b,t,a} I}{y_t} = \frac{30 \times (60.91 \times 10^6)}{134} = 13.64 \text{ kN.m}$$

Safe moment resisting capacity of the section is $M_r = 13.64$ kN.m.

Therefore, $\quad 0.625w_0 = 13.64 \times 10^3 \quad$ or $\quad w_0 = 21.82$ kN/m

Thus the maximum safe load is 21.82 kN/m.

EXAMPLE 8.11

A simply supported beam of inverted T cross-section has an over-hang at one end and carries a uniformly distributed load of intensity w_0 on its entire span as illustrated in Figures 8.13(a) and (c). Determine the dimension t of the section so that the factor of safety with respect to the failure by yielding must not be less than 2. Consider $a = 1.0$ m; $w_0 = 10$ kN/m and $\sigma_{y,t} = \sigma_{y,c} = \sigma_y = 350$ MPa.

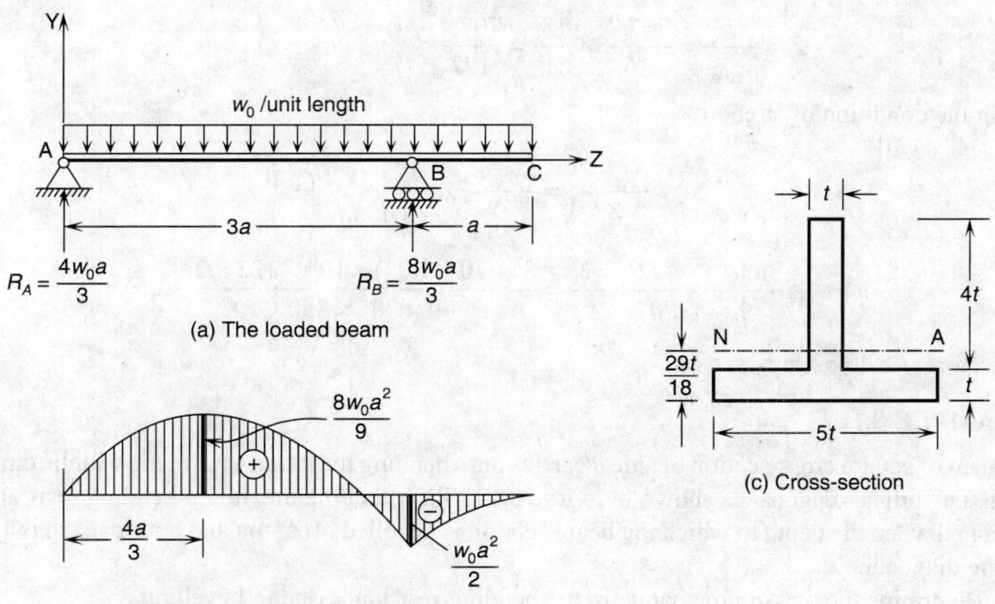

Figure 8.13 Simply supported beam of T-shaped cross-section subjected to a uniformly distributed load.

Solution: The reactions at the supports are,

$$\sum M_A = -(4w_0)(2a) + R_B(3a) = 0 \quad \text{or} \quad R_B = \frac{8w_0}{3}$$

$$\sum F_V = -(4w_0)(a) + R_A + R_B = 0 \quad \text{or} \quad R_A = 4w_0 - R_B = \frac{4w_0}{3}$$

The bending moment diagram is shown in Figure 8.13(b),

$$M_{\max} = \frac{8w_0 a^2}{9}$$

To determine the location of centroid of the cross-section consider top of section to be the reference axis;

$$\bar{y} = \frac{\sum_i A_i \bar{y}_i}{\sum_i A_i y_i} = \frac{A_1\bar{y}_1 + A_2\bar{y}_2}{A_1 + A_2} = \frac{(4t \times t) \times (2t) + (5t \times t) \times (4t + 0.5t)}{(4t \times t) + (5t \times t)} = \left(\frac{61}{18}\right)t$$

$$I = \frac{1}{3} \times t \times \left(\frac{61t}{18}\right)^3 + \frac{1}{3} \times \left[5t \times \left(\frac{29t}{18}\right)^3 - 4t \times \left(\frac{11t}{18}\right)^3\right] = \frac{707t^4}{36}$$

and
$$Z = \frac{I}{y_{\max}} = \frac{707t^4}{36} \times \frac{18}{61t} = \frac{707t^3}{122}$$

From the condition of strength

$$\frac{M_{\max}}{Z}\left(= \frac{8w_0 a^2}{9} \times \frac{122}{707t^3}\right) \le \frac{\sigma_y}{2}$$

$$t^3 \ge \frac{8w_0 a^2}{9} \times \frac{122}{707} \times \frac{2}{\sigma_y} = \frac{8 \times (10 \times 1.0^2) \times 10^6 \times 122 \times 2}{9 \times 707 \times 350}$$

Therefore, $t \ge 20.62$ mm.

EXAMPLE 8.12

A beam of square cross-section of side a carries pure bending moment in the plane which coincides with the vertical diagonal as shown in Figure 8.14. Show that the maximum bending resistance is obtained when the depth to which the beam section is bevelled off at top and bottom each is 1/18th of the diagonal.

Determine the maximum change in the bending resistance due to bevelling.

Solution: The bending resistance of a beam section is governed by its section modulus. The section moduli in two cases are:

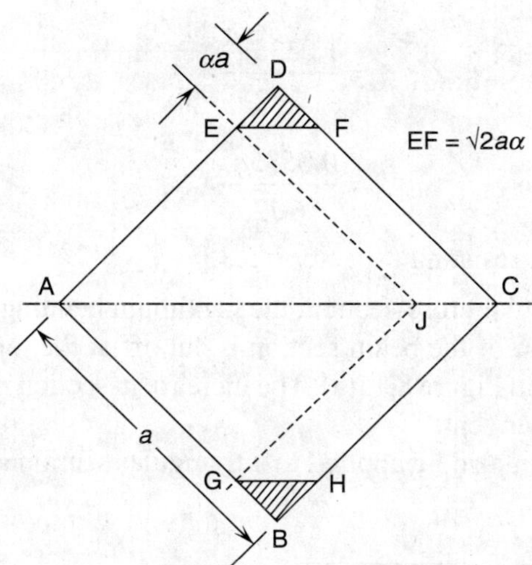

Figure 8.14 Bevelled square cross-section of Example 8.12.

Case (a):

$$y_{\max} = \frac{a}{\sqrt{2}}, I_x = 2 \times \frac{bh^3}{12} = 2 \times \frac{\sqrt{2}a \times (a/\sqrt{2})^3}{12} = \frac{a^4}{12} \text{ and } Z = \frac{a^3}{6\sqrt{2}}$$

Case (b): Let ED = αa, where α represents the fraction by which the side of the section is shortened. The resulting cross-section may be considered to consist of a square AEJG and two *parallelograms* CFEJ and CJGH. Therefore,

$$I_x' = \frac{a^4(1-\alpha)^4}{12} + 2 \times \frac{1}{3} \times (\alpha a\sqrt{2}) \times \left[\frac{a(1-\alpha)}{\sqrt{2}}\right]^3$$

$$= \frac{a^4(1+3\alpha)(1-\alpha)^3}{12}$$

$$y'_{\max} = \frac{(1-\alpha)a}{\sqrt{2}}$$

Therefore, $$Z' = \frac{(1+3\alpha)(1-\alpha)^2 a^3}{6\sqrt{2}} = \frac{(1+\alpha-5\alpha^2+3\alpha^3)a^3}{6\sqrt{2}}$$

For Z' to be maximum $dZ'/d\alpha = 0$, i.e.

$$1 - 10\alpha + 9\alpha^2 = 0 \quad \text{or} \quad \alpha^2 - \left(\frac{10}{9}\right)\alpha + \frac{1}{9} = 0 \quad \text{or} \quad \alpha \approx \frac{1}{9}$$

Therefore, the depth, up to which the beam section is cut off at top and bottom, each is:

$$\frac{\alpha a}{\sqrt{2}} = \frac{a}{9\sqrt{2}} = \frac{1}{18}(\sqrt{2}a) = \frac{\text{Length of diagonal}}{18}$$

For bevelled cross-section, $$Z'_{\max} = \frac{1.0535a^3}{6\sqrt{2}}$$

Increase in sectional modulus $$= \frac{0.0535a^3}{6\sqrt{2}}$$

Per cent increase in moment resisting capacity = 5.35.

In case of beams having circular cross-section, the maximum bending resistance is obtained when the shaded segmental portion of the beam section is cut off at the top and bottom each by 0.011 times the diagonal as shown in Figure 8.15(a). The increase in section modulus or moment resisting capacity in this case is 0.7 per cent.

The above procedure can also be applied to a triangular section as shown in Figure 8.15(b).

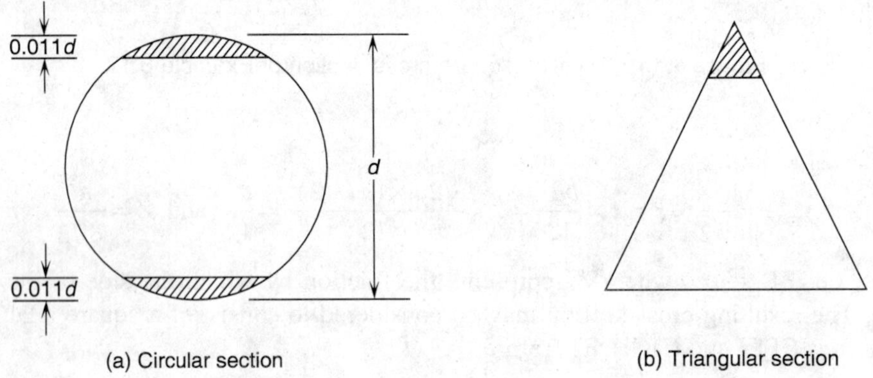

Figure 8.15 Bevelled cross-section to maximize the moment resisting capacities.

EXAMPLE 8.13

A rectangular timber beam of size 50 × 100 mm deep is subjected to a bending moment such that the maximum bending stress developed in the beam is 10 MPa. Determine the total force carried by the shaded areas shown in Figures 8.16(a) to (c) and the moment of these forces about the neutral axis of the cross-section.

Solution: The total force carried by the shaded areas is,

F = Stress at the centroid of the shaded area × Area under consideration

$$= \left(\frac{\sigma_{b,\max}}{y_{\max}} \times \bar{y}\right) \times A = \frac{\sigma_{b,\max}}{y_{\max}} \times A\bar{y}$$

The moment of the force about the neutral axis of the cross-section,

$$M = F \times \bar{y} = \left(\frac{\sigma_{b,\max}}{y_{\max}} \times A\bar{y}\right) \times \bar{y} = \frac{\sigma_{b,\max}}{y_{\max}} \times I_{NA}$$

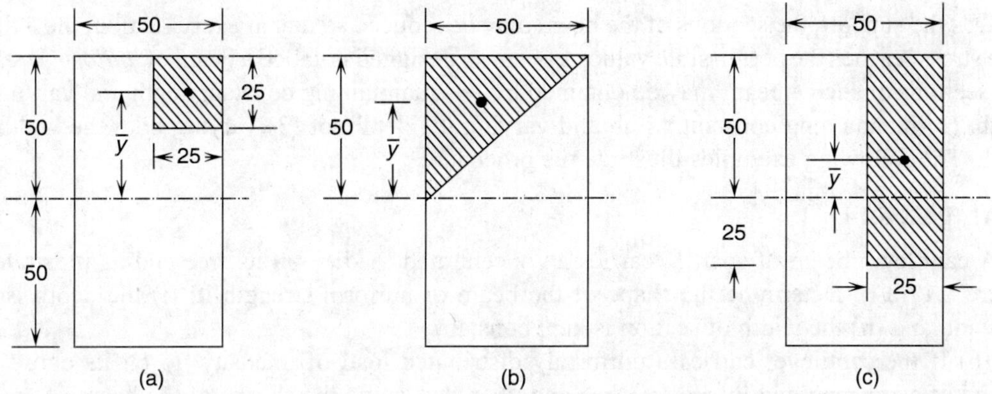

Figure 8.16 Portions of the cross-section for computation of total force and moment (all dimensions in mm).

(a)
$$F = \frac{10}{50} \times [(25 \times 25) \times (12.5 + 25)] = 4687.5 \text{ N} = 4.6875 \text{ kN}$$

$$I_{NA} = \frac{25 \times 25^3}{12} + (25 \times 25) \times (37.5)^2 = 911458 \text{ mm}^4$$

$$M = \left(\frac{10}{50} \times 911458\right) \times 10^{-3} = 182.3 \text{ Nm}$$

(b)
$$F = \frac{10}{50} \times \left(\frac{1}{2} \times 50 \times 50\right) \times \left(\frac{2}{3} \times 50\right) = 8333 \text{ N} = 8.333 \text{ kN}$$

$$I_{NA} = \frac{50 \times 50^3}{4} = 1562500 \text{ mm}^4$$

$$M = \left(\frac{10}{50} \times 1562500\right) \times 10^{-3} = 312.5 \text{ Nm}$$

(c)
$$F = \frac{10}{50} \times (25 \times 75) \times (37.5 - 25) = 4687.5 \text{ N} = 4.687 \text{ kN}$$

$$I_{NA} = \frac{25 \times 75^3}{12} + 25 \times 75 \times (12.5)^2 = 1171875 \text{ mm}^4$$

$$M = \left(\frac{10}{50} \times 1171875\right) \times 10^{-3} = 234.4 \text{ Nm}$$

8.7 THE BEAMS OF UNIFORM STRENGTH

In the conventional beams of uniform cross-section, the maximum fibre stress at a point where the bending moment is the maximum reaches permissible value. Hence the beam of uniform cross-section is uneconomical as all other sections are under-stressed. As the bending moment decreases towards the support, the sections of the beam may be reduced so that at every section, the extreme fibre stress reaches the permissible value. A beam so designed is called a beam of *uniform strength*. The sections of such a beam may be obtained by: (1) maintaining constant depth and varying the width; (2) maintaining constant width and varying the depth, or (3) varying both the width and depth. The following examples illustrate the procedure.

EXAMPLE 8.14

(a) A cantilever beam of span L, carries a concentrated load W on its free end as illustrated in Figure 8.17(a-i). Determine the shape of the beam of uniform strength if: (i) the depth is kept constant, and (ii) the width of section is kept constant.

(b) If the cantilever carries a uniformly distributed load of intensity w_0 on its entire span instead of a concentrated load on its free end, then determine the shapes of the beam of uniform strength for the cases (i) and (ii) in (a).

Solution: (a) When the cantilever supports a concentrated load W on its free end
The bending moment diagram is shown in Figure 8.17(a-i),

$$M_z = Wz$$

(i) If the depth of the beam d is kept constant and the width varies along the span, then the width b_z of the beam at a distance z from the free end A for uniform strength σ_b is obtained from,

$$\frac{b_z d^2}{6} \times \sigma_b = Wz \quad \text{or} \quad b_z = \frac{6W}{\sigma_b d^2} z = k_1 z$$

The variation of width along the span is shown in Figure 8.17(a-ii).

(ii) When the width of the beam b is kept constant and depth varies along the span, then the depth d_z of the beam at distance z from the free end A is given by

$$\frac{b d_z^2}{6} \times \sigma_b = Wz \quad \text{or} \quad d_z = \sqrt{\frac{6W}{\sigma_b b} z} = k_2 \sqrt{z}$$

The variation depth along the span is shown in Figure 8.17(a-iii).

(b) When the cantilever supports a uniformly distributed load of intensity w_0, the bending moment diagram is shown in Figure 8.17(b-i)

$$M_z = \frac{w_0 z^2}{2}$$

(i) Depth of the beam d is kept constant
The width b_z of the beam at distance z from the support A is given by

$$\frac{b_z d^2}{6} \times \sigma_b = \frac{w_0 z^2}{2} \quad \text{or} \quad b_z = \frac{3w_0}{\sigma_b d^2}(z^2) = k_3 z^2$$

The variation of width along the span is shown in Figure 8.17(b-ii).

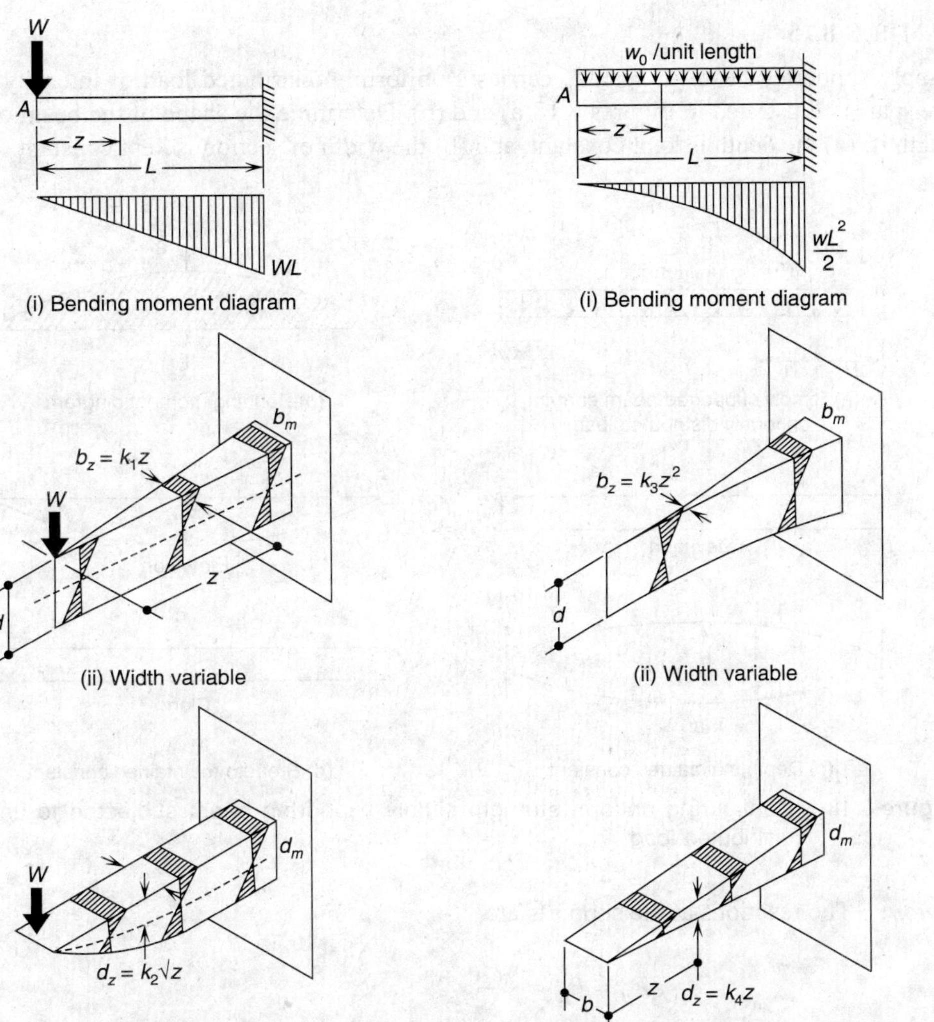

(a) Cantilever beam carrying concentrated load (b) Cantilever beam carrying distributed loading

Figure 8.17 Configuring the beam as per bending moment distribution (Example 8.14).

(ii) Width of the beam b is kept constant

The depth d_z of the beam at distance z from the free end of the cantilever is for uniform strength, σ_b is

$$\frac{bd_z^2}{6} \times \sigma_b = \frac{w_0}{2}(z^2) \quad \text{or} \quad d_z = \sqrt{\frac{3w_0}{\sigma_b b}}\, z = k_4 z$$

The variation depth along the span is shown in Figure 8.17(b-iii).

EXAMPLE 8.15

A simply supported beam of span L, carries a uniformly distributed load of intensity w_0 on its entire span as illustrated in Figures 8.18(a) and (b). Determine the shape of the beam of uniform strength if: (a) the depth is kept constant, and (b) the width of section is kept constant.

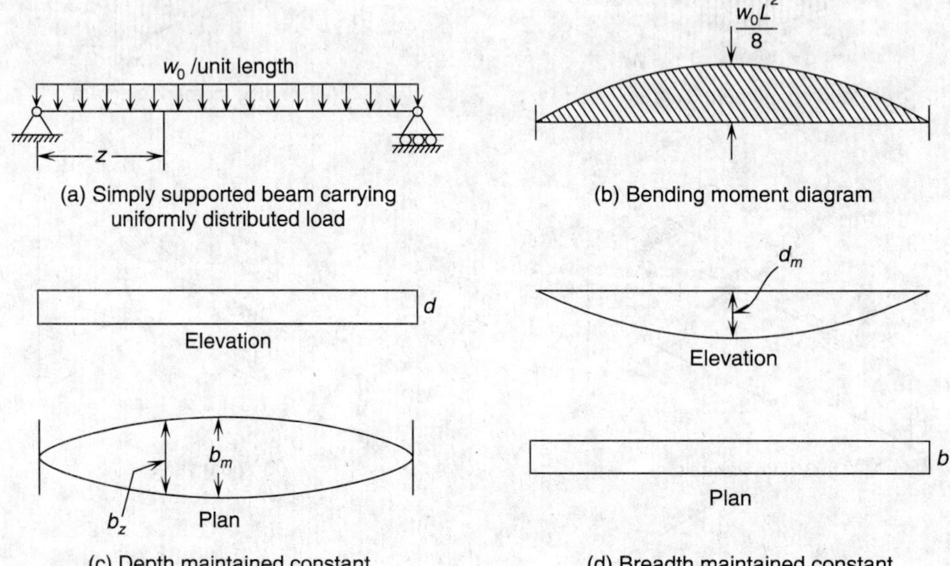

Figure 8.18 Configuring uniform strength simply supported beam subjected to uniformly distributed load.

Solution: The reactions at the supports are,

$$R_B = R_A = \frac{w_0 L}{2}$$

$$M_x = \left(\frac{w_0 L}{2}\right)z - \frac{w_0 z^2}{2} = \frac{w_0}{2}(Lz - z^2)$$

(a) When the depth of the beam d is kept constant and the width varies along the span.

Let the width of the beam at distance z from the support A is b_z. For uniform strength, σ_b is constant throughout. Therefore,

$$\frac{b_z d^2}{6} \times \sigma_b = \frac{w_0}{2}(Lz - z^2) \quad \text{or} \quad b_z = \frac{3w_0}{\sigma_b d^2}(Lz - z^2) \text{ i.e. } b_z \propto M_z$$

At the mid-span $z = L/2$, $\qquad b_m = \frac{3}{4} \times \left(\frac{w_0 L^2}{\sigma_b d^2}\right)$

Thus, $$b_z = \frac{4b_m}{L}\left[\frac{z}{L} - \left(\frac{z}{L}\right)^2\right]$$

The variation of width along the span is shown in Figure 8.18(c).

(b) When the width of the beam b is kept constant and depth varies along the span.

Let the depth of the beam at distance z from the support A is d_z. For uniform strength, σ_b is constant throughout. Therefore,

$$\frac{bd_z^2}{6} \times \sigma_b = \frac{w_0}{2}(Lz - z^2) \quad \text{or} \quad d_z = \sqrt{\frac{3w_0}{\sigma_b b}(Lz - z^2)} \quad \text{i.e.} \quad d_z \propto \sqrt{M_x}$$

At $z = L/2$, $$d_m = \frac{L}{2} \times \sqrt{\frac{3w_0}{\sigma_b b}}$$

Thus, $$d_z = 2d_m\left[\left(\frac{z}{L}\right) - \left(\frac{z}{L}\right)^2\right]^{1/2}$$

The variation depth along the span is shown in Figure 8.18(d).

EXAMPLE 8.16

A simply supported beam of rectangular cross-section carries a point load which can be placed anywhere on the span L. Determine the shape of the beam of uniform strength when the width of section b is maintained.

Solution: At a section at distance z from the support where the section is $b \times d_z$ the moment is

$$M_z = \frac{Wz(L - z)}{L}$$

For uniform strength, σ_b is constant throughout. Therefore,

$$\frac{bd_z^2}{6} \times \sigma_b = \frac{Wz(L - z)}{L} \quad \text{or} \quad d_z = \sqrt{\frac{6W}{\sigma_b b} z(L - z)}$$

At the mid-span $z = L/2$, $$d_m = L\sqrt{\frac{3w_0}{2\sigma_b b}}$$

Thus, $$d_z = \frac{2d_m}{L}\sqrt{z(L - z)}$$

It should be noted that the shapes derived in above three examples are bending moment and cross-sectional geometry specific. The effects of other important parameters have not been considered and are applicable to rectangular cross-sections only.

8.8 FLITCHED BEAMS

These are beams of compound or *composite sections* generally consisting of wooden beams strengthened by mild steel plates as shown in Figures 8.19(a), (b), (c), (d) and (e). The steel and wooden components are firmly secured throughout their length such that there is no slippages between them, i.e. two components act as integral unit and strain in two materials at a level is same. In other words at a level under consideration the stress in steel is m times that in wood where m $(= E_s/E_t)$ is the *modular ratio* between steel and wood.

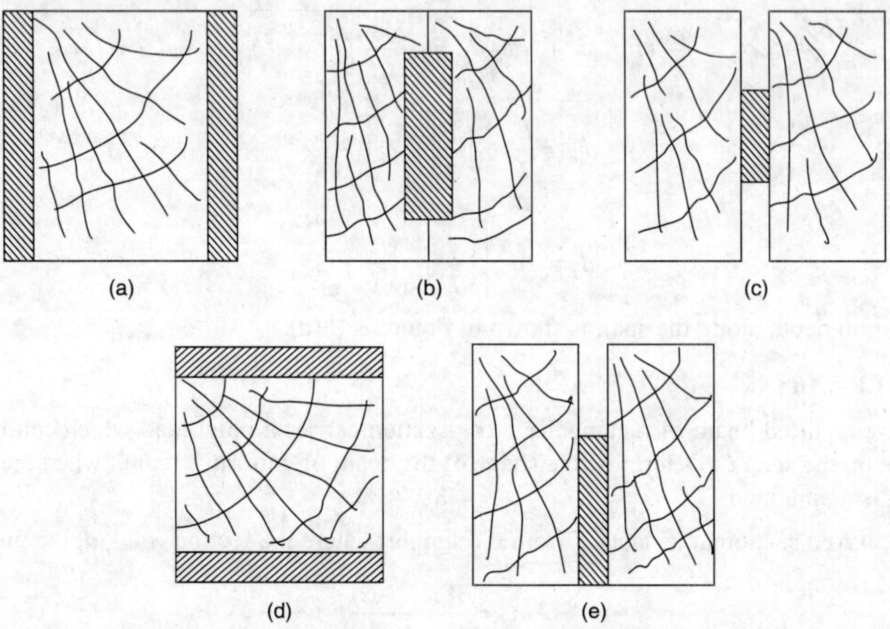

Figure 8.19 Typical composite sections of flitched beams.

The moment of resistance of the *compound beam* is determined by considering the equivalent wooden or steel section or alternatively moment of resistance of individual components may be computed and total *moment resisting capacity* is determined. The following examples illustrate the method for computation of moment of resistance of *flitched beams*.

EXAMPLE 8.17

A timber beam of size 150 mm × 80 mm deep is strengthened by bolting securely a 150 mm × 55 mm rolled steel channel as shown in Figure 8.20(a). Determine the maximum uniformly distributed load, the composite beam can support safely over a simply supported span of 3 m without exceeding bending stresses of 7.0 and 130 MPa in timber and steel, respectively. The elastic moduli for steel and timber are 200 and 10 GPa, respectively. The cross-sectional properties of channel section are:

$$A = 1265 \text{ mm}^2, I_x = 4.711 \times 10^6 \text{ mm}^4 \text{ and } I_y = 3.79 \times 10^5 \text{ mm}^4$$

The distance of centroid of channel from its back is 16.6 mm.

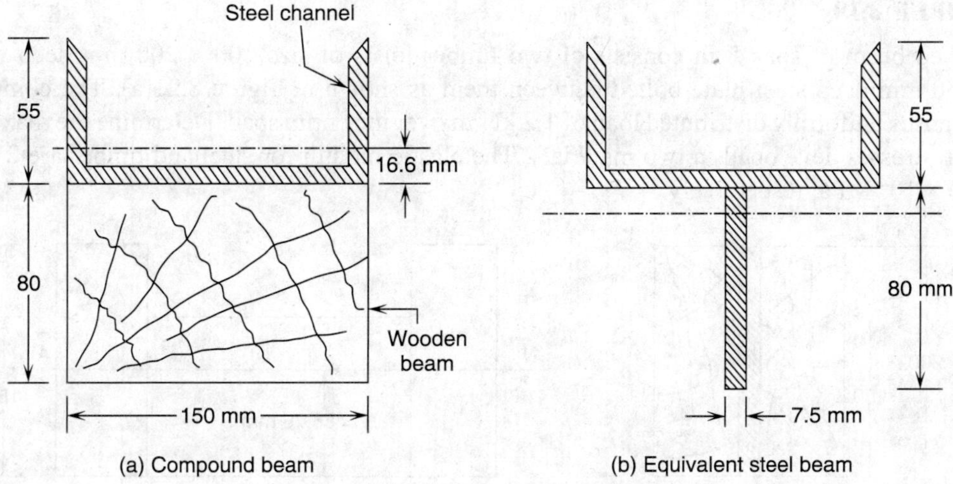

(a) Compound beam (b) Equivalent steel beam

Figure 8.20 Compound beam of Example 8.17.

Solution: The modular ratio of the two materials is,

$$m = \frac{E_s}{E_t} = \frac{200}{10} = 20$$

Therefore, the equivalent width of the wooden beam in terms of steel,

$$b_t = \frac{150}{20} = 7.5 \text{ mm}$$

For the equivalent steel beam (shown in Figure 8.20(b)), the distance of centroid of section from the bottom,

$$\bar{y} = \frac{(80 \times 7.5) \times 40 + (1265) \times (80 + 16.6)}{(80 \times 7.5) + 1265} = 78.4 \text{ mm}$$

$$I_x = 3.79 \times 10^5 + 1265 \times (96.6 - 78.4)^2 + \frac{7.5}{3} \times [(80 - 78.4)^3 + (78.4)^3]$$

$$= 2.0027 \times 10^6 \text{ mm}^4$$

The minimum (the safe) moment resisting capacity of the section in terms of allowable stresses is given by,

$$M_{r,b} = m \times \left(\frac{\sigma_{b,t,a} I_x}{y_b}\right) = 20 \times \left[\frac{7.0 \times (2.0027 \times 10^6)}{78.4}\right]$$

Equate this to the bending moment,

$$M_{\max} = M_{r,b} \text{ or } \frac{(w_0 \times 3.0^2) \times 10^6}{8} = 20 \times \frac{7.0 \times (2.0027 \times 10^6)}{78.4}$$

Therefore, $w_0 = 3.179$ kN/m

EXAMPLE 8.18

A flitched beam of span 8 m consists of two timber joists of size 100 × 200 mm deep with a 15 × 150 mm deep steel plate bolted between them as shown in Figure 8.21(a). The composite beam carries uniformly distributed load of 1.2 kN/m over the entire span. Determine the maximum bending stresses developed in two materials. The elastic moduli for steel and timber are 2×10^5 and 1.0×10^4 MPa, respectively.

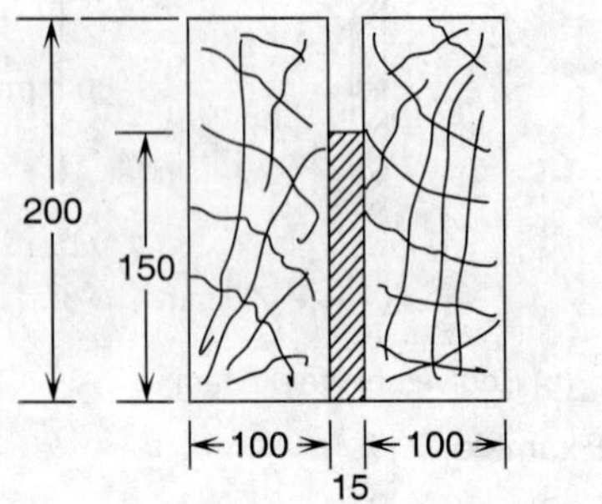

(a) Compound section of flitched beam

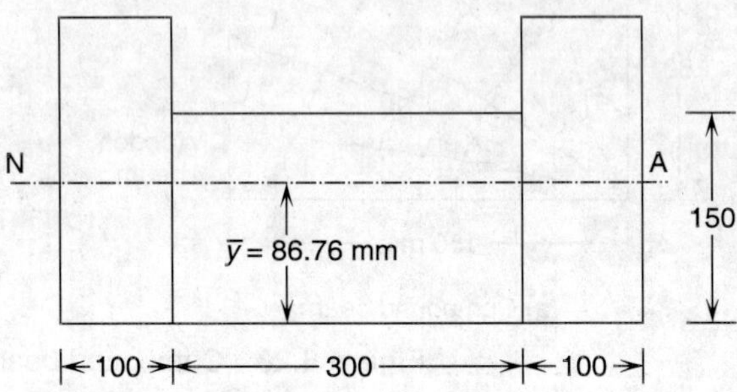

(b) Equivent timber section

Figure 8.21 Analysis of flitched beam of Example 8.18.

Solution: The modular ratio of the two materials is,

$$m = \frac{E_s}{E_t} = \frac{2.0 \times 10^5}{1.0 \times 10^4} = 20$$

The maximum bending moment acting on the beam is,

$$M_{\max} = \frac{1.2 \times 8^2}{8} = 9.6 \text{ kN.m}$$

The equivalent timber section is shown in Figure 8.21(b). The steel plate is replaced by an equivalent timber section of width 20 × 15 = 300 mm and depth 150 mm.

The height of neutral axis,

$$\bar{y} = \frac{2 \times (100 \times 200) \times 100 + (300 \times 150) \times 75}{2 \times (100 \times 200) + (300 \times 150)} = 86.76 \text{ mm}$$

$$I = 2 \times \frac{1}{3} \times (100 \times 113.24^3) + 2 \times \frac{1}{3} \times (100 \times 86.76^3) + \frac{1}{3} \times (300 \times 63.24^3)$$

$$+ \frac{1}{3} \times (300 \times 86.76^3)$$

$$= 230.9436 \times 10^6 \text{ mm}^4$$

Extreme stress in timber,

$$\sigma_{t,\text{timber}} = \frac{(9.6 \times 10^6) \times 86.76}{230.9436 \times 10^6} = 3.606 \text{ MPa}$$

$$\sigma_{c,\text{timber}} = \frac{113.24}{86.76} \times 3.606 = 4.707 \text{ MPa}$$

Extreme stress in steel,

$$\sigma_{t,\text{steel}} = 20 \times 3.606 = 72.12 \text{ MPa}$$

$$\sigma_{\text{c,steel}} = 20 \times \frac{63.24}{113.24} \times 4.707 = 52.57 \text{ MPa}$$

EXAMPLE 8.19

A flitched beam consists of two timber joists of rectangular section each having breadth B and depth D with a steel plate having breadth (i.e. thickness) b and depth d bolted between them such that the whole section is symmetrical about the horizontal axis as shown in Figure 8.22(a). The allowable stresses in steel and timber are 125 and 7 MPa, respectively, which reach simultaneously. Determine the ratios D/d and B/b in order that the moment of resistance of timber alone shall be equal to that of the steel alone. The elastic moduli for steel and timber are 200 and 10 GPa, respectively.

If the timber joists are 250 mm deep and 75 mm wide, determine the size of steel plate.

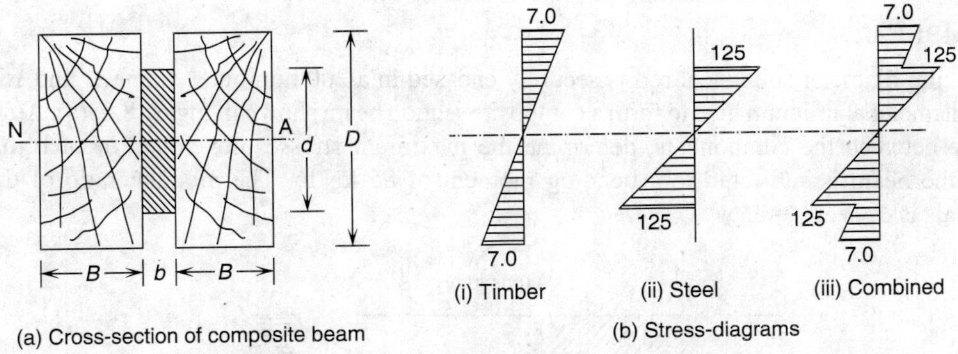

Figure 8.22 Analysis of composite beam of Example 8.19.

Solution: The modular ratio of the two materials is,

$$m = \frac{E_s}{E_t} = \frac{2.0 \times 10^5}{1.0 \times 10^4} = 20$$

For the given materials the allowable stresses, $\sigma_{\text{timber}} = 7$ MPa and $\sigma_{\text{steel}} = 125$ MPa. The stress in steel in terms of stress in timber is given by

$$\sigma_s = \left(\frac{E_s}{E_t}\right)\left(\frac{d}{D}\sigma_t\right) \quad \text{or} \quad \frac{D}{d} = \left(\frac{E_s}{E_t}\right)\left(\frac{\sigma_t}{\sigma_s}\right) = 20 \times \frac{7}{125} = \frac{28}{25}$$

Figure 8.22(b) shows the stress diagram.

The moments of resistance of timber and steel components are,

$$M_{r,\text{timber}} = 2 \times \frac{1}{6} \times 7 \times (B \times D^2) \quad \text{and} \quad M_{r,\text{steel}} = \frac{1}{6} \times 125 \times (b \times d^2)$$

If the moments of resistance of timber and steel components are equal,

$$2 \times \frac{1}{6} \times 7 \times (B \times D^2) = \frac{1}{6} \times 125 \times (b \times d^2) \quad \text{or} \quad \frac{B}{b} = \frac{125}{14} \times \left(\frac{d}{D}\right)^2$$

Therefore, $$\frac{B}{b} = \frac{125}{14} \times \left(\frac{25}{28}\right)^2 = 7.12$$

For the case $D = 250$ mm and $B = 75$ mm,

$$d = \frac{25}{28} \times D = \frac{25}{28} \times 250 = 232.2 \text{ mm}$$

and $$b = \frac{B}{7.12} = 10.5 \text{ mm.}$$

EXAMPLE 8.20

A 150 mm diameter solid steel rod is securely encased in a 200 mm outer diameter and 150 mm inner diameter aluminium tube to form a composite section beam shown in Figure 8.23(a). Assuming no slip between the components, determine the maximum stresses developed in each material when the beam is subjected to a bending moment of 85 kN.m. The modular ratio of the two materials is 3, i.e. $E_s = 3E_a$.

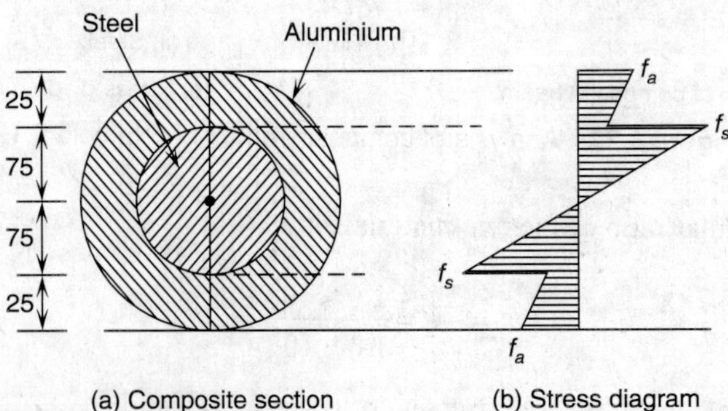

Figure 8.23 Composite beam of Example 8.20 (All dimensions in mm).

Solution: If σ_s is the maximum stress developed in steel, the stress induced in aluminium at a distance 75 mm from neutral axis,

$$\sigma_a' = \frac{\sigma_s}{m} = \frac{\sigma_s}{3}$$

Therefore, maximum stress in aluminium,

$$\sigma_a = \frac{100}{75} \times \frac{\sigma_s}{3} = \frac{4\sigma_s}{9}$$

The moments of resistance of steel and aluminium components are,

$$M_{r,s} = \sigma_s Z_s = \sigma_s \times \left(\frac{\pi \times 150^3}{32}\right) = 331340\sigma_s$$

$$M_{r,a} = \sigma_a Z_a = \frac{4\sigma_s}{9} \times \frac{\pi}{32}\left(\frac{200^4 - 150^4}{200}\right) = 238619\sigma_s$$

Total moment of resistance, $M_r = M_{r,s} + M_{r,a} = 569959\sigma_s$

Equate total moment of resistance to the applied bending moment,

$$569959\sigma_s = 85 \times 10^6 \quad \text{or} \quad \sigma_s = 149.13 \text{ MPa}$$

and $$\sigma_a = \frac{4}{9} \times 149.13 = 66.28 \text{ MPa}$$

8.9 PROBLEMS

8.1 A wire of diameter d is wound round a cylinder of diameter D. Determine the bending stresses developed in the wire.

$$\left[\textbf{Ans. } \sigma_{b,\max} = E\left(\frac{d}{D}\right)\right]$$

8.2 An inclined wooden beam shown in Figure 8.24 carries a 25 kN concentrated load at the middle of its span. Determine the maximum compressive stress developed in the beam.

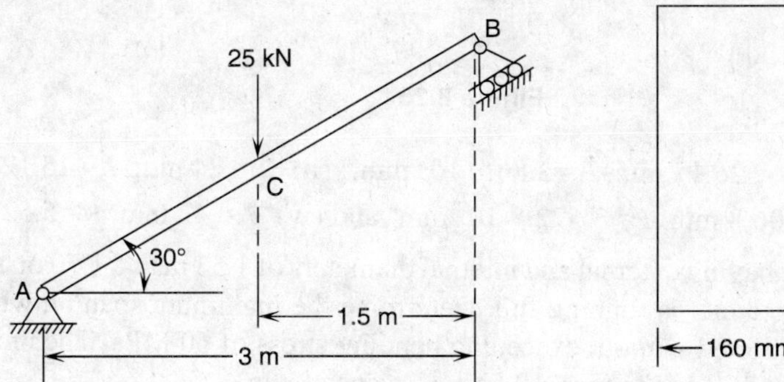

Figure 8.24

[**Ans.** $M_{\max}$ = 18.75 kN.m; $\sigma_{d,c}$ = 0.2604 MPa, $\sigma_{b,c}$ = 7.8125 MPa and $\sigma_{\max,c}$ = 8.0729 MPa]

8.3 Determine the moment of inertia I and section modulus Z for the special cross-sections produced by welding separate components as shown in Figure 8.25.

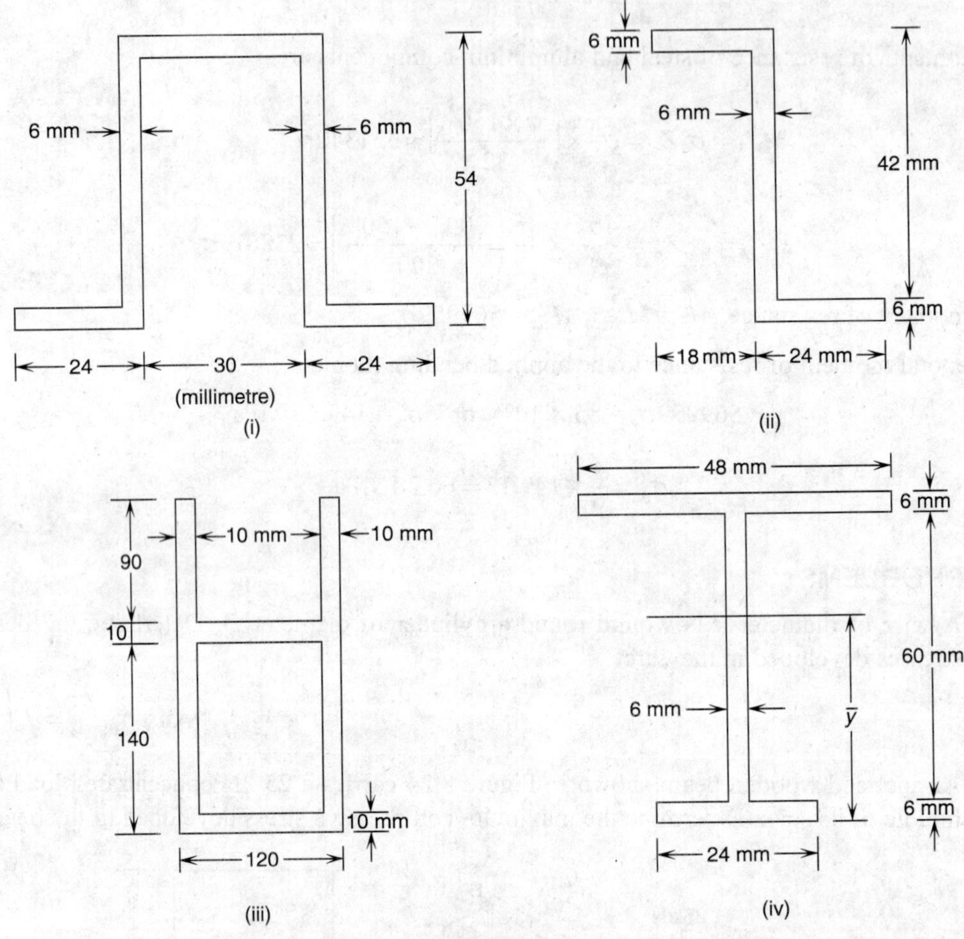

Figure 8.25

[**Ans.** (i) $\bar{y}$ = 26.17 mm, I = 3.86 × 10^5 mm^4; (ii) $\bar{y}$ = 24 mm, I = 15.12 × 10^4 mm^4; (iii) $\bar{y}$ = 106.8 mm, I = 3.572 × 10^7 mm^4 and (iv) $\bar{y}$ = 42 mm, I = 5.52 × 10^5 mm^4]

8.4 A main water pipe having external and internal diameters of 1.53 m and 1.50 m, respectively, supplying water to a city is running full. Determine the maximum span on which the pipe may be freely supported without exceeding bending stress of 60 MPa. The unit weights of steel and water are 76.4 kN/m^3 and 10 kN/m^3, respectively.

[**Ans.** I = 20485 × 10^6 mm^4; w_0 = 23.07 kN/m; and L = 23.6 m]

8.5 A simply supported beam of unsymmetrical I-section having a cross-section illustrated in Figure 8.26 carries a uniformly distributed load over a span of 5 m. Determine the maximum load the beam can support safely if the allowable tensile stress, $\sigma_{b,t}$ = 175 MPa.

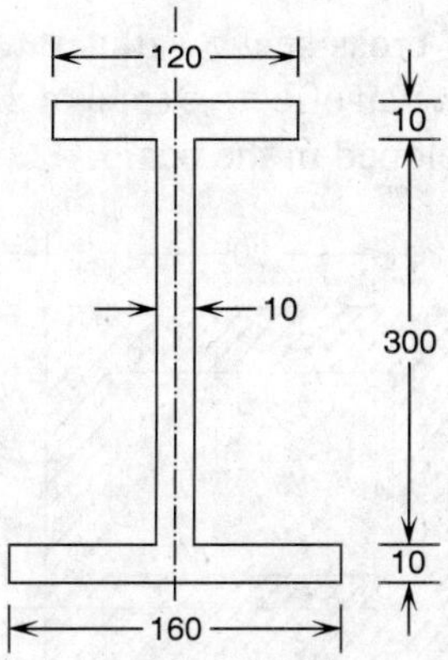

Figure 8.26

[**Ans.** $y_c = 170.7$ mm, $y_t = 149.3$ mm, $I = 8.913 \times 10^7$ mm^4 and $w_0 = 33.43$ kN/m]

8.6 A reinforced concrete aqueduct of channel cross-section shown in Figure 8.27 carries water over a simple span of 8 m. Determine the maximum bending stresses developed at the mid-span. The unit weights of concrete and water are 25 kN/m^3 and 10 kN/m^3, respectively.

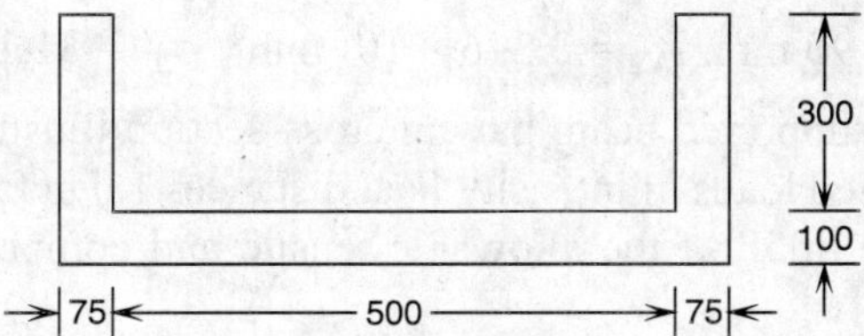

Figure 8.27

[**Ans.** $y_t = 132$ mm, $M_{max} = 34$ kNm, $I = 1460 \times 10^6$ mm^4 and $w_0 = 4.25$ kN/m;
$\sigma_{b,c} = 6.26$ MPa and $\sigma_{b,t} = 3.08$ MPa]

8.7 A right-angled bend of square cross-section of size 50 × 50 mm shown in Figure 8.28 supports a uniformly distributed load of 0.3 kN/unit horizontal length over a span of 5 m. Determine the maximum bending stress developed at the section X-X.

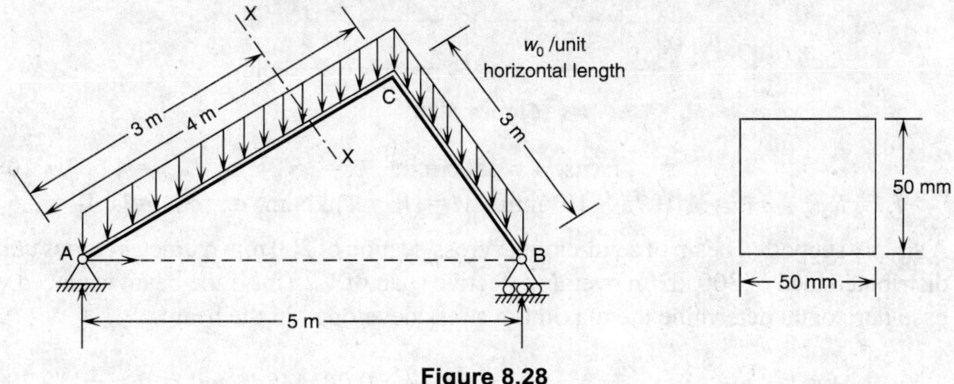

Figure 8.28

[**Ans.** $\sigma_b = 45.0$ MPa]

8.8 A simply supported beam of cross-section illustrated in Figure 8.29, carries uniformly distributed load of 30 kN/m over an effective span of 2.0 m. Determine the maximum tensile and compressive stresses developed in the beam.

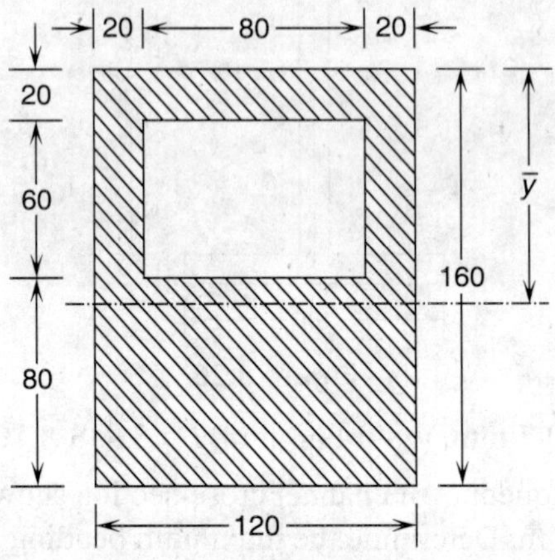

Figure 8.29

[**Ans.** $\overline{y}$ = 90 mm, I_{NA} = 3.376 × 10^7 mm^4, $\sigma_{b,t}$ = 31.1 MPa and $\sigma_{b,c}$ = 40.0 MPa]

8.9 A 3 m long simply supported beam having cross-section illustrated in Figure 8.30, carries two equal concentrated loads of intensity W at distances 1.0 m from the supports. Determine the maximum value of W, if the allowable tensile and compressive stresses are 120 and 100 GPa, respectively.

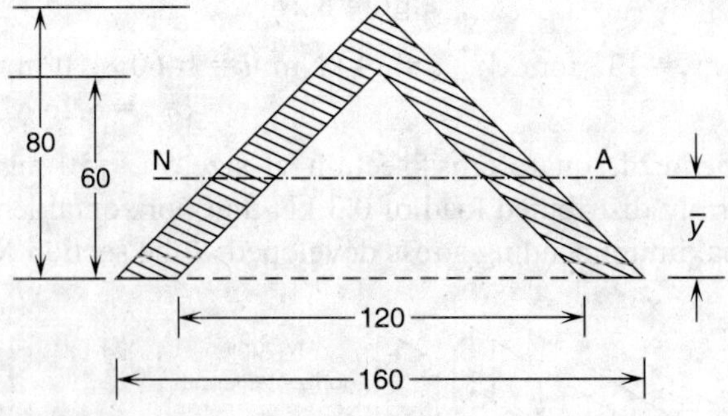

Figure 8.30

[**Ans.** A = 2800 mm^2, $\overline{y}$ = 35.2 mm, I_{base} = 4.667 × 10^6 mm^4, $I_{xx} = I_{\text{base}} - A\overline{y}^2$ = 1.1975 × 10^6 mm^4, M = (W × 1) kN.m, $\sigma_{b,c}$ controls; W = 2.673 kN]

8.10 A simply supported beam of semi-circular cross-section of 200 mm diameter carries uniformly distributed load of 300 kN/m over an effective span of 1.25 m. If the beam is placed with its base horizontal determine the maximum stress developed in the beam.

$$\left[\textbf{Ans. } I = \frac{d^4}{18\pi}, \overline{y} = \frac{4r}{3\pi}, y_{\max} = r - \frac{4r}{3\pi}, Z = 0.061445d^3 \text{ and } \sigma_{b,\max} = 119.20 \text{ MPa}\right]$$

8.11 Determine the size of the strongest rectangular beam that can be cut from a cylindrical wood log of diameter D shown in Figure 8.31.

[**Hint:** For the rectangular beam ABCD with OD at an angle θ with horizontal diameter, the size of the beam is $b = 2r\cos\theta$, and $h = 2r\sin\theta$. Therefore, $Z = \dfrac{4r^3\cos\theta\sin^2\theta}{3}$, for strongest beam, $\dfrac{dZ}{d\theta} = 0 = \sin\theta(2\cos^2\theta - \sin^2\theta)$ or $\tan\theta = \sqrt{2}$, $b = \dfrac{D}{\sqrt{3}}$ and $h = \dfrac{\sqrt{2}D}{\sqrt{3}}$]

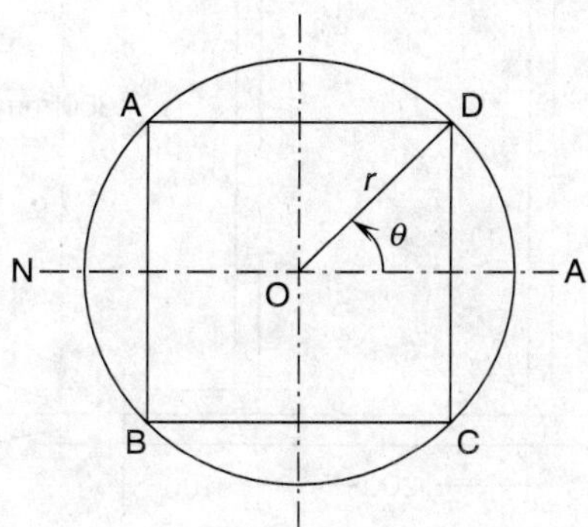

Figure 8.31

8.12 A uniformly tapering vertical post of height H having diameters D_2 and D_1 at the top and the base, respectively, is fixed at the base as shown in Figure 8.32. Determine the magnitude and location of the maximum bending stress developed in the post due to horizontal force P applied at the top of the post.

[**Hint.** At the section X–X: $D_x = D_2 + kx$

where $k = \dfrac{D_1 - D_2}{H}$; $Px = \sigma_b\left(\dfrac{\pi D_x^3}{32}\right)$ or

$$\sigma_b = \frac{32P}{\pi}\left[\frac{x}{(D_2 + kx)^3}\right];$$

for σ_b to be maximum, $\dfrac{\partial\sigma_b}{\partial x} = 0$, i.e. $x = \dfrac{D_2}{2k}$.

Therefore, $\sigma_{b,\max} = \dfrac{128}{27\pi}\left[\dfrac{PH}{(D_1 - D_2)D_2^2}\right]$]

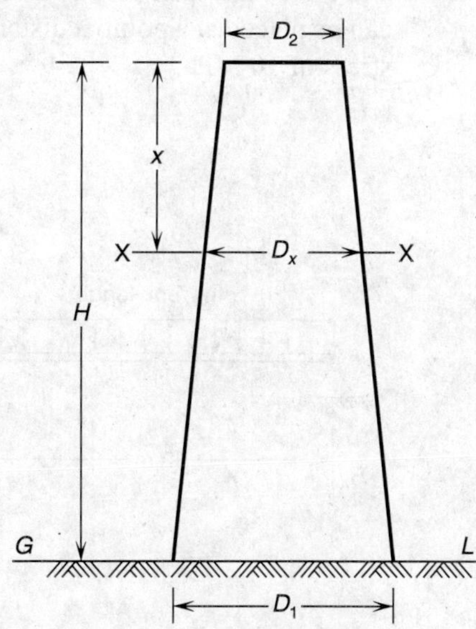

Figure 8.32

8.13 A beam of built-up cross-section illustrated in Figure 8.33 is to be used over an opening of 7.5 m. Determine the section modulus when the beam is placed with the web of the section: (a) vertical and (b) horizontal.

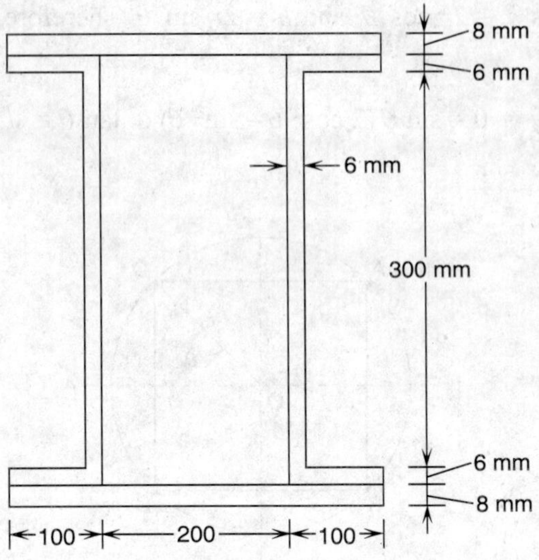

Figure 8.33

[**Ans.** $Z_v = 6.75 \times 10^5$ mm³ and $Z_h = 1.506 \times 10^6$ mm³]

8.14 A built-up box beam of timber consists of four planks each 50 × 200 mm in size, securely spiked together as shown in Figure 8.34. The beam supports a uniformly distributed load of 4 kN/m over a span of 4 m. Determine the maximum additional concentrated load W which can be placed at a point at distance 3 m from the left support without exceeding the bending stress of 10 MPa.

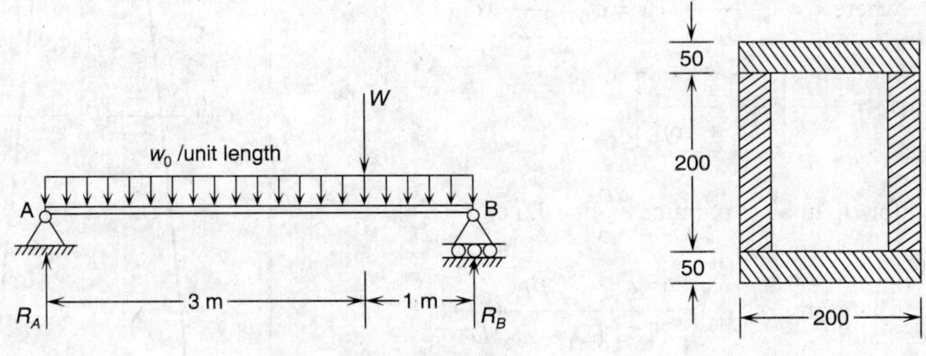

Figure 8.34

[**Ans.** $I = \{(23 \times 10^8)/6\}$ mm⁴ and $W = 26.075$ kN]

8.15 Beams of approximately equal cross-sectional area and of same material but of different shapes, rectangular and I-section are considered for the use in an application that has to resist a maximum moment of M_{max}. The I-section is 300 mm deep with flanges 150 mm wide and 25 mm thick, and web 12 mm thick. The rectangular section has a depth equal to twice its width. Compare the flexural strength of two beams.

[**Ans.** The ratio of flexural strength of I-beam to that of rectangular beam = 8.58]

8.16 A 200 mm wide rectangular timber beam used as a railway sleeper supported by a uniformly distributed reaction, carries two uniformly distributed loads each 50 kN as shown in Figure 8.35. Determine the depth of the section if the maximum stress is limited to 8 MPa.

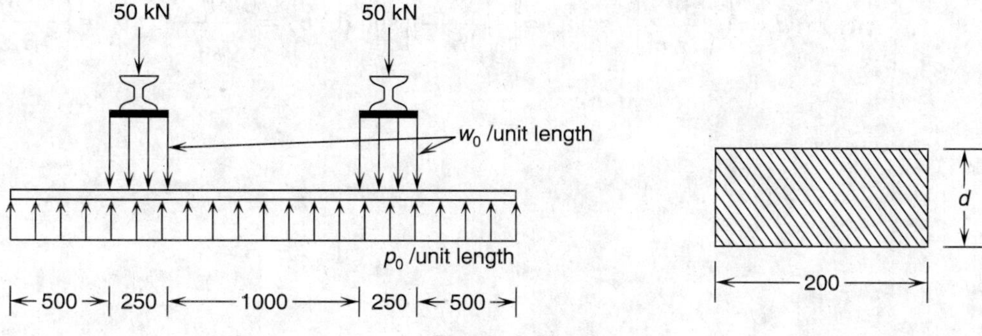

Figure 8.35

[**Ans.** M_{max} = 6.25 kN.m and d = 153.09 mm]

8.17 A simply supported beam carries a uniformly distributed load of 16 kN/m over an effective span of 10 m. Select the lightest wide flange beam so that maximum stress in the beam does not exceed 120 MPa. Determine the actual maximum stress developed in the beam.

[**Ans.** $W610 \times 82$; $\sigma_b = 113$ MPa]

8.18 A timber beam of size 150 × 200 mm deep is strengthened by bolting securely two steel flitches each 12.5 × 150 mm deep in section. Determine moment of resistance when: (a) the flitches are attached symmetrically at top and bottom; and (b) the flitches are attached symmetrically at the sides. The allowable bending stress in timber is 6.0 MPa. The elastic moduli for steel and timber are, respectively. What is the maximum stress in steel in each case?

[**Ans.** (a) M_r = 56.86 kN.m and (b) M_r = 14.4375 kN.m]

8.19 A timber beam of size 250 × 400 mm deep is reinforced by bolting securely a steel plate 250 mm wide and 10 mm thick to its lower face so that the overall section is 250 × 410 mm. Determine the maximum stresses developed in timber and steel when the beam section is subjected to a sagging bending moment of 75 kN.m. The elastic moduli for steel and timber are 200 and 10 GPa, respectively.

[**Hint.** For simplicity replace timber component by an equivalent steel component of width 250/20 = 12.5 mm resulting in inverted T-section; $\overline{y}$ from the bottom edge = 141.7 mm; I_e = 136729175 mm^4; $\sigma_{t,\text{steel}}$ =77.73 MPa; $\sigma_{c,\text{steel}}$ = 147.17 MPa and $\sigma_{c,\text{timber}}$ = 147.17/20 = 7.36 MPa]

8.20 A 25 mm diameter solid bronze bar is securely fitted in a steel tube of 35 mm external diameter to form a composite section beam. Assuming no slip between the components, determine the moment of resistance of composite section. The maximum stresses in bronze and steel are limited to 90 MPa and 120 MPa, respectively. The Young's modulus for steel is 1.75 times that of bronze, i.e. $E_s = 1.75E_b$.

[**Ans.** $M_r = 0.448759$ N.m]

8.21 A simply supported beam of rectangular cross-section carries a point load at the mid-span L. Determine the shape of the beam of uniform strength when: (a) the breadth b of section is maintained constant and (b) the depth of section d is maintained constant.

$$\left[\textbf{Ans.}\text{ (a) } d_x = \left(\sqrt{\frac{3W}{\sigma_b b}}\right)\sqrt{x}\,;\ d_m = \left(\sqrt{\frac{3WL}{2\sigma_b b}}\right) \text{ and (b) } b_x = \left(\frac{3W}{\sigma_b d^2}\right)x\,;\ b_m = \left(\frac{3W}{2\sigma_b d^2}\right)\right]$$

Chapter 9

Shearing Stresses

9.1 INTRODUCTION

This chapter deals with the elastic analysis of shearing stresses caused in beams due to transverse shear force. The expression relating the shear force and cross-sectional area of a beam to the shearing stress has been developed. The related problem of built-up beam wherein separate longitudinal parts of a beam are attached by means of bolts, gluing, or welding are also discussed.

As discussed earlier in Chapter 2, the equilibrium of vertical shear force V (see Figure 9.1(b)) acting on a cross-section of a loaded beam of Figure 9.1(a) is established through the development of *vertical shearing stresses* τ_v in the beam (see Figure 9.1(c)). The resulting force due to these stresses $V_R = \int_A \tau_v \, dA$, is equal in magnitude but opposite in sense to external shear force V as shown in Figure 9.1(c). Since the shear force in loaded beam varies along the length of the beam, there is corresponding variation in the shearing stresses developed in the beam.

It may be recalled that the shear force is essentially linked with the change in the bending moment at the adjoining sections throughout the beam. Thus, if a shear and a bending moment are present at one section, a different bending moment will exist at an adjoining section, although shear may remain constant as shown in Figure 9.2(b). This variation of moment and hence in bending stresses along the beam results in development of shear stresses on the imaginary longitudinal planes (horizontal plane) that are parallel to the axis of the member. Thus, in addition to vertical shear stresses, *horizontal shear stresses* are also created in the beam.

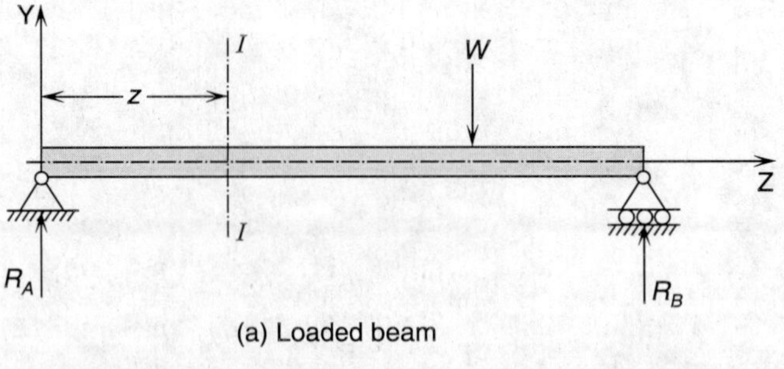

(a) Loaded beam

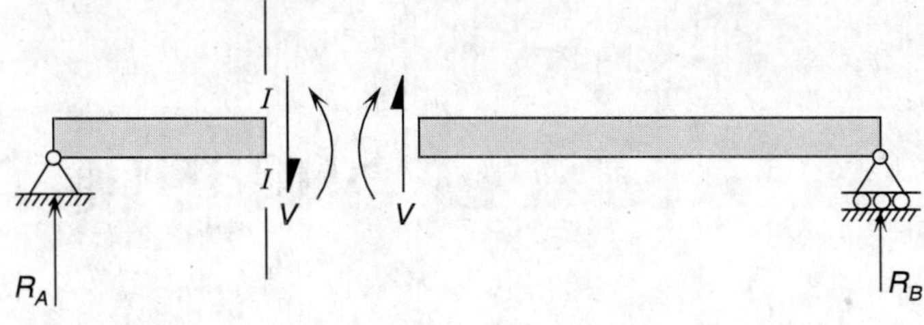

(b) Free-body diagrams of beam parts

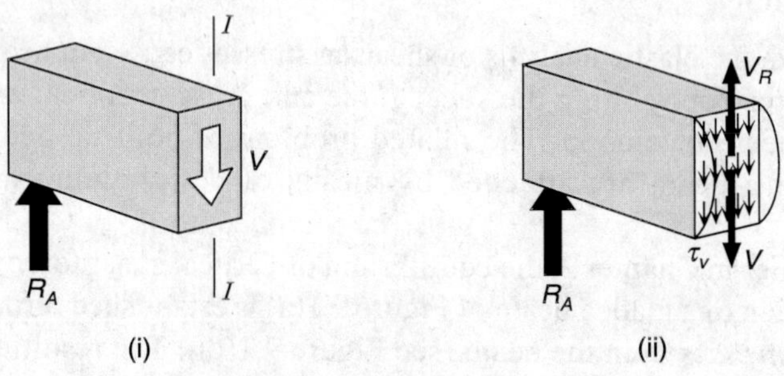

(c) Load balance mechanism

Figure 9.1 Basic shear-carrying mechanism.

As explained in section 9.2 the magnitude of these horizontal shear stresses can be determined by considering the equilibrium in the horizontal direction of an element of the beam. For rotational equilibrium of the element, the horizontal shearing stress is always equal to the vertical shearing stress. Thus, a horizontal shearing stress is always accompanied by an equal vertical shearing stress and the distribution of vertical shearing stress will be the same as that for the horizontal shearing stress. For the beam of Figure 9.2(a), the variations of the horizontal and vertical shearing stresses on a cross-section are as shown in Figure 9.2(c).

Since at a point, equal shearing stresses exist on the mutually perpendicular planes, the shearing stresses whose direction is coincident with the shear force at a section will be determined.

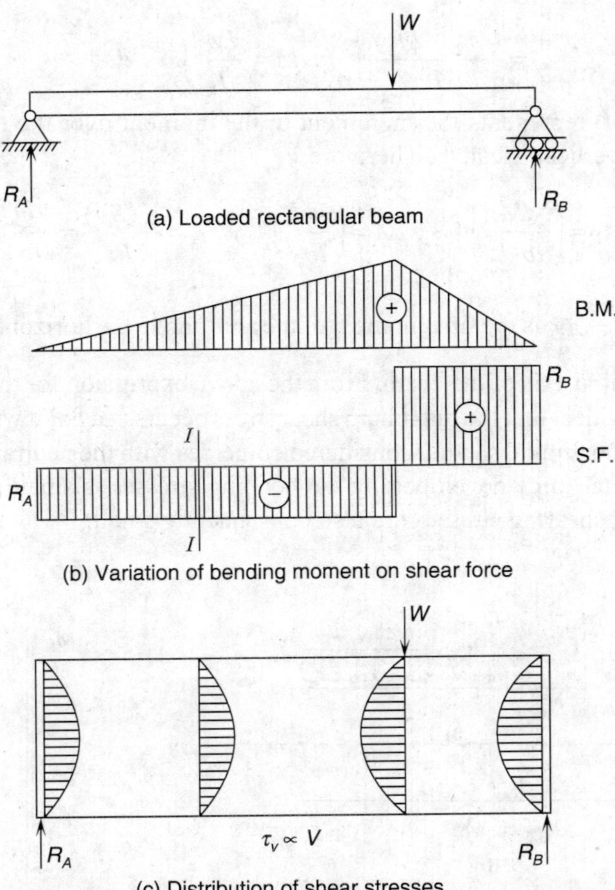

Figure 9.2 Distribution of shear and shear stresses.

9.2 SHEARING STRESS FORMULA

Consider two adjacent sections *I–I* and *II–II* in a loaded beam separated by infinitesimal distance *dx* as shown in Figure 9.3(a), and treat the part between the sections as an isolated free body as shown in Figure 9.3(b). Due to variation of bending moment along the length of a beam, the bending stress at one section, say *II–II*, will be larger than at the other. Therefore, the resultant horizontal thrust F_2 caused by the compressive force on the section *II–II* will be greater than the resultant horizontal thrust F_1 on the section *I–I*. For equilibrium in the horizontal direction, horizontal shear force dF must develop to balance the difference in forces produced by bending action $F_2–F_1$ as shown in Figures 9.3(c) and (d). If τ_h represents the average shears over the differential area of length dz and width b, then the equilibrium in horizontal direction yields,

$$dF = \tau_h (b\,dz) = F_2 - F_1 = \int_{y_1}^{y_t} \left(\frac{M_2 y}{I}\right) dA - \int_{y_1}^{y_t} \left(\frac{M_1 y}{I}\right) dA$$

$$= \frac{(M_2 - M_1)}{I} \int_{y_1}^{y_t} y \, dA = \frac{dM}{I} \int_{y_1}^{y_t} y \, dA$$

where $M_2 - M_1$ (= dM) represents the increment in the moment over the distance dx and thus, dM/dz represents the vertical shear, V. Therefore,

$$\tau_h = \left(\frac{dM}{Ib \, dz} \right) \int_{y_1}^{y_t} y \, dA = \left(\frac{V}{Ib} \right) \int_{y_1}^{y_t} y \, dA = \frac{V (A\bar{y})}{Ib} = \frac{VQ}{Ib} \tag{9.1}$$

The term $Q = \int_{y_1}^{y_t} y \, dA = A\bar{y}$ is the first moment of the area above the horizontal section considered with respect to the neutral axis of the beam. From the above expression for the horizontal shearing stress in a beam, it is evident that the maximum shear stress occurs at a point where $A\bar{y}$ is maximum. This occurs when the horizontal section considered coincides with the neutral axis. Near the top of the beam, the horizontal force developed by the bending stresses is small since they act over a small area. Horizontal shearing stresses are also consequently small.

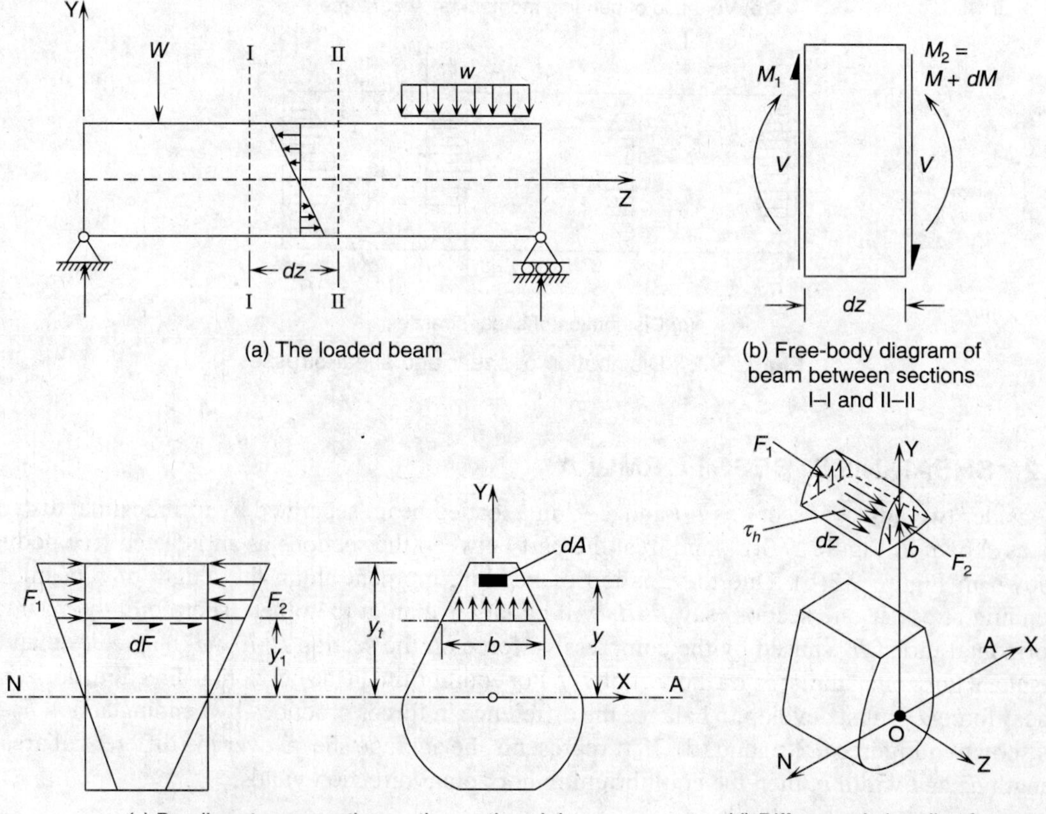

Figure 9.3 Development of horizontal shear in a loaded beam.

A horizontal shearing stress is always accompanied by an equal vertical shearing stress. To obtain relationship between τ_h and τ_v, consider the free-body diagram of a typical element *dz–dx–dy* in a beam subjected only to shear stresses as shown in Figure 9.4(a). For equilibrium of this element, the shearing force τ_h on the bottom face requires an equal balancing shearing stress

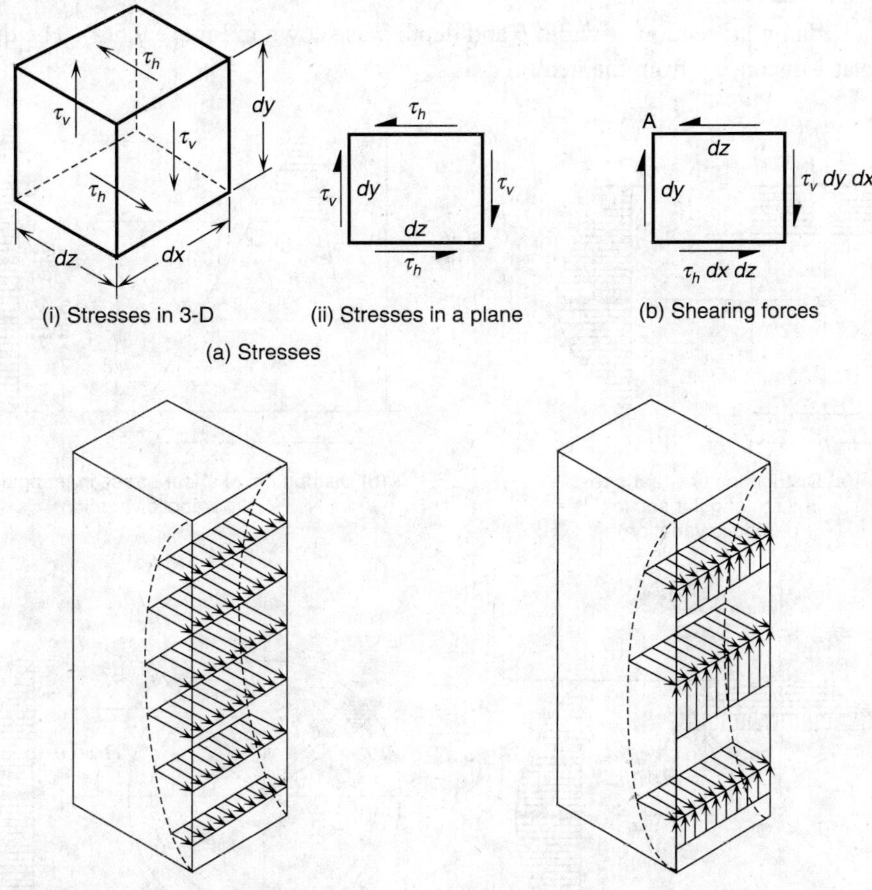

Figure 9.4 Magnitudes of the horizontal and vertical shear stressing at any point in a beam.

on the top face. These horizontal shear stresses acting on the upper and lower faces of the element produce a *rotational moment* which must be balanced by a rotational moment due to vertical shear stresses acting on the side (vertical) faces of the element (see Figure 9.4(b)). Thus,

$$\sum M = \tau_h \,(dx\; dz)\; dy - \tau_v \,(dy\; dx)\; dz = 0 \quad \text{or} \quad \tau_h = \tau_v \tag{9.2}$$

Hence, the horizontal and vertical shear stresses must be equal at the point as illustrated in Figure 9.4(d). The distribution of horizontal and vertical shear-stresses in a rectangular cross-section are shown in Figures 9.4(c) and (d).

9.3 SHEARING STRESS DISTRIBUTIONS

Shearing stress distributions in some of the commonly used cross-sections are derived in this section.

9.3.1 Rectangular Section

Consider a rectangular section of width b and depth h as shown in Figure 9.5(a). The quantity $A\overline{y}$ for a layer at a distance y from the neutral axis is given by,

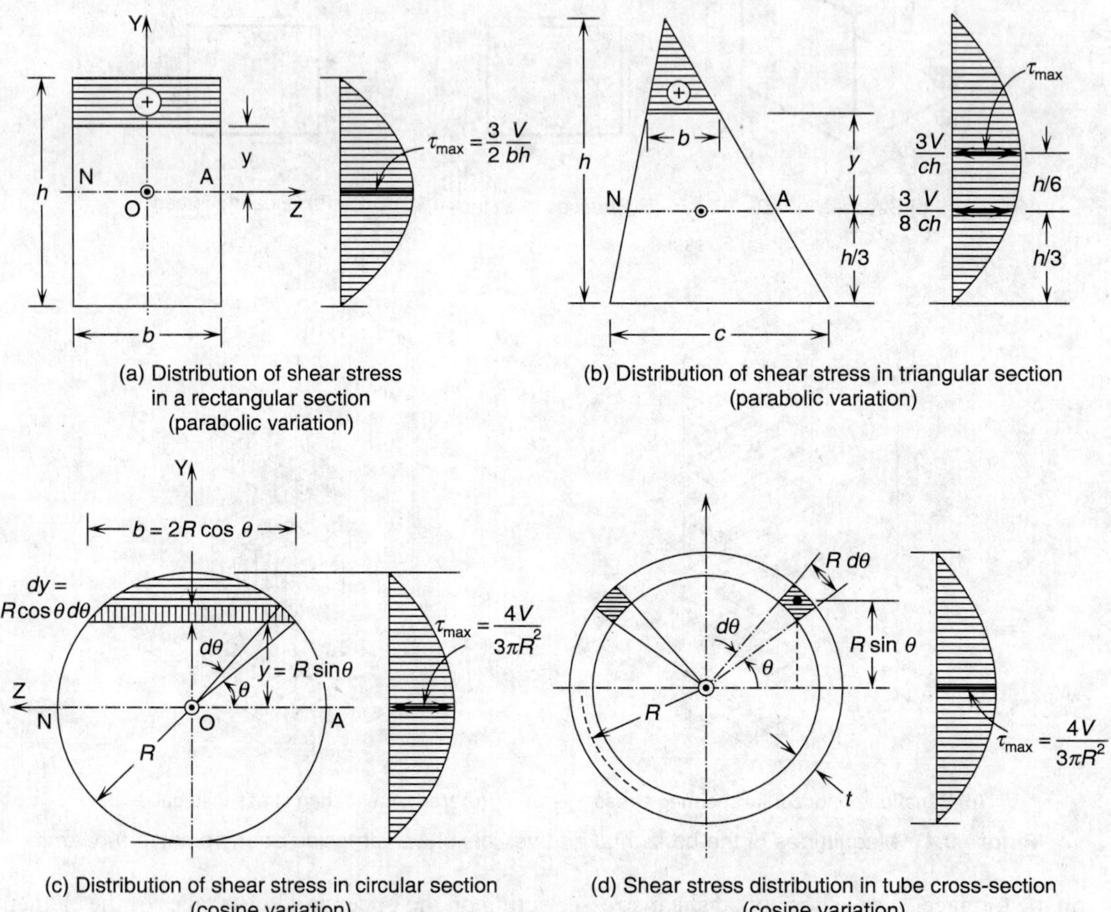

(a) Distribution of shear stress in a rectangular section (parabolic variation)

(b) Distribution of shear stress in triangular section (parabolic variation)

(c) Distribution of shear stress in circular section (cosine variation)

(d) Shear stress distribution in tube cross-section (cosine variation)

Figure 9.5 Shear stress distributions in different cross-sections.

$$A\overline{y} = \left[b\left(\frac{h}{2} - y\right)\right]\left[y + \frac{1}{2}\left(\frac{h}{2} - y\right)\right] = \frac{b}{2}\left(\frac{h^2}{4} - y^2\right)$$

Therefore,
$$\tau_h = \frac{V(A\overline{y})}{Ib} = \frac{V}{2I}\left(\frac{h^2}{4} - y^2\right) = \frac{6V}{bh^3}\left(\frac{h^2}{4} - y^2\right) \tag{9.3}$$

This relationship indicates that shearing stress varies parabolically down the depth of the section. The maximum shear stress occurs at neutral axis, i.e. at $y = 0$:

$$\tau_{h,\max} = \frac{6V}{bh^3}\left(\frac{h^2}{4}\right) = \frac{3}{2}\frac{V}{bh} = \frac{3}{2}\frac{V}{A} \tag{9.4}$$

It should be noted that $\tau_{av} = V/A$ represents the average shearing stress on the section. Thus, the maximum shear stress on a rectangular cross-section is 1.5 times higher than the average. The distribution of shear stress in a rectangular section is shown in Figure 9.5(b). Figures 9.5(c) and (d) show distribution of shear stresses in circular section (cosine variation) and tube cross-sections (cosine variation).

9.3.2 Triangular Section

Consider a member having a cross-section in the form of a triangle with base c and height h as shown in Figure 9.4(b), then

$$A\bar{y} = \frac{1}{2}(b)\left(\frac{2h}{3} - y\right)\left[\frac{1}{3}\left(\frac{2h}{3} - y\right) + y\right] = \frac{1}{3}(b)\left(\frac{2h}{3} - y\right)\left(\frac{h}{3} + y\right)$$

and

$$I = \frac{ch^3}{36}$$

Therefore,

$$\tau_v = \frac{V(A\bar{y})}{Ib} = \frac{12V}{ch^3}\left(\frac{2h}{3} - y\right)\left(\frac{h}{3} + y\right)$$

$$= \frac{12V}{ch^3}\left(\frac{2h^2}{9} + \frac{h}{3}y - y^2\right) \tag{9.5}$$

For maximum shear stress $\quad \dfrac{d\tau_v}{dy} = 0 = \dfrac{12V}{ch^3}\left(\dfrac{h}{3} - 2y\right) \quad$ or $\quad y = \dfrac{h}{6}$

The maximum shear stress occurs at a distance $y = \dfrac{h}{6}$ from the neutral axis, from Eq. (9.5):

$$\tau_{v,\max} = \frac{12V}{ch^3}\left(\frac{2h}{3} - \frac{h}{6}\right)\left(\frac{h}{3} + \frac{h}{6}\right)$$

$$= \frac{3V}{ch} = \frac{3}{2}\frac{V}{(ch/2)} = \frac{3V}{2A} \tag{9.6}$$

Thus, the maximum shear stress in a triangular cross-section is 3/2 times the average, located at the mid-point of the altitude. The distribution of shear stress is shown in Figure 9.4(b).

9.3.3 Circular Section

A procedure similar to the one used for rectangular section can be adopted to develop shearing stress formula for the circular shape. For a circular section of radius R, consider an element of width $2R\cos\theta$ and thickness $R\,d\theta\cos\theta$ at an angle θ from the diameter, i.e at distance $R\sin\theta$ as illustrated in Figure 9.4(c). The moment of the elemental (shaded) area is,

$$d(A\bar{y}) = (2R\cos\theta)(R\,d\theta\cos\theta) \times R\sin\theta$$

Therefore,

$$A\bar{y} = \int_{\theta}^{\pi/2} (2R\cos\theta)(R\,d\theta\cos\theta) \times R\sin\theta = -2R^3 \int_{\theta}^{\pi/2} (\cos^2\theta)(-\sin\theta)\,d\theta$$

$$= -\frac{2R^3}{3}\left[\cos^3\theta\right]_{\theta}^{\pi/2} = \frac{2R^3}{3}(\cos^3\theta)$$

Moreover, $I = \dfrac{\pi D^4}{64} = \dfrac{\pi R^4}{4}$

Therefore, $$\tau_v = \frac{V(A\bar{y})}{Ib} = \frac{4V}{\pi R^4 (2R\cos\theta)} \times \frac{2R^3}{3}(\cos^3\theta) = \frac{4V\cos^2\theta}{3\pi R^2} \tag{9.7}$$

The maximum shear stress occurs at neutral axis, i.e. at $\theta = 0$:

$$\tau_{v,\max} = \frac{4}{3}\frac{V}{\pi R^2} = \frac{4V}{3A} \tag{9.8}$$

Thus, the maximum shear stress on a circular cross-section is 4/3 times higher than the average. The distribution of shear stress is shown in Figure 9.5(c).

9.3.4 Hollow Circular Section

Removing material from the centre of circular cross-section tends to increase the local value of shear stress, especially near the diameter where the maximum shearing stress occurs. Consider a *thin-walled tube* of external and internal diameters of D and $d(= D - 2t)$, respectively, where t is the thickness of the tube.

$$I = \frac{\pi D^4}{64} - \frac{\pi d^4}{64} = \frac{\pi D^4}{64} - \frac{\pi (D-2t)^4}{64} = \frac{\pi D^4}{64}\left[1 - \left(1 - \frac{2t}{D}\right)^4\right]$$

$$= \frac{\pi D^4}{64}\left[1 - \left(1 - 4 \times \frac{2t}{D}\right)\right] = \frac{\pi D^3 t}{8} = \pi R^3 t \quad \text{for thin tubes} \quad t << r$$

where R is the radius of centre line of the tube. Consider an element $(R\,d\theta\,t)$ at an angle θ from the diameter as shown in Figure 9.4(d). The moment of the elemental (shaded) area,

$$d(A\bar{y}) = (R\,d\theta\,t)\,R\sin\theta$$

Therefore, $$A\bar{y} = 2\int_{\theta}^{\pi/2} (R\,d\theta\,t)\,R\sin\theta = 2R^2 t\cos\theta$$

$$\tau_v = \frac{V(A\bar{y})}{Ib} = \frac{V(2R^2 t\cos\theta)}{(\pi R^3 t)(2t)} = \frac{V\cos\theta}{\pi R t} \tag{9.9}$$

The maximum value occurs at $\theta = 0$, i.e. at the neutral axis, thus

$$\tau_{v,\max} = \frac{V}{\pi R t} = \frac{2V}{(2\pi R)t} = \frac{2V}{A} = 2\left(\frac{V}{A}\right) \tag{9.10}$$

Thus, the maximum shear stress in a *thin-walled tube* is approximately twice the average.

It should be noted that the multiplying factor has increased from 4/3 from the solid circular section to 2 for thin-walled circular tube.

9.3.5 Section with Abrupt Changes in Thickness

The stress distributions for some of the commonly used shapes with abrupt changes in thickness are shown in Figures 9.6(a) to (e).

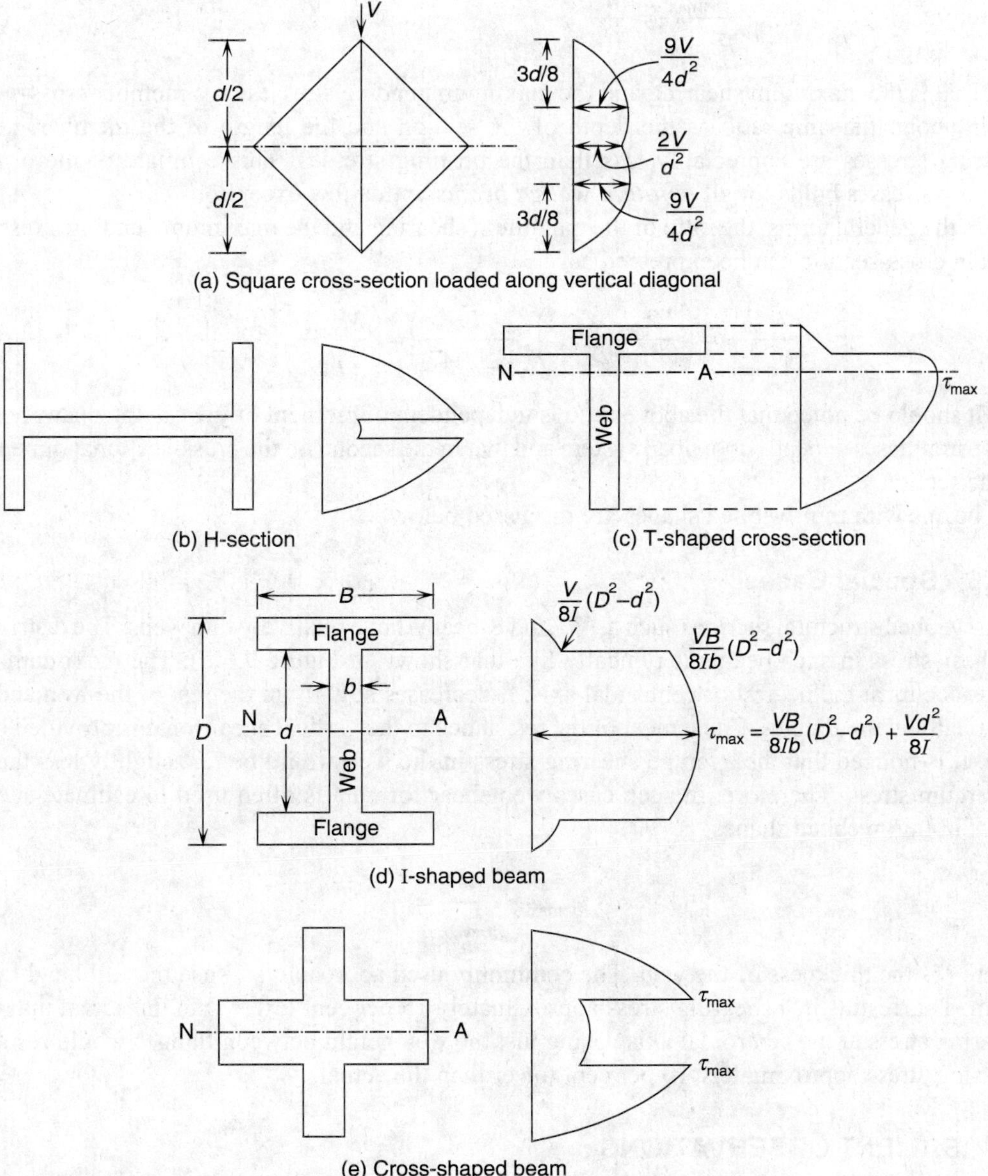

Figure 9.6 Shearing stress distributions for typical shapes with abrupt change in dimension.

To estimate the order of shearing stresses, consider the absolute values of the maximum bending and maximum shearing stresses on a member cross-section. For example, for a cantilever of length L having rectangular section of size $b \times h$, subjected to a concentrated load W at its free end,

$$\sigma_{b,\max} = \frac{M_{\max}}{Z} = \frac{6WL}{bh^2} \text{ and } \tau_{v,\max} = \frac{3}{2}\frac{W}{bh}$$

Therefore,
$$\frac{\tau_{v,\max}}{\sigma_{b,\max}} = \frac{h}{4L} \tag{9.11}$$

That is the maximum shearing and the maximum bending stresses on a member cross-section are in about the same ratio as the depth of the section and the length of the member, i.e. the shearing stresses are appreciably less than the bending stresses. This estimate of the order of shearing stresses holds for all *non-thin-walled beams* with a few exceptions.

In the general terms, the ratio of the maximum shearing and the maximum bending stresses on a beam cross-section can be expressed as,

$$\frac{\sigma_{b,\max}}{\tau_{v,\max}} = \left(\frac{M_{\max}}{I} y_{\max}\right)\left(\frac{Ib}{V_{\max} A\bar{y}}\right) = \left(\frac{M_{\max}}{V_{\max}}\right)\left(\frac{y_{\max} b}{A\bar{y}}\right) \tag{9.12}$$

It should be noted that the above ratio is independent of moment of inertia; the quantity in the first parentheses depends upon load system and that in the second on the cross-sectional dimensions of the section.

The beams with thin-webbed shapes are discussed below.

9.3.6 Special Cases

Thin-webbed structural sections such as W- and S-beams have relatively thin webs. The distribution of shear-stress in such beams is typically like that shown in Figure 9.6(d). The maximum shear stress occurs at the neutral or centroidal axis. It decreases slightly in the rest of the web and then drastically in the flanges. Thus, most of the resistance to the vertical shear force is provided by the web. It is noticed that the average shearing stress in the web would be just slightly less than the maximum stress. Therefore, in such cases web shear formula is often used to estimate shearing stress in thin-webbed shapes

$$\tau_{v,\max} \approx \frac{V}{A_{\text{web}}} = \frac{V}{th} \tag{9.13}$$

where t is the thickness of the web. The commonly used approach is to use the full height of the beam. This results in a shearing stress approximately 15 per cent lower than the actual maximum shearing stress at the centroidal axis. Using just the web height between flanges would result in a shearing stress approximately 10 per cent lower than the actual.

9.4 SALIENT OBSERVATIONS

From the foregoing discussions on the distribution of shearing stress in a cross-section, it is noted that:

1. The vertical shearing stress is zero on the extreme (i.e. topmost and bottommost) planes of the section. As it should be on the free surfaces, i.e. on boundaries.
2. The maximum shearing stress in the cross-section occurs at the centroidal axis provided the thickness there does not exceed that at some other section.
3. In the portion of the cross-section where the width b is constant, the shearing stress has a non-linear variation, decreasing with the increase in distance from the centroidal axis. The variation of shearing stress is usually parabolic.
4. At an axis where width changes abruptly, e.g. at the interface of flange and the web of a T- or an I-section, the shearing stress also change abruptly, being much smaller in the flange than in the thinner web.
5. A horizontal shearing stress is always accompanied by an equal vertical shearing stress.

9.5 LIMITATIONS OF SHEARING STRESS FORMULA

As the shearing stress formula for beams is based on the flexural formula, all the limitations imposed on flexural formula are applicable, i.e. member is assumed to be straight having symmetrical cross-section with material to be elastic having same modulus of elasticity in tension as in compression.

In certain cases the shearing stresses predicted by shear formula are incompatible with the boundary conditions. However, in practical situations the errors are within the permissible limits.

9.6 GENERAL PROCEDURE FOR ANALYSIS

The general procedure for computing shearing stress in the beams is outlined below:

1. Determine the vertical shearing force V at the section of interest.
2. Locate centroid of the entire cross-section and pass the neutral axis through the centroid.
3. Compute moment of inertia of the cross-section with respect to the neutral axis.
4. Identify the level at which the shearing stress is to be computed and determine the width of the section at that level, include all the parts of the section that are cut by the axis.
5. Compute $A\bar{y}$ with respect to neutral axis of the partial areas away from the axis of interest. If the partial area is a composite area consisting of several component parts, divide it into simple areas and locate centroid of each component part. Compute $A\bar{y}$ as follows,

$$A\bar{y} = A_1y_1 + A_2y_2 + A_3y_3 + \cdots \tag{9.14}$$

6. Calculate shearing stress using the shear formula.

The following examples will illustrate the procedure.

EXAMPLE 9.1

Determine the distribution of horizontal shear stress in a regular hexagon of side a as shown in Figure 9.7.

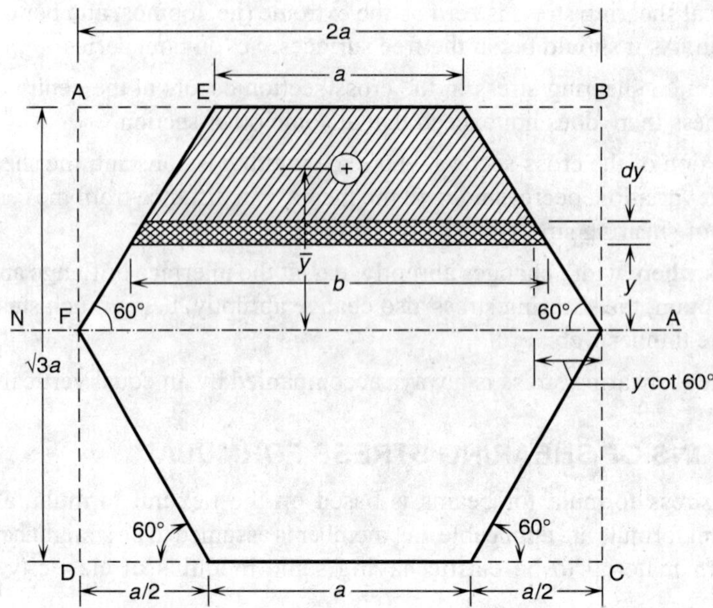

Figure 9.7 Regular hexagon cross-section of Example 9.1.

Solution: The area and moment of inertia of an hexagon about the neutral axis:

$$A = 2a \times (\sqrt{3}a) - 4 \times \left[\frac{1}{2} \times \left(\frac{a}{2} \times \frac{\sqrt{3}a}{2}\right)\right] = \frac{3\sqrt{3}a^2}{2}$$

Moment of inertia of the regular hexagon cross-section

I_{hex} = Moment of inertia of the rectangle ABCD – 4 × Moment of inertia of a triangle AEF about its apex

$$= \frac{2a \times (\sqrt{3}a)^3}{12} - 4 \times \left[\frac{1}{4} \times \frac{a}{2}\left(\frac{\sqrt{3}a}{2}\right)^2\right] = \frac{5\sqrt{3}a^4}{16}$$

Consider an element of width b and thickness dy at a distance y from the neutral axis as illustrated in Figure 9.7. Then

$$b = 2a - 2y \cot 60° = 2a - \frac{2}{\sqrt{3}}y$$

$$A\bar{y} = \int_y^{\sqrt{3}a/2} dA\, y = \int_y^{\sqrt{3}a/2} (b\, dy)\, y = \int_y^{\sqrt{3}a/2} by\, dy$$

$$= \int_y^{\sqrt{3}a/2} \left(2a - \frac{2}{\sqrt{3}}y\right) y\, dy = \int_y^{\sqrt{3}a/2} \left(2ay - \frac{2}{\sqrt{3}}y^2\right) dy$$

$$= \left| ay^2 - \frac{2}{3\sqrt{3}}y^3 \right|_y^{\sqrt{3}a/2} = \frac{a^3}{2} - ay^2 + \frac{2}{3\sqrt{3}}y^3$$

Therefore,
$$\tau_v = \frac{V(A\bar{y})}{I_{\text{hex}}b} = \frac{V\left(\dfrac{a^3}{2} - ay^2 + \dfrac{2}{3\sqrt{3}}y^3\right)}{\dfrac{5\sqrt{3}a^4}{16}\left(2a - \dfrac{2}{\sqrt{3}}y\right)}$$

The maximum shearing stress in the cross-section occurs at neutral axis, i.e. $y = 0$. Thus,

$$\tau_{\max} = \frac{V\left(\dfrac{a^3}{2}\right)}{\dfrac{5\sqrt{3}a^4}{16}(2a)} = \frac{4V}{5\sqrt{3}a^2} = \frac{0.4618V}{a^2}$$

Shearing stress at $y = a/2$, $\tau_{y=a/2} = 0.386V/a^2$

The average shearing stress

$$\tau_{\text{av}} = \frac{V}{A} = \frac{V}{3\sqrt{3}a^2/2} = \frac{2V}{3\sqrt{3}a^2}$$

Therefore,
$$\frac{\tau_{\max}}{\tau_{\text{av}}} = \frac{4V}{5\sqrt{3}a^2} \times \frac{3\sqrt{3}a^2}{2V} = \frac{6}{5} = 1.2$$

Thus, the maximum shear stress in a regular hexagon cross-section is 6/5 times the average.

EXAMPLE 9.2

A simply supported wide flange beam shown in Figure 9.8(a) carries a uniformly distributed load of 20 kN/m over an effective span of 9.0 m. At a section at distance 1.0 m from the support determine: (a) the maximum shearing stress in the cross-section, (b) the shearing stress at the junction of the flange and the web, and (c) draw the shearing stress distribution in the web and determine the percentage of shear carried by the web alone.

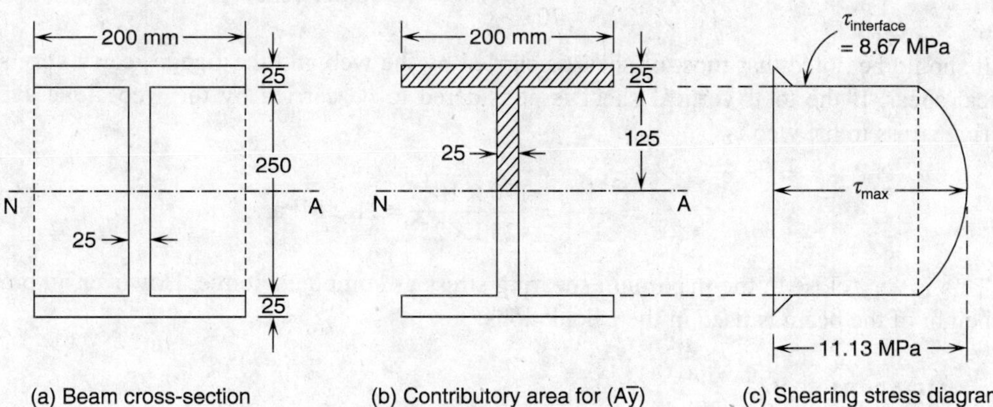

Figure 9.8 Wide-flange beam of Example 9.2.

Solution: The shearing force at a section at distance 1.0 m from the support is:

$$V_z = R_A - w_0\,(1.0) = 90 - 20 \times 1.0 = 70 \text{ kN}$$

The moment of inertia about the neutral axis:

$$I = \frac{200 \times 300^3}{12} - \frac{175 \times 250^3}{12} = 2.22 \times 10^8 \text{ mm}^4$$

The maximum shearing stress in the cross-section occurs at the neutral axis. Figure 9.8(b) shows the contributory area for $(A\bar{y})$. Thus,

$$\tau_{\max} = \frac{V_z(A\bar{y})}{Ib} = \frac{(70 \times 10^3) \times (200 \times 25 \times 137.5 + 25 \times 125 \times 62.5)}{(2.22 \times 10^8) \times 25}$$

$$= 11.13 \text{ MPa}$$

The shearing stress at the junction of the flange and the web,

$$\tau_{\text{interface}} = \frac{V_z(A\bar{y})}{Ib} = \frac{(70 \times 10^3) \times (200 \times 25 \times 137.5)}{(2.22 \times 10^8) \times 25} = 8.67 \text{ MPa}$$

These stresses vary parabolically from the top to the bottom of the web as shown in Figure 9.8(c). The total shearing force carried by the web is

$$V_{\text{web}} = \tau_{\text{interface}} \times (25 \times 250) + \frac{2}{3} \times (\tau_{\max} - \tau_{\text{interface}}) \times (250 \times 25)$$

$$= 8.67 \times (25 \times 250) + \frac{2}{3} \times (11.13 - 8.67) \times (250 \times 25) = 64.44 \text{ kN}$$

The percentage shear carried by the web alone

$$\frac{V_{\text{web}}}{V} \times 100 = \frac{64.44}{70} \times 100 = 92.06 \text{ per cent}$$

It should be noted that most of shear is carried by the web and the flanges carry very small vertical shear. If the total vertical shear is considered to be carried by the web alone average shearing stress in the web is

$$\tau_{\text{av,web}} = \frac{V}{A_{\text{web}}} = \frac{(70 \times 10^3)}{250 \times 25} = 11.2 \text{ MPa}$$

This is very close to the maximum shearing stress as computed above. However, in practice full height of the beam is used in the calculations.

9.7 SHEAR FLOW

In practice, built-up sections are generally fabricated from the available sections and standard plates as per the structural requirements. Due to the existence of horizontal shearing forces at the contact surface between the elements joined by the fasteners, i.e. the nails, bolts, and rivets, these fasteners are subjected to shear. Thus the built-up sections used for beams, must be analysed to determine the proper size and spacing of fasteners that would resist the sliding action between elements and ensure that all parts of the beam act together.

The quantity q obtained by multiplying the shearing stress at a level in the section by the width b of the section at that level is called *shear flow*. Thus the shear flow represents the longitudinal force per unit length transmitted across the section at the level y_1. Therefore, from Eq. (9.1):

$$q = \tau b = \frac{V(A\bar{y})}{I} \tag{9.15}$$

The units of q are *force per unit length*, e.g. newton per metre (N/m), N/mm, etc. *The shear flow is a measure of shearing force that is to be resisted at a particular section per unit length.* With shearing force capacity of a fastener predetermined, the safe spacing of the fasteners can be selected. If the maximum spacing of the fasteners is s_{max}, the force resisted by fastener is given by,

$$F_d = s_{max}\, q = s_{max} \frac{V(A\bar{y})}{I} = \frac{Vs_{max}}{I}(A\bar{y}) \tag{9.16}$$

Friction being neglected, this force is resisted by the shearing or bearing strength P_r of the fasteners, whichever is smaller. Thus equating P_r to F_d,

$$P_r = \frac{V(A\bar{y})}{I} s_{max} \quad \text{or} \quad s_{max} = \frac{P_r I}{V(A\bar{y})} \tag{9.17}$$

EXAMPLE 9.3

A simply supported timber beam of T cross-section carries a uniform distributed load of intensity 10 kN/m over an effective span of 4.0 m. The T cross-section is composed of two rectangular components 250 × 50 mm and 50 × 300 mm used as top flange element and web element, respectively, glued together at the interface as illustrated in Figure 9.6(b). Determine: (1) the maximum stress in the cross-section, (2) the shear stress developed at the interface which is to be carried by the glue and (3) the shear force to be carried by a connecting nail spaced 80 mm centre-to-centre.

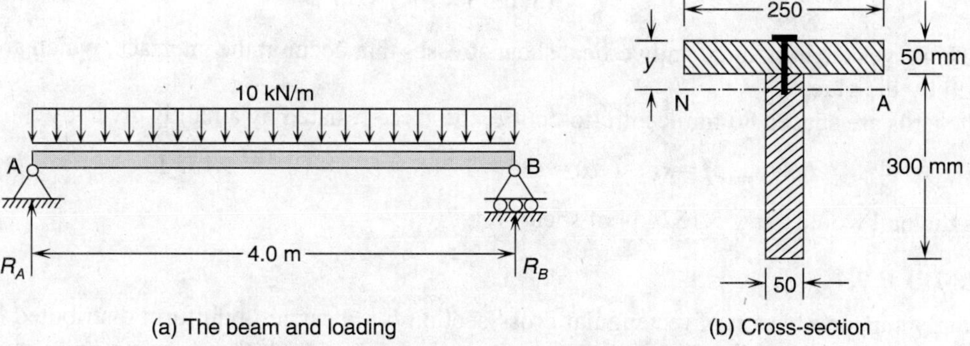

(a) The beam and loading

(b) Cross-section

Figure 9.9 T-beam of Example 9.3.

Solution: Consider top of the flange to be the reference axis;

$$\bar{y} = \frac{(250 \times 50) \times 25 + (50 \times 300) \times (50 + 150)}{(250 \times 50) + (50 \times 300)} = 120.45 \text{ mm}$$

Moment of inertia about the neutral axis:

$$I = I_1 + I_2 = \frac{250 \times 120.45^3}{3} - \frac{(250 - 50) \times (120.45 - 50)^3}{3} + \frac{50 \times (350 - 120.45)^3}{3}$$

$$= 3.7053 \times 10^8 \text{ mm}^4$$

The maximum shear stress occurs in the beam at the section where the shear force V is the maximum. The maximum shear force is:

$$V_{max} = \frac{10 \times 4.0}{2} = 20 \text{ kN}$$

In a section the shear stress at a layer located at a distance y from the neutral axis is given by:

$$\tau = \frac{V(A\bar{y})}{Ib}$$

where the term $(A\bar{y})$ represents the first moment of the area beyond the horizontal layer considered. The maximum shear stress in the section occurs at the neutral axis. In the present case, it is convenient to consider the area below the neutral axis in evaluating $(A\bar{y})$. Thus,

$$\tau = \frac{V(A\bar{y})}{Ib} = \frac{(20 \times 10^3) \times [50 \times (350 - 120.45)][(350 - 120.45)/2]}{(3.7053 \times 10^8) \times 50}$$

$$= 1.422 \text{ MPa}$$

The shear stress at the interface is evaluated by considering $(A\bar{y})$ for the area of cross-section between the interface and the top of flange.

$$\tau_{interface} = \frac{V(A\bar{y})}{Ib} = \frac{(20 \times 10^3) \times [(250 \times 50) \times (120.45 - 25)]}{(3.7053 \times 10^8) \times 50} = 1.288 \text{ MPa}$$

This gives the horizontal and vertical shear stresses that occur at the interface, which are to be carried by the glue.

As the nails are spaced 80 mm centre-to-centre, the force resisted by a nail is given by,

$$F_d = s_{max} q = s_{max}(\tau b) = (80 \times 1.288 \times 50) \times 10^{-3} = 5.152 \text{ kN}$$

Thus the nail would carry 5.152 kN of shear.

EXAMPLE 9.4

A simply supported beam of rectangular cross-section $b \times h$ carries a uniform distributed load of intensity w_0 over a span L. Determine the critical length at which the shearing stress τ and the flexural stress σ_b reach their ultimate values simultaneously.

Solution: The maximum shearing force, $V_{\max} = R_A = \dfrac{w_0 L}{2}$

The maximum bending moment, $$M_{\max} = \frac{w_0 L^2}{8}$$

Maximum load based on allowable shearing stress on a rectangular cross-section,

$$\tau_a = \tau_{\max} = \frac{3}{2}\frac{V}{A} = \frac{3}{2} \times \frac{(w_0 L/2)}{bh} = \frac{3 w_0 L}{4bh} \quad \text{or} \quad w_0 = \frac{4\tau_a bh}{3L}$$

Maximum load based on allowable bending stress,

$$\sigma_{b,a} = \sigma_{b,\max} = \frac{M_{\max}}{Z} = \frac{w_0 L^2}{8} \times \frac{6}{bh^2} = \frac{3 w_0 L^2}{4bh^2} \quad \text{or} \quad w_0 = \frac{4\sigma_{b,a} bh^2}{3L^2}$$

Therefore, $$\frac{4\tau_a bh}{3L} = \frac{4\sigma_{b,a} bh^2}{3L^2} \quad \text{or} \quad L = \frac{\sigma_{b,a} h}{\tau_a}$$

If the length exceeds the critical value, flexure will govern the design and for shorter lengths shear will govern.

EXAMPLE 9.5

A simply supported timber box-beam with an overhang, supports the load system shown in Figure 9.10(a). Determine the maximum value of load W at which the shearing stress τ and the flexural stress σ_b will not exceed their allowable values, 0.7 MPa and 7.0 MPa, respectively.

Solution: The shear force and bending moment diagrams are shown in Figures 9.10(b) and (c).

The maximum shearing force, $V_{\max} = (0.5W + 2)$ kN

The maximum bending moment in terms of W is at $z = 2$ m and its value is,

$$M_{\max} = (W - 4) \text{ kN.m}$$

Moment of inertia, $$I = \frac{200 \times 250^3}{12} - \frac{150 \times 200^3}{12} = 1.604 \times 10^8 \text{ mm}^4$$

Section modulus, $$Z = \frac{I}{y_{\max}} = \frac{(1.604 \times 10^8)}{(250/2)} = 1.28 \times 10^6 \text{ mm}^3$$

The maximum shear stress in the section occurs at the neutral axis,

$$\tau_{\max} = \frac{V_{\max}(A\bar{y})}{Ib}$$

where $A\bar{y} = (200 \times 25) \times 112.5 + 2 \times (100 \times 25) \times 50 = 8.125 \times 10^5 \text{ mm}^3$

Alternatively, $A\bar{y} = (200 \times 125) \times (125/2) - (150 \times 100) \times (100/2) = 8.125 \times 10^5 \text{ mm}^3$

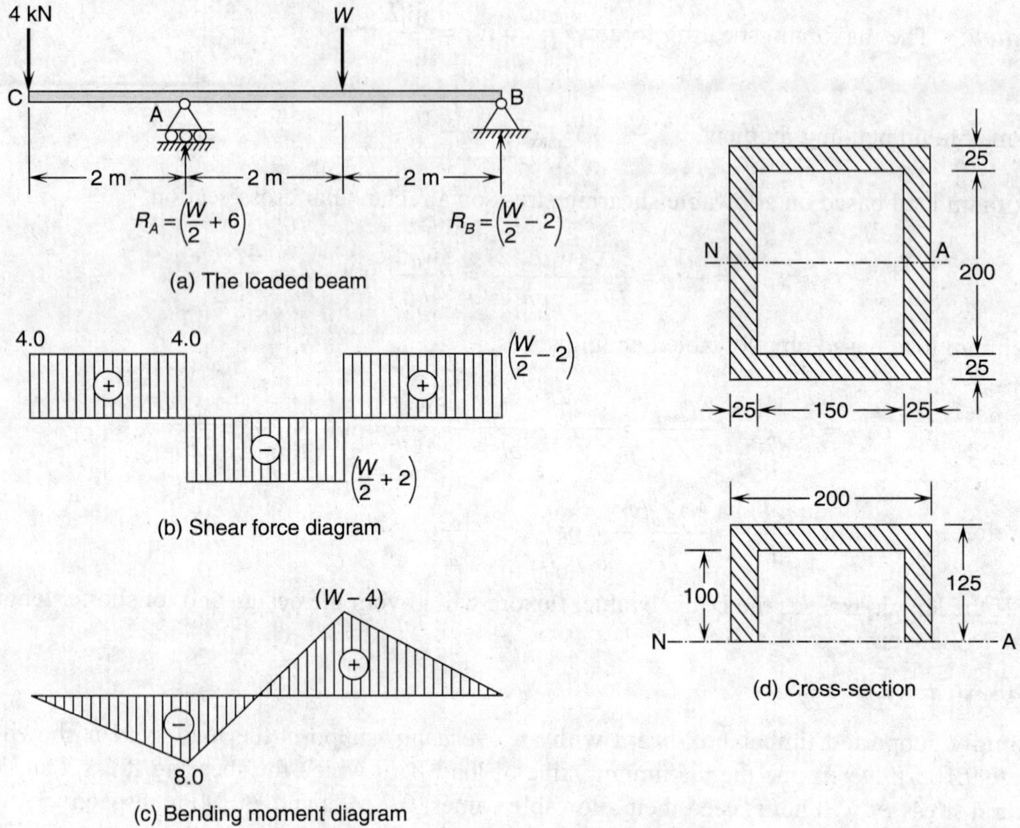

Figure 9.10 Box-beam of Example 9.5.

Therefore, the maximum load based on allowable shearing stress,

$$\tau_a = 0.7 = \frac{[(0.5P + 2) \times 10^3] \times (8.125 \times 10^5)}{(1.604 \times 10^8) \times 50}$$

or $$W = 2 \times \left(\frac{0.7 \times 1.604 \times 10^8 \times 50}{10^3 \times 8.125 \times 10^5} - 2 \right) = 9.819 \text{ kN}$$

The maximum load based on allowable bending stress,

$$\sigma_{b,a} = 7.0 = \frac{(W - 4) \times 10^6}{1.28 \times 10^6} = \frac{(W - 4)}{1.28}$$

or $$W = 7.0 \times 1.28 + 4 = 12.98 \text{ kN}$$

Therefore, the safe load which is smaller of the two is 9.819 kN.

EXAMPLE 9.6

Using the first principles, determine the ratio of maximum to the mean value of shear stress in the beam of hollow circular cross-section whose outside radius is thrice the inside radius.

Solution: Consider an elementary strip of width b and thickness dy at a distance y from the neutral axis of a circular section of external and internal radii R and r, respectively, as illustrated in Figure 9.11. Then,

$$y = R \sin \theta = r \sin \phi$$

Therefore,

$$dy = R \cos \theta \, d\theta = r \cos \phi \, d\phi$$

$$b = 2(R \cos \theta - r \cos \phi)$$

Thus,

$$dA = b\, dy = 2(R \cos \theta - r \cos \phi)\, dy = 2(R^2 \cos^2 \theta \, d\theta - r^2 \cos^2 \phi \, d\phi)$$

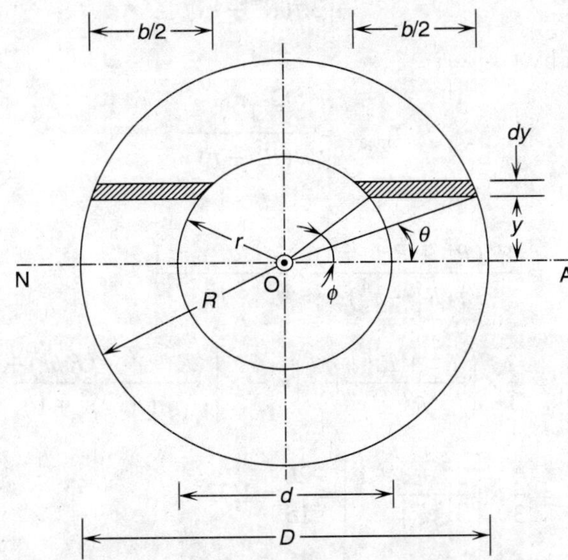

Figure 9.11 Hollow circular cross-section of beam of Example 9.6.

The moment of the elemental (shaded) area is,

$$d(A\bar{y}) = 2(R^2 \cos^2 \theta \, d\theta \times R \sin \theta - r^2 \cos^2 \phi \, d\phi \times r \sin \phi)$$

$$A\bar{y} = \int_{\theta}^{\pi/2} 2R^3 \cos^2 \theta \sin \theta \, d\theta - \int_{\phi}^{\pi/2} 2r^3 \cos^2 \phi \sin \phi \, d\phi$$

Therefore,

$$A\bar{y} = -\int_{\theta}^{\pi/2} 2R^3 \cos^2 \theta (-\sin \theta)\, d\theta + \int_{\phi}^{\pi/2} 2r^3 \cos^2 \phi (-\sin \theta)\, d\phi$$

$$= -\frac{2R^3}{3}\left[\cos^3 \theta\right]_{\theta}^{\pi/2} + \frac{2r^3}{3}\left[\cos^3 \phi\right]_{\phi}^{\pi/2} = \frac{2R^3}{3}(\cos^3 \theta) - \frac{2r^3}{3}(\cos^3 \phi)$$

Shear stress, $$\tau = \frac{V(A\bar{y})}{Ib} = \frac{V \times 2(R^3 \cos^3 \theta - r^3 \cos^3 \phi)}{3 \times [\pi(R^4 - r^4)/4] \times 2(R \cos \theta - r \cos \phi)}$$

$$= \frac{4V(R^3 \cos^3 \theta - r^3 \cos^3 \phi)}{3\pi \times (R^4 - r^4)(R \cos \theta - r \cos \phi)}$$

$$= \frac{4V[R^2 \cos^2 \theta + (R \cos \theta)(r \cos \phi) + r^2 \cos^2 \phi]}{3\pi \times (R^4 - r^4)}$$

Maximum shear stress occurs at $\theta = \phi = 0$, thus

$$\tau_{\max} = \frac{4V(R^2 + Rr + r^2)}{3\pi(R^4 - r^4)}$$

The mean stress is given by,

$$\tau_{\text{mean}} = \frac{V}{\pi(R^2 - r^2)}$$

Therefore,

$$\frac{\tau_{\max}}{\tau_{\text{mean}}} = \frac{4V(R^2 + Rr + r^2)}{3\pi(R^4 - r^4)} \times \frac{\pi(R^2 - r^2)}{V}$$

$$= \frac{4}{3} \times \left(\frac{R^2 + Rr + r^2}{R^2 + r^2} \right) = \frac{4}{3} \times \left[\frac{(R/r)^2 + (R/r) + 1}{(R/r)^2 + 1} \right]$$

For $\frac{R}{r} = 3$, $$\frac{\tau_{\max}}{\tau_{\text{mean}}} = \frac{4}{3} \times \left[\frac{3^2 + 3 + 1}{3^2 + 1} \right] = \frac{26}{15} = 1.73$$

EXAMPLE 9.7

A simply supported wooden box-beam is made of 150 × 25 mm wooden planks connected by mails as shown in Figure 9.12(a). The beam carries a uniformly distributed load of 2.5 kN/m over a span of 4 m. If each nail can safely transmit shear force of 1.25 kN, determine the maximum spacing of nails along the length of the beam. Sketch the corresponding shear stress distribution curve.

Solution: The maximum shear force transmitted by the cross-section,

$$V_{\max} = \frac{wL}{2} = \frac{2.5 \times 4}{2} = 5.0 \text{ kN}$$

Moment of inertia about the neutral axis:

$$I = \frac{150 \times 200^3}{12} - \frac{100 \times 150^3}{12} = 7.1875 \times 10^7 \text{ mm}^4$$

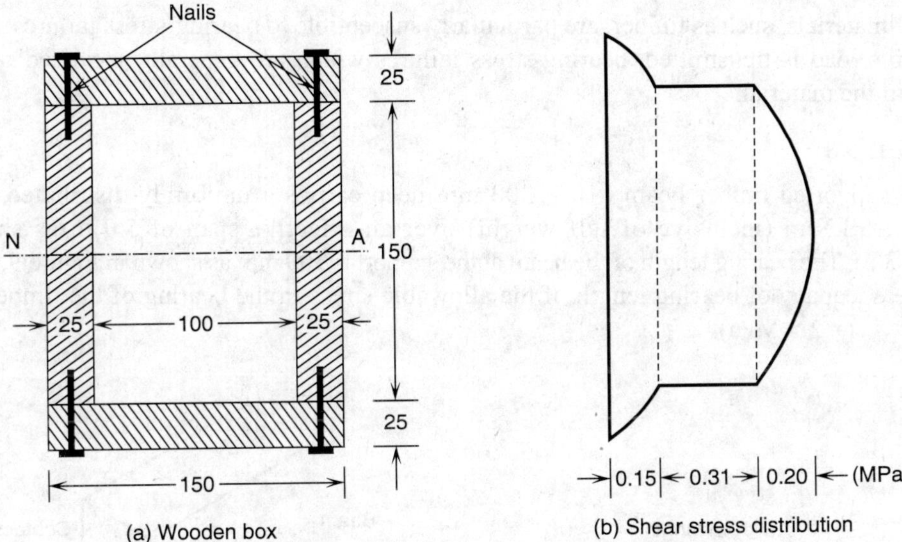

(a) Wooden box

(b) Shear stress distribution

Figure 9.12 Wooden box made of planks connected by nails.

Horizontal shear stress at the junction of flange and web,

$$\tau = \frac{V(A\bar{y})}{Ib} = \frac{(5\times10^3)\times[(150\times25)\times87.5]}{(7.1875\times10^7)\times(2\times25)} = 0.457 \text{ MPa}$$

Consider pitch p, the horizontal shear force per pitch length

$$0.457 \times [(2 \times 25)p] = 22.85p \text{ N}$$

Equate the horizontal shear force per pitch length to the shearing strength of the nails,

$$22.85p = 2 \times 1250 \quad \text{or} \quad p = 109.4 \text{ mm c/c.}$$

The shear stress distribution curve is shown in Figure 9.12(b).

9.8 BEARING STRESSES

The stresses developed at the point of contact between two loaded elements are called *bearing stresses*. Such stresses, for example, are developed at the ends of a simply supported beam where it rests on end supports having certain dimensions. The magnitudes of the stresses developed are dependent on the magnitude of the forces transmitted through the point of contact and the surface area of the contact between the two elements. The smaller the *contact area*, the greater the bearing stresses. The stresses produce deformation in both elements at the point of contact. These deformations typically extend into the elements only a small distance.

The magnitude of the bearing stress at a point is simply equal to the load transmitted, divided by the area of contact, or $\sigma_{bg} = P/A$. This relation is based on the crude but reasonable assumption that the bearing stresses are uniformly distributed over the contact area.

Many materials, such as timber, are particularly susceptible to bearing-stress failures. When a compressive load is transmitted, bearing-stress failures which are generally localized occur by crushing in the material.

EXAMPLE 9.8

A simply supported timber beam 110 × 300 mm deep carries a uniformly distributed load of intensity 7.5 kN/m (inclusive of self weight) over an effective span of 5.0 m as shown in Figure 9.13(a). The bearing length of the beam at the support is 100 mm as shown in Figure 9.13(b-ii). Check the adequacy of bearing length, if the allowable stress in the bearing of the timber in the beam is $\sigma_{a,bg}$ (= 2.5 MPa).

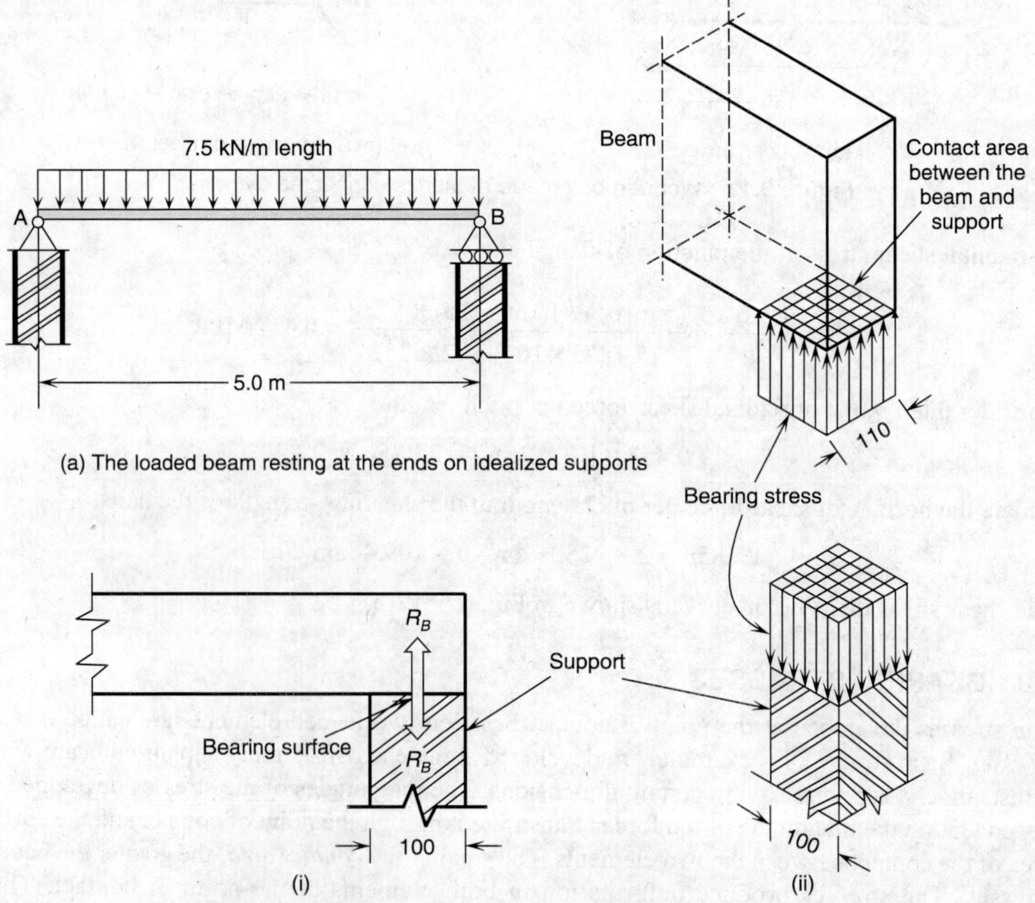

Figure 9.13 Bearing stress at a support of a beam.

Solution: The support reactions are,

$$R_A = R_B = \frac{1}{2} \times (7.5) \times (5.0) = 18.75 \text{ kN}$$

Bearing area, $A_{\text{bearing}} = 110 \times 100$ mm

Therefore, bearing stress is given by:

$$\sigma_{bg} = \frac{R_B}{A_{\text{bearing}}} = \frac{18.75 \times 10^3}{110 \times 100} = 1.705 \text{ MPa} \le F_{bg} \ (= 2.5 \text{ MPa})$$

Since the actual bearing stress is less than the allowable bearing stress, the bearing length is adequate. Figure 9.13(b) shows distribution of bearing stress (assuming uniformly distributed load).

Other important case in bearing is encountered in *bolted connections*, wherein the highly irregular pressure develops between a bolt and the plates. The average nominal intensity of the pressure is obtained by dividing the load transmitted by the projected area of the bolt onto the connecting plate. This is referred to as the bearing stress. The bearing stress developed in the connection of Figure 9.14(a) is $\sigma_{bg} = P/td$, where t is the thickness of the plate and d the diameter of the bolt. For the case in Figure 9.14(b), the bearing stresses in the middle and outer plates are $\sigma_{bg,1} = P/t_1 d$ and $\sigma_{bg,2} = P/2t_2 d$, respectively. This procedure is also applicable to riveted connections as illustrated in the following example.

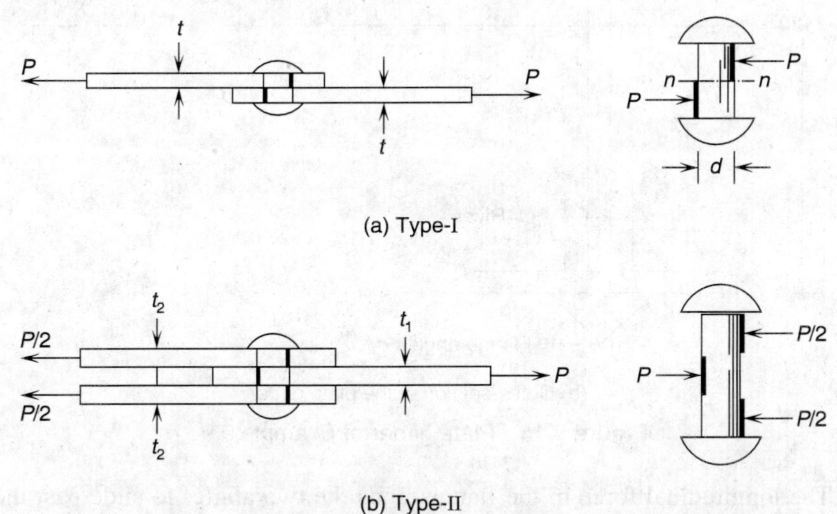

Figure 9.14 Rivetted or bolted connections.

EXAMPLE 9.9

A simply supported plate girder is fabricated by attaching the short legs of four 125 × 90 × 13 mm angles to a 1100 × 10 mm thick web plate to form a 1120 mm deep section as illustrated in Figures 9.15(a) and (b). The moment of inertia of the section about neutral axis is $I = 4.140 \times 10^9$ mm^4. The plate girder carries a uniformly distributed load 100 kN/m over a clear span of 9.0 m. Determine the centre-to-centre spacing of 19 mm rivets that fasten angles to the web plate. Use $\tau_a = 100$ MPa; $\sigma_{bg} = 200$ MPa for rivets in single shear and $\sigma_{bg} = 250$ MPa for rivets in double shear.

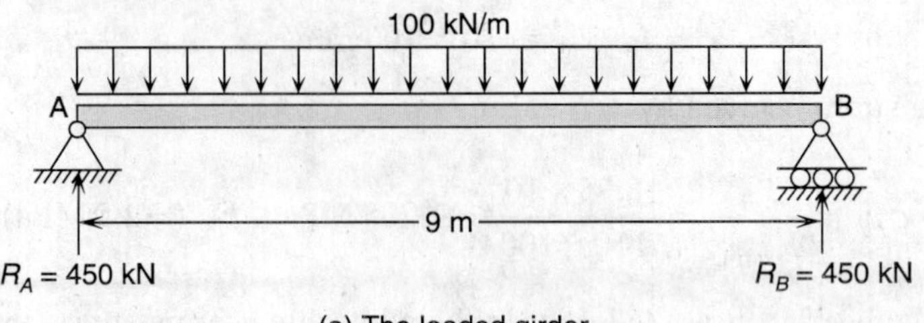

(a) The loaded girder

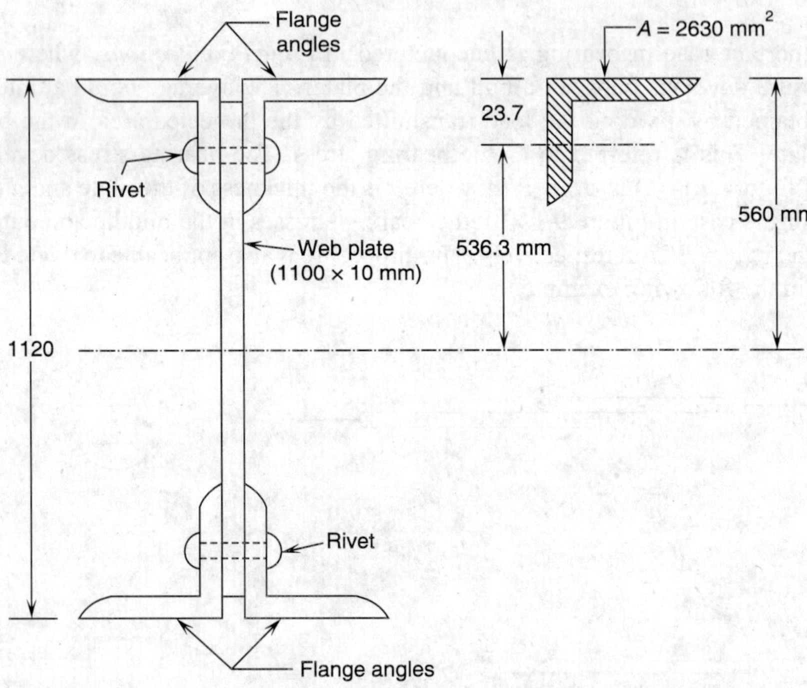

(b) Cross-section of the plate girder

Figure 9.15 Plate girder of Example 9.9.

Solution: The longitudinal force in the flange tends the two angles to slide past the web. This sliding tendency of angles is resisted by the rivets. Hence $A\bar{y}$ of these two flange angle must be considered in computation of shearing stress, therefore,

$$A\bar{y} = 2 \times (2630)(560 - 23.7) = 2.82 \times 10^6 \text{ mm}^3$$

The force resisting capacity of a rivet called rivet value is the minimum of the shear resistance and bearing resistance of a rivet against the web plate.

The shear resistance of a 19 mm rivet in double shear,

$$P_s = 2(a_s \tau_a) = 2 \times \frac{\pi}{4}(19)^2 \times 100 = 56700 \text{ N} = 56.7 \text{ kN}$$

The bearing resistance of the rivet against the web plate is,

$$P_b = (a_b\sigma_{bg}) = (dt\sigma_{bg}) = (19 \times 10) \times 250 = 47500 \text{ N} = 47.5 \text{ kN}$$

The lower of these values is the rivet value, i.e. $P_r = 47.5$ kN.

The maximum shear stress occurs in the girder at the section where the shear force V is a maximum. The maximum shear force is:

$$V_{\max} = \frac{100 \times 9.0}{2} = 450 \text{ kN}$$

Therefore, the rivet spacing or pitch is given by,

$$s_{\max} = \frac{P_r I}{V(A\bar{y})} = \frac{(47.5 \times 10^3)(4.140 \times 10^9)}{(450 \times 10^3)(2.82 \times 10^6)} = 154.96 \text{ mm} \quad \text{(say 154 mm)}.$$

EXAMPLE 9.10

The hoisting arrangement detailed in Figure 9.16(a) uses a 200 mm wide and 300 mm deep timber beam which is anchored by two 20 mm diameter threaded bolts with 16 mm diameter at the roots at the end A and rests on masonry block at B. Determine the load W that can be handled safely. The allowable stresses in various materials are:

Allowable bearing stress of the brick work = 1.5 MPa
Allowable tensile stress in bolts = 150 MPa

Allowable stresses in the bearing, bending and shearing in the timber are 2.5 MPa, 10.0 MPa and 1.0 MPa, respectively.

Solution: The free-body diagram of hoisting beam is shown in Figure 9.16(b). To calculate various reaction components consider equilibrium equations:

$$\sum F_z = R_{AH} = 0$$

$$\sum M_A = 0 = R_{BV} \times 1.25 - W \times 5.0 \quad \text{or} \quad R_{BV} = 4W(\uparrow)$$

$$\sum F_y = R_{AV} + R_{BV} - W = 0 \quad \text{or} \quad R_{AV} = -3W\,(\downarrow)$$

The tensile strength of two 20 mmϕ threaded bolts,

$$P_{\text{bolts}} = 2(a_s\tau_a) = 2 \times \frac{\pi}{4}(16)^2 \times 150 = 60300 \text{ N} = 60.3 \text{ kN}$$

Therefore, $R_{AV} = 3W = 60.3$ kN or $W = 20.1$ kN

The bearing resistance of the masonry block against the timber beam,

$$W_{bg,\text{masonry}} = (A_{bg}\sigma_{bg,m}) = (200 \times 200) \times 1.5 = 60000 \text{ N} = 60.0 \text{ kN}$$

Therefore, $R_{BV} = 4W = 60.0$ kN or $W = 15.0$ kN

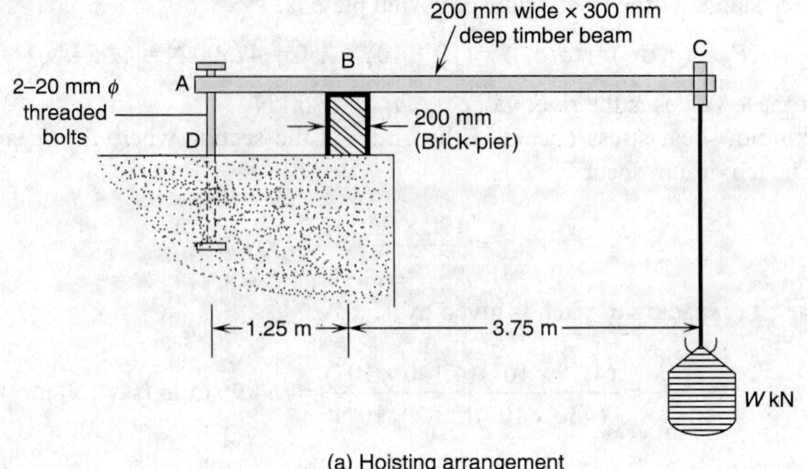

(a) Hoisting arrangement

$R_{AH} = 0$
W
A
B
C
1.25 m
3.75 m
$R_{AV} = 3W$
$R_B = 4W$

(b) Free-body diagram of the hoisting beam

Figure 9.16 Beam used for hoisting equipment of Example 9.10.

The bearing resistance of the timber beam against the masonry block,

$$W_{bg,\text{timber}} = (A_{bg}\sigma_{bg,t}) = (200 \times 200) \times 2.5 = 100000 \text{ N} = 100.0 \text{ kN}$$

Therefore, $R_{BV} = 4W = 100.0$ kN or $W = 25.0$ kN

The shearing resistance of the timber beam,

$$W_{\text{beam, shear}} = \frac{2}{3}(A_s\, \tau_a) = \frac{2}{3} \times (200 \times 300) \times 1.0 = 40000 \text{ N} = 40.0 \text{ kN}$$

Therefore, $V_{\max} = 3W = 40.0$ kN or $W = 13.33$ kN

The bending resistance of the timber beam,

$$M_{\text{beam, bending}} = Z\sigma_b = \frac{1}{6} \times (200 \times 300^2) \times 10.0 = 30 \times 10^6 \text{ N} = 30.0 \text{ kN.m}$$

$$M_{\max} = 3.75W = 30.0 \text{ kN.m} \quad \text{or} \quad W = 8.0 \text{ kN}$$

The lower of these values is the load that can be handled safely, i.e. $W = 8.0$ kN.

9.9 PROBLEMS

9.1 A simply supported beam of length L having a rectangular cross-section of width b and height h carries a concentrated load at the mid-span. Show that the ratio of the maximum shearing and the maximum bending stresses is $\tau_{max}/\sigma_{b,max} = h/2L$.

9.2 Show that the maximum horizontal shearing stress developed in a beam with a cross-section shown in Figure 9.17 occurs at a layer $h/8$ above or below the neutral axis.

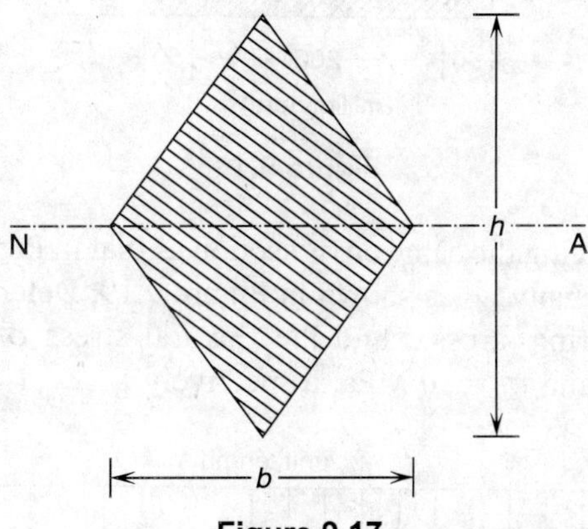

Figure 9.17

9.3 A beam of square cross-section of side a with one diagonal placed in a vertical position is subjected to a vertical shearing force V, show that (a) the horizontal shearing stress at the neutral axis is equal to the mean shearing stress, (b) the maximum horizontal shearing stress developed is $9V/8a^2$ and occurs at a layer $a/4\sqrt{2}$ above or below the neutral axis and (c) the ratio of the maximum and the average shearing stress over the cross-section is 9/8.

9.4 Show that the maximum shear stress in a thin-walled tube is approximately twice the average, i.e. $\tau_{max} \approx 2V/A$.

9.5 Prove that the maximum shear stress in a beam having equilateral triangle cross-section of side b with vertex uppermost is $\tau_{max} = 2\sqrt{3}V/b^2$, located at the height $\sqrt{3}b/4$.

9.6 A certain section of a beam of symmetrical triangular cross-section of width b and depth h carries a vertical shearing force V along the axis of symmetry. Determine (a) the shearing stress τ at any depth y from the vertex of the section and (b) the ratio of the maximum and the average shearing stresses over the cross-section.

[**Ans.** $\tau = [12V\,y(h-y)]/(bh^3)$ and ratio = 2/3]

9.7 A simply supported timber box-beam of span L with cross-section shown in Figure 9.18 carries a uniformly distributed load of 5 kN/m. Determine the limiting value of L to cause maximum flexural stress of 15 MPa.

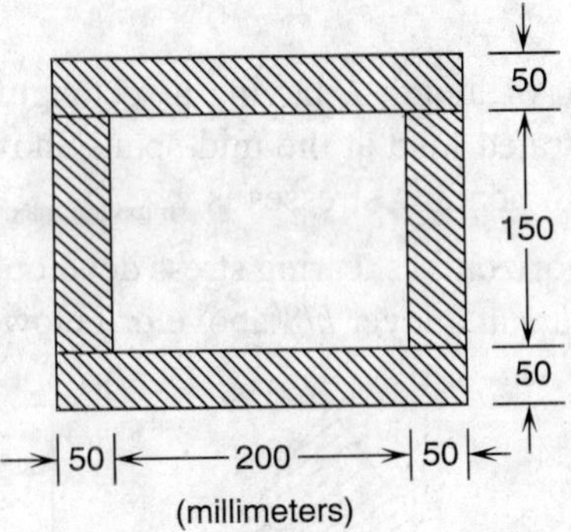

Figure 9.18

9.8 A timber box-beam having the same cross section as that in Problem 9.7 carries a uniformly distributed load of intensity w_0 as shown in Figure 9.19. Determine the maximum value of w_0 at which the shearing stress τ and the flexural stress σ_b will attain their allowable values, $\tau_a = 1.0$ MPa and $\sigma_b = 10$ MPa, respectively.

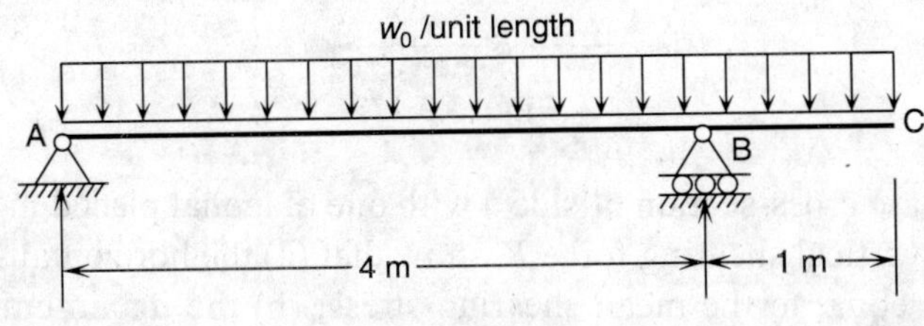

Figure 9.19

9.9 A 2.2 m long tube having cross-section of 100 mm outer diameter and 6 mm thickness shown in Figure 9.20 is used as a cantilever. Determine the maximum concentrated load it can carry at its free end without exceeding the allowable stresses in shear and bending of $\tau_a = 80$ MPa and $\sigma_{b,a} = 150$ MPa, respectively.
[**Hint.** For $t << r$, $I = \pi r^3 t$, $A\bar{y} = 2r^2 t \cos\theta$; $W_{\text{bending}} = 2.83$ kN and $W_{\text{shear}} = 70.8$ kN]

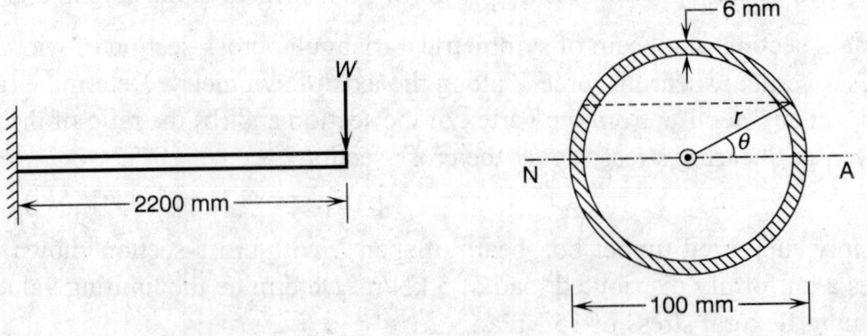

Figure 9.20

9.10 A simply supported wide-flange beam of length 10 m having cross-section shown in Figure 9.21 carries a uniformly distributed load of intensity 30 kN/m over its entire span. Determine the maximum and minimum shearing stresses in the web of the section at a distance 2 m from the support.

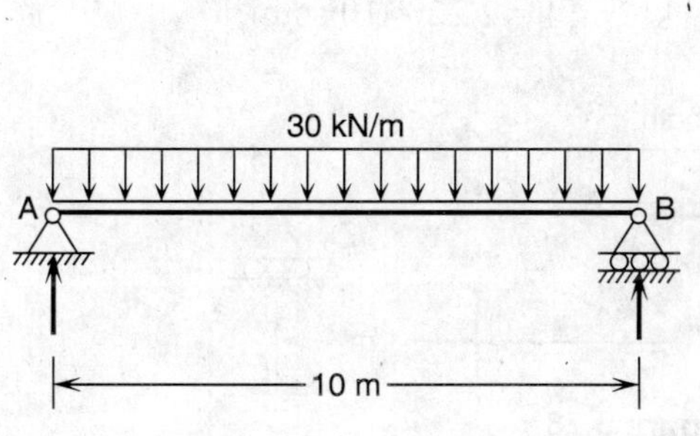

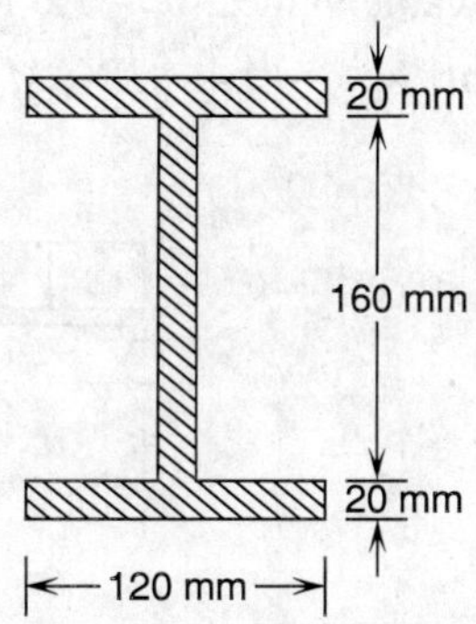

Figure 9.21

[**Ans.** τ_{max} = 27.45 MPa; τ_{min} = 21.15 MPa; vertical shear carried by the web = 90.2 per cent]

9.11 A simply supported I-beam of length L having cross-section shown in Figure 9.22 carries a uniformly distributed load of intensity w_0. Determine the ratio of the maximum bending and the maximum shearing stresses.

$$\left[\textbf{Hint.}\ \frac{\sigma_{b,\max}}{\tau_{\max}} = \left(\frac{M_{\max}}{V_{\max}}\right)\left(\frac{y_{\max} b}{A\bar{y}}\right) = \left(\frac{L}{4}\right)\left(\frac{y_{\max} b}{A\bar{y}}\right)\right]$$

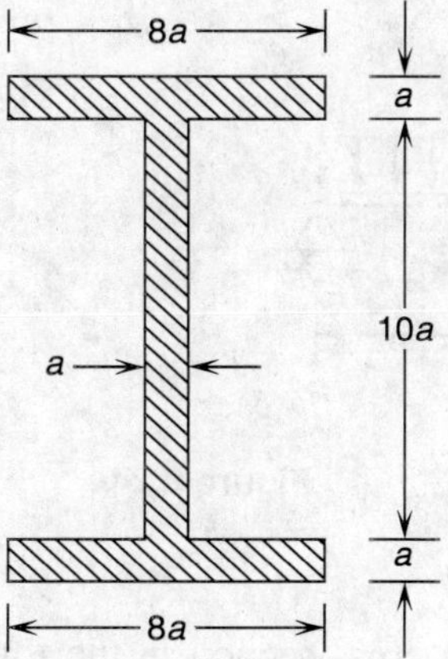

Figure 9.22

9.12 A simply supported I-beam of length 5 m having the same cross-section as that in Problem 9.11 with a = 25 mm, carries a uniformly distributed load of intensity w_0 kN/m over a length of 4 m from the left support as shown in Figure 9.23. Determine the maximum value of w_0 at which the shearing stress τ and the flexural stress σ_b will not exceed their allowable values, $\tau_a = 1.0$ MPa and $\sigma_{b,a} = 10$ MPa, respectively.

[**Hint.** $V_{max} = 2.4w_0$ kN; $M_{max} = 2.88w_0$ kN.m; I = 2.22×10^8 mm^4]

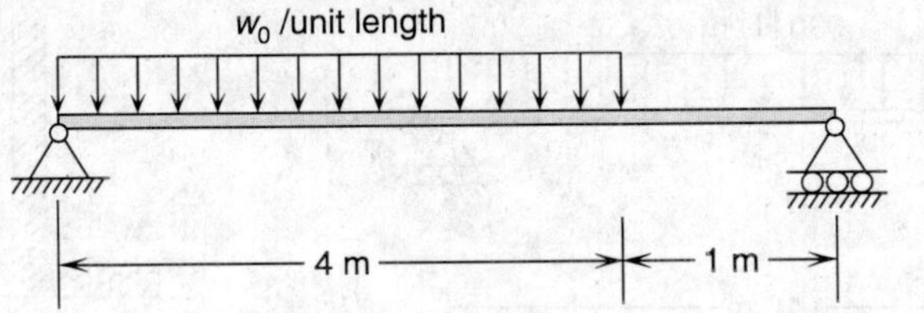

Figure 9.23

9.13 A simply supported T-beam is composed of two long wooden planks of size 200 mm wide by 50 mm thick. One plank is used as top flange element and other as web element, nailed together at the interface as illustrated in Figure 9.24. The beam carries a uniformly distributed load 2.5 kN/m over a span of 3.2 m. Determine the necessary spacing of the nails joining two planks to make them act as a unit. Consider that the allowable shear force carried by a connector nail is 880 N.

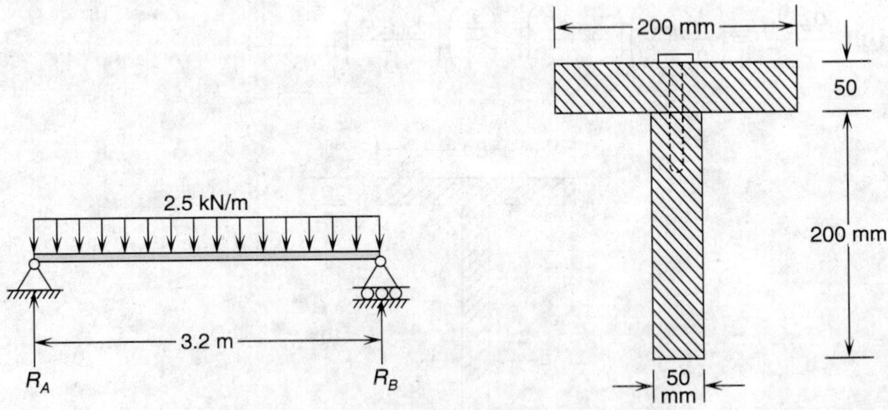

Figure 9.24

[**Ans.** s_{max} = 40 mm]

9.14 A thin walled element has a cross-section in the form of an isosceles triangle shown in Figure 9.25. Determine the shear stresses at the mid-height and centroidal levels of the cross-section developed due to vertical shear V = 120 kN.

[**Ans.** $\tau_{centroid}$ = 163.2 MPa; $\tau_{mid\text{-}height}$ = 171.6 MPa]

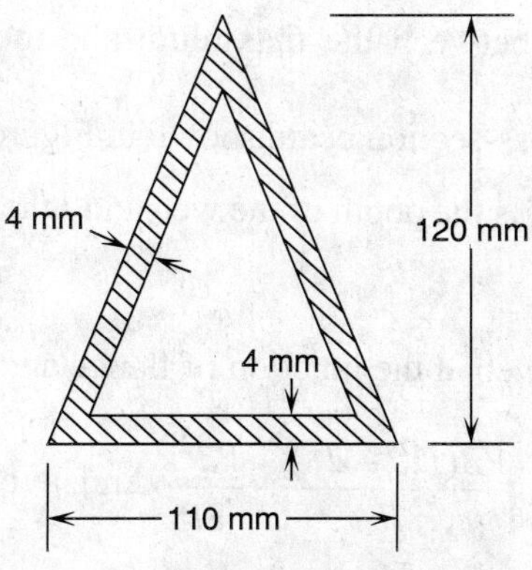

Figure 9.25

9.15 A simply supported beam with cross-section composed of seven timber pieces as illustrated in Figure 9.26 carries a uniformly distributed load 4.0 kN/m over a span of 6.5 m. Determine the minimum spacing of the 10 mm lag screws to be used to fasten the timber pieces to make them act as a unit. The shear force resisting capacity of a screw is specified as 2 kN.

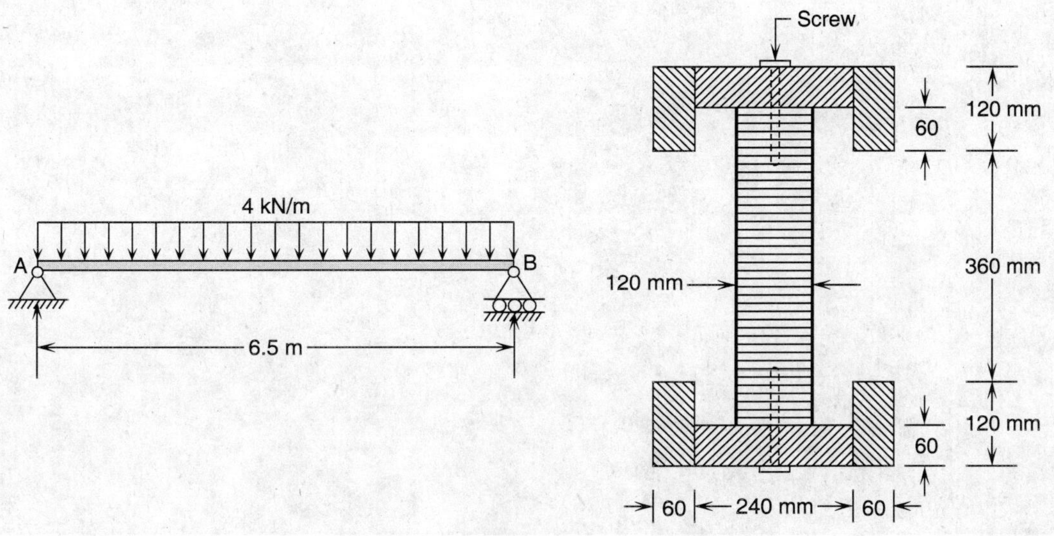

Figure 9.26

[**Ans.** $I = 4.90752 \times 10^9$ mm^4; $A\bar{y} = 7.344 \times 10^6$ mm^3 and $s_{max} = 102$ mm]

9.16 A I cross-section has flanges of width a and overall depth of $2a$. The flanges and the web are of uniform thickness t. Show that the ratio of the maximum and the average shearing stresses is 2.25.

[**Hint.** For small t: $A = 4bt$; $I = (8b^3t)/3$ and $\tau_{max} = 9V/(16bt)$]

9.17 Show that the difference between the maximum and mean shearing stresses down the rectangular web of an I cross-section beam shown in Figure 9.26 is $\left(\frac{Vd^2}{24I}\right)$, where V is the shear force at the section, d is the depth of the web and I the moment of inertia of the section about the neutral axis.

[**Hint:** Shear stress in the web at the junction of flange and web, $\frac{VB(D^2 - d^2)}{8Ib}$; maximum shear stress in the web, $\frac{VB(D^2 - d^2)}{8Ib} + \frac{Vd^2}{8I}$ and average shear stress in the web, $\frac{VB(D^2 - d^2)}{8Ib} + \frac{Vd^2}{12I}$ as shown in Figure 9.26(b)]

Chapter 10

Springs

10.1 INTRODUCTION

Springs are among the most common elastic elements in structural and mechanical engineering that are employed in various structures as *shock absorbers*, i.e. as *accumulators of elastic energy*, e.g. in wagons, valves, etc. Though they form mainly parts of machine elements, the derivation of fundamental design formulas are based on the principles of solid mechanics.

10.2 HELICAL SPRINGS

A helical spring may be regarded as a piece of rod bent in three dimensions with its axis line representing a *helix* in its simplest form. The geometrical shape of the axial line is determined by the diameter of the coil D, number of coils n and the angle of slope α. The slope of the helix axis of the coil is related to the pitch of the spring s as shown in Figure 10.1.

$$s = \pi D \tan \alpha \tag{10.1}$$

For the springs generally encountered in practice, the pitch s is very small as compared to πD, and hence the angle α may be considered small. It is commonly smaller than 5°. Such a spring with angle α being so small that the bending effects can be neglected, then the spring is called a *close-coiled spring*. The springs are generally wound of circular wire with its diameter denoted by d.

Based on the type of applied load the helical springs are divided into *tension springs*, *compression springs* and *torsion springs* as shown in Figure 10.2(a). In the first two cases, a spring is subjected to the forces having resultant acting along its axis. A torsion spring is subjected

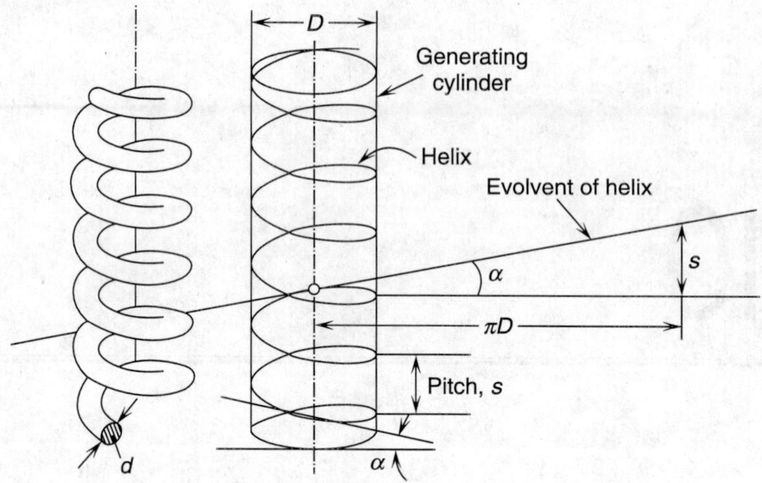

Figure 10.1 Generation of a helical spring.

to two moments in a plane perpendicular to the axis of spring. The end coils of a compression spring are turned down and ground off on the ends to provide bearing planes. The end coils of a tension and a torsion springs are unbent to provide means of its fastening to the adjacent parts. Figures 10.2(b) and (c) show conical spring and flat spiral spring, respectively.

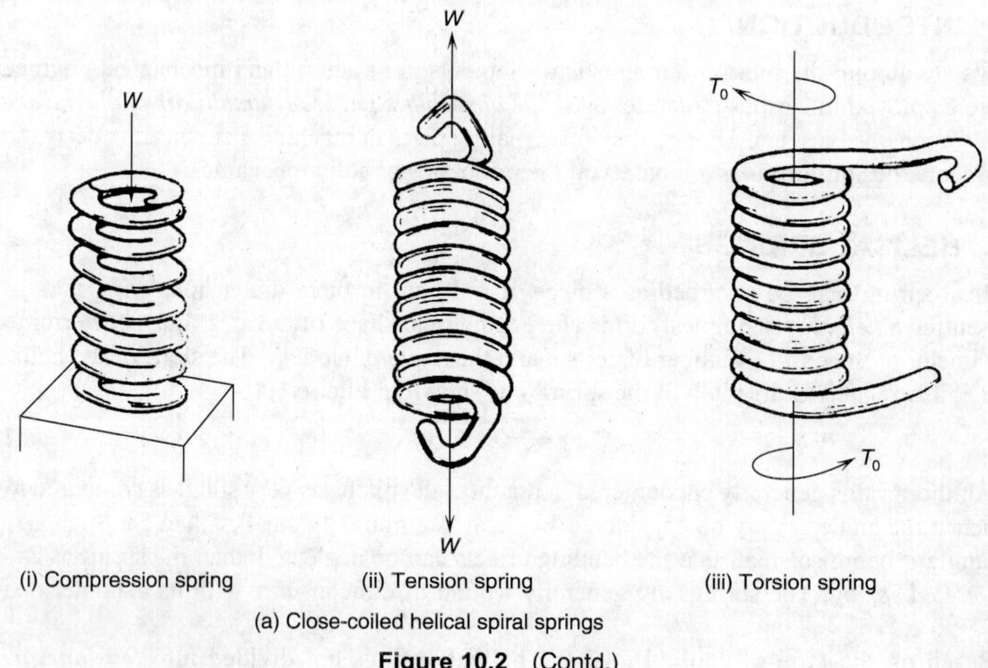

Figure 10.2 (Contd.)

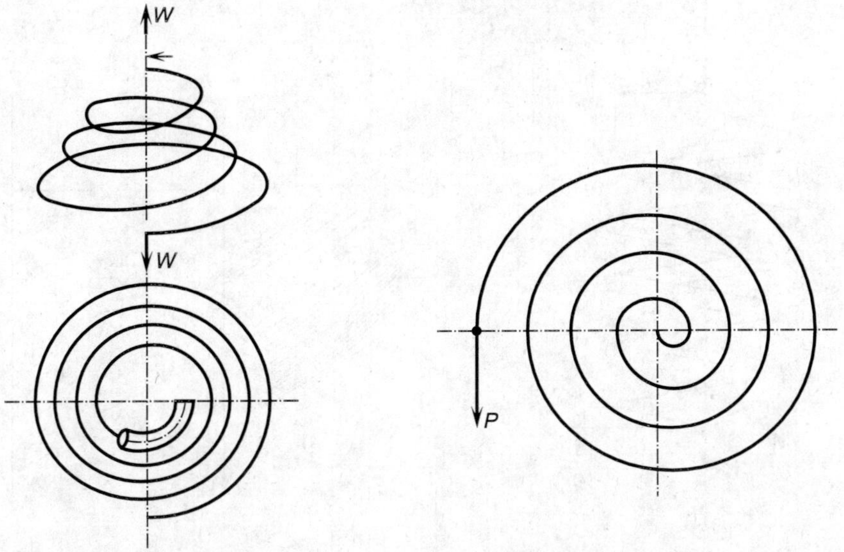

(b) Conical spring

(c) Flat spiral spring

Figure 10.2 Common types of spiral springs.

10.3 CLOSE-COILED HELICAL SPRINGS

10.3.1 Axial Load

Consider the close-coiled helical spring carrying load W as shown in Figure 10.3(a). The mean diameter, total number of active coils, the diameter of spring wire and shear modulus of the spring material are D, n, d and G, respectively. Consider the section of one of the coils of the spring obtained by a vertical plane containing the spring axis shown in Figure 10.3(b). It should be noted that for equilibrium a shear load W and a twisting moment $T(= WD/2)$ should act on this cross-section. The direct uniform shear stress due to W

$$\tau_d = \frac{4W}{\pi d^2} \tag{10.2}$$

Using torsion formula

$$\frac{T}{J} = \frac{\tau}{r} = \frac{G\theta}{L} \tag{10.3}$$

The maximum shear stress due to twisting moment which occurs at the surface is

$$\tau_{\max} = \frac{Tr}{J} = \frac{(WD/2)(d/2)}{(\pi d^4/32)} = \frac{8WD}{\pi d^3} \tag{10.4}$$

At the inner surface of the section, the two stresses are additive and the total stress is

$$\tau = \tau_d + \tau_{\max} = \frac{4W}{\pi d^2} + \frac{8WD}{\pi d^3} = \frac{4W}{\pi d^2}\left(1 + \frac{2D}{d}\right) \tag{10.5}$$

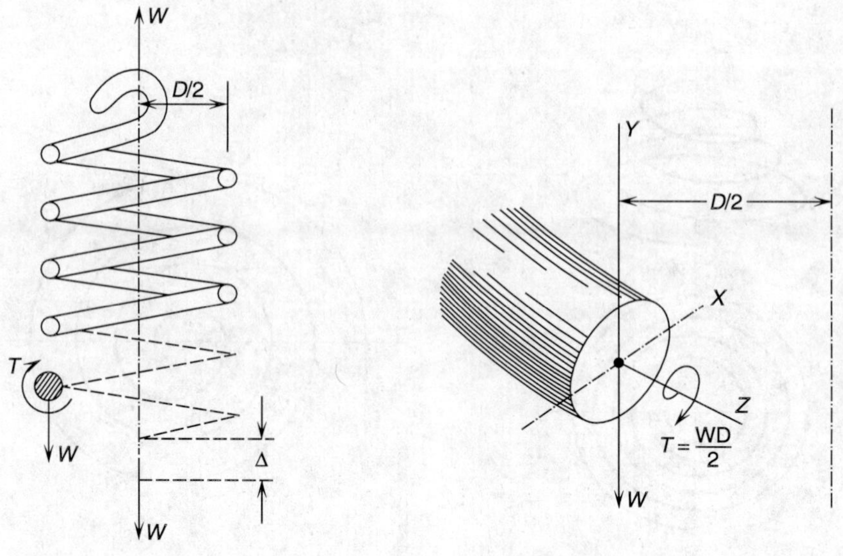

(a) Tension helical spring (b) Development of torque at a section

Figure 10.3 Close-coiled helical spring carrying axial force.

For normal springs the second term within brackets is generally very large compared to unity and hence unity can be neglected, i.e. the effect of pure shear being small in comparison to torsion shear can be safely ignored. Thus,

$$\tau \approx \tau_{\max} = \frac{8WD}{\pi d^3} \tag{10.6}$$

$$\theta = \frac{TL}{GJ} = \frac{(WD/2)(\pi Dn)}{G(\pi d^4/32)} = \frac{16nWD^2}{Gd^4} \tag{10.7}$$

Therefore, deflection $$\Delta = \frac{D\theta}{2} = \frac{8nWD^3}{Gd^4} \tag{10.8}$$

Alternatively, the deflection can be computed by equating external work done by the applied load to the strain energy stored in the spring.

Strain energy, $$U = \frac{T}{2}\theta = \frac{T}{2} \times \frac{TL}{GJ} = \frac{T^2 L}{2GJ} \tag{10.9}$$

Here $T = (WD/2)$; $L = (\pi D)n$ and $J = \pi d^4/32$. Thus

$$U = \frac{(WD/2)^2(\pi Dn)}{2G(\pi d^4/32)} = \frac{4nW^2 D^3}{Gd^4} \tag{10.10}$$

As the applied load gradually moves through the deflection Δ, the work done by the load W is $U_e = W\Delta/2$. Therefore,

$$U_e = U$$

$$\frac{W\Delta}{2} = \frac{4nW^2D^3}{Gd^4}$$

or
$$\Delta = \frac{8nWD^3}{Gd^4} \tag{10.11}$$

The ratio W/Δ, the load required per unit deflection is generally referred to as the *spring constant*, k.

Therefore,
$$k = \frac{W}{\Delta} = \frac{Gd^4}{8nR^3} \tag{10.12}$$

10.3.2 Strain Energy in an Axially Loaded Spring

From Eq. (10.10),
$$U = \frac{4nW^2D^3}{Gd^4}$$

To compute strain energy in terms of maximum shear stress, from Eq. (10.6)

$$W = \frac{\tau_{\max}\,\pi d^3}{8D} \tag{10.13}$$

Therefore,
$$U = \frac{4nD^3}{Gd^4} \times \left(\frac{\tau_{\max}\,\pi d^3}{8D}\right)^2$$

$$= \frac{\tau_{\max}^2}{4G} \times \left(\frac{\pi}{4}d^2\right)(\pi Dn)$$

$$= \frac{\tau_{\max}^2}{4G} \times \text{Volume of the spring} \tag{10.14}$$

The term $\frac{\tau_{\max}^2}{4G}$ is called *proof resilience* of the spring.

10.3.3 Spring Subjected to an Impact Load

Consider a weight W to fall on to a spring from a height H measured from the uncompressed state of spring. Let W_e is the equivalent static load and Δ the compression of the spring under W_e, then

Work done by the falling weight
$$= W(H + \Delta) \tag{10.15}$$

Work stored in the spring
$$= \frac{1}{2}W_e\,\Delta \tag{10.16}$$

From the *principle of conservation* of energy

$$W(H+\Delta)=\frac{1}{2}W_e\Delta \tag{10.17}$$

where $\Delta=\dfrac{8nW_eD^3}{Gd^4}$. Hence W_e and Δ can be determined.

10.3.4 Combinations of Springs

(a) *Springs in series.* When springs of different stiffness are joined end to end to support a common load W they are said to be connected in series. All the springs carry the same load W and the total deflection is

$$\frac{W}{k}=\frac{W}{k_1}+\frac{W}{k_2}+\frac{W}{k_3}+\cdots \tag{10.18}$$

where k is the combined or resultant stiffness.

(b) *Springs in parallel.* When springs of different stiffness are arranged with their axes parallel to each other to support a common load W, they are said to be connected in parallel. The springs share the load such that they all have a common deflection Δ

The common deflection, $\quad \Delta=\dfrac{W}{k}=\dfrac{W_1}{k_1}=\dfrac{W_2}{k_2}=\dfrac{W_3}{k_3}+\cdots$

Thus, $\quad W_1=W\dfrac{k_1}{k};\ W_2=W\dfrac{k_2}{k};\ W_3=W\dfrac{k_3}{k};\cdots$

The springs share the load in the ratios of their stiffness. The total load is

$$W=W_1+W_2+W_3+\cdots$$

$$=\frac{W}{k}(k_1+k_2+k_3+\cdots)$$

The combined stiffness, $\quad k=k_1+k_2+k_3+\cdots \qquad (10.19)$

10.3.5 Axial Torque or Torsion

A *torsion spring* is subjected to two couples in a plane perpendicular to the axis of spring. Consider a closely coiled helical spring subjected to an axial couple T_0 as illustrated in Figure 10.4(a). The couple T_0 produces bending in the coils of the spring as illustrated in Figure 10.4. Due to torsion the curvature of the coils increases or decreases. Let R_1 and R_2 be the initial and changed radius of curvatures of coils, respectively, and n_1 and n_2 are the corresponding number of the coils in the spring.

Change of curvature, $\quad \dfrac{1}{R_2}-\dfrac{1}{R_1}=\dfrac{T_0}{EI} \qquad (10.20)$

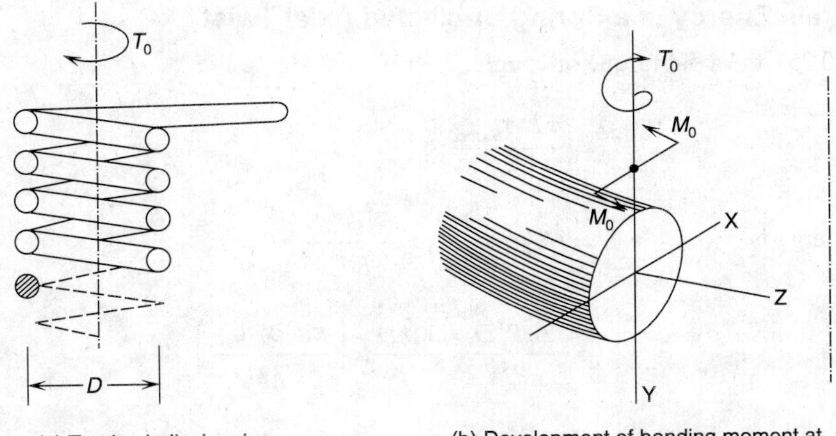

(a) Torsion helical spring

(b) Development of bending moment at a cross-section

Figure 10.4 Close-coiled helical spring under axial torque.

Since the total length of spring wire remains the same, i.e.

$$L = 2\pi R_1 n_1 = 2\pi R_2 n_2 \quad \text{or} \quad \frac{1}{R_2} - \frac{1}{R_1} = \frac{2\pi}{L}(n_2 - n_1)$$

Therefore,
$$T_0 = \frac{2\pi EI}{L}(n_2 - n_1) \tag{10.21}$$

The *wind up angle* or *angle of twist* in radians is,

$$\phi = 2\pi (n_2 - n_1) \tag{10.22}$$

Hence
$$T_0 = \frac{EI\phi}{L} \quad \text{or} \quad \phi = \frac{T_0 L}{EI} = \frac{T_0(2\pi R_1 n_1)}{EI} \tag{10.23}$$

For helical torsion spring made of round wire of diameter d, number of coils n, and mean coil diameter $D = 2R_1$,

$$\phi = \frac{T_0 \pi (2R_1) n}{E(\pi d^4/64)} = \frac{64DnT_0}{Ed^4} \tag{10.24}$$

The bending stress,
$$\sigma_b = \frac{T_0 y}{I}$$

Therefore, the maximum stress,

$$\sigma_{b,\max} = \frac{T_0(d/2)}{(\pi d^4/64)} = \frac{32}{\pi d^3} T_0 \tag{10.25}$$

10.3.6 Strain Energy in a Spring Subjected Axial Twist

From Eq. (10.25), the axial torque or couple,

$$T_0 = \frac{\pi d^3 \sigma_{b,\max}}{32} \tag{10.26}$$

The strain energy is

$$U = \frac{1}{2} T_0 \phi = \frac{1}{2} T_0 \times \frac{64 n T_0 D}{E d^4}$$

$$= \frac{32 n T_0^2 D}{E d^4} = \frac{32 n D}{E d^4} \times \left(\frac{\pi d^3 \sigma_{b,\max}}{32} \right)^2$$

$$= \frac{\sigma_{b,\max}^2}{8E} \times \left(\frac{\pi}{4} d^2 \right) (\pi D n)$$

$$= \frac{\sigma_{b,\max}^2}{8E} \times \text{Volume of the spring} \tag{10.27}$$

The quantity $\dfrac{\sigma_{b,\max}^2}{8E}$ is called the *proof resilience.*

EXAMPLE 10.1

A close-coiled helical spring has a stiffness of 10 N/mm. The length of the spring when fully compressed with adjacent coils in contact is 400 mm. Determine (a) the diameter of the spring wire and the mean diameter of the coils if their ratio is 1/10, (b) the maximum load that can be applied before adjacent coils touch, if the gap between any two adjacent coils is 2.5 mm and (c) the maximum shear stress in the spring. The modulus of rigidity of the material of the spring is 80 GPa.

Solution: The solid length of the spring, $nd = 400$ mm or $n = 400/d$

(a) Ratio of spring wire and coil diameter, $\dfrac{d}{D} = \dfrac{1}{10}$ or $D = 10d$

Stiffness,

$$k = \frac{W}{\Delta} = \frac{G d^4}{8 n D^3}$$

$$10 = \frac{(80 \times 10^3) d^4}{8(400/d)(10d)^3} \quad \text{or} \quad d^2 = \frac{10 \times 8 \times 400 \times 10^3}{80 \times 10^3} = 400$$

Therefore,

$$d = 20 \text{ mm}$$

$$D = 10d = 10 \times 20 = 200 \text{ mm}$$

$$n = \frac{400}{d} = \frac{400}{20} = 20 \text{ coils}$$

(b) Gap between adjacent coils, $\delta = 2.5$ mm

Maximum deflection $\Delta = 2.5 \times 20 = 50$ mm

$$k = \frac{W}{\Delta} \quad \text{or} \quad 10 = \frac{W}{50}$$

Therefore, $W = 10 \times 50 = 500$ N

(c)
$$\tau_{\max} = \frac{8WD}{\pi d^3} = \frac{8 \times 500 \times 200}{\pi \times 20^3} = 31.83 \text{ MPa}$$

EXAMPLE 10.2

A close-coiled helical spring of stiffness of 1 N/mm is to be designed for a maximum axial load of 45 N with maximum shear limited to 130 MPa. The length of the spring when the coils are in contact, i.e. solid length is to be 50 mm. Determine the diameter of the spring wire, the mean diameter of the coils and the number of coils required. The modulus of rigidity of the material is 45 GPa.

Solution: The solid length of the spring, $nd = 50$ mm or $n = 50/d$ (i)

Stiffness,
$$k = \frac{W}{\Delta} = \frac{Gd^4}{8nD^3}$$

Therefore,
$$1 = \frac{(45 \times 10^3) \times d^4}{8 \times n \times D^3} \quad \text{or} \quad d^4 = \frac{8 \times n \times D^3}{(45 \times 10^3)} \tag{ii}$$

The maximum shear is

$$\tau_{\max} = \frac{8WD}{\pi d^3} \quad \text{or} \quad 130 = \frac{8 \times 45 \times D}{\pi \times d^3}$$

Therefore,
$$D = \frac{130 \times \pi \times d^3}{8 \times 45} = 1.1345\, d^3 \tag{iii}$$

Substitute from Eqs. (i) and (iii) into (ii)

$$d^4 = \frac{8}{(45 \times 10^3)} \times \frac{50}{d} \times (1.1345 d^3)^3 \quad \text{or} \quad d^4 = \frac{(45 \times 10^3)}{8 \times 45 \times 1.1345^3}$$

Hence $d = 3.042$ mm

$D = 1.1345 d^3 = 1.1345 \times 3.042^3 = 31.94$ mm

$$n = \frac{50}{d} = \frac{50}{3.042} = 16.44$$

EXAMPLE 10.3

A close-coiled helical spring made of 12 mm round steel has eleven coils and the mean diameter of the coil is 160 mm. (a) If the spring is subjected to an axial load of 155 N, determine the elongation, intensity of torsional stress and strain energy stored per cubic metre of the spring. (b) If the axial load is removed and axial torque of 12 N.m is applied, determine the axial twist, intensity of bending stress, and the work stored per cubic metre of the spring under the loaded condition. The modulus of rigidity and elasticity are 80 GPa and 200 GPa, respectively.

Solution: (a) Under axial load, the deflection of the spring,

$$\Delta = \frac{8nWD^3}{Gd^4} = \frac{8 \times 11 \times 155 \times 160^3}{(80 \times 10^3) \times 12^4} = 33.67 \text{ mm}$$

The maximum shear stress is

$$\tau_{\max} = \frac{8WD}{\pi d^3} = \frac{8 \times 155 \times 160}{\pi \times 12^3} = 36.547 \text{ MPa}$$

Strain energy stored per cubic metre of the spring

$$U = \frac{\tau_{\max}^2}{4G} \times 10^9 = \frac{36.547^2 \times 10^9}{4 \times (80 \times 10^3)}$$

$$= 4.174 \times 10^6 \text{ N.mm/m}^3 = 4.174 \text{ kN.m/m}^3$$

(b) Under axial torque, axial twist in the spring ϕ

$$\phi = \frac{64nT_0D}{Ed^4} = \frac{64 \times 11 \times (12 \times 10^3) \times 160}{(200 \times 10^3) \times 12^4} = 0.3259 \text{ radian}$$

The maximum bending stress is

$$\sigma_{b,\max} = \frac{32T_0}{\pi d^3} = \frac{32 \times (12 \times 10^3)}{\pi \times 12^3} = 70.736 \text{ MPa}$$

Strain energy stored per cubic metre of the spring,

$$U = \frac{\sigma_{b,\max}^2}{8E} = \frac{70.736^2 \times 10^9}{8 \times (200 \times 10^3)}$$

$$= 3.127 \times 10^6 \text{ Nmm/m}^3 = 3.127 \text{ kNm/m}^3$$

EXAMPLE 10.4

A close-coiled helical spring of 150 mm free uncompressed length is designed to absorb strain energy equal to 50 Nm when fully compressed with coils in contact. The maximum shearing stress is limited to 140 MPa. If the mean diameter of the coil is 100 mm, determine the diameter of the steel wire required and the number of coils. The modulus of rigidity of the spring material is 80 GPa.

Solution: Strain energy stored,

$$U = \frac{\tau_{\max}^2}{4G} \times \text{Volume} = \frac{\tau_{\max}^2}{4G} \times \left(\frac{\pi}{4} d^2\right)(\pi D n)$$

Therefore, $$50 \times 10^3 = \frac{140^2}{4 \times (80 \times 10^3)} \times \left(\frac{\pi}{4} \times d^2\right)(\pi \times 100 \times n)$$

$$d^2 n = 3308.4 \qquad \text{(i)}$$

The maximum shear stress is given by

$$\tau_{\max} = \frac{8W_e D}{\pi d^3} \quad \text{or} \quad 140 = \frac{8 \times W_e \times 100}{\pi \times d^3}$$

Hence $$W_e = \frac{140 \times \pi}{8 \times 100} d^3 = 0.5498 d^3$$

Compression, $$\Delta = \text{Free length} - \text{Compressed length} = \frac{8 n W_e D^3}{G d^4}$$

or $$150 - nd = \frac{8 \times n \times (0.5498 d^3) \times 100^3}{(80 \times 10^3) \times d^4} = \frac{54.98 n}{d}$$

Thus, $$\left(\frac{54.98}{d} + d\right) n = 150 \qquad \text{(ii)}$$

Eliminate n between Eqs. (i) and (ii)

$$\left(\frac{54.98}{d} + d\right)\left(\frac{3308.4}{d^2}\right) = 150$$

$$d^3 - \frac{3308.4 \times (54.98 + d^2)}{150} = 0 \quad \text{or} \quad d^3 - 22.056 d^2 - 1212.64 = 0$$

By trial and modification $d = 24.1375$

Therefore, $$n = \frac{3308.4}{d^2} = \frac{3308.4}{24.1375^2} = 5.678$$

EXAMPLE 10.5

A vehicle weighing 30 kN and running at 1.8 m/s is brought to rest by a buffer spring. The buffer spring consists of close-coiled helical springs, each having 15 coils of 25 mm diameter steel rod coiled to a mean diameter of 200 mm. Determine the number of springs to absorb the energy of motion during compression of 250 mm. The modulus of rigidity and acceleration due to gravity are 95 GPa and 9.8 m/s^2, respectively.

Solution: Kinetic energy of the vehicle,

$$\text{K.E.} = \frac{mv^2}{2} = \frac{(W/g)v^2}{2} = \frac{(30 \times 10^3) \times 1.8^2}{2 \times 9.8} = 4959.2 \text{ Nm (joule)}$$

Consider P newton be the gradually applied force on one spring so as to compress it by 250 mm. Then

$$\Delta = \frac{8nPD^3}{Gd^4} \quad \text{or} \quad 250 = \frac{8 \times 15 \times P \times 200^3}{(95 \times 10^3) \times 25^4}$$

Therefore,

$$P = \frac{250 \times (95 \times 10^3) \times 25^4}{8 \times 15 \times 200^3} = 9663.9 \text{ N}$$

Strain energy stored by one spring,

$$U = \frac{1}{2} P\Delta = \frac{1}{2} \times 9663.9 \times 250 = 1207987.5 \text{ N mm} = 1207.987 \text{ joule}$$

Number of springs required,

$$N = \frac{4959.2}{1207.987} = 4.11 \text{ say 5 springs}$$

EXAMPLE 10.6

A horizontal rigid bar of negligible weight is suspended from three close-coiled helical springs of equal length as shown in Figure 10.5. The arrangement transmits a load W at a distance x from left support such that the bar remains horizontal. The springs are made out of the rods of equal diameters. The numbers of coils in the three springs are 10, 12 and 16, respectively, while the mean diameters of coils are in the proportion 1:1.2:1.42, respectively. Determine the distance x of point of application of the load.

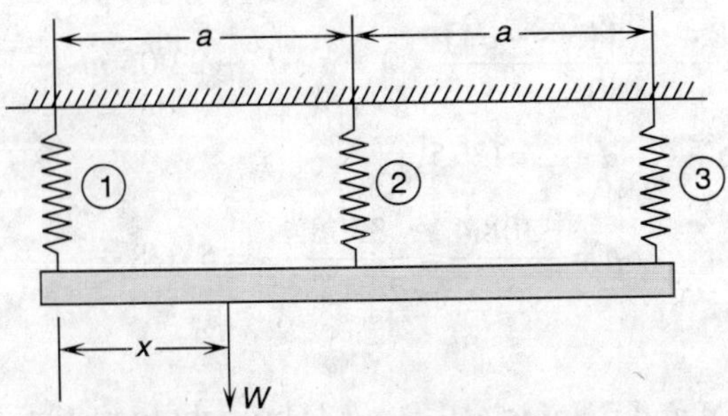

Figure 10.5 Horizontal rigid bar of Example 10.5.

Solution: Since the springs are made out of the same rod,

$$d_1 = d_2 = d_3 = d \quad \text{and} \quad G_1 = G_2 = G_3 = G$$

Numbers of coils in three springs $n_1 = 10$, $n_2 = 12$ and $n_3 = 16$

Mean diameters of the coils $D_1 : D_2 : D_3 :: 1.0 : 1.2 : 1.42$

As the bar remains horizontal after the application of the load, the extension in three springs is the same, i.e. $\Delta_1 = \Delta_2 = \Delta_3 = \Delta$. Then

$$\Delta_i = \frac{8n_i W_i D_i^3}{G_i d_i^4} \quad \text{or} \quad W_i = \frac{\Delta_i G_i d_i^4}{8n_i D_i^3} = \frac{C}{n_i D_i^3}$$

Where the constant $C = \Delta G d^4/8$.

Consider the loads carried by three springs to be W_1, W_2 and W_3, respectively. Then for moment equilibrium, the point of application of load W must coincide with the C.G. of the spring loads, i.e.

$$x = \frac{W_2 a + W_3(2a)}{(W_1 + W_2 + W_3)} = \left(\frac{W_2 + 2W_3}{W_1 + W_2 + W_3}\right) a$$

As the relative magnitude of the forces can be used, it is convenient to work in terms of ratios. Divide the numerator and denominator by W_1,

$$x = \frac{(W_2 / W_1) + 2(W_3 / W_1)}{1 + (W_2 / W_1) + (W_3 / W_1)} a$$

The various ratios are

$$\frac{W_2}{W_1} = \frac{n_1 D_1^3}{n_2 D_2^3} = \frac{10 \times 1.0^3}{12 \times 1.2^3} = 0.48225$$

$$\frac{W_3}{W_1} = \frac{n_1 D_1^3}{n_3 D_3^3} = \frac{10 \times 1.0^3}{16 \times 1.42^3} = 0.21828$$

Therefore,

$$x = \frac{0.48225 + 2 \times 0.21828}{1 + 0.48225 + 0.21828} a = 0.5403a$$

10.4 CONCENTRIC OR CLUSTER SPRINGS

Concentric or cluster springs are close-coiled helical springs placed one inside the other as shown in Figures 10.6(a) and (b). The clearance between the coils is

$$\delta = \frac{(D_2 - D_1)}{2} - \frac{(d_1 + d_2)}{2} \tag{10.28}$$

The arrangement has to satisfy the following conditions:

(a) For the springs made of same material and having equal free length, the maximum shear stress in them is equal,

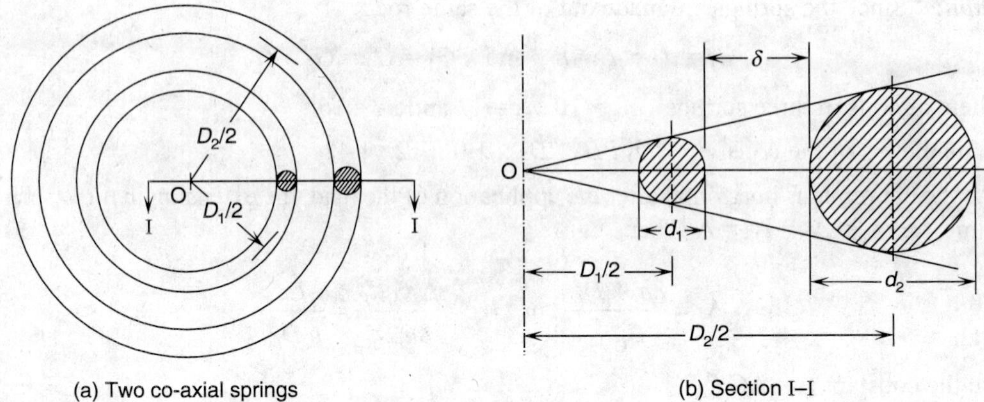

(a) Two co-axial springs

(b) Section I–I

Figure 10.6 Concentric or co-axial springs.

$$\frac{8W_1D_1}{\pi d_1^3} = \frac{8W_2D_2}{\pi d_2^3} \quad \text{or} \quad \frac{W_1}{W_2} = \left(\frac{D_2}{D_1}\right)\left(\frac{d_1}{d_2}\right)^3 \tag{10.29}$$

(b) If the deflection is same for both the springs,

$$\frac{8n_1W_1D_1^3}{G_1d_1^4} = \frac{8n_2W_2D_2^3}{G_2d_2^4} \quad \text{or} \quad \frac{W_1}{W_2} = \left(\frac{D_2}{D_1}\right)^3\left(\frac{d_1}{d_2}\right)^4\left(\frac{G_1}{G_2}\right)\left(\frac{n_2}{n_1}\right) \tag{10.30}$$

(c) If the deflection is unequal in the springs,

$$\frac{W_1}{W_2} = \left(\frac{\Delta_1}{\Delta_2}\right)\left(\frac{D_2}{D_1}\right)^3\left(\frac{d_1}{d_2}\right)^4\left(\frac{G_1}{G_2}\right)\left(\frac{n_2}{n_1}\right) \tag{10.31}$$

(d) If both the springs have the same solid length,

$$n_1d_1 = n_2d_2 \quad \text{or} \quad \frac{n_2}{n_1} = \frac{d_1}{d_2} \tag{10.32}$$

Comparing Eqs. (10.29) and (10.30):

$$1 = \left(\frac{D_2}{D_1}\right)^2\left(\frac{d_1}{d_2}\right)\left(\frac{G_1}{G_2}\right)\left(\frac{n_2}{n_1}\right) \tag{10.33}$$

Eliminating n_1 and n_2 between Eqs. (10.32) and (10.33):

$$1 = \left(\frac{D_2}{D_1}\right)^2\left(\frac{d_1}{d_2}\right)^2\left(\frac{G_1}{G_2}\right) \tag{10.34}$$

If $G_1 = G_2$, then

$$\left(\frac{D_1}{d_1}\right) = \left(\frac{D_2}{d_2}\right)$$

The ratio D/d is called the *spring index*, thus the spring indices are equal.

Therefore, Eq. (10.29) reduces to

$$\frac{W_1}{W_2} = \left(\frac{d_1}{d_2}\right)^4 \quad \text{and} \quad \tan\theta = \left(\frac{d_1}{D_1}\right) = \left(\frac{d_2}{D_2}\right) \tag{10.35}$$

The total load carried by the springs, $W = W_1 + W_2$.

EXAMPLE 10.7

In a compound spring consisting of two concentric close-coiled helical springs, the inner spring has 10 turns of 7.2 mm diameter steel bar coiled to a mean diameter of 100 mm, and the outer spring has 8 turns of 8 mm diameter bar coiled to a mean diameter of 125 mm. The uncompressed axial lengths of inner and outer springs are 80 mm and 75 mm, respectively. The compound spring is placed between a pair of two parallel plates and compressed until the distance between the two plates reduces to 65 mm. Determine (a) the load applied to the plates and (b) the maximum stress developed in each spring. The modulus of rigidity is 82 GPa.

Solution: For inner spring, $n_1 = 10$, $d_1 = 7.2$ mm, $D_1 = 100$ mm and $l_1 = 80$ nm

For outer spring, $n_2 = 8$, $d_2 = 8$ mm, $D_2 = 125$ mm and $l_2 = 75$

Compression of two springs,

$$\Delta_1 = 80 - 65 = 15 \text{ mm and } \Delta_2 = 75 - 65 = 10 \text{ mm}$$

The load carried by two springs

$$W_1 = \frac{\Delta_1 G_1 d_1^4}{8n_1 D_1^3} = \frac{15 \times (82 \times 10^3) \times 7.2^4}{8 \times 10 \times 100^3} = 41.32 \text{ N}$$

$$W_2 = \frac{\Delta_2 G_2 d_2^4}{8n_2 D_2^3} = \frac{10 \times (82 \times 10^3) \times 8^4}{8 \times 8 \times 125^3} = 27.87 \text{ N}$$

Therefore, $W = W_1 + W_2 = 68.19$ N

The stresses in the two springs are

$$\tau_{\max} = \frac{8W_1 D_1}{\pi d_1^3} = \frac{8 \times 41.32 \times 100}{\pi \times 7.2^3} = 28.19 \text{ MPa}$$

$$\tau_{\max} = \frac{8W_2 D_2}{\pi d_2^3} = \frac{8 \times 26.87 \times 125}{\pi \times 8^3} = 16.705 \text{ MPa}$$

EXAMPLE 10.8

In a compound spring consisting of two co-axial close-coiled helical springs, the inner spring has 20 turns of 10 mm diameter steel bar coiled to a mean diameter of 90 mm, and the outer spring has 15 turns coiled to a mean diameter of 120 mm. The springs are made of same material and have equal free length. Determine (a) the diameter of the steel wire required for the outer spring and (b) the stiffness of the compound spring. The modulus of rigidity is 80 GPa.

Solution: For inner spring, $n_1 = 20$, $d_1 = 10$ mm and $D_1 = 90$ mm
For outer spring, $n_2 = 15$ and $D_2 = 120$ mm
Since the maximum shear stress is same in both the springs

$$\frac{8W_1D_1}{\pi d_1^3} = \frac{8W_2D_2}{\pi d_2^3} \quad \text{or} \quad \frac{W_1}{W_2} = \left(\frac{D_2}{D_1}\right)\left(\frac{d_1}{d_2}\right)^3$$

Therefore,
$$\frac{W_1}{W_2} = \left(\frac{120}{90}\right)\left(\frac{10}{d_2}\right)^3 \quad \text{(i)}$$

Since both the springs have same initial length, they will have same deformation

$$\frac{8n_1W_1D_1^3}{G_1d_1^4} = \frac{8n_2W_2D_2^3}{G_2d_2^4} \quad \text{or} \quad \frac{W_1}{W_2} = \left(\frac{D_2}{D_1}\right)^3\left(\frac{d_1}{d_2}\right)^4\left(\frac{G_1}{G_2}\right)\left(\frac{n_2}{n_1}\right)$$

$$\frac{W_1}{W_2} = \left(\frac{120}{90}\right)^3\left(\frac{10}{d_2}\right)^4\left(\frac{15}{20}\right) \quad \text{(ii)}$$

Operate on Eqs. (i) and (ii) as: $\dfrac{\text{(ii)}}{\text{(i)}}$ or $\left(\dfrac{120}{90}\right)^2\left(\dfrac{10}{d_2}\right)\left(\dfrac{15}{20}\right) = 1$

Therefore,
$$d_2 = \left(\frac{120}{90}\right)^2 (10)\left(\frac{15}{20}\right) = 13.33 \text{ mm}$$

Since the springs are in parallel,

$$k = k_1 + k_2 = \frac{W_1}{\Delta} + \frac{W_2}{\Delta} = \frac{G_1d_1^4}{8n_1D_1^3} + \frac{G_2d_2^4}{8n_2D_2^3}$$

Stiffness
$$= \left[\frac{(80\times 10^3)\times 10^4}{8\times 20\times 90^3} + \frac{(80\times 10^3)\times 13.33^4}{8\times 15\times 120^3}\right] = 19.04 \text{ kN/m}$$

10.5 CONICAL SPRING

Consider the conical spring of minimum radius R_1, maximum radius R_2 and the number of turns n shown in Figure 10.7. At any point on the spring, the radius R from the axis of spring is given by

$$R = R_1 + \frac{(R_2 - R_1)\alpha}{2\pi n} \quad (10.36)$$

where α represents the angular distance of the point under consideration from the start of spring at smaller end as shown in Figure 10.7. Consider an element of length ds of the spring at radius R. The angle of twist produced by W in the length ds

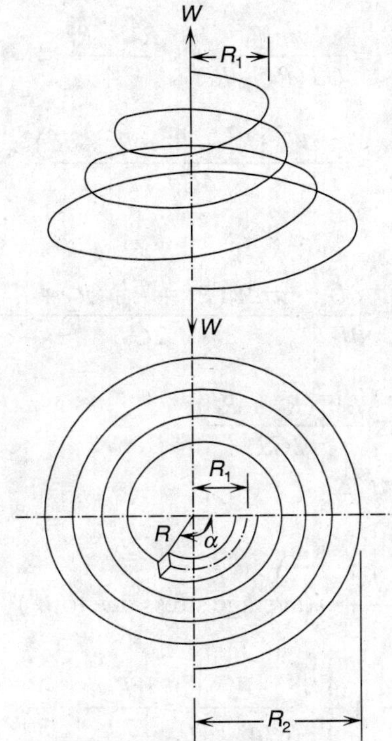

Figure 10.7 Conical spring.

$$d\theta = \frac{WR\,ds}{GJ} \tag{10.37}$$

Strain energy stored in the element

$$dU = \frac{1}{2}WR\,d\theta = \frac{1}{2}WR \times \frac{WR\,ds}{GJ} = \frac{W^2R^2\,(R\,d\alpha)}{2GJ}$$

$$= \frac{W^2R^3\,d\alpha}{2GJ} \tag{10.38}$$

From Eq. (10.36) $$dR = \frac{(R_2 - R_1)}{2\pi n}\,d\alpha \quad \text{or} \quad d\alpha = \frac{2\pi n}{(R_2 - R_1)}\,dR$$

Thus, $$dU = \frac{W^2R^3}{2GJ} \times \frac{2\pi n}{(R_2 - R_1)}\,dR = \frac{\pi n W^2}{GJ(R_2 - R_1)}\,R^3\,dR \tag{10.39}$$

Strain energy in the spring,

$$U = \frac{\pi n W^2}{GJ(R_2 - R_1)}\int_{R_1}^{R_2} R^3\,dR$$

$$= \frac{\pi n W^2}{GJ(R_2 - R_1)} \times \frac{R_2^4 - R_1^4}{4}$$

$$= \frac{\pi n W^2 (R_2^2 + R_1^2)(R_2 + R_1)}{4GJ} \tag{10.40}$$

Therefore, axial deflection

$$y_0 = \frac{\partial U}{\partial W} = \frac{\pi n W (R_2^2 + R_1^2)(R_2 + R_1)}{2GJ}$$

$$= \frac{LW(R_2^2 + R_1^2)}{2GJ} \tag{10.41}$$

where $L = n\pi(R_2 + R_1)$ and $J = \pi d^2/32$

Maximum shear stress

$$\tau_{max} = \frac{WR_2 d}{2J} + \frac{4}{3}(\text{average stress due to } W)$$

$$= \frac{WR_2 d}{2J} + \frac{4}{3} \times \frac{W}{(\pi d^2/4)} = \frac{WR_2 d}{2J} + \frac{4}{3} \times \frac{Wd^2}{8(\pi d^4/32)}$$

$$= \frac{WR_2 d}{2J} + \frac{1}{3} \times \frac{Wd^2}{2J} = \frac{Wd}{6J}(3R_2 + d) \tag{10.42}$$

EXAMPLE 10.9

A conical spring having 8 coils with minimum and maximum diameters of 60 mm and 120 mm, respectively, is made of 16 mm diameter steel wire. Determine the maximum load the spring can support without exceeding the shear stress of 200 MPa. Also determine the deflection of the spring. The modulus of rigidity of the material of the spring is $G = 84$ GPa.

Solution: From Eq. (10.42): $W = \dfrac{\tau_{max}(6J)}{d(3R_2 + d)}$

$$= \frac{200 \times 6 \times (\pi \times 16^4/32)}{16 \times (3 \times 60 + 16)}$$

$$= 2461.98 \text{ N} = 2.462 \text{ kN}$$

Deflection, $$y_0 = \frac{\pi n W (R_2^2 + R_1^2)(R_2 + R_1)}{2GJ}$$

$$= \frac{\pi \times 8 \times 2462 \times (60^2 + 30^2) \times (60 + 30)}{2 \times (84 \times 10^3) \times (\pi \times 16^4/32)} = 23.18 \text{ mm}$$

10.6 FLAT SPIRAL SPRINGS

This type of spring is made of wire with one end attached to a winding spindle at O and the other end is fixed to a pin C as shown in Figure 10.8. If a couple M is applied to the spindle, it will cause a reactive force F at C. For analysis, it is assumed that the shape of spring remains such that its C.G. is at the point O and spring at C remains normal to CO. Further it is assumed that during winding the coils do not come in contact with each other.

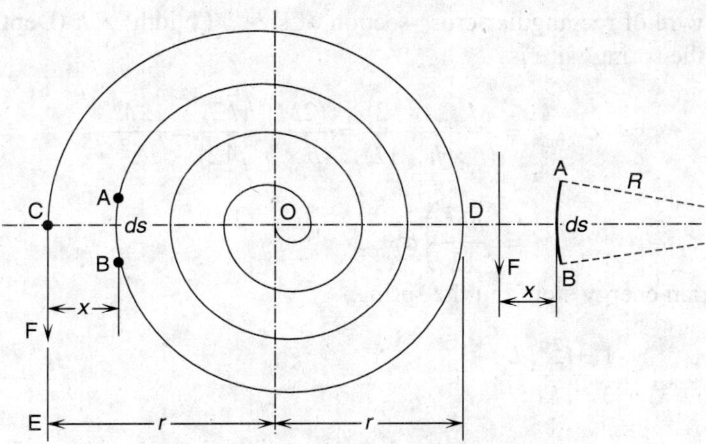

Figure 10.8 Flat spiral spring.

Consider an element of length ds of the spring wire at distance x from C. The element undergoes a change in angle between its two ends A and B during winding operation, i.e. before and after straining. The change in the curvature of the element $d\theta/ds$ is related to bending moment as

$$\frac{d\theta}{ds} = \frac{M_x}{EI} \quad \text{or} \quad d\theta = \frac{M_x\, ds}{EI} = \frac{(Fx)\, ds}{EI} \tag{10.43}$$

Total angle of winding up due to straining,

$$\theta = \int d\theta = \int \frac{(Fx)\, ds}{EI} = \frac{F}{EI}\int x\, ds \tag{10.44}$$

The quantity $\int x\, ds$ in the above expression is given by

$$\int x\, ds = \sum \text{Length of spring elements} \times \text{Their distance from fixed end}$$

$$= \text{Length of spring} \times \text{Distance between the winding spindle and the fixed end}$$

$$= L \times r \tag{10.45}$$

Hence from Eq. (10.44):
$$\theta = \frac{F}{EI} \times (Lr) = (Fr) \times \frac{L}{EI} = \frac{ML}{EI} \tag{10.46}$$

where $Fr = M$, the couple applied to the spring

Strain energy stored in the spring

$$U = \frac{1}{2}M\theta = \frac{1}{2}M \times \frac{ML}{EI} = \frac{1}{2}\frac{M^2 L}{EI} \tag{10.47}$$

The maximum moment occurs at the point D ($x = 2r$) and thus

$$M_{\text{max}} = F(2r) = 2(Fr) = 2M \tag{10.48}$$

For a spring wire of rectangular cross-section of size b (Width) × h (Depth), the maximum bending stress in the spring wire is

$$\sigma_{b,\text{max}} = \frac{M_{\text{max}}(h/2)}{(b \times h^3/12)} = \frac{(2M)(h/2)}{(b \times h^3/12)} = \frac{12M}{bh^2} \tag{10.49}$$

Therefore, $$M = \left(\frac{bh^2}{12}\right)\sigma_{b,\text{max}} \tag{10.50}$$

The maximum strain energy stored in the spring,

$$U_{\text{max}} = \frac{1}{2}\frac{M_{\text{max}}^2 L}{EI}$$

$$= \frac{1}{2} \times \left[\left(\frac{bh^2}{12}\right)\sigma_{b,\text{max}}\right]^2 \times \frac{L}{E} \times \frac{12}{bh^3}$$

$$= \frac{\sigma_{b,\text{max}}^2}{24E}(bhL) = \frac{\sigma_{b,\text{max}}^2}{24E} \times (\text{Volume of spring}) \tag{10.51}$$

Thus, proof resilience of the spring is $\dfrac{\sigma_{b,\text{max}}^2}{24E}$.

EXAMPLE 10.10

A flat spring is made of 12 mm wide × 0.6 mm thick special steel wire of 3.5 m length. Determine the torque applied on its central spindle, the amount of energy stored and the number of turns to wind up the spring, when the bending stress in the spring is limited to 600 MPa. The elastic modulus of the material of the spring is 200 GPa.

Solution: For the given spring,

$$b = 12\text{ mm};\ h = 0.6\text{ mm};\ L = 3.5\text{ m};\ \sigma_{b,\text{max}} = 600\text{ MPa and } E = 200\text{ GPa}$$

Torque applied on its central spindle, from Eq. (10.50);

$$M = \left(\frac{bh^2}{12}\right)\sigma_{b,\text{max}} = \frac{12 \times 0.6^2}{12} \times 600 = 216\text{ N}\cdot\text{mm}$$

Winding angle of the spring per turn,

$$\theta = \frac{ML}{EI} = \frac{216 \times (3.5 \times 10^3)}{(200 \times 10^3) \times (12 \times 0.6^3 / 12)} = 17.5 \text{ rad}$$

The number of turns to wind up

$$n = \frac{\theta}{2\pi} = \frac{17.5}{2\pi} = 2.785$$

The energy stored in the spring

$$U_{\max} = \frac{1}{2} \frac{M_{\max}^2 L}{EI}$$

$$= \frac{1}{2} \times \frac{216^2 \times (3.5 \times 10^3)}{(200 \times 10^3) \times (12 \times 0.6^3 / 12)} = 1890 \text{ N.mm}$$

10.7 LEAF SPRINGS

These are commonly used in carriages like cars, wagons, etc. A leaf spring consists of a number of plates of same width and thickness but of different lengths laminated one over the other. They are also called *laminated* or *built-up springs* or *carriage springs*. Based on the shape of initial curve sometimes they are also called *semi-elliptical* or *quarter-elliptic leaf springs*. The friction between plates is negligible and each plate can be considered to be free to slide over the neighbouring plate. These *plates are given a curvature, so that the initial deflection reduces under the load*. The plates are assumed to maintain the same radius. Hence each plate section has its own neutral axis.

Consider a semi-elliptical leaf spring supported at the ends and loaded at the centre as shown in Figure 10.9(a). Each of the plates is stressed to a maximum bending stress of $\sigma_{b,\max}$, and the number of plates at a given section is such that the moment resisted by them is equal to the applied bending moment. The load under which initial deflection vanishes and plates become straight is called the *proof load* of the leaf spring.

From the bending formula

$$\left[\frac{M}{I} = \frac{E}{R}\right] = \frac{\sigma_b}{y} \quad \text{or} \quad M = \frac{EI}{R} \tag{10.52}$$

where R is the radius of curvature of the deformed plate. With thickness of plates being negligible compared to R, all plates will have the same radius; hence the moment required to bend each plate into an arc of this radius is the same. Thus, *if a moment M is applied to a system of n plates, the moment is shared equally by all n plates*. Alternatively, for analysis, the leaf spring can be considered to be a beam of uniform depth equal to the thickness of one plate and width varying uniformly from zero at each end to nb at the centre, where n is the number of plates and b is the width of each plate.

For the leaf spring of Figure 10.9(b), the maximum bending moment which occurs at the centre is $WL/4$. If the spring consists of n plates with cross-section of each plate being $b \times t$, the moment resisted by each plates is $WL/4n$.

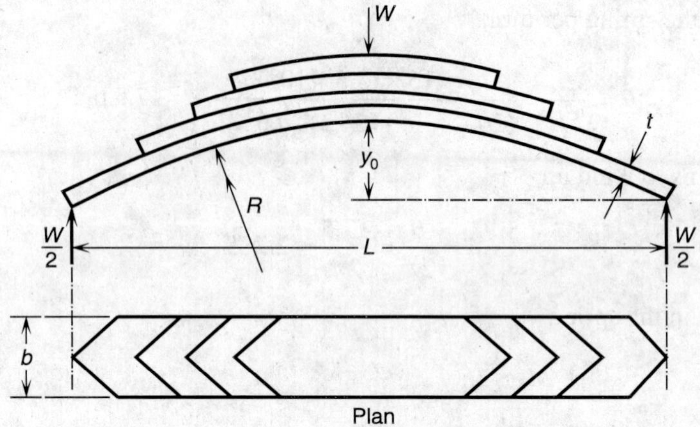

(a) Semi-elliptical laminated or leaf spring

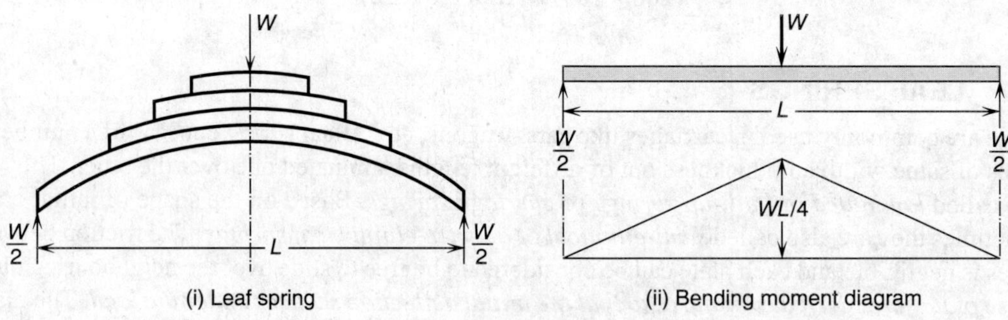

(b) Analysis of a laminated spring

Figure 10.9 Laminated or leaf springs.

The resulting bending stress $\sigma_{b,\max}$ is given by

$$\sigma_{b,\max} = \frac{M}{Z} = \frac{(WL/4n)}{bt^2/6} = \frac{3WL}{2nbt^2} \tag{10.53}$$

The initial central deflection, y_0 can be computed from the geometric properties of a segment of a circle

$$\left(\frac{L}{2}\right)^2 = R^2 - (R - y_0)^2 = 2Ry_0 - y_0^2 \tag{10.54}$$

Since y_0 is very small as compared to R, the term y_0^2 can be ignored in Eq. (10.54), thus

$$y_0 = \frac{L^2}{8R} \tag{10.55}$$

From Eq. (10.52):

$$\frac{1}{R} = \frac{\sigma_b}{Ey} = \frac{\sigma_{b,\max}}{E(t/2)} = \frac{2\sigma_{b,\max}}{Et} \tag{10.56}$$

Substitute the value of $\sigma_{b,\max}$ from Eq. (10.53)

$$\frac{1}{R} = \frac{2}{Et} \times \frac{3WL}{2nbt^2} = \frac{3WL}{nEbt^3} \tag{10.57}$$

Alternatively,
$$\frac{1}{R} = \frac{M}{EI} = \frac{WL}{4n} \times \frac{12}{Ebt^3} = \frac{3WL}{nEbt^3}$$

Therefore,
$$y_0 = \frac{L^2}{8R} = \frac{L^2}{8} \times \frac{2\sigma_{b,\max}}{Et} = \frac{\sigma_{b,\max} L^2}{4Et} \tag{10.58}$$

Substituting for $\sigma_{b,\max}$ from Eq. (10.53):

$$y_0 = \frac{3WL}{2nbt^2} \times \frac{L^2}{4Et} = \frac{3WL^3}{8nEbt^3} \tag{10.59}$$

The corresponding stress and deflection expressions for the *quarter-elliptic leaf spring* shown in Figure 10.10(a), can be obtained by substituting $W = 2W'$ and $L = 2L'$ in the expressions of semi-elliptical leaf spring, i.e.

$$\sigma_b = \frac{3(2W')(2L')}{2nbt^2} = \frac{6W'L'}{nbt^2} \tag{10.60}$$

$$y_0 = \frac{3(2W')(2L')^3}{8nEbt^3} = \frac{6W'L'^3}{nEbt^3} \tag{10.61}$$

For convenience, the terms W' and L' can be replaced by W and L in the above expressions, where these symbols would represent the applied load and length of the quarter-elliptic spring, respectively. Figure 10.10(b) shows the bending moment diagram.

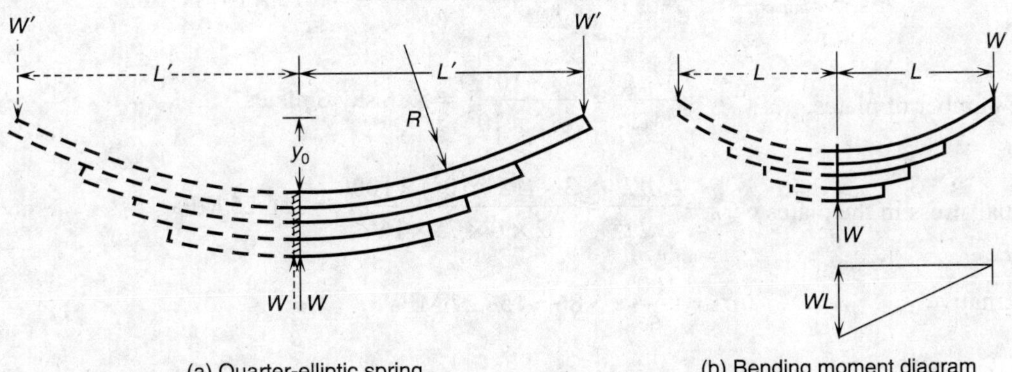

(a) Quarter-elliptic spring (b) Bending moment diagram

Figure 10.10 Analysis of quarter-elliptic leaf spring.

EXAMPLE 10.11

An 800 mm long leaf spring is required to support a central point load of 8 kN. Determine the thickness, width, initial radius of curvature and number of plates, when the central deflection and bending stress are limited to 20 mm and 200 MPa, respectively. Take the width of the plate to be 12 times its thickness. The elastic modulus of the material of the spring is 200 GPa.

Solution: From Eq. (10.58),

$$t = \frac{\sigma_{\max} L^2}{4Ey_0} = \frac{200 \times 800^2}{4 \times (200 \times 10^3) \times 20} = 8 \text{ mm}$$

Therefore, $b = 12t = 12 \times 8 = 96$ mm

From Eq. (10.53), $$n = \frac{3WL}{2\sigma_{\max} bt^2} = \frac{3 \times (8 \times 10^3) \times 800}{2 \times 200 \times 96 \times 8^2} = 7.81 \quad \text{i.e.} \quad 8 \text{ plates}$$

From Eq. (10.55), $$R = \frac{L^2}{8y_0} = \frac{800^2}{8 \times 20} = 4000 \text{ mm}$$

EXAMPLE 10.12

A one metre long carriage spring consisting of 75 × 16 mm (thick) plates supports a central point load of 12 kN. Determine the number of plates, central deflection and the initial radius of curvature to which the spring must be bent so that it becomes flat under the load. The bending stress in the plates is limited to 160 MPa. The elastic modulus of the material of the spring is 200 GPa.

Solution: The maximum bending moment induced in the spring

$$M_{\max} = \frac{WL}{4} = \frac{(12 \times 10^3) \times 1000}{4} = 3 \times 10^6 \text{ N} \cdot \text{mm}$$

Moment resistance of each plate in the spring

$$M_r = \sigma_{b,\max} Z = \sigma_{b,\max} \left(\frac{bt^2}{6} \right) = \frac{160 \times 75 \times 16^2}{6} = 5.12 \times 10^5 \text{ N} \cdot \text{mm}$$

The number of plates, $$n = \frac{M_{\max}}{M_r} = \frac{3 \times 10^6}{5.12 \times 10^5} = 5.86 \text{ say 6 plates}$$

Actual stress in the plates $$\sigma_b = \frac{3WL}{2nbt^2} = \frac{3 \times (12 \times 10^3) \times 1000}{2 \times 6 \times 75 \times 16^2} = 156.25 \text{ MPa}$$

Alternatively, $$\sigma_b = \frac{160}{6} \times 5.86 = 156.27 \text{ MPa}$$

$$R = \frac{Et}{2\sigma_b} = \frac{(200 \times 10^3) \times 16}{2 \times 156.25} = 10240 \text{ mm} = 10.24 \text{ m}$$

Therefore, $$y_0 = \frac{L^2}{8R} = \frac{1000^2}{8 \times 10240} = 12.21 \text{ mm}$$

EXAMPLE 10.13

A 1.4 m long laminated carriage spring has leaves of 100 mm width and 10 mm thickness. The spring has to absorb 125 N.m of energy when straightened without exceeding the bending stress of 160 MPa. Determine the number of leaves and the initial radius of curvature to which the spring must be bent so that it becomes flat under the load. The elastic modulus of the material of the spring is 200 GPa.

Solution: The energy stored in a spring,

$$U = \frac{\sigma_{b,\max}^2}{6E} \times \text{Volume of the spring}$$

$$(125 \times 10^3) = \frac{160^2}{6 \times (200 \times 10^3)} \times \frac{n \times 100 \times 10 \times (1.4 \times 10^3)}{2}$$

$$n = 8.37 \text{ say 9 leaves}$$

Actual stress in the plates

$$(125 \times 10^3) = \frac{\sigma_b^2}{6 \times (200 \times 10^3)} \times \frac{9 \times 100 \times 10 \times (1.4 \times 10^3)}{2}$$

$$\sigma_b = 154.3 \text{ MPa}$$

Alternatively, $$\sigma_b^2 = \frac{160^2}{9} \times 8.37 \quad \text{or} \quad \sigma_b = 154.3 \text{ MPa}$$

$$R = \frac{Et}{2\sigma_b} = \frac{(200 \times 10^3) \times 10}{2 \times 154.3} = 6480.88 \text{ mm} = 6.48 \text{ m}$$

EXAMPLE 10.14

A 500 mm long quarter-elliptic leaf spring has leaves of 55 mm width and 7 mm thickness. Determine the least number of leaves that are required if the deflection under a gradually applied load of 2.5 kN is not to exceed 60 mm.

If instead of being gradually applied the load of 2.5 kN falls from a height of 6 mm on to the un-deflected spring. Determine the maximum deflection and stress produced in the spring. The elastic modulus of the material of the spring is 200 GPa.

Solution:

$$y_0 = \frac{6WL^3}{nEbt^3} \quad \text{or} \quad 60 = \frac{6 \times (2.5 \times 10^3) \times 500^3}{n \times (200 \times 10^3) \times 55 \times 7^3}$$

Therefore, $$n = 8.28 \text{ say 9}$$

Consider W_e N to be the equivalent gradually applied static load which would produce the same deflection as caused by the impact load.

$$y_e = \frac{6W_eL^3}{nEbt^3} \quad \text{or} \quad y_e = \frac{6\times W_e \times 500^3}{9\times(200\times10^3)\times 55\times 7^3}$$

Hence, $W_e = 45.235y_e$ N

Work done by impact load W, $U_e = W(6 + y_e) = (2.5 \times 10^3)(6 + y_e)$

Strain energy stored in the spring, $U_i = \frac{1}{2}W_e y_e = \frac{45.235\,y_e^2}{2}$

From energy principles, $U_e = U_i$ i.e. $\frac{45.235\,y_e^2}{2} = (2.5\times10^3)(6 + y_e)$

or $y_e^2 - 110.534y_e - 663.2 = 0$ or $y_e = 116.24$ mm

Therefore, $W_e = 45.235 \times 116.24 = 5258.12$ N $= 5.258$ kN

Hence the maximum stress,

$$\sigma_{b,\max} = \frac{6WL}{nbt^2} = \frac{6\times(5.258\times10^3)\times500}{9\times55\times7^2} = 650.34 \text{ MPa}$$

10.8 PROBLEMS

10.1 Design a closely coiled helical spring which shall deflect 10 mm under an axial load of 120 N with shear stress limited to 100 MPa. The spring is to be made out of a round steel bar with mean coil diameter to be 10 times that of the bar. Determine the diameter and the length of the bar necessary to form the spring. The material of the bar has the modulus of rigidity of 80 GPa.

[**Ans.** $d = 5.53$ mm; $D = 55.28$ mm; $n = 4.61$ and $L = 800.61$ mm]

10.2 A railway wagon weighing 75 kN and moving with a speed of 2.78 m/s is to be stopped by four buffer springs in which the maximum compression allowed is 200 mm. Each buffer spring is made of a round steel rod of 20 mm diameter coiled to a mean diameter of 200 mm. Calculate the number of coils in each spring. The modulus of rigidity of the spring material and acceleration due to gravity are 84 GPa and 9.81 m/s^2, respectively.

[**Ans.** $n = 5.69$ say 6 coils]

10.3 A closely coiled compression helical spring having mean coil diameter of 180 mm and initial compression of 25 mm is used to hold a 100 mm diameter safety valve designed to blow off at a gauge pressure of 1.5 MPa. Determine the diameter of the spring wire and the number of turns required if the shear stress is limited to 100 MPa. The modulus of rigidity of the spring material is 80 GPa.

[**Ans.** $d = 37.8$ mm and $n = 7.43$ say 8 turns]

10.4 A 2.5 kN weight is dropped on to a closely coiled helical spring consisting of 16 turns of 30 mm diameter rod coiled to a mean diameter of 250 mm. Determine the height through which the weight is dropped before striking the spring such that the spring is compressed by 225 mm. The modulus of rigidity of the material is 84 GPa.

[**Ans.** $h = 119.45$ mm]

10.5 A closely coiled helical vertical spring having 10 coils is fixed at both the ends as shown in Figure 10.11. If a load W is applied to it at an intermediate point having 6 coils above it, determine the load carried by top and bottom portions of the spring.

[**Hint:** The load is distributed such that the deflections of two parts are the same]

[**Ans.** $P_t = 2W/5$ and $P_b = 3W/5$]

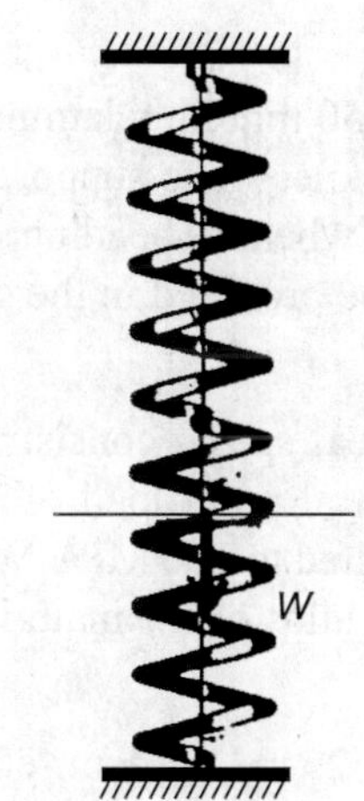

Figure 10.11 Closely coiled helical spring of Problem 10.6.

10.6 A closely coiled helical spring having n turns is made of a round steel bar of diameter d, coiled to a mean radius of $R = 5d$. Show that the stiffness of this spring can be expressed as $k = C(R/n)$, where C is a constant. If the modulus of rigidity of the material is 80 GPa, determine the value of the constant.

If the spring is required to support a load of 1 kN with an extension of 100 mm, and a maximum shearing stress of 360 MPa, determine the mean diameter of the coil and the number of turns. What is the weight of the material used in the spring if its unit weight is 7.75×10^{-5} N/mm^3.

[**Ans.** $C = 2$; $R = 42.05$ mm; $n = 8.41$ and weight = 9.568 N]

10.7 A compound spring consists of two close-coiled helical springs connected in series. The first spring has 12 turns of 3 mm diameter steel bar coiled to a mean diameter of 30 mm, and the second has 12 turns of a bar of same material again coiled to the same mean diameter of 30 mm. If the stiffness of the compound spring is 870 N/m, determine (a) the bar diameter of second spring, (b) the maximum load that can be carried by the spring and the corresponding extension for a maximum shearing stress of 210 MPa. The modulus of rigidity is 80 GPa.

[**Hint:** Being in series, springs carry same load; total deflection is the sum of individual deflections and the permissible load will be based on smaller bar diameter].

[**Ans.** $d = 2.5$ mm, $W = 42.95$ N and $\Delta = 49.37$ mm]

10.8 In a composite spring consisting of two concentric helical springs, the outer spring has 10 turns of 3 mm diameter spring wire coiled to a mean diameter of 30 mm, and the uncompressed axial length of inner spring having 8 turns is 10 mm shorter than the outer one. The radial clearance between the springs is 2 mm. (a) Determine the stiffness of the inner spring if an axial load of 100 N causes the composite spring to compress by 20 mm. (b) Determine the wire and mean coil diameters of the inner spring. The modulus of rigidity of the material of the springs is 80 GPa.

[**Ans.** $k_i = 4$ N/mm; $d_i = 2.3076$ mm and $D_i = 20.6924$ mm]

10.9 A flat spring is made of steel wire 12 mm wide × 0.5 mm thick and 3.5 m long. Determine the maximum bending stress induced and the amount of energy stored in the spring when the central spindle is wind up by three complete turns. The elastic modulus of the material of the spring is 200 GPa.

[**Hint:** Use eqns in order: $M = (E/\theta)/L$; $U = (1/2)M\theta$ and $f_{b,\max} = 12M/(bh^2)$]

[**Ans.** $\sigma_{\max} = 538.56$ MPa]

10.10 A 750 mm long laminated carriage spring consisting of 40 mm wide twelve leaves of same thickness is to support a central point load of 7.5 kN. Determine the thickness of the spring leaf when the bending stress is limited to 200 MPa. Also determine the initial central deflection to be provided in the spring? The elastic modulus of the material of the spring is 200 GPa.

[**Ans.** $t = 9.4$ mm and $y_0 = 14.88$ mm]

10.11 A leaf spring consisting of seven steel plates of size of 75 × 6 mm (thick) is to support a central point load of 3 kN. Determine the span of the spring when the bending stress is limited to 160 MPa. What is the initial central deflection provided in the spring? The elastic modulus of the material of the spring is 210 GPa.

[**Ans.** $L = 672$ mm and $y_0 = 14.336$ mm]

10.12 A 1.25 m long semi-elliptical leaf spring has ten leaves of 80 mm width and 9 mm thickness. Determine the initial radius of leaves if the proof stress in bending is 600 MPa.
Find the maximum height from which a load of 500 N can fall onto the centre of spring, if the bending stress produced is to be one half that of proof stress. The elastic modulus of the material of the spring is 200 GPa.

[**Ans.** $R = 1.5$ m; $W_e = 10.368$ kN; $y_0 = 65.1$ mm and $h = 609.86$ mm]

10.13 A conical spring having 10 coils with minimum and maximum diameters of 50 mm and 100 mm, respectively, is made of 15 mm diameter steel wire. Determine the maximum load, the spring can support without exceeding the shear stress of 180 MPa. Also determine the deflection of the spring. The elastic modulus of the material of the spring is $G = 80$ GPa.

[**Ans.** $W = 2169$ N and $y_0 = 20.07$ mm]

Chapter 11

Combined Stresses:

Elastic Bending of Members with Axial Loads

11.1 INTRODUCTION

In the preceding chapters, the stress state on an element due to the three fundamental loadings, namely the axial, flexural and torsional loadings were discussed. Each of these loadings was discussed on the assumption that only one of these loadings was acting on a structure at a time. The present chapter deals with the cases in which axial and flexural loadings act simultaneously on a structure which result in *unsymmetric* or *skew bending*. Both the axial and flexural loadings cause only normal or direct stresses and therefore form the simplest combination. The analysis employs the *method of superposition* in handling stresses due to elastic bending and axial forces obtained independently. The discussion is confined on the stress analysis of structures with doubly symmetric cross-sections. The stress analysis in structures like chimneys, dams and retaining walls where the resultant of vertical and horizontal loads has a large eccentricity at a section which results in the bending stresses σ_b dominating over the axial or normal stresses σ_o, and the vertical load W may produce tensile stresses on one face of the section, has been included.

11.2 ECCENTRIC TENSION AND COMPRESSION

In eccentric loading, the resultant of external forces does not coincide with the axis of the member, but is shifted relative to Z-axis and remains parallel to it as shown in Figure 11.1(a). Consider that the resultant of external forces W acts at the point A having eccentricities of e_x and e_y, with respect to the Y- and X-axes, respectively. The moments produced by the resultant force W about the principal axes are then

$$M_x = We_y \text{ and } M_y = We_x \tag{11.1}$$

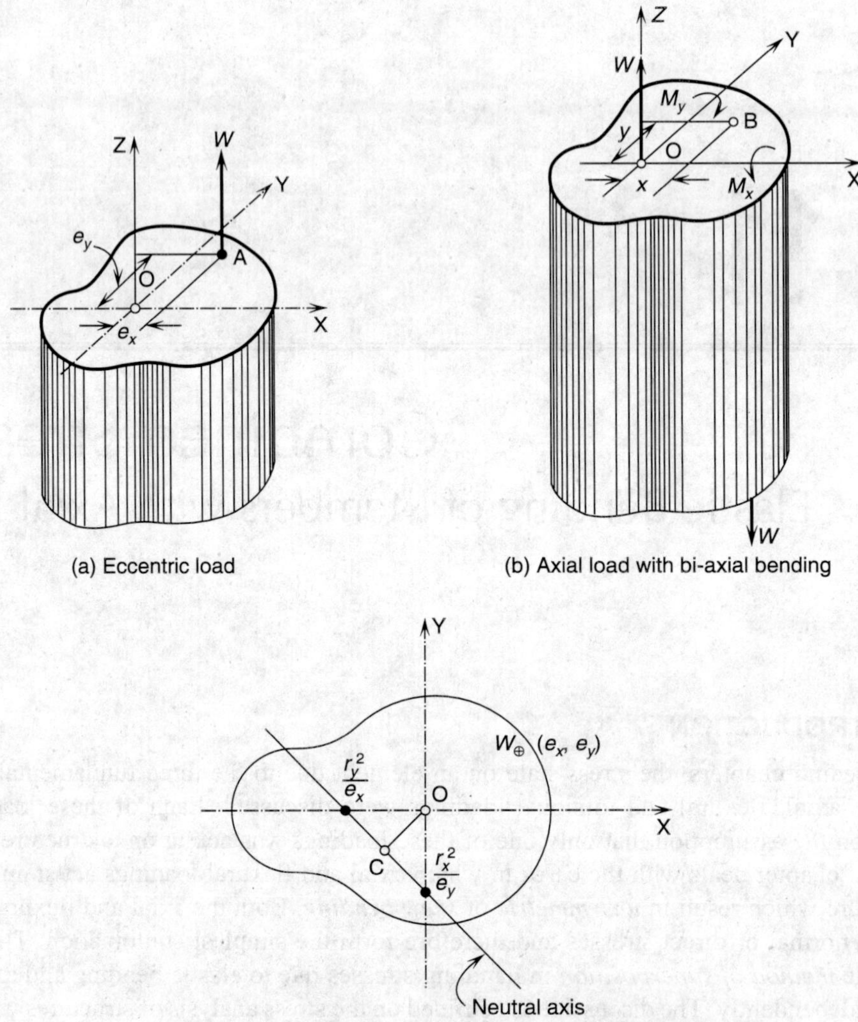

(a) Eccentric load

(b) Axial load with bi-axial bending

(c) Cross-section subjected to eccentric load

Figure 11.1 Analysis of a member subjected to eccentric load.

The resultant stress at a point can be obtained by superimposing the effects of the axial load W, and moments M_x and M_y (see Figures 11.1(b)). The resultant normal stress σ_r at an arbitrary point B with co-ordinates x and y is thus given by

$$\sigma_r = \sigma_o + \sigma_{bx} + \sigma_{by} = \frac{W}{A} \pm \frac{(We_y)y}{I_x} \pm \frac{(We_x)x}{I_y} \tag{11.2}$$

Therefore,

$$\sigma_r = \frac{W}{A}\left[1 \pm \frac{e_y y}{r_x^2} \pm \frac{e_x x}{r_y^2}\right] \tag{11.3}$$

where $I_x = Ar_x^2$ and $I_y = Ar_y^2$, the terms r_x and r_y represent the radii of gyration.

This relation represents a plane, and the position of line of zero stress is obtained by equating σ_r to zero

$$1 \pm \frac{e_y y}{r_x^2} \pm \frac{e_x x}{r_y^2} = 0$$

or

$$\frac{y}{(\mp r_x^2 / e_y)} + \frac{x}{(\mp r_y^2 / e_x)} = 1 \tag{11.4}$$

The form of Eq. (11.4) resembles the equation of straight line $y/b + x/a = 1$ with intercepts of $a = \mp(r_y^2 / e_x)$ and $b = \mp(r_x^2 / e_y)$, on X- and Y-axes, respectively. Thus the line of zero stress generally called *neutral axis* passes through the quadrant which is opposite to that containing the load W and the stress on the portion of the cross-section towards the centroid from the neutral axis has the same sign as the applied force.
The distance OC of straight line is given by analytical geometry as

$$OC = \frac{1}{\sqrt{(e_y / r_x^2)^2 + (e_x / r_y^2)^2}} \tag{11.5}$$

Consequently, as the point of application of force W approaches the centroid of the section the neutral axis moves away from it. In the limit when the force is applied centrally, $e_x = e_y = 0$ the neutral axis is at infinity, and stresses are uniformly distributed over the section. If, on the other hand e_x and e_y are relatively large, i.e. the point of application of force moves away from the centroid, the distance OC is reduced and the neutral axis approaches the centroid. In this case, the neutral axis lies inside the section and the area on one side of this axis will be in compression and that on the other side will be in tension, i.e. both tensile and compressive stresses develop on the section. Thus, in case of eccentric tension or compression, the stress at any point in the section varies linearly with its distance from the line of zero stress which is similar to the neutral axis of simple bending. However, the neutral axis does not pass through the centroid of the section unlike the *pure bending* since the member also carries an axial force.

In the design practice the main concern is the maximum and minimum stresses developed on the section. These stresses occur at the boundary of the section, i.e. at the points where the coordinates x and y have the maximum values. Moreover, in structures made of masonry and concrete, tensile are not allowed in the cross-section. It is then important to determine the maximum possible values for the eccentricities, i.e. the region within which W can be applied for a no tension condition. The region in the vicinity of the centroid where this condition is satisfied is called the *core* or *kern of the cross-section*. If the trace of the force W is inside the core of the section, the stresses are of the same sign at all the points in the section. If the force is applied outside the core of the section, the neutral axis crosses the section and the stresses on the section are both tensile and compressive. When the point of application of the force is on the boundary of the core, the neutral axis is tangential to contour of the section. In order to determine the *core of the section*, consider that the neutral axis is rolled round the section. The point of application of the force will then draw the contours of the core.

Rectangular section: For a rectangular cross-section of size $b \times h$, to obtain the boundary of the core, equate extreme values of σ_r given by Eq. (11.2) to zero

$$\sigma_r = \frac{W}{A} \pm \frac{(We_y)\,y_{\max}}{I_x} \pm \frac{(We_x)\,x_{\max}}{I_y}$$

or

$$0 = \frac{W}{bh} \pm \frac{(We_y)(h/2)}{bh^3/12} \pm \frac{(We_x)(b/2)}{b^3h/12}$$

Therefore,

$$\mp\frac{6e_y}{h} \mp \frac{6e_x}{b} = 1 \quad \text{or} \quad \mp\frac{e_y}{(h/6)} \mp \frac{e_x}{(b/6)} = 1 \tag{11.6}$$

This represents equation of straight lines with reference to e_x and e_y axes. The intercepts on e_x and e_y axes are $\mp(b/6)$ and $\mp(h/6)$, respectively. The plot of these four points define a rhombus shaped region whose diagonals are $b/3$ and $h/3$; is the core of the rectangular section. The load W must lie within this shaded region shown in Figure 11.2(a) if there is to be no reversal of stress in any part of the cross-section. Since the core region lies in the middle of the section and is one-third its size, it is called the *middle-third rule.* In case of a square cross-section of side a, the diagonals of the rhombus which are along the X and Y axes measure $a/3$ and its side is $\sqrt{2}a/6$.

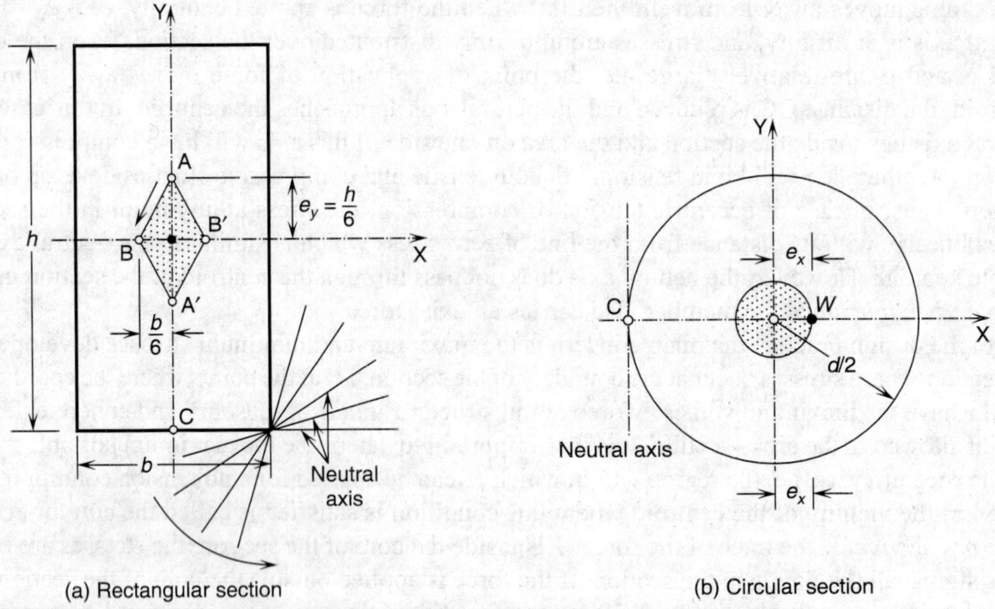

Figure 11.2 Cores of rectangular and circular cross-sections.

Circular section: Consider a circular cross-section of diameter d. Whatever be the position of the point of application of the load W, there exists an axis of symmetry passing through this point and the problem reduces to one of eccentricity about a single axis of symmetry. Thus, it is convenient to consider the diameter along the line joining the centre of the circle to the point of application of load as shown in Figure 11.2(b). Then e_x is the eccentricity of the load and $e_y = 0$, so that the minimum stress at a point C is given by

$$\sigma_{min} = \frac{W}{A} - \frac{(We_x)x_{max}}{I} \tag{11.7}$$

The boundary of the core may be obtained by equating this extreme value of stress to zero

$$0 = \frac{W}{(\pi d^2/4)} - \frac{(We_x)(d/2)}{(\pi d^4/64)}$$

Therefore, $e_x = d/8$.

Thus, the core shown in Figure 11.2(b) is the shaded circular area of diameter $d/4$ which is also referred to as the *middle quarter rule*.

It should be noted that the core of a cross-section depends on the geometric properties of the section and is independent of the magnitude of the load or the strength properties of the material.

EXAMPLE 11.1

Determine which of the bars with cut-outs shown in Figures 11.3(a) and (b) is capable of carrying a larger load without developing plastic deformations.

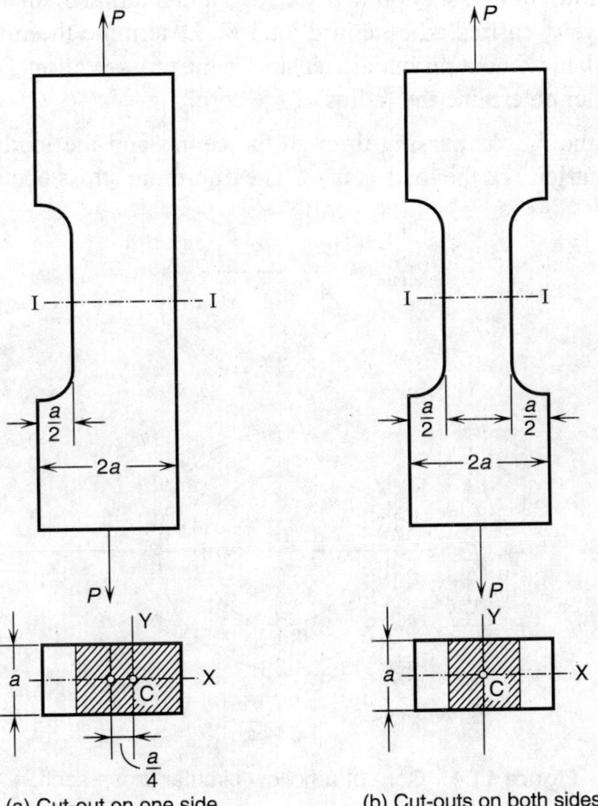

(a) Cut-out on one side (b) Cut-outs on both sides

Figure 11.3 Adverse effects of eccentricity.

Solution: The maximum stress developed will govern the load carrying capacity of the section. In case (a) bar has cut-out on one side and thus the force W is eccentric for the weakened section. The eccentricity with respect to Y-axis is (a/4). Consequently, the maximum tensile stress is

$$\sigma_{\max} = \frac{W}{A} + \frac{(We_x)x_{\max}}{I_y} = \frac{W}{4} + \frac{(We_x)}{Z_y}$$

$$= \frac{W}{a \times (3a/2)} + \frac{6 \times [W \times (a/4)]}{a \times (3a/2)^2} = \frac{4W}{3a^2}$$

On the other hand, in case (b) the bar has cut-outs on both the sides and the force W acts axially. Thus, the maximum tensile stress is

$$\sigma_{\max} = \frac{W}{A} = \frac{W}{a \times a} = \frac{W}{a^2}$$

The bar having cut-outs on both sides has smaller stress and thus is capable of carrying a larger load. This example illustrates the adverse effects of eccentricity.

EXAMPLE 11.2

A short column of hollow circular section with external and internal diameters of the cross-section of d_1 and d_2, respectively, carries an eccentric load W. Determine the maximum eccentricity at which load can be applied without producing tension in the cross-section. If inside diameter is half that of external diameter determine the radius of the core.

Solution: Consider the X-axis passing through the centre and the loading point, as shown in Figure 11.4. The eccentricity of the load W is e_x. The minimum stress occurs at the point C and is given by

$$\sigma_{\min} = \frac{W}{A} - \frac{(We_x)x_{\max}}{I}$$

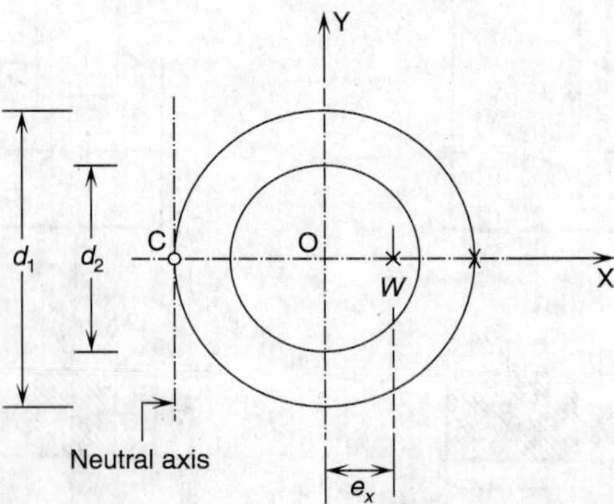

Figure 11.4 Core of a hollow circular cross-section.

If the tension is to be just avoided, equate this value of stress to zero

$$0 = \frac{W}{\pi(d_1^2 - d_2^2)/4} - \frac{(We_x)(d_1/2)}{\pi(d_1^4 - d_2^4)/64}$$

or

$$0 = 1 - \frac{e_x(d_1/2)}{(d_1^2 + d_2^2)/16}$$

Therefore,

$$e_x = \frac{d_1^2 + d_2^2}{8d_1} = \frac{1}{8}\left(d_1 + \frac{d_2^2}{d_1}\right)$$

This also indicates that the core is a circular area of diameter $2e_x$.
For the case $d_2 = d_1/2$; the radius of the core is

$$e_x = \frac{1}{8}\left(d_1 + \frac{d_2^2}{d_1}\right) = \frac{1}{8}\left[d_1 + \frac{(d_1/2)^2}{d_1}\right] = \frac{5d_1}{32}$$

EXAMPLE 11.3

Determine the core of the I-section shown in Figure 11.5. The geometric properties of the section are:

$$A = 3233 \text{ mm}^2;\ Z_x = 2.235 \times 10^5 \text{ mm}^3 \text{ and } Z_y = 3.0 \times 10^4 \text{ mm}^3$$

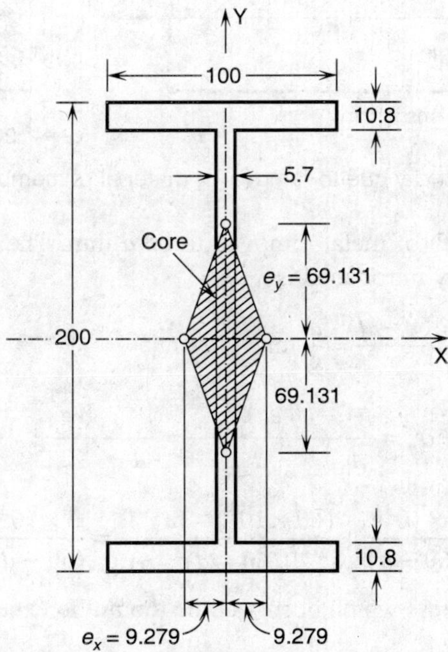

Figure 11.5 Core of I-section of Example 11.3 (All dimensions are in mm).

Solution: Consider the maximum eccentricities along X- and Y-axes to be e_x and e_y, respectively, which just avoid the development of tensile stresses on the section. If, W is the longitudinal load acting on the section, then for no stress point on Y-axis (i.e. $x = 0$)

$$\frac{W}{A} - \frac{(We_y)y_{\max}}{I_x} = \frac{W}{A} - \frac{We_y}{Z_x} = 0 \quad \text{or} \quad e_y = \frac{Z_x}{A}$$

Therefore, $$e_y = \frac{2.235 \times 10^5}{3233} = 69.131 \text{ mm}$$

Similarly for neutral point on X-axis (i.e. $y = 0$)

$$e_x = \frac{Z_y}{A} = \frac{3.0 \times 10^4}{3233} = 9.279 \text{ mm}$$

The core of the section is shown in Figure 11.5 by rhombus shaped shaded region.

EXAMPLE 11.4

A tie bar 20 mm wide and 60 mm deep subjected to an axial load of 60 kN as shown in Figure 11.6. Determine the depth of metal that may safely be removed from one of the narrow sides such that the maximum stress over the reduced depth may not exceed 90 MPa.

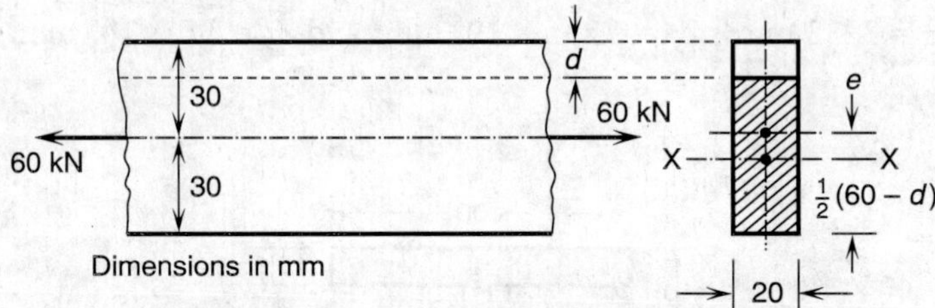

Figure 11.6 Eccentricity due to removal of material of member of Example 11.4.

Solution: Consider the depth of metal removed to be d mm. The removal of metal makes the load eccentric; the eccentricity introduced is

$$e_y = 30 - \frac{60 - d}{2} = \frac{d}{2} \text{ mm}$$

Then $$\sigma_{\max} = \sigma_o + \sigma_b = \frac{W}{A} + \frac{(We_y)y_{\max}}{I_x} = \frac{W}{A} + \frac{We_y}{Z_x}$$

$$= \frac{(60 \times 10^3)}{20(60 - d)} + \frac{(60 \times 10^3) \times (d/2)}{[20(60 - d)^2/6]} = \frac{3 \times 10^3}{(60 - d)} + \frac{9 \times 10^3 \times d}{(60 - d)^2}$$

For the maximum (tensile) stress which occurs at the top not to exceed 90 MPa,

$$\frac{3 \times 10^3}{(60 - d)} + \frac{9 \times 10^3 \times d}{(60 - d)^2} \le 90 \quad \text{or} \quad 100 \times (60 - d) + 300d \le 3(60 - d)^2$$

For the limiting case, $$3d^2 - 560d + 4800 = 0$$

Therefore, $$d = \frac{280}{3} - \sqrt{\left(\frac{280}{3}\right)^2 - 1600} = 9.006 \text{ mm}$$

Hence, 9 mm of metal may be safely removed from one of the narrow sides.

EXAMPLE 11.5

A 3 m long tie bar of 20 mm thick rectangular section is tapered from a depth of 40 mm to 120 mm at the ends as shown in Figure 11.7. A load of 30 kN is applied through the centroid of the smaller end and parallel to the edge AB. Determine the position and magnitude of the maximum tensile stress.

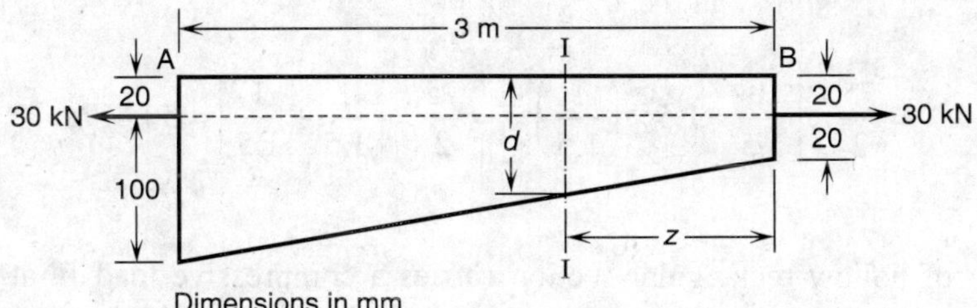

Figure 11.7 Eccentricity due to taper in the member of Example 11.5.

Solution: Consider the section I-I at a distance z mm from the smaller end B. The depth of the cross-section is

$$d = 40 + \left(\frac{100 - 20}{3}\right)z = 40\left(1 + \frac{2z}{3}\right) \text{ mm}$$

The eccentricity of the load e_y at the section I-I is

$$e_y = \frac{1}{2} \times \left[40\left(1 + \frac{2z}{3}\right)\right] - 20 = \frac{40z}{3} \text{ mm}$$

Properties of cross-section at I-I,

$$A_z = 20 \times 40\left(1 + \frac{2z}{3}\right) = 800\left(1 + \frac{2z}{3}\right) \text{ mm}^2$$

$$Z_z = \frac{1}{6} \times 20 \times \left[40\left(1 + \frac{2z}{3}\right)\right]^2 = \frac{16000}{3} \times \left(1 + \frac{2z}{3}\right)^2 \text{ mm}^3$$

The maximum tensile stress is

$$\sigma_{\max} = W\left(\frac{1}{A} + \frac{e_y}{Z_z}\right) = \frac{(30 \times 10^3)}{800}\left[\left(1 + \frac{2z}{3}\right)^{-1} + \frac{40z}{3} \times \frac{3}{20}\left(1 + \frac{2z}{3}\right)^{-2}\right]$$

$$= \frac{(30 \times 10^3)}{800}\left[\left(1 + \frac{2z}{3}\right)^{-1} + 2z\left(1 + \frac{2z}{3}\right)^{-2}\right]$$

For maximum stress, $\dfrac{d\sigma_{max}}{dz} = 0$. Ignoring the constant quantity outside the brackets

$$-\frac{2}{3}\times\left(1+\frac{2z}{3}\right)^{-2}+2\times\left(1+\frac{2z}{3}\right)^{-2}+2z\times\left(-\frac{4}{3}\right)\left(1+\frac{2z}{3}\right)^{-3}=0$$

$$-\frac{2}{3}+2-\frac{8z}{3}\left(1+\frac{2z}{3}\right)^{-1}=0 \quad \text{or} \quad \left(1+\frac{2z}{3}\right)-2z=0$$

Therefore, $\quad 3+2z-6z=0 \quad \text{or} \quad z=0.75 \text{ m}$

$$\sigma_{max}=\frac{(30\times10^3)}{800}\left[\left(1+\frac{2\times0.75}{3}\right)^{-1}+2\times0.75\times\left(1+\frac{2\times0.75}{3}\right)^{-2}\right]$$

$$=\frac{75}{2}\times\left[\frac{1}{1.5}+1.5\times\left(\frac{1}{1.5}\right)^2\right]=\frac{75}{2}\times\left[\frac{1}{1.5}+\frac{1}{1.5}\right]=50 \text{ MPa}$$

EXAMPLE 11.6

A masonry pier of hollow rectangular section carries a compressive load W at the point A as shown in Figure 11.8. Determine the maximum value of load W, if the maximum compressive stress in masonry is limited to 2.24 MPa. Ignore the self weight of the masonry.

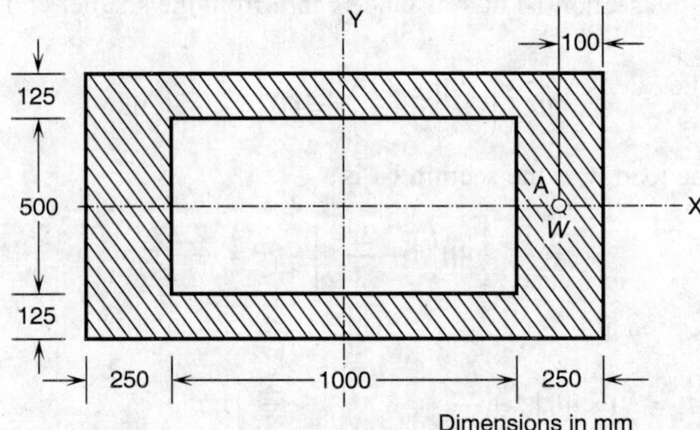

Figure 11.8 Hollow section of the masonry pier of Example 11.6.

Solution: Consider the maximum load the pier can support to be W kN. The geometric properties of the cross-section are,

Area of the cross-section, $A = 750 \times 1500 - 500 \times 1000 = 6.25 \times 10^5 \text{ mm}^2$

Moment of inertia, $\quad I_y=\dfrac{750\times1500^3}{12}-\dfrac{500\times1000^3}{12}=1.6927\times10^{11}\text{ mm}^4$

The maximum compressive stress on the section

$$\sigma_{\max} = \frac{W}{A} + \frac{(We_x)x_{\max}}{I_y}$$

$$= \frac{(W \times 10^3)}{6.25 \times 10^5} + \frac{(W \times 10^3) \times (500 + 150) \times 750}{1.6927 \times 10^{11}} = 4.48 \times 10^{-3} \text{ W}$$

or $\quad 4.48 \times 10^{-3}\, W = 2.24$ (given)

$\therefore \quad W = 500$ kN

EXAMPLE 11.7

A cylindrical cast iron shaft of external diameter 150 mm has an eccentric bore of 105 mm diameter such that the thickness of the shaft cross-section varies from 15 mm to 30 mm as shown in Figure 11.9. Determine the maximum longitudinal compressive load W acting along the axis of the bore the section can carry, if the maximum compressive stress in the material of the shaft is limited to 60 MPa.

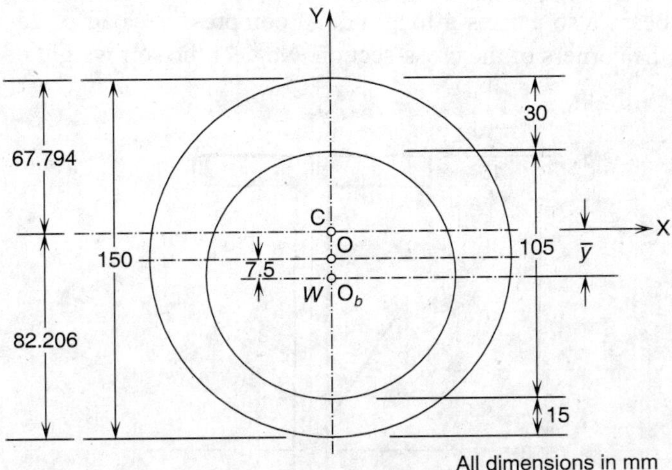

Figure 11.9 Shaft with eccentric hole.

Solution: Consider the maximum load the shaft can support to be W kN. The geometric properties of the cross-section are,

Area of the cross-section, $\quad A = \dfrac{\pi}{4}(150^2 - 105^2) = 9.0124 \times 10^3 \text{ mm}^2$

Let the centroid of the net section of the shaft is located at a distance $\bar{y}$ from the centre of bore O_b. Taking moment about O_b

$$\bar{y} = \frac{(\pi \times 150^2/4) \times 7.5 - (\pi \times 105^2/4) \times 0}{(\pi \times 150^2/4) - (\pi \times 105^2/4)} = \frac{150^2 \times 7.5}{150^2 - 105^2} = 14.706 \text{ mm}$$

Moment of inertia,

$$I_x = \left(\frac{\pi \times 150^4}{64} + \frac{\pi \times 150^2}{4} \times 7.206^2\right) - \left(\frac{\pi \times 105^4}{64} + \frac{\pi \times 105^2}{4} \times 14.706^2\right)$$

$$= 1.79288 \times 10^7 \text{ mm}^4$$

The maximum compressive stress

$$\sigma_{\max} = \frac{(W \times 10^3)}{9.0124 \times 10^3} + \frac{(W \times 10^3) \times 14.706}{1.79288 \times 10^7} \times 82.206 = 0.17839\,W$$

or $\quad 0.17839W = 60.0$ MPa (given)

$\therefore \quad W = 336.34$ kN

EXAMPLE 11.8

A 3.5 m long simply-supported I-section beam supports a concentrated load of 35 kN inclined at 30° to the web passing through the centroid of the section at the mid-span point as shown in Figure 11.10. The beam also carries a longitudinal compressive load of 15 kN. Determine the stresses induced at the corners of the cross-section. Neglect the self weight of the beam.

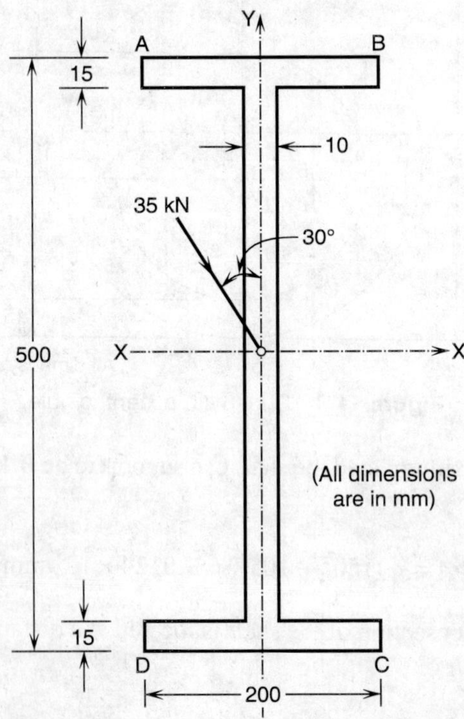

Figure 11.10 Symmetrical I-section subjected to inclined load.

Solution: The geometric properties of the cross-section are,
Area of the cross-section, $A = 2 \times (200 \times 15) + 470 \times 10 = 1.07 \times 10^4$ mm^2
Moments of inertia are,

$$I_x = \frac{200 \times 500^3}{12} - \frac{190 \times 470^2}{12} = 4.3947 \times 10^8 \text{ mm}^4$$

$$I_y = 2 \times \frac{15 \times 200^3}{12} + \frac{470 \times 10^3}{12} = 2.0039 \times 10^7 \text{ mm}^4$$

Vertical component of the applied load, $W_y = 35 \cos 30° = 30.31$ kN
Horizontal component of the applied load, $W_x = 35 \sin 30° = 17.5$ kN
The bending moments,

$$M_x = \frac{W_y L}{4} = \frac{30.31 \times 3.5}{4} = 26.521 \text{ kN.m}$$

$$M_y = \frac{W_x L}{4} = \frac{17.5 \times 3.5}{4} = 15.313 \text{ kN.m}$$

Direct stress due to longitudinal load,

$$\sigma_o = \frac{P}{A} = \frac{15 \times 10^3}{1.07 \times 10^4} = 1.402 \text{ MPa}$$

Extreme stresses due to transverse load components,

$$\sigma_{bx} = \pm \frac{M_x}{Z_x} = \pm \frac{(26.521 \times 10^6) \times 250}{4.3947 \times 10^8} = 15.087 \text{ MPa}$$

$$\sigma_{by} = \pm \frac{M_y}{Z_y} = \pm \frac{(15.312 \times 10^6) \times 100}{2.0039 \times 10^7} = \pm 76.41 \text{ MPa}$$

Then the resultant stresses at the corners are,

$\sigma_A = 1.402 + 15.087 + 76.41 = 92.899$ MPa (compressive)

$\sigma_B = 1.402 + 15.087 - 76.41 = -59.921$ MPa (tension)

$\sigma_C = 1.402 - 15.087 - 76.41 = -90.095$ MPa (tension)

$\sigma_D = 1.402 - 15.087 + 76.41 = 62.725$ MPa (compressive)

11.3 STRESSES IN CHIMNEYS

The free standing structures like chimneys and towers are subjected to horizontal forces in addition to their self-weight. Consider a chimney of dimensions shown in Figure 11.11 with self-weight W. The chimney is subjected to the design wind pressure of intensity p. Then the total horizontal load P acting on the chimney due to wind pressure is

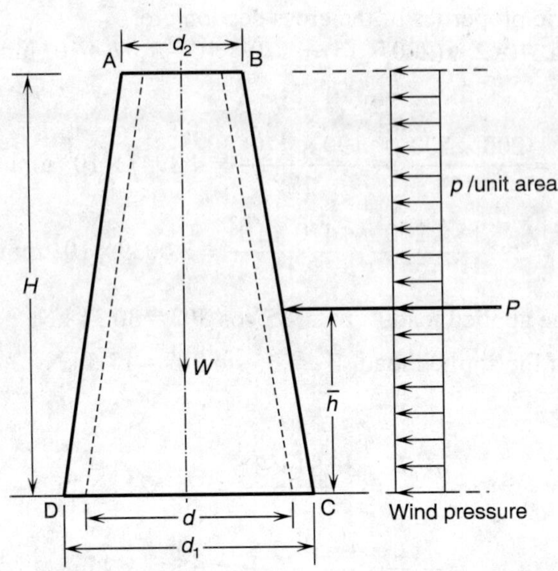

Figure 11.11 Forces acting on a chimney.

$$P = p \times \text{Projected area of chimney, i.e. area of trapezium ABCD}$$

$$= p \times \frac{1}{2}(d_1 + d_2)H \tag{11.8}$$

This wind load acts through the centroid of the projected area of chimney located at height $\bar{h}$ above the base. The height of the centroid is

$$\bar{h} = \left(\frac{d_1 + 2d_2}{d_1 + d_2}\right) \times \frac{H}{3} \tag{11.9}$$

Moment due to wind load about the base,

$$M = P\bar{h} = \left[p \times \frac{1}{2}(d_1 + d_2)H\right] \times \left(\frac{d_1 + 2d_2}{d_1 + d_2}\right) \times \frac{H}{3}$$

$$= \frac{1}{6}p(d_1 + 2d_2)H^2 \tag{11.10}$$

The resultant stresses on the base

$$\sigma_{\max} \text{ and } \sigma_{\min} = \sigma_o \pm \sigma_b = \frac{W}{A} \pm \frac{My_{\max}}{I}$$

$$= \frac{W}{[\pi(d_1^2 - d^2)/4]} \pm \frac{(P\bar{h}) \times (d_1/2)}{[\pi(d_1^4 - d^4)/64]} \tag{11.11}$$

For no tension in the base $\sigma_{min} = 0$,

$$\frac{W}{[\pi(d_1^2 - d^2)/4]} - \frac{(P\bar{h}) \times (d_1/2)}{[\pi(d_1^4 - d^4)/64]} = 0$$

Therefore,

$$W = \frac{8P\bar{h}d_1}{(d_1^2 + d^2)} = \left[\frac{4}{3}\frac{(d_1 + 2d_2)d_1}{(d_1^2 + d^2)}\right] pH^2 \qquad (11.12)$$

11.3.1 Wind Pressure

The design wind pressure p_z at any height z above mean ground level is computed as explained in the national codes and the horizontal load is equal to the wind pressure multiplied by the projected area of the structure normal to the direction on the wind. In case of circular chimney or circular tower, the projected area is equal to the external diameter times the height. For the cylindrical and inclined surfaces, the wind pressure is multiplied by a *shape factor* which accounts for the reduction in pressure due to the curvature or inclination of the surface.

EXAMPLE 11.9

A 20 metre high masonry chimney shaft of hollow square section tapers from 2.5 m external side at the base to 1.5 m external side at the top as shown in Figure 11.12(a). The thickness of the wall is 0.5 m throughout the height. The intensity of wind pressure acting on a surface is $p = 1.5$ kN/m^2. Determine the maximum and minimum stresses at the base when wind acts normal to: (a) a side face and (b) a diagonal. The unit weight of masonry is 22 kN/m^3. The shape factor or wind pressure reduction coefficient for the case (b) is 0.5.

Solution: Area at the base, $A_b = (2.5^2 - 1.5^2) = 4.0$ m^2
Self weight of the chimney,

$$W = \frac{1}{2}[A_b + A_t] \times H \times \gamma = \frac{1}{2}[(2.5^2 - 1.5^2) + (1.5^2 - 0.5^2)] \times 20 \times 22$$
$$= 1320 \text{ kN}$$

(a) Wind acting normal to a side face (see Figure 11.12(a)):
The moment of inertia section

$$I = (2.5^4 - 1.5^4)/12 = 2.83 \text{ m}^4$$

The bending moment at the base due to wind pressure on projected trapezoidal area,
From Eq. (11.10):

$$M = \frac{1}{6} \times 1.5 \times (2.5 + 2 \times 1.5) \times 20^2 = 550 \text{ kNm}$$

The resultant stresses on the base

$$\sigma_{max} \text{ and } \sigma_{min} = \frac{W}{A} \pm \frac{My_{max}}{I} = \frac{1320}{4.0} \pm \frac{550 \times (2.5/2)}{2.83}$$
$$= 330 \pm 242.93 = 572.9 \text{ kN/m}^2 \text{ and } 87.07 \text{ kN/m}^2$$

Both stresses are compressive.

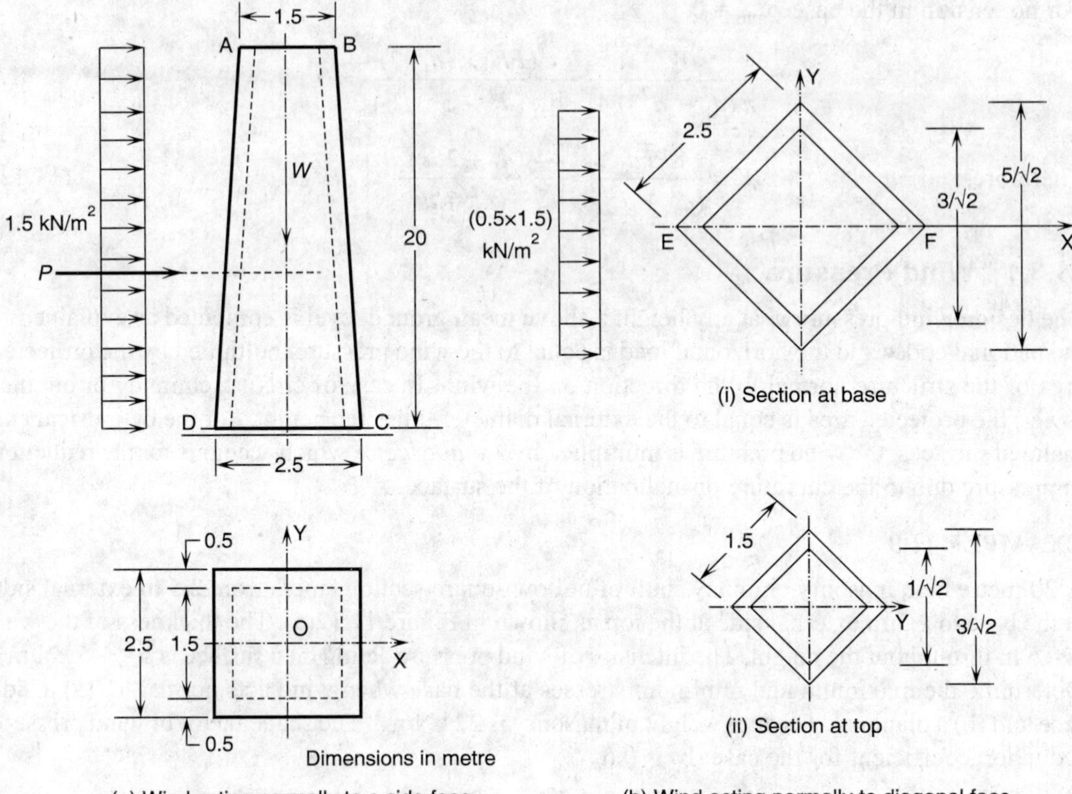

(a) Wind acting normally to a side face (b) Wind acting normally to diagonal face

Figure 11.12 Tapered masonry chimney subjected to wind pressure.

(b) Wind acts normal to the diagonal (see Figure 11.12(b)):

The area of cross-section at the base and self weight of the chimney are the same as in the previous case. The moment of inertia section

$$I = (2.5^4 - 1.5^4)/12 = 2.83 \text{ m}^4$$

The bending moment at the base due to wind pressure on projected trapezoidal area,
From Eq. (11.10):

$$M = \frac{1}{6} \times (1.5 \times 0.5) \times (2.5\sqrt{2} + 2 \times 1.5\sqrt{2}) \times 20^2 = 388.91 \text{ kNm}$$

The resultant stresses on the base

$$\sigma_{\max} \text{ and } \sigma_{\min} = \frac{W}{A} \pm \frac{My}{I} = \frac{1320}{4.0} \pm \frac{388.91 \times (2.5\sqrt{2}/2)}{2.83}$$

$$= 330 \pm 242.93 = 572.93 \text{ kN/m}^2 \text{ and } 87.07 \text{ kN/m}^2$$

EXAMPLE 11.10

A 20 metre high masonry chimney shaft is 2 metre square at the base tapers to 1.0 metre square at the top. The tapered central flue is circular in cross-section and has one metre diameter at the base. If the weight of chimney is 1.4 MN, determine the intensity of uniform wind pressure that may act on one face of the chimney such that the stress distribution across the base just ceases to be wholly compressive.

Solution: Consider the intensity of uniform wind pressure to be p.

Area at the base, $A = 2.0^2 - (\pi \times 1.0^2/4) = 3.215 \text{ m}^2$

The moment of inertia of the section at the base

$$I = \frac{2.0^4}{12} - \frac{\pi \times 1.0^4}{64} = 1.284 \text{ m}^4$$

The bending moment at the base due to wind pressure on projected trapezoidal area, From Eq. (11.10):

$$M = \frac{1}{6} \times p \times (2.0 + 2 \times 1.0) \times 20^2 = \frac{800p}{3} \text{ kNm}$$

For the stress distribution across the base, just to cease to be wholly compressive,

$$\sigma_{\min} = \frac{W}{A} - \frac{My_{\max}}{I}$$

$$= \frac{1400}{3.215} - \frac{800 \times (2.0/2)}{3 \times 1.284} p = 0$$

Therefore, $p = 2.0967 \text{ kN/m}^2$.

11.4 STRESSES IN DAMS

Figure 11.13 illustrates the trapezoidal cross-section ABCD of a masonry dam of height H which retains water on its vertical face to a depth h. The top and bottom widths of the section are represented by a and b, respectively.

(i) *Weight W of the dam:* For unit length of the dam its weight W is given by

$$W = \gamma_m \left(\frac{a+b}{2}\right) H \tag{11.13}$$

The weight W acts at a distance $\bar{x}$ from its vertical water face BC through the centroid of the dam cross-section and is given by,

$$\bar{x} = \frac{a^2 + ab + b^2}{3(a+b)} = \frac{b^3 - a^3}{3(b^2 - a^2)} \tag{11.14}$$

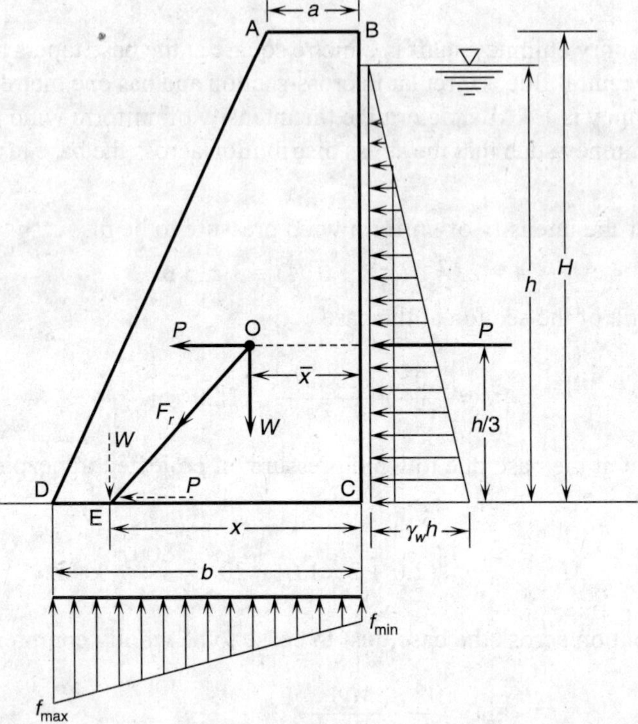

Figure 11.13 Masonry dam of trapezoidal section subjected to various forces.

(ii) *Horizontal water pressure:* At any depth y from the free surface of water the intensity of water pressure is $\gamma_w y$. Hence the horizontal water pressure increases linearly from zero at the free surface of water to $\gamma_w h$ at the base of the dam, i.e. the pressure is proportional to distance from the top. Thus the water pressure distribution is triangular in shape.

Total pressure on unit length of dam

$$P = \text{Average pressure} \times \text{Height of free surface}$$

$$= \frac{\gamma_w h}{2} \times h = \frac{\gamma_w h^2}{2} \tag{11.15}$$

The total pressure acts at height $h/3$ above the bottom or the base as shown in Figure 11.13.

(iii) *The reaction at the base of dam:* The lines of action of W and P intersect at the point O. The resultant F_r of the forces W and P intersects the base of the dam at point E at distance x from the face BC. Resolve the resultant force F_r acting at point E into its vertical and horizontal components W and P, respectively. The vertical component W acting at E is resisted by the normal reaction at the base. The horizontal component P of the resultant force F_r acting at point E is resisted by the friction between the bottom of the dam and the underlying soil.

The vertical component W acting at E is an eccentric load. The eccentricity of loading is

$$e = x - \frac{b}{2} \tag{11.16}$$

The location of the point E where the resultant strikes the base can be determined by taking moments about the point C while keeping in mind that the moment of a resultant force F_r is equal to that of its components W and P, i.e.

$$Wx = W\bar{x} + P\left(\frac{h}{3}\right) \quad \text{or} \quad x = \bar{x} + \left(\frac{P}{W}\right)\frac{h}{3} \tag{11.17}$$

From Eqs. (11.16) and (11.17),

$$e = \bar{x} + \left(\frac{P}{W}\right)\frac{h}{3} - \frac{b}{2} \tag{11.18}$$

The stresses across the base can be obtained as follows.

Direct stress,

$$\sigma_o = \frac{W}{A} = \frac{W}{b \times 1} \tag{11.19}$$

Moment due to water pressure about the base, $M = We$

Bending stress,

$$\sigma_b = \pm\frac{M}{Z} = \pm\frac{We}{(1 \times b^2)/6} = \pm\frac{6We}{b^2} \tag{11.20}$$

The extreme resultant stresses,

$$\sigma_{\max}, \sigma_{\min} = \sigma_o + \sigma_b = \frac{W}{b} \pm \frac{6We}{b^2} = \frac{W}{b}\left(1 \pm \frac{6e}{b}\right) \tag{11.21}$$

Therefore,

$$\sigma_{\max} = \frac{W}{b}\left(1 + \frac{6e}{b}\right) \text{ outer face and}$$

$$\sigma_{\min} = \frac{W}{b}\left(1 - \frac{6e}{b}\right) \text{ at the water face}$$

For no tension to occur at the base of the dam,

$$\sigma_{\min} = \frac{W}{b}\left(1 - \frac{6e}{b}\right) > 0 \quad \text{or} \quad \left(1 - \frac{6e}{b}\right) > 0$$

or

$$e < \frac{b}{6} \tag{11.22}$$

11.4.1 Stability of the Dam

In addition to the design requirements at the base that no tension should develop there and the compressive stress should not exceed the allowable value, a stable dam should neither slide over the soil on which it is resting nor it should overturn. To avoid sliding, the available frictional resistance should be greater than the horizontal water pressure P i.e. $\mu W > P$, where μ is the coefficient of friction between the masonry dam and the underlying soil. The ratio $\mu W/P$ represents the *factor of safety against sliding*. The factor of safety against sliding should be at least 1.5.

The overturning of the dam, if at all occurs, can be about the heel point D. To avoid overturning, the resultant F_r of the weight W and horizontal force P must intersect the base of the dam within its width. For the dam shown in Figure 11.13, consider the moments about D.

Overturning moment,
$$M_o = \frac{Ph}{3} \tag{11.23}$$

Available stabilizing or restoring moment,
$$M_r = W(b - \bar{x}) \tag{11.24}$$

The ratio of available restoring moment to the overturning moment is called the *factor of safety against overturning*. To prevent development of tensile stress and overturning, the dam section should be such that *the resultant force strikes the base within the middle third*, i.e. $x < \frac{2b}{3}$.

11.4.2 Minimum Base Width for a Dam

In order that the tension may not be developed in the base section, the bottom width of the dam section shall not be less than a certain limit. For the trapezoidal cross-section of a dam of height H having the top and bottom widths of a and b, respectively, as shown in Figure 11.13, the weight per unit length of dam is given by Eq. (11.13),

$$W = \gamma_m \left(\frac{a+b}{2}\right) H \tag{11.25}$$

If the water face is vertical, then the distance of centroid of the cross-section from the vertical water face is

$$\bar{x} = \frac{a^2 + ab + b^2}{3(a+b)} \tag{11.26}$$

Horizontal force due to water stored for depth h, $P = \frac{\gamma_w h^2}{2}$

Hence,
$$\frac{P}{W} = \frac{h^2}{H(a+b)}\left(\frac{\gamma_w}{\gamma_m}\right) \tag{11.27}$$

For the case when the tension at the base just vanishes, from Eq. (11.17),

$$x = \bar{x} + \frac{P}{W}\frac{h}{3} = \frac{2b}{3}$$

$$\frac{a^2 + ab + b^2}{3(a+b)} + \frac{h^2}{H(a+b)}\left(\frac{\gamma_w}{\gamma_m}\right)\frac{h}{3} = \frac{2b}{3}$$

or
$$a^2 + ab + b^2 + \left(\frac{\gamma_w}{\gamma_m}\right)\frac{h^3}{H} = 2b(a+b)$$

Therefore,
$$b^2 + ab - a^2 = \left(\frac{\gamma_w}{\gamma_m}\right)\frac{h^3}{H} = \frac{h^3}{SH}$$

or
$$b^2\left[1 + \frac{a}{b} - \left(\frac{a}{b}\right)^2\right] = \left(\frac{\gamma_w}{\gamma_m}\right)\frac{h^3}{H} = \frac{h^3}{SH} \tag{11.28}$$

For full reservoir condition, $h = H$.

Therefore,
$$b^2 + ab - a^2 = \left(\frac{\gamma_w}{\gamma_m}\right)H^2 = \frac{H^2}{S} \tag{11.29}$$

where $S = (\gamma_m/\gamma_w)$, the *specific gravity* of masonry. The above relation can be used to compute the minimum base width.

Special cases

(i) Triangular section $a = 0$, and Eq. (11.29) reduces to

$$b^2 = \left(\frac{\gamma_w}{\gamma_m}\right)H^2 \quad \text{or} \quad b = \sqrt{\left(\frac{\gamma_w}{\gamma_m}\right)}\,H \tag{11.30}$$

(ii) Rectangular section $a = b$, and Eq. (11.29) reduces to

$$b^2 = \left(\frac{\gamma_w}{\gamma_m}\right)H^2 \quad \text{or} \quad b = \sqrt{\left(\frac{\gamma_w}{\gamma_m}\right)}\,H \tag{11.31}$$

Minimum width to prevent sliding

It is given by: $\mu W > P$ i.e.

$$\mu\left[\frac{(a+b)}{2}\gamma_m H\right] > \frac{\gamma_w H^2}{2} \quad \text{or} \quad a + b > \frac{H}{\mu S} \tag{11.32}$$

For critical condition $a + b = \dfrac{H}{\mu S}$ or $b = \dfrac{H}{\mu S} - a$

Minimum width from the maximum stress consideration

$$\sigma_{\max} = \frac{W}{b}\left(1 + \frac{6e}{b}\right) = \sigma_a \tag{11.33}$$

Hence for safety and stability, the largest of the three values of b is adopted.

EXAMPLE 11.11

A 10 m high masonry dam of trapezoidal cross-section ABCD has the top and bottom widths of 1.0 m and 6.0 m, respectively as shown in Figure 11.14. The dam retains water on its vertical face to a depth h. Determine: (a) the stresses developed at the base when the dam retains water up to the top, (b) determine the depth h to which the water can be stored without developing tension at the base, also examine the stability of the dam against sliding and overturning. The unit weights of the masonry and water may be taken as 22 kN/m³ and 10 kN/m³, respectively. The coefficient of friction μ between the masonry dam and the underlying soil at the site is 0.7.

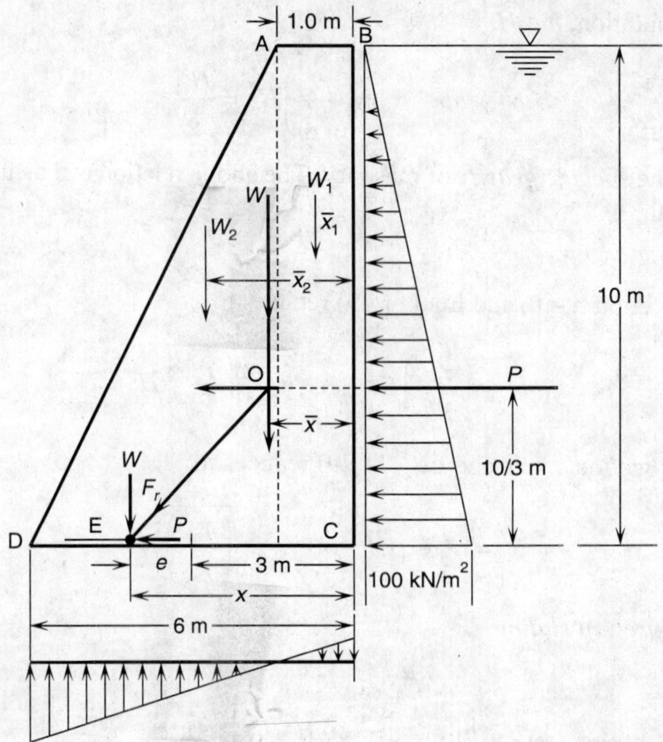

Figure 11.14 Forces acting on the masonry dam of Example 11.1.

Solution: Consider one-metre length of the dam.

(a) The dam retains water up to its top:

The horizontal water pressure increases linearly from zero at the free surface of water to $\gamma_w H = 10 \times 10 = 100$ kN/m² at the base of the dam. Total horizontal force acting on the dam is

$$P = \frac{\gamma_w H^2}{2} = \frac{10 \times 10^2}{2} = 500 \text{ kN}$$

This acts at height (10/3) m above the base as shown in Figure 11.14. The weight of the dam, $W = W_1 + W_2$ is given by

$$W_1 = \gamma_m aH = 22 \times 1.0 \times 10 = 220 \text{ kN}$$

which acts at $\bar{x}_1 = 0.5(= a/2)$ from the water face

$$W_2 = \gamma_m \left(\frac{b-a}{2}\right) H = 22 \times \left(\frac{6-1}{2}\right) \times 10 = 550 \text{ kN}$$

which acts at $\bar{x}_2 = 1.0 + (5/3) = 2.667$ m from the water face

Consider that the resultant F_r strikes the base at the point E located at distance x from the water face. x can be determined by taking moments about the point E while keeping in mind that the moment of a resultant force is equal to that of its components W and P, i.e.

$$P\left(\frac{H}{3}\right) - W_1 x_1 - W_2 x_2 = 0$$

or

$$500 \times \left(\frac{10}{3}\right) - 220 \times (x - 0.5) - 550 \times (x - 2.667) = 0$$

Therefore,

$$x = \frac{1}{(220 + 550)} \times \left[500 \times \left(\frac{10}{3}\right) + 220 \times 0.5 + 550 \times 2.667\right] = 4.2124 \text{ m}$$

Thus the resultant F_r strikes the base just outside the middle third of the rectangular base which is at $(2/3) \times 6 = 4$ m. Hence the tension will develop at the base. The load $W = 770$ kN acts at an eccentricity,

$$e = x - \frac{b}{2} = 4.2124 - \frac{6}{2} = 1.2124 \text{ m}$$

The extreme resultant stress,

$$\sigma_r = \frac{W}{b}\left(1 \pm \frac{6e}{b}\right) = \frac{770}{6} \times \left(1 \pm \frac{6 \times 1.2124}{6}\right)$$

Therefore, $\sigma_{max} = 283.925$ kN/m^2 (compression) at the outside face and

$\sigma_{min} = -27.258$ kN/m^2 (tension) at the water face

Alternatively, from Eq. (11.13)

$$W = 22 \times \left(\frac{1.0 + 6.0}{2}\right) \times 10 = 770 \text{ kN}$$

For a dam with vertical water face, the distance of centroid of the dam cross-section from the vertical face is given by Eq. (11.14),

$$\bar{x} = \frac{6^3 - 1^3}{3 \times (6^2 - 1^2)} = 2.0476 \text{ m}$$

Total horizontal force acting on the dam,

$$P = \frac{\gamma_w H^2}{2} = \frac{10 \times 10^2}{2} = 500 \text{ kN}$$

The resulting eccentricity of load W is given by Eq. (11.18)

$$e = \bar{x} + \frac{P}{W}\frac{H}{3} - \frac{b}{3} = 2.0476 + \frac{500}{770} \times \frac{10}{3} - \frac{6}{3} = 1.2121 \text{ m}$$

The stresses at the base can be computed as above.

(b) If the dam retains water up to a depth h, the total horizontal force acting on the dam is

$$P = \frac{\gamma_w h^2}{2} = 5h^2$$

This acts at height $h/3$ above the base. For no tension to occur at C, the resultant F_r strikes the base at E which lies within the middle third and in limiting case at $x = (2b/3)$, i.e. at $x = (2/3) \times 6 = 4$ m. Equate the moment of the resultant force to that of its components W and P about the point E, i.e.

$$5h^2 \times (h/3) = 220 \times (4 - 0.5) + 550 \times (4 - 2.67) \quad \text{or} \quad h^3 = 900.9$$

Therefore, $h = 9.658$ m.

Stability of the dam against sliding and overturning

Horizontal sliding force, $P = 5 \times 9.658^2 = 466.38$ kN

Resisting frictional force, $\mu W = 0.7 \times 770 = 539$ kN

Since $\mu W > P$, sliding will not occur. The factor of safety against sliding is given by

$$\frac{\mu W}{P} = \frac{539}{466.38} = 1.1557$$

To check the stability of dam against overturning, consider moments about the heel point D.

Overturning moment, $M_o = \frac{Ph}{3} = \frac{466.38 \times 9.658}{3} = 1501.14$ kNm

The available stabilizing or restoring moment,

$$M_r = W_1(b - \bar{x}_1) + W_2(b - \bar{x}_2)$$
$$= 220 \times (6 - 0.5) + 550 \times (6 - 2.67) = 3041.5 \text{ kNm}$$

The factor of safety against overturning is

$$\frac{M_r}{M_o} = \frac{3041.5}{1501.14} = 2.026$$

EXAMPLE 11.12

A 30 m high masonry dam of trapezoidal cross-section retains water on its vertical face to a depth of 28 m. Determine the top and bottom widths, if the normal pressure developed at the base varies from zero at one side to a value equal to the safe bearing capacity of the soil of 900 kN/m^2 at the other. The unit weights of the masonry and water may be taken as 22 kN/m^3 and 10 kN/m^3, respectively.

Solution: Consider one-metre length the dam with top and bottom widths of the section to be a and b, respectively. The weight of the structure is

$$W = \gamma_m \left(\frac{a+b}{2}\right) H = 22 \times \left(\frac{a+b}{2}\right) \times 30 = 330(a+b) \tag{i}$$

Since the tension at the base is just avoided, $e = \dfrac{b}{6}$

The maximum pressure is then given by

$$\sigma_{\max} = \frac{W}{b}\left(1 + \frac{6e}{b}\right) = \frac{W}{b}\left(1 + \frac{6}{b} \times \frac{b}{6}\right) = \frac{2W}{b} = 900$$

or

$$\frac{W}{b} = 450$$

Substituting for W from Eq. (i):

$$\frac{330(a+b)}{b} = 450 \quad \text{or} \quad \frac{a}{b} = \frac{450}{330} - 1 = \frac{4}{11}$$

Therefore,

$$a = \frac{4b}{11}$$

From Eq. (11.28), for the case when tension just vanishes,

$$b^2\left[1 + \left(\frac{4}{11}\right) - \left(\frac{4}{11}\right)^2\right] = \frac{10}{22} \times \frac{28^3}{30} \quad \text{or} \quad b^2 = 270.103$$

Therefore, $b = 16.435$ m and $a = 5.98$ m.

EXAMPLE 11.13

The base of a concrete dam of triangular cross-section is designed to be a minimum for no tension to be developed with 30 m depth of water. The design was subsequently modified before the completion of the dam to store water for a depth of 32 m and it was proposed that the top of dam be rectangular with width of 5 m. Determine the new height of the dam for no tension to be developed anywhere. The unit weights of the concrete and water may be taken as 22.5 kN/m^3 and 10 kN/m^3, respectively.

Solution: Specific gravity of concrete, $S = \dfrac{22.5}{10} = 2.25$

(i) For the original design with triangular cross-section ABC to store 30 m deep water

$$\text{Width of the base, } b = \frac{H}{\sqrt{S}} = \frac{30}{\sqrt{2.25}} = 20 \text{ m}$$

(ii) Modified design for storing water to a depth of 32 m.

Consider ADFG being the additional concrete section provided in the modified design and the additional height AG required is represented by h_1 m as shown in Figure 11.15. The depth AJ from the top of triangle where the width of dam is 5 m, i.e. the level above which the section is to be a rectangular is given by

$$\frac{AJ}{AB} = \frac{JD}{BC} \quad \text{or} \quad AJ = \frac{JD}{BC} \times AB = \frac{5}{20} \times 30 = 7.5 \text{ m}$$

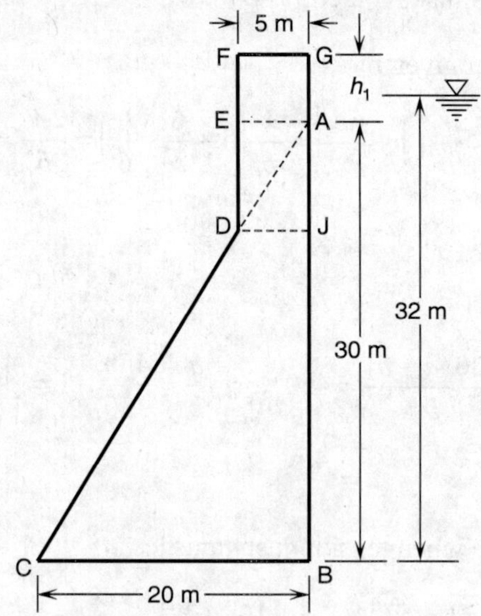

Figure 11.15 Design of concrete dam of Example 11.13 for increased depth of storage.

Consider one-metre length the dam. The weights of the various parts of the dam are,
Weight of trapezoidal part

$$W_1 = \gamma_m \left(\frac{a+b}{2}\right) H = 22.5 \times \left(\frac{5+20}{2}\right) \times 22.5 = 6328.125 \text{ kN}$$

this acts at a distance $\bar{x}$ from the vertical face of the dam,

$$\bar{x} = \frac{b^3 - a^3}{3(b^2 - a^2)} = \frac{20^3 - 5^3}{3(20^2 - 5^2)} = 7 \text{ m}$$

Weight of rectangular part

$$W_2 = \gamma_m aH = 22.5 \times 5 \times (h_1 + 7.5) = (112.5h_1 + 843.75) \text{ kN}$$

this acts at a distance of 2.5 from the water face.

The horizontal water pressure,

$$P = \frac{\gamma_w h^2}{2} = \frac{10 \times 32^2}{2} = 5120 \text{ kN}$$

For the tension to be just avoided, the distance x of the point from the water face at which the resultant F_r strikes the base should be equal to $(2b/3)$, i.e.

$$x = \frac{\text{Total moment about the point B}}{\text{Total vertical load}} = \frac{P(H/3) + W_1 x_1 + W_2 x_2}{W_1 + W_2} = \frac{2b}{3}$$

or
$$\frac{5120 \times (32/3) + 6328.125 \times 7 + (112.5h_1 + 843.75) \times 2.5}{6328.125 + (112.h_1 + 843.75)} = \frac{2 \times 20}{3}$$

$$112.5 \times (40 - 7.5)h_1 = 5120 \times 32 + 6328.125 \times 21 + 843.75 \times 7.5$$
$$- 40 \times (6328.125 + 843.75)$$

or
$$3656.25h_1 = 16183.75$$

Therefore,
$$h_1 = 4.4263 \text{ m}$$

Final height of the dam, $H = 30 + h_1 = 34.4263$ m.

11.4.3 Trapezoidal Dam Cross-section with Battered or Inclined Water Face

This case differs from that of a dam with vertical water face as the direction of pressure intensity being normal to the battered surface is inclined to the horizontal. The procedure discussed in the preceding section is also applicable to this case. However, it is convenient to consider the weight of wedge of water on the battered surface and horizontal pressure P_h instead of considering total water pressure P perpendicular to the battered surface in the calculations. As usual the horizontal pressure P_h acts at $h/3$ above the base and the weight of wedge of water on the battered surface acts at the centre of gravity of the wedge (see Figure 11.16).

EXAMPLE 11.14

A 3 m high concrete wall of trapezoidal cross-section is 800 mm thick at the base which reduces linearly to 250 mm at the top. Determine the pressures developed at the base when the wall retains water on its battered face up to a height of 2.85 m. The unit weights of the concrete and water may be taken as 25 kN/m^3 and 10 kN/m^3, respectively (see Figure 11.16).

Solution: Batter or inclination of the wall, $\alpha = \tan^{-1}(3/0.55) = 79.61°$

and
$$b_1 = 2.85 \cot 79.61° = 0.52256 \text{ m}$$

Inclined water face of the wall,

$$L = h \text{ cosec } \alpha = 2.85 \times \text{cosec } 79.61 = 2.8975 \text{ m}$$

Consider one-metre length of the wall. The weight of the wall is

(i) Weight of the rectangular portion of wall

$$W_1 = \gamma_m aH = 25 \times 0.25 \times 3 = 18.75 \text{ kN}$$

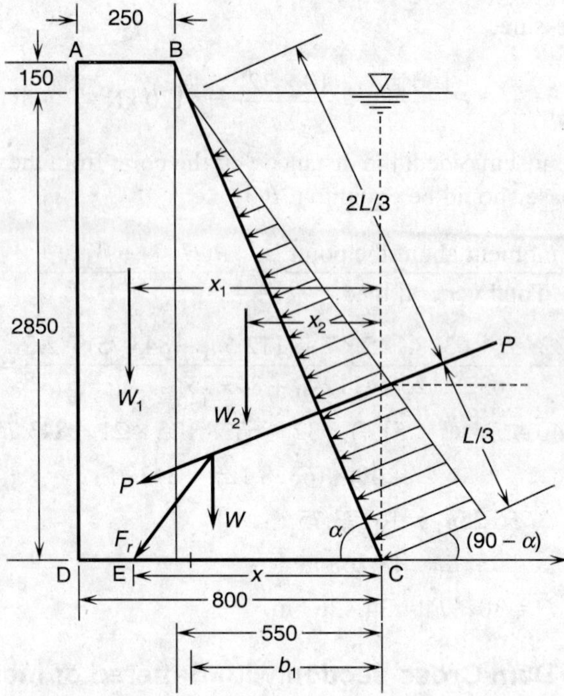

Figure 11.16 Water retaining wall of trapezoidal section of Example 11.14 (battered water face).

(ii) Weight of triangular portion of wall

$$W_2 = \frac{1}{2}(\gamma_m bH) = \frac{1}{2} \times 25 \times 0.55 \times 3 = 20.626 \text{ kN}$$

The total water pressure acting normal to the battered surface,

$$P = \frac{1}{2} \times (\gamma_w h) \times L = \frac{1}{2} \times (10 \times 2.85) \times 2.8975 = 41.2894 \text{ kN}$$

This acts at a point at distance $(L/3)$ measured along the surface from the base. The horizontal and vertical distances of the point are $b_1/3$ and $h/3$, respectively.

The horizontal and vertical components of water pressure are:

$$P_h = \frac{1}{2}\gamma_w h^2 = \frac{1}{2} \times 10 \times 2.85^2 = 40.6125 \text{ kN}$$

$$P_v = P \cos \alpha = 41.2894 \times \cos 79.61° = 7.4464 \text{ kN}$$

Alternatively, $$P_v = \frac{1}{2}(\gamma_w h) b_1 = \frac{1}{2} \times (10 \times 2.85) \times 0.52256 = 7.4465 \text{ kN}$$

Consider that the resultant F_r strikes the base at the point E located at distance x from the edge C. Since the moment of a resultant force is equal to that of its components W and P i.e.

$$P(L/3) + W_1x_1 + W_2x_2 = (P_v + W_1 + W_2)x$$

$$x = \frac{P(L/3) + W_1x_1 + W_2x_2}{P_v + W_1 + W_2}$$

Therefore,

$$x = \frac{41.2894 \times (2.8975/3) + 18.75 \times 0.675 + 20.625 \times (2/3) \times 0.55}{7.4464 + 18.75 + 20.625}$$

$$= 1.2835 \text{ m}$$

Eccentricity, $e = x - (b/2) = 1.2835 - (0.80/2) = 0.8835$ m

The pressures at the base are then given by

$$\sigma_{max} = \frac{W}{b}\left(1 \pm \frac{6e}{b}\right) = \frac{46.82}{0.80} \times \left(1 \pm \frac{6}{0.80} \times 0.8835\right)$$

Hence $\sigma_{max} = 446.33 \text{ kN/m}^2$ and $\sigma_{min} = -329.28 \text{ kN/m}^2$

11.5 RESULTANT THRUST STRIKING THE BASE AT POINT OUTSIDE THE MIDDLE THIRD

In order that no tension is produced at a section than $e \leq b/6$ i.e. the resultant force on the section must strike the section within the middle third. However, in practice normally the resultant force falls outside the middle third and a part of the section is subjected to tension. For the safety of the structure the tension should be within permissible limits of the material of the body.

On the other hand, in case of a base section resting on the ground a tension between the base and the supporting ground means that base has lifted up the ground. This would result in the reduction of contact area and hence in a change in pressure distribution. However, the pressure distribution is assumed to follow the linear law and pressure intensity on the base varies uniformly from σ_{max} at the edge D to zero at a certain point G as shown in Figure 11.17. For equilibrium, the downward load W over the length of l of structure is balanced by the total upward reaction on the area $b \times l$. Thus,

$$\text{Total upward reaction} = \text{Area of pressure diagram} = \frac{1}{2}(\sigma_{max} \times \text{DG}) \times l = W$$

Therefore,

$$\sigma_{max} = \frac{2W}{l \times \text{DG}} \tag{11.34}$$

Moreover, the line of action of W must pass through the centroid of the pressure diagram, i.e. DG = 3DE, then DG = $3(b - x)$

Therefore,

$$\sigma_{max} = \frac{2W}{3 \times l \times (b - x)} \tag{11.35}$$

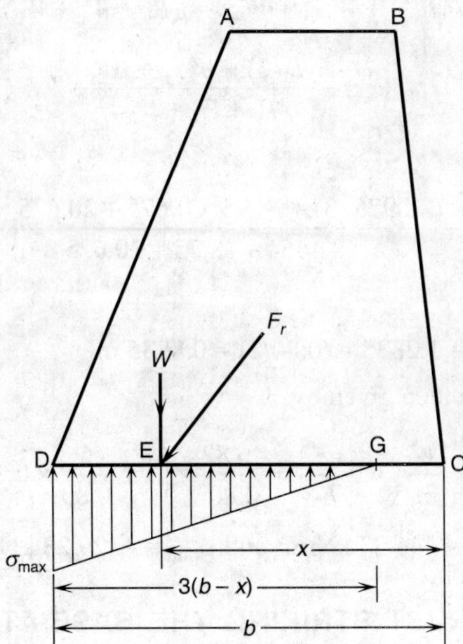

Figure 11.17 Resultant thrust striking the base at a point outside the middle-third.

EXAMPLE 11.15

A rectangular concrete block of the base size 0.8 × 1.5 m rests on the ground and supports a load of 150 kN at an eccentricity of 500 mm along the longer dimension. If no tension can be transferred between the contact surface of the block and the supporting ground, determine the maximum stress exerted on the surface.

Solution: As the load acts outside the middle-third, a part of the contact surface is in tension the base lifts up the ground. From Eq. (11.35):

$$\sigma_{max} = \frac{2W}{3l(b-x)}$$

Here,

$$W = 150 \text{ kN}; \; l = 0.8 \text{ m}; \; b = 1.5 \text{ m and } x = (L/2) + e = 0.75 + 0.5 = 1.25 \text{ m}$$

Therefore,
$$\sigma_{max} = \frac{2 \times 150}{3 \times 0.8 \times (1.5 - 1.25)} = 500 \text{ kN/m}^2$$

11.6 STRESSES IN RETAINING WALLS

A retaining wall is a structure constructed primarily to retain or support earth or some other material in a relatively vertical position on one or both sides of it at different heights. The common applications are in hill roads, bridges, canals, and swimming pools. The materials generally retained (e.g., earth, loose stone), when unsupported, attain a natural slope, the angle of which to the horizontal is known as the *angle of repose*. The angle of repose varies with the type of material and its moisture content. The granular materials such as sand or gravel behave differently than the cohesive materials such as clay. The material retained or supported by the retaining wall is called *backfill* which may have its top surface horizontal or inclined as shown in Figures 11.18(b-i). The portion of the backfill lying above a horizontal plane at the top of the wall is called *surcharge* and its inclination to the horizontal is called the *surcharge angle* δ. Various components of a retaining wall are shown in Figure 11.18.

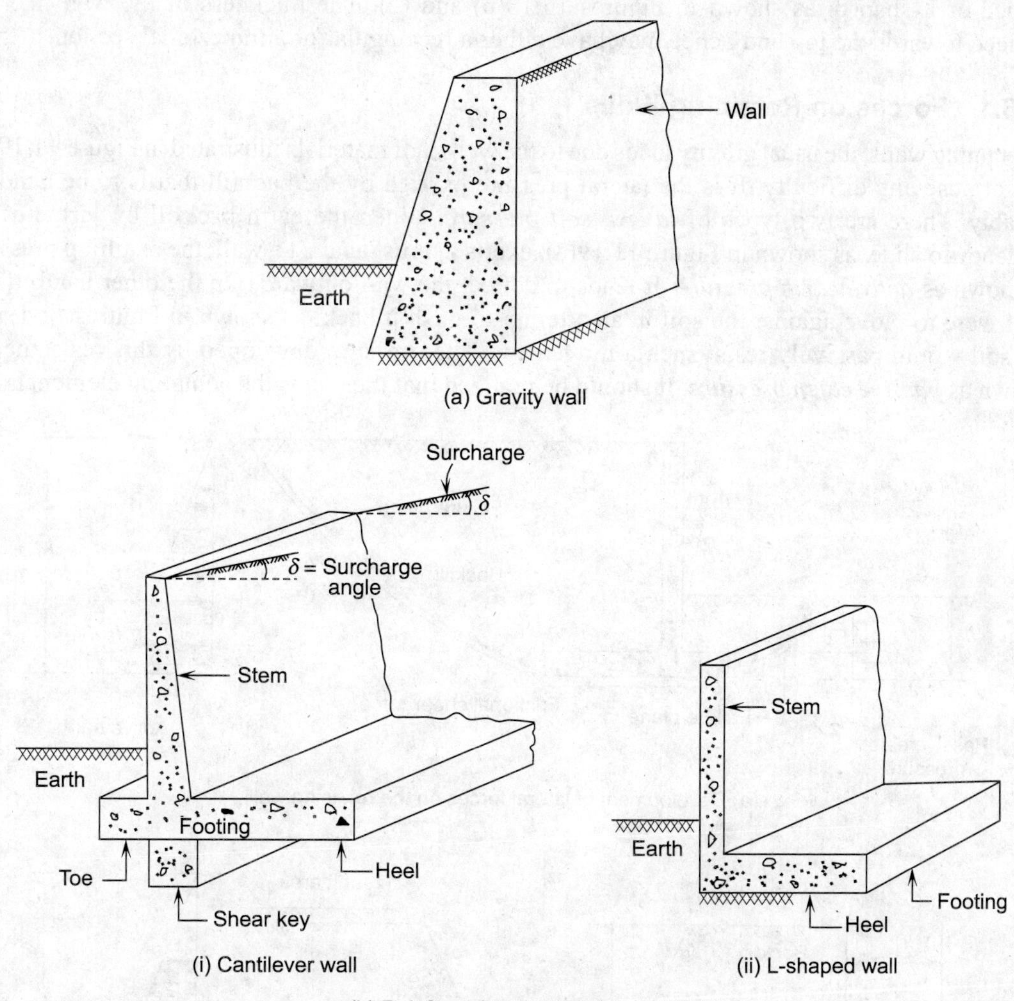

Figure 11.18 Commonly used types of retaining walls.

In its simplest form, a retaining wall is a thick plain concrete or masonry vertical cantilever supported at the base. The dead weight in such a structure is a major factor providing *stability against overturning* and *horizontal sliding* under the action of lateral earth pressure. This type of wall shown in Figure 11.18(a) is termed a *gravity retaining wall*. Its dimensions are so proportioned that there is no tension induced in the wall, i.e. the resultant of forces remains within the middle third of the cross-section.

In case of reinforced concrete, a relatively thin vertical cantilever with a horizontal base slab shown in Figure 11.18(b) is commonly used. This type of wall, called the cantilever retaining wall, utilizes the weight of backfill itself to provide the desired stability. The cantilever portion of the retaining wall is called the stem, the portion of the base slab under the backfill is called the heel and the front portion of the base is called the toe. The cantilever retaining walls may be either T-shaped or L-shaped, as shown in Figures 11.18(b) and (c). The thickness of the wall may be reduced towards the top and hence may have either a rectangular or a trapezoidal section.

11.6.1 Forces on Retaining Walls

In retaining walls, the usual gravity loads due to the weight of materials illustrated in Figure 11.19(b) do not pose any difficulty. It is the lateral pressure exerted by the backfill that is to be handled suitably. There are two types of *lateral soil pressure.* When the earth backfill by virtue of its tendency to slide, as shown in Figure 11.19(a), exerts a thrust against a wall, the resulting pressure is known as *active earth pressure.* It tends to deflect the wall outward. On the other hand, if the wall were to move against the soil in an attempt to push it back, as shown in Figures 11.19(a), the soil would passively resist such a movement. The pressure developed by this resistance is known as *passive earth pressure.* It should be realized that the soil is the actuating element in the

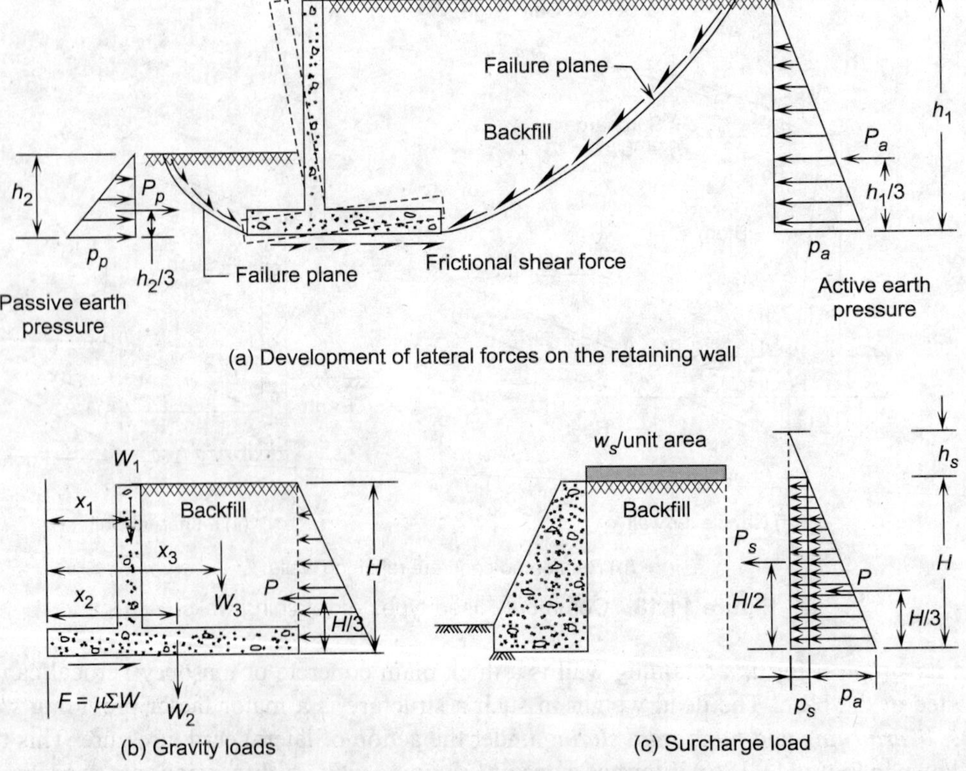

Figure 11.19 Forces acting on retaining walls.

development of active soil pressure, and the wall is the decisive factor in the development of passive pressure. In the analysis of retaining walls, a designer is concerned primarily with the active earth pressure, hereafter referred to as earth pressure.

The earth pressure on a wall is a function of the type of soil, the degree of saturation, the compaction of the backfill, and the surcharge. For calculating the lateral earth pressure, various theories have been proposed. In this text, Rankine's theory will be used. This theory neglects the friction between the retaining wall and the retained earth; as a result, the resultant pressure shall be parallel to the surface of the backfill which may be horizontal or inclined.

Consider a retaining wall supporting the earth as shown in Figure 11.20. The pressure, p_a, exerted by the soil on the wall is assumed to be proportional to the unit weight of soil γ_s and depth h of the section under consideration from the top. Thus for a retaining wall having the backfill whose surface is at an angle δ with the horizontal, p_a is given by:

$$p_a = k_a \gamma_s h \tag{11.36}$$

per metre length of wall. The typical pressure distribution for level backfill is shown in Figure 11.19(b). The total force, P_a per metre length of wall which is equal to the area under the pressure diagram is given by:

$$P_a = p_{max} \times (h/2) = k_a \gamma_s h (h/2) = k_a \gamma_s h^2/2 \tag{11.37}$$

where γ_s = Unit weight of backfill, kN/m^3,

h = Depth of the section under consideration, m,

and k_a is the coefficient of active earth pressure. For the walls with a vertical backface,

$$k_a = \cos\delta \left[\frac{\cos\delta - \sqrt{\cos^2\delta - \cos^2\phi}}{\cos\delta + \sqrt{\cos^2\delta - \cos^2\phi}} \right] \tag{11.38}$$

where ϕ is the angle of repose of the backfill and δ is the angle the surface of backfill makes with the horizontal.

The force P for the total height of the wall is $P_a = (k_a \gamma_s H^2)/2$ acting at $(H/3)$ from the base and is inclined at an angle δ to the horizontal.

If the backfill is level, i.e. $\delta = 0$, and

$$P_a = \frac{1}{2} \gamma_s H^2 \left(\frac{1 - \sin\phi}{1 + \sin\phi} \right) \tag{11.39}$$

which acts at a height $(H/3)$ from the base. The passive earth pressure due to passive movement of the wall towards the earth is given by:

$$P_p = (k_p \gamma_s H^2)/2 \tag{11.40}$$

where k_p, the coefficient of passive earth pressure is given by

$$k_p = \cos\delta \left[\frac{\cos\delta + \sqrt{\cos^2\delta - \cos^2\phi}}{\cos\delta - \sqrt{\cos^2\delta - \cos^2\phi}} \right] \tag{11.41}$$

and acts $H/3$ from the base and is *parallel to the backfill surface.*

For the horizontal backfill, i.e. $\delta = 0$,

$$P_p = \frac{\gamma_s H^2}{2}\left(\frac{1+\sin\phi}{1-\sin\phi}\right) \tag{11.42}$$

11.6.2 Surcharge

Loads on the earth behind a retaining wall impose an additional force to be resisted by the wall. Such a load is commonly referred to as *surcharge* (see Figure 11.20(a) and (b)). These loads may be due to railway tracks, roadways, cranes, vehicles and storage of materials, etc. For analysis and design, it is a common practice to transform the surcharge load into an equivalent height of earth backfill by dividing the surcharge pressure by the unit weight of soil, etc. When a surcharge load of intensity w_s acts on the horizontal surface of the backfill as shown in Figure 11.20(b), then

$$h_s = \frac{w_s}{\gamma_s} \tag{11.43}$$

where h_s = Equivalent height of soil for surcharge, m,
w_s = Intensity of surcharge, kN/m², and
γ_s = Unit weight of soil, kN/m³.

The effect of added height on the pressure exerted on the retaining wall is illustrated in Figure 11.20(b). The intensity of additional horizontal pressure p_s is:

$$p_s = k_a\,\gamma_s\,h_s = w_s\left(\frac{1-\sin\phi}{1+\sin\phi}\right) \tag{11.44}$$

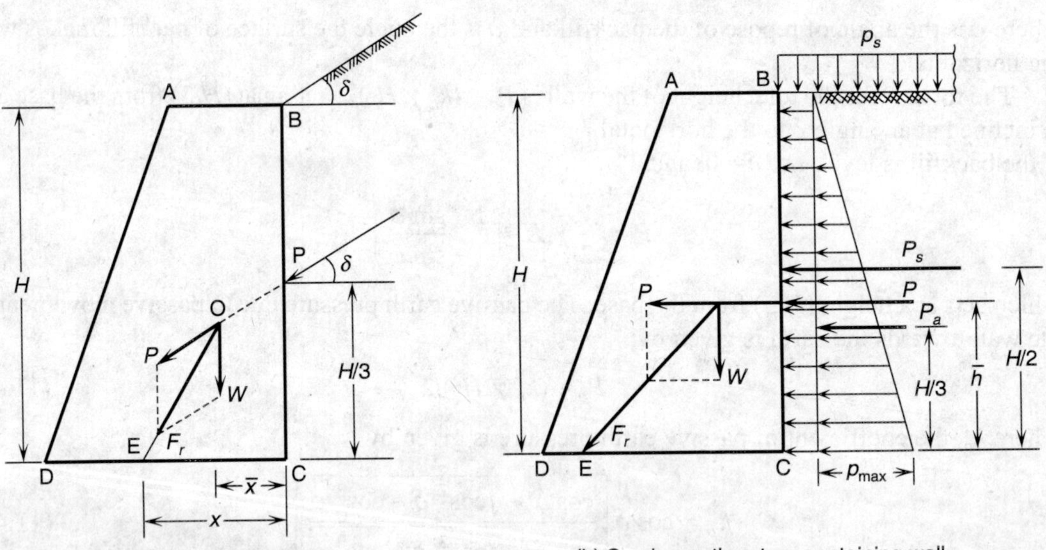

Figure 11.20 Surcharge on retaining walls.

It is seen that horizontal pressure due to surcharge is constant throughout the depth of the soil up to the base of the footing. Thus the horizontal earth pressure due to surcharge is given by

$$P_s = w_s H \left(\frac{1 - \sin \phi}{1 + \sin \phi} \right) \tag{11.45}$$

which acts at a height (H/2) from the base.

The total earth pressure is the sum of horizontal pressures due to backfill and surcharge. Thus,

Total earth pressure, $$P = P_b + P_s = H\left(p_s + \frac{1}{2}\gamma_s H \right) \times \left(\frac{1 - \sin \phi}{1 + \sin \phi} \right) \tag{11.46}$$

and acts at a height $\bar{h}$ from the base,

$$\bar{h} = \frac{P_a(H/3) + P_s(H/2)}{P_a + P_s} \tag{11.47}$$

11.6.3 Stability Requirements

A retaining wall is normally subjected to gravity loads due to self-weight W. The lateral pressure P is composed of backfill pressure P_a and surcharge pressure P_s. As in the case of dams, for stability of the wall, the following requirements should be satisfied.

(1) *Overturning*: To ensure stability against overturning about the outside edge of the wall, stabilizing moments due to gravity loads should be more than the overturning moments due to lateral forces, i.e. the resultant of the forces F_r should lay within the base width as shown in Figure 11.21. The Factor Of Safety, i.e. the ratio of stabilizing moments to the overturning moments should be in the range 1.5 to 2.0.

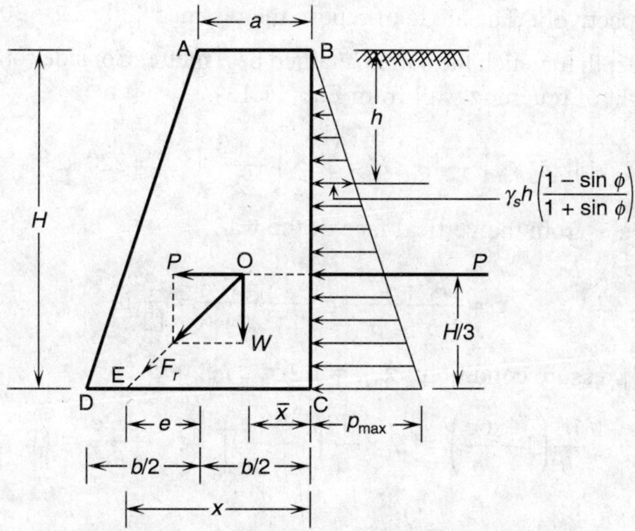

Figure 11.21 Earth pressure on a retaining wall.

(2) *Sliding*: The resisting force against sliding should be more than the sliding force so as to get a factor of safety of 1.5, i.e.

$$\left(\mu \sum W\right) / P_{ah} > 1.5 \tag{11.48}$$

where $\sum W$ = Total gravity load,

P_{ah} = Horizontal component of earth pressure,

μ (= tan ϕ) = Friction coefficient between base and ground, and

ϕ = Angle of repose or angle of internal friction of soil.

The passive earth pressure is not taken into account while calculating the resistance to sliding.

(3) *Base pressure*: The vertical pressure on the soil under the base should not exceed the *bearing capacity of soil*. There should be no tension under the base, since tension means that the edge of the wall will be lifted above the soil, which is not permissible. The latter condition can be accomplished by making the resultant of the forces to fall within the *middle third* of the base.

The procedure for analysis of gravity retaining wall is similar to one used for dams except that the linearly varying lateral pressure is due to backfill instead of water, hence the formulae derived for dams are valid for retaining walls.

EXAMPLE 11.16

A 5 m high masonry retaining wall of trapezoidal cross-section has the top and bottom widths of 1.0 m and 3.0 m, respectively. The soil is stored on the side of vertical face of retaining wall. Determine the depth h to which the soil can be stored so that the maximum pressure at the base is 1.2 times the minimum pressure at the base. The unit weights of the masonry and soil are as 22 kN/m^3 and 16.5 kN/m^3, respectively. The angle of repose the retained soil is $\phi = 30°$.

Solution: Let the depth to which the soil is retained be h metre. Consider one-metre length of the retaining wall. Weight of retaining wall from Eq. (11.13),

$$W = \gamma_m \left(\frac{a+b}{2}\right) H = 22 \times \left(\frac{1+3}{2}\right) \times 5 = 220 \text{ kN}$$

this acts at a distance $\overline{x}$ from the vertical face of the wall,

$$\overline{x} = \frac{b^3 - a^3}{3(b^2 - a^2)} = \frac{3^3 - 1^3}{3(3^2 - 1^2)} = \frac{13}{12} \text{ m}$$

From the stipulated pressure condition: $\sigma_{\max} = 1.2\sigma_{\min}$ i.e.

$$\frac{W}{b}\left(1 + \frac{6e}{b}\right) = 1.2 \times \frac{W}{b}\left(1 - \frac{6e}{b}\right) \quad \text{or} \quad \frac{13.2e}{b} = 0.2$$

Therefore, $$e = \frac{1}{22} \text{ m}$$

If the retaining wall stores soil up to a depth h, the total horizontal force acting on the wall is

$$P = \frac{1}{2}\gamma_s h^2 \left(\frac{1-\sin\phi}{1+\sin\phi}\right) = \frac{1}{2}\times 16.5\times h^2\left(\frac{1-\sin 30°}{1+\sin 30°}\right) = \frac{11}{4}h^2$$

This acts at a height $h/3$ above the base. From Eq. (11.18)

$$e = \bar{x} + \frac{P}{W}\frac{h}{3} - \frac{b}{2} \quad \text{or} \quad \frac{1}{22} = \frac{13}{12} + \frac{11h^2}{4\times 220}\times\frac{h}{3} - \frac{3}{2}$$

$$\frac{11h^3}{2640} = \frac{1}{22} - \frac{13}{12} + \frac{3}{2} = \frac{61}{132}$$

Therefore,
$$h^3 = \frac{61}{132}\times\frac{2640}{11} \quad \text{or} \quad h = 4.8046 \text{ m}$$

EXAMPLE 11.17

A 6 m high concrete gravity retaining wall of trapezoidal cross-section has the top width of 1.0 m and retains soil level with its top. The soil is stored on the side of vertical face of wall. There is a uniform surcharge of 27 kN/m^2. Determine: (a) The base width of wall for no tension, and (b) the maximum pressure developed at the base. The unit weights of the concrete and the soil are as 24 kN/m^3 and 18 kN/m^3, respectively. The angle of repose of the retained soil is $\phi = 30°$.

Solution: Let the width of the base is b m. Consider one-metre length of the retaining wall. Weight of retaining wall,

$$W = \gamma_m\left(\frac{a+b}{2}\right)H = 24\times\left(\frac{1+b}{2}\right)\times 6 = 72\,(1+b)\text{ kN}$$

which acts at a distance $\bar{x}$ from the vertical face of the dam,

$$\bar{x} = \frac{a^2+ab+b^2}{3(a+b)} = \frac{1+b+b^2}{3(1+b)}$$

Intensity of lateral pressure at the top of wall,

$$p_{\text{top}} = p_s = 27\times\left(\frac{1-\sin 30°}{1+\sin 30°}\right) = 9 \text{ kN/m}^2$$

Intensity of lateral pressure at the bottom of wall,

$$p_{\text{bottom}} = (27+18\times 6)\times\left(\frac{1-\sin 30°}{1+\sin 30°}\right) = 45 \text{ kN/m}^2$$

Total lateral pressure on the wall,

$$P = \frac{6}{2}\times(9+45) = 162 \text{ kN}$$

P acts at a height $\bar{h}$ above the base which is given by

$$\bar{h} = \frac{45 + 2 \times 9}{45 + 9} \times \frac{6}{3} = \frac{7}{3} \text{ m}$$

Consider that for no tension the resultant F_r strikes the base at a point located at distance $x = (2b/3)$ m from the soil face.

Take moments about the soil face, $P\bar{h} + W\bar{x} = Wx$ i.e.

$$162 \times \frac{7}{3} + 72(1 + b) \times \frac{1 + b + b^2}{3(1 + b)} = 72(1 + b) \times \frac{2b}{3}$$

$$\frac{21}{4(1 + b)} + \frac{1 + b + b^2}{3(1 + b)} = \frac{2b}{3} \quad \text{or} \quad 63 + 4 + 4b + 4b^2 = 8b + 8b^2$$

$$4b^2 + 4b - 67 = 0 \quad \text{or} \quad b^2 + b - 16.75 = 0$$

Therefore, $\qquad b = 3.623$ m

A reader should note that the procedure is exactly similar to the one used for dams.

11.7 PROBLEMS

11.1 A tie bar of T-section is subjected to a longitudinal pull P acting at a point on Y-axis but not at the centroid of the section as shown in Figure 11.22. Determine the magnitude of P and the position of its line of action if the stresses across the section vary from 10 MPa compression at the top to 120 MPa tension at the bottom.

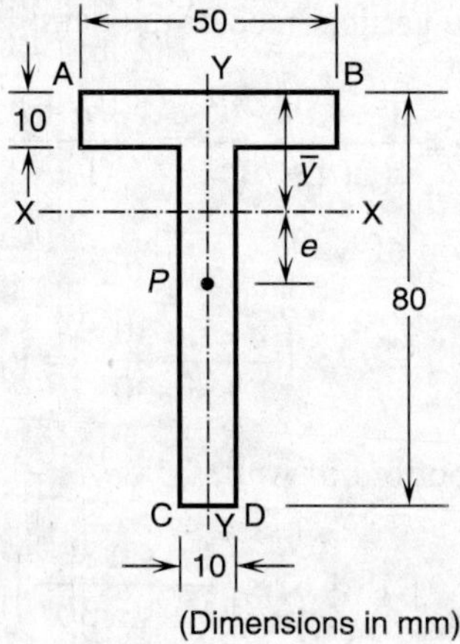

Figure 11.22

[**Ans.** $A = 1200$ mm^2; $\bar{y} = 28.23$ mm; $I_x = 7.5667 \times 10^5$ mm^4; $e_y = 28.42$ mm and $P = 43.25$ kN]

11.2 A short hollow square pier with outside and inside sides of 2.5 m and 1.5 m, respectively, as shown in Figure 11.23, supports a vertical point load of 100 kN located at the diagonal at distance of 1.4 m from the vertical axis of the pier. Ignoring the self weight of the pier, determine the normal stresses at the four outside corners on the horizontal section of pier.

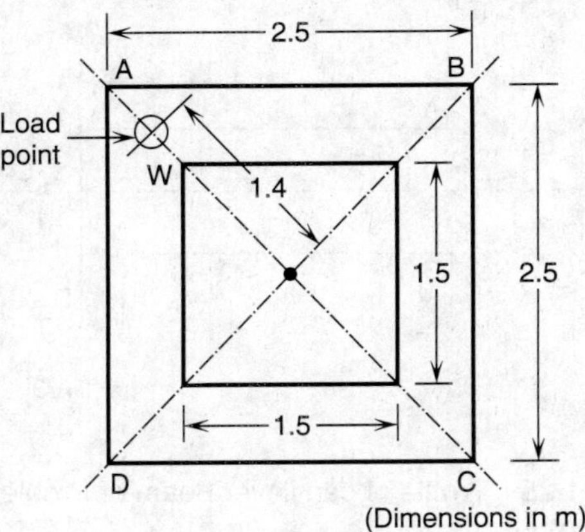

Figure 11.23

[**Ans.** $\sigma_A = 1.124$ MPa (C); $\sigma_C = 0.624$ MPa (T) and $\sigma_B = \sigma_D = 0.25$ MPa (C)]

11.3 A 150 mm wide and 20 mm thick steel plate is subjected to a longitudinal pull of 120 kN along its longitudinal centroidal axis. A hole of 40 mm diameter is drilled through the plate whose centre is 50 mm from the original longitudinal axis of the plate as shown in Figure 11.24. Determine the maximum stresses induced in the plate.

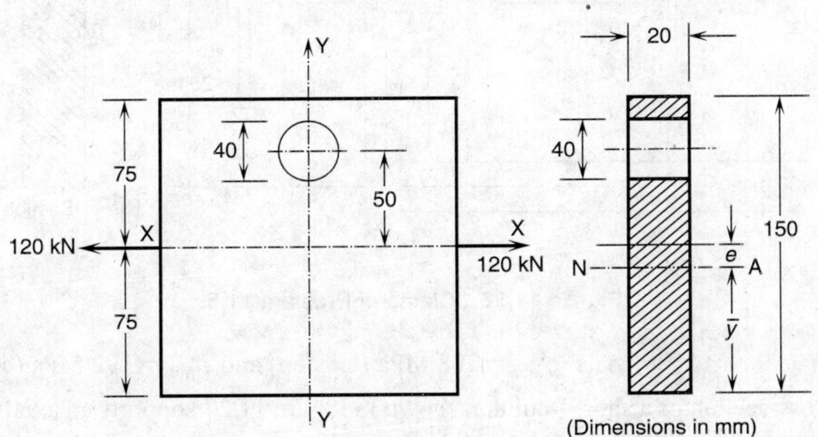

Figure 11.24

[**Ans.** $A = 2200$ mm^2; $\bar{y} = 56.82$ mm; $I_{NA} = 2.791 \times 10^6$ mm^4; $e_y = 18.18$ mm; $\sigma_{\text{top}} = 127.368$ MPa and $\sigma_{\text{bottom}} = 10.128$ MPa]

11.4 A cantilever beam of profile shown in Figure 11.25 is subjected to an inclined force $P = 30$ kN as shown in figure. Determine the resultant normal stresses at the points A and B at the fixed end with $a = 150$ mm, $b = 50$ mm and $h = 150$ mm.

[**Hint:** Transfer the horizontal component of the force to the axis of the cantilever as an axial force along with moment].

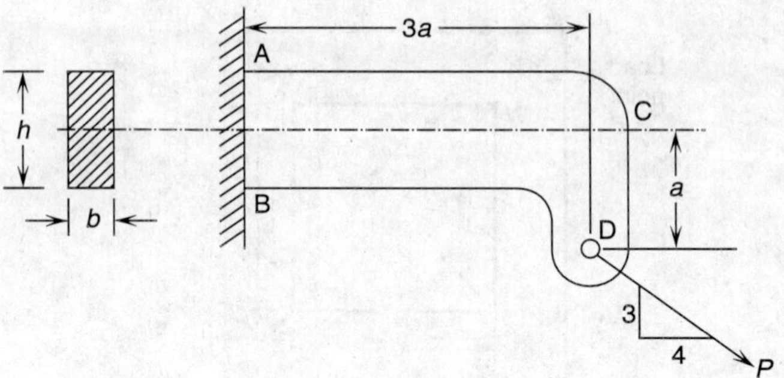

Figure 11.25 Profile of cantilever beam of Problem 11.4.

[**Ans.** $\sigma_A = 27.2$ MPa (tension) and $\sigma_B = 20.8$ MPa (compression)]

11.5 Determine the maximum stresses induced on the section I-I of the clamp shown in Figure 11.26 when a force of 2.4 kN is exerted by the screw. The cross-section at I-I is rectangular with size of 10 × 30 mm.

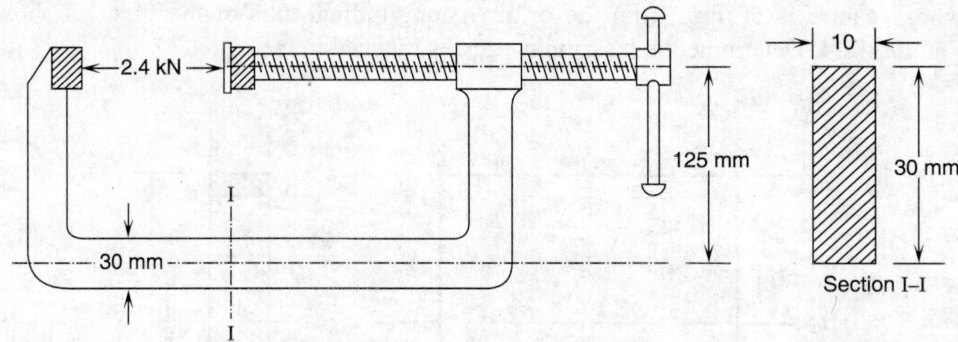

Figure 11.26 Clamp of Problem 11.5.

[**Ans.** $\sigma_{max} = 208$ MPa (tension) and $\sigma_{min} = 192$ MPa (compression)

11.6 The cross-section of a short I-column shown in Figure 11.27 supports an axial load W_1 and an eccentric W_2 at an eccentricity of 150 mm. If the stresses are to remain compressive throughout and vary from 15 MPa at one edge to 90 MPa at the other, determine the magnitudes of loads W_1 and W_2. The properties of the section are: $A = 3671$ mm^2 and $I = 2.624 \times 10^7$ mm^4.

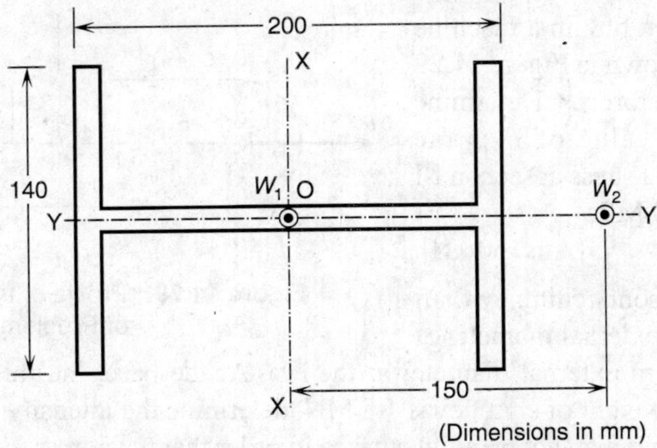

Figure 11.27 I-cross section of column of Problem 11.6.

[**Ans.** W_1 = 127.15 kN and W_2 = 65.577 kN]

11.7 The line of thrust in a test specimen of 15 mm diameter is parallel to its axis but displaced from it. Determine the distance of line of thrust from the axis of the specimen when the maximum stress is 20 per cent greater than the mean stress on a normal section.

$$\left[\textbf{Hint}: \sigma_{\max} = \frac{4P}{\pi d^2} + \frac{32Pe}{\pi d^3}; \sigma_o = \frac{4P}{\pi d^2} \text{ and } \sigma_{\max} = 1.20\sigma_o\right]$$

[**Ans.** e = 0.375 mm]

11.8 A 3 m long bar ABC of rectangular cross-section of size 80 × 150 mm (deep) fixed at the end A is cranked at 90° at the point B at a distance of 2.0 m from the fixed end as shown in Figure 11.28. The cranked bar is loaded by a force P, the direction of which passes through the centroids C and D of the cross-sections at C and D. Determine the maximum normal stress in the bar.

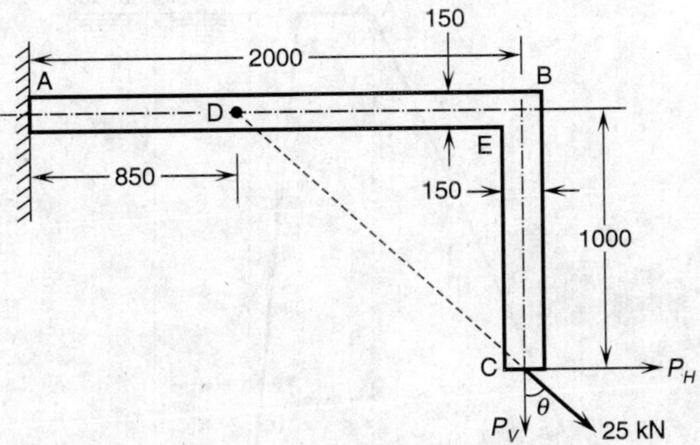

Figure 11.28 Cranked bar of Problem 11.8.

[**Ans.** θ = 49°; Critical points for moments are A and B; M_A = 13.93 kNm; $\sigma_{\max,A}$ = 48.00 MPa; M_E = 18.87 kNm; and $\sigma_{\max,E}$ = 65.84 MPa]

11.9 The 50 mm thick link in a machine with a profile shown in Figure 11.29 is subjected to a force P. Determine the maximum value of P if the maximum normal stress at section I-I is limited to 100 MPa.

[**Ans.** 50 kN].

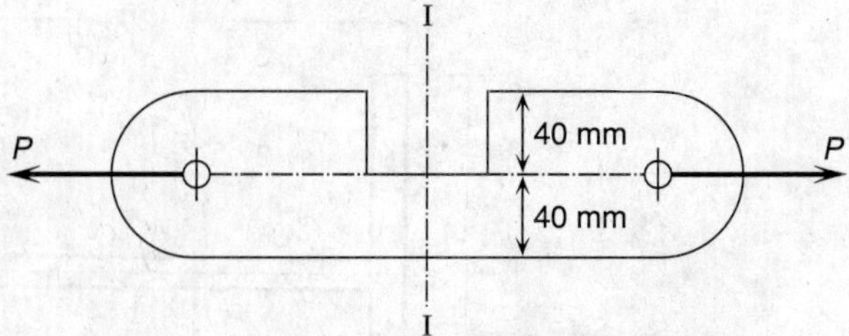

Figure 11.29 Profile of the link of machine of Problem 11.9.

11.10 A 25 m high masonry chimney shaft taper from 3 m external diameter at the base to 2.0 m external diameter at the top. At the base, the thickness of chimney is 500 mm. If the weight of chimney is 1.8 MN, determine the intensity of uniform horizontal wind pressure that may act on a flat surface in order that tension at the base is just avoided. The shape factor for the circular section is 0.6.

[**Ans.** $\sigma_o = 458.365$ kN/m^2; $A = 62.5$ m^2; $\bar{y} = (35/3)$ m and $p = 2.228$ kN/m^2]

11.11 A reinforced concrete chimney of uniform hollow circular cross-section of external and internal diameters of 3.0 and 2.5 m, respectively, is subjected to a wind pressure of intensity 1.25 kN/m^2 on a flat surface. Determine the maximum height of the chimney above the ground level if no tension is permitted in the base. The unit weight of concrete is 25 kN/m^3. The shape factor for the circular section is 0.6.

[**Ans.** $W = 53.996h$ kN; $P_h = 2.25h$ kN; $A = 2.1598$ m^2; $I = 2.0586$ m^4; and $h = 30.498$ m]

11.12 The 10 m high masonry dam shown in Figure 11.30 has trapezoidal cross-section with the top and bottom widths of 1.5 m and 6.5 m, respectively. The dam stores water to its full height. The water face of the dam has a batter of 1 in 10. Determine the maximum and minimum normal pressures at the base. The unit weights of the masonry and water are 22.5 kN/m^3 and 9.81 kN/m^3, respectively.

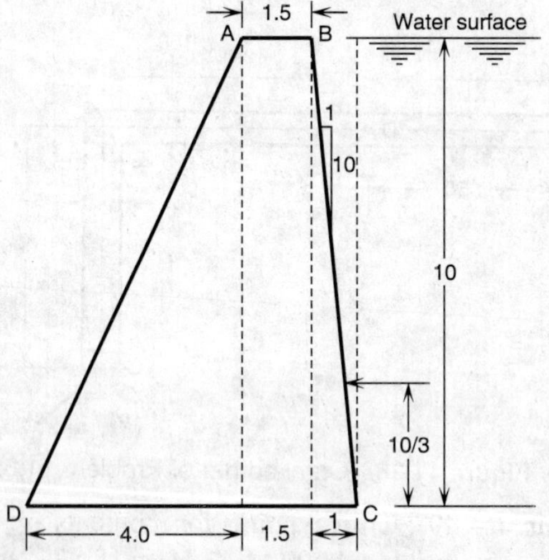

Figure 11.30 Trapezoidal masonry retaining wall of Problem 11.12.

[**Ans.** $W = 949.05$ kN; $P_h = 490.5$ kN; $x = 4.25$ m; $e = 1.0$ m; $\sigma_{max} = 280.79$ kN/m^2, $\sigma_{min} = 11.23$ kN/m^2]

11.13 (a) State the importance of *middle third rule* in gravity dams.
(b) Distinguish between *active* and *passive earth pressures.*

11.14 A 8 m high concrete retaining wall of trapezoidal cross-section has the top and the bottom widths of 1.0 m and 4.0 m, respectively. The wall retains soil to its full height against the vertical face. (a) Determine the pressures at the base. The unit weights of the plain concrete and the soil are as 24 kN/m^3 and 18 kN/m^3, respectively. The angle of repose the retained soil is $\phi = 30°$.

(b) Also determine the additional height h_1 of soil that the wall can retain without developing any tension at the base. Neglect the soil on the top of the wall.

[**Hint:** In part (b) earth pressure distribution is trapezoidal varying from $6h_1$ kN/m^2 at the top to $6(h_1 + 8)$ kN/m^2 at the bottom and total horizontal force acts at height $\bar{h}$ from the base]

[**Ans.** (a) $W = 480$ kN; $\bar{x} = 1.4$ m; $P = 192$ kN; $e = (7/15)$ m; $\sigma_{max} = 204$ kN/m^2, $\sigma_{min} = 36$ kN/m^2 and (b) $P = 48(h_1 + 4)$ kN; $\bar{h} = [4(3h_1 + 8)/3(h_1 + 4)]$; $h = 0.5$ m]

11.15 A 6 m high masonry retaining wall of trapezoidal cross-section with a vertical face on the backfill side has the top and bottom widths of 1.5 m and 3.5 m, respectively. The soil is stored level with the top of wall. Determine the maximum and the minimum pressure intensities at the base. The unit weight of the masonry is 25 kN/m^3. The unit weight of soil is 16 kN/m^3 for the top half the depth and 18 kN/m^3 below this level. The angle of repose the retained soil is 30° for both types of soils.

[**Ans.** $W = 345$ kN; $\bar{x} = 1.32$ m; $P = 99$ kN; $x = 1.89$ m; $e = 0.14$ m; $\sigma_{max} = 122.23$ kN/m^2 and $\sigma_{min} = 74.914$ kN/m^2]

Chapter 12

Rivetted and Bolted Connections

12.1 INTRODUCTION

In practice the metal structures are generally fabricated from built-up sections, made of plates, angles, rolled beams or channel sections. The various components of these members are connected by means of fasteners or are welded to enable them to act as single composite units. Where two or more members meet at a joint, the forces and moments are transferred from one member to the other. These connections are designed in such a way that the transfer of forces is accomplished safely. The connections are also required for extending the lengths of members, joining columns to footings, etc.

These connections are made by screws; pins and bolts; cotters; rivets and welds. Of these screws, pins, bolts and cotters are used as temporary fasteners, since components connected can be separated easily. On the other hand, rivets and welds are used as permanent fastenings where the components connected are not likely to require separation.

Since the behaviour of joints is very complex due to local effects, complicated geometry, and non-linear load deformation characteristics, a number of approximations and assumptions are made in the design of bolted, riveted and welded connections.

12.2 CONNECTOR PARAMETERS

The rivets are mild steel round rods with a head on one end. These are manufactured by heating the rivet rod and upsetting one end of the rod by running it into the rivet machine. The commonly used types of rivet heads are shown in Figures 12.1(a), (b) and (c). The components to be joined are initially placed exactly in the required position and riveted. In order to prevent dislocation from the required position during the process of riveting, the components are generally held together by bolts.

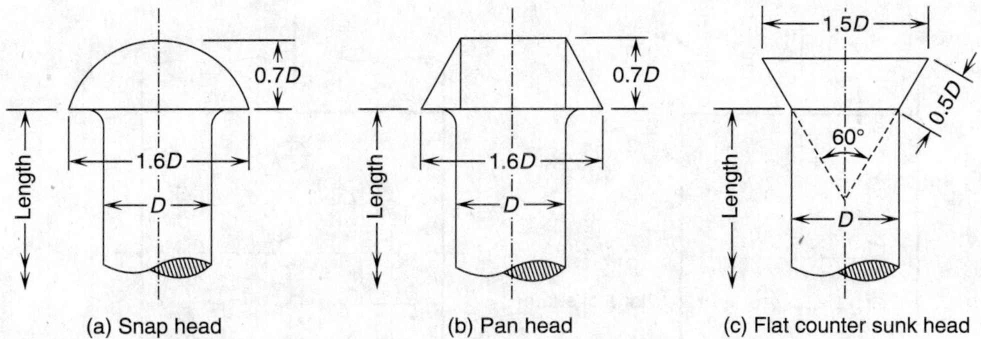

Figure 12.1 Commonly used rivet heads for structural connections.

12.2.1 Connector Diameter

The *nominal diameter* of a rivet is the diameter of the cold shank before driving. The diameter of the *hole* or *closed rivet* is called the *gross* or *effective diameter* of a rivet. The strength of a rivet is based on its gross diameter. The diameter of a rivet hole is taken as the nominal diameter of the rivet plus 1.5 mm for rivets of nominal diameter less than or equal to 25 mm, and 2 mm for rivets of nominal diameter exceeding 25 mm, unless specified otherwise.

The diameter of the rivets to suit the thickness t (mm) of the component (plate) to be connected can be calculated using the following empirical relations:

(i) *Unwin's formula* $d = 6.0\sqrt{t}$

(ii) *French formula* $d = 1.5t + 4$

(iii) *The German formula* $d = \sqrt{50t} - 2$

In structural design, the above relations are not much used. In practice, 20 and 22 mm rivets are commonly used.

12.2.2 Arrangement of Connectors

In a joint, according to requirement there can be one or more rows of connectors (rivets or bolts). The connectors in a multi-row connector joint may be arranged either in a *chain* or *zigzag* fashion as shown in Figure 12.2. If the rivets are arranged as shown in Figure 12.2(a), it is called the chain riveting. On the other hand, rivets arranged in the manner shown in Figure 12.2(b), are called *zigzag riveting* or *diamond riveting*. The latter type of riveting provides a much stronger joint of higher efficiency than the former one.

12.2.3 Pitch of the Connectors

The centre-to-centre distance between the adjacent connectors (rivets or bolt holes) in a given row is called *the pitch* and is denoted by p. The line of rivets which is parallel to the direction of load is called *gauge line.* The distance between consecutive rows of connectors is called the *back pitch or gauge* and is denoted g. When the connectors in consecutive rows are staggered, the distance between their centres is the *diagonal pitch* as shown in Figure 12.3.

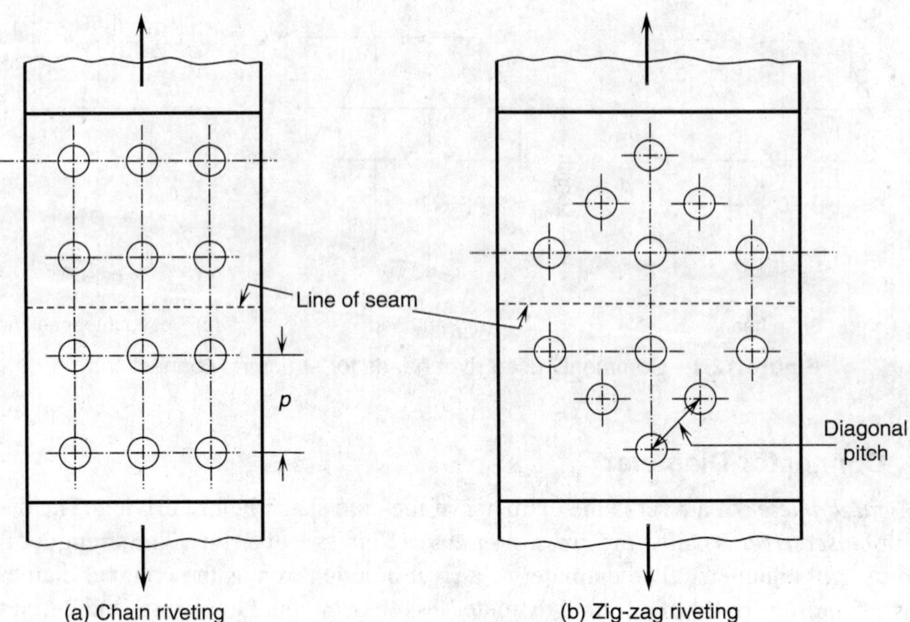

Figure 12.2 Types of rivet arrangements.

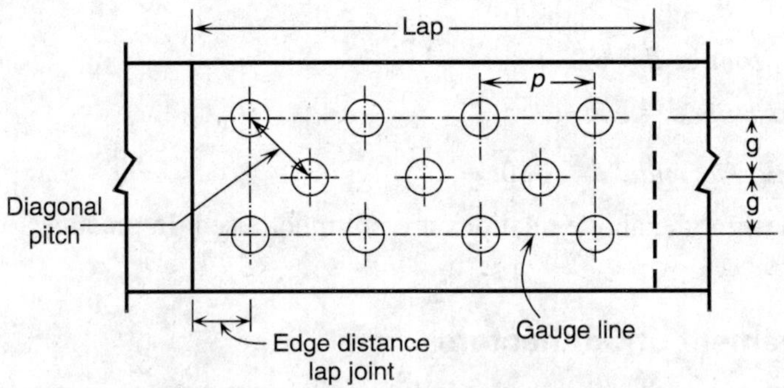

Figure 12.3 Definitions of some important terms.

As per IS:800 specifications the pitch shall not be less than 2.5 times the diameter of the connector hole. In practice, a pitch equal to three times the diameter of the connector is generally adopted. On the other hand, the distance between centres of any two adjacent rivets (including tacking rivets) connecting together the components of compression or tension members shall not exceed $32t$ or 300 mm, whichever is less, where t is the thickness of thinner outside plate.

The distance between centres of any two adjacent rivets in a row lying in the direction of stress shall not exceed $16t$ or 200 mm, whichever is less in tension member, and $12t$ or 200 mm, whichever is less in compression members. In case of compression members, in whom forces are transferred through butting faces this distance shall not exceed 4.5 times the diameter of the rivets for a distance from the abutting faces equal to 1.5 times the width of the member.

Tacking Rivets: Tacking rivets are additional rivets used, beside those required by strength considerations, to satisfy the maximum pitch requirements. The pitch of tacking rivets shall not exceed $32t$ or 300 mm, whichever is lower. Where exposed to weather, it should not exceed $16t$ or 200 mm whichever is less.

When a member is composed of two elements that are to act as one unit, the elements are stitched together by tacking rivets. The tacking rivets are also called as *stitching rivets*. For tension member the pitch of tacking rivets should not exceed 1000 mm and for compression member 600 mm.

12.2.4 Permissible Stresses in the Connectors

The permissible stresses in the rivets and bolts specified in IS:800 are listed in Table 12.1. For field rivets the permissible stresses are reduced by 10 per cent.

TABLE 12.1 Permissible stresses in the fasteners, i.e. rivets and bolts

Description of the rivets	*Permissible stress on gross area, MPa*		
	Shear stress, τ_f	*Bearing stress,* σ_{fp}	*Axial tensile stress,* σ_{ft}
Power-driven rivets	100	300	100
Hand-driven rivets	80	250	80
Close-tolerance and turned bolts	100	300	120
Bolts in clearance holes	80	250	120

12.2.5 Types of Connections

There are two types of riveted and bolted joints: *lap joints* and *butt joints*. In a lap joint, the plates to be connected are lapped over one another and fastened together by one or more rows of connectors on the overlap portion as shown in Figures 12.4(a) and (b). This form of joint has a disadvantage that the two lines of pull are not in alignment and hence the joint is subjected to bending action. As a result, the joint is distorted as shown in Figures 12.5(a) and (b).

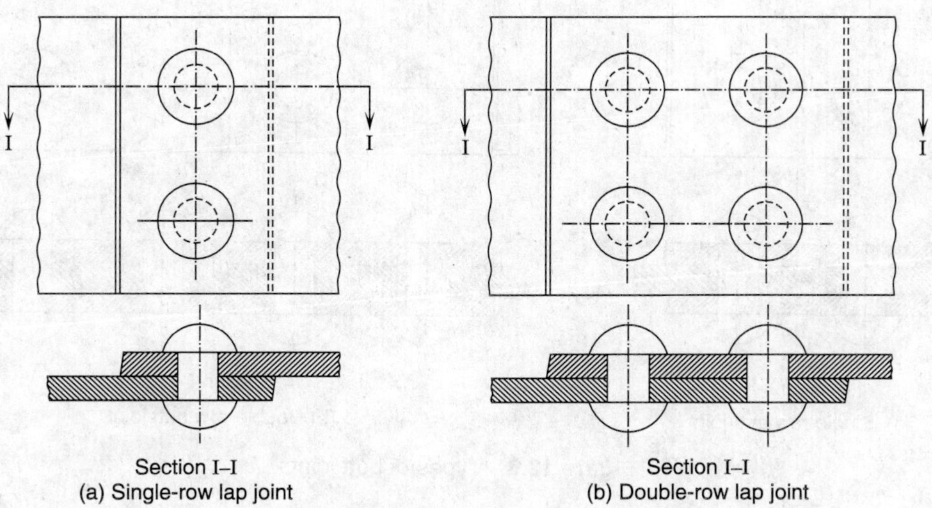

Figure 12.4 Types of lap joints.

Unlike in lap joint, in a *butt joint*, the connected ends of the plates lie in the same plane and are butted together. The abutting ends of the plates are covered by one or two *cover* or *strap plates* fastened to each of the main plates by means of a number of rows of connectors as shown in Figures 12.6(a) and (b). A butt joint with single cover plate is shown in Figure 12.7(a). This form of joint has the same disadvantage as the lap joint wherein the joint is subjected to bending action resulting in the distortion of the joint as shown in Figure 12.7(b). The foregoing defect of distortion is eliminated in butt joints with double cover plates. As shown in Figure 12.6, the pull being axial, no bending action acts on the joint.

The distance normal to the joint between edges of the overlapping plates in a lap joint or between the joint and the edge of cover plates in a butt joint is termed *lap*.

The number of rows of connectors that fasten the plates together (in a lap joint) or the cover plate to each of the main plates (in a butt joint) are used to identify the joint as *single-row*, *double-row*, etc., lap or butt joint. In case of lap joint, it is the total number of rows, and in case of a butt joint it is the number of rows on each side of the joint, which determine the type of joint as illustrated in Figures 12.4 and 12.6.

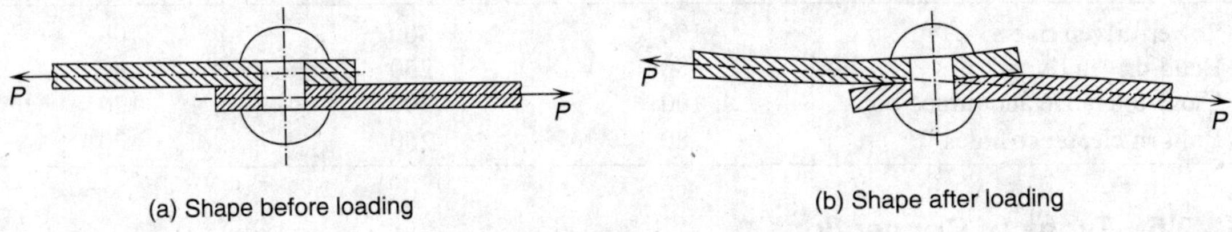

(a) Shape before loading

(b) Shape after loading

Figure 12.5 Distortion in a lap joint.

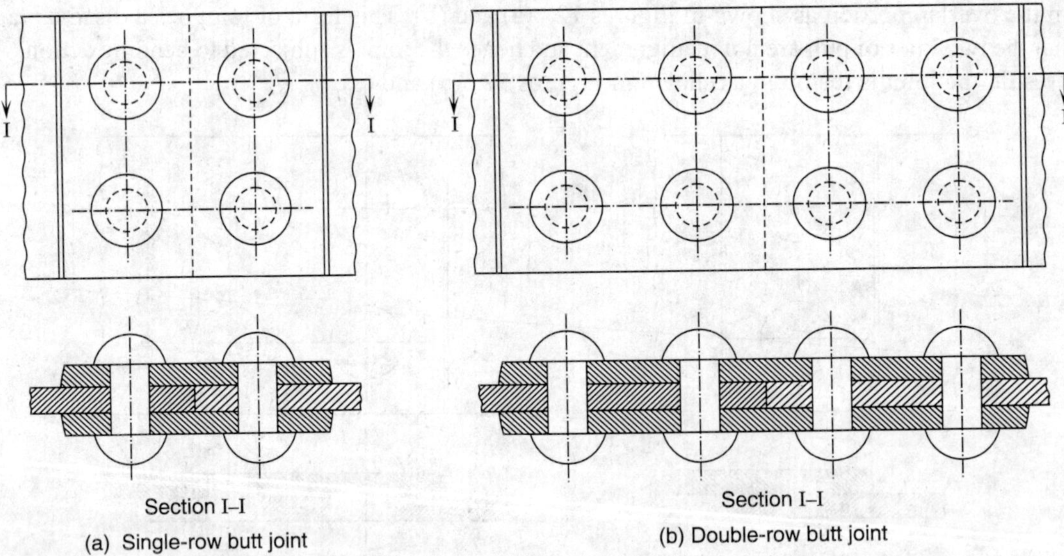

(a) Single-row butt joint

(b) Double-row butt joint

Figure 12.6 Types of butt joints.

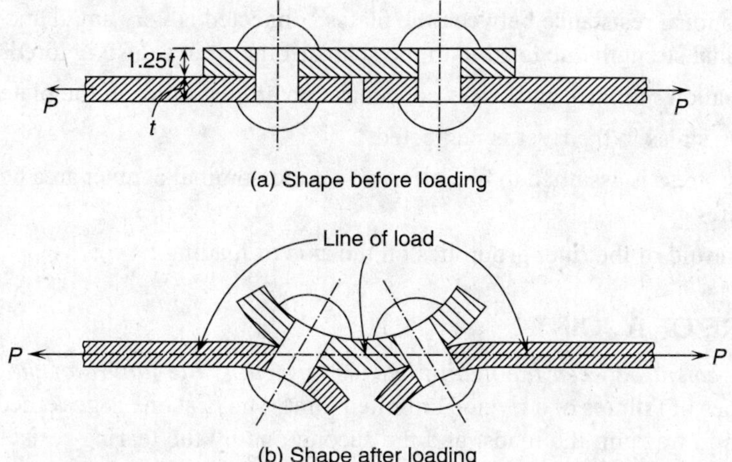

Figure 12.7 Distortion of a single cover plate butt joint.

The thickness of cover plates is generally taken as follows:

(i) $1\frac{1}{8}$ times the thickness of the main plate for single cover butt joint and

(ii) $\frac{5}{8}$ times the thickness of the main plate for double cover butt joint as shown in Figure 12.8.

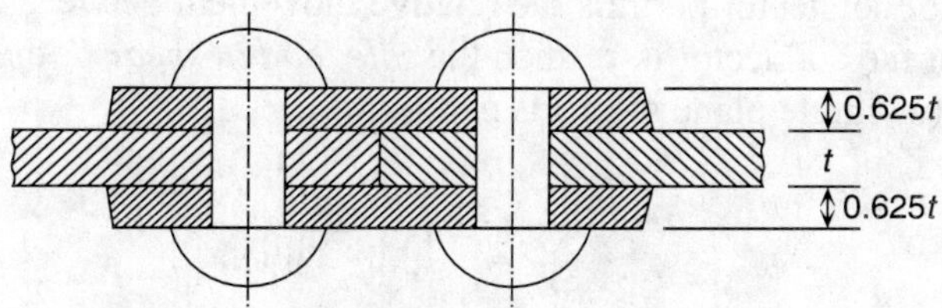

Figure 12.8 A double-cover plate butt joint.

12.3 ASSUMPTIONS MADE IN ANALYSIS

Due to the presence of many indeterminate factors, the exact analysis of riveted, bolted and welded connections is very difficult if not impossible. However, to obtain practical solutions, certain simplifying assumptions are made. The following assumptions are made in the analysis of simple riveted joints:

(i) All riveted connections are considered to be of *bearing type*, i.e. connected plates bear directly on the rivets for transmission of the loads.

(ii) When tightly fitted in position, the rivets plastically deform to fill the rivet holes completely and hence no slip between plates can occur. In calculations, therefore, the diameter of the driven rivet is considered equal to that of rivet hole.

(iii) The applied load is resisted equally by all rivets in the joint.

(iv) The frictional resistance between the plates connected is very small and can be ignored. Additional strength due tension in the shank of the rivet is also ignored.

(v) Deformation of plates and stress concentration around holes in the plates are neglected.

(vi) Bending stress in the rivet is neglected.

(vii) Bearing stress is assumed to be uniform over the nominal contact area between the rivets and plates.

(viii) The centroid of the rivet group lies on the axis of loading.

12.4 FAILURE OF A JOINT

Any relative movement between the main plates is termed as the failure of the joint. There are basically two types of failures of a riveted or bolted joint. The first one is governed by the capacity of rivets or bolts to transmit the loads, and the second one by the tearing resistance of the main plates.

(a) *Rivet capacity:* The strength of a rivet is determined by the lowest of the values of the load that a rivet can carry in single- or double-shear (as per the arrangement), load that will crush the rivet in the main plate, or in one or two cover plates (as the case may be).

(b) *Tearing capacity:* The load externally applied to the joint acts directly to tear the main plate.

To understand elemental types of failure, consider a single-row lap joint. As shown in Figure 12.9, shearing of connector permits the relative movement between the main plates. If the safe shearing stress for the connector is τ_a, then the *safe load in shear* or the *shear strength* of the joint per pitch length (in single plane shear) is given by

$$R_{ss} = \tau_a A_s = \tau_a \left(\frac{\pi d^2}{4} \right) \tag{12.1}$$

where d represents the diameter of both the connector and the connector hole. The friction between the plates is neglected.

In this case of lap joint, a connector is liable to fail by shear in only one plane. Hence, the connectors of a lap joint are said to be in *single-shear*. The strength of one connector in single shear is $\tau_a (\pi d^2/4)$. This is called *shear value of one connector*.

In case of a multiple-row lap joint, if n connectors are covered per pitch length, then *shearing strength* of joint per pitch length would be

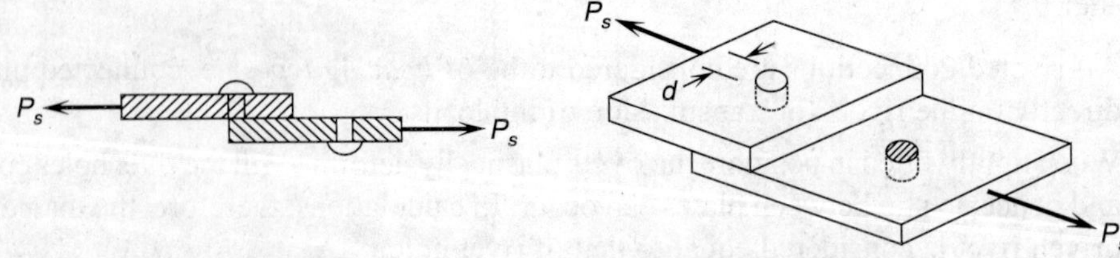

Figure 12.9 Failure by shear of rivet (single-shear).

$$R_{ss} = n\tau_a\left(\frac{\pi d^2}{4}\right) \tag{12.2}$$

In case of a butt joint with two cover plates, a connector failing in shear will shear along two planes simultaneously. Hence, connectors used in such a joint are said to be in *double-shear*. The safe load for a connector in double shear is

$$R_{ds} = 2 \times \tau_a\left(\frac{\pi d^2}{4}\right) \tag{12.3}$$

Some designers prefer to take factor $(1\tfrac{7}{8})$ instead of 2 in double shear, i.e.

$$R_{ds} = \frac{15}{8} \times \tau_a\left(\frac{\pi d^2}{4}\right) \tag{12.4}$$

In case of multiple-row butt joint where *n* rivets are covered per pitch length, the *shearing strength* of joint per pitch length is

$$R_{ds} = n\tau_a\left(2 \times \frac{\pi d^2}{4}\right) \tag{12.5}$$

In the second type of failure, called *bearing failure* shown in Figure 12.10, the relative movement between the main plates may result from a permanent deformation or enlargement of the connector hole caused by excessive *bearing pressure.* If these pressures are high, it is quite possible that the plate or connector itself may be crushed. The intensity of bearing pressure on the contact surface is non-uniform. However, in practice the *bearing stress* or *pressure* σ_p is assumed to be uniformly distributed over the projected area of the hole. Thus, the safe load in bearing or the *bearing strength of the joint* per pitch length is expressed as

$$R_p = \sigma_{ap}\, A_b = \sigma_{ap}\,(dt) \tag{12.6}$$

In case of multiple-row lap joint, if *n* rivets are covered per pitch length, then *bearing strength* per pitch length would be

$$R_p = n\sigma_{ap}\,(dt) \tag{12.7}$$

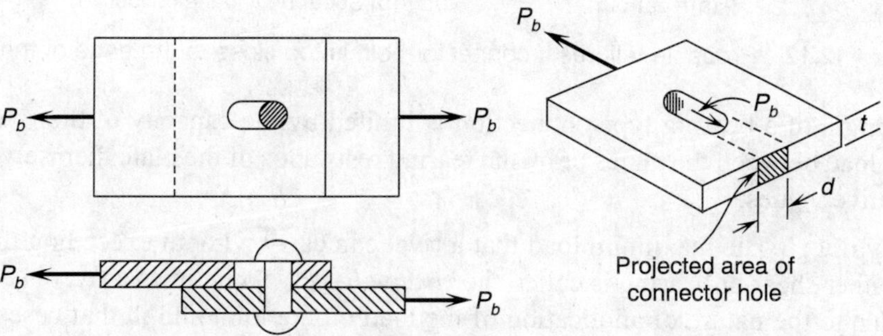

Figure 12.10 Bearing deformation of the upper plate.

Figure 12.11 represents a type of failure where the relative movement between the main plates is caused by the *tearing of the main plate*. This failure occurs on the section through the connector hole having minimum sectional area and hence minimum *tearing resistance*. The *resisting area* is the product of the net width of the plate $(p - d)$ and the thickness t. If the safe tearing stress for plate is σ_{at}, then the safe load in tearing or the *tearing strength of the joint* per pitch length is given by

$$R_t = \sigma_{at} A_t = \sigma_{at} (p - d)t \tag{12.8}$$

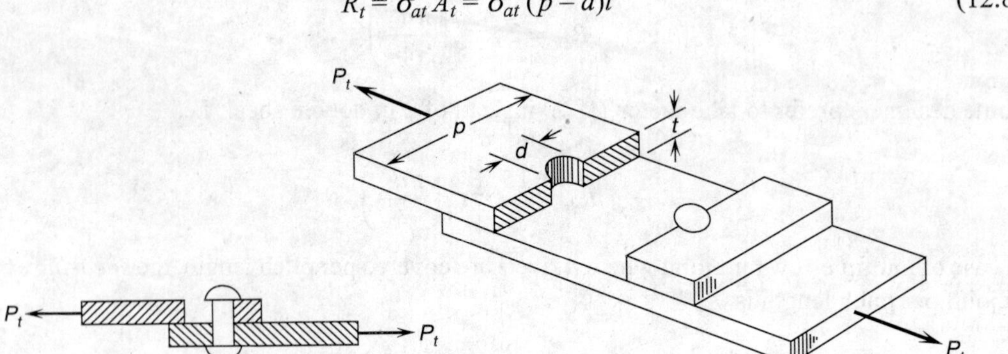

Figure 12.11 Failure caused by the tearing of main plate through connector.

Other types of elemental failures which permit the relative movement between the main plates include the bursting through the edge of the plate at the back of a connector hole, as shown in Figure 12.12(a). This type of failure of splitting of the plate at the edge may be caused by the internal pressure of an overdriven connector. The shear failure of plate behind a connector hole as shown in Figure 12.12(b) can occur alone or in a combination with the bursting through the edge. Such failures are prevented by providing an *edge distance*, i.e. distance from the edge of the plate to the centre of the connector of 1.75 to 2 times the diameter of the connector.

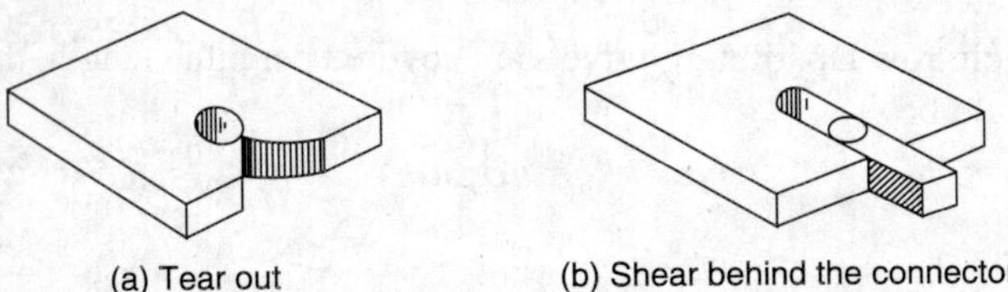

(a) Tear out (b) Shear behind the connector

Figure 12.12 Probable failures if connector hole is too close to the edge of the plate.

The strength of a bearing type connection is limited by the capacity of the rivets or bolts to transmit the load between the plates or by the tearing resistance of the plate themselves, depending on their relative values.

Connector value: The maximum load that a rivet can carry without exceeding the permissible stresses in either shear or bearing is called the *rivet value* R_v. Sometimes a rivet itself is subjected to tension due to the nature of application of the load on the joint and in that case its strength in tension must also be considered.

12.5 EFFICIENCY OF A JOINT

The efficiency of a riveted or bolted joint indicates how well a joint has been designed. It compares the strength of the joint with that of the solid plate and is expressed as

$$\text{Efficiency of the joint, } \eta = \frac{\text{Strength of joint}}{\text{Strength of solid plate}} \tag{12.9}$$

Consider one pitch length p of a joint with R_t, R_s and R_p being the safe loads per pitch length from tearing, shearing and bearing considerations. The strength of the solid plate of thickness t would be $P_t = \sigma_{at}\,(pt)$, then

$$\eta = \frac{\text{Least of } R_t,\ R_s \text{ and } R_p}{\sigma_{at}\,(pt)} \tag{12.10}$$

In practice, R_t is normally kept lower than R_s and R_p, thus

$$\eta = \frac{R_t}{P_t} = \frac{\sigma_{at}\,(p-d)\,t}{\sigma_{at}\,(pt)} = \frac{p-d}{p} = 1 - \frac{d}{p} \tag{12.11}$$

EXAMPLE 12.1

A double riveted butt-joint is used for connecting two 16 mm thick plates with two 12 mm cover plates. The 22 mm diameter rivets are arranged in the joint such that the pitch of rivets in outer rows is twice that in the inner rows. Determine the pitches of rivets in two rows, if the permissible stresses in tension, shear and bearing are 100, 75 and 150 MPa, respectively.

Solution: Consider the arrangement of the rivets in the joint as shown in Figures 12.13(a) and (b). If p is the pitch of rivets in the outer row, then there are three rivets per pitch.

The gross diameter of the rivet, $d = 22 + 1.5 = 23.5$ mm

The strength of rivets in double shear per pitch length

$$R_{ds} = 3 \times \left[2 \times 75 \times \left(\frac{\pi \times 23.5^2}{4} \right) \right] = 195181.26 \text{ N} = 195.181 \text{ kN}$$

Bearing strength of the rivets per pitch length

$$R_p = 3 \times [150 \times (23.5 \times 16)] = 169200 \text{ N} = 169.2 \text{ kN}$$

The rivet value, R_v = lesser of R_{ds} and R_p = 169.2 kN

Tearing strength of the plate with rivet holes per pitch length

$$R_t = 100 \times (p - 23.5) \times 16 = 1600 \times (p - 23.5) \text{ kN}$$

$$R_t = R_v \quad \text{i.e.} \quad 1600 \times (p - 23.5) = 169.2 \times 10^3$$

Therefore, the required pitch is

$$p = \frac{169200}{1600} + 23.5 = 129.25 \text{ mm}$$

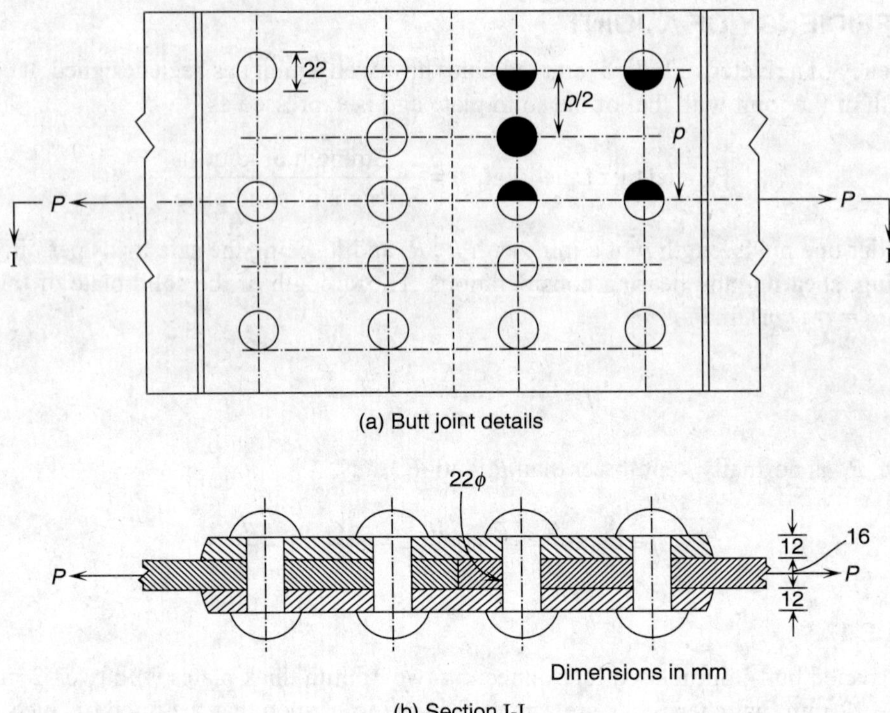

Figure 12.13 Double cover double riveted butt joint of Example 12.1.

However,

(i) p_{max} = 16t or 200 mm whichever is less

= 16 × 16 or 200 mm whichever is less = 200 mm

(ii) p_{min} generally lies between 2d (= 47 mm) and 3d (= 70.5 mm).

Keeping in view the inner row, a pitch of 120 mm may be adopted. The arrangement of the rivets in the joint is shown in Figure 12.13.

EXAMPLE 12.2

A cylindrical pressure vessel 5 m in diameter is fabricated of 16 mm steel plates. The longitudinal joint of the vessel is a double-row lap joint with no staggering between the rows as shown in Figures 12.14(a) and (b). The rivets used are of 20 mm nominal diameter with a pitch of 75 mm. Determine the efficiency of joint and the safe pressure in the vessel. The permissible stresses are 140 MPa in tension, 110 MPa in shear and 240 MPa in bearing.

Solution: The gross or effective diameter of the rivet,

$$d = 20 + 1.5 = 21.5 \text{ mm}$$

The strength of rivets in single shear per pitch length

$$R_{ss} = 2 \times 110 \times \left(\frac{\pi \times 21.5^2}{4} \right) = 79871.07 \text{ N} = 79.871 \text{ kN}$$

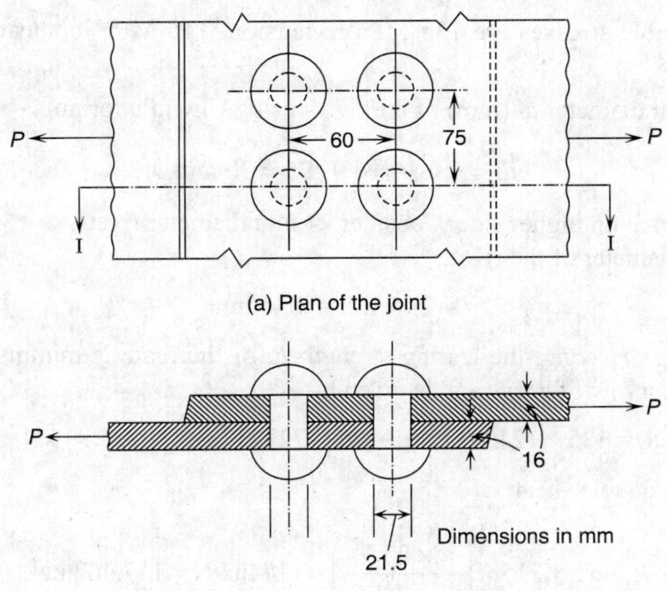

Figure 12.14 Double-row lap joint of cylindrical vessel of Example 12.2.

Bearing strength of the rivets per pitch length

$$R_p = 2 \times [240 \times (21.5 \times 16)] = 165120 \text{ N} = 165.12 \text{ kN}$$

Tearing strength of the plate with rivet holes per pitch length

$$R_t = 140 \times (75 - 21.5) \times 16 = 119840 \text{ N} = 119.84 \text{ kN}$$

Tearing strength of the solid plate per pitch length

$$P_t = 140 \times 75 \times 16 = 168000 \text{ N} = 168 \text{ kN}$$

The R_s is the minimum, therefore,

$$\eta = \frac{R_s}{P_t} \times 100 = \frac{79.871}{168} \times 100 = 47.54 \text{ per cent}$$

For a cylindrical vessel subjected to internal pressure;

$$\sigma_{at} = \frac{p_i D}{2t\eta} \quad \text{or} \quad p_i = \frac{2t\eta\sigma_{at}}{D}$$

Therefore, $$p_i = \frac{2 \times 16 \times (47.54/100) \times 140}{5 \times 10^3} = 0.426 \text{ MPa} = 426 \text{ kPa}$$

EXAMPLE 12.3

A double cover butt-joint, with rivets arranged in either diamond or chain riveting form, is to be provided for connecting two tie bars, each 240 mm wide and 25 mm thick. Design the joint and determine the working strength, efficiency of joint and the actual stresses in the plate and rivets. The permissible stresses are 135 MPa in tension, 115 MPa in shear and 200 MPa in bearing.

Solution: Nominal diameter d' (mm) of the rivets using Unwin's formula

$$d' = 6.0\sqrt{t} = 6.0\sqrt{25} = 30 \text{ mm}$$

This value is somewhat on higher side. Consider 24 mm diameter rivets.
Gross or effective diameter of the rivet,

$$d = 24 + 1.5 = 25.5 \text{ mm}$$

(a) *For diamond type riveting,* the tearing strength R_t of the plate is minimum at section II-II through the outermost row of rivets and is given by

$$R_t = 135 \times (240 - 25.5) \times 25 = 723937.5 \text{ N} = 723.94 \text{ kN}$$

Strength of rivets in double shear

$$R_{ds} = 2 \times 115 \times \left(\frac{\pi \times 25.5^2}{4}\right) = 117462 \text{ N} = 117.462 \text{ kN}$$

For computation of bearing strength of the rivets, t is taken as the thickness of the main plate or the total thickness of the cover plates whichever is less. In a double cover butt-joint, the thicknesses of the cover plates are usually taken as $0.625t$ mm where t is the thickness of the main plate, i.e. $0.625 \times 25 = 15.625$ mm. Consider 16 mm thick cover plates, with their total thickness of 32 mm which is more than the thickness of the main plate. Thus,

$$R_p = \sigma_{ap}\,(dt) = 200 \times (25.5 \times 25) = 127500 \text{ N} = 127.5 \text{ kN}$$

The rivet value, R_v = lesser of R_{ds} and R_p = 117.462 kN

The number of rivets required for the joint

$$N = \frac{R_t}{R_v} = \frac{723.94}{117.462} = 6.16$$

Adopting 6 rivets arranged in a zig-zag fashion as shown in Figure 12.15(a).

Design of cover plates

The critical section for cover plates is IV-IV where they are weakened by three rivets. If, t is the thickness of each cover plate in mm, then for two cover plates

$$2[\sigma_{at}\,(b - 3d)t] = R_t \quad \text{or} \quad 2 \times [135 \times (240 - 3 \times 25.5) \times t] = 723.94 \times 10^3$$

Therefore,
$$t = \frac{723.94 \times 10^3}{270 \times (240 - 3 \times 25.5)} = 16.4 \text{ mm}$$

Provide 18 mm thick cover plates.

Strength against failure of main plate by tearing at various sections:

Strength at II-II, $P_{II} = \sigma_{at}\,(b - d)t = 135 \times (240 - 25.5) \times 25 = 723.94$ kN

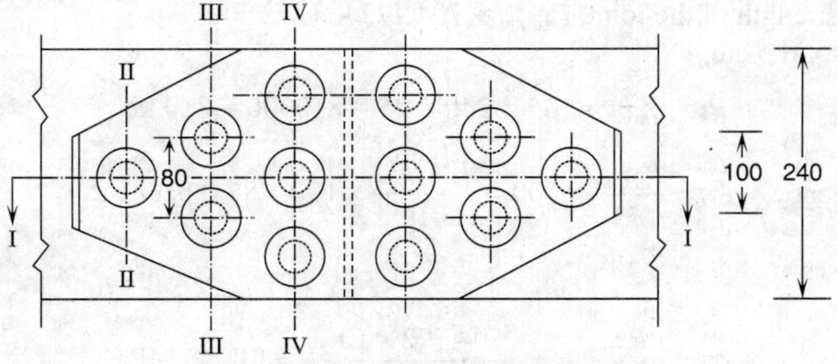

(i) Plan of zig-zag arrangement

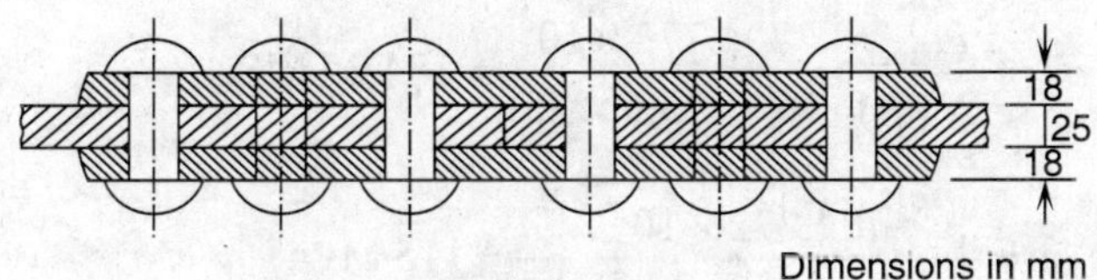

(ii) Section I–I

(a) Diamond riveting arrangement

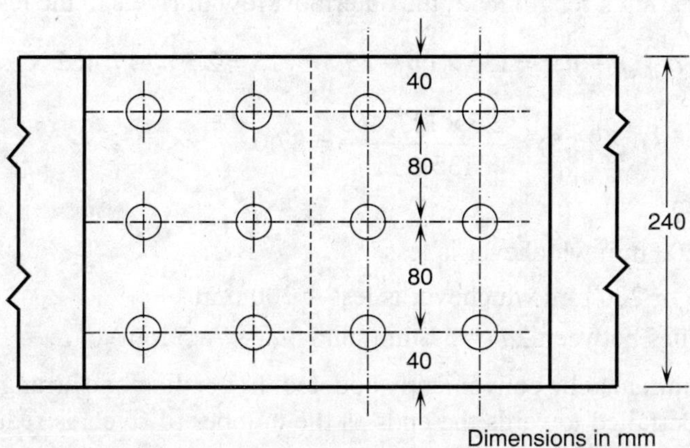

(b) Chain riveting arrangement

Figure 12.15 Zig-zag and chain reviting arrangement of the joint of Example 12.3.

Strength at III-III, $P_{III} = \sigma_{at}\,(b - 2d)t + R_v$

$= [135 \times (240 - 2 \times 25.5) \times 25] \times 10^{-3} + 117.462$

$= 755.34$ kN

Strength at IV-IV, $P_{IV} = \sigma_{at}\,(b - 3d)t + 3R_v$

$= [135 \times (240 - 3 \times 25.5) \times 25] \times 10^{-3} + 3 \times 117.462$

$= 904.2$ kN

Strength of the joint in shearing of all the rivets or crushing of all rivets or plate,

$$P_{\text{rivets}} = NR_v = 6 \times 117.462 = 704.772 \text{ kN}$$

Thus the actual strength of the joint, $P_{\text{actual}} = 704.772$ kN
Strength of the solid plate,

$$P_t = \sigma_{at} bt = 135 \times 240 \times 25 = 810000 = 810 \text{ kN}$$

Therefore, efficiency $\quad \eta = \dfrac{P_{\text{actual}}}{P_t} \times 100 = \dfrac{704.772}{810} \times 100 = 87.0$ per cent.

The actual stresses are

$$\sigma_t = \frac{P_{\text{actual}}}{(b-d)t} = \frac{704.772 \times 10^3}{(240 - 25.5) \times 25} = 131.43 \text{ MPa}$$

$$\sigma_p = \frac{P_{\text{actual}}}{N(dt)} = \frac{704.772 \times 10^3}{6 \times (25.5 \times 25)} = 184.25 \text{ MPa}$$

$$\tau = \frac{704.772 \times 10^3}{6 \times [2 \times (\pi \times 25.5^2 / 4)]} = 115 \text{ MPa}$$

Pitch: Consider one pitch length p of the joint. There are 2.5 rivets per pitch length. For economy equate the (minimum) tearing strength R_t' at the outermost row of rivets to the lesser of R_s and R_p.

$$R_t' = \sigma_{at}(p-d)t = 135 \times (p - 25.5) \times 25 = 2.5 \times 117{,}462 \text{ N}$$

or $$(p - 25.5) = \frac{2.5 \times 117{,}462}{135 \times 25} = 87.0$$

Therefore, $p = 112.5$ mm.

(i) $p_{\max} = 16t$ or 200 mm whichever is less
$= 16 \times 25$ or 200 mm whichever is less $= 200$ mm

(ii) $p_{\min}$ generally lies between $2d$ (= 51 mm) and $3d$ (= 76.5 mm).

Thus, a pitch of 80 mm may be conveniently adopted. In the diamond arrangement, the width of cover plates can be curtailed towards the ends as the number of rivets is reduced. A width of 100 mm has been kept at the end.

Edge distance $\quad p_{\text{edge}} = 1.5d = 1.5 \times 25.5 = 38.25 \text{ mm} \approx 40$ mm

Row distance $\quad p_{\text{row}} = 2d = 2 \times 25.5 \approx 50$ mm

(b) In an alternative arrangement chain riveting can be adopted instead of zig-zag riveting as shown in Figure 12.15(b). In this arrangement, 6 rivets have been arranged in two rows of three rivets each. Hence, tearing strength (minimum) at the outermost row

$$R_t = \sigma_t (b - 3d)t = 135 \times (240 - 3 \times 25.5) \times 25$$
$$= 551.813 \times 10^{-3} \text{ N} = 551.813 \text{ kN}$$

This is less than the least strength due to six rivets (= $6R_v = 6 \times 117.462 = 704.77$ kN). The efficiency of the joint is

$$\eta = \frac{551.813}{810} \times 100 = 68.125 \text{ per cent}$$

Therefore, for the same joint, with the same number of rivets, diamond arrangement is more efficient than chain arrangement and hence more economical.

12.6 ECCENTRICALLY LOADED CONNECTIONS

Generally in practice, the applied load does not pass through the centroid of the connector group of the joint, resulting in an *eccentric loading*. In such a case, load is not equally distributed over all the rivets or bolts. The effect of eccentric load is statically equivalent to a *concentric load*, and a twisting moment or a bending moment, which is equal to the product of the load and its eccentricity. If the load axis acts in the plane of connector group the resulting moment will be the *twisting moment* as shown in Figure 12.16(a). On the other hand, if the axis of the load does not lie in the plane of the connectors as shown in Figure 12.16(b), the applied load causes *bending moment* which is not in the plane of connectors and the connectors are either in compression or in tension.

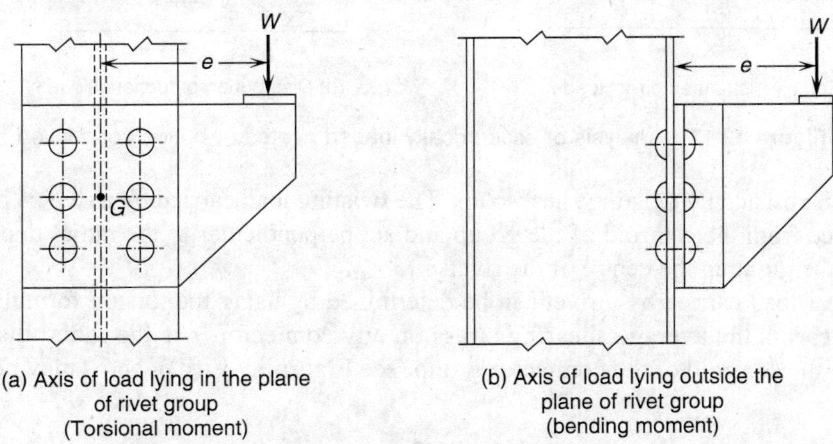

(a) Axis of load lying in the plane of rivet group (Torsional moment)

(b) Axis of load lying outside the plane of rivet group (bending moment)

Figure 12.16 Eccentrically loaded (moment resisting) connections.

12.6.1 Eccentric Load Causing Twisting Moment

In this case, the axis of the load lies in the plane of connectors, and only shear stress develops in the connectors, and problem consists of identifying the connector which carries the maximum shear. So if the joint is to be safe, the maximum value of shear stress should be within the permissible limits.

For illustration consider the connection shown in Figure 12.17(a) subjected to an eccentric force acting in the plane of connector group at distance e from its centroid C. By adding a pair of equal but oppositely directed collinear forces of magnitude W at the centroid C, the applied eccentric load is replaced by central load W and torsional couple or twisting moment $M_t = We$, as shown in Figure 12.17(b). The central load W is resisted as direct shear load $R_s = W/N$ equally distributed over all the rivets as shown in the free-body diagram of the plate in Figure 12.17(c). The twisting

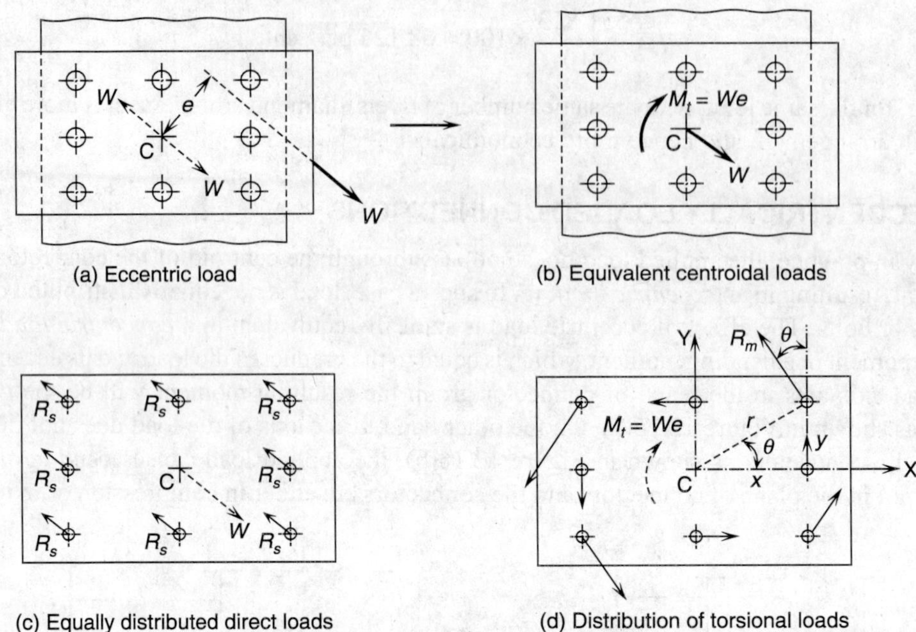

Figure 12.17 Analysis of eccentrically loaded riveted or bolted connection.

moment M_t is resisted by twisting/shear loads. The twisting load carried by a rivet is proportional to its distance from the centroid of the group and act perpendicular to the radial line joining the centroid of the group to the centre of the rivet.

The shear load carried by a rivet can be determined by using the torsion formula $\tau = M_t r/J$. Here τ represents the average shearing stress on any connector, r is the radial distance from its centre to the centroid of the connector group (see Figure 12.17(d)), and J may be expressed as:

$$J = \sum Ar^2 \tag{12.12}$$

If all the connectors have same area A with distance r of any connector expressed in terms of its co-ordinates as $r^2 = x^2 + y^2$, Eq. (12.12) can be rewritten

$$J = A\left(\sum x^2 + \sum y^2\right) \tag{12.13}$$

Therefore, the torsion formula reduces to

$$\tau A = \frac{M_t r}{\sum x^2 + \sum y^2} \quad \text{or} \quad R_m = \frac{M_t r}{\sum x^2 + \sum y^2} \tag{12.14}$$

The resultant load on a typical connector can be conveniently obtained by resolving R_s and R_m into x and y components. The components $R_{s,x}$ and $R_{s,y}$ of R_s are constant for all connectors, whereas $R_{m,x}$ and $R_{m,y}$ of R_m depend upon the components x and y of the rivet measured from the centroid of the group, hence

$$R_{m,x} = R_m \sin\theta = R_m\left(\frac{y}{r}\right) = \frac{M_t y}{\sum x^2 + \sum y^2} \tag{12.15}$$

$$R_{m,y} = R_m \cos\theta = R_m\left(\frac{x}{r}\right) = \frac{M_t x}{\sum x^2 + \sum y^2} \tag{12.16}$$

The maximum load occurs on a connector where all the components are additive, as at connector at the upper right hand corner

$$R_r = \sqrt{(R_{s,x} + R_{m,x})^2 + (R_{s,y} + R_{m,y})^2} \tag{12.17}$$

The angle θ between R_r and $R_{s,x}$ is given by

$$\tan\theta = \frac{R_{s,y} + R_{m,y}}{R_{s,x} + R_{m,x}} \tag{12.18}$$

12.6.2 Most Heavily Loaded Connector

R_s is same for all the connectors and R_m being proportional to r, will be maximum for the rivet which is at the maximum distance from centroid of connector group. The angle between two forces acting on the most heavily stressed connector will be the minimum. Sometimes reader may encounter a situation where a connector having maximum r may not have least angle. In such cases, a connector having the next lower value of r but the least value of θ may be considered.

Alternatively, as described earlier, the resultant load on a typical connector can be obtained conveniently by resolving R_s and R_m into x and y components and the most heavily loaded connector is the one where all the components are additive.

EXAMPLE 12.4

The gusset plate connection shown in Figure 12.18(a) consists of a row of five rivets with 20.5 mm diameter holes at a pitch of 75 mm. The allowable load W was based on permissible shear stress of 70 MPa. Determine the maximum shear stress in the rivets if the rivet at V is improperly driven so that it cannot carry any load.

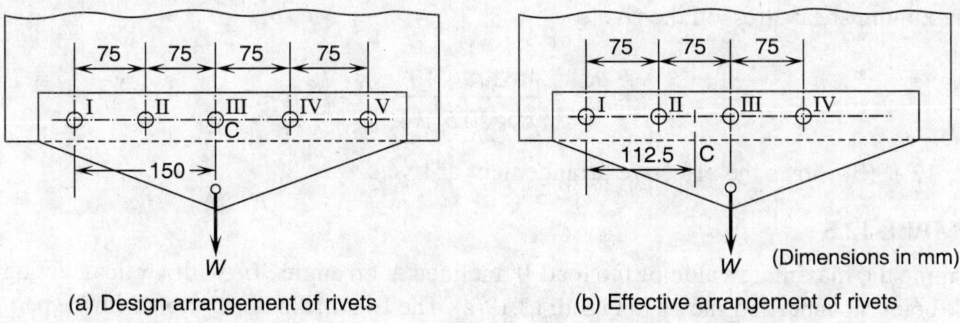

Figure 12.18 Gusset plate connection of Example 12.4.

Solution: In the design stage, the load was assumed to pass through the centroid of the connectors group and hence it is equally distributed over all the connectors. The design load is

$$W = 5\times\left(70\times\frac{\pi\times 20.5^2}{4}\right) = 115522.25\text{ N} = 115.52\text{ kN}$$

In the service condition due to malfunctioning of the rivet-V, the applied load becomes eccentric over the group of four active rivets and the joint will also be subjected to a twisting moment.

The distance of centroid of the group of four active rivets from the rivet-I is

$$\bar{x} = \frac{75\times(A+2A+3A)}{4\times A} = 112.5\text{ mm}$$

Therefore eccentricity, $e = 150 - 112.5 = 37.5$ mm

Therefore, twisting moment is

$$M_t = We = 115.52\times(37.5) = 4332\text{ kN.mm (clockwise)}$$

The torsion factor

$$\sum x^2 = 2\times(112.5^2 + 37.5^2) = 28125\text{ mm}^2$$

Direct shear load,

$$R_s = \frac{W}{N} = \frac{115.52}{4} = 28.88\text{ kN}$$

The maximum torsional loads are induced in the remotest rivets, i.e. rivet-I and rivet-IV. Therefore,

$$R_m = \frac{M_t x}{\sum x^2} = \frac{4332\times 112.5}{28125} = 17.328\text{ kN}$$

The rivet-IV is the most heavily loaded as both load components are additive. Thus, the maximum resultant load R_r on rivet-IV is

$$R_r = 28.88 + 17.328 = 46.208\text{ kN}$$

The maximum shear stress in the rivets is

$$\tau_{\max} = \frac{R_r}{A} = \frac{46.208\times 10^3}{\pi\times 20.5^2/4} = 140\text{ MPa}$$

Figure 12.18(b) shows the effective arrangement of rivets.

EXAMPLE 12.5

Determine the maximum value of the load W inclined at an angle 30° with vertical, acting on the double bracket connection shown in Figure 12.19(a). The 16 mm diameter rivets with the permissible shear stress of 100 MPa are used in the fabrication.

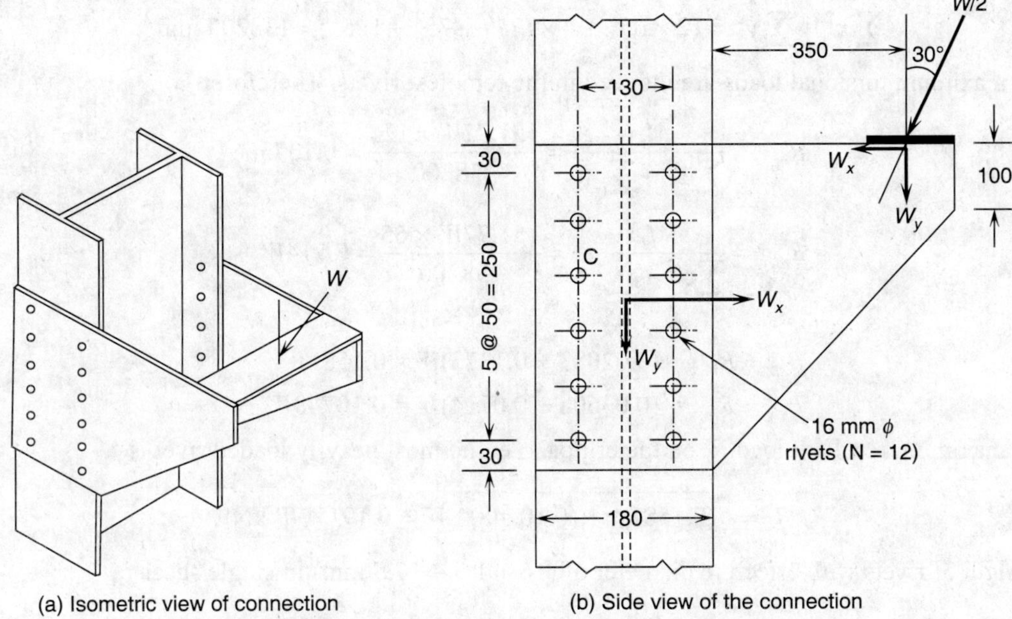

(a) Isometric view of connection

(b) Side view of the connection

Figure 12.19 Double bracket connection of the Example 12.5.

Solution: Due to symmetry, the applied load W kN is equally shared by two bracket plates. Thus load carried by each plate is $W/2$ (see Figure 12.19(b)). For each plate, the effect of applied load is equivalent to an equal central load acting through the centroid of the rivet group plus a twisting moment which is equal to the product of the load and its eccentricity. Noting that the moment of a force is equal to the sum of moments of its components. The components of the loads and the corresponding eccentricities are

$$W_x = (W/2) \sin 30° = 0.25W; \quad e_y = 125 + 30 = 155 \text{ mm}$$

$$W_y = (W/2) \cos 30° = 0.433W; \quad e_x = 90 + 350 = 440 \text{ mm}$$

Therefore, twisting moment is

$$M_t = W_x e_y + W_y e_x = (0.25W) \times (155) - (0.433W) \times (440)$$
$$= 151.77W \text{ kN.mm (clockwise)}$$

Number of rives connecting each plate, $N = 12$, therefore,

$$R_{s,x} = \frac{W_x}{N} = \frac{0.25W}{12} = 0.02083W$$

$$R_{s,y} = \frac{W_y}{N} = \frac{0.433W}{12} = 0.03608W$$

There are six rivets each having X coordinate of 65 mm and –65 mm, respectively. The Y coordinates of the rivets are two each of (125, 75, 25, –25, –75 and –125 mm). Thus the rotational factor

$$\sum x^2 + \sum y^2 = 12 \times 65^2 + 4 \times (25^2 + 75^2 + 125^2) = 138200 \text{ mm}^2$$

The maximum torsional loads are induced in the remotest rivets. Therefore,

$$R_{m,x} = \frac{M_t y}{\sum x^2 + \sum y^2} = \frac{151.77W \times 125}{138200} = 0.1373W$$

$$R_{m,y} = \frac{M_t x}{\sum x^2 + \sum y^2} = \frac{151.77W \times 65}{138200} = 0.714W$$

Thus,

$$R_{s,x} + R_{m,x} = (0.02083 + 0.1373)W = 0.1581W$$

$$R_{s,y} + R_{m,y} = (0.03608 + 0.0714)W = 0.10748W$$

The maximum resultant load R_r on the critical, i.e. the most heavily loaded rivet is

$$R_r = \sqrt{(0.1581)^2 + (0.10748)^2}\ W = 0.19117W \text{ kN}$$

Strength of rivets with effective diameter of (16 + 1.5 = 17.5 mm) in single shear

$$R_s = 100 \times \left(\frac{\pi \times 17.5^2}{4}\right) = 24052.82 \text{ N} = 24.05 \text{ kN}$$

Therefore, $0.19117W = 24.05$ or $W = 125.80$ kN

EXAMPLE 12.6

A riveted bracket connection supports a concentrated load of 50 kN at an eccentricity of 200 mm as shown in Figure 12.20. Determine the diameter of the rivets and the thickness of the bracket plate when the permissible stresses are 100 MPa in shear and 200 MPa in bearing.

Solution: The rotational factor with respect to the centroid of the group of rivets

$$\sum x^2 + \sum y^2 = 4 \times (80^2 + 100^2) = 65600 \text{ mm}^2$$

The twisting moment,

$$M_t = We_x = 50 \times (200) = 10000 \text{ kN.mm (clockwise)}$$

Direct shear load in a rivet

$$R_{s,y} = \frac{W}{N} = \frac{50}{4} = 12.5 \text{ kN}$$

The maximum torsional loads are induced in the remotest rivet A. Therefore,

$$R_{m,x} = \frac{M_t y}{\sum x^2 + \sum y^2} = \frac{10000 \times 80}{65600} = 12.195 \text{ kN}$$

$$R_{m,y} = \frac{M_t x}{\sum x^2 + \sum y^2} = \frac{10000 \times 100}{65600} = 15.244 \text{ kN}$$

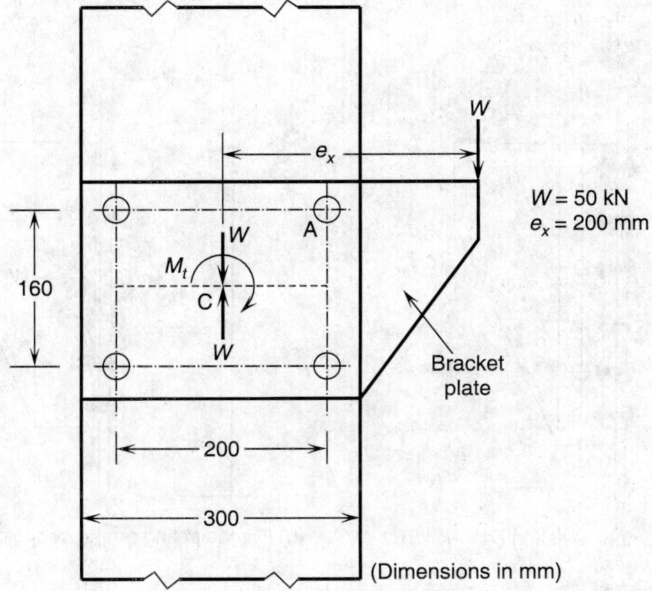

Figure 12.20 Eccentric riveted bracket connection of Example 12.6.

Thus, $R_{s,x} + R_{m,x} = 0 + 12.195 = 12.195$ kN

$$R_{s,y} + R_{m,y} = 12.5 + 15.244 = 27.744 \text{ kN}$$

The maximum resultant load R_r on the critical rivet, i.e. the most heavily loaded rivet is

$$R_r = \sqrt{(12.195)^2 + (27.744)^2} = 30.306 \text{ kN}$$

Consider the rivet value $R_r = R_s$, the single shear strength of the rivet. Therefore,

$$R_s = R_r \quad \text{i.e.} \quad 100 \times \left(\frac{\pi \times d^2}{4}\right) = 30.306 \times 10^3 \text{ N}$$

Hence, effective diameter of rivet, $d = 19.64$ mm

Nominal diameter = 19.64 – 1.5 = 18.14 mm. Use 19 mm nominal diameter rivets.

To determine the thickness of the bracket plate, equate the bearing strength of rivet to its shearing strength, i.e. $R_p = R_s$

$$200 \times (20.5 \times t) = 100 \times \left(\frac{\pi \times 20.5^2}{4}\right)$$

Therefore, $t = 8.05$ mm.

Provide bracket plate of thickness $t \geq 8.05$ mm.

12.6.3 Eccentric Load Causing Bending Moment

In this type of connections shown in Figure 12.21 the load does not lie in the plane of connectors and the external bending moment due to eccentric load tends to rotate the connection about an axis

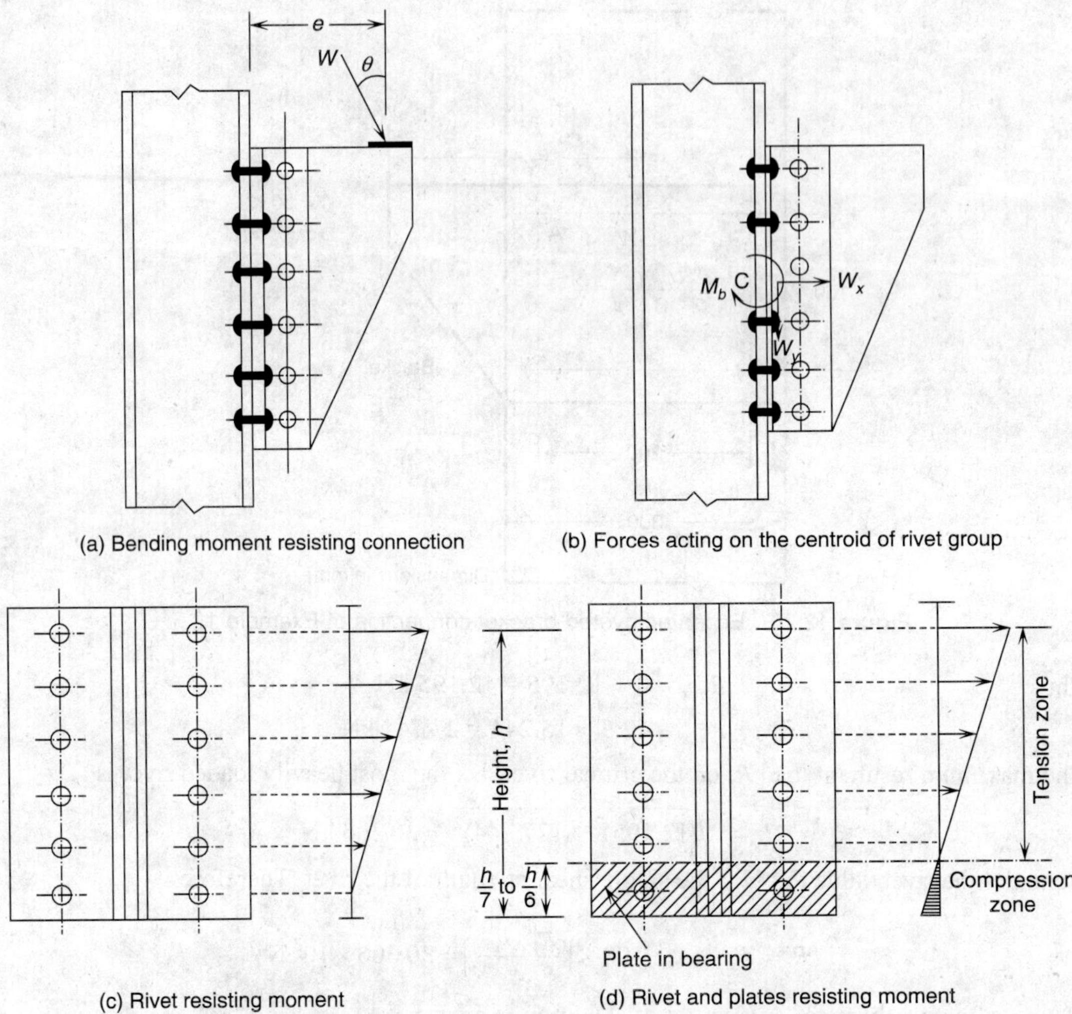

(a) Bending moment resisting connection

(b) Forces acting on the centroid of rivet group

(c) Rivet resisting moment

(d) Rivet and plates resisting moment

Figure 12.21 Analysis of bending moment resisting connection.

lying in the plane of connectors. The connectors below the axis of rotation will be compressed and those above it pulled. On tension side, there is separation between the connected elements and on the compression side elements press each other. The variation of strains and stresses is linear, i.e. varies from zero at the axis of rotation or neutral axis to a maximum value in the extreme connector.

Different codes stipulate different locations of axis of rotation or neutral axis. These stipulations are briefly described below:

1. Neutral axis is taken to pass through the centroid of the connector group. This assumption does not agree with the actual behaviour of the joint.
2. Neutral axis is taken to pass through the bottom of the bracket, thus all the rivets are subjected to tension. However, underlying assumptions are not justified in practice. British Constructional Steel Works Association stipulates the centre line of the bottommost rivet as the line of rotation.

3. The line of rotation is assumed to be at a distance 1/7th to 1/6th of the effective bracket depth from the bottom of the bracket. The effective depth is the distance from the centre line of the topmost connector to the bottom of the bracket. This stipulation assumes that the neutral axis passes through the centroid of the effective cross-section. The effective cross-section consists of circular areas of connector alone above the neutral axis and the net bearing or contact area below the neutral axis. This approach agrees with conditions of axial and eccentric connections. American code uses this assumption in its stipulations.

The effect of applied eccentric load is considered equivalent to central loads W_x and W_y acting through the centroid of the connector group in *X*- and *Y*-directions, respectively; plus a bending moment which is equal to the product of the load and its eccentricity. The bending moment acts about the major neutral axis of the connector group. Noting that the moment of a force is equal to the sum of moments of its components, i.e.

$$M_b = W_x e_y + W_y e_x \tag{12.19}$$

Here e_x and e_y are eccentricities of W_y and W_x with respect to the centroid of the connector group.

12.6.4 Maximum Tension in a Connector

The stipulation (1) above is not adopted in practice. The maximum tension developed in the connector using stipulations (2) and (3) are discussed below.

British practice: In this method, the centre line of the bottommost connector(s) is taken to be the line of rotation or neutral axis. Consider the connection shown in Figures 12.22(a) where the tensions in connectors marked 1, 2, 3 and 4 from the top are represented by T_1, T_2, T_3 and T_4. From Figure 12.22 (a), by similar triangles,

$$\frac{T_1}{r_1} = \frac{T_2}{r_2} = \frac{T_3}{r_3} = \frac{T_4}{r_4} \tag{12.20}$$

If M_1, M_2, M_3 and M_4 are moments shared by the connectors 1, 2, 3 and 4, respectively, the total bending moment M_b resisted by the connector group is given by

$$M_b = M_1 + M_2 + M_3 + M_4$$

$$= T_1 r_1 + T_2 r_2 + T_3 r_3 + T_4 r_4 = T_1 r_1 + \frac{T_1 r_2^2}{r_1} + \frac{T_1 r_3^2}{r_1} + \frac{T_1 r_4^2}{r_1}$$

$$= \frac{T_1}{r_1}\left(r_1^2 + r_2^2 + r_3^2 + r_4^2\right) = \frac{T_1}{r_1}\sum r^2$$

Therefore,
$$T_1 = \frac{M_b r_1}{\sum r^2} \tag{12.21}$$

In general the tension in *i*th connector is given by

$$T_i = \frac{M_b r_i}{\sum r^2} \tag{12.22}$$

The maximum tension always develops in the topmost connector. In this type of connection, the bending moment is entirely resisted by connectors if they are subjected to initial tension.

American practice: In this method, the connectors are considered to have no initial tension the, and moment is resisted by connectors in tension zone and the plate in bearing in compression zone of the connection. In this bearing type connection, the neutral axis or the line of rotation is assumed to be at a distance of 1/7th to 1/6th of the *effective depth h* of bracket from the bottom of the bracket as shown in Figure 12.22(b). The *effective depth h* is the distance from the centre line of the topmost connector to the bottom of the bracket. This stipulation assumes that the neutral axis passes through the centroid of the effective cross-section. The effective cross-section consists of circular areas of connector alone above the neutral axis and the net bearing or contact area below the neutral axis.

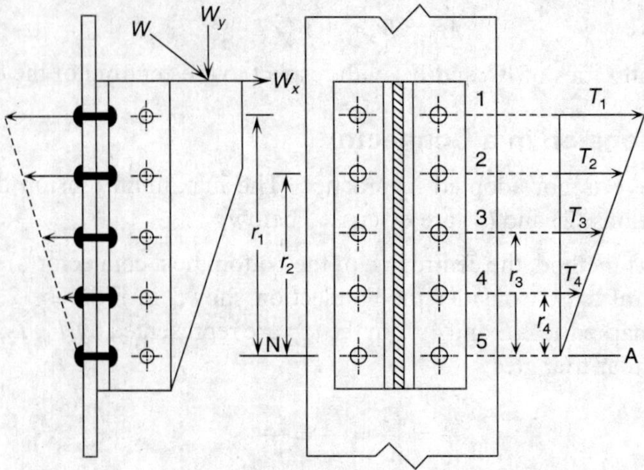

(a) Neutral axis passing through centre of bottommost connector

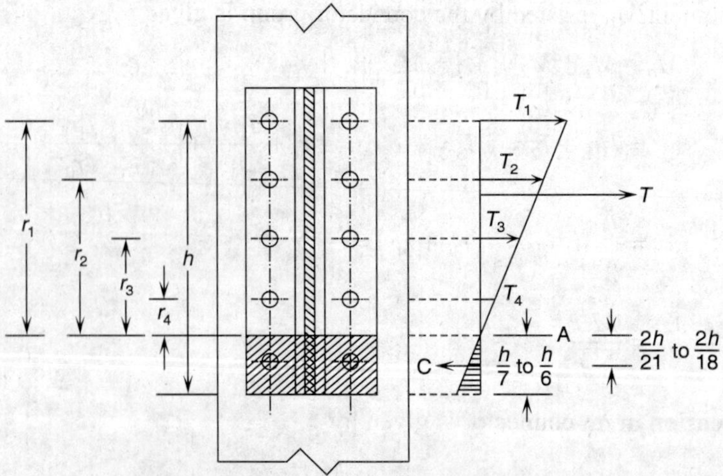

(b) Neutral axis at $h/7$ to $h/6$ from the bottom

Figure 12.22 Stress diagrams for the connection subjected to bending moment.

Consider the total tension and total compression developed in the connectors and plate to be T and C, respectively, and M_b' be the moment shared by the connectors in tension. The tension developed in the ith connector is

$$T_i = \frac{M_b' r_i}{\sum r^2}$$

Therefore,
$$T = T_1 + T_2 + T_3 + \cdots = \sum T_i = \frac{M_b' \sum r_i}{\sum r^2} \tag{12.23}$$

For internal equilibrium,
$$T = C = \frac{M_b' \sum r_i}{\sum r^2}$$

The moment of compressive force C about the line of rotation is

$$M_b'' = C\left(\frac{2h}{21 \text{ or } 18}\right) = \left(\frac{M_b' \sum r_i}{\sum r^2}\right)\frac{2h}{21} \tag{12.24}$$

Thus
$$M_b = M_b' + M_b'' = M_b' + \frac{M_b' \sum r_i}{\sum r^2}\frac{2h}{21} = M_b'\left(1 + \frac{\sum r_i}{\sum r^2}\frac{2h}{21}\right)$$

$$M_b' = M_b\left(1 + \frac{\sum r_i}{\sum r^2}\frac{2h}{21}\right)^{-1} \tag{12.25}$$

Hence the maximum tension is
$$T_1 = \frac{M_b' r_1}{\sum r^2} \tag{12.26}$$

If the approximate position of the neutral axis is known, it is possible to identify the connectors located in the tension zone. Then the exact position of the neutral axis can be determined using normal procedure.

12.6.5 Combined Tensile and Shear Stresses in Connectors

The additional forces, i.e. shear forces due to W_x and W_y are computed as usual. Thus a connector is subjected to shear and tension and its strength under this combination is adequate if the following interaction formula in terms of stresses is satisfied.

$$\left(\frac{\tau_r'}{\tau_a}\right)^2 + \left(\frac{\sigma'_{ft,m}}{\sigma_{at}}\right)^2 \le 1 \tag{12.27}$$

where τ_r', $\sigma'_{ft,m}$ are the calculated shear and tensile stresses, respectively; and τ_a, σ_{at} are the corresponding allowable stresses in shear and tension.

The forces on a connector are

$$R_{s,x} = \frac{W_x}{N} \quad \text{and} \quad R_{s,y} = \frac{W_y}{N} \tag{12.28}$$

The corresponding shear stresses are

$$\tau_x = \frac{R_{s,x}}{A} \quad \text{and} \quad \tau_y = \frac{R_{s,y}}{A} \tag{12.29}$$

The tension in a connector is given by

$$\sigma'_{ft,m} = \frac{M_b}{Z} = \frac{M_b y_j}{I} = \frac{M_b y_j}{A\sum y_i^2} \tag{12.30}$$

where A, y_i and y_j are the connector area, Y-coordinate of a typical connector with respect to the centroid axis of the group, and the distance of connector under consideration from the neutral axis, respectively.
The procedure may require a further trial. The following examples will illustrate the procedure.

EXAMPLE 12.7

Investigate the safety of the riveted joint connecting two angles with column flanges using 16 mm diameter rivets as per the arrangement shown in Figures 12.23(a) to (c). The joint supports a reaction of 80 kN at an eccentricity of 160 mm from the column face. The working stresses in the rivets used in the joint are: 80 MPa in shear and 110 MPa in tension. Consider the neutral axis to pass through the lowermost row of rivets.

Solution: Effective diameter of rivet = 16 + 1.5 = 17.5 mm

Shear force per rivet

$$R_{s,y} = \frac{W}{N} = \frac{80}{10} = 8 \text{ kN}$$

The corresponding shear stress

$$\tau_y = \frac{8 \times 10^3}{(\pi \times 17.5^2)/4} = 33.26 \text{ MPa}$$

Tension due to bending moment

The topmost rivets are heavily loaded. To determine tension compute I with respect to the neutral axis considered to be at the lowermost row of rivets

$$I = 2 \times A\sum y_i^2 = 2 \times A \times (60^2 + 120^2 + 180^2 + 240^2) = 216000 A \text{ mm}^4$$

Tension in the topmost rivet at distance y = 240 mm from the neutral axis

$$\sigma'_{ft,m} = \frac{M_b y}{I} = \frac{(We)y}{216000 A} = \frac{[(80 \times 10^3) \times 160] \times 240}{216000 \times (\pi \times 17.5^2/4)} = 59.13 \text{ MPa}$$

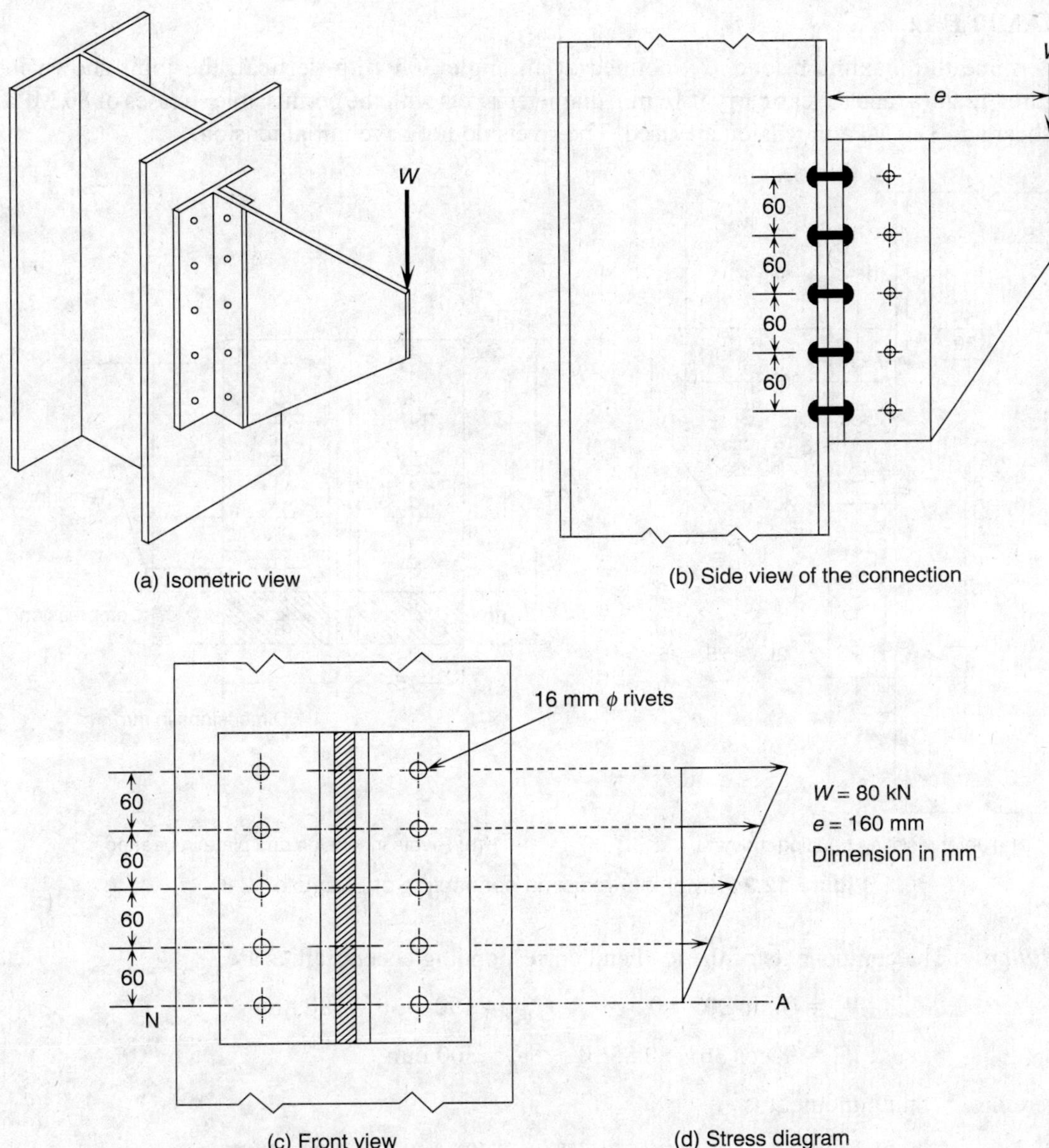

Figure 12.23 Bending moment resisting connection of Example 12.7.

To check safety, use interaction formula

$$\left(\frac{\tau_r'}{\tau_a}\right)^2 + \left(\frac{\sigma'_{ft,m}}{\sigma_{at}}\right)^2 = \left(\frac{33.26}{80}\right)^2 + \left(\frac{59.13}{110}\right)^2 = 0.4618 \leq 1.0$$

Hence the joint is safe.

EXAMPLE 12.8

Determine the maximum load W, inclined at an angle 30° with vertical, the joint shown in Figures 12.24(a) and (b) can carry if 16 mm diameter rivets with the permissible stresses of 80 MPa in shear and 120 MPa in tension are used. The rivets do not have initial tension.

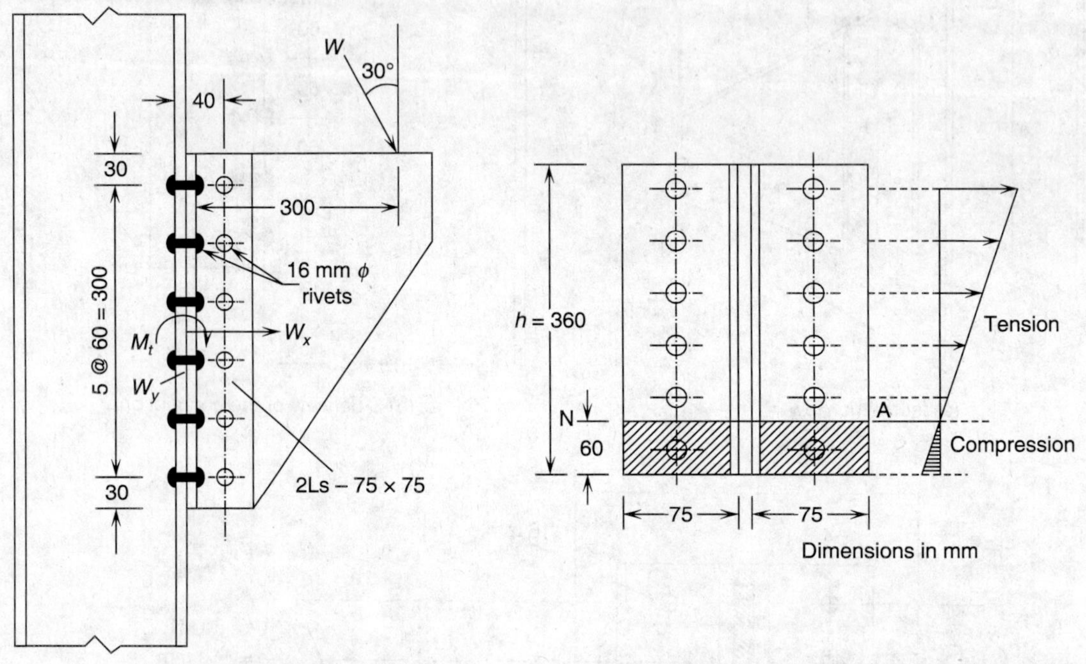

Figure 12.24 Moment resisting connection of Example 12.8.

Solution: The components of the load and corresponding eccentricities are

$$W_x = W \sin 30° = 0.5W; \qquad e_y = 150 + 30 = 180 \text{ mm}$$

$$W_y = W \cos 30° = 0.866W; \qquad e_x = 300 \text{ mm}$$

Therefore, bending moment is

$$M_b = W_x e_y + W_y e_x = (0.5W) \times (180) + (0.866W) \times (300)$$

$$= 349.8W \text{ kN.mm (clockwise)}$$

Total number of rivets connecting bracket, $N = 12$, therefore,

$$R_{s,x} = \frac{W_x}{N} = \frac{0.5W}{12} = 0.04167W$$

$$R_{s,y} = \frac{W_y}{N} = \frac{0.866W}{12} = 0.0722W$$

The corresponding shear stresses are

$$\tau_x{}' = \frac{0.04167W}{(\pi \times 16^2)/4} = 0.0002W \quad \text{and} \quad \tau_y{}' = \frac{0.0722W}{(\pi \times 16^2)/4} = 0.3591W$$

Therefore, $\tau_r{}' = \sqrt{(0.0002^2 + 0.3591)^2}\, W = 0.3591W$

Tension due to bending moment

Topmost rivet is the most heavily loaded. Depth of the connection is 360 mm, consider neutral axis at a height 360/6 = 60 mm from the bottom. With pitch of 60 mm, there are 10 rivets above the neutral axis which are in tension, and width of contact or bearing area is 150 (= 75 + 75) mm. Area of the rivet

$$A = (\pi \times 16^2)/4 = 201.062 \text{ mm}^2$$

$$I = 2 \times A \sum y_i^2 + \frac{1}{3} bh^3$$

$$= 2 \times 201.062 \times (30^2 + 90^2 + 150^2 + 210^2 + 270^2) + \frac{150 \times 60^3}{3}$$

$$= 7.0515 \times 10^7 \text{ mm}^4$$

$$Z = \frac{7.0515 \times 10^7}{300 - 30} = 2.612 \times 10^5 \text{ mm}^3$$

The tension in the topmost rivets is

$$\sigma'_{ft,m} = \frac{M_b}{Z} = \frac{349.8W \times 10^3}{2.612 \times 10^5} = 1.339W$$

Using interaction formula

$$\left(\frac{\tau_r{}'}{\tau_a}\right)^2 + \left(\frac{\sigma'_{ft,m}}{\sigma_{at}}\right)^2 = \left(\frac{0.3591W}{80}\right)^2 + \left(\frac{1.339W}{120}\right)^2 \le 1$$

or $\quad 1.4466 \times 10^{-4}\, W^2 \le 1 \quad$ i.e. $\quad W \le 83.14$ kN

12.7 PROBLEMS

12.1 A double cover triple-row riveted butt-joint is used for connecting two plates, each 75 mm wide × 12 mm thick. The cover plates are of unequal lengths and rivets of unequal diameters as shown in Figure 12.25. Determine the working strength and efficiency of the joint, if the permissible stresses are 133.33 MPa in tension, 100 MPa in shear and 200 MPa in bearing.

[**Hint:** Calculate the rivet values of three types of rivets. Compute strength of the plates of appropriate thickness at the sections passing through the rivets. The least of these values will give the strength of the joint].

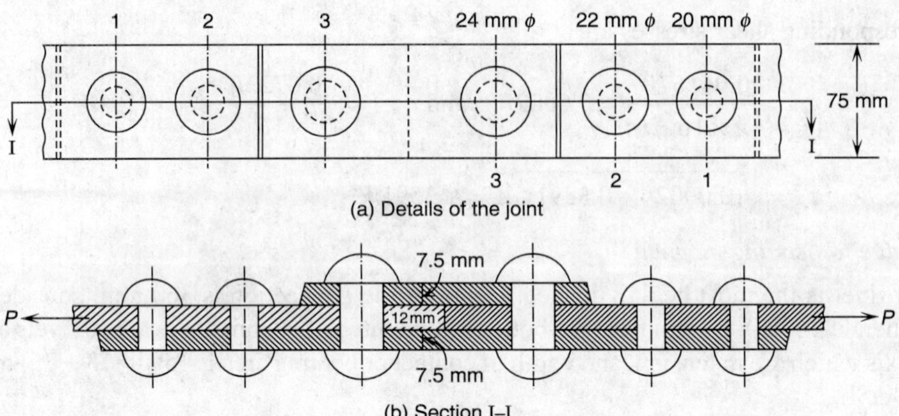

Figure 12.25 Double cover triple-row riveted butt joint of Problem 12.1.

[**Ans.** Strength of joint (minimum corresponds to that of main plate at section 1-1) P_{1-1} = 85.4 kN; P = 120 kN and η = 71.3 per cent]

12.2 Two plates are connected by four rivets of 25 mm in diameter as shown in Figure 12.26. Determine the working load P, if the permissible stresses are 112.5 MPa in tension, 100 MPa in shear and 150 MPa in bearing.

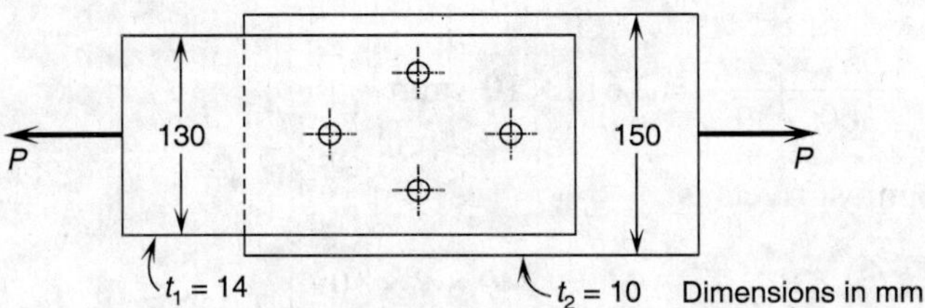

Figure 12.26 Plate connection of Problem 12.2.

[**Ans.** P = 138.75 kN]

12.3 Two 8 mm thick plates are connected by a diamond shape five-row lap joint using 19 mm diameter rivets as shown in Figure 12.27. Determine the safe load P if the working stresses are 168 MPa in tension, 114 MPa in shear and 264 MPa in bearing.

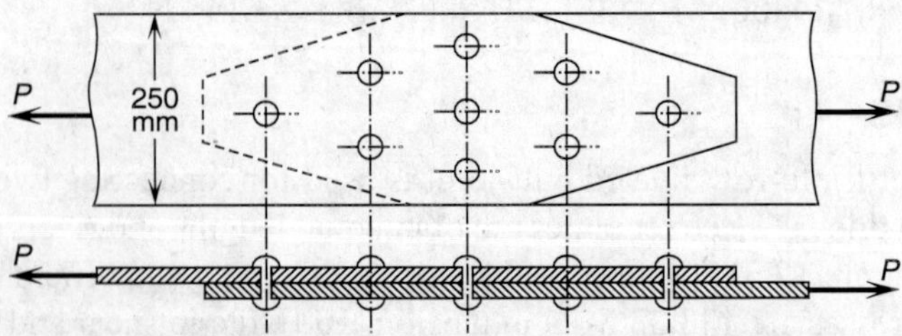

Figure 12.27 Diamond shape lap joint of Problem 12.3.

[**Ans.** P = 290.4 kN]

12.4 Design a riveted boiler of 2.5 m in diameter to sustain internal pressure of 600 kPa. The efficiency of longitudinal single-row riveted joint of the vessel is 72.5 per cent. The stress in the material of the shell is limited to 65 MPa. For the rivets diameter use the relation $d = 6.0\sqrt{t}$. Determine the thickness of the vessel and necessary pitch of the rivets on the joint.

[**Ans.** t = 16 mm; rivet diameter d = 24 mm and p = 92.73 mm (provide 92 mm)]

12.5 A boiler shell is fabricated from 14 mm thick plates having strength of 400 kN in a pitch length of 400 mm. The efficiency of longitudinal joint of the shell is 40 per cent and the tensile stress in the material of the shell is limited to 80 MPa. Determine the maximum diameter of the vessel if it is designed to sustain internal pressure of 1.6 MPa.

[**Ans.** D = 1.2 m]

12.6 A double-cover double-row butt-joint is used for connecting two 12 mm thick plates with 22 mm diameter, rivets. Determine the working strength and efficiency of the joint, if the permissible stresses are 150 MPa in tension, 100 MPa in shear and 300 MPa in bearing. [**Hint:** Number of rivet per pitch length = 2; R_r = 169.2 kN; put $P_t = R_r$]

[**Ans.** p = 117.5 mm and η = 80 per cent]

12.7 The rivet group in a joint consists of ten 20 mm diameter rivets arranged in two vertical rows of 5 rivets each at a pitch of 80 mm. The centre to centre horizontal distance between two rows is 100 mm. The joint carries a concentrated load of 100 kN at an eccentricity of 250 mm in the plan of rivets. Identify the most heavily loaded rivet and determine the stress induced in it.

[**Ans.** The topmost and the bottommost rivets of right hand vertical row are the most heavily loaded and τ_{max} = 73.7 MPa]

12.8 A gusset plate is riveted to a larger plate by four rivets with 22 mm diameter rivet holes arranged and loaded as shown in Figure 12.28. Determine the maximum and the minimum shear stress developed in the rivets.

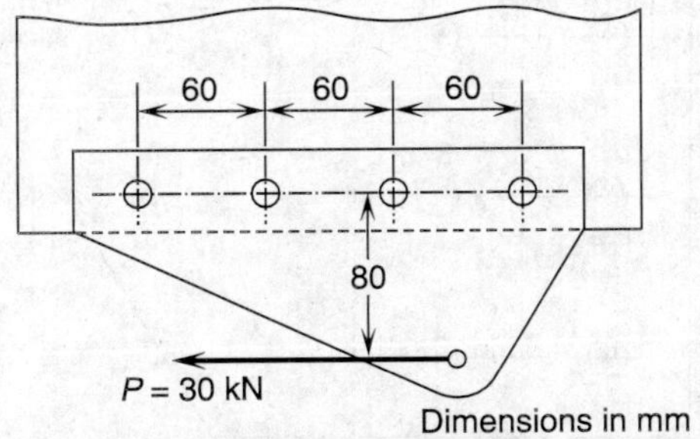

Figure 12.28 Gusset plate connection of Problem 12.8.

[**Ans.** τ_{max} = 37.2 MPa and τ_{min} = 22.4 MPa]

12.9 The connection shown in Figure 12.29 uses six rivets with 22 mm diameter holes. Determine the thickness of the plate required so that the joint carries a load of 80 kN without exceeding the bearing stress of 140 MPa.

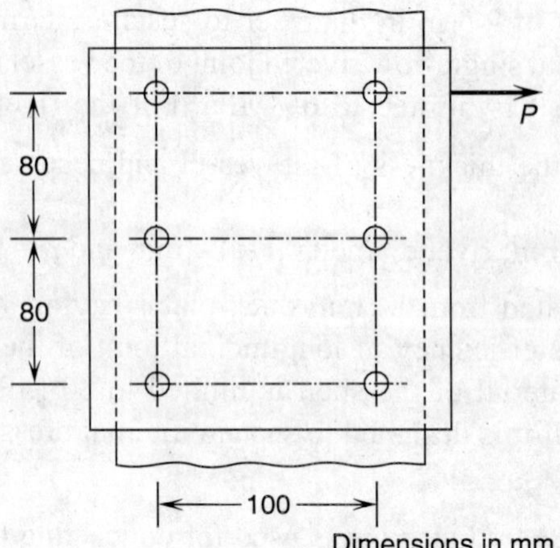

Figure 12.29 Connection for Problem 12.9.

[**Ans.** $t = 8.80$ mm]

12.10 For the riveted connection in Problem 12.9 compute the maximum value of load P that can be carried by the joint without exceeding the rivet load of 30 MPa if the lower left rivet is removed.

12.11 The bracket plate of a double-bracket connection shown in Figure 12.30 carry a load W inclined at an angle 45° with vertical. Determine the maximum value of load W, which can be allowed when the maximum stress in the rivets is limited to 100 MPa. The diameter of the rivets used in the fabrication is 20 mm.

[**Hint:** Right hand bottommost rivet is the most heavily loaded]

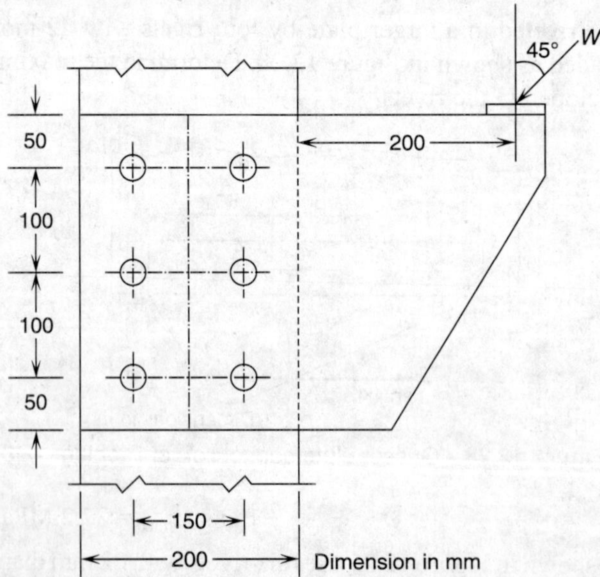

Figure 12.30 Double bracket connection of Problem 12.11.

[**Ans.** $M_t = 106$ kN mm; $\Sigma r^2 = 73750$ mm^2 and $W = 105.26$ kN]

12.12 The connection of a bracket to a column flange is shown in Figure 12.31. The joint supports a reaction of 200 kN at an eccentricity of 250 mm from the column face. Determine the maximum stresses in the rivets and check the safety of the joint, if 22 mm diameter rivets with the permissible stresses of 100 MPa in shear and 80 MPa in tension are used in the joint. The rivets do not have initial tension.
[**Hint:** Consider N.A. at height *h*/7 from the bottom]

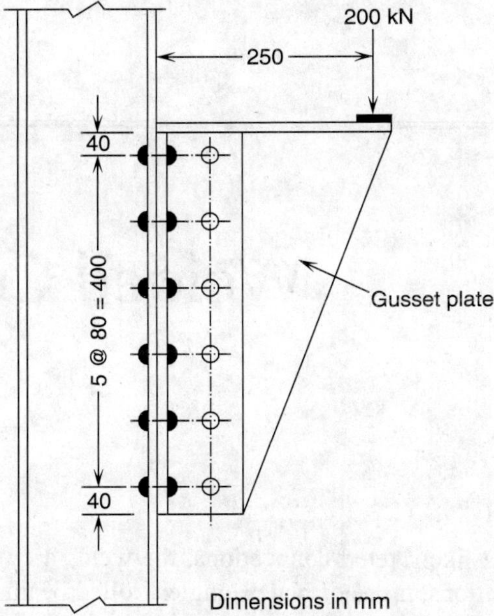

Figure 12.31 Column–bracket connection of Problem 12.12.

[**Ans.** τ = 38.4 MPa; $\sigma_{t,m}$ = 62.9 MPa and the design is safe.]

Chapter 13

Welded Connections

13.1 INTRODUCTION

As discussed in Chapter 12 like riveted connections, the welds are used in fabrication of metal structures from built-up sections, made of plates, angles, rolled beams or channel to enable them to act as single composite units. These connections are designed in such a way that the transfer of forces and moments from one member to the other is accomplished safely.

The major advantages of welded connections over riveted connections from structural point of view include: absence of driving holes make the entire area of welded member effective; higher strength of the joint to the order of parent metal and higher rigidity results in smaller bending moments; welded structures are relatively lighter and economical; repairs and new additions can be done more easily; the ease of welding the components complicated shapes; and lesser noise, better finish and appearance.

On the other hand, welding jobs require skilled manpower and supervision. The testing the quality of connection may require X-ray examination. Due to uneven heating and cooling warping at surfaces and internal stresses may develop. Not withstanding these disadvantages, the reliability of welded connections has increased to the point where they are used extensively to supplement or replace riveted or bolted connections in structural and machine design. In practice, it is economical to fabricate complicated members by welding simple component parts together rather than using complicated castings.

13.2 WELDING PROCESS

Welding is a method of *joining metals by fusion*. The welding process involves melting the metal at the joint and fusing it with additional metal from a *welding rod*. The melting of the metal is achieved by high temperature of the order of 3300°C with the heat from either an *electric arc* or an *oxyacetylene torch*. On cooling, the weld material and base metal form a continuous and almost homogeneous joint. To protect the weld from excessive oxidation, a heavily coated welding rod is used which releases an inert gas that envelops the arc stream, this technique is called *shielded arc process*.

The two principal types of welds used are the butt welds and the fillet welds illustrated in Figures 13.1.

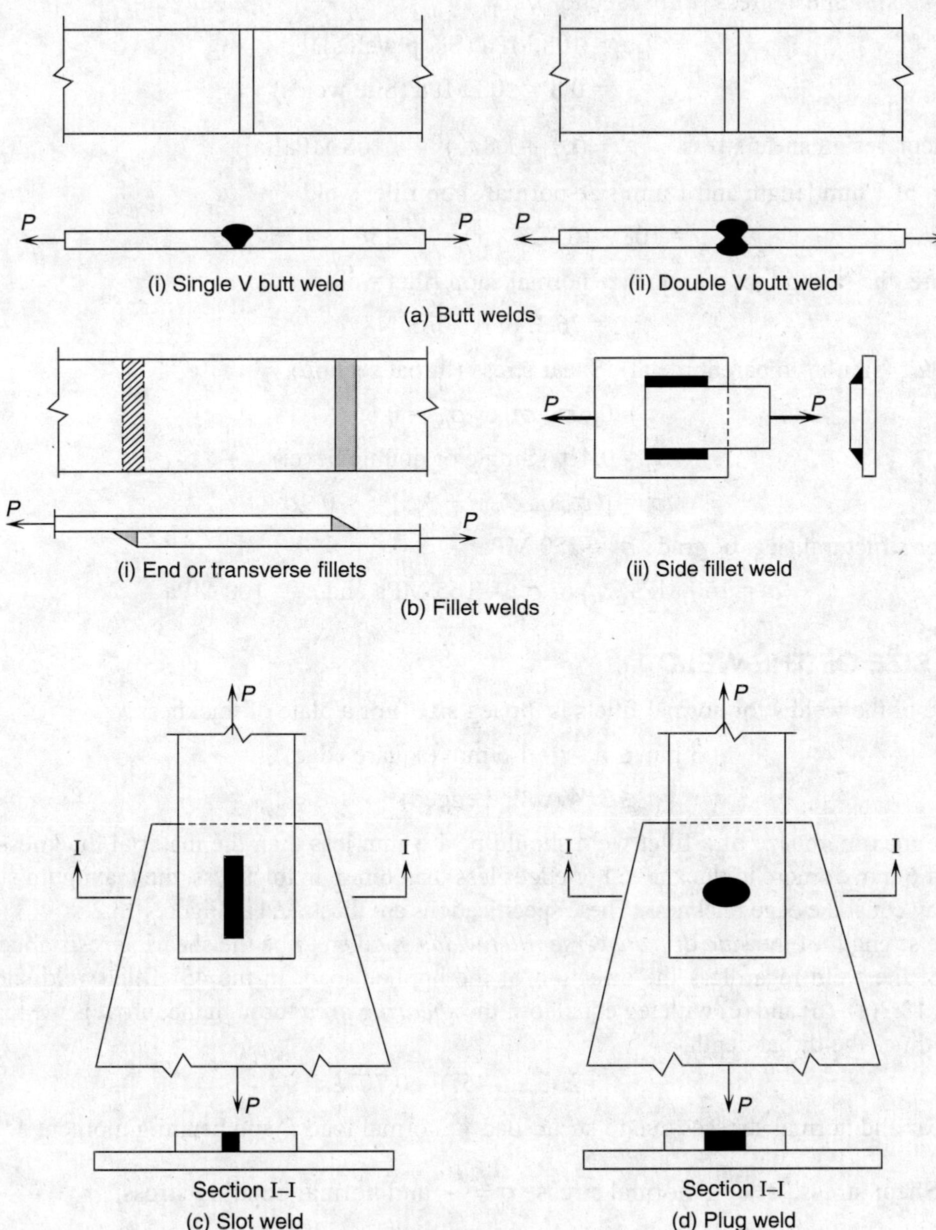

Figure 13.1 Types of welds.

13.3 ALLOWABLE STRESSES

The allowable stresses for fillet welds depend on the electrode used in the welding process and on the grade of steel being welded. The allowable shearing stress and the calculated shearing stresses are denoted by the symbols τ_a and τ, respectively. As per Indian welding code, the allowable stresses are given below:

Fillet Welds: Shear stress (throat section),

$$\tau_a = 108 \text{ MPa (Shop welds)}$$
$$= 0.8 \times 108 \text{ MPa (Site welds)} \tag{13.1}$$

Equivalent design shear stress, $\tau_e = (\sigma_b^2 + 1.8\tau^2)^{1/2} < 108 \text{ MPa}$ (13.2)

Strength of 1 mm length and 1 mm size normal shop fillet weld

$$q = 108 \times (0.707 \times 1) = 76.356 \text{ N} \approx 76 \text{ N}$$

Therefore, the strength of a t mm size normal shop fillet weld

$$q_s = 76.356t \text{ N} \approx 76t \text{ N} \tag{13.3}$$

Butt Welds (similar to parent metal): Shear stress (throat section),

$$\tau_t = 0.6\sigma_y, \ \sigma_{bt} \text{ or } \sigma_{bc} = 0.66\sigma_y \tag{13.4}$$
$$\tau_a = 0.4\sigma_y \text{ (Single or double bevel)}$$
$$\sigma_e = [(\sigma_{bt} \text{ or } \sigma_{bc})^2 + 3\tau^2]^{1/2} < 0.9\sigma_y \tag{13.5}$$

Thus, for structural steel of grade $\sigma_y = 250$ MPa:

$$\sigma_t = 150 \text{ MPa}; \ \sigma_{bt} \text{ or } \sigma_{bc} = 165 \text{ MPa and } \tau_a = 100 \text{ MPa}$$

13.4 SIZE OF THE WELD

The size of the weld s for normal fillets is the leg size. For a plate of thickness t,

$$3 \text{ mm} \le s \le t - 1.5 \text{ mm (square edge)}$$
$$\le 3t/4 \text{ (rolled edged)} \tag{13.6}$$

The maximum size of a fillet weld should be 1.5 mm less than the material thickness along edges of 6 mm or more in thickness. For edges less than 6 mm in thickness, the maximum size of a weld may equal the edge thickness. These specifications are illustrated in Figures 13.2.

The strength of the *side* or *transverse fillet welds* is taken to be the shearing resistance of the throat of the weld regardless the direction of the applied load. In the 45° fillet weld shown in Figures 13.3(a), (b) and (c) with leg equal to s, the *shearing area* through the throat is the length of weld L times the throat depth,

$$A = L\,(s \sin 45°) = 0.707Ls \tag{13.7}$$

The shear and normal stresses due to shear load V, normal load P and bending moment M are:

Shear stress, $\tau = \dfrac{V}{A}$; normal stress, $\sigma = \dfrac{P}{A}$ and normal bending stress, $\sigma_b = \dfrac{M}{Z}$

where Z is the sectional modulus of the throat area at the point where stress is required.

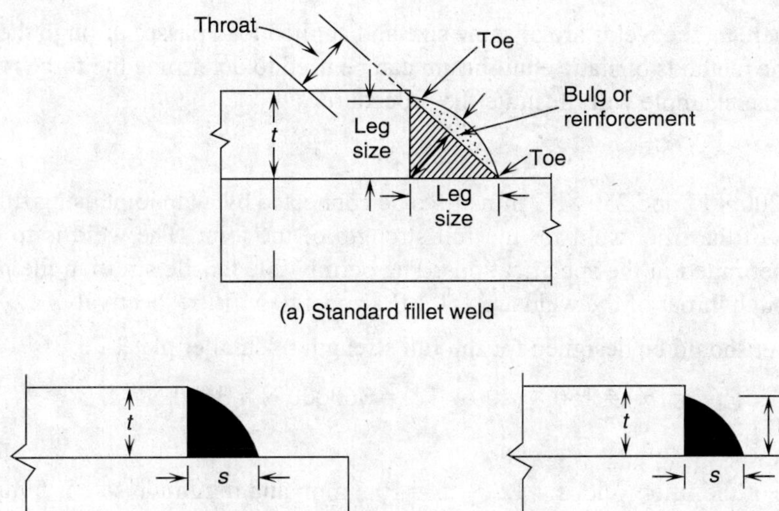

(a) Standard fillet weld

(i) $t < 6$ mm

(ii) $t > 6$ mm

(b) Maximum size of fillet weld

Thickness of thicker plate, mm	*Minimum size* mm
$t \le 10$	3
$10 < t \le 20$	5
$20 < t \le 32$	6
$32 < t \le 50$	8

(c) Minimum size of fillet weld

Figure 13.2 Specification for fillet welds.

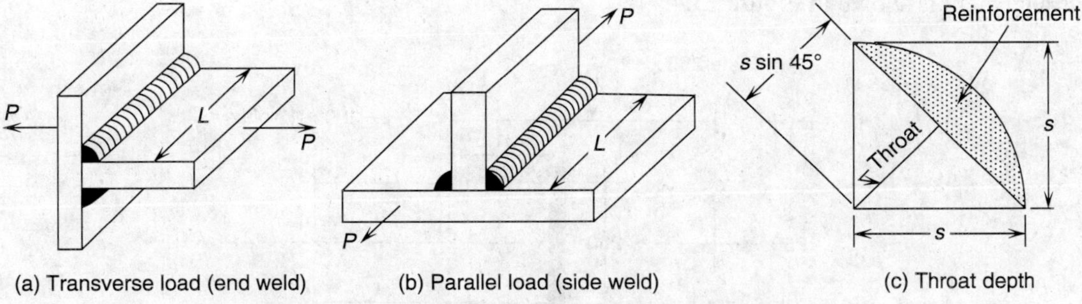

(a) Transverse load (end weld) (b) Parallel load (side weld) (c) Throat depth

Figure 13.3 Strength of side and transverse fillet welds.

The strength of a butt weld is equal to the allowable stress multiplied by the product of the length of the weld times the thickness of the thinner plate of the joint. The allowable stress is taken to be the same as that of the base metal.

In the cases where all the welds are of same size and applied load passes through the centroid of the weld lines, the methods of static equilibrium can be used to determine the force resisted by a weld. The following example will illustrate the procedure.

EXAMPLE 13.1

Two plates of size 200 × 12 and 250 × 12 mm are to be connected by a lap joint using fillet welds. Determine the size of the fillet weld for the full strength of the joint. The weld is to be of the maximum size as permitted in the specifications. The permissible tensile stress in the plates and shearing stress through throat of the weld are 150 MPa and 110 MPa, respectively.

Solution: The joint should be designed for the full strength of smaller plate, i.e.

$$P = \sigma_t\,(bt) = 150 \times (200 \times 12) = 360000 \text{ N} = 360 \text{ kN}$$

Size of the weld:

The maximum size of the fillet weld, $s = 12 - 1.5 = 10.5$ mm; and minimum size is 5 mm. Hence 10 mm fillet welds can be used.

Strength of weld per unit length,

$$q = (0.707 \times 10) \times 110 = 777.7 \text{ N}$$

Total effective length of the weld required,

$$L = \frac{P}{q} = \frac{360 \times 10^3}{777.7} = 462.9 \text{ mm}$$

This length is to be distributed on the two sides and on the end of top smaller plate such that: (i) for non-eccentric connection the lengths of weld on the sides should be equal and (ii) if weld is provided on two sides only, the transverse distance between them should not exceed $16t$ (= 160 mm).

Since the width of the plate 200 mm is greater than $16t$, the weld have to be provided at the end also. Thus provide 200 mm at the end and 131.5 mm on each side of the plate. The weld arrangement is shown in Figure 13.4.

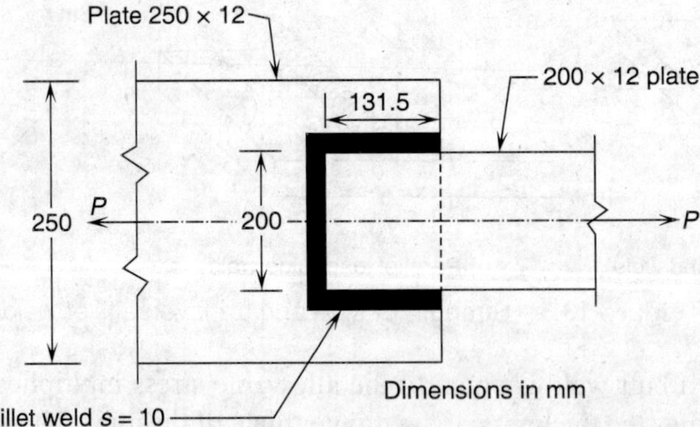

Figure 13.4 Two plates connected by fillet welds of Example 13.1.

EXAMPLE 13.2

A circular penstock of one metre diameter is fabricated from 12 mm thick plate by lapping over and securing it by transverse fillet welds on the inside and outside of the penstock as shown in Figure 13.5. Determine the safe internal pressure, if the permissible tensile stress in the plates and shearing stress through throat of the weld are 150 MPa and 110 MPa, respectively. Use the welds of the maximum size as permitted in the specifications.

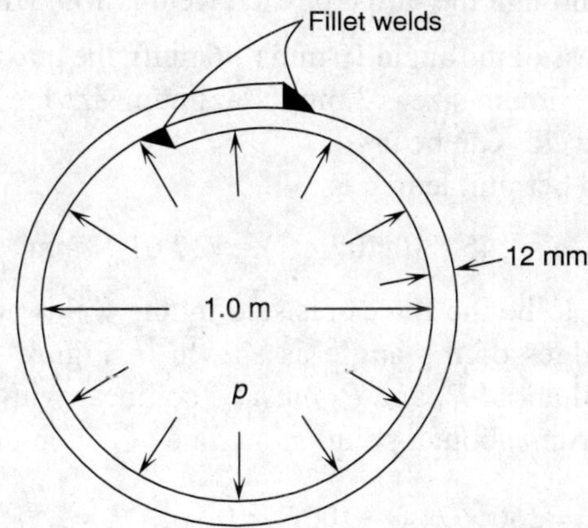

Figure 13.5 Welded penstock of Example 13.2.

Solution: Since the thickness of the plate 12 mm (> 6 mm), the maximum size of the fillet weld, $s = 12 - 1.5 = 10.5$ mm and minimum size is 5 mm. Hence 10 mm fillet welds can be used. The strength of the weld per unit length is

$$q = 110 \times (0.707 \times 10) = 777.7 \text{ N/mm}$$

The resistance offered by the welds per unit length of the penstock

$$q_{\text{resisting}} = 2 \times q = 1555.4 \text{ N}$$

The bursting force per unit length in penstock due to internal pressure p,

$$q_{\text{bursting}} = \sigma t = \left(\frac{pD}{2t}\right)t = \frac{pD}{2}$$

For $q_{\text{bursting}} = q_{\text{resisting}}$: $$\frac{pD}{2} = 1555.4$$

Therefore, safe pressure $$p = \frac{1555.4 \times 2}{D} = \frac{1555.4 \times 2}{(1 \times 10^3)} = 3.1108 \text{ MPa}$$

Check for stress in penstock plate:

Stress developed in the plate due to the internal pressure

$$\sigma = \frac{pD}{2t} = \frac{3.1108 \times (1 \times 10^3)}{2 \times 12} = 129.617 \text{ MPa} < 150 \text{ MPa}$$

EXAMPLE 13.3

A 100 × 100 × 10 mm angle which is to be welded to a gusset plate, carries a centroidal load of 360 kN. Determine the lengths of side fillet welds required at the heel and toe of the angle for non-eccentric connection when (a) no weld is provided at the end of the angle and (b) a fillet weld is added along the entire length of the end of the angle. All the welds are to be of the same size as the maximum permitted in the specifications. The distance of centroid from the heel is 28.4 mm. The permissible shearing stress through the throat of each weld is 145 MPa.

Solution: Since the thickness of the angle 10 mm (> 6 mm), the maximum size of the fillet weld, $s = 10 - 1.5 = 8.5$ mm and minimum size is 3 mm. Maximum size for rounded toe $s = 10 \times (3/4) = 7.5$ mm. Hence 7 mm fillet welds can be used.

The strength of the weld per unit length is

$$q = 145 \times (0.707 \times 7) = 717.61 \text{ N/mm}$$

(a) Consider P_1 and P_2 to be the forces resisted by the welds at heel and toe, respectively, acting along the edges of the angle as shown in Figure 13.6(a). For non-eccentric connection, the resultant of P_1 and P_2 must be collinear with the applied force. Consider the moment equilibrium about a point on the line of action of P_2

$$\sum M_{P_2} = 360 \times 71.6 - 100P_1 = 0 \quad \text{or} \quad P_1 = 257.76 \text{ kN}$$

Similarly, considering a point on the line of action of P_1

$$\sum M_{P_1} = 360 \times 28.4 - 100P_2 = 0 \quad \text{or} \quad P_2 = 102.24 \text{ kN}$$

Therefore, the required lengths of the weld are

$$L_1 = \frac{P_1}{q} = \frac{257.76 \times 10^3}{717.61} = 359.19 \text{ mm}$$

$$L_2 = \frac{P_2}{q} = \frac{102.24 \times 10^3}{717.61} = 142.47 \text{ mm}$$

(b) When a transverse fillet weld of 7 mm size is added along the entire length of the end of the angle, the force P_3 resisted by this weld acting at its midpoint is

$$P_3 = qL = 717.61 \times 100 = 71761 \text{ N} = 71.76 \text{ kN}$$

For non-eccentric connection, the resultant of P_1, P_2 and P_3 must be collinear with the applied force. As in the previous case, consider moment equilibrium about points on the line of action of P_1 and P_2

$$\sum M_{P_2} = 360 \times 71.6 - 71.76 \times 50 - 100P_1 = 0 \quad \text{or} \quad P_1 = 221.88 \text{ kN}$$

$$\sum M_{P_1} = 100P_2 - 71.76 \times 50 - 360 \times 28.4 \quad \text{or} \quad P_2 = 138.12 \text{ kN}$$

The required lengths of the weld are

$$L_1 = \frac{P_1}{q} = \frac{221.88 \times 10^3}{717.61} = 309.19 \text{ mm}$$

$$L_2 = \frac{P_2}{q} = \frac{138.12 \times 10^3}{717.61} = 192.47 \text{ mm}$$

The lengths are generally increased by a small amount to provide for starting and stopping of the weld. Figure 13.6(b) shows the side fillet and the end fillet.

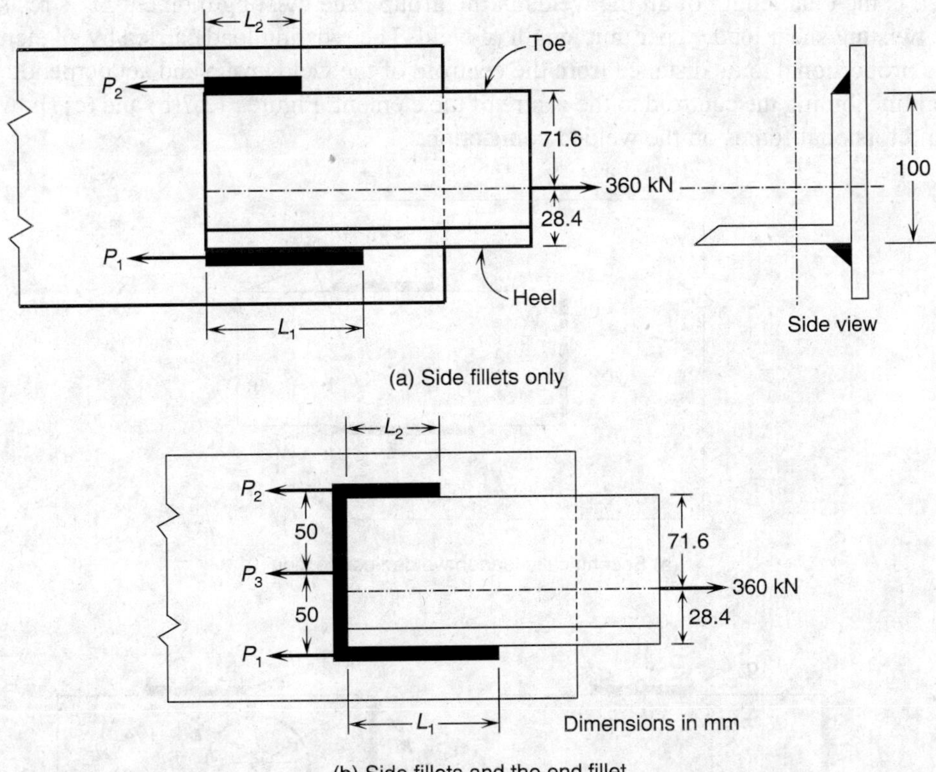

Figure 13.6 An angle tension member welded to a gusset plate of Example 13.3.

13.5 ECCENTRICALLY LOADED WELDED CONNECTIONS

If the line of action of the applied load does not pass through the centroid of the weld lines, the load per unit length of weld is not constant. Hence, the connection is called an *eccentric welded connection*. As in case of riveted connection, the effect of eccentric load is statically equivalent to a central load W and a twisting moment or a bending moment M_t $(M_b) = We$, as shown in the free-body diagram of the plate in Figure 13.7(a). If the load axis acts in the plane of weld lines, the resulting moment will be in the plane of weld and have the *twisting effect*. On the other hand, if the axis of the load does not lie in the plane of the weld lines, the applied load will cause bending moment in a plane normal to the plane of weld.

13.5.1 Eccentric Load Causing Twisting Moment

In this case, the axis of the load lies in the plane of weld lines, and only shear stress develops in the welds. The central load W is resisted as direct shear load q_d per unit length which is uniformly distributed over all the welds. The magnitude of q_d is

$$q_d = W / \sum L \tag{13.8}$$

where ΣL is the total length of all the welds in the group. The twisting moment M_t is resisted by variable twisting/shear load q_m per unit length of weld. The twisting load carried by an element of a weld is proportional to its distance from the centroid of the weld group and act perpendicular to the radial line joining the centroid to the centre of the element. Figures 13.7(b) and (c) show direct forces and torsional forces on the welded connection.

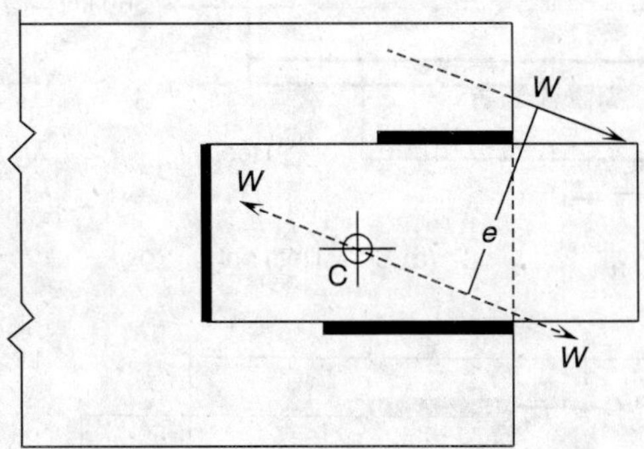

(a) Eccentrically loaded welded connection

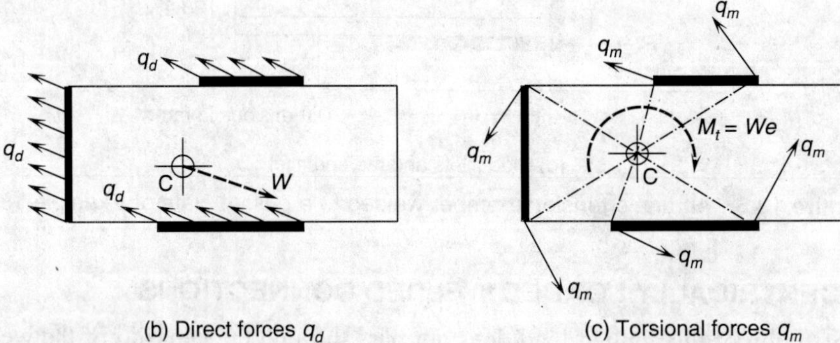

(b) Direct forces q_d

(c) Torsional forces q_m

Figure 13.7 Analysis of eccentrically loaded welded connection.

The shear load carried by an element can be determined by using the torsion formula $\tau = M_t r/J$ with modified value of J. Here τ represents average shearing stress on the element; r is the radial distance from its centre to the centroid of the welds. To formulate expression for J consider the arrangement of straight welds shown in Figure 13.8. For any weld of unit width and length L_i, the centroial value $\bar{J}_i$ is the sum of the rectangular moment of inertia with respect to axes through the centre directed along and perpendicular to its length. This may be expressed as:

$$\bar{J}_i = I_x + I_y = 0 + (L_i^3/12) \tag{13.9}$$

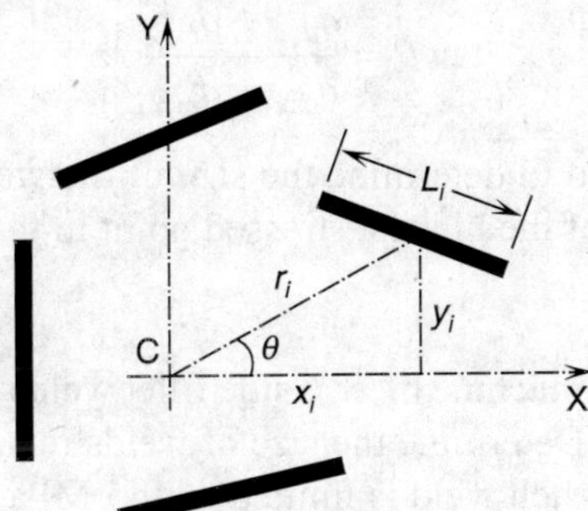

Figure 13.8 Determination of J.

The value J_i of the weld with respect to the centroid C of the weld group is

$$J_i = \bar{J}_i + A_i r_i^2 = \bar{J}_i + (L_i \times 1) r_i^2 = (L_i^3/12) + L_i(x_i^2 + y_i^2) \tag{13.10}$$

where x_i and y_i are the coordinates of the centre of each weld relative to the common centroid C. For all the welds of the group, the modified J of the torsion formula which is the sum of J_i for each weld is given by

$$J = \sum J_i = \frac{1}{12}\left(\sum L_i^3\right) + \sum L_i(x_i^2 + y_i^2) = \sum L_i\left(\frac{1}{12}L_i^2 + x_i^2 + y_i^2\right) \tag{13.11}$$

Torsion force q_m acting perpendicular to the radial location r of any point on the weld is obtained by using the torsion formula

$$q_m = \frac{M_t r}{\sum L_i\left(\frac{1}{12}L_i^2 + x_i^2 + y_i^2\right)} \tag{13.22}$$

As in the case of riveted connections it is convenient to express the torsion force q_m in its X- and Y-components as follows:

$$q_{m,x} = q_m \sin\theta = q_m \frac{y}{r} = \frac{M_t y}{\sum L_i\left(\frac{1}{12}L_i^2 + x_i^2 + y_i^2\right)} \tag{13.33}$$

Similarly,

$$q_{m,y} = \frac{M_t x}{\sum L_i\left(\frac{1}{12}L_i^2 + x_i^2 + y_i^2\right)} \tag{13.14}$$

The maximum intensity of the weld force occurs at the point where all the components are additive,

$$q_r = \sqrt{(q_{d,x} + q_{m,x})^2 + (q_{d,y} + q_{m,y})^2} \tag{13.15}$$

The angle θ between q_r and $q_{d,x}$ is given by

$$\tan\theta = \frac{q_{d,y} + q_{m,y}}{q_{d,x} + q_{m,x}} \tag{13.16}$$

Frequently this value of q_r is used to determine the size of the weld, but occasionally the size of each weld is based on the value of the highest stressed point in that weld.

EXAMPLE 13.4

A plate attached to the frame of a machine by two side fillet welds as shown in Figure 13.9(a) is to support a vertical load of 50 kN. Determine the size of welds required to resist the load when the shearing stress through throat of each weld is limited to 145 MPa.

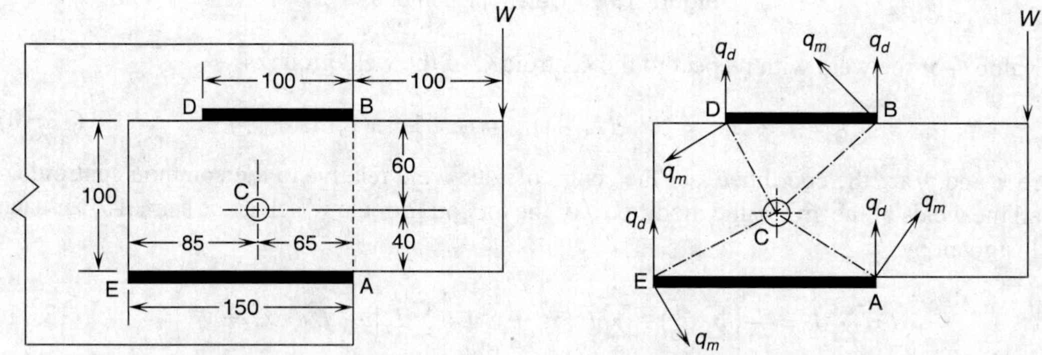

Figure 13.9 The plate with two side fillet welds of Example 13.4.

Solution: The centroid $C(\overline{x}, \overline{y})$ of weld lines with respect to origin (say point A) can be computed as

$$\overline{x} = \frac{\sum L_i x_i}{L} \quad \text{i.e.} \quad \overline{x} = \frac{100 \times 50 + 150 \times 75}{(100 + 150)} = 65 \text{ mm}$$

$$\overline{y} = \frac{\sum L_i y_i}{L} \quad \text{i.e.} \quad \overline{y} = \frac{100 \times 100 + 150 \times 0}{(100 + 150)} = 40 \text{ mm}$$

Figure 13.9(b) shows the free-body diagram for direct and torsional loads.

Therefore, torsional moment about centroid C is

$$M_t = We_x = 50 \times (65 + 100) = 8250 \text{ kN.mm (clockwise)}$$

The modified J for the weld lines of unit width is

$$J = \sum L_i \left(\frac{1}{12} L_i^2 + x_i^2 + y_i^2 \right)$$

$$= 150 \times \left[\frac{150^2}{12} + 10^2 + 40^2 \right] + 100 \times \left[\frac{100^2}{12} + 15^2 + 60^2 \right]$$

$$= 1002083.33 \text{ mm}^4$$

Components of direct load are

$$q_{d,x} = \frac{W_x (= 0)}{\sum L} = 0 \quad \text{and} \quad q_{d,y} = \frac{W}{\sum L} = \frac{50 \times 10^3}{(150 + 100)} = 200 \text{ N/mm}$$

The components of the torsional forces at the points of the maximum torsional intensity A and B are:

Horizontal direction

At point B: $$q_{m,x} = \frac{M_t y}{J} = \frac{(8250 \times 10^3) \times 60}{1002083.33} = 493.97 \text{ N/mm}$$

At point A: $$q_{m,x} = \frac{M_t y}{J} = \frac{(8250 \times 10^3) \times 40}{1002083.33} = 329.31 \text{ N/mm}$$

Vertical direction

At points B and A: $$q_{m,y} = \frac{M_t x}{J} = \frac{(8250 \times 10^3) \times 65}{1002083.33} = 535.14 \text{ N/mm}$$

The resultant of the direct and torsional force components at the two points are

$$q_{r,B} = \sqrt{(493.97)^2 + (200 + 535.14)^2} = 885.68 \text{ N/mm}$$

$$q_{r,A} = \sqrt{(329.31)^2 + (200 + 535.14)^2} = 805.53 \text{ N/mm}$$

The strength of the weld per unit length

$$q = 145 \times (0.707 \times s) = 102.5s \text{ N/mm}$$

Therefore, the size of the welds based on the maximum load at point B is

$$s = \frac{885.68}{102.5} = 8.64 \text{ mm}$$

Provide 9 mm thick welds.

EXAMPLE 13.5

The eccentric welded connection shown in Figure 13.10(a) consists of two bracket plates welded to the flanges of a column. The plates are welded to the flanges by shop fillet welds on three sides. Determine the maximum load the joint can support if 8 mm welds are used. The permissible shearing stress through the throat of each normal shop fillet weld is 108 MPa.

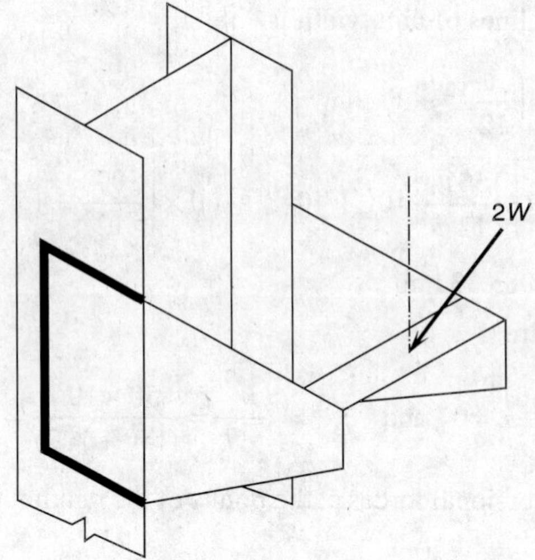

(a) Isometric view of the connection

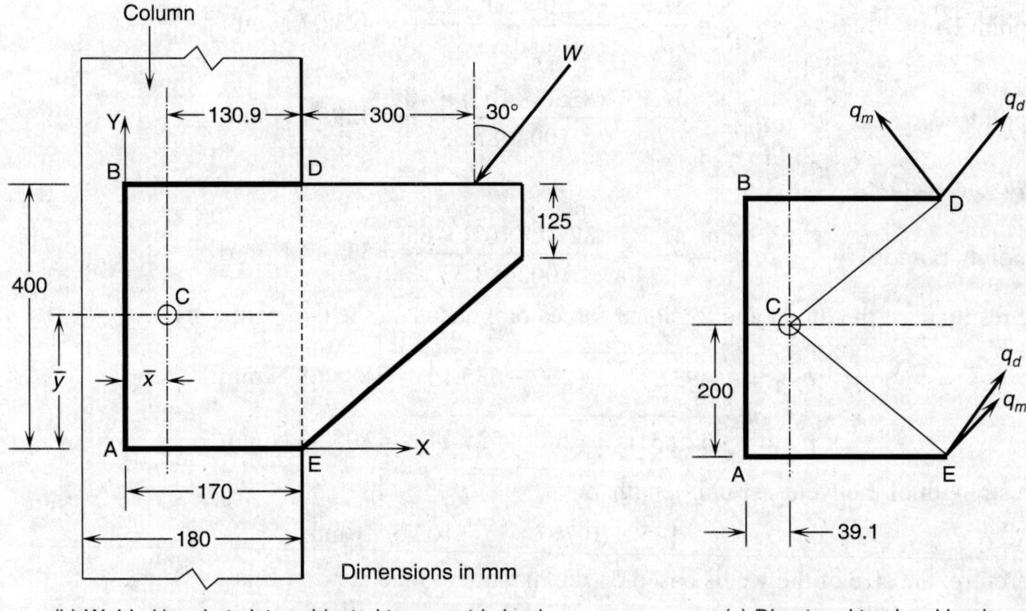

(b) Welded bracket plate subjected to eccentric load (c) Direct and torsional loads

Figure 13.10 Bracket plates with three side fillet welds of Example 13.5.

Solution: The total effective length of the weld, $L = 400 + 2 \times 170 = 740$ mm

Centroid $C(\bar{x}, \bar{y})$ of weld lines with respect to origin (say point A) can be computed as

$$\bar{x} = \frac{2 \times 170 \times 85}{740} = 39.1 \text{ mm} \quad \text{and} \quad \bar{y} = \frac{400}{2} = 200 \text{ mm}$$

The load components and corresponding eccentricities are

$$W_x = W \sin 30° = 0.5W; \qquad e_y = 200 \text{ mm}$$

$$W_y = W \cos 30° = 0.866W; \qquad e_x = 300 + (170 - 39.1) = 430.9 \text{ mm}$$

Therefore, torsion moment is

$$M_t = W_x e_y + W_y e_x = -(0.5W) \times (200) + (0.866W) \times (430.9)$$

$$= 273.2W \text{ kN mm (clockwise)}$$

Components of direct load are

$$q_{d,x} = \frac{W_x}{\sum L} = \frac{0.5W}{740} = 0.6757 \times 10^{-3} W \text{ kN/mm}$$

$$q_{d,y} = \frac{W_y}{\sum L} = \frac{0.866W}{740} = 1.1703 \times 10^{-3} \text{ kN/mm}$$

The J for the welds of unit width is

$$J = \sum L_i \left(\frac{1}{12} L_i^2 + x_i^2 + y_i^2 \right)$$

$$= 400 \times \left[\frac{400^2}{12} + 39.1^2 \right] + 2 \times 170 \times \left[\frac{170^2}{12} + (85 - 39.1)^2 + 200^2 \right]$$

$$= 21080006 = 21.08 \times 10^6 \text{ mm}^4$$

The pointh E has the maximum stress intensity since it is at the maximum distance from the centroid of weld group and the angle between the direct and torsional forces is the minimum. Figure 13.10(b) shows the welded bracket plate subjected to eccentric load and Figure 13.10(c) shows direct and torsional loads. Components of the torsional force at E are:

$$q_{m,x} = \frac{M_t y}{J} = \frac{(273.2W) \times 200}{21.08 \times 10^6} = 2.5920 \times 10^{-3} W \text{ N/mm}$$

$$q_{m,y} = \frac{M_t x}{J} = \frac{(273.2W) \times 130.9}{21.08 \times 10^6} = 1.6965 \times 10^{-3} W \text{ N/mm}$$

The resultant of the direct and torsional force components is

$$q_{r,E} = \left[\sqrt{(0.6757 + 2.5920)^2 + (1.1703 + 1.6965)^2} \right] \times 10^{-3} W$$

$$= 4.347 \times 10^{-3} W \text{ N/mm}$$

The strength of the weld per unit length

$$q = 108 \times (0.707 \times s) = 76.356 \times 8 = 610.848 \text{ N}$$

For $q = q_{r,E}$: $$W = \frac{610.848}{4.347 \times 10^{-3}} = 140.52 \times 10^3 \text{ N} = 140.52 \text{ kN}$$

The maximum load $= 2 \times 140.52 = 281.04$ kN.

13.5.2 Eccentric Load Causing Bending Moment

As in case of other connections, the effect of an eccentric load on the weld line is statically equivalent to direct load at the centroid and a bending moment $M_b = We$. This external bending moment due to eccentric load tends to rotate the connection about an axis lying in the plane of weld lines and causes normal stresses, i.e. tensile and compressive stresses at the throat section of the fillet weld. The portion of the weld below the axis of rotation will be compressed and the one above it pulled. On the other hand, the direct load causes direct or shear stresses. The variation of strains and stresses is linear, i.e. varies from zero at the *axis of rotation* or *neutral axis* to a maximum value in the extreme points.

Consider the bracket shown in Figures 13.11(a) and (b) consisting of a plate abutting the flange of the column and welded to the flange by fillet welds on both the sides of the plate for the full depth of the bracket. The direct load per unit length of the weld is

$$q_d = \frac{W}{\sum L} \tag{13.17}$$

where $\sum L$ is the total length of all the welds in the group. The normal stress developed in an element can be determined by using the bending formula:

$$\sigma_b = M_b y/I$$

The normal load carried by the element per unit length,

$$q_b = \sigma_b (1 \times s) = \left(\frac{M_b y}{I}\right) s \tag{13.18}$$

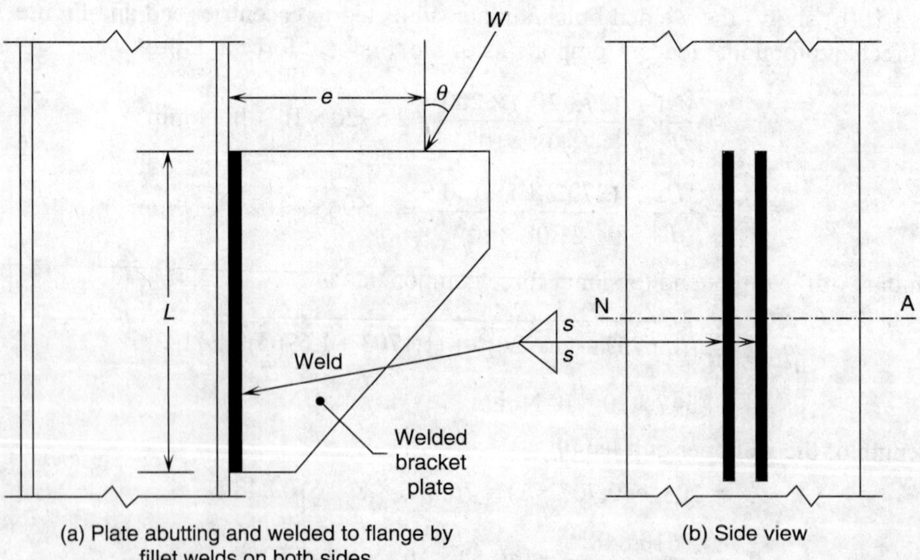

Figure 13.11 Welded plate bracket subjected to bending.

For a weld of unit width, i.e. for $s = 1$, $$q_b = \frac{M_b y}{I} \tag{13.19}$$

where q_b represents the bending stress on the element; y is the distance of the centre of element from the neutral axis passing through the centroid of the welds. To determine I for any arrangement of straight welds, consider a typical weld of unit width and length L_i having the centroial value $\overline{I}_{x,i}$ as the rectangular moment of inertia with respect to an axes passing through the centre of the element and parallel to the axis of rotation or the neutral axis.

The value $I_{x,i}$ of the weld with respect to the centroid C of the weld group is

$$I_{x,i} = \overline{I}_{x,i} + A_i y_i^2 = \overline{I}_{x,i} + (L_i \times 1)\, y_i^2 = (\overline{I}_{x,i} + L_i y_i^2) \tag{13.20}$$

For all the welds of the group, the I of the bending formula which is the sum of $I_{x,i}$ for each weld is given by

$$I = \sum I_{x,i} = \sum\left(\overline{I}_{x,i} + L_i y_i^2\right) \tag{13.21}$$

For a weld perpendicular to neutral axis, $\overline{I}_{x,i} = (L_i^3/12)$ and thus $I_{x,i} = L_i\left(\frac{1}{12}L_i^2 + y_i^2\right)$.

On the other hand when a weld is parallel to neutral axis, $\overline{I}_{x,i} \approx 0$ and $I_{x,i} = L_i y_i^2$.
Since q_d and q_b are perpendicular to each other, the resultant load per unit length of weld is

$$q_{es} = (q_b^2 + q_d^2)^{1/2} \quad \text{or} \quad \tau_e = (\sigma_b^2 + \tau^2)^{1/2} \quad \text{as } (\tau = q_d/s) \tag{13.22}$$

However, in practice the combined effect of these normal and shear stresses at the throat section is treated as an *equivalent shear stress* from the strength point of view. The equivalent shear stress is given by

$$\tau_e = (\sigma_b^2 + 1.8\tau^2)^{1/2} \tag{13.23}$$

where σ_b and τ are the normal bending stress and the shear stresses, respectively, and $\tau_e \le \tau_a$ (108 MPa).
In case of butt welds, the equivalent shear stress is given by

$$\tau_e = (\sigma_b^2 + 3\tau^2)^{1/2} \le 0.9\tau_a \tag{13.24}$$

The procedure involves the following steps:

1. Locate the centroid of the weld lines group.
2. Determine the bending moment and direct load at the centroid.
3. Identify the critical weld point where the normal tensile stress is the maximum.
4. Determine the equivalent shear stress at the critical weld point.
5. Compare the equivalent shear stress with the fillet weld strength.

EXAMPLE 13.6

The welded bracket connection shown in Figure 13.12 consists of a plate welded to the flange of a column by 10 mm shop fillet welds on two sides. Determine the maximum load W the bracket

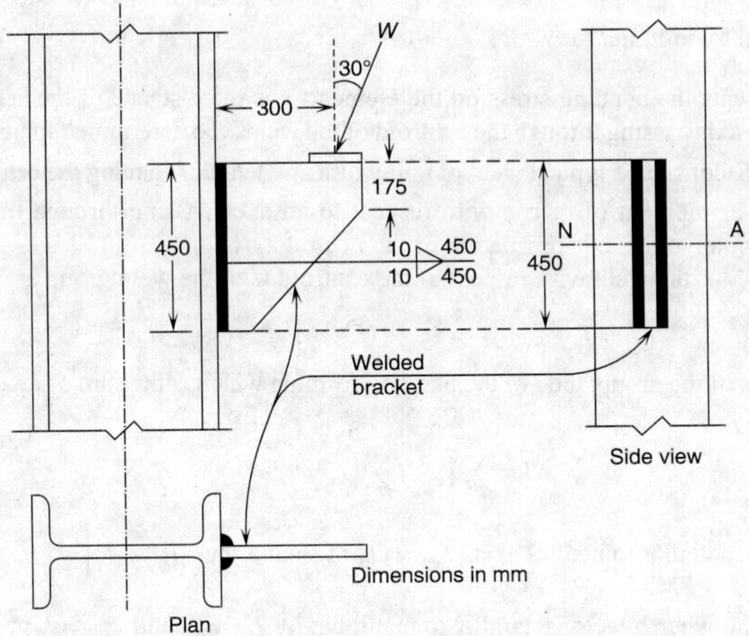

Figure 13.12 Welded plate bracket connection of Example 13.6.

can support at a distance of 300 mm from the face of the column if the permissible stress in the weld is limited to 110 MPa.

Solution: The total effective length of the weld, $\sum L = 2 \times 450 = 900$ mm

Moment of inertia of weld lines,

$$I_x = 2 \times \left(\frac{450^3}{12} \right) = 1.5187 \times 10^7 \text{ mm}^3$$

The load components and corresponding eccentricities are

$$W_x = W \sin 30° = 0.5W; \qquad e_y = 225 \text{ mm}$$

$$W_y = W \cos 30° = 0.866W; \qquad e_x = 300 \text{ mm}$$

Therefore, the bending moment is

$$M_t = W_x e_y + W_y e_x = -(0.5W) \times (225) + (0.866W) \times (300)$$

$$= 147.3W \text{ kN/mm (clockwise)}$$

Components of direct load are

$$q_{d,x} = \frac{W_x}{\sum L} = \frac{0.5W}{900} = 5.556 \times 10^{-4} W \text{ kN/mm}$$

$$q_{d,y} = \frac{W_y}{\sum L} = \frac{0.866W}{900} = 9.622 \times 10^{-4} W \text{ kN/mm}$$

Components of the normal force at the point of the maximum bending stress intensity,

$$q_{b,x} = \frac{M_b y}{I_x} = \frac{(147.3W)\times 225}{1.5187\times 10^7} = 2.1823\times 10^{-3} W \text{ kN/mm}$$

The resultant of the direct and normal force components is

$$q_r = \left(\sqrt{(0.5556+2.1823)^2 + (0.9622)^2}\right)\times 10^{-3} W$$

$$= 2.902\times 10^{-3} W \text{ kN/mm}$$

The equivalent shear load

$$q_{es} = (q_b^2 + 1.8 q_d^2)^{1/2} = \left(\sqrt{(0.5556+2.1823)^2 + 1.8\times(0.9622)^2}\right)\times 10^{-3} W$$

$$= 3.027\times 10^{-3} W \text{ N/mm}$$

The strength of the weld per unit length

$$q = 110 \times (0.707 \times 10) = 777.7 \text{ N}$$

For $q = q_{es}$:
$$W = \frac{777.7}{3.027\times 10^{-3}} = 237.829\times 10^3 \text{ N} = 256.92 \text{ kN}$$

The maximum load is 256.92 kN.

EXAMPLE 13.7

The bracket connection shown in Figures 13.13(a) and (b) consist of a T-section connected to the flange of a column. The T-section is welded to the face of flange by shop fillet welds provided along the perimeter of the section. Determine the size of the weld required if the bracket is to support a load of 100 kN at a distance of 150 mm from the face of the column. The permissible shearing stress through the throat of each normal shop fillet weld is 108 MPa.

Solution: The total effective length of the weld,

$$\sum L = 125 + 2\times 10 + 115 + 2\times 190 + 10 = 650 \text{ mm}$$

Centroid $C(\bar{x}, \bar{y})$ of weld lines with respect to section I-I can be computed as

$$\bar{y} = \frac{125\times 0 + 2\times 10\times 5 + 115\times 10 + 2\times 190\times 105 + 10\times 200}{650}$$

$$= \frac{43150}{650} = 66.38 \text{ mm}$$

The moment of inertia for the welds of unit width,

$$I_x = [0 + 125\times(66.38)^2] + 2\times\left(\frac{10^3}{12} + 10\times(66.38-5)^2\right) + [0 + 115\times(56.38)^2]$$

$$+ 2\times\left(\frac{56.38^3}{3} + \frac{133.62^3}{3}\right) + [0 + 10\times(133.62)^2] = 2.8803\times 10^6 \text{ mm}^3$$

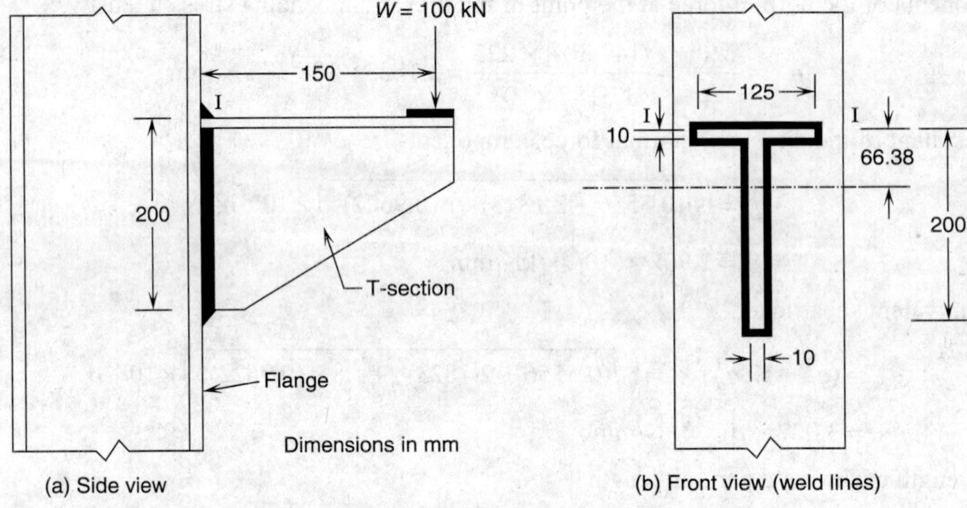

Figure 13.13 T-section eccentric welded bracket connection of Example 13.7.

The distance of the remotest point from the neutral axis

$$y = 200 - \bar{y} = 200 - 66.38 = 133.62 \text{ mm}$$

Therefore, bending moment is

$$M_b = We = 100 \times (150) = 15000 \text{ kN.mm (clockwise)}$$

The direct load per unit length of the weld is

$$q_d = \frac{100 \times 10^3}{650} = 153.85 \text{ N/mm}$$

The normal load per unit length due to bending is given by

$$q_b = \frac{(15000 \times 10^3) \times 133.62}{2.8803 \times 10^6} = 695.87 \text{ N/mm}$$

The equivalent shear load

$$q_{es} = (q_b^2 + 1.8 q_d^2)^{1/2} = (695.87^2 + 1.8 \times 153.85^2)^{1/2} = 725.84 \text{ N/mm}$$

The strength of the weld per unit length

$$q = 108 \times (0.707 \times s) = 76.356s \text{ N}$$

For $q = q_{es}$:

$$s = \frac{725.84}{76.356} = 9.51 \text{ mm}$$

Hence welds of size 10 mm are required.

EXAMPLE 13.8

The bracket connection shown in Figures 13.14(a) and (b) consist of a joist cutting welded to the flange of a column by shop fillet welds 8 mm in size on the flanges and 6 mm on the web. Determine the safe load W the bracket can support at a distance of 200 mm from the face of the column if the permissible stress in the weld is limited to 110 MPa.

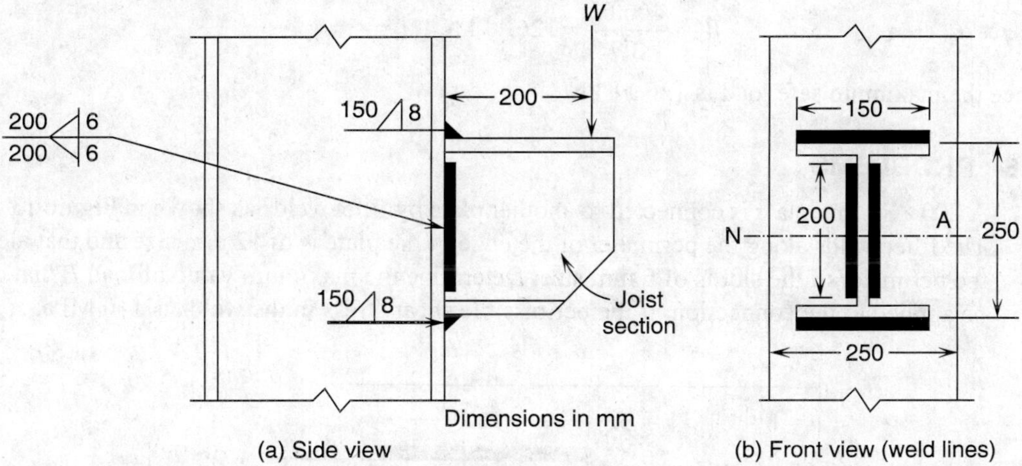

Figure 13.14 Eccentrically loaded joist bracket connection of Example 13.8.

Solution: The strength of flange weld is 8/6 i.e. 4/3 times that of web welds. Thus the welds can be analysed in terms of 6 mm weld size. The total effective length of the weld,

$$\sum L = 2 \times 150 \times (4/3) + 2 \times 200 = 800 \text{ mm}$$

The moment of inertia for the group of welds of unit width,

$$I_x = 2 \times \left(\frac{200^3}{12}\right) + 2 \times [0 + 150 \times (4/3) \times (125)^2] = 7.583 \times 10^6 \text{ mm}^4$$

The bending moment is

$$M_b = We = W \times 200 = 200W \text{ kN.mm (clockwise)}$$

The direct load per unit length of the weld is

$$q_d = \frac{W \times 10^3}{800} = 1.25W \text{ N/mm}$$

The normal load per unit length due bending is given by

$$q_b = \frac{(200W \times 10^3) \times 125}{7.583 \times 10^6} = 3.297W \text{ N/mm}$$

The equivalent shear load

$$q_{es} = (q_b^2 + 1.8q_d^2)^{1/2} = W(3.297^2 + 1.8 \times 1.25^2)^{1/2} = 3.7W \text{ N/mm}$$

The strength of the weld per unit length

$$q = 110 \times (0.707 \times 6) = 466.62 \text{ N}$$

For $q = q_{es}$:

$$W = \frac{466.62}{3.7} = 126.11 \text{ kN}$$

Hence the maximum safe load is 126.11 kN.

13.6 PROBLEMS

13.1 A 200 × 14 mm plate is connected to another plate by fillet welds as shown in Figure 13.15. The fillet welds along the perimeter of the edge of the plate is of 12 mm size and that along the perimeter of the slot is of 6 mm size. Determine the maximum value of load P that can be applied to the connection, if the permissible shear stress in the welds is 110 MPa.

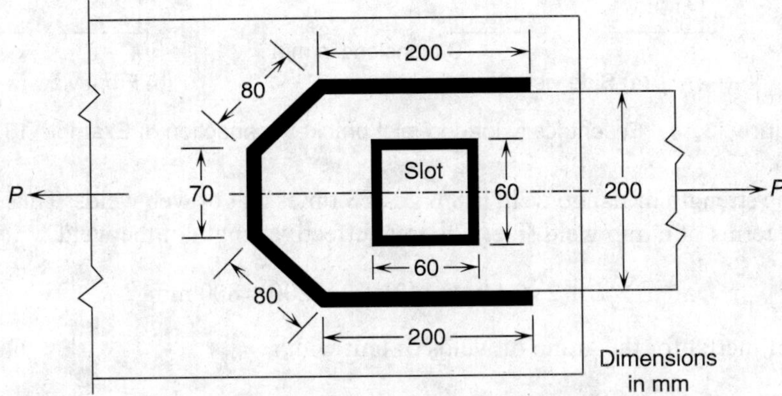

Figure 13.15 Welded connection of Problem 13.1.

[**Ans.** P = 69.3 kN]

13.2 A 150 × 115 × 8 mm angle carrying centroidal tensile load of 250 kN is to be welded to a gusset plate, with the 150 mm leg against the plate. Determine the lengths of 6 mm side fillet welds required at the heel and toe of the angle for non-eccentric connection. The distance of centroidal axis of angle from the heel is 44.6 mm. The allowable shearing stress through the throat of each weld is 110 MPa.

[**Ans.** L_1 = 160.88 mm and L_2 = 380.25 mm]

13.3 A 100 × 75 × 10 mm angle carrying a centroidal load of P = 247.5 kN is to be welded to a 12 mm thick gusset plate with 100 mm leg against the plate as shown in Figures 13.16(a) and (b). Determine the lengths of side fillet welds required at the heel and toe of the angle for non-eccentric connection when (a) no weld is provided at the end of the angle and (b) a fillet weld is added along the entire length of the end of the angle. All the welds are to be of the same size as the maximum permitted in the specifications. The distance of centroid from the heel is 31.9 mm. The permissible shearing stress through throat of each weld is 110 MPa. [**Hint:** $s = 7$ mm]

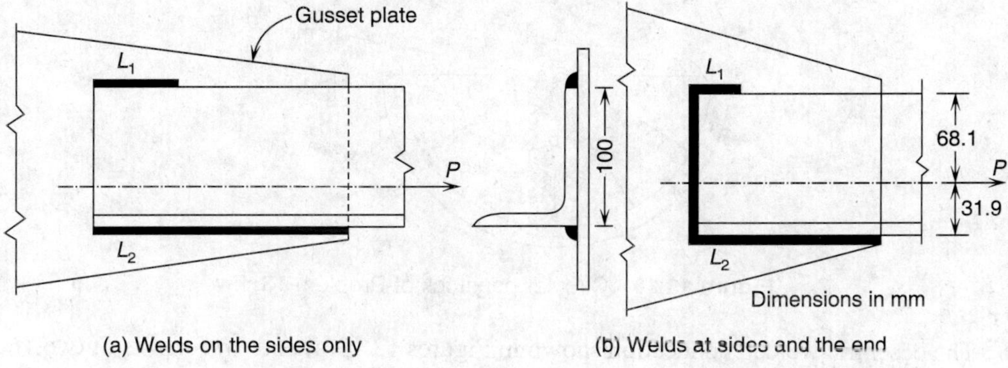

Figure 13.16 Design of connection of angle with a gusset of Problem 13.3.

[**Ans.** (a) $L_1 = 147$ mm; $L_2 = 313$ mm and (b) $L_1 = 97$ mm; $L_2 = 263$ mm]

13.4 An angle carrying load P acting through its centroid was to be welded to a plate with specified lengths of 8 mm welds as shown in Figure 13.17(a), but a welder applies them as shown in Figure 13.17(b). Using the design load computed from (a) determine the maximum load per millimetre of weld in (b). The allowable shearing stress through throat of each weld is 110 MPa.

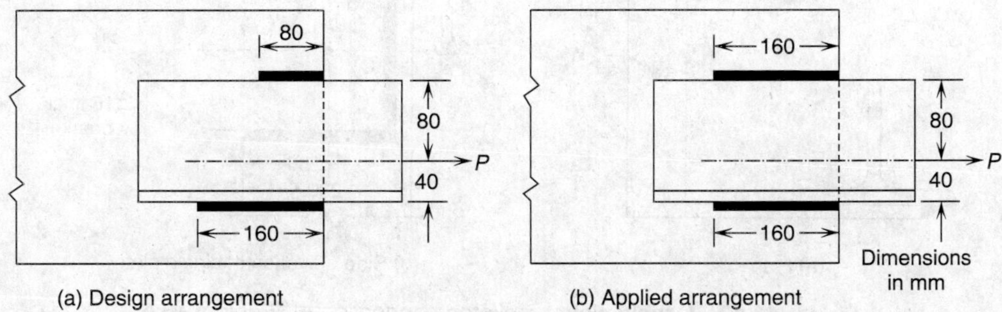

Figure 13.17 Angle welded to plate Problem 13.4.

[**Ans.** (a) $P = 149.318$ kN; (b) Find maximum load per mm due to eccentric load]

13.5 A circular penstock of 1.5 diameter is fabricated from 16 mm thick plate by lapping over and securing it by transverse fillet welds on the inside and outside of the penstock as shown in Figure 13.18. Determine the safe internal pressure, if the permissible tensile stress in the plates and shearing stress through throat of the weld are 150 MPa and 110 MPa, respectively. Use the welds of the maximum size as permitted in the specifications.

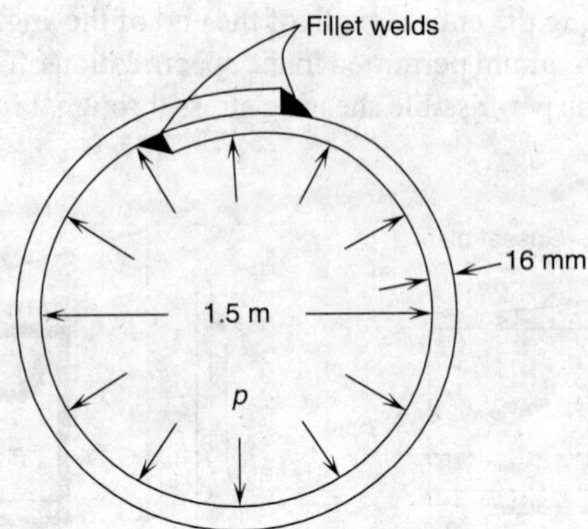

Figure 13.18 Circular penstock of Problem 13.5.

13.6 The eccentric welded connection shown in Figures 13.19(a) and (b) consists of two 16 mm thick bracket plates welded to the flanges of a column. The plates are welded to the flanges by shop fillet welds on three sides. The bracket has to transmit an end reaction of 360 kN to the column. Determine the size of the welds to be used. The permissible shearing stress through throat of each normal shop fillet weld is 108 MPa.

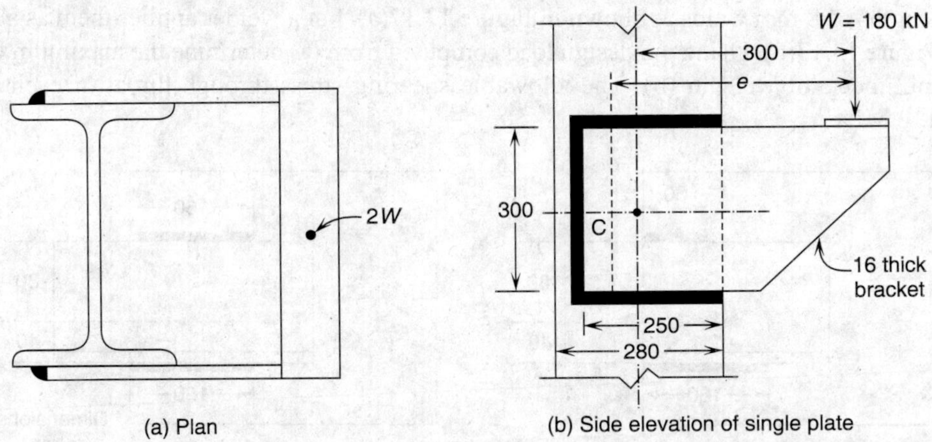

Figure 13.19 Double plate bracketed connection of Problem 13.6.

[**Ans.** $e = 331.9$ mm; $J = 1.903 \times 10^7$ mm^4; $s = 12$ mm]

13.7 A plate bracket of depth h and thickness t is butt welded to the flange of a column. The bracket transmits an end reaction of W at a distance e from the face of the column as shown in Figure 13.20. Show that the equivalent design load per unit length of the weld is

$$\frac{W(36e^2 + 3h^2)^{1/2}}{h^2}.$$

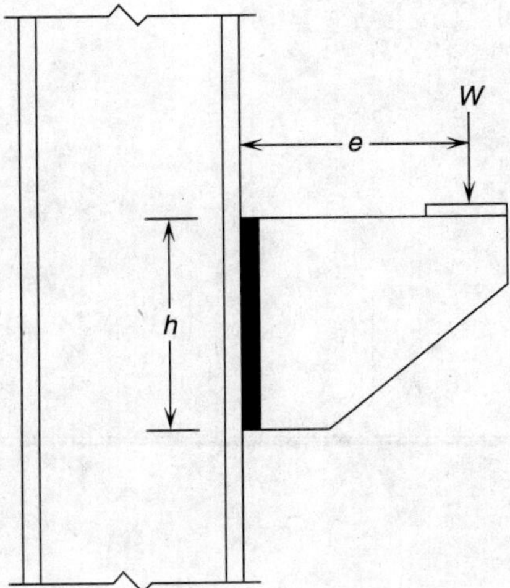

Figure 13.20 Eccentrically loaded butt-welded plate bracket of Problem 13.7.

[**Ans.** The loads are, $q_d = (W/h)$; $q_b = (6We/h^2)$ and $q_{es} = (q_b^2 + 3q_d^2)^{1/2}$]

Chapter 14

Deflections

(Direct Integration and Geometrical Methods)

14.1 INTRODUCTION

A structure when loaded, its material is subjected to internal stresses which cause it to deform. These deformations in the materials result into certain movements in the structure, called *deflections*. The deformations in the structure are caused by *bending moments*, *axial forces*, and *shearing forces* developed due to externally applied loads. In beams and frames the deformations are caused primarily by bending moments, the deformations caused by axial and shearing forces are generally small and ignored. In *trusses*, however, deformations are mainly due to axial extensions or shortenings of members which cause distortion in the shape of truss. In this chapter, the deformations caused by shearing forces are neglected as they are quite small in almost all beam-like structures. For the commonly used beam with depth-to-span ratio of 1/12 to 1/6, the shear deflections are in the range of 1 per cent to 8 per cent of bending deflections. These deformations or deflections disappear when the load is removed, provided, the elastic limit of the material is not exceeded. Such deflections are referred to as *elastic*. The deflections of the structure due to settlement of supports, lack of fit of members, shrinkage and creep in concrete structures which are permanent in nature, are referred to as *inelastic deformations*.

The deflections in a structure are evaluated to check that they satisfy the *limit state of serviceability*, i.e. deflections do not exceed the design limitations for satisfactory performance of structure. For example, the deflection of the floor of a building is limited so as to minimize cracking of plaster in the ceiling. Most of the national codes have placed limitations on horizontal and vertical deflections in various structural elements. The deflections in a structure are also computed for the analysis of statically indeterminate structures, as the static equilibrium equations alone are not sufficient to determine the unknown redundant forces. The deflections provide additional conditions of *consistent deformations* or *compatibility*. Knowledge of deflections in a structure is also required for the analysis of dynamic characteristics of a structure and vibration control.

There are a number of methods for computing deflections, but in this chapter, only geometric and energy or virtual work or unit load methods are discussed to develop the concept and through understanding. In geometric methods, the deformations are obtained directly from the strains in the structure, whereas, the energy methods are based on the principle of *conservation of energy*. However, both the methods provide identical results. These methods are generally used to obtain the displacements and rotations at isolated points on the structure. In this chapter, both the linear displacements of points (deflections) and rotational displacements of the lines (slopes) have been discussed.

14.2 DEFORMED SHAPES OF THE STRUCTURES

The practice of drawing qualitative sketch for the expected deformed shape of the structure before making actual calculations is beneficial in understanding the displacement behaviour of the structure under the applied loads. In drawing the qualitative expected deformed shape certain basic requirements must be fulfilled. For example, the deflected shape of an elastic member is a continuous smooth curve called *elastic curve*, which must satisfy the support conditions and connection between members.

The following simple rules are useful in obtaining a reasonable qualitative indication of the deflection response of flexural members, i.e. beams and frames.

14.2.1 Members

- A member deforms in the direction of the external load acting on it.
- The member end and the connecting joint rotate by the same amount and in the same direction, i.e. the elastic curve is continuous, unless there is a hinge between the member and the joint.
- The deflections are small and members retain their original lengths.
- The members with higher *stiffness* (EI/L) undergo smaller deformations as compared to the members with lower stiffness, e.g. the short stocky members will undergo smaller deformations than the long *slender members*.
- The deflected shapes of the loaded members are sketched first, followed in order by that of joints and unloaded members.

14.2.2 Joints

- A joint in a structure is assumed to be rigid, i.e. it does not change its size or shape as it undergoes displacement. The whole joint displaces as a unit, during displacement the joint rigidity causes the members to retain their initial angular relation to one another. The amount of rotation that occurs depends on the *joint stiffness* that in turn depends on the *relative stiffness* of the members meeting at the joints.
- The displacement of a joint depends on the external support conditions. A joint at a fixed support can neither translate nor rotate, whereas a joint at hinged support can rotate but cannot translate in any direction. On the other hand, a roller connection allows rotation freely but does not allow movement perpendicular to the surface on which the roller is mounted. However, the roller can translate parallel to the mounting surface.

The sketch of the elastic curve is drawn carefully keeping in mind the above simple rule that as no deflection is possible at a pinned or a roller support, the elastic must pass through such supports. At a fixed support, neither displacement nor rotation of the tangent to the elastic curve is permitted; elastic curve must be drawn tangent to the direction of the unloaded axis. The qualitative deflected shapes of some loaded beams are drawn in Figure 14.1. The following examples will illustrate the procedure for drawing the qualitative deflected shape of structures.

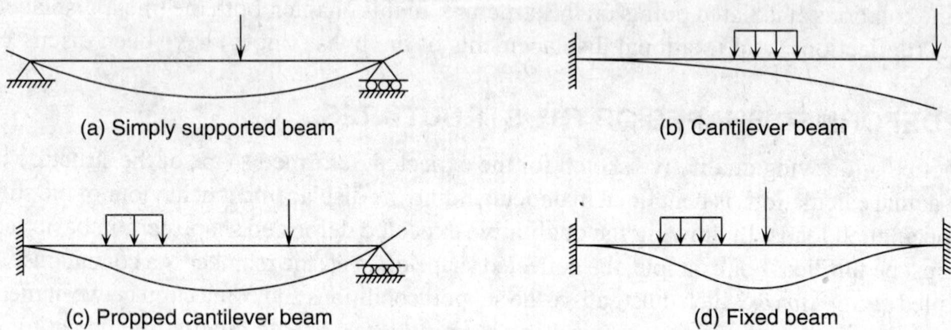

Figure 14.1 Qualitative deformed shapes of typical loaded beams.

EXAMPLE 14.1

Draw the qualitative deflected shape for the three-span continuous beam subjected to concentrated and distributed loads as shown in Figure 14.2(a).

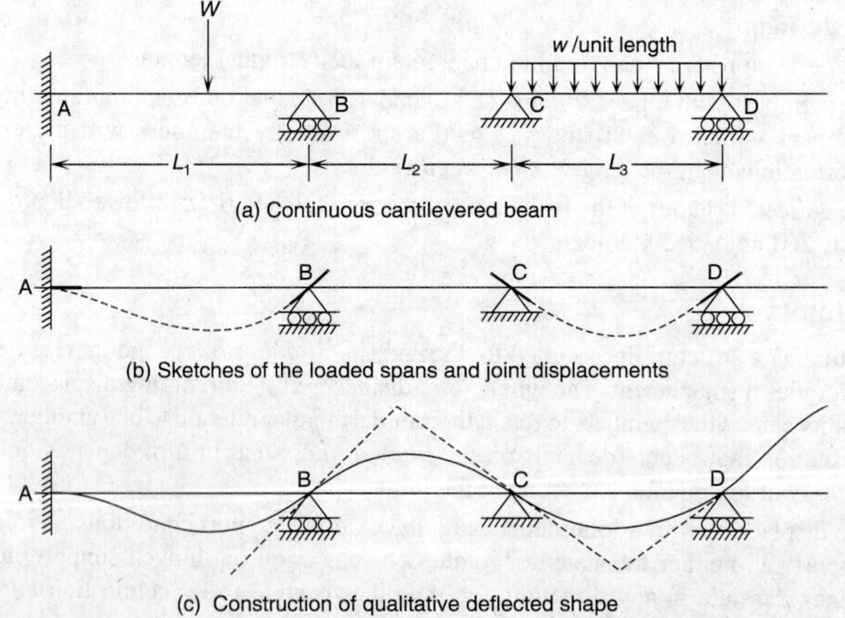

Figure 14.2 Qualitative deflected shape of the beam of Example 14.1.

Solution: An inspection of the structure indicates the salient features of the deflection curve as:

The spans A–B and C–D will deflect downward due to externally applied downward loads. The end A of member A–B can neither rotate nor translate due to fixity at A; however; the right end B can rotate but cannot translate vertically. Both the ends of loaded span C–D can rotate but can not translate vertically.

With the above information sketch the deformed shapes of loaded spans followed by the sketch of displacements of joints A, B, C and D as illustrated in Figure 14.2(b). Now sketch the deformed shape of the middle span B–C. Since the member has no external load acting on it, it deflects only in response to the displacements of the joints B and C to which it is connected. While sketching the deformed shapes of the members meeting at a joint, it should be insured that they have the same slope or slope continuity as shown in Figure 14.2(c).

EXAMPLE 14.2

Draw the qualitative deflected shapes for the braced or non-sway and unbraced rigid frames subjected to external loads as shown in Figures 14.3(a) and 14.4(a), respectively.

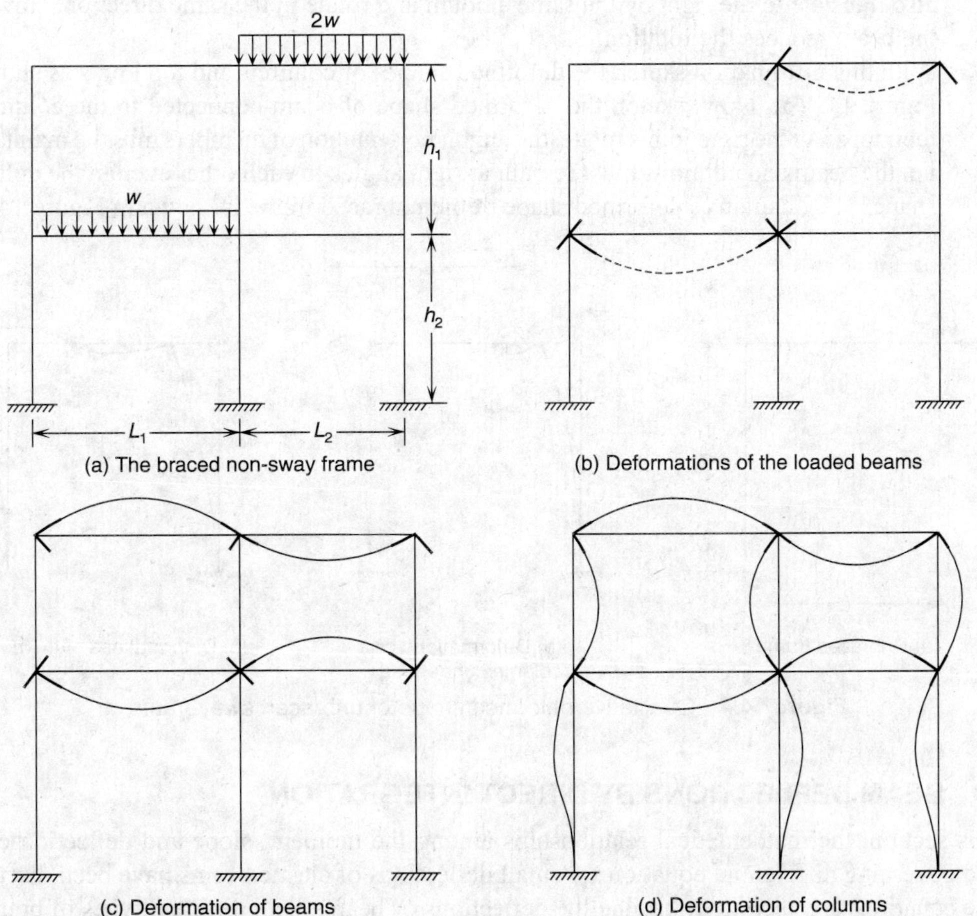

Figure 14.3 Qualitaive deflected shape of the non-sway (braced) frame.

Solution: Though the columns carry axial loads, the axial deformations are ignored.

(a) *Braced rigid frame*: In braced frames, the joints will not translate relative to each other. As usual the first step is to draw the defected shapes of loaded members which deform in the direction of loads. Based on these defected shapes, indicate the directions of rotation of connecting end joints as shown in Figure 14.3(b). Now draw the defected shapes of other members connected to the joints with known displacement directions while maintaining the continuity of slope as shown in Figure 14.3(c).

Since the columns have no external load acting on them, they deflect only in response to the displacements of the joints of the beams to which they are connected. The deformed shapes of the columns connected to the beams through the joints are sketched while maintaining the continuity of slope at the joints. The resulting qualitative deformed shape of the braced rigid frame is shown in Figure 14.3(d).

(b) *Unbraced rigid frame*: In this frame, the joints translate relative to one another. Due to the applied lateral load, the top of left column will translate to the right and will rotate. Since the beam is considered to maintain its original length the top of right column will also translate to the right by the same amount and rotate in the same direction. However, the beam reduces the rotation.

With this information sketch the deformed shapes of columns and top joints as shown in Figure 14.4(b). Now, sketch the deformed shape of beam connected to the columns at their tops. As the rigid joints rotate, the angular orientation of members must be maintained, i.e. the beam and columns must remain at right angles to each other even in the deflected shape. The qualitative deformed shape of the unbraced frame is shown in Figure 14.4(c).

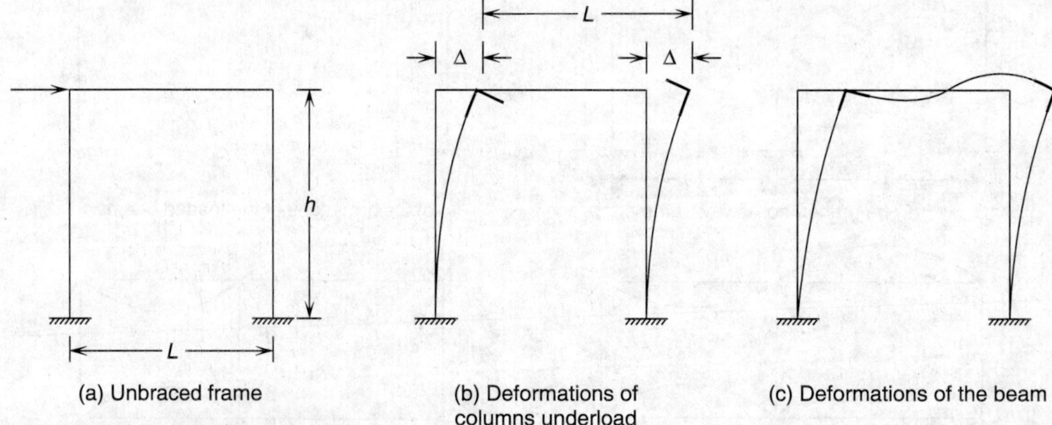

(a) Unbraced frame (b) Deformations of columns underload (c) Deformations of the beam

Figure 14.4 Qualitative deformed shape of unbraced sway-frame.

14.3 BEAM DEFLECTIONS BY DIRECT INTEGRATION

In this section, the mathematical relationships among the moment, slope and deflection curves called governing differential equation for small deflections of elastic beams have been discussed. These equations are used to determine the deflections of beams with different types of boundary and load conditions.

The general approach of *successive integration* presented in this section, allows the determination of deflection at any point on the beam. This approach has the following advantages:

1. The approach results in a set of equations for deflection at all parts of the beam, from which the deflection at any point can be determined.
2. The equations for deflection can be plotted to obtain the shape of elastic curve of deflection.
3. The equations for deflection emphasize the fundamental relationships among loads, type of support, flexural stiffness of beam, slope, and deflections. These relationships would help in evolving efficient designs.
4. The maximum deflection and its location can be found directly from the resulting equations.

14.3.1 Governing Differential Equation

The differential relations among the applied loads, shearing force and moment are usually combined to obtain various types of governing differential equations. The sequence of equations generally followed is,

$$y(z) = \text{Deflection at point on the elastic curve} \tag{14.1}$$

$$\theta(z) = \frac{dy(z)}{dz} = y'(z) = \text{ Slope at a point on elastic curve} \tag{14.2}$$

$$EI\frac{d^2y(z)}{dz^2}[= EIy''(z)] = M(z) = \text{ Bending moment at a section} \tag{14.3}$$

$$\frac{dM(z)}{dz} = EI\frac{d^3y(z)}{dz^3}[= EIy'''(z)] = V(z) = \text{ Shear at a section} \tag{14.4}$$

$$\frac{dV(z)}{dz} = EI\frac{d^4y(z)}{dz^4}[= EIy''''(z)] = q(z)$$

$$= \text{Load intensity at a section} \tag{14.5}$$

The following three alternative *governing differential equations* for determining the deflection of loaded beams with constant flexural rigidity (EI) are generally used:

Second-order differential equation, $EI\dfrac{d^2y(z)}{dz^2} = M(z)$

Third-order differential equation, $EI\dfrac{d^3y(z)}{dz^3} = V(z)$

Fourth-order differential equation, $EI\dfrac{d^4y(z)}{dz^4} = q(z)$

The choice of the equation depends upon the ease with which an expression for load, shearing force or moment can be written. The second- and fourth-order governing differential equations are commonly used, since it is more convenient to start a solution either with load function $q(z)$ or the moment function $M(z)$. A lower-order differential equation needs fewer constants of integration.

14.3.2 Boundary Conditions

For the solution of beam-deflection problems, the boundary conditions must be prescribed. There are various types of *boundary conditions*. The boundary conditions pertaining to the force quantities are termed *static boundary conditions* and those describing geometrical or deformational behaviour of an end are called *kinematic boundary conditions*. Non-homogeneous boundary conditions involving a given shear, moment, rotation, or displacement are also encountered in practice. These boundary conditions apply both to statically determinate and indeterminate beams. The boundary conditions generally encountered in practice are;

1. *Clamped* or *fixed support*: At such a support, the displacement y and the slope dy/dz vanish. Hence, at the end where $z = a$,

$$y(a) = 0 \quad \text{and} \quad \left(\frac{dy}{dz}\right)_{z=a} = y'(a) = 0 \tag{14.6}$$

   where, as usual, a prime ($'$) denotes differentiation with respect to z.

2. *Roller* or *pinned support*: In this case, neither deflection nor moment M can exist. Hence,

$$y(a) = 0 \quad \text{and} \quad M(a) = EI\,y''(a) = 0 \tag{14.7}$$

3. *Free end*: Such an end is free of moment and shear. Hence,

$$M(a) = EI\,y''(a) = 0 \quad \text{and} \quad V(a) = (EI\,y''')_{z=a} = 0 \tag{14.8}$$

4. *Guided support*: In such a support, free movement in guided direction is permitted, but the rotation of the end is prevented. The support is not capable of resisting any shear. Therefore,

$$y'(a) = 0 \text{ and } V(a) = (EI\,y''')_{z=a} = 0 \tag{14.9}$$

The above boundary conditions for beams with constant EI are summarized in Figure 14.5. In Figure 14.5(a) both conditions are kinematic; in Figure 14.5(c) both are static; whereas in Figures 14.5(b) and (d) the conditions are mixed.

14.3.3 Procedure

The general procedure for computation of slopes and deflections in the beams using the method of successive integration involves the following steps:

1. Determine the reactions at the supports of the beam.
2. Draw the bending moment diagram and identify the magnitude at the critical points.
3. If the loading, shearing force, and moment functions are continuous and flexural rigidity EI is constant, the evaluation of integration constants is direct. When discontinuities occur, divide the beam into segments in which the bending moment functions are continuous by designating the points where abrupt changes occur with the letters A, B, C etc.

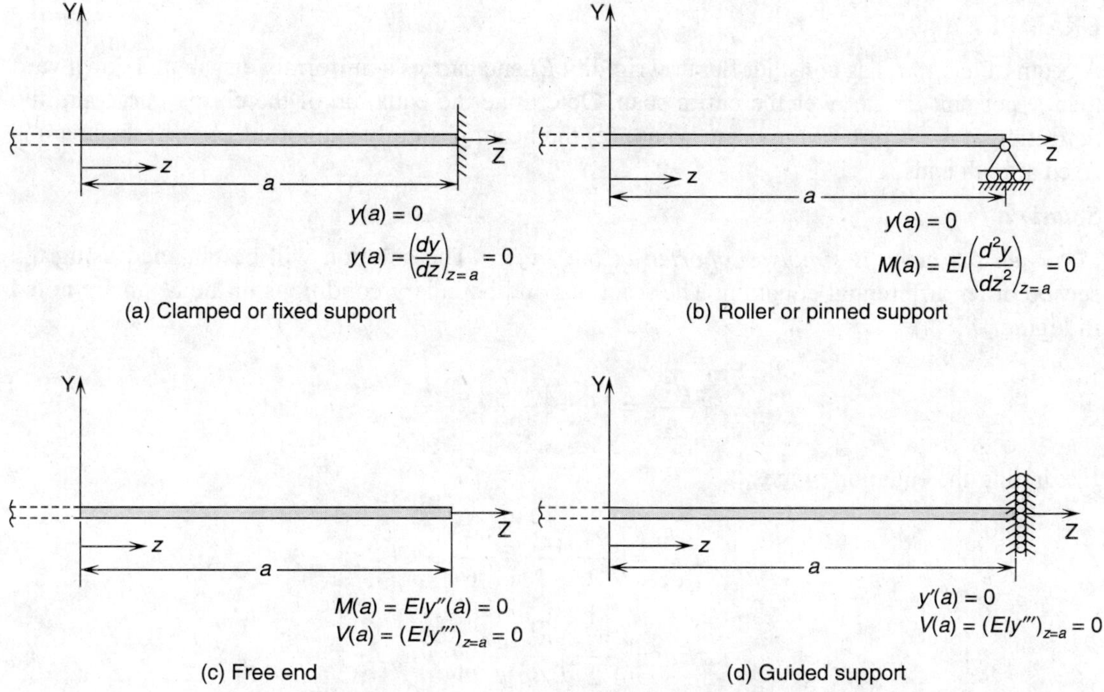

Figure 14.5 Common boundary conditions.

4. Formulate relevant functions, e.g. bending moment expression for the second-order governing differential equations and load variations for the fourth-order governing differential equations, in each segment.
5. Integrate the governing differential equations twice in case of the second-order and four times in case of the fourth-order differential equations for each segment; identify the constants of integration separately by subscripted letters such as C_1, C_2, C_3 and C_4.
6. Establish the boundary conditions for the slope and deflection diagrams. The number of boundary conditions must be equal to the number of unknown constants. When the beam is divided into segments due the presence of discontinuities, the continuity or compatibility conditions at the common boundaries of the beam segments are enforced, i.e. for the continuity of elastic curve, the slope/deflection of the beam at the end of one segment must be equal to the slope/deflection of the beam at the beginning of the next segment.
7. Evaluate the constants of integration by using the boundary and compatibility conditions. This may involve the solution of a set of simultaneous equations.
8. Substitute back the values of the constants of integration into the displacement relations to obtain the elastic curve equations.
9. Determine the desired quantities like the maximum deflection and its location, etc.

Several illustrative examples of statically determinate and statically indeterminate beam problems are given below.

EXAMPLE 14.3

A beam of length L has constant flexural rigidity EI and carries a uniformly distributed downward load w per unit length over the entire span. Determine the equation of the elastic curve and the deflection at the centre of the beam when: (a) the beam is simply supported, and (b) the beam is fixed at both ends.

Solution:

(a) *When the beam is simply supported at the ends:* The solution will be obtained using the second order differential equation. The reactions and boundary conditions of the beam are noted in Figure 14.6(a).

$$EI\frac{d^2y}{dz^2} = M_z = \left(\frac{wL}{2}\right)z - \frac{wz^2}{2}$$

Integrating the equation twice,

$$EI\frac{dy}{dz} = \left(\frac{wL}{2}\right)\frac{z^2}{2} - \frac{wz^3}{6} + C_1$$

$$EIy = \left(\frac{wL}{2}\right)\frac{z^3}{6} - \frac{wz^4}{24} + C_1 z + C_2 \qquad \text{(i)}$$

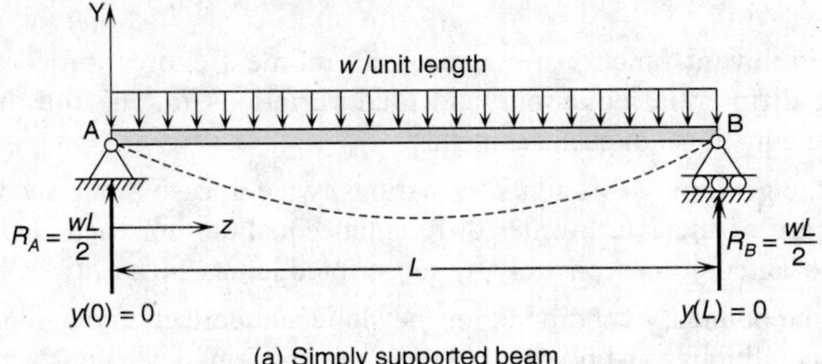

(a) Simply supported beam

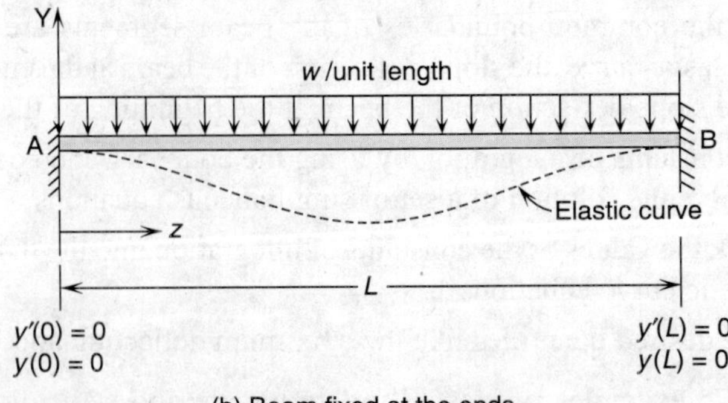

(b) Beam fixed at the ends

Figure 14.6 Loading and boundary conditions of the beams of Example 14.3.

Boundary conditions:

$$EIy(0) = EIy_A = 0 = C_2 = 0$$

$$EIy(L) = EIy_B = \left(\frac{wL}{2}\right)\frac{L^3}{6} - \frac{wL^4}{24} + C_1 L = 0 \quad \text{or} \quad C_1 = -\frac{wL^3}{24}$$

Therefore, the equation of the elastic curve is

$$EIy = \left(\frac{wL}{2}\right)\frac{z^3}{6} - \frac{wz^4}{24} - \frac{wL^3}{24}z = -\frac{w}{24}(L^3 z - 2Lz^3 + z^4)$$

or

$$y = -\frac{w}{24El}(L^3 z - 2Lz^3 + z^4) = -\frac{wL^4}{24El}(\xi - 2\xi^3 + \xi^4)$$

where $\xi = z/L$. Because of symmetry, maximum deflection occurs at $z = L/2$ or $\xi = 1/2$.

Therefore,

$$y_{\max} = -\frac{wL^4}{24EI}\left[\left(\frac{1}{2}\right) - (2)\left(\frac{1}{2}\right)^3 + \left(\frac{1}{2}\right)^4\right] = -\frac{5wL^4}{384EI}\downarrow$$

(a) *When the beam is fixed at the ends*: The reactions and boundary conditions of the beam are noted in Figure 14.6(b). The solution will be obtained using the fourth-order differential equation given below:

$$EI\frac{d^4 y}{dz^4} = q(z) = -w$$

Integrating the equation four times,

$$EI\frac{d^3 y}{dz^3} = -wz + C_1 \tag{i}$$

$$EI\frac{d^2 y}{dz^2} = -\frac{wz^2}{2} + C_1 z + C_2 \tag{ii}$$

$$EI\frac{dy}{dz} = -\frac{wz^3}{6} + C_1\frac{z^2}{2} + C_2 z + C_3 \tag{iii}$$

$$EIy = -\frac{wz^4}{24} + C_1\frac{z^3}{6} + C_2\frac{z^2}{2} + C_3 z + C_4 \tag{iv}$$

Boundary conditions: Four kinematic boundary conditions are available for determining the coefficients of integration.

$$EIy(0) = EIy_A = 0 \quad \Rightarrow \quad C_4 = 0$$

$$EIy'(0) = EIy_A' = 0 \quad \Rightarrow \quad C_3 = 0$$

$$EIy(L) = EIy_B = -\frac{wL^4}{24} + C_1\frac{L^3}{6} + C_2\frac{L^2}{2} = 0 \tag{v}$$

$$EIy'(L) = EIy_B' = -\frac{wL^3}{6} + C_1\frac{L^2}{2} + C_2L = 0 \quad \text{(vi)}$$

By solving the Eqs. (v) and (vi) simultaneously,

$$C_1 = \frac{wL}{2} \quad \text{and} \quad C_2 = -\frac{wL^2}{12}$$

By substituting these constants into the Eq. (iv) for elastic curve,

$$EIy = -\left(\frac{wz^4}{24}\right) + \left(\frac{wL}{2}\right)\left(\frac{z^3}{6}\right) + \left(-\frac{wL^2}{12}\right)\left(\frac{z^2}{2}\right) = -\left(\frac{wz^2}{24}\right)(z^2 - 2Lz + L^2)$$

$$= -\left(\frac{wz^2}{24}\right)(L-z)^2 = -\left(\frac{wL^4}{24}\right)[\xi^2(1-\xi)^2]$$

where $\xi = z/L$. It should be noted that since EI times the second and the third derivative of deflection $y(z)$ represent, respectively, $M(z)$ and $V(z)$; at $z = 0$, these relations define the reactions at the support A. Hence, C_1 is the vertical reaction and C_2 is the moment at the support.

Because of symmetry maximum deflection occurs at $z = L/2$ or $\xi = 1/2$.

Therefore, $$y_{\max} = \frac{wL^4}{24EI}\left(\frac{1}{2}\right)^2\left(1-\frac{1}{2}\right)^2 = -\frac{wL^4}{384EI}\downarrow$$

EXAMPLE 14.4

A two-span continuous beam ABC with each span of length L carries a uniformly distributed downward load w per unit length over the entire length as shown in Figure 14.7(a). The flexural rigidity EI is constant. Determine the equation of the elastic curve of the beam.

Solution: The solution can be obtained using the second order differential equation. Since the structure has both geometric and load symmetry, the deflection and slope at the middle support are zero. Thus, the problem reduces to the one-degree statically indeterminate problem. The unknown reaction at the end support A is the redundant action. The boundary conditions of the beam are noted in Figure 14.7(b).

$$EI\frac{d^2y}{dz^2} = M_z = R_Az - \frac{wz^2}{2}$$

Integrating the equation twice,

$$EI\frac{dy}{dz} = R_A\frac{z^2}{2} - \frac{wz^3}{6} + C_1$$

$$EIy = R_A\frac{z^3}{6} - \frac{wz^4}{24} + C_1z + C_2 \quad \text{(i)}$$

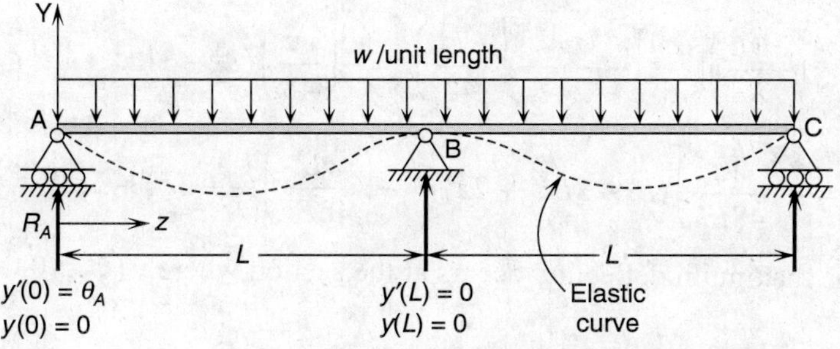

(a) Two-span continuous beam

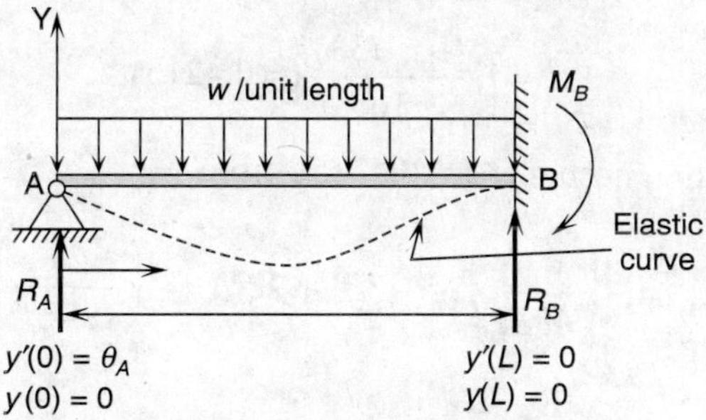

(b) Idealized beam for analysis

Figure 14.7 Analysis of continuous beam of Example 14.4.

Boundary conditions:

$$EIy(0) = EIy_A = 0 = C_2 = 0$$

$$EIy'(L) = R_A \frac{L^2}{2} - \frac{wL^3}{6} + C_1 = 0 \tag{ii}$$

$$EIy(L) = EIy_B = R_A \frac{L^3}{6} - \frac{wL^4}{24} + C_1 L = 0 \tag{iii}$$

By solving Eqs. (ii) and (iii):

$$C_1 = \frac{wL^3}{48} \quad \text{and} \quad R_A = \frac{3wL}{8}$$

Therefore,

$$R_C = R_A = \frac{3wL}{8} \quad \text{and} \quad R_B = 2wL - (R_C + R_A) = 2wL - \frac{2 \times 3wL}{8} = \frac{5wL}{4}$$

The equation of the elastic curve of the beam is obtained by substituting these values in Eq. (i),

$$EIy = R_A \frac{z^3}{6} - \frac{wz^4}{24} + C_1 z + C_2$$

$$= \left(\frac{3wL}{8}\right)\left(\frac{z^3}{6}\right) - \frac{wz^4}{24} - \left(\frac{wL^3}{48}\right)z = -\left(\frac{wz}{48}\right)(L^3 - 3Lz^2 + 2z^3)$$

Therefore, $$y = -\left(\frac{wz}{48EI}\right)(L^3 - 3Lz^2 + 2z^3) = -\left(\frac{wL^4}{48EI}\right)(\xi - 3\xi^3 + 2\xi^4)$$

where $\xi = z/L$. The maximum deflection occurs at the section where $y'(\xi) = 0$, i.e.

$$\frac{d}{d\xi}(\xi - 3\xi^3 + 2\xi^4) = 1 - 9\xi^2 + 8\xi^3 = 0$$

or $$\xi = 1, \frac{(1 \pm \sqrt{33})}{16} (= 0.4215)$$

Thus, the maximum deflection occurs at $z = 0.4215L$ from A or C.

Therefore, $$y_{\max} = -\left(\frac{wL^4}{48EI}\right)(\xi - 3\xi^3 + 2\xi^4) = -\left(\frac{wL^4}{185EI}\right)$$

EXAMPLE 14.5

A simply supported beam ACB of length L carries anticlockwise moment at its left support A. The flexural rigidity EI varies along the length as shown in Figure 14.8. Determine the equation of the elastic curve of the beam.

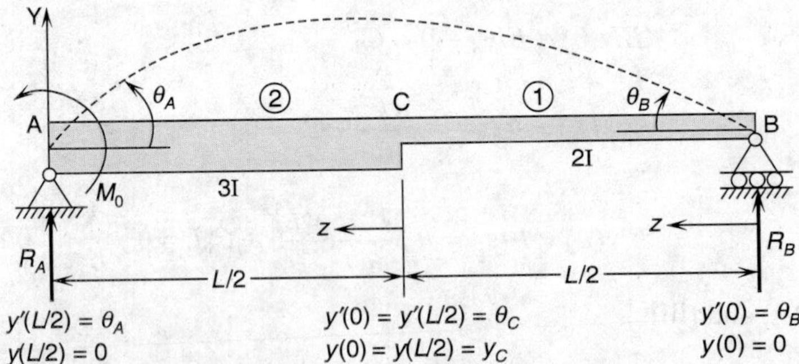

Figure 14.8 Boundary condition for the beam of variable section of Example 14.5.

Solution: In this problem, different M/EI expressions are applicable to segments AC and CB of the beam. The solution can be obtained by considering beam in two segments and enforcing the continuity conditions at the common boundary point C, in addition to usual boundary conditions. The origin is selected at one end of the beam for the first segment and for the second and subsequent segments at every junction of the segments and all z's are to be positive in the same direction. The second order differential equation has been used for the solution of the current problem.

Segment BC *with origin at* B, $0 < z < L/2$

For a section at distance z from B,

$$E(2I)\frac{d^2y}{dz^2} = M_z = \frac{M_0 z}{L}$$

Integrating the governing equation twice,

$$\frac{dy}{dz} = \frac{M_0 z^2}{2E(2I)L} + C_1$$

$$y = \frac{M_0 z^3}{6E(2I)L} + C_1 z + C_2$$

At $z = 0$; $\quad y(0) = y_B = 0$ and $\theta(0) = \theta_B$

Hence, $\quad C_1 = \theta_B$ and $C_2 = 0$

At $z = \left(\frac{L}{2}\right)$; $\quad y_C = \left(\frac{M_0 L^2}{96EI}\right) + \theta_B\left(\frac{L}{2}\right)$ and $\theta_C = \left(\frac{M_0 L}{16EI}\right) + \theta_B$

Segment CA *with origin at* C, $\quad 0 < z < L/2$

For a section at distance z from C,

$$E(3I)\frac{d^2y}{dz^2} = M_z = \frac{M_0}{L}\left(z + \frac{L}{2}\right)$$

Integrating the equation twice,

$$\frac{dy}{dz} = \frac{M_0}{E(3I)L}\left(\frac{z^2}{2} + \frac{Lz}{2}\right) + C_3$$

$$y = \frac{M_0}{E(3I)L}\left(\frac{z^3}{6} + \frac{Lz^2}{4}\right) + C_3 z + C_4$$

At $z = 0$; $\quad y(0) = y_C$ and $\theta(0) = \theta_C$

Hence, $\quad C_3 = \theta_C$ and $C_4 = y_C$

At $z = \left(\frac{L}{2}\right)$; $\quad y_A = \left(\frac{M_0 L^2}{36EI}\right) + \theta_C\left(\frac{L}{2}\right) + y_C$ and $\theta_A = \left(\frac{M_0 L}{8EI}\right) + \theta_C$

Apply boundary conditions at A,

$$y_A = \left(\frac{M_0 L^2}{36EI}\right) + \theta_C\left(\frac{L}{2}\right) + y_C$$

$$= \left(\frac{M_0 L^2}{36EI}\right) + \left[\left(\frac{M_0 L}{16EI}\right) + \theta_B\right]\left(\frac{L}{2}\right) + \left(\frac{M_0 L^2}{96EI}\right) + \theta_B\left(\frac{L}{2}\right)$$

or

$$= \left(\frac{5M_0 L^2}{72EI}\right) + \theta_B(L) = 0 \quad \text{or} \quad \theta_B = -\frac{5M_0 L}{72EI}$$

Therefore, $\theta_C = \left(\dfrac{M_0 L}{16EI}\right) + \theta_B = \left(\dfrac{M_0 L}{16EI}\right) - \dfrac{5M_0 L}{72EI} = -\dfrac{M_0 L}{144EI}$

$$\theta_A = \left(\frac{M_0 L}{8EI}\right) - \frac{M_0 L}{144EI} = \frac{17M_0 L}{144EI}$$

$$y_C = \left(\frac{M_0 L^2}{96EI}\right) + \theta_B\left(\frac{L}{2}\right) = \left(\frac{M_0 L^2}{96EI}\right) - \left(\frac{5M_0 L}{72EI}\right)\left(\frac{L}{2}\right) = -\frac{7M_0 L^2}{288EI}$$

For the part of the beam from B to C: $0 < z < L/2$

$$y = \frac{M_0 z^3}{6E(2I)L} - \frac{5M_0 L}{72EI} z = \frac{M_0 L^2}{72EI}(6\xi^3 - 5\xi)$$

For the part of the beam from C to A: $0 < z < L/2$

$$y = \frac{M_0}{E(3I)L}\left(\frac{z^3}{6} + \frac{Lz^2}{4}\right) - \frac{M_0 L}{144EI} z - \frac{7M_0 L^2}{288EI}$$

$$= \frac{M_0}{288EI}\left(16\frac{z^3}{L} + 24z^2 - 2Lz - 7L^2\right)$$

$$= \frac{M_0 L^2}{288EI}(16\xi^3 + 24\xi^2 - 2\xi - 7)$$

where $\xi = z/L$.

14.3.4 Macaulay Procedure

When discontinuities occur in loading, i.e. when the point loads are applied or sudden change in magnitude of uniformly distributed load occurs, the method of successive integration described above cannot be applied directly. To handle such situations, Macaulay introduced a mathematical procedure which consists of expressing the bending moment at a section as a function of $f\langle z - a\rangle$ where a is constant depending upon the position of a given load. The expression within the square bracket is ignored if its value is negative. This expression can be integrated in a usual manner by treating it as a single quantity, e.g.

$$\int \langle z - a\rangle\, dx = \frac{\langle z - a\rangle^2}{2} \tag{14.10}$$

In using this method, the following rules are adopted:

1. The origin for z is taken at the left end of the beam.
2. An expression for bending moment is written at the section just to the left of the right hand end of the beam by considering moment of all the forces to the left of the section.
3. In case of uniformly distributed load, it must extend to the right hand end of the beam. This can always be achieved by applying upward acting uniformly distributed load over extended portion of the load.

The following examples will illustrate the procedure.

EXAMPLE 14.6

A simply supported beam AB of length L carries a concentrated downward load W at a distance a from the left support A as shown in Figure 14.9. The flexural rigidity EI is constant. Find the equation of the elastic curve of the beam.

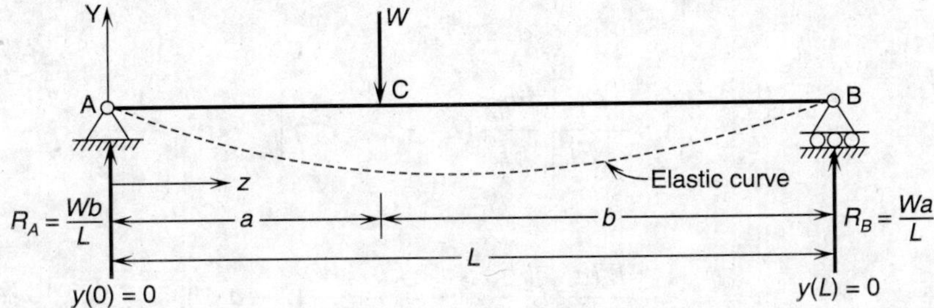

Figure 14.9 Simply supported beam subjected to a concentrated load of Example 14.6.

Solution: The solution will be obtained using the second order differential equation. The reactions and boundary conditions of the beam are noted in Figure 14.9. There is a discontinuity in M_z at $z = a$, requiring different functions for the two parts. However, the situation can be handled using Macaulay procedure. The bending moment at a point just to the left of B at a distance z from A is given by:

$$M_z = EI\frac{d^2y}{dz^2} = Wbz - W(z-a)$$

Integrating the equation twice,

$$EI\frac{dy}{dz} = \frac{Wb}{2L}z^2 - \frac{W}{2}\langle z-a\rangle^2 + C_1 \qquad \text{(i)}$$

$$EIy = \frac{Wb}{6L}z^3 - \frac{W}{6}\langle z-a\rangle^3 + C_1 z + C_2 \qquad \text{(ii)}$$

Boundary conditions: Substituting the boundary conditions in Eq. (ii):

$$EIy(0) = EIy_A = 0 = C_2 = 0$$

$$EIy(L) = \frac{Wb}{6L}L^3 - \frac{W}{6}\langle L-a\rangle^3 + C_1 L + C_2$$

$$= \frac{WbL^2}{6} - \frac{Wb^3}{6} + C_1 L = 0$$

or

$$C_1 = -\frac{WbL^2}{6L} + \frac{Wb^3}{6L}$$

Therefore, the equation of the elastic curve is

$$EIy = \frac{Wb}{6L}z^3 - \frac{W}{6}\langle z-a\rangle^3 + \left(-\frac{WbL^2}{6L} + \frac{Wb^3}{6L}\right)z$$

or
$$y = -\frac{1}{EI}\left[\frac{Wbz}{6L}(L^2 - z^2 - b^2) + \frac{W}{6}\langle z-a\rangle^3\right]$$

For segment AC:

$$y = -\frac{1}{EI}\left[\frac{Wbz}{6L}(L^2 - z^2 - b^2)\right] \tag{iii}$$

For segment CB:

$$y = -\frac{1}{EI}\left[\frac{Wbz}{6L}(L^2 - z^2 - b^2) + \frac{W}{6}(z-a)^3\right] \tag{iv}$$

Deflection at the point of applied force,

$$y_C = -\frac{1}{EI}\left[\frac{Wbz}{6L}(L^2 - z^2 - b^2)\right]_{z=a} = -\frac{1}{EI}\left[\frac{Wba}{6L}(L^2 - a^2 - b^2)\right]$$

$$= -\frac{1}{EI}\left[\frac{Wba}{6L}(2ab)\right] = -\frac{Wa^2b^2}{3EIL}\downarrow \quad \text{since } L = a + b$$

The maximum deflection occurs in the longer segment of the beam where slope equals zero. Therefore, for $a > b$

$$\frac{dy}{dz} = -\frac{1}{EI}\frac{d}{dz}\left[\frac{Wbz}{6L}(L^2 - z^2 - b^2)\right]$$

$$= -\frac{1}{EI}\left[\frac{Wb}{6L}(L^2 - 3z^2 - b^2)\right] = 0$$

or
$$L^2 - 3z_m^2 - b^2 = 0 \text{ or } z_m^2 = \frac{L^2 - b^2}{3} = \sqrt{\frac{a(a+2b)}{3}}$$

$$y_{\max} = -\frac{1}{EI}\left[\frac{Wbz}{6L}(L^2 - z^2 - b^2)\right]_{z=z_m} = -\frac{Wbz_m}{6EIL}(L^2 - 3z_m^2 - b^2 + 2z_m^2)$$

$$= -\frac{Wbz_m}{6EIL}(2z_m^2) = -\frac{Wbz_m^3}{3EIL} = -\frac{Wb}{3EIL}\left(\frac{L^2 - b^2}{3}\right)^{3/2}$$

$$= -\frac{Wb(L^2 - b^2)^{3/2}}{9\sqrt{3}EIL}$$

If the load is applied at the middle of span, when $a = b = L/2$, by substitution in the equation

$$y_{\max} = \frac{WL^3}{48EI}$$

EXAMPLE 14.7

A simply supported beam AB of length $4a$ has a constant flexural rigidity EI and carries a uniformly distributed downward load w per unit length over the second quarter of the span from the left support. Determine the deflection at the centre of the beam.

Solution: The solution will be obtained using the second order differential equation. The boundary conditions of the beam are noted in Figure 14.10(a). There is a discontinuity in M_z at $z = a$, and $z = 2a$ requiring different functions for the three parts. However, the situation can be handled easily by using Macaulay procedure. Consider the overall equilibrium of the beam to determine reactions,

$$\sum M_B = R_A(4a) - (wa)\left(\frac{5a}{2}\right) = 0 \quad \text{or} \quad R_A = \left(\frac{5wa}{8}\right)$$

Therefore, $$R_B = wa - R_A = \left(\frac{3wa}{8}\right)$$

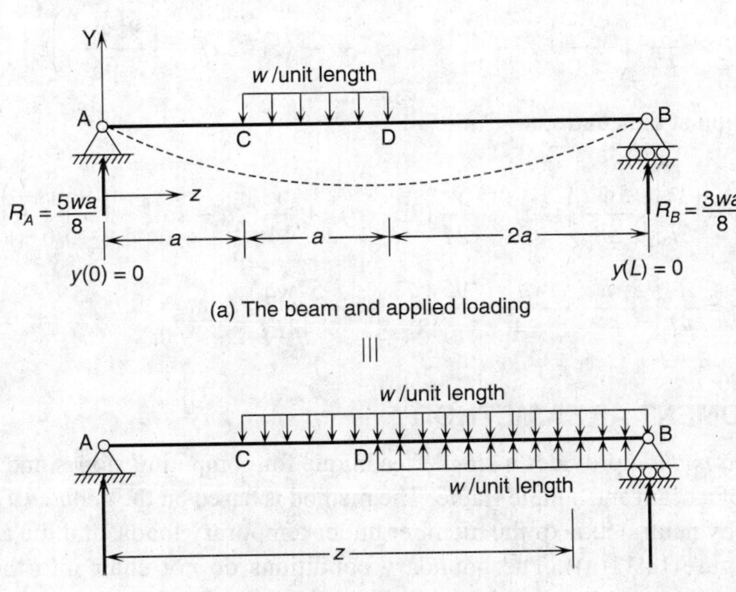

Figure 14.10 Analysis of partially loaded beam of Example 14.7 by Macaulay's method.

Using Macaulay procedure, the uniformly distributed downward acting load is extended to the right hand support B and to balance this additional load over the length DB an upward acting uniformly distributed load w is applied on the length from D to B as shown in Figure 14.10(b). Then bending moment at a point just to the left of B at a distance z from A is given by:

$$EI\frac{d^2y}{dz^2} = M_z = \frac{5wa}{8}z - \frac{w}{2}\langle z-a\rangle^2 + \frac{w}{2}\langle z-2a\rangle^2$$

Integrating the equation twice,

$$EI\frac{dy}{dz} = \frac{5wa}{16}z^2 - \frac{w}{6}\langle z-a\rangle^3 + \frac{w}{6}\langle z-2a\rangle^3 + C_1 \quad \text{(i)}$$

$$EIy = \frac{5wa}{48}z^3 - \frac{w}{24}\langle z-a\rangle^4 + \frac{w}{24}\langle z-2a\rangle^4 + C_1z + C_2 \quad \text{(ii)}$$

Boundary conditions: Substituting the boundary conditions in Eq. (ii):

$$EIy(0) = EIy_A = 0 = C_2 = 0 \text{ and}$$

$$EIy(4a) = \frac{5wa}{48}(4a)^3 - \frac{w}{24}(3a)^4 + \frac{w}{24}(2a)^4 + C_1(4a) = 0 \text{ or } C_1 = \frac{95wa^3}{96}$$

Therefore, from Eq. (ii) the expression of elastic curve is

$$y = \frac{1}{EI}\left[\left(\frac{5wa}{48}\right)z^3 - \frac{w}{24}\langle z-a\rangle^4 + \frac{w}{24}\langle z-2a\rangle^4 - \left(\frac{95wa^3}{96}\right)z\right]$$

Thus, for deflection at the centre, $z = 2a$

$$y = \frac{1}{EI}\left[\left(\frac{5wa}{48}\right)(2a)^3 - \frac{w}{24}(2a-a)^4 + \frac{w}{24}(2a-2a)^4 - \left(\frac{95wa^3}{96}\right)(2a)\right]$$

$$(y)_{z=2a} = \frac{1}{EI}\left[\frac{5wa^4}{6} - \frac{wa^4}{24} - \frac{95wa^4}{48}\right] = -\frac{57wa^4}{48EI}\downarrow$$

14.4 THE MOMENT-AREA METHOD

The *moment-area method* provides a simple technique for computing slopes and deflections at a particular point of a beam and simple frame. The method is based on the *geometry of elastic curve*, the shape taken by neutral axis of the member under temporary loads, and the associated M/EI diagram (see Figure 14.11(a)). The boundary conditions do not enter into the derivation of relationships since these are based only on the interpretation of definite integrals.

Consider an element of length dz along the neutral axis of the beam segment AB subjected to external loads, being on the neutral axis its length does not change on deformation. The bending moments developed due to external loads are positive and cause shortening of the beam fibres above the neutral axis and lengthening of those below the neutral axis.

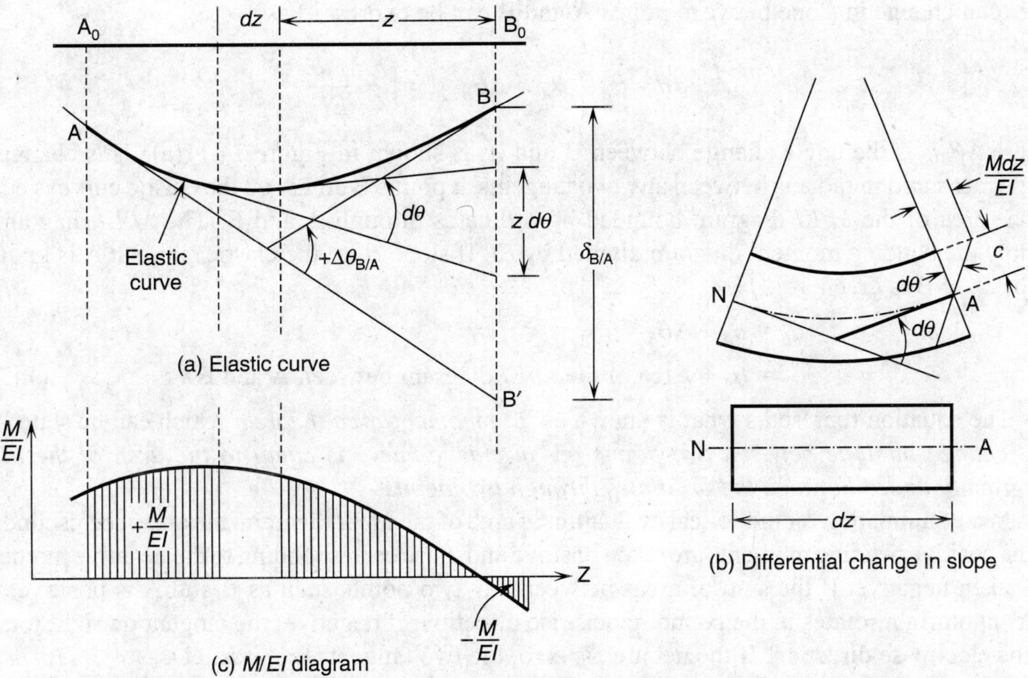

Figure 14.11 Relationship between the *M/EI* diagram and elastic curve.

In the elemental length *dz* the neutral axis changes direction by an amount $d\theta$ due to the effect of curvature as shown in Figure 14.11(b). The change in slope $d\theta$ can also be expressed as

$$\tan d\theta = d\theta = \frac{\text{Shortening of fibre}}{c}$$

$$= \frac{\text{Strain} \times dz}{c} = \frac{\varepsilon\, dz}{c} = \frac{(\sigma / E)\, dz}{c}$$

where the stress σ is given by flexural equation,

$$\sigma = \frac{Mc}{I}$$

Therefore,
$$d\theta = \frac{(Mc / EI)\, dz}{c} = \frac{M\, dz}{EI} \tag{14.11}$$

This change in slope $d\theta$ over a differential length *dz* can be obtained from the differential equation of the elastic curve of a flexural member, i.e.

$$\frac{d^2 y}{dz^2}\left[= \frac{d}{dz}\left(\frac{dy}{dz}\right) = \frac{d\theta}{dz}\right] = \frac{M}{EI} \quad \text{or} \quad d\theta = \frac{M\, dz}{EI}$$

since the slope of the elastic curve is expressed as $\tan\theta \approx \theta = \dfrac{dy}{dz}$.

The total change in slope between points A and B can be expressed as

$$\int_A^B d\theta = \theta_B - \theta_A = \Delta\theta_{B/A} = \int_A^B \frac{M\,dz}{EI} \tag{14.12}$$

where $\Delta\theta_{B/A}$ is the angle change between B and A as shown in Figure 14.11(a). This change in angle measured in radians between any two tangents at points B and A on the elastic curve is equal to the area of the *M/EI* diagram bounded by ordinates through A and B. The *M/EI* diagram is simply the bending moment diagram divided by *EI*. If slope θ_A of the elastic curve at A is known, slope θ_B at B is given as

$$\begin{aligned}\theta_B &= \theta_A + \Delta\theta_{B/A}\\ &= \theta_A + \text{Area of the } M/EI \text{ diagram between A and B}\end{aligned} \tag{14.13}$$

The equation represents what is known as *first moment-area theorem* which can be stated as: *The change in slope between two points on an elastic curve is equal to the area of the M/EI diagram bounded between the ordinates through two points.*

In the summation for numerical evaluation of area of the *M/EI* diagram, the areas corresponding to the positive bending moments are taken positive and those corresponding to the negative moments are taken negative. If the sum of areas between any two points such as B and A is positive, the tangent on right rotates in the counterclockwise direction; if negative; the tangent on right rotates in the clockwise direction. If the net area is zero, the two tangents are parallel.

The change of slope between the tangents to the elastic curve may be used for computing deflection between the tangents. The change in the neutral axis direction of elemental length due to effect of curvature results in a vertical displacement. The vertical displacement is obtained by multiplying $d\theta$ by a distance z from an arbitrary origin to the same element. Thus, the deflection between the tangents drawn at two points due to this angle change is the equal to

$$d\delta = z\,d\theta \tag{14.14}$$

where z is generally the distance from the point at which the deflection is desired to the point at which tangent is drawn. Thus, substituting the relation of Eq. 14.11 into Eq. 14.14 gives

$$d\delta = z\frac{M\,dz}{EI} = \frac{Mz\,dz}{EI} \tag{14.15}$$

By summing the effect for all elements from A to B, vertical distance BB′ is obtained. Geometrically, this distance represents the displacement of a point B from a tangent to the elastic curve at A. This displacement which is termed tangential deviation of a point B from a tangent at A is designated by $\delta_{B/A}$.

$$\delta_{B/A} = \int_A^B \frac{Mz\,dz}{EI} \tag{14.16}$$

The above equation is a mathematical statement of the *second moment-area theorem* which is stated as:

The deflection or *tangential deviation of a point* A *on the elastic curve of a beam from a tangent at another point* B *on the elastic curve is equal to the statical* (or *first*) *moment of the area of M/EI diagram between the two points about the point at which deflection is desired.*

For convenience of numerical applications Eq. 14.16 can be rewritten in terms of centre of gravity (C.G.) of the area as

$$\delta_{B/A} = \int_A^B \frac{Mz\,dx}{EI} = A_M \bar{z} \tag{14.17}$$

where A_M is the total area of the *M/EI* diagram between two points under consideration and $\bar{z}$ is the horizontal distance to the centroid of the area from B. Similarly, the tangential deviation of a point A on the elastic curve from a tangent at another point B is

$$\delta_{A/B} = A_M \bar{z}_1 \tag{14.18}$$

The sign of the tangential deviation depends on the sign of bending moments. A positive value for the tangential deviation indicates that a given point lies above a tangent to the elastic curve drawn through the other point, and vice versa.

14.4.1 Application of Moment-Area Theorems

When applying the moment-area theorems, it should be realized that the theorems are applicable between any two points on a *continuous curve* of any beam for any loading. It must be emphasized that the theorems do not always give directly the slope or deflection at a point in the beam with respect to the horizontal. Rather, only relative rotations of the tangents and tangential deviations are obtained directly, i.e. the theorems give the change in slope of the elastic curve from one point to another or the deflection of the tangent at one point with respect to the tangent at another point on a beam. Therefore, the method is very convenient for determining slopes and deflections of beams when the slope of the elastic curve at one point or more points is known. Such beams include cantilevers and symmetrically loaded beams, where the slopes at the fixed end and axis of symmetry, respectively, do not change.

In applying the moment-area method, a carefully drawn sketch of the elastic curve is required. Since no deflection is possible at a pinned or a roller support, the elastic curve passes through such supports. At a fixed support, neither displacement nor rotation of the tangent to the elastic curve is permitted; elastic curve must be drawn tangent to the direction of the unloaded axis. The deflection of a point on the beam is usually referred to as being above or below its initial position, i.e. upward or downward without emphasis on the signs.

The beams loaded with concentrated loads can be handled easily because the moment diagrams consist of straight lines. When the distributed and concentrated loads are applied simultaneously, the calculations may be simplified by drawing a separate *M/EI* diagram for each of the loads and computing the slopes and deflections caused by each load separately and adding the values obtained for the individual loads for the final results.

14.4.2 Procedure for Computing Deflection by Moment-Area Method

The following procedure for computing the deflection of any point on a beam uses the two moment-area theorems developed in the preceding section.

1. Draw the load and bending moment diagrams.
2. Draw the *M/EI* diagram by dividing the bending moment values by the beam stiffness, *EI*. The dimension for the quantity *M/EI* is $(\text{length})^{-1}$.

3. Compute the area of the *M/EI* diagram and locate its centroid. If the diagram is not of simple shape, break it into parts and compute the area and centroid of each part separately.

   In case of a cantilever, if the deflection at the end of the cantilever is to be determined, the area of the entire *M/EI* diagram is used, if the deflection at some other point is desired; only the area of the *M/EI* diagram between the support and the point under consideration is used.

   In case of a symmetrically loaded simply supported beam, if the maximum deflection at the centre of span is desired, the area of the *M/EI* diagram between the centre and one of the supports, i.e. half of the diagram, is used.

4. (i) *A cantilever beam*: Use Theorem 2 to compute the vertical deflection at the point of interest from the tangent to the beam at the support, because the tangent at the support coincides with the original position of the beam, the deflection thus found is the actual deflection of the beam at the point of interest. If all loads are in the same direction, the maximum deflection occurs at the end of cantilever.

   (ii) *A symmetrically loaded simply supported beam*: Use Theorem 2 to compute the maximum vertical deviation of the point at one of the supports from the tangent to the beam at its centre. Because the tangent to the beam at its centre is horizontal and since the deflection at the support is actually zero, the deviation thus computed is the actual maximum deflection of the beam at its centre.

   To compute the vertical deflection at some other point, the area of the *M/EI* diagram between the centre of the beam and the point under consideration is used. Theorem 2 is used to compute the vertical deviation of the point of interest from the tangent to the beam at its centre. Subtract this deviation from the maximum deflection at the centre.

   (iii) *An asymmetrically loaded simply supported beam*: Sketch the probable shape of elastic curve. Then, draw the tangent to the elastic curve at one of the supports. In accordance with Theorem 2, take moment of the entire *M/EI* diagram with respect to the second support to compute its vertical deviation from the tangent line at the other support.

   Compute the vertical distance at the point interest from the horizontal-axis to the tangent line by using proportions. From this distance, subtract the vertical deviation of the point of interest from the tangent. The latter deviation is computed by taking the moment of area of the *M/EI* diagram between the first support and the point of interest.

Several examples of the beam problems are given below to illustrate the application of the moment-area theorems.

EXAMPLE 14.8

Using the moment-area theorems, compute the slope and deflection at the free end of the cantilever beam AB of span L subjected to: (a) a concentrated load W at the free end B, (b) moment M at the free end and (c) uniformly distributed load over the entire span. The beam is of uniform cross-section with constant EI.

Solution: The tangent to the elastic curve at the fixed end is horizontal, i.e. the slope and deflection at the fixed end are zero. Therefore, the changes in the slope and deflection of the tangent at the free end, with respect to a tangent at the fixed end are the slope and vertical deflection of that point. Therefore,

$$\theta_B = \theta_A + \text{Area of the } M/EI \text{ diagram between the points A and B}$$

$$\delta_{B/A} = \text{Deflection at point B w.r.t. the tangent at point A}$$

$$= \text{Moment of the } M/EI \text{ diagram between A and B about B}$$

where $\theta_A = 0$. The bending moment and the corresponding M/EI diagrams for the cantilever subjected to various load conditions are shown in Figures 14.12(a), (b) and (c).

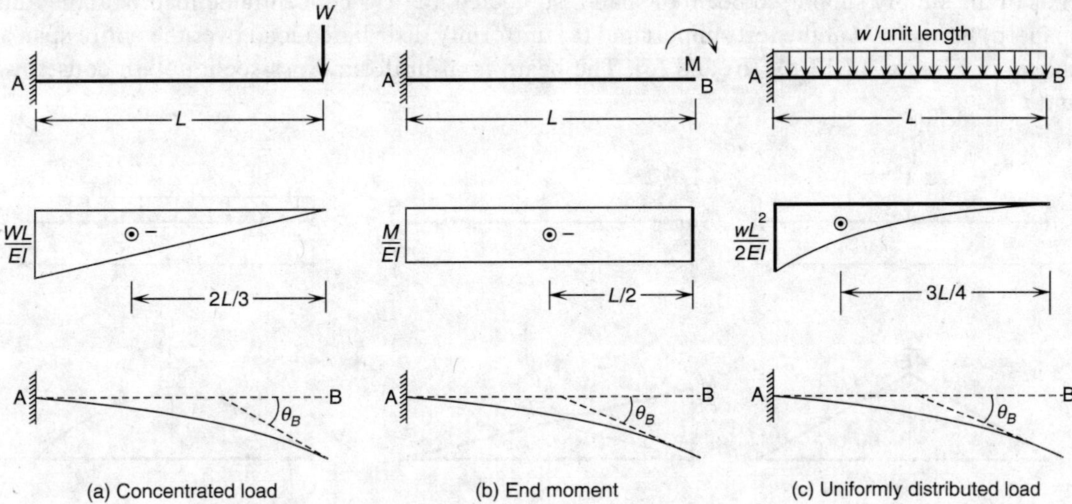

Figure 14.12 Cantilever beam subjected to different types of loadings of Example 14.8.

(a) *A concentrated load W at the free end*

$$\theta_B = \left(\frac{1}{2}\right) L \left(-\frac{WL}{EI}\right) = -\frac{WL^2}{2EI} \text{ (clockwise)}$$

The negative sign indicates that the second tangent rotates clockwise w.r.t. the first tangent.

$$\delta_{B/A} = \left(\frac{1}{2}\right) L \left(-\frac{WL}{EI}\right)\left(\frac{2L}{3}\right) = -\frac{WL^3}{3EI} \downarrow$$

The negative sign indicates that the point B moves down w.r.t. the first tangent.

(b) *Moment M at the free end*

$$\theta_B = L\left(-\frac{M}{EI}\right) = -\frac{ML}{EI} \text{ (clockwise)}$$

and

$$\delta_{B/A} = L\left(-\frac{M}{EI}\right)\left(\frac{L}{2}\right) = -\frac{ML^2}{2EI} \downarrow$$

(c) *Uniformly distributed load over the entire span*

$$\theta_B = \left(\frac{1}{3}\right) L \left(-\frac{wL^2}{2EI}\right) = -\frac{wL^3}{6EI} \text{ (clockwise)}$$

and
$$\delta_{B/A} = \left(-\frac{wL^3}{6EI}\right)\left(\frac{3L}{4}\right) = -\frac{wL^4}{8EI} \downarrow$$

EXAMPLE 14.9

Using the moment-area theorems, compute the deflection at the centre line and slopes at the ends of the simply supported beam of span L subjected to: (a) a concentrated load W at the mid-span, (b) moment M at the left support and (c) uniformly distributed load over the entire span as shown in Figures 14.13(a), (b) and (c). The beam is of uniform cross-section with constant E and I.

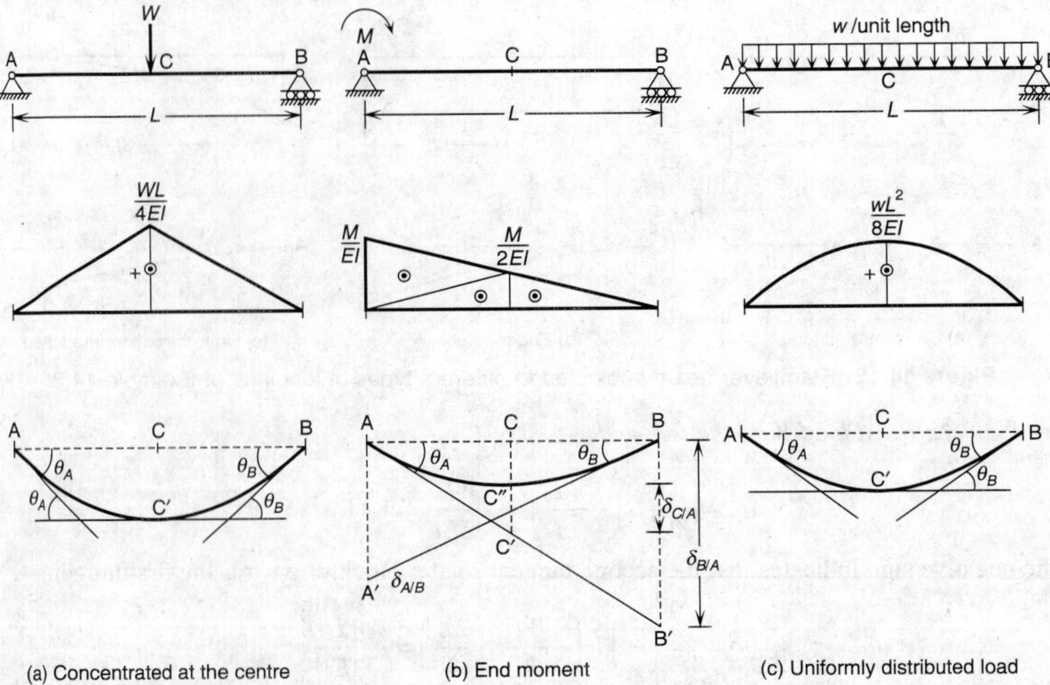

Figure 14.13 Simply supported beam subjected to different types of loads of Example 14.9.

Solution: The deformations of simply supported beam for different load conditions are computed as below.

(a) *A concentrated load W at the mid-span* C:

Because of symmetry in geometry and loading, the tangent at C is horizontal. The end slopes are equal to the area of the *M/EI* diagram between A and C,

$$\theta_B = -\theta_A = \left(\frac{1}{2}\right)\left(\frac{L}{2}\right)\left(\frac{WL}{4EI}\right) = \frac{WL^2}{16EI}$$

The minus sign is due to the fact that second tangent at A makes a clockwise rotation with respect to the first tangent at C.

The deflection of the point A w.r.t. the tangent at C is equal to the moment of the *M/EI* diagram betweenA and C about A.

$$\delta_{C/A} = \left(\frac{1}{2}\right)\left(\frac{L}{2}\right)\left(\frac{WL}{4EI}\right)\left[\frac{2}{3}\left(\frac{L}{2}\right)\right] = \frac{WL^3}{48EI} \uparrow$$

(b) *Moment M acting at the left support*

In the absence of a horizontal tangent to the elastic curve, the end slopes and deflection at the centre line cannot be computed directly. The indirect procedure used is illustrated in this example.

The deflection of the point B w.r.t. the tangent at A is equal to the moment of the *M/EI* diagram between A and B about B,

$$\delta_{B/A} = \left(\frac{1}{2}\right)(L)\left(\frac{M}{EI}\right)\left[\frac{2L}{3}\right] = \frac{ML^2}{3EI}$$

Since θ_A is very small, consider AB′ = AB = L, therefore,

$$\theta_A = \frac{\delta_{B/A}}{L} = \frac{ML}{3EI}$$

Similarly, by taking moment of the *M/EI* diagram between A and B about A,

$$\delta_{A/B} = \left(\frac{1}{2}\right)(L)\left(\frac{M}{EI}\right)\left[\frac{L}{3}\right] = \frac{ML^2}{6EI}, \quad \text{and} \quad \theta_B = \frac{\delta_{A/B}}{L} = \frac{ML}{6EI}$$

The deflection at the mid-point C, from Figure 14.13(b):

$$\delta_C = CC' - C'C'' = \frac{\delta_{B/A}}{2} - \delta_{C/A}$$

where $$\delta_{C/A} = \left(\frac{1}{2}\right)\left(\frac{L}{2}\right)\left(\frac{M}{EI}\right)\left[\frac{2}{3}\left(\frac{L}{2}\right)\right] + \left(\frac{1}{2}\right)\left(\frac{L}{2}\right)\left(\frac{M}{2EI}\right)\left[\frac{1}{3}\left(\frac{L}{3}\right)\right] = \frac{5ML^2}{48EI}$$

Thus, $$\delta_C = \frac{ML^2}{6EI} - \frac{5ML^2}{48EI} = \frac{ML^2}{16EI}$$

(c) *Uniformly distributed load over the entire span*

Because of symmetry in geometry and loading, the tangent at C is horizontal. The end slopes are equal to the area of the *M/EI* diagram between A and C,

$$\theta_B = -\theta_A = \left(\frac{2}{3}\right)\left(\frac{L}{2}\right)\left(\frac{wL^2}{8EI}\right) = \frac{wL^3}{24EI}$$

The deflection of the point A w.r.t. from the tangent at C is equal to the moment of the *M/EI* diagram between A and C about A,

$$\delta_{A/C} = \left(\frac{wL^3}{24EI}\right)\left[\frac{5}{8}\left(\frac{L}{2}\right)\right] = \frac{5wL^4}{384EI} \uparrow$$

EXAMPLE 14.10

Using moment-area theorems, compute the slope and the deflection at the point C and free end B of the cantilever beam AB of variable moment of inertia subjected to loads as shown in Figures 14.14(a) and (b).

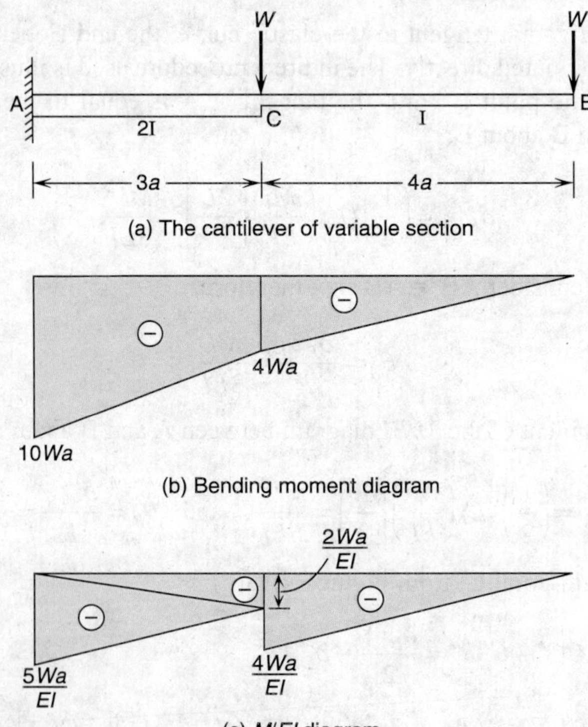

Figure 14.14 Computation of slope and deflection of the cantilever of variable section of Example 14.10.

Solution: The *M/EI* diagram of beam shown in Figure 14.14(c) is divided into triangles to facilitate computations. The tangent to the elastic curve at the fixed end is horizontal.

The slope and deflection at the point C:

$$\theta_C = \theta_A + \text{Area of the } M/EI \text{ diagram between A and C}$$

$$\theta_C = \left(\frac{1}{2}\right)\left(-\frac{Wa}{EI}\right)[5 \times 3a + 2 \times 3a] = -\frac{21Wa^2}{2EI}$$

$\delta_{C/A}$ = Deflection at point C w.r.t. the tangent at A

= Moment of the *M/EI* diagram between A and C about C

$$\delta_{C/A} = \left(\frac{1}{2}\right)\left(-\frac{Wa}{EI}\right)[5 \times 3a \times (2a) + 2 \times 3a \times (a)] = -\frac{18Wa^3}{EI}$$

The slope and deflection at the free-end:

$\theta_B = \theta_A$ + Area of the *M/EI* diagram between A and B

$$\theta_B = \left(\frac{1}{2}\right)\left(-\frac{Wa}{EI}\right)[5 \times 3a + 2 \times 3a + 4 \times 4a] = -\frac{37Wa^2}{2EI}$$

$\delta_{B/A}$ = Deflection at point B w.r.t. the tangent at A

= Moment of the *M/EI* diagram between A and B about B

$$\delta_{B/A} = \left(\frac{1}{2}\right)\left(-\frac{Wa}{EI}\right)\left[5 \times 3a \times (2+4)a + 2 \times 3a \times (1+4)a + 4 \times 4a \times \left(\frac{2}{3} \times 4a\right)\right]$$

$$= -\frac{244Wa^3}{3EI}$$

EXAMPLE 14.11

Using moment-area theorems, compute the slopes at the ends and the mid-point; and deflection at the mid-point of a simply supported beam loaded as shown in Figure 14.15(a). Also determine the maximum deflection and its location in the beam.

Solution: The *M/EI* diagram of the beam is shown in Figure 14.15(b).

(a) *The slopes*

In the absence of a horizontal tangent to the elastic curve, the end slopes and deflections are computed by indirect procedure. As in the case (b) of Example 14.9:

$$\delta_{B/A} = \left(\frac{1}{2}\right)\left(\frac{2Wa}{3EI}\right)\left[(a)\left(2a + \frac{a}{3}\right) + (2a)\left(\frac{4a}{3}\right)\right] = \frac{5Wa^3}{3EI}$$

Therefore, $$\theta_A = \frac{\delta_{B/A}}{3a} = \frac{5Wa^2}{9EI}$$

and $$\delta_{A/B} = \left(\frac{1}{2}\right)\left(\frac{2Wa}{3EI}\right)\left[(a)\left(\frac{2a}{3}\right) + (2a)\left(a + \frac{2a}{3}\right)\right] = \frac{4Wa^3}{3EI}$$

Therefore, $$\theta_B = \frac{\delta_{A/B}}{3a} = \frac{4Wa^2}{9EI}$$

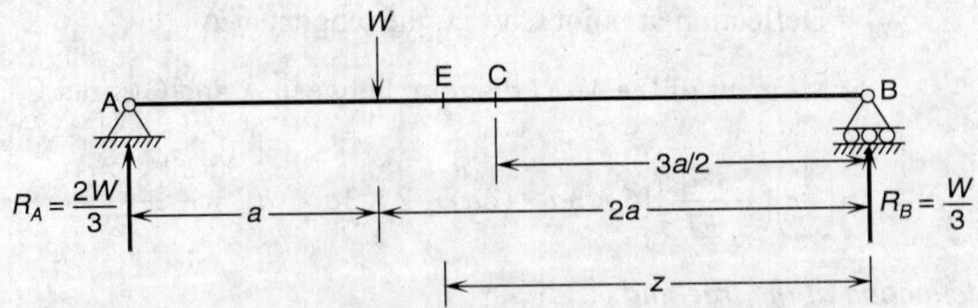

(a) Simply supported beam subjected to a concentrated load

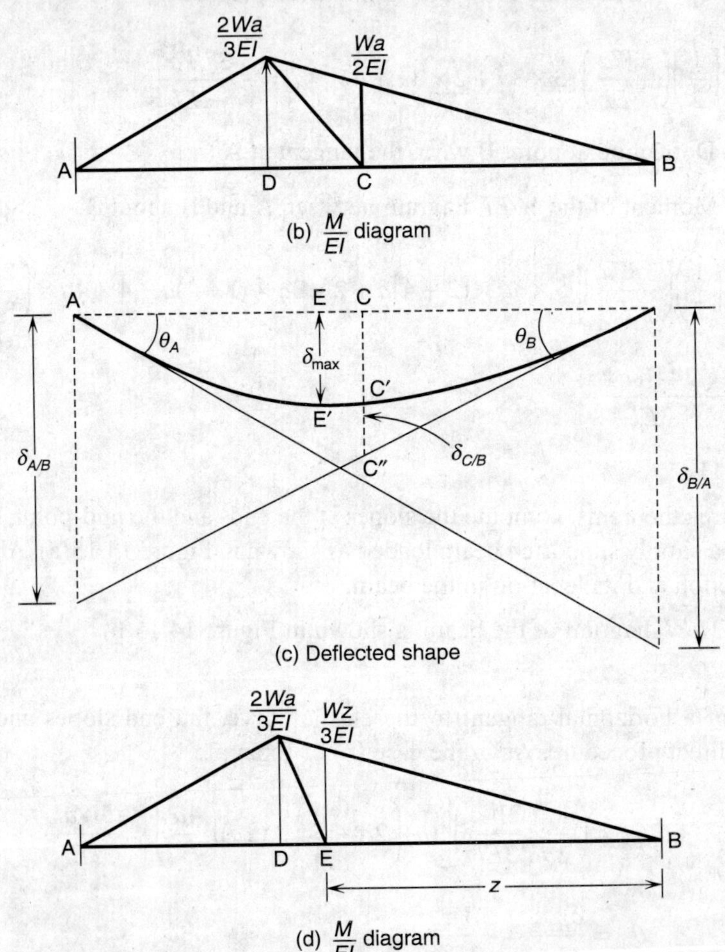

Figure 14.15 Slopes and deflections at various locations in a simply supported beam.

For the right hand support,

$$\theta_B = \theta_C + \text{Area of the } M/EI \text{ diagram between C and B}$$

Thus,

$$\theta_C = \theta_B - \text{Area of the } M/EI \text{ diagram between C and B}$$

$$\theta_C = \frac{4Wa^2}{9EI} - \left(\frac{1}{2}\right)\left(\frac{3a}{2}\right)\left(\frac{Wa}{2EI}\right) = \frac{5Wa^2}{72EI}$$

The deflection at the mid-point C,

$$\delta_C = CC' = CC'' - C'C'' = \frac{\delta_{A/B}}{2} - \delta_{C/B}$$

where $$\delta_{C/B} = \left(\frac{1}{2}\right)\left(\frac{3a}{2}\right)\left(\frac{Wa}{2EI}\right)\left[\frac{1}{3}\left(\frac{3a}{2}\right)\right] = \frac{3Wa^3}{16EI}$$

Thus, $$\delta_C = \frac{2Wa^3}{3EI} - \frac{3Wa^3}{16EI} = \frac{23Wa^3}{48EI}$$

(b) *The maximum deflection*

The maximum deflection of beam will occur in the beam segment A–D or D–B whichever is greater than span/2 because it is the larger portion in which the sign of slope changes from positive to negative or vice versa and therefore at some point within this length, slope will be zero, i.e. the tangent to the elastic curve will be horizontal. Consider the point E between D and B at distance z from B where deflection is the maximum. Then,

$$\theta_B = \theta_E\ (= 0) + \text{Area of the } M/EI \text{ diagram between E and B}$$

Therefore, $$\frac{4Wa^2}{9EI} = \left(\frac{1}{2}\right)(z)\left(\frac{Wz}{3EI}\right) = \frac{Wz^2}{6EI} \quad \text{or} \quad z = \sqrt{\frac{8}{3}}a = 1.633a$$

The deflection of the point B w.r.t. the tangent at E is equal to the moment of the *M/EI* diagram between E and B about B,

$$\delta_{\max} = \delta_{B/E} = \frac{4Wa^2}{9EI}\left[\frac{2z}{3}\right] = \frac{8Wa^2z}{27EI} = \frac{(8 \times 1.633)Wa^3}{27EI}$$

The maximum deflection of beam always occurs near the middle of the beam (see Figures 14.15(a) and (b)).

EXAMPLE 14.12

A beam A–B–D hinged at A and supported at the same level at B with lengths AB and BD of $4a$ and a, respectively, carries a concentrated load W at the overhanging end D as shown in Figure 14.16(a). Using moment-area theorems, compute the deflection at the mid-point C of the supported span. Also compute the deflection at the free end D when the concentrated load W is applied at the point C as shown in Figures 14.16(a). The beam is of uniform cross-section with constant E and I.

Solution: The elastic curve and *M/EI* diagram of the beam are shown in Figures 14.16(b) and (c), respectively.

Case-I: The deflection at the mid-point C when concentrated load W is at D,

$$\delta_C = CC'' - C'C'' = \frac{\delta_{B/A}}{2} - \delta_{C/A}$$

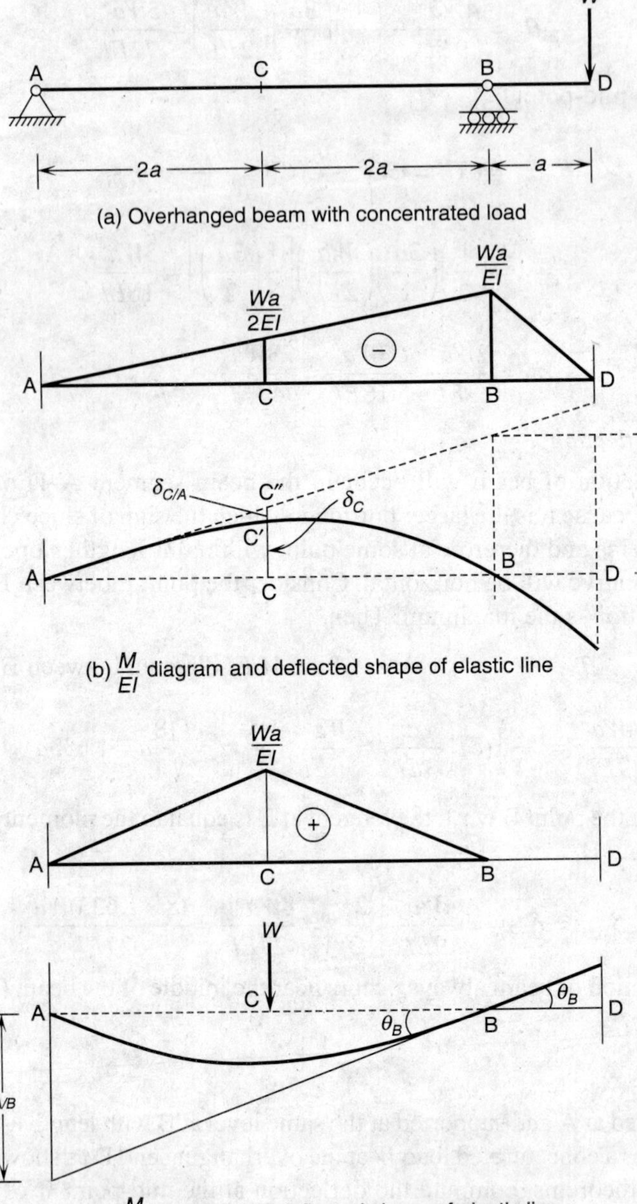

Figure 14.16 Computation of slopes in overhanged beam of Example 14.12.

where
$$\delta_{B/A} = \left(\frac{1}{2}\right)(4a)\left(\frac{Wa}{EI}\right)\left[\frac{1}{3}(4a)\right] = \frac{8Wa^3}{3EI}$$

$$\delta_{C/A} = \left(\frac{1}{2}\right)(2a)\left(\frac{Wa}{2EI}\right)\left[\frac{1}{3}(2a)\right] = \frac{Wa^3}{3EI}$$

Thus, $$\delta_C = \frac{4Wa^3}{3EI} - \frac{Wa^3}{3EI} = \frac{Wa^3}{EI}$$

Case-II: The deflection at the point D when the concentrated load W is at C,

$$\delta_D = a \times \theta_B = a \times \left(\frac{\delta_{A/B}}{4a}\right) = \frac{\delta_{A/B}}{4}$$

where $$\delta_{A/B} = \left(\frac{1}{2}\right)(4a)\left(\frac{Wa}{EI}\right)\left[\frac{(4a+2a)}{3}\right] = \frac{4Wa^3}{EI}$$

Therefore, $$\delta_D = \frac{4Wa^3}{4EI} = \frac{Wa^3}{EI}$$

It should be noted that a load W at D causes a deflection at C which is equal to the deflection at D caused the load W acting at C. This is precisely the Maxwell's reciprocal theorem of deflections.

EXAMPLE 14.13

A bracket BC of length $3a$ connected to the top B of a column AB of length $5a$ as shown in Figure 14.17(a) carries a concentrated load W at its free end C. Using moment-area theorems, compute the slope, horizontal and vertical deflections at the point C. Both the beams are of uniform cross-section with constant E and I.

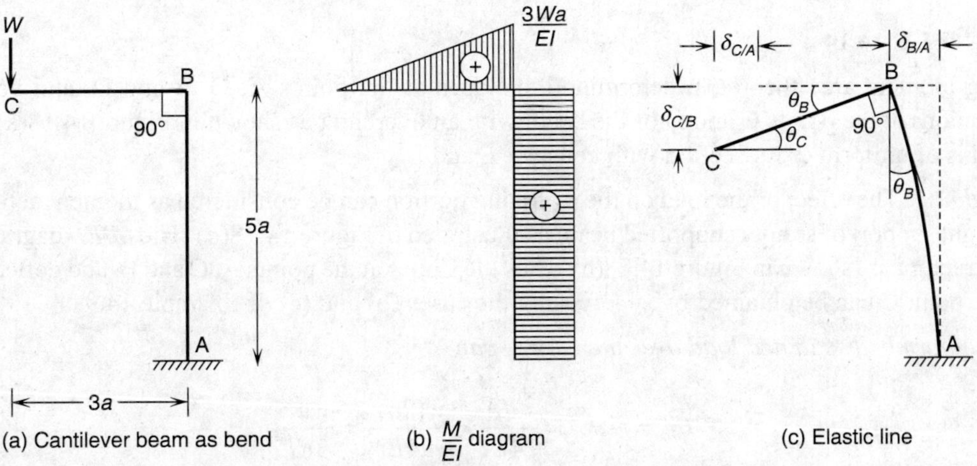

(a) Cantilever beam as bend (b) $\frac{M}{EI}$ diagram (c) Elastic line

Figure 14.17 Column with bracket of Example 14.13.

Solution: The *M/EI* diagram and elastic curve of the structure are shown in Figures 14.17(b) and (c), respectively. The axial deformation in the member AB is neglected.

1. *Slope at* C:

$$\theta_C = \theta_B + \text{Area of the } \frac{M}{EI} \text{ diagram between } B \text{ and } C$$

$$\theta_C = (5a)\left(\frac{3Wa}{EI}\right) + \frac{1}{2}(3a)\left(\frac{3Wa}{EI}\right) = \left(\frac{39Wa}{2EI}\right)$$

2. *Horizontal deflection at* C:

$$\delta_{C,H} = \delta_{B/A} = (5a)\times\left(\frac{3Wa}{EI}\right)\times\left(\frac{5a}{2}\right) = \frac{75Wa}{2EI}$$

3. *Vertical deflection at* C:

$$\delta_{C,V} = (3a)\,\theta_B + \delta_{C/B}$$

$$= (3a)\left[(5a)\left(\frac{3Wa}{EI}\right)\right] + \frac{1}{2}(3a)\left(\frac{3Wa}{EI}\right)\left[\frac{2}{3}(3a)\right]$$

$$= \frac{54Wa^3}{EI}$$

As discussed earlier in Chapter 1, the principle of superposition can advantageously be used in the analysis of linear elastic structures. According to this principle, the displacements resulting from each of a number of loads acting independently may be summed up to obtain the resultant displacements due to the combined action of all loads. The following examples illustrate the application of the principle of superposition in the computation of slopes and displacements using area-moment theorems.

EXAMPLE 14.14

Using moment-area theorems, determine the slopes at the points A, C, B and D, and vertical deflections at the points C and D of the beam with an overhang as shown in Figure 14.18(a). The beam is of uniform cross-section with constant E and I.

Solution: The effect of the load on the overhang portion can be considered as moment acting on the right support of simply supported beam as illustrated in Figure 14.18(b). The *M/EI* diagrams of the structure are shown in Figure 14.18(c). Thus the slopes at the points A, C and B, and deflections at the point C can be obtained by superposing the cases (b) and (c) of Example 14.9.

For uniformly distributed load over the entire span

$$\theta_B' = -\theta_A' = \frac{wL^3}{24EI} = \frac{w(2a)^3}{24EI} = \frac{wa^3}{3EI}$$

For moment M (= wa²) acting at the right support

$$\theta_A'' = \frac{ML}{6EI} = \frac{(wa^2)(2a)}{6EI} = \frac{wa^3}{3EI} \quad \text{and}$$

$$\theta_B'' = -\frac{ML}{3EI} = -\frac{(wa^2)(2a)}{3EI} = -\frac{2wa^3}{3EI}$$

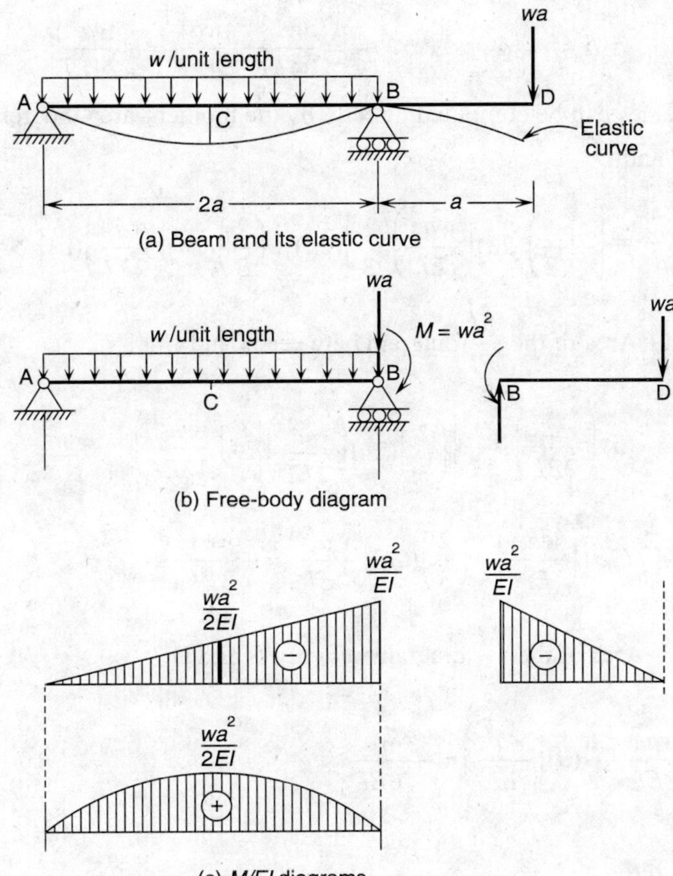

Figure 14.18 Method of superposition for area-moment theorem Example 14.14.

Therefore,
$$\theta_A = \theta_A' + \theta_A'' = -\frac{wa^3}{3EI} + \frac{wa^3}{3EI} = 0$$

$$\theta_B = \theta_B' + \theta_B'' = \frac{wa^3}{3EI} - \frac{2wa^3}{3EI} = -\frac{wa^3}{3EI}$$

Deflections at C:

For uniformly distributed load

$$\delta'_{A/C} = \frac{5wL^4}{384EI} = \frac{5w(2a)^4}{384EI} = \frac{5wa^4}{24EI}$$

For moment M (= $-wa^2$) at the right support

$$\delta''_{C/A} = \frac{ML^2}{16EI} = \frac{(-wa^2)(2a)^2}{16EI} = -\frac{wa^4}{4EI}$$

Therefore, $$\delta_{A/C} = \delta'_{A/C} + \delta''_{A/C} = \frac{5wL^4}{24EI} - \frac{wa^4}{4EI} = -\frac{wa^4}{24EI}$$

Alternatively, the values can be computed directly by the moment-area theorems.

1. *Slopes at* A, C, B and D:

$$\theta_A = \frac{\delta_{B/A}}{2a} = \left\{-\left(\frac{1}{2}\right)(2a)\left(\frac{wa^2}{EI}\right)\left(\frac{1}{3}\right)(2a) + \left(\frac{2}{3}\right)(2a)\left(\frac{wa^2}{2EI}\right)(a)\right\} \times \frac{1}{2a} = 0$$

$$\theta_C = \theta_A + \text{Area of the } \frac{M}{EI} \text{ diagram between A and C}$$

$$= 0 - \frac{1}{2}(a)\left(\frac{wa^2}{2EI}\right) + \frac{1}{2}\left\{\left(\frac{2}{3}\right)(2a)\left(\frac{wa^2}{2EI}\right)\right\} = \left(\frac{wa^3}{12EI}\right)$$

$$\theta_B = 0 - \frac{1}{2}(2a)\left(\frac{wa^2}{EI}\right) + \left(\frac{2}{3}\right)(2a)\left(\frac{wa^2}{2EI}\right) = -\frac{wa^3}{3EI}$$

$$\theta_D = \theta_B + \text{Area of the } \frac{M}{EI} \text{ diagram between B and D}$$

$$\theta_D = -\frac{wa^3}{3EI} - \frac{1}{2}(a)\left(\frac{wa^2}{EI}\right) = -\frac{5wa^3}{6EI}$$

2. *Deflection at* C

As the slope at A is zero,

$$\delta_C = -\frac{1}{2}(a)\left(\frac{wa^2}{2EI}\right)\left(\frac{a}{3}\right) + \frac{1}{2}\left\{\left(\frac{2}{3}\right)(2a)\left(\frac{wa^2}{2EI}\right)\right\}\left(\frac{3a}{8}\right) = \frac{wa^4}{24EI}$$

3. *Deflection at* D

$$\delta_D = (a)\theta_B + \delta_{D/B} = (a)\left(-\frac{wa^3}{3EI}\right) - \left(\frac{1}{2}\right)(a)\left(\frac{wa^2}{EI}\right)\left(\frac{2a}{3}\right) = -\frac{2wa^4}{3EI}$$

Statically indeterminate beams can readily be analysed for the unknown reactions using moment-area theorems together with superposition. The procedures developed consider the bending-moment diagrams in two parts — one due to applied loading on the beam considered to be determinate, the other part for the redundant reactions. The known and the unknown parts of the bending-moment diagram together give a complete bending-moment diagram of the structure. This diagram can then be used in applying the moment-area theorems to the elastic curve of the beam to compute the rotations and deflections. The procedure for computation of the redundant reactions, slopes and displacements are illustrated in the following examples.

EXAMPLE 14.15

Using moment-area theorems and superposition, determine the maximum deflection of the beam shown in Figure 14.19(a) due to applied load. The beam is of uniform cross-section with constant *E* and *I*.

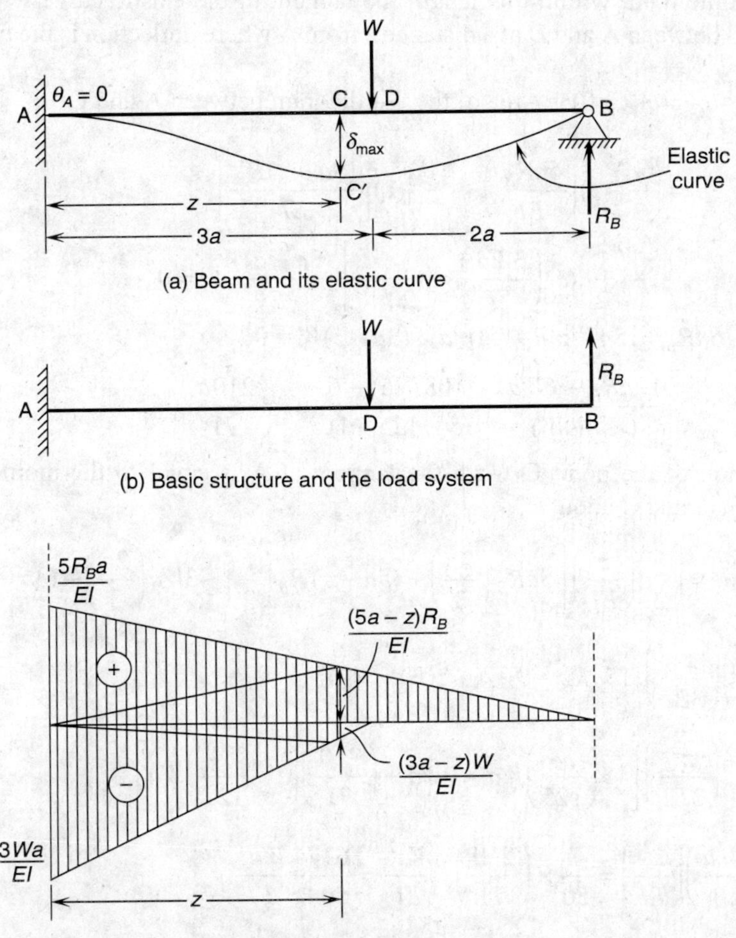

Figure 14.19 Computation of maximum deflection of propped cantilever of Example 14.15.

Solution: Consider that the beam is released from the redundant reaction R_B. The *M/EI* diagram of both the parts are drawn on the same base line are shown in Figure 14.19(c). Since the tangent at the built-in end remains horizontal after the application of force *W*, the geometrical or compatibility condition to be satisfied is $\delta_{B/A} = 0$. Basic structure and the load system. Therefore, from moment-area theorem,

$$\delta_{B/A} = \left(\frac{1}{2}\right)(5a)\left(\frac{5aR_B}{EI}\right)\left[\left(\frac{2}{3}\right)(5a)\right] - \left(\frac{1}{2}\right)(3a)\left(\frac{3Wa}{EI}\right)\left[2a + \left(\frac{2}{3}\right)(3a)\right]$$

$$= \frac{125a^3 R_B}{3EI} - \frac{18Wa^3}{EI} = 0 \quad \text{or} \quad R_B = \frac{54W}{125}$$

The maximum deflection of beam will occur in the beam segment A–D which is greater than span/2, and at some point within this length the tangent to the elastic curve will be horizontal. Consider point C between A and D at a distance z from A where deflection is the maximum. Then,

$$\theta_C = \theta_A(=0) + \text{Area of the } \frac{M}{EI} \text{ diagram between A and C}$$

$$= \left(\frac{1}{2}\right)(z)\left(\frac{5aR_B}{EI}\right) + \left(\frac{1}{2}\right)(z)\left[\frac{(5a-z)R_B}{EI}\right]$$

$$- \left(\frac{1}{2}\right)(z)\left(\frac{3Wa}{EI}\right) - \left(\frac{1}{2}\right)(z)\left[\frac{(3a-z)W}{EI}\right] = 0$$

or

$$5aR_B + (5a-z)R_B - 3Wa - (3a-z)W = 0$$

$$z = \frac{10aR_B - 6Wa}{(R_B - W)} = \frac{(108/25) - 6}{(54/125 - 1)}(a) = \frac{210a}{71}$$

The deflection of the point C w.r.t. the tangent at A is equal to the moment of the *M/EI* diagram between A and C about C.

$$\delta_{\max} = \delta_{C/A} = \left(\frac{1}{2}\right)\left(\frac{z}{EI}\right)\left[5aR_B\left(\frac{2z}{3}\right) + (5a-z)R_B\left(\frac{z}{3}\right) - 3Wa\left(\frac{2z}{3}\right) - (3a-z)W\left(\frac{z}{3}\right)\right]$$

$$= \left(\frac{1}{2}\right)\left(\frac{z^2}{3EI}\right)[15aR_B - 9Wa + z(W - R_B)]$$

$$= \left(\frac{1}{2}\right)\left(\frac{z^2}{3EI}\right)\left[15\left(\frac{54}{125}\right)Wa - 9Wa + \left(\frac{210}{71}\right)\left(1 - \frac{54}{125}\right)Wa\right]$$

$$= \left(\frac{7Wa}{50}\right)\left(\frac{z^2}{EI}\right) = \frac{7}{50} \times \left(\frac{210}{71}\right)^2 \frac{Wa^2}{EI} = \frac{6174}{5041}\frac{Wa^3}{EI}$$

EXAMPLE 14.16

Using moment-area theorems and superposition, determine the maximum deflection of the beam built-up at the ends shown in Figure 14.20(a) due to uniformly distributed load w per unit length. The beam is of uniform cross-section with constant E and I.

Solution: The structure is indeterminate to the second degree. However, due to geometric and loading symmetry, the fixed end moments and vertical reactions are equal, therefore, single geometric condition is sufficient to yield the redundant. From the geometry of elastic curve $\delta_{B/A} = 0$ or $\theta_{B/A} = 0$. Therefore,

$$R_B = R_A = \frac{wL}{2} \quad \text{and} \quad M_B = M_A = M$$

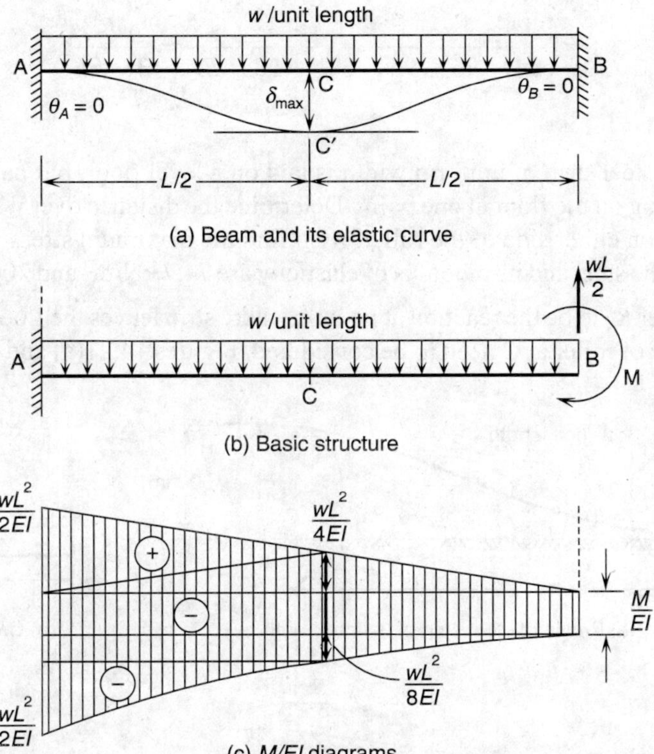

Figure 14.20 Computation of maximum deflection of beam of Example 14.16.

Consider that the beam is released from the redundant reactions at the support B. The basic structure is shown in Figure 14.20(b). The *M/EI* diagrams for the three parts of the load are drawn on the same base line as shown in Figure 14.20(c). Since the tangents at the built-in ends remain horizontal after the application of forces, the geometrical or compatibility condition $\theta_{B/A} = 0$ can be used. Therefore, from moment-area theorem,

$$\theta_{B/A} = \left(\frac{1}{2}\right)(L)\left(\frac{wL^2}{2EI}\right) + (L)\left(-\frac{M}{EI}\right) + \left(\frac{1}{3}\right)(L)\left(-\frac{wL^2}{2EI}\right) = 0$$

Thus,
$$M = \frac{wL^2}{4} - \frac{wL^2}{6} = \frac{wL^2}{12}$$

The maximum deflection of beam will occur at a point where the tangent to the elastic curve is horizontal. Due to geometric and loading symmetry the slope at the centre of the beam is zero, hence maximum deflection will occur at the centre. Therefore,

$$\delta_{max} = \delta_{C/A} = \left(\frac{1}{2}\right)\left(\frac{L}{2}\right)\left(\frac{wL^2}{2EI}\right)\left(\frac{L}{3}\right) + \left(\frac{1}{2}\right)\left(\frac{L}{2}\right)\left(\frac{wL^2}{4EI}\right)\left(\frac{L}{6}\right) + \left(\frac{L}{2}\right)\left(-\frac{wL^2}{12EI}\right)\left(\frac{L}{4}\right)$$
$$+ \left(\frac{L}{2}\right)\left(-\frac{wL^2}{8EI}\right)\left(\frac{L}{4}\right) + \left(\frac{1}{3}\right)\left(\frac{L}{2}\right)\left(-\frac{3wL^2}{8EI}\right)\left[\left(\frac{3}{4}\right)\left(\frac{L}{2}\right)\right]$$

$$= \left(\frac{1}{24} + \frac{1}{96} - \frac{1}{96} - \frac{1}{64} - \frac{3}{128} \right) \frac{wL^4}{EI} = \frac{wL^4}{384EI}$$

EXAMPLE 14.17

A long 3 mm thick steel strip of uniform width is laid on a level floor, but passes over a roller of 50 mm diameter lying on the floor at one point. Determine the distance over which the strip will be clear of the ground on either side of the roller. Also find the maximum stress induced in the strip. The unit weight of the steel and its modulus of elasticity are 74.7 kN/m^3 and 200 GPa, respectively.

Solution: Consider R_A to be the reaction at point A where strip leaves the floor. Due to symmetry, the strip on one side of roller at C need to be considered. Figures 14.21(a) and (b) show the profile

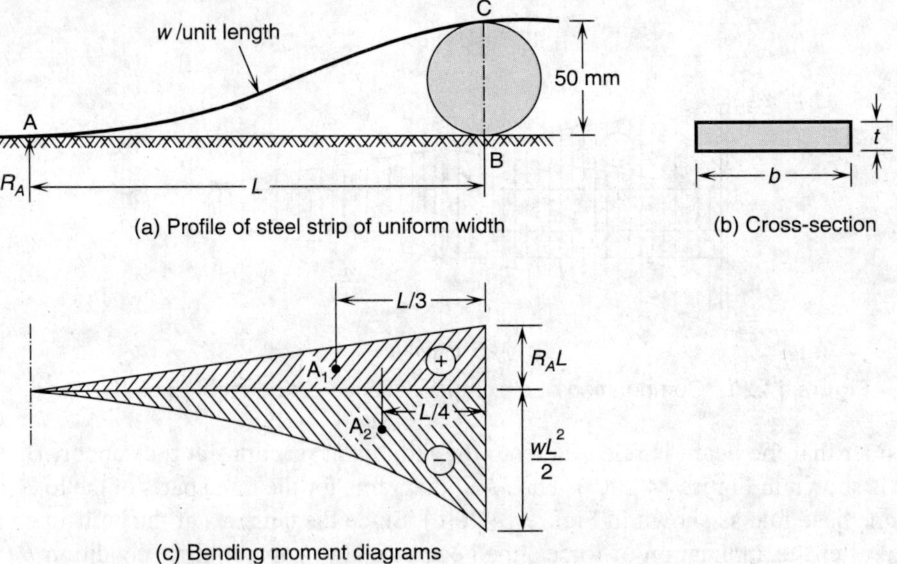

Figure 14.21 Geometry of the steep strip lying on the floor of Example 14.17.

of steel-strip of uniform width and the cross-section. The bending moment diagrams due to reaction R_A and self weight of strip per unit length w are shown in Figure 14.21(c). As there is no change in slope between points A and C,

$$A_1 = A_2 \quad \text{or} \quad \frac{1}{2} \times (R_A L)(L) = \frac{1}{3} \times \left(\frac{wL^2}{2} \right)(L) \quad \text{or} \quad R_A = \frac{wL}{3}$$

The vertical distance between A and C, $\delta_{A/C} = \delta_{C/A} = 50$ mm. Therefore,

$$\delta_{C/A} = \frac{1}{EI}(A_1 \bar{x}_1 - A_2 \bar{x}_2)$$

$$= \frac{1}{EI} \left[\frac{1}{2} \times (R_A L)(L) \left(\frac{L}{3} \right) - \frac{1}{3} \times \left(\frac{wL^2}{2} \right)(L) \left(\frac{L}{4} \right) \right]$$

$$= \frac{1}{EI}\left[\frac{1}{2}\times\left(\frac{wL^2}{3}\right)(L)\left(\frac{L}{3}\right) - \frac{1}{3}\times\left(\frac{wL^2}{2}\right)(L)\left(\frac{L}{4}\right)\right] = \frac{wL^4}{72EI}$$

Therefore, $$L^4 = \frac{72EI\delta_{C/A}}{w} = \frac{72\times(200\times10^3)\times(b\times3^3/12)\times50}{[(3\times b\times1000\times10^{-9})\times74.7]\times10^3} = 7.369\times10^{12}$$

Thus, $L = 1640$ mm

$$M_{\max} = R_A L - \frac{wL^2}{2} = \frac{wL^2}{3} - \frac{wL^2}{2} = \frac{wL^2}{6}$$

Maximum bending stress, $$\sigma_b = \frac{6M_{\max}}{bt^2} = \frac{wL^2}{bt^2}$$

$$\sigma_b = \frac{[(3\times b\times1000\times10^{-9})\times74.7\times10^3]\times1640^2}{b\times3^2} = 65.7 \text{ MPa}$$

14.5 ELASTIC LOAD METHOD

A reader shall note that the form of the moment-area procedure, for computation of the change in slope between two points A and B on an elastic curve, which involve taking static moment about an axis through B of the area under the *M/EI* curve between A and B divided *L*, is very much similar to computation of vertical reactions of a loaded simple beam supported at points A and B. If the distributed vertical load applied to a simple beam were *M/EI* diagram then the computation of vertical reaction at point A would involve taking moments about an axis through point B and as a matter of fact, this vertical reaction at point A would be $\theta_{A/B}$ (= θ_A) as computed in the forgoing section.

This analogy may be developed further if we review the form of the computation of slope and displacement θ_C and δ_C. It should be noted that θ_C at point C gives the slope of the tangent with reference to the direction of the chord AB of the elastic curve and δ_C gives the deflection of point C from the same chord.

Consider an imaginary beam loaded with *M/EI* diagram as shown in Figure 14.22(b) and reactions R_A and R_B are determined,

$$R_B = \left(\frac{1}{L}\right)\int_A^B \frac{(Mdz)z}{EI} = \left(\frac{1}{L}\right)A_M\bar{z} = \theta_B$$

where A_M is the total area of the *M/EI* diagram between A and B, and $\bar{z}$ is the horizontal distance to the centroid of this area from A. Similarly, the R_A is

$$R_A = \left(\frac{1}{L}\right)A_M(L-\bar{z}) = \theta_A$$

Now, consider point C,

$$\theta_C = \theta_A - \Delta\theta_{C/A} = R_A - \int_A^C \frac{M}{EI}dz$$

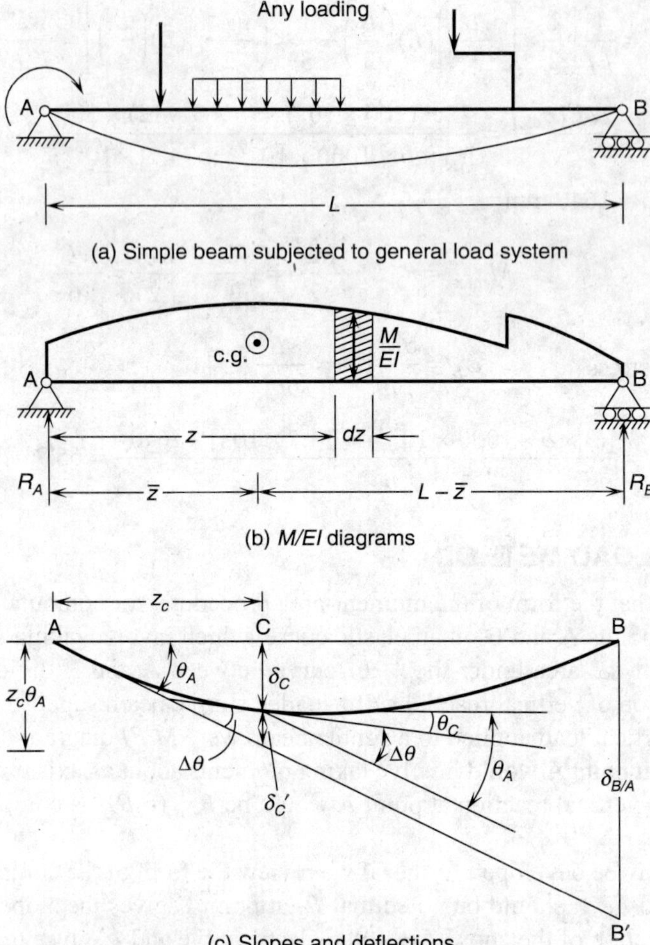

Figure 14.22 Development of elastic load method.

Likewise,
$$\delta_C = \theta_A(z_c) - \int_A^C \frac{Mz\,dz}{EI}$$

From these considerations, it may be concluded that the elastic curve of a beam AB is exactly the same as the bending moment diagram for an imaginary simply supported beam of same span AB, which is loaded with a distributed transverse load equal to *M/EI* diagram of the actual beam AB.

The procedure of applying the elastic-load method may be summarized in the following statement:

The slopes and displacements of an elastic curve measured with reference to one of its chords AB are equal, respectively, to the (elastic) shears and (elastic) moments of an imaginary simply supported beam of span AB loaded with distributed (elastic) load consisting of the *M/EI* diagram for that portion AB.

The sign convention used in drawing regular load, shear, and bending-moment diagrams are followed, thus, positive *M/EI* diagram indicates upward loads. The positive bending-moment on the imaginary beam therefore indicates deflections above the chord, and negative values indicate deflections below the chord. Similarly, the positive shear on the imaginary beam indicates that the elastic curve slopes upward, proceeding from left to right, and negative value indicates that it slopes downward.

EXAMPLE 14.18

Using the elastic-load method, compute the slopes and deflection at the salient points of a simply supported beam loaded as shown in Figure 14.23(a). Also determine the maximum deflection and its location in the beam. *EI* is constant throughout.

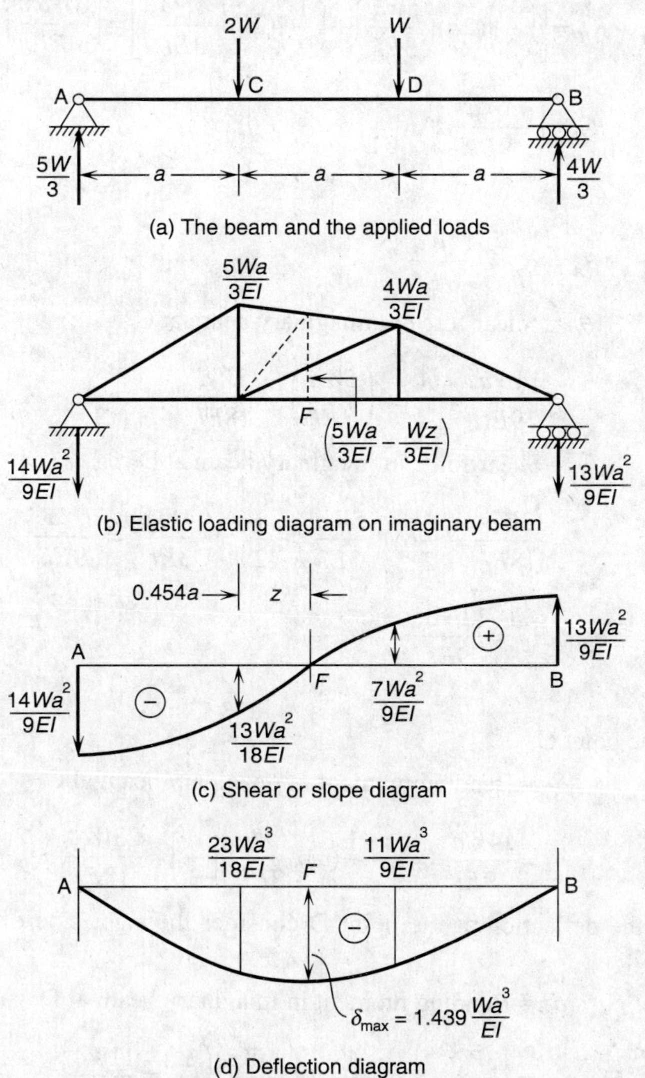

Figure 14.23 Slopes and deflection of a simple beam by elastic-load method.

Solution: The elastic-load curve of the beam is shown in Figure 14.23(b). The elastic reactions of the imaginary beam are:

$$R_B = \frac{1}{3a}\sum M_A = \left(\frac{1}{3a}\right)\left[\frac{1}{2}(a)\left(\frac{5Wa}{3EI}\right)\left(\frac{2a}{3}\right) + \frac{1}{2}(a)\left(\frac{5Wa}{3EI}\right)\left(a + \frac{a}{3}\right)\right.$$

$$\left. + \frac{1}{2}(a)\left(\frac{4Wa}{3EI}\right)\left(a + \frac{2a}{3}\right) + \frac{1}{2}(a)\left(\frac{4Wa}{3EI}\right)\left(2a + \frac{a}{3}\right)\right]$$

Therefore, $$R_B = \left(\frac{13Wa^2}{9EI}\right)$$

$$R_A + R_B = \left[\left(\frac{1}{2}\right)(2a)\left(\frac{5wa}{3EI}\right) + \left(\frac{1}{2}\right)(2a)\left(\frac{4Wa}{3EI}\right)\right] = \left(\frac{3Wa^2}{EI}\right)$$

Thus, $$R_A = \left(\frac{14Wa^2}{9EI}\right)$$

(a) *The slopes*

$$\theta_A = R_A = \frac{14Wa^2}{9EI}$$

θ_C = Shear force in imaginary beam at C

$$= \frac{14Wa^2}{9EI} - \frac{1}{2}(a)\left(\frac{5Wa}{3EI}\right) = \frac{13Wa^2}{18EI}$$

θ_D = Shear force in imaginary beam at D

$$= \frac{13Wa^2}{18EI} - \frac{1}{2}(a)\left(\frac{5Wa}{3EI}\right) - \frac{1}{2}(a)\left(\frac{4Wa}{3EI}\right) = \frac{7Wa^2}{9EI}$$

and $$\theta_B = R_B = \frac{13Wa^2}{9EI}$$

(b) *The deflections*

The deflection at the point C,

δ_C = Bending moment in imaginary beam at C

$$= \frac{14Wa^2}{9EI} \times a - \frac{1}{2}(a)\left(\frac{5Wa}{3EI}\right)\left(\frac{a}{3}\right) = \frac{23Wa^2}{18EI}$$

To determine the deflection at the point D consider the right side of point D for ease in computations, thus,

δ_D = Bending moment in imaginary beam at D

$$= \frac{13Wa^2}{9EI} \times a - \frac{1}{2}(a)\left(\frac{4Wa}{3EI}\right)\left(\frac{a}{3}\right) = \frac{11Wa^2}{9EI}$$

(c) *The maximum deflection*

The maximum deflection of beam will occur at the point where slope, i.e. the shear is zero on the imaginary beam. Consider this point F between C and D at distance z from C where deflection is the maximum. Then,

$$\frac{13Wa^2}{18EI} - (z)\left(\frac{5Wa}{3EI}\right) + \left(\frac{1}{2}\right)(z)\left(\frac{Wz}{3EI}\right) = z^2 - 10z + \frac{13a}{3} = 0$$

or $$z = 0.454a$$

At this point z:

$$\delta_{max} = \text{Bending moment in imaginary beam at F}$$

$$= 1.439\frac{Wa^3}{EI}$$

The slope and deflection curves of the beam are shown in Figures 14.23(c) and (d).

14.6 APPLICATION OF MOMENT-AREA THEOREM AND ELASTIC-LOAD METHOD

The elastic-load method cannot be applied directly to the fictitious beam span which includes an intermediate hinge, because at such hinges, there may be sharp change (or discontinuity) in the slope of the elastic curve. However, in some cases based on the study of elastic curve of the beam the elastic-load method can be combined with moment-area theorem to obtain the solution. Similarly, in case of cantilevers and overhangs, the sketch of elastic curve of the structure may be helpful.

EXAMPLE 14.19

Using the elastic-load method, compute the maximum deflection for beam loaded as shown in Figure 14.24(a) is constant throughout.

Solution: In this problem, the elastic-load method cannot be applied directly to the fictitious beam span CB because of the presence of a hinge at C. However, a study of elastic curve of the beam shown in Figure 14.24(e) reveals that the method can be combined with moment-area theorem to obtain the solution. The second moment-area theorem can be applied to the portion AC to determine the deflection δ_C. This deflection will establish the position of the chord CB. Now the elastic-load method can be used to compute the slopes and deflections with reference to the chord CB. Free body diagram and *M/EI* diagram are shown in Figures 14.24(b) and (c).

1. Using second moment-area theorem, the deflection at the hinge C is,

$$\delta_C = \frac{1}{2}(2a)\left(\frac{Wa}{EI}\right)(a) + \frac{1}{2}(a)\left(\frac{3Wa}{EI}\right)\left(a + \frac{2a}{3}\right) = \frac{7Wa^3}{2EI} \downarrow$$

Rotation of chord CB, $$\theta_C' = \frac{\delta_C}{6a} = \left(\frac{1}{6a}\right)\left(\frac{7Wa^3}{2EI}\right) = +\frac{7Wa^2}{12EI}$$

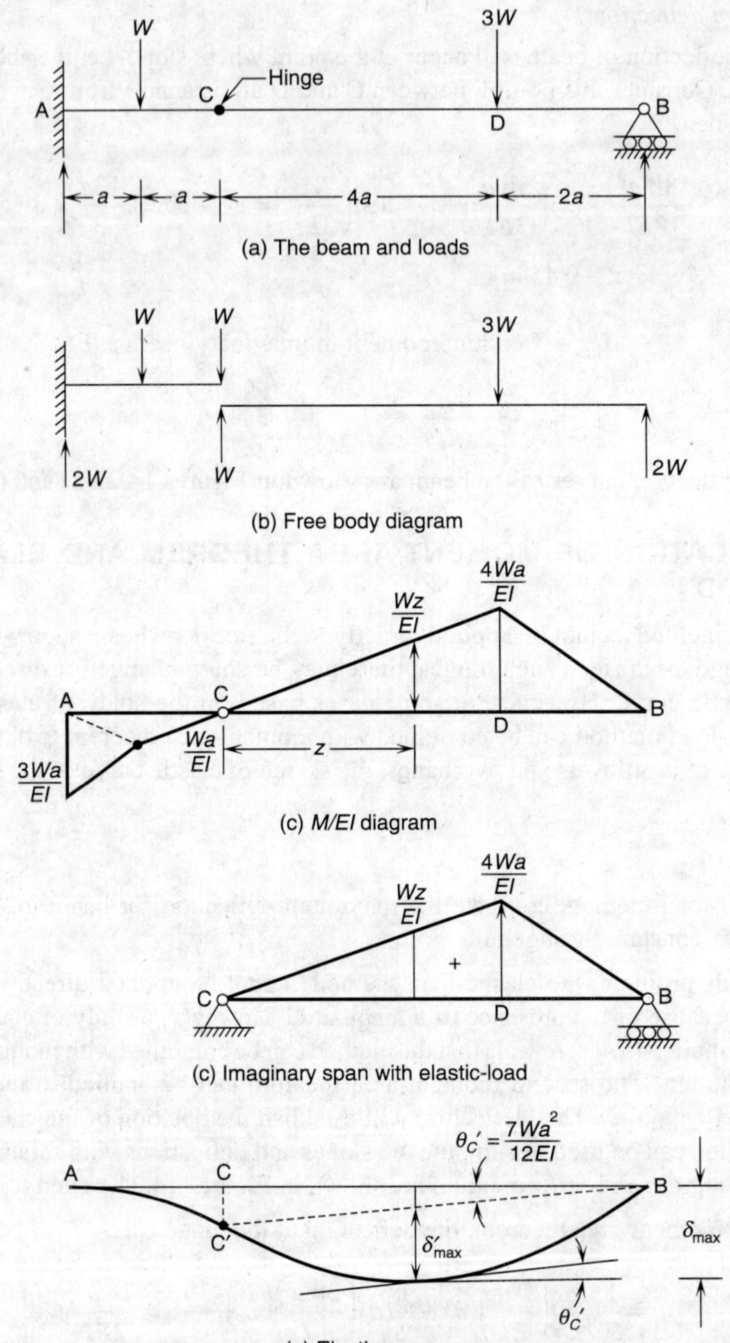

Figure 14.24 Area-moment theorem and elastic-load method for computation of slopes and deflections.

2. Using elastic-load method,

The elastic reactions of the fictitious beam CB of Fig. 14.24(d) are:

$$R_B = \frac{\sum M_C}{6a} = \left(\frac{1}{6a}\right)\left[\frac{1}{2}(6a)\left(\frac{4Wa}{EI}\right)\left(\frac{4a+6a}{3}\right)\right] = \frac{20Wa^2}{3EI}\downarrow$$

and
$$R_C = \frac{\sum M_B}{6a} = \left(\frac{1}{6a}\right)\left[\frac{1}{2}(6a)\left(\frac{4Wa}{EI}\right)\left(\frac{2a+6a}{3}\right)\right] = \frac{16Wa^2}{3EI}\downarrow$$

(a) *The slopes*: The actual slopes with respect to the horizontal are,

$$\theta_C = R_C + \theta_C' = -\frac{16Wa^2}{3EI} + \left(\frac{7Wa^2}{12EI}\right) = -\frac{57Wa^2}{12EI}$$

$$\theta_B = R_B + \theta_C' = \frac{20Wa^2}{3EI} + \left(\frac{7Wa^2}{12EI}\right) = \frac{87Wa^2}{12EI}$$

(b) *The deflections*

$$\delta_D = \frac{71Wa^3}{6EI}\downarrow$$

(c) *The maximum deflection*

The maximum deflection will occur at the point where the tangent to elastic curve is horizontal, i.e. the tangent slopes down to right with respect to the chord CB by an amount $\frac{7Wa^2}{12EI}$ i.e.

$$-\frac{16Wa^2}{3EI} + \left(\frac{1}{2}\right)(z)\left(\frac{Wz}{EI}\right) = -\frac{7Wa^2}{12EI} \quad \text{or} \quad z^2 = \left(\frac{171}{18}\right)a^2$$

Therefore, $z = 3.082a$ and deflection with respect to the chord C′B

$$\delta'_{\max} = -\left(\frac{16Wa^2}{3EI}\right)(3.082a) - \left(\frac{1}{2}\right)(3.082a)\left(\frac{3.082Wa}{EI}\right)\left(\frac{3.082a}{3}\right)$$

$$= -\frac{11.558Wa^3}{EI}\downarrow$$

The actual maximum deflection with respect to the horizontal is,

$$\delta_{\max} = \delta'_{\max} - \left(\frac{7Wa^2}{12EI}\right)(6a - 3.082a) = -\frac{13.2902Wa^3}{EI}\downarrow$$

14.7 CONJUGATE-BEAM METHOD

The conjugate-beam method is an extension and slight variation of the elastic-load method. The procedure consists in selecting a conjugate beam, i.e. a corresponding fictitious beam, which has the same length as the actual beam but is supported in such a manner that, *when the conjugate beam is loaded by M/EI diagram of the real beam as an elastic load, the (elastic) shear at a location is equal to the slope of the real beam at the corresponding location, and (elastic) bending-moment in the conjugate beam is equal to the corresponding deflection of the real beam.*

The support conditions of the conjugate beam can be selected based on the known characteristics of the elastic curve of the actual beam at its supports or any other special feature such as hinge. The conjugate substitutes of actual supports and connections are illustrated in Table 14.1. For example, consider the case of fixed end or support of an actual beam. The support has neither slope nor deflection; therefore, there must be neither shear nor moment at this point of conjugate-beam, i.e. the point must be free and unsupported as illustrated in column 4 of the table. Similarly, at the simply supported end the beam can rotate but cannot deflect; therefore, there must be shear but no moment at this point of conjugate-beam, i.e. the point must be provided with the vertical reaction of a roller support. This analogy can be extended to the interior hinge where the real beam can deflect but has discontinuity in slope, i.e. there are two different slopes on either side of the hinge; therefore, conjugate-beam, at this point must be provided with the vertical reaction to create the sudden change in shear but with capability of resisting bending moment. The objective can be achieved by providing a roller support at this point of conjugate-beam. At an intermediate support, the beam can rotate but cannot deflect, hence at the corresponding point of conjugate-beam, a shear force must exist and bending moment must be zero. This objective can be achieved by providing an interior hinge at the corresponding point. Thus the reactions, shears, and bending moments are now computed as for a statically determinate beam. The (elastic) shears and (elastic) bending-moment diagrams for the conjugate beam represent the variations of slope and deflection measured with respect to original position of real beam along its span; i.e. they represent the true slopes and deflections.

TABLE 14.1 Conjugate substitutes of actual support and connector

Real beam		*Conjugate beam*	
Type of support or *connection*	*Kinematic* or *displacement boundary*	*Static* or *force boundary condition*	*Type of support* or *connection required*
Hinged or roller supported end	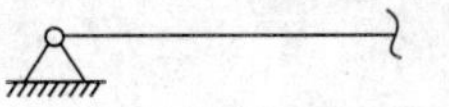 Or  $\theta \neq 0$, $y = 0$	$V \neq 0$ $M = 0$	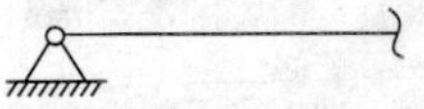 Hinged or roller supported end 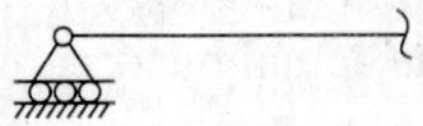
Fixed end	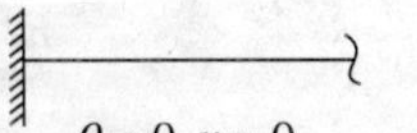 $\theta = 0$, $y = 0$	$V = 0$ $M = 0$	
Free or overhang end	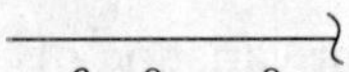 $\theta \neq 0$, $y \neq 0$	$V \neq 0$ $M \neq 0$	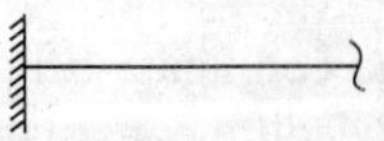

(Contd.)

(*Contd.*)

Real beam		Conjugate beam	
Type of support or *connection*	*Kinematic* or *displacement boundary*	*Static* or *force boundary condition*	*Type of support* or *connection required*
Overhang support	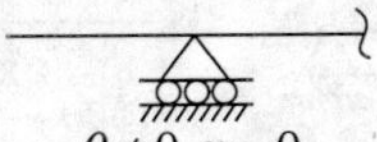 $\theta \neq 0, y = 0$	$V \neq 0$ $M = 0$	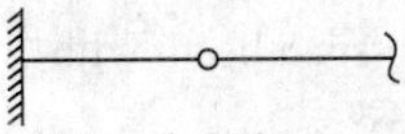 Interior hinge
Interior support (hinged or roller)	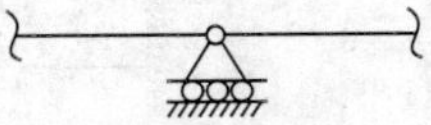 $\theta \neq 0, y = 0$	$V \neq 0$ $M = 0$	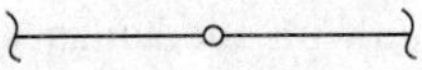
Interior hinge	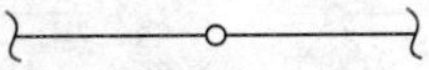 θ is discontinuous $y \neq 0$	V must be discontinuous $M \neq 0$	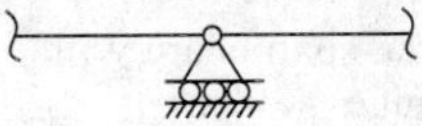

The convention used for signs of the (elastic) load, (elastic) shear and (elastic) bending-moment is the usual beam convention as described in Art. 14.5 regarding elastic-load method.

It should be noted that the conjugate beams corresponding to the statically determinate real beams are always statically determinate. The conjugate beams corresponding to the statically indeterminate real beams appear to be unstable. However, such conjugate beams are in equilibrium since they are stabilized by *M/EI* loading diagram for the corresponding real beam. The conjugate beams corresponding to some typical statically determinate and indeterminate real beams are illustrated in Table 14.2.

TABLE 14.2 Actual beam and corresponding conjugate beam

Type of beam	*Actual or real beam*	*Conjugate beam*
	I. Statically determinate beams	
Simply supported	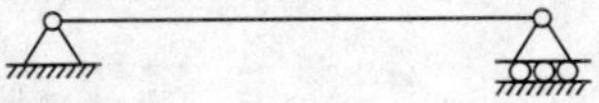	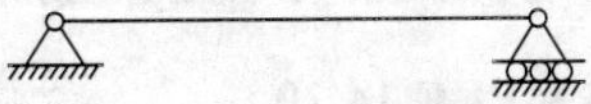
Cantilever	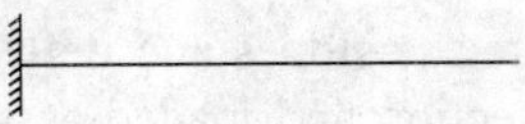	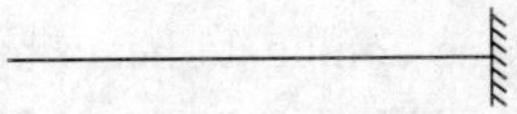
Beam with an overhang	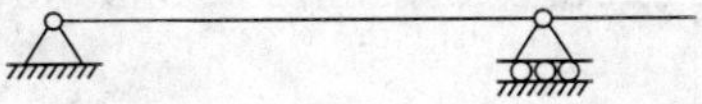	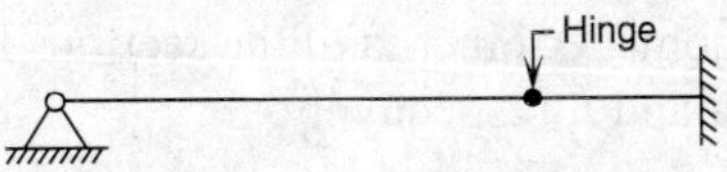 Hinge
Beam with an hinge	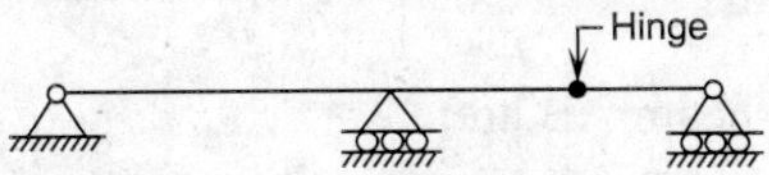 Hinge	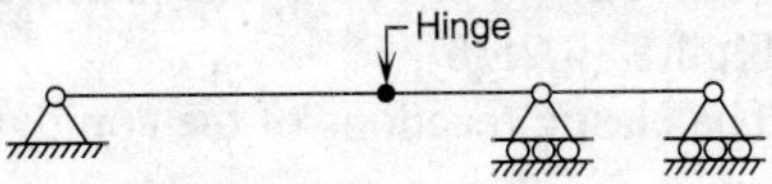 Hinge
Propped cantilever with a hinge	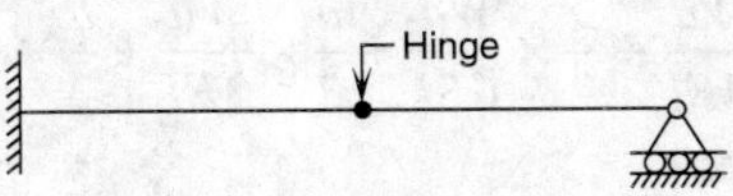 Hinge	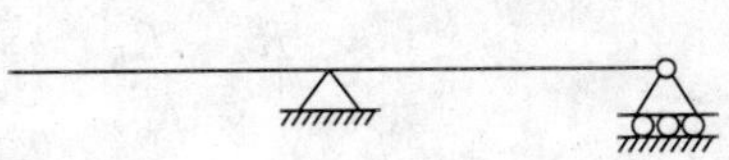

(*Contd.*)

(*Contd.*)

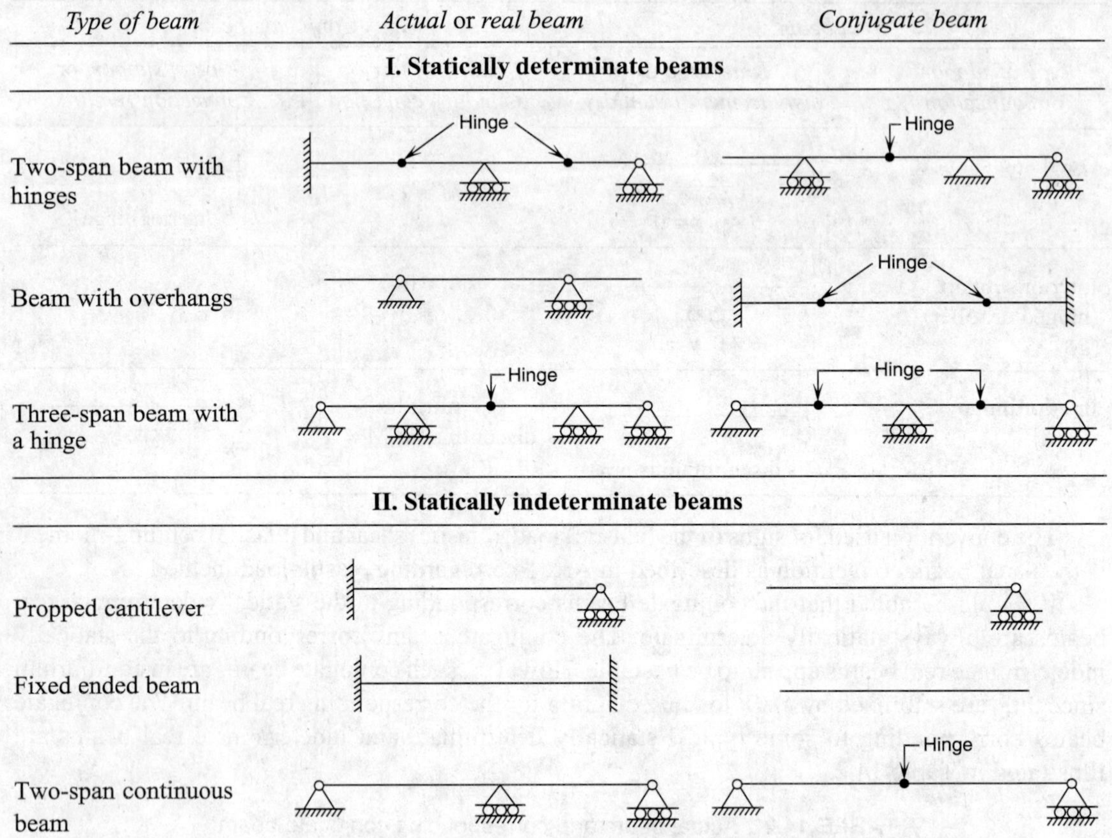

Type of beam	*Actual* or *real beam*	*Conjugate beam*
I. Statically determinate beams		
Two-span beam with hinges		
Beam with overhangs		
Three-span beam with a hinge		
II. Statically indeterminate beams		
Propped cantilever		
Fixed ended beam		
Two-span continuous beam		

Though, the conjugate-beam method involves almost the same computations as using the combination of the moment area theorems and the elastic load method, however, conjugate-beam method provides a straightforward procedure with well-established sign convention.

EXAMPLE 14.20

Using the conjugate-beam method, compute the slopes and deflection at the salient points of a simply supported beam of variable cross-section and loaded as shown in Figure 14.25(a). The moment of inertia of the sections for the left one-third, middle one-third and right one-third are $3I$, $2I$ and I, respectively.

Solution: The conjugate-beam carrying elastic load corresponding to the real beam is shown in Figure 14.25(b).

The elastic reactions of the conjugate-beam AB are:

$$R_A + R_B = \left[\frac{a}{2} \times \frac{Wa}{9EI} + \frac{a}{2} \times \frac{Wa}{6EI} + \frac{a}{2} \times \frac{Wa}{3EI} + \frac{a}{2} \times \frac{2Wa}{3EI}\right]$$

$$= \frac{23Wa^2}{36}$$

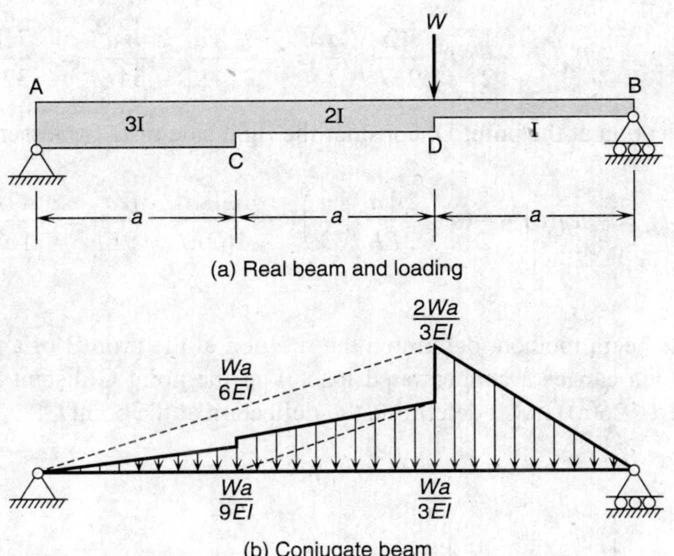

Figure 14.25 Loaded conjugate beam of Example 14.20.

$$R_B = \frac{\sum M_A}{3a} = \frac{1}{3a} \times \frac{1}{2}(a)\left[\left(\frac{Wa}{9EI}\right)\left(\frac{2a}{3}\right) + \left(\frac{Wa}{6EI}\right)\left(\frac{4a}{3}\right) + \left(\frac{Wa}{3EI}\right)\left(\frac{5a}{3}\right) + \left(\frac{2Wa}{3EI}\right)\left(\frac{7a}{3}\right)\right] = \frac{65Wa^2}{162}$$

$$R_A = \frac{23Wa^2}{36} - R_B = \frac{23Wa^2}{36} - \frac{65Wa^2}{162} = \frac{77Wa^2}{324}$$

(a) *The slopes*: The actual slopes with respect to the horizontal are given by the elastic shears in the conjugate-beam at the corresponding points. Therefore,

$$\theta_A = V_A = R_A = -\frac{77Wa^2}{324EI} \quad \text{and} \quad \theta_B = V_B = R_B = \frac{65Wa^2}{162EI}$$

$$\theta_C = V_C = R_A + \frac{1}{2}(a)\left(\frac{Wa}{9EI}\right) = -\frac{77Wa^2}{324EI} + \frac{Wa^2}{18EI} = -\frac{59Wa^2}{324EI}$$

For determine the slope at the point D, consider the right side of D for lesser calculations,

$$\theta_D = V_D = R_B - \frac{1}{2}(a)\left(\frac{2Wa}{3EI}\right) = \frac{65Wa^2}{162EI} - \frac{Wa^2}{3EI} = \frac{11Wa^2}{162EI}$$

(b) *The deflections*: The actual deflection of any point on the real beam with respect to the horizontal is equal to the elastic bending moment in the conjugate-beam at the corresponding point. Therefore,

$$\delta_C = M_C = -R_A(a) + \frac{1}{2}(a)\left(\frac{Wa}{9EI}\right)\left(\frac{a}{3}\right) = -\frac{77Wa^3}{324EI} + \frac{Wa^3}{54EI} = -\frac{71Wa^3}{324EI}\downarrow$$

For finding the deflection at the point D, consider the right side of D for lesser calculations,

$$\delta_D = M_D = -R_B(a) + \frac{1}{2}(a)\left(\frac{2Wa}{3EI}\right)\left(\frac{a}{3}\right) = -\frac{65Wa^3}{162EI} + \frac{Wa^3}{9EI} = -\frac{47Wa^3}{162EI}\downarrow$$

EXAMPLE 14.21

Using the conjugate-beam method, determine the reaction at the prop B of a propped cantilever AB of length L which carries a concentrated load W at the point C distant αL from the fixed support (see Figure 14.26(a)). Also determine the deflection at the point C.

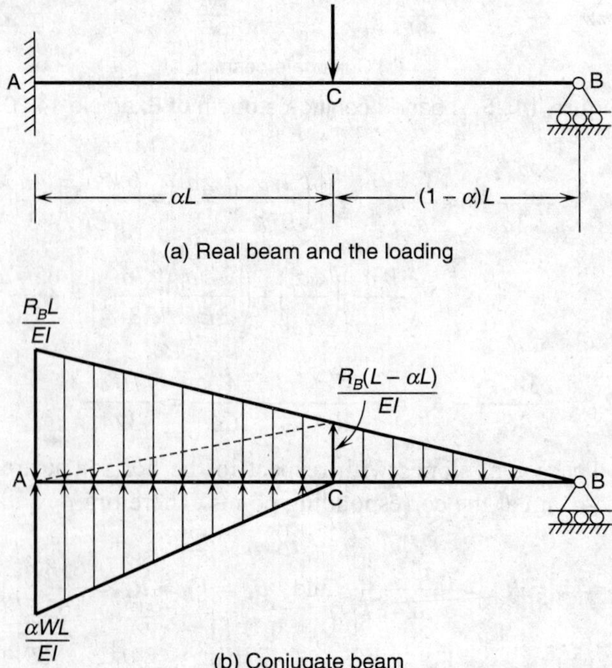

Figure 14.26 Propped cantilever of Example 14.21.

Solution: Consider that the beam is released from the redundant reaction R_B, which is treated as a part of external load system. The conjugate-beam corresponding to the real beam carrying elastic loading M/EI due to both the parts of the load drawn on the same baseline is shown in Figure 14.26(b).

(1) *Reaction at the support* B: For zero deflection at the support B of real beam, the bending moment in the conjugate-beam at point B should be zero. Thus,

$$\delta_B = M_B = -\frac{1}{2}(\alpha L)\left(\frac{\alpha WL}{EI}\right)\left(L - \frac{\alpha L}{3}\right) + \frac{1}{2}(L)\left(\frac{R_B L}{EI}\right)\left(\frac{2L}{3}\right) = 0$$

or
$$R_B = \frac{(3\alpha^2 - \alpha^3)W}{2}$$

(2) *The deflection at the point* C: The deflection at C is equal to the elastic bending moment in the conjugate-beam at the point C. Therefore,

$$\delta_C = M_C = -\frac{1}{2}(\alpha L)\left(\frac{\alpha WL}{EI}\right)\left(\frac{2\alpha L}{3}\right)$$

$$+\frac{1}{2}(\alpha L)\left(\frac{R_B L}{EI}\right)\left(\frac{2\alpha L}{3}\right) + \frac{1}{2}(\alpha L)\left[\frac{R_B(1-\alpha)L}{EI}\right]\left(\frac{\alpha L}{3}\right)$$

$$= -\frac{W(\alpha L)^3}{3EI} + \frac{R_B(\alpha L)^2 L}{6EI}[2 + (1-\alpha)] = -\frac{W(\alpha L)^2 L}{6EI}\left[2\alpha - \left(\frac{R_B}{W}\right)(3-\alpha)\right]$$

$$= -\frac{W(\alpha L)^2 L}{6EI}\left[2\alpha - \frac{(3\alpha^2 - \alpha^3)}{2}(3-\alpha)\right] = -\frac{WL^3\alpha^3}{12EI}[4 - \alpha(3-\alpha)^2]$$

EXAMPLE 14.22

A beam AB of length 2*a* is fixed at A and simply supported at B. For the left half of the beam, the moment of inertia of section is 2*I* and for the right half it is *I*. The beam is subjected to a clockwise moment *M* at the simple support (see Figure 14.27(a)). Using the conjugate-beam method, determine the reaction and slope at the prop B; and bending moment at the end A.

Solution: As in Example 14.21 consider the beam to be released from the redundant reaction R_B and treat it as a part of external load system. The conjugate-beam with elastic loading *M/EI* due to both the redundant and the applied load drawn on the same base line is shown in Figure 14.27(b). For zero deflection at the end B of real beam, the bending moment in the conjugate-beam at point B should be zero. Thus,

$$\delta_B = M_B = (a)\left(\frac{M}{2EI}\right)\left(a + \frac{a}{2}\right) + (a)\left(\frac{M}{EI}\right)\left(\frac{a}{2}\right)$$

$$-\frac{1}{2}(a)\left(\frac{R_B a}{EI}\right)\left(a + \frac{2a}{3}\right) - \frac{1}{2}(a)\left(\frac{R_B a}{2EI}\right)\left(a + \frac{a}{3}\right) - \frac{1}{2}(a)\left(\frac{R_B a}{EI}\right)\left(\frac{2a}{3}\right) = 0$$

or
$$R_B = \frac{5M}{6a}$$

$$\theta_B = V_B = (a)\left(\frac{M}{2EI}\right) + (a)\left(\frac{M}{EI}\right) - \frac{1}{2}(a)\left(\frac{R_B a}{EI}\right) - \frac{1}{2}(a)\left(\frac{R_B a}{2EI}\right) - \frac{1}{2}(a)\left(\frac{R_B a}{EI}\right)$$

$$= \frac{3Ma}{2EI} - \frac{5R_B a^2}{4EI} = \frac{3Ma}{2EI} - \frac{25Ma}{24EI} = \frac{11Ma}{24EI}$$

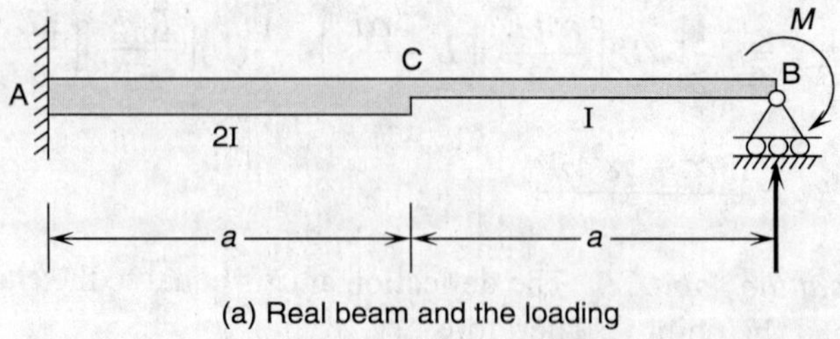

(a) Real beam and the loading

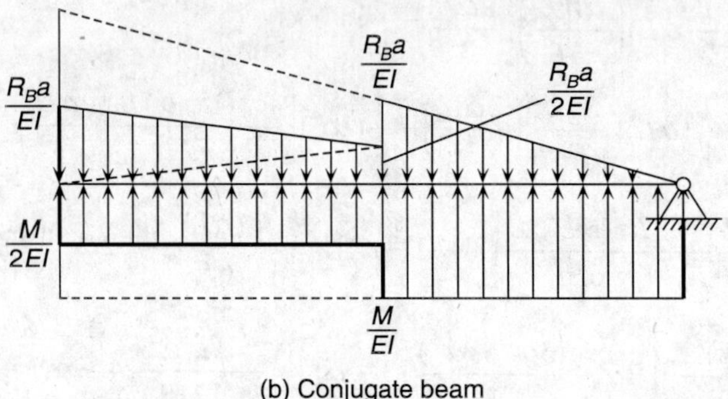

(b) Conjugate beam

Figure 14.27 Propped cantilever of variable section of Example 14.22.

Bending moment at A,

$$M_A = R_B(2a) - M = \frac{5M}{6a} \times (2a) - M = \frac{2M}{3}$$

EXAMPLE 14.23

Using the conjugate-beam method, compute: (i) the slopes at the sections C and C′ and (ii) the deflections at the sections C and D of the beam of variable cross-section and loaded as shown in Figure 14.28(a).

Solution: The behaviour of the link is similar to an interior hinge and as such it can be considered as an interior support. The conjugate-beam carrying elastic load corresponding to the real beam is shown in Figure 14.28(c).

(i) *Reactions*: The elastic reactions (see Figure 14.28(b)) of the conjugate-beam ACBD are:

$$R_{C'} = \frac{\sum M_B}{4a} = \left(\frac{1}{4a}\right)\left[\left(\frac{1}{2}\right)(4a)\left(\frac{3Wa}{EI}\right)\left(4a + \frac{8a}{3}\right) - \left(\frac{1}{2}\right)(4a)\left(\frac{3Wa}{2EI}\right)\left(\frac{4a}{3}\right)\right]$$

$$= \frac{9Wa^2}{EI} \uparrow$$

$$V_{B,\text{left}} = \left(\frac{1}{2}\right)(4a)\left(\frac{3Wa}{EI}\right) - \left(\frac{1}{2}\right)(4a)\left(\frac{3Wa}{2EI}\right) - \frac{9Wa^2}{EI} = -\frac{6Wa^2}{EI} \uparrow$$

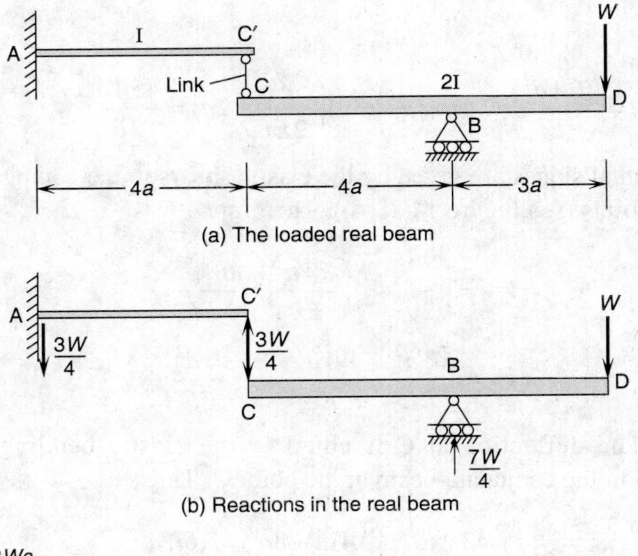

(a) The loaded real beam

(b) Reactions in the real beam

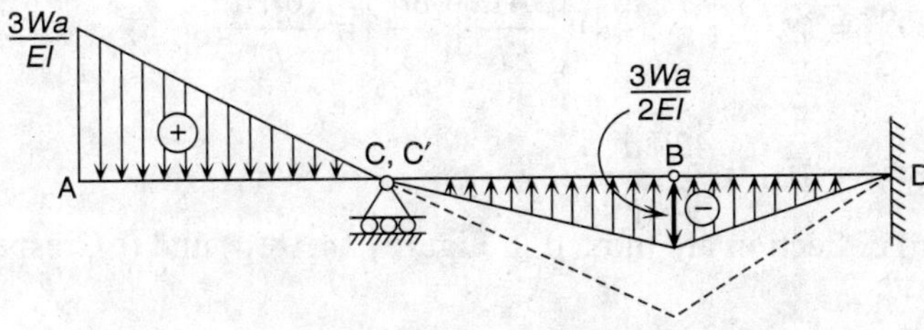

(c) Conjugate beam with elastic loading

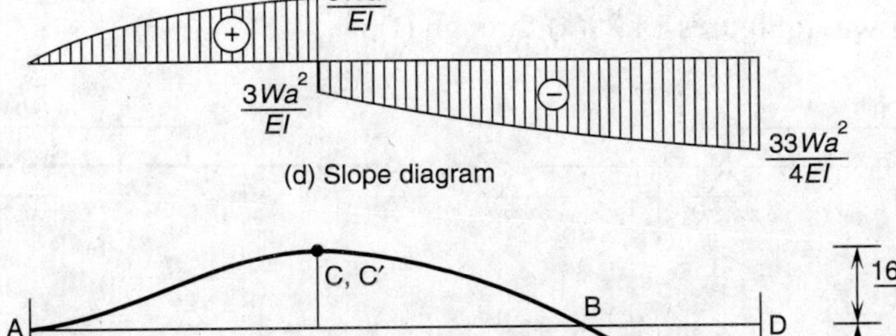

(d) Slope diagram

C, C′ 16Wa³/EI

A B D 45Wa³/2EI

(e) Deflection diagram

Figure 14.28

$$R_D = \frac{6Wa^2}{EI} + \left(\frac{1}{2}\right)(3a)\left(\frac{3Wa}{2EI}\right) = \frac{33Wa^2}{4EI} \downarrow$$

$$M_D = \left(\frac{1}{2}\right)(4a)\left(\frac{3Wa}{EI}\right)\left(7a + \frac{8a}{3}\right) - \left(\frac{1}{2}\right)(7a)\left(\frac{3Wa}{2EI}\right)\left(\frac{7a+3a}{3}\right) - \frac{9Wa^2}{EI}(7a) = \frac{45Wa^2}{2EI}$$

Alternatively,

$$M_D = \left(\frac{6Wa^2}{EI}\right)(3a) + \left(\frac{1}{2}\right)(3a)\left(\frac{3Wa}{2EI}\right)(2a) = \frac{45Wa^3}{2EI}$$

(ii) *Slopes*: The actual slopes are given by the elastic shears in the conjugate-beam at the corresponding points (see Figure 14.28(d)). Therefore,

$$\theta_{C'} = V_{C'} = \left(\frac{1}{2}\right)(4a)\left(\frac{3Wa}{EI}\right) = \frac{6Wa^2}{EI}$$

$$\theta_C = V_C = \frac{6Wa^2}{EI} - \frac{9Wa^2}{EI} = -\frac{3Wa^2}{EI}$$

(iii) *Deflections*: The deflection at C is equal to the elastic bending moment (see Figure 14.28(e)) in the conjugate-beam at the point C. Thus,

$$\delta_C = M_C = \left(\frac{1}{2}\right)(4a)\left(\frac{3Wa}{EI}\right)\left(\frac{8a}{3}\right) = \frac{16Wa^3}{EI}$$

Similarly, $$\delta_D = M_D = \frac{45Wa^3}{2EI}$$

The variations of slope and deflection are plotted in Figures 14.18(d) and (e), respectively.

14.8 PROBLEMS

14.1 Draw the qualitative deflected shapes for the beams subjected to concentrated and distributed loads as shown in Figures 14.29(a) through (f).

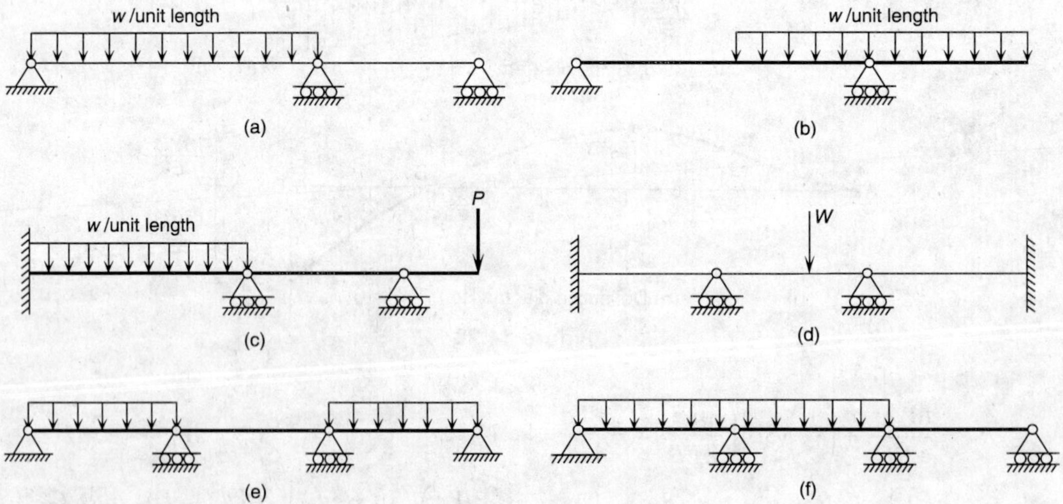

Figure 14.29

14.2 Draw the qualitative deflected shapes for the braced or non-sway and un-braced rigid frames subjected to external loads as shown in Figures 14.30(a) through (f).

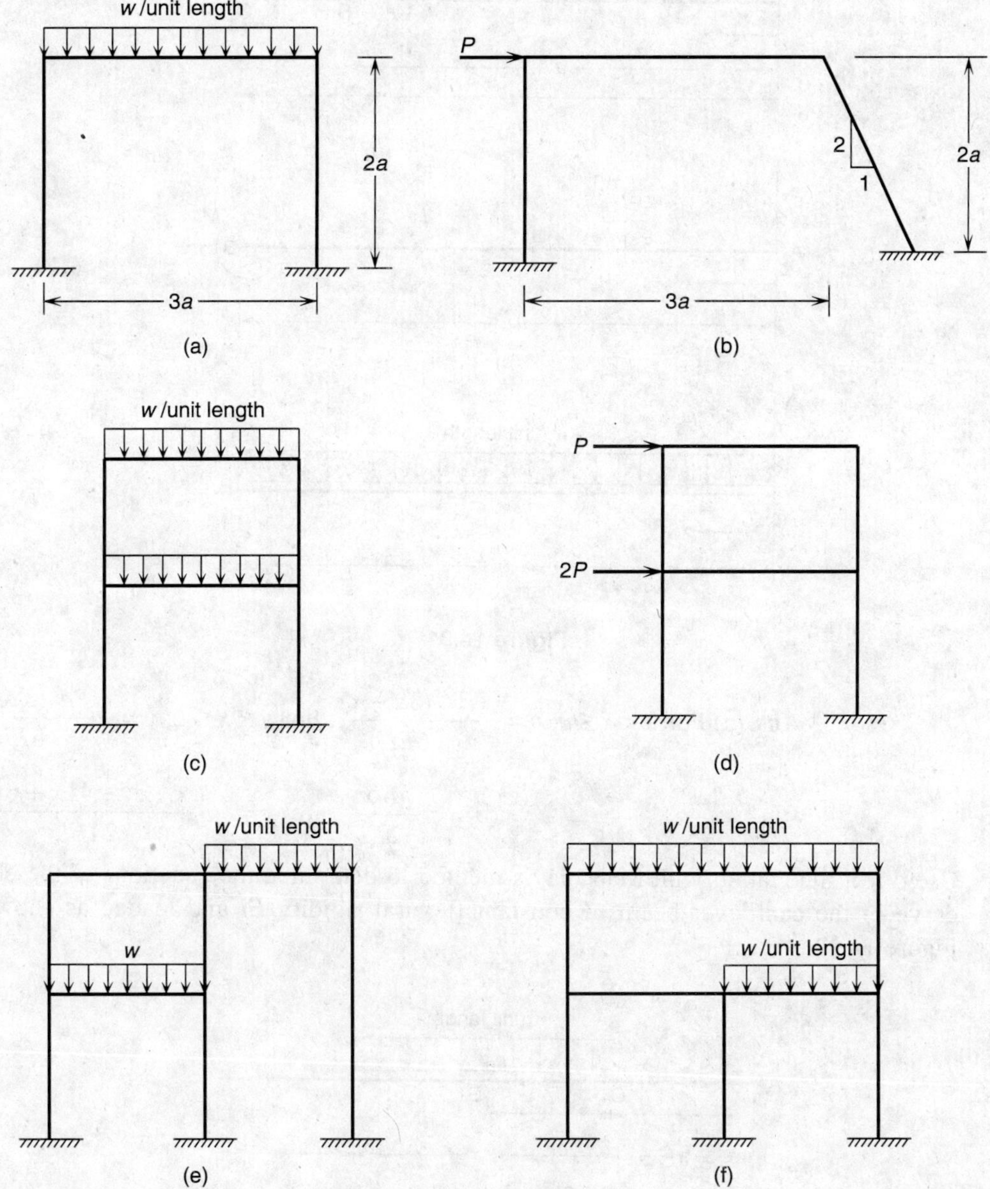

Figure 14.30

14.3 Use direct integration to determine the equations of the elastic curve of the cantilever beams of constant flexural rigidity EI and loaded as shown in Figures 14.31(a) through (c).

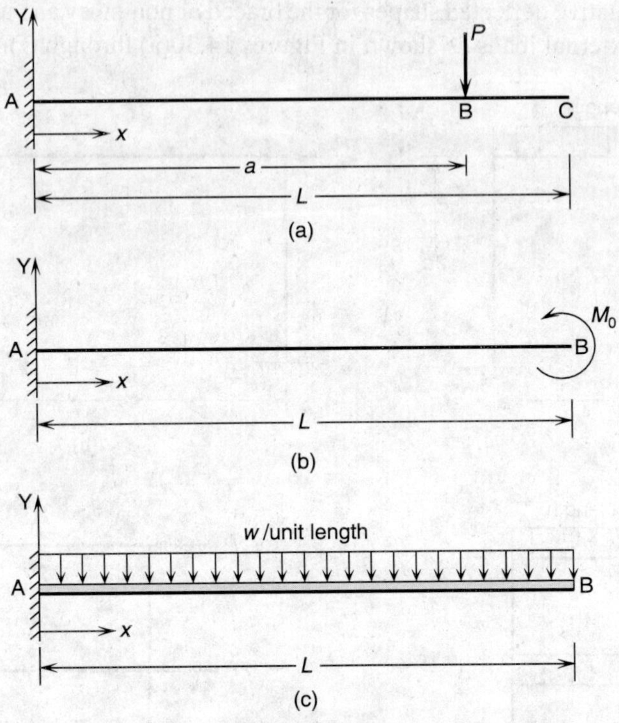

Figure 14.31

$$\left[\textbf{Ans.} \text{ (a) For } 0 < x \le a;\ y = -\frac{Px^2(3a - x)}{6EI}; \text{ For } a \le x \le L;\ y = -\frac{Pa^2(3x - a)}{6EI} \right.$$

$$\left. \text{(b) } y = \frac{M_0 x^2}{2EI}; \quad \text{(c) } y = -\frac{wx^2(x^2 - 4Lx + 6L^2)}{24EI} \right]$$

14.4 Use direct integration with Macaulay's method to determine the equations of the elastic curve of the cantilever beam of constant flexural rigidity EI and loaded as shown in Figure 14.32.

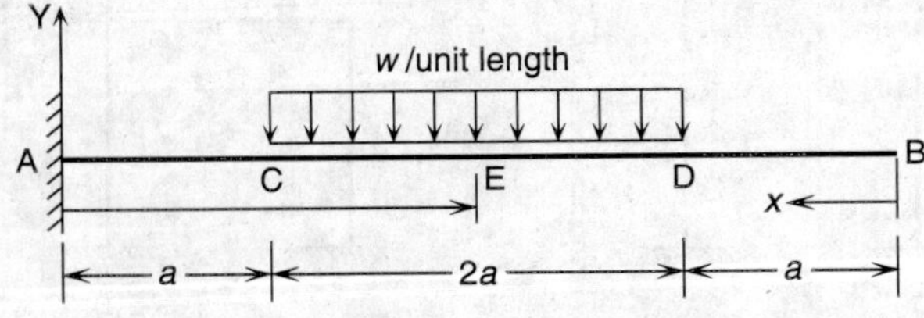

Figure 14.32

$$\left[\textbf{Ans.} \ y = -\frac{wa^4}{24EI}\left[\langle x - a\rangle^4 - \langle x - 3a\rangle^4 - 104\xi + 336\right] \right.$$

$$\left. \text{where } \xi = \frac{x}{a}, \frac{\theta_E}{\theta_B} = \frac{25}{26} \text{ and } \frac{y_E}{y_B} = \frac{43}{112} \right]$$

14.5 Use direct integration method to determine the equations of the elastic curve of the simply supported beams of constant flexural rigidity EI and loaded as shown in Figures 14.33(a) through (c).

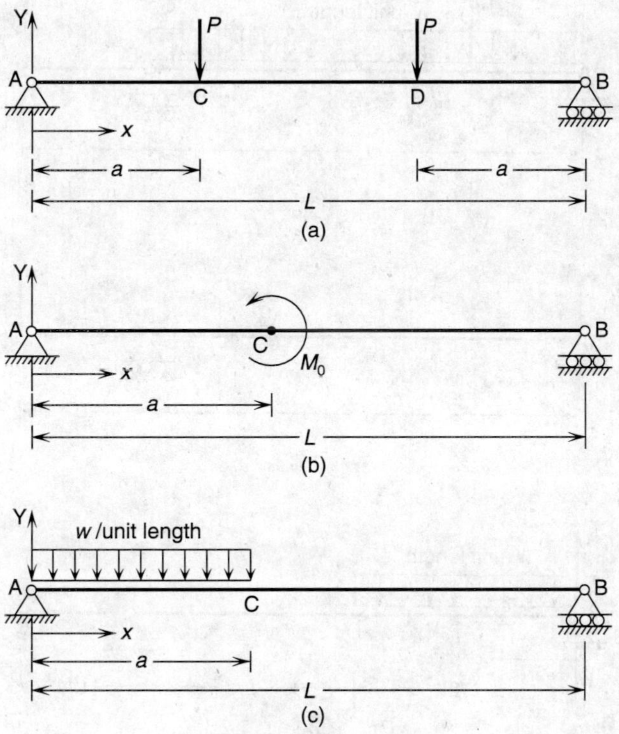

Figure 14.33

$$\left[\textbf{Ans. } \text{(a) For } 0 \le x \le a;\ y = -\frac{Px(3aL - 3a^2 - x^2)}{6EI};\right.$$

$$\text{For } a \le x \le L - a;\ y = -\frac{Pa}{6EI}(3Lx - 3x^2 - a^2)$$

$$\text{(b) For } 0 \le x \le a;\ y = -\frac{M_0}{6EIL}(6Lax - 3a^2x - 2L^2x - x^3)$$

$$\text{For } a \le x \le L;\ y = \frac{M_0}{6EIL}(3La^2 + 3Lx^2 - x^3 - 2L^2x - 3a^2x)$$

$$\text{(c) For } 0 \le x \le a;\ y = -\frac{wx}{24EIL}\{a^2(2L - a)^2 - 2ax^2(2L - a) + Lx^2\}$$

$$\left.\text{For } a \le x \le L;\ y = -\frac{wa^2(L - x)}{24EIL}(4Lx - 2x^2 - a^2)\right]$$

14.6 Determine the equations of the elastic curve of the beams of constant flexural rigidity EI with an overhang at one end and loaded as shown in Figures 14.34(a) through (c).

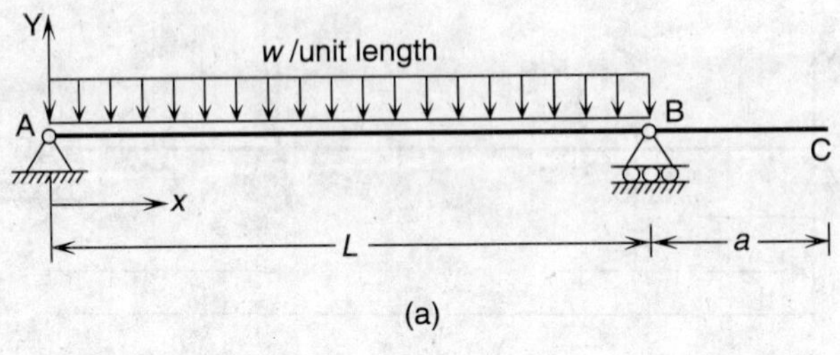

(a)

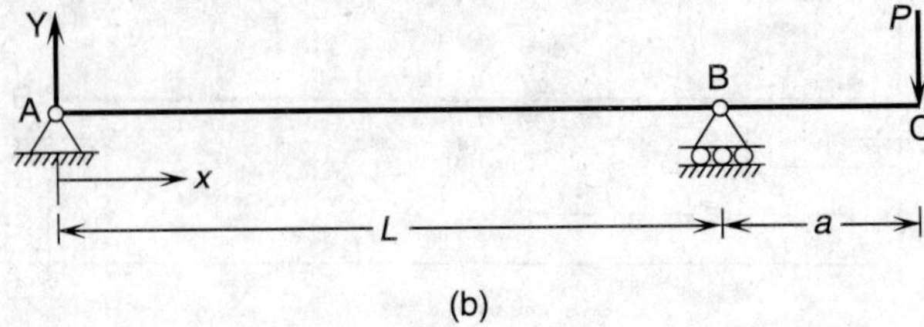

(b)

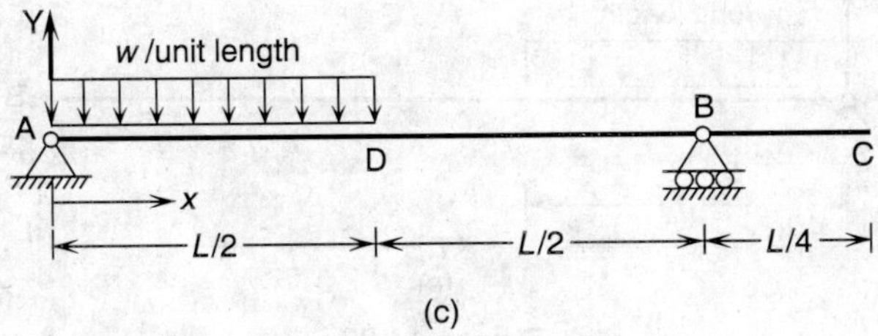

(c)

Figure 14.34

[**Ans.** (a) For $0 \le x \le L$; $y = -\dfrac{wx}{24EI}(L^3 - 2Lx^2 + x^3)$ and $y_C = \dfrac{wL^3 a}{24EI}$

(b) $y_c = -\dfrac{Pa^2(L+a)}{3EI}$; For max. deflection, $x_m = 0.577L$ and $y_{max} = \dfrac{0.06415PaL^2}{EI} \uparrow$

(c) $y = \dfrac{wL^4}{384EI}[16\xi^4 - 24\xi^3 - \langle 2\xi - 1\rangle^4 - 8\langle \xi - 1\rangle^4 + 9\xi]$ where $\xi = \dfrac{x}{L}$]

14.7 Determine the equations of the elastic curve of the beams of constant flexural rigidity EI and loaded as shown in Figures 14.35(a) through (c).

14.8 Determine the equations of the elastic curve of the beams of variable flexural rigidity EI and loaded as shown in Figure 14.36.

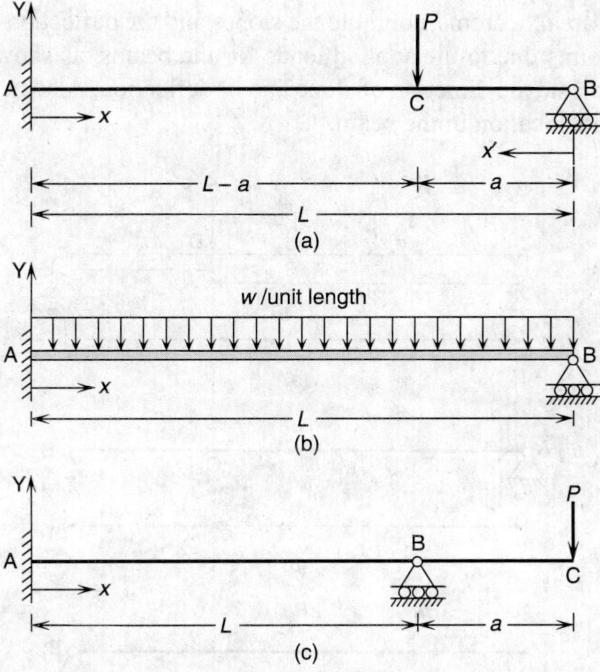

Figure 14.35

$$\left[\textbf{Ans.}\ \text{(a) For } 0 < x < L - a;\ y = -\frac{Pax^2}{12EIL^3}(3C_1 - C_2 x) \text{ where } C_1 = L(L^2 - a^2);\right.$$

$$C_2 = (3L^2 - a^2),\ \text{For } 0 < x' \le a;\ y = -\frac{P(L-a)^2 x'}{12EIL^3}[3L^2 a - (2L + a)x'^2]$$

$$\left.\text{(b) } y = -\frac{wx^2(L-x)(3L-2x)}{48EI};\ \text{(c) } R_A = -\frac{3Pa}{2L},\ M_A = \frac{Pa}{2} \text{ and } y_C = -\frac{Pa^2}{12EI}(3L + 4a)\right]$$

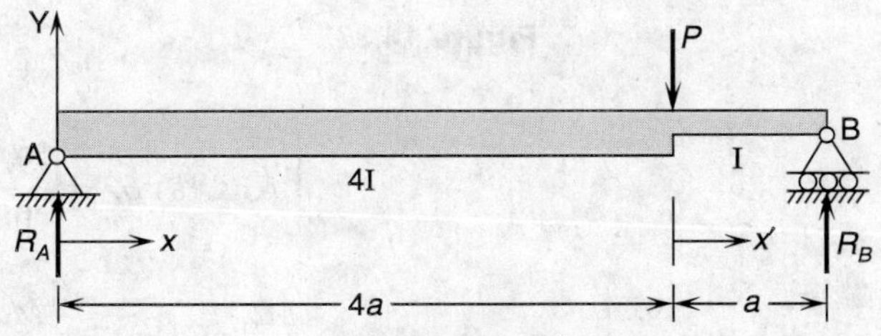

Figure 14.36

$$\left[\textbf{Ans.}\ \text{For } 0 \le x \le 4a;\ y = \frac{Pa^3}{600EI}(5x^3 - 144x);\ x_m = 3.1a,\ y_{\max} = \frac{1.99Pa^3}{4EI}\right.$$

$$\left.\text{For } 0 \le x' \le a;\ y = \frac{2Pa^3}{75EI}(15x'^2 - 5x'^3 + 6x' - 16)\right]$$

14.9 Using moment-area theorems, compute the slopes and the deflection of the elastic curves at the specified points due to the applied loads for the beams, as shown in Figures 14.37(a) through (d). Indicate the direction of slope and of deflection. Also determine the maximum deflection and its location in the beam.

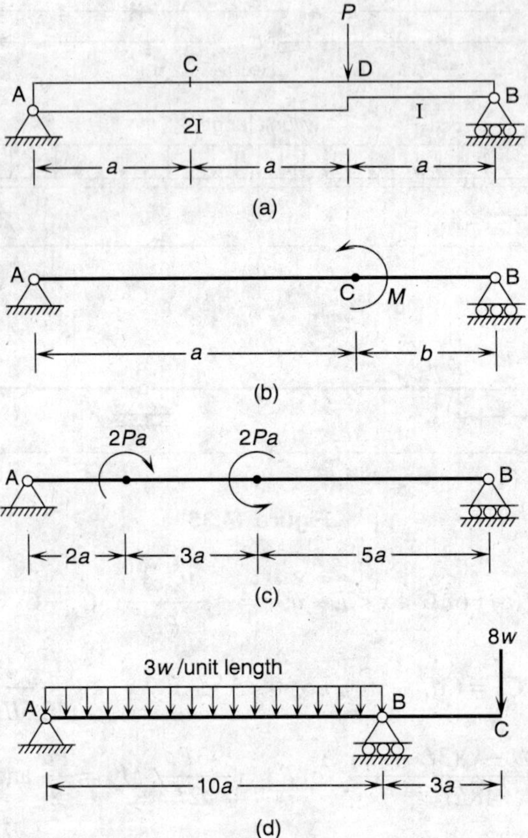

Figure 14.37

[**Ans.** (a) $\theta_C = \dfrac{19Pa^2}{108EI}$ and $\delta_C = \dfrac{25Pa^3}{108EI}$;

(b) $\theta_C = -\dfrac{Ma}{3EI}$ and $\delta_C = \dfrac{4Ma^2}{9EI}$; (d) $\delta_C = \dfrac{62wa^4}{EI}$]

14.10 A beam A–B–C–D is subjected to an end moment M_0 at A and an unknown moment M_C at the mid-point of the overhanging portion as shown in Figure 14.38. Using moment-area theorems, determine the magnitude of the bending moment M_C, so that the deflection at the point D will be equal to zero.

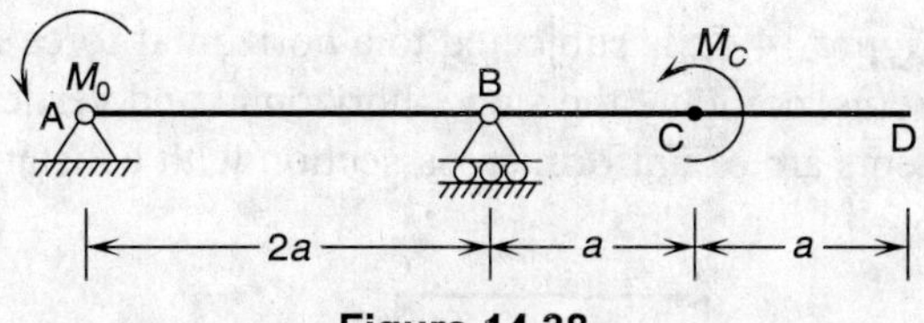

Figure 14.38

[**Ans.** For $\delta_D = 0$; $M_C = 0.235M_0$]

14.11 Use moment-area theorems to determine the slopes and deflections at the salient points of the beams loaded as shown in Figures 14.39(a) and (b). The beams are of variable cross-section as indicated.

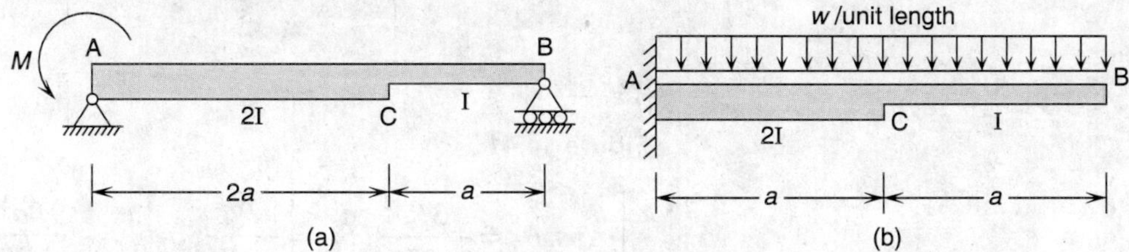

Figure 14.39

$$\left[\textbf{Ans.}\ \text{(a)}\ \theta_A = \frac{14Ma}{27EI}, \theta_B = -\frac{17Ma}{54}, \theta_C = -\frac{4Ma}{27EI} \text{ and } \delta_C = \frac{7Ma^2}{27EI} \uparrow\right.$$

$$\left.\text{(b)}\ \theta_B = -\frac{3wa^3}{4EI}, \theta_C = -\frac{7wa^3}{12EI}, \delta_B = -\frac{17wa^4}{16EI} \downarrow \text{ and } \delta_C = -\frac{17wa^4}{48EI} \downarrow\right]$$

14.12 A cantilever BC of length $2a$ connected to the top B of a column AB of length $3a$ as shown in Figure 14.40 carries a concentrated load W at its free end C. Using moment-area theorems, compute the slope, horizontal and vertical deflections at the point C. Both the beams are of uniform cross-section with constant E and I.

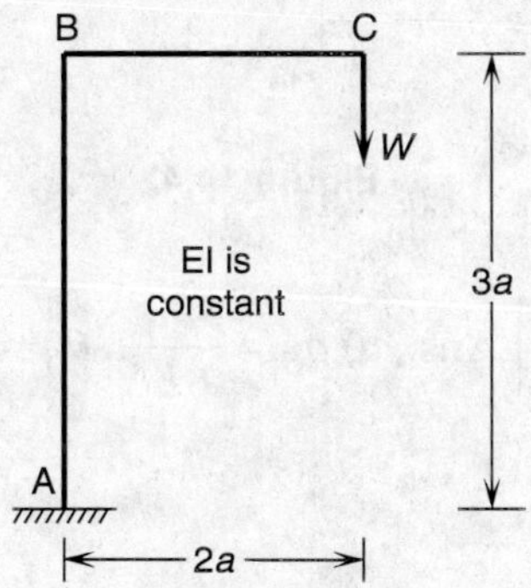

Figure 14.40

$$\left[\textbf{Ans.}\ \theta_B = \frac{6Wa^2}{EI}, \delta_{C,V} = -\frac{44Wa^3}{3EI} \downarrow\right]$$

14.13 A bent shown in Figure 14.41 is subjected to a horizontal force P at its free end D. Using moment-area theorems, compute the slope, horizontal and vertical deflections at the point D. All the components are of uniform cross-section with constant E and I.

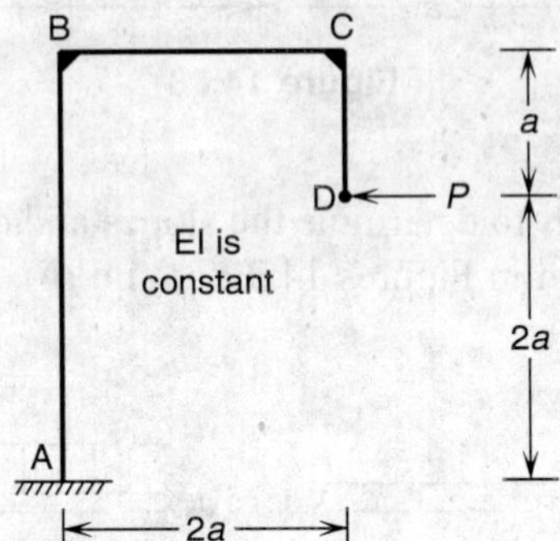

Figure 14.41

$$\left[\textbf{Ans. } \theta_B = -\frac{3Pa^2}{2EI}, \theta_C = -\frac{Pa^2}{2EI}, \delta_{B,H} = -\frac{9Pa^3}{2EI} \leftarrow \right.$$
$$\left. \text{and } \delta_{C,V} = \delta_{D,V} = \frac{Pa^3}{EI} \uparrow, \delta_{D,H} = -\frac{16Pa^3}{3EI} \leftarrow \right]$$

14.14 Using the elastic-load method, compute the slope and deflection as specific below for beams loaded as shown in Figures 14.42(a) and (b). EI is constant throughout. (a) θ_B, θ_C, δ_B and δ_C; (b) θ_C and δ_C.

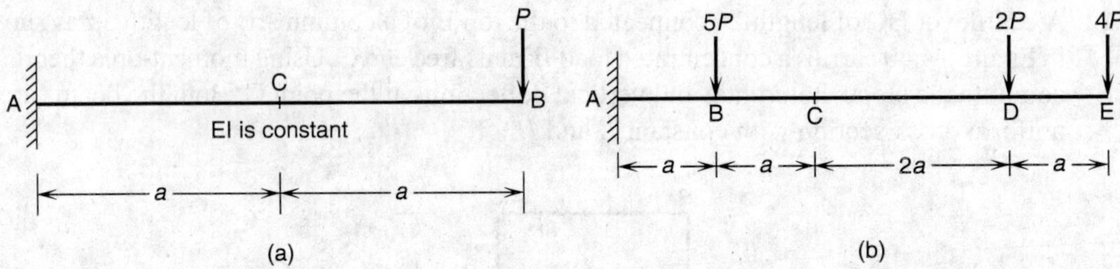

Figure 14.42

$$\left[\textbf{Ans. } (a)\ \delta_C = \frac{5Pa^3}{6EI}, \theta_C = \frac{3Pa^2}{2EI}, \theta_B = \frac{2Pa^2}{EI} \text{ and } \delta_B = \frac{8Pa^3}{3EI}\right.$$
$$\left. (b)\ \theta_C = \frac{93Pa^2}{2EI}, \delta_C = \frac{313Pa^3}{6EI}\right]$$

14.15 Using the elastic-load method, compute the slopes and deflections for the beams shown in Figures 14.43(a) to (b) as specified below.

(a) θ_B and δ_B; (b) θ_A, θ_B, δ_B and δ_C

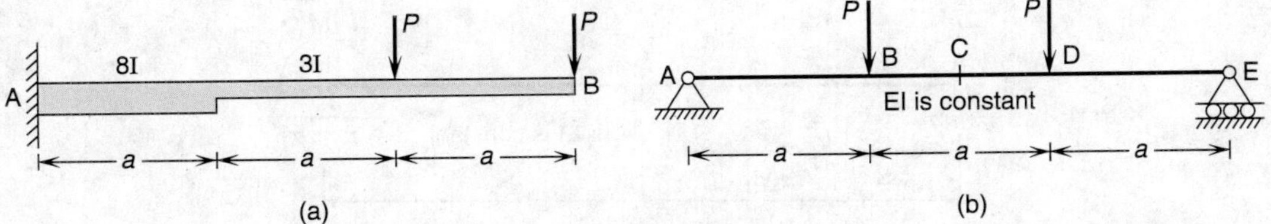

Figure 14.43

$$\left[\textbf{Ans.}\ (a)\ \theta_B = \frac{4Pa^2}{3EI} \text{ and } \delta_B = \frac{39Pa^3}{16EI}\right.$$

$$\left.(b)\ \theta_A = \frac{Pa^2}{EI}, \theta_B = \frac{Pa^2}{2EI}, \delta_B = \frac{5Pa^3}{6EI} \text{ and } \delta_C = \frac{23Pa^3}{24EI}\right]$$

14.16 Using the elastic-load method, compute the slopes at salient points and deflection at the location of hinge for beams loaded as shown in Figures 14.44(a) and (b). EI is constant throughout.

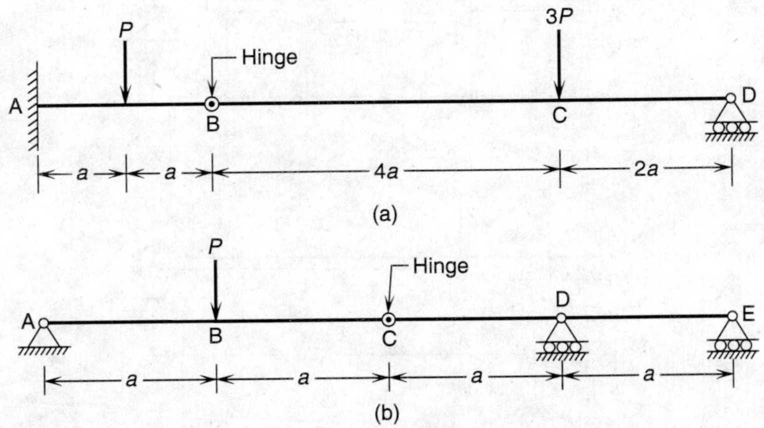

Figure 14.44

$$\left[\textbf{Ans.}\ (a)\ \theta_A = 0, \theta_{B,\text{left}} = -\frac{5Pa^2}{2EI}, \theta_{B,\text{right}} = -\frac{19Pa^2}{4EI}, \theta_D = \frac{29Pa^2}{4EI}, \text{ and } \delta_B = \frac{7Pa^3}{2EI}\downarrow\right.$$

$$\left.(b)\ \theta_A = -\frac{5Pa^2}{12EI}, \theta_B = -\frac{Pa^2}{6EI}, \theta_{C,\text{left}} = \frac{Pa^2}{12EI}, \theta_{C,\text{right}} = -\frac{5Pa^2}{12EI}, \text{ and } \delta_B = \delta_C = \frac{Pa^3}{3EI}\right]$$

14.17 Using moment-area method, show that the maximum deflection of a beam built-in at both the ends and supporting a uniformly distributed load is one-fifth the maximum deflection of the same beam simply supported. EI is constant for the beam.

14.18 Using conjugate-beam method, compute the slopes at the ends and deflection at the mid-point of a simply supported beam loaded as shown in Figure 14.45.

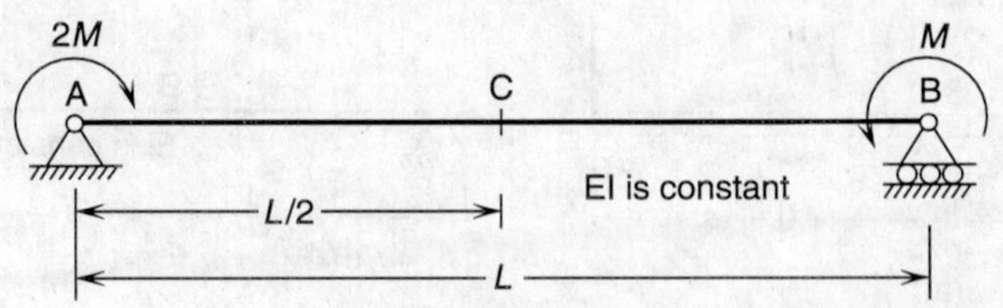

Figure 14.45

$$\left[\textbf{Ans. } \theta_A = \frac{5ML}{6EI}, \theta_B = \frac{2ML}{3EI} \text{ and } \delta_C = \frac{3ML^2}{EI}\right]$$

14.19 Using conjugate-beam method, compute the slopes and deflections for the beams of variable cross-section shown in Figures 14.46(a) to (b) as specified below.

(a) θ_A, θ_B, θ_C, θ_D, δ_C and δ_D; (b) θ_A, θ_B, δ_B and δ_{max}

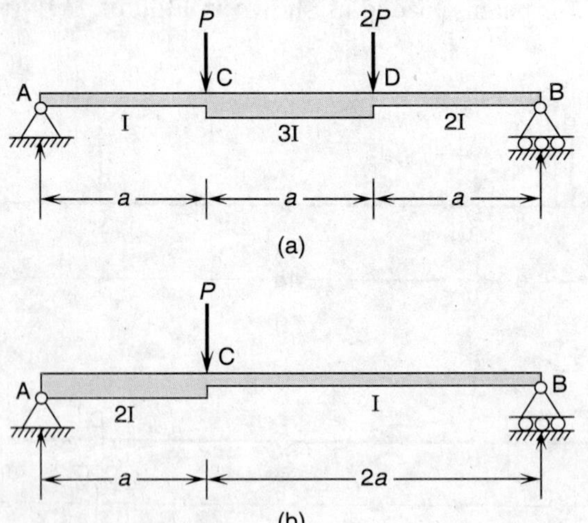

Figure 14.46

$$\left[\textbf{Ans. } \text{(a) } \theta_A = \frac{139Pa^2}{162EI}, \theta_B = \frac{235Pa^2}{324EI}, \theta_C = \frac{31Pa^2}{162EI},\right.$$

$$\theta_D = \frac{25Pa^2}{81EI}, \delta_C = \frac{103Pa^3}{162EI} \text{ and } \delta_D = \frac{95Pa^3}{162EI}$$

$$\left.\text{(b) } \theta_A = \frac{23Pa^2}{54EI}, \theta_B = \frac{11Pa^2}{27EI}, \delta_C = \frac{10Pa^3}{EI} \text{ and } \delta_{max} = \frac{0.425Pa^3}{EI}\right]$$

14.20 Using conjugate-beam method, compute the slopes and deflections for the beams shown in Figures 14.47(a) to (d) as specified. *EI* is constant for the beams.

(a) θ_A, θ_C, θ_D, δ_B and δ_C; (b) θ_C, θ_D, δ_B and δ_C; (c) δ_{max} and (d) θ_C and δ_C

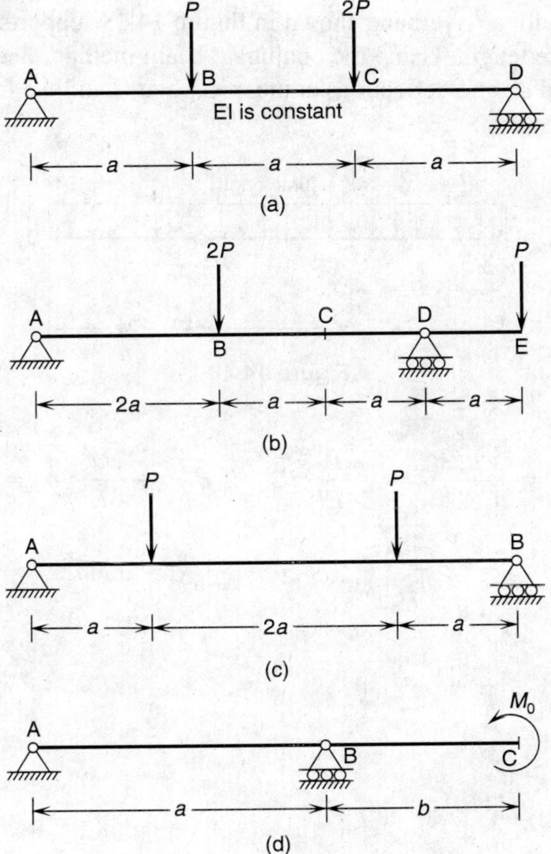

Figure 14.47

[**Ans.** (a) $\theta_A = -\dfrac{13Pa^2}{9EI}$, $\theta_C = \dfrac{13Pa^2}{18EI}$, $\theta_D = \dfrac{14Pa^2}{9EI}$, $\delta_B = \dfrac{11Pa^3}{9EI} \downarrow$ and $\delta_C = \dfrac{23Pa^3}{18EI} \downarrow$

$\delta_{\max} = \dfrac{1.44Pa^3}{EI}$ occurs at 1.55a from the end A

(b) $\theta_C = \dfrac{25Pa^2}{24EI}$, $\theta_D = \dfrac{2Pa^2}{3EI}$, $\delta_B = \dfrac{5Pa^3}{3EI} \downarrow$ and $\delta_C = \dfrac{Pa^3}{3EI} \uparrow$

(c) Maximum deflection $\delta_{\max} = \dfrac{11Pa^3}{6EI}$

(d) $\theta_C = \dfrac{M_0}{3EI}(a+3b)$ and $\delta_C = \dfrac{2M_0 b}{6EI}(2a+3b)$]

14.21 The beam ABD with an overhang shown in Figure 14.48 supports a uniformly distributed load over its entire length. Using the conjugate-beam method, determine the slopes at the points A, B, C and D; and deflections at the mid-point C and the free end D. *EI* is constant for the beam.

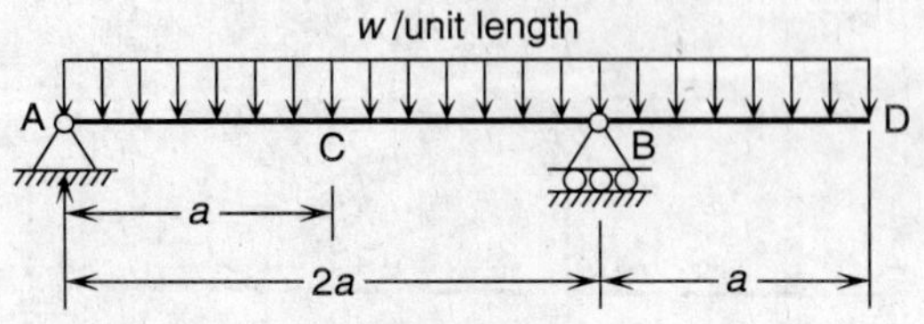

Figure 14.48

$$\left[\textbf{Ans. } \theta_A = -\frac{wa^3}{6EI}, \theta_C = \frac{wa^3}{24EI}, \theta_B = 0, \theta_D = -\frac{wa^3}{6EI}, \delta_C = \frac{wa^4}{12EI} \downarrow \text{ and } \delta_D = \frac{wa^4}{8EI} \right]$$

Chapter 15

Deflections

(Energy Methods)

15.1 INTRODUCTION

In the preceding chapter, the slopes and deflections of the structures have been computed using various forms of the direct integration and geometrical methods. These procedures are satisfactory for many structures, including some that were rather complicated. However, methods are applicable to the structures subjected mainly to flexural action like beams and frames; they cannot be used for the structures where axial forces are predominant, e.g. trusses.

In this chapter, the methods for computation of the slopes and deflections of structures, based on the principle of *conservation of energy* are discussed. The two commonly used methods, namely the *virtual work* and *Catigliano's theorems* are introduced. Because of the simplicity of the formulation and their applicability to a wide range of structures these methods are popular.

15.2 ENERGY METHODS

All energy methods use the concept of work. When a structure deforms, the external loads move along the deformations and do some work. The internal stresses developed due to applied loads also act through their respective strains and do internal work.

15.2.1 Work of Externally Applied Forces

The work is defined as the product of a force and its displacement in the direction in which the force is acting. If the force is constant and acts in the direction of displacement, the work done is expressed as:

$$W_e = F\Delta \tag{15.1}$$

On the other hand, if the force F varies during displacement or deformation, consider the force F moving through a very small but finite distance $d\Delta$ along its direction of action. The force will not change in magnitude appreciably during this small movement $d\Delta$ and the elementary work is defined as

$$dW_e = F\,d\Delta \tag{15.2}$$

If the force has moved a total distance Δ, the work done could be calculated by dividing the distance Δ into a number of arbitrary small distances $d\Delta$ and the work would be approximated as,

$$W_e = \int_0^{\Delta} F\,d\Delta \tag{15.3}$$

In structural analysis problems force F varies during the deformation, e.g. consider the very mportant case of linear relationship between the force F and the displacement Δ, i.e. F is linear nction of Δ. Thus,

$$F = c\Delta \tag{15.4}$$

re c is a constant. Substituting this into Eq. (15.3)

$$W_e = \int_0^{\Delta} (c\Delta)\,d\Delta = \frac{c\Delta^2}{2} \tag{15.5}$$

lenotes the displacement or deflection of a point, and since $F = c\Delta$, then

$$W_e = \frac{1}{2}F\Delta \tag{15.6}$$

Iere the loading F represents a single force, P or moment M, and the deformation Δ represents corresponding displacement and rotation, respectively. Thus F and Δ are frequently referred to generalized force and generalized deformation. The curve $F = f(\Delta)$ would be in general some ion-linear continuous curve as shown in Figure 15.1(a). The work done expression $\int_0^{\Delta} F\,d\Delta$

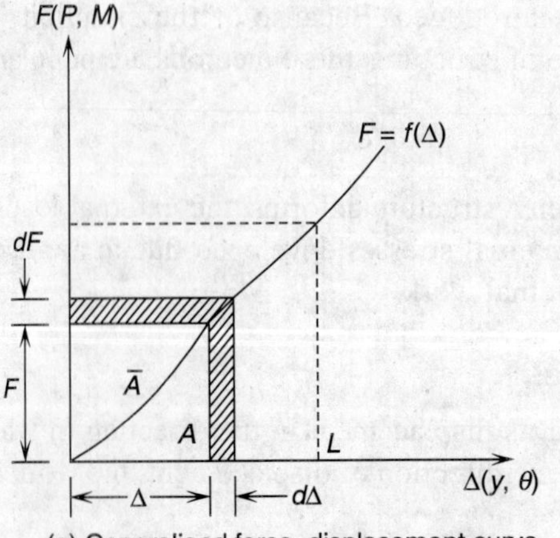

(a) Generalised force–displacement curve

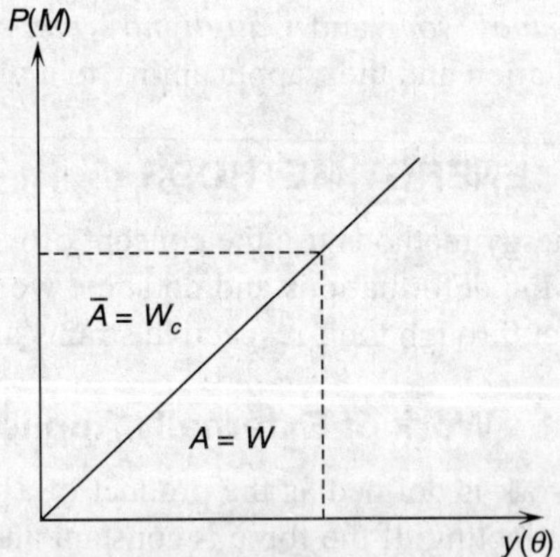

(b) Load–deformation work diagram

Figure 15.1 Geometric representation of work w.r.t. load deformation diagram.

represents the area under the curve. In other words, the work can be interpreted geometrically as area under load-deformation diagram. If the curve $F = f(\Delta)$ becomes straight line, i.e. there exists a linear relationship between force and deformation, the area renders a triangle as shown in Figure 15.1(b) and we obtain

$$W_e = \frac{1}{2} P\Delta \quad \text{or} \quad \frac{1}{2} M\theta \tag{15.7}$$

The area $\overline{A}$ which together with A forms rectangle is obviously numerically identical to A in the case of linear force-deformation relationship. The area $\overline{A}$ is termed *complementary work* W_c. Thus the complementary work is defined as

$$W_c = \int dP\Delta \quad \text{or} \quad \int dM\,\theta \tag{15.8}$$

15.2.2 Eigen Work and Displacement Work

The total work done by a force comprises of two parts namely *the* eigen work and the displacement work. *Eigen work is defined as the work done by a force moving along the deformation caused by it*. In contrast to eigen work, *the displacement work is the work done by a force along the displacement caused by another force*.

As an example, consider the beam shown in Figure 15.2(a) carrying multiple loads. In the double subscripted quantities term δ_{ij} represents a displacement at the point i due to force P_j acting at the point j. Consider the force P_1 acting at point 1 which produces eigen work $W_{11} = (P_1\delta_{11})/2$. Now, suppose that another load P_2 is applied to the beam at point 2. This load will cause additional displacements δ_{22} and δ_{12} at the points 2 and 1, respectively. Thus the eigen work of load P_2, $W_{22} = (P_2\delta_{22})/2$.
The eigen work of the external forces (loading) of the system is

$$W_{\text{eig}} = \sum_{i=1}^{2} W_{ii} = W_{11} + W_{22} = \frac{1}{2} P_1\delta_{11} + \frac{1}{2} P_2\delta_{22} \tag{15.9}$$

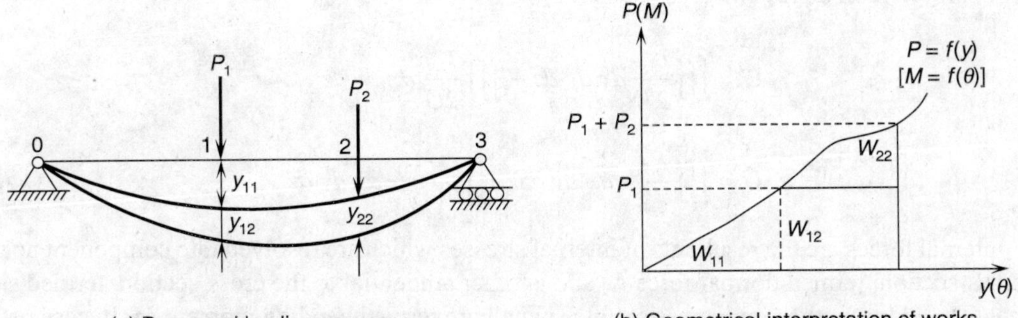

(a) Beam and loading sequence

(b) Geometrical interpretation of works

Figure 15.2 Loading the beam and geometrical interpretation of the eigen work and displacement works.

while the displacement work is

$$W_{\text{dis}} = \sum_{i,j}^{2} W_{ij} = W_{12} = P_1\delta_{12} \tag{15.10}$$

Figure 15.2(b) presents the geometrical interpretation of eigen and displacement works. The displacement work wherein the load remains constant could be interpreted as *virtual work* which *is the product of a constant load and an imaginary very small displacement* (*virtual displacement*). This concept of virtual work will be discussed latter in this chapter.

15.2.3 Work of Internal Forces: Strain Energy

When a body is gradually loaded, it deforms and the internal stresses developed act through their respective displacements (strains) and do work equal to one-half of the product of the corresponding quantities. In case of a general state of stress on an element, the stored elastic strain energy density U_0 per unit volume is expressed as

$$U_0 = \frac{1}{2}(\alpha_x\varepsilon_x + \alpha_y\varepsilon_y + \alpha_z\varepsilon_z + \tau_{xy}\gamma_{xy} + \tau_{yz}\gamma_{yz} + \tau_{zx}\gamma_{yz}) \tag{15.11}$$

Therefore, the general expression for the total internal strain energy in a linearly elastic body is

$$U = \frac{1}{2}\iiint(\alpha_x\varepsilon_x + \alpha_y\varepsilon_y + \alpha_z\varepsilon_z + \tau_{xy}\gamma_{xy} + \tau_{yz}\gamma_{yz} + \tau_{zx}\gamma_{yz})\,dx\,dy\,dz \tag{15.12}$$

The integration extends over the volume of a body. In an engineering structure, it is generally customary to work with the *stress resultants* causing *axial*, *bending*, *shearing* and *torsion actions*. Thus, ignoring out of plane deformations for a one-dimensional elastic structure, a less general expression given below is sufficient,

$$U = \frac{1}{2}\iiint(\alpha_z\varepsilon_z + \tau_{xy}\gamma_{xy})\,dx\,dy\,dz \tag{15.13}$$

For the linearly elastic materials, for uniaxial stress, $\varepsilon_z = \alpha_z/E$, and for pure shear, $\tau_{xy} = \gamma_{xy}/G$, the Eq. (15.13) can be rewritten as

$$U = \iiint\frac{\alpha_z^2}{2E}\,dx\,dy\,dz + \iiint\frac{\tau_{xy}^2}{2G}\,dx\,dy\,dz$$

or

$$U = \iiint\frac{E\varepsilon_z^2}{2}\,dx\,dy\,dz + \iiint\frac{G\tau_{xy}^2}{2}\,dx\,dy\,dz \tag{15.14}$$

Internal forces are the resultants of internal stresses which are resolved into component normal to cross-section, termed normal force N, and another tangential to the cross-section, termed shear force V. In addition, there are stress couples which are termed bending moment M. In case of 3-D or space structures, there is another kind of internal force, called torsion moment T. All these internal forces are double action forces. They are different from the *fixation forces* or *reactions at a support*. The equations for internal strain energy are specially tailored for the solution of problems encountered in engineering structures. For stress resultants N, V, M, and T the triple integrals are reduced to single integrals. One-dimensional continuous elastic bodies will be discussed in detail in the following sections. Considering that elastic parameters E and G are constant, the above expression can be used to derive the corresponding expression for these stress resultants.

15.3 STRAIN ENERGY FOR DEFORMED MEMBERS

15.3.1 Axially Loaded Members

In this type of structure $\alpha_z = N_z/A_z$ and at any section $\iint dx\, dy = \iint dA = A_z$, the axial force N_z and cross-sectional area A_z are the functions of z. Therefore, elongation elastic strain energy is,

$$U_n = \iiint_V \frac{\alpha_z^2}{2E}\, dx\, dy\, dz = \iiint_V \frac{N_z^2}{2A_z^2 E}\, dx\, dy\, dz$$

$$= \int_L \frac{N_z^2}{2A_z^2 E}\left(\iint_A dx\, dy\right) dz = \int_L \frac{N_z^2}{2A_z E}\, dz \tag{15.15}$$

where, EA_z is the *axial stiffness*. Integration along the member length L gives the elastic strain energy stored in the member. In a prismatic member where N (= P), A and E are constant, the elongation of the member $\Delta = PL/AE$. The work done which is stored as strain energy in the body, is the product of average of externally applied force $P/2$ and the deformation Δ. Hence

$$U_n = \frac{P^2 L}{2AE} = \frac{AE\Delta^2}{2L} \tag{15.16}$$

15.3.2 Member Subjected to Bending

The elastic strain energy in a member subjected to pure bending about one of its principal axes can be obtained by integration along the beam length L, i.e.

$$U_b = \iiint_V \frac{\alpha_z^2}{2E}\, dx\, dy\, dz = \iiint_V \frac{1}{2E}\left(\frac{M}{I}y\right)^2 dx\, dy\, dz$$

$$= \int_L \frac{M^2}{2EI^2}\left(\iint_A y^2\, dy\, dx\right) dz = \int_L \frac{M^2}{2EI}\, dz \tag{15.17}$$

where, EI is the *flexural stiffness* of the member and M is the bending moment. In Eq. 15.17 bending moment M and the moment of inertia I can vary along the length of the member.

15.3.3 Member Subjected to Shear

For a member with regular cross-section and length L subjected to shear V, the elastic strain energy is given by,

$$U_s = \iiint_V \frac{\tau_{xy}^2}{2G}\, dx\, dy\, dz = k \int_L \frac{V^2}{2GA}\, dz \tag{15.18}$$

where, the factor k depends on the cross-sectional properties of the member. In Eq. (15.18) shearing force V and area A can vary along the length of the member.

15.3.4 Members Subjected to Torsion

In this case, the *torsion shear* is analogous to bending shear, thus the strain energy of torsion can be expressed as

$$U_t = \iiint_V \frac{\tau^2}{2G}\, dx\, dy\, dz = \int_L \frac{T^2}{2GJ}\, dx \tag{15.19}$$

where, G is modulus of shear and J is polar moment of inertia, GJ is the *torsion rigidity* of the member and T is the twisting moment.

For a prismatic member of length L, the total elastic strain energy due to axial force, shearing force, bending moment and torsion can be expressed as:

$$U = U_n + U_b + U_s + U_t$$

$$= \int_L \frac{N^2}{2AE}\, dz + \int_L \frac{M^2}{2EI}\, dz + k \int_L \frac{V^2}{2GA}\, dz + \int_L \frac{T^2}{2GJ}\, dz \tag{15.20}$$

Integrating the expression over the whole length L of each member constituting the structure and summing up the results, the following expression is obtained for the whole structure.

$$U = \sum \int_L \frac{N^2}{2AE}\, dz + \sum \int_L \frac{M^2}{2EI}\, dz + \sum k \int_L \frac{V^2}{2GA}\, dz + \sum \int_L \frac{T^2}{2GJ}\, dz \tag{15.21}$$

It may be noted from the above expression that:

1. The *strain energy is always positive* as the expression contains stress resultants N, V, M, and T in the second power.
2. The strain energy is expressed by a homogeneous equation of the stresses or strains in the second power. The strains are directly proportional to the stresses.
3. The strain energy stored under the action of a certain system of forces is not equal to the sum of the strain energies of each of these separate forces. Therefore, the *principle of superposition is no longer valid*. This follows from the fact that the strain energy is a function of the second power of stress resultants N, V, M, and T; and that the square of a sum is not equal to the sum of squares.
4. The strain energy stored in a body is independent of the sequence in which external forces are applied, because the strain energy depends only on the final values of stress resultants N, V, M, and T which are independent of the sequence.

15.4 THE WORK EQUATION

According to the *law of conservation of energy*, there is no gain or loss of energy in a conservative system during deformation under external loads. It means that the change in energy of applied load, i.e. the work done by the externally applied forces during loading process W_{ex} is equal to the increase in strain energy stored in the system, i.e. the work done by the internal forces, i.e. $W_{ex} = U$.

In the other words, all the work of external forces is transformed into strain energy. The latter is stored in the body during the period of increasing strains and deformations. In a perfectly elastic closed system the work done by a load (stored as strain energy) will be released back completely when the load is removed in the absence of thermal dissipation or frictional or damping losses. Systems not following these conditions are commonly referred to as *non-conservative systems*. Thus the total or net work W must be zero, and

$$W = W_{\text{ex}} + W_{\text{in}} = 0$$

Therefore,
$$U = -W_{\text{in}} = W_{\text{ex}} \tag{15.22}$$

where W_{in} has a negative sign because the deformations are opposed by the internal forces. This *work balance equation* is equivalent to *static equilibrium conditions*. Using the principle of virtual work which is an axiom of equilibrium,

$$\delta W = \delta W_{\text{ex}} + \delta W_{\text{in}} = 0$$

or
$$\delta W - \delta W_{\text{in}} = \delta W_{\text{ex}} \tag{15.23}$$

which means that change in the energy of applied loads is equal to the increase in the stored energy. The work balance Eq. (15.22) can be used directly in solving problems in structural analysis.

15.5 COMPUTATION OF DISPLACEMENTS USING WORK EQUATION

15.5.1 Member Subjected to Bending

For most of the *beams* and *frames*, axial and shear deformations can be ignored in determining the deflection of central axis. On the other hand, large axial force deformations in arches, suspension bridges and trusses have to be taken into account. Similarly, for deep beams, shear deformation can reach large values.

EXAMPLE 15.1

Determine the order of magnitude of contribution of shear force in the maximum deflection/rotaion at the load point in a: (i) cantilever with load P applied at its free end and (ii) simply supported beam due to a moment M applied at its left hand support. Both the beams are of rectangular cross-section with k value of 1.2. Consider that the cross-section and magnitude of modulus of elasticity are constant along the entire length of the beams of Figure 15.3.

Solution: The eigen work of internal forces namely bending moment and shear force is

$$W_{\text{in}} = \int_0^L \frac{M_z^2\, dz}{2EI} + k\int_0^L \frac{V_z^2}{2GA}\, dz$$

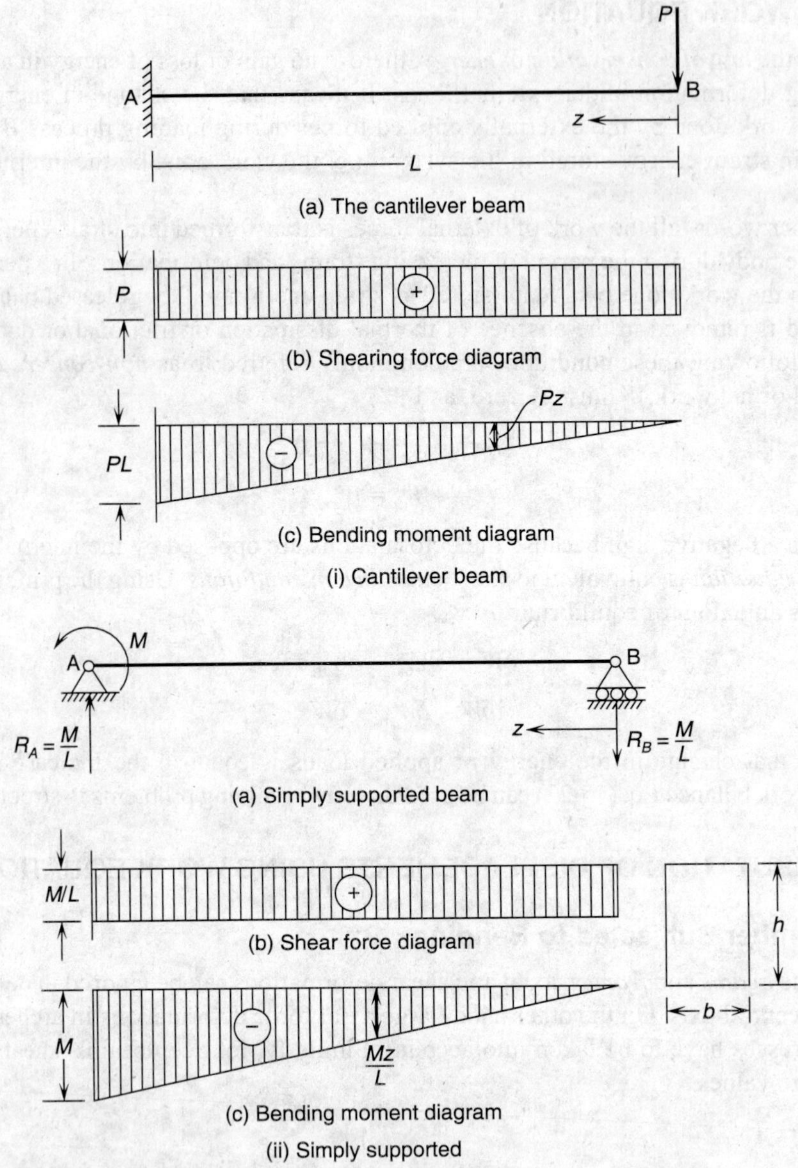

Figure 15.3 Computation of deflections in cantilever and simply supported beams of Example 15.1.

(i) *Cantilever beam with load P applied at its free end* (see Figure 15.3(i) (a–c))

$$-W_{\text{in}} = \int_0^L \frac{(-Pz)^2\,dz}{2EI} + 1.2\int_0^L \frac{P^2\,dz}{2GA} = \frac{P^2L^3}{6EI} + \frac{6}{10}\frac{P^2L}{GA}$$

The external eigen work done during the displacement δ_{max} is

$$W_{\text{ex}} = P\delta_{\text{max}}/2$$

Equating external work to the eigen work of internal forces, i.e. $W_{ex} = -W_{in}$

$$\frac{1}{2}P\delta_{max} = \frac{P^2L^3}{6EI} + \frac{6}{10}\frac{P^2L}{GA}$$

$$\delta_{max} = \left(\frac{PL^3}{3EI}\right) + \left(\frac{6PL}{5GA}\right) = \delta_{bend} + \delta_{shear}$$

$$= \frac{PL^3}{3EI}\left[1 + \frac{18EI}{5GAL^2}\right] = \frac{PL^3}{3EI}\left[1 + \frac{18}{5}\left(\frac{E}{G}\right)\left(\frac{bh^3/12}{bhL^2}\right)\right]$$

$$= \frac{PL^3}{3EI}\left[1 + \frac{3}{10}\left(\frac{E}{G}\right)\left(\frac{h}{L}\right)^2\right]$$

where b and h are the width and height of the rectangular cross-section, respectively.

(ii) *Simply supported beam subjected to a moment M at its supports* (see Figure 15.3(ii) (a–c))

The magnitude of bending moment and shearing force are,

$$M_z = \left(\frac{M}{L}\right)z \quad \text{and} \quad V_z = \frac{M}{L}$$

Thus,
$$-W_{in} = \int_0^L \frac{1}{2EI}\left(\frac{Mz}{L}\right)^2 dz + 1.2\int_0^L \frac{1}{2GA}\left(\frac{M}{L}\right)^2 dz = \frac{M^2L}{6EI} + \frac{6}{10}\frac{M^2}{GAL}$$

The external eigen work done during the rotation θ_{max} is,

$$W_{ex} = \frac{M\theta_{max}}{2}$$

Equating external work to the eigen work of internal forces, i.e. $W_{ex} = -W_{in}$

$$\frac{M\theta_{max}}{2} = \frac{M^2L}{6EI} + \frac{6}{10}\frac{M^2}{GAL}$$

or
$$\theta_{max} = \frac{ML}{3EI} + \frac{6}{5}\frac{M}{GAL} = \frac{ML}{3EI}\left(1 + \frac{18}{5} \times \frac{EI}{GAL^2}\right)$$

$$= \frac{ML}{3EI}\left[1 + \frac{3}{10}\left(\frac{E}{G}\right)\left(\frac{h}{L}\right)^2\right]$$

The relationship between material constants E and G is given by:

$$\frac{E}{G} = 2(1 + \nu)$$

Taking $\nu = 0.2$ a typical value for concrete, the ratio E/G becomes 2.4. The total deflection and rotation of the beams are:

$$\delta_{\max}(\theta_{\max}) = \left[1 + 0.72\left(\frac{h^2}{L^2}\right)\right]\delta_{\text{bend}}(\theta_{\text{bend}})$$

For a very short or deep beam say $L/h = 1$, the total deflection is 1.72 times that due to bending alone. Hence, shear deformations are important. On the other hand, for long or slender typical beam with $L/h = 15$, the deflection due to shear is 0.32 per cent. However, it should be realized that this is not always the case. For an I-beam with strong flanges and very thin web, the shape factor can become up to ten times larger than rectangular section and shear force V could not be ignored.

EXAMPLE 15.2

A beam AB of length $2a$ is simply supported at the ends A and B. For the left half of the beam, the moment of inertia of the section is $2I$ and for the right half it is I as shown in Figure 15.4. The beam is subjected to a concentrated load at the mid-span point. Using the work balance equation, determine the deflection at the load point C.

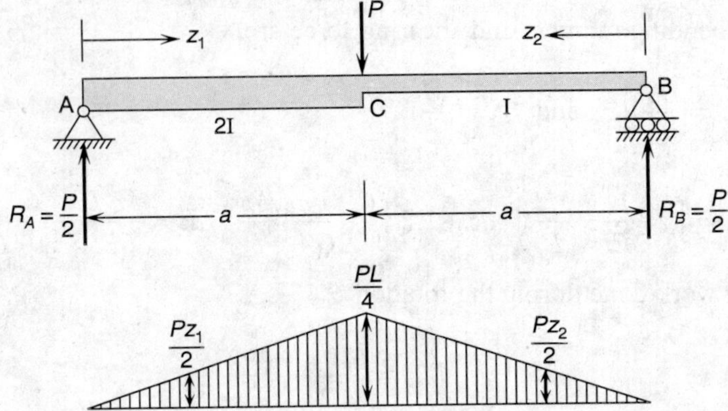

Figure 15.4 Beam of variable flexural rigidity of Example 15.2.

Solution: Let δ_C be the deflection at the point C, then

$$W_{\text{ex}} = \frac{1}{2}P\delta_C$$

The increase in strain energy of the system, i.e. the work done by the internal forces,

$$-W_{\text{in}} = U = U_{AC} + U_{CB}$$

$$= \left(\frac{1}{2}\right)\int_0^a \left(\frac{1}{2EI}\right)\left(\frac{Pz_1}{2}\right)^2 dz_1 + \left(\frac{1}{2}\right)\int_0^a \left(\frac{1}{EI}\right)\left(\frac{Pz_2}{2}\right)^2 dz_2$$

$$= \left(\frac{P^2}{16EI}\right)\left[\int_0^a z_1^2\,dz_1 + 2\int_0^a z_2^2\,dz_2\right]$$

$$= \left(\frac{P^2}{16EI}\right)\left[\left(\frac{a^3}{3}\right) + 2\left(\frac{a^3}{3}\right)\right] = \left(\frac{P^2a^3}{16EI}\right)$$

It should be noted that for the computation of internal work in the part CB of the beam, the point B has been taken as the origin for simplification. Therefore, from the work equation, $W_{ex} = -W_{in}$:

$$\frac{1}{2}P\delta_C = \left(\frac{P^2a^3}{16EI}\right) \quad \text{or} \quad \delta_C = \left(\frac{Pa^3}{8EI}\right)$$

EXAMPLE 15.3

A simply supported beam of length $L(= a + b)$ is subjected to a concentrated load P at a distance a from the left support. Using the work balance equation, determine the deflection at the load point. The flexural rigidity EI is constant.

Solution: Let δ_C be the deflection at the location of point load, then external work is:

$$W_{ex} = \frac{1}{2}P\delta_C$$

The internal work can be computed in two parts A–C and C–B since the bending moment diagram is discontinuous at the load point C as shown in Figure 15.5. Therefore,

$$-W_{in} = \int_0^L \frac{M_z^2}{2EI}\,dz = \int_0^a \frac{1}{2EI}\left(\frac{Pb}{L}z_1\right)^2 dz_1 + \int_0^b \frac{1}{2EI}\left(\frac{Pa}{L}z_2\right)^2 dz_2$$

Thus, $$-W_{in} = \frac{P^2b^2a^3}{6EIL^2} + \frac{P^2a^2b^3}{6EIL^2} = \frac{P^2a^2b^2}{6EIL^2}(a+b) = \frac{P^2a^2b^2}{6EIL}$$

Finally from work equation:

$$\frac{1}{2}P\delta_C = \frac{P^2a^2b^2}{6EIL} \quad \text{or} \quad \delta_C = \frac{Pa^2b^2}{3EIL}$$

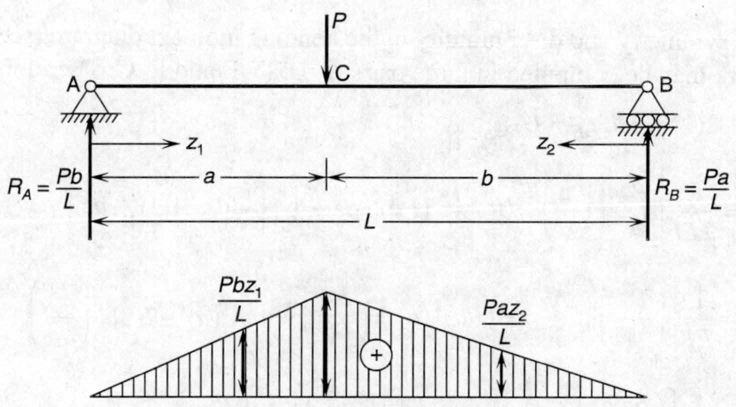

Figure 15.5 Simply supported beam subjected to concentrated load of Example 15.3.

EXAMPLE 15.4

Using the work balance method, determine the vertical deflection at the mid-point of the bent beam due to a concentrated force acting there as shown in Figure 15.6(a). The flexural rigidity EI is constant.

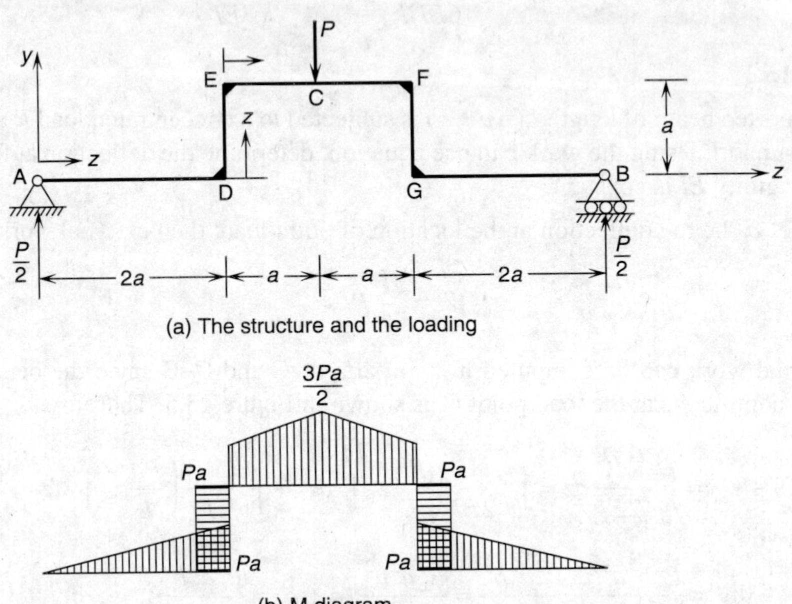

Figure 15.6 Computation of deflection in a cranked beam of Example 15.4.

Solution: Let y_C be the deflection at the location of point load, then external work is:

$$W_{\text{ex}} = \frac{1}{2} P\delta_C$$

Because of symmetry and discontinuity in the bending moment diagram (see Figure 15.6(b)), the internal work may be computed in three parts A–D, D–E and E–C. Therefore,

$$-W_{\text{in}} = 2(U_{AD} + U_{DE} + U_{EC})$$

$$= \frac{2}{2EI}\left[\int_0^{2a}\left(\frac{Pz}{2}\right)^2 dx + \int_0^a (Pa)^2\, dz + \int_0^a \left\{\frac{P}{2}(2a+z)\right\}^2 dz\right]$$

$$= \frac{1}{EI}\left[\left(\frac{P^2}{4}\right)\left(\frac{8a^3}{3}\right) + (Pa)^2(a) + \left(\frac{P^2}{4}\right)\left\{(4a^2)(a) + 2(2a)\left(\frac{a^2}{2}\right) + \frac{a^3}{3}\right\}\right]$$

Thus, $$-W_{\text{in}} = \frac{1}{EI}\left[\left(\frac{2P^2a^3}{3}\right) + (P^2a^3) + \left(\frac{19P^2a^3}{12}\right)\right] = \frac{39P^2a^3}{12EI}$$

Finally from work equation:

$$\frac{1}{2}P\delta_C = \frac{39P^2a^3}{12EI} \quad \text{or} \quad \delta_C = \frac{13Pa^3}{2EI}$$

15.5.2 Axially Loaded Members

The work balance method is equally applicable to structures carrying axial loads.

EXAMPLE 15.5

Using the work balance equation, determine the deflection at the load point 1 of the truss shown in Figure 15.7(a).

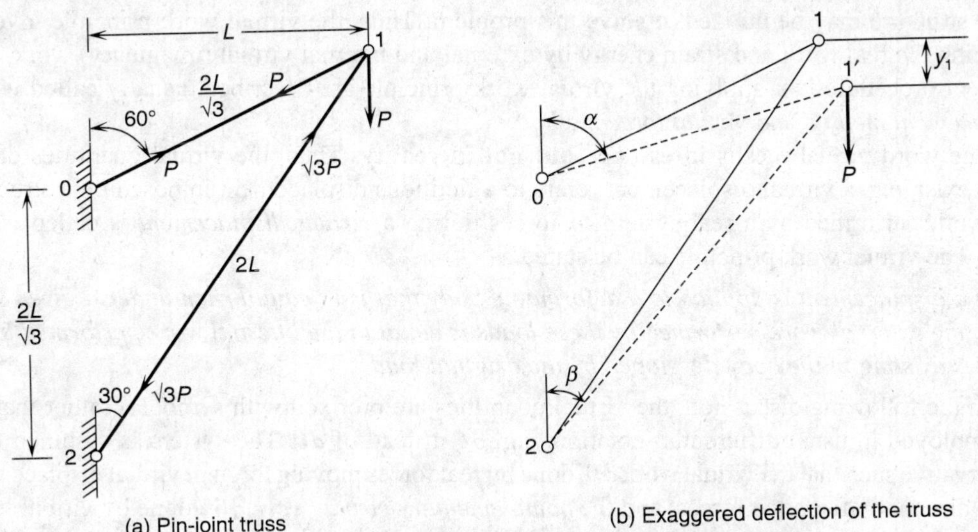

Figure 15.7 The pin-jointed truss and its deflected shape.

Solution: An exaggerated deflection of the truss is shown in Figure 15.7(b) for clarity. An assumption of small deflection theory for load-deflection relationship implies $\alpha \cong 60°$ and $\beta \cong 30°$. The external work done is,

$$W_{\text{ex}} = \frac{1}{2}P\delta_C$$

The strain energy stored in the members of axial stiffness EA is given by

$$-W_{\text{in}} = \sum_i \frac{1}{2}\frac{N_i^2 L_i}{(AE)_i} = \frac{1}{2AE}\left[P^2\left(\frac{2L}{\sqrt{3}}\right) + (\sqrt{3}P)^2\,(2L)\right]$$

$$= \frac{P^2L}{AE}\left(\frac{1}{\sqrt{3}} + 3\right)$$

Therefore, from the work equation,

$$W_{\text{ex}} = -W_{\text{in}}$$

$$\frac{1}{2}P\delta_C = \frac{P^2L}{AE}\left(\frac{1}{\sqrt{3}}+3\right) \quad \text{or} \quad \delta_C = \left(\frac{2}{\sqrt{3}}+6\right)\left(\frac{PL}{AE}\right)$$

15.6 VIRTUAL WORK METHODS

The work balance method described above fails if several deflections and / or rotations are required to be determined in a deformed body subjected to one or more forces. However, the *virtual work principle* which relates a *system of forces in equilibrium* to a *compatible system of displacements* in the structure may be utilized to solve this problem. Thus, the virtual work principle involves replacing the real work and strain energy by external and internal virtual (imaginary) work. Two different procedures for applying the virtual work principle are described, usually called *virtual displacement method* and *virtual force method.*

The word virtual means **in essence but not in reality**. Thus, the virtual quantities do not really exist, e.g. a virtual displacement refers to a fictitious displacement imposed on a structure. The work performed by a real system of forces during a *virtual displacement* is called *virtual work.* The virtual work principle can be stated as

If a displacement is applied to a deformable body that is in equilibrium under a given set of loads, the external work performed by these loads is equal to the internal work performed by the stresses existing in the body developed by the original loads.

In the following discussion, the virtual quantities are prefixed with symbol ∂ rather than the '*d*' employed in usual differential notation, e.g. $\partial\theta$ instead of $d\theta$. The system is assumed to be conservative such that the virtual work ∂W done by real forces moving through virtual displacements in the direction of applied forces; and the *complementary virtual work* ∂W_c done by virtual forces moving through real displacements are zero. Such works are positive when the directions of forces and displacements are consistent. The virtual displacements and virtual forces are perfectly arbitrary and bear no relation to the applied forces and displacements, respectively.

15.6.1 Virtual Work and Complementary Virtual Work

The equilibrium condition states that for a body to be in equilibrium, the sum of all the forces acting on the body must be zero $\left(\sum P_i = 0\right)$. Now suppose that a rigid body acted upon by several forces P_i has moved an arbitrary small virtual distance $\partial\Delta$ which is compatible with the constraints on the body. Then the work done by this force system would be given by the vector equation

$$\partial W = (P_1 \cos\theta_1 + P_2 \cos\theta_2 + P_3 \cos\theta_3 + \dots)\,\partial\Delta = \Sigma P_i \partial\Delta \tag{15.24}$$

which is simply multiplying an equilibrium condition by $\partial\Delta$. This formulation of *equilibrium conditions* has several computational advantages, e.g. the reactions of fixed supports drop out from the equations because they do not work.

In the preceding calculation, the virtual work has been defined as the product of a real force moving through a virtual displacement. In this form, the principle is referred to more accurately as *principle of virtual displacement*. However, the principle of virtual work can be stated in another form known as *principle of complementary virtual work*. Here, the virtual work is defined as the product of virtual force moving through a real displacement. This principle thus states that if the system is in equilibrium, then the sum of all virtual complementary works is zero for *compatibility* or *geometric continuity* (e.g. zero slope at the fixed point in case of fixed support, and members meeting at a rigid joint have the same absolute rotation). Thus, the principle of virtual work or virtual displacement assumes compatibility and leads to equations of equilibrium while the principle of complementary virtual work or virtual force assumes equilibrium and leads to equations of compatibility.

To illustrate the dual character of these two fundamental principles of mechanics, consider the discrete frame structure consisting of rigid-bars supported by rotational springs capable of activating reaction moments at the supports as shown in Figure 15.8.

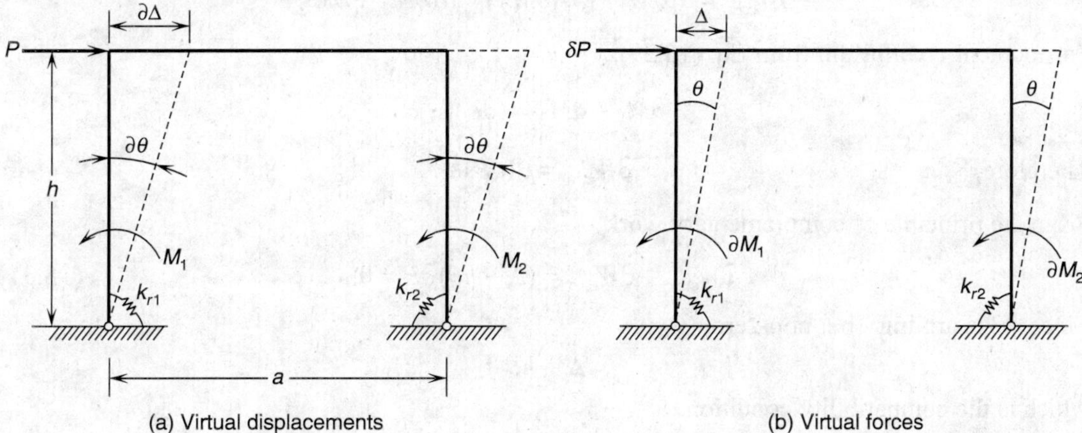

Figure 15.8 Concept of virtual displacements and virtual forces.

(1) *Principle of virtual work*

Give the system a virtual rotation as shown in Figure 15.8(a). The virtual works of various forces are

$$\partial W_{ex} = P(\partial \Delta) \tag{15.25}$$

$$-\partial W_{in} = M_1\, \partial \theta_1 + M_2\, \partial \theta_2 \tag{15.26}$$

For compatibility $\partial \theta_1 = \partial \theta_2 = \partial \theta = \partial \Delta / h$, then

$$-\partial W_{in} = M_1 \left(\frac{\partial \Delta}{h} \right) + M_2 \left(\frac{\partial \Delta}{h} \right) = \frac{1}{h}(M_1 + M_2)\, \partial \Delta \tag{15.27}$$

From the principle of virtual work using Eqs. (15.25) and (15.27)

$$\partial W_{ex} + \partial W_{in} = \left[P - \frac{1}{h}(M_1 + M_2) \right] \partial \Delta = 0 \tag{15.28}$$

Since $\delta\Delta$ is perfectly arbitrary but non-zero, the expression in parentheses must be zero, i.e.

$$\left[P - \frac{1}{h}(M_1 + M_2)\right] = 0 \quad \text{or} \quad Ph = (M_1 + M_2) \tag{15.29}$$

which is equilibrium condition. Noting that $M_i = k_i\theta_i$

$$\theta = (Ph)/(k_{r1} + k_{r2}) \tag{15.30}$$

(2) *Principle of complementary virtual work*

Apply on the system a virtual force ∂P. The virtual work done by various forces (see Figure 15.8(b)) are

$$\partial M_{\text{c,ex}} = \Delta(\partial P)$$

$$-\partial W_{\text{c,in}} = \theta(\partial M_1) + \theta(\partial M_2) = \theta(\partial M_1 + \partial M_2)$$

For moment equilibrium from Eq. (15.29):

$$\partial M_1 + \partial M_2 = (\partial P)\, h,$$

Therefore,

$$-\partial W_{\text{c,in}} = \theta(\partial P)h$$

From the principle of complementary work

$$\partial W_{\text{c,ex}} + \partial W_{\text{c,in}} = (\Delta - h\theta)\partial P = 0 \tag{15.31}$$

Since ∂P is arbitrary but non-zero,

$$\Delta = h\theta$$

which is the compatibility condition.

15.6.2 Applications of Virtual Work Methods

Virtual work method is one of the most important methods used to calculate displacement of elastic structures and also forms the basis of *force* or *flexibility method* for the analysis of statically indeterminate systems. In this method, a virtual load is considered to act at the point and in the direction in which displacement is required. For illustration, consider the simply supported beam shown in Figure 15.9(a), carrying load P at point 1. It is required to determine vertical displacement at point 2. Let the load P be removed temporarily from the beam and virtual load ∂P of arbitrary magnitude be applied at point 2 in the direction of desired deflection as shown in Figure 15.9(b).

Now load the beam with the real load P, producing additional displacements δ_1 and δ_2 under loads P and ∂P, respectively, as shown in Figure 15.9(c). From the principle of energy balance that the external virtual work is equal to internal virtual work, we have

$$-\partial W_{\text{in}} = \partial W_{\text{ex}} = -\partial P \delta_2$$

$$= \text{Virtual force } \partial P \times \text{Real displacement} \tag{15.31a}$$

If M is the bending moment induced in the beam due to the applied load system and δM the moment due to the virtual load ∂P. The total internal strain energy of the beam carrying applied load system is given by

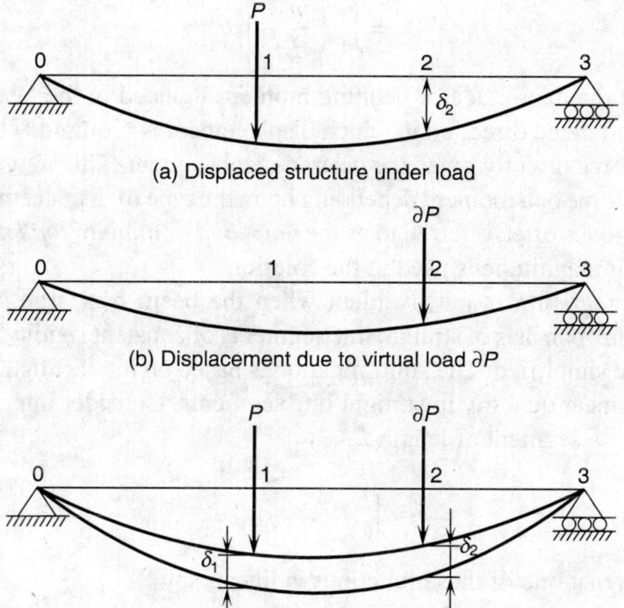

Figure 15.9 Simply supported beam subjected to different load sequences.

$$U_1 = \int_0^L \frac{M^2}{2EI}\,dz \tag{15.32}$$

If virtual load ∂P is now added, M increases to $M + \partial M$ and the total strain energy of the beam is given by

$$U_2 = \int_0^L \frac{(M+\partial M)^2}{2EI}\,dz = \int_0^L \frac{M^2}{2EI}\,dz + \int_0^L \frac{M \times \partial M}{EI}\,dz + \int_0^L \frac{(\partial M)^2}{2EI}\,dz$$

The change in strain energy

$$U_2 - U_1 = \int_0^L \frac{M \times \partial M}{EI}\,dz + \int_0^L \frac{(\partial M)^2}{2EI}\,dz \tag{15.33}$$

Equating external work done by ∂P to the change of strain energy

$$\partial P \delta_2 = \int_0^L \frac{M \times \partial M}{EI}\,dz + \int_0^L \frac{(\partial M)^2}{2EI}\,dz$$

or

$$\delta_2 = \int_0^L \frac{M}{EI}\left(\frac{\partial M}{\partial P}\right) dz + \int_0^L \frac{\partial M}{2EI}\left(\frac{\partial M}{\partial P}\right) dz \tag{15.34}$$

Since ∂P is arbitrary and moment ∂M is a linear function of ∂P, the quantity $(\partial M/\partial P)$ can be replaced by m i.e. ∂P can be replaced by a unit load and ∂M by m. The second term on the right hand side of Eq. 15.34 being product of two virtual quantities is put to zero, therefore,

$$\delta_2 = \int_0^L \left(\frac{mM}{EI}\right) dz \tag{15.35}$$

Numerically m is the magnitude of bending moment induced in the beam due to a unit load applied at the point and in the direction in which displacement is required. Thus, an application of unit external virtual force directly gives the desired displacement. This unit external force can be in the form of either a force or a moment depending upon the type of displacement to be determined. The virtual work method is often referred to as the *unit-load* or *dummy-load* method, as a fictitious or dummy load of unit magnitude is used in the solution.

The numerical integration is inconvenient when the beam has large number of segments. However, if the structure consists of straight line segments of constant rigidity within each segment, the integration may be simplified. The simplification is based on the fact that the diagram m due to unit load is always a linear quantity in straight line segments. Consider integral of product of two functions $f_1(z)\, f_2(z)$ for a segment of length L,

$$I = \int_0^L f_1(z)\, f_2(z)\, dz \tag{15.36}$$

subject to a condition that one of these functions is linear, say

$$f_2(z) = b + kz \tag{15.37}$$

The Eq. (15.36) then becomes

$$I = b\int_0^L f_1(z)\, dz + k\int_0^L z f_1(z)\, dz$$

The first of the integrals represents the area bounded by the curve $f_1(z)$, i.e.

$$\int_0^L f_1(z)\, dz = A$$

The second integral represents the static moment of the above area with respect to the origin of the segment, i.e.

$$\int_0^L z f_1(z)\, dz = A\bar{z} \tag{15.38}$$

where $\bar{z}$ is the coordinate of the centroid of the first diagram, thus the integrand becomes

$$I = \int_0^L f_1(z)\, f_2(z)\, dz = bA + k(A\bar{z})$$

$$= (b + k\bar{z})A = f_2(\bar{z}) \int_0^L f_1(z)\, dz \tag{15.39}$$

Thus the integration of the product of two functions is equal to the multiplication of the area under the first diagram by the ordinate of the second (linear) diagram at the location of the centroid of the first diagram. Therefore, the virtual work integral may be represented by the following simple expression:

$$\delta = \int \frac{mM}{EI}\, dz = \bar{m} \int \left(\frac{M}{EI}\right) dz$$

$$= \bar{m} \text{ (area of } M / EI \text{ diagram)} \tag{15.40}$$

where $\bar{m}$ is the ordinate of diagram at the location of centre of gravity of corresponding continuous part of (M/EI) diagram as shown in Figures 15.10(a) and (b). The method also called *area centre-of-gravity method* or *graphical integration* and is applicable to structures made up of straight members. In the cases when both the functions $f_1(z)$ and $f_2(z)$ are linear, the operation of multiplication is commutative, i.e. it is immaterial whether the area under the first diagram be multiplied by the ordinate of the second diagram or the area under the second diagram be multiplied by the ordinate of the first diagram. The graphical integration is extensively used for computation of stiffness coefficients.

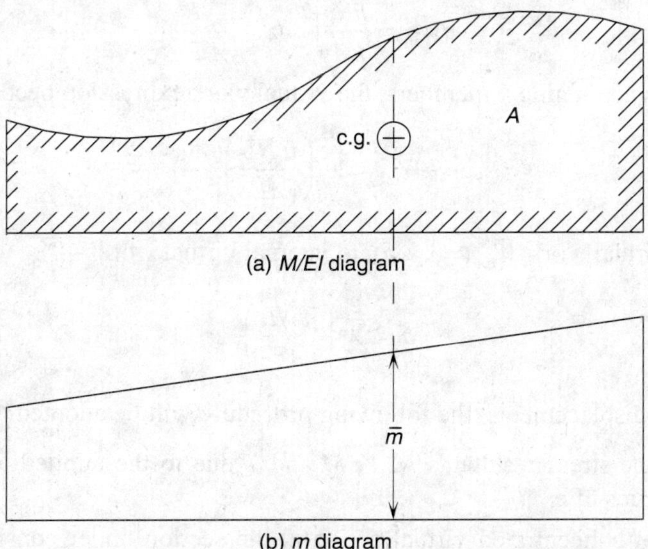

Figure 15.10 Computation of deflection by area centre of gravity method.

For a rigid frame consisting of a number of discrete elements or members

$$\delta_b = \frac{A_1\bar{m}_1}{E_1I_1} + \frac{A_2\bar{m}_2}{E_2I_2} + \cdots = \sum_i \left(\frac{A\bar{m}}{EI}\right)_i \tag{15.41}$$

By an analogous treatment, it can be shown that deflections due to axial forces, shear forces and torsion moments are

$$\delta_n = \int \frac{nN}{EA}\, dz, \quad \delta_q = \int \frac{vV}{GA}\, dz \quad \text{and} \quad \delta_{tor} = \int \frac{tT}{GJ}\, dz \tag{15.42}$$

respectively, where G is modulus of shear and J is polar moment of inertia.

Since, bending moment diagrams encountered in practice may be divided into simple figures; rectangle, triangle, parabola and conic parabola for which the area and the position of the centroid are known, the graphical procedure considerably simplifies the integration. The axial force, shearing force and torsion diagrams are still simpler; they are linear as a rule and consist of rectangles and triangles in various combinations.

The virtual work or unit-load method can also be used for determining the deflection of a truss. Since bending moments in a pin-jointed truss are by definition all zero, the virtual work formula in this case is

$$-d\overline{W}_{\text{in}} = \int \frac{nN}{EA}\, dz \tag{15.43}$$

Noting that N and n are always constant over the whole length, the internal virtual work of a discrete member of length L is

$$-d\overline{W}_{\text{in}} = \frac{\overline{n}N}{EA}\int_0^L dz = \frac{\overline{n}NL}{AE} \tag{15.44}$$

For a complete truss containing k members, the virtual work expression becomes

$$-\overline{W}_{\text{in}} = \sum_{i=1}^{k}\left(\frac{\overline{n}\,NL}{AE}\right)_i \tag{15.45}$$

Equating external virtual work $\overline{W}_{\text{ex}}\ (=1\times\delta)$ to internal virtual work $-\overline{W}_{\text{in}}$

$$\delta = \sum_{i=1}^{k}\left(\frac{\overline{n}\,NL}{AE}\right)_i \tag{15.46}$$

For computation of displacements, the following procedure will be adopted

1. Determine the stress resultants N, V, M and T due to the applied loads at an arbitrary section in terms of z.
2. Apply a unit concentrated virtual action at the section under consideration and in the direction in which deformation is required.
3. Compute the stress resultants n, v, m and t due to the unit action at the same section as in step (1) in terms of z.
4. Substitute the stress resultants obtained in steps (1) and (3) in the appropriate equation and integrate along all the elements of the entire structure.

When the computed value of the displacement is positive, its direction coincides with that adopted for the unit action; and when it is negative, it is opposite to the one adopted for the unit action.

EXAMPLE 15.6

A simply supported beam AB of span L is subjected to a concentrated load P at the point C at a distance a from the left support. Using the virtual work method, determine the deflection at the load point D at a distance z_1 from the right support. The flexural rigidity EI is constant.

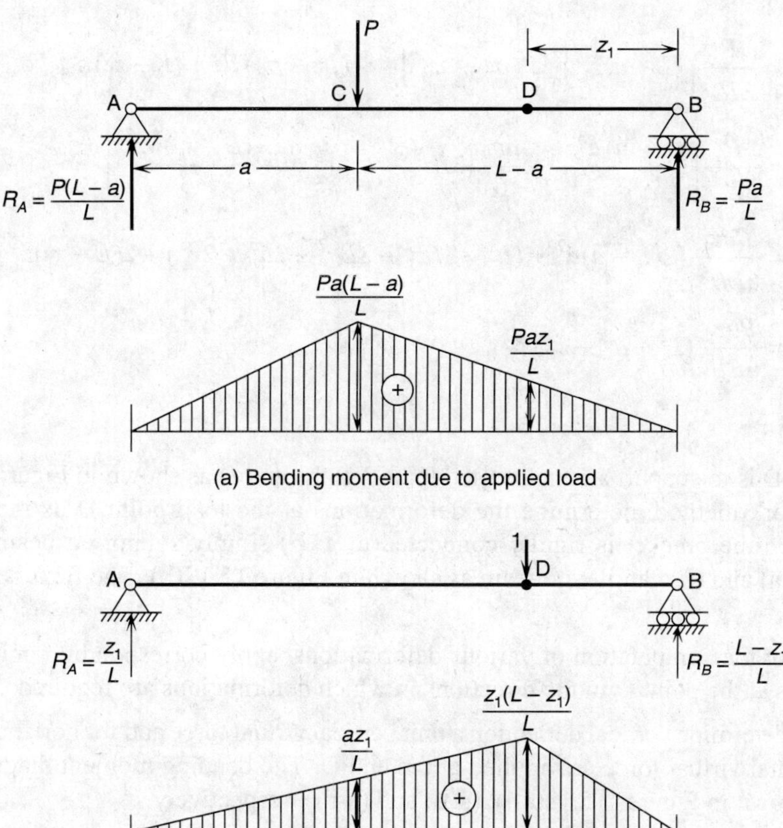

Figure 15.11 Beam of Example 15.6.

Solution: The bending moment diagrams of M and m are shown in Figures 15.11(a) and (b), respectively. Due to discontinuities in M and m diagrams at the point of application of load P and unit load, the integration $\int_0^L (mM/EI)\, dz$ is evaluated by summing up $\int_s (mM/EI)\, dz$ separately for the regions A–C, C–D and D–B, i.e.

$$\delta_D = \int_0^a m\left(\frac{M}{EI}\right) dz + \int_a^{L-z_1} m\left(\frac{M}{EI}\right) dz + \int_0^{z_1} m\left(\frac{M}{EI}\right) dz$$

By numerical integration:

$$\delta_D = \int_0^a \left(\frac{z_1 z}{L}\right)\left\{\frac{P(L-a)z}{EIL}\right\} dz + \int_a^{L-z_1} \left(\frac{z_1 z}{L}\right)\left\{\frac{Pa(L-z)}{EIL}\right\} dz$$
$$+ \int_0^{z_1} \left\{\frac{(L-z_1)z}{L}\right\}\left(\frac{Paz}{EIL}\right) dz$$

$$= \frac{P}{EIL^2}\left[(L-a)z_1 \int_0^a z^2\, dz + az_1 \int_a^{L-z_1} (Lz - z^2)\, dz + (L-z_1)a \int_0^{z_1} z^2\, dz\right]$$

$$= \frac{P}{EIL^2}\left[\frac{(L-a)a^3 z_1}{3} + \frac{az_1}{6}\left(3Lz^2 - 2z^3\right)_a^{L-z_1} + \frac{(L-z_1)az_1^3}{3}\right]$$

$$= \frac{Paz_1}{6EIL^2}[2(L-a)a^2 + (L^3 - 3Lz_1^2 - 3La^2 + 2a^3 + 2z_1^3) + 2(L-z_1)z_1^2]$$

$$= \frac{Paz_1}{6EIL}(L^2 - a^2 - z_1^2)$$

EXAMPLE 15.7

A bracket BCD is to support a concentrated load P at the end D as shown in Figure 15.12. Using the virtual work method, determine the deformations at the load point D as indicated on the diagram, when the bracket is rigidly connected to a: (i) simply supported beam as shown in Figure 15.12(a) and (ii) cantilever beam as shown in Figure 15.12(b). The flexural rigidity EI is constant.

Solution: For the computation of various deformations, apply corresponding unit concentrated virtual actions at the point D in the directions in which deformations are required.

Case-I: To determine vertical deflection a unit vertical virtual force and for horizontal deflection, a unit horizontal virtual force are applied at the end D. The bending moment diagrams of M, m_1 and m_2 are shown in Figures 15.12(a-i), (a-ii) and (a-iii), respectively.

Therefore,
$$\delta_{DV} = \frac{1}{EI}\int_S m_1 M\, dz$$

Using geometrical integration or the area centre-of-gravity method,

$$\delta_{DV} = \left(\frac{1}{2}\times 3a \times \frac{2Pa}{EI}\times\frac{4a}{3}\right) + \left(a \times \frac{2Pa}{EI}\times 2a\right) + \left(\frac{1}{2}\times 2a \times \frac{2Pa}{EI}\times\frac{4a}{3}\right) = \frac{32Pa^3}{3EI}$$

$$\delta_{DH} = \left(\frac{1}{2}\times 3a \times \frac{2Pa}{EI}\times\frac{2a}{3}\right) + \left(a \times \frac{2Pa}{EI}\times\frac{a}{2}\right) = \frac{3Pa^3}{EI}$$

Case-II: To determine vertical deflection and rotation at the end D, a virtual unit force and unit moment, respectively, are applied at the section D. The bending moment diagrams of M, m_1 and m_2 are shown in Figures 15.12(b-i), (b-ii) and (b-iii), respectively.

Therefore,
$$\delta_{DV} = \frac{1}{EI}\int_S m_1 M\, dx$$

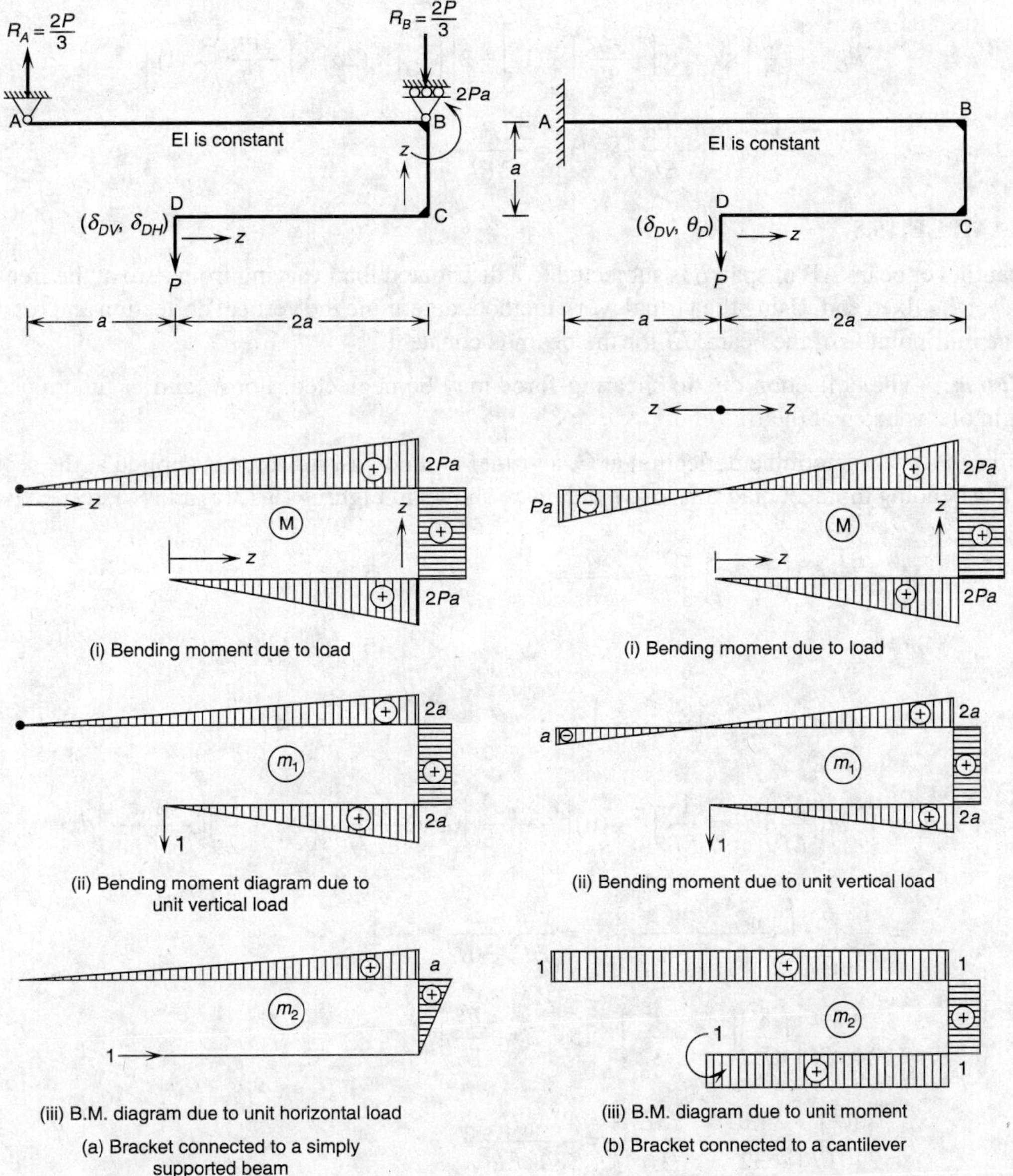

Figure 15.12 Computation of deformations in the frames of Example 15.7.

Using geometrical integration or the area centre-of-gravity method using Table 15.2,

$$\delta_{DV} = \left(\frac{1}{3} \times a \times \frac{Pa}{EI} \times a\right) + 2\left(\frac{1}{3} \times 2a \times \frac{2Pa}{EI} \times 2a\right)$$

$$+\left(a \times \frac{2Pa}{EI} \times 2a\right) = \frac{29Pa^3}{3EI}$$

$$\theta_D = \left[\left(\frac{1}{2}\right)\times(a)\times\left(-\frac{Pa}{EI}\right)\times(1)\right] + 2\left[\left(\frac{1}{2}\right)\times(2a)\times\left(\frac{2Pa}{EI}\right)\times(1)\right]$$

$$+\left[(a)\times\left(\frac{2Pa}{EI}\right)\times(1)\right] = \frac{11Pa^2}{2EI}$$

EXAMPLE 15.8

A cantilever beam AB of span L is subjected to a distributed load varying from zero at the free end to w_o at the fixed end. Using the virtual work method, determine the vertical deflection and rotation at the mid-point C of the beam. EI for the beam is constant.

Solution: The deflection due to shearing force may be neglected. For M and m functions, the origin of z is taken at the free end.

Deflection: To determine deflection at C, a virtual unit downward force is applied at the section C. The bending moment diagrams of M and m are shown in Figures 15.13(a) and (b), respectively.

$$M = -\frac{1}{2}(z)\left(\frac{w_0 z}{L}\right)\left(\frac{z}{3}\right) = -\frac{w_0 z^3}{6L} \qquad 0 \le z \le L$$

$$m = 0 \qquad 0 \le z \le L/2$$

$$m = -(1)\left(z - \frac{L}{2}\right) = \left(-z + \frac{L}{2}\right) \qquad L/2 \le z \le L$$

$$\delta_C = \int_0^L m\left(\frac{M}{EI}\right)dz = \frac{1}{EI}\int_0^{L/2}(0)\left(-\frac{w_0 z^3}{6L}\right)dz + \frac{1}{EI}\int_{L/2}^{L}\left(-z+\frac{L}{2}\right)\left(-\frac{w_0 z^3}{6L}\right)dz$$

$$= \frac{1}{EI}\int_{L/2}^{L}\left(\frac{w_0 z^4}{6L} - \frac{w_0 z^3}{12}\right)dz = \frac{w_0}{EI}\left(\frac{z^5}{30L} - \frac{z^4}{48}\right)_{L/2}^{L}$$

$$= \frac{w_0}{240EI}\left[\left(\frac{8}{L}\right)\left(L^5 - \frac{L^5}{32}\right) - (5)\left(L^4 - \frac{L^4}{16}\right)\right]$$

$$= \frac{w_0}{240EI}\left[\frac{248L^4}{32} - \frac{75L^4}{16}\right] = \frac{49w_0 L^4}{3840EI}$$

Rotation: To determine rotation at C, a virtual unit moment is applied at the section C. The corresponding m diagram is shown in Figures 15.13(c). The virtual moment $m = 0$ for z between 0 and $L/2$, and $m = 1$ for z between $L/2$ and L. Therefore,

$$\theta_C = \int_0^L m\left(\frac{M}{EI}\right)dz = \frac{1}{EI}\int_0^{L/2}(0)\left(-\frac{w_0 z^3}{6L}\right)dz + \frac{1}{EI}\int_{L/2}^{L}(1)\left(-\frac{w_0 z^3}{6L}\right)dz$$

$$= -\frac{w_0}{6EIL}\int_{L/2}^{L}(z^3)\,dz = -\frac{w_0}{6EIL}\left(\frac{z^4}{4}\right)_{L/2}^{L} = -\frac{15w_0 L^3}{384EI}$$

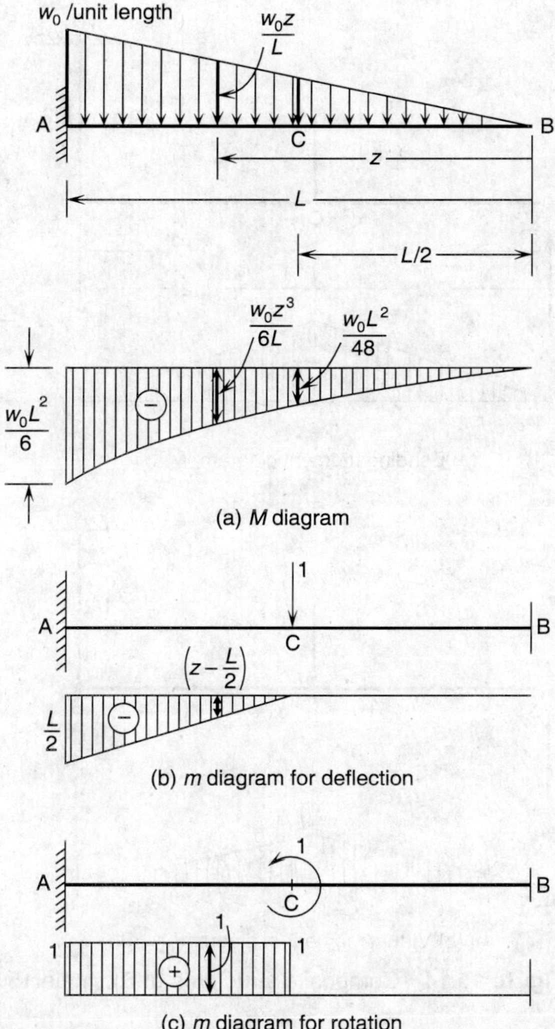

Figure 15.13 Cantilever beam of example 15.8 subjected to uniformly varying load.

EXAMPLE 15.9

In the structure of Figure 15.14, the left end A of the beam AB of span L is hinge-supported while the end B is suspended by a rod of length h. The moment of inertia of the beam and cross-sectional area of the rod are I and A, respectively. The Young's modulus of the materials of both the members is E. Determine the vertical deflection at the load point by using: (i) the work balance method and (ii) the virtual work method. EI for the beam is constant.

Solution: (i) *The work balance method*:

Let δ_C be the deflection at the point C, then

$$W_{ex} = \frac{1}{2} P \delta_C$$

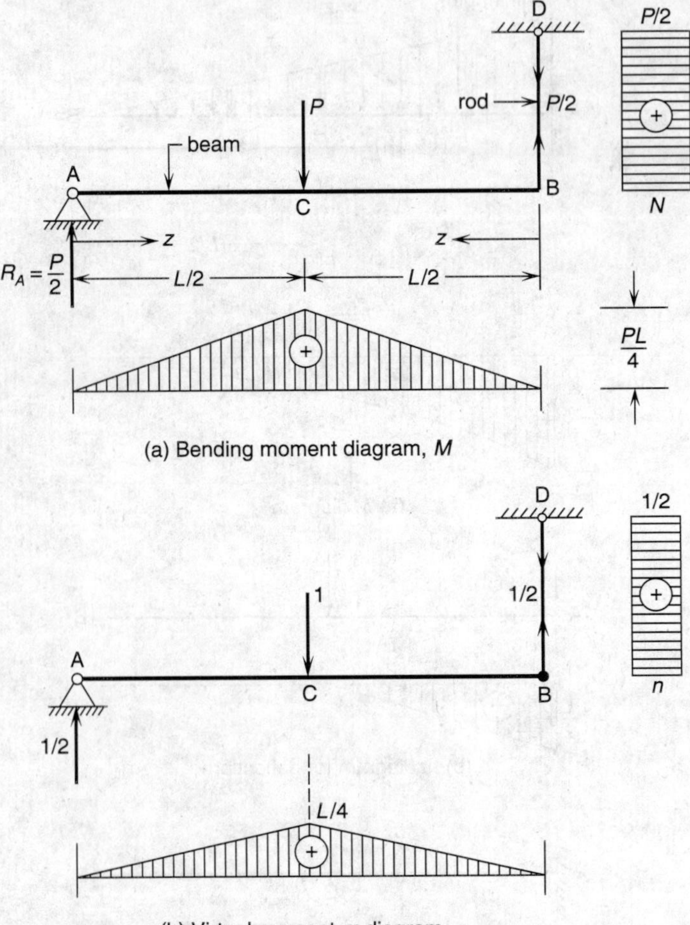

Figure 15.14 Composite structures of Example 15.9.

The increase in strain energy of the system, i.e. the work done by the internal forces,

$$-W_{\text{in}} = U = U_{AB} + U_{BD}$$

$$= 2\left[\int_0^{L/2}\left(\frac{1}{2EI}\right)\left(\frac{Pz}{2}\right)^2 dz\right] + \frac{N^2 h}{2AE} = \left(\frac{P^2}{4EI}\right)\int_0^{L/2} z^2\, dx + \frac{(P/2)^2 h}{2AE}$$

$$= \left(\frac{P^2}{4EI}\right)\frac{(L/2)^3}{3} + \frac{P^2 h}{8AE} = \left(\frac{P^2 L^3}{96EI}\right) + \frac{P^2 h}{8AE}$$

From work equation:

$$\frac{1}{2}P\delta_C = \left(\frac{P^2 L^3}{96EI}\right) + \frac{P^2 h}{8AE} \quad \text{or} \quad \delta_C = \frac{PL^3}{48EI} + \frac{Ph}{4AE}$$

(ii) *The virtual work method*: To determine deflection at C, a virtual unit downward force is applied at the section C. The bending moment diagrams of M and m are shown in Figures 15.14(a) and (b), respectively.

$$\delta_C = \frac{1}{EI}\int_0^L mM\,dx + \frac{\bar{n}NL}{AE}$$

Using geometrical integration or the area centre-of-gravity method,

$$\delta_C = (2)\left[\frac{1}{EI}\left(\frac{2}{3}\times\frac{L}{4}\right)\times\left(\frac{1}{2}\times\frac{PL}{4}\times\frac{L}{2}\right)\right] + \frac{1}{AE}\left(\frac{1}{2}\times\frac{P}{2}\times h\right)$$

$$= \frac{PL^3}{48EI} + \frac{Ph}{4AE}$$

EXAMPLE 15.10

In the structure of Figure 15.15(a), use the virtual work method to determine the relative distance by which the points A will move under the action of forces P. EI for the structure is constant.

Solution: To determine the mutual displacement or separation of two points A, apply at these points equal and opposite virtual unit forces acting along the straight line joining the points. The bending moment diagrams M and m are shown in Figures 15.15(b) and (c), respectively, and δ_A is given by:

$$\delta_A = \sum \frac{1}{EI}\int_0^L mM\,dz$$

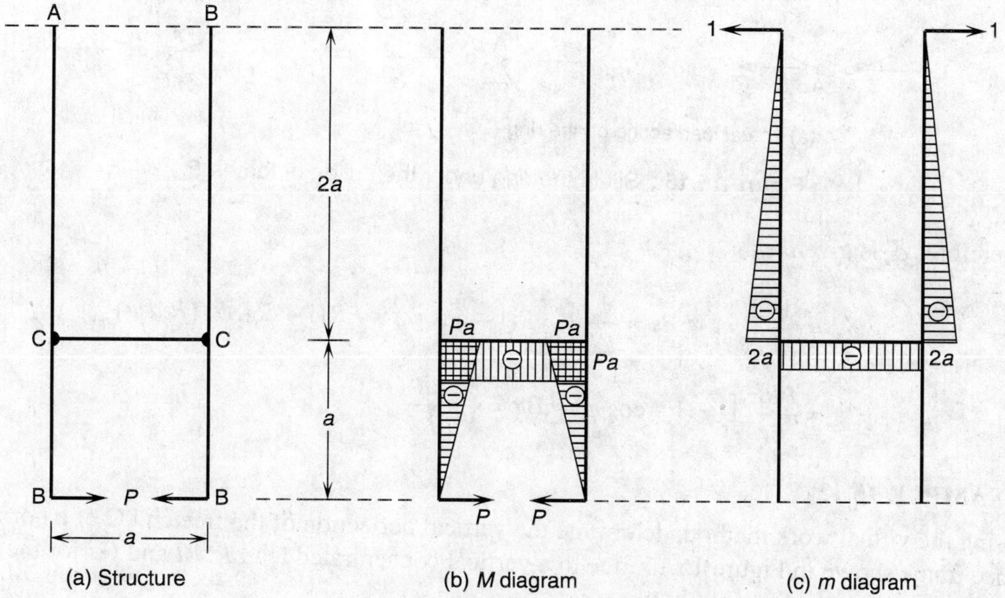

(a) Structure (b) M diagram (c) m diagram

Figure 15.15 The rigid frame system of Example 15.10.

It is seen from M and m diagrams that the result of multiplication of diagrams in the vertical portions is zero. Using geometrical integration or the area centre-of-gravity method for the horizontal portion,

$$\delta_A = \left[\frac{1}{EI}(-2a)\times(-Pa\times a)\right] = \frac{Pa^3}{EI}$$

EXAMPLE 15.11

Using the virtual work method, determine the distance by which the clearance in a cut ring opens under the action of forces P. EI for the ring is constant.

Solution: The bending moment M due to given forces at a point C at an angle ψ from the vertical as shown in Figure 15.16(a), is:

$$M_1 = PR(1 - \cos\psi)$$

To determine the mutual displacement of two points A, apply at these points equal and opposite virtual unit forces acting along the straight line joining the points as shown in Figure 15.16(b). The virtual moment m at the point C, due to unit virtual force is,

$$m_z = R(1 - \cos\psi)$$

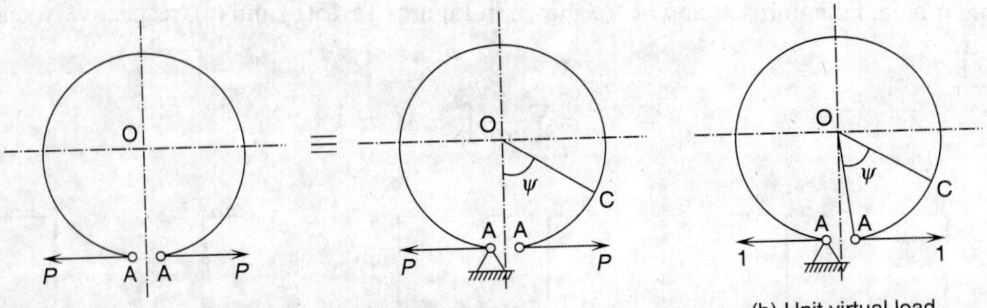

(a) Actual load acting on the ring

(b) Unit virtual load

Figure 15.16 Slit in the ring under the action of forces P.

Therefore, δ_A is given by:

$$\delta_A = \frac{1}{EI}\int_S mM\,ds = \frac{1}{EI}\int_0^{2\pi} R(1-\cos\psi)\times PR(1-\cos\psi)\,(R\,d\psi)$$

$$= \frac{PR^3}{EI}\int_0^{2\pi}(1-\cos\psi)^2\,d\psi = \frac{3\pi PR^3}{EI}$$

EXAMPLE 15.12

Using the virtual work method, determine the vertical deflection of the free end C of a horizontal knee frame shown in Figure 15.17, due to a vertical concentrated load P. EI and GJ for the frame member are constant.

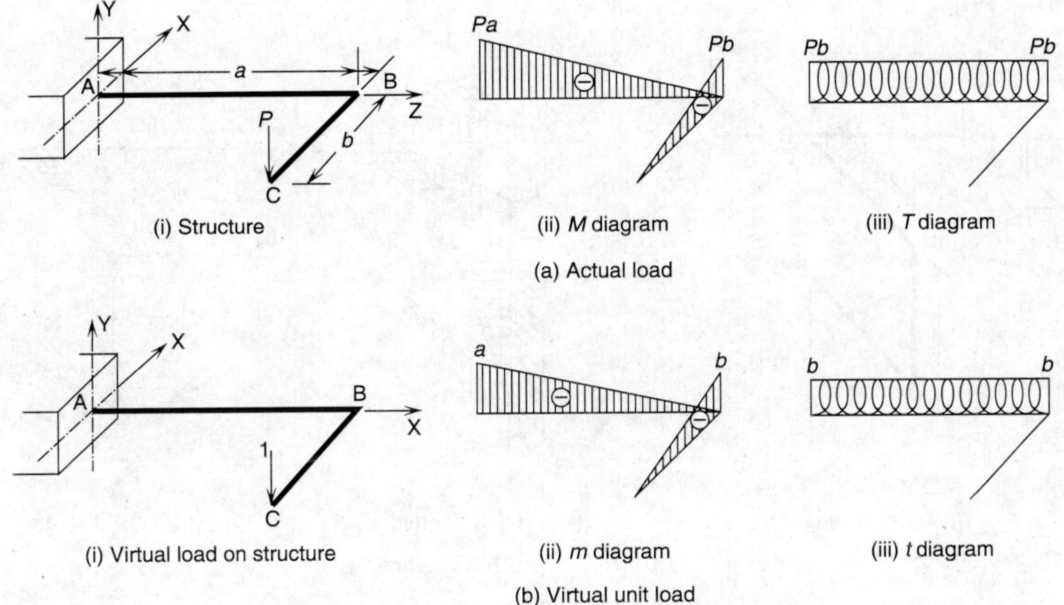

Figure 15.17 Horizontal knee frame of Example 15.12.

Solution: The basic displacements in the horizontal frame are due to bending and twisting of the members. The stress resultant diagrams of *M*, *T*, *m* and *t* are shown in Figures 15.17(a) and (b). The deflection at point C is given by,

$$\delta_C = \frac{1}{EI}\int_S m\,M\,ds + \frac{1}{GJ}\int_S t\,T\,ds$$

Using graphical integration for computation of deflection at point C,

$$\delta_C = \frac{1}{EI}\left[\left(\frac{1}{2}\right)(a)(Pa)\left(\frac{2a}{3}\right)\right] + \frac{1}{EI}\left[\left(\frac{1}{2}\right)(b)(Pb)\left(\frac{2b}{3}\right)\right] + \frac{1}{GJ}(a)(Pb)(b)$$

$$= \frac{P(a^3 + b^3)}{3EI} + \frac{Pab^2}{GJ}$$

EXAMPLE 15.13

Using the virtual work method, determine the *z*-component of deflection of the free end *E* of three-dimensional frame due to the load system shown in Figure 15.18. The flexural rigidity of the members in both the planes is *EI*. The torsional rigidity is *GJ*.

Solution: The basic displacements in the 3D system are due to bending and twisting of the members. The stress resultant diagrams of *M*, *T*, *m* and *t* are shown in Figures 15.18(a) and (b). The deflection at point *E* is given by,

$$\delta_E = \frac{1}{EI}\int_S m\,M\,ds + \frac{1}{GJ}\int_S t\,T\,ds$$

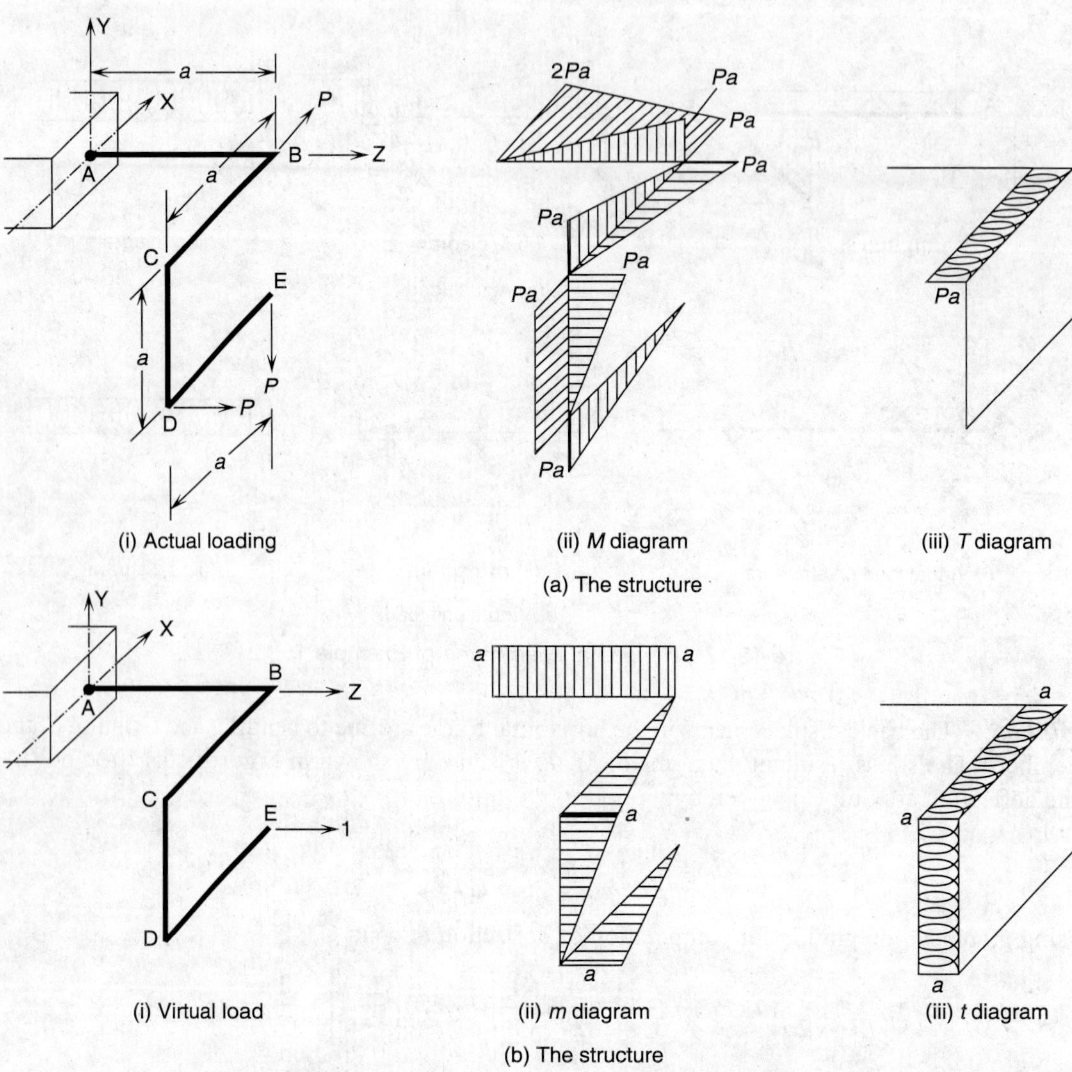

Figure 15.18 Three-dimensional system of Example 15.13.

The graphical integration has been used for the computation of deflection at point E. In the process, only the diagrams lying in the same plane are multiplied under the integral signs. Therefore,

$$\delta_E = \frac{1}{EI}\left[\left(\frac{1}{2}\right)(a)(Pa)(a)\right] + \frac{1}{EI}\left[\left(\frac{1}{2}\right)(a)(Pa)\left(\frac{a}{3}\right)\right]$$

$$+\frac{1}{EI}\left[\left(\frac{1}{2}\right)(a)(Pa)\left(\frac{2a}{3}\right)\right] + \frac{1}{GJ}[(a)(Pa)(a)]$$

$$= \frac{Pa^3}{EI} + \frac{Pa^3}{GJ}$$

15.7 ANALYSIS OF TRUSSES

The following examples illustrate the procedure for evaluating deflections of the truss joints.

EXAMPLE 15.14

Using the virtual work method, determine the vertical component of deflection of the joint 2 of the truss shown in Figure 15.19. All the members of the truss are of same cross-section and material.

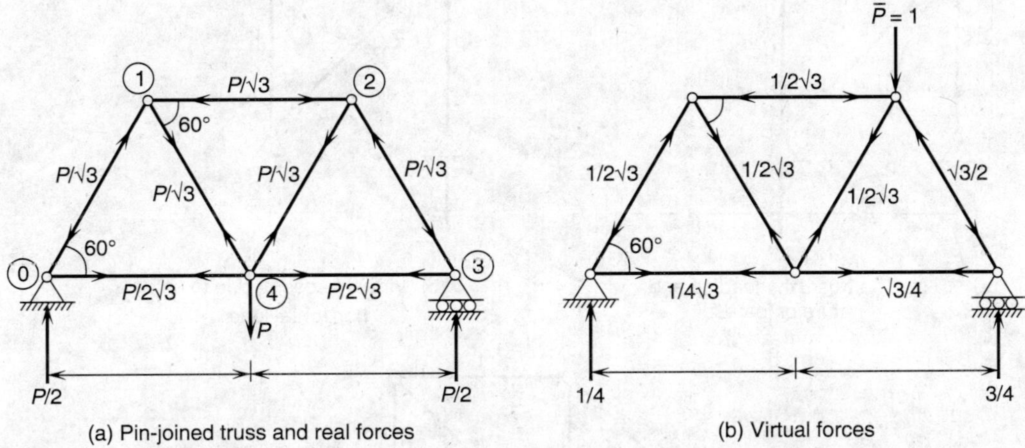

Figure 15.19 Pin-joined truss with real and virtual forces.

Solution: The basic displacements in the truss are due to axial deformation of the members. The stress resultant diagrams of N and n are shown in Figures 15.19(a) and (b), respectively. The vertical component of deflection at point 2 is given by,

$$\delta_2 = \frac{a}{AE}\sum_{i=1}^{m}(N\bar{n})_i$$

$$= \frac{a}{AE}\left[\left(\frac{P}{\sqrt{3}}\cdot\frac{1}{2\sqrt{3}}\right)+\left(\frac{P}{2\sqrt{3}}\cdot\frac{1}{4\sqrt{3}}\right)+\left(\frac{P}{\sqrt{3}}\cdot\frac{1}{2\sqrt{3}}\right)\right.$$

$$\left.+\left(\frac{P}{\sqrt{3}}\cdot\frac{1}{2\sqrt{3}}\right)+\left(\frac{P}{\sqrt{3}}\cdot\frac{1}{2\sqrt{3}}\right)+\left(\frac{P}{2\sqrt{3}}\cdot\frac{\sqrt{3}}{4}\right)+\left(\frac{P}{\sqrt{3}}\cdot\frac{\sqrt{3}}{2}\right)\right]$$

$$= \frac{32Pa}{24AE} = \frac{4Pa}{3AE}$$

EXAMPLE 15.15

Using the virtual work method, determine the horizontal and the vertical components of deflection of the joint D of the truss shown in Figure 15.20. The truss members have been designed such that the stress in the tension members is 12σ while that in the compression members is 7σ. E is same for all the members.

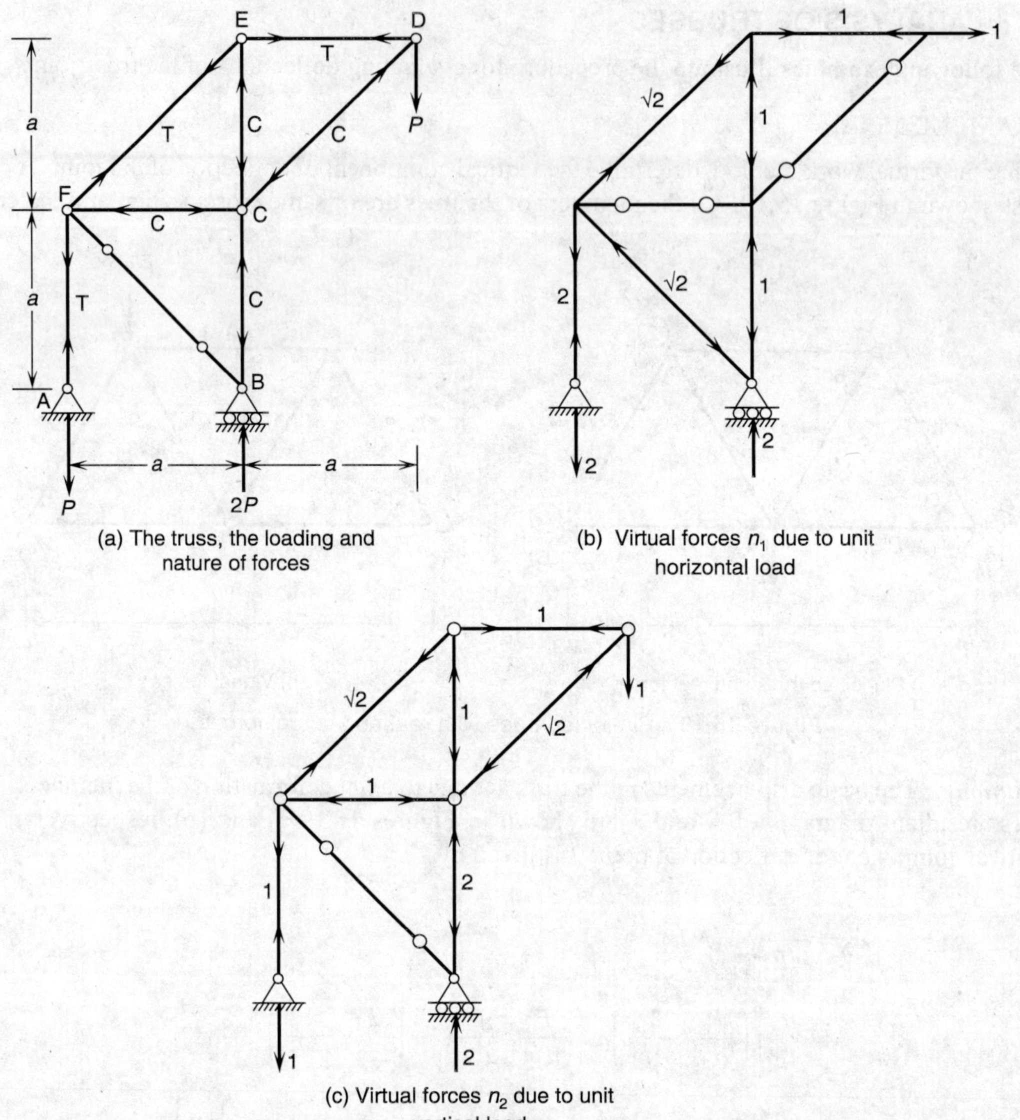

(a) The truss, the loading and nature of forces

(b) Virtual forces n_1 due to unit horizontal load

(c) Virtual forces n_2 due to unit vertical load

Figure 15.20 Nature of real forces and magnitudes of virtual forces in the truss of Example 15.15.

Solution: The real and virtual forces in the member of truss are shown in Figures 15.20(a) and (b) and (c), respectively. The horizontal component of deflection at point D is given by,

$$y_{DH} = \sum_{i=1}^{m}\left(\frac{\bar{n}_1 NL}{AE}\right)_i = \frac{1}{E}\sum_{i=1}^{m}\left(\frac{N}{A}\times\bar{n}_1 L\right)_i = \frac{1}{E}\sum_{i=1}^{m}(p\,\bar{n}_1 L)_i$$

For convenience, the computation may be performed in tabular form (Table 15.1) as follows:

TABLE 15.1 Computations for deflection in the truss of Example 15.15

Member	*Length,* L	*Stress,* p	*Force due to unit virtual load* Horizontal, $\bar{n}_1$	Vertical, $\bar{n}_2$	$p\bar{n}_1L$	$p\bar{n}_2L$
AF	a	12σ	2	1	$24\sigma a$	$12\sigma a$
FB	$\sqrt{2}a$	0	$-\sqrt{2}$	0	0	0
FC	a	-7σ	0	-1	0	$7\sigma a$
FE	$\sqrt{2}a$	12σ	$\sqrt{2}$	$\sqrt{2}$	$24\sigma a$	$24\sigma a$
EC	a	-7σ	-1	-1	$7\sigma a$	$7\sigma a$
ED	a	12σ	1	1	$12\sigma a$	$12\sigma a$
DC	$\sqrt{2}a$	-7σ	0	$-\sqrt{2}$	0	$14\sigma a$
CB	a	-7σ	-1	-2	$7\sigma a$	$14\sigma a$
				$\frac{1}{E}\sum_{i=1}^{m}(pn_1L)_i$	$\frac{74\sigma a}{E}$	$\frac{90\sigma a}{E}$

Therefore,

$$\delta_{DH} = \frac{74\sigma a}{E} \quad \text{and} \quad \delta_{DV} = \frac{90\sigma a}{E}$$

15.8 THE PRINCIPLE OF SUPERPOSITION OF MECHANICAL WORK

An important property of linear deformation is the validity of principle of superposition which means that: if a force F_1 produces a deformation r_1 and F_2 produces another deformation r_2, then deformation due to $F(= F_1 + F_2)$ is $r\ (= r_1 + r_2)$. However, as far as mechanical work is concerned the principle of superposition can be applied to the *displacement work* component but it is not valid for eigenwork component. It can be noticed that in the case of eigenwork relationship between work W_{ii} and displacement δ_{ii} is parabolic and therefore the principle of superposition does not hold good, while in case of displacement work, the relationship is linear and the principle of superposition is thus valid. Since the virtual work method is based on displacement work, the principle of superposition can be applied. The following example illustrates the procedure.

EXAMPLE 15.16

Using the virtual work method, determine the vertical deflection at the mid-point of the supported span due to the load system shown in Figure 15.21(a). The flexural rigidity EI is constant.

Solution: The stress resultant diagram for uniformly distributed load M_1, for concentrated load at the mid-span M_2, for concentrated load at the end of the overhang M_3, and virtual load diagramare shown in Figures 15.21(b), (c), (d) and (e), respectively. Using the principle of superposition, the deflection at point C is given by,

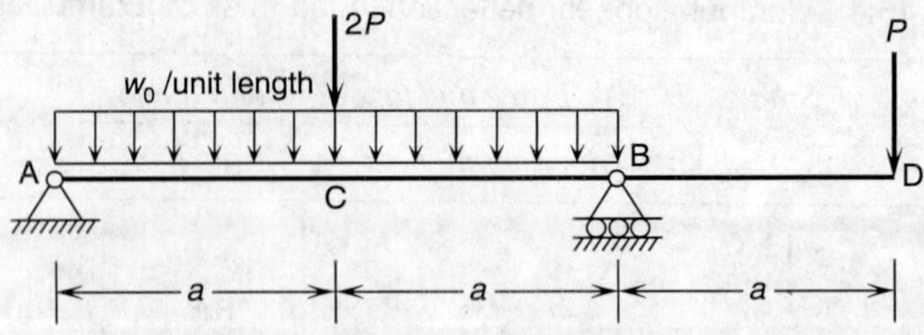

(a) The overhang beam and the loading

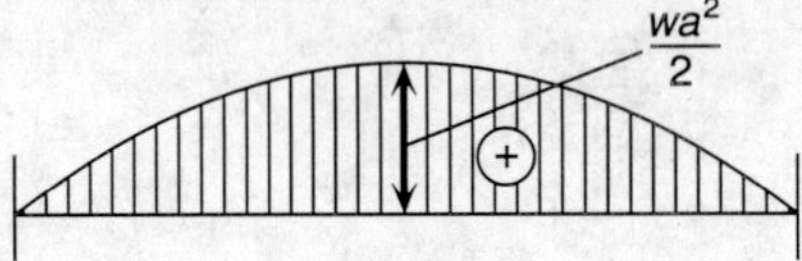

(b) M_1 diagram for uniformly distributed load

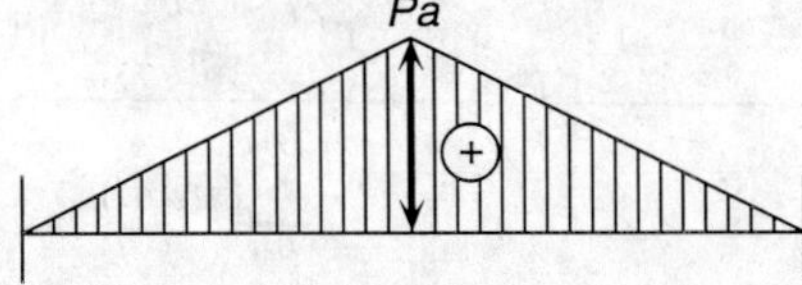

(c) M_2 diagram for concentrated load at the centre

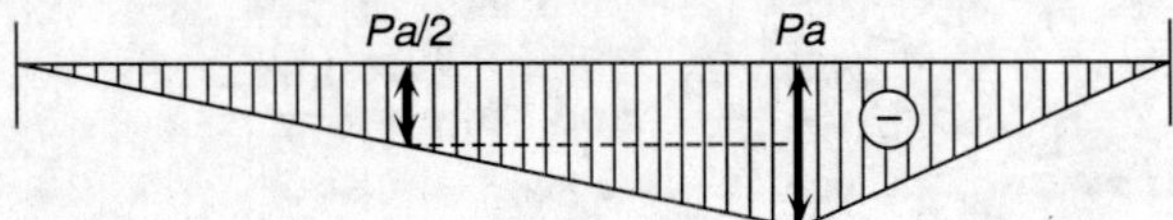

(d) M_3 diagram for concentrated load at the overhang end

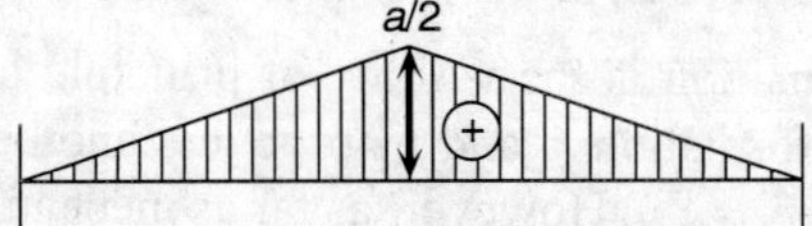

(e) *m* diagram due to unit virtual load at *C*

Figure 15.21 Computation of deflection by superposition.

$$\delta_C = \delta_1 + \delta_2 + \delta_3$$

where

$$\delta_1 = 2 \times \frac{1}{EI}\left[\left(\frac{2}{3}\right)(a)\left(\frac{wa^2}{2}\right)\left(\frac{5}{8} \times \frac{a}{2}\right)\right] = \frac{5wa^4}{24EI} \downarrow$$

$$\delta_2 = 2 \times \frac{1}{EI}\left[\left(\frac{1}{2}\right)(a)(Pa)\left(\frac{2}{3} \times \frac{a}{2}\right)\right] = \frac{Pa^3}{3EI} \downarrow$$

$$\delta_3 = \frac{1}{EI}\left[\left(\frac{1}{2}\right)(a)\left(-\frac{Pa}{2}\right)\left(\frac{2}{3} \times \frac{a}{2}\right) + (a)\left(-\frac{Pa}{2}\right)\left(\frac{1}{2} \times \frac{a}{2}\right)\right.$$

$$\left. + \left(\frac{1}{2}\right)(a)\left(-\frac{Pa}{2}\right)\left(\frac{1}{3} \times \frac{a}{2}\right)\right] = -\frac{Pa^3}{4EI} \uparrow$$

Therefore,
$$\delta_C = \frac{5wa^4}{24EI}\downarrow + \frac{Pa^3}{3EI}\downarrow - \frac{Pa^3}{4EI}\uparrow$$
$$= \left(\frac{5wa^4}{24EI} - \frac{Pa^3}{12EI}\right)\downarrow$$

15.9 ENERGY THEOREMS OF ELASTIC SYSTEMS

In this section, some of the important theorems of elastic systems are discussed.

15.9.1 Theorems of Betti and Maxwell

Consider the beam shown in Figure 15.22(a) subjected to load systems P_i and P_j at points i and j, respectively. Consider that the load system P_i is applied first and then subsequently the load system P_j. The work done by the forces is

$$W_{\text{I}} = W_{ii} + (W_{jj} + W_{ij}) \tag{15.47}$$

where W_{ii} and W_{jj} are eigenworks of P_i and P_j, respectively, and W_{ij} is the displacement work of P_i due to P_j. Now let the order of loading is reversed by bringing the load P_j first and then the load P_i as shown in Figure 15.22(a-ii). The work done by the forces in the second case is

$$W_{\text{II}} = W_{jj} + (W_{ii} + W_{ji}) \tag{15.48}$$

where W_{ji} is the displacement work done by P_j due to P_i. Since the total work done is independent of sequence of loading, W_{I} must be equal to W_{II}.

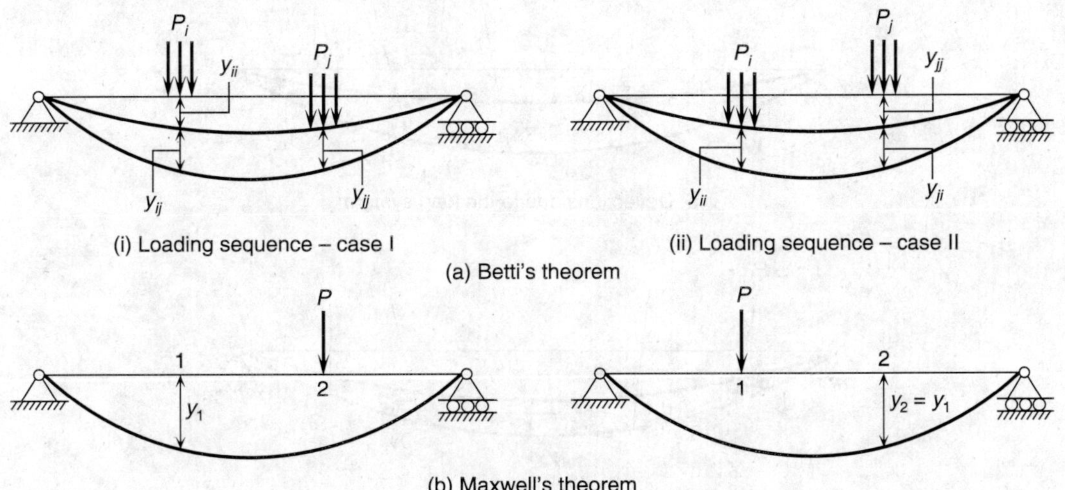

Figure 15.22 Basic features of Betti's and Maxwell's theorem.

Thus
$$W_{ii} + W_{jj} + W_{ij} = W_{jj} + W_{ii} + W_{ji}$$

or
$$W_{ij} = W_{ji} \quad \text{or} \quad \sum P_i\,\delta_{ij} = \sum P_j\,\delta_{ji} \tag{15.49}$$

This theorem is known as *Betti's theorem* and may be stated: *for a linearly elastic structure, the work done by a set of external force P_i acting through the displacement δ_{ij} produced by another set of force P_j is equal to the work done by the latter set of external force P_j acting through the displacements δ_{ji} produced by force P_i* (see Figure 15.22(a)).

Consider both the load systems P_i and P_j to be consisting of single load P (as shown in Figure 15.22(b) having the same magnitude but not necessarily in the same direction), then

$$P_i\delta_{ij} = P_j\delta_{ji} \quad \text{or} \quad \delta_{ij} = \delta_{ji} \tag{15.50}$$

This is known as *Maxwell's theorem of reciprocal deflection* and states that: the deflection of point j due to force P at point i is numerically equal to the deflection of point i due to force P applied at point j. It should be noted that deflections are measured in the direction of the forces. Here force means a generalized force (including moment).

15.9.2 Cotterill–Castigliano's Theorems

In the earlier sections, the eigen work has been used in conjunction with the work equation to determine the deflection at the point of application of a load. However, eigen work could be generalized to calculate deflections at any section. For illustration, consider a simply supported beam shown in Figure 15.23(a) subjected to a system of loads $f(P_1, P_2, P_3, ..., P_n) = f(P)$ applied gradually. The beam undergoes deformation and strain energy, which is a function of external loads and is equal to external work done, is stored in the system. Therefore,

$$W_{ex} = f(P_1, P_2, P_3, ..., P_n) = U \tag{15.51}$$

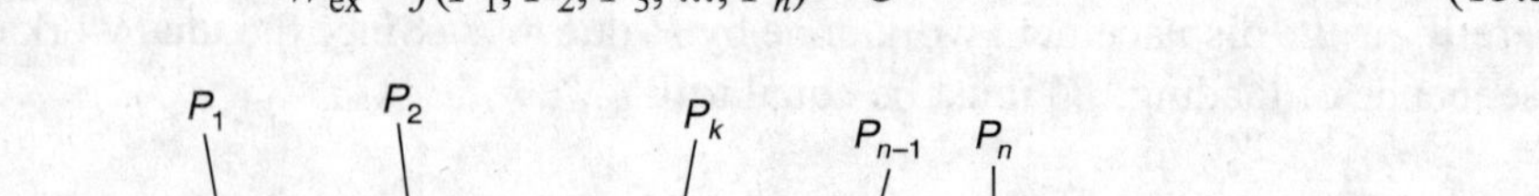

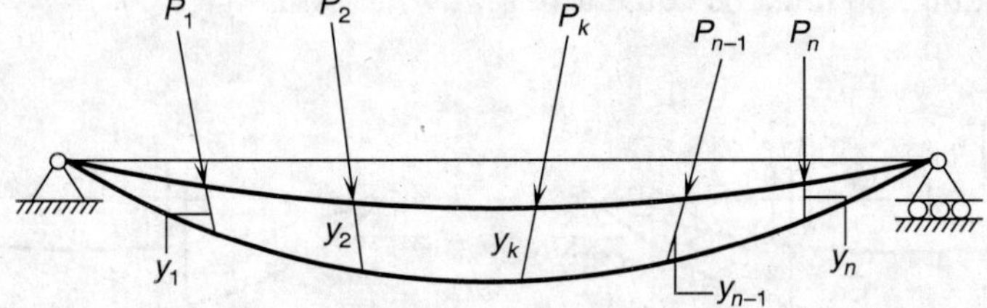

(a) Deflections due to the load system

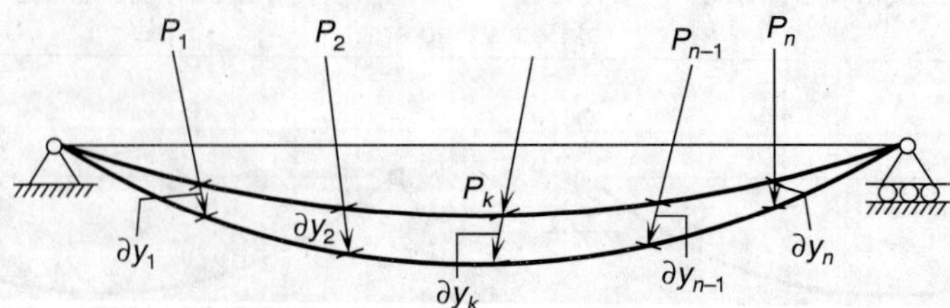

(b) Additional deflections due to δP_n

Figure 15.23 Simply supported beam subjected to a given load system.

Now, if any one of the loads, say P_k is increased by a differential amount dP_k, the strain energy of the system will change by an amount $(\partial U/\partial P_k)\, dP_k$. The expression for total strain energy becomes,

$$U_t = U + \left(\frac{\partial U}{\partial P_k}\right) dP_k$$

$$= W_{\text{ex,t}} = U + \sum_{i=1}^{n} P_i \, d\delta_i + \frac{1}{2} dP_k \, d\delta_k$$

Neglecting the last term as being the product of two differential quantities, we have

$$U_t = U + \sum_{i=1}^{n} P_i \, d\delta_i \tag{15.52}$$

If the sequence of loading is reversed, i.e. dP_k is applied first and then the system of loads $f(P)$ as shown in Figure 15.23(b) the total work done is

$$W_{\text{ex,t}} = U_t = U + \left(\frac{1}{2} dP_k \, d\delta_k\right) + dP_k \, \delta_k$$

The term $(dP_k \, d\delta_k/2)$ being of second order can be neglected. Consequently expression of work done reduces to

$$U_t = U + dP_k \, \delta_k \tag{15.53}$$

Since the order of application of loads is immaterial, the total work done or the total internal strain energy in both the loading sequences must be equal. Therefore,

$$U + \frac{\partial U}{\partial P_k} dP_k = U + dP_k \, \delta_k \tag{15.54}$$

or

$$\frac{\partial U}{\partial P_k} = \delta_k \quad \text{or} \quad \delta_k = \frac{\partial U}{\partial P_k} = \frac{\partial (W_{\text{ex}})}{\partial P_x} = \frac{\partial (-W_{\text{in}})}{\partial P_k}$$

The above expression can be generalized to:

$$\delta_k = \frac{\partial (-W_{\text{in}})}{\partial P_k} \quad \text{and} \quad \theta_k = \frac{\partial (-W_{\text{in}})}{\partial M_k} \tag{15.55}$$

where, W_{in} is the internal work of the system. In other words, for a linearly elastic structure, partial derivative of the total strain energy with respect to a typical load P_k gives the deflection due to this load in its direction. This is *Cotterill–Castigliano's second theorem.*

Now, if one of the displacements, say δ_k, is changed by an infinitesimal amount $d\delta_k$ while all other displacements are kept unchanged, the corresponding change in the strain energy would be $(\partial U/\partial \delta_k) \, d\delta_k$. During such a change, the force P_k is the only one which does work amounting to $P_k \, d\delta_k$ Equating the change in the internal energy to the additional work done

$$\left(\frac{\partial U}{\partial \delta_k}\right) d\delta_k = P_k \, dy_k \quad \text{i.e.} \quad \frac{\partial U}{\partial \delta_k} = P_k \tag{15.56}$$

This is known as *Cotterill–Castigliano's first theorem*. It states that the partial derivative of strain energy of a system with respect to any one of the displacements at a certain point gives singular force at the same point. It should be noted that this theorem does not place any restriction on the relationship between deformation and force being linear.

To apply Castigliano's second theorem for determining deflections, the strain energy must be expressed in terms of external loads. Consider, for example, the flexural system where the internal strain energy is due to bending. The expression for deflection is

$$\delta_k = \frac{\partial U}{\partial P_k} = \frac{\partial}{\partial P_k} \int \frac{(M_z)^2}{2EI} dz$$

It is much easier to first differentiate the quantity under integral sign and then evaluate the integral, i.e.

$$\delta_k = \int_0^L \left(\frac{M_z}{EI}\right)\left(\frac{\partial M_z}{\partial P_k}\right) dz \tag{15.57}$$

Similar expression can be developed for trusses where internal energy is due to axial strains. The expression for deflection of a truss point is

$$\delta_k = \sum_{i=1}^{n} \left(\frac{N}{EA}\right)_i \left(\frac{\partial N}{\partial P_k}\right)_i L_i \tag{15.58}$$

When the frame is subjected to all types of stress resultants,

$$\delta_k = \int_0^L \left(\frac{M_z}{EI}\right)\left(\frac{\partial M_z}{\partial P_k}\right) dz + \sum_{i=1}^{n} \left(\frac{N}{EA}\right)_i \left(\frac{\partial N}{\partial P_k}\right)_i L_i + \int_0^L \left(\frac{T_z}{GJ}\right)\left(\frac{\partial T_z}{\partial P_k}\right) dz + \int_0^L \left(\frac{V_z}{GA}\right)\left(\frac{\partial V_z}{\partial P_k}\right) dz$$

It must be noted that if a deflection component is required at a point where no action is applied or if an action exists at that point but not in the direction of desired deflection, then an imaginary action is assumed until the partial derivative for the total strain energy has been computed. In the resulting expression, the imaginary action is then reduced to zero.

The Cotterill–Castigliano's second theorem can be advantageously used for the calculation of redundant actions in *statically indeterminate systems*. The procedure consists in making the system statically determined by removing the redundant supports and replacing them with unknown redundant actions. Thus, the system under the given loading would have deflections at the location and in the direction of support reactions. The work is expressed in terms of external known loads and unknown redundant actions. The partial derivatives with respect to redundant actions R_i give deflection at their location and directions. However, the deflections are suppressed by the support that makes the system statically indeterminate in the first place. Thus

$$-\frac{\partial W_{\text{in}}}{\partial R_i} = 0, \quad i = 1, 2, \ldots, n \tag{15.59}$$

where n is the number of redundant actions. Suppose R_i is equal to unity, then we would have a unit deflection δ_i. Thus

$$\delta_{Ri} = R_i\,\delta_i$$

Since δ_R must be equal to the prescribed deflection δ_{pi}, thus the net deflection

$$\delta_{Ri} = R_i\,\delta_i - \delta_{pi}$$

However, from Cotterill–Castigliano's second theorem

$$\left(-\frac{\partial W_{\text{in}}}{\partial R_i}\right) = R_i\,\delta_i - \delta_{pi}$$

Differentiating the equation with respect to R_i, we obtain

$$-\frac{\partial^2 W_{\text{in}}}{\partial R_i^2} = \delta_i > 0 \tag{15.60}$$

that is δ_i must be positive. Thus Eq. (15.58) is the necessary condition that $-W_{\text{in}}$ is an extremum. While Eq. (15.59) says that extremum is minimum. Thus *statically indeterminate redundancy takes a value that makes the work of internal forces* $-W_{\text{in}}$ *a minimum, i.e.*

$$\partial(-W_{\text{in}}) = 0 \quad \text{and} \quad \partial^2(-W_{\text{in}}) > 0$$

The Cotterill–Castigliano's first and second theorems lead to the formulation of *stiffness* and *flexibility methods*. It should be noted that if a structure behaves elastically, there is no difference between virtual work method and Castigliano's second theorem for deflections. However, it should be realized that while Castigliano's theorem is valid only for structures with a linear load-deflection relation but the virtual work method may be adopted for any type of non-elastic structure.

EXAMPLE 15.17

A simply supported beam AB of span $2a$ is strengthened by a simply supported cross (secondary) beam DE of length a placed symmetrically to its mid-span as shown in Figure 15.24(a). If the main beam carries a uniformly distributed load w_0 per unit length, determine the vertical deflection at the mid-point of the beam. The flexural rigidity of the main beam is twice that of secondary beam.

Solution: Consider the flexural rigidity of the main beam to be EI. The secondary beam behaves as an elastic spring, i.e. the secondary beam carries load which is k times the deflection where k is spring constant or the stiffness of the cross beam. Figures 15.24(b) and (c) show free-body diagram of the structure and the equivalent spring supported system. Let the reaction at the secondary beam is R. Then the reaction at the each end of the main beam due to symmetry will be,

$$R_A = R_B = \frac{2w_0 a - R}{2}$$

Bending moment at a section at distance z from A

$$M_z = \left(\frac{2w_0 a - R}{2}\right)z - \frac{w_0 z^2}{2} \quad \text{and} \quad \frac{\partial M_z}{\partial R} = -\frac{z}{2}$$

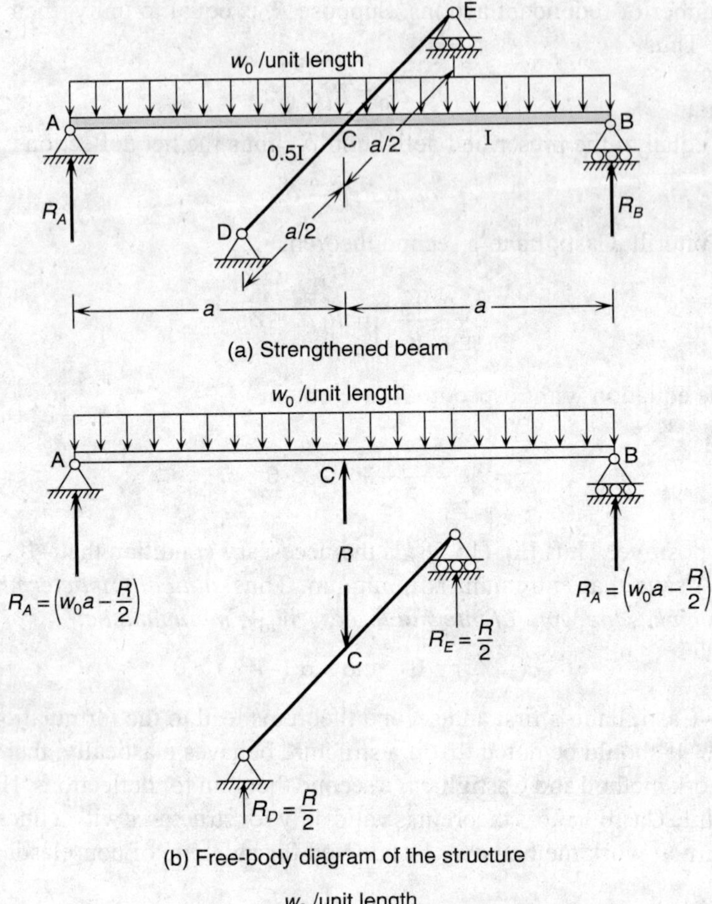

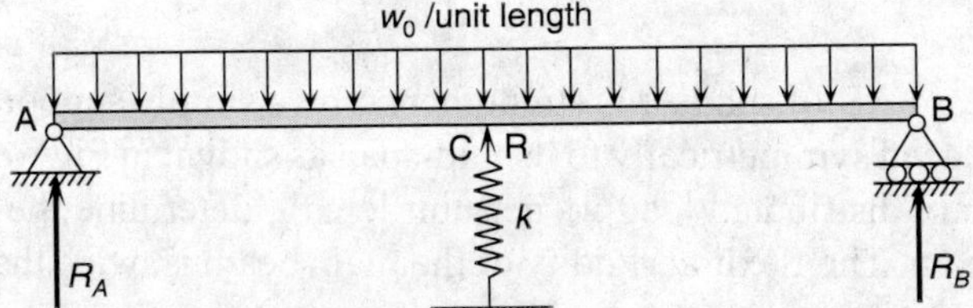

(c) Equivalent spring supported system

Figure 15.24 Analysis of strengthened beam of Example 15.17.

The deflection at C in the secondary beam,

$$\delta_C = \frac{Ra^3}{48(EI/2)}$$

The spring constant of the beam,

$$\frac{R}{\delta_C} = k = \frac{24EI}{a^3} \quad \text{or} \quad \delta_C = \frac{R}{k}$$

The deflection at C in the main beam is given by:

$$\delta_C = \int_S \left(\frac{M_z}{EI}\right)\left(\frac{\partial M_z}{\partial R}\right) dz$$

Therefore, for compatibility of displacements of the main and secondary beams

$$\delta_C = 2 \times \left[\frac{1}{EI}\int_0^a \left\{\left(\frac{2w_0 a - R}{2}\right)z - \frac{w_0 z^2}{2}\right\}\left(-\frac{z}{2}\right)dz\right] = -\frac{R}{k}$$

It should be noted that the deflection in secondary beam is opposite to the direction of R

$$\frac{1}{EI}\int_0^a \left\{-\left(\frac{2w_0 a - R}{2}\right)z^2 + \frac{w_0 z^3}{2}\right\} dz = \frac{1}{EI}\left\{\left(-\frac{2w_0 a}{2} + \frac{R}{2}\right)\frac{z^3}{3} + w_0 \frac{z^4}{8}\right\}_0^a = -\frac{R}{k}$$

$$\frac{1}{EI}\left\{\left(-\frac{2w_0 a^4}{6} + \frac{a^3 R}{6}\right) + \frac{w_0 a^4}{8}\right\} = -\frac{R}{k}$$

or

$$\frac{1}{EI}\left\{-\frac{2w_0 a^4}{6} + \frac{w_0 a^4}{8}\right\} = -\left(\frac{1}{k} + \frac{a^3}{6EI}\right)R$$

$$-\frac{5w_0 a^4}{24EI} = -\left(\frac{1}{k} + \frac{a^3}{6EI}\right)R \quad \text{or} \quad R = \frac{\left(\frac{5w_0 a^4}{24EI}\right)}{\left(\frac{1}{k} + \frac{a^3}{6EI}\right)}$$

Substitute the value of k,

$$R = \frac{\left(\frac{5w_0 a^4}{24EI}\right)}{\left(\frac{a^3}{24EI} + \frac{a^3}{6EI}\right)} = w_0 a$$

The deflection at C,

$$\delta_C = (w_0 a)\left(\frac{a^3}{48EI}\right) = \frac{w_0 a^4}{48EI}$$

EXAMPLE 15.18

A simply supported beam AB of span $2a$ is strengthened by two struts placed symmetrically to its mid-span as shown in Figure 15.25(a). If the beam carries a uniformly distributed load w_0 per unit length, determine the vertical deflection at the mid-point of the beam. The flexural rigidities of the beam and struts are EI and AE, respectively.

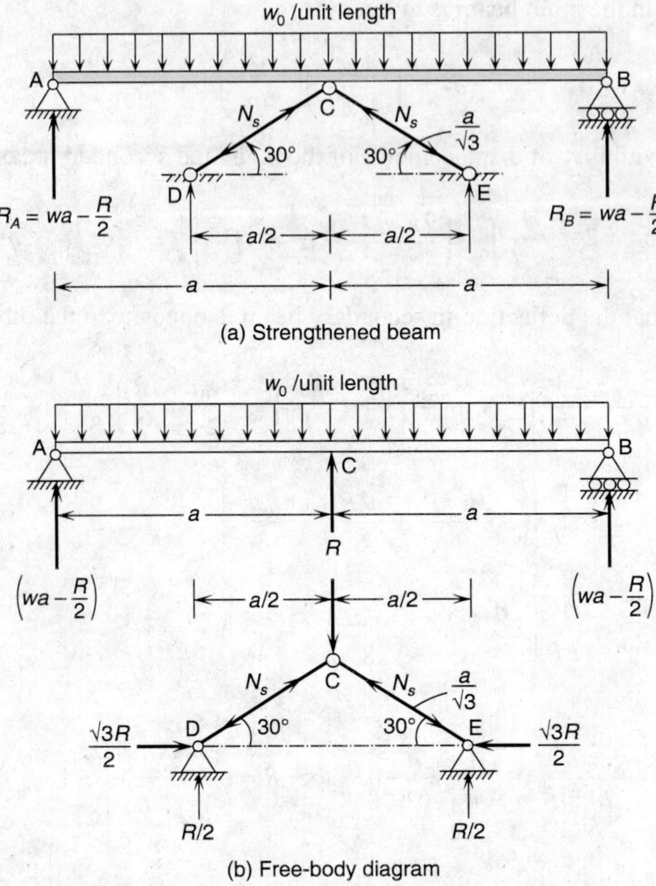

Figure 15.25 Strengthened beam of Example 15.18.

***Solution*:** As in the preceding example the strut assembly behaves like an elastic spring, i.e. it deflects by an amount R/k where R and k are the vertical reaction and spring constant or stiffness of the strut assembly, respectively. Free-body diagram is shown in Figure 15.25(b). The reaction at the each end of main beam due to symmetry is,

$$R_A = R_B = w_0 a - \frac{R}{2}$$

The bending moment in the main beam at a section at distance z from A

$$M_z = \left(w_0 a - \frac{R}{2}\right)z - \frac{w_0 z^2}{2} \quad \text{and} \quad \frac{\partial M_z}{\partial R} = -\frac{z}{2}$$

The axial force in each strut is determined by considering vertical equilibrium at point C,

$$2N_s \sin 30° = R \quad \text{or} \quad N_s = R \quad \text{and} \quad \frac{\partial N_s}{\partial R} = 1$$

Length of strut, $$L = \frac{a}{2} \sec 30° = \frac{a}{2} \times \frac{2}{\sqrt{3}} = \frac{a}{\sqrt{3}}$$

As the relative displacement between the main beam and the strut assembly at C is zero. Therefore, for compatibility of displacements of the main beam and strut assembly,

$$\delta_{C,\text{beam-strut}} = \int_S \left(\frac{M_z}{EI}\right)\left(\frac{\partial M_z}{\partial R}\right) dz + \sum\left(\frac{N_s L}{EA}\right)\left(\frac{\partial N}{\partial R}\right) = 0$$

or

$$2\times\left[\frac{1}{EI}\int_0^a \left\{\left(w_0 a - \frac{R}{2}\right)z - \frac{w_0 z^2}{2}\right\}\left(-\frac{z}{2}\right)dz\right] + 2\times\frac{RL}{EA} = 0$$

$$\frac{1}{EI}\left\{\left(-w_0 a + \frac{R}{2}\right)\frac{z^3}{3} + w_0\frac{z^4}{8}\right\}_0^a + \frac{2RL}{EA} = 0$$

$$\frac{1}{EI}\left\{\left(-\frac{w_0 a^4}{3} + \frac{a^3 R}{6}\right) + \frac{w_0 a^4}{8}\right\} = -\frac{2R(a/\sqrt{3})}{AE}$$

or

$$\frac{1}{EI}\left\{-\frac{w_0 a^4}{3} + \frac{w_0 a^4}{8}\right\} = -\left(\frac{2u}{\sqrt{3}AE} + \frac{a^3}{6EI}\right)R$$

$$-\frac{5w_0 a^4}{24EI} = -\left(\frac{2a}{\sqrt{3}EA} + \frac{a^3}{6EI}\right)R \quad \text{or} \quad R = \frac{\left(\dfrac{5w_0 a^4}{24EI}\right)}{\left(\dfrac{2a}{\sqrt{3}EA} + \dfrac{a^3}{6EI}\right)}$$

The deflection at C,

$$\delta_C = \frac{R(a/\sqrt{3})}{EA} = \frac{\left(\dfrac{5w_0 a^4}{24EI}\right)}{\left(2 + \dfrac{EAa^2}{2\sqrt{3}EI}\right)}$$

EXAMPLE 15.19

A pin-joined system of five bars is loaded as shown in Figures 15.26(a) and (b). The geometrical properties of the bars are indicated in the figure. Determine the relative displacement of the ends A and B.

Solution: The axial forces in the members are determined by the method of joints. Considering the symmetry of construction and loading,

Joint A: $\sum F_z = 0$: $\quad N_{AC}\cos\theta_1 - N_{AD}\cos\theta_2 - P = 0$

or

$$N_{AC}\left(\frac{4}{5}\right) - N_{AD}\left(\frac{4}{2\sqrt{13}}\right) - P = 0 \qquad \text{(i)}$$

$\sum F_y = 0$: $\quad N_{AC}\sin\theta_1 - N_{AD}\sin\theta_2 = 0$

or

$$N_{AC}\left(\frac{3}{5}\right) - N_{AD}\left(\frac{6}{2\sqrt{13}}\right) = 0 \qquad \text{(ii)}$$

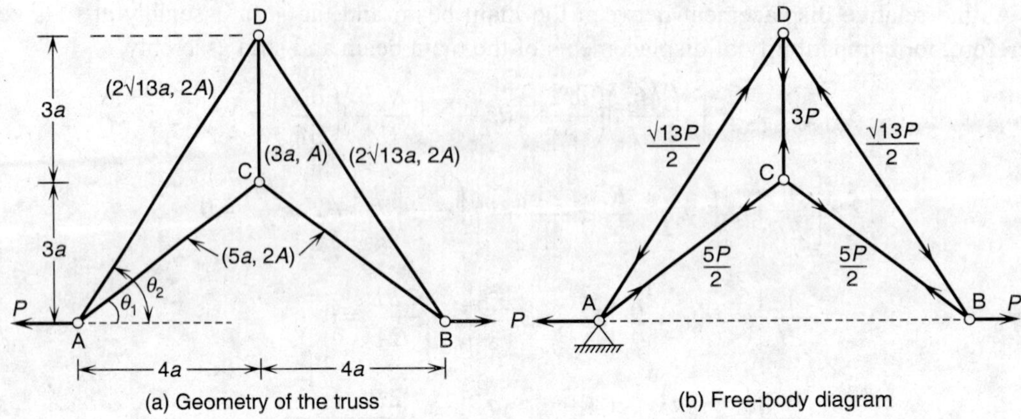

Figure 15.26 Analysis of the truss of Example 15.19.

Solving Eqs. (i) and (ii):

$$N_{AC} = \left(\frac{5P}{2}\right) \quad \text{and} \quad N_{AD} = \left(\frac{\sqrt{13}P}{2}\right)$$

Joint D: $\sum F_y = 0: \quad N_{AD} \sin \theta_2 + N_{BD} \sin \theta_2 - N_{CD} = 0$

as $N_{BD} = N_{AD}, \quad N_{CD} = 2N_{AD} \sin \theta_2 = 2\left(\frac{\sqrt{13}P}{2}\right)\left(\frac{6}{2\sqrt{13}}\right) = 3P$

Computations of displacement:

$$\delta_k = \sum_{i=1}^{5} \left(\frac{N}{EA}\right)_i \left(\frac{\partial N}{\partial P_k}\right)_i L_i = 2\left[\frac{1}{E(2A)}\left(\frac{5P}{2}\right)\left(\frac{5}{2}\right)(5a)\right]$$

$$+ 2\left[\frac{1}{E(2A)}\left(\frac{\sqrt{13}P}{2}\right)\left(\frac{\sqrt{13}}{2}\right)(2\sqrt{13}\,a)\right] + \left[\frac{1}{E(A)}(3P)(3)(3a)\right]$$

$$= \frac{233 + 26\sqrt{13}}{4}\left(\frac{Pa}{AE}\right)$$

EXAMPLE 15.20

A bar is bent into a quadrant of a circle of radius a and supported in a horizontal plane as shown in Figure 15.27. Determine the deflection at the end A caused by a vertical load W applied there.

Solution: Consider an element of length $ds = a\,d\theta$ at angular distance θ from the free end. The element is subjected to torsion and bending caused by the vertical load W over its entire length. The moment about moment axis EO is

$$M_\theta = \text{Load} \times \text{Moment arm AC} = W \times a \sin \theta = Wa \sin \theta$$

and

$$\frac{\partial M_\theta}{\partial W} = a \sin \theta$$

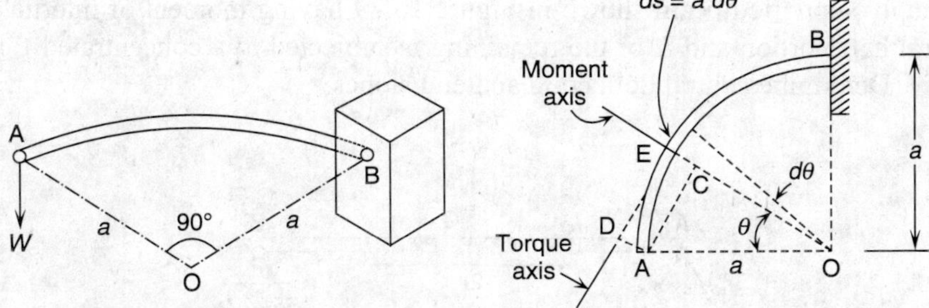

Figure 15.27 Bar bend to a quadrant of a circle of Example 15.20.

The twisting moment about torque axis DE is

$$T_\theta = \text{Load} \times \text{Moment arm AD} = W \times a\,(1 - \cos\theta) = Wa\,(1 - \cos\theta)$$

and
$$\frac{\partial T_\theta}{\partial W} = a(1 - \cos\theta)$$

Therefore,

$$\delta_A = \frac{1}{EI}\int M_\theta\left(\frac{\partial M_\theta}{\partial W}\right) ds + \frac{1}{GJ}\int T_\theta\left(\frac{\partial T_\theta}{\partial W}\right) ds$$

$$= \frac{1}{EI}\int_0^{\pi/2}(Wa\sin\theta)(a\sin\theta)\,a\,d\theta + \frac{1}{GJ}\int_0^{\pi/2}[Wa(1-\cos\theta)][a(1-\cos\theta)]\,a\,d\theta$$

$$= \frac{Wa^3}{EI}\int_0^{\pi/2}\sin^2\theta\,d\theta + \frac{Wa^3}{GJ}\int_0^{\pi/2}(1-\cos\theta)^2\,d\theta$$

$$\delta_A = \frac{Wa^3}{EI}\left(\frac{\pi}{4}\right) + \frac{Wa^3}{GJ}\left(\frac{3\pi}{4} - 2\right)$$

15.10 PROBLEMS

15.1 A simply supported beam AB shown in Figure 15.28 is subjected to an end moment M_0 at the end B. Determine the end slopes and maximum deflection.

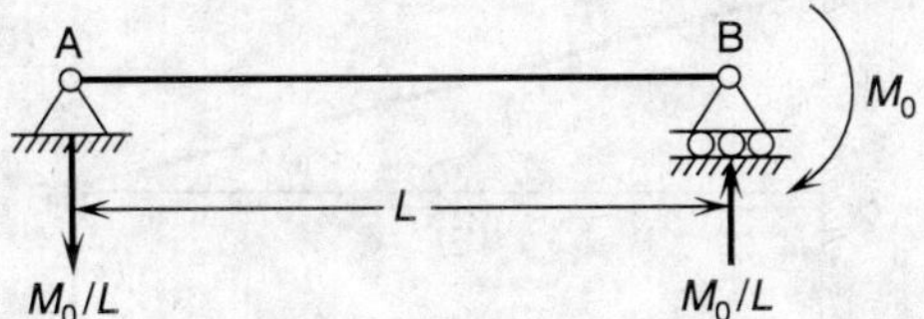

Figure 15.28

[**Ans.** $\theta_A = M_0L/6EI$ and $\theta_B = M_0L/3EI$. The deflection at distance z from the end A is given by: $\delta_z = M_0Lz\,(1 - z^2/L^2)/6EI$. Maximum deflection occurs at $z = L/\sqrt{3}$ and its magnitude is: $\delta_{max} = M_0L^2/(9\sqrt{3}EI)$]

15.2 A simply supported beam shown in Figure 15.29 having moment of intertia of $2I$ at the central half portion and I for the remaining is subjected to a concentrated force P at the centre. Determine central deflection and end slopes.

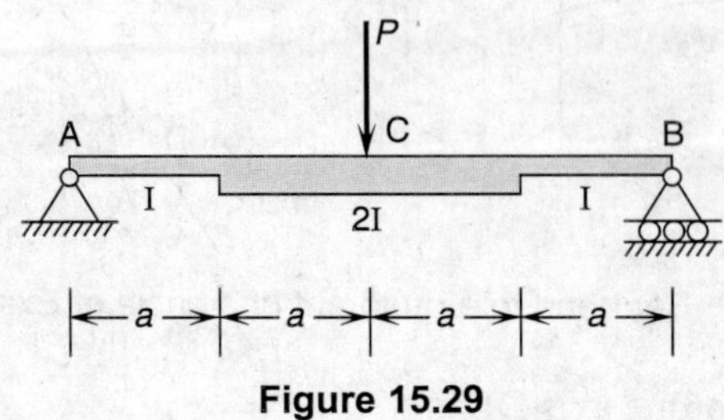

Figure 15.29

[**Ans.** $\delta_C = 3Pa^3/4EI$; $\theta_B = \theta_A = 5Pa^2/8EI$]

15.3 Use Cotterill–Castigliano's theorem to compute vertical deflection at the point D of the beam shown in Figure 15.30.

[**Hint:** Apply fictitious load W at the point D. The deflection at the point D is:

$\delta_D = \partial U/\partial W = \int[M_z(\partial M_z/\partial W)\,dz]/EI = Pa^3/4EI(\uparrow)$]

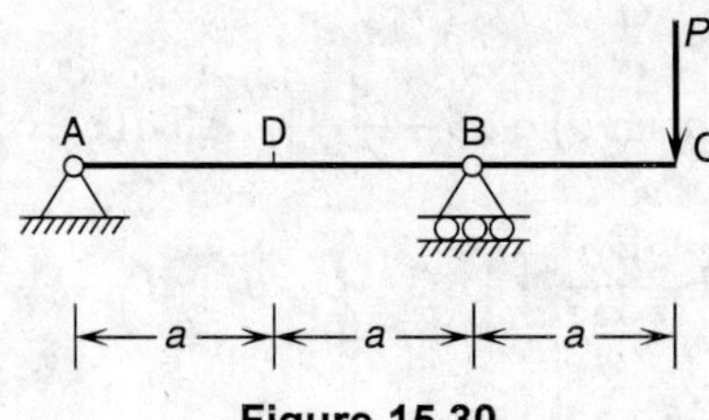

Figure 15.30

15.4 The free end of cantilever beam shown in Figure 15.31 is supported by an inclined tie rod. The cross-sectional area of tie rod is A and the flexural rigidity of cantilever beam is EI. Use strain energy method to determine vertical displacement at the joint B and tension in the tie rod due to concentrated load P acting at B. E is same for the beam and tie rod.

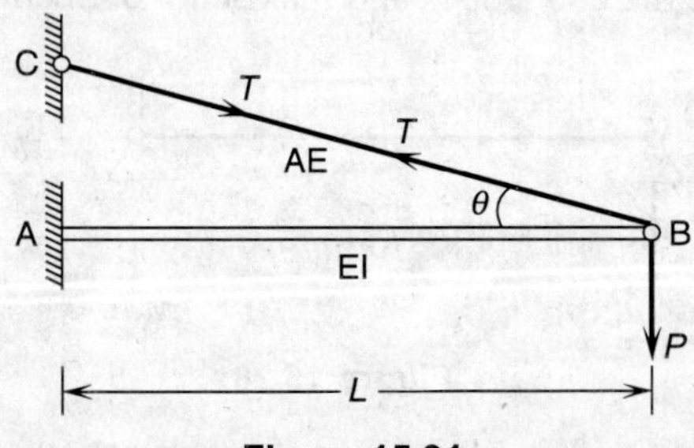

Figure 15.31

[**Ans.** $U = \{AE \sin^2\theta \,.\, \cos\theta/2L\} + (8EI/2L^3)\}\ \delta^2$; where δ is downward deflection at the point B. $\delta = P/\{AE \sin^2\theta \,.\, \cos\theta/L + (3EI/L^3)\}$].

15.5 The rigid cantilever frame shown in Figure 15.32 has constant flexural rigidity *EI* and carries a concentrated load *P* at the free end. Using different versions of energy approach determine displacements at the points B, C and D slope at the point D.

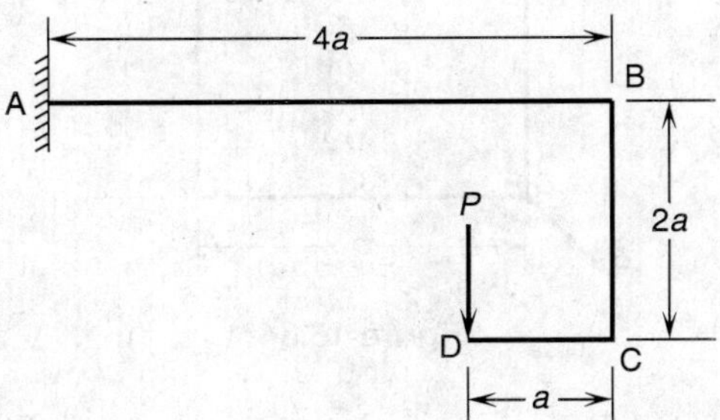

Figure 15.32

$$[\textbf{Ans. } \delta_{BV} = 40P\alpha^3/3EI\,(\downarrow),\ \delta_{BH} = 0;\ \delta_{CV} = \frac{40Pa^3}{3EI}(\downarrow),\ \delta_{CH} = \frac{6Pa^3}{EI}(\leftarrow);$$

$$\delta_{DV} = \frac{35Pa^3}{3EI},\ \delta_{DH} = \frac{6Pa^3}{EI}(\leftarrow)\ \text{and}\ \theta_D = \frac{3Pa^2}{2EI}\text{(anticlockwise)}]$$

15.6 In the rigid frame shown in Figure 15.33 determine the distance by which the points A move closer under the action of force *P* acting at the points *P*. *EI* is constant throughout.

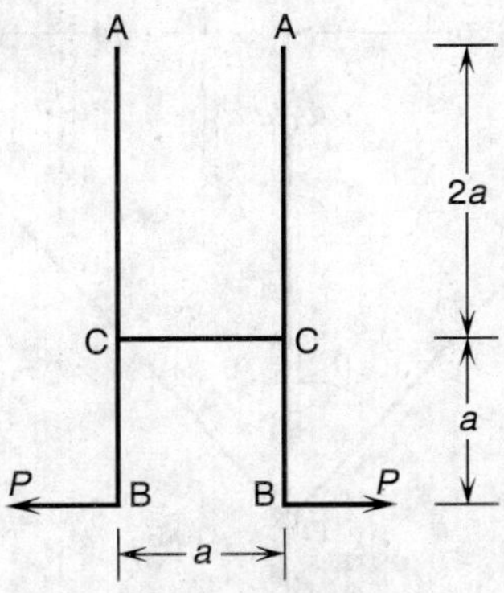

Figure 15.33

[**Ans.** $\delta_A = 2Pa^3/EI$].

15.7 The square rigid frame with uniform cross-section shown in Figure 15.34 is subjected to diagonally opposite forces *P* at the points B and D. Ignoring axial deformations, determine the distance by which the points A and C move closer.

[**Hint:** Due to symmetry only half the frame ABD carrying load *P*/2 at the ends B and D need be considered with roller supports such that movement is allowed only along the diagonal BD, i.e. presume the reference coordinate system along the diagonal $\delta_{AC} = Pa^3/24EI$].

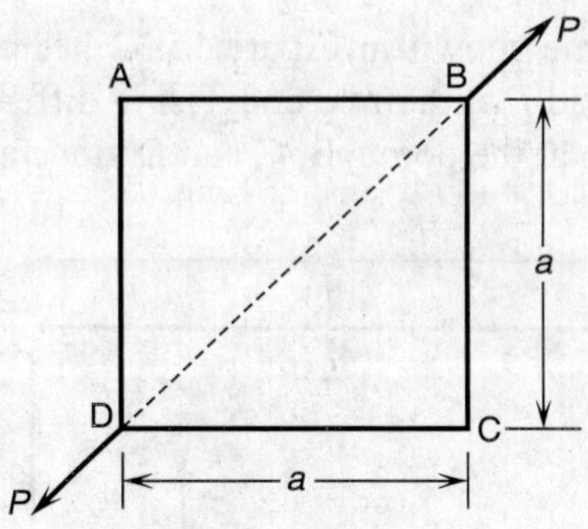

Figure 15.34

15.8 The three-bar pin-jointed frame in a vertical plane shown in Figure 15.35 is subjected to a vertical load P at the common point D. Use energy method to analyze the frame. The bars are of constant cross-section and are of same material.

[**Hint:** Consider BD as the redundant member with force T_1. The forces in the other bars are T_2, and $\cos\theta = 4/5$, $\sum V_D = T_1 + 2T_2 \cos\theta = P$ and $U = \dfrac{5a}{AE}\left\{\dfrac{2}{5}T_1^2 + \dfrac{25}{64}(P - T_1)^2\right\}$

Therefore, from the theorem of least work: $T_1 = 125P/253$].

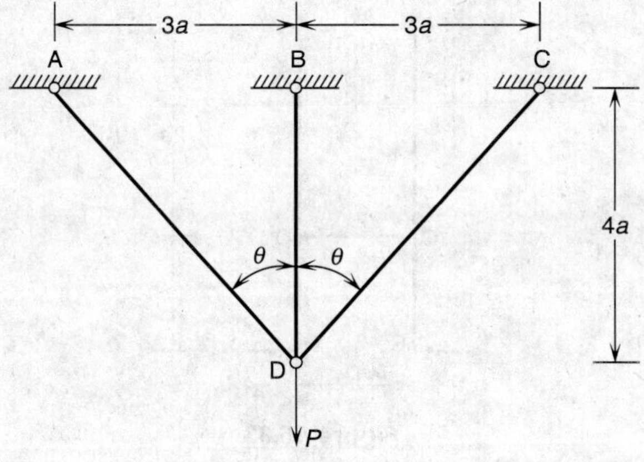

Figure 15.35

15.9 While fabricating the pin-jointed plane frame shown in Figure 15.36, it was found that member AC is fabricated Δ too short. Determine the forces in the members after assembly. AE is same for all members.

$$\left[\textbf{Ans. } F_6 = \frac{3}{5+3\sqrt{3}}\left(\frac{\Delta AE}{L}\right)(T);\ F_1 = F_2 = F_3 = F_4 = F_5 = F_6/\sqrt{3}\ (\mathrm{C})\right]$$

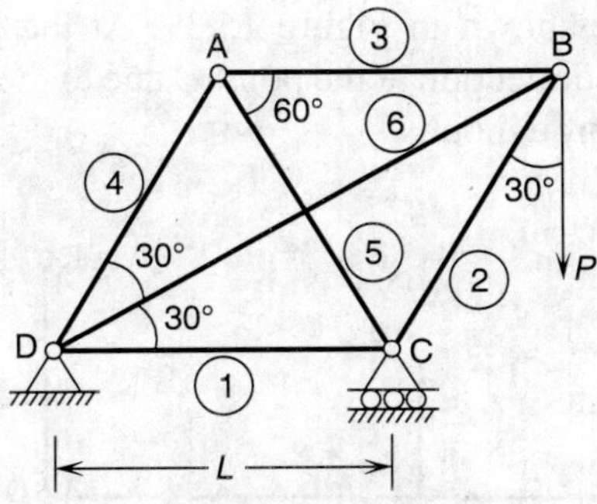

Figure 15.36

15.10 All members of the truss shown in Figure 15.37 are of same cross-section and material. Compute force in each of the members due to opposite forces P acting at B and D.
[**Hint:** Consider only half the frame ABD as in Problem 15.7].

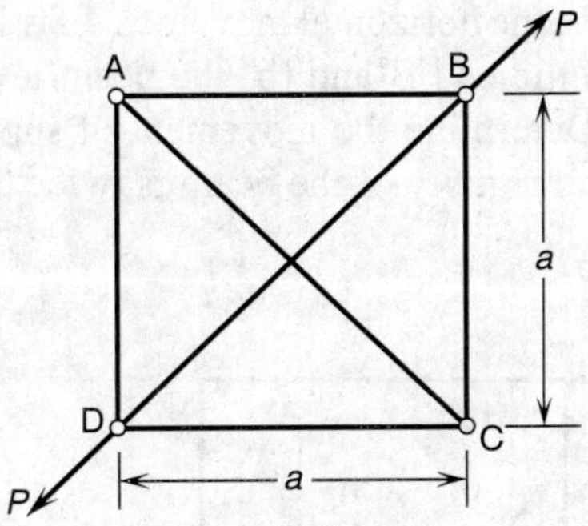

Figure 15.37

15.11 Analyze the continuous beam ABCD shown in Figure 15.38. The beam is fixed at the end A and supported at B and C, and free at the end D. The beam carries a concentrated load P at the free end.

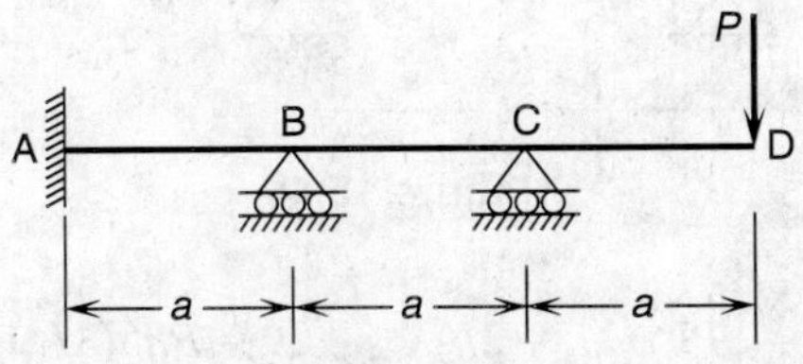

Figure 15.38

$$\left[\textbf{Ans. } R_A = \frac{3P}{7}(\uparrow);\ R_B = \frac{12P}{7}(\downarrow);\ R_C = \frac{16P}{7}(\uparrow) \text{ and} \right.$$

$$\left. M_{AB} = -\frac{Pa}{7};\ M_{BA} = M_{BC} = \frac{2Pa}{7} \text{ and } M_{CB} = M_{CD} = -Pa \right]$$

15.12 Analyze the rigid frame shown in Figure 15.39. At the point C a frictionless hinge is provided. Also calculate deflection at the point C due to the concentrated load acting at the point B. *EI* is constant throughout.

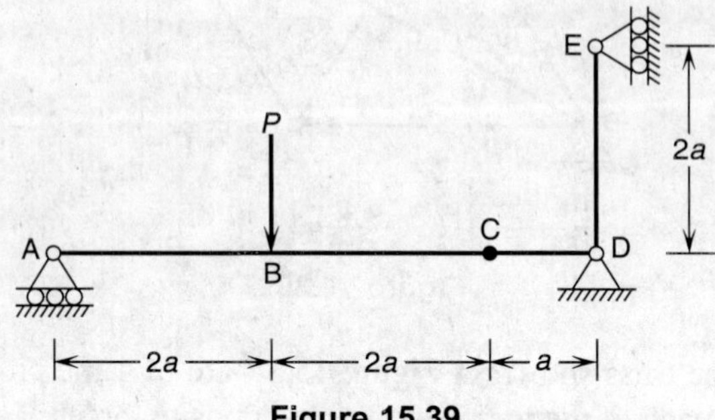

Figure 15.39

15.13 A portal frame ABCD has its end A hinged while the end B is placed on rollers as shown in Figure 15.40. Determine the horizontal movement of rollers support B due to a horizontal force P is applied at: (a) the end B and (b) the beam level. All members have the same flexural rigidity *EI*. (c) Determine the movement of support B for the load condition of case (a), when the flexural rigidity of the beam is twice that of columns.

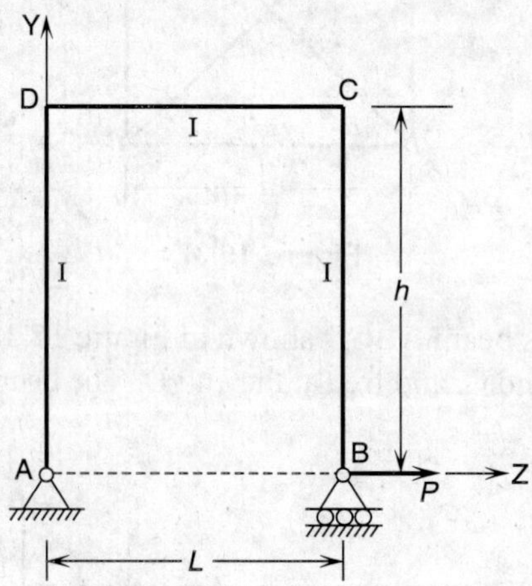

Figure 15.40

$$\left[\textbf{Ans.}\ (a)\ \delta_B = \frac{Ph^2(3L+2h)}{3EI};\ (b)\ \delta_B = \frac{Ph^2(3L+2h)}{6EI};\ (c)\ \delta_B = \frac{Ph^2(3L+4h)}{12EI}\right]$$

15.14 Use the virtual work method to determine the distance by which the point C in the structures of Figures 15.41 and 15.42 will move away from point A under the action of forces *P*. *EI* for the structure is constant.

$$\left[\textbf{Ans.}\ \delta_H = \frac{3375Pa^3}{2EI},\ \delta_V = \frac{18125Pa^3}{8EI};\ \delta_{AC} = \frac{Pa^3}{24EI}\right]$$

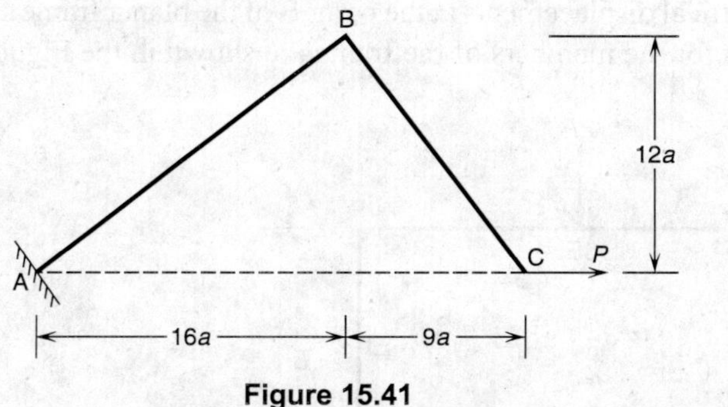

Figure 15.41

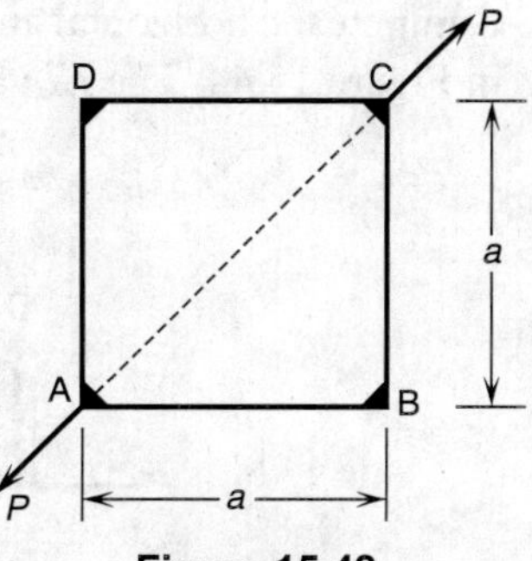

Figure 15.42

15.15 Compute the slopes and deflections for the planer frames shown in Figure 15.43(a) to (d) as specified. *EI* is constant for all the members of the frames.

(a) θ_C, δ_{CH} and δ_{CV}; (b) θ_D; (c) θ_C; and (d) θ_D, δ_{DH} and δ_{DV}.

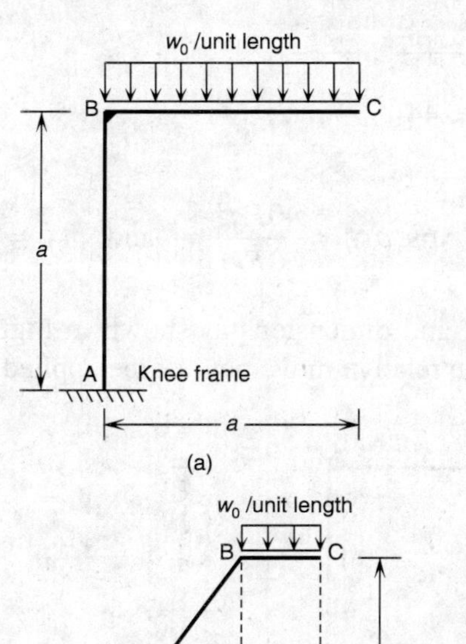

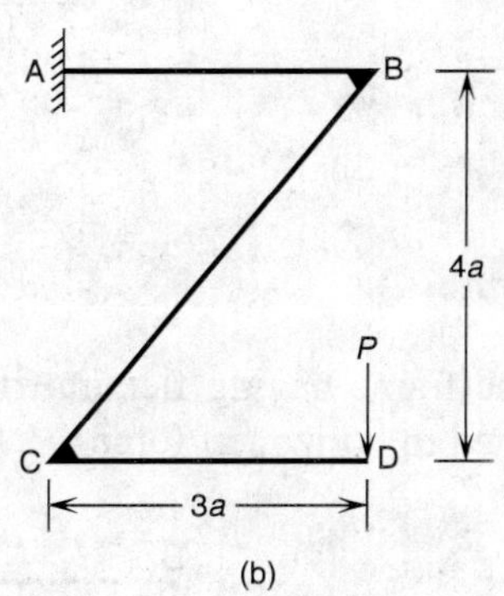

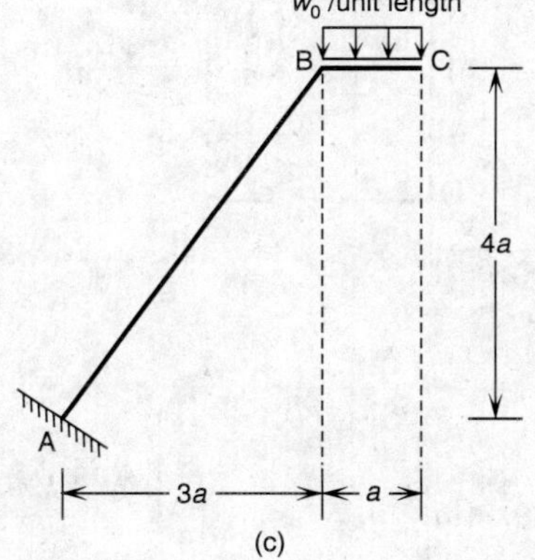

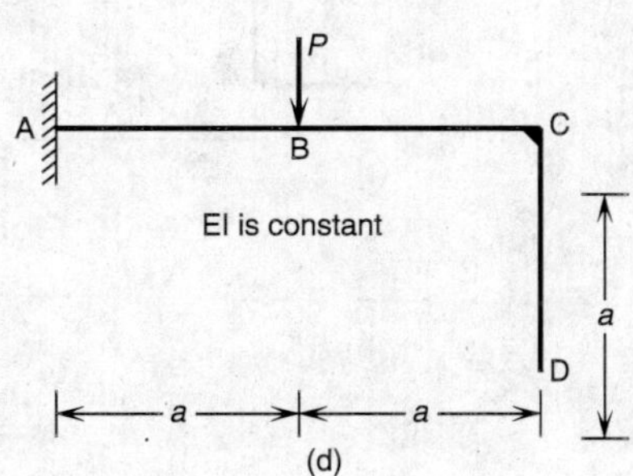

Figure 15.43

$$\left[\textbf{Ans.}\ (a)\ \theta_C = \frac{2w_0a^3}{3EI}, \delta_{CH} = \frac{w_0a^4}{4EI}, \delta_{CV} = \frac{5w_0a^4}{8EI};\quad (b)\ \theta_D = \frac{33Pa^2}{4EI}\right.$$

$$\left.(c)\ \theta_C = \frac{61w_0a^3}{6EI};\quad (d)\ \delta_{DH} = \frac{Pa^3}{2EI}\leftarrow, \delta_{DV} = \frac{5Pa^3}{6EI}\downarrow \text{ and } \theta_D = \frac{Pa^2}{2EI}\circlearrowleft\right]$$

15.16 Compute the horizontal and vertical displacements at the point B of the planer frame shown in Figure 15.44. The EI values for the members of the frame are shown in the Figure.

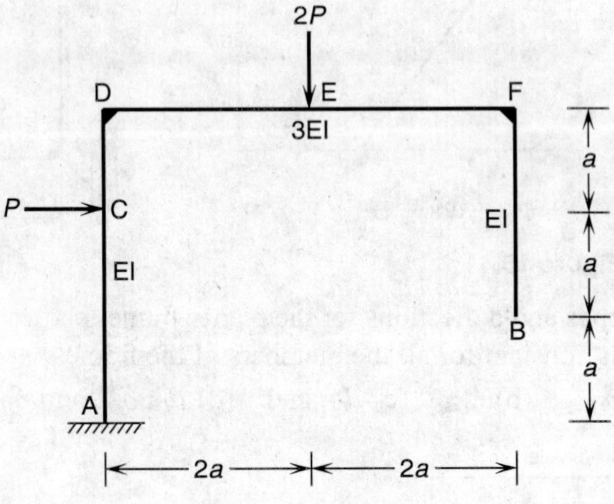

Figure 15.44

$$\left[\textbf{Ans. } \delta_{BH} = \frac{40Pa^3}{3EI} \leftarrow \text{ and } \delta_{BV} = \frac{208Pa^3}{3EI} \downarrow\right]$$

15.17 A plane frame having flexural rigidity EI and dimensions as shown in Figure 15.45 is subjected to horizontal forces P. Determine relative movement of the applied forces.

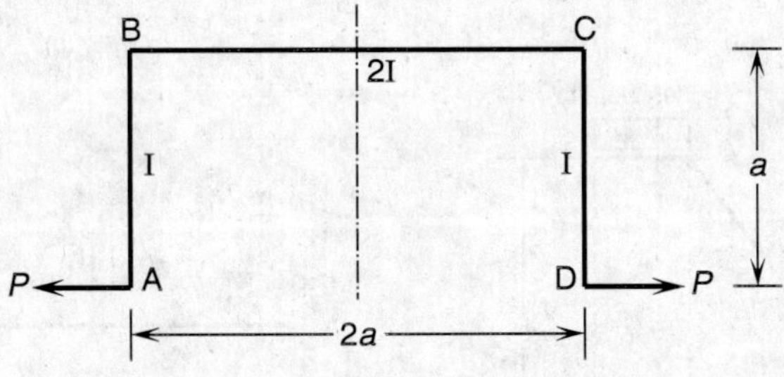

Figure 15.45

$$\left[\textbf{Ans. } \delta_{AD} = \frac{5Pa^3}{3EI}\right]$$

15.18 A U-shaped member of constanthas the dimensions shown in Figure 15.46. Determine relative displacement of the load points.

$$\left[\textbf{Ans. } \delta_{AE} = \frac{P}{EI}\left\{\frac{2b^3}{3} + a\left(\pi b^2 + \frac{\pi a^2}{2} + 4ab\right)\right\}\right]$$

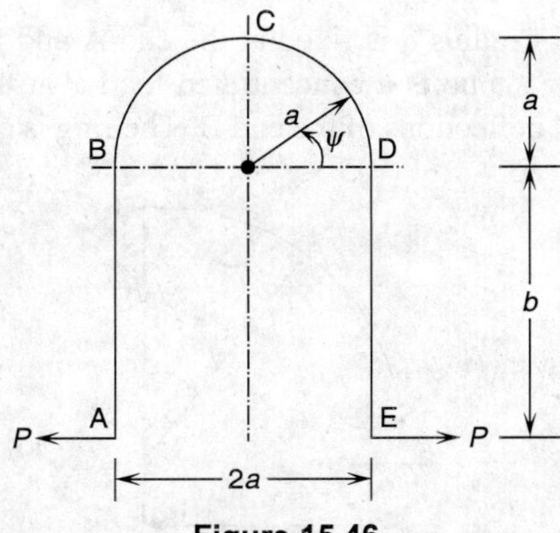

Figure 15.46

15.19 A semi-circular arch ACD shown in Figure 15.47 is subjected to a horizontal force P applied at the end B. Determine the horizontal movement of the end B when (a) the end A is fixed and the end B is free, and (b) the end A is hinged and the end B is on rollers. The flexural rigidity EI is uniform throughout.

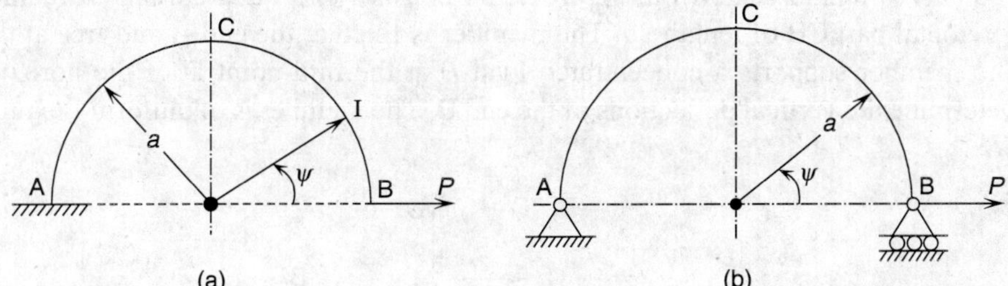

Figure 15.47

$$\left[\textbf{Ans. } \text{(a) } \delta_{BH} = \frac{\pi Pa^3}{2EI}, \delta_{BV} = \frac{2Pa^3}{EI} \text{ and } \theta_B = \frac{2Pa^2}{EI}; \text{(b) } \delta_{BH} = \frac{\pi a^3 P}{2EI}\right]$$

15.20 A lamp post ABC shown in Figure 15.48 consists of a vertical part AB of length b and a quadrant BC of radius a at its top. The lower end of the post is fixed. It carries a concentrated vertical force P at C. Determine the horizontal and vertical deflections of the end C of the lamp post. The structure is of uniform flexural rigidity EI.

$$\left[\textbf{Ans. } \delta_{CV} = -\frac{(\pi a + 4b)Pa^2}{4EI}\downarrow, \delta_{CH} = \frac{Pa(a+b)^2}{2EI}\right]$$

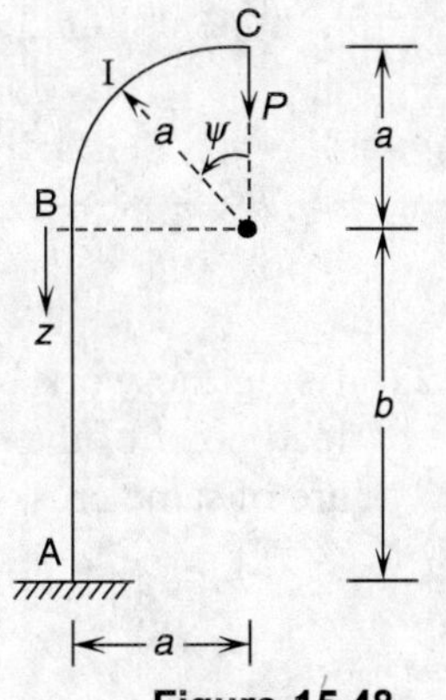

Figure 15.48

15.21 A quadrant ring AB of radius a is fixed at the end A and free at the end B as shown in Figure 15.49. The ring supports a concentrated load P at the free end B. Determine the horizontal and vertical deflections of the end B. The ring is of uniform flexural rigidity EI.

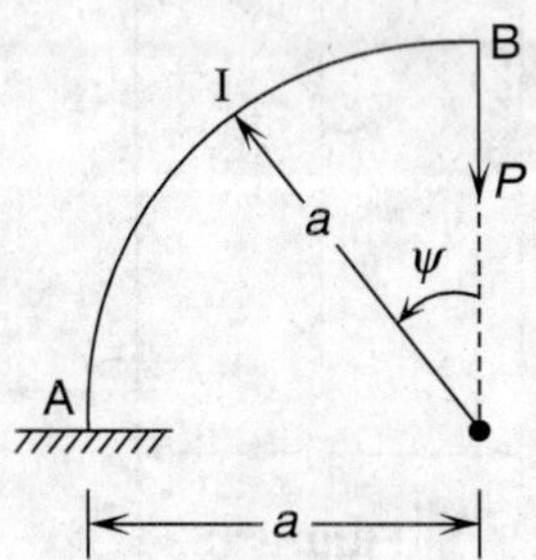

Figure 15.49

$$\left[\textbf{Ans. } \delta_{BV} = -\frac{\pi P a^3}{4EI} \text{ and } \delta_{BH} = -\frac{Pa^3}{2EI} \leftarrow\right]$$

15.22 The curved member shown in Figure 15.50 consists of a quadrant AC of radius a and a horizontal part CB of length $2a$. The member is fixed at the end A and free at the end B. The member supports a concentrated load P at the mid-point D of the horizontal part. Determine the vertical deflections of the end B. The member is of uniform flexural rigidity EI.

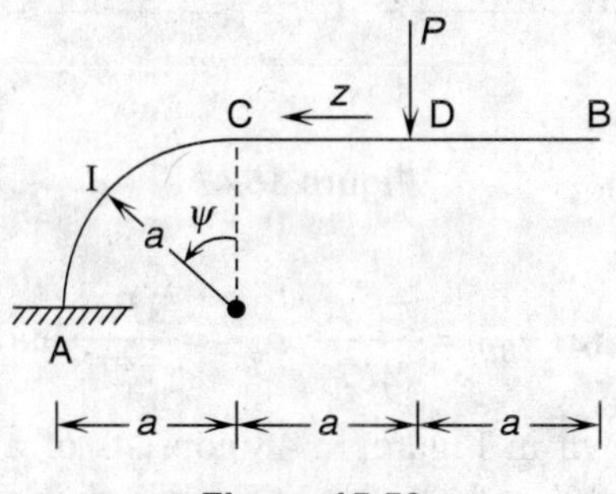

Figure 15.50

$$\left[\textbf{Ans. } \delta_{BV} = -\frac{(\pi - 1)Pa^3}{2EI} \downarrow\right]$$

15.23 Using the work balance method, determine the vertical component of deflection at the load point of the trusses shown in Figures 15.51(a) and (b). All the members of the truss are of same cross-section and material.

$$\left[\textbf{Ans. } \text{(a) } \delta_{CV} = \frac{(7 + 4\sqrt{2})Pa}{AE} \downarrow, \text{ (b) } \delta_{DV} = \frac{(80 + 16\sqrt{3})Pa}{AE}\right]$$

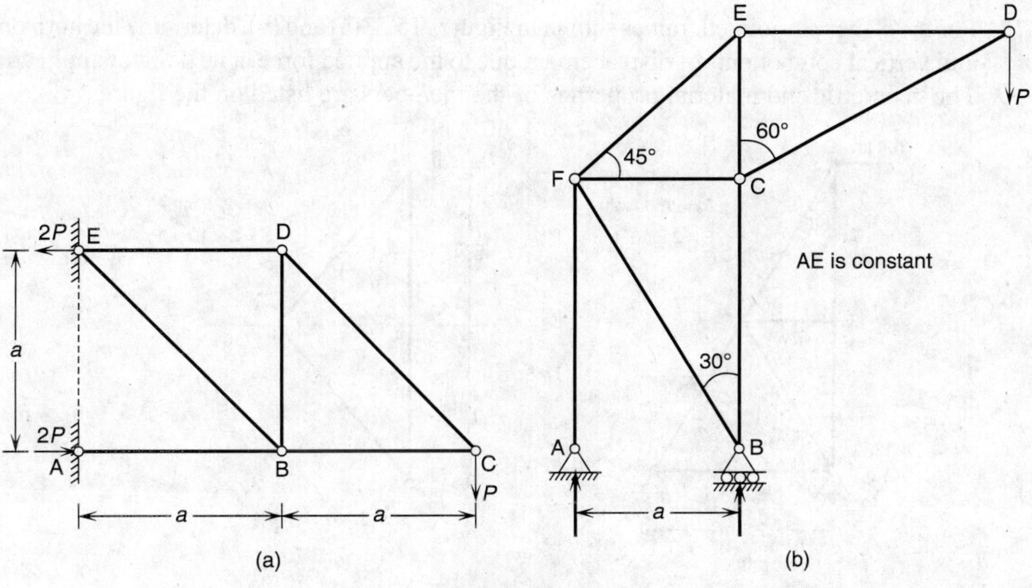

Figure 15.51

15.24 Using the virtual work method determine the horizontal component of deflection of the loaded joint B of the truss shown in Figure 15.52. All the members of the truss are of same cross-section and material.

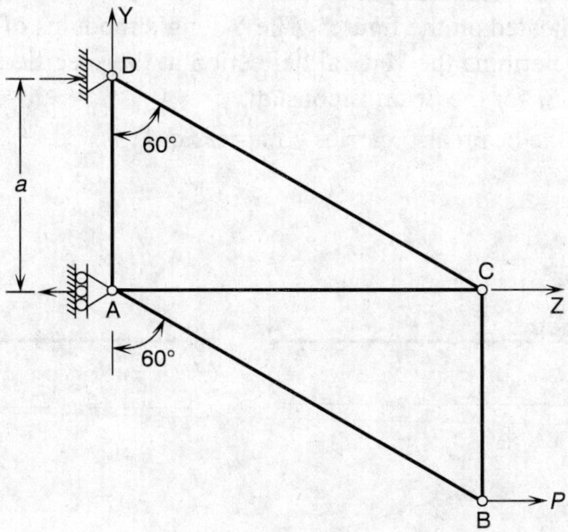

Figure 15.52

$$\left[\textbf{Ans. } \delta_{BH} = \frac{(6+\sqrt{3})Pa}{AE} \right]$$

15.25 For the planer pin-jointed frames shown in Figures 15.53(a) and (b), determine the horizontal and vertical components of displacement due to the applied force at its point of application. The geometric and material properties of the members are listed on the figures.

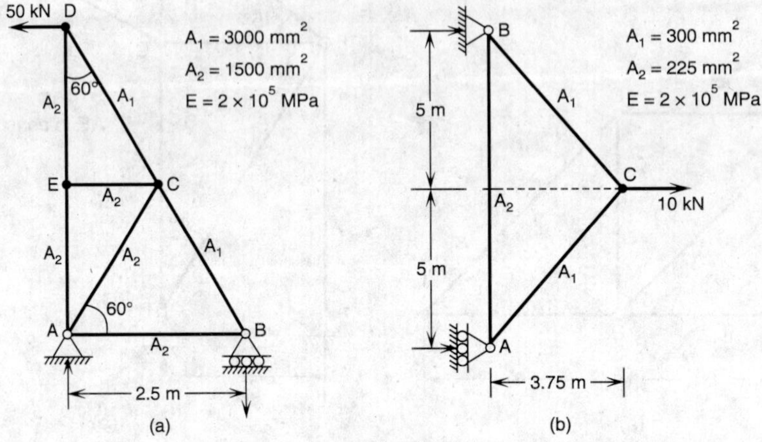

Figure 15.53

[**Ans.** (a) δ_{DH} = 4.25 mm and δ_{DV} = 1.25 mm; (b) δ_{CH} = 2.43 mm and δ_{CV} = 0.74 mm]

15.26 In the structures of Figures 15.54(a) and (b), the beam AB is hinge-supported at the left end A and is suspended by a rod as shown in the respective figures while in the beam of Figure 15.54(c) end A is fixed. The moment of inertia of the beam and cross-sectional area of the rod are indicated on the figures. The Young's modulus of the materials of both the members is *E*. Determine the vertical deflection at the specified point by using the work balance method. *EI* for the beam is constant.

[**Hint:** A part of the beam also carries compression]

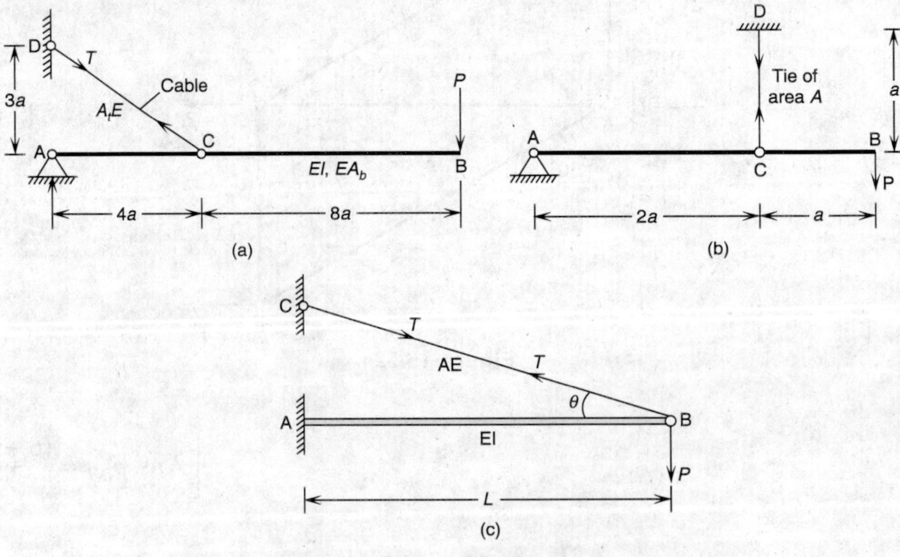

Figure 15.54

$$\left[\textbf{Ans.} \text{ (a) } \delta_{CV} = \frac{125Pa}{EA_t} + \frac{64Pa}{EA_b} + \frac{256Pa^3}{EI}; \text{ (b) } \delta_{BV} = \frac{Pa^3}{EI} + \frac{9Pa}{4EA}\right.$$

$$\left.\text{and (c) } \delta_{BV} = \frac{P}{(AE\sin^2\theta \,.\cos\theta/L) + (3EI/L^3)}\right]$$

15.27 In the structure of Figure 15.55 the beam AB is hinge-supported at the end A and is propped C at by an inclined strut as shown in the figure. The moment of inertia of the beam and cross-sectional area of the strut are indicated on the figure. The Young's modulus of the materials of both the members is E. Determine the vertical deflection at the point B by using the virtual work method. *EI* for the beam is constant.

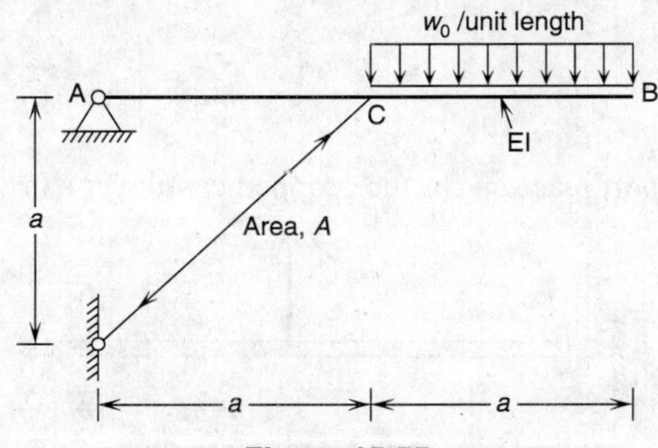

Figure 15.55

[**Hint:** Ignore tension in the beam] $\left[\textbf{Ans. } \delta_{BV} = \frac{7w_0a^4}{24} + \frac{6\sqrt{2}\,wa^2}{EA}\right]$

15.28 A bar bent 90° at two points is fixed at one end as shown in Figures 15.56(a) and (b). Application of force *P* at the free end causes an axial force, direct shear, bending, and torsion in the three segments. Determine the deflection of the free end. The rigidities are constant through the bar.

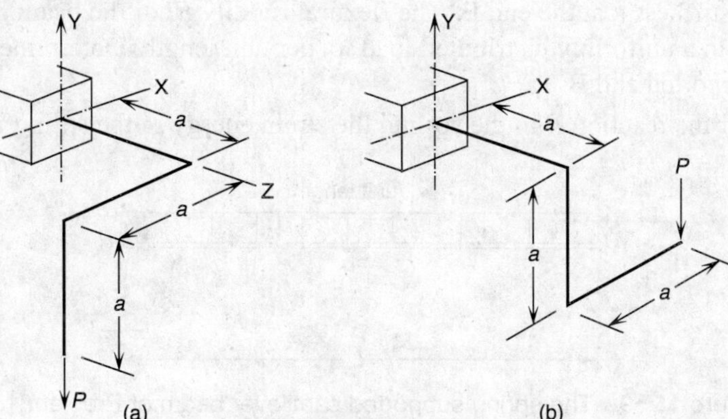

Figure 15.56

$$\left[\textbf{Ans. } \delta_V = \frac{Pa}{AE} + \frac{2Pa^3}{3EI} + \frac{20Pa}{9GA} + \frac{Pa^3}{GJ}\right]$$

15.29 For the stepped cantilever ACB shown in Figure 15.57, determine the vertical deflection at the end B.

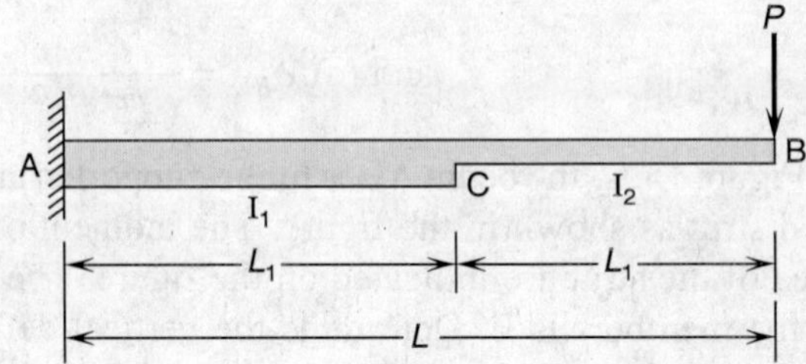

Figure 15.57 Stepped cantilever of Problem 15.29.

$$\left[\textbf{Ans. } \delta_B = \frac{P}{3EI_1}[L^3 - L_2^3] + \frac{PL_2^3}{3EI_2}\right]$$

15.30 Determine the support reaction for the propped cantilever shown in Figure 15.58.

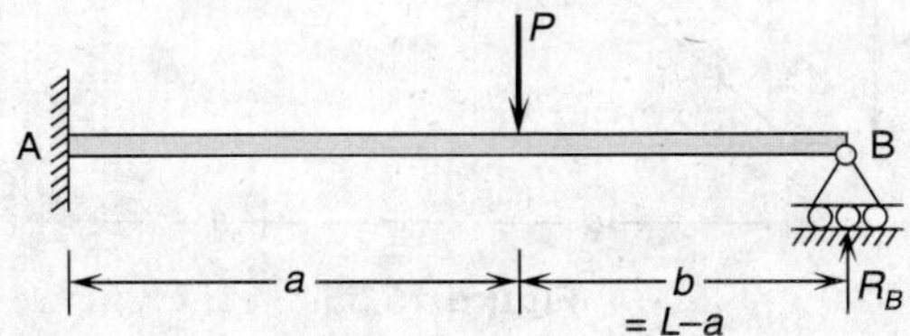

Figure 15.58 Propped cantilever of Problem 15.30

$$\left[\textbf{Ans. } R_B = \frac{Pa^2}{2} \times \frac{2a+3b}{(a+b)^3} = P\left(\frac{a}{L}\right)^2 \left\{\frac{3}{2} - \frac{1}{2}\left(\frac{a}{L}\right)\right\}\right]$$

15.31 The uniform beam AB of span L of Figure 15.59 is fixed at the end A and supported by a spring of stiffness k at the end B. The flexural rigidity EI of the beam is constant and it is subjected to a uniformly distributed load w_0 per unit length. Determine the reaction at the spring supported end B.

[**Hint:** For the reaction R in the spring, the strain energy in the spring is $R^2/2k$]

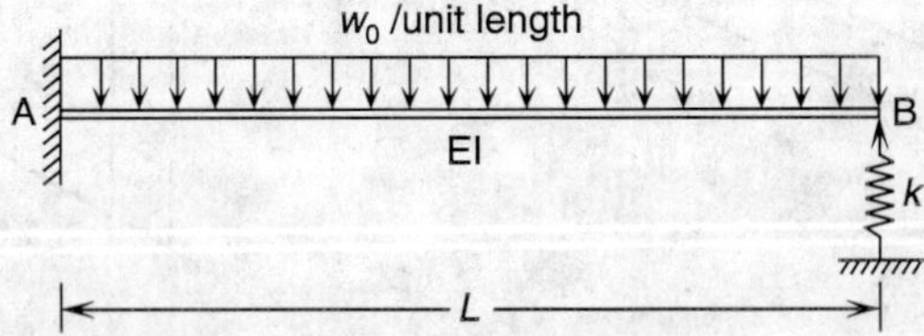

Figure 15.59 The spring supported cantilever beam of Problem 15.31.

$$\left[\textbf{Ans. } R = \frac{3w_0L}{8}\left\{\frac{1}{(3EI/kL^3)+1}\right\}\right]$$

15.32 A load $P = 7.0$ kN acting at point X of a simply supported beam shown in Figure 15.60 produces vertical deflection at the points A, B and C of the beam as:

$$\delta_A = 30 \text{ mm}, \quad \delta_B = 80 \text{ mm} \text{ and } \delta_C = 50 \text{ mm}$$

Determine the deflection of point X when the beam carries loads of $P_A = 6.5$ kN; $P_B = 3.5$ kN and $P_C = 4.5$ kN, at A, B and C respectively.

[**Hint:** Use reciprocal theorem, $\delta_X = \delta_{XA} P_A + \delta_{XB} P_B + \delta_{XC} P_C$
$= (\Delta_A/P_X)P_A + (\Delta_B/P_X)P_B + (\Delta_C/P_X)P_C$]

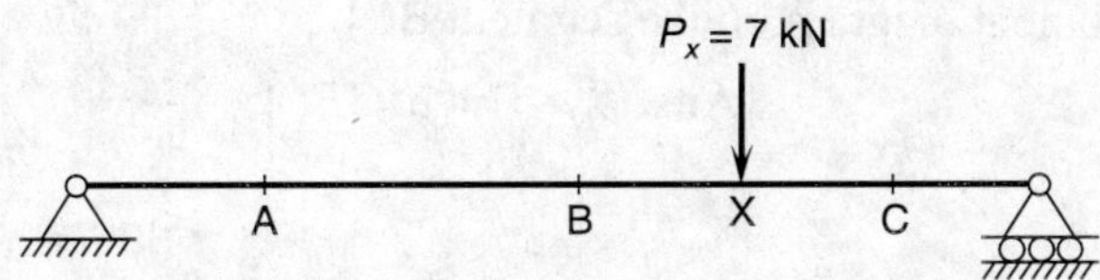

Figure 15.60 Simply supported beam of Problem 15.32.

[**Ans.** $\delta_X = 100$ mm]

15.33 For the horizontal beam shown in Figure 15.61 a vertical displacement of support B of 6 mm causes a reaction $R_A = 9$ kN at A. Determine the reaction R_B at B due to a vertical displacement 8 mm at support A.

[**Hint:** Use reciprocal theorem $\delta_{BA} = \Delta_B/R_A = \Delta_A/R_B$]

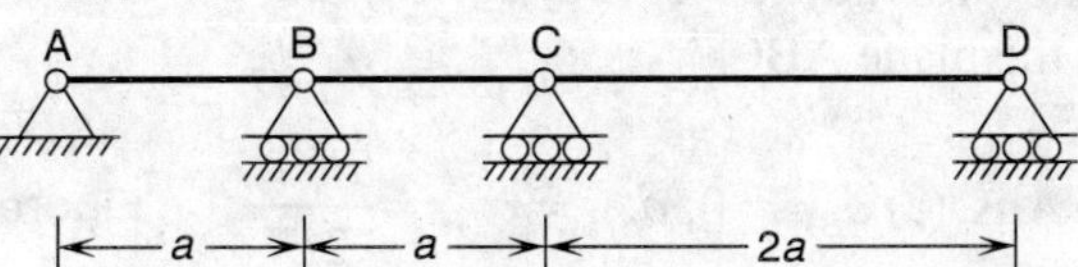

Figure 15.61 Beam of Problem 15.33.

[**Ans.** $R_B = 12$ kN]

15.34 A bar of circular cross-section is bent into a shape of a half ring and supported in a vertical plane as shown in Figure 15.62. Determine: (a) the horizontal movement of point B and the vertical movement of point C under the action of horizontal force acting at the free end B in the plane of ring; and (b) the horizontal movement of point B in the direction of the horizontal force acting there perpendicular to the plane of the ring.

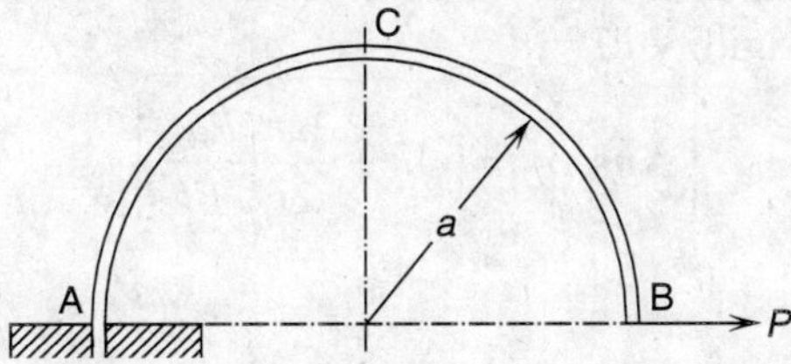

Figure 15.62 Half ring of Problem 15.34.

$$\left[\textbf{Ans.} \text{ (a) } \delta_{BH} = \frac{Pa^3(\pi)}{2EI}, \ \delta_{CV} = \frac{Pa^3}{2EI}; \text{ (b) } \delta_{BH} = \frac{3Pa^3(\pi)}{2JG} + \frac{Pa^3(\pi)}{2EI}\right]$$

15.35 A rigid frame of constant EI throughout shown in Figure 15.63 is hinged at A and supported on rollers at D. The frame carries uniformly distributed load of intensity w_0 on the beam BC. Neglecting the effect of axial deformations, determine the horizontal deflection at the roller D.

[**Hint:** As no moment exists on the columns AB and CD due to actual loading, hence area-moment method is applied to moment diagram on the segment BC].

[**Ans.** $\delta_D = 7w_0a^4/3EI$]

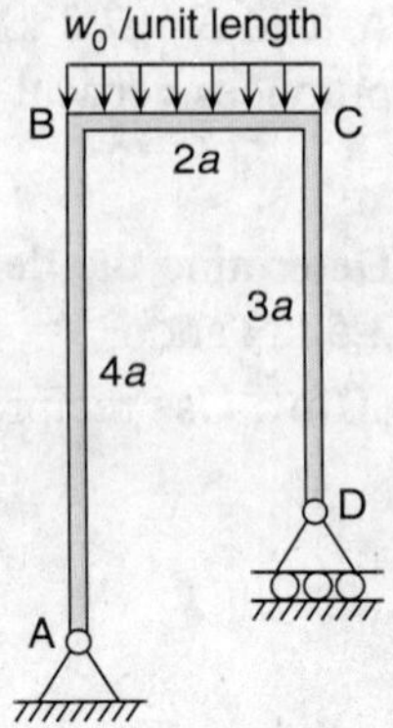

Figure 15.63 Rigid frame of Problem 15.35.

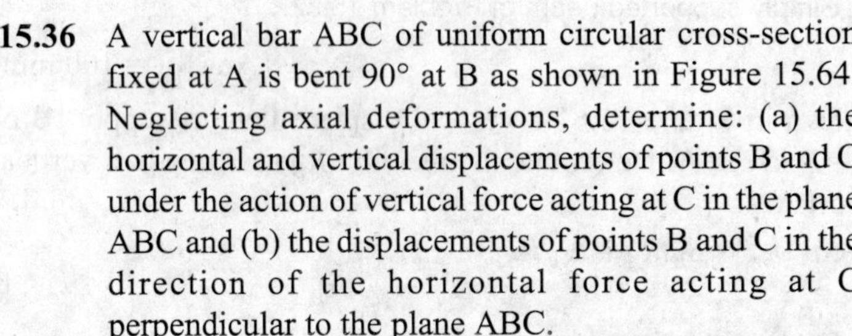

15.36 A vertical bar ABC of uniform circular cross-section fixed at A is bent 90° at B as shown in Figure 15.64. Neglecting axial deformations, determine: (a) the horizontal and vertical displacements of points B and C under the action of vertical force acting at C in the plane ABC and (b) the displacements of points B and C in the direction of the horizontal force acting at C perpendicular to the plane ABC.

$$\left[\textbf{Ans.}\text{ (a) } \delta_{BV} = 0;\ \delta_{BH} = \delta_{CH} = \frac{Pba^2}{2EI};\right.$$

$$\left.\delta_{CV} = \frac{Pb^2}{EI}\left(a + \frac{b}{3}\right);\text{ (b) } \delta_C = \frac{Pab^2}{JG} + \frac{P}{3EI}(a^3 + b^3)\right]$$

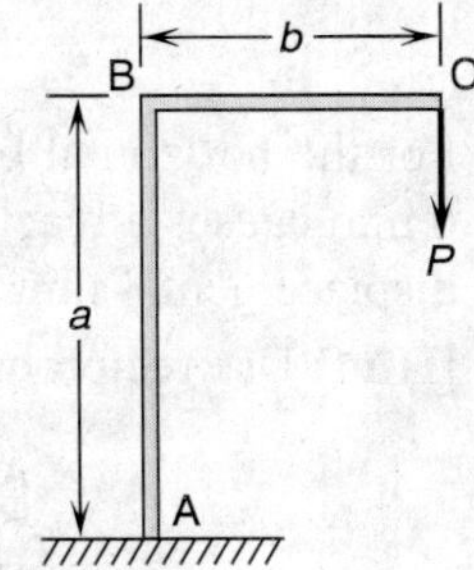

Figure 15.64 Bent bar of Problem 15.36.

15.37 The structure shown in Figure 15.65 consists of a semi-circle BCD of radius a and a vertical part AB of length h. The member is fixed at the end A and free at the end D. The member supports a vertical load P at the end D. Determine the vertical deflection at D. The member is of uniform flexural rigidity EI.

$$\left[\textbf{Ans.}\ \delta_D = \left(4h + \frac{3\pi}{2}\right)\frac{Pa^2}{EI}\right]$$

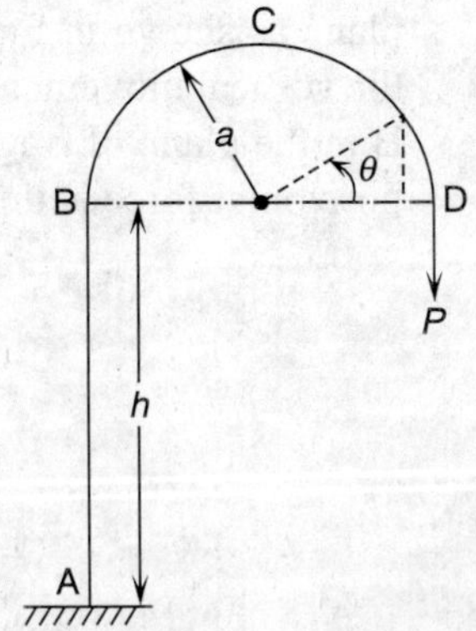

Figure 15.65 Structure of Problem 15.37.

15.38 A thin-walled three-quarter circle ring ABC supported in vertical plane is loaded as shown in Figure 15.66. Determine the vertical displacement of the point C which is free to rotate but can move in vertical direction only. Consider only bending energy.

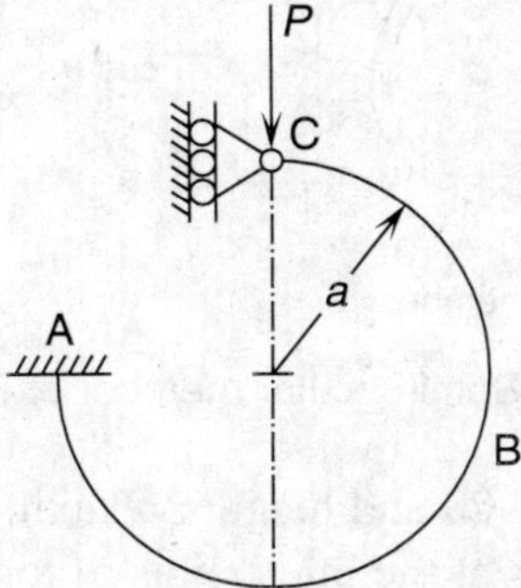

Figure 15.66 Thin-walled three-quarter ring of Problem 15.38.

$$\left[\textbf{Ans. } \delta = \frac{Pa^3}{EI} \times \left(\frac{3\pi}{4} - \frac{1}{9\pi + 8}\right)\right]$$

15.39 A thin-walled circular ring with a slit is loaded as shown in Figure 15.67. Determine the relative displacement of the points A and B. Consider only bending energy.

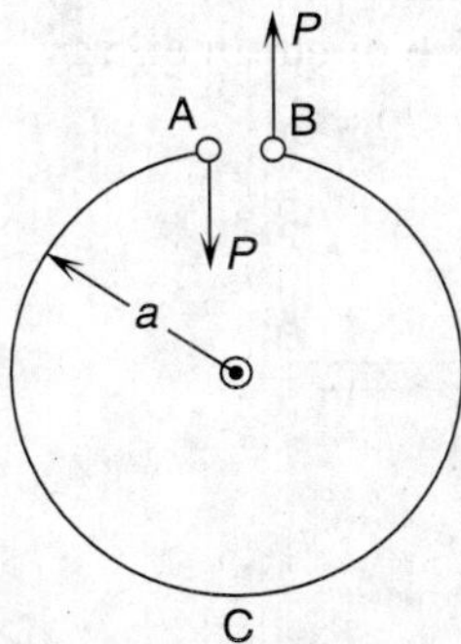

Figure 15.67 Circular ring with a slit of Problem 15.39.

$$\left[\textbf{Ans. } \delta_{AB} = \frac{\pi P a^3}{EI}\right]$$

15.40 A semi-circular member of radius r and having circular cross-section of radius a lying in the horizontal plane is subjected to a torque T at its free end as shown in Figures 15.68(a) and (b). Determine the twist at the free end A.

$$\left[\textbf{Hint}: J = 2I = \frac{\pi a^4}{2}\right]$$

$$\left[\textbf{Ans. } \theta = \frac{T}{a^3} \times \left(\frac{1}{2G} + \frac{1}{E}\right)\right]$$

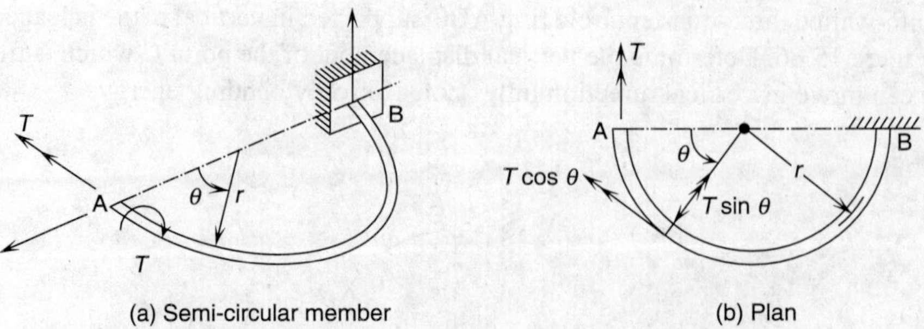

Figure 15.68 Semi-circular member of Problem 15.40.

15.41 A rectangular frame of width $2a$ and height $2b$ having uniform cross-section is subjected to concentrated vertical loads at the mid-points of top and bottom horizontal members as shown in Figure 15.69(a). Determine the moment at the section C and also the decrease in the distance between the two points of application of loads P.

[**Hint:** Due to the symmetry of geometry and loading, the top and bottom members deform in such a way that the tangents at the load points remain horizontal and there is no shear across the section C–C. Hence consider only quarter of the frame as shown in Figure 15.69(b) and calculate its strain energy. Since due to symmetry, the slope at C is zero, equate the partial derivative of strain energy w.r.t. M to zero and obtain the value of M_C. Equate the partial derivative of strain energy w.r.t. P to zero and obtain the distance by which the load points come closer].

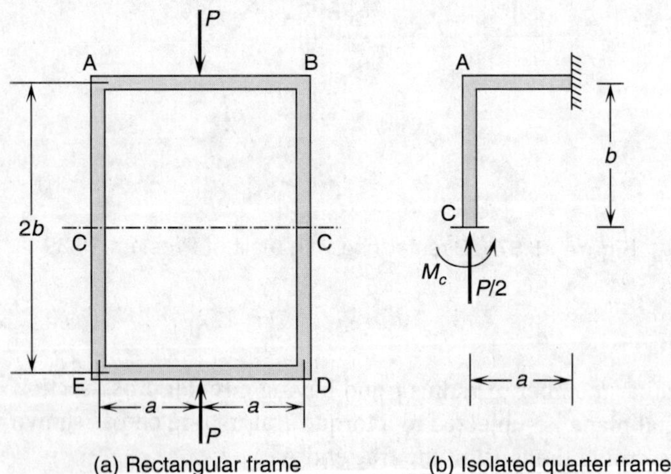

Figure 15.69 Rectangular frame of Problem 15.41.

$$\left[\textbf{Ans. } M_C = \frac{Pa^2}{4(a+b)} \text{ and } \delta = \frac{Pa^3}{12EI} \times \left(\frac{a+4b}{a+b}\right)\right]$$

15.42 A ring composed of two semi-circular members of radius a and two straight members of length $2a$ as shown in Figures 15.70(a) and (b) subjected to a load system shown in the figure. Determine the moment at the section C and change in distance between the load points. Consider only the bending energy.

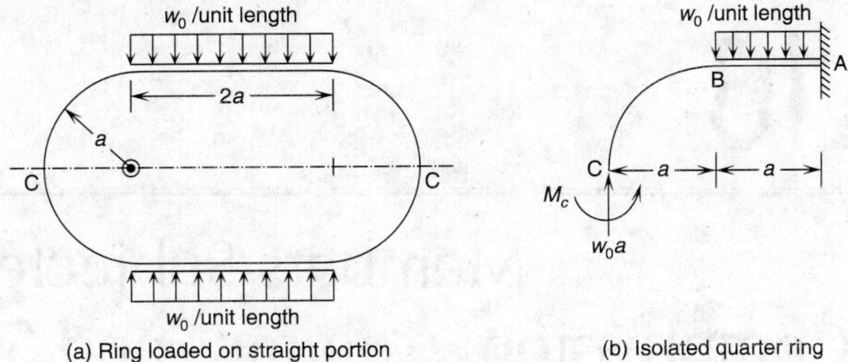

Figure 15.70 Ring of Problem 15.42.

$$\left[\textbf{Ans. } M_C = \left(\frac{3\pi + 2}{3(\pi + 2)}\right) w_0 a^2 \text{ and } \delta_V = 1.72\left(\frac{w_0 a^4}{EI}\right)\right]$$

15.43 A thin-walled rectangular ring with a slit is loaded as shown in Figure 15.71. Determine the relative displacement of the points A and B. Consider only bending energy.

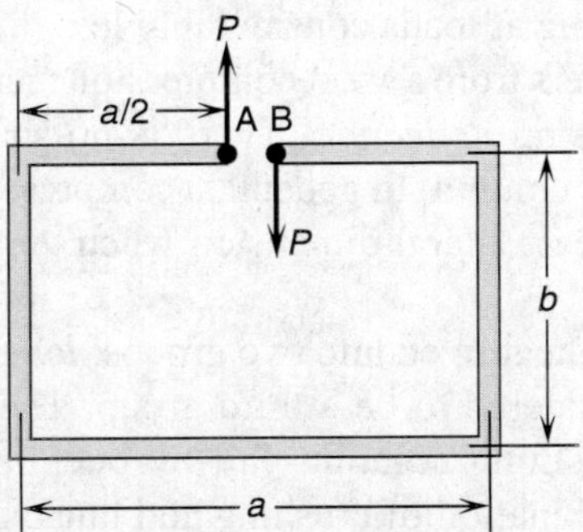

Figure 15.71 Thin-walled rectangular ring with a slit of Problem 15.43.

$$\left[\textbf{Ans. } \delta_{AB} = \frac{(a + 3b)Pa^2}{6EI}\right]$$

Chapter 16

Members Subjected to Compression: Columns and Struts

16.1 INTRODUCTION

A column is a *compression member* that is so slender compared to its length that under gradually increasing loads it fails in buckling at loads considerably less than those required to cause failure by crushing. In this respect, it differs from a short compression member, which, even if eccentrically loaded, undergoes negligible lateral *deflection*. There is no sharp line of demarcation between short compression member and a column, In general, a compression member with its *unsupported length* more than 10 times the least lateral dimension when vertical is considered to be column, when inclined a *strut*.

The columns are generally categorized into two groups: *long* and *intermediate*, sometimes the short compression block is considered to be a third group. The distinction between the three is determined by their behaviour. Long columns fail by buckling or excessive lateral bending; intermediate columns, by a combination of crushing and buckling; short compression blocks, by crushing.

An *ideal column* is considered to be a member of constant cross-section and constant *EI*, made of homogeneous and isotropic material, which is initially perfectly straight and is subjected to axial compressive loads. However, *actual columns* always have small imperfections of material and fabrication, as well as unavoidable accidental eccentricities of the load. The initial crookedness of the column, together with unknown eccentricities caused by the placement of load, results in a direct compressive stress and a flexure stress.

If the eccentricity is small and the member is short, the lateral deflection is negligible and flexural stress is insignificant compared with direct compressive stress. On the other hand, a long member is quite flexible, since its lateral deflection is proportional to the cube of its length; hence a relatively low value of P may cause a large flexural stress accompanied by a negligible direct compressive stress. Thus at these two extremes, a *short column* carries principally direct compressive, and a *long column* is subjected primarily to flexural stress. As the length of column increases, the importance of direct compressive stress decreases, and that of flexural stress increases. Unfortunately, in the *intermediate column* range it is not possible to determine the contributions of each stress in the resultant stress. This has lead to many formulas for intermediate columns.

In practice, the columns are generally distinguished in terms of actual length. The member in which the least cross-sectional dimension is appreciable relative to its length is termed *short column*, it tends to fail by crushing of materials, a *strength failure*. On the other hand, the member in which the length is relatively large as compared to its least lateral dimension is called long column. The long columns tend to fail by *buckling*, which is an *instability failure* rather than a strength failure where material ruptures only after a member has buckled. The potential for buckling tends to limit the *load-carrying capacity* of the long columns.

The *ultimate load-carrying capacity* of a short column is independent of its length; it depends primarily on the strength of materials used and its cross-sectional area. As the length of compression member is increased progressively, the relative proportions of the member change to the extent that it can be described as a long column and its behaviour changes dramatically under the load.

16.2 SHORT COLUMNS

A compression member is called short when its length does not exceed ten times its smallest lateral dimension; the flexural deflection is so small that its effect can be neglected. The compression members of this category fail primarily by crushing or direct stresses and ultimate load-carrying capacity is consequently independent of member lengths.

Axial loads: When the load P is applied at the centroid of the cross-section of the loaded element, uniform compressive stresses of magnitude $\sigma_o = P/A$ are developed. The failure occurs when the actual stress exceeds the crushing strength of the material, i.e. $\sigma_o \geq \sigma_y$. The *crushing load* is given by $P_y = A\sigma_y$ where A is the cross-sectional area of the column and σ_y is the yield or crushing strength of the material.

Eccentric loads: When the loads are eccentrically applied, i.e. applied away from the centroid of the cross-section, the resultant stress distribution will be non-uniform. The effect of eccentric loads is to produce bending stresses in the member, which in turn interact with the direct compressive stresses. If the load eccentricity is large, tensile stresses may be developed over a part of the cross section.

Consider a member subjected to a compressive load P acting at a distance e_y from the centroidal X-axis, i.e. along the principal axis Y of the member as shown in Figures 16.1(a) and (b). The stresses produced by this load at any section I-I can be obtained by resolving the eccentric load into a statically equivalent axial force P producing only uniform stresses σ_o and a moment Pe_y which produces only bending stresses σ_b as shown in Figures 16.1(c) and (d). The final stresses are obtained by adding the two stress distributions as shown in Figure 16.1(e). Therefore,

Uniform stress, $$\sigma_o = \frac{P}{A} \tag{16.1}$$

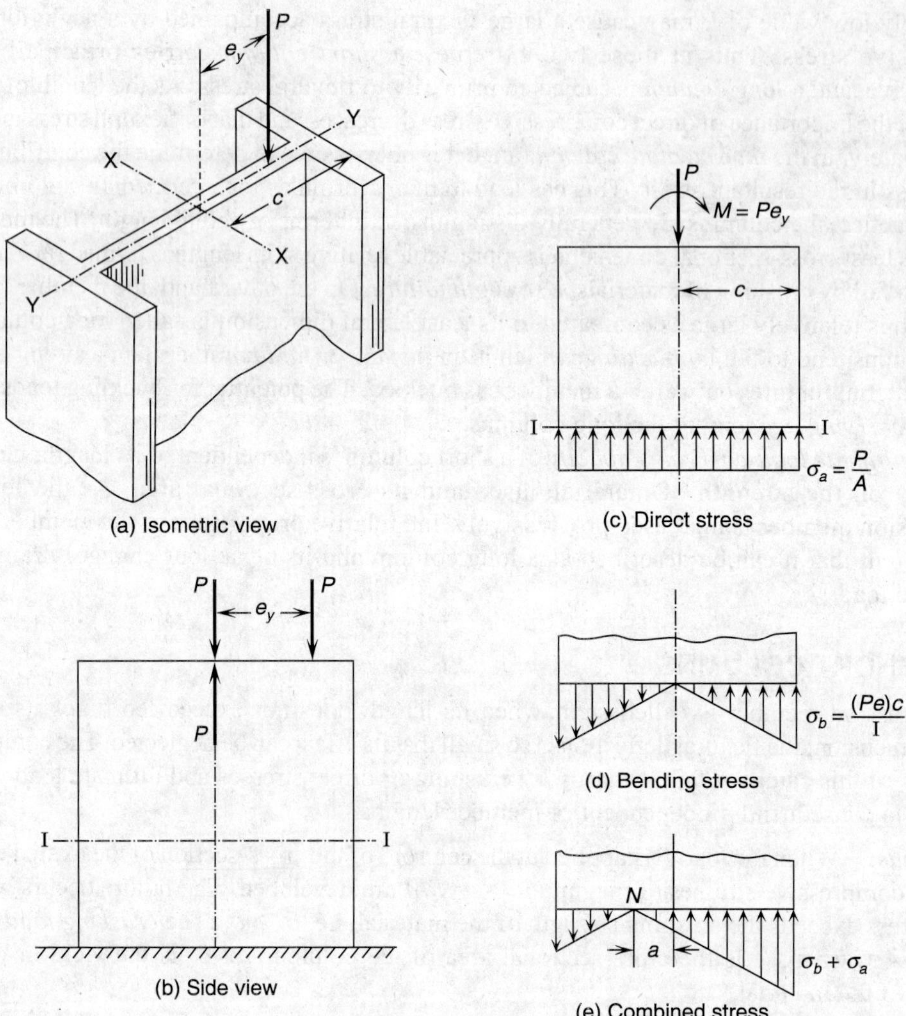

Figure 16.1 Compressive load applied at an eccentricity.

$$\text{Bending stress,} \qquad \sigma_b = \pm\frac{M_x y}{I} \quad \text{where} \quad M_x = Pe_y \tag{16.2}$$

$$\text{The combined stress,} \qquad \sigma_o + \sigma_b = \frac{P}{A} \pm \frac{M_x y}{I_x} = \frac{P}{A} \pm \frac{(Pe_y)y}{I_x} \tag{16.3}$$

The stress distribution illustrated in Figure 16.1(d) reveals that the magnitude of the bending stresses is proportional to the eccentricity e_y of the load. In the case of large eccentricity, the bending stresses σ_b dominate over the axial or normal stresses σ_o, and the vertical load P may produce tensile stresses on one face of the section. When $e_y = 0$, only compressive stresses will exist. To have only compressive stresses in the member, the eccentricity of the load must be restricted such that on the tension side resultant stress is zero, i.e. $\sigma_o + \sigma_b = 0$. For a rectangular section,

$$\sigma_o + \sigma_b = \frac{P}{A} - \frac{(Pe_y)y}{I_x} = \frac{P}{bh} - \frac{(Pe_y)(h/2)}{(bh^3/12)} = \frac{P}{bh} - \frac{6Pe_y}{bh^2}$$

$$= 0$$

Therefore, $$e_y = h/6 \tag{16.4}$$

Thus, if the load were placed within this maximum value on either side of X-axis, the stresses produced would all be compressive. Placing the load exactly at this point i.e. $e_y = h/6$, will produce zero stresses on the opposite face. If the eccentricity exceeds this value, it will cause tensile stresses to develop on the tension face. This location is referred to as the *Kern point*. For a load which can have eccentricity in either direction, Kern points exist on either side of the centroidal axis. Since these locations are at one-third points on the cross-section, this load placement concept is called the *middle-third rule*.

Similarly by considering eccentricity e_x about Y-axis, it can be shown that if the load were placed within the maximum value of $e_x = b/6$ on either side of Y-axis, the stresses produced would all be compressive. If the eccentricity exceeds this value, it will cause tensile stresses to develop on the part of the section.

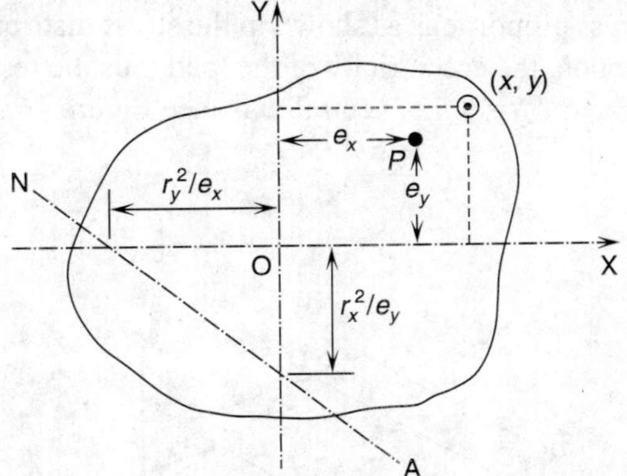

Figure 16.2 Neutral axis for a column carrying a load with biaxial eccentricity.

For a column carrying a load with biaxial eccentricity, i.e. the eccentricity of the load is about both the axes as shown in Figure 16.2, the resultant stress at any point (x, y) is given by

$$\sigma_r = \frac{P}{A} + \frac{(Pe_y)y}{I_x} + \frac{(Pe_x)x}{I_y} = \frac{P}{A}\left[1 + \frac{e_y y}{r_x^2} + \frac{e_x x}{r_y^2}\right] \tag{16.5}$$

where $I_x = Ar_x^2$ and $I_y = Ar_y^2$. To determine the *neutral axis* or line of zero stress in the cross-section, set $\sigma_r = 0$

$$0 = \frac{P}{A}\left[1 + \frac{e_y y}{r_x^2} + \frac{e_x x}{r_y^2}\right] \quad \text{or} \quad -\frac{e_y y}{r_x^2} - \frac{e_x x}{r_y^2} = 1$$

$$\frac{y}{(-r_x^2/e_y)} + \frac{x}{(-r_y^2/e_x)} = 1 \tag{16.6}$$

This equation represents a straight line, which makes intercepts $(-r_y^2/e_x)$ and $(-r_x^2/e_y)$ on X- and Y-axes, respectively. Thus the neutral axis passes through the quadrant which is opposite to that containing the load P and in general it is not perpendicular to the direction OP as in Figure 16.2.

For illustration, consider a rectangular section carrying a load which is eccentric to both the axes, i.e. has biaxial eccentricity as shown in Figure 16.3(a). The resultant stresses at the critical points A, B, C and D of the section are

$$\sigma_A = \frac{P}{A}\left[1 + \frac{6e_y}{h} + \frac{6e_x}{b}\right]; \qquad \sigma_B = \frac{P}{A}\left[1 - \frac{6e_y}{h} + \frac{6e_x}{b}\right]$$

$$\sigma_C = \frac{P}{A}\left[1 - \frac{6e_y}{h} - \frac{6e_x}{b}\right] \text{ and } \sigma_D = \frac{P}{A}\left[1 + \frac{6e_y}{h} - \frac{6e_x}{b}\right] \tag{16.7}$$

The intersections of the neutral axis with BC and CD or their extensions are determined by stress proportions as shown in the stress distribution of Figure 16.3(b). For no tension in the cross-section, the eccentricity of the load must be restricted such that on the tension side resultant stress is zero $\sigma_C = 0$, i.e. neutral axis (see Figure 16.3(c)) passes through the corner C. Then

$$\sigma_C = \frac{P}{A}\left[1 - \frac{6e_y}{h} - \frac{6e_x}{b}\right] = 0$$

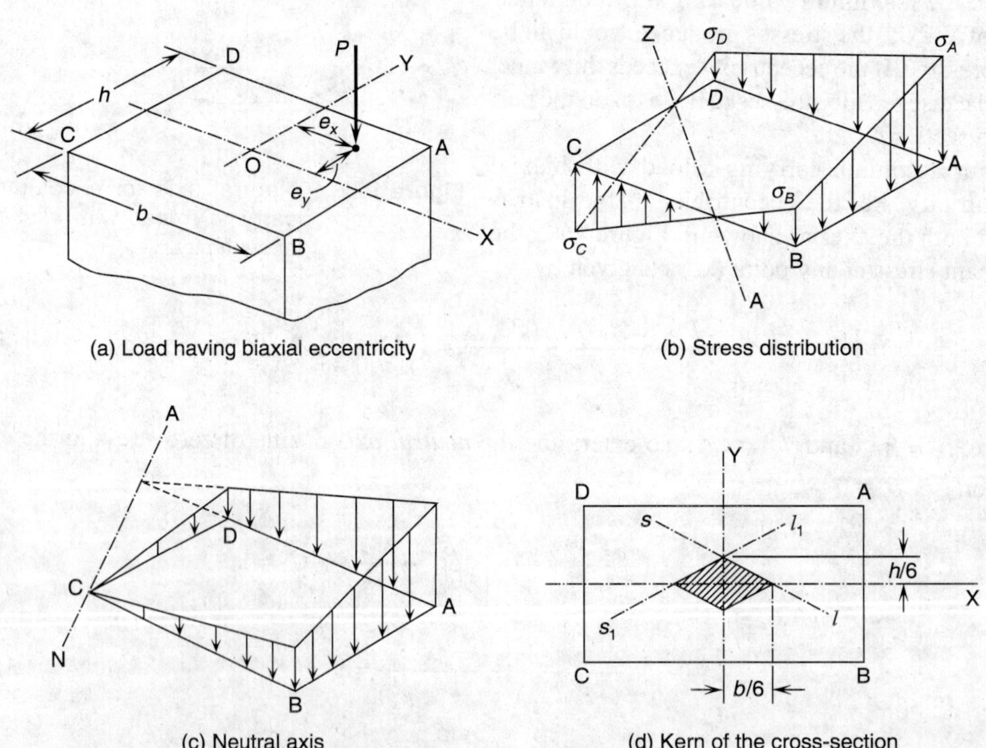

Figure 16.3 Neutral axes for eccentrically applied load P for a rectangular column.

or

$$\frac{6e_y}{h} + \frac{6e_x}{b} = 1 \quad \text{or} \quad \frac{e_x}{(b/6)} + \frac{e_y}{(h/6)} = 1 \tag{16.8}$$

This equation represents a straight line sI, which makes intercepts (b/6) and (h/6) on X and Y axes, respectively, as shown in Figure 16.3(d). This line is the *locus of points* of application of P, for which corner C has zero stress. Any compressive load to the right of this line will cause tension at C. Similarly, the line s_1I_1 is the locus of load point P, that will cause zero stress at the B. Continuing the procedure it will be seen that no corner and no part of the cross-section will be in tension if the resultant compressive load lies on or within the rhombus or diamond shaped shaded area. This shaded figure in the centre is called the *core* or *kern* of the column section.

The kern of a circular section is a circle whose diameter is one-quarter of the diameter of the section.

16.3 LONG COLUMNS

A long compression member subjected to an increasing load maintains its straight configuration when the load is small and continues to do so as the load is increased. As a particular load level is reached, the member suddenly becomes unstable and deforms into a curved shape. This phenomenon is termed *buckling*. When the member has buckled, it is no longer capable of carrying any additional load. Any further increase in load will immediately cause additional deformation, which will eventually cause the member to *snap*. *A structure in buckled mode is not serviceable*.

The *buckling* phenomenon is associated with the stiffness characteristics of the member reflected by the *modulus of elasticity*, E representing specific properties of material and *moment of inertia*, I which represents the amount of material and its distribution in the cross-section. A member of low stiffness will buckle before one with high stiffness. Increasing member length reduces stiffness of the member. The member tends to buckle about the axis having lesser moment resisting capacity (lesser I value or about the weak axis). The *ultimate-load-carrying capacity* of a long compression member varies inversely as the square of its length.

16.3.1 Critical Load

Consider a long vertical beam hinged at the ends so that it is free to bend in any direction. A central horizontal or lateral load Q is applied to the beam to cause bending in its weaker plane as shown in Figure 16.4(a). Since flexural stress is proportional to deflection, there will be no change in stress if an axial load P is added gradually at each end as shown in Figure 16.4(b), with Q being decreased simultaneously as P increases so that the midspan deflection Δ is unaltered. The midspan bending moment reduces from

$$M = (Q/2) \times (L/2) + P\Delta \quad \text{to} \quad M = P_{cr}\Delta \tag{16.9}$$

when Q has been reduced to zero. The load P_{cr} is the *critical load* required to maintain the column in its deflected position without any side thrust. Any increase in P beyond this value increases the deflection Δ, thereby increasing M, and hence Δ, and so on, until the column fails by *buckling*. On the other hand, if P is decreased slightly below this critical value, deflection is decreased, thereby decreasing the bending moment, thence the deflection, and so on, and column straightens out. A critical load, therefore, can be interpreted as an axial load that is just sufficient to hold the column in a slightly deformed configuration, although in such an unstable condition that a slight lateral disturbance will cause it to bow out as shown in Figure 16.4(c).

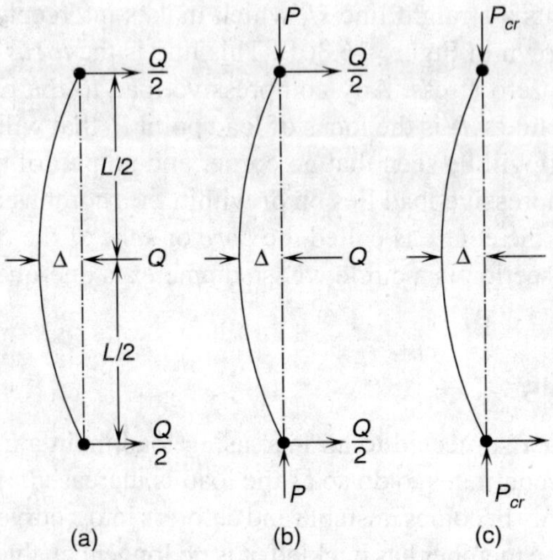

Figure 16.4 Beam and column sustaining equal deflection.

16.3.2 Critical Load by Euler's Formula

Leonard Euler a Swiss mathematician proposed an analysis based on differential equation of elastic curve $EI(d^2y/dz^2) = M$. Such an analysis is valid for an *ideal column* and only up to the *stress at the proportionality limit*. The critical load can be obtained by considering the state of equilibrium of the column in its bend form under the action of axial compressive load.

Consider the center line of an ideal column having *hinged ends*, also called *pinned* or *pivoted ends*, subjected to an axial compressive load P as shown in Figure 16.5(a). The column is assumed to be restrained against lateral movement at the ends. The column deflects in to a curved shape such that the lateral deflection at a point at distance z from one end of the column is defined by the distance y. The maximum Δ is so small that the difference between the original length of the column and the projection of its deflected shape on its vertical plane is not appreciable. Under these conditions, the slope of bend shape is very small. The external bending moment at any cross-section in the bend configuration of the column is

$$M = P(y) = Py \tag{16.10}$$

The approximate differential equation of the elastic or deflection curve is:

$$M = -EI\frac{d^2y}{dz^2} = Py \quad \text{or} \quad \frac{d^2y}{dz^2} + \left(\frac{P}{EI}\right)y = 0$$

$$\frac{d^2y}{dz^2} + \alpha^2 y = 0 \quad \text{where} \quad \alpha^2 = \frac{P}{EI} \tag{16.11}$$

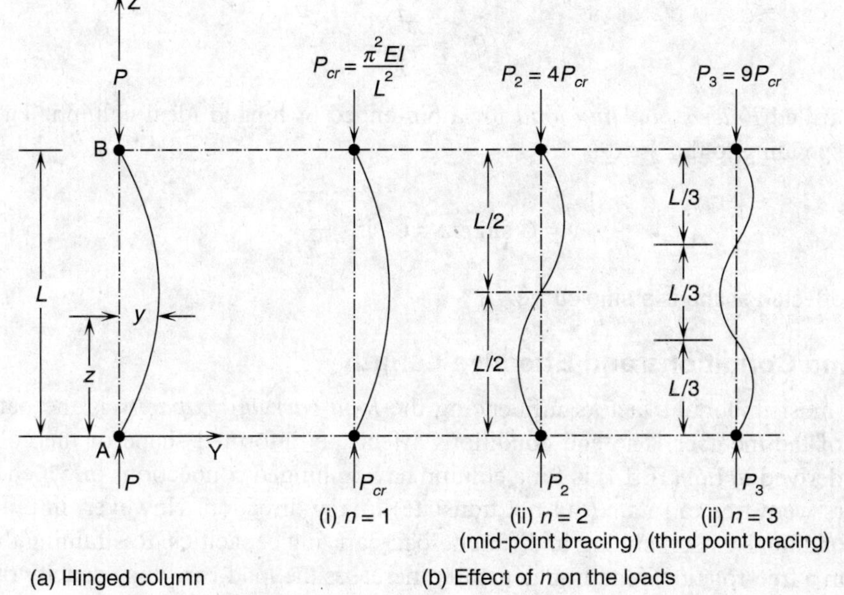

Figure 16.5 Buckling of a hinged column.

This *second order homogenous differential equation* can be solved by standard method wherein the solution is given by

$$y = C \sin \alpha z + D \cos \alpha z \tag{16.12}$$

The constants C and D are determined from the fact that a correct solution has to satisfy the boundary conditions.

At the end A of the bar, $y = 0$ at $z = 0$ i.e. $0 = 0 + D$ or $D = 0$

At the end B, $y = 0$ at $z = L$ i.e. $0 = C \sin \alpha L + D \cos \alpha L$

Substituting $D = 0$ in the second condition gives:

$$0 = C \sin \alpha L \quad \text{i.e.} \quad \text{either } C = 0 \quad \text{or} \quad \sin \alpha L = 0$$

If $C = 0$, then y is zero everywhere which is the trivial case of un-deformed straight bars, and is of no interest. Thus, valid solution is $\sin \alpha L = 0$ which gives

$$\alpha L = n\pi \text{ radians} \quad \text{where} \quad n = 1, 2, 3, \ldots$$

Therefore,
$$\alpha^2 = \left(\frac{n\pi}{L}\right)^2 = \frac{P}{EI} \quad \text{or} \quad P = \frac{n^2 \pi^2 EI}{L^2} \tag{16.13}$$

The value $n = 0$ is meaningless as it makes P zero. For other values of n the column bends into shapes shown in Figure 16.5(b). Of these, (i) is the critical since the other occurs with larger loads and are possible only if the column is braced at the middle or third points, respectively. Thus the smallest value of this load occurs when $n = 1$, and the *critical load* is given by

$$P_{cr} = \frac{\pi^2 EI}{L^2} \tag{16.14}$$

This is called *Euler's buckling load* for a pin-ended or hinged ideal column. The deflected shape corresponding to this load is

$$y = C \sin \alpha z = C \sin \sqrt{\frac{P}{EI}}\, z \tag{16.15}$$

Thus, the deflected shape is a sine curve.

16.3.3 End Conditions and Effective Length

One of the most important factors influencing the *load-carrying capacity* is the nature of end conditions of the member. The end conditions affect the deformed shape of the columns. The expression derived in Eq. (16.14) is for a column having hinged connections in which the ends of the members were free to rotate (but not translate) in any direction. However, in practice many other end conditions exist resulting in different load-carrying capacities. Restraining the ends of a column from a free-rotation condition generally increases the load-carrying capacity of a column. Allowing translations as well as rotations at the ends of a column generally reduces its load-carrying capacity.

Four columns that are identical in all respects except for their end conditions are illustrated in Figures 16.6(i) to (iv). The end conditions shown in the figure represent *idealized* or *theoretical extremes*, since end conditions in actual practice are often a combination of these primary conditions. The effects of end conditions on critical or buckling loads of columns are shown in the figure.

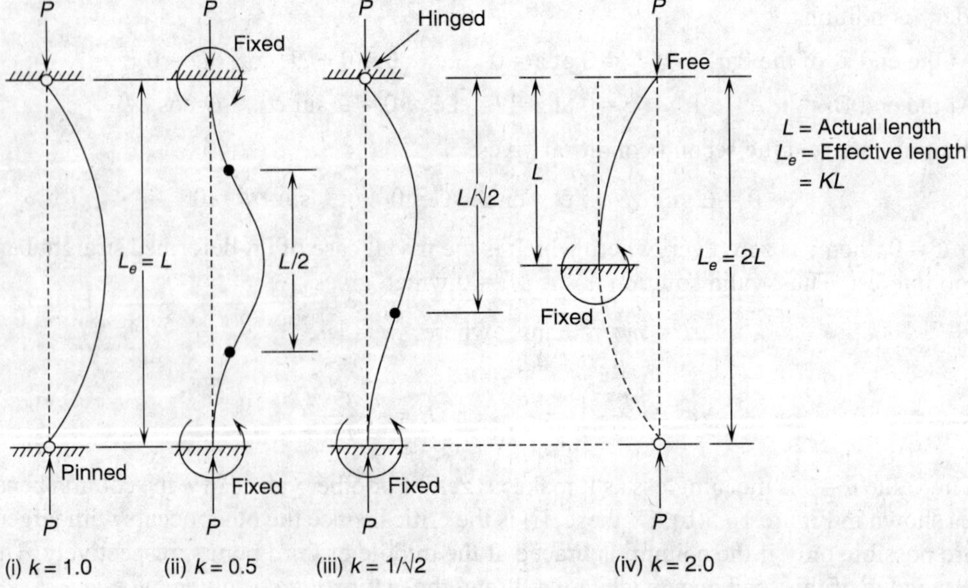

Figure 16.6 Effective lengths of columns with different end conditions.

The theoretical buckling load, for columns with the other end conditions, is computed analytically in a way similar to that for the hinged column but with appropriate boundary conditions. The buckling load for each of these columns can be expressed in terms of the critical load for a hinged column, which is taken as the *primary* or *fundamental case*. Alternatively, the load-carrying of these columns is estimated by considering the effect of end conditions on the *deflected shape* of the columns. The buckling load of a column is controlled by the portion of the deflected column between the *points of inflection* which are analogous to the *hinged connections*. This distance between the points of inflection or imaginary hinged-connection is called the *effective length*, L_e. The buckling load of this portion is given by $P_{cr} = \pi^2 EI/(L_e)^2$. Thus, the concept of effective length is useful in the analysis of columns with different end conditions as it provides an easy way to estimate the load-carrying capacities of the columns. The numerical value that modifies the unsupported length L is called the k factor of column, i.e. $L_e = kL$. However, while sketching the deformed shape of a column, it should be ensured that the curve is continuous and the tangent to the member at the fixed support remains vertical. At the point of change of sense of curvature an *inflection point* must exist. In case of fixed-ended column in Figure 16.7(a), due to symmetry the points of inflection are at quarter points of its unsupported lengths, i.e. at $L/4$ from each end. Since *the bending moment is zero at the point of inflection*, the *free-body diagrams* (Figure 16.7(b)) show that the middle half of the fixed-ended column is equivalent to a hinged column having effective length $L_e = L/2$. Thus the critical load on a fixed-ended column is

$$P_{cr} = \frac{\pi^2 EI}{L_e^2} = \frac{\pi^2 EI}{(L/2)^2} = 4\left(\frac{\pi^2 EI}{L}\right) \tag{16.16}$$

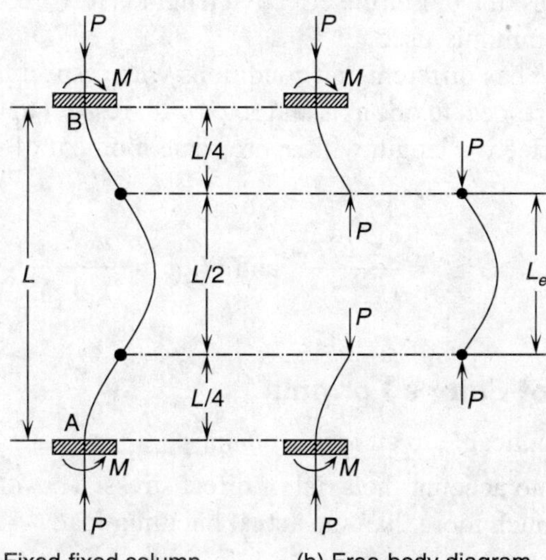

(a) Fixed-fixed column (b) Free-body diagram

Figure 16.7 Behaviour of fixed-fixed column.

This is four times the strength of the column if its ends were hinged. Similarly, for a column hinged at one end and built-in or fixed at the other, as in Figure 16.6(iii), the point of inflection is at $L/\sqrt{2}$ (nearly $0.7L$) from the hinged end, i.e. the effective length $L_e = 0.7L$. Thus the critical load on this column is

$$P_{cr} = \frac{\pi^2 EI}{L_e^2} = \frac{\pi^2 EI}{(L/\sqrt{2})^2} = 2\left(\frac{\pi^2 EI}{L^2}\right) \tag{16.17}$$

This is twice the critical load on a hinged column of same length.

Figure 16.6(iv) illustrates the method for determining the load capacity of a column built in a one end and free at the other, i.e. the cantilever or flagpole type of column. The critical loads on it and on the pin-ended column are equal, provided the pin-ended column is two times as long as this column. In other words, the effective length is twice its unsupported length, i.e. $L_e = 2L$. Thus the critical load on this column is

$$P_{cr} = \frac{\pi^2 EI}{L_e^2} = \frac{\pi^2 EI}{(2L)^2} = \frac{1}{4}\left(\frac{\pi^2 EI}{L^2}\right) \tag{16.18}$$

The load is one-quarter of the critical load on a hinged column of same length.

The effect of end conditions on critical load can, therefore, be expressed in terms of the critical length for the fundamental type of hinged column of same length. The k factor for pined-ended column is 1.0, hence $L_e = kL = 1.0L$; for a fixed-ended column, $k = 0.5$ and $L_e = kL = 0.5L$; for a fixed-hinged column, $k = 0.7$ and $L_e = kL = 0.7L$ and for a flagpole type column, $k = 2.0$ and $L_e = kL = 2L$ as illustrated in Figure 16.6.

The effects of imperfect end conditions in field applications are taken into account by modifying the effective length of the member. Theoretically the effective length of a column supported at the ends lies between $0.5L$ to $1.0L$. The more nearly the full restraint, the closer is the effective length to $0.5L$. The more it rotates, the closer the effective length is to $1.0L$. In practice, the values in the range $0.6L$ to $0.7L$ are commonly used.

Quite often, a column has different end conditions with respect to one axis than another, e.g. it may be pin-ended with respect to one axis and fixed with respect to the other, thus it is necessary to correlate the correct effective length with appropriate moment of inertia or radius of gyration. Thus,

$$P_{cr,z} = \frac{\pi^2 EI_z}{(k_z L_z)^2} \quad \text{and} \quad P_{cr,y} = \frac{\pi^2 EI_y}{(k_y L_y)^2} \tag{16.19}$$

16.3.4 Limitations of Euler's Formula

(i) This formula is applicable to an *ideal column*.

(ii) It does not take into account the axial or direct stress. Thus the critical load given by this formula may be much more than the actual buckling load.

(iii) It considers that the assumptions of theory of bending are also valid for the buckling theory.

(iv) It is evident from formula that the critical load for a column does not depend on the strength of material, but only on its *geometric configuration* (moment of inertia) and *modulus of elasticity*. For this reason, two geometrically identical slender struts, one of high-strength steel and the other of ordinary structural steel, will buckle at the same critical load since they have the same modulus of elasticity.

A column always tends to buckle in its weaker direction, i.e. about the axis of least moment of inertia of the cross-section, thus the value of moment of inertia I in the column formula is always the least moment of inertia of the cross-section.

Thus, for a good design with a given area, the material should be distributed as far away as possible from the centroid and in such a manner that the moment of inertia about the principal axes are equal or as nearly equal as possible.

(v) Euler's formula is valid only if the bending stress developed during buckling, i.e. the *critical stress* is within the *proportional limit*.

16.3.5 Critical Stress

For design purposes the critical load is generally expressed in terms of critical stress. The critical stress is computed by substituting $I = Ar^2$, where A is the cross-sectional area; and r is the *radius of gyration* of the section, which is defined by $r = \sqrt{I/A}$. Thus, from Eq. (16.14) for the hinged column,

$$\sigma_{cr} = \frac{P_{cr}}{A} = \frac{\pi^2 EI}{AL^2} = \frac{\pi^2 E}{L^2}\left(\frac{I}{A}\right) = \frac{\pi^2 Er^2}{L^2} = \frac{\pi^2 E}{(L/r)^2} \tag{16.20}$$

Here P_{cr}/A is the average stress in the column when subjected to its critical load, often called the *critical stress*. Its limiting value is obtained at the proportional limit. The ratio L/r is called the *slenderness ratio* of the column. The critical stress is inversely proportional to the square of the slenderness ratio. The higher the slenderness ratio, the lower is the critical stress at which buckling occurs. Since an axially loaded column tends to buckle about the axis of least moment of inertia, the least radius of gyration is used in the slenderness ratio. For a more general case

$$\frac{P_{cr,x}}{A} = \frac{\pi^2 E}{(kL_x/r_x)^2} \quad \text{and} \quad \frac{P_{cr,y}}{A} = \frac{\pi^2 E}{(kL_y/r_y)^2} \tag{16.21}$$

Here, *the critical stress will be governed not by the least radius of gyration but by the higher value of the slenderness ratio.*

A *long column* is generally defined as one for which Euler's formula is applicable. The limiting slenderness ratio for a column of a given material can be estimated from the known values of the proportional limit and modulus of elasticity of the material. For example, consider a column of the structural steel having the proportional limit and modulus of elasticity of σ_{pl} = 200 MPa and E = 200 GPa, respectively. The limiting slenderness ratio is

$$\left(\frac{L}{r}\right)^2 = \frac{\pi^2 E}{\sigma_{pl}} = \frac{\pi^2 \times (200 \times 10^3)}{200} \approx 10000 \quad \text{or} \quad \frac{L}{r} = 100$$

It is evident from the Euler's curve between L/r and $\sigma = P/A$ shown in Figure 16.8 that below this value of slenderness ratio, i.e. in the dashed portion of the curve, the Euler's stress exceeds the proportional limit. Hence for $L/r \leq 100$, Euler's formula is not valid and the proportional limit is taken to be the critical stress. The curve also shows that the critical or allowable stress on the

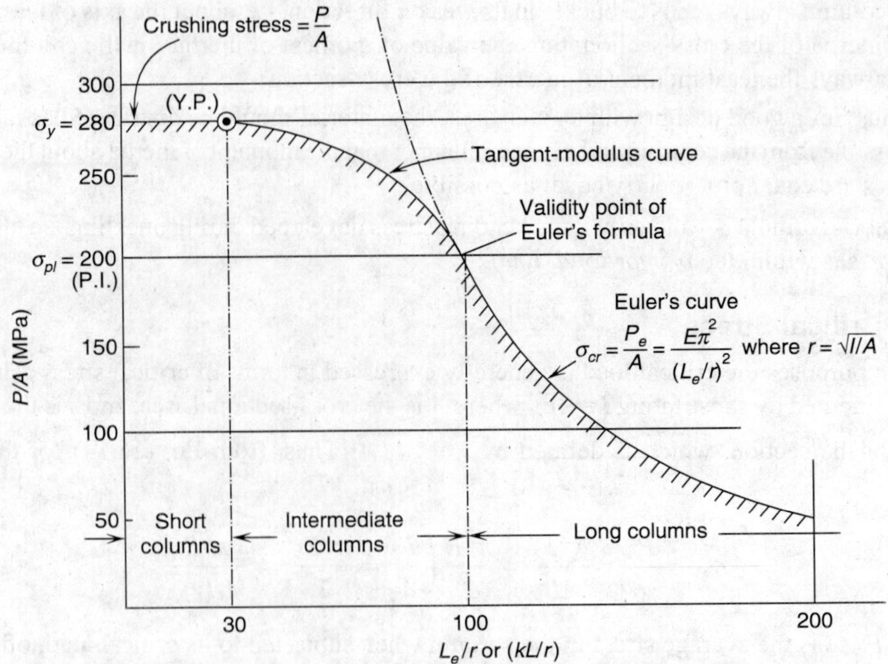

Figure 16.8 Critical or allowable stress is given by solid line.

column decreases rapidly as the slenderness ratio increases, hence in a good design slenderness ratio is kept as small as possible. The allowable stress can be obtained by dividing the critical stress by a suitable *factor of safety*, usually 1.7 to 2.5, depending on the material of the column.

The critical load for columns with other end or boundary conditions can also be computed analytically by considering the state of equilibrium of the column in its bent form as in the case of a hinged column but with appropriate boundary conditions.

16.4 CANTILEVER COLUMN

Consider the cantilever column, i.e. a column fixed at one end and free at the other, carrying a compressive force P at its free end as shown in Figure 16.9. The external bending moment at any cross-section in the bend configuration of the column is

$$M = P(\Delta - y) \tag{16.22}$$

The approximate differential equation of the elastic or deflection curve is:

$$M = -EI\frac{d^2y}{dz^2} = P(\Delta - y) \quad \text{or} \quad \frac{d^2y}{dz^2} + \left(\frac{P}{EI}\right)y = \left(\frac{P}{EI}\right)\Delta$$

$$\frac{d^2y}{dz^2} + \alpha^2 y = \alpha^2\Delta \quad \text{where} \quad \alpha^2 = \frac{P}{EI} \tag{16.23}$$

This second order differential equation can be solved by standard method wherein the general solution is given by

$$y = C \sin \alpha z + D \cos \alpha z + \Delta \tag{16.24}$$

The constants C and D can be determined from the fact that a correct solution has to satisfy the prescribed boundary conditions at the ends.

At the fixed end $A(z = 0)$ of the bar,

(i) $y = 0$ i.e. $0 = 0 + D + \Delta$ or $D = -\Delta$ and

(ii) $\dfrac{dy}{dz} = 0$ i.e. $0 = [C\alpha \cos \alpha x - D\alpha \sin \alpha x]_{z=0}$ or $C = 0$

Substitution of these values in the expression for y gives:

$$y = \Delta(1 - \cos \alpha z) \tag{16.25}$$

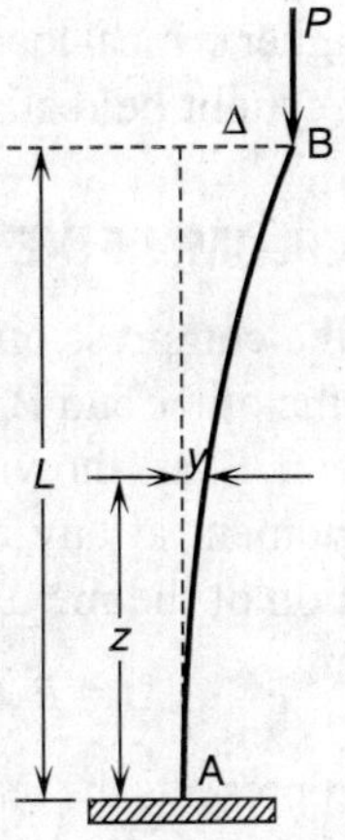

Figure 16.9 Cantilever or flagpost column.

At the free end $B(z = L)$ of the column, $y = \Delta$ i.e.

$$\Delta = \Delta(1 - \cos \alpha L) \quad \text{or} \quad \Delta \cos \alpha L = 0$$

If $\Delta = 0$, then y is zero everywhere which is the trivial case of undeformed straight column. Thus, valid solution is $\cos \alpha L = 0$ which gives

$$\alpha L = (2n - 1)\frac{\pi}{2} \text{ radians where } \quad n = 1, 2, 3, \ldots$$

Therefore, $$\alpha^2 = \left[\frac{(2n-1)\pi}{2L}\right]^2 = \frac{P}{EI} \quad \text{or} \quad P = \frac{(2n-1)^2 \pi^2 EI}{4L^2}$$

The smallest value of P corresponds to $n = 1$; thus the critical load is given by

$$P_{cr,1} = \frac{\pi^2 EI}{4L^2} = \frac{\pi^2 EI}{(2L)^2} = \frac{\pi^2 EI}{L_e^2} \tag{16.26}$$

Thus, equivalent length is $L_e = 2L$.

The deflected shape shown in Figure 16.9 corresponding to this load is

$$y = \Delta\left(1 - \cos \frac{\pi}{2L} z\right) \tag{16.27}$$

For $n = 2, 3$ the corresponding values of critical load are

$$P_{cr,2} = \frac{9\pi^2 EI}{4L^2} \quad \text{and} \quad P_{cr,3} = \frac{25\pi^2 EI}{4L^2} \tag{16.28}$$

The corresponding deflected shapes are,

$$y = \Delta\left(1 - \cos \frac{3\pi}{2L} z\right) \quad \text{and} \quad y = \Delta\left(1 - \cos \frac{5\pi}{2L} z\right) \tag{16.29}$$

The higher critical loads and their corresponding deflected shapes have no practical significance since they cannot be realized.

16.5 COLUMN FIXED AT ONE END AND HINGED AT THE OTHER

Consider the compression member AB fixed at end A and hinged at the other end B and carrying a compressive load P at the end B as shown in Figure 16.10. The external bending moment at any cross-section in the slightly bent configuration of the member is

$$M = R_B(L - z) + Py \tag{16.30}$$

where R_B represents the reaction at the end B. The governing differential equation of the elastic or deflection curve is:

$$M = -EI\frac{d^2y}{dz^2} = R_B(L - z) + Py$$

or

$$\frac{d^2y}{dz^2} + \left(\frac{P}{EI}\right)y = -\left(\frac{R_B}{EI}\right)(L - z)$$

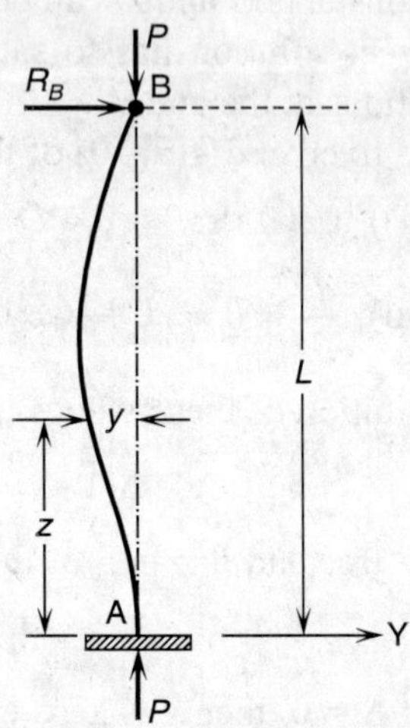

Figure 16.10 Fixed–hinged column.

Differentiate the above equation twice with respect to z

$$\frac{d^4y}{dz^4} + \alpha^2\frac{d^2y}{dz^2} = 0 \quad \text{where} \quad \alpha^2 = \frac{P}{EI} \tag{16.31}$$

The standard solution of this fourth order differential equation is given by

$$y = C_1 \sin \alpha z + C_2 \cos \alpha z + C_3 z + C_4 \tag{16.32}$$

The prescribed boundary conditions which are to be satisfied:
At the fixed end A($z = 0$) of the member,

(i) $y = 0$ i.e. $0 = 0 + C_2 + 0 + C_4 \rightarrow C_4 = -C_2$

(ii) $\frac{dy}{dz} = 0$ i.e. $0 = [C_1\alpha \cos az - C_2\alpha \sin az + C_3]_{z=0}$

or $\quad 0 = \alpha C_1 + 0 + C_3 \rightarrow C_3 = -\alpha C_1$

At the hinged end B($z = L$) of the member,

(iii) $y = 0$ i.e. $0 = C_1 \sin \alpha L + C_2 \cos \alpha L + C_3 L + C_4$

(iv) $\frac{d^2y}{dz^2} = 0$ i.e. $0 = [-C_1\alpha^2 \sin \alpha z - C_2\alpha^2 \cos \alpha z]_{z=L}$

or $\quad 0 = C_1\alpha^2 \sin \alpha L + C_2\alpha^2 \cos \alpha L \rightarrow C_2 = -C_1 \tan \alpha L$

The constants C_1, C_2, C_3 and C_4 can be determined by solving these four linear homogeneous algebraic equations. Substitute the values of C_2, C_3 and C_4 in terms of C_1 in (iii):

$$C_1 \sin \alpha L - C_1 \sin \alpha L - \alpha C_1 L + C_1 \tan \alpha L = 0$$

or
$$\tan \alpha L - \alpha L = 0 \tag{16.33}$$

This transcendental equation is known as *characteristic equation*, which can be solved graphically or numerically. The smallest root of this equation will provide the lowest critical load. Using trial and modification procedure,

αL, radians	π	4	4.5	4.4	4.45	4.48	4.49	4.495	4.493	4.494
$\tan \alpha L - \alpha L$	–3.1	–2.8	0.14	–1.30	–0.73	–0.25	–0.07	0.03	–0.008	0.012

The smallest root is 4.493, thus

$$\alpha L = 4.493 \quad \text{or} \quad \left(\frac{P}{EI}\right) L^2 = (4.493)^2$$

Therefore,
$$P_{cr} = \frac{20.19EI}{L^2} \approx \frac{2\pi^2 EI}{L^2} = \frac{\pi^2 EI}{(L/\sqrt{2})^2} \tag{16.34}$$

Thus, equivalent length, $L_e = L/\sqrt{2} \approx 0.7L$.

16.6 FIXED-FIXED COLUMN

Consider the compression member AB fixed at both the ends and carrying a compressive load P as shown in Figure 16.11. The external bending moment at any section in the slightly bent shape of the member is

$$M = R_B(L - z) + Py - M_B \tag{16.35}$$

where R_B and M_B represent the reaction and bending moment at the end B, respectively. The governing differential equation of the elastic or deflection curve is:

$$M = -EI\frac{d^2y}{dz^2} = R_B(L - z) + Py - M_B$$

or
$$\frac{d^2y}{dz^2} + \left(\frac{P}{EI}\right)y = -\left(\frac{R_B}{EI}\right)(L - z) + \frac{M_B}{EI}$$

Differentiate this equation twice with respect to z

$$\frac{d^4y}{dz^4} + \alpha^2\frac{d^2y}{dz^2} = 0 \quad \text{where} \quad \alpha^2 = \frac{P}{EI} \tag{16.36}$$

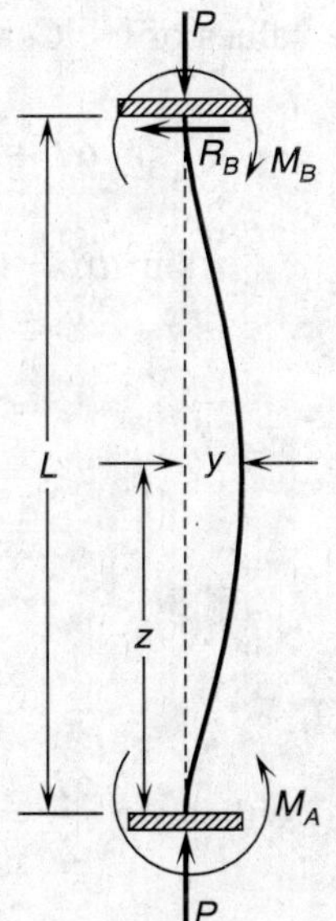

Figure 16.11 Fixed-fixed column.

It should be noted that this equation is exactly the same as Eq. (16.31). However, the boundary conditions are different. The standard solution of this fourth order differential equation as usual is given by

$$y = C_1 \sin \alpha z + C_2 \cos \alpha z + C_3 z + C_4 \tag{16.37}$$

The prescribed boundary conditions to be satisfied are:
At the fixed end A($z = 0$) of the member,

(i) $y = 0$ i.e. $0 = 0 + C_2 + 0 + C_4 \quad \rightarrow C_4 = -C_2$

(ii) $\dfrac{dy}{dz} = 0$ i.e. $0 = [C_1\alpha \cos \alpha z - C_2\alpha \sin \alpha z + C_3]_{z=0}$

or $0 = \alpha C_1 + 0 + C_3 \quad \rightarrow C_3 = -\alpha C_1$

At the fixed end B($z = L$) of the member,

(iii) $y = 0$ i.e. $0 = C_1 \sin \alpha L + C_2 \cos \alpha L + C_3 L + C_4$

(iv) $\dfrac{dy}{dz} = 0$ i.e. $0 = [C_1\alpha \cos \alpha z - C_2\alpha \sin \alpha z + C_3]_{z=L}$

or $0 = C_1\alpha \cos \alpha L - C_2\alpha \sin \alpha L + C_3$

These four linear homogeneous algebraic equations provide values of the constants C_1, C_2, C_3 and C_4.
From (ii) and (iv):

$$C_1\alpha (\cos \alpha L - 1) - C_2\alpha \sin \alpha L = 0 \quad \rightarrow C_2 = \frac{(\cos \alpha L - 1)}{\sin \alpha L} C_1$$

Substitute values of C_2, C_3 and C_4 in terms of C_1 in (iii):

$$C_1 \sin \alpha L + \frac{(\cos \alpha L - 1)}{\sin \alpha L} C_1 \cos \alpha L - \alpha C_1 L - \frac{(\cos \alpha L - 1)}{\sin \alpha L} C_1 = 0$$

$$\sin^2 \alpha L + (\cos \alpha L - 1)^2 - \alpha L \sin \alpha L$$

$$= (\sin^2 \alpha L + \cos^2 \alpha L) - 2 \cos \alpha L + 1 - \alpha L \sin \alpha L$$

$$= 2(1 - \cos \alpha L) - \alpha L \sin \alpha L$$

$$= 2\left(2 \sin^2 \frac{\alpha L}{2}\right) - \alpha L\left(2 \sin \frac{\alpha L}{2} \cos \frac{\alpha L}{2}\right) = 0$$

or

$$\sin \frac{\alpha L}{2}\left(\sin \frac{\alpha L}{2} - \frac{\alpha L}{2} \cos \frac{\alpha L}{2}\right) = 0 \tag{16.38}$$

Thus, either $\sin \dfrac{\alpha L}{2} = 0$ or $\left(\sin \dfrac{\alpha L}{2} - \dfrac{\alpha L}{2} \cos \dfrac{\alpha L}{2}\right) = 0$

(i) $\sin \dfrac{\alpha L}{2} = 0$ gives: $\dfrac{\alpha L}{2} = n\pi$ radians where $n = 1, 2, 3, \ldots$

Therefore, $\alpha^2 = \left(\dfrac{2n\pi}{L}\right)^2 = \dfrac{P}{EI}$ or $P = \dfrac{4n^2\pi^2 EI}{L^2}$

The least critical load corresponding to $n = 1$ is

$$P_{cr} = \frac{4\pi^2 EI}{L^2}$$

And corresponding deflected shape is given by, $y = \left(1 - \cos\frac{\alpha L}{2}\right)$

(ii) $$\left(\sin\frac{\alpha L}{2} - \frac{\alpha L}{2}\cos\frac{\alpha L}{2}\right) = 0 \quad \text{or} \quad \tan\frac{\alpha L}{2} = \frac{\alpha L}{2} \tag{16.39}$$

This equation is similar to that for fixed-hinged column, thus

$$\frac{\alpha L}{2} = 4.493 \quad \text{i.e.} \quad \alpha^2 = \left(\frac{2 \times 4.493}{L}\right)^2 = \frac{P}{EI} \quad \text{or} \quad P = \frac{80.75EI}{L^2}$$

Thus, the lowest critical load,

$$P_{cr} = \frac{4\pi^2 EI}{L^2} = \frac{\pi^2 EI}{(L/2)^2} = \frac{\pi^2 EI}{L_e^2} \tag{16.40}$$

Thus, equivalent length,

$$L_e = L/2 = 0.5L.$$

16.7 BRACINGS

Bracing a column changes its effective length and consequently its buckling mode. These are frequently used at one or more levels along the height of columns to reduce the column lengths and thus increase their load-carrying capacities. In practice, bracings generally form the part of the structural framework and also serve other functions. In Figure 16.12(a), the hinged column has no bracings. As illustrated in Figure 16.12(b) a hinged column braced at mid-height will deform into an S-shaped curve. Since the shape of the member between the bracing and the point of inflection is analogous to that of a pin-ended column, the effective length of column is equal to this distance. In the column shown in Figure 16.12(b) effective length is halved, which increases its load-carrying capacity by a factor of 4. Figure 16.12(c) shows where bracings are set at every one-third length (or third-point). In case of asymmetrically placed bracings the longer unbraced portion would buckle prior to the shorter portion, thus effective length of the longer portion in the direction of bracing will govern the buckling load as shown in Figure 16.12(d). The more a column is braced, the shorter its effective length becomes and higher the buckling load. However, symmetrically placed bracings are usually more effective. It should be noted that in case of a column braced about one axis but not the other, it has a tendency to buckle in the direction associated with the highest *slenderness ratio*. All the columns must be checked for buckling about both the axes by calculating $P_{cr,y}$ and $P_{cr,x}$ as shown in Figure 16.13. The smaller of the loads is the *critical load* at which the column will buckle first. When a column (see Figure 16.14(a)) with asymmetric cross-section is braced in only one plane, it can potentially buckle in two modes as shown in Figure 16.14(b). The column will buckle in the mode associated with the higher slenderness ratio as shown in Figure 16.14(c).

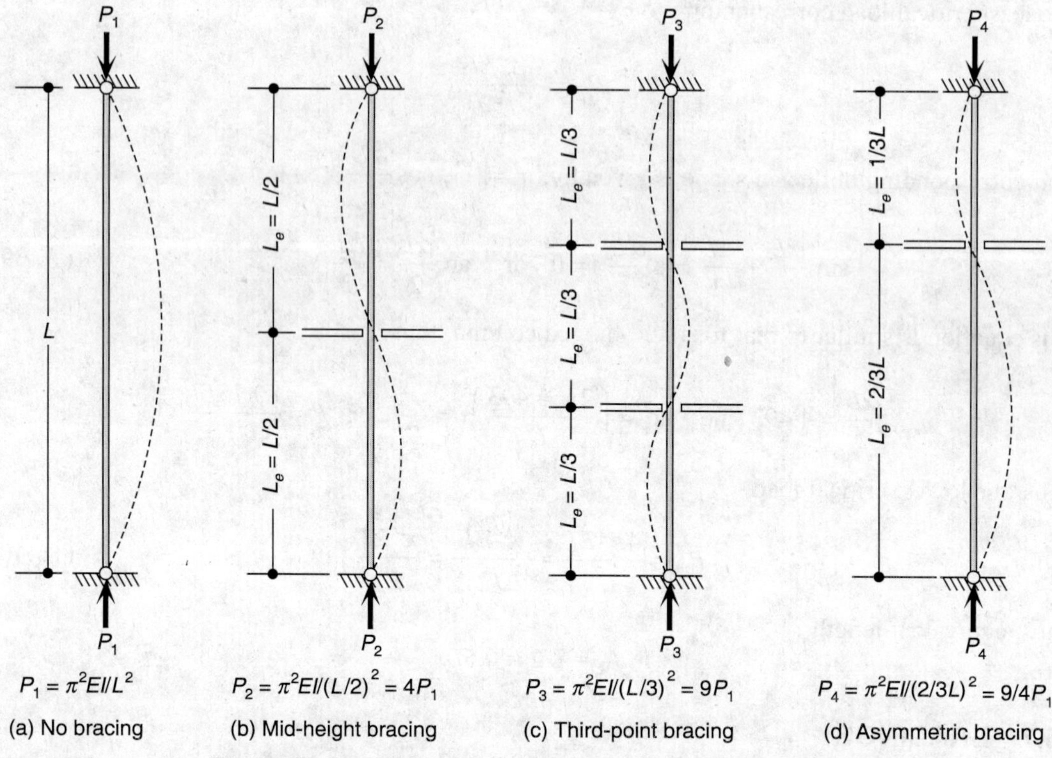

Figure 16.12 Effect of lateral bracing on column buckling.

EXAMPLE 16.1

A 2 m long round steel member of 15 mm diameter is subjected to a gradually increasing axial compressive load. Determine: (a) the Euler's buckling load for the member with pin-ended conditions, and (b) the maximum lateral deflection corresponding to the buckling conditions of the member in (a). The Young's modulus and yield stress for steel are 200 GPa and 240 MPa, respectively.

Solution: The geometric properties of the member are,

$$\text{Area, } A = \frac{\pi d^2}{4} = \frac{\pi \times (15)^2}{4} = 176.71 \text{ mm}^2$$

Moments of inertia,

$$I = \frac{\pi d^4}{64} = \frac{\pi \times (15)^4}{64} = 2485 \text{ mm}^4$$

Therefore,

$$P_e = \frac{\pi^2 EI}{L^2} = \frac{\pi^2 \times (200 \times 10^3) \times 2485}{(2000)^2} = 1226.3 \text{ N} = 1.2263 \text{ kN}$$

Direct compressive stress

$$\sigma_o = \frac{P_e}{A} = \frac{1226.3}{176.71} = 6.94 \text{ MPa}$$

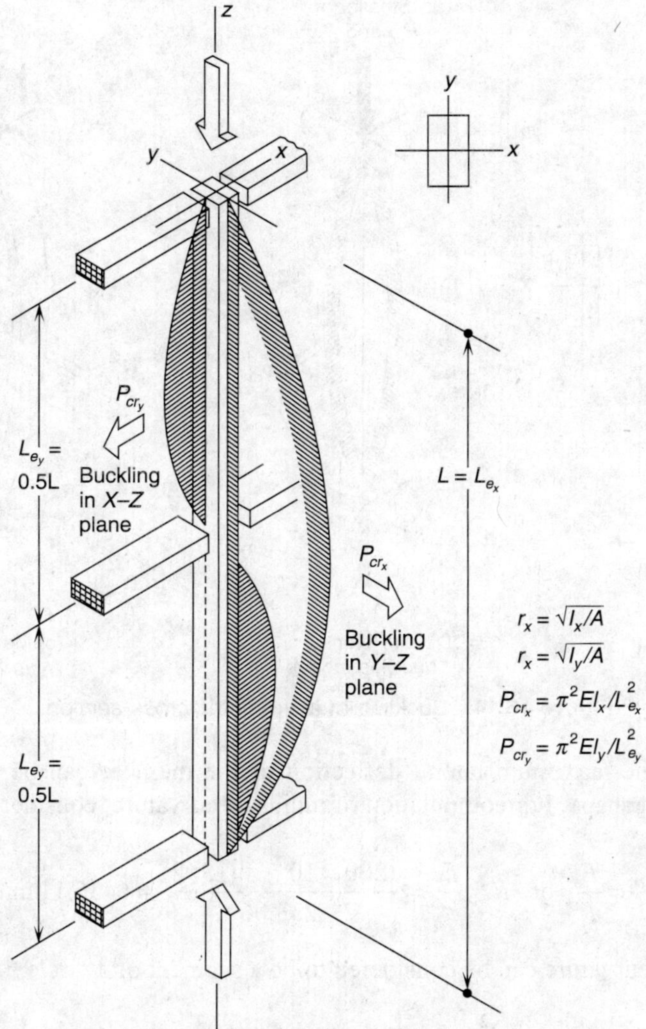

Figure 16.13 Column bracing in X–Z plane.

Consider the maximum bending stress corresponding to the buckling condition to be σ_b

$$\sigma_o + \sigma_b = \sigma_y \quad \text{or} \quad \sigma_b = \sigma_y - \sigma_o = 240 - 6.94 = 233.06 \text{ MPa}$$

The maximum bending moment at the centre,

$$M_{\max} = P_e y_m = \sigma_b Z = \sigma_b \frac{1}{(d/2)}$$

Therefore,

$$y_m = \frac{\sigma_b}{P_e} \frac{I}{(d/2)} = \frac{233.06}{1226.3} \times \frac{2485}{(15/2)} = 62.97 \text{ mm}$$

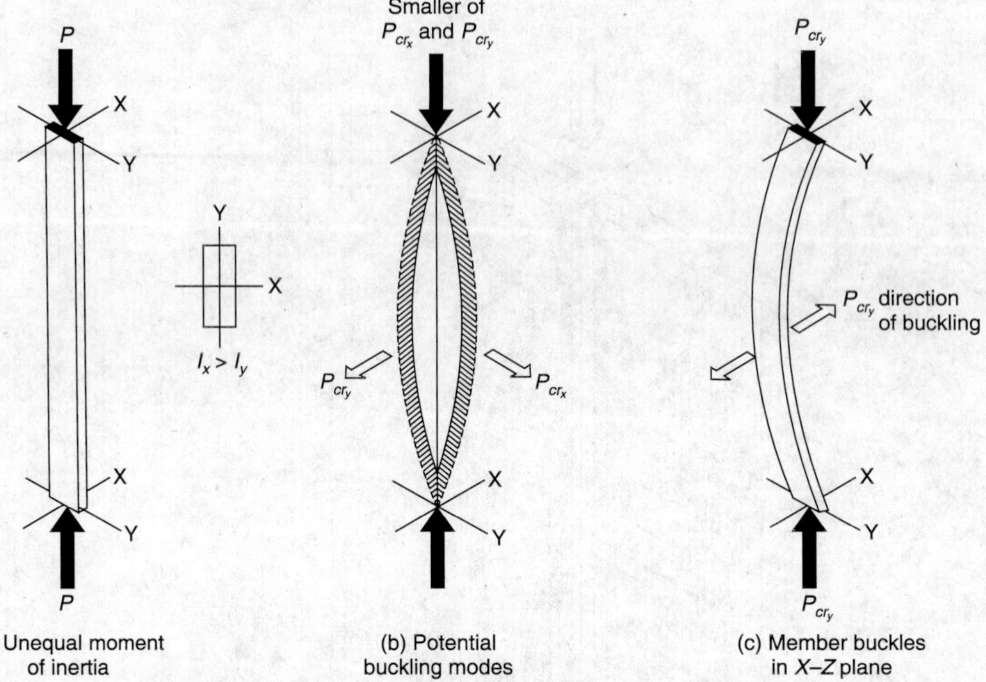

Figure 16.14 Buckling of asymmetric cross-section.

Alternatively, the maximum lateral deflection of the member can be estimated from the geometry of the bent shape. For computation of radius of curvature, consider flexural relation:

$$\frac{\sigma_b}{c} = \frac{E}{R} \quad \text{or} \quad R = \frac{Ec}{\sigma_b} = \frac{(200 \times 10^3) \times (15/2)}{233.06} = 6436.11 \text{ mm}$$

A curve with small curvature can be considered to be a segment of a circle, i.e.

$$y_m (2R - y_m) = \left(\frac{L}{2}\right)^2 \quad \text{or} \quad y_m^2 - 2Ry_m + \left(\frac{L}{2}\right)^2 = 0$$

Therefore,
$$y_m = R - \sqrt{R^2 - \left(\frac{L}{2}\right)^2} = 6436.11 - \sqrt{6436.11^2 - \left(\frac{2000}{2}\right)^2}$$
$$= 78.16 \text{ mm}$$

EXAMPLE 16.2

The cross-section of a steel compression member is built-up of two unequal ISA 7550 × 6 mm angles placed with longer legs back to back and connected at regular intervals along the length so that they act together. Determine the maximum compressive load the member can carry with a length of 3 m. Consider effective length to be 0.85L. The net cross-sectional area of each angle is

716 mm^2, the least radius of gyration for the whole cross-section is 18.5 mm, and modulus of elasticity of the material 200 GPa.

Solution: The slenderness ratio,

$$\frac{L}{r} = \frac{0.85 \times (3 \times 10^3)}{18.5} = 137.84$$

Thus critical load can be determined by the Euler's formula

$$P = \frac{\pi^2 EI}{L^2} = \frac{\pi^2 EA}{(L/r)^2} = \frac{\pi^2 \times (200 \times 10^3) \times (2 \times 716)}{(137.84)^2}$$

$$= 1.48772 \times 10^5 \text{ N} = 148.772 \text{ kN}$$

EXAMPLE 16.3

A rectangular column of 80 mm × 100 mm size is braced at the mid-point along its weaker direction. Determine the critical axial load the column can support with pin-ended conditions over a length of 3.25 m. The modulus of elasticity of the material is 20 GPa.

Solution: Moments of inertia,

$$I_x = \frac{bd^3}{12} = \frac{80 \times (100)^3}{12} = 6.67 \times 10^6 \text{ mm}^4$$

$$I_y = \frac{b^3 d}{12} = \frac{80^3 \times 100}{12} = 4.267 \times 10^6 \text{ mm}^4$$

The slenderness ratios,

$$\left(\frac{L}{r}\right)_x = \frac{(3.25 \times 10^3)}{[(6.67 \times 10^6)/(80 \times 100)]^{1/2}} = 112.56$$

$$\left(\frac{L}{r}\right)_y = \frac{(1.625 \times 10^3)}{[(4.267 \times 10^6)/(80 \times 100)]^{1/2}} = 70.36$$

Thus the column will buckle in the stronger, i.e. the unbraced, direction and the critical load is

$$P = \frac{\pi^2 EA}{(L/r)^2} = \frac{\pi^2 \times (20 \times 10^3) \times (80 \times 100)}{(112.56)^2}$$

$$= 1.2464 \times 10^5 \text{ N} = 124.64 \text{ kN}$$

EXAMPLE 16.4

A column is to support an axial service load of 500 kN with a factor of safety of 2.75. Determine the lightest wide flange rolled steel section which can safely be used when: (a) the length is 5.0 m and both ends are hinged, and (b) the length is 7.0 m with one end fixed and the other hinged.

The stress at proportional limit and modulus of elasticity are σ_{pl} = 200 MPa and E = 210 GPa, respectively.

Solution: For steel with σ_{pl} = 200 MPa and E = 210 GPa, the minimum slenderness ratio for validity of Euler's formula

$$\frac{L_e}{r} = \left(\frac{\pi^2 E}{\sigma_{pl}}\right)^{1/2} = \sqrt{\frac{\pi^2 \times (210 \times 10^3)}{200}} = 101.8$$

Critical Euler's load, P_{cr} = 500 × 2.75 = 1375 kN

(a) *Euler's criteria*-I: For L = 5.0 m and $L_e/r \geq 101.8$

The least moment of inertia

$$I \geq \frac{P_{cr} L_e^2}{\pi^2 E} = \frac{(1375 \times 10^3) \times (5 \times 10^3)^2}{\pi^2 \times (210 \times 10^3)} = 1.659 \times 10^7 \text{ mm}^4$$

For $L/r \geq 101.8$, the least r is

$$r \leq \frac{L_e}{101.8} = \frac{5 \times 10^3}{101.8} = 49.12 \text{ mm}$$

Thus the section must have a least $I \geq 1.659 \times 10^7$ mm^4 and a least $r \leq 49.12$ mm. From the table of structural steel sections, the section ISWB 450 × 79.4 kg having the least $I = 1.7067 \times 10^7$ mm^4 and r = 41.1 mm satisfies the conditions.

Euler's criteria-II: The section is based on proportional limit, i.e. for $L_e/r < 101.8$

$$\text{Minimum area of the section, } \frac{P_{cr}}{\sigma_{pl}} = \frac{1375 \times 10^3}{200} = 6875 \text{ mm}^2$$

The least $r > 49.12$ mm

Section satisfying these conditions is ISWB 500 × 95.2 kg having the least A = 12122 mm^2 and r = 49.6 mm. Therefore, the lightest suitable section is ISWB 500 × 95.2 kg.

(b) For L = 7.0 m and critical Euler's load, P_{cr} = 1375 kN

For a fixed-hinged column, the effective length of equivalent hinged column

$$L_e = 0.7L = 0.7 \times 7 = 4.9 \text{ m}$$

Using *Euler's criteria*-I: For $L_e/r \geq 101.8$

$$I \geq \frac{P_{cr} L_e^2}{\pi^2 E} = \frac{(1375 \times 10^3) \times (4.9 \times 10^3)^2}{\pi^2 \times (210 \times 10^3)} = 1.5923 \times 10^7 \text{ mm}^4$$

and $$\text{least } r \leq \frac{L}{101.8} = \frac{4.9 \times 10^3}{101.8} = 48.13 \text{ mm}$$

The lightest section that satisfies these conditions is the section ISWB 450 × 79.4 kg having least $I = 1.707 \times 10^7$ mm^4 and $r = 41.1$ mm.

Euler's criteria-II: The section based on proportional limit, i.e. for $L/r < 101.8$

$$\text{Minimum area of the section, } A \geq \frac{1375 \times 10^3}{200} = 6875 \text{ mm}^2$$

The least $r \geq 48.13$ mm.

For which the lightest section available is ISWB 500 × 95.2 kg with the least area $A = 12122$ mm^2 and $r = 49.6$ mm. A comparison of the two sections obtained above indicates that the appropriate section is ISWB 500. It should be noted that in both the parts the section is governed by Euler's formula for the elastic stability, i.e. $L/r \geq 101.8$.

EXAMPLE 16.5

A rectangular steel member of 25 mm × 100 mm size is to support axial compressive load. Determine: (a) the limiting length of the member to behave as a short column, (b) the critical buckling stress for the member with pin-ended conditions over a length of 4.5 m, and (c) the critical buckling load for the member in (b) above if it is laterally braced in weaker direction at depth from 1.5 m from the top. The modulus of elasticity of the material $E = 200$ GPa. The grade of steel used is Fe250.

Solution: Moments of inertia of the cross-section,

$$I_x = \frac{bd^3}{12} = \frac{25 \times (100)^3}{12} = 2.083 \times 10^6 \text{ mm}^4$$

$$I_y = \frac{b^3 d}{12} = \frac{25^3 \times 100}{12} = 0.1302 \times 10^6 \text{ mm}^4$$

(a) *Load for failure by crushing*:

$$P_{\max} = \sigma_y A = 250 \times (25 \times 100) = 625 \times 10^3 \text{ N} = 625 \text{ kN}$$

Buckling length for $P_{\max}$:

The buckling will occur about the weaker axis (i.e. Y-axis), thus

$$P_{cr,y} = \frac{\pi^2 EI_y}{L^2} = P_{\max}$$

or

$$L = \sqrt{\frac{\pi^2 EI_y}{P_{\max}}} = \sqrt{\frac{\pi^2 \times (200 \times 10^3) \times (0.1302 \times 10^6)}{625 \times 10^3}} = 641.25 \text{ mm}$$

Thus, the limiting length for the member to behave as a short column is 640 mm. However, it should be noted that in practice such exactly defined transitions do not occur.

(b) *Critical buckling loads*:

$$P_{cr,x} = \frac{\pi^2 EI_x}{L_x^2} = \frac{\pi^2 \times (200 \times 10^3) \times (2.083 \times 10^6)}{(4500)^2} = 2.03 \times 10^5 \text{ N} = 203 \text{ kN}$$

$$P_{cr,y} = \frac{\pi^2 EI_y}{L_y^2} = \frac{\pi^2 \times (200 \times 10^3) \times (0.1302 \times 10^6)}{(4500)^2} = 0.1269 \times 10^5 \text{ N} = 12.69 \text{ kN}$$

Since the buckling load about Y-axis (i.e weaker axis) is smaller than that about X-axis (i.e. stronger axis), the member will buckle in the direction of least dimension at the load of 12.69 kN.

Critical buckling stress:

$$\sigma_{cr,y} = \frac{P_{cr,y}}{A} = \frac{12.69 \times 10^3}{25 \times 100} = 5.076 \text{ MPa}$$

(c) The bracing of column changes its buckling mode and consequently its effective length. Therefore,

$$L_x = L = 4.5 \text{ m} \quad \text{and} \quad L_y = 4.5 - 1.5 = 3.0 \text{ m}$$

$$P_{cr,x} = \frac{\pi^2 EI_x}{L_x^2} = \frac{\pi^2 \times (2 \times 10^5) \times (2.083 \times 10^6)}{(4500)^2} = 2.03 \times 10^5 \text{ N} = 203 \text{ kN}$$

$$P_{cr,y} = \frac{\pi^2 EI_y}{L_y^2} = \frac{\pi^2 \times (2 \times 10^5) \times (0.1302 \times 10^6)}{(3000)^2}$$

$$= 0.2856 \times 10^5 \text{ N} = 28.56 \text{ kN}$$

Since, $P_{cr,y} < P_{cr,x}$, the column will buckle at 28.56 kN about its weak axis in the plane of bracing in an S-shaped buckling mode, and there is no danger of buckling out of the bracing.

EXAMPLE 16.6

A straight column of length L having uniform cross-sectional area A and flexural stiffness EI_c is heated so that its temperature varies linearly from $t/4$ at one end to t at the other end. One end is pinned to a rigid foundation; the other end is pin-jointed to the centre of a horizontal elastic beam of length $2a$ and flexural stiffness EI_b as shown in Figure 16.15(i), the thermal expansion the column is resisted by the beam. If there is no load in the beam when $t = 0$, show that the column buckles in flexure when

$$t = \frac{8\pi^2 I_c}{5\lambda AL^2}\left(1 + \frac{EA}{K_t L}\right)$$

where $K_t = 6EI_b/a^3$ and λ is the coefficient of linear expansion.

Solution: In restraining the expansion of the column the beam behaves as a spring of stiffness K_t which is calculated as

$$\Delta = \frac{P(2a)^3}{48EI_b} \quad \text{or} \quad K_t = \frac{P}{\Delta} = \frac{6EI_b}{a^3}$$

The elastically restrained column is shown in Figure 16.15(ii).

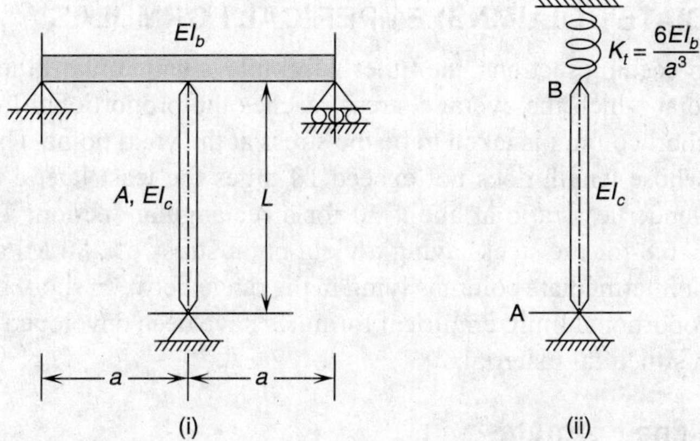

Figure 16.15 Elastically restrained column of Example 16.6.

Average rise in the temperature of column $= \frac{1}{2}\left(\frac{t}{4} + t\right) = \frac{5t}{8}$

Free expansion of the column $= \frac{5}{8}\lambda t L$

Consider the force exerted by the beam on the column to be P, then

Decrease in the length of the column $= \frac{PL}{AE}$

Deflection in the beam $\Delta = \frac{P}{K_t}$

Net expansion of the column $\delta L = \frac{5}{8}\lambda t L - \frac{PL}{AE}$

For compatibility net expansion of the column must be equal to the deflection in the beam, i.e.

$$\frac{5}{8}\lambda t L - \frac{PL}{AE} = \frac{P}{K_t}$$

Therefore,
$$P = \frac{5\lambda t L}{8\left(\frac{L}{AE} + \frac{1}{K_t}\right)}$$

For buckling of the column $P = P_e = \frac{\pi^2 EI_c}{L^2} = \frac{5\lambda t L}{8\left(\frac{L}{AE} + \frac{1}{K_t}\right)}$

Thus,
$$t = \frac{8\pi^2 EI_c}{5\lambda L^3}\left(\frac{L}{AE} + \frac{1}{K_t}\right) = \frac{8\pi^2 I_c}{5\lambda AL^2}\left(1 + \frac{AE}{K_t L}\right)$$

16.8 INTERMEDIATE COLUMNS; EMPERICAL FORMULAS

As discussed in the preceding sections, the Euler's formula is valid only if the slenderness ratio is larger than the value at which the average stress reaches the proportional limit. In practice, the limiting stress on a short column is taken to be the stress at the yield point. The definition of short column as the one whose length does not exceed 10 times the least lateral dimension fixes the upper limit of the slenderness ratio at about 30 for a rectangular section. These conditions are illustrated in Figure 16.8 for the steel having a yield point stress of 280 MPa and a proportional limit of 200 MPa. For intermediate columns lying in the range between short and long columns the stress exceeds the proportional limit, empirical formulas have been developed mainly for steel due to its being common structural material.

16.8.1 The Rankine Formula

For intermediate compression members falling in the range between short and long members, the failure of the member is due to the combined effect of direct and bending stresses. For this category of members, the following empirical formula was proposed by Rankine. The critical load for a column can be computed as

$$\frac{1}{P_{cr}} = \frac{1}{P_c} + \frac{1}{P_e} \tag{16.41}$$

where P_{cr} = Critical load

P_c = Crushing load = $\sigma_y A$

P_e = Euler's load = $\dfrac{\pi^2 EI}{L_e^2}$

σ_y = Ultimate compressive stress for column material

A = Area of cross-section

For a member of given material and cross-section, the crushing load $P_c (= \sigma_y A)$ is constant. However, the Euler's load P_e depends upon the effective length of the member. If the member is short, P_e will be large and $1/P_e$ will be small enough to be ignored as compared to $1/P_c$. Thus as the length of a member decreases

$$\frac{1}{P_{cr}} \to \frac{1}{P_c} \quad \text{or} \quad P_{cr} \to P_c$$

On the other hand for a long member P_c will be large and $1/P_c$ will be small enough to be ignored as compared to $1/P_e$. Thus as the length of a member increases

$$\frac{1}{P_{cr}} \to \frac{1}{P_e} \quad \text{or} \quad P_{cr} \to P_e$$

Thus the Rankine formula is assumed to provide satisfactory results for compression members.

Therefore,
$$P_{cr} = \frac{P_e \times P_c}{P_c + P_e} = \frac{P_c}{1 + (P_c / P_e)}$$

$$= \frac{\sigma_c A}{1 + \dfrac{\sigma_c A L_e^2}{\pi^2 EI}} = \frac{\sigma_c A}{1 + \dfrac{\sigma_c}{\pi^2 E}\left(\dfrac{L_e}{r}\right)^2}$$

$$= \frac{\sigma_c A}{1 + a\left(\dfrac{L_e}{r}\right)^2} \tag{16.42}$$

where $a = \dfrac{\sigma_c}{\pi^2 E}$, a constant for the given material of the member and L_e/r is the slenderness ratio. However, a is determined experimentally.

The values of a and σ_c for some important materials are listed in the table.

Material	*Compressive Stress*, σ_c MPa	*Elastic modulus* E, GPa	*Constant*, a
Cast iron	550	90	1/1600
Wrought iron	250	215	1/9000
Mild steel	320	210	1/7500
Medium carbon steel	415	210	1/5000
Aluminium	120	70	1/5000
Strong timber	50	95	1/750

The effective length of the column at which the critical loads by Euler's and Rankine's formulae would be equal is given by

$$\frac{\pi^2 EI}{L_e^2} = \frac{\sigma_c A}{1 + a\left(\dfrac{L_e}{r}\right)^2} \quad \text{or} \quad 1 + a\left(\frac{L_e}{r}\right)^2 = \frac{\sigma_c A L_e^2}{\pi^2 EI} = \frac{\sigma_c L_e^2}{\pi^2 E r^2}$$

or

$$\frac{L_e^2}{r^2}\left(\frac{\sigma_c}{\pi^2 E} - a\right) = 1$$

Therefore,

$$L_e^2 = \frac{r^2}{\left(\dfrac{\sigma_c}{\pi^2 E}\right) - a} \tag{16.43}$$

To take into account the eccentricities of the load Rankine's formula can be modified as

$$P = \frac{\sigma_c A}{\left\{1 + a\left(\dfrac{L_e}{r}\right)^2\right\}\left[1 + \dfrac{e_y y}{r_x^2} + \dfrac{e_x x}{r_y^2}\right]} \tag{16.44}$$

where x and y are distances of the fibre from the neutral axis.

16.8.2 Gordon's Formula

According to the Gordon's formula, the crippling load is given by

$$P_{cr} = \frac{\sigma_c A}{1 + b\left(\frac{L_e}{d}\right)^2} \tag{16.45}$$

where P_{cr} = critical load

d = least lateral dimension

$$b = \text{constant} = \frac{\sigma_c A d^2}{\pi^2 EI}$$

16.8.3 Other Empirical Formulae

The formulae discussed so far presume ideal geometric, ends and load conditions which generally do not exist in practice. Hence a number of empirical formulae have been proposed from time to time for practical design. Some of them are listed below:

(1) *The Straight-line Formulae*: These straight-line formulae provide linear relationships between allowable average stress and slenderness ratio. The general form is

$$\frac{P_{cr}}{A} = a - b\left(\frac{L_e}{r}\right) \tag{16.46}$$

Here again a and b are experimentally determined constants and L_e is the effective length. Some of the typical relations are listed below.

(i) Critical stress for a compression member of structural steel

$$\sigma_{cr} = \frac{P_{cr}}{A} = 367.5 - 2.0\left(\frac{L_e}{r}\right) \text{ MPa} \tag{16.47}$$

(ii) Critical stress for a compression member of cast iron

$$\sigma_{cr} = 23.8 - 0.6\left(\frac{L_e}{r}\right) \text{ MPa} \tag{16.48}$$

(iii) Safe stress for a compression member of mild steel

$$\sigma = 150\left[1 - 0.0038\frac{L_e}{r}\right] \text{ MPa} \tag{16.49}$$

(2) *The Parabolic Formula*: This formula relates allowable average stress and slenderness ratio. The general form is

$$\frac{P_{cr}}{A} = a - b\left(\frac{L_e}{r}\right)^2 \tag{16.50}$$

The experimentally determined constants a and b are usually selected in such a way that the parabola becomes tangential to the Euler's curve in the long column range. A typical relation called Johnson's formula is given below

$$P_{cr} = \sigma_c A\left[1 - c\left(\frac{L_e}{r}\right)^2\right]$$

where $$c = \frac{\sigma_c}{4\pi^2 E} \tag{16.51}$$

(3) *Fidler's formula*

$$P = \frac{A}{c}\left[(\sigma_e + \sigma_c) - \sqrt{(\sigma_e + \sigma_c)^2 - 2(\sigma_e \times \sigma_c)}\right]$$

where $$\sigma_e = \frac{\pi^2 E r^2}{L_e^2}. \tag{16.52}$$

EXAMPLE 16.7

A short length of a tube having outer and inner diameters of 50 and 40 mm, respectively, failed in compression at a load of 245 kN. When a 2 m length of the same tube was tested as strut with fixed ends, the load at failure was 160 kN. Assuming that σ_c in Rankine's formula is given by the first test; determine (a) the value of the constant a in the same formula and (b) the crippling load for a 2.5 m long strut of this tube with one end fixed and the other end hinged using Rankine's formula.

Solution: The geometric properties of the tube section are:

Area of cross-section, $$A = \frac{\pi}{4}(50^2 - 40^2) = 706.86 \text{ mm}^2$$

Moment of inertia, $$I = \frac{\pi}{64}(50^4 - 40^4) = 181132.45 \text{ mm}^4$$

Therefore, $$r^2 = \frac{I}{A} = \frac{d_o^2 + d_i^2}{16} = \frac{(50^2 + 40^2)}{16} = 256.25 \text{ mm}^2$$

(a) Crushing strength as per first test,

$$\sigma_c = \frac{245 \times 10^3}{706.86} = 346.6 \text{ MPa}$$

As per second test $$L_e = L/2 = 2000/2 = 1000 \text{ mm}$$

$$P_{cr} = 160 \times 10^3 \text{ N}$$

Therefore,

$$P_{cr} = \frac{\sigma_c A}{1 + a\left(\dfrac{L_e^2}{r^2}\right)} = \frac{346.6 \times 706.86}{1 + a \times \dfrac{1000^2}{256.25}} = 160 \times 10^3 \text{ N}$$

$$1 + a \times \frac{1000^2}{256.25} = \frac{346.6 \times 706.86}{160 \times 10^3} = 1.5312$$

or $$a = \frac{(1.5312 - 1) \times 256.25}{1000^2} = \frac{1}{1000^2/(0.5312 \times 256.25)} = \frac{1}{7346}$$

(b) $$L_e = L/\sqrt{2} = 2500/\sqrt{2} = 1767.8 \text{ mm}$$

$$P_{cr} = \frac{\sigma_c A}{1 + a\left(\frac{L_e^2}{r^2}\right)} = \frac{346.6 \times 706.86}{1 + \frac{1}{7346} \times \frac{1767.8^2}{256.25}}$$

$$= 92.099 \times 10^3 \text{ N} = 92.1 \text{ kN}$$

EXAMPLE 16.8

A 2.25 m long tubular steel strut with hinged ends having outer and inner diameters of 37.5 and 32.5 mm, respectively, is subjected to a compressive load which is parallel to its axis. (a) Compare the crippling loads obtained by Rankine's and Euler's formulae, and (b) If the elastic limit for the material of the strut is 200 MPa, determine the length of strut beyond which Euler's formula applies. The yield stress and the modulus rigidity of the material of the strut are 315 GPa and 200 GPa, respectively. The constant for Rankine's formula is 1/7500.

Solution: The geometric properties of the strut section are:
Area of cross-section,

$$A = \frac{\pi}{4}(37.5^2 - 32.5^2) = 274.89 \text{ mm}^2$$

Moment of inertia,

$$I = \frac{\pi}{64}(37.5^4 - 32.5^4) = 42307.19 \text{ mm}^4$$

Therefore,

$$r^2 = \frac{I}{A} = \frac{d_o^2 + d_i^2}{16} = \frac{(37.5^2 + 32.5^2)}{16} = 153.91 \text{ mm}^2$$

Euler's load

$$P_e = \frac{\pi^2 EI}{L_e^2} = \frac{\pi^2 \times (200 \times 10^3) \times 42307.19}{(2.25 \times 10^3)^2} = 16496.01 \text{ N}$$

According to Rankine's formula

$$P_{cr} = \frac{\sigma_c A}{1 + a\left(\frac{L_e^2}{r^2}\right)} = \frac{315 \times 274.89}{1 + \frac{1}{7500} \times \frac{2250^2}{153.91}} = 16077.89 \text{ N}$$

Thus, the ratio

$$\frac{P_e}{P_{cr}} = \frac{16496.01}{16077.89} = 1.026$$

The limiting length for validity of Euler's formula

$$P_e = \frac{\pi^2 EI}{L_e^2} = \sigma_{pl} A \quad \text{or} \quad L_e^2 = \frac{\pi^2 EI}{\sigma_{pl} A} = \frac{\pi^2 Er^2}{\sigma_{pl}}$$

$$L_e^2 = \frac{\pi^2 \times (200 \times 10^3) \times 153.91}{200} = 1519030.8 \text{ mm}^2$$

Therefore, $L = 1232.49 \text{ mm} = 1.232 \text{ m}$

EXAMPLE 16.9

A 1.5 m long straight steel bar of 5 × 20 mm cross-section is compressed longitudinally until it buckles. Assuming Euler's formula to apply to this case, estimate the maximum central deflection before the steel passes the yield point stress of 320 MPa. The modulus of elasticity of the material $E = 210$ GPa.

Solution: The minimum moment of inertia

$$I_{\min} = \frac{b^3 d}{12} = \frac{(5)^3 \times 20}{12} = 208.33 \text{ mm}^4$$

Euler's load, $$P_e = \frac{\pi^2 EI}{L_e^2} = \frac{\pi^2 \times (210 \times 10^3) \times 208.33}{1500^2} = 191.9 \text{ N}$$

If the central deflection is y_m mm, then
Maximum bending moment,

$$M_{\max} = P_e y_m = 101.9 y_m \text{ N mm},$$

Direct stress,

$$\sigma_o = \frac{P}{A_e} = \frac{191.9}{5 \times 20} = 1.92 \text{ MPa}$$

Bending stress,

$$\sigma_b = \frac{M_{\max} y}{I} = \frac{(191.9 y_m) \times (5/2)}{(5^3 \times 20/12)} = 2.3028 y_m \text{ MPa}$$

Resultant maximum stress,

$$\sigma_y = \sigma_o + \sigma_b = 1.92 + 2.3028 y_m = 320 \text{ MPa}$$

Therefore,

$$y_m = \frac{320 - 1.92}{2.3028} = 138.13 \text{ mm}$$

EXAMPLE 16.10

A 200 mm long straight steel link bar of solid circular cross-section of 9 mm diameter is compressed longitudinally until it buckles. The two ends of the link are hinged. Using the Johnson's parabolic formula, estimate the critical load the link can support. The yield point stress and the modulus of elasticity of the material are 400 MPa and 200 GPa, respectively.

Solution: The geometric properties of the link are

Area of cross-section

$$A = \frac{\pi}{4}(9)^2 = 63.617 \text{ mm}^2$$

Therefore,

$$r^2 = \frac{I}{A} = \frac{(\pi d^4/64)}{(\pi d^2/4)} = \frac{d^2}{16} \quad \text{or} \quad r = \frac{d}{4} = 2.25 \text{ mm}$$

Johnson's parabolic formula

$$P_{cr} = \sigma_c A\left[1 - c\left(\frac{L_e}{r}\right)^2\right]$$

where

$$c = \frac{\sigma_c}{4\pi^2 E} = \frac{400}{4\pi^2 \times (200 \times 10^3)} = 5.066 \times 10^{-5}$$

Therefore,

$$P_{cr} = 400 \times 63.617 \times \left[1 - (5.066 \times 10^{-5}) \times \left(\frac{200}{2.25}\right)^2\right]$$

$$= 15261 \text{ N} = 15.261 \text{ kN}$$

EXAMPLE 16.11

(a) Two rolled steel channels ISLC 200 are latticed to form the cross-section of a column as shown in Figure 16.16. How far apart should these channels be placed back to back, so that the column may carry the maximum load? The properties of one channel are:

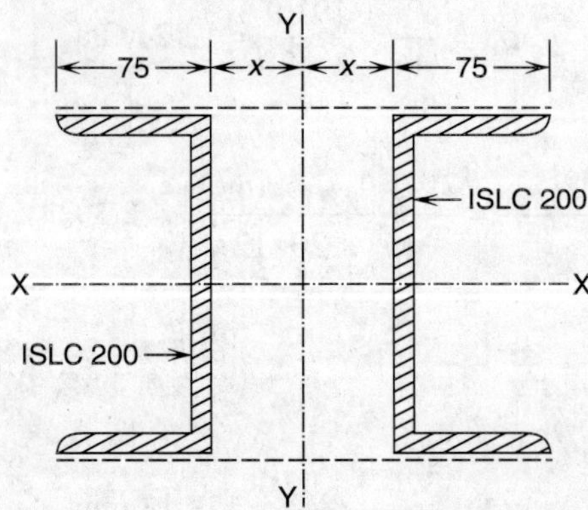

Figure 16.16 Latticed cross-section of the column of Example 16.11.

$b = 75$ mm; Area $A = 2622$ mm^2; $I_{xx} = 1.7255 \times 10^7$ mm^4;

$I_{yy} = 1.469 \times 10^6$ mm^4 and distance of C.G. from back of channel = 23.5 mm

(b) The channels of the above column are spaced 150 mm apart back to back. Determine: the safe load the column can carry with fixed-ends conditions over a length of 6 m, taking factor of safety 2.5. The constants for Rankine's formula are 320 MPa and 1/7500.

Solution: (a) For the maximum load carrying capacity of the column the moments of inertia of the combined section about two principal axes must be equal, i.e. $I_x = I_y$

$$I_x = 2I_{xx} = 2 \times 1.7255 \times 10^7 = 3.451 \times 10^7 \text{ mm}^4$$

Consider the back to back distance between the channels to be $2x$, then

$$I_y = 2 \times [1.469 \times 10^6 + 2622 \times (x + 23.5)^2] = 3.451 \times 10^7 \text{ mm}^4$$

$$(x + 23.5)^2 = \frac{(3.451 \times 10^7/2) - 1.469 \times 10^6}{2622} = 6020.595$$

$$x + 23.5 = 77.59 \quad \text{or} \quad x = 54.09$$

Therefore, back to back distance between the channels is 108.18 mm.

(b) For $2x = 150$ mm

$$I_{\text{min}} = I_x = 3.451 \times 10^7 \text{ mm}^4$$

$$A = 2 \times 2622 = 5244 \text{ mm}^2$$

$$r^2 = \frac{I_{\text{min}}}{A} = \frac{3.451 \times 10^7}{5244} = 6580.9 \text{ mm}^2$$

Therefore, from Rankine's formula

$$P_{cr} = \frac{\sigma_c A}{1 + a\left(\frac{L_e^2}{r^2}\right)} = \frac{320 \times 5244}{1 + \frac{1}{7500} \times \frac{3000^2}{6580.9}} = 1419280.12 \text{ N}$$

The safe load, $$P_w = \frac{P_{cr}}{m} = \frac{1419280.12}{2.5} = 567712.05 \text{ N} = 567.71 \text{ kN}$$

EXAMPLE 16.12

(a) Four rolled steel equal angles ISA 110 × 110 × 10 mm are latticed to form the cross-section of a column as shown in Figure 16.17. The properties of each angle are:

$A = 2106$ mm^2; $I_{xx} = I_{yy} = 2.384 \times 10^6$ mm^4; and distance of C.G. from back = 30.8 m

Determine the safe load the column can carry with fixed-ends condition over a length of 8 m. The safe compressive stress is given the straight line formula:

$$\sigma_c = 150\left[1 - 0.0038\frac{L_e}{r}\right] \text{ MPa}$$

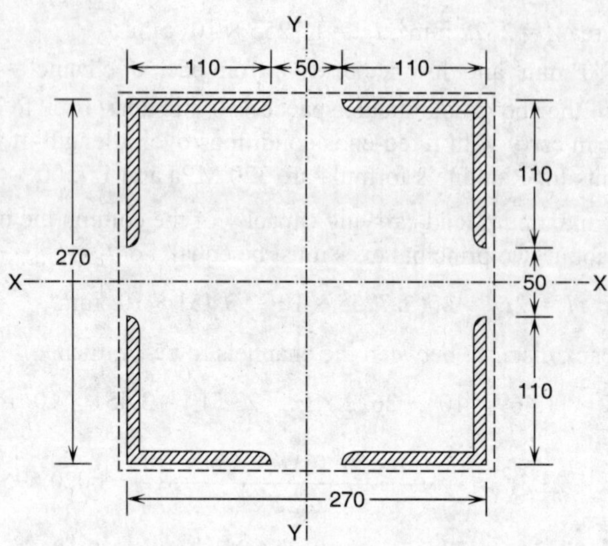

Figure 16.17 Built-up cross-section of the column of Example 16.12.

Solution: The properties of combined section are

$$\text{Area, } A = 4 \times 2106 = 8424 \text{ mm}^2$$

The moments of inertia of the combined section

$$I_x = 4 \times [2.384 \times 10^6 + 2106 \times (135 - 30.8)^2] = 1.01 \times 10^8 \text{ mm}^4$$

Therefore, $$r = \sqrt{\frac{I}{A}} = \sqrt{\frac{1.01 \times 10^8}{8424}} = 109.5 \text{ mm}$$

Effective length $$L_e = 8/2 = 4 \text{ m}$$

The slenderness ratio, $$\frac{L_e}{r} = \frac{4000}{109.5} = 36.53$$

Thus, safe compressive stress

$$\sigma_c = 150 \times [1 - 0.0038 \times 36.53] = 129.177 \text{ MPa}$$

Therefore, safe working load, $$P_w = \sigma_c A = 129.177 \times 8424$$

$$= 1088191.7 \text{ N} = 1088.19 \text{ kN}$$

16.9 ECCENTRICALLY LOADED COLUMNS

In the preceding sections, the ideal columns that are perfectly straight, with compressive load acting exactly along the axis have been discussed. In practice, there is always a certain amount of *eccentricity* in the application of the compressive loads. In addition, the columns subjected to loads having a definite eccentricity are commonly encountered, e.g. the case of a beam connected to the column flange in a building. This eccentricity introduces an additional bending moment that changes the column's load-deflection behaviour.

Consider an initially straight, hinged column subjected to a compressive force P applied at an eccentricity e as shown in Figure 16.18. The differential equation of the deflected configuration is

$$M = -EI\frac{d^2y}{dz^2} = P(y) = Py \quad \text{or} \quad \frac{d^2y}{dz^2} + \left(\frac{P}{EI}\right)y = 0$$

$$\frac{d^2y}{dz^2} + \alpha^2 y = 0 \quad \text{where} \quad \alpha^2 = \frac{P}{EI} \tag{16.53}$$

The solution of this second order homogenous differential equation is given by

$$y = C\sin\alpha z + D\cos\alpha z$$

The constants C and D can be determined from the end or boundary conditions.

At the end B of the bar,

$$y = e \text{ at } z = L/2$$

$$e = C\sin\alpha(L/2) + D\cos\alpha(L/2) \qquad \text{(i)}$$

and at the end A of the bar,

$$y = e \text{ at } z = -L/2$$

$$e = -C\sin\alpha(L/2) + D\cos\alpha(L/2) \qquad \text{(ii)}$$

Operate on Eqs. (i) and (ii) as;

(i) + (ii) gives, $D = e\sec(\alpha L/2)$

(i) – (ii) gives, $C = 0$

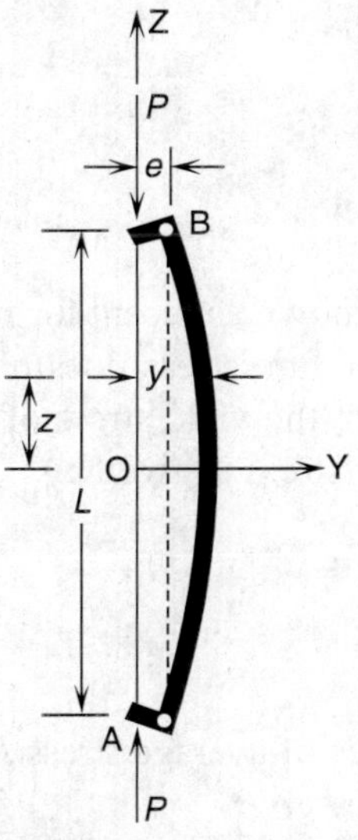

Figure 16.18 Eccentrically loaded strut.

Thus the deflected shape of the column is

$$y = e\sec\left(\frac{\alpha L}{2}\right)\cos\alpha z \tag{16.54}$$

The maximum deflection is $y_{max} = e\sec(\alpha L/2)$ and hence the maximum bending moment in the column is

$$M_{max} = Py_{max} = Pe\sec(\alpha L/2) \tag{16.55}$$

The maximum compressive stress due to the maximum bending moment and compressive force in the column is

$$\sigma_{max} = \frac{P}{A} + \frac{M_{max}c}{I} = \frac{P}{A} + \frac{[Pe\sec(\alpha L/2)]c}{Ar^2}$$

$$= \frac{P}{A}\left(1 + \frac{ec}{r^2}\sec\frac{\alpha L}{2}\right) = \frac{P}{A}\left(1 + \frac{ec}{r^2}\sec\frac{L}{2}\sqrt{\frac{P}{EI}}\right) \tag{16.56a}$$

Therefore, $$\sigma_{\max} = \frac{P}{A}\left[1 + \frac{ec}{r^2}\sec\sqrt{\frac{PL^2}{4EI}}\right] = \frac{P}{A}\left[1 + \frac{ec}{r^2}\sec\sqrt{\frac{\pi^2}{4}\frac{P}{(\pi^2 EI/L^2)}}\right]$$

$$= \frac{P}{A}\left[1 + \frac{ec}{r^2}\sec\sqrt{\frac{\pi^2}{4}\frac{P}{P_e}}\right] = \frac{P}{A}\left[1 + \frac{ec}{r^2}\sec\frac{\pi}{2}\sqrt{\frac{P}{P_e}}\right] \tag{16.56b}$$

where c is the distance of the outermost fibre from the neutral axis, and r is the least radius of gyration. For other end conditions, substitute effective length L_e for unsupported length L. The Eq. (16.56a) can be expressed in a more general form as

$$\sigma_{\max} = \frac{P}{A}\left[1 + \frac{ec}{r^2}\sec\left(\frac{PL_e^2}{4EI}\right)^{1/2}\right]$$

$$= \frac{P}{A}\left[1 + \frac{ec}{r^2}\sec\left\{\frac{L_e}{r}\left(\frac{P}{4EA}\right)^{1/2}\right\}\right] \tag{16.57}$$

This is known as secant formula. The expression is applicable to columns of any length so long as the maximum stress does not exceed the elastic limit, i.e. the yield stress. If m is the factor of safety, σ_y the yield stress of the material and P_w the safe working load, then $P = mP_w$. The safe working stress is given by

$$\frac{P_w}{A} = \frac{\sigma_y / m}{1 + \frac{ec}{r^2}\sec\left\{\frac{L_e}{2r}\left(\frac{mP_w}{EA}\right)^{1/2}\right\}} \tag{16.58}$$

The value of average stress P_w/A can be determined by trial and modification for the given values of L/r and ec/r^2.

The term ec/r^2 is *called the eccentricity factor*. The relationship is plotted between P_w/A and slenderness ratio L/r for different values of ec/r^2 for a given material of known yield point in Figure 16.19. The curves indicate that in very slender columns, those with $L/r \geq 150$, Euler's formula appears to provide a good estimate, as the curves for different eccentricities move closer towards merging with Euler's curve. In this range, lie the so called long *columns*, whose buckling load can be determined by the Euler's formula. For the ratios between 0 and 40 where the stress is almost independent of slenderness ratios, lay the short columns that behave like compression blocks with very little deflection. A slenderness ratio in the range 40 and 150 gives intermediate columns, for which the limiting average stress can be determined only by the secant formula.

Perry modified the secant formula given by Eq. (16.56b) to a more useful form by approximating secant term as follows:

$$\sec\frac{\pi}{2}\sqrt{\frac{P}{P_e}} \cong \frac{1.2P_e}{P_e - P}$$

Therefore, $$\sigma_{\max} = \frac{P}{A}\left[1 + \frac{ec}{r^2}\left(\frac{1.2P_e}{P_e - P}\right)\right] = \sigma_o\left[1 + \frac{ec}{r^2}\left(\frac{1.2\sigma_e}{\sigma_e - \sigma_o}\right)\right]$$

where σ_e (= P_e/A) and σ_o (= P/A) are Euler and direct stresses, respectively.

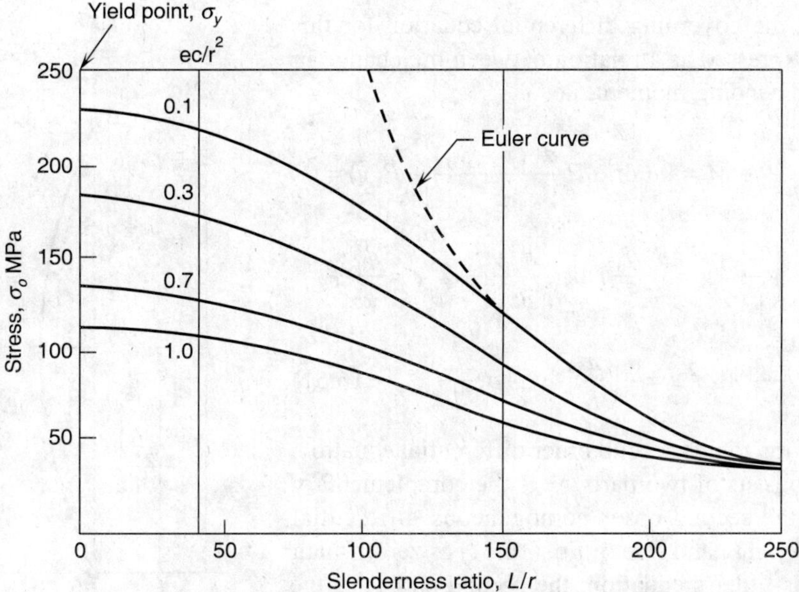

Figure 16.19 Relationship between σ_o (= P_w/A) and L/r.

$$\frac{\sigma_{\max}}{\sigma_o} - 1 = \frac{ec}{r^2}\left(\frac{1.2\sigma_e}{\sigma_e - \sigma_o}\right) = \frac{ec}{r^2}\left[\frac{1.2}{1-(\sigma_o/\sigma_e)}\right]$$

Therefore,

$$\left(\frac{\sigma_{\max}}{\sigma_o} - 1\right)\left(1 - \frac{\sigma_o}{\sigma_e}\right) = 1.2\left(\frac{ec}{r^2}\right) \tag{16.59}$$

This is known as Perry's formula. For predetermined or known $\sigma_{\max}$, and σ_e determined from Euler's load, σ_o and hence $P = \sigma_o A$ can be calculated. The maximum value of $\sigma_{\max}$ may be taken as σ_y, the yield stress.

16.10 STRUTS WITH INITIAL GEOMETRIC CURVATURE

The stress developed due to load is not much affected by the precise form of the curvature, but depends on the amount of deviation from the straightness. Consider a strut, hinged at both ends and with an initial geometric imperfection represented by $y_o(z)$ loaded along the axis joining end points as shown in Figure 16.20.

Since the initial shape of strut $y_o(z)$ having end conditions $y_o(0) = y_o(L) = 0$ is continuous and has a finite number of maxima and minima in the range $0 \le z \le L$, it can be represented by a sine series. For the ease of analysis a single term of the sine series has been taken. Thus the initial shape of the bent strut can be considered to be a sine curve of the form

$$y_o = y_m \sin\frac{\pi z}{L} \tag{16.60}$$

As usual the governing differential equation for this case is also expressed as a relation between the change in curvature and bending moment, i.e.

$$EI\frac{d^2y_1}{dz^2} + M = 0 \quad \text{or} \quad EI\frac{d^2(y - y_o)}{dz^2} + P(y) = 0$$

$$\text{or} \quad \frac{d^2y}{dz^2} + \left(\frac{P}{EI}\right)y = \frac{d^2y_o}{dz^2} \quad \text{or} \quad \frac{d^2y}{dz^2} + \alpha^2 y = \frac{d^2y_o}{dz^2}$$

Therefore,
$$\frac{d^2y}{dz^2} + \alpha^2 y = -\left(\frac{\pi}{L}\right)^2 y_m \sin\frac{\pi z}{L} \tag{16.61}$$

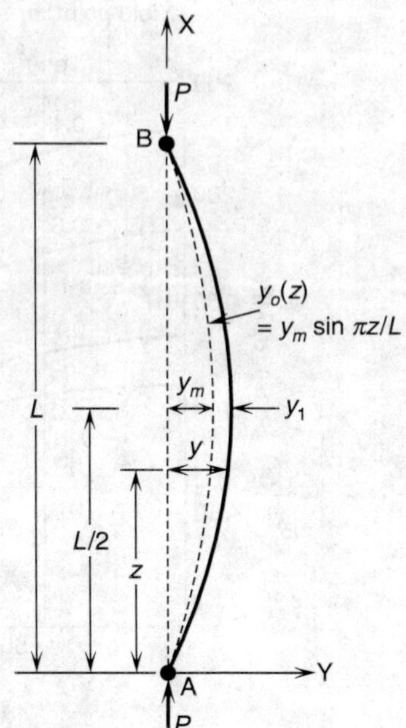

Figure 16.20 Strut with initial geometric curvature.

The solution to this second order differential equation, $y = y_c + y_p$ consists of two parts, y_c is the complementary solution for the second order homogeneous differential equation with right hand side of Eq. (16.61) as zero, which is the same as Euler's equation; the second part y_p is the particular solution which is of interest. Since the end conditions of deflected shape for this case are $y(0) = y(L) = 0$, and there is no moment at the ends; the particular solution may be taken as

$$y = Cy_m \sin\frac{\pi z}{L} \tag{16.62}$$

From Eq. (16.61),
$$-\left(\frac{\pi}{L}\right)^2 Cy_m \sin\frac{\pi z}{L} + \alpha^2\left(Cy_m \sin\frac{\pi z}{L}\right) = -\left(\frac{\pi}{L}\right)^2 y_m \sin\frac{\pi z}{L}$$

or
$$\left[\frac{\pi^2}{L^2} - \alpha^2\right]C = \frac{\pi^2}{L^2} \quad \text{or} \quad \left[\frac{\pi^2}{L^2} - \frac{P}{EI}\right]C = \frac{\pi^2}{L^2}$$

or
$$C\left[\frac{\pi^2 EI}{L^2} - P\right] = \frac{\pi^2 EI}{L^2} \quad \text{or} \quad C[P_e - P] = P_e$$

Hence, the deflected shape of the member is

$$y = \frac{P_e}{P_e - P}y_m \sin\frac{\pi z}{L} = \frac{1}{(1 - P/P_e)}y_m \sin\frac{\pi z}{L} \tag{16.63}$$

when $y_m = 0$, i.e. the strut is initially straight and $P = P_e$. The maximum deflection y_{max} will be at the midspan, i.e. at $z = L/2$, thus

$$y_{max} = \frac{1}{(1 - P/P_e)}y_m \tag{16.64}$$

The quantity $\dfrac{1}{(1-P/P_e)}$ is termed the *deflection magnification factor* caused by the axial force P. Hence the maximum bending moment in the column is

$$M_{\max} = Py_{\max} = \frac{P}{(1-P/P_e)}\,y_m \tag{16.65}$$

The maximum compressive stress due to the maximum bending moment and compressive force in the strut is

$$\sigma_{\max} = \frac{P}{A} + \frac{M_{\max}c}{I} = \frac{P}{A} + \left(\frac{P}{(1-P/P_e)}\,y_m\right)\frac{c}{Ar^2}$$

$$= \frac{P}{A}\left(1 + \frac{1}{(1-P/P_e)}\frac{cy_m}{r^2}\right) = \sigma_o\left(1 + \frac{\sigma_e}{\sigma_e - \sigma_o}\frac{cy_m}{r^2}\right)$$

Therefore, rearrangement of terms results in the following expression

$$\left(\frac{\sigma_{\max}}{\sigma_o} - 1\right)\left(\frac{\sigma_e - \sigma_o}{\sigma_e}\right) = \frac{cy_m}{r^2} \tag{16.66}$$

16.11 COMBINED ECCENTRIC LOADING AND INITIAL CURVATURE

The Eqs. (16.59) and (16.66) for maximum compressive stresses due to eccentric loading and initial curvature, respectively, are similar if maximum deviation y_c from the line joining two supports due to both causes be specified as $y_c = y_m + 1.2e$.

Then a strut initially bent and also eccentrically loaded may be considered as equivalent to an axially loaded column with increased initial curvature given by y_c. Thus maximum stress is given by

$$\left(\frac{\sigma_{\max}}{\sigma_o} - 1\right)\left(\frac{\sigma_e - \sigma_o}{\sigma_e}\right) = \frac{cy_c}{r^2} \tag{16.67}$$

For limited $\sigma_{\max}$, with σ_e determined from Euler's load, σ_o and hence $P = \sigma_o A$ can be calculated. The maximum value of $\sigma_{\max}$ may be taken as σ_y, the yield stress.

EXAMPLE 16.13

A 2.5 m long tubular steel strut with hinged ends has external and internal diameters of 65 and 50 mm, respectively. The strut is subjected to a load which is parallel to the axis but eccentric. Determine the maximum eccentricity, when the crippling stress is 0.75 of the Euler stress. The yield stress and the elastic modulus of the material of the strut are 320 GPa and 210 GPa, respectively.

Solution: The geometric properties of the strut section are:

Area of cross-section, $\quad A = \dfrac{\pi}{4}(65^2 - 50^2) = 1354.8\ \text{mm}^2$

Moment of inertia, $I = \dfrac{\pi}{64}(65^4 - 50^4) = 569444.3 \text{ mm}^4$

Therefore, $r^2 = \dfrac{I}{A} = \dfrac{5.694 \times 10^5}{1.355 \times 10^3} = 420.32$

Euler's load, $P_e = \dfrac{\pi^2 EI}{L_e^2} = \dfrac{\pi^2 \times (210 \times 10^3) \times 5.694 \times 10^5}{(2.5 \times 10^3)^2} = 188823.7 \text{ N}$

$$\sigma_e = \frac{188823.7}{1.3548 \times 10^3} = 139.37 \text{ MPa}$$

Eccentric load stress $\sigma_o = 0.75\sigma_e = 0.75 \times 139.37 = 104.53$ MPa

The maximum stress given by secant formula is

$$\sigma_{\max} = \sigma_o\left[1 + \frac{ec}{r^2}\sec\frac{\pi}{2}\sqrt{\frac{P}{P_e}}\right] = \sigma_o\left[1 + \frac{ec}{r^2}\sec\frac{\pi}{2}\sqrt{\frac{\sigma_o}{\sigma_e}}\right]$$

Therefore, $\dfrac{\sigma_{\max} - \sigma_o}{\sigma_o} = \dfrac{ec}{r^2}\sec\dfrac{\pi}{2}\sqrt{\dfrac{\sigma_o}{\sigma_e}} = \dfrac{ec}{r^2}\sec\dfrac{\pi}{2}\sqrt{0.75}$

or $\dfrac{320 - 104.53}{104.53} = \dfrac{e(65/2)}{420.32}\sec(1.36035)$

hence $e = \dfrac{(320 - 104.53) \times 420.32 \times \cos(1.36035)}{104.53 \times (65/2)} = 5.569 \text{ mm}$

EXAMPLE 16.14

A 4 m long circular cast iron column with fixed ends has 200 mm external diameter and 20 mm thickness. The column carries a load of 130 kN at an eccentricity of 30 mm from the axis of the column. Determine (a) the extreme stresses on the cross-section and (b) the maximum eccentricity when there is no tension anywhere on the section. The elastic modulus of the material of the column is 80 GPa.

Solution: The geometric properties of the strut section are:

Internal diameter $d = 200 - 2 \times 20 = 160$ mm

Area of cross-section $A = \dfrac{\pi}{4}(200^2 - 160^2) = 11310 \text{ mm}^2$

Moment of inertia $I = \dfrac{\pi}{64}(200^4 - 160^4) = 4.637 \times 10^7 \text{ mm}^4$

Section modulus $Z = \dfrac{I}{d/2} = \dfrac{4.637 \times 10^7}{100} = 4.637 \times 10^5 \text{ mm}^3$

Effective length of column $L_e = \frac{L}{2} = \frac{4000}{2} = 2000$ mm

Maximum displacement, $y_{\max} = e \sec(\alpha L/2) = e \sec \frac{L_e}{2}\sqrt{\frac{P}{EI}}$

The angle, $\frac{L_e}{2}\sqrt{\frac{P}{EI}} = \frac{2000}{2}\sqrt{\frac{130 \times 10^3}{(80 \times 10^3) \times (4.637 \times 10^7)}} = 0.1872$ radian

Maximum bending moment

$$M_{\max} = Py_{\max} = Pe \sec \frac{L_e}{2}\sqrt{\frac{P}{EI}}$$

$$= (130 \times 10^3) \times 30 \times \sec 0.1872 \text{ rad} = 3969347.7 \text{ N}$$

Extreme stresses

$$\sigma_{\max} = \frac{P}{A} \pm \frac{M_{\max}}{Z} = \frac{(130 \times 10^3)}{11310} \pm \frac{3969347.7}{4.637 \times 10^5}$$

$$= 11.49 \pm 8.56 = 20.05;\ 2.93 \text{ MPa (compression)}$$

When tension is just avoided, even with maximum eccentricity

$$\frac{P}{A} = \frac{M_{\max}}{Z} = \frac{Pe \sec \frac{L_e}{2}\sqrt{\frac{P}{EI}}}{Z}$$

$$e = \frac{Z}{A} \cos \frac{L_e}{2}\sqrt{\frac{P}{EI}} = \frac{(4.637 \times 10^5)}{11310} \cos 0.1872 \text{ rad} = 40.28 \text{ mm}$$

EXAMPLE 16.15

A 2 m long steel strut of rectangular cross-section of size 60 × 80 mm fixed at both ends is initially curved with sinusoidal centre line having a maximum deviation of 6 mm in the weaker direction. The strut is subjected to a compressive load of 60 kN at an eccentricity of 15 mm in the plane of curvature. Determine the maximum stress induced in the cross-section. The elastic modulus of the material of the strut is 200 GPa.

Solution: The effective eccentricity taking in to account the initial curvature in the strut is

$$y_c = y_m + 1.2e = 6 + 1.2 \times 15 = 24 \text{ mm}$$

Euler's load, $P_e = \frac{\pi^2 EI}{L_e^2} = \frac{\pi^2 \times (200 \times 10^3) \times (80 \times 60^3/12)}{(0.5 \times 2 \times 10^3)^2} = 2842446$ N

$$\sigma_e = \frac{P_e}{A} = \frac{2842446}{60 \times 80} = 592.18 \text{ MPa}$$

$$\sigma_o = \frac{P}{A} = \frac{60 \times 10^3}{60 \times 80} = 12.5 \text{ MPa}$$

$$r^2 = \frac{I}{A} = \frac{(80 \times 60^3 / 12)}{60 \times 80} = 300 \text{ mm}^2$$

The maximum stress for this case is given by the secant formula

$$\left(\frac{\sigma_{\max}}{\sigma_o} - 1\right)\left(\frac{\sigma_e - \sigma_o}{\sigma_e}\right) = \frac{c y_c}{r^2}$$

Hence
$$\left(\frac{\sigma_{\max}}{12.5} - 1\right)\left(\frac{592.18 - 12.5}{592.18}\right) = \frac{30 \times 24}{300}$$

$$\sigma_{\max} = 12.5 \times \left[\frac{30 \times 24 \times 592.18}{300 \times (592.18 - 12.5)} + 1\right] = 43.147 \text{ MPa}$$

16.12 COLUMN DESIGN

Due to the presence of all types of indeterminate imperfections in compression members, national design codes have provided empirical formulas-based theoretical analysis, experimental results and field tested factors of safety. The commonly used formulas for the intermediate columns have been discussed earlier in section 16.8.3.

EXAMPLE 16.16

A steam engine cylinder of 400 mm internal diameter is subjected to a steam pressure of 0.72 MPa to move a piston having rod length between piston and cross-head of 1.2 m. Determine the diameter of piston rod taking a factor of safety of 4, consider the piston rod to be firmly fixed to the piston and cross-head. The constants for Rankine's formula are 320 MPa and 1/7500.

Solution: The compressive load on the piston rod,

$$P_w = \text{Steam pressure} \times \text{Area of cylinder} = 0.72 \times (\pi \times 400^2/4) = 90477.9 \text{ N}$$

The crippling load on the cylinder

$$P_{cr} = m \times P_w = 4 \times 90477.9 = 361911.6 \text{ N}$$

Effective length of the piston rod

$$L_e = L/2 = 1200/2 = 600 \text{ mm}$$

$$r^2 = \frac{I}{A} = \frac{(\pi d^4 / 64)}{(\pi d^2 / 4)} = \frac{d^2}{16}$$

Therefore, from the Rankine's formula

$$P_{cr} = \frac{\sigma_c A}{1 + a\left(\dfrac{L_e^2}{r^2}\right)} \quad \text{or} \quad \frac{320 \times (\pi d^2 / 4)}{1 + \dfrac{1}{7500} \times \dfrac{600^2 \times 16}{d^2}} = 361911.6$$

$$251.33d^2 = 361911.6 \times \left(1 + \frac{768}{d^2}\right) \quad \text{or} \quad d^2 = 1440 \times \left(1 + \frac{768}{d^2}\right)$$

$$d^4 - 1440d^2 - 1105920 = 0$$

$$d^2 = (1440/2) + \sqrt{(1440/2)^2 + 1105920} = 1994.49 \text{ mm}^2$$

Therefore, $d = 44.66$ mm

EXAMPLE 16.17

An aluminium-alloy compression member of rectangular cross-section with the depth equal to twice its width, i.e. $d = 2b$ is 600 mm long. Determine the dimensions of the cross-section if the member is to support a load of 7.5 kN with a factor of safety of 2. For the slenderness ratios in the range 18 to 64, the permissible stress is given by the following straight-line formula

$$\frac{P}{A} = 308.9 - 2.158\left(\frac{L}{r}\right) \text{MPa}$$

The modulus of elasticity of the material 75 GPa.

Solution: The minimum moments of inertia of the cross-section is $b^3d/12$, thus the least radius of gyration

$$r = \left(\frac{1}{A}\right)^{1/2} = \left(\frac{b^3 d}{12 \times bd}\right)^{1/2} = \frac{b}{\sqrt{12}}$$

Thus the slenderness ratio

$$\frac{L}{r} = \frac{600}{b/\sqrt{12}} = \frac{600\sqrt{12}}{b}$$

Consider L/r to be in the range 18 to 64, then

$$\frac{P}{A} = \frac{mP_w}{A} = \frac{\text{F.O.S.} \times \text{Load to be carried}}{\text{Area}}$$

$$= \frac{2 \times (7.5 \times 10^3)}{b \times 2b} = \frac{7.5 \times 10^3}{b^2}$$

Therefore, $$\frac{7.5 \times 10^3}{b^2} = 308.9 - 2.158\left(\frac{600\sqrt{12}}{b}\right)$$

or $$b^2 - \frac{2.158 \times 600\sqrt{12}}{308.9} b - \frac{7.5 \times 10^3}{308.9} = b^2 - 14.52b - 24.28 = 0$$

Therefore, $b = 16.03$ mm. Hence the slenderness ratio

$$\frac{L}{r} = \frac{600\sqrt{12}}{b} = \frac{600\sqrt{12}}{16.03} = 129.66$$

This value of slenderness ratio is outside the valid range, thus Euler's formula for the long column can be used,

$$\frac{mP_w}{A} = \frac{\pi^2 E}{(L/r)^2} \quad \text{or} \quad \frac{2 \times (7.5 \times 10^3)}{b \times 2b} = \frac{\pi^2 \times (75 \times 10^3) \times b^2}{(600\sqrt{12})^2}$$

or $$b^4 = \frac{2 \times (7.5 \times 10^3) \times (600\sqrt{12})^2}{2 \times \pi^2 \times (75 \times 10^3)} = 4.3771 \times 10^4 \text{ mm}^4$$

Therefore, $b = 14.46$ mm. The section of the column is 14.46 × 28.93 mm.

16.13 ELASTICALLY RESTRAINED PERFECT COLUMNS

In practice, most of the columns are essentially structural members connected to other structural members which provide elastic types of supports or restraints at the ends. These *elastic supports* or restraints are similar to *spring supports* of rotational and translational types. The following simple examples would illustrate the concept.

EXAMPLE 16.18

A perfect strut of length L is elastically supported at two ends with rotational springs of spring constant k_r i.e., the restraining moment at each end is k_r times the slope at each end. Show that the buckling load P_{cr} of the column is given by the characteristic equation,

$$\tan\frac{\alpha L}{2} = -\frac{P_{cr}}{\alpha k_r} \quad \text{where} \quad \alpha^2 = \frac{P}{EI}$$

Solution: Consider the compression member AB elastically supported at two ends and carrying a compressive load P as shown in Figure 16.21. The restraining moments at the ends are

$$M_B = M_A = M_o = k_r\theta$$

where θ is the slope at each end. The external bending moment at any section in the slightly bent shape of the member is

$$M = Py - M_o$$

The governing differential equation of the elastic or deflection curve is:

$$M = -EI\frac{d^2y}{dz^2} = Py - M_o$$

or $$\frac{d^2y}{dz^2} + \left(\frac{P}{EI}\right)y = \frac{d^2y}{dz^2} + \alpha^2 y = \frac{M_o}{EI}$$

where $\alpha^2 = \dfrac{P}{EI}$.

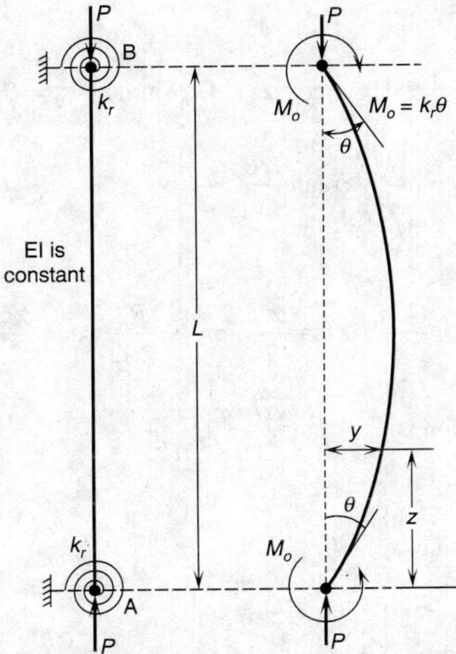

Figure 16.21 Elastically restrained column (rotational spring).

The standard solution of this second order differential equation is given by

$$y = y_c + y_p = (C \sin \alpha z + D \cos az) + \frac{M_o}{P_{cr}}$$

Thus $$\frac{dy}{dz} = \alpha C \cos \alpha z - \alpha D \sin \alpha z$$

The prescribed boundary conditions to be satisfied are

(i) At the end A ($z = 0$) of the member,

$$y = 0 \quad \text{i.e.} \quad 0 = 0 + D + \frac{M_o}{P_{cr}} \quad \rightarrow D = -\frac{M_o}{P_{cr}}$$

(ii) At the mid point ($z = L/2$) of the member,

$$\frac{dy}{dz} = 0 \quad \text{i.e.} \quad 0 = (\alpha C \cos \alpha z - \alpha D \sin \alpha z)_{z = L/2}$$

$$= \left[\left(\alpha \cos \frac{\alpha L}{2}\right) C + \frac{M_o}{P_{cr}} \alpha \sin \frac{\alpha L}{2}\right]$$

Therefore, $$C = -\frac{M_o}{P_{cr}} \tan \frac{\alpha L}{2}$$

(iii) At the end A ($z = 0$)

$$\frac{dy}{dz} = \theta = [C\alpha \cos \alpha z - D\alpha \sin \alpha z]_{z=0} = C\alpha$$

Therefore,
$$\theta = -\frac{M_o}{P_{cr}} \alpha \tan \frac{\alpha L}{2}$$

where $M_o = k_r \theta$.

Therefore,
$$\frac{M_o}{k_r} = -\frac{M_o}{P_{cr}} \alpha \tan \frac{\alpha L}{2} \quad \text{or} \quad -\frac{P_{cr}}{\alpha k_r} = \tan \frac{\alpha L}{2}$$

Thus the characteristic equation is,
$$\tan \frac{\alpha L}{2} = -\frac{P_{cr}}{\alpha k_r}$$

Special cases:

(1) when both ends are fixed, i.e. $k_r \to 0$

$$\tan \frac{\alpha L}{2} \left(= \frac{\sin \frac{\alpha L}{2}}{\cos \frac{\alpha L}{2}} \right) = \infty \quad \text{i.e.} \quad \cos \frac{\alpha L}{2} = 0$$

or
$$\alpha L = 2n\pi \text{ radians} \quad \text{where } n = 1, 2, 3, \ldots$$

Therefore,
$$\text{for } n = 1, \quad P_{cr} = \frac{4\pi^2 EI}{L^2}$$

(2) when both ends are hinged, i.e. $k_r = 0$

$$\tan \frac{\alpha L}{2} \left(= \frac{\sin \frac{\alpha L}{2}}{\cos \frac{\alpha L}{2}} \right) = \infty \quad \text{i.e.} \quad \cos \frac{\alpha L}{2} = 0$$

or
$$\alpha L = (2n - 1)\pi \text{ radians} \quad \text{where } n = 1, 2, 3, \ldots$$

Therefore,
$$\text{for } n = 1, \quad P_{cr} = \frac{\pi^2 EI}{L^2}$$

EXAMPLE 16.19

A perfect strut of length L is built-in at its lower end and elastically supported at its upper end against lateral or transverse deflection with a normal spring of spring constant k_t i.e. the restraining force is k_t times the end deflection. Show that the crippling load P_{cr} of the column is given by the characteristic equation,

$$\frac{\tan \alpha L}{\alpha L} = 1 - \frac{P_{cr}}{k_t L} \quad \text{where} \quad \alpha^2 = \frac{P}{EI}$$

Solution: Consider the cantilever compression member AB elastically supported at its upper end and carrying a compressive load P as shown in Figure 16.22. The restraining force at the upper end is $k_t\Delta$. Here Δ is the lateral deflection at the upper end. The external bending moment at any section in the slightly bent shape of the member is

$$M = P(\Delta - y) - k_t\Delta(L - z)$$

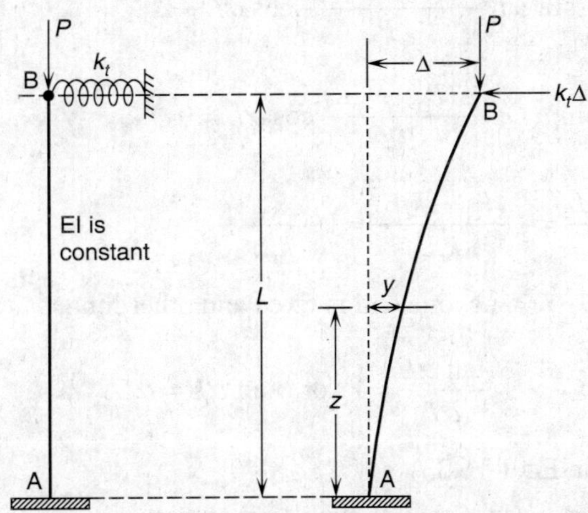

Figure 16.22 Elastically restrained column (linear spring).

The governing differential equation of the elastic or deflection curve is:

$$M = -EI\frac{d^2y}{dz^2} = -[P(\Delta - y) - k_t\Delta(L - z)]$$

or

$$\frac{d^2y}{dz^2} + \left(\frac{P}{EI}\right)y = \frac{d^2y}{dz^2} + \alpha^2 y = \frac{P\Delta - k_t\Delta(L - z)}{EI}$$

The standard solution of this differential equation is

$$y = y_c + y_p = (C \sin \alpha z + D \cos \alpha z) + \frac{P_{cr}\Delta - k_t\Delta(L - z)}{P_{cr}}$$

Thus

$$\frac{dy}{dz} = \alpha C \cos \alpha z - \alpha D \sin \alpha z + \frac{k_t\Delta}{P_{cr}}$$

The prescribed boundary conditions to be satisfied are

(i) At the end $A(z = 0)$ of the member,

$$y = 0 \quad \text{i.e.} \quad 0 = 0 + D + \frac{P_{cr}\Delta - k_t\Delta L}{P_{cr}} \quad \rightarrow \quad D = -\frac{\Delta(P_{cr} - k_t L)}{P_{cr}}$$

$$\frac{dy}{dz} = 0 = C\alpha + \frac{k_t \Delta}{P_{cr}} \rightarrow C = -\frac{k_t \Delta}{P_{cr}\alpha}$$

(ii) At the upper end B ($z = L$) of the member, $y = \Delta$ i.e.

$$\Delta = \left(-\frac{k_t \Delta}{P_{cr}\alpha} \sin \alpha z - \frac{\Delta(P_{cr} - k_t L)}{P_{cr}} \cos \alpha z + \frac{P_{cr}\Delta - k_t \Delta(L - z)}{P_{cr}} \right)_{z=L}$$

$$= -\frac{k_t \Delta}{P_{cr}\alpha} \sin \alpha L - \frac{\Delta(P_{cr} - k_t L)}{P_{cr}} \cos \alpha L + \Delta$$

Therefore,
$$-\frac{k_t \Delta}{P_{cr}\alpha} \sin \alpha L - \frac{\Delta(P_{cr} - k_t L)}{P_{cr}} \cos \alpha L = 0$$

$$\frac{\tan \alpha L}{\alpha L} = -\frac{(P_{cr} - k_t L)}{k_t L} = 1 - \frac{P_{cr}}{k_t L}$$

Special cases: (1) when $k_t \rightarrow \infty$, i.e. one end is fixed and other hinged

$$\frac{\tan \alpha L}{\alpha L} = 1 \quad \text{or} \quad \tan \alpha L = \alpha L$$

Same as obtained earlier in Eq. (16.33).

(2) When $k_t = 0$, the case reduces to that of cantilever column

$$\frac{\tan \alpha L}{\alpha L} = \left(\frac{\sin \alpha L}{\alpha L} \right) \frac{1}{\cos \alpha L} = \infty$$

Thus, valid solution is $\cos \alpha L = 0$ which gives

$$\alpha L = (2n - 1)\frac{\pi}{2} \text{ radians} \quad \text{where} \quad n = 1, 2, 3, \ldots$$

Therefore,
$$\text{for } n = 1, \quad P_{cr} = \frac{\pi^2 EI}{4L^2}$$

16.14 BEAM-COLUMNS OR LATERALLY LOADED COLUMNS

In practice, the structural members are generally acted upon simultaneously by the axial and transverse forces or moments causing bending. Such members are referred to as *beam-columns*. The linearized analyses of elastic beam-columns, where deflections are small to moderate employ the usual linear differential equation for elastic deflection of beams. However, for this procedure the equilibrium equation must be written in the deformed state. The following simple examples would illustrate the procedure.

16.14.1 Hinged Strut Subjected to a Transverse or Lateral Concentrated Load

Consider the hinged strut AB carrying an axial thrust P and subjected to a transverse point load W at its centre as shown in Figure 16.23. The bending moment at a section distant z from the end A is given by

$$M = -Py - \frac{W}{2}z \qquad 0 \le z \le L/2 \tag{16.68}$$

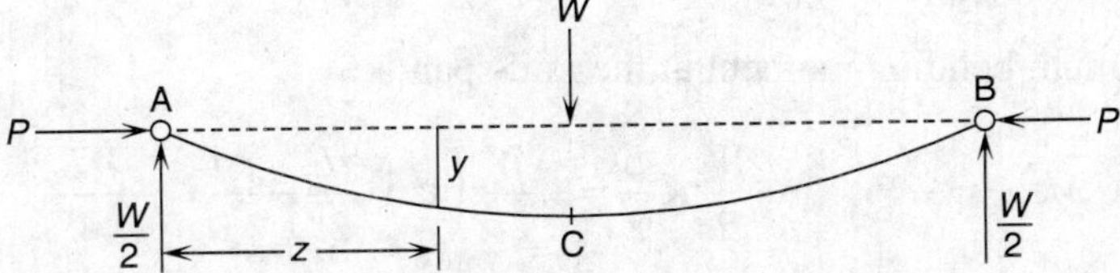

Figure 16.23 Beam-column carrying concentrated load at mid-span.

The governing differential equation of the elastic or deflection curve is:

$$M = -EI\frac{d^2y}{dz^2} = Py + \frac{W}{2}z \quad \text{or} \quad \frac{d^2y}{dz^2} + \frac{P}{EI}y = -\frac{Wz}{2EI}$$

Thus governing differential equation is

$$\frac{d^2y}{dz^2} + \alpha^2 y = -\frac{Wz}{2EI} \tag{16.69}$$

The solution to this differential equation is

$$y = (C\sin\alpha z + D\cos\alpha z) - \frac{Wz}{2P} \tag{16.70}$$

Thus, the slope at any section is,

$$\frac{dy}{dz} = \alpha C\cos\alpha z - \alpha D\sin\alpha z - \frac{W}{2P}$$

The prescribed boundary conditions to be satisfied are:

(i) At the end A ($z = 0$) of the strut,

$$y = 0 \quad \text{i.e.} \quad 0 = 0 + D + 0 \quad \rightarrow D = 0$$

(ii) At the centre C($z = L/2$), due to symmetry, $\frac{dy}{dz} = 0$

Therefore, $\quad 0 = \alpha C\cos\frac{\alpha L}{2} - \alpha D\sin\frac{\alpha L}{2} - \frac{W}{2P} \quad C \rightarrow \frac{W}{2\alpha P}\sec\frac{\alpha L}{2}$

Thus for the region $0 < z < L/2$

$$y = \frac{W}{2\alpha P}\sec\frac{\alpha L}{2}\sin\alpha z - \frac{Wz}{2P} = \frac{W}{2\alpha P}\left(\sec\frac{\alpha L}{2}\sin\alpha z - \alpha z\right)$$

The maximum deflection occurs at the centre, i.e. $z = L/2$,

$$y_{\max} = \frac{W}{2\alpha P}\left(\sec\frac{\alpha L}{2}\sin\frac{\alpha L}{2} - \frac{\alpha L}{2}\right) = \frac{W}{2\alpha P}\left(\tan\frac{\alpha L}{2} - \frac{\alpha L}{2}\right)$$

The absolute maximum bending moment at the mid-span is

$$M_{\max} = -Py_{\max} - \frac{W}{2}\times\frac{L}{2} = -\frac{W}{2\alpha}\left(\tan\frac{\alpha L}{2} - \frac{\alpha L}{2}\right) - \frac{WL}{4}$$

$$= -\frac{W}{2\alpha}\tan\frac{\alpha L}{2} \tag{16.71}$$

At buckling deflection $y_{\max}$ and moment $M_{\max}$ become infinite, i.e.

$$\tan\frac{\alpha L}{2} \to \infty \quad \text{or} \quad \frac{\alpha L}{2} = \frac{n\pi}{2} \qquad n = 1, 3, 5, \ldots$$

Setting $n = 1$, the critical buckling load is obtained

$$\alpha L = \pi \quad \text{or} \quad \sqrt{\frac{P_{cr}}{EI}}\,L = \pi \;\rightarrow P_{cr} = \frac{\pi^2 EI}{L^2}$$

An approximate solution to this problem can be obtained by expanding $\tan\theta\left(=\frac{\alpha L}{2}\right)$ in to Taylor series:

$$\tan\theta = \theta + \frac{\theta^3}{3} + \frac{2\theta^5}{15} + \frac{17\theta^7}{315} + \ldots = \ldots \quad \text{or} \quad \tan\theta = \theta + \frac{\theta^3}{3} \quad \text{where } \theta \text{ is small}$$

where $$\theta = \frac{\alpha L}{2} = \frac{L}{2}\sqrt{\frac{P}{EI}} \quad \text{and} \quad \theta^2 = \frac{PL^2}{4EI} = \frac{\pi^2}{4}\frac{P}{P_e} = 2.4674\,\frac{P}{P_e}$$

$$y_{\max} = \frac{W}{2\alpha P}(\tan\theta - \theta) = \frac{W}{2\alpha P}\left(\frac{\theta^3}{3} + \frac{2\theta^5}{15} + \frac{17\theta^7}{315} + \ldots\right)$$

$$= \frac{W}{2\alpha P}\frac{\theta^3}{3}\left(1 + \frac{2\theta^2}{5} + \frac{17\theta^4}{105} + \ldots\right) = \frac{W}{2\alpha P}\left(\frac{\alpha^3 L^3}{3\times 8}\right)\left(1 + \frac{2\theta^2}{5} + \frac{17\theta^4}{105} + \ldots\right)$$

$$= \frac{WL^3}{48EI}\left(1 + \frac{2\theta^2}{5} + \frac{17\theta^4}{105} + \ldots\right)$$

By substituting the value of θ^2 and simplifying

$$y_{\max} = \frac{WL^3}{48EI}\left[1 + 0.9870\left(\frac{P}{P_e}\right) + 0.9857\left(\frac{P}{P_e}\right)^2 + \dots\right]$$

$$\approx \frac{WL^3}{48EI}\left[1 + \left(\frac{P}{P_e}\right) + \left(\frac{P}{P_e}\right)^2 + \left(\frac{P}{P_e}\right)^3 + \dots\right] = \frac{WL^3}{48EI}\left(\frac{1}{1 - P/P_e}\right) \qquad (16.72)$$

The bracketed quantity gives the deflection magnification factor caused by the axial force P. When the axial force reaches P_{cr}, the deflection becomes infinite. This magnification factor is same for any kind of transverse loading. The approximate bending moment can be calculated directly as

$$M_{\max} = -\frac{W}{2\alpha}\left[\frac{\alpha L}{2} + \frac{1}{3}\left(\frac{\alpha L}{2}\right)^3 + \dots\right]$$

$$= -\left[\frac{WL}{4} + \frac{WL^3}{48}\alpha^2\right] = -\left[\frac{WL}{4} + \frac{WL^3}{48EI}P_{cr}\right] \qquad (16.73)$$

where the first term is due to transverse loading, and second to axial force.

16.14.2 Hinged Strut Subjected to a Transverse Uniformly Distributed Load

Consider the hinged strut AB carrying an axial thrust P and subjected to a transverse uniformly distributed load w per unit length as shown in Figure 16.24. The bending moment at a section distant z from the end A is given by

$$M = Py + \frac{wL}{2}z - \frac{wz^2}{2} = Py + \frac{wz(L-z)}{2}$$

The governing differential equation of the deflection curve is:

$$\frac{d^2y}{dz^2} + \frac{P}{EI}y = -\frac{wz(L-z)}{2EI} \qquad (16.74)$$

The solution to the differential equation is

$$y = (C\sin\alpha z + D\cos\alpha z) - \frac{wz(L-z)}{2P} - \frac{wEI}{P^2}$$

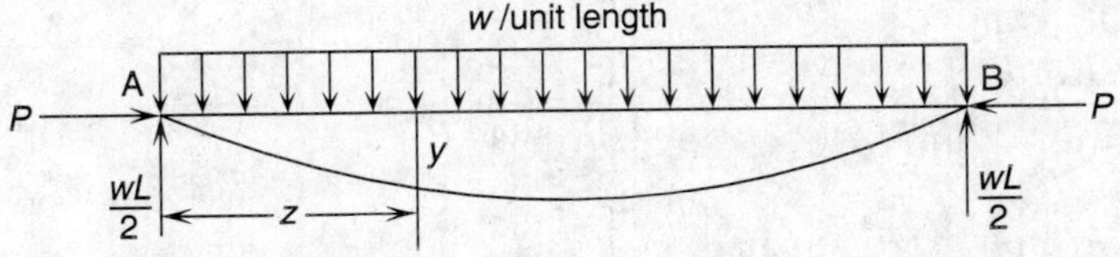

Figure 16.24 Beam-column carrying uniformly distributed load.

The slope at any section is,

$$\frac{dy}{dx} = \alpha C \cos \alpha z - \alpha D \sin \alpha z + \frac{wz}{P} - \frac{wL}{2P}$$

The prescribed boundary conditions are:
(i) At the end A($z = 0$) of the strut,

$$y = 0 \quad \text{i.e.} \quad 0 = 0 + D + 0 - \frac{wEI}{P^2} \quad \rightarrow D = \frac{wEI}{P^2}$$

(ii) At the centre C($z = L/2$), due to symmetry, $\dfrac{dy}{dz} = 0$

Therefore, $$0 = \alpha C \cos \frac{\alpha L}{2} - \alpha D \sin \frac{\alpha L}{2} + \frac{wL}{2P} - \frac{wL}{2P} \quad C \rightarrow D \tan \frac{\alpha L}{2}$$

For the region $0 < z < L/2$, substitute values of C and D

$$y = \left(\frac{wEI}{P^2} \tan \frac{\alpha L}{2} \sin \alpha z + \frac{wEI}{P^2} \cos \alpha z \right) - \frac{wz(L - z)}{2P} - \frac{wEI}{P^2}$$

The maximum deflection occurs at the centre, i.e. $z = L/2$,

$$y_{\max} = \left(\frac{wEI}{P^2} \tan \frac{\alpha L}{2} \sin \frac{\alpha L}{2} + \frac{wEI}{P^2} \cos \frac{\alpha L}{2} \right) - \frac{wL^2}{8P} - \frac{wEI}{P^2}$$

$$= \frac{wEI}{P^2} \left(\sin^2 \frac{\alpha L}{2} + \cos^2 \frac{\alpha L}{2} \right) \frac{1}{\cos \frac{\alpha L}{2}} - \frac{wL^2}{8P} - \frac{wEI}{P^2}$$

$$= \frac{wEI}{P^2} \left(\sec \frac{\alpha L}{2} - 1 \right) - \frac{wL^2}{8P} \tag{16.75}$$

An approximate solution to this problem can be obtained by expanding $\sec \theta \left(= \dfrac{\alpha L}{2} \right)$ into the Taylor series:

$$\sec \theta = 1 + \frac{\theta^2}{2!} + \frac{5\theta^4}{4!} + \frac{61\theta^6}{6!} + \ldots \quad \text{or} \quad \sec \theta = 1 + \frac{\theta^2}{2!} + \frac{5\theta^4}{4!} \quad \text{when } \theta \text{ is small}$$

where $$\theta = \frac{\alpha L}{2} = \frac{L}{2} \sqrt{\frac{P}{EI}} \quad \text{and} \quad \theta^2 = \frac{PL^2}{4EI} = \frac{\pi^2}{4} \frac{P}{P_e} = 2.4674 \frac{P}{P_e}$$

$$y_{\max} = \frac{wEI}{P^2} \left[\frac{\theta^2}{2} + \frac{5\theta^4}{24} + \frac{61\theta^6}{720} + \ldots \right] - \frac{wL^2}{8P}$$

$$= \frac{wEI}{P^2}\left(\frac{\theta^2}{2}\right)\left[1 + \frac{5\theta^2}{12} + \frac{61\theta^4}{360} + \ldots\right] - \frac{wL^2}{8P}$$

$$= \frac{wL^2}{8P}\left[1 + \frac{5\theta^2}{12} + \frac{61\theta^4}{360} + \ldots\right] - \frac{wL^2}{8P} = \frac{wL^2}{8P}\left[\frac{5\theta^2}{12} + \frac{61\theta^4}{360} + \ldots\right]$$

$$= \frac{wL^2}{8P}\left(\frac{5\theta^2}{12}\right)\left[1 + \frac{12}{5} \times \frac{61\theta^2}{360} + \ldots\right] = \frac{5wL^4}{384EI}\left[1 + \frac{12}{5} \times \frac{61\theta^2}{360} + \ldots\right]$$

As in the previous case, it can be shown that

$$y_{\max} = \frac{5wL^4}{384EI}\left[1 + 1.0034\left(\frac{P}{P_e}\right) + 1.032\left(\frac{P}{P_e}\right)^2 + \ldots\right]$$

$$\approx \frac{5wL^4}{384EI}\left[1 + \left(\frac{P}{P_e}\right) + \left(\frac{P}{P_e}\right)^2 + \left(\frac{P}{P_e}\right)^3 + \ldots\right]$$

$$= \frac{5wL^4}{384EI}\left(\frac{1}{1 - P/P_e}\right) \tag{16.76}$$

The bracketed quantity represents the deflection magnification factor caused by the axial force P. The maximum bending moment at the centre can be calculated directly

$$M_{\max} = -Py_{\max} - \frac{wL^2}{8}$$

$$= -P\left[\frac{wEI}{P^2}\left(\sec\frac{\alpha L}{2} - 1\right) - \frac{wL^2}{8P}\right] - \frac{wL^2}{8} = -\frac{wEI}{P}\left(\sec\frac{\alpha L}{2} - 1\right)$$

$$= -\frac{wEI}{P}\left[1 + \frac{1}{2}\left(\frac{\alpha L}{2}\right)^2 + \frac{5}{24}\left(\frac{\alpha L}{2}\right)^4 + \ldots - 1\right]$$

$$= -\frac{w}{\alpha^2}\left[\frac{\alpha^2 L^2}{8} + \frac{5\alpha^4 L^4}{384}\right] = -\left[\frac{wL^2}{8} + \frac{5wL^4}{384EI}P_{cr}\right] \tag{16.77}$$

16.15 PROBLEMS

16.1 A solid circular compression member of 60 mm diameter is to be replaced by a hollow circular section of the same material. Determine the size of the hollow section if its internal diameter is 0.75 times the external diameter.

[**Ans.** d_e = 65.985 mm and d_i = 49.489 mm]

16.2 Two circular columns one hollow and other solid, have the same length, cross-sectional area and end conditions, are made of the same material. If the internal diameter of the hollow column is β times its external diameter, show that the ratio of the buckling strengths of two columns is

$$\frac{P_h}{P_s} = \frac{1+\beta^2}{1-\beta^2}$$

16.3 A 4 m long circular bar when freely supported gives a mid-span deflection of 12.5 mm under a concentrated load of 100 N acting at its centre. If the same bar is used as a strut with pin-jointed ends, determine the critical load it can support.

[**Ans.** P_e = 12.85 kN]

16.4 A hollow tube of length L with outer and inner diameters of d_o and d_i, respectively, was found to extend by Δ under an axial tensile load of W. Show that with a factor of safety of m, the safe load for the tube when it is used as a strut with hinged ends is

$$P_w = \frac{\pi^2 (d_o^2 + d_i^2) W}{16 m \Delta L}$$

16.5 A 5 m long hollow tube with outer and inner diameters of 48 mm and 30 mm, respectively, was found to extend by 6.5 mm under an axial tensile load of 90 kN. Determine: (a) the buckling load for the tube when it is used as a strut with hinged ends and (b) safe compressive load with a factor of safety of 4.5.

[**Ans.** P_e = 5.472 kN and P_w = 1.216 kN]

16.6 A simply supported bar of length L subjected to a uniformly distributed load of intensity w over its entire span deflects Δ at its centre. Show that with a factor of safety of m, the safe load for the bar when it is used as a strut with one end fixed and the other hinged end is

$$P_w = \frac{5\pi^2 w L^2}{192 m \Delta}$$

16.7 Determine the buckling loads if the above beam is used as a column with both ends fixed.

16.8 A round steel hinged strut of length L and diameter d is subjected to a gradually increasing axial load. Assuming that Euler's formula is applicable to this case, show that the maximum central deflection of the strut before the steel passes the yield point stress σ_y is

$$y_m = \left(\frac{16\sigma_y L^2}{\pi^2 d^2 E} - 1\right)\frac{d}{8}$$

where E is the modulus of elasticity of the material.

16.9 A 2.25 m long straight round steel rod of 16 mm diameter is subjected to a gradually increasing axial compressive load until it buckles. Both the ends of the bar may be taken as hinged. Using Euler's formula, estimate the maximum central deflection before the steel passes the yield point stress of 250 MPa. The modulus of elasticity of the material is 200 GPa.

[**Ans.** y_m = 74.33 mm]

16.10 A bar of uniform cross-section is hinged at both the ends in such a way that its length is unaltered during buckling. If the bar is heated uniformly to a temperature t. Show that the bar will buckle when temperature attains a value of $t_{cr} = \frac{1}{\lambda}\left(\frac{\pi r}{L}\right)^2$, where λ is the coefficient of thermal expansion, r the least radius of gyration and L is the length.

16.11 A 50 × 100 mm rectangular timber strut with fixed ends is used to support an axial compressive load. Determine: (a) the minimum length at which the Euler's formula can be applied and (b) the safe load the strut can carry with a factor of safety of 2, for a length of 2.5 m. The stress at proportional limit and modulus of elasticity are $\sigma_{pl} = 30$ MPa and $E = 10$ GPa, respectively.

[**Ans.** $L = 1.66$ m and $P = 32.9$ kN]

16.12 Two 250 × 10 mm plates are welded one to each flange of two rolled steel channel sections ISJC 200 to form a composite column section as shown in Figure 16.25. Determine: (a) the safe load the column can carry with fixed-ends conditions over a length of 6 m with a factor of safety 3.5 and (b) if the maximum permissible compressive stress is 80 MPa, determine the maximum eccentricity of a 400 kN load supported by the above column from the Y-axis of the column. The load line lies in the vertical pane through the X-axis. The properties of one channel are:

$$b = 70 \text{ mm : Area } A = 1.777 \times 10^3 \text{ mm}^2; I_{xx} = 1.1612 \times 10^7 \text{ mm}^4;$$

$$I_{yy} = 8.42 \times 10^5 \text{ mm}^4 \text{ and distance of C.G. from back of channel} = 19.7 \text{ mm}$$

The constants for Rankine's formula are 320 GPa and 1/7500. The modulus of elasticity is 200 GPa.

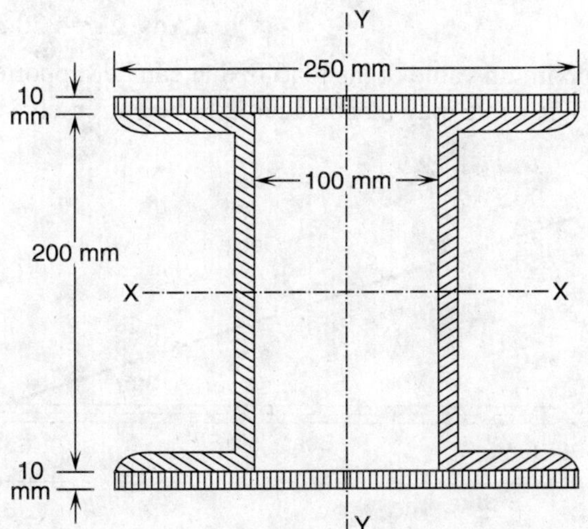

Figure 16.25 Built-up section of Problem 16.12.

[**Ans.** (a) $P = 63.68$ kN and $e = 29.5$ mm]

16.13 (a) Two rolled steel channels ISLC 350 are latticed to form the cross-section of a column as shown in Figure 16.26. How far apart should these channels be placed back to back, so that the column may carry the maximum load? The properties of one channel are:

$b = 100$ mm : Area $A = 4947$ mm^2 ; $I_{xx} = 9.3126 \times 10^7$ mm^4 ;

$I_{yy} = 3.946 \times 10^6$ mm^4 and distance of C.G. from the back of channel = 24.1 mm

(b) The channels of the above column are spaced 235 mm apart back to back. Determine the safe load the column can carry with fixed-ends conditions over a length of 7.5 m, taking factor of safety 3. The modulus of elasticity is 200 GPa.

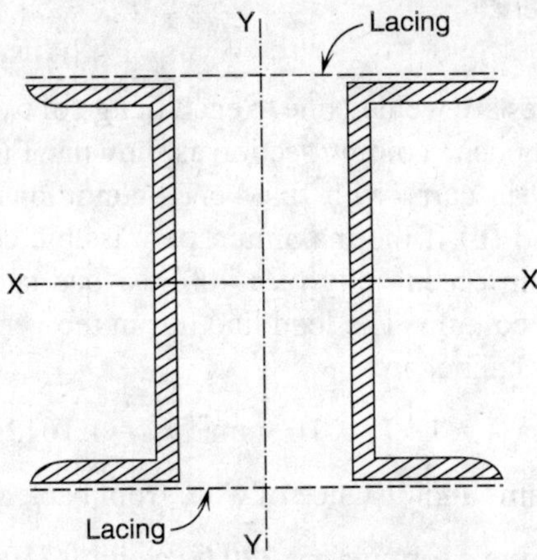

Figure 16.26 Cross-section of Problem 16.13.

[**Ans.** $2x = 220.33$ and $P_w = 2178.65$ kN]

16.14 Determine the maximum value of the load W that can be supported by the structure shown in Figure 16.27. The cable may be considered to be extremely strong in tension.

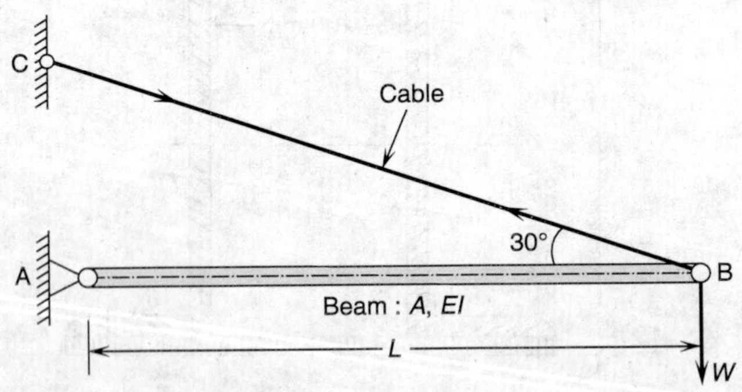

Figure 16.27 Cable supported beam of Problem 16.14.

$\left[\textbf{Ans. } \sigma_o A/\sqrt{3} \text{ or } \pi^2 EI/\sqrt{3}L^2 \text{ whichever is smaller}\right]$

16.15 A straight rod of length L having uniform cross-sectional area A and flexural stiffness EI is heated so that its temperature varies linearly from $t/3$ at one end to t at the other end. One end is pinned to a rigid foundation; the other end is pin-jointed so that it can slide in the direction of the length of the bar, the thermal expansion the bar is resisted by a compression spring of stiffness k. If there is no load in the spring when $t = 0$, show that the rod buckles in flexure when

$$t = \frac{3\pi^2 I}{2\lambda LA}\left(1 + \frac{EA}{kL}\right)$$

where λ is the coefficient of linear expansion.

16.16 A simply supported built-up steel beam of cross-section shown in Figure 16.28 is subjected to a uniformly distributed load of intensity 50 kN/m over its entire span. Determine: (a) the length of the beam when the deflection at its centre is 10 mm, and (b) safe load if the above beam is used as a column with both ends fixed. The factor of safety is 5. The elastic modulus of the material of the column is 210 GPa.

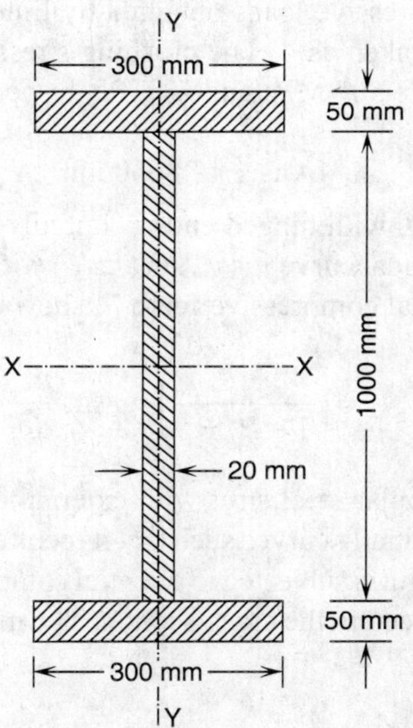

Figure 16.28 Built-up beam section of Problem 16.16.

[**Ans.** $L = 13.38$ m and $P_w = 2089.49$ kN]

16.17 A mild steel rod of 25 × 25 mm cross-section is to be used as a compression member with one end fixed and the other hinged to carry a service load of 50 kN. Determine maximum length of the rod which can be used when a factor of safety of 3 is allowed. The constants for Rankine's formula are 320 GPa and 1/7500.

[**Ans.** $L = 510.29$ mm]

16.18 A 2 m long hollow cylindrical cast-iron column with both ends hinged is to support an axial compressive load of 300 kN. Use Rankine's formula and adopt a factor of safety of 4, to determine the size of the column with internal diameter being 0.8 times the external diameter. The constants for Rankine's formula are 550 GPa and 1/1600.

[**Ans.** d_e = 135.58 mm and d_i = 108.47 mm]

16.19 A 6 m long hollow cylindrical cast-iron column with hinged ends having outer diameter of 150 mm and thickness of 20 mm, is subjected to an axial compressive load. (a) Compare the crushing loads obtained by Euler's and Rankine's formulae, and (b) If the elastic limit for the material of the strut is 200 MPa, determine the length of strut below which Euler's formula ceases to apply. The modulus of elasticity of the material of the column is 80 GPa. The constants for Rankine's formula are 550 GPa and 1/1600.

[**Hint:** r^2 = 2162.5 mm^2 and assume $\sigma_{pl} = \sigma_c$ = 550 MPa]

[**Ans.** P_e = 387.4 kN; P_r = 393.9 kN and L = 176.1 mm]

16.20 A 4 m long hollow cylindrical cast-iron column with hinged ends having outer and inner diameters of 300 mm and 220 mm, respectively, is subjected to an axial compressive load. Determine the safe compressive loads obtained by Euler's and Rankine's formulae. The factor of safety may be taken as 4. The crushing stress and modulus of elasticity of the material of the column are 567 MPa and 70 GPa, respectively. The Rankine's constant is 1/1600.

[**Ans.** r^2 = 8650 mm^2; P_{ew} = 3050 kN and P_{rw} = 2148 kN]

16.21 An elastic rod of length L with hinged ends is initially curved such that its centre line is represented by the sinusoidal curve $y = y_m \sin(\pi z/L)$ with a maximum deviation of y_m. The rod is subjected to an axial compressive force P, show that the total deflection is

$$y = \left(\frac{1}{1 - P/P_{cr}}\right) y_m \sin\frac{\pi z}{L}$$

16.22 A 2.5 m long hinged tubular steel strut with external and internal diameters of 100 and 88 mm, respectively, is initially curved such that its centre line is sinusoidal with a maximum deviation of 5 mm. The strut is subjected to a central compressive load of 12.5 kN. Determine the maximum stress induced in the cross-section. The modulus of elasticity of the material of the strut is 210 GPa.

[**Ans.** $\sigma_{\max}$ = 8.675 MPa]

16.23 Show that the equation for the bending moment for an elastic beam-column of constant EI carrying uniformly distributed load of intensity w_0 is

$$M = -\frac{w_0}{\alpha^2}\left(\frac{\cos\alpha L - 1}{\sin\alpha L}\sin\alpha z - \cos\alpha L + 1\right)$$

16.24 Show that the equation of elastic curve for an hinged strut of constant EI shown in Figure 16.29 subjected to a transverse distributed load of sinusoidal variation $w = w_0 \sin(\pi z/L)$ is

$$y = \left(\frac{1}{1 - P/P_{cr}}\right)\left(\frac{w_0 L^4}{\pi^4 EI}\right)\sin\frac{\pi z}{L}$$

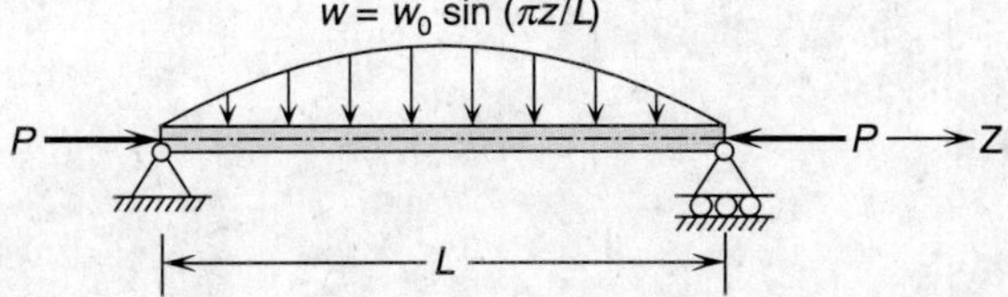

Figure 16.29 Beam-column subjected to sinusoidal load.

16.25 Show that for the hinged strut of length L, which is simultaneously subjected to moment M_o at each end causing single curvature and axial force P, the maximum deflection is

$$y_{\max} = \frac{M_o}{P}\left(\sec\frac{\alpha L}{2} - 1\right)$$

and the absolute maximum bending moment is $M_{\max} = M_o \sec(\alpha L/2)$.

16.26 Show that the equation for the bending moment for an elastic beam-column of constant EI, subjected to a load uniformly increasing from zero at left end to an intensity w_0 at the right end as shown in Figure 16.30 is

$$M = -\frac{w_0}{\alpha^2}\frac{\sin\alpha z}{\sin\alpha L}$$

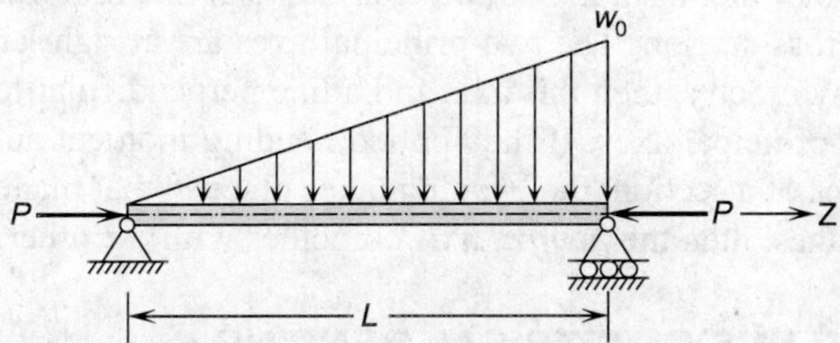

Figure 16.30 Beam-column subjected to uniformly varying load.

16.27 Show that for the hinged strut of length L, simultaneously loaded by an end moment M_o and axial force P, the deflection is

$$y = \frac{M_o}{P}\left(\frac{\sin\alpha z}{\sin\alpha L} - \frac{z}{L}\right)$$

and the bending moment is $M = -M_o \sin\alpha z / \sin\alpha L$.

Chapter 17

Unsymmetrical (Skew) Bending

17.1 INTRODUCTION

All the cross-sections have two axes passing through the centroid such that the moment of inertia of the section about one of them is a maximum and the other is a minimum. These are called the *principal axes* of the cross-section. The two principal axes are at right angles to each other. If a section has an axis of symmetry, then this axis and a line perpendicular to it and passing through the centroid constitute principal axes. If the applied bending moment acts in a plane containing one of the principal axes of a section, the bending takes place in that plane. If a moment M acts in one of the principal planes, then the *neutral axis* coincides with the other principal axis.

17.2 OBLIQUE OR UNSYMMETRICAL BENDING

In the cases where the plane of forces or bending moment does not coincide with the principal axis of the section, the bending is termed as *oblique* or *unsymmetrical bending*. Sometimes it is also called *skew* or *inclined bending*. Generally, it occurs when the beam cross-section is unsymmetrical with respect to the plane of bending. However, the assumption that the line of action of the loads passes through the point called *shear centre of the cross-section* so that the *twisting* is not developed.

Figure 17.1(a) illustrates an unsymmetrical section with its centroid at the point O and subjected to a moment M in the vertical plane containing the O–Y axis. Since the section is not symmetrical and thus the bending moment does not act in any plane of symmetry, i.e. it is a case of oblique bending. The principal axes of the cross-section are represented by the lines U–U and V–V at right angles to each other.

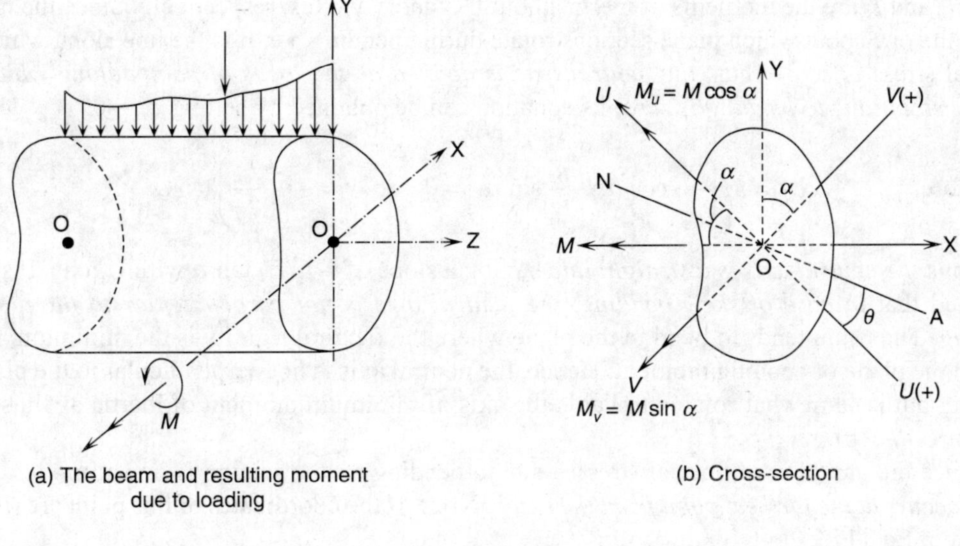

(a) The beam and resulting moment due to loading

(b) Cross-section

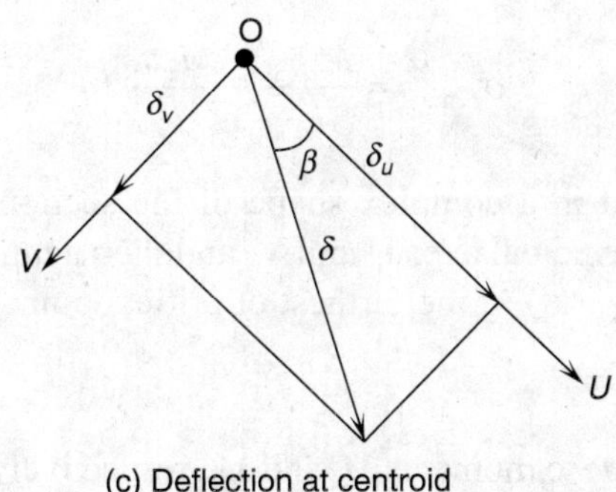

(c) Deflection at centroid

Figure 17.1 Principal axes and principal planes of a section.

The case of inclined loading or skew bending can most conveniently be treated as the simultaneous bending of the beam in two principal planes Z–U and Z–V shown in Figure 17.1(b). This can be accomplished by resolving the bending moment into two component moments about the U- and V-axes,

$$M_u = M \cos \alpha \text{ and } M_v = M \sin \alpha$$

or

$$\tan \alpha = \frac{M_v}{M_u} \tag{17.1}$$

where α *is the angle of inclination of the load with the V-axis*. The normal bending stress σ_b at a point having coordinates u and v is given by the sum of stresses due to moments M_u and M_v considered to be acting independently, i.e.

$$\sigma_b = \frac{M_u v}{I_u} + \frac{M_v u}{I_v} = M\left(\frac{v}{I_u} \cos \alpha + \frac{u}{I_v} \sin \alpha\right) \tag{17.2}$$

where I_u and I_v are the moments of inertia about UU and VV axes, respectively. Since the neutral axis is the one about which plane sections rotate during bending, i.e. it is the line along which the flexural stress is zero. Thus, *the neutral axis is defined as the locus of points along which the normal stress due to bending is zero*, its equation can be obtained by setting $\sigma_b = 0$.

Therefore, $$\sigma_b = M\left(\frac{v}{I_u}\cos\alpha + \frac{u}{I_v}\sin\alpha\right) = 0 \quad \text{or} \quad v = -u\left(\frac{I_u}{I_v}\right)\tan\alpha \tag{17.3}$$

Thus *the neutral axis is a straight line* having a slope of $(-I_u/I_v)$ tan α with U-axis. It should be noted that *in case of skew bending, the neutral axis is not perpendicular to the plane of bending*. The beam tends to bend in the plane where the flexural rigidity is the minimum, rather than in the plane of bending moment. Hence, the neutral axis is not perpendicular to the plane of bending but is somewhat rotated towards the axis of minimum moment of inertia as illustrated in Figure 17.1(b).

Since the variation of normal stresses due to bending on the section is linear, *the maximum stress occurs at the farthest point from the neutral axis*. If the coordinates of this point are (u_1, v_1), then from Eq. (17.2),

$$\sigma_{b,\max} = \frac{M_u v_1}{I_u} + \frac{M_v u_1}{I_v} \tag{17.4}$$

To locate the critical point in a complex shape of the section, it is convenient to draw the section to scale and mark the principal axes U and V, and then neutral axis. Then using a ruler and a set square, the coordinates (u_1, v_1) of the farthest or critical point can be recorded directly.

17.2.1 Deflections

If δ_x and δ_y are the deflections due to moments M_u and M_v, respectively as illustrated in Figure 17.1(c) the resultant deflection is given by

$$\delta = \sqrt{\delta_u^2 + \delta_v^2} \tag{17.5}$$

The inclination β with the UU axis is given by

$$\beta = \tan^{-1}\frac{\delta_v}{\delta_u} \tag{17.6}$$

17.2.2 Symmetrical Section

For a beam having symmetrical section, the principal axes U and V coincide with X- and Y-axes.

EXAMPLE 17.1

A simply supported wooden rectangular beam of size $a \times 2a$ (deep), is subjected to forces P and $2P$, in the X- and Y-directions, respectively, as shown in Figure 17.2(a). Determine the maximum stress developed in the beam section.

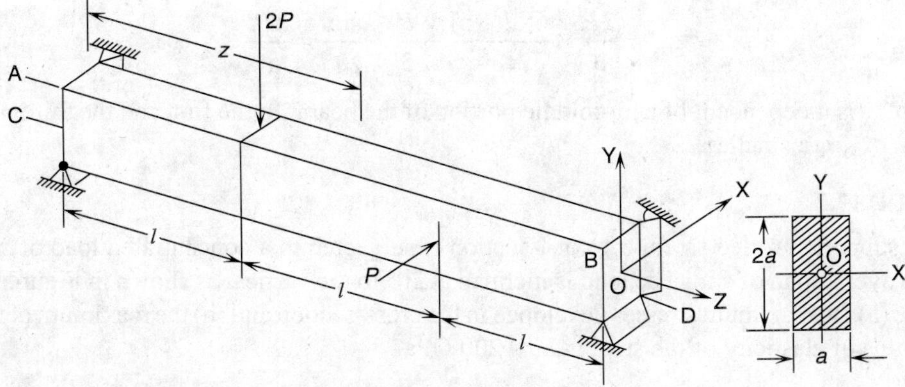

(a) The simply supported beam and the load system

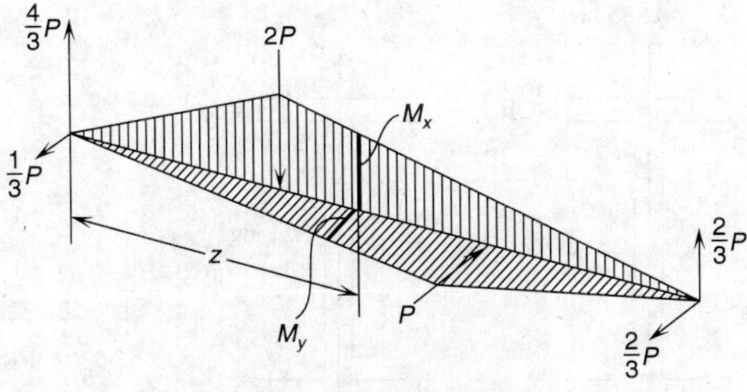

(b) Bending moment diagrams

Figure 17.2 Simply supported rectangular beam of Example 17.1.

Solution: For the given symmetrical section with total depth = $2a$ and width = a, the principal axes U and V coincide with X- and Y-axes. Thus

$$Z_x = \frac{a(2a)^2}{6} = \frac{4a^3}{6} \quad \text{and} \quad Z_y = \frac{(2a)a^2}{6} = \frac{2a^3}{6}$$

The bending moment diagrams are shown in Figure 17.2(b).

As the external forces are applied along the principal axes of the section, it is convenient to consider separately the bending due to two forces. The critical points are located at the edge AB where the maximum compressive stresses are summed up or at the edge CD where the maximum tensile stresses are summed up. Consider a section at a distance z from the left support in the middle portion of the beam, the bending moments are

$$M_x = \frac{2P}{3}(3l - z) \quad \text{and} \quad M_y = \frac{P}{3}(z)$$

For a point at the edge CD

$$\sigma_{\max} = \frac{2P}{3}\frac{(3l - z)}{Z_x} + \frac{P}{3}\frac{z}{Z_y}$$

Therefore, $$\sigma_{max} = \frac{2P}{3} \times \frac{6(3l - z)}{4a^3} + \frac{P}{3} \times \frac{6z}{2a^3} = \frac{3Pl}{a^3}$$

The σ_{max} is independent of z for middle portion of the beam. In the first and the third portions, the values σ_{max} are smaller.

EXAMPLE 17.2

A simply supported rolled steel beam of I-section is subjected to a concentrated load of 50 kN at the centre over a span of 4 m. The load is inclined at 20° to the vertical as shown in Figure 17.3(a). Determine (a) the maximum stresses developed in the cross-section and (b) the maximum deflection. The modulus of elasticity of the steel is $E_s = 200$ GPa.

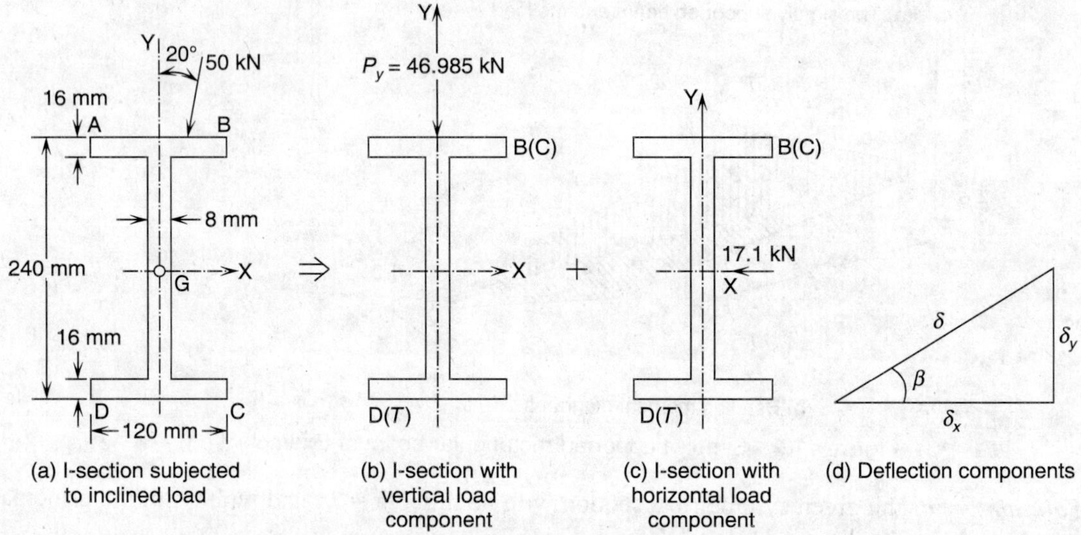

Figure 17.3 Unsymmetrical bending of I-section of Example 17.2.

Solution: For the given section,

$$I_u = I_x = \frac{1}{12}[120 \times (240)^3 - 112 \times (208)^3] = 5.42502 \times 10^7 \text{ mm}^4$$

$$I_v = I_y = \frac{1}{12}[2 \times (16 \times 120^3) + 208 \times 8^3] = 4.616875 \times 10^6 \text{ mm}^4$$

The components of the load in the X- and Y-directions are,

$$P_x = 50 \times \sin 20° = 17.1 \text{ kN} \quad \text{and} \quad P_y = 50 \times \cos 20° = 46.985 \text{ kN}$$

The bending moments due to these load components are,

$$M_x = \frac{P_y L}{4} = \frac{46.985 \times 4}{4} = 46.985 \text{ kN.m}$$

$$M_y = \frac{P_x L}{4} = \frac{17.1 \times 4}{4} = 17.1 \text{ kN.m}$$

The points of the maximum compressive stresses are located at the point B where the maximum compressive stresses are summed up or at the edge D where the maximum tensile stresses are summed up as shown in Figures 17.3(b) and (c). Thus, the maximum stress at point B (compression) and the maximum stress at point D (tension) are

$$\sigma_{b,\max} = \frac{(46.985 \times 10^6) \times 120}{5.42502 \times 10^7} + \frac{(17.1 \times 10^6) \times 60}{4.616875 \times 10^6} = 326.16 \text{ MPa}$$

The central vertical deflection is

$$\delta_y = \frac{P_y L^3}{48 E I_x} = \frac{(46.985 \times 10^3) \times (4 \times 10^3)^3}{48 \times (200 \times 10^3) \times (5.42502 \times 10^7)} = 5.774 \text{ mm}$$

The central horizontal deflection is

$$\delta_x = \frac{P_x L^3}{48 E I_y} = \frac{(17.1 \times 10^3) \times (4 \times 10^3)^3}{48 \times (200 \times 10^3) \times (4.616875 \times 10^6)} = 24.69 \text{ mm}$$

Therefore, the resultant deflection is given by

$$\delta = \sqrt{\delta_x^2 + \delta_y^2} = \sqrt{(5.774)^2 + (24.69)^2} = 25.356 \text{ mm}$$

The inclination β with the X-axis as shown in Figure 17.3(c) is given by

$$\beta = \tan^{-1} \frac{\delta_y}{\delta_x} = \tan^{-1} \frac{5.774}{24.69} = 13.16°$$

EXAMPLE 17.3

A cantilever beam of symmetrical cross-section is bent by a couple in the axial plane I-I as shown in Figure 17.4(a). Determine the equation of the curve traced by the end of the cantilever when the angle of inclination with vertical α, defining the plane of bending moment, varies from 0 to 2π.

Solution: The components of the bending moment M acting in the principal planes U and V are $M_u = M \cos \alpha$ and $M_v = M \sin \alpha$, respectively, as shown in Figure 17.4(b). The deflection components δ_u and δ_v at the end of cantilever due to moments M_u and M_v, respectively, are given by:

$$\delta_u = \frac{M_u L^2}{2 E I_v} = \frac{(M \cos \alpha) L^2}{2 E I_v}$$

and

$$\delta_v = \frac{M_v L^2}{2 E I_u} = \frac{(M \sin \alpha) L^2}{2 E I_v}$$

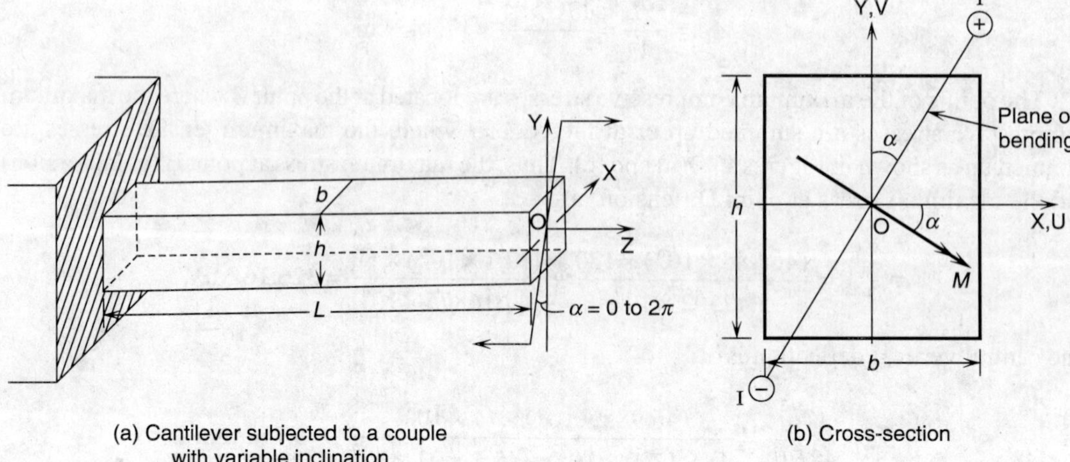

(a) Cantilever subjected to a couple with variable inclination

(b) Cross-section

Figure 17.4 Cantilever of a symmetrical section bend by a couple.

Eliminate the variable angle of inclination α,

$$\frac{\delta_u^2}{\left(\dfrac{ML^2}{2EI_v}\right)^2}+\frac{\delta_v^2}{\left(\dfrac{ML^2}{2EI_u}\right)^2}=\cos^2\alpha+\sin^2\alpha=1$$

Hence, the end of the cantilever traces an ellipse with the semi-axes as $\dfrac{ML^2}{2EI_v}$ and $\dfrac{ML^2}{2EI_u}$.

EXAMPLE 17.4

A cantilever beam of I-section is subjected to two concentrated load of 6 kN and 2 kN inclined at 30° and 45° to the vertical or V-axis as shown in Figure 17.5(a). Locate the neutral axis and determine the maximum stresses developed in the cross-section.

Solution: The geometric properties of I-section (see Figure 17.5(b)),

$$I_u=I_x=\frac{1}{12}[120\times(160)^2-104\times(128)^3]=2.27847\times10^7\ \text{mm}^4$$

$$I_v=I_y=2\times\frac{1}{12}(16\times120^3)+\frac{1}{12}(128\times6^3)=4.65170\times10^6\ \text{mm}^4$$

The maximum moments occur at the fixed support of the cantilever. The moments causing tension on positive sides of the axes are considered positive. Thus

M_u due to vertical or Y-direction component of the loads,

$$M_u=(6\times\cos30°)\times1.5+(2\times\cos45°)\times2.5=11.33\ \text{kN and}$$

M_v due to horizontal or X-direction component of the loads,

$$M_v=(6\times\sin30°)\times1.5-(2\times\sin45°)\times2.5=0.9645\ \text{kN}$$

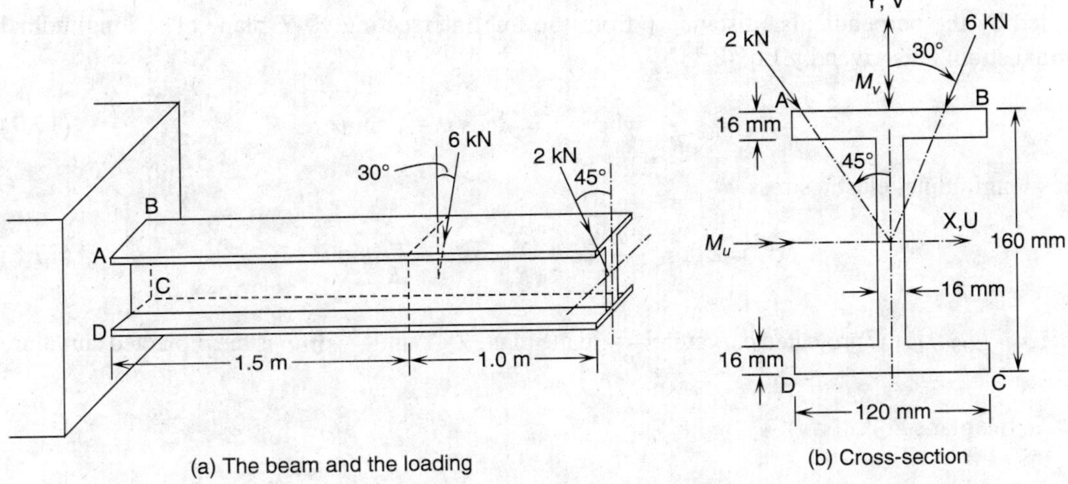

Figure 17.5 Unsymmetrical bending of cantilever beam of Example 17.4.

For the given section, the inclination θ of the neutral axis with the U-U axis is

$$\tan\theta = \left(\frac{I_u}{I_v}\right)\tan\alpha = \left(\frac{I_u}{I_v}\right)\left(\frac{M_v}{M_u}\right) = \left(\frac{2.27847\times 10^7}{4.6517\times 10^6}\right)\times\left(\frac{0.9645}{11.33}\right)$$

$$= 0.417$$

Therefore, $\theta = 22.63°$ and the bending stress is

$$\sigma_b = \frac{M_u v_1}{I_u} + \frac{M_v u_1}{I_v} = \frac{11.33\times 10^6}{2.27847\times 10^7}v_1 + \frac{0.9645\times 10^6}{4.65170\times 10^6}u_1$$

$$= 0.49726v_1 + 0.20734u_1$$

The maximum *tensile stress* occurs at the farthest point B (u_1 = 60 mm, v_1 = 80 mm). The coordinates of the farthest point D where *compressive stress* is the maximum of (u_3 = –60 mm, v_3 = –80 mm). Thus,

At point 1, $\sigma_{b,\text{max}}$ = 0.49726 × 80 + 0.20734 × 60 = 52.22 MPa (tension)

At point 3, $\sigma_{b,\text{max}}$ = 0.49726 × (–80) + 0.20734 + (–60) = –52.22 MPa (compression)

17.2.3 Unsymmetrical Section

As discussed earlier, the bending moment M can be resolved into two components $M_u = M\cos\alpha$ and $M_v = M\sin\alpha$ in two principal planes Z–U and Z–V, respectively, and bending can be treated as the simultaneous bending of the beam in the principal planes.

However, a *generalized flexural formula* for computation of elastic bending stress at any beam cross-section with arbitrarily directed coordinate axes can be derived using basic hypothesis that a plane section through a beam would rotate and intersect X–Y plane at an angle ϕ with X-axis. Consider an infinitesimal element dA in positive quadrant of X, Y axes. This element is located by the perpendicular distance v from the line intersecting X–Y plane. The longitudinal normal strain ε_z is given by Eq. (8.1):

$$\varepsilon_z = -\frac{v}{R} \quad \text{where } v = y\cos\theta - x\sin\theta \tag{17.7}$$

Thus longitudinal elastic stress

$$\sigma_z = E\varepsilon_z = -\frac{Ev}{R} = -\frac{Ey}{R}\cos\theta + \frac{Ex}{R}\sin\theta \tag{17.8}$$

where $\frac{1}{R}\cos\theta$ is the projected curvature $\frac{1}{R_x}$ in the plane Z–Y and $\frac{1}{R}\sin\theta$ is the projected curvature $\frac{1}{R_y}$ in the plane Z–X.

Therefore,
$$\sigma_z = -\frac{Ey}{R_x} + \frac{Ex}{R_y} \tag{17.9}$$

For internal equilibrium at a section,

$$\int \sigma_z \, dA = -\frac{E}{R_x}\int y\, dA + \frac{E}{R_y}\int x\, dA = 0$$

This equation is identically satisfied provided co-ordinate axes are taken with their origin at the centroid of the cross-section.

For moment equilibrium at a section, two moment component equations are required to balance the externally applied moment around either axis by internal system of stresses,

$$M_x = \int -\sigma_z y\, dA = \frac{E}{R_y}\int y^2\, dA - \frac{E}{R_x}\int xy\, dA$$

$$M_y = \int \sigma_z x\, dA = -\frac{E}{R_y}\int xy\, dA + \frac{E}{R_x}\int x^2\, dA$$

The constants are taken outside the integrals in the expressions on the right hand side. These integrals define moments and product of inertia. Thus

$$\frac{E}{R_y}I_x - \frac{E}{R_x}I_{xy} = M_x \quad \text{and} \quad -\frac{E}{R_y}I_{xy} + \frac{E}{R_x}I_y = M_y \tag{17.10}$$

$$\begin{bmatrix} I_x & -I_{xy} \\ -I_{xy} & +I_y \end{bmatrix}\begin{bmatrix} E/R_y \\ E/R_x \end{bmatrix} = \begin{bmatrix} M_x \\ M_y \end{bmatrix} \tag{17.11}$$

Solving these equation simultaneously

$$\frac{E}{R_y} = \frac{M_x I_y + M_y I_{xy}}{I_x I_y - I_{xy}^2} \quad \text{and} \quad \frac{E}{R_x} = \frac{M_y I_x + M_x I_{xy}}{I_x I_y - I_{xy}^2}$$

Thus the elastic bending stress with respect to any arbitrarily directed coordinate axes is

$$\sigma_b = -\frac{Ey}{R_x} + \frac{Ex}{R_y} = -\left(\frac{M_x I_y + M_y I_{xy}}{I_x I_y - I_{xy}^2}\right) y + \left(\frac{M_y I_x + M_x I_{xy}}{I_x I_y - I_{xy}^2}\right) x \tag{17.12}$$

This is generalized flexural formula. If the principal axes are used $I_{xy} = 0$. By setting this equation to zero, the angle θ for the location of neutral axis in arbitrary coordinate system is obtained,

$$\tan\theta = \frac{y}{x} = \frac{M_y I_x + M_x I_{xy}}{M_x I_y + M_y I_{xy}} \tag{17.13}$$

EXAMPLE 17.5

A 200 × 200 × 20 angle section beam AB shown in Figures 17.6(a) and (b) carries two concentrated loads of 20 kN each at one-third points over a simply supported span of 3.75 m as shown in Figure 17.6(a). The angle section is placed with one of its legs vertically upward. Determine: (a) the direction of neutral axis at a section between the loads and (b) the stresses developed in the cross-section at the points 1, 2 and 3.

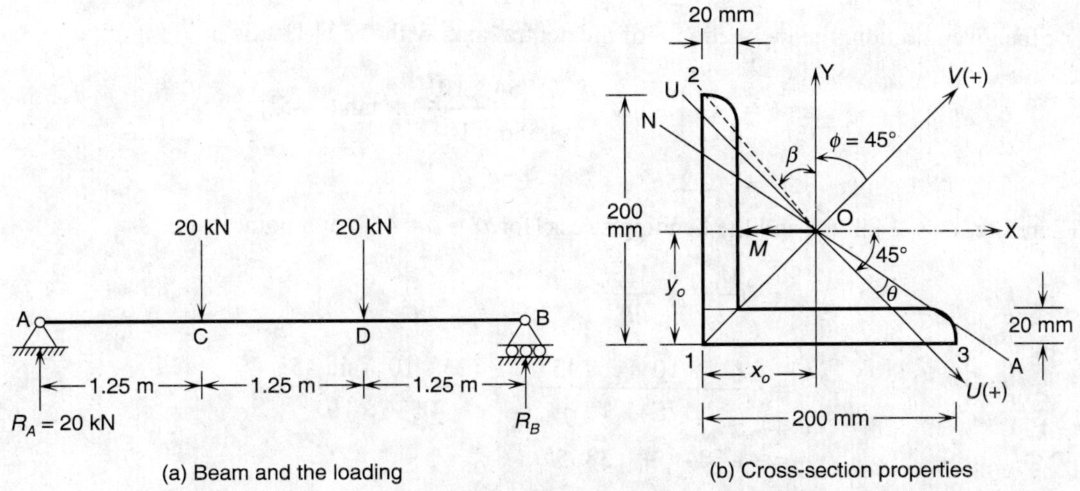

(a) Beam and the loading

(b) Cross-section properties

Figure 17.6 Bending of beam of angle section of Example 17.5.

Solution: The maximum bending moment acting on the section between the loads is

$$M = R_A \times \frac{L}{3} = 20 \times 1.25 = 25 \text{ kN.m}$$

The geometrical properties of the angle section are:

$$x_o = \frac{(200 \times 20) \times 100 + (180 \times 20) \times 10}{200 \times 20 + 180 \times 20} = \frac{436000}{7600} = 57.37 \text{ mm}$$

$$y_o = \frac{(200 \times 20) \times 100 + (180 \times 20) \times 10}{200 \times 20 + 180 \times 20} = \frac{436000}{7600} = 57.37 \text{ mm}$$

$$I_x = I_y = \frac{1}{3} \times [20 \times (200 - 57.37)^3 + 200 \times 57.37^3 + 180 \times 37.37^3] = 2.8807 \times 10^7 \text{ mm}^4$$

$$I_{xy} = 4000 \times 42.63 \times (-47.37) + 3600 \times (-47.37) \times 52.63 = -1.7053 \times 10^7 \text{ mm}^4$$

From Eq. (A.17): $\tan 2\phi = \dfrac{2I_{xy}}{I_y - I_x} = -\infty$ or $\phi = -45°$

Therefore, from Eq. (A.23) the principal moment of inertia are

$$I_u = I_x \cos^2 \phi + I_y \sin^2 \phi - I_{xy} \sin 2\phi$$
$$= 2.8807 \times 10^7 - 1.7053 \times 10^7 = 1.1754 \times 10^7 \text{ mm}^4$$

$$I_u + I_v = I_x + I_y = 2 \times 2.8807 \times 10^7 = 5.7614 \times 10^7$$

Thus, $I_v = 5.7614 \times 10^7 - I_u = 5.7614 \times 10^7 - 1.1754 \times 10^7 = 4.586 \times 10^7 \text{ mm}^4$

Alternatively, from Eq. (A.28):

$$I_u, I_v = \frac{1}{2}(I_x + I_y) \pm \sqrt{\frac{1}{2}(I_x - I_y)^2 + I_{xy}^2}$$

For the given section, the inclination θ of the neutral axis with the U-U axis is

$$\tan \theta = \left(\frac{I_u}{I_v}\right) \tan \alpha = \left(\frac{1.1754 \times 10^7}{4.586 \times 10^7}\right) \times \tan(-45°)$$
$$= -0.2563$$

Therefore, $\theta = -14.376°$, and the bending stress (for $\alpha = \phi = 45°$) at a point is

$$\sigma_b = -\frac{M_u v_i}{I_u} + \frac{M_v u_i}{I_v}$$
$$= -\frac{(25 \times 10^6) \cos 45°}{1.1754 \times 10^7} v_i + \frac{(25 \times 10^6) \sin 45°}{4.586 \times 10^7} u_i$$
$$= -1.504 v_i + 0.3855 u_i$$

The coordinates of the points 1, 2 and 3 are

For point 1: $u_1 = 0,\ v_1 = \sqrt{(57.37)^2 + (57.37)^2} = 81.133 \text{ mm}$

For point 2: Distance $0\text{–}2 = \sqrt{(57.37)^2 + (142.63)^2} = 153.736 \text{ mm}$

Inclination of 0–2 with Y-axis, $\beta = \tan^{-1}\left(\dfrac{57.37}{142.63}\right) = 21.91°$

Therefore, inclination of 0–2 with U-axis, i.e. $\angle 2 - O - U$

$$\beta' = 45 - \beta = 45 - 21.91 = 23.09°$$

Thus, $u_2 = -153.736 \times \cos 23.09° = -141.42$ mm and

$$v_2 = -153.736 \times \sin 23.09° = 60.29 \text{ mm}$$

For point 3: $u_3 = 141.42$ mm and $v_3 = 60.29$ mm

Thus, the bending stresses at various points are,

$$\sigma_{b,1} = -1.504 \times (-81.133) + 0.3855 \times 0 = 122.024 \text{ MPa (tension)}$$

$$\sigma_{b,2} = -1.504 \times (60.29) + 0.3855 \times (-141.42) = -145.19 \text{ MPa (compression)}$$

$$\sigma_{b,3} = -1.504 \times (60.29) + 0.3855 \times (141.42) = -36.159 \text{ MPa (compression)}$$

EXAMPLE 17.6

A rolled steel beam of unequal angle section ISA 200 × 100 × 10 shown in Figures 17.7(a) and (b) is subjected to a concentrated load W (kN) at its centre over a simply supported span of 2.0 m. The angle section is placed with its longer leg vertically downward. Determine the load W, if the maximum allowable stress developed in the cross-section is limited to 145 MPa. The geometrical properties of the angle section are:

$I_u = 1286.7 \times 10^4$ mm^4; $I_v = 132.5 \times 10^4$ mm^4; tan $\alpha = 0.27$, and the distances of the centroid of the section from the short and long legs are 69.6 mm and 20.1 mm, respectively.

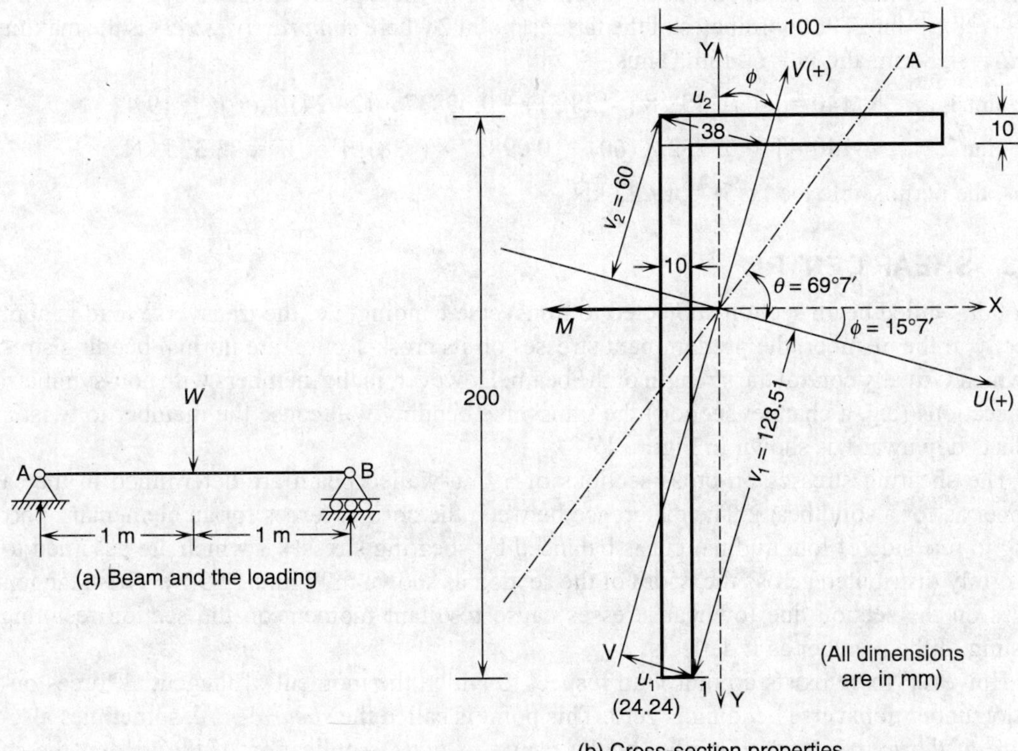

Figure 17.7 Bending of unequal angle section beam of Example 17.6.

Solution: For the given section, $x_o = 20.1$ mm, $y_o = 69.6$ mm and $\phi = \alpha = 15.11°$

The maximum bending moment acting on the section is

$$M = \frac{WL}{4} = \frac{(W \times 10^3) \times (2 \times 10^3)}{4} = (5 \times 10^5)\, W \text{ (N.mm)}$$

For the given section, the inclination θ of the neutral axis with the U-U axis is

$$\tan\theta = \left(\frac{I_u}{I_v}\right) \tan\alpha = \left(\frac{1286.7 \times 10^4}{132.5 \times 10^4}\right) \times 0.27 = 2.622$$

Therefore, $\theta = 69.12$, and the bending stress is

$$\sigma_{b,\max} = -\frac{M_u v_i}{I_u} + \frac{M_v u_i}{I_v} = -\frac{(5 \times 10^5)\, W \cos 15.11°}{1286.7 \times 10^4} v_i + \frac{(5 \times 10^5)\, W \sin 15.11°}{132.5 \times 10^4} u_i$$

$$= -0.03752 W v_i + 0.09837 W u_i$$

The coordinates of the *critical* or remotest points 1 and 2 are determined as the previous example. The maximum *tensile stress* occurs at the farthest point 1, where $u_1 = 24.24$ mm and $v_1 = -128.51$ mm. The coordinates of the farthest point 2 where *compressive stress* is the maximum are $u_2 = -38$ mm and $v_2 = 60$ mm. Thus,

At point 1, $140 = [-0.03752 \times (-128.51) + 0.09837 \times (24.24) W$ or $W = 19.42$ kN

At point 2, $-140 = [-0.03752 \times (60) + 0.09837 \times (-38)] W$ or $W = 23.375$ kN

Thus, the permissible load is $W = 19.42$ kN.

17.3 SHEAR CENTRE

In a thin-walled beam section subjected to transverse bending, i.e. the transverse load is applied directly to the member, the predominant stresses on its cross-section are normal bending stresses σ_b which virtually control the strength of the beam. However, in the members with non-symmetrical thin sections (e.g. a channel section) the transverse bending will cause the member to twist as it deflects downward as shown in Figure 17.8.

The shearing stresses on cross-sections of a thin-walled beam are determined in the same manner as for a solid beam. The difference between the normal forces for an elementary portion lying to one side of longitudinal cut is balanced by shearing stresses τ which are assumed to be uniformly distributed across the width of the section as shown in Figure 17.9(b). These tangential forces on the section due to shear stresses cause resultant moment on the section resulting in twisting of the member as it deflects.

However, there exists a point with respect to which the moment of tangential forces on the section under transverse bending is zero. This point is called the *shear centre*, sometimes also the *centre of flexure*, or *flexural centre*, or the *centre of twist*. Application of the load at the shear centre of the member will cause the member to deflect downward without twisting. The shear centre of many non-symmetric members often lies outside the member.

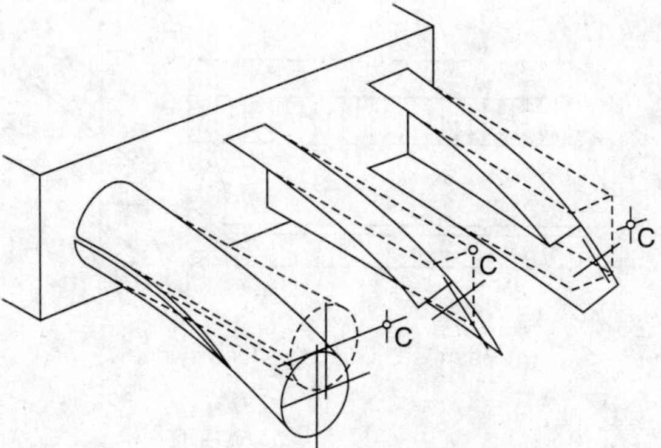

Figure 17.8 Twisting of thin walled sections while bending and location of shear centres.

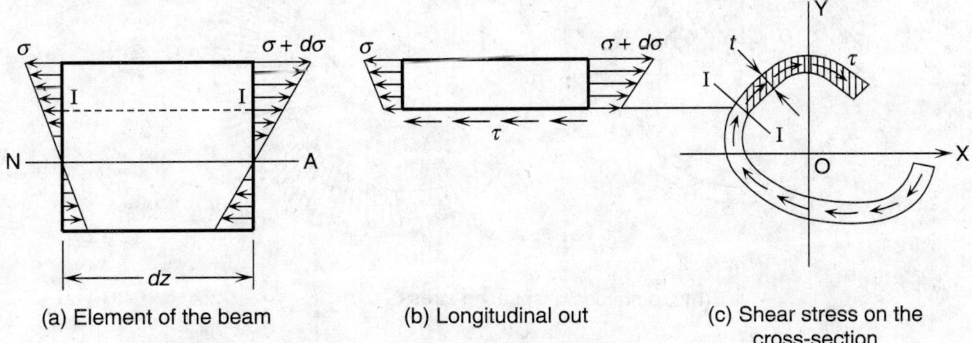

Figure 17.9 Development of shear stress in thin-walled beam section.

The laws of mechanics enable a system of forces lying in the plane of a cross-section to be transferred to any point in the plane in the form of a resultant force and a moment. The magnitude of the resultant force does not depend on the point of transfer or reduction and is always equal to the shearing force V. However; the resultant moment at the section depends on the position of the point of transfer of the forces. This principle can be used for locating the shear centre of the cross-section of a beam. Thus, the problem of locating the shear centre of a cross-section of a beam consists in locating the line of action of the resultant shearing forces on the section such that the resultant moment vanishes. The shear centre has no meaning in case of pure bending as no resultant shear force exists in this case.

The concept is illustrated by the example of circular open section shown in Figure 17.10(a). The moment of inertia of the section with respect to the X-axis is,

$$I_x = \frac{\pi}{4}R^4 - \frac{\pi}{4}(R-t)^4 = \frac{\pi R^4}{4}\left[1-\left(1-\frac{t}{R}\right)^4\right]$$

$$= \frac{\pi R^4}{4}\left[1-\left(1-4\frac{t}{R}+\cdots\right)\right] = \pi R^3 t \qquad (17.14)$$

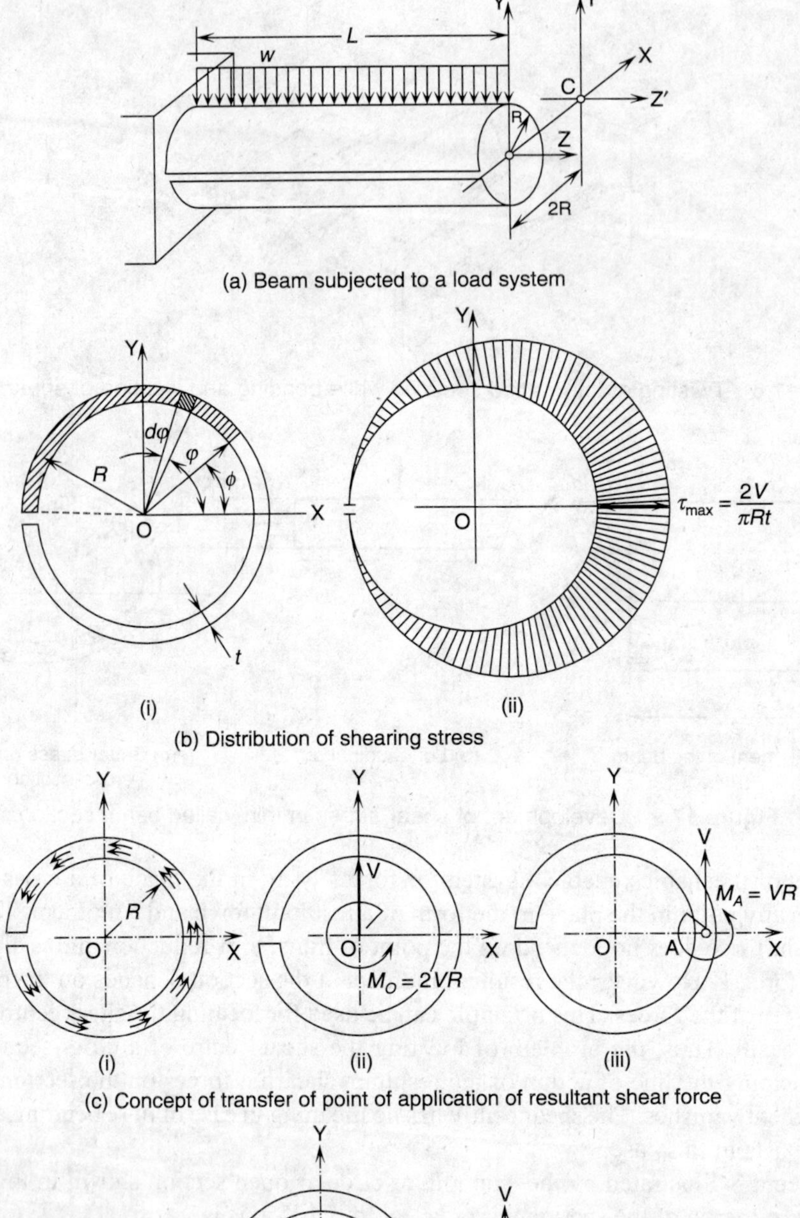

(a) Beam subjected to a load system

(i) (ii)

(b) Distribution of shearing stress

(i) (ii) (iii)

(c) Concept of transfer of point of application of resultant shear force

(d) Shear centre

Figure 17.10 Circular open section.

The shear stress acting vertically can be calculated in the usual manner,

$$\tau = \frac{VA\bar{y}}{I_x t} \tag{17.15}$$

where the static moment of shaded part of the section, i.e. the portion beyond the section at ϕ is given by

$$A\bar{y} = \int_{\phi}^{\pi} (tR\,d\varphi)(R \sin \varphi) = R^2 t \int_{\phi}^{\pi} \sin \varphi \, d\varphi$$

$$= R^2 t\,(1 + \cos \phi)$$

Accordingly,

$$\tau = \frac{VR^2 t}{I_x t}(1 + \cos \phi) = \frac{V}{\pi R t}(1 + \cos \phi) \tag{17.16}$$

The shear stress diagram is shown in Figure 17.10(b–i and ii).
The resultant of the tangential forces along the Y-axis is given by

$$\int_A \tau \cos \varphi \, dA = \int_{-\pi}^{\pi} \frac{V}{\pi R t}(1 + \cos \varphi) \cos \varphi \,(tR\,d\varphi)$$

$$= \frac{V}{\pi}\int_{-\pi}^{\pi} (1 + \cos \varphi) \cos \varphi \, d\varphi = V \tag{17.17}$$

Thus the magnitude of the resultant of the tangential forces is always equal to the shear forces and is independent of the point of transfer. However, so far the resultant moment at the section is concerned; it depends in the first place on the position of the point of transfer of the forces. For example, consider the centre of the circle as the transfer point and then the resultant moment of shear forces at the section is

$$M_o = \int_A (\tau\, dA)\, R = \int_{-\pi}^{\pi} \frac{V}{\pi R t}(1 + \cos \varphi)\,(tR\,d\varphi)\, R$$

$$= \frac{VR}{\pi}\int_{-\pi}^{\pi} (1 + \cos \varphi)\, d\varphi = 2VR \tag{17.18}$$

When the moment is passed or reduced to another point (say) at distance a from the first point of reduction O, it changes by an amount Va. Thus, if the forces are reduced to a point A shown in Figure 17.10(c-iii), then

$$M_A = M_o - VR = 2VR - VR = VR \tag{17.19}$$

Hence the point with respect to which the moment of tangential forces on the section is zero occurs at a distance $s = 2R$ from the centre of the circle as shown in Figure 17.10(d).

17.3.1 A Channel Section

Consider the channel section shown in Figure 17.11(a) used as a cantilever beam subjected to external vertical loading. Figure 17.11(b) illustrates a short length dz of the beam subjected to

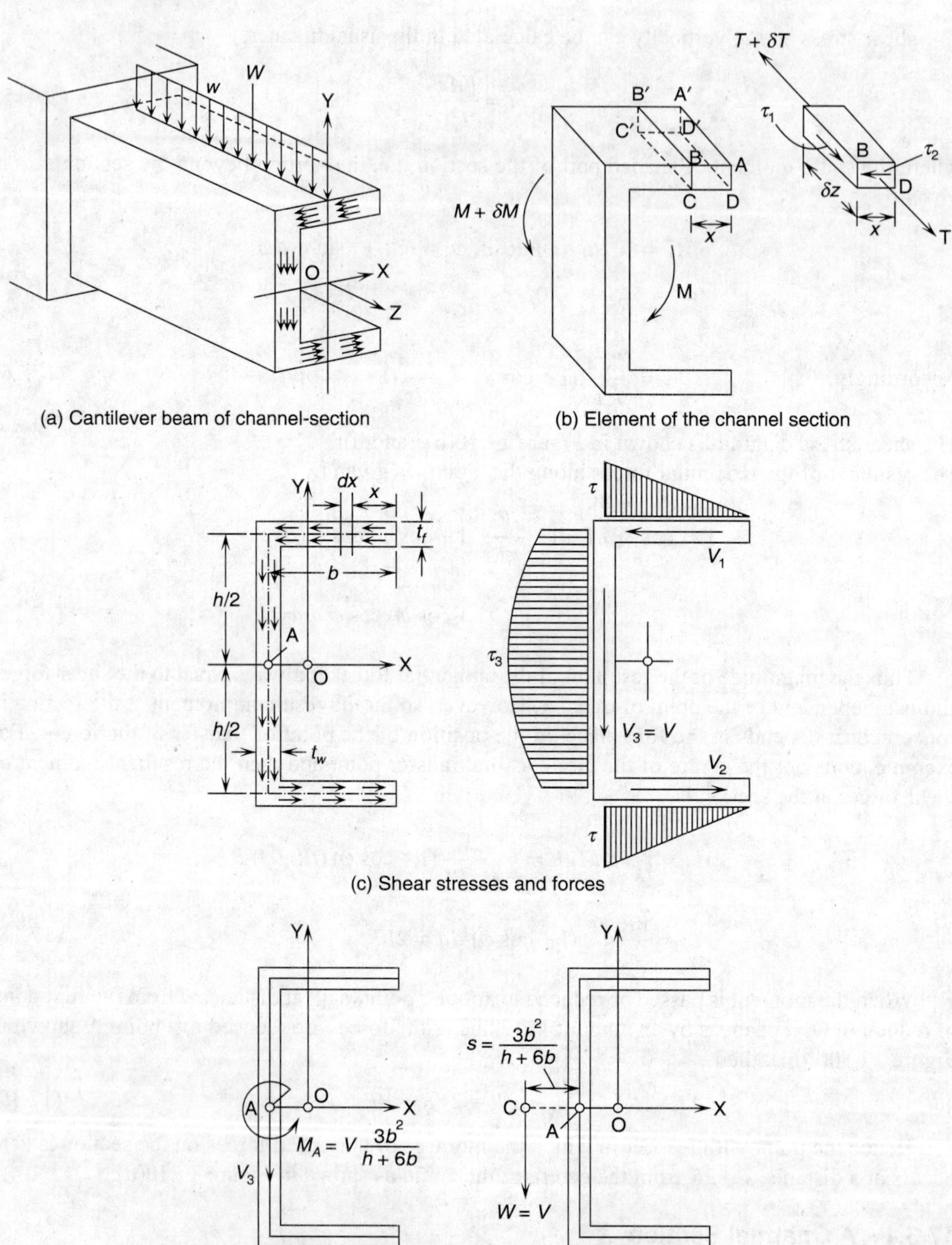

(a) Cantilever beam of channel-section

(b) Element of the channel section

(c) Shear stresses and forces

(d) Resultant moment of shear forces on the section

Figure 17.11 Shear centre for a channel section.

bending moments increasing from M to $M + \delta M$. The thickness of the flange, the depth of the channel section from the centre of the top flange to the centre of the bottom flange and the moment of inertia of the section are t_f, h and I, respectively. Since the thickness of the flange is small, the bending stress is taken to be constant over the flange.

Consider a small rectangular element ABCDA′B′C′D′ of length δz cut from the top flange through the section BCC′B′ as shown in Figure 17.11(b). The normal forces T and $T + \delta T$ acting on the faces ABCD and A′B′C′D′ of the element due to bending are,

$$T + \delta T = \left(\frac{M + \delta M}{I} \times \frac{h}{2}\right)(t_f\,x) \text{ and } T = \left(\frac{M}{I} \times \frac{h}{2}\right)(t_f\,x) \tag{17.20}$$

Therefore, the net normal force acting on the rectangular element

$$\delta T = \left(\frac{\delta M}{I} \times \frac{h}{2}\right)(t_f\,x) \tag{17.21}$$

This net force δT is balanced by the shear force $\tau_1(\delta z\, t_f)$ acting on the face BCC′B′:

$$\tau_1 \delta z t_f = \left(\frac{\delta M}{I} \times \frac{h}{2}\right)(t_f\,x) \quad \text{or} \quad \tau_1 = \frac{\delta M}{\delta z}\left(\frac{hx}{2I}\right) \tag{17.22}$$

In the limit as δz tends to zero, $\dfrac{\delta M}{\delta z} \Rightarrow \dfrac{dM}{dz} = V$, the shear force

Therefore,
$$\tau_1 = \left(\frac{Vh}{2I}\right)x \tag{17.23}$$

The shear stress τ_1 is accompanied by an equal complementary shear stress, τ_2 acting along the flange, i.e.

$$\tau_2 = \left(\frac{Vh}{2I}\right)x \tag{17.24}$$

The shear stress τ_2 varies directly with the distance x from right to left with a maximum value of $\tau = \dfrac{Vhb}{2I}$ as shown in Figure 17.11(c). Similarly, at any distance x from the edge, a shear stress of the same intensity will be induced in the bottom flange. However, since the bending stress in the bottom flange is compressive, the direction of shear stress is opposite to that in the top flange. The variation shear stresses acting on the cross-section in the flanges and the web are shown in Figure 17.11(c).

The resultant horizontal shear force in each flange

$$V_1 = \text{Average shear stress in flange} \times \text{Area of flange}$$
$$= \text{Half the maximum shear in flange} \times \text{Area of flange}$$

$$= \frac{1}{2} \times \left(\frac{Vhb}{2I}\right) \times (bt_f) = \frac{Vhb^2 t_f}{4I} \tag{17.25}$$

The shear stress acting vertically in the web can be calculated in the usual manner,

$$\tau_3 = \frac{VA\bar{y}}{It_w} \tag{17.26}$$

The vertical resultant, $$V_3 = \int_{-h/2}^{h/2} \tau_3 \, t_w \, dy$$

The vertical resultant, V_3 balances the external shear force V, i.e. $V_3 = V$ while horizontal shear resultants V_1 in the flanges do not balance but results in a counter-clockwise torque $M_o = V_1 h$ on the cross-section. The cross-section can be prevented from twisting by applying an external load W through a point C so that W and V form a couple equal and opposite to M_o as shown in Figure 17.11(d). Thus, if s is the distance of the point C from the median of the web of the section, then for the resultant moment of shear forces on the section to be zero,

$$M_o = V_1 h = Vs \quad \text{where } V = W$$

Therefore, $$s = \frac{V_1 h}{V} = \frac{h^2 b^2 t_f}{4I} \tag{17.27}$$

Alternatively, the location of shear centre can be obtained more directly by using the following procedure. Consider the shear or tangential forces V_1, V_2 and V_3 developed in the top and bottom flanges, and the web of the channel cross-section shown in Figure 17.5(c), respectively. These shear forces are given by

$$V_1 = \int_{A=t_f x} \tau \, dA = \int_0^b \frac{VA\bar{y}}{It_f}(t_f \, dx) = \left(\frac{V}{It_f}\right)\int_0^b A\bar{y}\,(t_f \, dx)$$

$$= \left(\frac{V}{It_f}\right)\int_0^b (t_f \, x) \times \left(\frac{h}{2}\right) \times (t_f \, dx)$$

$$= \frac{Vt_f h}{2I}\int_0^b x \, dx = \frac{Vt_f hb^2}{4I} \tag{17.28}$$

Therefore, $$V_1 = \frac{Vt_f hb^2}{4I} = V_2$$

Consider the shear centre to be at a distance s from the median of the web. Then, for the moment of shear forces on channel section to be zero with respect to the shear centre C,

$$V_R \times s = V_1 \times \frac{h}{2} + V_2 \times \frac{h}{2} = V_1 h = \frac{Vt_f h^2 b^2}{4I} \tag{17.29}$$

$$s = \frac{Vt_f h^2 b^2}{V_R 4I} = \frac{t_f h^2 b^2}{4I} \quad (\text{since } V_R = V) \tag{17.30}$$

Ignoring the moment of inertia of the flanges about their axes,

$$I = 2A_f \times \left(\frac{h}{2}\right)^2 + \frac{t_w h^3}{12} = \frac{A_f h^2}{2} + \frac{A_w h^2}{12} = \frac{h^2}{12}(A_w + 6A_f) \tag{17.31}$$

where A_w and A_f are the areas of web and flange, respectively.

Therefore,

$$s = \frac{3b^2 t_f}{A_w + 6A_f} = \frac{3b^2 t_f}{ht_w + 6(bt_f)} = \frac{3b}{(t_w/t_f)h + 6b} \tag{17.32}$$

If $t_w = t_f = t$,

$$s = \frac{3b^2}{h + 6b} \tag{17.33}$$

It follows that the shear centre is at a distance $\dfrac{3b^2}{h+6b}$ from the median line of the web as shown in Figure 17.11(d). Thus, s is independent of the value of V and is a function of the dimensions of the cross-section only.

The shear stress τ acting on the flange or web is considered to be constant over the thickness t, since it is small compared to the overall dimensions of the cross-section. The shear force τt acting over a unit length of the flange or web is called *shear flow*, and is denoted by q. As t is constant over the unit length of the flange or web, q will have a variation similar to τ.

17.3.2 Unsymmetrical I-section

Consider a beam having I-section with unequal flanges placed vertically as shown in Figure 17.12. Since the vertical shear taken by the horizontal web is relatively very small, the total shear V can be assumed to be carried as the forces V_1 and V_2 in the left and right flanges of the I-section, respectively.

$$V_1 = \int_{A=t_1 y} \tau\, dA = 2\int_0^{b_1/2} \frac{VA\bar{y}}{It_1}(t_1 b_1\, dy) = 2\left(\frac{V}{I}\right)\int_0^{b_1/2} A\bar{y}\,(dy)$$

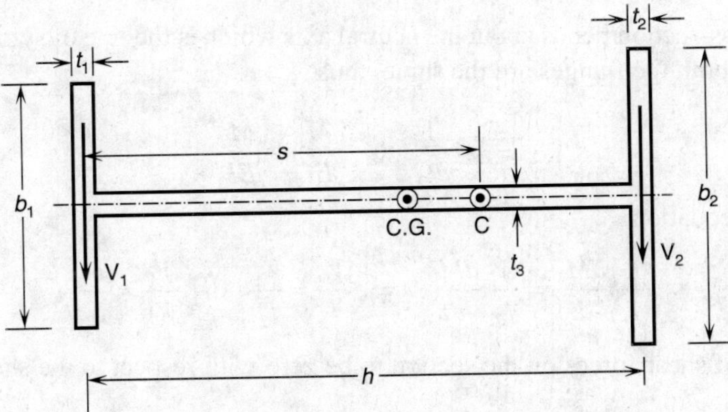

Figure 17.12 Shear centre of unsymmetrical I-section.

The quantity $A\bar{y}$ for a layer at distance y from the neutral axis is given by,

$$A\bar{y} = \left[t_1\left(\frac{b_1}{2} - y\right)\right]\left[y + \frac{1}{2}\left(\frac{b_1}{2} - y\right)\right] = \frac{t_1}{2}\left(\frac{b_1^2}{4} - y^2\right)$$

Therefore,

$$V_1 = 2\left(\frac{V}{I}\right)\int_0^{b_1/2} \frac{t_1}{2}\left(\frac{b_1^2}{4} - y^2\right)(dy) = \left(\frac{Vt_1}{I}\right)\int_0^{b_1/2}\left(\frac{b_1^2}{4} - y^2\right)(dy)$$

$$= \left(\frac{Vt_1}{I}\right)\left|\frac{b_1^2}{4}y - \frac{y^3}{3}\right|_0^{b_1/2}$$

$$= \left(\frac{Vt_1}{I}\right)\times\frac{b_1^3}{12} = \frac{V}{I}\times\left(\frac{t_1 b_1^3}{12}\right) = \frac{VI_1}{I} \tag{17.34}$$

Similarly,

$$V_2 = \frac{VI_2}{I}, \text{ where } I_2 = \frac{t_2 b_2^3}{12} \tag{17.35}$$

Consider the shear centre to be at a distance from the left flange. Then, for the moment of shear forces on the section to be zero with respect to the shear centre,

$$V_1 s = V_2(h - s) \text{ or } \frac{VI_1}{I}s = \frac{VI_2}{I}(h - s)$$

Therefore,

$$\frac{h - s}{s} = \frac{I_1}{I_2} = \frac{t_1 b_1^3}{t_2 b_2^3} \text{ or } s = \left(\frac{t_2 b_2^3}{t_1 b_1^3 + t_2 b_2^3}\right)h \tag{17.36}$$

Alternatively, consider the bending moments shared by the two flanges M_1 and M_2 which are related to the shear forces V_1 and V_2 as follows,

$$V_1 = \frac{dM_1}{dx} \text{ and } V_2 = \frac{dM_2}{dx} \tag{17.37}$$

Since the cross-section bends about the neutral axis which is the line of symmetry in this case, the curvatures of both the flanges are the same, thus

$$\frac{1}{R_1} = \frac{1}{R_2} \text{ or } \frac{M_1}{EI_1} = \frac{M_2}{EI_2} \tag{17.38}$$

Differentiate the equation

$$\frac{1}{I_1}\frac{dM_1}{dz} = \frac{1}{I_2}\frac{dM_2}{dz} \Rightarrow \frac{V_1}{I_1} = \frac{V_2}{I_2} \text{ or } \frac{V_1}{V_2} = \frac{I_1}{I_2} \tag{17.39}$$

For the moment of shear forces on the section to be zero with respect to the shear centre,

$$V_1 s = V_2(h - s) \text{ or } \frac{V_1}{V_2} = \frac{h - s}{s} \tag{17.40}$$

Therefore, comparing Eqs. (17.39) and (17.40):

$$\frac{h-s}{s} = \frac{I_1}{I_2} = \frac{t_1 b_1^3}{t_2 b_2^3}$$

Therefore,
$$s = \left(\frac{t_2 b_2^3}{t_1 b_1^3 + t_2 b_2^3}\right) h \tag{17.41}$$

Thus, s is independent of the value of V and is a function of the dimensions of the cross-section only. It should be noted that the distance of shear centre from the flanges is inversely proportional to the moment of inertia of the flanges.

17.3.3 Sections Having Two Axes of Symmetry

As a typical example of the sections having two axes of symmetry, consider a beam having I-section shown in Figure 17.13(a). Using the same procedure as for the unsymmetrical channel section consider an element cut in the right-hand portion of the top flange of the I-section to deduce that the shear stress on this part acts from right to left. On the left-hand portion the shear stress acts from left to right as shown in Figure 17.13(b). These shear stresses have linear variation from zero at the free ends to a maximum value in the middle of the flange. The shear stress variations on the cross-section in the flanges and the web are shown in Figure 17.13(b) while the resultant forces due to these stresses in Figure 17.13(c). Their magnitudes are

$$V_1 = \text{Average shear stress in flange} \times \text{Area of flange part}$$

$$= \frac{\tau}{2} \times \left(t_f \frac{b}{2}\right) = \frac{\tau t_f b}{4} \tag{17.42}$$

The vertical resultant due to the shear stress acting vertically in the web is,

$$V_3 = \int_{-h/2}^{h/2} \tau_3 \, t_w \, dy \tag{17.43}$$

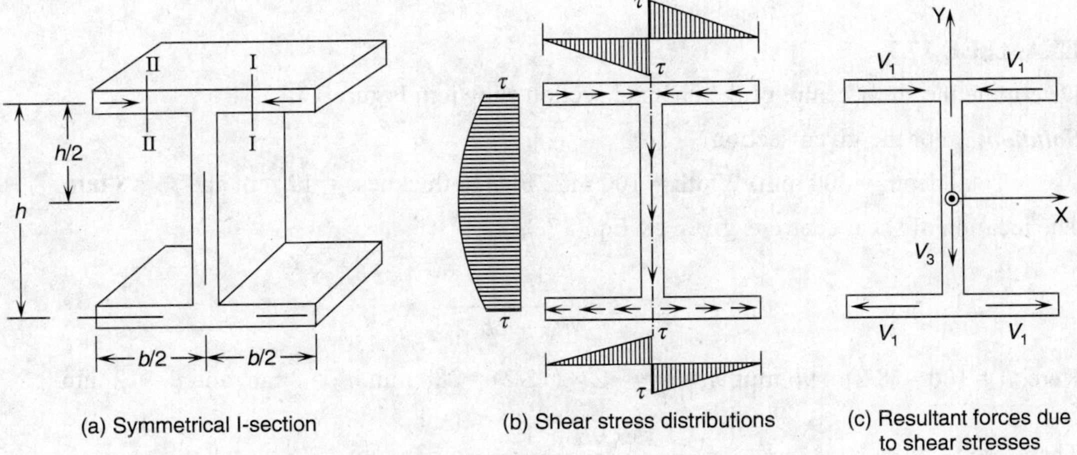

(a) Symmetrical I-section

(b) Shear stress distributions

(c) Resultant forces due to shear stresses

Figure 17.13 I-section having two axes of symmetry.

The vertical resultant, V_3 balances the external shear force V, while the horizontal shear resultants V_1 in the flanges balance each other on the left- and right-hand portions of the flanges. Thus, the net effect of horizontal resultants on a symmetrical cross-section like the I-section vanishes and the shear centre is coincident with the centroid.

From the forgoing discussion it is evident that the distribution of tangential forces on a thin-walled beam subjected to transverse bending depends on the conditions of application of external forces.

As the moment of tangential forces at the section with respect to an axis passing through the shear centre is zero, the resultant of external forces must also pass through the shear centre of the section, i.e. the external forces must also produce a zero moment with respect to this axis. Otherwise, the transverse bending of a thin-walled beam is accompanied by twisting.

For the sections having one axis of symmetry, the shear centre does not coincide with the centroid but lies on the axis of symmetry. In some of the simple cases the position of the shear centre can be located by inspection only. For example, for a T-section and the angle sections shown in Figures 17.14(a) and (b), the shear centre is at the point of intersection of the median lines of the web and flange. The moment of tangential forces at the section with respect to this point is always zero.

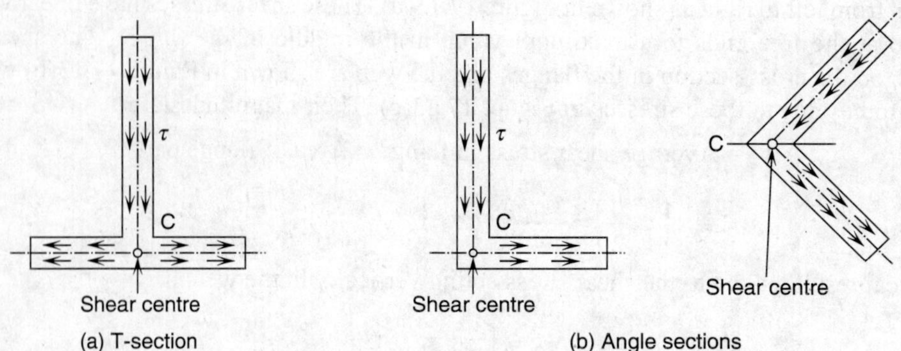

(a) T-section

(b) Angle sections

Figure 17.14 Shear centres for T- and angle-sections used as beams.

EXAMPLE 17.7

Determine the shear centre of the channel section shown in Figure 17.15.

Solution: For the given section,

Total depth = 300 mm, Width = 100 mm, Flange thickness = 12 mm and t_w = 8 mm

The location of shear centre is given by Eq. (17.32);

$$s = \frac{3b^2 t_f}{ht_w + 6bt_f}$$

Here, $b = 100 - (8/2) = 96$ mm, $h = 300 - 2 \times (12/2) = 288$ mm, $t_w = 8$ mm and $t_f = 12$ mm.

Therefore,
$$s = \frac{3 \times 96^2 \times 12}{288 \times 8 + 6 \times 96 \times 12} = 36 \text{ mm}$$

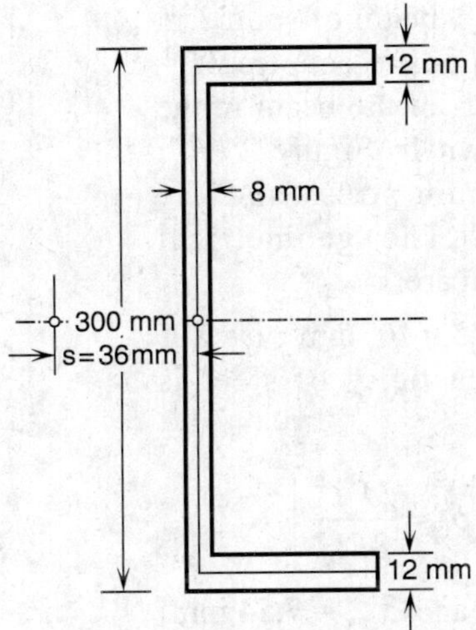

Figure 17.15 Channel section of Example 17.7.

17.4 PROBLEMS

17.1 A beam having unequal angle section 120 × 80 × 10 mm shown in Figure 17.16 is subjected to a concentrated load 6 kN at its centre over a simply supported span of 4 m. The angle section is placed with its longer leg vertically upward. The concentrated load lies in the plane making angle $\alpha = 100°$ with the X-axis. The load passes through the shear centre. Determine the maximum tensile and compressive stresses developed in the cross-section at the point of application of load.

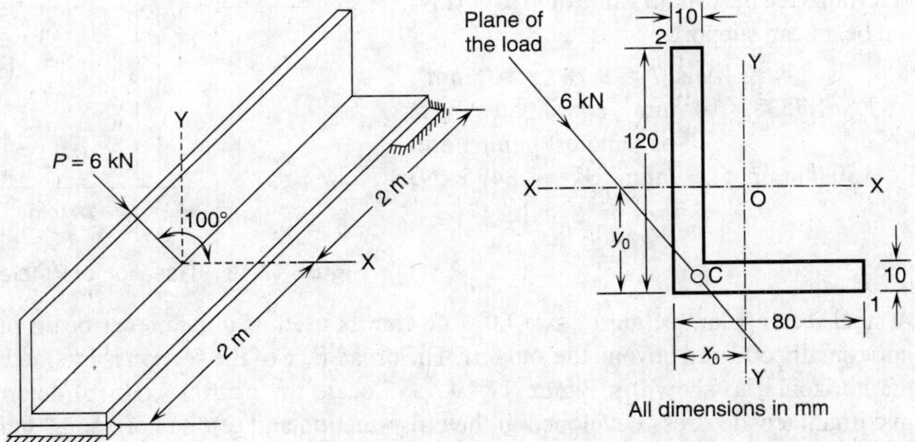

Figure 17.16 Unequal angle section of Problem 17.1.

[**Ans.** $I_u = 3.2115 \times 10^6$ mm^4; $I_v = 0.5749 \times 10^6$ mm^4; coordinates (u, v) of points 1 and 2 are: (34.09 mm, 28.41 mm) and (–23.45 mm, –77.37 mm); $\sigma_{b,1} = 136.22$ MPa and $\sigma_{b,2} = -198.64$ MPa]

17.2 A standard rolled steel I-beam of span 3 m is bent by equal and opposite couples M = 4 kN.m acting at the ends of the beam in the axial plane I-I as shown in Figure 17.17. Determine the maximum stress and the maximum deflection. The geometrical properties of the I-section are:

$I_u = 2.4 \times 10^7$ mm^4; $I_v = 1.5 \times 10^6$ mm^4; $\alpha = 30°$ and the modulus of elasticity of steel is $E = 200$ GPa.

$$\left[\textbf{Hint}: \delta_u = \frac{M_u l^2}{8EI_v} \text{ and } \delta_v = \frac{M_v l^2}{8EI_u}\right]$$

[**Ans.** $\sigma_{max} = 81.1$ MPa and $\delta_{max} = 7.54$ mm]

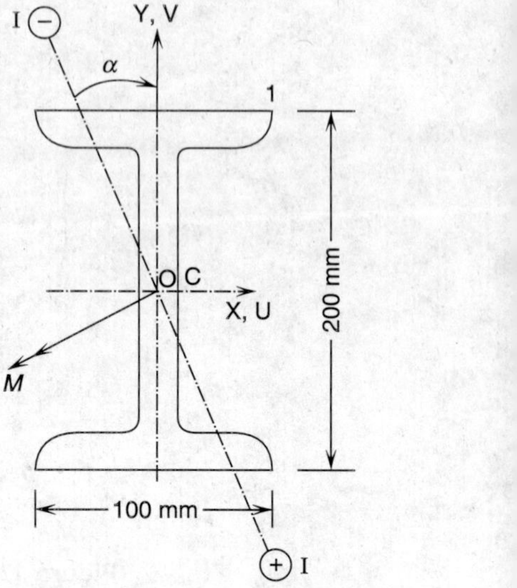

Figure 17.17 I-section of Problem 17.2.

17.3 A 3.0 m long simply supported beam of T-section is subjected to a concentrated load W (kN) at its centre. The load is inclined at angle $\alpha = 30°$ with the Y-axis and passes through the centroid of the section as shown in Figure 17.18. If the maximum compressive and tensile stresses in bending are restricted to 75 MPa and 35 MPa, respectively, determine the maximum value of load W (kN) the beam can support.

[**Ans.** $I_u = 9.787 \times 10^4$ mm^4; $I_v = 1.7533 \times 10^4$ mm^4; coordinates (u, v) of remotest point 1 are: (50.00 mm, 52.39 mm); $W = 4.8495$ kN]

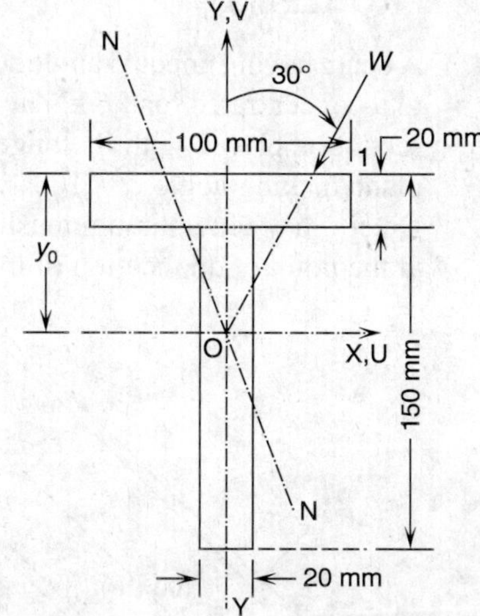

Figure 17.18 T-section of Problem 17.3.

17.4 A steel rectangular section of size 60 × 40 mm is used as a cantilever beam projecting horizontally 600 mm beyond the support. The broad face of the section is inclined at 30° to the horizontal as shown in Figure 17.19. (a) Locate the neutral axis; (b) determine the maximum tensile stress developed in the cross-section and (c) the horizontal and vertical deflections at the free end when the beam is subjected to a concentrated load 0.25 kN at its free end. The modulus of elasticity of steel is $E = 200$ GPa.

$$\left[\textbf{Hint}: \delta = \frac{WL^3}{3EI}\right]$$

[**Ans.** $\sigma_{max} = 11.244$ MPa; $\phi = 14.3916°$; $\delta_h = 0.0678$ mm and $\delta_v = 0.242$ mm]

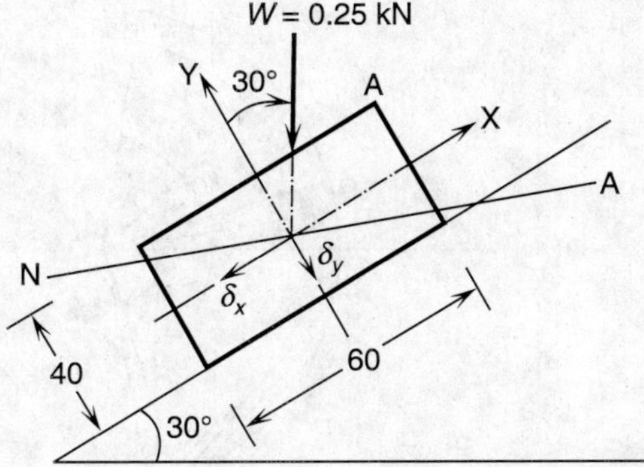

Figure 17.19 Rectangular cross-section of beam of Problem 17.4.

17.5 A standard rolled steel channel section shown in Figure 17.20 is used as a 3.0 m long simply supported beam (purlin) in a roof truss. Determine the maximum compressive and tensile stresses in bending in the beam when it carries a uniformly distributed load of 1.2 kN/m. The principal moment of inertias of the channel section are:

$$I_u = 8.87 \times 10^6 \text{ mm}^4; I_v = 40.3 \times 10^4 \text{ mm}^4$$

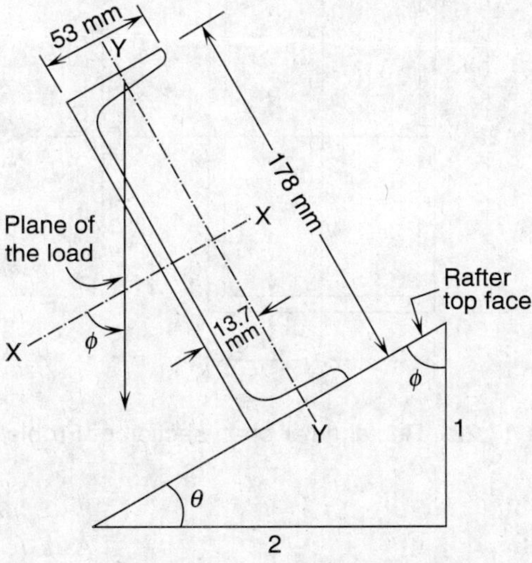

Figure 17.20 Channel section of roof purlin of Problem 17.5.

[**Ans.** σ_{max} = 32.67 MPa and –71.01 MPa]

17.6 Determine the shear centre of a section having the shape of a segment of a circle as shown in Figure 17.21. What is the location of shear centre for a semi-circular section?

$$\left[\textbf{Ans. } s = \frac{2R(\sin\phi - \phi\cos\phi)}{(\phi - \sin\phi\cos\phi)} \text{ from the centre towards convex side; } s = \frac{4R}{\pi}\right]$$

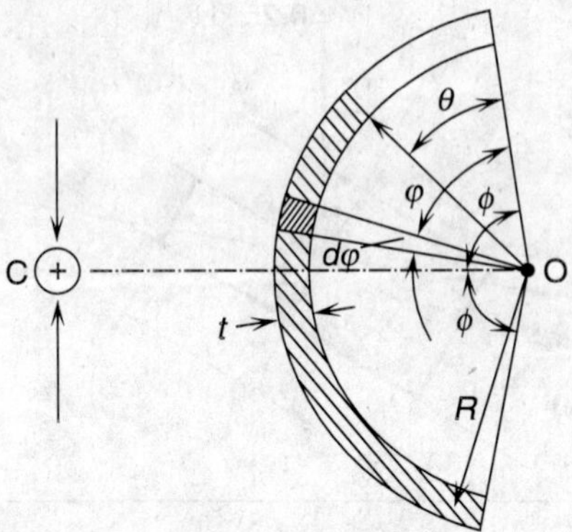

Figure 17.21 Circular arc section of Problem 17.6.

17.7 Determine the location of the shear centre for a rectangular thin section of size a (width) × $2a$ (height) having a cut at the X-axis as shown in Figure 17.22.

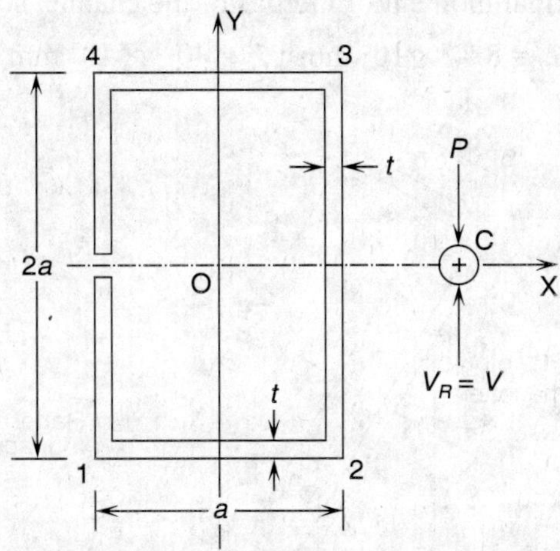

Figure 17.22 Rectangular open-section of Problem 17.7.

$$\left[\textbf{Ans. } I_x = \frac{10a^3t}{3};\ s_x = \frac{17}{6}a;\ s_y = 0\right]$$

17.8 Determine the location of the shear centre for a rectangular thin section of size a (width) × $2a$ (height) having a cut at the lower left corner as shown in Figure 17.23.

$$\left[\textbf{Ans. } I_x = \frac{10a^3t}{3};\ I_y = \frac{7a^3t}{6} \text{ and } s_x = \frac{3}{5}a;\ s_y = \frac{12}{7}a \text{ are the distances from the centroid towards upper right corner}\right]$$

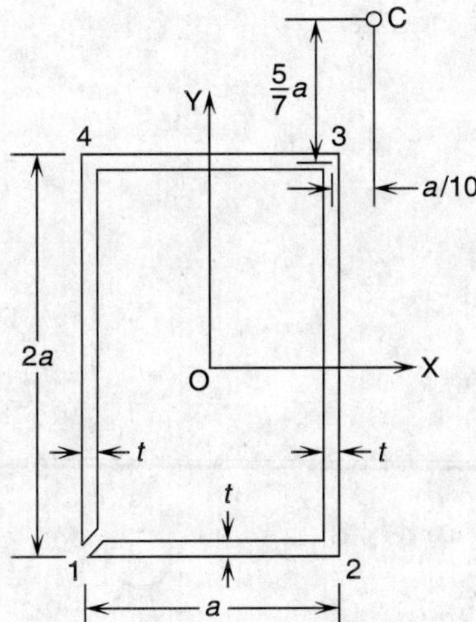

Figure 17.23 Rectangular open-section of Problem 17.8.

17.9 Determine the location of the shear centre for a thin I-section of Figure 17.24 which is unsymmetrical about Y-axis.

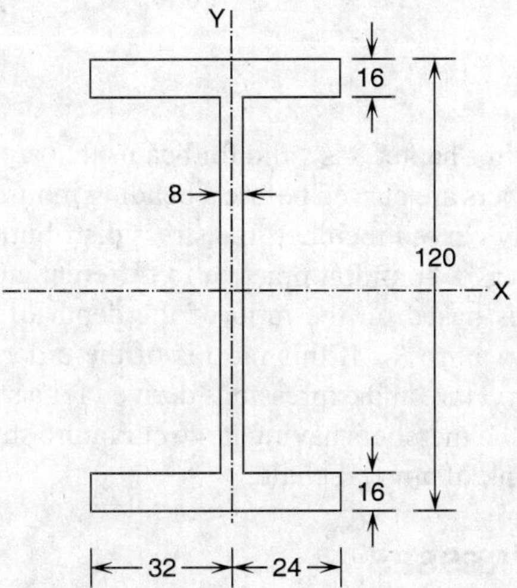

Figure 17.24 Unsymmetrical I-section of Problem 17.9.

[**Ans.** $s_x = 3.63$ mm; $s_y = 0$]

Chapter 18

Curved Members Subjected to Bending

18.1 INTRODUCTION

The flexural formula derived in Chapter 8 is valid for beams that are initially straight. However, in practice, sometimes the members are curved before a bending moment is applied in the plane of its curvature. In case of a sharply curved member, the stress distribution is markedly different from that for a straight member. It is a common practice to differentiate members of small and large curvatures. This distinction is based on the ratio of the depth of the section h in the plane of curvature to the radius of curvature R_c. If this ratio is of the order of 0.2 or less, the member is considered to have a small curvature and expressions derived for a straight member can be applied.

The typical examples of the members having large curvature shown in Figures 18.1(a) and (b) include a crane hook and a link of a metal chain.

18.2 STRAINS AND STRESSES

The following assumptions are made in the analysis of a curved member subjected to bending action:

1. The radial plane sections before bending remain plane after bending.
2. The material is considered isotropic and elastic which is strained within the elastic limit.
3. The radial strain is considered to be negligible.

Consider the portion of a curved member shown in Figure 18.2(a) having uniform cross-section between two adjacent sections ab and cd making angle $d\theta$ before bending as shown in Figure 18.2(b). Let under the action of pure moment the sections ab and cd rotate relative to each other by the amount $d\phi$ as shown in Figure 18.2(b), but remain straight and undistorted. As the member element

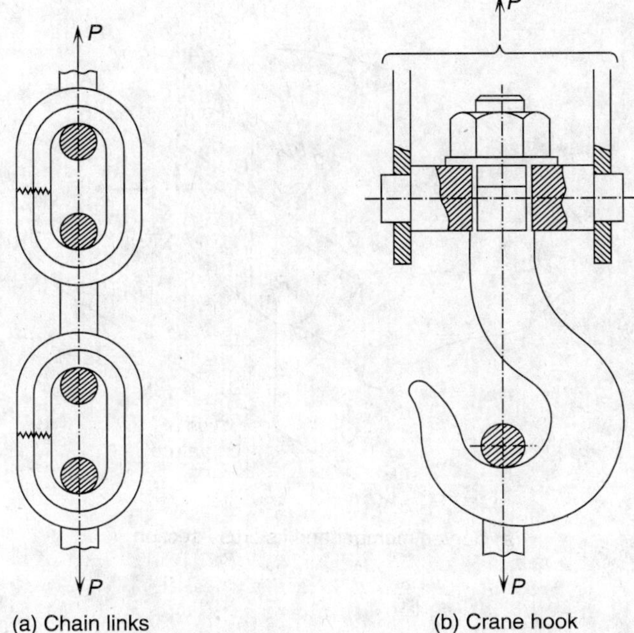

(a) Chain links (b) Crane hook

Figure 18.1 Common examples of curved members subjected to bending.

bends, the fibres on upper side of element are elongated and those on the lower side shortened. The deformations vary linearly from the maximum elongation on one face to the maximum shortening on the other. There exists a layer somewhere near thc middle of the member section where the fibres ef are neither shortened nor elongated. The plane containing fibres like ef is called *neutral surface* because such fibres remain unchanged in length and hence carry no stress. However, although the fibres equidistant from the neutral surface have equal deformations, but due to unequal lengths they will have unequal strains and consequently unequal stresses. This results in *nonlinear stress distribution* as shown in Figure 18.2(d).

As a consequence of the nonlinear stress distribution, there can be no balance between the compressive and tensile forces over the section if the neutral surface passes through the centroid of the section; the neutral surface must shift from the centroid of the section towards the axis of curvature O. This shift of neutral surface results in increased stress at outer fibres and decreased stress at inner fibres as compared with the stresses computed from flexure formula.

Consider a typical element dA at distance y from the neutral axis, which is at a distance e from the centroid of the section. The total deformation (elongation) of a typical fibre gh located at distance y from the neutral surface is $\delta = y\, d\phi$. The initial length of this fibre is $(R_c - e + y)\, d\theta$. Thus, the corresponding strain is

$$\varepsilon = \frac{\delta}{dL} = \frac{y\, d\phi}{(R_c - e + y)\, d\theta} \tag{18.1}$$

where R_c represents the radius of curvature of the centroidal axis. For a homogeneous and elastic material the stress in the fibre is given by

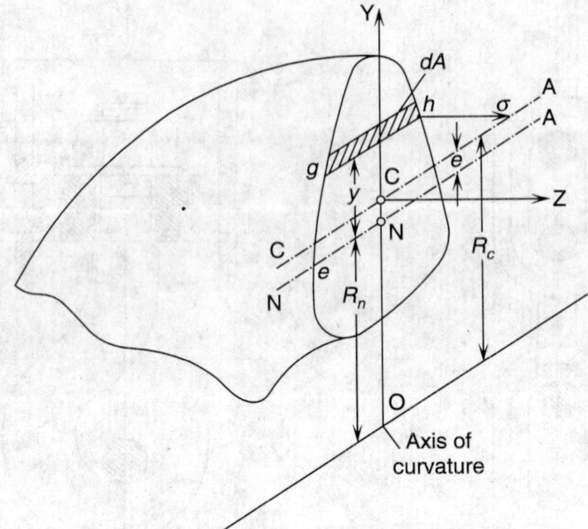

(a) Curved member and its cross-section

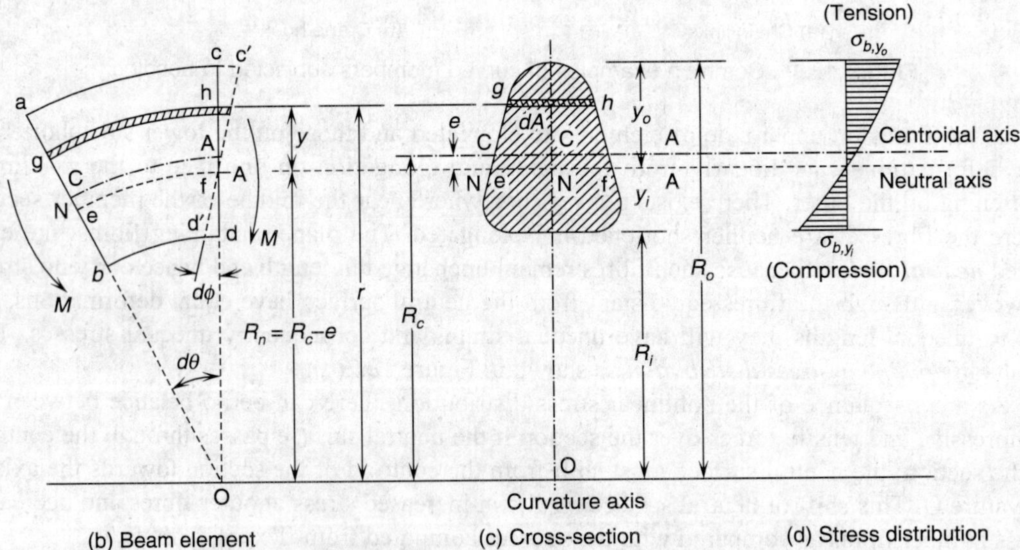

(b) Beam element (c) Cross-section (d) Stress distribution

Figure 18.2 Bending of a curved beam.

$$\sigma_{b,y} = E\varepsilon = \frac{Ey\,d\phi}{(R_c - e + y)\,d\theta} = \left(\frac{E\,d\phi}{d\theta}\right)\frac{y}{(R_c - e + y)} \tag{18.2}$$

Thus, the deformations caused in the member develop normal stresses which are compressive on one part of the cross-section and tensile on the other. The effects of the compressive and tensile stresses are represented by the corresponding resultant forces. These two resultant forces separated by *moment arm* provide the *internal resisting moment*.

18.3 EQUILIBRIUM CONDITIONS

If the member is loaded in pure bending, the conditions of equilibrium require that the sum of normal forces over the cross-section be equal to zero and the internal resisting couple M_r produced by tensile and compressive *stress fields* balance the externally applied bending moment M_z. In accordance with the first equilibrium condition, the force summation over the entire area must vanish

$$\int \sigma_{b,y}\, dA = \left(\frac{E\, d\phi}{d\theta}\right) \int \frac{y\, dA}{(R_c - e + y)} = 0 \tag{18.3}$$

Since $\dfrac{E\, d\phi}{d\theta}$ cannot be zero, then

$$\int \frac{y\, dA}{(R_c - e + y)} = 0 \tag{18.4}$$

Equation (18.4) may be used to determine unknown quantity e. Consider the distance from the axis of curvature to the element dA to be represented by r i.e. $r = R_c - e + y$, then

$$\int \frac{y}{(R_c - e + y)}\, dA = \int \frac{[r - (R - e)]}{r}\, dA$$

$$= \int dA - (R_c - e) \int \frac{dA}{r} = A - (R_c - e) \int \frac{dA}{r} = 0$$

Therefore,
$$\frac{A}{R_c - e} = \int \frac{dA}{r} \quad \text{or} \quad \frac{A}{R_n} = \int \frac{dA}{r} \tag{18.5}$$

The Eq. (18.5) can also be expressed as

$$e = R_c - \frac{A}{\int \frac{dA}{r}} = R_c - R_n \tag{18.6}$$

where R_n is the radius of curvature of the neutral axis. The integral appearing in the denominator represents a geometrical characteristic of the section similar to the *static moment* or *moment of inertia*.

The computation of e as the difference between R_c and R_n involves difficulties because the difference of large quantities R_c and R_n is very small, but need to be evaluated accurately since this directly affects the result of computation of stress. Thus the quantity R_n should be computed to a large number of decimal places.

For rotational equilibrium, the sum of the moments produced by the elemental forces about the neutral axis provides internal resisting moment, M_r which balances the applied external moment, M_z is given by

$$M_z = \int (\sigma_{b,y}\, dA)\, y = \left(\frac{E\, d\phi}{d\theta}\right) \int \frac{y^2\, dA}{(R_c - e + y)} \tag{18.7}$$

The quantity under integration can be simplified as follows

$$\int \frac{y^2\,dA}{(R_c - e + y)} = \int \frac{y[(R_c - e + y) - (R_c - e)]\,dA}{(R_c - e + y)}$$

$$= \int y\,dA - (R_c - e)\int \frac{y\,dA}{(R_c - e + y)} \tag{18.8}$$

The first integral on the right hand side of Eq. (18.8) is the moment of entire cross-section about the neutral axis and equals Ae, and the second is equal to zero as a consequence of Eq. (18.4). Thus, Eq. (18.7) can be rewritten as

$$\frac{M_z}{Ae} = \left(\frac{E\,d\phi}{d\theta}\right) \tag{18.9}$$

From Eqs. (18.2) and (18.9)

$$\sigma_{b,y} = \frac{M_z}{Ae}\frac{y}{(R_c - e + y)} = \frac{M_z}{Ae}\frac{y}{r} \tag{18.10}$$

Equations (18.6) and (18.10) can be used to determine the stresses in the curved members. The analysis of curved members described above is called Winker–Bach theory. It is applicable when the member is subjected to pure bending. If the direct stresses are also present, the resulting normal stress at a point is obtained by superposing the uniform direct stress algebraically on the stresses produced by bending. The shearing stresses induced by the transverse shear V may be calculated as for a straight member, using relation $\tau = (V/Ib)Q$.

18.4 NEUTRAL AXIS AND BENDING STRESS DISTRIBUTION

In the following sections, the procedure for determining the position of neutral axis and bending stress distribution in the members having different cross-section shapes has been illustrated.

18.4.1 Rectangular Section

Consider the rectangular section of a curved member having width b and height h with other details illustrated in Figure 18.3(a). To determine the position of neutral axis, consider an elementary area dA of depth dr at a distance r from the initial curvature axis. Then $dA = b\,dr$, the area of cross-section $A = bh$ and integral of Eq. (18.5) is

$$\frac{A}{R_n} = \int_A \frac{dA}{r} = \int_{R_i}^{R_o} \frac{b\,dr}{r} = b(\log_e R_o - \log_e R_i) = b\log_e(R_o/R_i)$$

or

$$R_n = \frac{A}{b\log_e(R_o/R_i)} = \frac{bh}{b\log_e(R_o/R_i)} = \frac{h}{\log_e(R_o/R_i)}$$

$$e = R_c - R_n = R_c - \frac{h}{\log_e(R_o/R_i)} \tag{18.11}$$

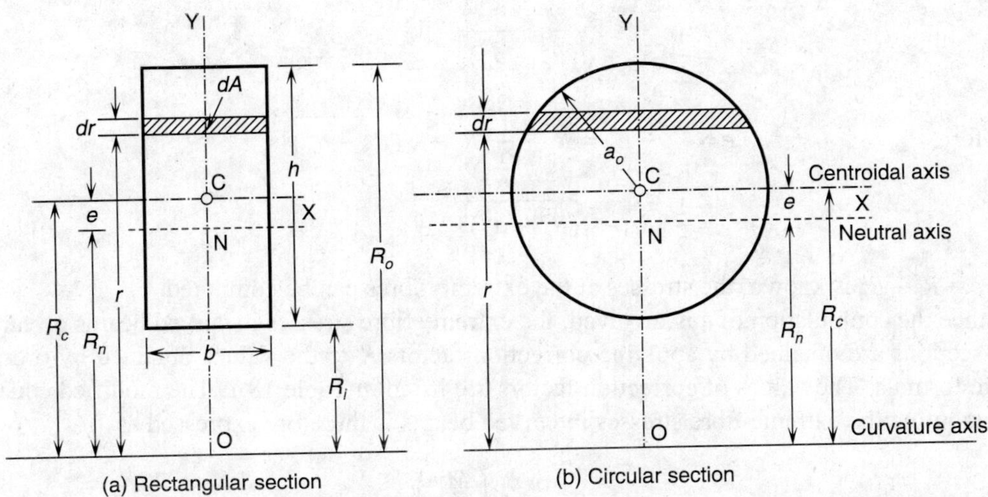

Figure 18.3 Properties of rectangular and circular sections used in curved members subjected to bending moment.

Since $R_o = R_c + (h/2)$ and $R_i = R_c - (h/2)$, Eq. (18.11) reduces to

$$e = R_c - \frac{h}{\log_e \left[\dfrac{R_c + (h/2)}{R_c - (h/2)}\right]} \tag{18.12}$$

Once the distance e is known, the stresses at the extreme fibres can be computed.

18.4.2 Circular Section

Consider the circular section of radius a_o of a curved beam of initial radius R_c with other dimension shown in Figure 18.3(b). To determine the position of neutral axis, consider an elementary area dA of depth dr at a distance r from initial curvature axis. Then

$$dA = 2\sqrt{a_o^2 - (R_c - r)^2}\ dr$$

The area of cross-section is $A = \pi a_o^2$ and the integral

$$\frac{A}{R_n} = \int_A \frac{dA}{r} = 2\int_{R_i}^{R_o} \sqrt{a_o^2 - (R_c - r)^2}\ \frac{dr}{r}$$

$$= 2\pi\left[R_c - \sqrt{(R_c^2 - a_o^2)}\right]$$

$$R_n = \frac{a_o^2}{2[R_c - \sqrt{(R_c^2 - a_o^2)}]}$$

or

$$= \frac{a_o^2[R_c + \sqrt{(R_c^2 - a_o^2)}]}{2[R_c - \sqrt{(R_c^2 - a_o^2)}][R_c + \sqrt{(R_c^2 - a_o^2)}]}$$

$$= \frac{1}{2}[R_c + \sqrt{(R_c^2 - a_o^2)}]$$

Therefore,
$$e = R_c - R_n = R_c - \frac{1}{2}[R_c + \sqrt{(R_c^2 - a_o^2)}]$$

$$= \frac{1}{2}[R_c - \sqrt{(R_c^2 - a_o^2)}] \tag{18.13}$$

Once $e = R_c - R_n$ is known, the stresses at the extreme fibres can be computed.

Since the computation of e is involved, the extreme fibre stresses in curved beams of various cross-sections are obtained by applying correction factors K to the values obtained by ordinary flexure formula. The values of correction factors are listed in Table 18.1. The modified equation for computing the extreme fibre stresses in curved beams is therefore expressed as

$$\sigma_{b,y} = K\left(\frac{My}{I}\right) \tag{18.14}$$

where y is the distance from the neutral axis to the fibre closest (at $r = R_i$) or farthest (at $r = R_o$) to it, i.e. the innermost or the outermost fibre. The correction factor K varies with the ratio $R_c/\{(h/2)$ or $a_o\}$, where R_c is the radius of curvature of the centroidal axis. The correction factors are greater than unity for inner fibres and less than unity for outer fibres. For values of R_c/R_i greater than 20, the factors approach unity and flexure formula may be applied directly to such slightly curved beams.

TABLE 18.1 Correction factors K for curved beams

Cross-section	*Rectangle*		*Circle* or *ellipse*		*Other sections* (average values)	
$R_c/\{(h/2)$ or $a_o\}$	*Inside*	*Outside*	*Inside*	*Outside*	*Inside*	*Outside*
1.2	2.89	0.57	3.41	0.54	When section is	
1.4	2.13	0.63	2.40	0.60	unsymmetrical, R_c/R_i	
1.6	1.79	0.67	1.96	0.65	refer to the inner fibre	
1.8	1.63	0.70	1.75	0.68	–	–
2.0	1.52	0.73	1.62	0.71	1.63	0.74
3.0	1.30	0.81	1.33	0.79	1.36	0.81
4.0	1.20	0.85	1.23	0.84	1.25	0.86
6.0	1.12	0.90	1.14	0.89	1.16	0.90
8.0	1.09	0.92	1.10	0.91	1.12	0.93
10.0	1.07	0.94	1.08	0.93	1.10	0.94
20.0	1.04	0.96	1.03	0.97	1.05	0.95

EXAMPLE 18.1

A circular link having rectangular cross-section 100 mm wide by 50 mm thick, is subjected to a diametrical compressive load of 60 kN as shown in Figure 18.4(a). (a) Determine the stresses at the points I, II, III and IV by using the correction factors. (b) Apply the expressions developed by Winker–Bach theory to check the stresses at I and II.

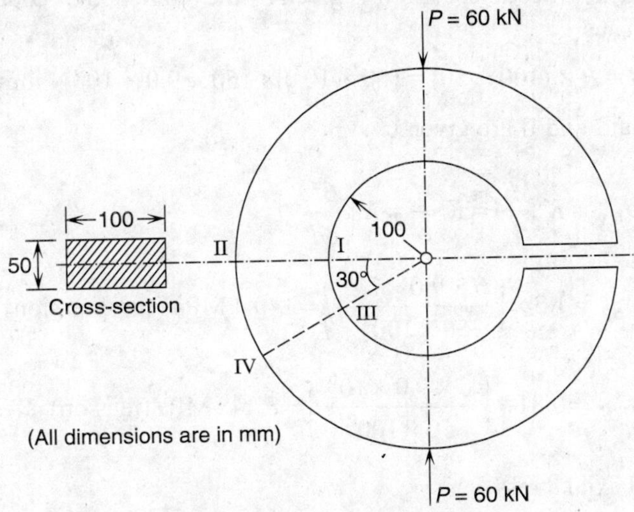

(a) Circular link and its cross-section

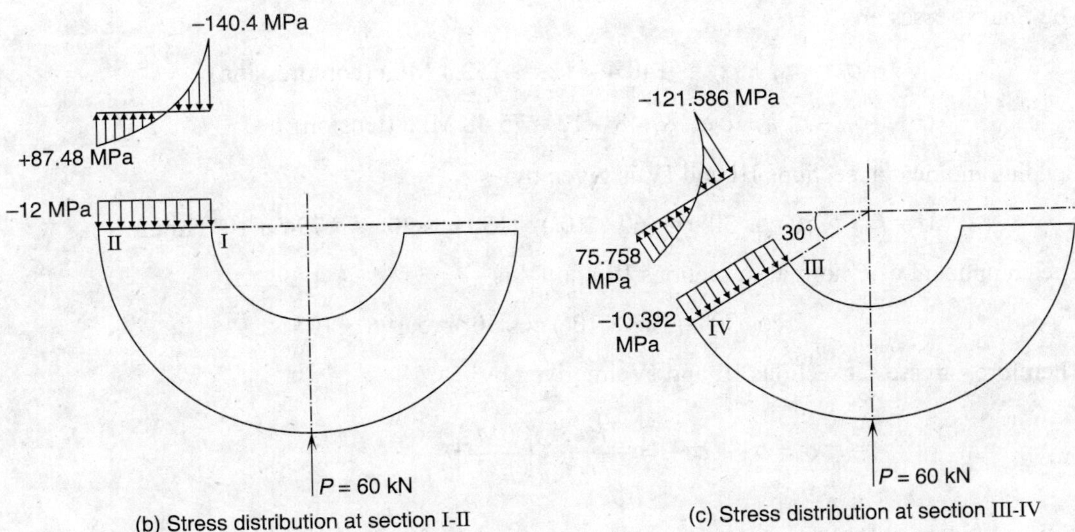

(b) Stress distribution at section I-II

(c) Stress distribution at section III-IV

Figure 18.4 Stresses at specified points on the circular link of Example 18.1.

Solution: The radius of curvature of the centroidal axis of the cross-section is $R_c = 100 + 50 = 150$.
(a) Stresses using the correction factors
For the ratio $R_c/(h/2) = 150/50 = 3$, from Table 18.1

$$K_i = 1.3 \text{ and } K_o = 0.81$$

The bending moment at sections I and II is given by the moment of P about the centroidal axis of the cross-section. Thus

$$M = P \times (100 + 50) = (60 \times 10^3) \times 150 = 9.0 \times 10^6 \text{ N.mm}$$

The bending stresses at I and II are given by:

$$\sigma_{b,y} = K\frac{My}{I} = K\frac{M}{Z} = K\left(\frac{6M}{bd^2}\right)$$

$$\sigma_{b,I} = 1.3 \times \left(\frac{6 \times 9.0 \times 10^6}{50 \times 100^2}\right) = 140.4 \text{ MPa (compression)}$$

$$\sigma_{b,II} = 0.81 \times \left(\frac{6 \times 9.0 \times 10^6}{50 \times 100^2}\right) = 87.48 \text{ MPa (tension)}$$

Uniformly distributed axial stress

$$\sigma_a = \frac{P}{A} = \frac{60 \times 10^3}{50 \times 100} = 12 \text{ MPa (compression)}$$

The final stresses are:

$$\sigma_I = \sigma_{b,I} + \sigma_a = -140.4 - 12 = -152.4 \text{ MPa (compression)}$$
$$\sigma_{II} = \sigma_{b,II} + \sigma_a = 87.48 - 12 = 75.48 \text{ MPa (tension)}$$

Bending moment at sections III and IV is given by:

$$M = P \times (150 \cos 30°) = (60 \times 10^3) \times 150 \cos 30° = 7.794 \times 10^6 \text{ N.mm}$$

The component of P normal to sections III and IV is

$$P \cos 30° = (60 \times 10^3) \cos 30° = 5.196 \times 10^4 \text{ N}$$

Therefore, stresses at sections III and IV are given by:

$$\sigma = \sigma_a + \sigma_{b,y} = -\frac{P}{A} \mp K\left(\frac{6M}{bd^2}\right)$$

$$\sigma_{III} = -\frac{5.196 \times 10^4}{50 \times 100} - 1.3 \times \left(\frac{6 \times 7.794 \times 10^6}{50 \times 100^2}\right)$$

$$= -10.392 - 121.586 = -131.978 \text{ MPa (compression)}$$

$$\sigma_{IV} = -\frac{5.196 \times 10^4}{50 \times 100} + 0.81 \times \left(\frac{6 \times 7.794 \times 10^6}{50 \times 100^2}\right)$$

$$= -10.392 + 75.758 = 65.366 \text{ MPa (tension)}$$

The stress distributions are shown in Figures 18.4(b) and (c).

(b) From Eqs. (18.12) and (18.10):

$$e = 150 - \frac{100}{\log_e \left[\dfrac{150 + 50}{150 - 50}\right]} = 150 - \frac{100}{\log_e 2} = 5.73 \text{ mm}$$

$$\sigma_{b,y_i}; \sigma_{b,y_o} = \frac{M}{Ae}\frac{y}{r} = \frac{M}{Ae}\left[\frac{(h/2) \mp e}{R_c \mp (h/2)}\right]$$

$$\sigma_{b,I} = \frac{M}{Ae}\frac{y_i}{R_i} = \frac{(9.0 \times 10^6)}{(50 \times 100) \times 5.73} \times \frac{50 - 5.73}{150 - 50} = 139.068 \text{ MPa (compression)}$$

$$\sigma_{b,II} = \frac{M}{Ae}\frac{y_o}{R_o} = \frac{(9.0 \times 10^6)}{(50 \times 100) \times 5.73} \times \frac{50 + 5.73}{150 + 50} = 87.534 \text{ MPa (tension)}$$

The values of axial stress are added to obtain the final stresses. The values agree closely with those obtained by using correction factors given in Table 18.1.

EXAMPLE 18.2

A circular ring of internal diameter of 100 mm has a circular cross-section of 80 mm diameter. The ring is subjected to a diametrical compressive load of 25 kN as shown in Figure 18.5. Determine the stresses at the points I and II by using: (a) the expressions developed by Winker–Bach theory and (b) the correction factors.

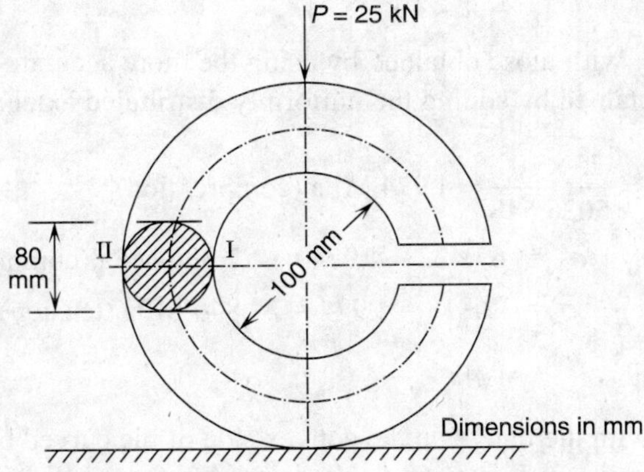

Figure 18.5 Circular open ring with circular cross-section of Example 18.2.

Solution: The radii of curvature of the centroidal axis, inner edge and outer edge of the cross-section are $R_c = 50 + 40 = 90$ mm, $R_i = 50$ mm and $R_o = 130$ mm, respectively.

Cross-sectional area, $A = \pi \times 40^2 = 5026.548\ \text{mm}^2$

The bending moment at section I-II is given by

$$M = P \times (40 + 50) = (25 \times 10^3) \times 90 = 2.25 \times 10^6\ \text{N.mm}$$

(a) From Eqs. (18.13) and (18.10):

$$e = \frac{1}{2}[R_c - \sqrt{(R_c^2 - a_o^2)}] = \frac{1}{2}[90 - \sqrt{(90^2 - 40^2)}]$$

$$= 4.6887\ \text{mm}$$

$$\sigma_{b,I} = \frac{M}{Ae}\frac{y_i}{R_i} = \frac{(2.25 \times 10^6)}{5026.548 \times 4.6887} \times \frac{40 - 4.6887}{50} = 67.422\ \text{MPa (compression)}$$

$$\sigma_{b,II} = \frac{M}{Ae}\frac{y_o}{R_o} = \frac{(2.25 \times 10^6)}{5026.548 \times 4.6887} \times \frac{40 + 4.6887}{130} = 32.818\ \text{MPa (tension)}$$

(b) The ratio $R_c/a_o = 90/40 = 2.25$; hence from Table 18.1

$$K_i = 1.5475 \text{ and } K_o = 0.73$$

The bending stresses at I and II are given by:

$$\sigma_{b,y} = K\frac{My}{I} = K\frac{M}{Z} = K\left(\frac{4M}{\pi R^3}\right)$$

$$\sigma_{b,I} = 1.5475 \times \left[\frac{4 \times (2.25 \times 10^6)}{\pi \times 40^3}\right] = 69.38\ \text{MPa (compression)}$$

$$\sigma_{b,II} = 0.73 \times \left[\frac{4 \times (2.25 \times 10^6)}{\pi \times 40^3}\right] = 32.68\ \text{MPa (tension)}$$

The values agree closely with those obtained by using the more accurate Winker–Bach method. The final stresses are obtained by adding the uniformly distributed axial stress,

$$\sigma_a = \frac{P}{A} = \frac{25 \times 10^3}{5026.548} = 4.974\ \text{MPa (compression)}$$

Thus, $\sigma_I = \sigma_{b,I} + \sigma_a = -67.422 - 4.974 = -72.396$ MPa (compression)

$\sigma_{II} = \sigma_{b,II} + \sigma_a = 32.818 - 4.974 = 27.844$ MPa (tension)

18.4.3 T-Section

Consider the T-section with the flange on the convex side of the curved beam of initial radius R_c. Other dimensions are shown in Figure 18.6. To determine the position of neutral axis consider elementary area dA of depth dr at a distance r from the initial curvature axis in flange and web portions as shown in Figure 18.6 (a). Then

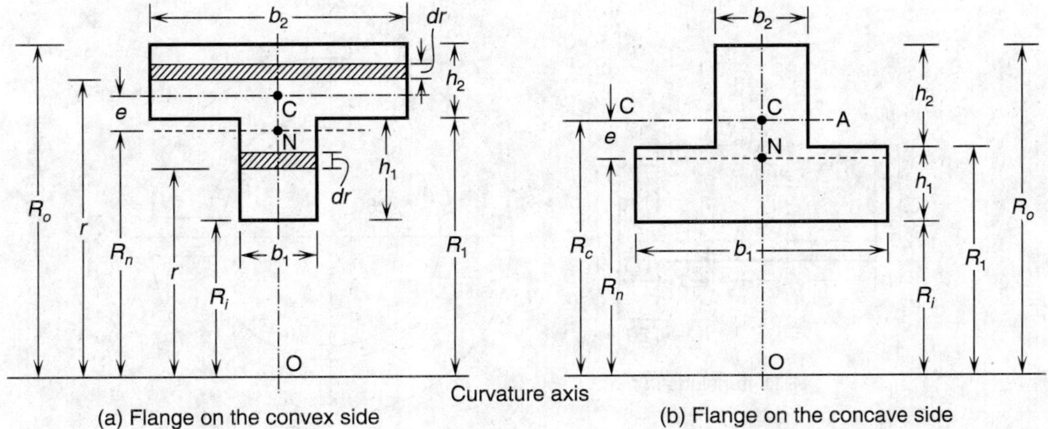

(a) Flange on the convex side (b) Flange on the concave side

Figure 18.6 Different arrangements of T-section used in a curved member subjected to a bending moment.

$$\frac{A}{R_n} = \int_A \frac{dA}{r} = \int_{R_i}^{R_1} \frac{b_1\, dr}{r} + \int_{R_1}^{R_o} \frac{b_2\, dr}{r} = b_1 \log_e \frac{R_1}{R_i} + b_2 \log_e \frac{R_o}{R_1}$$

where $A = b_1 h_1 + b_2 h_2$. Therefore,

$$R_n = \frac{b_1 h_1 + b_2 h_2}{b_1 \log_e \dfrac{R_1}{R_i} + b_2 \log_e \dfrac{R_o}{R_1}} \tag{18.15}$$

With the notation given in Figure 18.6 (b), this result is valid also for the section when the flange is on the concave side. The result can be extended to the sections composed of a series of rectangles as

$$R_n = \frac{b_1 h_1 + b_2 h_2 + \cdots + b_k h_k}{b_1 \log_e \dfrac{R_1}{R_i} + b_2 \log_e \dfrac{R_2}{R_1} + \cdots + b_k \log_e \dfrac{R_o}{R_{k-i}}} \tag{18.16}$$

where k is the number of rectangles, and R_i and R_o are distances from the curvature axis to the innermost and outermost fibres in respective rectangles.

EXAMPLE 18.3

An open circular ring of internal radius of 180 mm has T-section with flange on the concave side. The ring is subjected to a diametrical compressive load of 120 kN as shown in Figure 18.7. Determine the stresses at the points I and II.

Solution: Properties of the T-cross section are,

$$b_1 = 100 \text{ mm};\ h_1 = 20 \text{ mm};\ b_2 = 20 \text{ mm and } h_2 = 140 \text{ mm}$$

Area of the section, $A = 100 \times 20 + 140 \times 20 = 4800 \text{ mm}^2$

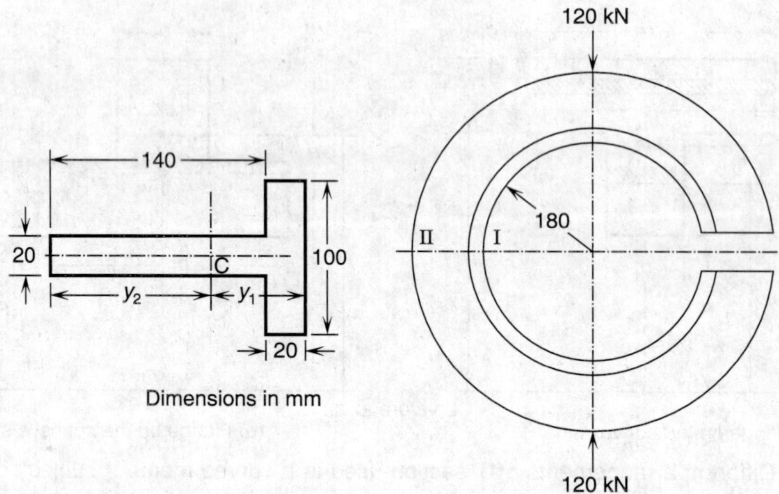

Figure 18.7 Open circular ring with T-cross section of Example 18.3.

The position of centroid of the cross-section with respect to inner edge of the section

$$y_o = \frac{(100 \times 20) \times 10 + (140 \times 20) \times 90}{4800} = 56.67 \text{ mm}$$

Thus, $y_1 = y_o = 56.67$ mm and $y_2 = 160 - 56.67 = 103.33$ mm

And $R_c = 236.67$ mm; $R_i = 180$ mm; $R_1 = 200$ mm; and $R_o = 340$ mm

$$R_n = \frac{(100 \times 20) + (20 \times 140)}{100 \times \log_e \dfrac{200}{180} + 20 \times \log_e \dfrac{340}{200}} = 226.965 \text{ mm}$$

$$e = R_c - R_n = 236.67 - 226.965 = 9.705$$

The bending moment at sections I and II is given by

$$M = (120 \times 10^3) \times (180 + 56.67) = 2.84 \times 10^7 \text{ N.mm}$$

$$\sigma_{b,I} = \frac{M}{Ae}\frac{y_i}{R_i} = \frac{2.84 \times 10^7}{4800 \times 9.705} \times \frac{56.67 - 9.705}{180} = 159.071 \text{ MPa (compression)}$$

$$\sigma_{b,II} = \frac{M}{Ae}\frac{y_o}{R_o} = \frac{2.84 \times 10^7}{4800 \times 9.705} \times \frac{103.33 + 9.705}{340} = 202.685 \text{ MPa (tension)}$$

Uniformly distributed axial stress,

$$\sigma_a = \frac{P}{A} = \frac{120 \times 10^3}{4800} = 25.000 \text{ MPa (compression)}$$

Thus, $\sigma_I = \sigma_{b,I} + \sigma_a = -159.071 - 25.000 = -184.071$ MPa (compression)

$\sigma_{II} = \sigma_{b,II} + \sigma_a = 202.685 - 25.000 = 177.685$ MPa (tension)

18.4.4 Triangular-Section

Figure 18.8(a) illustrates the triangular section of a curved beam of initial radius R_c with its base on the concave side. To determine the position of neutral axis consider an elementary area dA of depth dr at a distance r from initial curvature axis. Then

$$dA = \frac{b}{h}(R_o - r)\,dr$$

$$\frac{A}{R_n} = \int_A \frac{b}{h}\frac{(R_o - r)}{r}\,dr = \frac{b}{h}\left[R_o \int_{R_i}^{R_o} \frac{1}{r}\,dr - \int_{R_i}^{R_o} dr\right]$$

$$= \frac{b}{h}\left[R_o \log_e \frac{R_o}{R_i} - (R_o - R_i)\right] = b\left[\frac{R_o}{h}\log_e \frac{R_o}{R_i} - 1\right]$$

where $A = bh/2$. Therefore,

$$R_n = \frac{h}{2\left(\dfrac{R_o}{h}\log_e \dfrac{R_o}{R_i} - 1\right)} \tag{18.17}$$

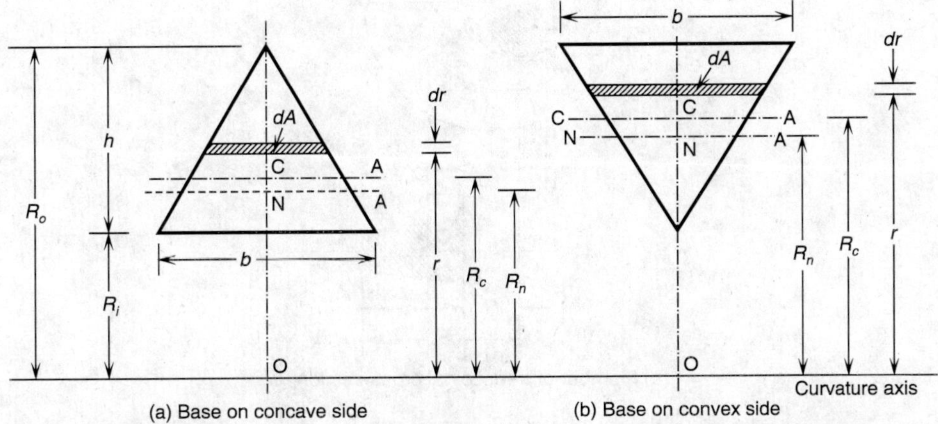

Figure 18.8 Bending of a curved beam with triangular cross-section.

If the section is so arranged that the base of the triangle is on the convex side (see Figure 18.8(b)),

$$dA = \frac{b}{h}(r - R_i)\,dr$$

$$\frac{A}{R_n} = \int_A \frac{b}{h}\frac{(r - R_i)}{r}\,dr = \frac{b}{h}\left(\int_{R_i}^{R_o} dr - R_i \int_{R_i}^{R_o} \frac{1}{r}\,dr\right)$$

$$= \frac{b}{h}\left[(R_o - R_i) - R_i \log_e \frac{R_o}{R_i}\right] = b\left[1 - \frac{R_i}{h}\log_e \frac{R_o}{R_i}\right]$$

where $A = bh/2$. Therefore,

$$R_n = \frac{h}{2\left(1 - \dfrac{R_i}{h}\log_e \dfrac{R_o}{R_i}\right)} \tag{18.18}$$

18.4.5 Trapezoidal-Section

Figures 18.9(a) and (b) shows the parameters of the trapezoidal section with its wider base on the concave side. To determine the position of neutral axis consider an elementary area *dA* of depth *dr* at a distance *r* from initial curvature axis. Then

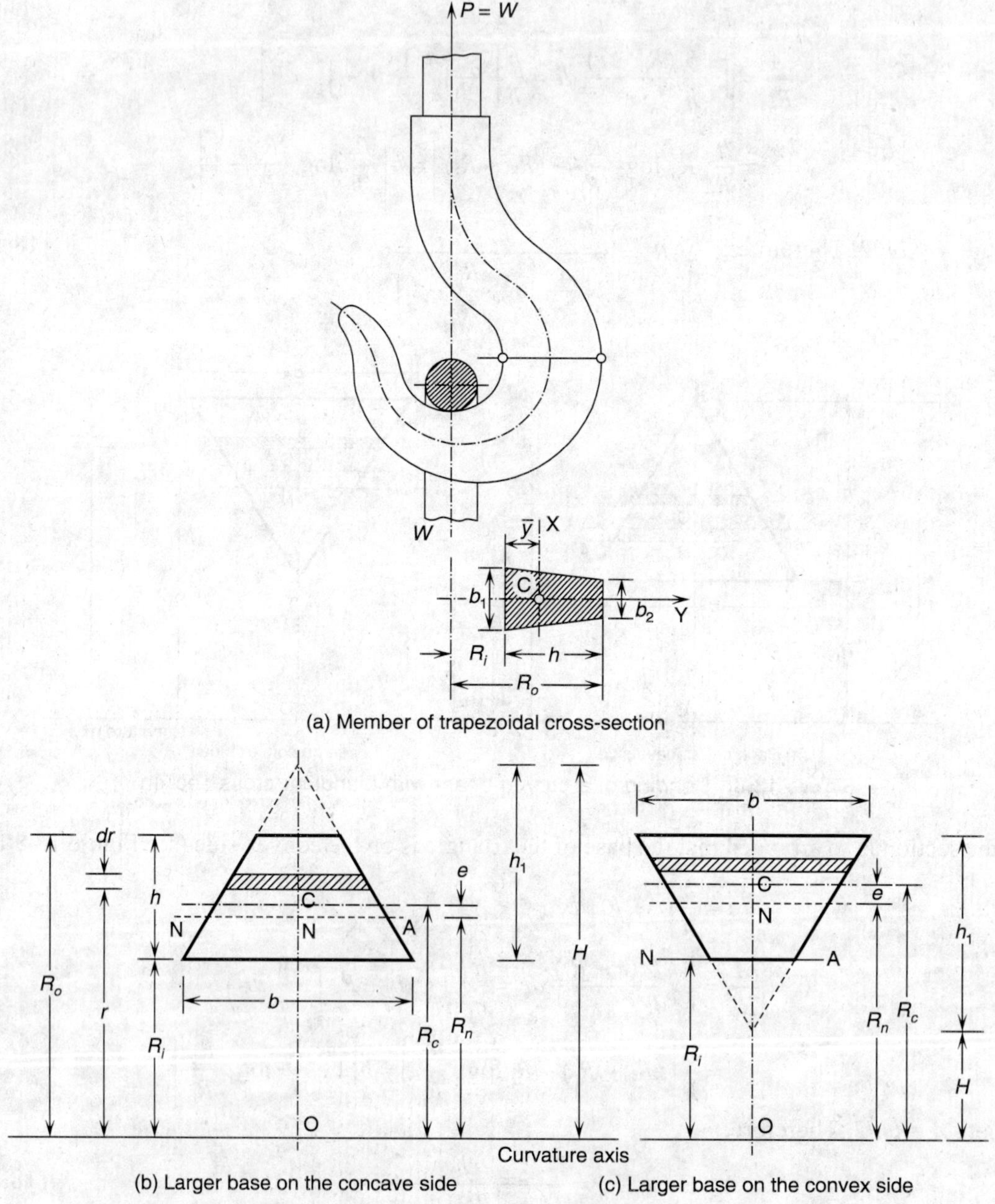

Figure 18.9 Bending of a curved member having trapezoidal cross-section.

$$dA = \frac{b}{h_1}(H - r)\, dr$$

$$\frac{A}{R_n} = \int_A \frac{b}{h_1}\frac{(H-r)}{r}\, dr = \frac{b}{h_1}\left[H\int_{R_i}^{R_o}\frac{1}{r}\, dr - \int_{R_i}^{R_o} dr\right]$$

$$= \frac{b}{h_1}\left[H \log_e \frac{R_o}{R_i} - (R_o - R_i)\right] = \frac{b}{h_1}\left[H \log_e \frac{R_o}{R_i} - h\right]$$

Therefore,

$$R_n = \frac{Ah_1}{bh\left(\frac{H}{h}\log_e \frac{R_o}{R_i} - 1\right)} \tag{18.19}$$

If the wider base of the trapezoidal section is on the convex side (see Figure 18.9(c)).

$$dA = \frac{b}{h_1}(H - r)\, dr$$

Proceeding as in the above case,

$$R_n = \frac{Ah_1}{bh\left(1 - \frac{H}{h}\log_e \frac{R_o}{R_i}\right)} \tag{18.20}$$

EXAMPLE 18.4

A crane hook curved to an internal diameter of 60 mm has a trapezoidal cross-section with base widths of 40 mm on concave side and 10 mm on the convex side, and a height of 70 mm as shown in Figure 18.10. Determine the capacity of the hook for a maximum allowable stress of 100 MPa in the material of the hook.

Solution: The distance y_o from the wider base of trapezium to its centroid is

$$y_o = \frac{b_1 + 2b_2}{b_1 + b_2}\frac{h}{3} = \frac{40 + 2\times 10}{40 + 10} \times \frac{70}{3} = 28 \text{ mm}$$

The radius of curvature of the centroidal axis of the cross-section is

$$R_c = 30 + 28 = 58 \text{ mm}$$

Radii of inner and outer fibres are:

$$R_i = 30 \text{ mm and } R_o = 30 + 70 = 100 \text{ mm}$$

As the wider base of the cross-section is on the concave side, from Eq. (18.19)

$$R_n = \frac{Ah_1}{bh\left(\frac{H}{h}\log_e \frac{R_o}{R_i} - 1\right)}$$

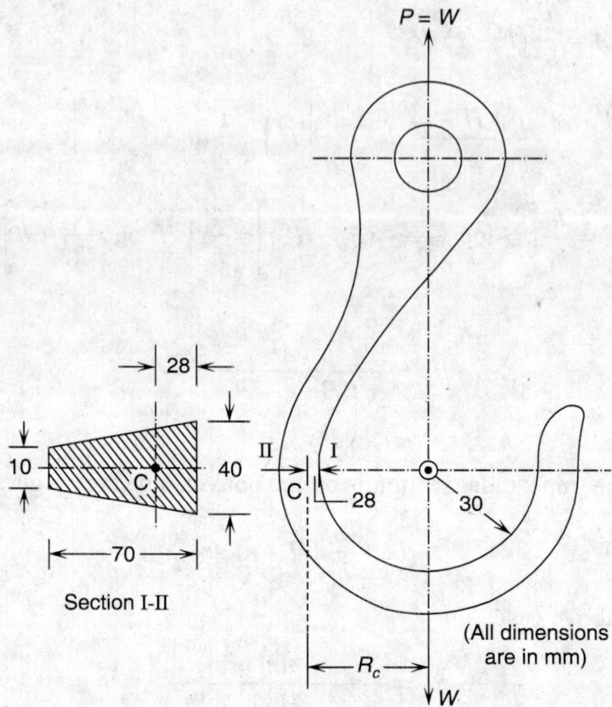

Figure 18.10 Crane hook of Example 18.4.

where $b = b_1 = 40$ mm; $h = 70$ mm;

$$A = \left(\frac{b_1 + b_2}{2}\right) h = \frac{40 + 10}{2} \times 70 = 1750 \text{ mm}^2$$

$$h_1 = \frac{b_1}{b_1 - b_2} h = \frac{40}{40 - 10} \times 70 = 93.33 \text{ mm}$$

And $$H = R_i + h_1 = 30 + 93.33 = 123.33 \text{ mm}$$

Therefore, $$R_n = \frac{1750 \times 93.33}{40 \times 70 \times \left(\frac{123.33}{70} \log_e \frac{100}{30} - 1\right)} = 52.0244 \text{ mm}$$

Hence, $$e = R_c - R_n = 58 - 52.0244 = 5.9756 \text{ mm}$$

The bending moment at a section about the centroidal axis of the cross-section is

$$M = W \times R_c = (W \times 10^3) \times 58 = (5.8 \times 10^4) W \text{ N.mm}$$

$$\sigma_{b,y_i} = \frac{M}{Ae} \frac{y_i}{R_i} = \frac{(5.8 \times 10^4) W}{(1750) \times 5.9756} \times \frac{28 - 5.9756}{30} = 4.074W \text{ MPa (tension)}$$

$$\sigma_{b,y_o} = \frac{M}{Ae} \frac{y_o}{R_o} = \frac{(5.8 \times 10^4) W}{(1750) \times 5.9756} \times \frac{42 + 5.9756}{100} = 2.661W \text{ MPa (compression)}$$

Uniformly distributed direct tensile stress

$$\sigma_a = \frac{W}{A} = \frac{W \times 10^3}{1750} = 0.571W \text{ MPa (tension)}$$

As the maximum stress is limited to 100 MPa,

$$(4.074 + 0.571)W = 100 \quad \text{or} \quad W = 21.52 \text{ kN}$$

18.5 PROBLEMS

18.1 A curved beam has a square cross-section of side 480 mm. If the radius of curvature of the centroidal axis is 600 mm, determine the ratio of the maximum to minimum bending stresses when the beam is subjected to a bending moment.

[**Ans.** 1.762]

18.2 A bar of circular cross-section of radius 200 mm is bent to a circular arc of radius 400 mm. Determine the extreme-fibre stresses in the bent bar, if it is subjected to an external bending moment of 600 kN.m.

[**Ans.** e = 26.795 mm; σ_t = 154.32 MPa and σ_c = 67.355 MPa]

18.3 A crane hook shown in Figure 18.11 is curved to an internal radius of 50 mm and has a circular cross-section of 100 mm diameter. Determine the maximum load W that may be supported without exceeding a stress of 100 MPa in the material of the hook.

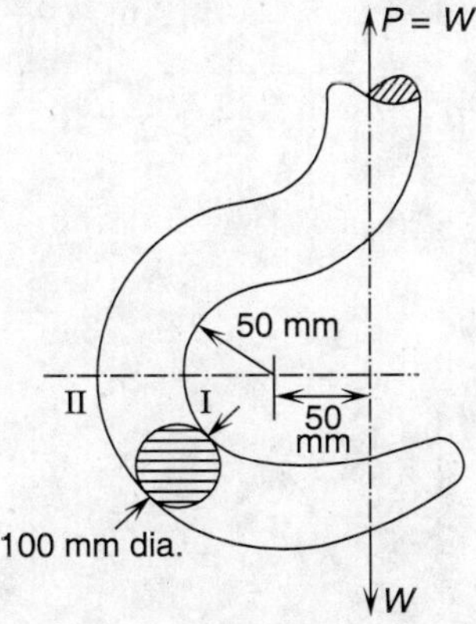

Figure 18.11 Crane hook of Problem 18.3.

[**Ans.** 39.25 kN]

18.4 A crane hook curved to an internal diameter of 50 mm has a trapezoidal cross-section with base widths of 40 mm on concave side and 20 mm on the convex side and a height of 100 mm as shown in Figure 18.12. Determine the capacity of the hook for a maximum allowable stress of 120 MPa in the material of the hook.

[**Ans.** 40.38 kN]

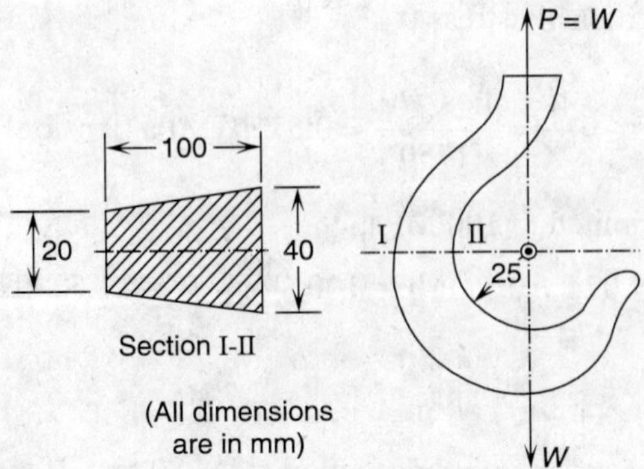

Figure 18.12 Crane hook of Problem 18.4.

18.5 The cross-section of a curved beam is a triangle of base 200 mm and height 300 mm. The radius of curvature of centroidal axis is 400 mm. If the beam is subjected to a bending moment of 120 kN.m which tries to open it. Determine the maximum stresses in the cross-section when the base of the triangle is on the: (a) convex side and (b) on the concave side.

[**Ans.** (a) $\sigma_t = 255.84$ MPa; $\sigma_c = 63.12$ MPa; (b) $\sigma_t = 100.632$ MPa; and $\sigma_c = 120.624$ MPa]

18.6 A crane hook curved to an internal diameter of 50 mm has a T-cross-section with a flange of 100 × 20 mm and a web of 80 × 20 mm. The flange is on the concave side. Determine the maximum stress in the cross-section when the hook carries a load of 100 kN.

[**Ans.** $\sigma_t = 153.43$ MPa and $\sigma_c = 107.58$ MPa]

Chapter 19

Plastic Bending

19.1 INTRODUCTION

In the preceding chapters, the materials were considered to be stressed within the elastic range; that stress was proportional to strain. The maximum permissible stress was the *proportionality limit*, even though the ultimate stress of the material was higher. However, in practice loads producing *inelastic strains* may safely be applied inspite of the fact that in such cases permanent deformations produced in the material do not vanish even when the loads are removed. Although these deformations are large compared to elastic deformations, but not so large to result in the collapse of the structure. These loads are designated as *limit loads*. *A limit is defined as the maximum load that can be applied to the structure before it collapses*. With a *ductile material*, this condition will not occur until the *yield point* has been reached throughout the most highly stressed section in a statically *determinate structure*. The application of limit loads to indeterminate structures is called *limit analysis*. This concept is applicable only to the materials having some ductility.

The idealized stress–strain relation for a ductile material is shown in Figure 19.1(a). The initial straightline portion of the diagram having slope E, the modulus of elasticity of the material, represents elastic region. The plastic region is also a straight line starting at the yield stress σ_{yp}, and having a slope C which is smaller than E. Thus, within plastic region the strain increases at faster rate than that in elastic region. Such a behaviour of the material is called *strain-hardening*. A material with C equal to zero is called *elastic-perfectly plastic*, i.e. *plastic flow* or *infinite strain* can occur with increase in the stress in the material beyond the yield point as shown in Figure 19.1(b).

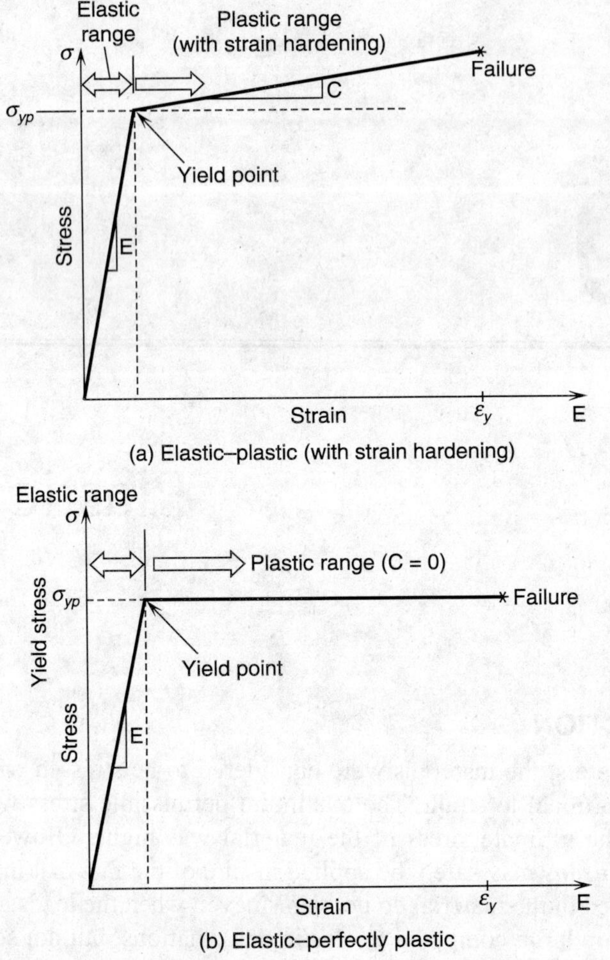

Figure 19.1 Idealized stress–strain relations for a ductile material.

19.2 LIMIT MOMENT

In dealing with inelastic action in flexure, the assumptions made in pure bending remain valid, except that the stresses need not be proportional to the strain. The transverse sections are still considered to remain plane; hence strains are proportional to their distance from the *neutral axis*. However, in a beam of elastic-perfectly plastic material, the stresses remain constant at the yield stress σ_{yp} whenever the strain exceeds the yield strain ε_{yp}. This chapter deals with the regions of the structures where the cross-section behaves plastically.

Consider a cantilever beam of rectangular cross-section shown in Figure 19.2(a) subjected to a concentrated load W acting at its free end. The bending moment diagram is shown in Figure 19.2(b). When the load W is increased beyond elastic limit, i.e. the structure is overloaded, the stresses in the material exceed the elastic or proportional limit σ_y and strains in some parts of the member may exceed ε_y (= σ_y/E). In all such parts where $\varepsilon > \varepsilon_y$, the material behaves plastically, i.e. it deforms at a constant yield stress of σ_y. The shaded area in Figure 19.2(a) denotes that portion of the beam that has been strained into plastic range. At section I-I, the stresses on the extreme fibres have just reached the yield stress, strain and stress distributions on the section are still elastic or linear as shown in Figures 19.2(c) and (d).

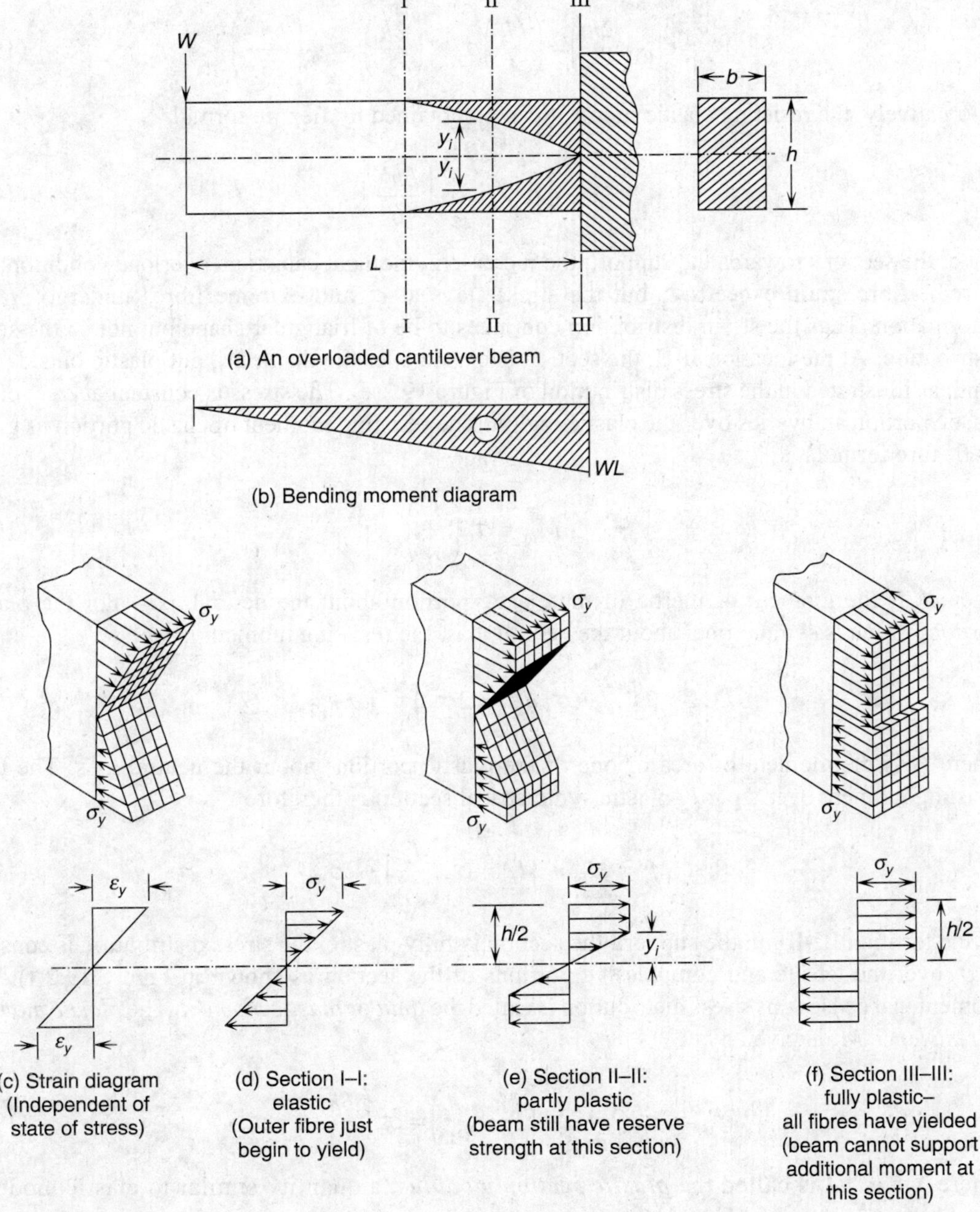

Figure 19.2 Variation of flexural stress distribution with the increase in the moment.

For internal force equilibrium,

$$C = T = \left(\frac{bh}{4}\right)\sigma_y \tag{19.1}$$

The limiting elastic resisting moment,

$$M_e = C \times \left(\frac{2h}{3}\right) = \left(\frac{bh}{4}\right)\sigma_y \times \left(\frac{2h}{3}\right) = \sigma_y\left(\frac{bh^3}{6}\right) \tag{19.2}$$

Alternatively, the resisting elastic moment can be obtained by flexure formula

$$M_e = \sigma_y Z = \sigma_y\left(\frac{bh^3}{6}\right) \tag{19.3}$$

But at the sections towards the support, the increase in moment causes an overload condition; the extreme-fibre strain exceeds ε_y but the stress stays at σ_y and extreme fibres undergo plastic deformation. Thus the strain distribution continues to be of triangular shape but not so the stress distribution. At the location II-II, the section is elastic over the depth $2y_i$, but plastic outside this depth as illustrated in the stress distribution of Figure 19.2(e). The stress is constant at σ_y over the plastic portion and varies over the elastic region. The resisting moment of elastic portion as given by flexure formula is

$$M_{ie} = \sigma_y\left(\frac{I_i}{y_i}\right) \tag{19.4}$$

where I_i is the moment of inertia of the elastic portion about the neutral axis. For the plastic portion, which is symmetrical about the neutral axis, the resisting moment is

$$M_{ip} = 2\int_{y_i}^{h/2} y(\sigma_y\, dA) = 2\sigma_y \int_{y_i}^{h/2} y\, dA = 2\sigma_y Q \tag{19.5}$$

where Q is the moment of area of one of the plastic portions about the neutral axis. The total resisting moment over a partly plastic symmetrical section is therefore

$$M_i = M_{ie} + M_{ip} = \sigma_y\left(\frac{I_i}{y_i}\right) + 2\sigma_y Q \tag{19.6}$$

At the location III-III on the support, the section is fully plastic; the stress distribution is constant at σ_y over the tensile and compressive portions of the section as shown in Figure 19.2(f). The moment that causes this stress distribution is called the *limit* or *plastic moment*. The *plastic moment of resistance* M_p is given by

$$M_p = 2\sigma_y Q = 2\sigma_y\left(\frac{bh}{2} \times \frac{h}{4}\right) = \sigma_y\left(\frac{bh^2}{4}\right) = \sigma_y S \tag{19.7}$$

where $S = bh^2/4$ is called the ***plastic section modulus***, a quantity similar to elastic modulus $Z = bh^2/6$.

19.3 THE SHAPE FACTOR

It is evident that $M_p > M_e$ as in fully plastic stage all the fibres are stressed to their full capacity σ_y, whereas at $M = M_e$ only the extreme fibres are stressed to σ_y.
From Eqs. (19.2) and (19.7):

$$\frac{M_p}{M_e} = \frac{S\sigma_y}{Z\sigma_y} = \frac{S}{Z} = s_f \tag{19.8}$$

The ratio M_p/M_e is called *the shape factor* of the cross-section. It is a constant which varies with the shape of the cross-section.

19.3.1 Rectangular Section

$$s_f = \frac{S}{Z} = \frac{bh^2/4}{bh^2/6} = 1.5$$

19.3.2 Circular Section

Consider a circular section beam with diameter D as shown in Figure 19.3(a). The stress distribution across the section is shown in Figure 19.3(b). In fully plastic stage, the compression and tension acting on the cross-section are,

$$C = T = A\sigma_y = \left(\frac{\pi D^2}{8}\right)\sigma_y \tag{19.9}$$

These forces act at the centre of gravities of the areas above and below the neutral axis which are at distance of $2D/3\pi$ from the neutral axis. Since for the fully plastic section σ_y is constant throughout,

$$M_p = 2\sigma_y Q = 2\sigma_y\left(\frac{\pi D^2}{8} \times \frac{2D}{3\pi}\right) = \sigma_y\left(\frac{D^3}{6}\right)$$

or $\quad S = D^3/6 \quad$ (19.10)

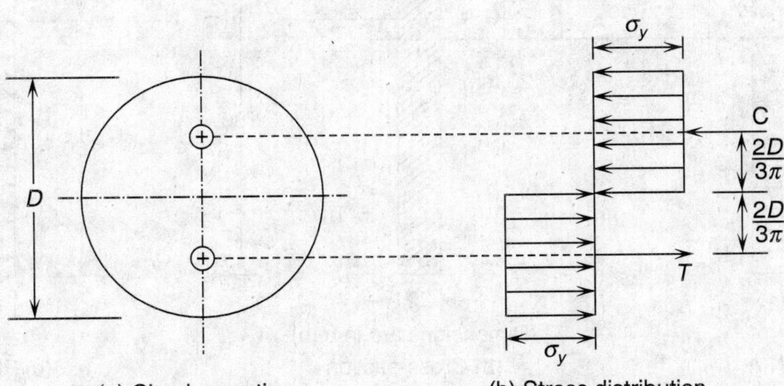

Figure 19.3 Computation of plastic moment of resistance.

For circular section, the shape factor is,

$$s_f = \frac{S}{Z} = \frac{D^3/6}{\pi D^3/32} = \frac{16}{3\pi} \approx 1.7$$

The values of this ratio for some of the shapes are listed in Table 19.1. It should be noted that the *limit moment* for rectangular and circular sections is 50 to 70 per cent over the yield moment, whereas beam sections are only about 10 per cent stronger if inelastic action is permitted.

TABLE 19.1 Ratio of the limit moment to the yield moment

Cross-section	M_p/M_e
Solid rectangle	1.5
Solid diamond	2.0
Solid circle	1.7
Thin walled circular tube	1.27
Wide flange beams (average)	1.14

For beams of unsymmetrical sections such as the T-beam shown in Figure 19.4(b), the neutral axis changes its location as the section enters the plastic range. In fully plastic case where the yield stress is constant over the section, the equilibrium condition that the total axial force on the section be zero requires that the areas subjected to tension and compression be equal, i.e.

$$T = C \quad \text{or} \quad \sigma_y A_t = \sigma_y A_c \quad \text{or} \quad A_t = A_c$$

For illustration, consider the T-section of Figure 19.4(a). The position of elastic neutral axis (see Figure 19.4(a)) from the bottom of the section is

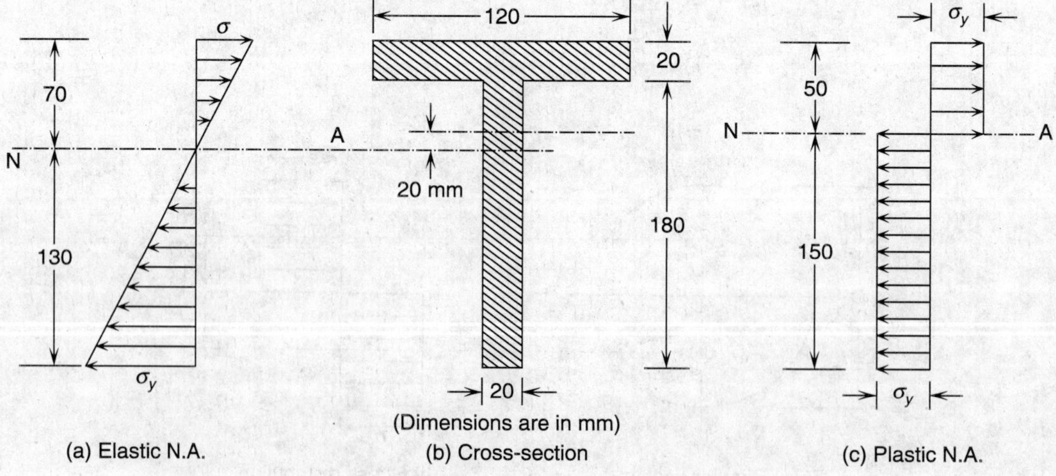

Figure 19.4 Shift in neutral axis in unsymmetrical section due to overload.

$$\bar{y}_e = \frac{A_1 y_1 + A_2 y_2}{A_1 + A_2} = \frac{(20 \times 180) \times 90 + (120 \times 20) \times 190}{(20 \times 180) + (120 \times 20)} = 130 \text{ mm}$$

The position of plastic neutral axis (see Figure 19.4(c)) from the bottom of the section is

$$\bar{y}_p = \frac{(A_1 + A_2)/2}{20} = \frac{[(20 \times 180) + (120 \times 20)]/2}{20} = 150 \text{ mm}$$

Thus the neutral axis for the fully plastic case is 20 mm above the neutral axis for the elastic case.

19.4 MOMENT-CURVATURE DIAGRAM

The moment of resistance M and curvature φ at different stages of overload can be determined, and *moment-curvature diagram* can be plotted as shown in Figure 19.5(d). The neutral axis is always determined from the condition $T = C$. Consider the case of a rectangular section where due to symmetry the neutral axis is located at mid-height, i.e. at $h/2$. (see Figures 19.5(a)–(c)).

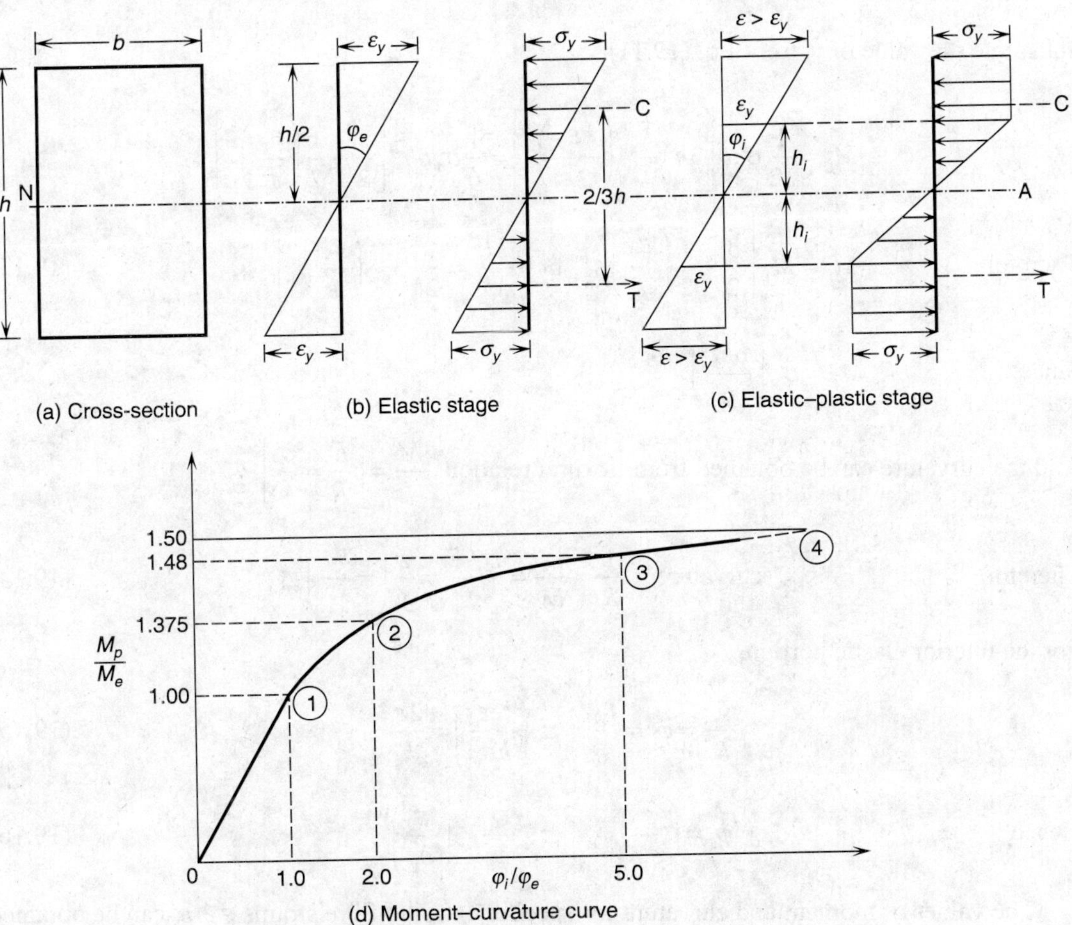

Figure 19.5 Moment–curvature diagram of a rectangular section.

Consider the overload case when the extreme-fibre strain $\varepsilon \geq \varepsilon_y$ and stress varies from 0 at neutral axis to σ_y at y_i and remains constant thereafter. From strain distribution,

$$\frac{\varepsilon}{(h/2)} = \frac{\varepsilon_y}{y_i} \quad \text{or} \quad y_i = \frac{h\varepsilon_y}{2\varepsilon} \tag{19.11}$$

The resisting moment over a partly plastic rectangular (symmetrical) section is given by Eq. (19.6)

$$M_i = \sigma_y\left[\frac{b(2y_i)^2}{6}\right] + 2\sigma_y\left[b\left(\frac{h}{2} - y_i\right)\left(\frac{1}{2}\right)\left(\frac{h}{2} + y_i\right)\right]$$

$$= \sigma_y b\left[\frac{2y_i^2}{3} + \left(\frac{h^2}{4} - y_i^2\right)\right] = \sigma_y b\left[\left(\frac{h^2}{4} - \frac{y_i^2}{3}\right)\right]$$

$$= \frac{1}{12}\sigma_y b(3h^2 - 4y_i^2) \tag{19.12}$$

Substitute the value of y_i from Eq. (19.11)

$$M_i = \frac{1}{12}\sigma_y b\left[3h^2 - 4\left(\frac{h\varepsilon_y}{2\varepsilon}\right)^2\right] = \frac{1}{6}\sigma_y bh^2\left[\frac{3}{2} - \frac{1}{2}\left(\frac{\varepsilon_y}{\varepsilon}\right)^2\right]$$

Therefore,
$$M_i = M_e\left[\frac{3}{2} - \frac{1}{2}\left(\frac{\varepsilon_y}{\varepsilon}\right)^2\right] \quad \text{or} \quad \frac{M_i}{M_e} = \frac{1}{2}\left[3 - \left(\frac{\varepsilon_y}{\varepsilon}\right)^2\right] \tag{19.13}$$

Since
$$M_e = \sigma_y\left(\frac{bh^3}{6}\right)$$

And the curvature can be obtained from flexural relation, $\dfrac{M}{I} = \left[\dfrac{E}{R} = \dfrac{\sigma}{y}\right]$

Therefore,
$$\text{curvature} = \frac{1}{R} = \frac{\sigma}{Ey} = \frac{\sigma/E}{y} = \frac{\varepsilon}{y}\left(= \frac{\varepsilon_y}{y_i}\right) \tag{19.14}$$

For the interior elastic portion,

$$\frac{1}{R_i} = \varphi_i = \frac{\varepsilon_y}{y_i} = \varepsilon_y\left(\frac{2\varepsilon}{h\varepsilon_y}\right) = \frac{2\varepsilon}{h} \tag{19.15}$$

Hence
$$\varphi_i = \left(\frac{\varepsilon}{\varepsilon_y}\right)\varphi_e \quad \text{or} \quad \frac{\varphi_i}{\varphi_e} = \left(\frac{\varepsilon}{\varepsilon_y}\right) \tag{19.16}$$

The values of moment and curvature for specific extreme-fibre strains $\varepsilon \geq \varepsilon_y$ can be obtained from Eqs. (19.13) and (19.16) for various stages of overloading:

(i) $\varepsilon = \varepsilon_y$ (linear elastic range): $\dfrac{M_i}{M_e} = 1.0$ and $\dfrac{\varphi_i}{\varphi_e} = 1.0$

(ii) $\varepsilon = 2\varepsilon_y$ (elasto-plastic range): $\dfrac{M_i}{M_e} = \dfrac{11}{48} = 1.375$ and $\dfrac{\varphi_i}{\varphi_e} = 2.0$

(iii) $\varepsilon = 5\varepsilon_y$ (elasto-plastic range): $\dfrac{M_i}{M_e} = \dfrac{74}{50} = 1.48$ and $\dfrac{\varphi_i}{\varphi_e} = 5.0$

(iv) $\varepsilon >> \varepsilon_y$ (fully-plastic range): $\dfrac{M_p}{M_e} = \dfrac{3}{2} = 1.50$ and $\dfrac{\varphi_i}{\varphi_e} = \infty$

The resulting moment-curvature diagram is plotted in Figure 19.5(d).

EXAMPLE 19.1

Determine the shape factor and plastic moment of resistance M_p of I-section shown in Figure 19.6(a). Also compute the working moment a beam having this cross-section can support if the factor of safety with respect to the failure by yielding is 2. Consider $\sigma_{y,t} = \sigma_{y,c} = \sigma_y = 250$ MPa.

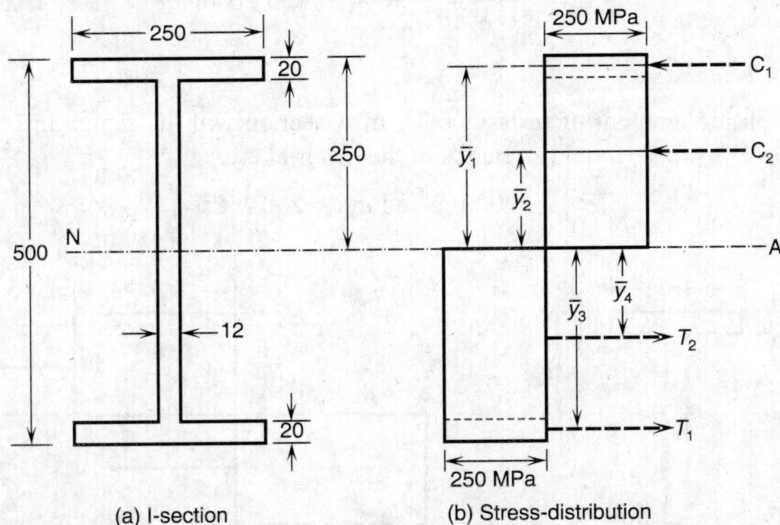

Figure 19.6 Computation of plastic moment of resistance of I-section.

Solution: Due to symmetry of the cross-section and for $\sigma_{y,t} = \sigma_{y,c}$, the neutral axis is located at the mid-height of the cross-section. The stress distribution is shown in Figure 19.6(b). The plastic moment of resistance is computed by taking moments of C_1, C_2, T_1 and T_2 about the neutral axis

$$M_p = (C_1\bar{y}_1 + C_2\bar{y}_2) + (T_1\bar{y}_3 + T_2\bar{y}_4)$$

$$= [(A_1\bar{y}_1 + A_2\bar{y}_2)] \times \sigma_y + [(A_3\bar{y}_3 + A_3\bar{y}_4)] \times \sigma_y$$

$$= 2 \times [(250 \times 20) \times 240 + (230 \times 12) \times 115] \times 250$$

$$= 758700000 \text{ N.mm} = 758.7 \text{ kN.m}$$

Therefore, plastic section modulus is

$$S = \frac{M_p}{\sigma_y} = \frac{758.7 \times 10^6}{250} = 3.0348 \times 10^6 \text{ mm}^3$$

The elastic section modulus is

$$Z = \frac{1}{y_t} = \frac{1}{y_c} = \frac{1}{12}[250 \times 500^3 - (250 - 12) \times (500 - 2 \times 20)^3] \times \frac{1}{250}$$

$$= 2.694677 \times 10^6 \text{ mm}^3$$

The shape factor

$$s_f = \frac{S}{Z} = \frac{3.0348 \times 10^6}{2.694677 \times 10^6} = 1.126$$

From the condition of strength, working moment is

$$M = \frac{M_p}{F.O.S.} = \frac{758.7}{2} = 379.35 \text{ kN.m}$$

EXAMPLE 19.2

Determine the plastic moment of resistance M_p of T-section with its flange in compression as shown in Figure 19.7(a). The yield strengths of the material are:

$$\sigma_{y,t} = 300 \text{ MPa and } \sigma_{y,c} = 250 \text{ MPa.}$$

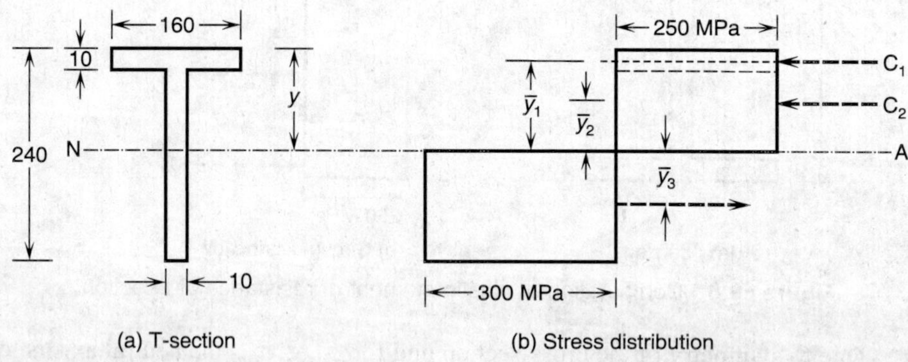

Figure 19.7 Computation of plastic moment of resistance of T-section of Example 19.2.

Solution: Consider that the neutral axis of the cross-section is located at depth y from the top of the flange. From the condition $C = T$:

$$[(160 \times 10) + (y - 10) \times 10] \times 250 = [(240 - y) \times 10] \times 300$$

Therefore, $\quad 110y = 6900 \quad \text{or} \quad y = 62.73 \text{ mm}$

The stress distribution is shown in Figure 19.7(b). Take moments of C_1, C_2 and T_1 about the neutral axis

$$M_p = (C_1\bar{y}_1 + C_2\bar{y}_2) + (T\bar{y}_3) = [(A_1\bar{y}_1 + A_2\bar{y}_2)] \times \sigma_{y,c} + (A_3\bar{y}_3) \times \sigma_{y,t}$$

$$= [(160 \times 10) \times (62.73 - 5) + (62.73 - 10) \times 10 \times (62.73 - 10)/2] \times 250$$

$$+ [(240 - 62.73) \times 10 \times (240 - 62.73)/2] \times 300$$

$$= 73704545 \text{ N.mm} = 73.70 \text{ kN.m}$$

EXAMPLE 19.3

A bracket is mounted on a rigid wall by three identical bolts as shown in Figure 19.8. The bracket is considered to be rigid so that the deformation of the bolts is caused by rotation of the bracket about its bottom point O. Determine: (a) the maximum load W that can be supported by the bracket at a distance $a/2$ from the face of the wall and (b) the ratio of the limit load to the yield load.

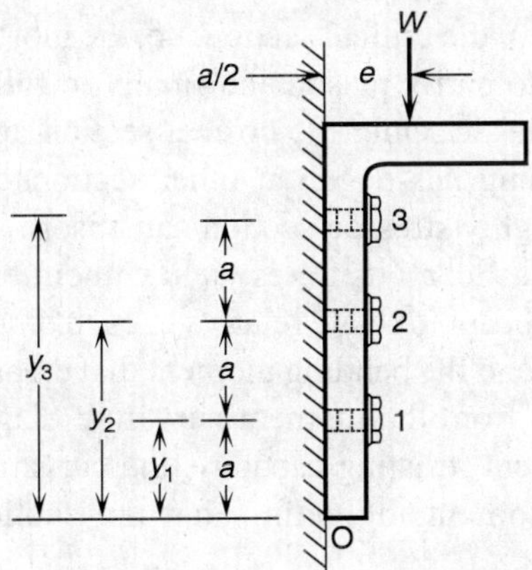

Figure 19.8 Bracket mounted on wall by bolts.

Solution: Consider the material of the bolts to be elastic-perfectly plastic. The deformations of the bolts are proportional to their distance from the point O and the bolts carry axial forces. The capacity of each bolt is given by the load at which yield begins, i.e. $F_p = A\sigma_y$. While any bolt remains in elastic range, rotation of the bracket will remain small. Hence excessive rotation will not occur until all the three bolts have reached their limit force. The limit force is determined by the moment equilibrium at point O,

$$\sum M_O = 0 \quad \text{or} \quad We - F_p(y_1 + y_2 + y_3) = 0 \quad \text{or} \quad We = F_p(y_1 + y_2 + y_3)$$

or

$$W_p(a/2) = (A\sigma_2)(a + 2a + 3a)$$

Therefore,

$$W_p = 12A\sigma_y$$

(b) In the elastic solution also the deformations of the bolts are proportional to their distance from the point O. Here all the bolts must remain in elastic range; the maximum stress in bolt 3 is σ_y, and the stresses in the bolts 2 and 1 will be $2\sigma_y/3$ and $\sigma_y/3$, respectively. The maximum elastic force is given by: $\sum M_O = 0$ i.e.

$$W_e\left(\frac{a}{2}\right)=A\left(\frac{\sigma_y}{3}\right)(a)+A\left(\frac{2\sigma_y}{3}\right)(2a)+A\sigma_y(3a)=\left(\frac{14}{3}\right)A\sigma_y a$$

Therefore,
$$W_e=\left(\frac{28}{3}\right)A\sigma_y$$

Thus, the ratio of the limit load to the yield load is

$$\frac{W_p}{W_e}=\frac{9}{7}=1.286$$

19.5 LIMIT ANALYSIS OF BEAMS

19.5.1 Concept of Plastic Hinges

As the load on the beam increases, the critical section, i.e. the most highly stressed section changes successively from fully elastic to partly plastic and finally to fully plastic. While other sections, remain fully elastic or partly plastic. Only the critical section approaches the fully plastic state because, although ductile yielding has begun at other sections also, the beam will not deflect uncontrollably as long as the highly stressed section can absorb an increase in bending moment. Once a section becomes fully plastic, all its fibres yield without further increase of stress, thereby permitting the other parts of the beam to rotate relative to each other. For this reason, a fully plastic section is called a *plastic hinge*, and the bending moment that creates it is considered to be the *limit moment* M_p. It should be noted from the moment curvature diagram of Figure 19.5, that as the section reaches its plastic moment resisting capacity, the curvature keeps on increasing without any further increase in the moment indicating that the section has acquired plastic hinge characteristics.

Collapse of statically determinate beam is considered to be synonymous with the formation of plastic hinge. The configuration of a beam when uncontrolled deformation can occur is called a *collapse mechanism*. Collapse mechanisms for a beam with typical ends and load conditions are shown in Figures 19.9(a), (b) and (c). In each case, the collapse mechanism is represented by dashed lines.

For illustration consider a simply-supported beam carrying a uniformly distributed load of intensity w /unit length. The plastic hinge will form at the highly stressed point, i.e. the centre when the load reaches its maximum or limit intensity w_p. At this (collapse) stage, the plastic hinge will cause the curvature to increase uncontrollably at the central section resulting in an uncontrollable deflection. The simply-supported beam with a hinge at the centre virtually behaves like a *mechanism* with no possibility of further increase in load beyond w_p.

The magnitude of w_p can be determined by equating the bending moment at the highly stressed section (hinge) to its plastic moment of resistance M_p;

$$\frac{w_p L^2}{8}=M_p \quad \text{or} \quad w_p=\frac{8M_p}{L^2} \tag{19.17}$$

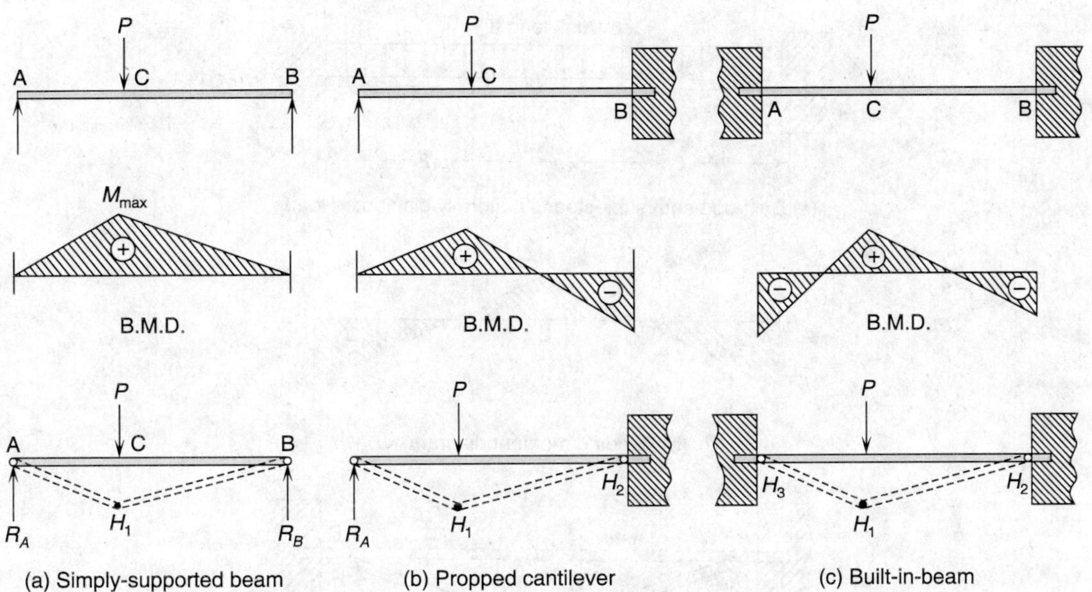

(a) Simply-supported beam (b) Propped cantilever (c) Built-in-beam

Figure 19.9 Collapse mechanisms for a beam with different end conditions.

This is known as *plastic-limit load*. The plastic design of structures is based on computing the working load by dividing plastic-limit load by a load factor.

In general, a plastic hinge forms at a section of zero shear, i.e. where the bending moment is maximum. The location of plastic hinges is obvious for beams subjected to concentrated loads. For indeterminate beams carrying distributed loads, the location of plastic hinges is not obvious; there is a possibility of existence of more than one collapse mechanism. In this case, the limit load is calculated for each possible mechanism and smallest value is taken as the limit load.

EXAMPLE 19.4

A beam built-in (fully restrained) at the ends carries a uniformly distributed load of intensity w /unit length as shown in Figure 19.10(a). Determine the limit load the beam can support and compare it with the maximum elastic load.

Solution: In view of symmetry, it is evident that the sections of maximum moment occur at mid-span and at the ends (see Figure 19.10(b)). The beam reduces to a collapse mechanism shown in Figure 19.10(c) when plastic hinges form at A, B and C as the moment at each of these positions approaches the limit moment M_p.

The redundant support reactions can now be obtained by applying equilibrium equations to the collapse mechanism with the moments $-M_A = -M_B = M_C = M_p$ i.e. M_A and M_B are negative.

$$\sum M_C = 0:\ M_p - R_{AV}\left(\frac{L}{2}\right) + M_p + \left(\frac{w_p L}{2}\right)\left(\frac{L}{4}\right) = 0$$

or

$$2M_p - R_{AV}\left(\frac{L}{2}\right) + \left(\frac{w_p L^2}{8}\right) = 0 \qquad \text{(i)}$$

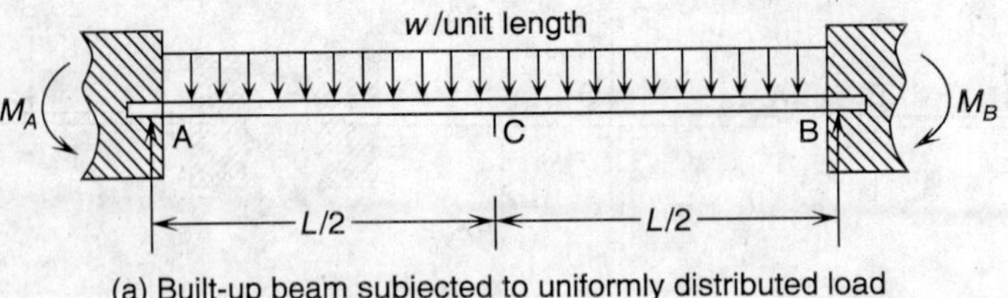

(a) Built-up beam subjected to uniformly distributed load

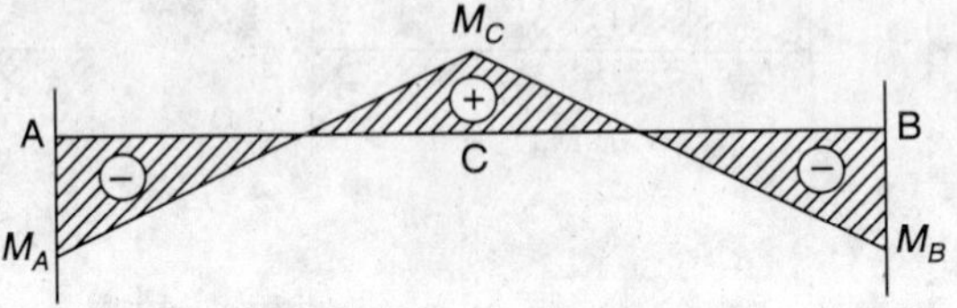

(b) Bending moment diagram

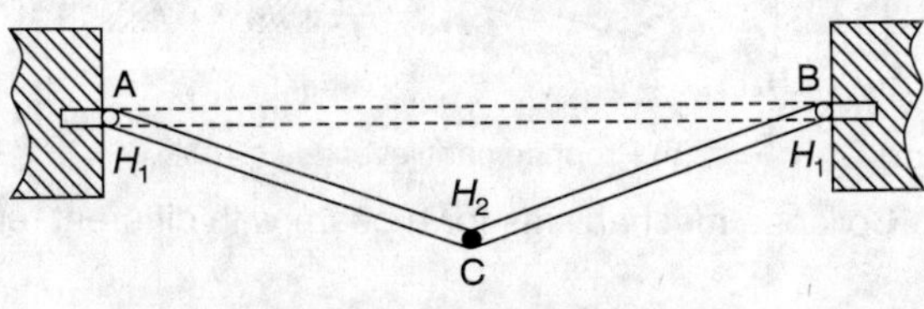

(c) Collapse mechanism

Figure 19.10 Collapse mechanism of a beam built-up at ends.

$$\sum M_B = 0: \; M_p - R_{AV}(L) + M_p - (w_p L)\left(\frac{L}{2}\right) = 0$$

or
$$-R_{AV}(L) + \left(\frac{w_p L^2}{2}\right) = 0 \tag{ii}$$

From Eq. (ii): $R_{AV} = \dfrac{w_p L}{2}$ and then from Eq. (i): $M_p = \dfrac{w_p L^2}{16}$

Thus, the limit load is,
$$w_p = \frac{16 M_p}{L^2}$$

The elastic moment at the support A is $M_A = -(w_e L^2/12)$. The limit load is given by,

$$w_e = \frac{12 M_e}{L^2}$$

The relation between the limit load and the elastic load can expressed as

$$\frac{w_p}{w_e} = \frac{4}{3}\left(\frac{M_p}{M_e}\right)$$

EXAMPLE 19.5

A propped beam carries a uniformly distributed load of intensity *w*/unit length as shown in Figure 19.11(a). Determine the relation between the limit load and the limit moment.

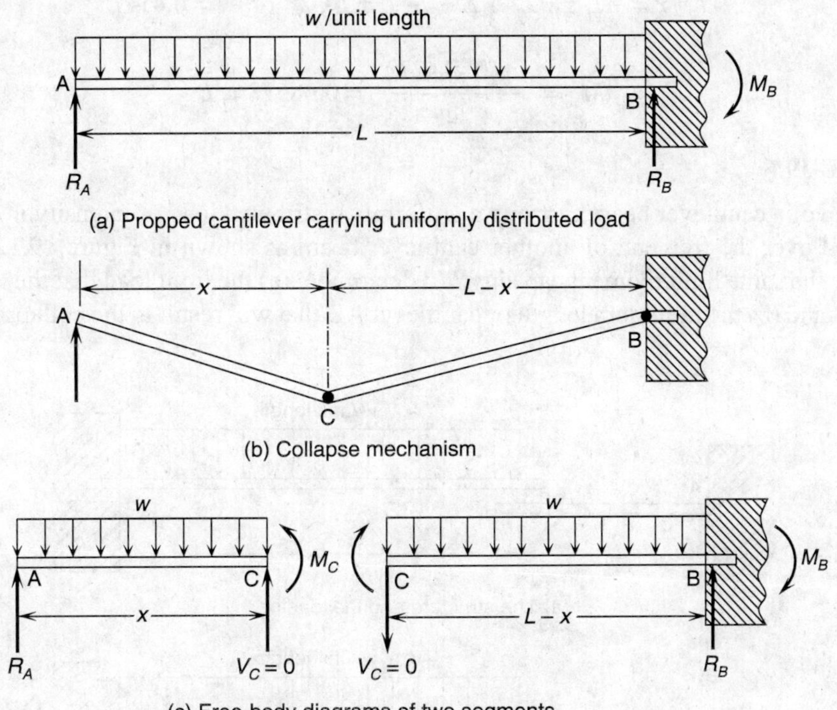

Figure 19.11 Collapse of propped cantilever of Example 19.5.

Solution: The collapse mechanism is shown in Figure 19.11(b). The moments at plastic hinge points C and B are the limit moments. The location of the point C of maximum moment that is of zero shear can be determined from the free-body diagrams of the beam segments AC and CB as illustrated in Figure 19.11(c). Consider the moment equilibrium about hinge points A and B (this will eliminate determination of unknown vertical reactions).

The segment AC: $$\sum M_A = 0 \quad \text{i.e.} \quad M_p - (w_p x)\left(\frac{x}{2}\right) = 0$$

or $$M_p = \frac{w_p x^2}{2}$$

The segment CB: $$\sum M_B = 0 \quad \text{i.e.} \quad -M_p - M_p + w_p(L - x)\left(\frac{L - x}{2}\right) = 0$$

or $$M_p = \frac{w_p (L - x)^2}{4}$$

Therefore, $$\frac{w_p x^2}{2} = \frac{w_p (L-x)^2}{4} \quad \text{or} \quad x^2 + 2Lx - L^2 = 0$$

$$x = -L \pm \sqrt{L^2 + L^2} = -L \pm L\sqrt{2} \quad \text{or} \quad x = 0.414L$$

Hence, $$M_p = \frac{w_p (0.4142L)^2}{2} = 0.08578 w_p L^2$$

EXAMPLE 19.6

The free end of a cantilever beam carrying a uniformly distributed load of intensity w /unit length, is supported over the free end of another cantilever beam as shown in Figure 19.12(a). If both beams have the same limit moment capacity M_p, determine: (a) the limit load that the arrangement can support and (b) the minimum length of cantilever AB that will result in the collapse of both the cantilevers.

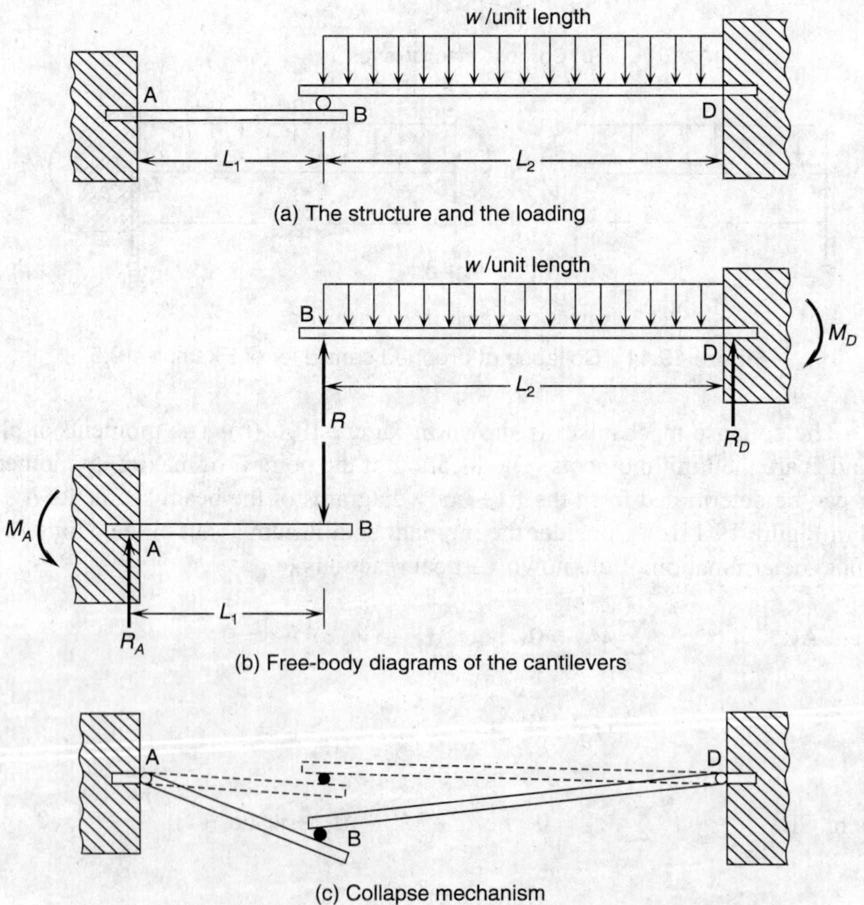

Figure 19.12 Two cantilever beams supported by a roller.

Solution: The free-body diagrams of the two cantilevers AB and BD along with the contact reactions R are illustrated in Figure 19.12(b). There are two possible collapse mechanisms; the first mechanism involves the collapse of cantilever BD only while the cantilever AB acts as a prop support. In this mechanism hinges form at a intermediate point C in the cantilever BD and its fixed support D as in Example 19.5. Whereas, in the second mechanism plastic hinges form at the support points A and D of the two cantilevers and structure collapses by rotation about these hinges. The second collapse mechanism is shown in Figure 19.12 (c).

(a) The correct collapse mechanism is the one which supports the smaller limit load.

First Mechanism: From Example 19.5,

$$M_p = 0.08578 w_{p,1} L_2^2 \quad \text{or} \quad w_{p,1} = \frac{M_p}{0.0857 L_2^2} = \frac{11.658 M_p}{L_2^2}$$

Second Mechanism: Consider the equilibrium at the hinge points (this will eliminate involvement of unknown vertical support reactions):

$$\sum M_A = 0 \quad \text{i.e.} \quad M_p - (R)(L_1) = 0 \quad \text{or} \quad R = \frac{M_p}{L_1}$$

$$\sum M_D = 0 \quad \text{i.e.} \quad -M_p - (R)(L_2) + (w_{p,2} L_2)\left(\frac{L_2}{2}\right) = 0$$

or

$$R = -\frac{M_p}{L_2} + \left(\frac{w_{p,2} L_2}{2}\right)$$

Eliminate the unknown reaction R,

$$\frac{M_p}{L_1} = -\frac{M_p}{L_2} + \left(\frac{w_{p,2} L_2}{2}\right) \quad \text{or} \quad w_{p,2} = \frac{2}{L_2}\left(\frac{M_p}{L_1} + \frac{M_p}{L_2}\right)$$

As a typical example consider $L_2 = 2L_1 = L$, the limit loads obtained from two mechanisms are:

$$w_{p,1} = \frac{11.658 M_p}{L^2} \quad \text{and} \quad w_{p,2} = \frac{6 M_p}{L^2}$$

The structure will collapse at the smaller limit load $w_{p,2}$ before the larger load $w_{p,1}$ is reached and hence determines the desired limit load.

(b) As the length L_1 is progressively reduced, the limit load $w_{p,2}$ increases until it attains the value $w_{p,1}$. Thus the limiting length of cantilever AB will be the one that will ensure the collapse of both the cantilevers and thus can be obtained by the following inequality

$$\frac{2}{L_2}\left(\frac{M_p}{L_1} + \frac{M_p}{L_2}\right) \le \frac{11.658 M_p}{L_2^2} \quad \text{or} \quad \left(\frac{L_2}{L_1} + 1\right) \le \frac{11.658}{2}$$

Therefore, $\left(\dfrac{L_2}{L_1}\right) \le \dfrac{9.658}{2} = 4.829$ or $L_1 \ge \dfrac{L_2}{4.829} = 0.207L_2$

The minimum length of AB at which collapse must occur with second mechanism is $0.207L_2$.

19.6 LIMIT TWISTING MOMENT OR TORQUE

In dealing with the torsion of circular bars stressed into plastic range, the assumptions made in elastic torsion remain valid, except that the strains may exceed the shear yield strain and stresses need not be proportional to the strain. The transverse sections are still considered rigid and does not warp but remain plane; consequently, the shearing strain remains proportional to the radial distance of the point under consideration from the centre of the bar.

Consider a circular shaft made of an elastic-perfectly plastic material that is twisted progressively through the elastic into the fully plastic range. When the stress on the extreme fibre of a section has just reached the shear yield point stress, strain and stress distributions on the section are still elastic or linear as shown in Figure 19.13(a).

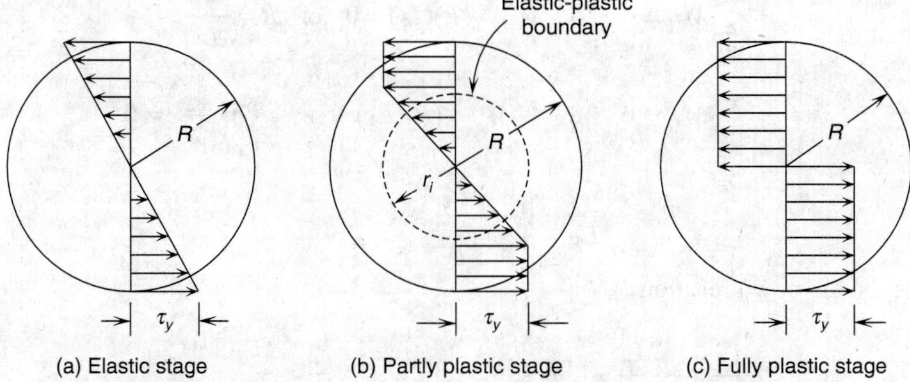

Figure 19.13 Shear stress distribution as twisting moment is increased.

The limiting elastic torque,

$$T_e = \tau_y\left(\frac{\pi R^3}{2}\right) \tag{19.18}$$

If the shaft is twisted beyond this point, the shearing strains continue to increase but the shear stresses remain constant at the yield stress τ_y. Thus at some intermediate radius r_i in the partly plastic case shown in Figure 19.13(b), the outer portion will be subjected to the constant yield stress τ_y; the inner core will remain elastic. The torque carried by the elastic core is

$$T_i = \frac{J_i}{r_i}\tau_y = \left(\frac{\pi r_i^3}{2}\right)\tau_y \tag{19.19}$$

The torque carried by the plastic (outer) portion is

$$T_o = \int_{r_i}^{R} r(\tau_y\, dA) = \int_{r_i}^{R} r(\tau_y\, 2\pi r\, dr)$$

$$= (2\pi\tau_y)\int_{r_i}^{R} r^2\,dr = \frac{2\pi}{3}(R^3 - r_i^3)\tau_y \tag{19.20}$$

The total torque carried by the shaft is

$$T = T_i + T_o = \left(\frac{\pi r_i^3}{2}\right)\tau_y + \frac{2\pi}{3}(R^3 - r_i^3)\tau_y$$

Therefore,

$$T = \frac{\pi R^3}{6}\left[4 - \left(\frac{r_i}{R}\right)^3\right]\tau_y \tag{19.21}$$

The fully plastic stage shown in Figure 19.13(c) cannot be reached because it requires an infinite angle of twist. The limiting torque T_p required to develop fully plastic condition is obtained by putting in $r_i = 0$ in Eq. (19.21):

$$T_p = \left(\frac{2\pi R^3}{3}\right)\tau_y = \frac{4}{3}T_e \quad \text{or} \quad \frac{T_p}{T_e} = \frac{4}{3}$$

Which is one-third more than the maximum elastic torque.

If the shaft section is hollow with external and internal diameters of D and d, respectively, the limiting torque T_p required developing fully plastic condition is obtained by putting $r_i = d$ and ignoring the elastic contribution in Eq. (19.21):

$$T_p = T_o = \frac{2\pi}{3}\left[\left(\frac{D}{2}\right)^3 - \left(\frac{d}{2}\right)^3\right]\tau_y = \frac{\pi}{12}(D^3 - d^3)\tau_y \tag{19.22}$$

EXAMPLE 19.7

A 64 mm diameter 1.6 m long aluminium shaft is rigidly connected to a 48 mm diameter 2.2 m long steel shaft, and the compound shaft which is rigidly fixed at the ends carries a torque T at the junction of aluminum and steel segments. Determine the maximum torque T that can be applied before uncontrolled rotation begins. The yield shear stresses in aluminium and steel are 155 and 140 MPa, respectively. The corresponding elastic moduli in two materials are 27 and 80 GPa, respectively.

Solution: Consider the two materials to be elastic-perfectly plastic. While either segment remains in elastic range, angular deformation will remain small. Hence excessive yielding will not occur until both segments have reached their limit torques. The limit torque is given by Eq. (19.20) with $r_i = 0$

$$T_p = \left(\frac{2\pi R^3}{3}\right)\tau_y$$

Limit torque T_p that can be carried at the junction of two segments is

$$T_p = T_{p,a} + T_{p,s} = \left(\frac{2\pi}{3}\right)(R_a^3\tau_{y,a} + R_s^3\tau_{y,s})$$

$$T_p = \left(\frac{2\pi}{3}\right)(32^3 \times 155 + 24^3 \times 140) = 14.691 \times 10^6 \text{ N.mm} = 14.691 \text{ kN.m}$$

EXAMPLE 19.8

A shaft of 50 mm diameter is made of a material which has bi-linear shear-stress vs. shear-strain curve shown in Figure 19.14(a). Determine the maximum torque T that can be applied when the maximum shear stress is restricted to 250 MPa.

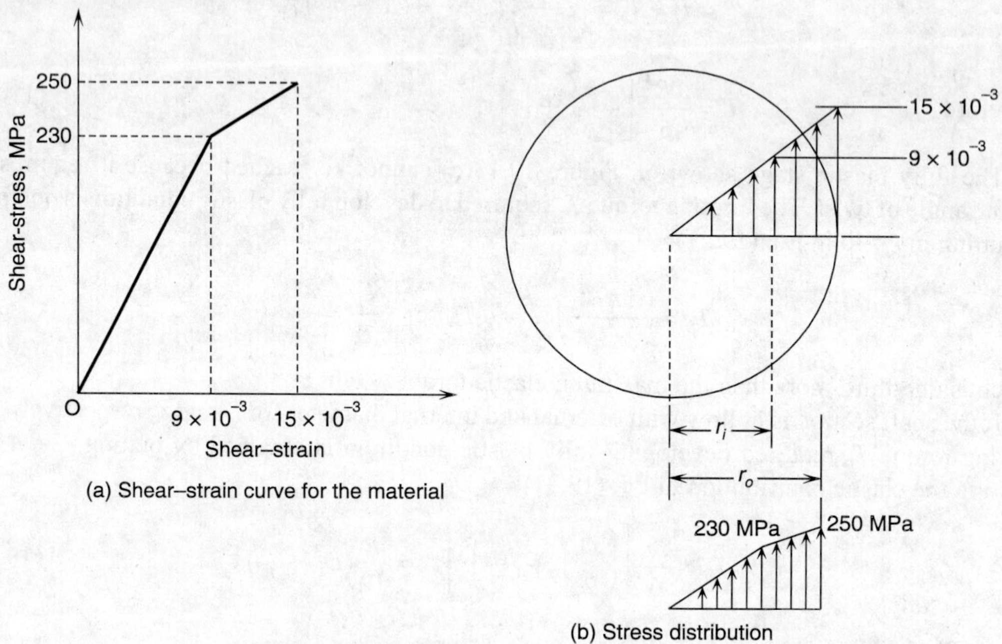

Figure 19.14 Stress distribution in shaft of Example 19.8.

Solution: As the maximum shear stress of 250 MPa develops at the outer periphery of the shaft, the strain varies linearly from zero at the centre to 9×10^{-3} radian at r_i and further to 15×10^{-3} radian at $r = 25$ mm, while the stress variation is obtained from the shear stress and strain curve of the material. The yield point strain 9×10^{-3} radian is developed at the radius

$$r_i = \frac{9.0 \times 10^{-3}}{15.0 \times 10^{-3}} \times 25 = 15 \text{ mm}$$

Thus, shear stress varies linearly from zero at the centre ($r = 0$) to 230 MPa at $r = 15$ mm and further to 250 MPa at $r = 25$ mm as shown in Figure 19.14(b).
The shear stress at any point in the range $0 < r < r_i$ is

$$\tau = \tau_i \times \frac{r}{r_i} = 230 \times \frac{r}{15} = (46/3)r \text{ MPa}$$

The shear stress at any point in the range $r_i < r < r_0$ is

$$\tau = \tau_i + \frac{\tau_o - \tau_i}{r_o - r_i}(r - r_i)$$

$$= 230 + \frac{250 - 230}{25 - 15}(r - 15) = (200 + 2r) \text{ MPa}$$

The torque carried by the inner core is

$$T_i = \int_0^{r_i} r(\tau\, dA) = \int_0^{r_i} r[(46r/3)(2\pi r\, dr)]$$

$$= (92\pi/3)\int_0^{r_i} r^3\, dr = \frac{92\pi r_i^4}{3 \times 4} = \left(\frac{23}{3}\right)\pi r_i^4$$

The torque carried by the cover (outer) portion is

$$T_o = \int_{r_i}^{r_o} r(\tau\, dA) = \int_{r_i}^{r_o} r[(200 + 2r)(2\pi r\, dr)]$$

$$= 2\pi \int_{r_i}^{r_o} (200r^2 + 2r^3)\, dr = 2\pi\left[200\left(\frac{r^3}{3}\right) + 2\left(\frac{r^4}{4}\right)\right]_{r_i}^{r_o}$$

$$= (400/3)\pi(r_o^3 - r_i^3) + \pi(r_o^4 - r_i^4)$$

$$T = T_i + T_o = (400/3)\pi(r_o^3 - r_i^3) + \pi(r_o^4 - r_i^4) + (23/3)\pi r_i^4$$

$$= (400/3)\pi(r_o^3 - r_i^3) + \pi[r_o^4 + (20/3)r_i^4]$$

$$= (400/3) \times \pi \times (25^3 - 15^3) + \pi \times [25^4 + (20/3) \times 15^4]$$

$$= 7.418740 \times 10^6 \text{ N.mm} = 7.41874 \text{ kN.m}$$

19.7 RESIDUAL STRESSES

A ductile material strained in to plastic range (curve OAB) when unloaded follows elastic path BC that is parallel to the initial elastic path OA. On reloading, a slight *hysteresis loop* is observed and material behaves elastically up to the previously loaded point B, beyond which it regains plastic behaviour (curve CBD). However, the material undergoes a *permanent set*, represented by OC. The curves representing loading, unloading, and reloading cycles of the actual and idealized materials are shown in Figures 19.15(a) and (b), respectively. In Figure 19.15(b) the unloading and reloading lines actually coincide but are shown slightly separated for clarity.

The permanent set mentioned above creates a system of *self balancing internal stresses* known as *residual stresses*. The magnitude and distribution of these stresses may be determined by superposing the stress pattern caused by the original loading on to the stress pattern caused by an elastically applied load which is equal in magnitude but opposite in sense to the original loading. The effect of applying a load equal but opposite to the original load is equivalent to the unloading the structure. This unloading creates a stress pattern that is considered to be elastic. However, the method of superposition cannot be used if the residual stresses so obtained exceed the yield stress.

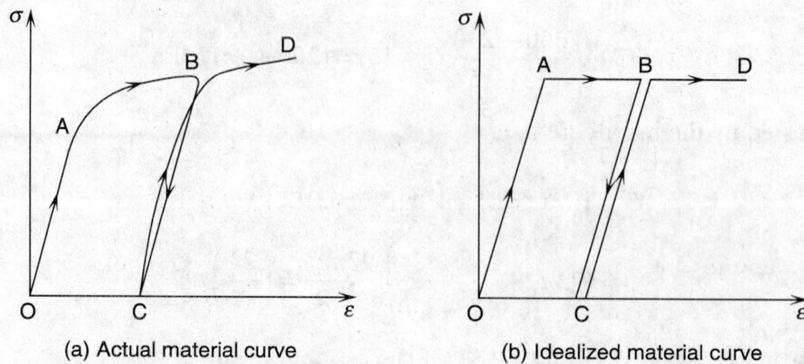

Figure 19.15 Loading, unloading and reloading of actual and idealized elastic and perfectly plastic materials.

As an illustration consider a rectangular section of a beam (see Figure 19.16(a-i)) that is strained into fully plastic state by application of limit moment, which is (3/2) times the yield moment and results in a stress distribution shown in Figure 19.16(a-ii). The unloading process is equivalent to elastically adding an equal but reversed loading shown in Figure 19.16(a-iii) to Figure 19.16(a-ii). This results in an unloaded member having the *residual stress distribution* shown in Figure 19.16(a-iv).

It is interesting to note that the unloaded member behaves elastically if the original limit moment is reapplied as shown in Figure 19.16(b-ii). Superposing parts (b-i) and (b-ii) of Figure 19.16(b) results in original plastic state shown in Figure 19.16 (b-iii). On the other hand, with reversed reloading (see Figures 19.16(c-i) and (c-ii)), the residual stresses are unfavorable and no more than three-fourth of the yield moment can be reapplied in the opposite sense before the member reaches its plastic limit as shown in Figure 19.16(c-iii). Further plastic yielding occurs if the sum of the original and the reversed loading exceeds twice the value of maximum yield moment. On application of load cycles additional yield after each cycle results in *rupture*, just like in bending a metal strip back and forth plastically for a few times leads to it rupture.

To illustrate the application of above concepts to the members subjected to torsion, consider a circular shaft strained into fully plastic state by the limit torque, which is (4/3) times the yield torque, and the resulting stress distribution is shown in Figure 19.17(a). The unloading of the shaft is achieved by adding elastically an equal but opposite torque as shown in Figure 19.17(b) to Figure 19.17(a). This superposition of loadings and stress patterns of parts (a) and (b) results in an unloaded member having the residual stress distribution shown in Figure 19.17(c). The elastic unloading produces the linear stress distribution shown in Figure 19.17(b).

Like a flexural member, a shaft unloaded from the fully plastic state if reloaded in the same sense, behaves elastically until original limit torque is reached. On the other hand, for reversed reloading, the residual stresses are unfavourable and no more than two-thirds of the yield torque can be reapplied in the opposite sense before the member reaches its plastic limit.

It should be noted that although residual stresses are self-balancing, if some of the material is removed, an unbalance is created. This phenomenon is observed in cold-formed members that get distorted after machining.

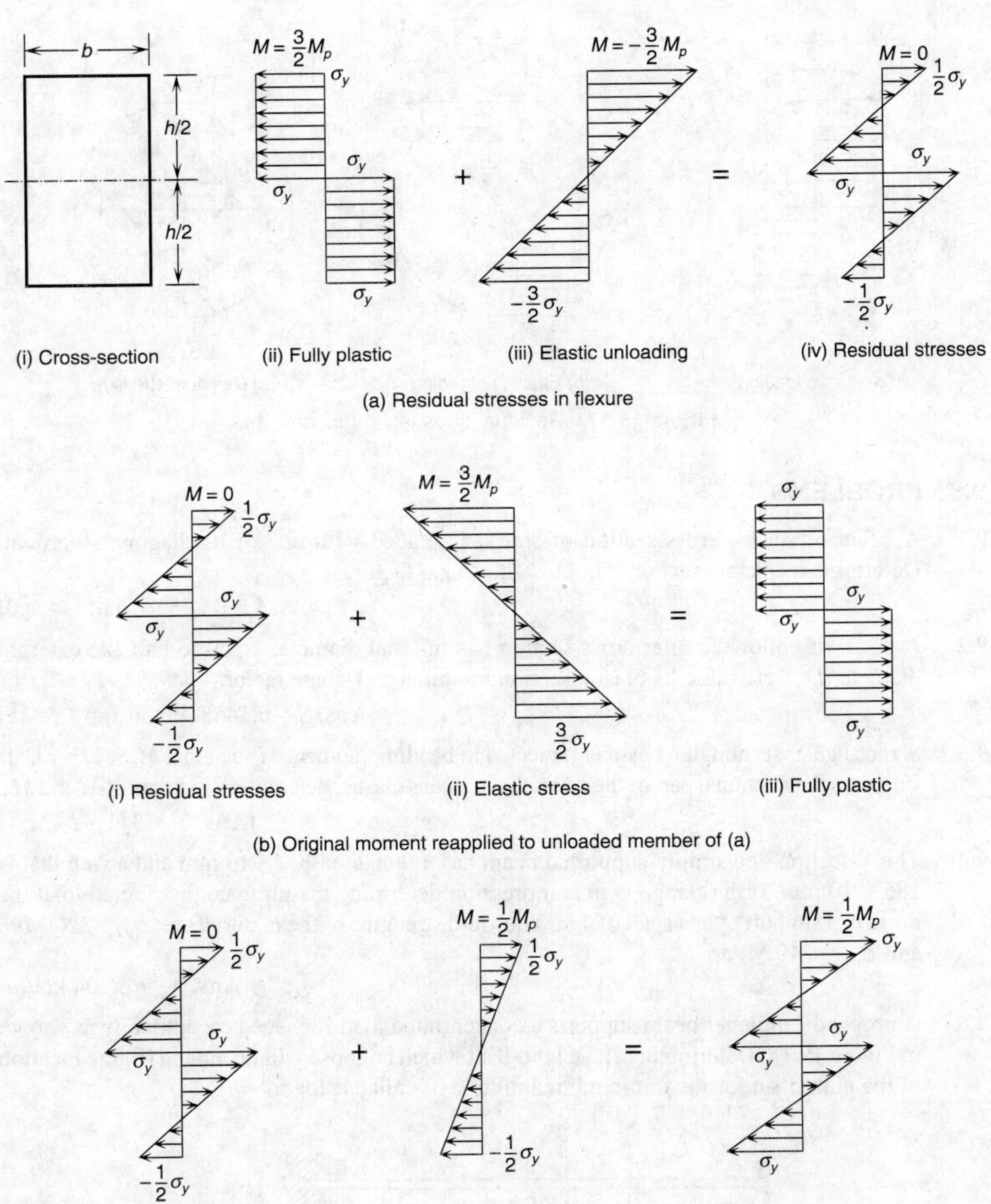

Figure 19.16 Residual stresses in a member strained into fully plastic state and effect of reloading.

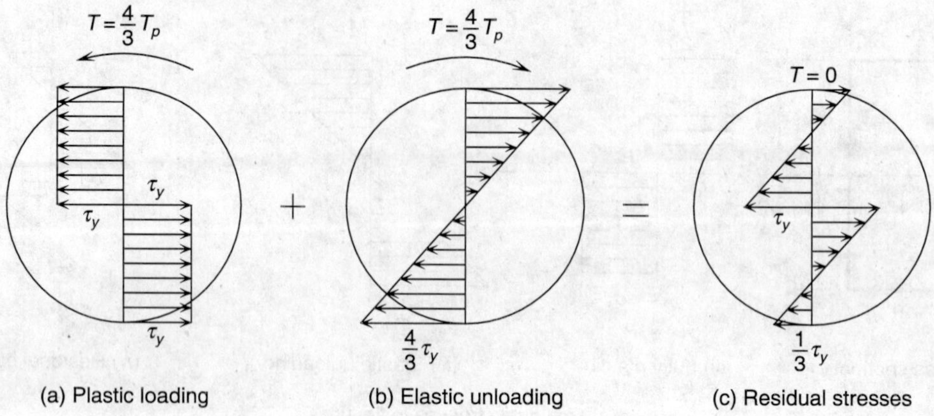

Figure 19.17 Residual stresses in torsion.

19.8 PROBLEMS

19.1 A beam of square cross-section of side a is placed with one of its diagonals vertical. Determine its plastic section modulus and shape factor.

[**Ans.** $S = 0.2357a^3$ and $s_f = 2.0$]

19.2 A beam of hollow circular cross-section has internal diameter equal to half the external diameter D. Determine its plastic section modulus and shape factor.

[**Ans.** $S = 0.1458D^3$ and $s_f = 1.5845$]

19.3 A rectangular section of a beam is subjected to bending moment M such that $M_e < M < M_p$. If k denotes the fractional part of the depth that remains elastic, determine k in terms of M and M_p.

[**Ans.** $k^2 = 3\{1 - M/M_p\}$]

19.4 The T-section of a simply-supported beam has a flange of 200 × 15 mm and a web that is 285 × 10 mm. If the flange is in compression determine the ultimate distributed load the beam can support over a span of 4 m. The yield strengths of the material are: $\sigma_{y,t} = 320$ MPa and $\sigma_{y,c} = 240$ MPa.

[**Ans.** $w_u = 66.04$ kN/m]

19.5 A propped cantilever beam supports a concentrated load W placed eccentrically as shown in Figure 19.18. Determine: (a) the load W at which collapse will occur and (b) the location of the point load for the minimum magnitude of collapse load.

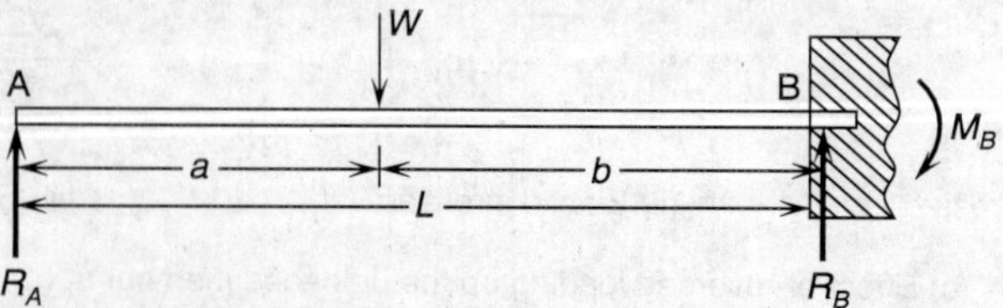

Figure 19.18 Propped cantilever of Problem 19.5.

[**Ans.** (a) $W_p = [(2a + b)M_p]/ab$ and (b) $a = 0.414L$; $W_{p,\min} = 5.828M_p/L$]

19.6 A beam fully restrained at the ends carries a concentrated load W placed eccentrically as shown in Figure 19.19. Determine the collapse load.

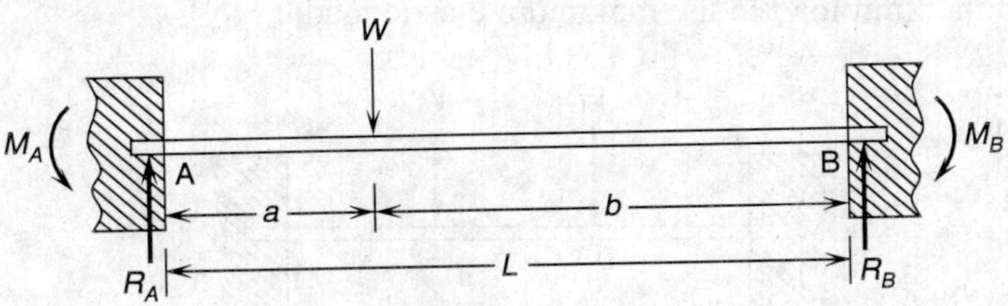

Figure 19.19 Restrained beam of Problem 19.6.

[**Ans.** $W_p = (2L/ab)M_p$]

19.7 Two cantilever beams are separated by a roller as shown in Figure 19.20. If both beams have the same limit moment capacity M_p, determine the limit concentrated load W, at which collapse will occur.

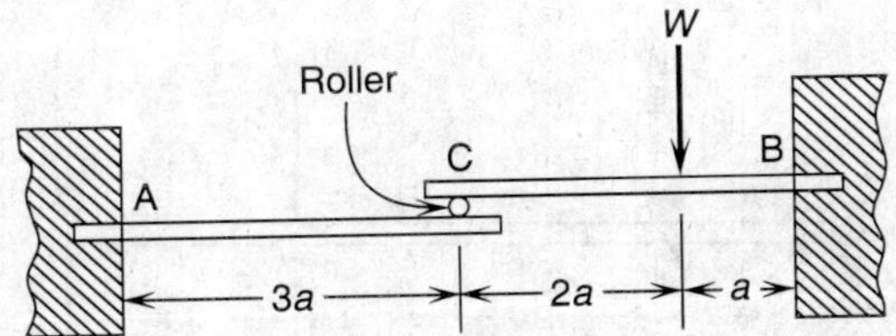

Figure 19.20 Cantilever beams of Problem 19.7.

[**Ans.** $W = 2M_p/a$]

19.8 A cantilever beam with its free end resting on a simply-supported beam supports a concentrated load W as shown in Figure 19.21. If the limit moment of the simple beam is three-quarters that of the cantilever beam, determine the load W, at which collapse will occur.

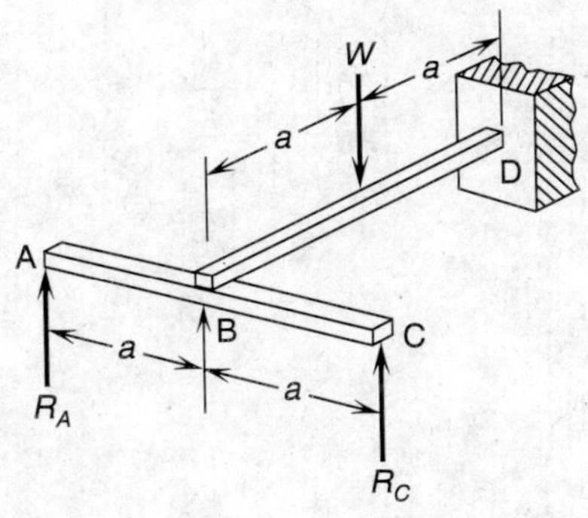

Figure 19.21 Cantilever beam propped by a simply supported beam.

[**Ans.** $W = 3M_p/a$]

19.9 A beam fully restrained at the ends carries two concentrated loads W at one-third points as shown in Figure 19.22. If the ratio of the limit moment to the yield moment is K, determine the ratio of the limit load to the maximum elastic load.

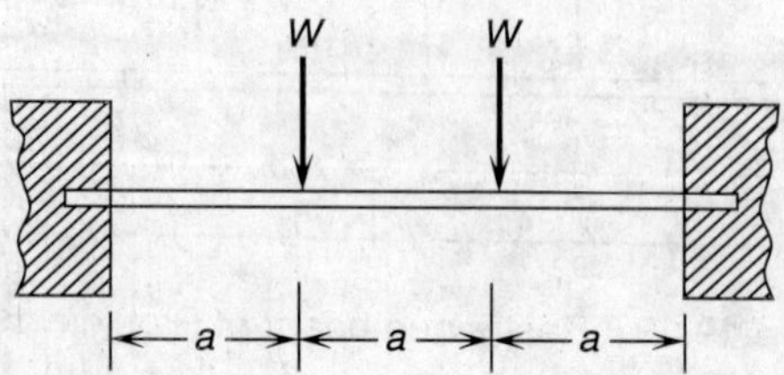

Figure 19.22 Fully restrained beam of Problem 19.9.

[Ans. $W_p/W_e = 4K/3$]

19.10 If both the spans of the continuous beam shown in Figure 19.23 have the same limit moment M_p, determine the limit concentrated load W.

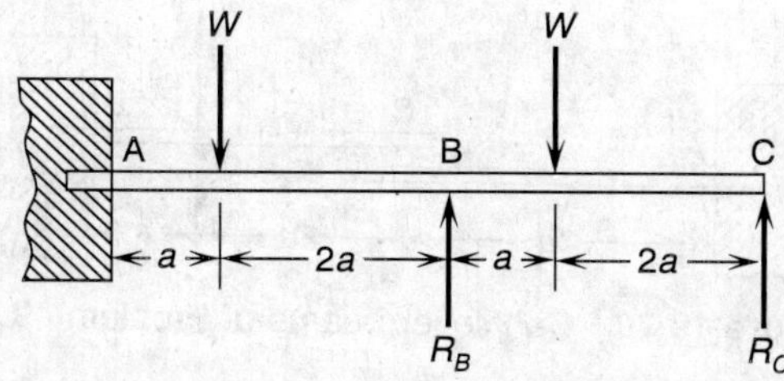

Figure 19.23 Continuous beam of Problem 19.10.

[Ans. $W = 5M_p/2a$]

Chapter 20

Principal and Complex Stresses

20.1 INTRODUCTION

In Chapter 11 on combined stresses the discussion was confined to the structures in which axial and flexural loadings acted simultaneously. As this system of loading combined only normal stresses it constituted the simplest combination and was considered first. The other three combinations of fundamental loadings, i.e. axial and torsional; flexural and torsional; and axial, torsional and flexural acting simultaneously are discussed in the present chapter. All the three systems of loadings discussed here combine shearing and normal stresses, and result in a complex set of stresses.

For design, it is required to know the orientation of the element on which the maximum normal stress exists, and its magnitude; also the orientation of the element on which maximum shearing stress exists, and its magnitude. This aspect of design has been discussed in detail in this chapter.

20.2 STATE OF STRESS

In a loaded body, the stresses on a plane passing through a given point depend on the orientation of the plane in space. The stresses vary in a certain manner with rotation of plane. The set of stresses occurring on the planes passing through the point under consideration is called the *state of stress* at the point.

In order to define the state of stress at the point *A* several cutting planes are imagined to have passed through this point and values of stresses occurring on them are determined. Consider an elementary volume in the form of a right parallelepiped in the vicinity of the point under consideration being isolated by six sections as shown in Figure 20.1. If the dimensions of the

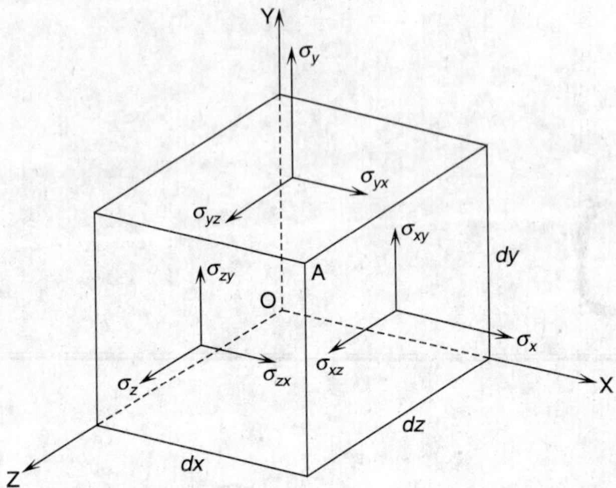

Figure 20.1 State of stress at a point.

parallelepiped are reduced, it will contract into that point. In the limit all faces of the parallelepiped pass through the point *A*, and the stresses on the corresponding cutting planes may be regarded as the stresses at the point under consideration.

The total stress acting on a cutting plane can be resolved into three components; one acting along the normal to the plane and two in the plane of the section. The normal stress is denoted by σ_i, with the subscript corresponding to the appropriate axis (*x*, *y* or *z*) as shown in Figure 20.1. The shearing stress is denoted by the symbol σ_{ij} with two subscripts: the first corresponds to the axis perpendicular to the plane, and second to the axis along which σ_{ij} vector is directed. However, shearing stress σ_{ij} is generally denoted by the symbol τ_{ij}. The orientation of axes is arbitrary.

20.2.1 Sign Convention

Normal tensile stresses σ_x are considered positive, and *compressive stresses* negative. A shearing stress τ_{ij} is considered positive if its moment about any point inside an element is clockwise; it is taken negative if this moment is anti-clockwise. The positive angle of inclination α is measured anti-clockwise from the plane normal to the stress σ_n.

20.2.2 Representation of State of Stress at a Point

The stresses acting on three faces of the element (on three mutually perpendicular planes passing through the point) are shown in Figure 20.1. The stresses of same magnitude but opposite in sense occur on the hidden faces of the element. Thus, at any point in a body there can be nine stress components

$$[\sigma] = \begin{bmatrix} \sigma_x & \tau_{yx} & \tau_{zx} \\ \tau_{xy} & \sigma_y & \tau_{zy} \\ \tau_{xz} & \tau_{yz} & \sigma_z \end{bmatrix} \tag{20.1}$$

The state of stress at any point in a body must be in equilibrium, i.e. the system of forces acting on the element must satisfy the *equilibrium equations*. Since the forces acting on the opposite faces are of different sign the equilibrium conditions $\sum F_x = 0, \sum F_y = 0$ and $\sum F_z = 0$ are identically satisfied and the sums of the projections of all forces on the *X*, *Y* and *Z* axes are zero. For moment equilibrium, the sums of the moments of all forces about the *X*, *Y* and *Z* axes must vanish. For the sum of the moments of all forces about the *X*-axis to vanish,

$$\sum M_x = (\tau_{yz}\, dx\, dz)\, dy - (\tau_{zy}\, dx\, dy)\, dz = 0 \quad \text{i.e.} \quad \tau_{yz} = \tau_{zy} \tag{20.2}$$

where the moment $\sum M_x$ represents moment about an axis perpendicular to *YZ*-plane. Similarly, $\tau_{zx} = \tau_{xz}$ and $\tau_{xy} = \tau_{yx}$.

Thus, *on two planes at right angles to each other the components of shearing stresses perpendicular to the common edge are equal and directed either both towards the edge or both away from the edge*. This universal law is called the *law of equality of shearing stresses*. Since the shearing stresses are pair-wise equal there are only six rather than nine independent stress components on the faces of the isolated element as shown in Figure 20.1. For analysis of design stress resultants in a loaded body subjected to combined forces the first step is to determine the normal and shear or tangential stresses on the faces of an isolated element.

EXAMPLE 20.1

A metal rod of square cross-section of size *a* carrying an axial tension of *P* is subjected to twisting moment *T* as shown in Figure 20.2(a). Determine the state of stress at the points A and B located at centroidal axis and surface, respectively.

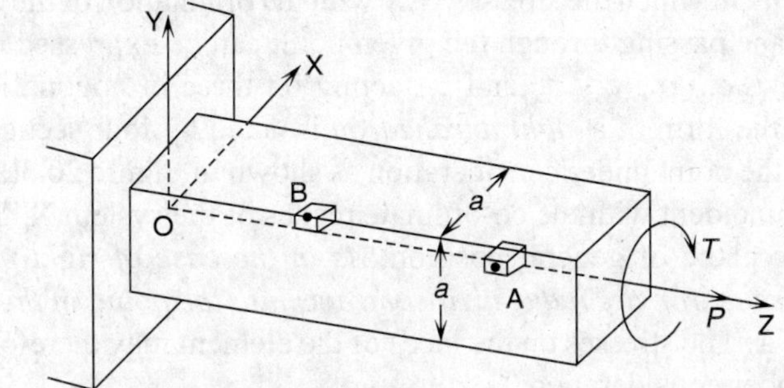

(a) The member subjected to torsion and axial load

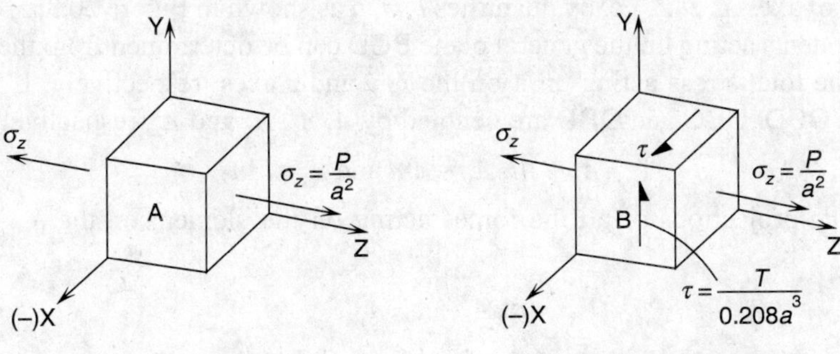

(b) Stress state at point A (c) Stress state at point B

Figure 20.2 Stress states in the member of Example 20.1.

Solution: Consider elementary volumes by cutting planes in the vicinity of the given points. For the convenience in computation of stresses, the planes are orientated along and across the axis of the rod. The enlarged isolated elements are shown in Figures 20.2(b) and (c).

The force P induces normal stress $\sigma_n = \frac{P}{a^2}$ at the cross-section of the rod. The stresses are shown on the faces of the elements. The twisting moment T develops shearing stresses on transverse and longitudinal planes. At the point B the shearing stress induced (see Figure 20.2(c)) is

$$\tau_{\max} = \frac{T_x}{\alpha(a)(t^2)} = \frac{T}{0.208(a)(a^2)} = \frac{T}{0.208a^3}$$

And at the point A the stress $\tau = 0$. Thus,

At the point A: $\sigma_x = \sigma_y = 0, \sigma_z = \frac{P}{a^2}$ and $\tau_{xy} = \tau_{yz} = \tau_{zx} = 0$

At the point B: $\sigma_x = \sigma_y = 0, \sigma_z = \frac{P}{a^2}, \tau_{zy} = 0, \tau_{yx} = \frac{T}{0.208a^3}$ and $\tau_{xz} = 0$

The states of stress are shown in Figures 20.2(b) and (c).

20.3 STRESSES ON A GENERAL PLANE

In this section the manner in which the stresses vary with the orientation of the element is discussed. The stresses on any plane passing through the given point can be expressed in terms of six stress components, namely σ_x, σ_y, σ_z, τ_{yz}, τ_{zx} and τ_{xy} acting on three orthogonal planes. Consider an elementary volume in the form of a *right tetrahedron* isolated by four sections from the stressed body in the vicinity of the point under consideration as shown in Figure 20.3(a). Three faces of the isolated element are coincident with the co-ordinate planes of the system X, Y, Z. The fourth face is formed by a cutting plane of general position. *As in the case of right parallelepiped, if the dimensions of right tetrahedron are reduced, it contracts into the point, in the limit all of the faces pass through the point* A. The stresses on the faces of the element may therefore be regarded as the stresses at the point under consideration.

The orientation of cutting plane BCD in space is defined by the direction cosines of the normal n in the system of axes X, Y, Z, i.e. by quantities l, m, n as shown in Figure 20.3(b).The normal and shearing components acting on the general plane BCD can be determined from the projections F_X, F_Y and F_Z of the total stress acting on it on the x, y and z axes, respectively. If the areas of the triangles BCD, OCD, OBC and OBD are denoted by A, A_x, A_y and A_z, respectively, then,

$$A_x = Al,\ A_y = Am \text{ and } A_z = An \tag{20.3}$$

Consider the projections of all the forces acting on the element on the X, Y and Z axes in succession,

$$F_X = \sigma_x A_x + \tau_{yx} A_y + \tau_{zx} A_z$$
$$F_Y = \tau_{xy} A_x + \sigma_y A_y + \tau_{zy} A_z$$
$$F_Z = \tau_{xz} A_x + \tau_{yz} A_y + \sigma_z A_z$$

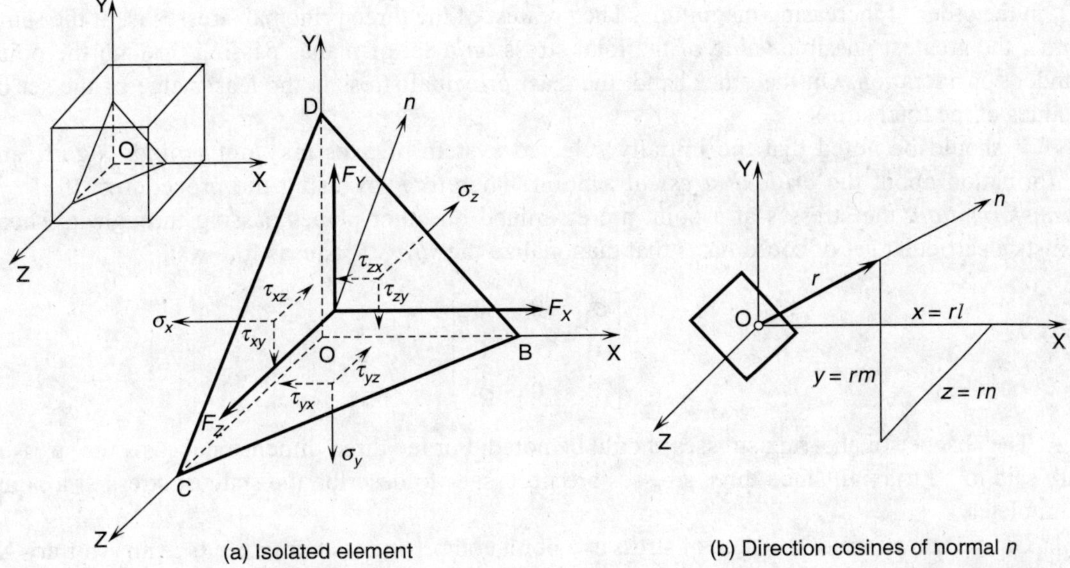

(a) Isolated element (b) Direction cosines of normal n

Figure 20.3 An element in the form of right tetrahedron.

Therefore, substituting from Eq. (20.3)

$$\begin{aligned} F_X &= \sigma_x\, l + \tau_{yx}\, m + \tau_{zx}\, n \\ F_Y &= \tau_{xy}\, l + \sigma_y\, m + \tau_{zy}\, n \\ F_Z &= \tau_{xz}\, l + \tau_{yz}\, m + \sigma_z\, n \end{aligned} \tag{20.4}$$

or

$$\begin{Bmatrix} F_X \\ F_Y \\ F_Z \end{Bmatrix} = \begin{bmatrix} \sigma_x & \tau_{yx} & \tau_{zx} \\ \tau_{xy} & \sigma_y & \tau_{zy} \\ \tau_{xz} & \tau_{yz} & \sigma_z \end{bmatrix} \begin{Bmatrix} l \\ m \\ n \end{Bmatrix} \tag{20.5}$$

The matrix in the above equation represents a *tensor* with quantities having no simple geometric interpretation, while single column matrices are the vector quantities. If a new system is selected in place of the original system of axes X, Y, Z the components of tensor will change, i.e. the values of σ_x, σ_y ... will be different. However, the tensor of state of stress remains the same.

The normal stress σ_n, on an inclined plane in terms of F_X, F_Y and F_Z is expressed as:

$$\sigma_n = F_X l + F_Y m + F_Z n \tag{20.6}$$

Substituting from Eq. (20.4),

$$\sigma_n = \sigma_x l^2 + \sigma_y m^2 + \sigma_z n^2 + 2\tau_{xy} lm + 2\tau_{yz} mn + 2\tau_{zx} nl \tag{20.7}$$

At a point under consideration in a stressed body, there exists such a system of axes X, Y, Z with respect to which the shearing stresses, τ_{xy}, τ_{yz} and τ_{zx} are zero. These axes are called *principal axes*. The corresponding mutually perpendicular planes are called *shear-less* or *principal planes*, and normal stresses on them are called *principal stresses*. These stresses are denoted by σ_3, σ_2 and

σ_1 in the order of increasing magnitude. The greatest of the three principal stresses is, at the same time, the greatest possible value of the total stress on a set of planes passing through the point under consideration. On the other hand, the least principal stress is the least value of the set of values of the total stress.

It should be noted that the initially selected system of axes may not provide significant information about the *critical stress* at a point. Therefore, by using the procedures of *stress transformation*, the stresses at a point are examined on other planes passing through it. There exists a particular set of coordinates that diagonalize the *stress tensor* as follows:

$$\begin{bmatrix} \sigma_1 & 0 & 0 \\ 0 & \sigma_2 & 0 \\ 0 & 0 & \sigma_3 \end{bmatrix}$$

The absence of shearing stresses should be noted. For the three-dimensional case, the stresses are said to be triaxial, since three stresses are necessary to describe the state of stress at a point completely.

Consider the case of the state of stress at a point comprising six components in an arbitrary X, Y, Z references system. Let the inclined plane be a principal plane, then the total stress S on this plane, which is also principal, is in the direction of the normal n. Therefore,

$$F_X = Sl,\ F_Y = Sm,\ F_Z = Sn \tag{20.8}$$

From Eq. (20.4):

$$\begin{aligned} (\sigma_x - S)\,l + \tau_{yx}\,m + \tau_{zx}\,n &= 0 \\ \tau_{xy}\,l + (\sigma_y - S)\,m + \tau_{zy}\,n &= 0 \\ \tau_{xz}\,l + \tau_{yz}\,m + (\sigma_z - S)\,n &= 0 \end{aligned} \tag{20.9}$$

The Eq. (20.9) represents a set of homogeneous system of equations in terms of unknowns l, m and n which define the orientation of the principal plane in the given X, Y, Z reference system. It must give a non-zero solution for l, m and n as direction cosines cannot all be zero simultaneously since

$$l^2 + m^2 + n^2 = 1 \tag{20.10}$$

For non-zero solution the determinant of this system must vanish, i.e.

$$\begin{vmatrix} \sigma_x - S & \tau_{yx} & \tau_{zx} \\ \tau_{xy} & \sigma_y - S & \tau_{zy} \\ \tau_{xz} & \tau_{yz} & \sigma_z - S \end{vmatrix} = 0 \tag{20.11}$$

Thus,
$$S^3 - J_1 S^2 + J_2 S - J_3 = 0 \tag{20.12}$$

where
$$J_1 = \sigma_x + \sigma_y + \sigma_z \tag{20.13}$$

$$J_2 = \sigma_y\sigma_z + \sigma_z\sigma_x + \sigma_x\sigma_y - \tau_{yz}^2 - \tau_{zx}^2 - \tau_{xy}^2 \tag{20.14}$$

$$J_3 = \begin{vmatrix} \sigma_x & \tau_{yx} & \tau_{zx} \\ \tau_{xy} & \sigma_y & \tau_{zy} \\ \tau_{xz} & \tau_{yz} & \sigma_z \end{vmatrix} \tag{20.15}$$

The three roots of the equation which are real give three values for the principal stresses σ_1, σ_2 and σ_3. The principal stresses which depend on the nature of state of stress are the same whatever be the reference axes system. Consequently, the coefficients J_1, J_2 and J_3 in Eq. (20.12) must remain unaltered as the X, Y, Z axes are rotated. These coefficients are called the *invariants of the state of stress*.

In some cases the invariants may take zero values, e.g. if $J_3 = 0$, one of the roots of Eq. (20.12) is also zero, the state of stress is said to be *biaxial* or *plane stress state.* In particular, the state of stress of pure shear represents a biaxial state of stress for which $\sigma_1 = -\sigma_3$ and $\sigma_2 = 0$. Such stresses occur, for example, in thin sheets stressed in two mutually perpendicular directions. The pressure vessel provides a typical example of this case.

If the second and third invariants are zero simultaneously, i.e. $J_2 = J_3 = 0$, then Eq. (20.12) has two zero roots and only one of the principal stresses is non-zero; such a state of stress is referred to as *uniaxial*, e.g. the cases of tension, compression and pure bending are included in this category.

EXAMPLE 20.2

Determine the principal stresses when: (a) all components of the state stress are equal, and (b) the state stress is

$$\begin{bmatrix} 0 & \tau & \tau \\ \tau & 0 & \tau \\ \tau & \tau & 0 \end{bmatrix}$$

Solution: (a) From Eqs. (20.13) to (20.15):

$$J_1 = 3\sigma \text{ and } J_2 = J_3 = 0$$

Therefore, $$S^3 - 3\sigma S^2 = 0 \quad \text{i.e.} \quad S_1 = 3\sigma,\ S_2 = S_3 = 0$$

Thus, the given state of stress represents uniaxial tension as shown in Figures 20.4(b). This problem represents the case of a tension bar with the element isolated in any manner. When the three cutting planes are equally inclined to the axis of tensile bar, the element has equal components of the state of stress on its faces as illustrated in Figure 20.4(c).

(b) From Eqs. (20.13) to (20.15):

$$J_1 = 0, J_2 = -3\tau^2 \text{ and } J_3 = 2\tau^3$$

Therefore, $$S^3 - 3\tau^2 S - 2\tau^3 = 0$$

By trial and modification,

$$S_1 = 2\tau \text{ and } S_2 = S_3 = -\tau$$

The state stress is thus triaxial as illustrated in Figures 20.5(b).

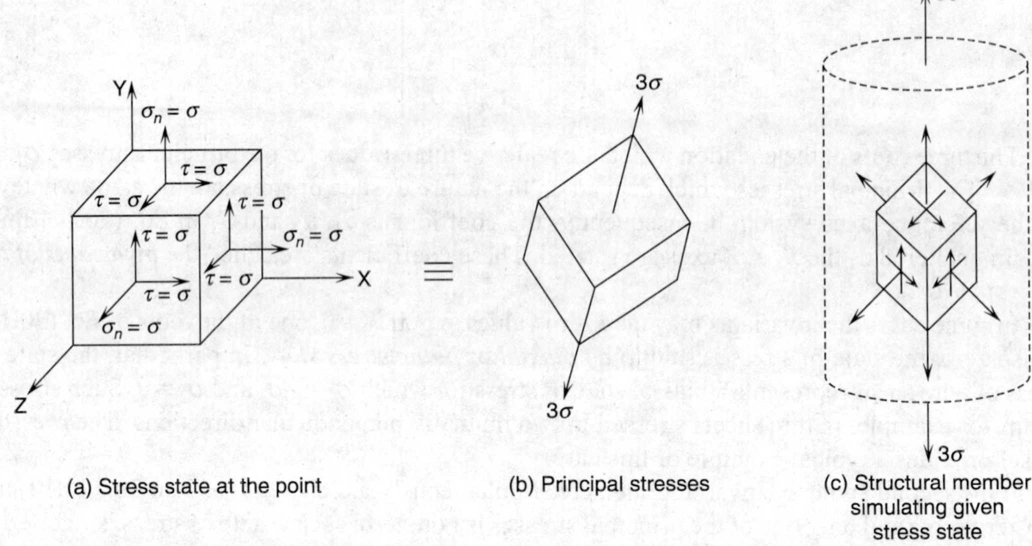

(a) Stress state at the point (b) Principal stresses (c) Structural member simulating given stress state

Figure 20.4 The principal stresses when components of the state of stress are equal.

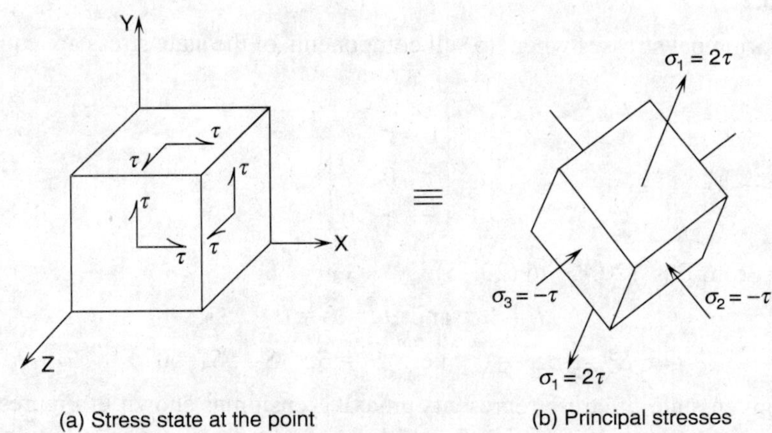

(a) Stress state at the point (b) Principal stresses

Figure 20.5 Triaxial state of stress of Example 20.2.

20.4 BIAXIAL STRESSES COMBINED WITH SHEAR

In the case of two dimensional problems, one of the principal stresses is zero and out of other two, one is the largest and the other is the least. Consider the state of stress at a point subjected to the stresses σ_y, σ_z and τ_{zy} be represented by a right parallelepiped with $\delta x = 1$ shown as a rectangle ABCD in Figure 20.6(a). For the general plane DE inclined at an angle θ with the major principal plane, i.e. the normal to the plane is inclined at an angle θ with the direction of σ_z, the normal and tangential or shearing stress components acting on the plane are represented by σ_n and τ, respectively. All the forces acting on the triangle ADE are shown in Figure 20.6(a).

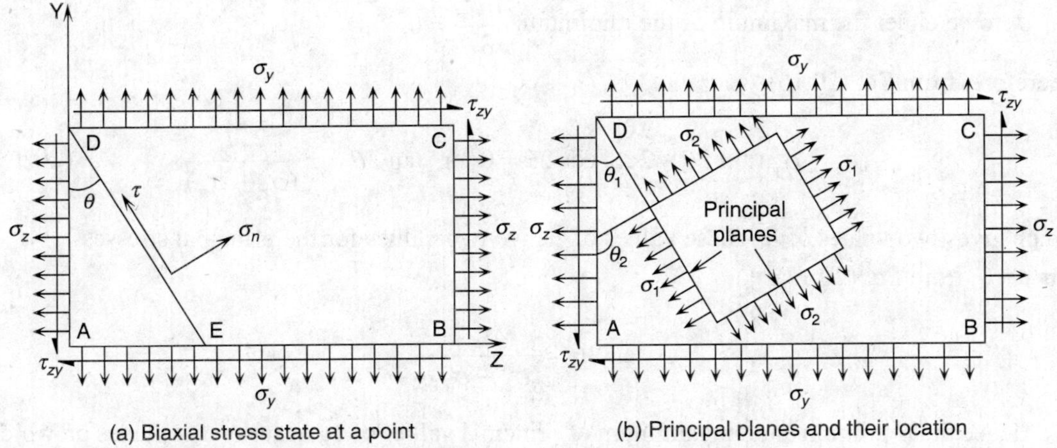

(a) Biaxial stress state at a point (b) Principal planes and their location

Figure 20.6 Principal planes for the biaxial state of stress at a point.

Project the forces acting on the triangle ADE on the direction of σ_n,

$$\sigma_n \text{ DE} - \sigma_y \text{ AE} \sin\theta - \tau_{zy} \text{ AD} \sin\theta - \sigma_z \text{ AD} \cos\theta - \tau_{zy} \text{ AE} \cos\theta = 0$$

where $\text{AD} = \delta y$; $\text{AE} = \delta y \tan\theta$; and $\text{DE} = \delta y \sec\theta$

Therefore,

$$\sigma_n\, \delta y \sec\theta - \sigma_y\, \delta y \tan\theta \sin\theta - \tau_{zy}\, \delta y \sin\theta - \sigma_z\, \delta y \cos\theta - \tau_{zy}\, \delta y \tan\theta \cos\theta = 0$$

i.e.

$$\sigma_n \sec\theta - \sigma_y \tan\theta \sin\theta - \tau_{zy} \sin\theta - \sigma_z \cos\theta - \tau_{zy} \tan\theta \cos\theta = 0$$

or

$$\sigma_n = \sigma_y \sin^2\theta + \tau_{zy} \sin\theta\cos\theta + \sigma_z \cos^2\theta + \tau_{zy} \sin\theta\cos\theta$$

$$= \sigma_y\left(\frac{1-\cos 2\theta}{2}\right) + \tau_{zy} \sin 2\theta + \sigma_z\left(\frac{1+\cos 2\theta}{2}\right)$$

$$= \left(\frac{\sigma_z + \sigma_y}{2}\right) + \left(\frac{\sigma_z - \sigma_y}{2}\right)\cos 2\theta + \tau_{zy} \sin 2\theta \tag{20.16}$$

Project the forces acting on the triangle ADE on the direction of τ,

$$\tau \text{ DE} - \sigma_y \text{ AE} \cos\theta - \tau_{zy} \text{ AD} \cos\theta + \sigma_z \text{ AD} \sin\theta + \tau_{zy} \text{ AE} \sin\theta = 0$$

$$\tau \sec\theta - \sigma_y \tan\theta \cos\theta - \tau_{zy} \cos\theta + \sigma_z \sin\theta + \tau_{zy} \tan\theta \sin\theta = 0$$

or

$$\tau = \sigma_y \sin\theta\cos\theta + \tau_{zy} \cos^2\theta - \sigma_z \sin\theta\cos\theta - \tau_{zy} \sin^2\theta$$

$$= \frac{\sigma_y}{2}\sin 2\theta + \tau_{zy}\cos^2\theta - \frac{\sigma_z}{2}\sin 2\theta - \tau_{zy}\sin^2\theta$$

$$= \left(\frac{\sigma_y - \sigma_z}{2}\right)\sin 2\theta + \sigma_{zy} \cos 2\theta \tag{20.17}$$

For σ_n to be either the maximum or the minimum, $\dfrac{\partial \sigma_n}{\partial \theta} = 0$

Therefore, from Eq. (20.16):

$$-(\sigma_z - \sigma_y) \sin 2\theta + 2\tau_{zy} \cos 2\theta = 0 \text{ or } \tan 2\theta = \pm \frac{2\tau_{zy}}{(\sigma_z - \sigma_y)} \quad (20.18)$$

which gives two values of θ. These values of θ give two values for the principal stresses.
For $\tau = 0$, from Eq. (20.17):

$$\tan 2\theta = \frac{2\tau_{zy}}{(\sigma_z - \sigma_y)}$$

Hence the maximum and the minimum or principal values of σ_n occur on the planes on which $\tau_{zy} = 0$, these principal planes are at right angles to each other as shown in Figure 20.6(b).

20.4.1 Shear Stress

Consider the state of stress at a point subjected to the simple complementary shear stresses σ_{zy} be represented by a right parallelepiped with $\delta x = 1$ shown as a rectangle ABCD in Figure 20.7(a). For the general plane DE inclined at an angle θ with the shear plane, the normal and tangential stress components acting on the plane are represented by σ_n and τ, respectively. All the forces acting on the triangle ADE are shown in Figure 20.7(a).

Project the forces acting on the triangle ADE on the direction of σ_n,

$$\sigma_n \text{DE} - \tau_{zy} \text{AD} \sin \theta - \tau_{zy} \text{AE} \cos \theta = 0$$

where AD = δy; AE = $\delta y \tan \theta$; and DE = $\delta y \sec \theta$

Therefore,

$$\sigma_n\, \delta y \sec \theta - \tau_{zy}\, \delta y \sin \theta - \tau_{zy}\, \delta y \tan \theta \cos \theta = 0$$

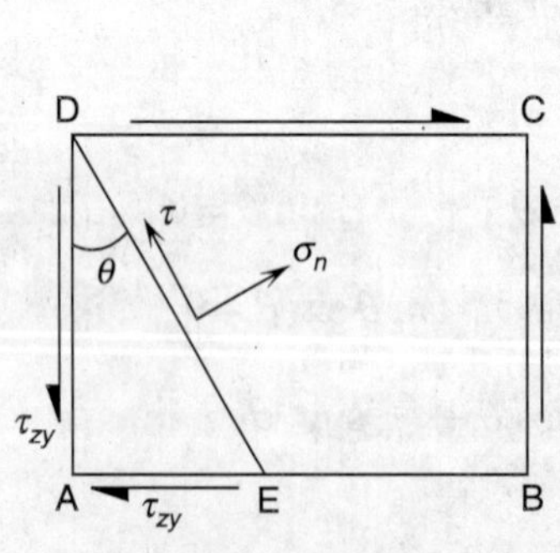

(a) State of simple shear at a point

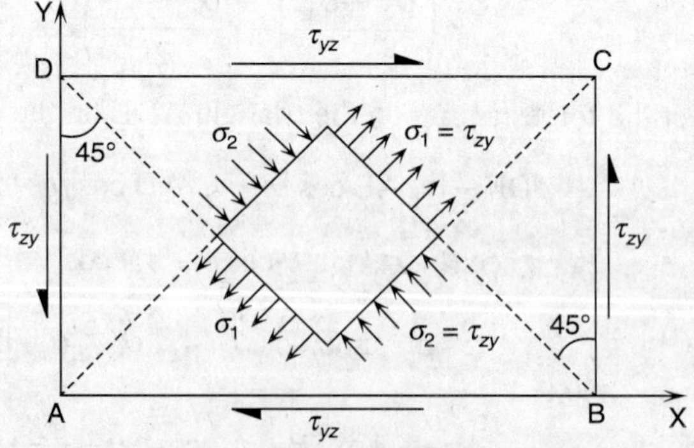

(b) The principal planes and their location

Figure 20.7 Complementary shear stresses acting at a point.

or $$\sigma_n = \tau_{zy} \sin\theta\cos\theta + \tau_{zy} \sin\theta\cos\theta = 2\tau_{zy} \sin\theta\cos\theta = \tau_{zy} \sin 2\theta \qquad (20.19)$$

Project the forces acting on the triangle ADE on the direction of τ,

$$\tau \text{DE} - \tau_{zy} \text{AD} \cos\theta + \tau_{zy} \text{AE} \sin\theta = 0$$

or $$\tau \sec\theta - \tau_{zy} \cos\theta + \tau_{zy} \tan\theta \sin\theta = 0$$

or $$\tau = \tau_{zy} \cos^2\theta - \tau_{zy} \sin^2\theta = \tau_{zy} \cos 2\theta \qquad (20.20)$$

For σ_n to be either the maximum or the minimum, $\dfrac{\partial \sigma_n}{\partial \theta} = 0$

Therefore, from Eq. (20.19):

$$2\tau_{zy} \cos 2\theta = 0 \quad \text{or} \quad \cos 2\theta = 0 \qquad (20.21)$$

Therefore, $$2\theta = 90° \text{ and } 270° \quad \text{or} \quad \theta = 45° \text{ and } 135°$$

For these two values of θ, $\tau = 0$. Hence, the maximum and the minimum values of σ_n occur on the planes on which $\tau_{zy} = 0$, i.e. two principal planes are at 45° and 135° with the plane AD. The principal planes are at right angles to each other. The corresponding principal stresses are:

$$\sigma_1 = \tau_{zy} \sin 2\theta = \tau_{zy} \sin 90° = \tau_{zy} \quad \text{and}$$

$$\sigma_2 = \tau_{zy} \sin 2\theta = \tau_{zy} \sin 270° = -\tau_{zy} \qquad (20.22)$$

i.e. one of the principal planes carry tensile stress and the other the compressive stress of magnitude equal to that of shear stress. These principal stresses are called *diagonal tensile* and *diagonal compressive stresses*.

Thus, two complementary shear stresses on planes right angles to each other are equivalent to tensile and compressive stresses of magnitude equal to that of shear stress acting on the planes inclined at 45° to the shear stress planes as shown in Figure 20.7(b).

20.5 PRINCIPAL STRESSES AND PRINCIPAL PLANES

As discussed earlier that within a stressed body there exist three mutually perpendicular planes on each of which the resultant stress is normal. These mutually perpendicular planes are called *principal planes*, and the *resultant normal stresses acting on them are called principal stresses*.

For σ_n to be either the maximum or the minimum from Eq. (20.18),

$$\tan 2\theta = \pm \frac{2\tau_{zy}}{(\sigma_z - \sigma_y)} \qquad (20.23)$$

Therefore, $$\sin 2\theta = \pm \frac{2\tau_{zy}}{\sqrt{(\sigma_z - \sigma_y)^2 + (2\tau_{zy})^2}} \qquad (20.24)$$

$$\cos 2\theta = \pm \frac{(\sigma_z - \sigma_y)}{\sqrt{(\sigma_z - \sigma_y)^2 + (2\tau_{zy})^2}} \qquad (20.25)$$

Substituting these values in Eq. (20.16)

$$\sigma_n = \left(\frac{\sigma_z + \sigma_y}{2}\right) \pm \left(\frac{\sigma_z - \sigma_y}{2}\right)\left[\frac{(\sigma_z - \sigma_y)}{\sqrt{(\sigma_z - \sigma_y)^2 + (2\tau_{zy})^2}}\right]$$

$$\pm \tau_{zy}\left[\frac{2\tau_{zy}}{\sqrt{(\sigma_z - \sigma_y)^2 + (2\tau_{zy})^2}}\right]$$

$$= \left(\frac{\sigma_z + \sigma_y}{2}\right) \pm \frac{1}{2}\sqrt{(\sigma_z - \sigma_y)^2 + (4\tau_{zy})^2}$$

Therefore,
$$\sigma_1, \sigma_2 = \left(\frac{\sigma_z + \sigma_y}{2}\right) \pm \sqrt{\left(\frac{\sigma_z - \sigma_y}{2}\right)^2 + \tau_{zy}^2} \tag{20.26}$$

One of the two values of σ_n is the maximum and the other minimum. As the maximum and the minimum values occur on the planes on which $\tau = 0$. From Eq. (20.17)

$$\tau = \left(\frac{\sigma_y - \sigma_z}{2}\right)\sin 2\theta + \tau_{zy}\cos 2\theta = 0 \quad \text{or} \quad \tan 2\theta = -\left(\frac{2\tau_{zy}}{\sigma_y - \sigma_z}\right) \tag{20.27}$$

For the shear stress to be the maximum and the minimum, $\frac{\partial \tau}{\partial \theta} = 0$, from Eq. (20.17)

$$(\sigma_y - \sigma_z)\cos 2\theta - 2\tau_{zy}\sin 2\theta = 0 \quad \text{or} \quad \tan 2\theta = -\left(\frac{\sigma_z - \sigma_y}{2\tau_{zy}}\right) \tag{20.28}$$

The two values of 2θ given by Eq. (20.28), differ by 180° and hence the values of θ differ by 90°. A comparison of Eqs. (20.27) and (20.28), indicates that the planes of principal stresses are inclined at 45° to the planes of maximum shear stress (see Figure 20.7(b)). Substitute Eq. (20.28) in Eq. (20.17):

$$\tau_{\max} = -\left(\frac{\sigma_y - \sigma_z}{2}\right)\left[\frac{(\sigma_z - \sigma_y)}{\sqrt{(\sigma_z - \sigma_y)^2 + (2\tau_{zy})^2}}\right] + \tau_{zy}\left[\frac{2\tau_{zy}}{\sqrt{(\sigma_z - \sigma_y)^2 + (2\tau_{zy})^2}}\right]$$

$$= \frac{1}{2}\frac{(\sigma_z - \sigma_y)^2 + (2\tau_{zy})^2}{\sqrt{(\sigma_z - \sigma_y)^2 + (2\tau_{zy})^2}} = \frac{1}{2}\sqrt{(\sigma_z - \sigma_y)^2 + (2\tau_{zy})^2} = \frac{\sigma_1 - \sigma_2}{2} \tag{20.29}$$

EXAMPLE 20.3

At a point in a material, the stresses acting on two planes right angles to each other are:

$$\sigma_z = 120 \text{ MPa}, \ \sigma_y = -200 \text{ MPa} \ \text{ and } \ \tau_{zy} = -80 \text{ MPa}$$

Determine: (a) the magnitude and direction of the principal stresses, (b) the maximum shear stress, (c) the tensile stress which, while acting alone, would develop the same maximum shear stress, and (d) the shear stress which, while acting alone, would develop the same principal stress.

Solution: The state of stress is shown in Figure 20.8

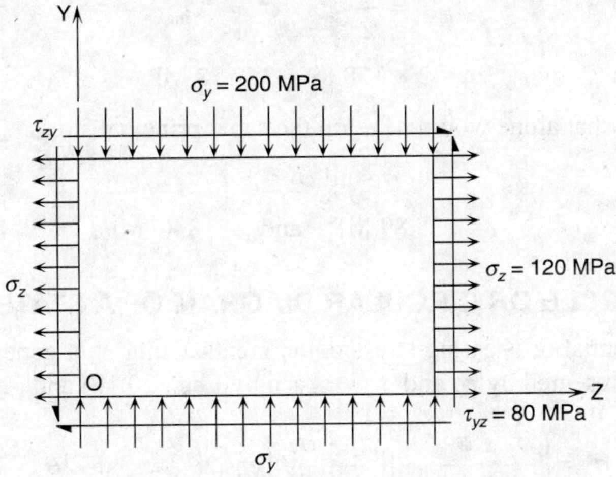

Figure 20.8 State of stress for the Example 20.3.

(a) At a point in a stressed material, the principal stresses are given by

$$\sigma_1, \sigma_2 = \left(\frac{\sigma_z + \sigma_y}{2}\right) \pm \sqrt{\left(\frac{\sigma_z - \sigma_y}{2}\right)^2 + \tau_{zy}^2}$$

$$= \left(\frac{120 - 200}{2}\right) \pm \sqrt{\left(\frac{120 + 200}{2}\right)^2 + (-80)^2}$$

$$= -40 \pm 178.89 = 138.89 \text{ MPa}, -218.89 \text{ MPa}$$

Therefore, the principal stresses are σ_1 = 138.89 MPa, and σ_2 = –218.89 MPa

Angular location of principal planes

$$\tan 2\theta = \frac{2\tau_{zy}}{\sigma_z - \sigma_y} = \frac{2 \times (-80)}{120 + 200} = -0.5$$

$$2\theta = -26.57° \quad \text{i.e.} \quad \theta_1 = -13.28° \quad \text{and} \quad \theta_2 = 76.72°$$

(b) The maximum shear stress

$$\tau_{\max} = \left(\frac{\sigma_1 - \sigma_2}{2}\right) = \left(\frac{138.89 + 218.89}{2}\right) = 178.89 \text{ MPa}$$

Angular location of planes carrying the maximum shear stress

$$\theta_3 = \theta_1 + 45° = 31.72° \text{ and } \theta_4 = \theta_2 + 45° = 121.72°$$

(c) The tensile stress σ_z that alone would develop the same maximum shear stress

$$\tau_{max} = \frac{\sigma_z}{2} \text{ or } \sigma_z = 2\tau_{max}$$

Therefore, $$\sigma_z = 2 \times 178.89 = 357.78 \text{ MPa}$$

(d) The shear stress τ that alone would develop the same principal stress

$$\sigma_1 \text{ and } \sigma_2 = \pm\, \tau$$

Therefore, $$\tau = 138.89 \text{ MPa and } -218.89 \text{ MPa}$$

20.6 MOHR'S CIRCLE OR CIRCULAR DIAGRAM OF A STATE OF STRESS

The normal and tangential or shearing stress components acting on a general plane as shown in Figure 20.9(a) are represented by σ_n and τ are given by Eqs. (20.16) and (20.17)

$$\sigma_n = \left(\frac{\sigma_z + \sigma_y}{2}\right) + \left(\frac{\sigma_z - \sigma_y}{2}\right)\cos 2\theta + \tau_{zy}\sin 2\theta$$

or $$\sigma_n - \left(\frac{\sigma_z + \sigma_y}{2}\right) = \left(\frac{\sigma_z - \sigma_y}{2}\right)\cos 2\theta + \tau_{zy}\sin 2\theta \tag{20.30}$$

$$\tau = \left(\frac{\sigma_y - \sigma_z}{2}\right)\sin 2\theta + \tau_{zy}\cos 2\theta \tag{20.31}$$

To eliminate the angle θ transpose $\left(\frac{\sigma_z + \sigma_y}{2}\right)$ and add up the squares of all terms of Eqs. (20.30) and (20.31):

$$\left(\sigma_n - \frac{\sigma_z + \sigma_y}{2}\right)^2 + \tau^2 = \left(\frac{\sigma_z - \sigma_y}{2}\right)^2 + \tau_{zy}^2$$

Therefore, $$\left(\sigma_n - \frac{\sigma_z + \sigma_y}{2}\right)^2 + \tau^2 = \frac{1}{4}\left[(\sigma_z - \sigma_y)^2 + 4\tau_{zy}^2\right] \tag{20.32}$$

With σ_n, τ taken as co-ordinates system as illustrated in Figure 20.9(b), Eq. (20.32) represents a circle whose centre is on σ_n axis at a distance $\left(\frac{\sigma_z + \sigma_y}{2}\right)$ from the origin. The radius of the circle is:

$$R = \frac{1}{2}\sqrt{(\sigma_z - \sigma_y)^2 + 4\tau_{zy}^2} \tag{20.33}$$

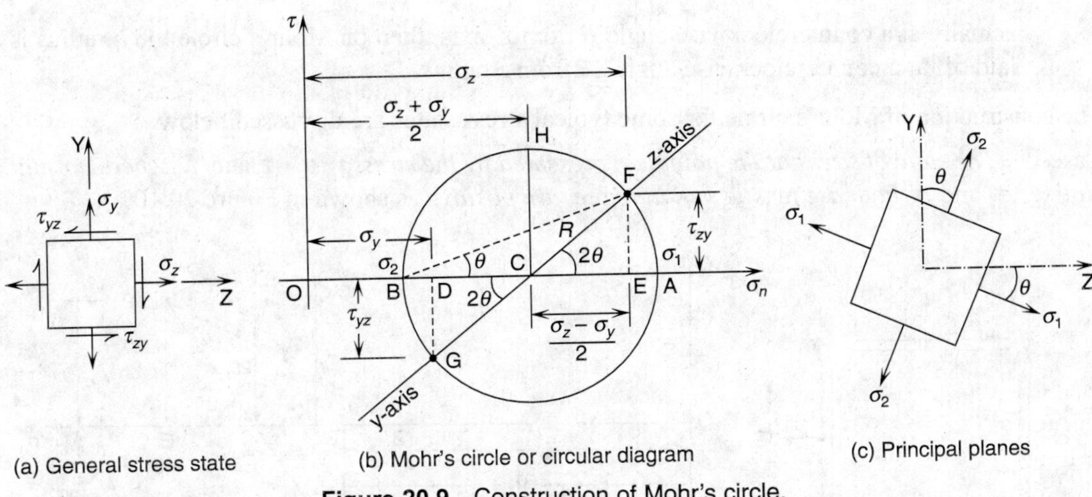

(a) General stress state

(b) Mohr's circle or circular diagram

(c) Principal planes

Figure 20.9 Construction of Mohr's circle.

In the other words, the circle is constructed on the segment $\sqrt{(\sigma_z - \sigma_y)^2 + 4\tau_{zy}^2}$ as diameter as illustrated in Figure 20.9(b). This circle is called *Mohr's circle* or *a circular diagram of the state of stress*. It should be noted that the angle θ sets up a correspondence between a point on the circle and cutting plane. Thus, corresponding to each cutting plane, there is a definite point on the Mohr's circle. Figure 20.9(c) shows the principal planes.

20.6.1 Construction of Mohr's Circle for the State of Stress

General rules

1. With orthogonal σ_n–τ axes as co-ordinates system, the points with coordinates (σ_z, τ_{zy}) and (σ_y, τ_{yz}) are plotted. These points represent the state of stress, i.e. the known normal and shearing stresses acting on the Z and Y faces of the element under consideration shown in Figure 20.9(a). For plotting these points the tension is assumed positive and compression negative, whereas the shearing stress is taken positive when its moment about the centre of the element is clockwise.
2. The points plotted above are joined by straight line which is the diameter of the circle having centre on the σ axis.
3. As different planes may pass through the point under consideration, the normal and shearing stress components on these planes are represented by the coordinates of the points whose position shifts around the circumference of Mohr's circle.
4. The radius of circle to any point on its circumference represents the axis directed normal to the plane whose stress components are given by the coordinates of that point.
5. The angle between the radii to the selected points on Mohr's circle is twice the angle between the normals to the actual planes represented by these points, or to twice the space angularity between the planes so represented. The rotational sense of this angle corresponds to the rotational sense of the actual angle between the normals to the planes, i.e. if *n* axis is

actually at a counterclockwise angle θ from Z-axis, then on Mohr's circle the n radius is laid off at a counterclockwise angle 2θ from z radius.

The construction of Mohr's circle for some typical stress states are discussed below.

Case-I: *The state of stress at the point is represented by the stresses* σ_z, σ_y and τ_{zy} *where normal stresses* σ_z and σ_y *both are tensile in nature, i.e. are positive* as shown in Figure 20.10(a).

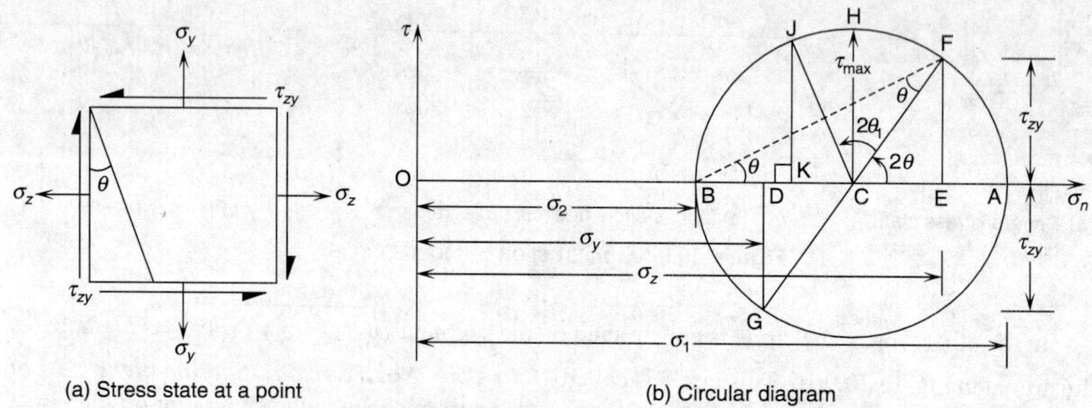

(a) Stress state at a point (b) Circular diagram

Figure 20.10 Mohr's stress circle.

Mark OE and OD representing unequal normal stresses σ_z and σ_y. Draw perpendicular offsets EF = DG representing shear stress. With C, the mid-point of DE as centre, draw a circle with CF or CG as radius and obtain the points A and B as shown in Figure 20.10(b).

The lengths OA and OB represent the major and minor principal stresses σ_1 and σ_2, respectively. The position of the principal plane is given by the angle ABF. The length CH represents the maximum shear stress $\tau_{max} = R = \dfrac{\sigma_1 - \sigma_2}{2}$.

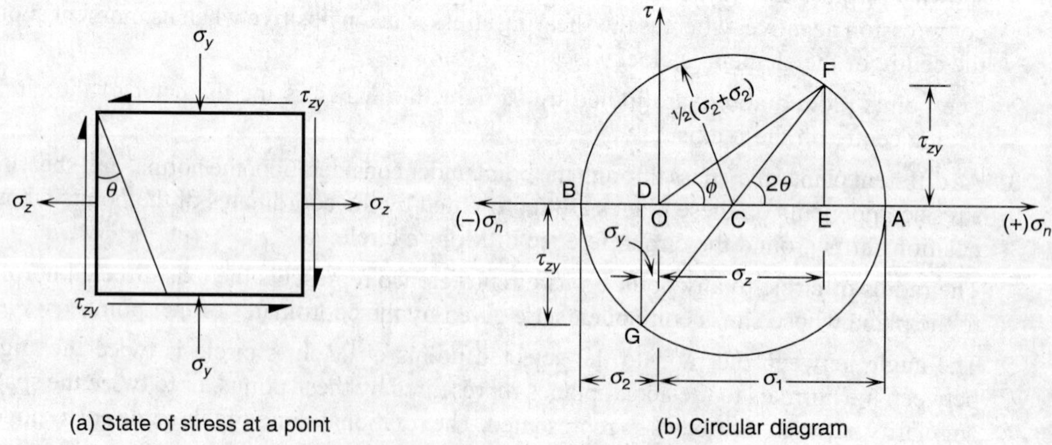

(a) State of stress at a point (b) Circular diagram

Figure 20.11 Mohr's stress circle.

To determine the normal and shear stresses on a plane inclined at an angle θ_1 with the principal plane, draw CJ so that the angle JCF = $2\theta_1$. Draw JK perpendicular to OC, the lengths OK and JK will provide the normal and tangential stresses on the plane.

Case-II: *The state of stress at the point is represented by the stresses σ_z, σ_y and τ_{zy} where one of the normal stresses, say σ_z is tensile and other σ_y compressive in nature* as shown in Figure 20.11(a).

The Mohr's circle for this state of stress is drawn as shown in Figure 20.11(b). It should be noted that one of the principal stresses is negative. The maximum shear stress $\tau_{\max} = R = \dfrac{\sigma_1 + \sigma_2}{2}$.

Case-III: *The state of stress at the point is represented by the principal stresses σ_1 and σ_2.*

In most of the cases encountered in practice, the position of one of the principal planes at the point under consideration is known and the other two principal planes can be easily determined in a set of planes perpendicular to the first one. Consider the conditions of equilibrium of the triangular prism formed by cutting an elementary parallelepiped by an inclined plane which remains parallel to one of the principal axes as shown in Figure 20.12(a). In this case, whatever may be the angle of inclination θ the inclined plane is parallel to principal Z-axis.

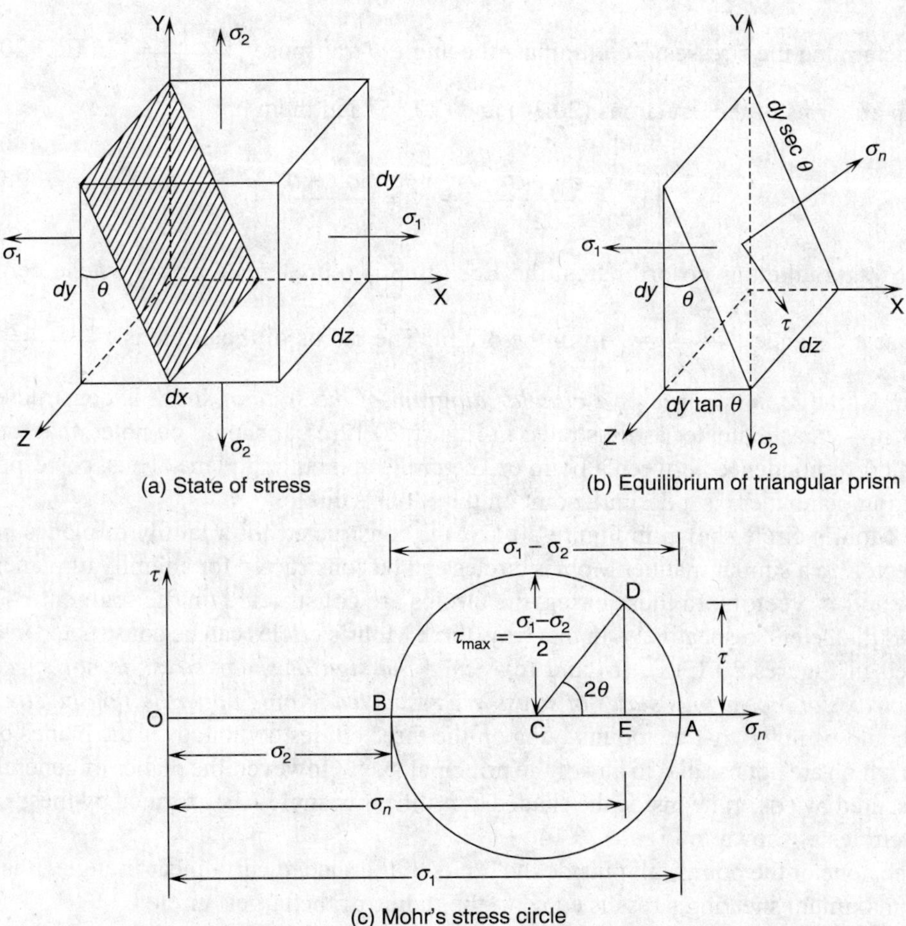

(a) State of stress

(b) Equilibrium of triangular prism

(c) Mohr's stress circle

Figure 20.12 Analysis when state of stress is represented by principal stresses.

Project all forces acting on the cut off triangular prism on the axes coincident with σ_n and τ vectors of inclined plane illustrated in Figure 20.12(b).

$$\sigma_n\, dz\,(dy \sec\theta) = \sigma_1\, dz\,(dy \cos\theta) + \sigma_2\, dz\, dy \tan\theta \sin\theta$$

or

$$\sigma_n = \sigma_1 \cos^2\theta + \sigma_2 \sin^2\theta$$

$$= \left(\frac{\sigma_1 + \sigma_2}{2}\right) + \left(\frac{\sigma_1 - \sigma_2}{2}\right)\cos 2\theta \tag{20.34}$$

and

$$\tau\, dz\,(dy \sec\theta) = \sigma_1\, dz\,(dy \sin\theta) - \sigma_2\, dz\, dy \tan\theta \cos\theta$$

$$\tau = (\sigma_1 - \sigma_2)\sin\theta\cos\theta$$

$$= \left(\frac{\sigma_1 - \sigma_2}{2}\right)\sin 2\theta \tag{20.35}$$

These expressions represent a set of planes parallel to one of the principal axes and can be used to determine the stresses. To eliminate the angle θ transpose $\left(\frac{\sigma_1 + \sigma_2}{2}\right)$ in Eq. (20.34) and square up all terms of the equations (20.34) and (20.35) add them up:

$$\left(\sigma_n - \frac{\sigma_1 + \sigma_2}{2}\right)^2 + \tau^2 = \left(\frac{\sigma_1 - \sigma_2}{2}\right)^2 \tag{20.36}$$

With σ_n, τ taken as co-ordinates, the Eq. (20.36) represents a circle whose centre is on σ_n–axis at a distance $\left(\frac{\sigma_1 + \sigma_2}{2}\right)$ from the origin. The radius of the circle is $\left(\frac{\sigma_1 - \sigma_2}{2}\right)$. In the other words, the *Mohr's circle* or *a circular diagram of the state of stress* is constructed on the segment $\sigma_1 - \sigma_2$ as diameter as illustrated in Figure 20.12(c). It should be noted that the angle θ sets up a correspondence between a point on the circle and cutting plane. Thus, corresponding to each cutting plane there is a definite point on the Mohr's circle.

The Mohr's circle shown in Figure 20.12(c) is constructed for a family of planes parallel to the σ_3 vector. In a similar manner Mohr's circles can be constructed for a family of planes parallel to the σ_1 and σ_2 vectors. In these cases, the circles are constructed on the segment $\sigma_2 - \sigma_3$ and $\sigma_1 - \sigma_3$ as diameters, respectively. In this way, three Mohr's circles can be constructed in a manner illustrated in Figures 20.13(a), (b) and (c). *Since the sign of shear stress is not specified, the Mohr's circles of the various states of stress are restricted to only upper half of the circle.*

It should be noted that the points lying on the three circles do not cover the planes of general position which are not parallel to any of the principal axes. However, the planes of general position K represented by (σ_n, τ) lie inside the shaded curvilinear triangle ABD formed by these combined Mohr's circles as shown in Figure 20.14.

Since none of the points fall outside the limits of the shaded curvilinear triangle, it is apparent that the maximum shearing stress is equal to the radius of the largest circle

$$\tau_{max} = \frac{\sigma_1 - \sigma_3}{2} \tag{20.37}$$

This stress occurs on a plane which is equally inclined to the planes of maximum and minimum principal stresses.

Thus the Mohr's circle enables the shear and normal stresses to be determined from the principal stresses and conversely enables principal stresses to be obtained if the shear; and normal and shear stresses on a given plane are known.

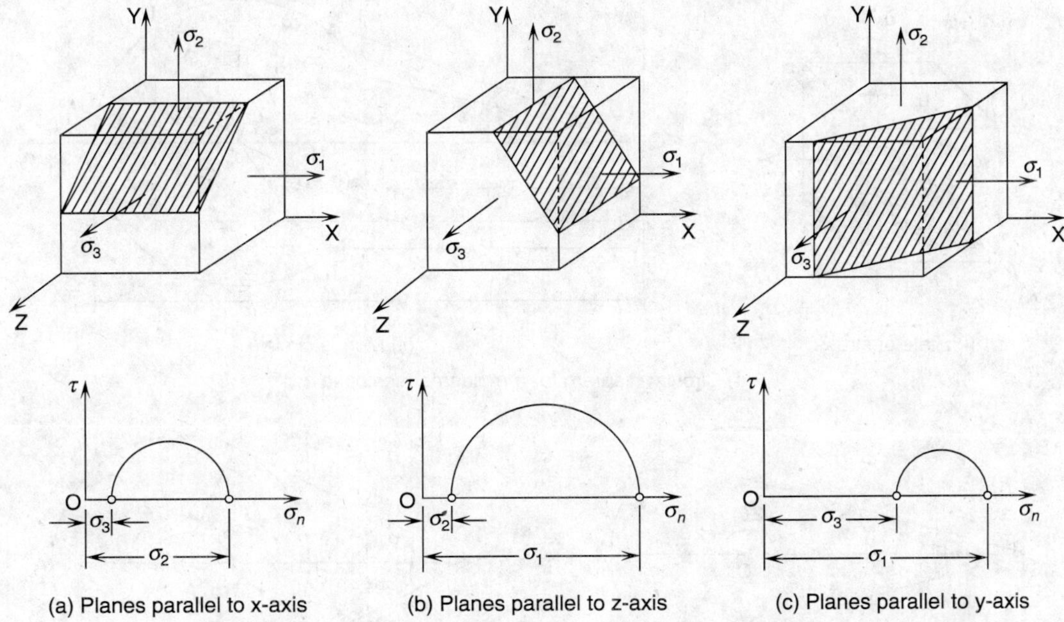

Figure 20.13 Mohr's stress circles for planes parallel to various axes.

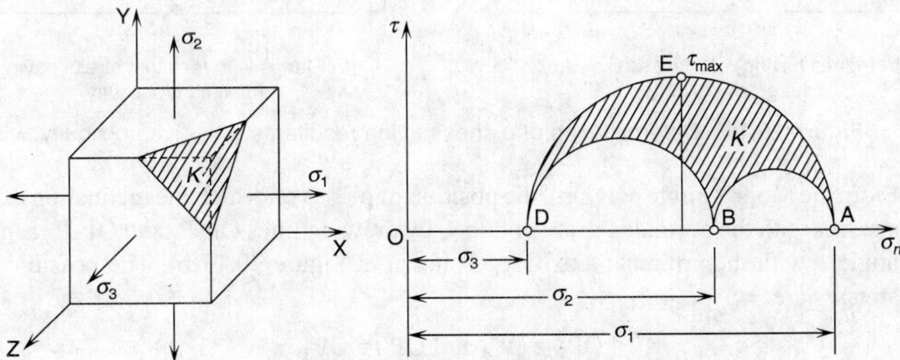

Figure 20.14 Mohr's stress circle for a plane inclined to the three reference axes.

20.6.2 Applications

(a) From the Mohr's circle diagram representing the state of stress at a point in a strained material shown in Figure 20.15(a), the angular position of a plane (say I) can be determined in terms of inclination ϕ of resultant stresses with the normal by drawing lines O–I from O to the point representing the plane on the diagram as shown in Figure 20.15(a-ii).

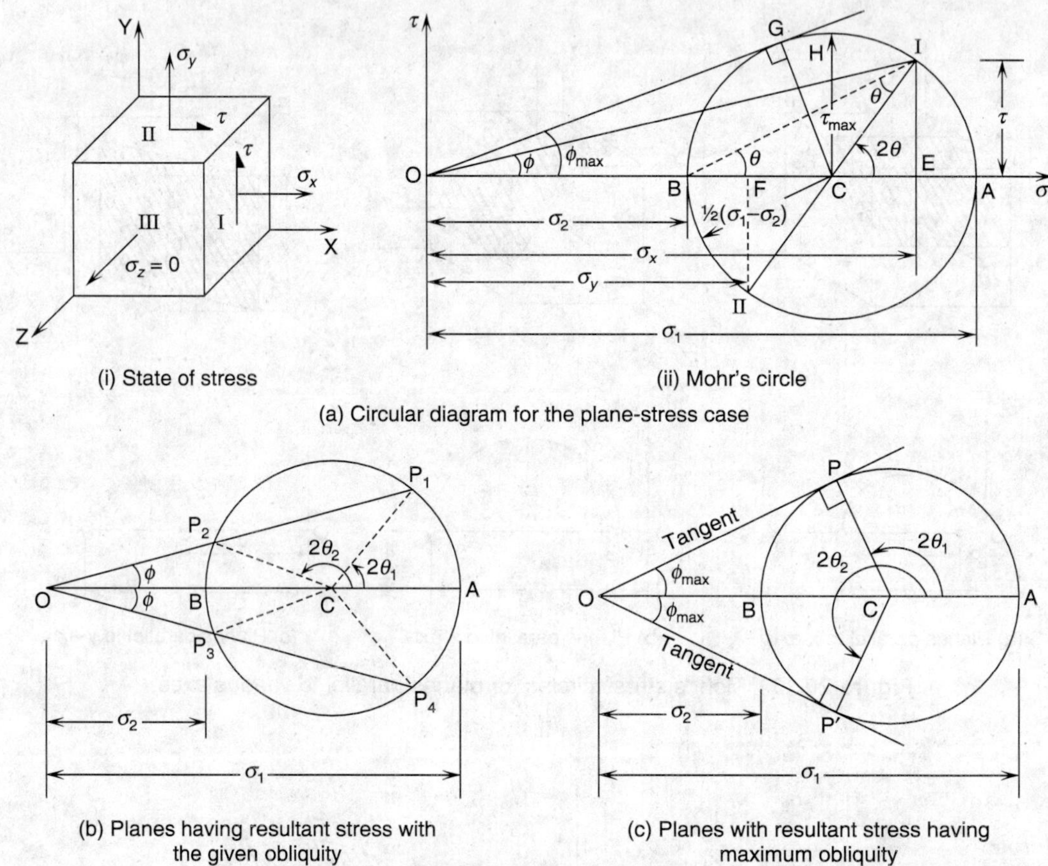

Figure 20.15 Determination of planes having resultants with given obliquity.

(b) From the Mohr's circle diagram, the position of planes on which the inclination of resultant stresses with the normal is ϕ, are obtained by drawing lines OP_2P_1 and OP_3P_4 from O at an angle ϕ with the principal axis OA as shown in Figure 20.15(b). The possible resultant stresses are:

$$OP_1 = OP_4 \text{ and } OP_2 = OP_3,$$

and their angular positions are given by:

(i) $\angle ACP_1 = 2\theta_1$, (ii) $\angle ACP_2 = 2\theta_2$, (iii) $\angle ACP_3 = 2\theta_3$, and (iv) $\angle ACP_4 = 2\theta_4$

(c) The Mohr's circle diagram can be used to locate the position of plane on which the inclination or obliquity ϕ of resultant stress is the maximum. This is obtained by drawing tangent lines OP and OP′ from O as shown in Figure 20.15(c). The inclination or obliquity of the resultant is given by ϕ_{max} and the angular positions of the planes are: (i) $\angle ACP = 2\theta_1$ and (ii) $\angle ACP' = 2\theta_2$.

EXAMPLE 20.4

The principal stresses at a point in a strained material subjected to two-dimensional stress are σ_1 and σ_2. Show that the: (a) the sum of the normal stresses on any two planes right angles to each other is constant, (b) the resultant stress σ_r on the plane carrying the maximum shear stress is given by $\sigma_r = \sqrt{\dfrac{\sigma_1^2 + \sigma_2^2}{2}}$ and (c) if the normal components of stresses σ_n and σ'_n in (a) are alike and the shear stress is τ, show that one of the principal stresses is zero if $\tau^2 = \sigma_n \sigma'_n$.

Solution: (a) For a plane inclined at an angle θ with the major principal plane, the normal stress is given by

$$\sigma_n = \left(\frac{\sigma_1 + \sigma_2}{2}\right) + \left(\frac{\sigma_1 - \sigma_2}{2}\right) \cos 2\theta$$

And for the plane inclined at an angle $\theta + 90°$ with the major principal plane,

$$\sigma_n' = \left(\frac{\sigma_1 + \sigma_2}{2}\right) + \left(\frac{\sigma_1 - \sigma_2}{2}\right) \cos(180 + 2\theta) = \left(\frac{\sigma_1 + \sigma_2}{2}\right) - \left(\frac{\sigma_1 - \sigma_2}{2}\right) \cos 2\theta$$

Therefore, $\qquad \sigma_n + \sigma_n' = \sigma_1 + \sigma_2 = \text{constant}$

(b) The plane carrying the maximum shear stress is inclined at an angle $\theta = 45°$ with the major principal plane. On this plane

$$\sigma_r = \sqrt{\sigma_1^2 \cos^2 \theta + \sigma_2^2 \sin^2 \theta}$$

$$= \sqrt{\sigma_1^2 \cos^2 45 + \sigma_2^2 \sin^2 45} = \sqrt{\frac{\sigma_1^2 + \sigma_2^2}{2}}$$

Alternatively use Eqs. (20.16) and (20.17) in $\sigma_r = \sqrt{\sigma_n^2 + \tau^2}$.

(c) At a point in a stressed material the principal stresses are given by

$$\sigma_1, \sigma_2 = \left(\frac{\sigma_n + \sigma_n'}{2}\right) \pm \sqrt{\left(\frac{\sigma_n - \sigma_n'}{2}\right)^2 + \tau^2}$$

If $\tau^2 = \sigma_n \sigma'_n$ is true, then

$$\sigma_1, \sigma_2 = \left(\frac{\sigma_n + \sigma_n'}{2}\right) \pm \sqrt{\left(\frac{\sigma_n - \sigma_n'}{2}\right)^2 + \sigma_n \sigma_n'}$$

$$= \left(\frac{\sigma_n + \sigma_n'}{2}\right) \pm \sqrt{\left(\frac{\sigma_n + \sigma_n'}{2}\right)^2} = (\sigma_n + \sigma_n') \text{ and } 0$$

In general a circular diagram can be drawn if the stresses on any two planes of a family of planes parallel to a principal axis are known. For example, consider the state of stress at a point shown in Figure 20.16, wherein Z-axis is principal axis and there are two planes I and II parallel to a principal axis on which the stresses are known. Consequently, the two corresponding points can be located on the opposite ends of a diameter of the circular diagram. The principal stresses from the circular can be computed as follows:

$$\sigma', \sigma'' = \left(\frac{\sigma_x + \sigma_y}{2}\right) \mp R$$

where R, the radius of circle $= \sqrt{\left(\frac{\sigma_x - \sigma_y}{2}\right)^2 + \tau^2}$

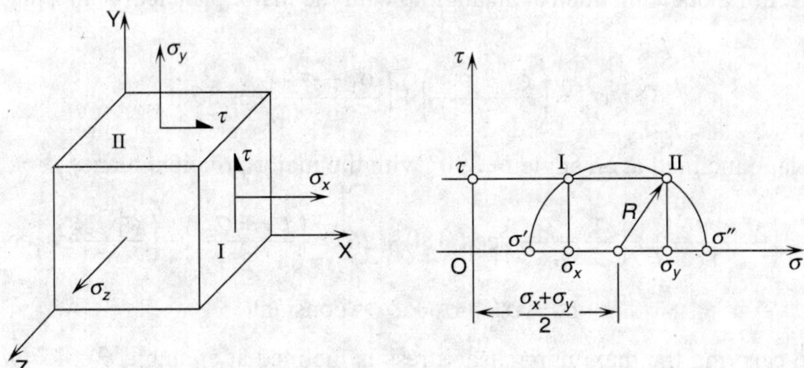

Figure 20.16 Circular diagram with two principal planes parallel to a principal axis.

Therefore, $$\sigma', \sigma'' = \left(\frac{\sigma_x + \sigma_y}{2}\right) \mp \sqrt{\left(\frac{\sigma_x - \sigma_y}{2}\right)^2 + \tau^2}$$

The stresses σ' and σ'' are compared with the magnitude of σ_x and σ_y, the stresses are redesignated by σ_1, σ_2 and σ_3 in the increasing order.

EXAMPLE 20.5

Determine the principal stresses for the state of stress shown in Figures 20.17(a) and (b). The stresses are given in arbitrary units.

Solution: (a) In the state of stress shown in Figure 20.17(a) one of the principal planes and one of the principal stresses are given. Consequently, the remaining principal stresses can be determined from Mohr's circle for family of planes parallel to the X-axis. Plot the points corresponding to planes I and II, and construct a circular diagram.

The stress state is: $\sigma_x = 500$, $\sigma_y = 400$ and $\tau = 300$

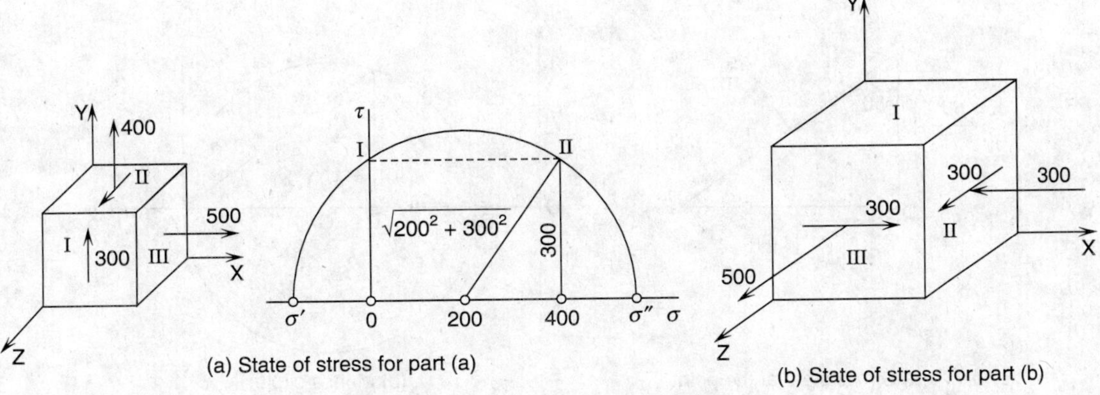

(a) State of stress for part (a)

(b) State of stress for part (b)

Figure 20.17 Stress states for Example 20.5.

The plane III is a principal plane. Thus,

$$\sigma' = \frac{0+400}{2} - \sqrt{\left(\frac{0-400}{2}\right)^2 + 300^2} = -160$$

$$\sigma'' = \frac{0+400}{2} + \sqrt{\left(\frac{0-400}{2}\right)^2 + 300^2} = 560$$

Consequently the principal stresses are $\sigma_1 = 560$, $\sigma_2 = 500$ and $\sigma_3 = -160$.

(b)The state of stress shown in Figure 20.17(b) is a plane stress case. The plane I is a principal plane. The other two principal planes are in a family of planes perpendicular to the first one. Consider Y-axis perpendicular to the principal plane. Then

$$\sigma_x = -300,\ \sigma_z = 500 \text{ and } \tau = 300$$

Plot the points corresponding to planes II and III, and construct a circular diagram. Thus,

$$\sigma' = \frac{-300+500}{2} - \sqrt{\left(\frac{-300-500}{2}\right)^2 + 300^2} = -400$$

$$\sigma'' = \frac{-300+500}{2} + \sqrt{\left(\frac{-300-500}{2}\right)^2 + 300^2} = 600$$

Therefore, the principal stresses are $\sigma_1 = 600$, $\sigma_2 = 0$ and $\sigma_3 = -400$.

EXAMPLE 20.6

A plane at a point in a strained material subjected to two-dimensional stress system transmits a normal tensile stress of 55 MPa along with an unknown shear stress as shown in Figure 20.18(a). The maximum principal stress in the material is 125 MPa tension acting on the plane inclined at 50° with the plane under consideration. (a) Determine analytically the second principal stress and (b) Construct the Mohr's circle diagram. Also calculate the strain normal to the plane. The elastic modulus and Poisson's of the material are 200 GPa and 0.3, respectively.

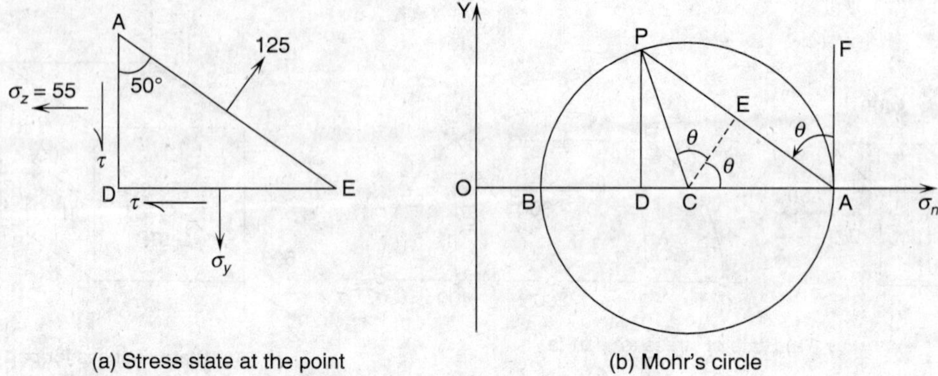

Figure 20.18 Stress state and corresponding Mohr's circle for Example 20.6.

Solution: The state of stress shown in Figure 20.18(a) is represented by,

$$\sigma_z = 55 \text{ MPa}, \sigma_y = ?, \tau = ? \text{ and } \sigma_1 = 125 \text{ MPa}, \theta_1 = 50°$$

(a) *Analytical solution*:

Project the forces acting on the element in the Z-direction:

$$\sigma_z \text{ AD} + \tau \text{DE} - 125\text{AE} \cos 50° = 0$$

$$55 \cos 50° + \tau \sin 50° - 125 \cos 50° = 0$$

$$\tau \sin 50° = 70 \cos 50° \quad \text{or} \quad \tau = 70 \times \cot 50° = 58.74 \text{ MPa}$$

In the Y-direction:

$$\sigma_y \text{ DE} + \tau \text{AD} - 125 \text{ AE} \sin 50° = 0$$

$$\sigma_y \sin 50° + \tau \cos 50° - 125 \sin 50° = 0$$

or $$(\sigma_y - 125) \sin 50° = -\tau \cos 50° \quad \text{i.e.} \quad \sigma_y = 125 - \tau \cot 50°$$

Therefore, $$\sigma_y = 125 - 49.29 = 75.71 \text{ MPa}$$

From the relation, $\sigma_1 + \sigma_2 = (\sigma_z + \sigma_y)$ or $\sigma_2 = (\sigma_z + \sigma_y) - \sigma_1 = 5.71$ MPa

(b) *Construction of the Mohr's circle diagram*:

Each cutting plane has a corresponding definite point on the Mohr's circle. In this problem, the position of one principal plane is known, whereas the position of the other plane carrying normal stress is known only on the axis as the magnitude of the shear stress is unknown, hence the point cannot be located directly on the circle. However, as its inclination with principal plane being known, the point can be located on the circle as follows.

Mark OA and OD representing the major principal stress σ_1 = 125 MPa and normal stress σ_z = 55 MPa, respectively. Draw perpendicular offset DP representing direction of shear stress. Draw the line AP at an angle θ_1 = 50° to AF meeting the line DP at P, thus fixing the position of second plane. It should be noted that

$$\angle\text{PCA} = \angle\text{PCE} + \angle\text{ECA} = 2 \times \angle\text{PAF} = 2\theta_1$$

Thus ∠PCA corresponds to $2\theta_1$.

Thus two points A and P are now available on the Mohr's circle. Draw a line EC bisecting AP at right angles, cutting the axis OA at C. With C as centre, draw a circle with CA or CP as radius, cutting the axis OA at B as shown in Figure 20.18(b). OB represents minor principal stress, $\sigma_2 = 5.71$ MPa.

The strain normal to the planes:

Strain in the direction of σ_1 :

$$\varepsilon_1 = \frac{(\sigma_1 - \nu\sigma_2)}{E} = \frac{(125 - 5.71 \times 0.3)}{200 \times 10^3} = 6.164 \times 10^{-4}$$

Strain in the direction of σ_2 :

$$\varepsilon_2 = \frac{(\sigma_2 - \nu\sigma_1)}{E} = \frac{(5.71 - 125 \times 0.3)}{200 \times 10^3} = -1.589 \times 10^{-4}$$

Strain on a plane inclined at an angle $\theta = 90 - \theta_1 = 90 - 50° = 40°$ to the direction of σ_1 :

$$\varepsilon = \varepsilon_1 \cos^2\theta + \varepsilon_2 \sin^2\theta = 0.413\varepsilon_1 + 0.587\varepsilon_2 = 1.613 \times 10^{-4}$$

20.7 TYPES OF STATES OF STRESS

To find the solutions to the strength design problems, it is important to know the state of stress under combined load conditions. Generally, the states of stress are divided into triaxial, biaxial and uniaxial, which are further classified according to the sign of the principal stresses.

Consider the case of principal stresses resulting from the interaction of bending and shearing stresses. In a beam, the result of the interaction between bending and shear stresses is to produce a set of resultant tensile and compressive stresses, as usual called the *principal stresses* that act in different directions from either the bending or shear stresses individually.

As usual the maximum and the minimum principal tensile and compressive stresses occur on the planes of zero shearing stress and maximum shearing stresses occur on planes at 45° angles to the planes of principal tensile and compressive stresses.

In the cantilever beam illustrated in Figure 20.19(a-i), a series of elements have been considered where both shear and bending stresses are determined and equivalent principal tensile and compressive stresses are found. The principal stresses have an inclination depending upon the relative magnitudes of the shear and bending stresses. For example, for the element at point B, Mohr's circle of Figure 20.19(a-ii) indicates the direction of the principal compressive stress makes a clockwise angle θ with the Z-axis, the principal tensile stress is at right angles. The value of θ is given by $\tan 2\theta = 2\tau/\sigma$ i.e. θ varies with the ratio τ/σ. On the extreme fibres A of the section τ is zero, an element carries only bending stresses in tension or compression and the principal stress directions are horizontal and vertical. At the neutral plane where σ is zero, the principal stresses are at 45° to the horizontal.

The solid and dashed lines on the beam shown in Figure 20.19(a-iii) are often called *stress trajectories* and *depict the direction of principle stresses* in the member. They consist of two systems of orthogonal curves with tangents at each point in the direction of principal stresses at that point. The solid lines indicate the direction of the maximum compressive stresses, and the dashed lines indicate the direction of the maximum tensile stresses. It should be noted that the stress trajectories are not the lines of constant stress but are lines of principle stress direction of variable stress intensity.

The *directions of principal stresses* in a simply supported beam are shown in Figure 20.19(b-i). Figure 20.19(b-ii) illustrates the load carrying mechanism in the beam.

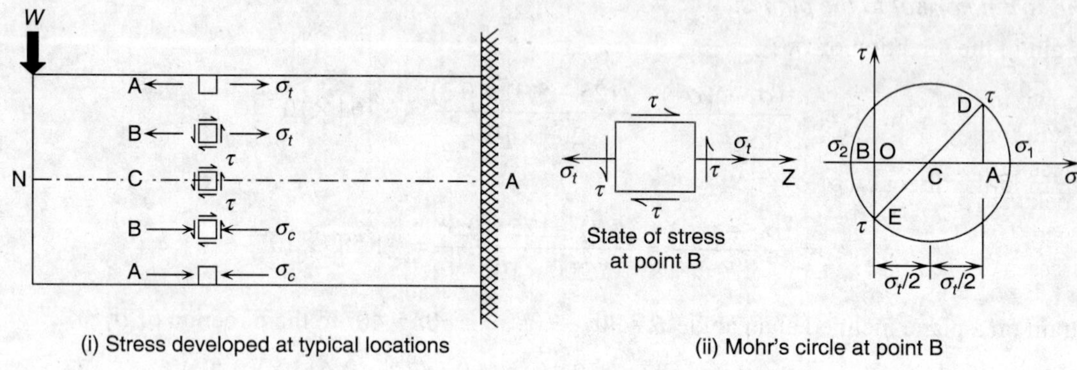

(i) Stress developed at typical locations

(ii) Mohr's circle at point B

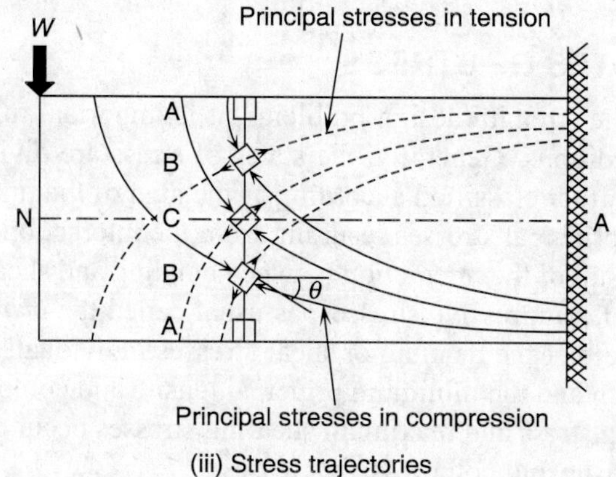

(iii) Stress trajectories

(a) Trajectories of principal stresses in a cantilever beam

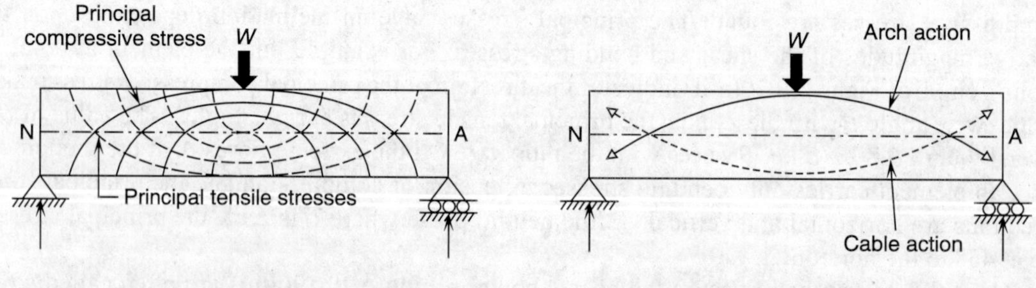

(i) Direction of principal stresses

(ii) Load-carrying mechanism

(b) Trajectories of principal stresses in a simply supported beam

Figure 20.19 Trajectories of principal stresses and load support mechanisms in beams.

It should be noted that at the neutral axis of the beam, where bending stresses are zero, only shear stresses exist which can be resolved into equivalent principal tensile and compressive stresses inclined at 45° angles to the neutral axis. On the other hand, at the extreme faces of the beam, an element carries only bending stresses in tension or compression and the principal tensile and compressive stresses become aligned with the corresponding bending stress and intersect the faces at 90°angles.

To study the pure shear state of stress, consider an element on the surface of a circular shaft shown in Figure 20.20(a-i). The element is subjected to the torsional shear stress and the Mohr's circle for this state of stress is shown in Figure 20.20(a-ii). The radius OD represents the shear stress on the element and the maximum tensile stress is represented by the point A which is 90° clockwise from OD. Thus the normal to the plane of maximum tensile stress is 45° clockwise from Z-axis. The *stress trajectories* which follow the directions of principal compressive and tensile stresses depicted by solid and dashed lines on the shaft shown in Figure 20.20(b) are 45° helices. In brittle materials which are weak in tension, the failure will occur along a 45° helix such as I-I.

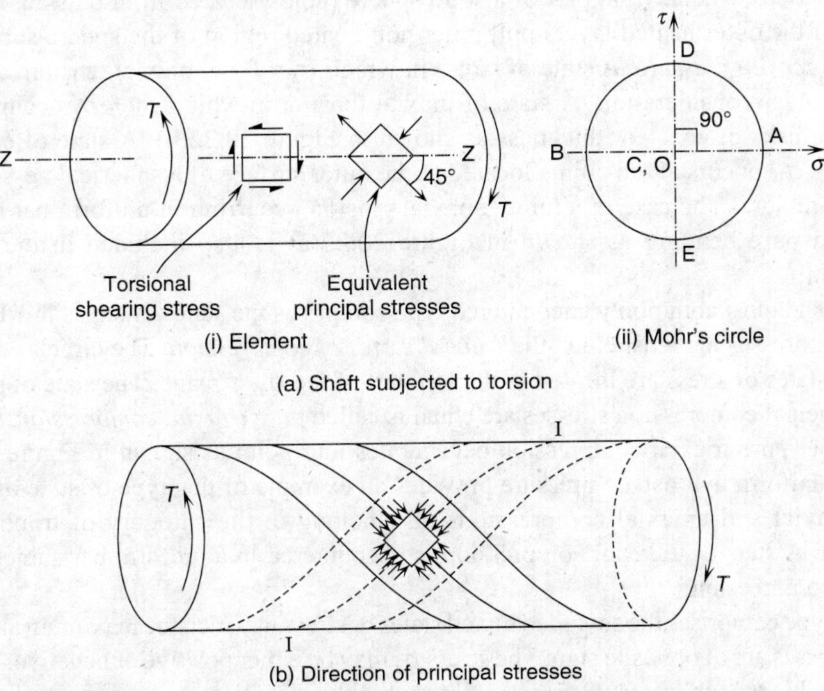

Figure 20.20 Stress trajectories of principal stresses due to torsion.

In a more general triaxial state of stress (see Figure 20.21), different types of stress conditions develop according to the sign of the principal stresses. The first type comprises *triaxial state of tension* i.e. the state in which none of the principal stresses is compressive. The circular diagrams for this type of states of stress are located in the right hand part of the σ_n, τ plane as shown in Figures 20.21(b) and (c). In the particular case, where all three principal tensile stresses are equal, the state of stress is called *pure triaxial tension.* The circular diagram under pure triaxial

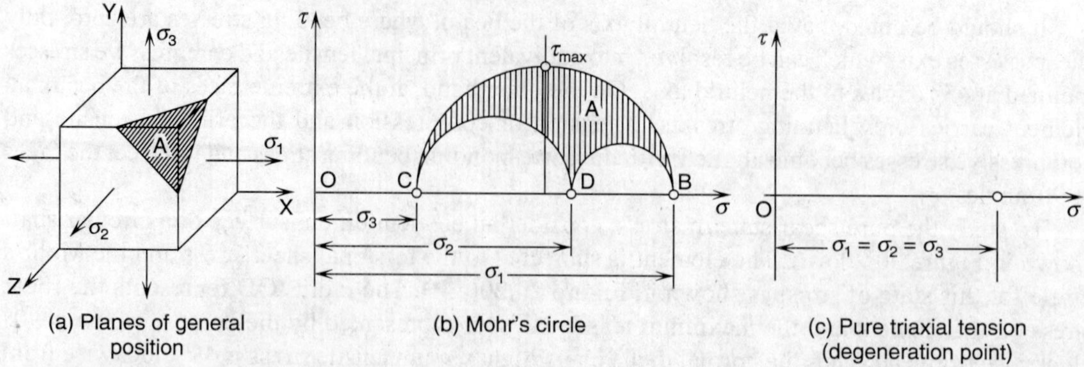

Figure 20.21 Combined Mohr's circle for the planes of general position in triaxial tension case.

tension degenerates into point as shown in Figure 20.21(c). An example of this type of state of stress is provided by the central part of a solid sphere rapidly heated from outside, where in the expansion of the outer heated layers pull inner non-heated region of the sphere subjecting it to an all-round *tensile pressure*. A state of stress in which $\sigma_3 = 0$ i.e. biaxial tension, also belongs to the type under consideration. A state of biaxial tension in which $\sigma_1 \neq \sigma_2$ occurs in rapidly rotating thin disks of constant thickness as shown in Figure 20.22(b). A state of equal biaxial tension ($\sigma_1 = \sigma_2$) occurs at the points located at the outer surface of a spherical vessel subjected to internal pressure. The case of simple uniaxial tension occurring in uniform bar subjected to stretching or pure bending as shown in Figure 20.22(a) is also included in the type under consideration.

The second most commonly encountered type comprises the states of stress in which none of the principal stresses are tensile, i.e. the *triaxial states of compression*. The circular diagrams for this type of states of stress are located in the left part of the σ_n, τ plane. The state of stress where all three principal compression stresses are equal is called *pure triaxial compression*. The circular diagram under pure triaxial compression degenerates into point as shown in Figure 20.22(e). A body under uniform hydrostatic pressure provides an example of this type of state of stress. The states of biaxial and uniaxial compressions also belong to the category of triaxial states of compression. A state of uniaxial compression is encountered in a uniform bar subjected to pure bending or compression.

A third type comprises the so-called mixed states of stress in which the maximum and minimum principal stresses are of opposite sign. The stress σ_2 may be either positive or negative. The circular diagrams for this type of states of stress are located in the middle part of the σ_n, τ plane as shown in Figure 20.22(c). For example, a mixed triaxial state of stress is set up in a thick-walled cylinder under internal pressure, and a biaxial mixed state of stress is set up in a rod subjected simultaneously to bending and torsion as illustrated in Figure 20.22(c). The pure shear also represents a mixed biaxial state of stress as shown in Figure 20.22(c-ii).

A number of typical states of stresses are shown in Figures 20.22(a) to 20.22(e). Since the sign of shear stress is not specified, the Mohr's circles of the various states of stress are restricted to only upper half of the circle.

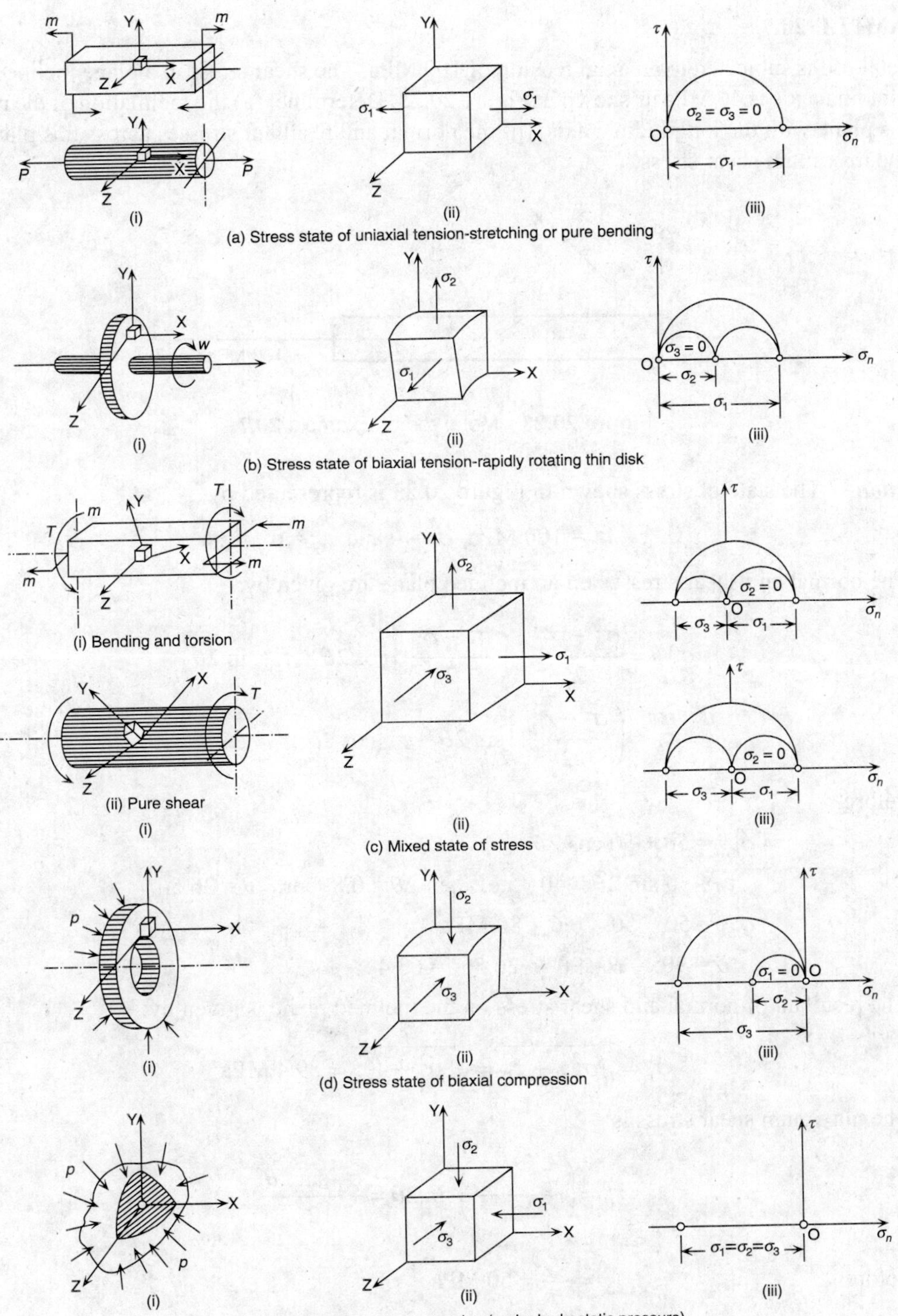

Figure 20.22 Circular diagrams of typical states of stress.

EXAMPLE 20.7

A metal rod is subjected to an axial tension of 100 MPa. The shear stress in a plane inclined to its longitudinal axis is 40 MPa as shown in Figure 20.23. Determine: (a) the inclination of the normal to this plane with the longitudinal axis, (b) the normal and resultant stresses across this plane and (c) the maximum shear stress.

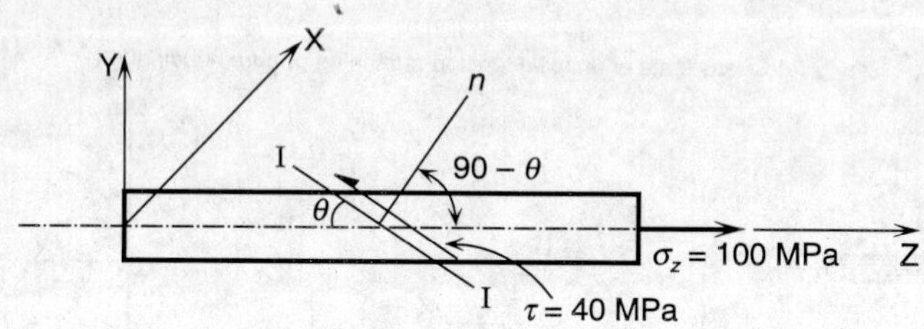

Figure 20.23 Metal rod of Example 20.7.

Solution: The state of stress shown in Figure 20.23 is represented by,

$$\sigma_z = 100 \text{ MPa}, \ \sigma_y = 0 \text{ and } \tau_{zy} = 0$$

(a) The normal and shear stresses on an inclined plane are given by:

$$\sigma_{n,\theta} = \frac{\sigma_z + \sigma_y}{2} + \left(\frac{\sigma_z - \sigma_y}{2}\right)\cos 2\theta + \tau_{zy} \sin 2\theta$$

$$\tau_\theta = \left(\frac{\sigma_z - \sigma_y}{2}\right)\sin 2\theta - \tau_{zy} \cos 2\theta$$

Therefore,

$$\sigma_{n,\theta} = 50 + 50 \cos 2\theta$$

$$\tau_\theta = 50 \sin 2\theta = 40 \quad \text{i.e.} \quad \sin 2\theta = 0.8 \quad \text{or} \quad \theta = 26.56°$$

Thus, $\sigma_{n,\theta} = 50 + 50 \times 0.6 = 80$ MPa

and $\alpha = 90° - \theta = 90° - 26.56° = 63.44°$

(b) The resultant of normal and shear stress on the inclined plane is given by:

$$\sigma_r = \sqrt{\sigma_{n,\theta}^2 + \tau_\theta^2} = \sqrt{80^2 + 40^2} = 89.4 \text{ MPa}$$

(c) The maximum shear stress is:

$$\tau_\theta = \left(\frac{\sigma_z - \sigma_y}{2}\right)\sin 2\theta = \frac{\sigma_z \sin 2\theta}{2}$$

Therefore, $$\tau_{\max} = \frac{\sigma_z}{2} = 50 \text{ MPa}$$

The stresses can also be determined from Mohr's circle.

EXAMPLE 20.8

A closed cylindrical tank of 1.0 m diameter made of 10 mm thick mild steel plate is subjected to an internal pressure of 6.0 MPa. Calculate the maximum normal and shear stresses developed in the wall of the tank.

Solution: For the given tank, $d = 1000$ mm, $t = 10$ mm, and $p = 6$ MPa.

At a point in the closed cylindrical tank wall, the material is subjected to hoop tension, and longitudinal stress.

Hoop or *circumferential tension*:

From Eq. (5.3):

$$\sigma_x = \frac{pd}{2t} = \frac{6 \times 1000}{2 \times 10} = 300 \text{ MPa}$$

Longitudinal stress:

From Eq. (5.6):

$$\sigma_z = \frac{pd}{4t} = \frac{6 \times 1000}{4 \times 10} = 150 \text{ MPa}$$

The state of stress is represented by,

$$\sigma_z = 150 \text{ MPa}, \ \sigma_x = 300 \text{ MPa and } \tau_{zx} = 0$$

The maximum normal stress,

$$\sigma_{n,\max} = \sigma_1 = \left(\frac{300 + 150}{2}\right) + \sqrt{\left(\frac{300 - 150}{2}\right)^2} = 300 \text{ MPa } (=\sigma_x)$$

The maximum shear stress is given by

$$\tau_{\max} = \left(\frac{\sigma_x - \sigma_z}{2}\right) = \left(\frac{300 - 150}{2}\right) = 75 \text{ MPa}$$

EXAMPLE 20.9

At a point in a strained material, the direct stresses acting on the two perpendicular planes are:

$$\sigma_z = 160 \text{ MPa (tension) and } \sigma_x = -120 \text{ MPa (compression)}$$

The direct stresses are accompanied by the shear stresses on the planes. The magnitude of resulting greater principal stress is $\sigma_1 = 150$ MPa. Determine: (a) the magnitude of shear stresses acting on the planes and (b) the maximum shear stress.

Solution: The state of stress at the point under consideration is represented by,

$$\sigma_z = 160 \text{ MPa}, \ \sigma_x = -120 \text{ MPa and } \sigma_1 = 200 \text{ MPa}$$

(a) The magnitude of shear stresses acting on the planes

From the condition that the sum of the normal stresses on any two planes right angles to each other is constant, i.e.

$$\sigma_n + \sigma_n' = \sigma_1 + \sigma_2 = \text{constant}$$

$$200 + \sigma_2 = 160 - 120 \quad \text{i.e.} \quad \sigma_2 = -160 \text{ MPa}$$

Also,

$$\sigma_2 = \left(\frac{\sigma_n + \sigma_n'}{2}\right) - \sqrt{\left(\frac{\sigma_n - \sigma_n'}{2}\right) + \tau^2}$$

or

$$200 = \left(\frac{160 - 120}{2}\right) - \sqrt{\left(\frac{160 + 120}{2}\right)^2 + \tau^2}$$

or

$$180 = \sqrt{140^2 + \tau^2} \quad \text{i.e.} \quad \tau^2 = (180)^2 - (140)^2$$

Therefore, $\tau = 113.14$ MPa

(b) The maximum shear stress

$$\tau_{max} = \left(\frac{\sigma_1 - \sigma_2}{2}\right) = \left(\frac{200 + 160}{2}\right) = 180 \text{ MPa}$$

EXAMPLE 20.10

At a point in a strained material the normal and tangential stresses acting on two planes at 40° are:

$$\sigma_{n,1} = 65 \text{ MPa (tension)}, \ \tau_1 = 60.62 \text{ MPa and } \sigma_{n,2} = 23.62 \text{ MPa (compression)}$$

Determine the principal stresses.

Solution: Consider the plane with greater normal stress to be inclined at an angle θ with the major principal plane. The normal and tangential stresses acting on the plane

$$\sigma_{n,1} = \left(\frac{\sigma_1 + \sigma_2}{2}\right) + \left(\frac{\sigma_1 - \sigma_2}{2}\right)\cos 2\theta = 65 \text{ MPa} \tag{i}$$

$$\tau_1 = \left(\frac{\sigma_1 - \sigma_2}{2}\right)\sin 2\theta = 60.62 \text{ MPa} \tag{ii}$$

The normal stress acting on the second plane inclined at an angle (θ + 40°) with the major principal plane is:

$$\sigma_{n,2} = \left(\frac{\sigma_1 + \sigma_2}{2}\right) + \left(\frac{\sigma_1 - \sigma_2}{2}\right)\cos (2\theta + 80°) = -23.62 \text{ MPa} \tag{iii}$$

Subtract Eq. (iii) from Eq. (i)

$$\left(\frac{\sigma_1 - \sigma_2}{2}\right)[\cos 2\theta - \cos (2\theta + 80°)] = 88.62 \tag{iv}$$

From Eqs. (ii) and (iv)

$$\frac{\cos 2\theta - \cos (2\theta + 80°)}{\sin 2\theta} = \frac{88.62}{60.62} = 1.46189$$

$$\frac{\cos 2\theta - (\cos 2\theta \times \cos 80° + \sin 2\theta \times \sin 80°)}{\sin 2\theta} = 1.46189$$

or $$(1 - \cos 80°) \cot 2\theta = (1.46189 - \sin 80°)$$

or $$\tan 2\theta = \frac{(1 - \cos 80°)}{(1.46189 - \sin 80°)} = 1.7320951$$

Therefore, $$2\theta = 60° \quad \text{i.e.} \quad \theta = 30°$$

From Eq. (ii):

$$\left(\frac{\sigma_1 - \sigma_2}{2}\right) \sin 60° = 60.62 \text{ or } \sigma_1 - \sigma_2 = \frac{60.62 \times 4}{\sqrt{3}} = 140 \text{ MPa} \quad \text{(v)}$$

From Eq. (i):

$$\left(\frac{\sigma_1 + \sigma_2}{2}\right) + \left(\frac{140}{2}\right) \cos 60° = 65 \text{ or } \sigma_1 + \sigma_2 = 60 \text{ MPa} \quad \text{(vi)}$$

From Eqs. (v) and (vi):

$$\sigma_1 = 100 \text{ MPa} \quad \text{and} \quad \sigma_2 = -40 \text{ MPa}$$

The tangential stress acting on the second plane,

$$\tau_2 = \left(\frac{100 + 40}{2}\right) \sin (60° + 80°) = 45 \text{ MPa}$$

EXAMPLE 20.11

A closed cylindrical pressure vessel of 1.0 m diameter made of 12 mm thick mild steel plate is subjected to an internal pressure of 2.4 MPa. Calculate the largest torque that can also be applied if the shearing stress in the wall of the vessel is limited to 60 MPa. Assume that the wall of the vessel is adequately braced against buckling.

Solution: For the given tank, $d = 1000$ mm, $t = 12$ mm, and $p = 2.4$ MPa.

The state of stress at a point on the surface of closed pressure vessel is represented by the hoop tension, and longitudinal stress which are given by

Hoop or tangential stress, $$\sigma_x = \sigma_t = \frac{pd}{2t} = \frac{2.4 \times 1000}{2 \times 12} = 100 \text{ MPa}$$

Longitudinal stress, $$\sigma_z = \sigma_l = \frac{pd}{4t} = \frac{2.4 \times 1000}{4 \times 12} = 50 \text{ MPa}$$

If the shearing stress due to torque is denoted by τ, the state of stress is represented by,

$$\sigma_z = 50 \text{ MPa}, \sigma_x = 100 \text{ MPa and } \tau_{zx} = \tau$$

The problem can conveniently be handled by constructing state of stress diagram as shown in Figure 20.24(a) and Mohr's circle as shown in Figure 20.24(b). The principal stresses are: $\sigma_1, \sigma_2 = \text{OC} \pm R$

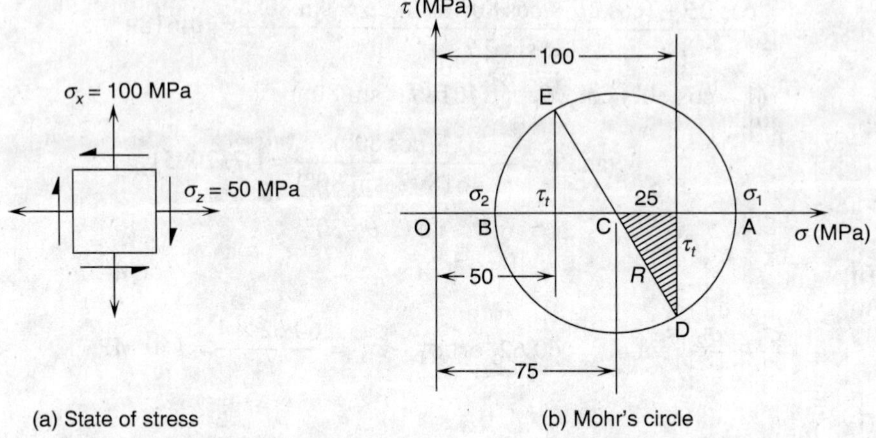

Figure 20.24 Stress state and Mohr's circle for the pressure vessel of Example 20.11.

$$\sigma_1 = \left(\frac{100 + 50}{2}\right) + R = (75 + R) \text{ MPa}$$

$$\sigma_2 = \left(\frac{100 + 50}{2}\right) - R = (75 - R) \text{ MPa}$$

Hence the problem reduces to determination of the largest radius R for which the shearing stress does not exceed 60 MPa. The absolute maximum shearing stress is the smallest of the following three values:

$$\frac{|\sigma_1 - \sigma_2|}{2} = R = 60 \text{ MPa} \qquad \text{or} \qquad R = 60 \text{ MPa}$$

$$\frac{|\sigma_1|}{2} = \frac{|75 + R|}{2} = 60 \text{ MPa} \qquad \text{or} \qquad R = 45 \text{ MPa}$$

$$\frac{|\sigma_2|}{2} = \frac{|75 - R|}{2} = 60 \text{ MPa} \qquad \text{or} \qquad R = 195 \text{ MPa}$$

Therefore, radius of Mohr's circle is 45 MPa. The in-plane shearing stress is given by

$$\tau_t = \sqrt{R^2 - 25^2} = \sqrt{45^2 - 25^2} = 37.42 \text{ MPa}$$

From torsion formula:

$$\tau_{max} = \frac{Tr}{J} \quad \text{or} \quad T = \frac{\tau_{max} J}{r}$$

Therefore,
$$T = \frac{37.42}{500} \times \left[\frac{\pi}{32} \times (1000^4 - 976^4)\right]$$

$$= 680.3 \times 10^6 \text{ N.mm} = 680.3 \text{ kN.m}$$

20.8 COMPLEX STRESS SYSTEMS

The states of stress discussed in the preceding sections result from the interaction of axial, bending, direct shearing and torsion stresses. The interaction between bending and shear stresses in a beam; bending and torsion in a shaft results in a set of principal stresses that act in different directions from either the bending or shear or torsion stresses individually. Their inclinations depend upon the relative magnitudes of the shear and bending stresses. The directions of principle stresses in a member are depicted by stress trajectories.

20.8.1 Combined Bending and Direct Shearing Stresses

It should be recalled that at the neutral axis of the beam, where bending stresses are zero, only shear stresses exist which are resolved into equivalent principal tensile and compressive stresses inclined at 45° angles to the neutral axis. On the other hand, at the extreme faces of the beam, an element carries only bending stresses in the form of tension or compression which become aligned with the corresponding bending stress and intersect the faces at 90° angles.

EXAMPLE 20.12

A built-up I-section beam carries concentrated loads of 125 kN at 0.8 m from each support over an effective simply supported span of 4.0 m. The flanges of the section are 150 mm wide and 25 mm thick, the overall depth of the section is 400 mm. The thickness of the web is 12.5 mm. Determine: (a) the principal stresses, and (b) the maximum shear stress in a section of the beam, where the bending moment and shearing force both have the maximum values.

Solution: The sections of the beam where the bending moment and shearing force both have the maximum values lie at 0.8 m from each support. The maximum bending moment and shearing force at this section are,

$$M_{max} = 125 \times 0.8 = 100 \text{ kN.m} \text{ and } V_{max} = 125 \text{ kN}$$

The moment of inertia of the section about neutral axis

$$I = \frac{1}{12}(150 \times 400^3 - 137.5 \times 350^3) = 3.087 \times 10^8 \text{ mm}^4$$

The maximum bending stress in the web

$$\sigma_b = \frac{M_z y}{I_x} = \frac{100 \times 10^6 \times 175}{3.087 \times 10^8} = 56.69 \text{ MPa (compressive)}$$

The maximum shearing stress in the web

$$\tau = \frac{V_z a\bar{y}}{I_x b} = \frac{125 \times 10^3 \times (150 \times 25) \times 187.5}{3.087 \times 10^8 \times 12.5} = 22.78 \text{ MPa}$$

(a) The principal stresses are,

$$\sigma_1, \sigma_2 = \left(\frac{\sigma_b}{2}\right) \pm \sqrt{\left(\frac{\sigma_b}{2}\right)^2 + \tau^2}$$

$$= \left(\frac{-56.69}{2}\right) \pm \sqrt{\left(\frac{-56.69}{2}\right)^2 + 22.78^2}$$

$$= 8.02 \text{ MPa and } -64.71 \text{ MPa}$$

(b) The maximum shear stress in a section,

$$\tau_{\max} = \left(\frac{\sigma_1 - \sigma_2}{2}\right) = \left(\frac{8.02 + 64.71}{2}\right) = 36.365 \text{ MPa}$$

EXAMPLE 20.13

A cantilever rod of circular cross-section of 50 mm diameter and 200 mm length is welded at its support to a plate by a 15 mm fillet weld all around the circumference of the rod. The section modulus of the weld is 22,000 mm^3. Determine the resultant stress in the weld when the cantilever carries 10 kN load at its free end.

Solution: The weld section is subjected to direct shear and bending stresses.

The bending moment and shearing force at the weld section are,

$$M = (10 \times 10^3) \times 200 = 2 \times 10^6 \text{ N.mm and } V = 10 \text{ kN}$$

Throat area for circular fillet weld,

$$A_f = \pi d\, \tau_w = p \times 50 \times (15/\sqrt{2}) = 1666.08 \text{ mm}^2$$

Therefore, Direct stress, $\tau = \dfrac{V}{A_f} = \dfrac{10 \times 10^3}{1666.08} = 6.00 \text{ MPa}$

$$\text{Bending stress, } \sigma_b = \frac{M}{Z} = \frac{2 \times 10^6}{22000} = 90.91 \text{ MPa}$$

(a) The resultant principal stresses are,

$$\sigma_1, \sigma_2 = \left(\frac{\sigma_b}{2}\right) \pm \sqrt{\left(\frac{\sigma_b}{2}\right)^2 + \tau^2}$$

$$= \left(\frac{90.91}{2}\right) \pm \sqrt{\left(\frac{90.91}{2}\right)^2 + 6.0^2}$$

$$= 45.455 \text{ MPa}, \pm 45.849 \text{ MPa}$$

Therefore, $\sigma_1 = 91.304 \text{ MPa}$

(b) The maximum shear stress in the section,

$$\sigma_{\max} = \sqrt{\left(\frac{\sigma_b}{2}\right)^2 + \tau^2} = \sqrt{\left(\frac{90.91}{2}\right)^2 + 6.0^2} = 45.849 \text{ MPa}$$

20.8.2 Combined Bending and Torsion Shearing Stresses

Consider a shaft of diameter D shown in Figure 20.25(a) subjected to bending moment M_z and twisting moment T_z at a section. At a point in the section at a radius r and distance y from the neutral axis the bending and torsion shear stresses are given by:

$$\sigma_b = \frac{M_z y}{I_x} \quad \text{and} \quad \tau = \frac{T_z r}{J_z} \tag{20.38}$$

where I_x and J_z are the moment of inertia about neutral axis and the polar moment of inertia of the section, respectively.

The location of the principal planes through a point is given by

$$\tan 2\theta = \frac{2\tau}{\sigma_n} \tag{20.39}$$

and principal stresses are,

$$\sigma_1, \sigma_2 = \left(\frac{\sigma_b}{2}\right) \pm \sqrt{\left(\frac{\sigma_b}{2}\right)^2 + \tau^2} \tag{20.40}$$

The effect of bending and twisting moments will be the maximum at the farthest points on the surface of the section (see Figure 20.25(b)). At these points the maximum bending stress is,

$$\sigma_{max} = \frac{M_z}{Z_x} = \frac{M_z}{(\pi D^3/32)} = \frac{32 M_z}{\pi D^3} \tag{20.41}$$

The maximum shear stress

$$\tau'_{max} = \frac{T_z}{Z_{p,z}} = \frac{T_z}{(\pi D^3/16)} = \frac{16 T_z}{\pi D^3} \tag{20.42}$$

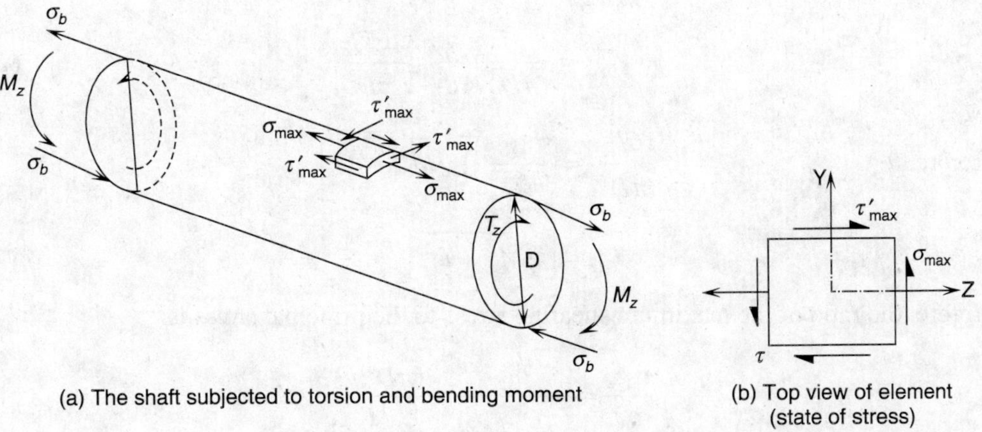

(a) The shaft subjected to torsion and bending moment

(b) Top view of element (state of stress)

Figure 20.25 Stresses caused by simultaneous action of flexure and torsion.

Therefore, the location of the principal planes through the point is given by:

$$\tan 2\theta = \frac{2\tau'_{\max}}{\sigma_{\max}} = \frac{T_z}{M_z} \tag{20.43}$$

and corresponding principal stresses are,

$$\sigma_1, \sigma_2 = \left(\frac{\sigma_{\max}}{2}\right) \pm \sqrt{\left(\frac{\sigma_{\max}}{2}\right)^2 + \tau'^2_{\max}} = \left(\frac{16}{\pi D^3}\right)\left[M_z \pm \sqrt{M_z^2 + T_z^2}\right] \tag{20.44}$$

The maximum shear stress is given by:

$$\tau_{\max} = \frac{\sigma_1 - \sigma_2}{2} = \left(\frac{16}{\pi D^3}\right)\left(\sqrt{M_z^2 + T_z^2}\right) \tag{20.45}$$

Equivalent bending and twisting moments

Consider $M_{e,z}$ to be the equivalent bending which acting alone will develop the maximum tensile stress equal to σ_1, then

$$\sigma_1 = \frac{M_{e,z}}{(\pi D^3/32)} = \frac{32M_{e,z}}{\pi D^3} \tag{20.46}$$

$$\frac{32M_{e,z}}{\pi D^3} = \left(\frac{16}{\pi D^3}\right)\left[M_z \pm \sqrt{M_z^2 + T_z^2}\right]$$

or

$$M_{e,z} = \left(\frac{1}{2}\right)\left[M_z \pm \sqrt{M_z^2 + T_z^2}\right] \tag{20.47}$$

Similarly, if $T_{e,z}$ is the equivalent twisting moment which acting alone develops the maximum shearing stress equal to $\tau_{\max}$, then

$$\tau_{\max} = \frac{T_{e,z}}{(\pi D^3/16)} = \frac{16T_{e,z}}{\pi D^3} \tag{20.48}$$

Therefore,

$$\frac{16T_{e,z}}{\pi D^3} = \left(\frac{16}{\pi D^3}\right)\left[\sqrt{M_z^2 + T_z^2}\right]$$

or

$$T_{e,z} = \left[\sqrt{M_z^2 + T_z^2}\right] \tag{20.49}$$

Therefore, the ratio of the maximum shearing stress to the principal stress is:

$$\frac{\tau_{\max}}{\sigma_1} = \frac{T_{e,z}}{(\pi D^3/16)} \times \frac{(\pi D^3/32)}{M_{e,z}}$$

$$= \frac{T_{e,z}}{2M_{e,z}} = \frac{\sqrt{M_z^2 + T_z^2}}{M_z + \sqrt{M_z^2 + T_x^2}} \tag{20.50}$$

For $M_{e,z} = \alpha T_{e,z}$ the above ratio reduces to:

$$\frac{\tau_{\max}}{\sigma_1} = \frac{\sqrt{\alpha^2 + 1}}{\alpha + \sqrt{\alpha^2 + 1}} \tag{20.51}$$

For illustration, consider $M_{e,z} = 1.5T_{e,z}$

$$\frac{\tau_{\max}}{\sigma_1} = \frac{\sqrt{1.5^2 + 1}}{1.5 + \sqrt{1.5^2 + 1}} = 0.546$$

The theory of combined stresses is mainly used in the design of components subjected to combined loadings or in the determination of safe loads the components can carry. In the location of critical element and computation of the maximum stresses to which it is subjected, the Mohr's circle plays an important role in visualizing the critical conditions and thus is advantageous over analytical calculations. The usual procedure is to consider the stress state on an element due to the three fundamental loadings, namely the axial, flexural and torsional loadings. A study of Mohr's circle for the element under consideration will indicate the design criteria. The analytical technique and the graphical method based on Mohr's circle are illustrated in the following examples.

EXAMPLE 20.14

A section of a shaft of 100 mm diameter is subjected to bending and twisting moments of 5.0 kN.m and 7.5 kN.m, respectively. Determine: (a) the maximum direct stress developed on the section, and specify the position of the plane on which it acts, (b) the stress which acting alone would develop the same maximum strain and (c) the maximum shear stress induced on the section, and specify the position of the plane on which it acts with respect to the axis of shaft. The Poisson's ratio of the material of the shaft is 0.26.

Solution: For the given problem,

$$M_z = 5.0 \text{ kN.m}, \; T_z = 7.5 \text{ kN.m}, \; D = 100 \text{ mm and } \nu = 0.26$$

(a) The principal stresses are,

$$\sigma_1, \sigma_2 = \left(\frac{16}{\pi D^3}\right)\left[M_z \pm \sqrt{M_z^2 + T_z^2}\right]$$

$$= \left(\frac{16}{\pi \times 100^3}\right) \times \left[5.0 \pm \sqrt{5.0^2 + 7.5^2}\right] \times 10^6$$

$$= 71.37, -20.44 \text{ MPa}$$

The location of the major principal plane through the point is given by:

$$\tan 2\theta = \frac{T_z}{M_z} = \frac{7.5}{5.0} = 1.5 \quad \text{or} \quad \theta = 28.155°$$

i.e. the principal plane makes an angle $90° - \theta = 90° - 28.155° = 61.845°$ with the axis of the shaft.

(b) The maximum strain is,

$$\varepsilon_1 = \frac{\sigma_1}{E} - \nu\frac{\sigma_2}{E} = \frac{1}{E}(71.37 + 0.26 \times 20.44) = \frac{76.68}{E}$$

If σ_a is the stress which acting alone would develop the same maximum strain

$$\frac{\sigma_a}{E} = \frac{76.68}{E} \quad \text{or} \quad \sigma_a = 76.68 \text{ MPa}$$

(c) The maximum shear stress is,

$$\tau_{\max} = \left(\frac{16}{\pi D^3}\right)\left[\sqrt{M_z^2 + T_z^2}\right] = \left(\frac{16}{\pi \times 100^3}\right) \times \left[\sqrt{5.0^2 + 7.5^2}\right] \times 10^6$$

$$= 45.91 \text{ MPa}$$

The plane of maximum shear makes an angle 61.845° ± 45 = 106.845° or 16.845° with the axis of the shaft.

EXAMPLE 20.15

A shaft 100 mm in diameter rotating at 28 Hz is subjected to a set of bending loads that produce a maximum flexural stress at extreme fibres of 100 MPa. Determine: (a) the torque that can also be applied simultaneously on the shaft if the shearing and normal stresses in the shaft are limited to 80 MPa and 120 MPa, respectively; and (b) the maximum power that can be transmitted by the shaft under the stipulated conditions of (a).

Solution: The applied torque produces maximum in-plane torsional shearing stress τ_t at the periphery resulting in the stress state shown in Figure 20.26(a). The problem can be handled conveniently by constructing Mohr's circle as shown in Figure 20.26(b). Since σ_1 and σ_2 are of opposite signs, the maximum shearing stress is equal to the in-plane shearing stress.

The proper radius of the circle R is the one which do not allow the shearing and normal stresses to exceed maximum permitted values.

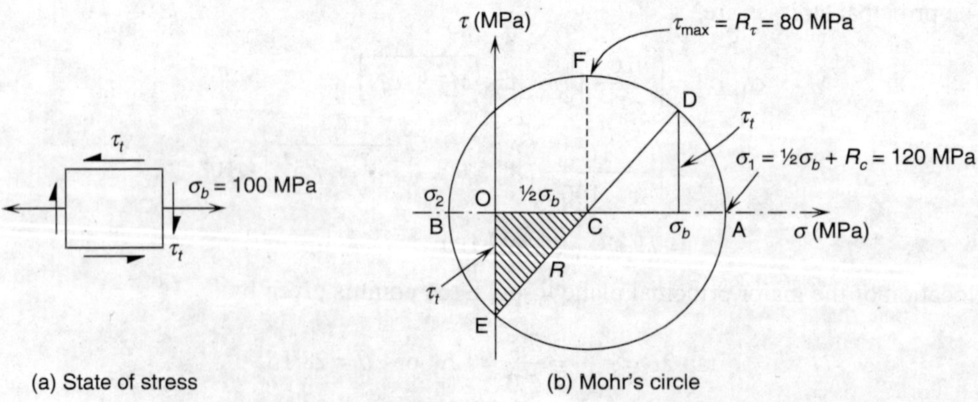

Figure 20.26 Stress state and Mohr's circle for shaft of Example 20.15.

(i) To develop the maximum permitted shearing stress, $R_\tau = 80$ MPa

(ii) The radius that will produce the maximum permitted normal stress must satisfy the condition

$$\sigma = 120 = OC + R_\sigma = 50 + R_\sigma \quad \text{or} \quad R_\sigma = 70 \text{ MPa}$$

The proper radius of the circle R is the smaller of the two values, i.e. $R = 70$ MPa which do not allow the permitted values to be exceeded.

The maximum in-plane torsional shearing stress τ_t that can be combined with the flexural stress is given by

$$\tau_t = \sqrt{R^2 - 50^2} = \sqrt{70^2 - 50^2} = 48.99 \text{ MPa}$$

The torque required to produce this torsional shearing stress is obtained from torsion formula:

$$T = \tau_t Z_p = \tau_t (\pi D^3/16)$$

Therefore, $$T = 48.99 \times \left[\frac{\pi}{16} \times 100^3\right] = 9.619 \times 10^6 \text{ N.mm} = 9.619 \text{ kN.m}$$

For the frequency $f = 28$ Hz (cps), the power P in watts (W) is given by Eq. (6.22):

$$P = (2\pi f)\, T_{\text{mean}}$$
$$= 2 \times \pi \times 28 \times (9.619 \times 10^3) = 1692.3 \times 10^3 \text{ N.m/s (W)} = 1655 \text{ kW}$$

EXAMPLE 20.16

A hollow shaft whose internal diameter is half the external diameter is subjected to bending and twisting moments of 54.0 kN.m and 72.0 kN.m, respectively. If the maximum shear stress induced on the section is not to exceed 96 MPa, determine the diameter of the shaft.

Solution: For the given problem, $M_z = 54.0$ kN.m and $T_z = 72.0$ kN.m.

Consider the external and internal diameters of the shaft to be D and $d(= 0.5D)$, respectively, then bending and torsion section moduli are,

$$Z_x = \frac{I_x}{D/2} = \frac{\pi D^3}{32}\left(1 - \frac{d^4}{D^4}\right) = \frac{\pi D^3}{32}\left[1 - \left(\frac{1}{2}\right)^4\right] = \frac{15\pi D^3}{512}$$

$$Z_{p,x} = \frac{J_x}{D/2} = \frac{\pi D^3}{16}\left(1 - \frac{d^4}{D^4}\right) = \frac{\pi D^3}{16}\left[1 - \left(\frac{1}{2}\right)^4\right] = \frac{15\pi D^3}{256}$$

The maximum shear stress is given by

$$\tau'_{\max} = \frac{T_{e,z}}{Z_{p,x}} = \left(\frac{256}{15\pi D^3}\right) T_{e,z}$$

where $$T_{e,z} = \sqrt{M_z^2 + T_z^2} = \sqrt{54^2 + 72^2} = 90 \text{ kN.m}$$

Therefore,

$$\tau'_{\max} = 96 = \left(\frac{256}{15\pi D^3}\right) \times (90 \times 10^6)$$

i.e. $$D^3 = \frac{256}{15\pi} \times \frac{(90 \times 10^6)}{96} = 5092958 \text{ or } D = 172.05 \text{ mm}$$

Therefore, the external and internal diameters of the hollow shaft are 172.05 and 86.03 mm, respectively.

EXAMPLE 20.17

A thin aluminium alloy cantilever tube of 50 mm external diameter and 120 mm length is subjected to an axial tension $P = 9.0$ kN, a torque $T = 72.0$ N.m and a bending load $W = 1.75$ kN at the free end. Determine the thickness of the tube so as to ensure a factor of safety of 4. The yield strength aluminium alloy is 276 MPa.

Solution: The force system acting on the cantilever is,

$$P = 9.0 \text{ kN}, T = 72.0 \text{ N.m and } M = (1.75 \times 10^3) \times 120 = 21 \times 10^4 \text{ N.mm}$$

The corresponding stresses are,

$$\sigma_a = \frac{P}{A} = \frac{P}{\pi D t} = \frac{9 \times 10^3}{\pi \times 50 \times t} = \frac{57.3}{t}$$

$$\tau = \frac{2T}{\pi D^2 t} = \frac{2 \times (72 \times 10^3)}{\pi \times 50^2 \times t} = \frac{18.33}{t}$$

$$\sigma_b = \frac{4M}{\pi D^2 t} = \frac{4 \times (21 \times 10^4)}{\pi \times 50^2 \times t} = \frac{106.95}{t}$$

The resultant direct stress, $\sigma_n = \sigma_a + \sigma_b = \frac{57.30}{t} + \frac{106.95}{t} = \frac{164.25}{t}$

The maximum principal stress is:

$$\sigma_1 = \left(\frac{\sigma_n}{2}\right) + \sqrt{\left(\frac{\sigma_n}{2}\right)^2 + \tau^2}$$

$$= \left(\frac{164.25}{2t}\right) + \sqrt{\left(\frac{164.25}{2t}\right)^2 + \left(\frac{18.33}{t}\right)^2} = \frac{166.27}{t}$$

Therefore, to ensure a factor of safety of 4,

$$\frac{166.27}{t} = \frac{276}{4} \text{ or } t = 2.4 \text{ mm}$$

EXAMPLE 20.18

A solid circular line shaft shown in Figure 20.27(a) is subjected to a load system such that the belts on the pulleys D and E pull the shaft vertically and that on pulley C horizontally. Determine the diameter of the shaft when the masses of the pulleys and shaft are neglected. The shearing and normal stresses in the shaft are limited to 84 MPa and 144 MPa, respectively.

Solution: The loads acting on the shaft produce bending in both horizontal and vertical planes. The bending moment diagrams in these planes and the torque distribution along the shaft are shown in Figures 20.27(b), (c) and (d), respectively. The resultant moment at any section is given by:

$$M = \sqrt{M_h^2 + M_v^2}$$

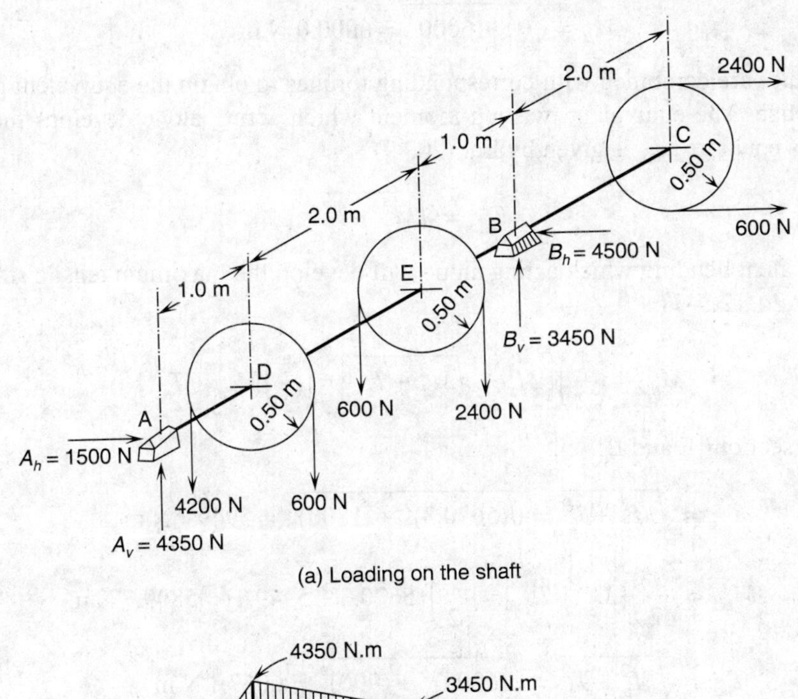

(a) Loading on the shaft

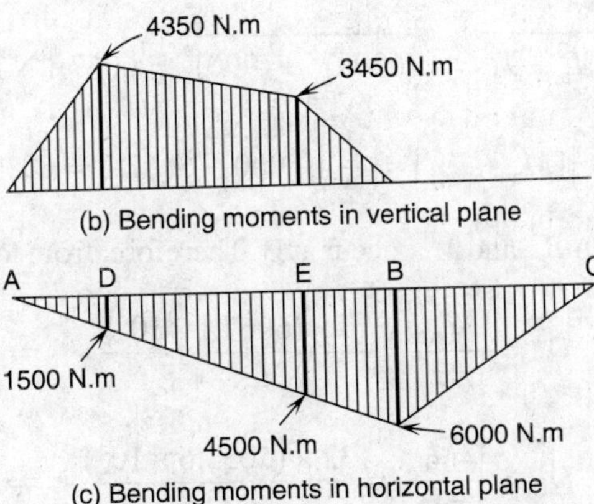

(b) Bending moments in vertical plane

(c) Bending moments in horizontal plane

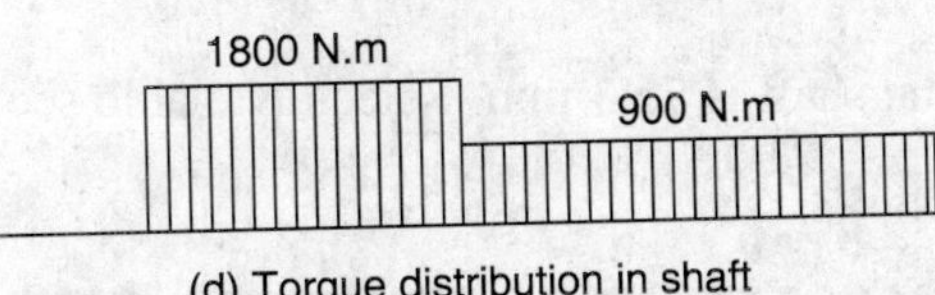

(d) Torque distribution in shaft

Figure 20.27 Loading, bending moment diagram and torsion distribution in the line shaft of Example 20.18.

The resultant moments at the salient points are:

$$M_D = \sqrt{4350^2 + 1500^2} = 4601.4 \text{ N.m}$$

$$M_E = \sqrt{3450^2 + 4500^2} = 5670.3 \text{ N.m}$$

$$M_B = \sqrt{0^2 + 6000^2} = 6000.0 \text{ N.m}$$

These values are combined with corresponding torques to obtain the equivalent moment and equivalent torque. The equivalent twisting moment which acting alone develops the maximum shearing stress equal to $\tau_{\max}$ is given by Eq. (20.49):

$$T_{e,z} = \sqrt{M_z^2 + T_z^2}$$

The equivalent bending which acting alone will develop the maximum tensile stress equal to σ_1 is given by Eq. (20.47):

$$M_{e,z} = \left(\frac{1}{2}\right)\left[M_z + \sqrt{M_z^2 + T_z^2}\right] = \left(\frac{1}{2}\right)[M_z + T_{e,z}]$$

At the critical sections E and B:

At E: $$T_{e,z} = \sqrt{M_z^2 + T_z^2} = \sqrt{(5670.3)^2 + (1800)^2} = 5949.1 \text{ N.m}$$

$$M_{e,z} = \left(\frac{1}{2}\right)[M_z + T_{e,z}] = \left(\frac{1}{2}\right)[5670.3 + 5949.1] = 5809.7 \text{ N.m}$$

At B: $$T_{e,z} = \sqrt{M_z^2 + T_z^2} = \sqrt{(6000)^2 + (900)^2} = 6067.1 \text{ N.m}$$

$$M_{e,z} = \left(\frac{1}{2}\right)[M_z + T_{e,z}] = \left(\frac{1}{2}\right)[6000 + 6067.1] = 6033.6 \text{ N.m}$$

The maximum values of both $T_{e,z}$ and $M_{e,z}$ occur at B. Therefore, from torsion and flexure formulae,

$$\tau = \frac{16T_{e,z}}{\pi D^3} \quad \text{or} \quad D^3 = \frac{16T_{e,z}}{\pi\tau} = \frac{16 \times (6067.1 \times 10^3)}{\pi \times 84} \quad \Rightarrow \quad D = 71.7 \text{ mm}$$

$$\sigma = \frac{32M_{e,z}}{\pi D^3} \quad \text{or} \quad D^3 = \frac{32M_{e,z}}{\pi\sigma} = \frac{32 \times (6033.6 \times 10^3)}{\pi \times 144} \quad \Rightarrow \quad D = 75.3 \text{ mm}$$

The required diameter of the shaft is 75.3 mm. Select available section of size greater than this value.

20.9 STRAIN ENERGY IN TERMS OF PRINCIPAL STRESSES

The maximum and the minimum direct strains in the material subjected to complex stresses are produced in the directions of the principal stresses.

Consider a strained element subjected to the principal stresses σ_1 and σ_2. The principal strains are,

$$\varepsilon_1 = \frac{\sigma_1}{E} - \nu\frac{\sigma_2}{E} = \frac{1}{E}(\sigma_1 - \nu\sigma_2)$$

and
$$\varepsilon_2 = \frac{\sigma_2}{E} - \nu\frac{\sigma_1}{E} = \frac{1}{E}(\sigma_2 - \nu\sigma_1) \quad (20.52)$$

where ν is the poisson's ratio. The strain energy stored per unit volume is

$$U = \frac{1}{2}(\sigma_1\varepsilon_1 + \sigma_2\varepsilon_2) = \frac{\sigma_1}{2}\left(\frac{\sigma_1}{E} - \nu\frac{\sigma_2}{E}\right) + \frac{\sigma_2}{2}\left(\frac{\sigma_2}{E} - \nu\frac{\sigma_1}{E}\right)$$

$$= \frac{1}{2E}(\sigma_1^2 + \sigma_2^2 - 2\nu\sigma_1\sigma_2) \quad (20.53)$$

In a general case of a strained element, three principal stresses σ_1, σ_2 and σ_3 acting on three principal planes may exist. The principal strains are given by:

$$\varepsilon_1 = \frac{\sigma_1}{E} - \nu\left(\frac{\sigma_2 + \sigma_3}{E}\right)$$

$$\varepsilon_2 = \frac{\sigma_2}{E} - \nu\left(\frac{\sigma_3 + \sigma_1}{E}\right)$$

and
$$\varepsilon_3 = \frac{\sigma_3}{E} - \nu\left(\frac{\sigma_1 + \sigma_2}{E}\right) \quad (20.54)$$

The strain energy stored per unit volume is:

$$U = \frac{1}{2}(\sigma_1\varepsilon_1 + \sigma_2\varepsilon_2 + \sigma_3\varepsilon_3)$$

$$= \frac{\sigma_1}{2}\left(\frac{\sigma_1}{E} - \nu\frac{\sigma_2 + \sigma_3}{E}\right) + \frac{\sigma_2}{2}\left(\frac{\sigma_2}{E} - \nu\frac{\sigma_3 + \sigma_1}{E}\right) + \frac{\sigma_3}{2}\left(\frac{\sigma_3}{E} - \nu\frac{\sigma_1 + \sigma_2}{E}\right)$$

$$= \frac{1}{2E}\left[\sigma_1^2 + \sigma_2^2 + \sigma_3^2 - \nu(2\sigma_1\sigma_2 + 2\sigma_2\sigma_3 + 2\sigma_3\sigma_1)\right] \quad (20.55)$$

EXAMPLE 20.19

A metallic plate specimen is tested under direct stresses acting in two mutually perpendicular directions. The strains in the two directions have been recorded as: $\varepsilon_1 = 7.81 \times 10^{-4}$ and $\varepsilon_2 = 1.94 \times 10^{-4}$. Determine: (a) the maximum direct stress developed on the section and (b) the inclination of the plane on which the normal stress is 65 MPa. The modulus of elasticity and Poisson's ratio of the material are 7.2×10^4 and 0.33, respectively.

Solution: For the given problem,

$$\varepsilon_1 = 7.81 \times 10^{-4},\ \varepsilon_2 = 1.94 \times 10^{-4},\ E = 7.2 \times 10^4 \text{ MPa and } \nu = 0.33$$

Consider a strained element subjected to the principal stresses σ_1 and σ_2. The principal strains from Eq. (20.52) are,

$$\sigma_1 - \nu\sigma_2 = E\varepsilon_1 \quad \text{and} \quad \sigma_2 - \nu\sigma_1 = E\varepsilon_2$$

and on addition and subtraction, these equations give,

$$(\sigma_1 + \sigma_2)(1 - \nu) = E(\varepsilon_1 + \varepsilon_2)$$

or
$$(\sigma_1 + \sigma_2) = \frac{(7.2 \times 10^4) \times (7.81 + 1.94) \times 10^{-4}}{(1 - 0.33)} = 104.78 \text{ MPa}$$

$$(\sigma_1 - \sigma_2)(1 + \nu) = E(\varepsilon_1 - \varepsilon_2)$$

or
$$(\sigma_1 - \sigma_2) = \frac{(7.2 \times 10^4) \times (7.81 - 1.94) \times 10^{-4}}{(1 + 0.33)} = 31.78 \text{ MPa}$$

Therefore, $\sigma_1 = 68.28$ MPa and $\sigma_2 = 36.50$ MPa.

The normal stress on a plane inclined at an angle θ is given by:

$$\sigma_n = \left(\frac{\sigma_1 + \sigma_2}{2}\right) + \left(\frac{\sigma_1 - \sigma_2}{2}\right)\cos 2\theta = 65$$

or
$$\left(\frac{104.78}{2}\right) + \left(\frac{31.37}{2}\right)\cos 2\theta = 65$$

i.e.,
$$52.39 + 15.685 \cos 2\theta = 65 \quad \text{or} \quad 2\theta = 36.49^\circ$$

Thus the inclination of the plane on which the normal stress is 65 MPa is 18.245° with the major principal plane.

EXAMPLE 20.20

A shaft of 90 mm diameter is subjected to bending and twisting moments of 8.0 kN.m and 12.0 kN.m, respectively. Determine the direct stress which acting alone will make the shaft store the same maximum strain energy per unit volume. The Poisson's ratio of the material is 0.33.

Solution: For the given problem, $M_z = 8.0$ kN.m and $T_z = 12.0$ kN.m

The maximum strain energy per unit volume is given by

$$U = \frac{1}{2E}(\sigma_1^2 + \sigma_2^2 - 2\nu\sigma_1\sigma_2)$$

where the principal stresses are,

$$\sigma_1 = \left(\frac{16}{\pi D^3}\right)\left[M_z + \sqrt{M_z^2 + T_z^2}\right]$$

$$\sigma_2 = \left(\frac{16}{\pi D^3}\right)\left[M_z - \sqrt{M_z^2 + T_z^2}\right]$$

Therefore, $$U=\left(\frac{1}{2E}\right)\left(\frac{16}{\pi D^3}\right)^2\left[4M_z^2+2T_z^2(1+\nu)\right]$$

Direct stress σ_a which acting alone will produce the same maximum strain energy per unit volume is given by:

$$U=\frac{\sigma_a^2}{2E}=\left(\frac{1}{2E}\right)\left(\frac{16}{\pi D^3}\right)^2\left[4M_z^2+2T_z^2(1+\nu)\right]$$

$$\sigma_a=\left(\frac{16}{\pi D^3}\right)\sqrt{4M_z^2+2T_z^2(1+\nu)}$$

Therefore, $$\sigma_a=\left(\frac{16}{\pi\times 90^3}\right)\sqrt{4\times 8^2+2\times 12^2\times(1+0.33)}=176.61\text{ MPa}$$

20.10 THE CRITERION FOR FAILURE

The failure means either yielding of material (which results in excessive permanent deformation) or actual rupture, whichever occurs first. Failure resulting from local buckling or elastic instability is not considered here. When a material is subjected to several types of stresses due to combined loading, the yielding depends on some combination of these stresses. There is no theoretical method available that correlates yielding in the uniaxial tensile test with yielding in more complex loadings. The commonly used failure theories for predicting onset of yielding are:

(i) *Rankine's maximum principal stress theory*

According to this theory failure occurs when the maximum principal stress on an element reaches its limiting value. The limiting stress is the yield stress in a simple tension test (or ultimate strength, if the material is brittle). Thus for a safe design, maximum principal stress shall not exceed the safe stress for the material. However, this theory disregards the effect of other principle stresses and of the shearing stresses on other planes through the element.

(ii) *St. Venant's maximum strain theory*

According to this theory a ductile material fails, i.e. begins to yield when the maximum principal strain on an element reaches the strain at which yielding occurs in a simple tension test or when the minimum principal strain, i.e. compressive strain equals the yield point strain in a simple compression test. Thus for a safe design the equivalent direct stress acting alone which produces a strain equal to maximum principal strain shall not exceed the safe stress for the material.

Consider Hooke's law for biaxial stress state of a strained element subjected to the maximum and minimum principal stresses σ_{max} and σ_{min}. The maximum principal strain is,

$$\varepsilon_1=\frac{\sigma_{max}}{E}-\nu\frac{\sigma_{min}}{E}=\frac{1}{E}(\sigma_{max}-\nu\sigma_{min}) \tag{20.56}$$

If the equivalent direct stress σ_{eq} acting alone produces the strain ε_1, then

$$\varepsilon_1=\frac{\sigma_{eq}}{E}=\frac{1}{E}(\sigma_{max}-\nu\sigma_{min}) \text{ i.e. } \sigma_{eq}=\sigma_{max}-\nu\sigma_{min} \tag{20.57}$$

This direct stress σ_{eq} shall not exceed the allowable stress for the material.

(iii) *Guest–Coulomb's maximum shear stress theory*

The maximum shear stress theory is based on the assumption that yielding begins when maximum shearing stress due to applied loading equals the maximum shearing stress developed at yielding in the simple tension test. Thus according to this theory, the maximum shear stress under applied loading shall not exceed the limiting shear stress for the material. In other words, the equivalent direct stress which acting alone produces the same maximum shear stress shall not exceed the limiting shear stress for the material.

In a strained element subjected to extreme principal stresses σ_{max} and σ_{min}, the maximum shear stress is given by,

$$\tau_{max} = \frac{\sigma_{max} - \sigma_{min}}{2} \tag{20.58a}$$

If the equivalent direct stress σ_{eq} acting alone produces the shear stress τ_{max}, then

$$\tau_{max} = \frac{\sigma_{max} - \sigma_{min}}{2} = \frac{\sigma_{eq}}{2} \quad \text{or} \quad \sigma_{eq} = \sigma_{max} - \sigma_{min} \tag{20.58b}$$

The direct stress σ_{eq} or maximum shear stress τ_{max} shall not exceed the corresponding safe limit. For the material having different yield points in tension and compression but equal factors of safety *n*, the condition of equivalence can be expressed as

$$\sigma_{eq} = \sigma_{max} - k\sigma_{min} \quad \text{where } k = \sigma_{y,t}/\sigma_{y,c}$$

For the brittle material, the ratio $\sigma_{y,t}/\sigma_{y,c}$ replaced by the ratio $\sigma_{u,t}/\sigma_{u,c}$.

(iv) *Beltrami and Haigh's maximum strain energy theory*

According to this theory, the *equivalent direct stress* acting alone which produces strain energy equal to maximum principal strain energy shall not exceed the safe stress for the material.

For a strained element subjected to the principal stresses σ_1 and σ_2, the strain energy stored per unit volume is

$$U_1 = \frac{1}{2E}\left(\sigma_1^2 + \sigma_2^2 - 2\nu\sigma_1\sigma_2\right) \tag{20.59}$$

If the direct stress σ_a' acting alone produces the same strain energy U_1, then

$$U_1 = \frac{\sigma_a'^2}{2E} = \frac{1}{2E}\left(\sigma_1^2 + \sigma_2^2 - 2\nu\sigma_1\sigma_2\right)$$

or

$$\sigma_a' = \sqrt{\sigma_1^2 + \sigma_2^2 - 2\nu\sigma_1\sigma_2} \tag{20.60}$$

This direct stress σ_a' shall not exceed the allowable shear stress for the material.

(v) *Maximum shear strain energy* or *distortion energy* or *Mises-Hencky yield theory*

According to this theory the equivalent direct stress acting alone which produces shear strain energy equal to the maximum shear strain energy shall not exceed the safe stress for the material.

For a strained element subjected to the principal stresses σ_1, σ_2 and σ_3, the strain energy stored per unit volume is

$$U_1 = \frac{1}{12G}[(\sigma_1 - \sigma_2)^2 + (\sigma_2 - \sigma_3)^2 + (\sigma_3 - \sigma_1)^2] \tag{20.61}$$

If the direct stress σ_a' acting alone produces the same shear strain energy U_1, then

$$U_1 = \frac{1}{12G}[(\sigma_a' - 0)^2 + (0 - 0)^2 + (0 - \sigma_a')^2] = \frac{\sigma_a'^2}{6G}$$

Therefore, $$2\sigma_a'^2 = (\sigma_1 - \sigma_2)^2 + (\sigma_2 - \sigma_3)^2 + (\sigma_3 - \sigma_1)^2$$

or $$\sigma_a' = \sqrt{\frac{[(\sigma_1 - \sigma_2)^2 + (\sigma_2 - \sigma_3)^2 + (\sigma_3 - \sigma_1)^2]}{2}} \tag{20.62}$$

This direct stress σ_a' shall not exceed the safe stress for the material limit.

The Mises-Hencky yield theory is preferable for ductile materials. For such materials, the maximum shearing stress theory also gives good agreement with experimental results. For rupture in brittle materials such as cast iron, the maximum principal stress theory is generally preferred.

EXAMPLE 20.21

A metal sheet is to be used to fabricate a closed cylindrical vessel of 800 mm diameter which is to be subjected to an internal pressure of 2.0 MPa. The stipulated factor of safety (FOS), elastic limit under pure tension and the Poisson's ratio of the metal are: 3, 240 MPa and 0.3, respectively. Determine the necessary thickness of the sheet based on: (i) Maximum principal stress theory, (ii) Maximum shear stress theory, and (iii) Maximum strain energy theory.

Solution: Safe axial tensile stress $\sigma_a = \dfrac{240}{3} = 80$ MPa

At a point in the closed cylindrical vessel, the material is subjected to major principal stress σ_1 i.e. hoop tension, and minor principal stress σ_2 i.e. longitudinal stress.

Therefore, $$\sigma_1 = \frac{pd}{2t} = \frac{2 \times 800}{2t} = \frac{800}{t} \text{ MPa (tension)}$$

$$\sigma_2 = \frac{pd}{4t} = \frac{2 \times 800}{4t} = \frac{400}{t} \text{ MPa (tension)}$$

(i) *Maximum principal stress theory*

Maximum principal stress, $$\sigma_1 = \frac{800}{t}$$

In the limiting case the maximum principal stress is equal to the safe stress for the material, i.e.

$$\frac{800}{t} = 80 \quad \text{or} \quad t = 10 \text{ mm}$$

(ii) *Maximum shear stress theory*

Safe shear stress $$\tau_a = \frac{\sigma_a}{2} = \frac{80}{2} = 40 \text{ MPa}$$

The maximum shear stress is,

$$\tau_{max} = \frac{\sigma_1 - \sigma_2}{2} = \frac{1}{2}\left(\frac{800}{t} - \frac{400}{t}\right) = \frac{200}{t}$$

In the limiting case the computed maximum shear stress τ_{max} shall be equal to the safe stress τ_a i.e. $\frac{200}{t} = 40$ or $t = 5$ mm.

(iii) *Maximum strain energy theory*

The computed direct stress σ_a' is given by:

$$\sigma_a' = \sqrt{\sigma_1^2 + \sigma_2^2 - 2\nu\sigma_1\sigma_2} = \left(\frac{pd}{2t}\right)\sqrt{1 + \left(\frac{1}{2}\right)^2 - \nu}$$

In the limiting case the computed equivalent direct stress σ_a' shall be equal to the safe stress σ_a i.e.

$$\left(\frac{pd}{2t}\right)\sqrt{1 + \left(\frac{1}{2}\right)^2 - \nu} = \sigma_a$$

or $$\left(\frac{800}{t}\right)\sqrt{1 + (0.5)^2 - 0.3} = 80 \quad \text{i.e.} \quad t = 9.75 \text{ mm}$$

EXAMPLE 20.22

A cube of 5 mm side is subjected to a load system as shown in Figure 20.28. (a) Determine the principal stresses and (b) will the cube yield according to Von-Mises yield criteria of the maximum shear strain energy theory, if the yield strength of the material is 75 MPa?

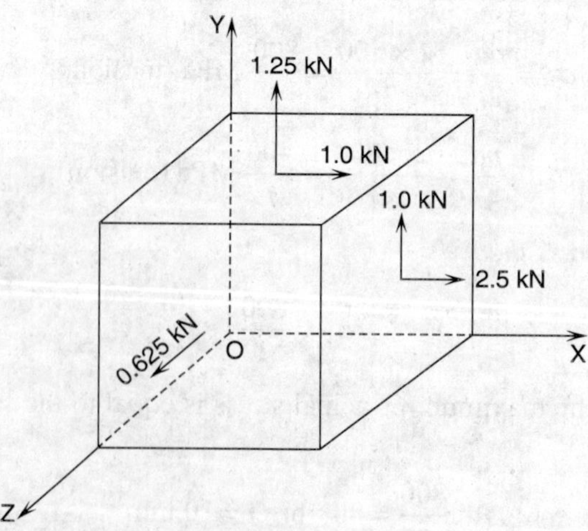

Figure 20.28 Loaded cube of Example 20.22.

Solution: The stresses developed under the given load system are:

$$\sigma_x = \frac{2500}{25} = 100 \text{ MPa (tension)}, \quad \sigma_y = \frac{1250}{25} = 50 \text{ MPa (tension)}$$

$$\sigma_z = \frac{625}{25} = 25 \text{ MPa (tension) and } \tau_{xy} = \frac{1000}{25} = 40 \text{ MPa}$$

The principal stresses are,

$$\sigma_{1,2} = \left(\frac{\sigma_x + \sigma_y}{2}\right) - \sqrt{\left(\frac{\sigma_x - \sigma_y}{2}\right)^2 + \tau_{xy}^2}$$

or
$$= \left(\frac{100 + 50}{2}\right) - \sqrt{\left(\frac{100 - 50}{2}\right)^2 + 40^2}$$

or
$$= 75 \pm 47.17 = 122.17 \text{ MPa and } 27.83 \text{ MPa}$$

and
$$\sigma_3 = \sigma_z = 25 \text{ MPa}$$

According to Von-Mises yield criteria, the equivalent direct yield stress σ_y' is given by:

$$\sigma_y' = \sqrt{\frac{[(\sigma_1 - \sigma_2)^2 + (\sigma_2 - \sigma_3)^2 + (\sigma_3 - \sigma_1)^2]}{2}}$$

$$= \sqrt{\frac{[(122.17 - 27.83)^2 + (27.83 - 25)^2 + (25 - 122.17)^2]}{2}}$$

$$= 95.79 \text{ MPa}$$

As direct equivalent yield stress σ_y' exceeds σ_y, the safe limit, the cube yields.

EXAMPLE 20.23

A 1 mm diameter steel wire is wound around an open ended copper tube with internal and external diameters of 120 mm and 140 mm, respectively, to increase the strength of the tube against internal pressure. Determine the initial tension which must be given to the wire, so that the maximum allowable stresses for the tube and wire materials, namely 90 MPa and 200 MPa, respectively, are reached simultaneously. Consider Poisson's ratio for the two materials to be 0.3 and the wire winding as a thin cylinder. The modulus of elasticity for the copper and steel are 100 GPa and 200 GPa, respectively.

Solution: Mean diameter of the tube, $d_m = \dfrac{120 + 140}{2} = 130$ mm

Thickness of the tube, $t = 10$ mm

The equivalent internal pressure p_e to develop maximum allowable stresses in the tube and wire materials, can be computed by considering equilibrium of the half tube of unit length, i.e.

Total tearing force on half tube due to internal pressure

= Total resisting force due to hoop stress in the tube and wire

i.e.
$$d_m p_e = \sigma_{c,a}(2t) + \left[\sigma_{w,a}\left(2\times\frac{\pi}{4}d_w^2\right)\right]\times\frac{1}{d_w}$$

or
$$130 p_e = 90\times 20 + \left[200\times\left(2\times\frac{\pi}{4}\right)\right]\times 1 \quad \text{or} \quad p_e = 16.262 \text{ MPa}$$

Against an internal pressure in the closed tube at a point the material is subjected to 2-dimensional principal stresses σ_1 and σ_2 i.e. hoop and longitudinal stresses, respectively, whereas the wire is subjected to direct stress only. For the initial tension T given to the wire, the circumferential force developed in the tube is obtained from,

Force in the wire = Force in the tube in the length d_w

i.e.
$$\left(\frac{\pi}{4}d_w^2\right)\times T = \sigma_t\times(t\,d_w) \quad \text{or} \quad \sigma_t = \left(\frac{\pi}{4}\frac{d_w}{t}\right)\times T$$

For the circumferential strain compatibility at the interface of wire and tube

$$\varepsilon_\theta = \frac{\sigma_1 - \nu\sigma_2}{E_c} = \frac{\sigma_{w,a} - T}{E_w}$$

where
$$\sigma_1 = \sigma_{c,a} - \sigma_t$$

$$\sigma_2 = \frac{p_e\,d}{4t} = \frac{p_e\times 130}{4\times 10} = 3.25\,p_e$$

Therefore,
$$\frac{1}{(100\times 10^3)}\times\left[\left(90 - \frac{\pi}{4}\times\frac{1}{10}\times T\right) - 0.3\times 3.25\times 16.262\right] = \frac{200 - T}{(200\times 10^3)}$$

i.e.
$$T = 61.347 \text{ MPa}$$

EXAMPLE 20.24

A thick spherical pressure vessel with internal diameter of 300 mm is subjected to an internal pressure of 80 MPa. Determine the wall thickness of pressure vessel based upon the: (a) maximum principal stress theory, (b) Maximum shear stress theory and (c) The principal strain theory. The yield strength and Poisson's ratio of the material of the vessel are 300 MPa and 0.30, respectively.

Solution: For the problem,

$$r_1 = 150 \text{ mm};\ p = 80 \text{ MPa};\ \sigma_y = 300 \text{ MPa and } \nu = 0.30$$

The maximum principal stresses in a thick spherical pressure vessel subjected to an internal pressure of p are given by Eq. (5.93):

$$\sigma_1 = \sigma_\theta = \frac{p}{2}\left(\frac{k^3+2}{k^3-1}\right) \quad \text{where } k = \frac{r_2}{r_1}$$

$$\sigma_2 = \sigma_\theta \quad \text{and} \quad \sigma_3 = \sigma_r = -p$$

(a) *The maximum principal stress theory*

$$\sigma_1 = \sigma_y \quad \text{or} \quad \frac{p(k^3+2)}{2(k^3-1)} = \sigma_y$$

$$k^3 = \frac{2(\sigma_y + p)}{(2\sigma_y - p)} = \frac{2(300+80)}{(2\times 300 - 80)} = 1.4616$$

$$k = \frac{r_1 + t}{r_1} = 1.135$$

$$t = r_1(1.135 - 1) = 150 \times (1.135 - 1) = 20.25 \text{ mm}$$

(b) *Maximum shear stress theory*

Safe shear stress, $\tau_a = \dfrac{\sigma_y}{2}$

The maximum shear stress, $\tau_{max} = \dfrac{\sigma_1 - \sigma_3}{2}$

In the limiting case, the computed maximum shear stress τ_{max} shall be equal to the safe stress τ_a i.e. $\sigma_1 - \sigma_3 = \sigma_y$

Therefore, $\dfrac{p(k^3+2)}{2(k^3-1)} + p = \sigma_y$

$$k^3 = \frac{2\sigma_y}{(2\sigma_y - 3p)} = \frac{2\times 300}{(2\times 300 - 3\times 80)} = 1.667$$

$$k = \frac{r_1 + t}{r_1} = 1.1856$$

$$t = r_1(1.1856 - 1) = 150 \times (1.1856 - 1) = 27.84 \text{ mm}$$

(c) *The principal strain theory*

$$\sigma_1 - \nu(\sigma_2 + \sigma_3) = \sigma_y$$

or
$$\frac{p(k^3+2)}{2(k^3-1)} - \nu p\left[\frac{(k^3+2)}{2(k^3-1)} - 1\right] = \sigma_y$$

or
$$\frac{p(k^3+2)}{2(k^3-1)}(1-\nu) + \nu p = \sigma_y \quad \text{or} \quad \frac{p(k^3+2)}{2(k^3-1)} = \left(\frac{\sigma_y - \nu p}{1-\nu}\right) = \sigma_y{}'$$

Thus the effective yield strength is,

$$\sigma_y{}' = \frac{\sigma_y - \nu p}{1-\nu} = \frac{300 - 0.3\times 80}{1 - 0.3} = 394.286 \text{ MPa}$$

Therefore, $$k^3 = \frac{2(\sigma_y' + p)}{(2\sigma_y' - p)} = \frac{2(394.286 + 80)}{(2 \times 394.286 - 80)} = 1.3387$$

$$k = \frac{r_1 + t}{r_1} = 1.102$$

$$t = r_1(1.102 - 1) = 150 \times (1.102 - 1) = 15.32 \text{ mm}$$

EXAMPLE 20.25

A thick cylindrical vessel with internal and external diameters of 500 mm and 700 mm, respectively, is closed at ends by rigid end plates welded to it. It is subjected to an axial pull of 1000 kN through the end plates, a torque of 100 kN.m and an internal pressure of 60 MPa. Determine the principal stresses and the absolute maximum shear stress at the inside surface of the vessel, ignoring the end effects.

Solution: Direct stress,

$$\sigma_d = \frac{P}{A} = \frac{1000 \times 10^3}{\pi(700^2 - 500^2)/4} = 5.305 \text{ MPa}$$

Shear stress,

$$\tau = \frac{16Td_2}{\pi(d_2^4 - d_1^4)} = \frac{16 \times (100 \times 10^6) \times 700}{\pi(700^4 - 500^4)} = 2.007 \text{ MPa}$$

The circumferential stress developed in the vessel is,

$$\sigma_\theta = p\left(\frac{k^2 + 1}{k^2 - 1}\right) \quad \text{where } k = \frac{r_2}{r_1} = \frac{350}{250} = 1.4$$

or $$= 60 \times \left(\frac{1.4^2 + 1}{1.4^2 - 1}\right) = 185 \text{ MPa}$$

$$\sigma_1, \sigma_2 = \frac{1}{2}(\sigma_\theta + \sigma_d) \pm \sqrt{(\sigma_\theta - \sigma_d)^2 + 4\tau^2}$$

$$= \frac{1}{2}(185 + 5.305) \pm \sqrt{(185 - 5.305)^2 + 4 \times (2.007)^2}$$

$$= 95.1525 \pm 89.8699 = 185.0224 \text{ MPa}, 5.2826 \text{ MPa}$$

The absolute maximum shear stress at the inside surface of the vessel

$$\tau_{\max} = \frac{\sigma_1 - \sigma_2}{2} = \frac{185.0224 - 5.2826}{2} = 89.87 \text{ MPa}$$

EXAMPLE 20.26

A 30 mm square cast iron rod with a horizontal bend and fixed to the wall as shown in Figure 20.29(a), breaks at the fixed end at certain load applied at the free end. Determine the maximum load the rod would have carried safely. The ultimate strengths of the material in tension and compression are 150 MPa and 330 MPa, respectively. The required FOS is 3.

Solution: From the bending and twisting moment diagrams shown in Figures 20.29(b), it is evident that the critical section lies at the support

Bending stress, $$\sigma_b = \frac{M}{Z} = \frac{6PL}{a^3}$$

Shear stress, $$\tau = \frac{T}{Z_p} = \frac{PL}{0.208a^3}$$

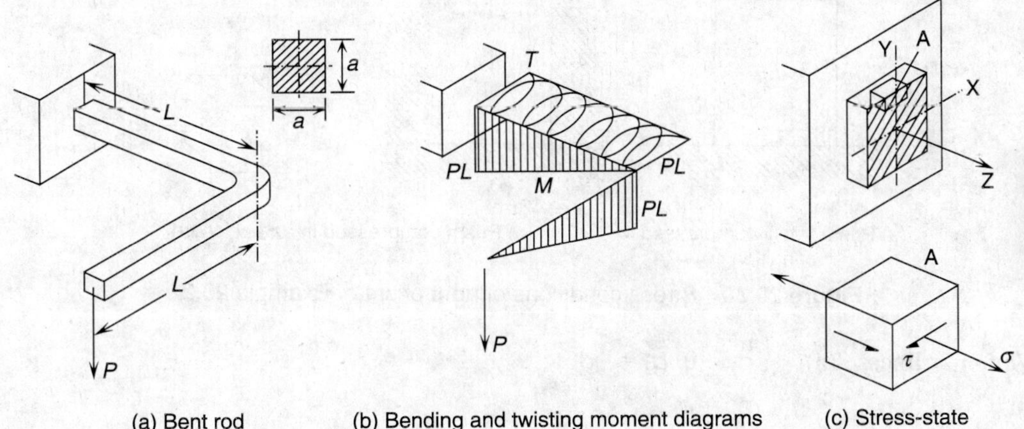

Figure 20.29 Cast iron rod with horizontal bend of Example 20.26.

Since the material is brittle and has different limiting stress points in tension and compression but with equal factors of safety n, the factor k is

$$k = \frac{\sigma_{u,t}}{\sigma_{u,c}} = \frac{150}{330} = 0.455$$

$$\sigma_{eq} = \sigma_1 - k\sigma_3 = \frac{1-k}{2}\sigma_b + \frac{1+k}{2}\sqrt{\sigma_b^2 + 4\tau^2}$$

$$= \left(\frac{PL}{a^3}\right)\left[3(1-k) + \frac{1+k}{2}\sqrt{6^2 + \frac{4}{(0.208)^2}}\right]$$

$$= \left(\frac{PL}{a^3}\right)\left[3(1-0.455) + \left(\frac{1+0.455}{2}\right)\sqrt{6^2 + \frac{4}{(0.208)^2}}\right]$$

$$= 9.88\left(\frac{PL}{a^3}\right) = 9.88 \times \frac{P \times 300}{30^3} = 0.1098P$$

Taking into account the FOS of 3, the safe load the rod could have carried

$$P = \frac{150}{3 \times 0.1098} = 455.37\,\text{N}$$

EXAMPLE 20.27

Compare the equivalent stresses in a right prism in two cases of loading: (a) the prism is compressed freely (see Figure 20.30(a)) the prism is compressed in a rigid enclosure which prevents its expansion laterally as shown in Figure 20.30(b).

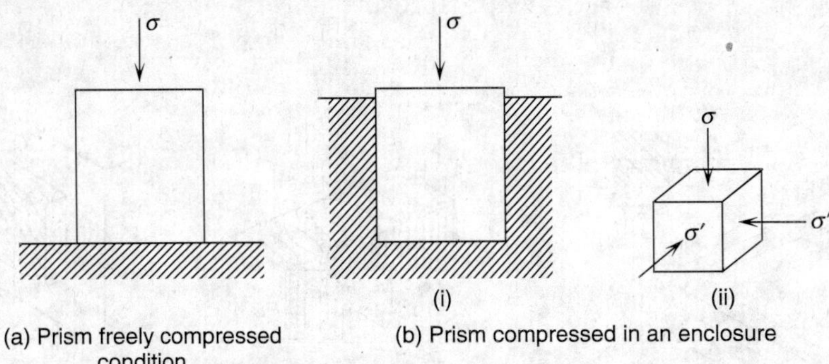

Figure 20.30 Stress conditions of right prism of Example 20.27.

Solution: In case (a) $\sigma_1 = 0,\ \sigma_3 = -\sigma.$

Thus, $\sigma_{eq} = \sigma$

(b) In this case, since lateral strains are zero, and hence lateral compressive stresses σ' develop as shown in Figure 20.30(b). The lateral compressive stresses σ' are determined using Hooke's law:

$$\varepsilon_{\text{lat}} = \frac{1}{E}[\sigma' - \nu(\sigma + \sigma')] = 0$$

whence lateral compressive stress is

$$\sigma' = \frac{\nu}{1-\nu}\sigma$$

For the resulting state of stress

$$\sigma_1 = -\frac{\nu}{1-\nu}\sigma, \quad \sigma_3 = -\sigma \quad \text{and} \quad \sigma_{eq} = \left(1 - \frac{\nu}{1-\nu}\right)\sigma$$

It should be noted that σ_{eq} is reduced due to lateral restraint.

20.11 PROBLEMS

20.1 A tension member carrying an axial force of 225 kN is comprised of two wooden joist 50 × 100 mm sglued together along a joint inclined at an angle of 40° to the axis of the member. Determine the normal and shearing stresses developed in the glue.

[**Ans.** σ = 18.56 MPa]

20.2 At a point in a strained body direct tensile stresses of 120 MPa and 70 MPa act on mutually perpendicular planes. Determine the shearing stress that can be applied so that the major principal tensile stress in the material is limited to 135 MPa. Also determine the minor principal stress and maximum shear stress developed in the material.

[**Ans.** $\sigma_a = \sigma_1 = 135$ MPa gives $\tau = 31.225$ MPa; $\sigma_2 = 55$ MPa and $\tau_{max} = 40$ MPa]

20.3 At a point in a strained material direct tensile stresses of 60 MPa and 100 MPa act on planes right angles to each other with shear stress τ. If the maximum tensile stress in the material is limited to 200 MPa, determine the shear stress developed in the material.

[**Ans.** $\tau_{max} = 355/3$ MPa]

20.4 At a point in a stressed body, the state of stress on two planes inclined at 45° is shown in Figure 20.31. Determine the principal stresses developed in the material.

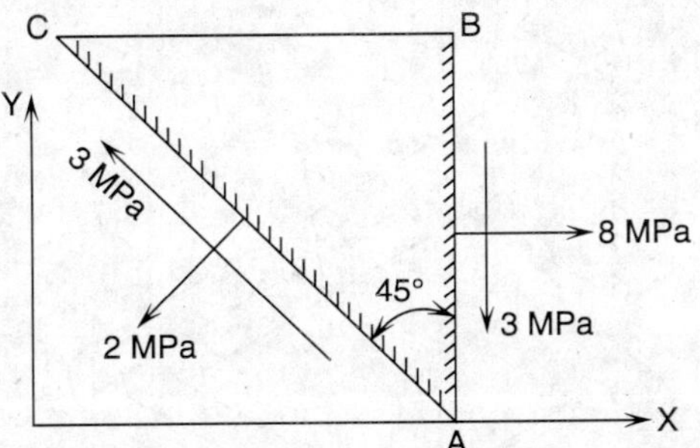

Figure 20.31 State of stress on two planes of Problem 20.4.

[**Ans.** Resolve vertically; $\sigma_x = 8$ MPa, $\sigma_y = 2$ MPa, $\tau_{xy} = 3$; $\sigma_{1,2} = 5 \pm 4.243$ MPa]

20.5 A two-dimensional state of stress at a point in a stressed body is given by:

$$\sigma_x = 60 \text{ MPa}, \ \sigma_y = 20 \text{ MPa and } \tau_{xy}$$

Determine the (a) the maximum permissible shear stress, τ_a, if the larger of the principal stresses is limited to 75 MPa, (b) the other principal stress and (c) maximum shear stress developed in the material.

[**Ans.** $\tau_a = 28.723$ MPa, $\sigma_2 = 5$ MPa, and $\tau_{max} = 35$ MPa]

20.6 Explain why the stress trajectories in a: (a) symmetrically loaded simply supported beam becomes horizontal at the mid-span, and (b) cantilever beam tends to become horizontal as they approach the support. What are the stress trajectories for axial tension and compression?

20.7 A solid shaft 100 mm in diameter is subjected simultaneously to an axial compressive force of 800 kN and to a torque that twists the shaft through an angle of 2° in a length of 8 m. If $G = 80$ GPa, determine the maximum normal and shearing stresses in the shaft.

[**Ans.** $\tau = 53.87$ MPa]

20.8 A solid cylindrical shaft is used to transmit simultaneously a maximum bending moment of 3.0 kN.m and a torque of 3.75 kN.m. Determine the smallest radius of the shaft that can be used if maximum normal and shearing stresses in the shaft are restricted to 120 MPa and 60 MPa, respectively.

[**Ans.** $D = 74.11$ mm]

20.9 A cylindrical shaft of 80 mm diameter is subjected to bending and twisting moments of 2.5 kN.m and 4.2 kN.m, respectively. If the yield stress of the shaft material is 300 MPa, determine the FOS of the shaft according to the maximum shearing stress theory of failure.

[**Hint & Ans.** $T_e = 4.8877$ kN.m, $\tau = 48.62$ MPa, $\sigma_a = 150$ MPa; FOS = 3.0855]

20.10 A 100 mm diameter bar with a built-in bracket is fixed to the wall and loaded as shown in Figure 20.32. Determine the principal stresses at the top extremity of the vertical diameter for the section marked I.

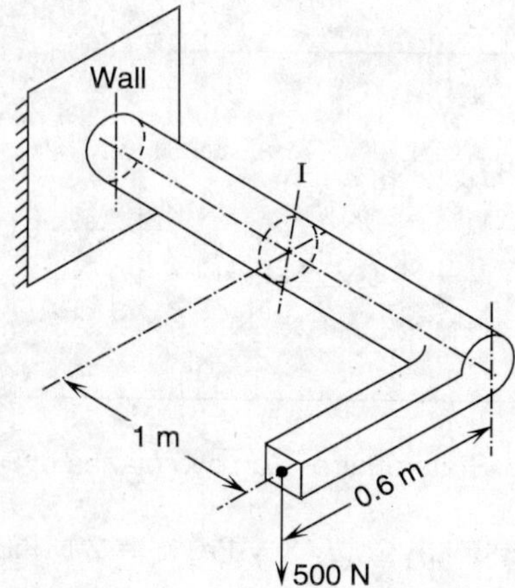

Figure 20.32 A bar with bracket of Problem 20.10.

[**Ans.** $M = 500$ N.m, $T = 300$ N.m, $M_e = 541.55$ N.m and $\sigma_1 = 5.516$ MPa, $\sigma_2 = -0.423$ MPa]

20.11 A plane at a point in a strained material subjected to two-dimensional stress system transmits a shear stress of 50 MPa along with an unknown normal tensile stress. The maximum principal stress in the material is 120 MPa (tension) acting on the plane inclined at 55° with the plane under consideration as shown in Figure 20.33(a). (a) Determine analytically the second principal stress and (b) construct the Mohr's circle diagram. Also calculate the strain normal to the plane. The elastic modulus and Poisson's of the material are 200 GPa and 0.3, respectively.

[**Hint:** To construct the Mohr's circle diagram, mark OA representing major principal stress $\sigma_1 = 120$ MPa and draw a line O′P parallel to the axis OA with OO′ representing shear stress $\tau = 50$ MPa. Draw the line AP at an angle $\theta_1 = 55°$ to the offset AF meeting the line O′P at P, thus fixing the position of second plane. Draw a line EC bisecting AP at right angles, cutting the axis OA at C. With C as centre, draw a circle with CA or CP as radius, cutting the axis OA at B as shown in Figure 20.33(b). OB represents minor principal stress].

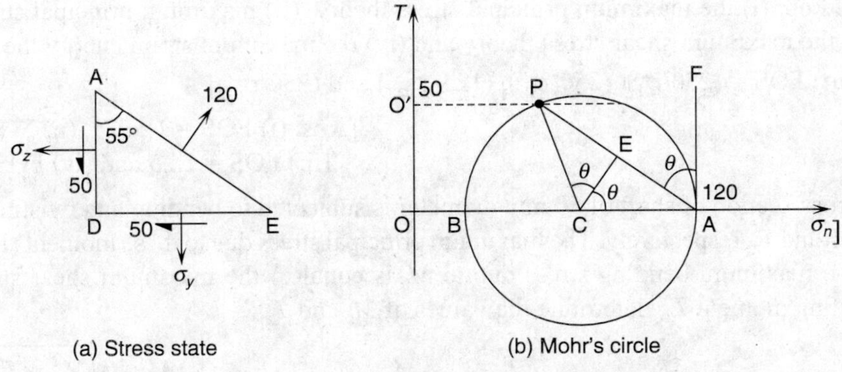

Figure 20.33 Stress state and Mohr's circle for Problem 20.11.

20.12 At a point in a material, the stresses acting on two planes right angles to each other are: $\sigma_x = 120$ MPa, $\sigma_y = -200$ MPa and $\tau = -80$ MPa.
Determine: (a) the magnitude and direction of the principal stresses, (b) the maximum shear stress, (c) the tensile stress which, while acting alone, would develop the same maximum shear stress, and (d) the shear stress which, while acting alone, would develop the same principal stress.

[**Ans.** (a) $\sigma_1 = 124.92$ MPa, $\sigma_2 = -204.93$ MPa, $\theta_1 = -3.5°$; (b) $\tau_{max} = 164.92$ MPa; (c) $\sigma_a = 329.84$ MPa and (d) $\tau = 124.92$ MPa, 204.92 MPa]

20.13 A metallic bar of circular cross-section is subjected to a longitudinal pull of 180 kN and the maximum shear stress on any oblique plane is limited to 120 MPa, determine the diameter of the bar.

[**Ans.** $d = 30.9$ mm]

20.14 An aircraft component is modelled as a cylindrical bar subjected to an axial compressive of force 8 kN, a bending moment of 3.5 kN.m and twisting moment of 30 N.m. A 20 mm diameter solid bar of aluminium having $\sigma_y = 591$ MPa, $\sigma_{yt} = 542$ MPa and $\sigma_{ys} = 271$ MPa is recommended for its use. Determine the factor of safety available based on: (i) the maximum principal stress theory and (ii) the maximum shear stress theory.

[**Ans.** $\sigma_1 = 73.71$ MPa; $\tau_{max} = 51.43$ MPa; (i) FOS = 7.35, and (ii) FOS = 5.269]

20.15 A bolt is subjected to the direct tension and transverse shear of 10 kN and 3.75 kN, respectively. Determine the necessary diameter of the bolt based on: (i) the maximum principal stress theory, (ii) the maximum shear stress theory, and (iii) the maximum strain energy theory. The elastic limit under pure tension, the Poisson's ratio and the factor of safety of the material are: 240 MPa, 0.3 and 3, respectively.

[**Ans.** (i) $d = 13.38$ mm, (ii) $d = 14.10$ mm and (iii) $d = 13.64$ mm]

20.16 The principal stresses σ_1, σ_2 and σ_3 at a point in a steel specimen subjected to three-dimensional stress system are 160 MPa, 80 MPa and −40 MPa, respectively. If the proportionality limit for the steel specimen is 250 MPa, determine the factor of safety

based on: (i) the maximum principal stress theory, (ii) maximum principal strain theory, (iii) the maximum shear stress theory and (iv) the maximum strain energy theory.
[**Hint:** FOS = $(250/\sigma_1)$, $(250/\sigma_a')$, $(125/\tau_{max})$ and $(250/\sigma_a')$]

[**Ans.** (i) FOS = 1.5625, (ii) FOS = 1.6892
(iii) FOS = 1.25 and (iv) FOS = 1.4046]

20.17 A cross-section of shaft of 50 mm diameter is subjected to bending and twisting moments of M_z and T_z, respectively. The maximum principal stress due to these moments is 150 MPa. If the maximum bending stress due to M_z is equal to the maximum shear stress due to twisting moment T_z, determine the values of M_z and T_z.

$$\left[\textbf{Hint}: \sigma_{max} = \frac{M_z}{(Z_p/2)}; \tau_{max} = \frac{T_z}{Z_p} \text{ i.e. } \frac{M_z}{T_z} = \frac{1}{2}; \sigma_1 = \frac{M_e}{Z} = \frac{(M_z + \sqrt{M_z^2 + T_z^2})/2}{(Z_p/2)}\right]$$

20.18 Three exactly similar mild steel tube specimens have the external and internal diameters 37.5 mm and 31.25 mm, respectively. One of these specimens was tested in pure tension and the limit of proportionality was recorded to be 70 kN. The second specimen was tested in torsion whereas the third was tested in torsion with a superimposed bending moment of 350 N.m. If the failure criterion is the maximum shear stress, determine the torque at which two specimens would fail.

[**Ans.** σ_a = 207.42 MPa; T_e = 309117.6 N.m and T_1, T_2 = 555.98 N.m and 432 N.m]

20.19 Draw Mohr's circle for a 2-dimensional stress field subjected to: (i) pure shear, (ii) pure biaxial tension, (iii) pure uniaxial compression, and (iv) pure uniaxial tension.

20.20 At a point in an elastic material, the stresses on three mutually perpendicular planes are:

First plane:	50 MPa (tensile) and 40 MPa (shear)
Second plane:	30 MPa (compressive) and 40 MPa (shear)
Third plane:	No stress

Determine: (a) the position of principal planes and magnitude of principal stresses, (b) the position of planes on which maximum shear stress acts and calculate the normal and shear stresses on them.

[**Ans.** (a) σ_1 = 66.57 MPa, σ_2 = −46.57 MPa, θ_1 = 22.5° and θ_2 = 112.5°; (b) θ_3 = 67.5°, σ_n = 10 MPa, τ = −56.57 MPa; θ_4 = 157.5°, σ_n = 10 MPa and τ = 56.57 MPa]

20.21 A thin cylindrical vessel with hemispherical ends is subjected to an internal pressure. Determine the ratio of thicknesses of cylindrical and hemispherical portions so that the junction section remains free from unequal deformation.

$$\left[\textbf{Ans.} \frac{t_1}{t_2} = \frac{2-\nu}{1-\nu}\right]$$

20.22 A thin walled tube is subjected to an internal pressure and twisting moment. The stresses developed in the walls are: (i) 60 MPa (tensile), (ii) 30 MPa (tensile) in a direction at right angles to (i), and complementary shear stress of 45 MPa in the directions of (i) and (ii). Determine the normal and tangential stresses on two planes which are equally inclined to (i) and (ii).

[**Hint:** $\sigma_x = 30$ MPa, $\sigma_y = 60$ MPa and $\sigma_{xy} = 45$ MPa]

[**Ans.** (a) $\sigma_1 = 92.43$ MPa and $\sigma_2 = -2.43$ MPa; at $\theta = 45°$ $\sigma_n = 0$, $\tau = 15$ MPa and $\theta = 135°$, $\sigma_n = 90$ MPa, $\tau = -45$]

20.23 A seamless spherical steel shell of 900 mm internal diameter and 10 mm thickness is filled with a fluid under pressure until the volume increases by 1.5×10^5 mm^3. Determine the pressure exerted by the fluid on the shell. The modulus of elasticity and Poisson's ratio for the material of the shell are 200 GPa and 0.3, respectively.

$$\left[\textbf{Hint}: \sigma_1 = \sigma_2 = \sigma_3 = \sigma = \frac{pd}{4t}; \varepsilon_\theta = \frac{\sigma}{E}(1-\nu); \varepsilon_v = 3\varepsilon_\theta; \text{and } \delta V = \varepsilon_v V\right]$$

[**Ans.** $p = 1.6634$ MPa]

20.24 A thin cylindrical vessel of 50 mm internal diameter and 2.5 mm wall thickness which is closed at ends is subjected to an axial pull of 10 kN and a torque of 500 N.m while under an internal pressure of 6.0 MPa. (a) Determine the principal stresses, (b) Express the state of stress at a point by means of square element, and (c) Sketch the Mohr's stress circle.

[**Hint:** σ_l $(= pd/4t) = 30$ MPa; σ_θ $(= pd/2t) = 60$ MPa; σ_d $(= P/\pi dt) = 25.465$ MPa]

[**Ans.** $\sigma_x = 30 + 25.465$ MPa; $\sigma_y = \sigma_\theta = 60$ MPa, σ_{xy} $(= 2T/\pi d^2 t) = 50.93$ MPa; $\sigma_1 = 108.61$ MPa; $\sigma_2 = 6.75$ MPa and $\tau_{max} = 50.98$ MPa]

20.25 A circular shaft is subjected to bending and twisting moments of M_z and T_z, respectively. With the help of Mohr's circle diagram represent the stresses on an element at the surface of the shaft. From this diagram or otherwise determine the maximum shear stress due to the combined effect of gradually applied loads of M_z and T_z.

[**Hint:** σ_b $(= 32\ M/\pi d^3)$; τ $(= 16T/\pi d^3)$; $\tau_{max} = \dfrac{1}{2}\sqrt{\sigma_b^2 + 4\tau^2}$

Mohr's circle diagram representing the state of stress on an element at the surface of the shaft is shown in Figure 20.34]

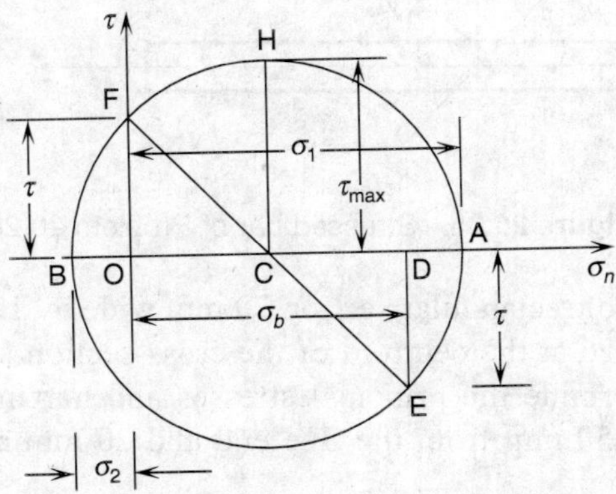

Figure 20.34 Mohr's circle for the compound stress of Problem 20.25.

20.26 A circular shaft of 50 mm diameter projects horizontally 250 mm from the fixed support. The shaft is subjected at its free end to vertical load of 3 kN, pure torque of 1.0 kN.m and axial pulling force of 15 kN. With the help of Mohr's circle diagram represent the stresses on the elements at the surface of the shaft at the topmost and bottommost points at the support.

[**Hint:** $\sigma_d (= 4P/\pi d^2 = 7.64$ MPa); $\sigma_b (= 32\,M/\pi d^3 = 61.12$ MPa); $\tau (= 16T/\pi d^3 = 40.74$ MPa)]

[**Ans.** At top point: $\sigma_x = 68.76$ MPa, $\sigma_1 = 87.69$ MPa, $\sigma_2 = -18.93$ MPa; At bottom point: $\sigma_x = -53.48$ MPa, $\sigma_1 = 0.42$ MPa, $\sigma_2 = -53.89$ MPa]

20.27 A thin cylindrical vessel of 80 mm mean diameter and 5 mm wall thickness is closed at ends and subjected to a torque of 1.5 kN.m while under an internal pressure of 5.0 MPa. If the yield strength of the material of the vessel is 240 MPa, determine the factor of safety according to each of the following theories of failure: (a) maximum normal or principal stress theory, (b) maximum normal strain theory and (c) maximum shear stress theory. The Poisson's ratio for the material is 0.3.

[**Hint:** $\sigma_x = 20$ MPa; $\sigma_\theta = 40$ MPa; $\tau = 29.84$ MPa; $\sigma_1 = 61.47$ MPa; $\sigma_2 = -1.47$ MPa; $\sigma_3 = 0$; $\tau_{max} = 31.47$ MPa]

$$\left[\textbf{Ans.}\ (a)\ n = \frac{\sigma_y}{61.47} = 3.9;\ (b)\ n = \frac{\sigma_y}{61.91} = 3.87;\ (c)\ n = \frac{\sigma_y/2}{31.47} = 3.81 \right]$$

20.28 A beam of rectangular cross-section of width b and height h is subjected to a pure bending moment M. Determine the bending and shearing stresses at a point (z, y) on a plane inclined at 45° with the horizontal plane as shown in Figure 20.35. The Z axis is along the axis of the member.

$$\left[\textbf{Hint}: \sigma_b = \tau = \frac{6M_z}{bh^2} \right]$$

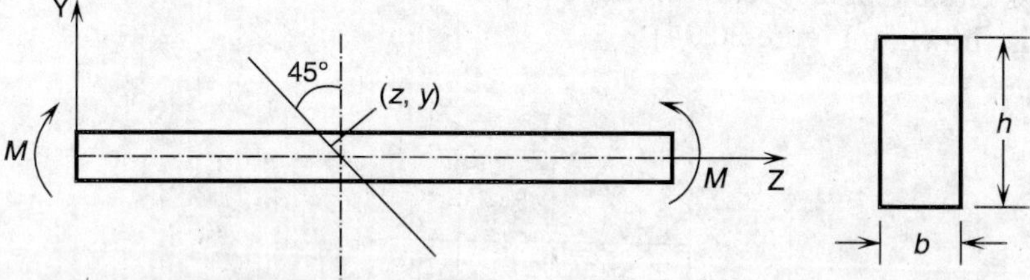

Figure 20.35 Stressed bar of Problem 20.28.

20.29 A cantilever beam of rectangular section 20 mm wide by 120 mm deep is subjected to a 60 kN load applied at the centroid of the cross-section at the free end as shown in Figure 20.36. Determine the principal stresses and maximum shearing stress at the point A which is 250 mm from the free end and 20 mm above the centerline of the beam.

[**Ans.** $\sigma_1 = 87.1$ MPa and $\theta = -12.9°$]

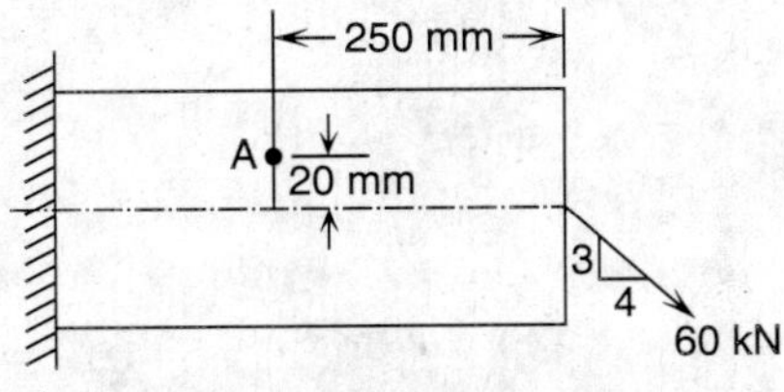

Figure 20.36

20.30 A solid circular line shaft 60 mm in diameter shown in Figure 20.37 is subjected to a load system such that the belts pull on the pulley C horizontally and pulley D vertically. Determine the maximum normal and shearing stresses developed in the shaft.

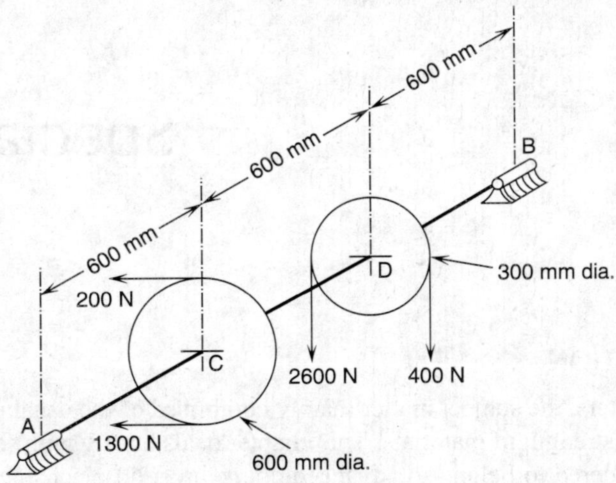

Figure 20.37

20.31 A cantilever beam of I-section 200 mm wide by 300 mm deep (overall) with 20 mm thickness throughout as shown in Figure 20.38 is subjected to a 50 kN vertical load applied at the free end and passing through the centroid of the cross-section. Determine the principal stresses at the point *A* which is at distance 1.0 m from the free end and just below the flange.

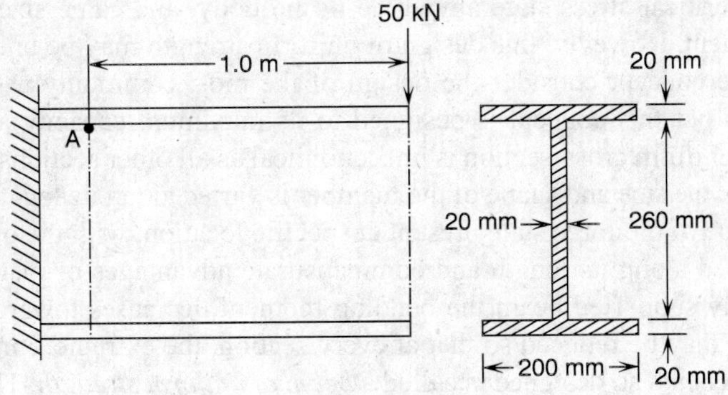

Figure 20.38

[**Ans.** $\sigma_1 = 36.5$ MPa]

Chapter 21

Special Topics

21.1 INTRODUCTION

In the preceding chapters, the subject matter mainly comprised of the usual undergraduate course in solid mechanics or strength of materials. This chapter deals briefly with some additional topics that are usually considered to belong to advanced course in solid mechanics. Each of the topics discussed here is complete in itself so that they may be studied independently of the others.

21.2 DESIGN CONCEPTS

A design of an element must meet specified strength and stiffness criteria for *safety* and *serviceability*. Generally, in the commonly used approach, the size and shape of the element is based on the most critical stress state anywhere in the body, and same size and shape is used throughout the element. However, this design of uniform strength may be uneconomical.

To illustrate the concept consider the design of the most commonly used element, i.e. the beam. The beam as per this concept is designed to its maximum capacity at one section only. Hence the beam of uniform cross-section is uneconomical as all other sections are under-stressed. In another approach, the size and shape of the member is varied along its length in response to the nature and magnitude of the force state present at specific location, with the objective to have the beam equally stressed along its length and commensurate advantages of material economy. For example, in a simply supported beam the bending moment decreases towards the supports, the section of the beam may be reduced so that at every section, the extreme fibre stress reaches the permissible value. A beam so designed is called a *beam of uniform strength*. The section of such a beam may be obtained by (i) maintaining constant depth and varying the width; (ii) maintaining constant width and varying the depth, or (iii) varying both the width and depth. The general principles are outlined in the following sections.

21.2.1 Strength and Stiffness Control

The element must be so proportioned that it is strong enough to carry the applied forces without distress in the material or excessive deformations. In most of the beam design problems for a given span, loading, and material, the design procedure is highly limited and it is confined to finding beam proportions and sectional properties for the member such that the actual stresses generated in the beam are limited to the predetermined safe levels dependent on the properties of the materials used. For example, in a beam given span, loading and support conditions provide maximum design moment M_{max} which is used to estimate the initial member size using the concept of required sectional modulus,

$$Z_{required} = M_{max}/\sigma_{b,a} \tag{21.1}$$

A member with sectional modulus equal to or greater than this value is selected as trial section checked for its adequacy for other design parameters.

21.2.2 Cross-sectional Shapes

The objective of design is to select a cross-sectional configuration that has the smallest possible area with material arranged in such a manner that I or Z has the maximum value, thus allowing the beam to carry the maximum possible external moment, hence the loading. From the definition of moment of inertia $I_x = \int y^2\, dA$ it is evident that the contribution of an elemental area (dA) to the total moment of inertia of a section depends on the square of the distance of this elemental area from the neutral axis of the section. Therefore, for a given cross-sectional area, i.e. the materials, a section with material arranged in the space as far as practically possible from the neutral axis of the section will provide an economical section. Consequently, beams with a high depth-to-width ratio are usually more efficient than those with shallower proportions. The steel *wide-flange* and *plate girder sections*, where a thin web connects two widely separated thick flanges, are the examples of highly efficient cross-sectional shapes where the material is moved away from the neutral axis to maximize the moment of inertia and hence the resistance to bending. However, it should be noted as discussed earlier, it is not the moment of inertia of the section alone but the ratio I_x/y_{max} is actually of primary importance. In case of symmetrical sections, an increase in I_x value is accompanied by an increase in the sectional modulus, Z_x. In case of non-symmetrical sections such as T-beams, the design must be based on the condition that maximum bending stress at any point on the beam is limited to the allowable stress.

21.2.3 Shaping a Beam along its Length

As the shears and moments vary along the length of the beam it is possible to improve the overall efficiency of a beam by appropriately shaping the beam along its length as discussed in Section 8.7. However, it should be noted that the shapes derived there are bending moment and cross-sectional geometry specific, i.e. shear stresses and deflection effects have been ignored and are applicable to rectangular cross-sections only. *These shapes are very much idealized responses to the loadings considered and may not have much practical utility.*

21.2.4 Effect of Support Locations and Boundary Conditions

The magnitudes of design moments can be reduced or their distribution altered by manipulating the support conditions for enhanced structural efficiency and economy. A classic way of reducing design moments is to provide *cantilever overhangs* on beams. Cantilevering one end of a simply supported beam carrying uniformly distributed load results in reduction in the positive moment present while a negative moment develops over the cantilever support. The greater the cantilever, the higher the negative moment and lower the positive moment. The overhang can be adjusted such that the overall moment in the beam is the minimum. This occurs when the numerical value of the positive moment is exactly equal to that of the negative moment. Consider the simple beams with variable length, cantilever(s) on one and both ends. In the absence of overhangs, i.e. simply supported beam the maximum moment is $wL^2/8$ (= $0.125wL^2$) at the mid-span. If the ends are cantilevered by moving in both the supports by an amount z, then

Negative moment at the support, $$M_{\text{negative}} = \frac{wz^2}{2}$$

Positive moment at the mid-span, $$M_{\text{positive}} = \frac{wL}{8}(L - 4z)$$

For optimum support locations, $$M_{\text{negative}} = M_{\text{positive}}$$

$$\frac{wz^2}{2} = \frac{wL}{8}(L - 4z) \quad \text{or} \quad z^2 + zL - \frac{L^2}{4} = 0$$

or $$z = \left(\frac{\sqrt{2} - 1}{2}\right)L = 0.207L \tag{21.2}$$

Therefore, $$M_{\text{design}} = \frac{w(0.207L)^2}{2} = 0.02145\,wL^2 \tag{21.3}$$

The optimum overhang is approximately one-fifth of the overall span. If the beam is cantilevered at one end by moving in one support by an amount z and for a balance between the positive and negative moments

$$\frac{wz^2}{2} = \frac{wL^2}{8}\left(\frac{L - 2z}{L - z}\right)^2 \quad \text{or} \quad z^2 - 2zL + \frac{L^2}{2} = 0$$

or $$z = \left(1 - \frac{1}{\sqrt{2}}\right)L = 0.2929\,L \tag{21.4}$$

Thus, $$M_{\text{design}} = \frac{w(0.2929L)^2}{2} = 0.0429\,wL^2 \tag{21.5}$$

The optimum overhang at one end is approximately three-tenth of the span. It should be noted that the values of overhang obtained in two cases sum up to 0.5L. There is significant reduction in moments and consequent saving in material will result. However, the optimum overhangs obtained above should be adjusted taking into consideration other factors like amount of deflection at the end of the cantilever, type of construction, etc.

21.2.5 Plastic or Ultimate Strength Design of Steel Beams

An idealized stress–strain curve for a ductile or *elastic-perfectly plastic* material where *plastic flow* or *infinite strain* can occur with no increase in the stress in the material beyond the yield point is illustrated in Figure 19.1(b). As explained in Chapter 19, at a certain stress level σ_{yp}, the material begins to undergo increased deformations without any additional increase in stress level. Actual rupture in the material does not occur until relatively large deformations occur. These are the *plastic deformations*. In the initial stages of loading when the external loads on the beam are low, the material in the beam is in the elastic range and bending stresses are linearly distributed across the cross-section. As loads increase, the bending stresses increase until the material at the extreme fibres reaches the yield point σ_{yp} where it begins to yield and enter the plastic range. In case of a rectangular beam of size $b \times d$ at the yield point, i.e. at the stage when outermost fibres just begin to yield, the resisting moment in the beam corresponds to $M_y = \sigma_{yp}(bd^2/6)$. Further increase in loads causes increased deformations in the beam fibres without a corresponding increase in the stress level in the material in the region where the beam fibres are deformed into the plastic region. At this stage, some fibres near the neutral axis are still below σ_{yp} and beam is still capable of carrying loads. Increasing the external load further causes increased deformations until all fibres in the cross-section begin to yield. *The load that finally causes the fibres nearest the neutral axis to yield is the maximum load the beam can support.* This load corresponds to a maximum resisting moment of $M_p = \sigma_{yp}(bd^2/4)$. The term $bd^2/4$ is generally termed *plastic sectional modulus, S*. Beyond this point, the beam has no further capacity to resist the external loads, and the further increase in loads would cause additional deformations until it ruptures and collapses. A *plastic hinge* is said to develop when all fibres are fully yielded at a cross-section.

For a rectangular section, the ratio s_f between the moment associated with the formation of a plastic hinge and that associated with the initial yielding,

$$s_f = M_p / M_y = [\sigma_{yp}(bd^2/4)]/[\sigma_{yp}(bd^2/6)] = 1.5 \tag{21.6}$$

This ratio often called *shape factor* for a beam, implies that the load W_f required to make the beam fail is 1.5 times the load W_y at which initial yielding to occur in the member. For a wide-flange beam, the shape factors vary but average around 1.14. Shape factors for a number of sections are listed in Table 19.1.

In *statically indeterminate structures*, the formation of a single plastic hinge may not lead directly to beam collapse. A number of hinges must form until a *collapse mechanism* is created. For illustration, consider the fixed-ended beam of Example 19.4 carrying uniformly distributed load of intensity *w*. As the load *w* is increased, the beam begins yielding at the ends where moments are the maximum. With the formation of plastic hinges at these points, the beam reduces to a simply supported beam with a fixed moment resistance at each end. It is seen that the formation of plastic hinges at the ends of the beam do not cause collapse to occur in the beam, since the structure still has moment-carrying capacity at mid-span. Further beam loading would eventually cause a plastic hinge to develop at mid-span. A collapse mechanism would then exist and beam would carry no additional load and fail. The failure load w_u of the beam is greater than would be implied by a consideration of just the shape factor of the beam alone.

21.2.6 Deflections

The deflection δ depends on W or w, L and EI and can be expressed as:

$$\delta = C_1\left(\frac{wL^4}{EI}\right) \text{ or } C_2\left(\frac{WL^3}{EI}\right) \tag{21.7}$$

where C_1 and C_2 are the constants depending upon supports conditions. For a simply supported beam carrying a uniformly distributed load, C_1 is 5/384 and the maximum deflection becomes: $\delta = \left(\frac{5wL^4}{384EI}\right)$, the deflections are highly sensitive to the length of the beam. If the ends this member are fixed rather than simply supported, the deflection is given by: $\delta = \left(\frac{wL^4}{384EI}\right)$. Thus, fixing the ends of a simply supported beam carrying a uniformly distributed load reduces the mid-span deflection by a factor of 5. The increased rigidity associated with fixed ended beams, or also continuous beams, is one of the primary reasons for using these members extensively.

Empirical guidelines are often used to control the magnitude of deflection of the beam to the prescribed limit. The limit on deflection is based on the criteria that it should neither cause discomfort to the occupant nor interfere with or impair functioning of another building element, i.e. allowable deflection could be based on acceptable tolerance for other systems. A commonly used criterion to control the deflection is that the deflection of floor should not exceed 1/360 of its span subjected to live load only. For the total gravity load consisting of both live and dead loads deflections are limited to $L/240$ of the span. If a member deflects more than that prescribed by these guidelines, it is usually considered not acceptable and other measures need be taken such as adopting a member of increased stiffness. The stiffness required of a member can be determined by equating the appropriate theoretical deflection to the allowable maximum deflection, e.g. the stiffness of a simply supported beam carrying uniformly distributed load for the allowable deflection of $L/240$ is given by,

$$\delta_{\text{allowable}} = \frac{L}{240} = \frac{5wL^4}{384EI} \text{ or } EI = \frac{5\times 240\times wL^4}{384L} = 3.125wL^3 \tag{21.8}$$

EXAMPLE 21.1

A simply supported rectangular timber beam of 230 mm width carries a uniformly distributed load of 5.0 kN/m over an effective span of 6.0 m. The beam is initially cambered upward so that it is level under dead load. Determine the depth of the beam so that the maximum live-load deflection does not exceed $L/360$. The modulus of elasticity of the timber $E = 1.080 \times 10^4$ MPa.

Solution: Allowable deflection under live-load,

$$\delta_{\text{allowable}} = \frac{L}{360} = \frac{5wL^4}{384EI}$$

or $$I = \frac{5 \times 360 \times wL^3}{384E} = \frac{5 \times 360 \times (5 \times 10^3 / 10^3) \times (6000)^3}{384 \times (1.080 \times 10^4)} = 4.6875 \times 10^8 \text{ mm}^4$$

Therefore, $$I = \frac{bd^3}{12} = \frac{230 \times d^3}{12} = 4.6875 \times 10^8 \quad \text{or} \quad d = 290.27 \text{ mm}$$

The following example illustrates the application the stress parameters discussed above.

EXAMPLE 21.2

Design timber floor beams of rectangular cross-section spaced 400 mm centre-to-centre to support uniformly distributed dead and live loads of 0.75 kN/m² (inclusive of weight of flooring deck system and self-weight of the beam, etc) and 2.5 kN/m², respectively. The allowable stresses in bending, shear, and bearing are 8.5, 1.0 and 2.75 MPa, respectively. The allowable deflections for live-load and live-plus-dead loads are $L/360$ and $L/240$, respectively. Determine appropriate depth of the 40 mm wide beams when their span is 5.0 m. The timber available in stock is 45 × 290 mm. The elastic modulus for the timber used is $E = 1.1 \times 10^4$ MPa.

Solution: For the timber used for the floor beam the stress parameters are:

$$\sigma_{b,a} = 8.5 \text{ MPa}; \ \tau = 1.0 \text{ MPa}; \ \sigma_{bg} = 2.75 \text{ MPa and } E = 1.1 \times 10^4 \text{ MPa}$$

Loads: w = Distributed load × Width of floor deck carried by one beam

$$= (2.5 + 0.75) \times 0.4 = 1.3 \text{ kN/m}$$

Design for bending stresses:

$$M_{max} = \frac{wL^3}{8} = \frac{1.3 \times 5^2}{8} = 4.0625 \text{ kN.m}$$

$$Z = \frac{bd^2}{6} = \frac{M_{max}}{\sigma_{b,a}}$$

or $$bd^2 = \frac{6M_{max}}{\sigma_{b,a}} = \frac{6 \times 4.0625 \times 10^6}{8.5} = 2.868 \times 10^6 \text{ mm}^3$$

For $b = 40$ mm, $d = 267.76$ mm. A beam of 40 × 270 mm size is adequate. However, keeping in view the availability of timber, provide 45 × 290 mm over-sized beams.

Check for shearing stresses: Maximum shearing force which occurs at the supports,

$$V_{max} = \frac{wL}{2} = \frac{1.3 \times 5}{2} = 3.25 \text{ kN}$$

Maximum shearing stress in the beam

$$\tau_{max} = \frac{3}{2}\left(\frac{V_{max}}{bd}\right) = \frac{3}{2}\left(\frac{3.25 \times 10^3}{45 \times 290}\right) = 0.374 \text{ MPa} < 1.0 \text{ MPa}$$

Thus, the beam size is adequate from shear considerations.

Check for bearing stresses:

$$\text{Length of bearing required} = \frac{V_{max}}{b\sigma_{bg}} = \frac{3.25 \times 10^3}{45 \times 2.75} = 26.26 \text{ mm}$$

Thus, bearing area of 45 × 30 mm is adequate. However, from practical considerations provide 45 × 50 mm as the bearing area and beam is under-stressed at the reactions.

Check for deflection:

Live-load deflection,

$$\delta_l = \frac{5w_l L^4}{384EI} = \frac{5 \times (2.5 \times 0.4) \times 5000^4}{384 \times (1.1 \times 10^4)} \left(\frac{12}{45 \times 290^3} \right)$$

$$= 8.09 \text{ mm} < \frac{L}{360} (= 13.89 \text{ mm})$$

Live- and dead-load deflection,

$$\delta = \frac{5wL^4}{384EI} = \frac{5 \times 1.3 \times 5000^4}{384 \times (1.1 \times 10^4)} \left(\frac{12}{45 \times 290^3} \right)$$

$$= 10.52 \text{ mm} < \frac{L}{240} (= 20.83 \text{ mm})$$

Thus, the beam meets all the deflection criteria.

Check for lateral stability:

Thin and deep beams are susceptible to lateral buckling and failure may occur much before the strength related failures. The beams with depth-to-width ratios of between 5:1 and 7:1 fall in this category. It is general practice in such cases to provide lateral bracing of appropriate type to ensure lateral stability. In the current example, the depth-to-width ratio is 6.44 which necessitates the provision of appropriate lateral bracing.

21.3 IMPACT OR DYNAMIC LOADING

The deformations produced in elastic bodies by impact loads cause them to act as springs as shown in Figure 21.1(a). As discussed earlier, the *equivalent spring constant* for such a body can be taken as the load to cause a unit deformation. For illustration, consider the problem of impact as analogous to that of a falling body stopped by a spring with constant k as shown in Figure 21.1(b). The mass m has zero velocity when first dropped from height h, and also when the spring is deflected through the maximum dynamic deflection δ. Since the work done by the externally applied load during loading process W_{ex} is equal to the increase in strain energy stored in the system, i.e. the work done by the internal forces,

$$W_{ex} = U = -W_{in}$$

where W_{in} has a negative sign because the deformations are opposed

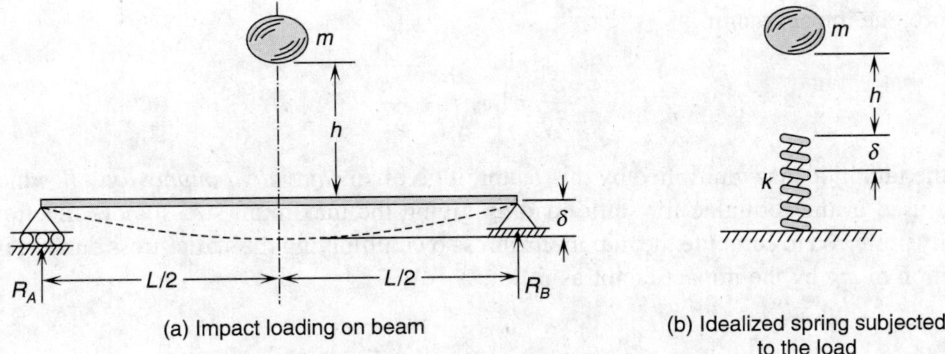

(a) Impact loading on beam

(b) Idealized spring subjected to the load

Figure 21.1 Impact loading on a structure.

$$mg(h+\delta) - \frac{1}{2}k\delta^2 = 0 \tag{21.9}$$

where $mg(h + \delta)$ is the work done by falling load on the body, and $(k\delta^2)/2$ is the strain energy stored by the equivalent spring. Rearranging Eq. (21.9) in the form

$$\delta^2 - 2\frac{mg}{k}\delta - 2\frac{mg}{k}h = 0$$

and replacing the term mg/k with δ_{st}, which is the static deformation produced by a gradual application of the load mg, Eq. (21.9) reduces to

$$\delta^2 - 2\delta_{st}\delta - 2\delta_{st}h = 0 \tag{21.10}$$

Thus the general value of dynamic deflection δ, is given by

$$\delta = \delta_{st} + \sqrt{(\delta_{st})^2 + 2\delta_{st}h} \tag{21.11}$$

Now two cases arise: (a) when h is large compared to δ_{st}, the work term $\delta_{st}\delta$ in Eq.(21.10) can be ignored, i.e.

$$\delta = \sqrt{2\delta_{st}h} \tag{21.12}$$

(b) when $h = 0$ (i.e., the load is suddenly applied) and Eq. (21.10) reduces to

$$\delta = 2\delta_{st} \tag{21.13}$$

Thus due to a suddenly applied load, the deflection and consequently the stress, which is directly proportional to it, are twice that caused by the same load applied gradually.

The ratio of the *maximum dynamic deformation* δ to the *static deformation* δ_{st} is called the *impact factor*. From Eq. (21.11)

$$\delta = \delta_{st} + \delta_{st}\sqrt{1+\frac{2h}{\delta_{st}}} = \delta_{st}\left(1+\sqrt{1+\frac{2h}{\delta_{st}}}\right)$$

Therefore, the impact factor is

$$\frac{\delta}{\delta_{st}} = 1 + \sqrt{1 + \frac{2h}{\delta_{st}}} \tag{21.14}$$

When the falling load is multiplied by this factor, it gives an *equivalent impact load W* which may then be used in the formulae for static loading giving the maximum stress and deflection. It is generally preferred to compute the maximum stress by multiplying the static stress due to a gradual application of *mg* by the impact factor as follows:

$$\sigma_{\max} = \sigma_{st}\left(1 + \sqrt{1 + \frac{2h}{\delta_{st}}}\right) \tag{21.15}$$

The above results can be applied to various types of impact loadings.

21.3.1 Tension

Consider the most commonly encountered case of impact loading where a weight *mg* drops freely through a height *h* before striking a stop on the end of the rod as shown in Figure 21.2. Assuming that the stress remains within the elastic range and that δ is very small compared to *h*, substitute $\delta_{st} = mgL/AE$, in Eq. (21.12)

$$\delta = \sqrt{2\delta_{st}h} = \sqrt{2\left(\frac{mgL}{AE}\right)h} \tag{21.16}$$

The corresponding stress in the rod is,

$$\sigma = \varepsilon E = \left(\frac{\delta}{L}\right)E = \sqrt{\left(\frac{2E}{AL}\right)mgh} \tag{21.17}$$

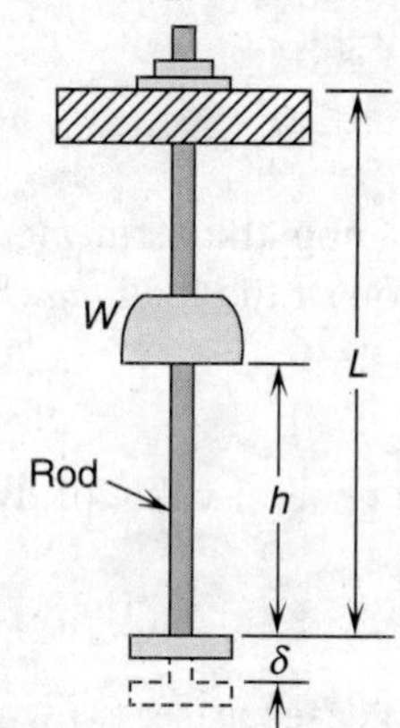

Figure 21.2 Impact loading on a rod.

Since the potential energy *mgh* is equal to the kinetic energy $mv^2/2$, Eq. (21.17) can also be used to determine the *shock stress* caused by the sudden stopping of mass *m* that is moving with a velocity *v*:

$$\sigma = \sqrt{\left(\frac{2E}{AL}\right) \times \left(\frac{mv^2}{2}\right)} \tag{21.18}$$

It should be noted from Eq. (21.17) that *the stress due to impact can be reduced by using a material with a lower value of E* or by increasing the area *A* or the length *L* of the rod. This is quite different from the case of static tension where the stress is independent of *E* or *L*.

In the foregoing discussion, it has been assumed that the stress remains below the proportionality limit. However, the case where the stress is above the proportionality limit can still be treated by considering that the shape of tensile test diagram does not depend on the rate with which the bar is strained, the area OABC in Figure 21.3(a) represents the work done on the bar to produce an elongation δ; this must be equal to the work $mg(h + \delta)$ done by the falling weight mg. When $mg(h + \delta)$ equals or exceeds the total area OADE, the falling weight will rupture the bar.

The resistance of a bar to impact also depends on the ductility of the material. In Figure 21.3(b) the tensile test diagram of a high strength steel of low *ductility* is superimposed on the diagram for a steel of lower strength but high ductility. The horizontally shaded area A_I is much larger than the vertically shaded area A_{II}, indicating that the *steel with higher ductility will absorb more energy before rupture than the less ductile steel*. For this reason, ductile materials are usually selected for members subject to *impact* or *shock loading*.

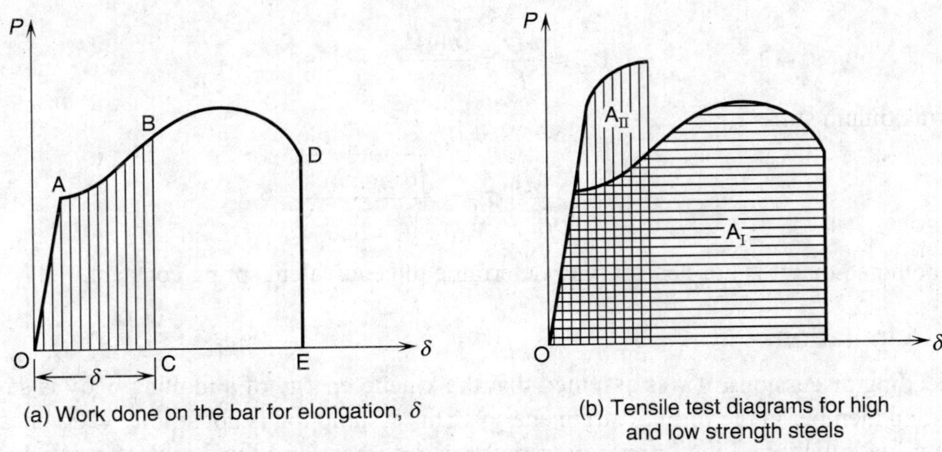

(a) Work done on the bar for elongation, δ

(b) Tensile test diagrams for high and low strength steels

Figure 21.3 Work done on the bars of different strengths and ductility.

The total area of a stress–strain diagram is called the *modulus of toughness;* it represents the energy absorbed per unit volume and its value is approximately equal to

$$U_t = \left(\frac{\sigma_y + \sigma_u}{2}\right) \times \varepsilon_u \tag{21.19}$$

where σ_y and σ_u are, respectively, the yield point and the ultimate strength, and ε_u is the ultimate strain. The partial area of a stress–strain diagram up to the stress σ_e at the elastic limit is called the *modulus of resilience;* it represents the energy that can be absorbed per unit volume without creating a permanent distortion. For a material with a linear stress–strain relation, its value is

$$U = \frac{1}{2}(\sigma_e \varepsilon_e) = \frac{1}{2}(\sigma_e)\left(\frac{\sigma_e}{E}\right) = \frac{\sigma_e^2}{2E} \tag{21.20}$$

21.3.2 Flexure

For illustration, consider the simply supported beam subjected to the impact of a load mg falling freely through the height h before striking the mid-point of the beam as shown in Figure 21.1(a). If the proportionality limit is not exceeded and h is large compared with δ_{st}, the dynamic deflection δ can be obtained from Eq. (21.12):

$$\delta = \sqrt{2h\delta_{st}}$$

Therefore, the *impact factor* for a simply supported beam carrying concentrated load *mg* at its mid-point ($\delta_{st} = mgL^3/48EI$) is given by

$$\frac{\delta}{\delta_{st}} = \sqrt{\frac{2h}{\delta_{st}}} = \sqrt{\frac{96EIh}{mgL^3}} \tag{21.21}$$

The static stress obtained from the flexure formula is

$$\sigma_{st} = \frac{My}{I} = \frac{(mg)L}{4} \times \frac{c}{I}$$

and the maximum stress is

$$\sigma_{\max} = \left(\frac{\delta}{\delta_{st}}\right)\sigma_{st} = \sqrt{\frac{6\,mgh\,Ec^2}{LI}} \tag{21.22}$$

It must be noted that it is not necessary to determine the equivalent spring constant.

21.3.3 Limitations

The preceding discussions, it was assumed that the kinetic energy of a moving body is stored in the resisting member in the form of strain energy. This assumption is not strictly valid because in the following situations kinetic energy transforms into heat and local inelastic deformations of both the moving body and the resisting member if: (i) the velocity of impact is high and the de-acceleration of the moving body is rapid, (ii) the velocity of impact is low and the resisting member possesses high stiffness, and (iii) the mass of the resisting member is large compared to that of the moving body, resulting in high inertia of the resisting member.

21.4 OTHER MECHANICAL PROPERTIES OF MATERIALS

21.4.1 Ductility

The materials that exhibit large plastic deformations before they rupture are generally referred to as *ductile materials*. Steel is a classic example of ductile material. The measure of ductility is the percentage elongation δ at rupture. The greater the δ, the more ductile is the material. Highly ductile materials include annealed copper, aluminium, brass, low carbon steel, etc. The low ductility materials include many alloy steels. On the other hand, a material which does not exhibit plastic behaviour under load but rather ruptures with little evident deformation is said to be a *brittle*. For such materials, the amount of elongation at rupture generally does not exceed 2 to 5 per cent. The typical examples of brittle materials are cast iron, high carbon tool steel, glass, brick, stone, plain concrete, etc. The stress–strain curves shown in Figures 21.3(a) and (b) illustrate the behaviour of the two materials under-loads.

The ductility of a material is an important consideration for structural design since the plastic region of the stress–strain curve represents a measure of *reserve strength*. Design or allowable stress levels are invariably set below the *yield stress* of the material and are well within the elastic range. For example, if a structure of a ductile material is loaded beyond its anticipated design levels, the stresses and strains increase until the yield point of the material is reached. At this point, the material yields but does not physically rupture but structure undergoes large deformations associated with the moving of the material into the plastic range. These deformations are visible compared to design deflections, and can serve as a visual warning of the impending failure. However, the member will fail only after the *ultimate strength* of the material is reached.

The structural members made of brittle materials do not exhibit plastic behaviour and do not visibly deform to any great degree prior to failure and thus give no advance warning of impending collapse. Though the concrete is a brittle material, the overall member can be designed to have a measure of ductility when used in conjunction with ductile steel.

Low temperatures and higher rate of load application on a structure decrease the plastic deformations in many ductile materials. The proportionality limits and yield point stresses often increase with increasing strain rates.

Under rapid and *impact loading*, *resilience* and *toughness* are important material properties. The *resilience defines the ability of the materials to absorb energy without undergoing plastic strains*. The *area of the elastic region under a stress–strain diagram represents the density of strain energy that can be absorbed without any permanent damage to the material*. This area is called modulus of resilience. *Toughness* on the other hand *defines the ability of the material to absorb energy prior to fracture*, thus the area under the stress–strain diagram represents the density of strain energy absorbed by material prior to fracture. The area under the complete monotonic stress–strain diagram is called modulus of toughness.

21.4.2 Creep

The term *creep refers to the time dependent continued permanent deformation of material under sustained stress level*. The plain concrete and plastics have large creep deformations whereas steel does not creep at all. Large plastic strains and, finally, rupture can result from creep. The creep can cause unfavourable stress redistribution in reinforced concrete members and long-term deflections increase significantly.

21.4.3 Repeated Loading: Fatigue

Many structural components are subjected to varying stresses caused by repeated loading and unloading. Components subjected to repeated stress cycles may fail at relatively low stress level, even less than the elastic strength of the material as determined by a static test. Failures of this type are known as *fatigue failures*.

Testing to determine these values is called *fatigue testing*. The simplest example of repeated loading and unloading is that of reverse bending. For example, consider a beam subjected to bending moments by a load W applied at its mid-point. As the load is reversed, a fiber that was originally on top passes from compression to tension and back to compression as the load is reapplied in the initial direction, thereby undergoing a complete reversal of stress for each revolution. The number of revolutions is noted until failure occurs. A typical result obtained in this stress reversal test is shown in the σ–N diagram in Figure 21.4, where stress vs. the number of

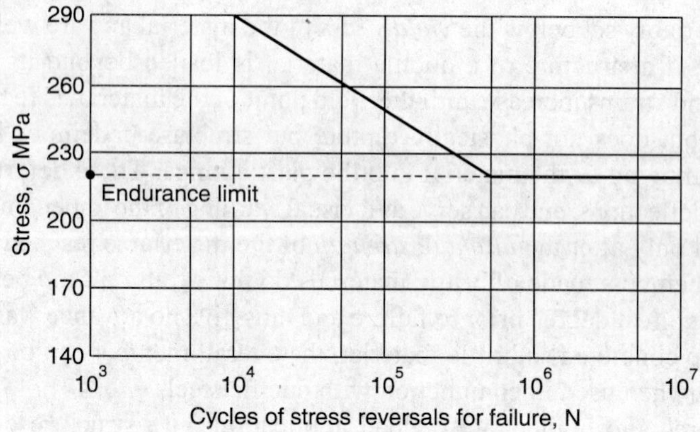

Figure 21.4 Stress versus number of cycles diagram on semi-logarithmic scales.

cycles is plotted with semi-logarithmic scales. The point at which this diagram levels off is called the endurance limit which is generally between 40 to 50 per cent of the ultimate strength. The endurance limit is denoted by σ_e. The *endurance limit* of a material is the maximum unit stress the material can withstand for an infinite number of cycles without failure. Most ferrous materials such as steel have well-defined endurance limits whereas, nonferrous materials such as aluminum do not.

When a ductile steel specimen is subjected to a gradually increasing load, yielding of the specimen is evident before actual failure occurs. But a specimen of the same material that is subjected to stress reversals fails suddenly, without any plastic deformation or any other warning. Thus the *fatigue failure of ductile steel is similar to* the *static failure of a brittle material*.

Fatigue failure can be explained satisfactorily by the *localized stress theory*, which is based on the stress concentrations that occur (a) inside a material because of discontinuities in the material itself and (b) at the surface of a material because of abrupt changes in section. These stress concentrations are not serious when a ductile material is subjected to a static load; but when the load is repeatedly applied, they cause minute cracks that spread with each repetition until the member suddenly fractures.

21.4.4 Stress Concentration

Sometimes in a structure where micro-cracks or flaws exist very high stresses often develop over a small area. These are called *stress concentrations*. In brittle materials, the stress concentrations cracks continue to propagate until the materials fail. In ductile materials local deformations at the stress concentration points relieve stresses by *redistribution*, and possibility of *crack propagation* is reduced.

It has been noticed during experimental investigations that stress concentrations occur inside a material because of discontinuities in the crystalline structure of the material and also at the surface of a material due to abrupt changes in section. For illustration, consider the effect of abrupt change in section on the stress distribution by providing a small circular hole of diameter d in a rectangular plate subjected to a uniform tensile stress σ_o as shown in Figure 21.5(a). The presence of hole changes the stress distribution across a section through the hole as shown by the shaded area. The stress at distance r from the centre of the hole is expressed by

$$\sigma_o' = \frac{\sigma_o}{2}\left[2 + \frac{1}{4}\left(\frac{d}{r}\right)^2 + \frac{3}{16}\left(\frac{d}{r}\right)^4\right] \tag{21.23}$$

From Eq. (21.23) for stress distribution across a section through the hole it should be noted that the stresses at points *I* and *II* to be 3σ. Because of bending action around the hole, compressive stresses of magnitude σ_o are created at the tip and bottom points of the hole.

The stress concentration caused by the small elliptical hole is shown in Figure 21.5(b). The maximum stress at the ends of the horizontal axis of the hole is given by

$$\sigma_{\max} = \sigma_o\left[1 + 2\left(\frac{b}{a}\right)\right] \tag{21.24}$$

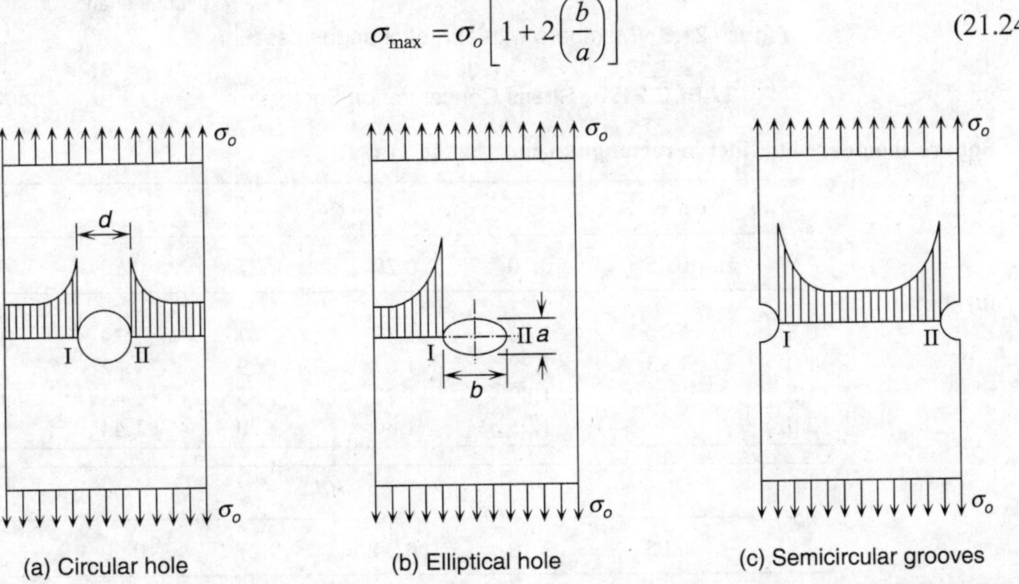

Figure 21.5 Stress concentration in members of rectangular cross-section with various types of discontinuities.

It should be noted that stress increases with the ratio b/a; hence a very high stress concentration is produced by a narrow hole or crack perpendicular to the direction of the tensile stress, and these cracks tend to spread. This spreading may be stopped by drilling small holes at the ends of the crack, thus replacing a high-stress concentration by a relatively smaller one.

Small semicircular grooves in a plate shown in Figure 21.5(c) produce stress concentrations at points *I* and *II* that are about three times the average stress σ_o applied at the ends of the plate.

The maximum stresses due to axial, torsional and flexural loads are generally given by

$$\sigma_o = k\frac{P}{A}, \quad \tau = k\frac{Tr}{J}, \quad \sigma_b = k\frac{Mc}{I} \tag{21.25}$$

where k represents the stress concentration factor. The values of stress concentration factors for several cases of abrupt change in section shown in Figure 21.6 are listed in Table 21.1.

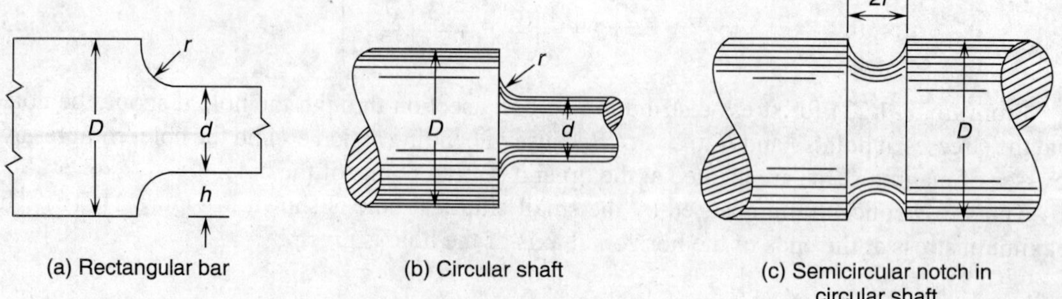

Figure 21.6 Members with abrupt change in section.

TABLE 21.1 Stress Concentration Factors

I. Square shoulder with fillet in rectangular bar [Figure 21.6(a)]

(a) Tension

h/r	r/d 0.05	0.10	0.20	0.27	0.50	1.0
0.5	1.70	1.60	1.53	1.47	1.39	1.21
1.0	1.93	1.78	1.67	1.59	1.42	1.22
1.5	—	1.89	1.72	1.65	1.43	1.23
2.0	—	1.95	1.80	1.70	1.44	1.23

(b) Bending

h/r	r/d 0.05	0.10	0.20	0.27	0.50	1.0
0.5	1.61	1.49	1.39	1.34	1.22	1.07
1.0	1.91	1.70	1.48	1.38	1.22	1.08
1.5	2.00	1.73	1.50	1.39	1.23	1.08
2.0	—	1.74	1.52	1.39	1.23	1.09

II. Square shoulder with fillet in circular shaft [Figure 21.6(b)]

(a) Tension: Approximately same as case I(a)

(b) Bending: Approximately same as case I(b)

(c) Torsion

D/d	r/d 0.005	0.01	0.02	0.03	0.04	0.10
2.00	—	3.0	2.25	2.00	1.82	1.44
1.33	—	2.7	2.16	1.91	1.76	1.40
1.20	3.00	2.5	2.00	1.75	1.62	1.34
1.09	2.20	1.88	1.53	1.40	1.30	1.15

III. Semicircular notch in circular shaft [Figure 21.6(c)]

(a) Tension

$\dfrac{r}{D-2r}$	0.05	0.15	0.30	0.40	0.52	0.75
k	2.57	2.16	1.81	1.65	1.51	1.36

(b) Bending

$\dfrac{r}{D-2r}$	0.05	0.10	0.20	0.30	0.50	0.75
k	2.20	1.86	1.59	1.45	1.30	1.18

(c) Torsion $k = \dfrac{2D}{D+2r}$

21.5 FINITE-ELEMENT ANALYSES

The types of stress and deformation analyses discussed in the preceding chapters are the basic tools commonly used for analyzing typical structures. However, many solid structures having complex geometries or boundary conditions or subjected to difficult loading conditions demand more sophisticated treatments. Such cases are usually handled easily by using computer-based *finite-element methods*. In this approach a continuous solid structure is replaced by a conforming meshed network of interconnected discrete elements of varying shapes and sizes. Forces are applied at connecting or nodal points. Energy-based structural analysis techniques and various compatibility requirements are then used to predict forces and displacements. Finite-element analysis results include various kinds of stress analyses (typically principal stresses), strain distributions, and displacement analyses. The finite-element approach is highly valuable in treating the complex volumetric solid structures, thin-shells and hybrid structures. Though versatile and highly valuable the finite element method is outside the scope of this text.

21.6 PROBLEMS

21.1 A simply supported beam of length L having a rectangular cross-section of width b and depth d is struck at the mid-span by a weight W falling through a height h. Show that the maximum bending stress developed in the beam is given by, $\sigma_{b,\max}^2 = 18WhE/bdL$.

21.2 A 4 m long simply supported beam having a rectangular cross-section 40 mm wide and 80 mm deep is struck onto a point 1.0 m from one end by a weight 9 kN falling through a height of 2.5 m. Compute the impact factor when the modulus of elasticity of the material of the beam is 200 GPa. Neglect the mass of the beam.

[**Ans.** $\delta/\delta_{st} \approx 17.0$]

Geometrical Characteristics of Cross-sections

A.1 INTRODUCTION

While dealing with the engineering problems relating to the strength of beams namely bending, torsion and shear; columns and deflection of beams, the knowledge of geometrical characteristics of the cross-sections of the members is essential.

A.2 CENTRE OF GRAVITY OR CENTROID

The concept of centre of gravity in plane areas (figures) is based on the centroid of a uniform lamina, a thin plate of uniform thickness, considered to consist of an infinite number of particles lying in the plane of lamina. Let the masses of the various particles be m_1, m_2, ... and their co-ordinates be (x_1, y_1), (x_2, y_2), ..., respectively, referred to a set of two rectangular reference axes OX and OY as shown in Figure A.1. The weights of these particles m_1g, m_2g, ... form a system of parallel forces and their resultant passing through a point C represents the weight of whole lamina (of mass M) Mg.

For rotational equilibrium the sum of the moments of the system of forces equals the moment of the resultant about the same axis. The point C in the plane of lamina through which the resultant passes is called the *centre of gravity* and its co-ordinates are denoted by $(\overline{x}, \overline{y})$.

Consider the moments of the system of forces about the axis OY

$$m_1gx_1 + m_2gx_2 + m_3gx_3 + ... = (m_1g + m_2g + m_3g + ...)\,\overline{x}$$

Therefore,

$$\overline{x} = \frac{m_1gx_1 + m_2gx_2 + m_3gx_3 + \cdots}{m_1g + m_2g + m_3g + \cdots} \tag{A.1}$$

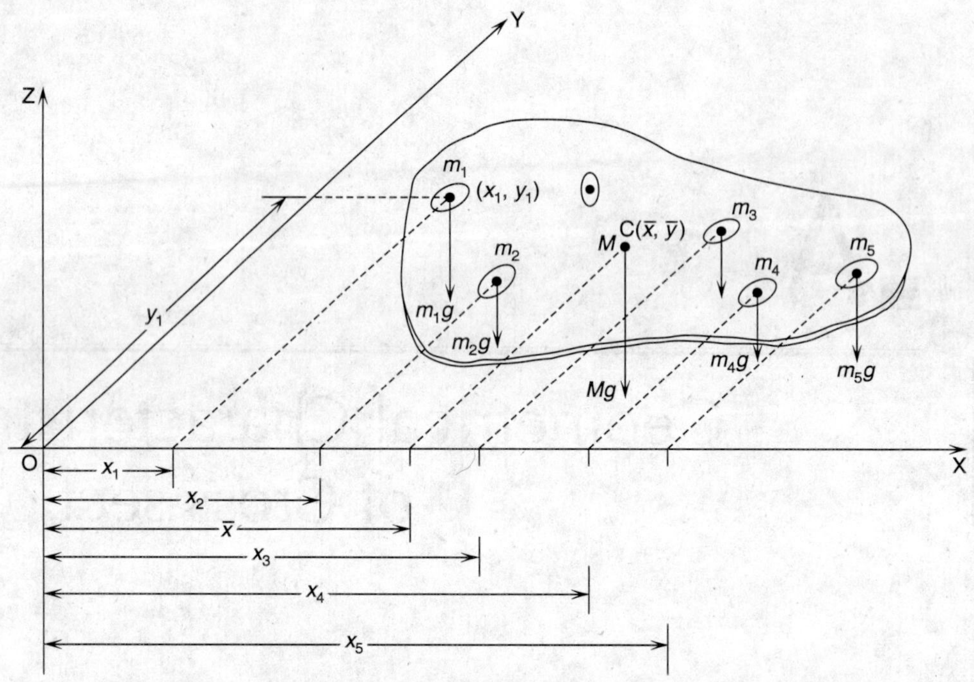

Figure A.1 A lamina of uniform thickness consisting of infinite particles.

In case of uniform lamina, the masses of the elements are proportional to their areas. If the surface density is w per unit area, then

$$\overline{x} = \frac{A_1 wgx_1 + A_2 wgx_2 + A_3 wgx_3 + \cdots}{A_1 wg + A_2 wg + A_3 wg + \cdots}$$

$$= \frac{A_1 x_1 + A_2 x_2 + A_3 x_3 + \cdots}{A_1 + A_2 + A_3 + \cdots} = \frac{\sum_i A_i x_i}{\sum_i A_i} = \frac{\sum_i A_i x_i}{A} \tag{A.2}$$

Similarly,

$$\overline{y} = \frac{A_1 y_1 + A_2 y_2 + \cdots}{A_1 + A_2 + \cdots} = \frac{\sum_i A_i y_i}{\sum_i A_i} = \frac{\sum_i A_i y_i}{A} \tag{A.3}$$

where A is the total area of the figure. For a continuous area, the co-ordinates of centroid can be written as

$$\overline{x} = \frac{\int_A x\, dA}{\int_A dA} \quad \text{and} \quad \overline{y} = \frac{\int_A y\, dA}{\int_A dA} \tag{A.4}$$

In the above mathematical expressions x and y represent the perpendicular distances from an element area dA to the reference axes. Thus, the centroid of a plane figure is that point in the figure at which all the area (mass) may be assumed to be concentrated or distributed. The point representing centroid of a plane figure is denoted by C.

It should be noted that the moment of an area has no real meaning when considered in isolation. However, when used in combination with mass of elements, it assumes significance.

The Eqs. (A.4) are used to compute the co-ordinates of the centre of gravity $(\overline{x}, \overline{y})$ of an area. However, for certain shapes like semi-circle in which the centre of gravity lies on its axis of symmetry at a distance from its diameter, the Pappus–Guldinus theorem can be used to obtain the distance of centre of gravity. According to this theorem, rotate the semi-circle about its diameter to obtain a solid of revolution, i.e. sphere whose volume is equal to $2\pi\overline{y}$ multiplied by the area of semi-circle

$$2\pi\overline{y} \times \frac{\pi R^2}{2} = \frac{4}{3}\pi R^3 \quad \text{or} \quad \overline{y} = \frac{4R}{3\pi} \tag{A.5}$$

A.3 MOMENT OF INERTIA

The engineering problems related to strength of beams (both bending and shear), columns and deflection of beams, involve the use of mathematical expression of the form $\int_A \rho^2 \, dA$, where ρ is the perpendicular distance from an element area dA to the reference axis. Based on the similarity of this expression to that for moment of inertia of the mass $\int_M \rho^2 \, dM$, it is termed as moment of inertia of the area. As in the case of the moment of an area, the moment of inertia applied to areas has no real meaning when considered by itself, it is merely a mathematical expression usually denoted by the symbol I. However, when used in combination with other terms, e.g. in the bending stress formula for beam $\sigma_b = \dfrac{M\rho}{I}$, it begins to have significance.

The mathematical definition of moment of inertia, $I = \int_A \rho^2 \, dA$ suggests that an area is divided into small elements as dA and each elemental area is multiplied by the square of the moment arm about the reference axis.

Consider cross-section of a beam shown in Figure A.2 referred to a co-ordinate system (X, Y). If the co-ordinates of the centre of the differential area dA are (x, y), then the moment of inertia I_x of the figure about the X-axis is the summation of the product of the each area and the square of the distance of the centroid from the reference axis or moment arm y. Thus,

$$I_x = \sum_i dA\, y^2 = \int_A y^2 \, dA \tag{A.6}$$

Similarly, the moment of inertia I_y of a figure about the Y-axis is

$$I_y = \sum_i dA\, x^2 = \int_A x^2 \, dA \tag{A.7}$$

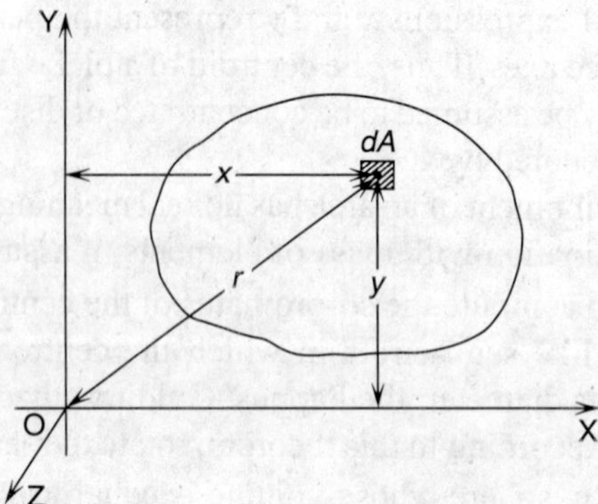

Figure A.2 Reference axes and symbols for moment and product of inertia.

The moment of inertia (of an area), I is sometimes called the *second moment of area*, as each differential area multiplied by its moment arm gives the moment of the area which when multiplied a second time by its moment arm, gives the second moment of the area.

It should be noted that the term $\int_A \rho^2 \, dA$ defining the moment of inertia is four-dimensional, as it consists of a distance squared multiplied by an area. Thus, if the dimensions of a figure are given in millimeters, then the unit of the moment of inertia is mm^4. Further, the moment of inertia is independent of the sign of moment arm as its square automatically makes it plus, it depends entirely on the sign of area. The positive area is taken as the one that adds to the area of the figure, and the negative area is the one that reduces the area of the figure. For the net area, the moment of inertia is always positive.

A.3.1 Radius of Gyration

The term *radius of gyration*, which is most frequently used in column problems, describes another mathematical property. The radius of gyration is denoted either by symbol k (in structural specifications) or the symbol r (in column theory) and is defined by the relation

$$I = \int \rho^2 \, dA = r^2 \int dA = r^2 \sum dA = Ar^2$$

Therefore, the radius of gyration is

$$r = \sqrt{(I / A)} \tag{A.8}$$

where A is the total area of the figure. The radius of gyration as defined above implies that the areas of all elements are at uniform distance from the reference axis, i.e. the entire area of the figure is assumed to be distributed/concentrated at distance r from the reference axis.

A.3.2 Product of Inertia

The *product of inertia* of an area is not used as often as the moment of inertia but is required for determination of the maximum and minimum of the moment of inertia, unsymmetrical bending of beams, and analysis of indeterminate structures.

The product of inertia of an area with respect to any pair of orthogonal reference axes may be defined as the sum of the products obtained by multiplying each element of area by the product of the two coordinates of the element with respect to reference axes and is mathematically expressed as

$$I_{xy} = \sum_i x\,y\,dA = \int x\,y\,dA \tag{A.9}$$

where dA is the area of an element of the given figure and x, y are the coordinates of the element with respect to two orthogonal reference axes. The product of inertia may be positive, negative or zero. Unlike the moment of inertia, the sign of the product of inertia depends on the location of the area relative to the axes, being positive if the area lies principally in the first or third quadrants and negative if the area lies principally in the second or fourth quadrants. The product of inertia of an area with respect to two orthogonal axes is zero when either of the two axes is an axis of symmetry.

A.3.3 Parallel Axis Theorem

It is often necessary to transfer the moment of inertia from one axis to another parallel axis. Transfer formulas which do not require further integration can be developed to achieve the objectives. For illustration consider the moment of inertia of the cross-section shown in Figure A.3(a) with respect to a centroidal $\overline{x}$ axis which is given by $I_{\overline{x}} = \int_A y'^2\, dA$. The moment of inertia of the same area with respect to a parallel axis X located at a distance $\overline{y}$ from the centroidal axis is

$$I_x = \int_A (y' + \overline{y})^2\, dA = \int_A y'^2\, dA + 2\overline{y}\int_A y'\, dA + \int_A \overline{y}^2\, dA \tag{A.10}$$

where $\overline{y}$ is the constant representing the distance separating the axes. The second term on right hand side of Eq. (A.10) represents the moment of area relative to the centroidal axes. Since the moment of area relative to the centroidal axes is zero, Eq. (A.10) finally reduces to

$$I_x = I_{\overline{x}} + A\overline{y}^2 \tag{A.11}$$

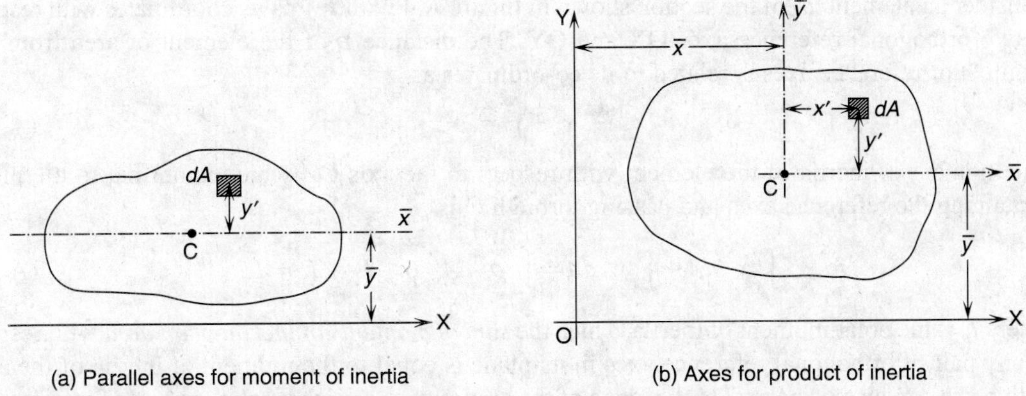

Figure A.3 Transfer of moment of inertia and product of inertia between axes.

Therefore, the moment of inertia of an area about any axis in its plane is equal to the total sum of the moment of inertia of the figure about a parallel axis through the centroid of the figure plus the product of the total area of the figure and the square of the distance between the parallel centroidal axis and the reference axis. Similarly, the other moment of inertias of an area can be expressed as

$$I_y = I_{\bar{y}} + A\bar{x}^2 \tag{A.12}$$

The above procedure is equally applicable to the product of inertia of an area. From the fundamental definition of product of inertia of an area with respect to a set of parallel axes X and Y located a distance $\bar{x}$ and $\bar{y}$ from the centroidal axes shown in Figure A.3(b) is

$$\begin{aligned} I_{xy} &= \int_A (x' + \bar{x})(y' + \bar{y})\, dA \\ &= \int_A x'y'\, dA + \bar{x}\int_A y'\, dA + \bar{y}\int_A x'\, dA + \int_A \bar{x}\,\bar{y}\, dA \end{aligned} \tag{A.13}$$

The second and third terms on right hand side of Eq. (A.13) represent the moments of area relative to the centroidal axes. Since the moments of area relative to the centroidal axes are zero, the Eq. (A.13) finally reduces to

$$I_{xy} = I_{\bar{x}\bar{y}} + A\,\bar{x}\,\bar{y} \tag{A.13a}$$

Thus, the *product of inertia of an area* with respect to any pair of orthogonal reference axes in its plane is equal to the product of inertia of the area with respect to a pair of parallel centroidal rectangular axes plus the product of area of the figure and the coordinates of the centroid of the area with respect to the given pair of axes.

It should be noted the least moment of inertia for any given direction of an axis is the centroidal moment of inertia. Further, the centroidal axis involved in the transfer formula is always the centroidal axis of the area being considered in the transfer term $A\bar{y}^2$.

A.3.4 Perpendicular Axes Theorem

Consider an element dA of the section shown in Figure A.4 with x, y as its coordinates with respect to two orthogonal reference axes OX and OY. The distance ρ of the element of area from the origin O of reference axes is related to its co-ordinates as

$$x^2 + y^2 = \rho^2 \tag{A.14}$$

The moment of inertia of the element with respect to the axis OZ which is normal to the plane containing the reference axes and passing through O is

$$\int_A x^2\, dA + \int_A y^2\, dA = \int_A \rho^2\, dA \quad \text{or} \quad I_x + I_y = I_p \tag{A.15}$$

where I_p is the polar moment of inertia. Thus, the *sum of moments of inertia of an area* with respect to any pair of orthogonal reference axes in its plane is equal to the moment of inertia of the area with respect to an axis normal to the plane of the area and passing through the point of intersection O of the axes OX and OY.

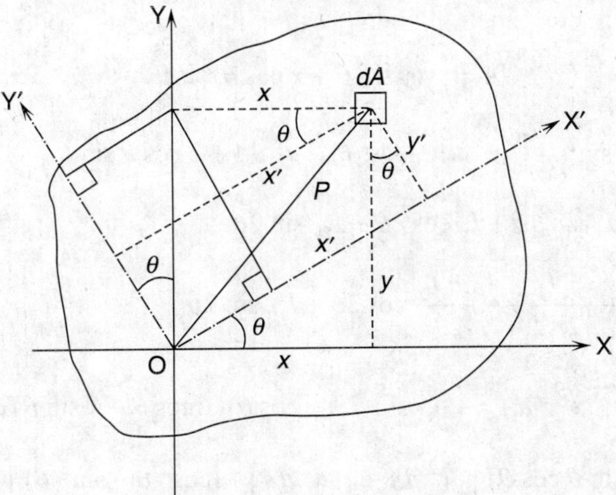

Figure A.4 Rotation of axes.

A.4 TRANSFORMATION LAWS

The moments of inertia change with the rotation of co-ordinate axes. Consider that the moments of inertia I_x, I_y and I_{xy} of a section are known with respect to orthogonal X–Y co-ordinate system (not necessarily centroidal). It is required to compute the moments of inertia $I_{x'}$, $I_{y'}$ and $I_{x'y'}$ with respect to another orthogonal X′–Y′ system of axes which is obtained after the X–Y axes have been rotated through an angle θ in anticlockwise direction as shown in Figure A.4. The co-ordinates of a point on an element with respect to the two co-ordinate systems are related as

$$x' = y \sin\theta + x \cos\theta$$

$$y' = y \cos\theta - x \sin\theta \tag{A.16}$$

Therefore,

$$I_{x'} = \int_A (y')^2\, dA = \int_A (y\cos\theta - x\sin\theta)^2\, dA$$

$$= \cos^2\theta \int_A y^2\, dA + \sin^2\theta \int_A x^2\, dA - 2\cos\theta\sin\theta \int_A x\,y\,dA$$

$$= I_x \cos^2\theta + I_y \sin^2\theta - I_{xy}\sin 2\theta \tag{A.17}$$

Since $2\cos\theta\sin\theta = \sin 2\theta$

Using the relations, $\cos^2\theta = \dfrac{1+\cos 2\theta}{2}$ and $\sin^2\theta = \dfrac{1-\cos 2\theta}{2}$

$$I_{x'} = \frac{I_x + I_y}{2} + \frac{I_x - I_y}{2}\cos 2\theta - I_{xy}\sin 2\theta \tag{A.18}$$

Similarly,

$$I_{y'} = \int_A (x')^2\, dA = \int_A (y \sin\theta + x\cos\theta)^2\, dA$$

$$= \sin^2\theta \int_A y^2\, dA + \cos^2\theta \int_A x^2\, dA + 2\cos\theta\sin\theta \int_A xy\, dA$$

$$= I_x \sin^2\theta + I_y \cos^2\theta + I_{xy}\sin 2\theta$$

$$= \frac{I_x + I_y}{2} - \frac{I_x - I_y}{2}\cos 2\theta + I_{xy}\sin 2\theta \tag{A.19}$$

and

$$I_{x'y'} = \int_A x'y'\, dA = \int_A (y\sin\theta + x\cos\theta)(y\cos\theta - x\sin\theta)\, dA$$

$$= \sin\theta\cos\theta\left(\int_A y^2\, dA - \int_A x^2\, dA\right) + (\cos^2\theta - \sin^2\theta)\int_A xy\, dA$$

$$= \frac{I_x - I_y}{2}\sin 2\theta + I_{xy}\cos 2\theta \tag{A.20}$$

Adding Eqs. (A.18) and (A.19)

$$I_{x'} + I_{y'} = I_x(\cos^2\theta + \sin^2\theta) + I_y(\sin^2\theta + \cos^2\theta) = I_x + I_y \tag{A.21}$$

Thus the sum of moments of inertia with respect to any set of two mutually perpendicular axes through the same point is independent of the angle θ and remains constant when the co-ordinate axes are rotated.

The product of inertia vanishes for doubly or singly symmetric section.

For $I_{x'y'} = 0$:

$$\frac{I_x - I_y}{2}\sin 2\theta + I_{xy}\cos 2\theta = 0 \quad \text{or} \quad \tan 2\theta = \frac{-2I_{xy}}{(I_x - I_y)}$$

Therefore,

$$2\theta = \tan^{-1}\left(\frac{2I_{xy}}{I_y - I_x}\right) \tag{A.22}$$

The co-ordinate axes with respect to which the product of inertia is 0 are called *principal axes.*

A.5 PRINCIPAL AXES

Since each of these $I_{x'}$ and $I_{y'}$ changes with the angle of rotation of the axes θ but their sum remains unchanged. Consequently, there exists such an angle θ at which one of the moments of inertia attains its maximum value while the other a minimum value. To obtain this value of angle θ, the expression for $I_{x'}$ is differentiated with respect to θ and the derivative is equated to zero. Thus

$$\frac{dI_{x'}}{d\theta} = -(I_x - I_y)\sin 2\theta - 2I_{xy}\cos 2\theta = 0$$

Therefore,

$$\tan 2\theta = \frac{-2I_{xy}}{(I_x - I_y)} = \frac{2I_{xy}}{(I_y - I_x)} \tag{A.23}$$

As seen earlier, for the same angle θ the product of inertia $I_{x'y'}$ vanishes. Since the function $\tan 2\theta$ is periodic, i.e. $\tan 2\theta = \tan(180° + 2\theta)$, corresponding values of angle θ differ by 90°. For one value of the angle θ, one of the moments of inertia will be a maximum I_u and for the other a minimum I_v. Thus, the axes with respect to which the product of inertia is zero and moments of inertia attain extreme values are called *principal axes u, v*. If the axes are centroidal, they are called *principal centroidal axes*.

If a section has an axis of symmetry, the product of inertia of the part of the section lying to one side of the axis is equal to the product of inertia of the part of the lying to the other but is of opposite sign, i.e. $I_{xy} = 0$; $\theta = 0$ and the axis is principal. Thus, the X and Y axes are the principal axes. The moments of inertias with respect to principal axes are called *principal moments of inertia*.

The moment of inertia and product of inertia of a plane figure represented by $I_{x'}$ and $I_{x'y'}$ expressed by Eqs. (19.18) and (19.20) can be rewritten as follows

$$I_{x'} - \frac{I_x + I_y}{2} = \frac{I_x - I_y}{2} \cos 2\theta - I_{xy} \sin 2\theta \tag{A.24}$$

$$I_{x'y'} = \frac{I_x - I_y}{2} \sin 2\theta + I_{xy} \cos 2\theta \tag{A.25}$$

Adding the squares of Eqs. (A.24) and (A.25):

$$\left(I_{x'} - \frac{I_x + I_y}{2}\right)^2 + (I_{x'y'})^2 = \left(\frac{I_x - I_y}{2}\right)^2 + I_{xy}^2 \tag{A.26}$$

With $I_{x'}$ and $I_{x'y'}$ taken as co-ordinates of the system the Eq. (A.26) represents a circle whose centre is on $I_{x'}$ axis at a distance $\left(\frac{I_x + I_y}{2}\right)$ from the origin. The radius of the circle is:

$$R = \sqrt{\left(\frac{I_x - I_y}{2}\right)^2 + I_{xy}^2} \tag{A.27}$$

The circle, called *Mohr's circle*, can be constructed as the Mohr's circle of stress. This construction gives a visual representation of all the possible values $I_{x'}$ and $I_{x'y'}$ with respect to all the axes passing through a specified point in an area.

The maximum and minimum values of moments of inertia or principal moments of inertia are

$$I_{max}, I_{min} = I_1, I_2 = \frac{I_x + I_y}{2} \pm \sqrt{\left(\frac{I_x - I_y}{2}\right)^2 + I_{xy}^2} \tag{A.28}$$

A.6 DETERMINATION OF MOMENT OF INERTIA BY INTEGRATION

For simplification, consider the differential area such that: (1) all parts of the differential area are at the same distance from the reference axis or (2) the moments of inertia of the differential area with respect to the reference axis is known.

A.6.1 Moment of Inertia of Simple Figures/Shapes

(1) Rectangle: To determine the centroidal moment of inertia of a rectangle of width b and height h, consider the differential strip element as shown in Figure A.5(a). All parts of the element are at the same distance from the centroidal $\bar{x}$ axis. From Eq. (A.6):

$$I_{\bar{x}} = \int_A y^2\, dA = \int_{-h/2}^{h/2} y^2\,(b\,dy) = b\int_{-h/2}^{h/2} y^2\, dy$$

$$= b\left[\frac{y^3}{3}\right]_{-h/2}^{h/2} = \frac{bh^3}{12} \tag{A.29}$$

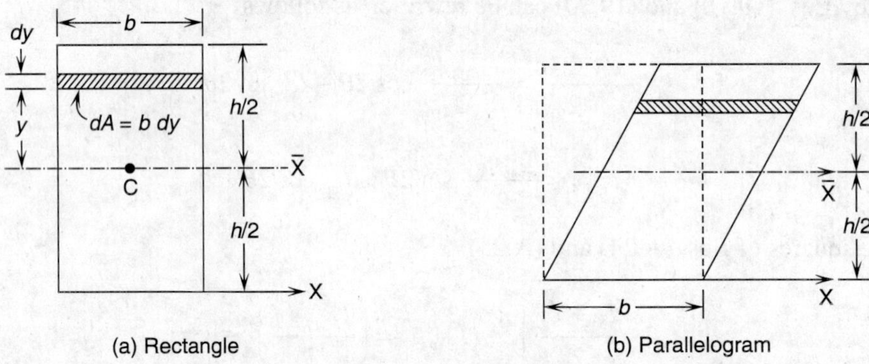

Figure A.5 Computation of moment of inertia of a rectangle.

The moment of inertia with respect to an axis coinciding with the base can be obtained by transferring the centroidal moment of inertia through the distance $h/2$ to the parallel base axis by using the transfer formula given by Eq. (A.11):

$$I_x = I_{\bar{x}} + A\bar{y}^2 = \frac{bh^3}{12} + (bh)\left(\frac{h}{2}\right)^2 = \frac{bh^3}{12} + \frac{bh^3}{4} = \frac{bh^3}{3} \tag{A.30}$$

The moment of inertia of parallelogram of Figure A.5(b) has the identical value as that for the rectangle because the elemental strips composing the parallelogram have merely shifted their position laterally from the rectangle of corresponding dimensions but have the same distance from the reference axis.

(2) Triangle: The moment of inertia of the triangle of base width b and altitude h with respect to an axis coinciding with the base can be determined by considering the differential strip element as shown in Figure A.6.

From similar triangles,
$$x = \left(\frac{b}{h}\right)(h - y) \tag{A.31}$$

From Eq. (A.6):

$$I_x = \int_A y^2\, dA = \int_0^h y^2\,(x\,dy) = \frac{b}{h}\int_0^h y^2\,(h - y)\, dy$$

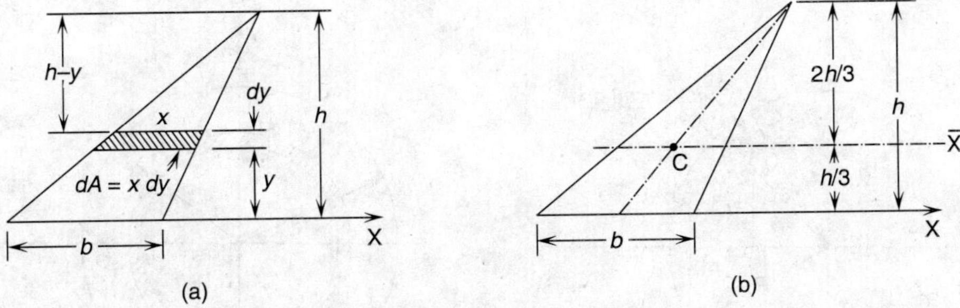

Figure A.6 Computation of moment of inertia of a triangle.

$$= \frac{b}{h}\left[\frac{hy^3}{3} - \frac{y^4}{4}\right]_0^h = \frac{b}{h}\left[\frac{h^4}{3} - \frac{h^4}{4}\right] = \frac{bh^3}{12} \tag{A.32}$$

The centroidal moment of inertia can be obtained by transferring the moment of inertia with respect the base through the distance $h/3$ to the parallel centroidal axis by using the transfer formula given by Eq. (A.11):

$$I_x = I_{\bar{x}} + A\bar{y}^2 \quad \text{or} \quad \frac{bh^3}{12} = I_{\bar{x}} + \left(\frac{bh}{2}\right)\left(\frac{h}{3}\right)^2$$

Therefore,

$$I_{\bar{x}} = \frac{bh^3}{12} - \frac{bh^3}{18} = \frac{bh^3}{36} \tag{A.33}$$

(3) Circular area of radius *R*: For a simple solution of circular areas, polar co-ordinates system is the most suitable. Using polar co-ordinates, consider the differential element as shown in Figure A.7(a). The moment of inertia with respect to the diameter is

$$I_{\bar{x}} = \int_A y^2\, dA = \int_0^R \int_0^{2\pi} (r \sin\theta)^2\, (r\, d\theta\, dr) = \int_0^R \left(\int_0^{2\pi} \sin^2\theta\, d\theta\right) r^3\, dr$$

$$= \int_0^R r^3\, dr \left(\int_0^{2\pi} \sin^2\theta\, d\theta\right) = \int_0^R r^3\, dr \int_0^{2\pi} \left(\frac{1-\cos 2\theta}{2}\right) d\theta$$

$$= \left[\frac{r^4}{4}\right]_0^R \times \frac{1}{2}\left[\theta - 2\sin 2\theta\right]_0^{2\pi} = \frac{R^4}{4} \times \pi \tag{A.34}$$

Therefore,

$$I_{\bar{x}} = \frac{\pi R^4}{4} = \frac{\pi D^4}{64} \tag{A.35}$$

where $D(= 2R)$ is the diameter of the circle.

For a complete circle, an alternate and simple solution is to consider the differential ring element as shown in Figure A.7(b). The polar moment of inertia is obtained from Eq. (A.15):

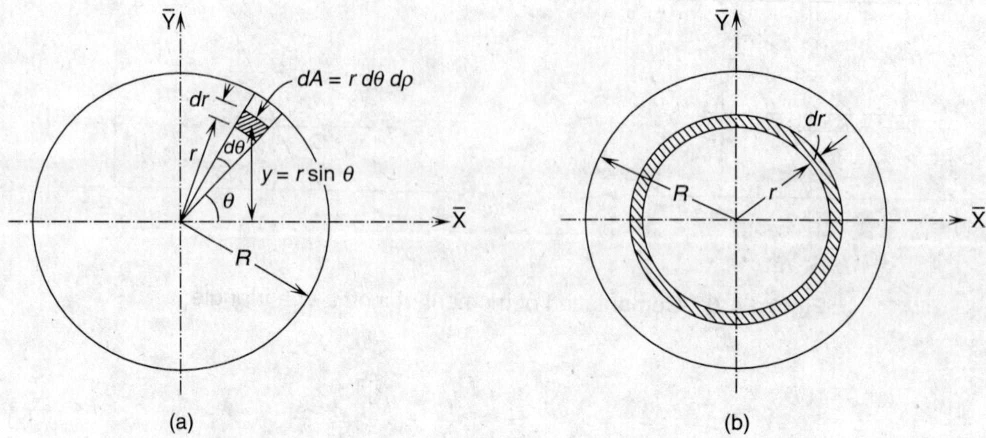

Figure A.7 Computation of moment of inertia of a circle.

$$I_p = \int_A r^2\, dA = \int_0^R r^2 (2\pi r\, dr)$$

$$= 2\pi \int_0^R r^3\, dr = 2\pi \left[\frac{r^4}{4} \right]_0^R = \frac{\pi R^4}{2} = \frac{\pi D^4}{32} \tag{A.36}$$

The diametric or rectangular moments of inertia $I_{\bar{x}}$ and $I_{\bar{y}}$ are equal due to the symmetry of the figure, thus from Eq. (A.15):

$$I_{\bar{x}} + I_{\bar{y}} = I_p = \frac{\pi R^4}{2} \quad \text{or} \quad I_{\bar{x}} = I_{\bar{y}} = \frac{\pi R^4}{4} \left(= \frac{\pi D^4}{64} \right) \tag{A.37}$$

A.6.2 Moment of Inertia of Composite Figures

The moment of inertia of a composite figure may be determined by sub-dividing it into a number simple geometric elements (rectangles, triangles, etc) for which centroids and moments of inertia are known. The moment of inertia of the composite figure is then obtained by summing up the moments of inertia of individual elements. Before the moments of inertia of separate elements are added, they must all be computed with respect to the same common reference axis. The centroidal moments of inertia of various elements of the figure can be transferred to the common reference axis using transfer formula.

Consider a figure to be sub-divided into simple figures of areas $A_1, A_2, \ldots$ and the perpendicular distances of the centroids of these areas from the two reference axes X and Y to be $y_1, y_2, \ldots$ and $x_1, x_2, \ldots$, respectively. The coordinates of the centroid (x_o, y_o) of the whole figure are given by

$$\bar{x} \text{ or } x_o = \frac{A_1 x_1 + A_2 x_2 + \cdots}{A_1 + A_2 + \cdots} = \frac{\sum_i A_i x_i}{\sum_i A_i} = \frac{\sum_i A_i x_i}{A} \tag{A.38}$$

$$\overline{y} \text{ or } y_o = \frac{A_1 y_1 + A_2 y_2 + \cdots}{A_1 + A_2 + \cdots} = \frac{\sum_i A_i y_i}{\sum_i A_i} = \frac{\sum_i A_i y_i}{A} \tag{A.39}$$

where A is the total area of the figure.

EXAMPLE A.1

Determine the moments of inertia of I-beam section shown in Figure A.8(a) with respect to the centroidal X and Y axes.

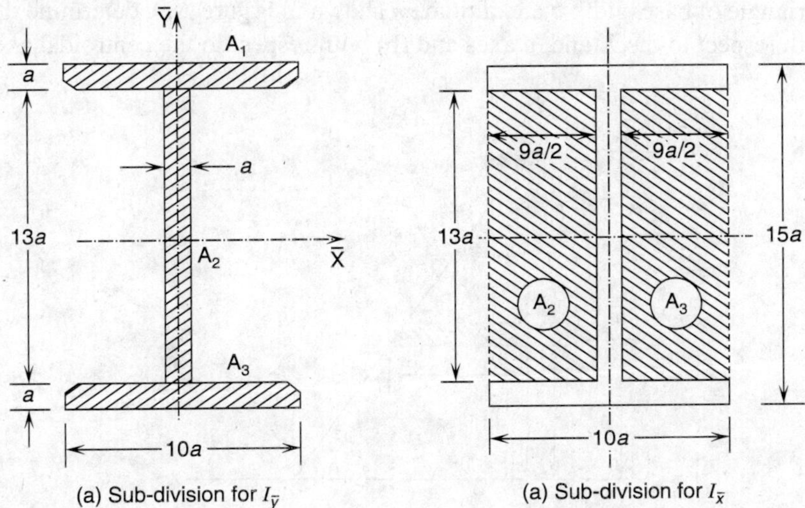

(a) Sub-division for $I_{\overline{y}}$

(a) Sub-division for $I_{\overline{x}}$

Figure A.8 I-beam section of Example A.1.

Solution: The moment of inertia of a composite figure is the sum of the moments of inertia of the various parts, computed with respect to the same common reference axis. With respect to the $\overline{x}$ axis, the simplest sub-division of the figure is to consider it to be made up of a large rectangular area A_1 ($10a \times 15a$) from which two smaller rectangles A_2 and A_3 each of size ($4.5a \times 13a$) are subtracted as shown in Figure A.8(b). The centroidal axis for each of these parts coincides with $\overline{x}$ or the reference axis of the figure, hence no transfer of centroidal moments of inertia of the parts is required, i.e. the expression $I_{\overline{x}} = \dfrac{bd^3}{12}$ is to be used for each part. Thus

$$I_{\overline{x}} = \frac{(10a)(15a)^3}{12} - 2 \times \frac{(4.5a)(13a)^3}{12} = \frac{13977a^4}{12} = 1164.75a^4$$

With respect to the $\overline{y}$ axis, the figure is considered to be made up of three rectangular areas as $A_1(a \times 10a)$, $A_2(13a \times a)$ and $A_3(a \times 10a)$ as shown in Figure A.8(a). Here, again the centroidal axis for each of these parts coincides with the $\overline{y}$ axis of the figure, hence this sub-division of the area eliminates the need for using the transfer formula, i.e. the expression $I_{\bar{y}} = bd^3/12$ can be used for each part. Thus

$$I_{\bar{x}} = \frac{(a)(10a)^3}{12} + \frac{(13a)(a)^3}{12} + \frac{(a)(10a)^3}{12}$$

$$= \frac{2013a^4}{12} = 167.75a^4$$

EXAMPLE A.2

For the right triangle of base width b and altitude h shown in Figure A.9, determine the product of inertia: (a) with respect to the X and Y axes and (b) with respect to the centroidal axes.

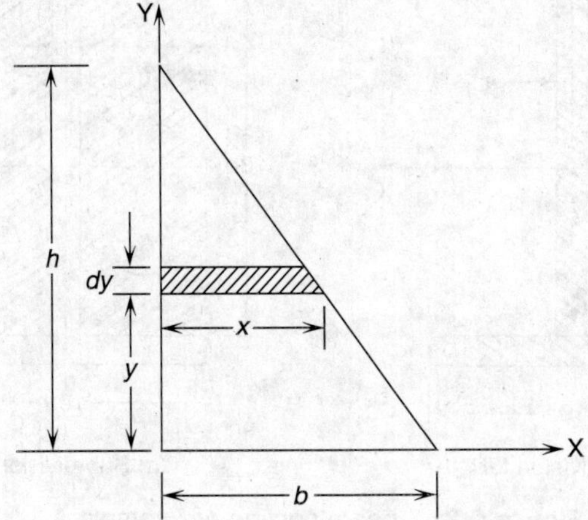

Figure A.9 Product of inertia of right triangle of Example A.2.

Solution: The product of inertia $I_{xy} = \int x\,y\,dA$ of the right triangle with respect to X, Y axes can be determined by considering the differential strip element as shown in Figure A.9. The area of the strip is $dA = x\,dy$ and the co-ordinates of its centroid are $x/2$ and y, then from similar triangles,

$$x = \left(\frac{b}{h}\right)(h - y)$$

Hence $$dA = x\,dy = \left(\frac{b}{h}\right)(h - y)\,dy$$

Therefore, the product of inertia is given by

$$I_{xy} = \int_0^h \left[\frac{1}{2}\left(\frac{b}{h}\right)(h - y)\right] y \left[\left(\frac{b}{h}\right)(h - y)\,dy\right] = \frac{b^2}{2h^2}\int_0^h y(h - y)^2\,dy$$

$$= \frac{b^2}{2h^2}\int_0^h (h^2 y - 2hy^2 + y^3)\,dy = \frac{b^2}{2h^2}\left[h^2\frac{y^2}{2} - 2h\frac{y^3}{3} + \frac{y^4}{4}\right]_0^h$$

$$= \frac{b^2}{2h^2}\left(\frac{h^4}{2} - \frac{2h^4}{3} + \frac{h^4}{4}\right) = +\frac{b^2h^2}{24}$$

The centroidal product of inertia can be obtained by transferring the product of inertia with respect the base through the distance $h/3$ to the parallel centroidal axis by using the transfer formula given by Eq. A.13:

$$I_{xy} = I_{\bar{x}\bar{y}} + A\bar{x}\bar{y} \quad \text{or} \quad \frac{b^2h^2}{24} = I_{\bar{x}\bar{y}} + \left(\frac{bh}{2}\right)\left(\frac{b}{3}\right)\left(\frac{h}{3}\right)$$

Therefore, $$I_{\bar{x}\bar{y}} = \frac{b^2h^2}{24} - \frac{b^2h^2}{18} = -\frac{b^2h^2}{72}$$

EXAMPLE A.3

Two standard rolled steel ISMC200 channels are laced together to form a section shown in Figure A.10. Determine how far apart the channels should be placed so that $I_{\bar{x}\bar{x}} = I_{\bar{y}\bar{y}}$ for the built-up section. Neglect the contribution of lattice bars indicated by the dotted lines in the figure.

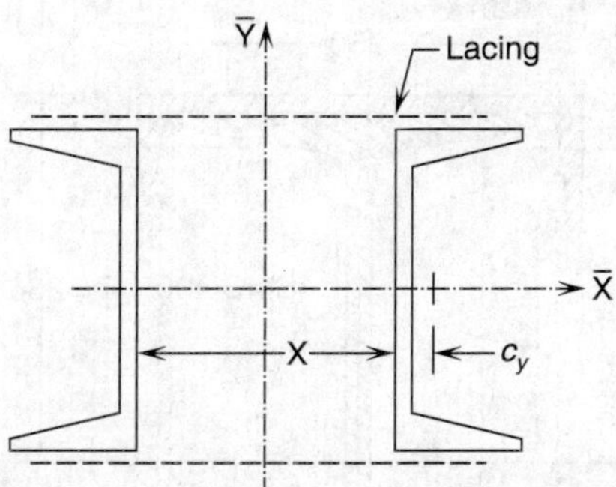

Figure A.10 Built-up section of Example A.3.

The geometrical properties of the channel-section are:

$$I_{\bar{x}} = 1.825 \times 10^7 \text{ mm}^4;\ I_{\bar{y}} = 1.404 \times 10^6 \text{ mm}^4;$$

$$A = 2.828 \times 10^3 \text{ mm}^2 \text{ and } c_y = 21.8 \text{ mm}$$

Solution: If the channels are placed at distance X apart, the moments of inertia of the built-up section are

$$I_{\bar{x}\bar{x}} = 2I_{\bar{x}} \text{ and } I_{\bar{y}\bar{y}} = 2\left[I_{\bar{y}} + A\left(\frac{X}{2} + c_y\right)^2\right]$$

For the condition $I_{\bar{x}\bar{x}} = I_{\bar{y}\bar{y}}$: $\quad 2I_{\bar{x}} = 2\left[I_{\bar{y}} + A\left(\dfrac{X}{2} + c_y\right)^2 \right]$

Therefore,

$$\left(\frac{X}{2} + c_y\right)^2 = \frac{I_{\bar{x}} - I_{\bar{y}}}{A} \quad \text{or} \quad X = 2\left(\sqrt{\frac{I_{\bar{x}} - I_{\bar{y}}}{A}} - c_y\right)$$

Substitute values of geometric values,

$$X = 2\left[\sqrt{\frac{(1.825 \times 10^7 - 1.404 \times 10^6)}{2.828 \times 10^3}} - 21.8\right] = 110.76 \text{ mm}$$

EXAMPLE A.4

A girder is composed of an ISWB 600 I-section and a ISMC 400 channel section. The channel section is connected to the top flange of the I-section as shown in Figure A.11. Compute the moments of inertia with respect to the centroidal axes.

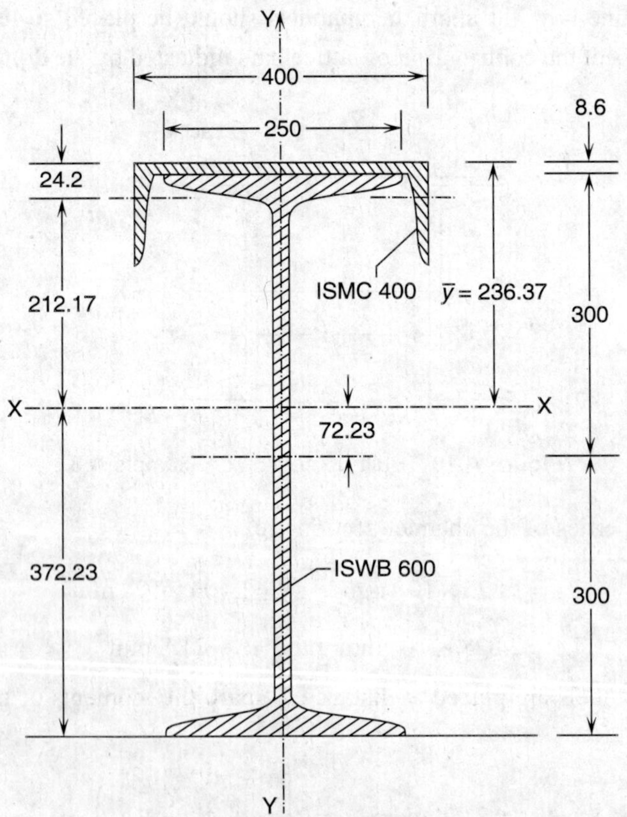

Figure A.11 Built-up girder of Example A.4.

The geometrical properties of the section are:

(1) I-section

$b = 250$ mm; $t_f = 23.6$ mm; $t_w = 11.8$ mm and $A = 1.8486 \times 10^4$ mm^2

$$I_{\bar{x}} = 115.627 \times 10^7 \text{ mm}^4 \text{ and } I_{\bar{y}} = 5.298 \times 10^7 \text{ mm}^4$$

(2) The channel-section

$b = 100$ mm; $t_f = 15.3$ mm; $t_w = 8.6$ mm; $c_y = 24.2$ mm and $A = 6.293 \times 10^3$ mm^2

$$I_{\bar{x}} = 15.083 \times 10^7 \text{ mm}^4 \text{ and } I_{\bar{y}} = 5.048 \times 10^6 \text{ mm}^4$$

Solution: Consider the depth of the centroidal axis from the top of the section to be $\bar{y}$

$$\bar{y} = \frac{(1.8486 \times 10^4) \times (300 + 8.6) + (6.293 \times 10^3) \times 24.2}{1.8486 \times 10^4 + 6.293 \times 10^3} = 236.37 \text{ mm}$$

The moments of inertia are

$$I_{\bar{x}\bar{x}} = I_{\bar{x}\bar{x}} \text{ of the I-section} + I_{\bar{x}\bar{x}} \text{ of the channel-section}$$

$$= [115.627 \times 10^7 + (1.8486 \times 10^4) \times (308.6 - 236.37)^2]$$

$$+ [5.048 \times 10^6 + (6.293 \times 10^3) \times (236.37 - 24.2)^2]$$

$$= 154.1049 \times 10^7 \text{ mm}^4$$

$$I_{\bar{y}\bar{y}} = I_{\bar{y}\bar{y}} \text{ of the I-section} + I_{\bar{y}\bar{y}} \text{ of the channel-section}$$

$$I_{\bar{y}\bar{y}} = 5.298 \times 10^7 + 15.083 \times 10^7 = 20.381 \times 10^7 \text{ mm}^4$$

EXAMPLE A.5

The hollow circular cross-section of external radius R shown in Figure A.12 has an eccentric circular hole of radius of $0.25R$ located at distance $0.125R$ from the edge. Determine the centroidal moments of inertia.

Solution: The geometrical properties of the section are:

Area of outer circle, $A_1 = \pi R^2$

Areas of inner circle, $A_2 = \pi \times (0.25R)^2 = 0.0625\pi R^2$

Centroidal moment of inertia of outer circle,

$$I_o = \frac{\pi R^4}{4} = 0.25\pi R^4$$

Centroidal moment of inertia of inner circle,

$$I_i = \frac{\pi \times (0.25R)^4}{4} = (9.7656 \times 10^{-4})\pi R^4$$

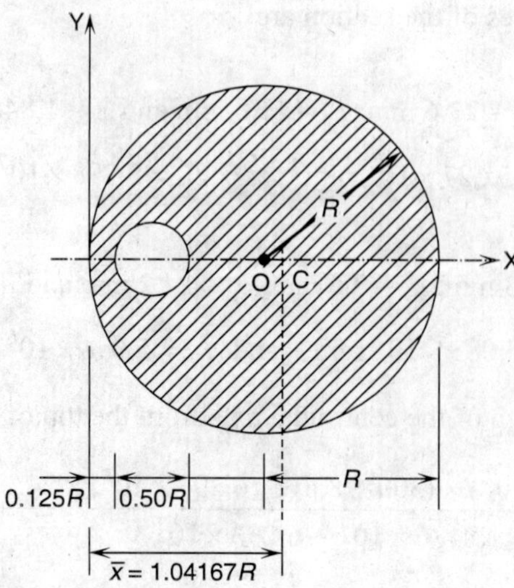

Figure A.12 Hollow circular cross-section of Example A.5.

The distance of the c.g. of the section from the edge is

$$\bar{x} = \frac{(\pi R^2) \times R - (0.0625\pi R^2) \times (0.375R)}{\pi R^2 - 0.0625\pi R^2} = 1.04167R$$

The moments of inertia are

$$I_{\bar{x}\bar{x}} = 0.25\pi R^4 - (9.7656 \times 10^{-4})\pi R^4 = 0.24902\,\pi R^4$$

$$I_{\bar{y}\bar{y}} = [0.25\pi R^4 + (\pi R^2) \times (1.04167R - R)^2]$$
$$- [(9.7656 \times 10^{-4})\pi R^4 + (0.0625\pi R^2) \times (1.04167R - 0.375\,R)^2]$$
$$= 0.222982\pi R^4$$

A.7 PROBLEMS

A.1 Determine moment of inertia of a triangle of base width b and altitude h with respect to an axis passing through the apex and parallel to the base.
[**Hint:** Consider an elementary strip of thickness dy at distance y from the apex, width of strip is by/h. Or alternatively transfer the centroidal moment of inertia.]

[**Ans.** $I = bh^3/4$]

A.2 The base b of an equilateral triangle is horizontal. Prove that the centroidal moments of inertia with respect to the horizontal and vertical axes are equal.

A.3 Determine the moment of inertia of an area enclosed by a regular hexagon of side a with respect to an axis passing through its two opposite apexes.

[**Ans.** $I = (5\sqrt{3}/16)a^4$]

A.4 The area of the shaded cross-section shown in Figure A.13 is 50×10^3 mm^2. If the moment of inertia $I_{x1} = 300 \times 10^6$ mm^4, determine the moment of inertia I_{x2}.

[**Hint:** $I_{x2} - I_{x1} = A(y_2^2 - y_1^2)$]

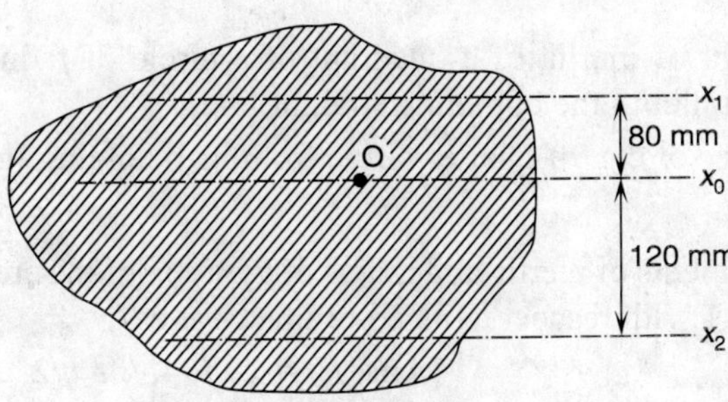

Figure A.13

[**Ans.** $I_{x2} = 700 \times 10^6$ mm^4]

A.5 Determine moment of inertia of a semi-circle of radius R shown in Figure A.14 with respect to its diameter.

[**Hint:** Use limits of θ from 0 to π in the Eq. (A.34)]

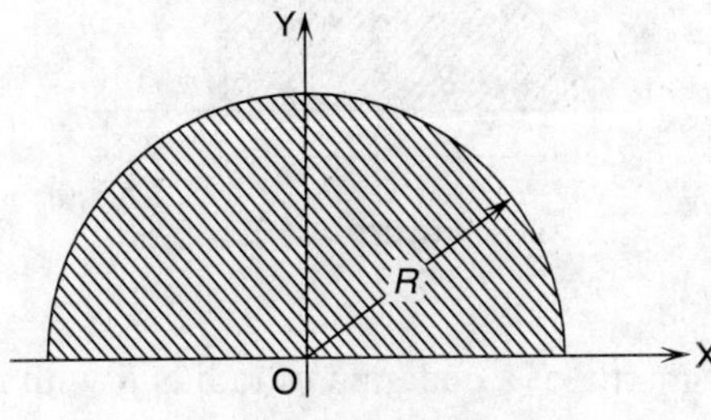

Figure A.14

[**Ans.** $I = \pi R^4/8$]

A.6 Determine moment of inertia of a quadrant of a circle of radius R with respect to one of its bounding straight lines as shown in Figure A.15.

[**Hint:** Use limits of θ from 0 to $\pi/2$ in the Eq. (A.34)]

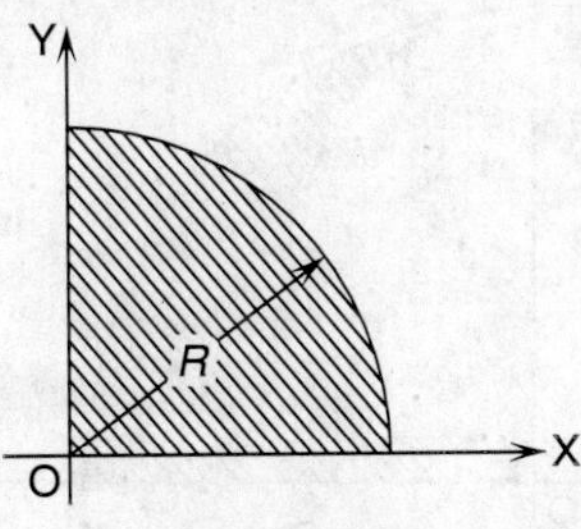

Figure A.15

[**Ans.** $I = \pi R^4/16$]

A.7 Determine moment of inertia of a semi-circle of radius R with respect to the centroidal axis parallel to its diameter.

[**Hint:** The distance of c.g. of semi-circle from its diameter is $4R/(3\pi)$]

[**Ans.** $I \approx 0.11R^4$]

A.8 Determine moment of inertia of a quadrant of a circle of radius R with respect to the centroidal axis parallel to the bounding straight line.

[**Hint:** The distance of c.g. of quarter-circle from its bounding line is $2R/(3\pi)$]

[**Ans.** $I \approx 0.055R^4$]

A.9 Determine the moments of inertia and radius of gyration of the area enclosed by the ellipse $(x^2/a^2) + (y^2/b^2) = 1$ with respect to its co-ordinates axes.

[**Ans.** $I_{xo} = \pi ab^3/4$, $k = b/2$, and $I_{yo} = \pi a^3 b/4$, $k = a/2$]

A.10 Determine the moments of inertia of the parabolic area enclosed by the curve $x = my^2$ as shown in Figure A.16 with respect to its co-ordinate axes.

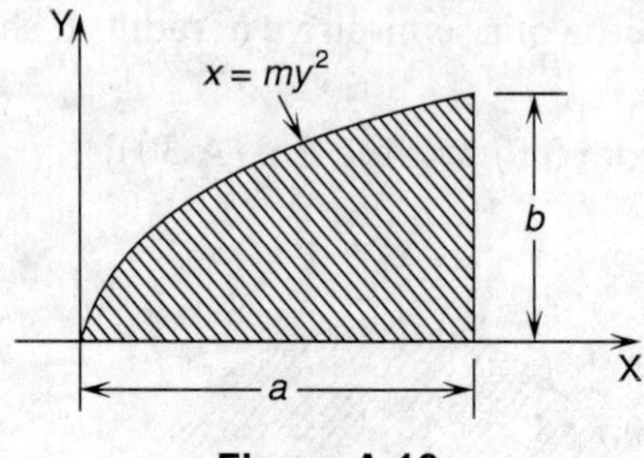

Figure A.16

[**Ans.** $I_x = 2ab^3/15$ and $I_y = 2a^3b/7$]

A.11 Determine product of inertia of a quadrant of radius R with respect to the X and Y axes.

[**Ans.** $I_{xy} = R^4/8$]

A.12 Determine product of inertia of the shaded area shown in Figure A.17 with respect to the X and Y axes.

[**Hint:** Use the result of Problem A.11]

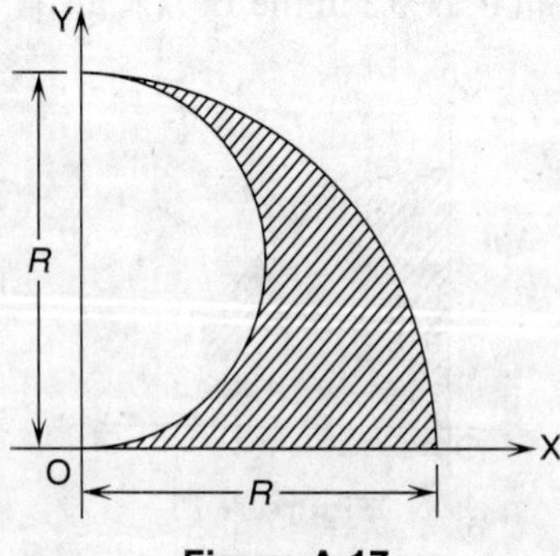

Figure A.17

[**Ans.** $I_{xy} = R^4/12$]

A.13 Determine the moments of inertia of the shaded area shown in Figure A.18 with respect to the axis X-X.

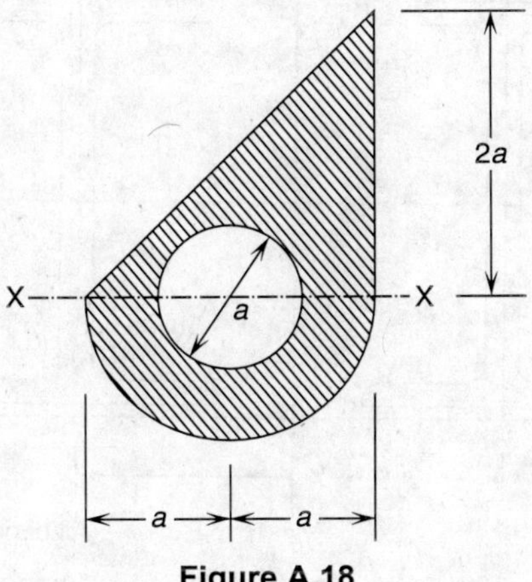

Figure A.18

[**Ans.** $I_{XX} = (21\pi + 256)a^4/192$]

A.14 Determine the moments of inertia of the shaded area of plane figure shown in Figure A.19 with reference to the X-axis.

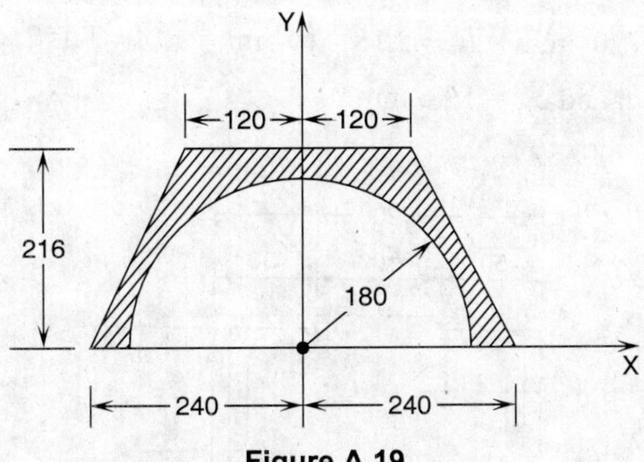

Figure A.19

[**Ans.** $\bar{y} = 133.14$ mm]

A.15 Two standard rolled steel ISWB 350 I-sections are laced together to form a section shown in Figure A.20. Determine how far apart the I-sections should be placed so that $I_{\bar{x}\bar{x}} = I_{\bar{y}\bar{y}}$ for the built-up section. Neglect the contribution of lattice bars indicated by the dotted lines in the figure.
The geometrical properties of the I-section are:

$$I_{\bar{x}} = 1.5522 \times 10^8 \text{ mm}^4;\ I_{\bar{y}} = 1.1759 \times 10^7 \text{ mm}^4 \text{ and } A = 7.25 \times 10^3 \text{ mm}^2$$

[**Ans.** $X = 280.6$ mm]

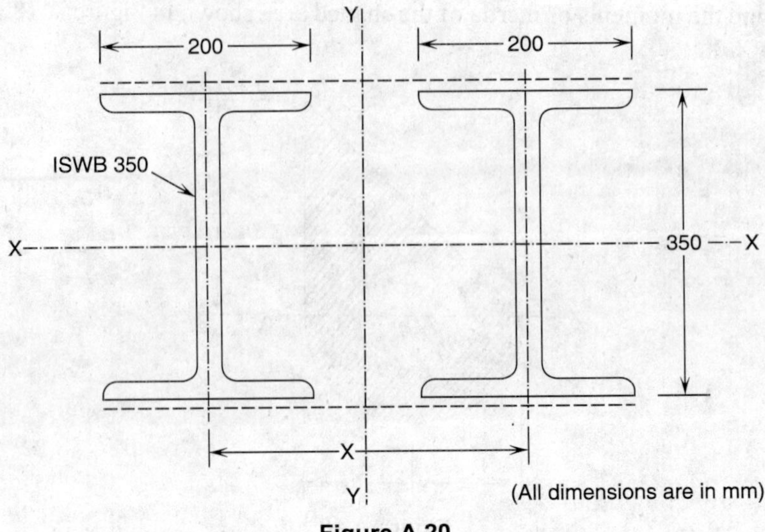

Figure A.20

A.16 The section shown in Figure A.21 is built-up of four ISA75 50 × 10 mm angles and two 250 × 10 mm thick cover plates. Determine the moments of inertia $I_{\bar{x}\bar{x}}$ and $I_{\bar{y}\bar{y}}$ of the built-up section. Neglect the contribution of lattice bars indicated by the dotted lines in the figure. The geometrical properties of the angle section are:

$$I_{\bar{x}} = 6.23 \times 10^5 \text{ mm}^4; I_{\bar{y}} = 2.18 \times 10^5 \text{ mm}^4; A = 1.152 \times 10^3 \text{ mm}^2;$$

$c_{yy} = 26$ mm and $c_{xx} = 13.6$ mm

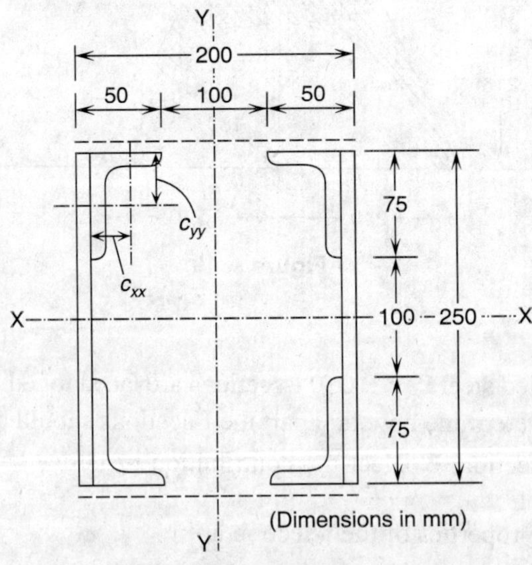

Figure A.21

[**Ans.** $I_{\bar{x}\bar{x}} = 7.365 \times 10^7 \text{ mm}^4$ and $I_{\bar{y}\bar{y}} = 9.044 \times 10^7 \text{ mm}^4$]

A.17 Two rolled steel ISMC 150 channel sections and one 275 × 10 mm thick cover plate are used to form a compound section shown in Figure A.22. Determine the moments of inertia $I_{\bar{x}\bar{x}}$ and $I_{\bar{y}\bar{y}}$ of the section. Neglect the lattice bars indicated by the dotted line in the figure. The geometrical properties of the channel section are:

$$I_{\bar{x}} = 7.794 \times 10^6 \text{ mm}^4; I_{\bar{y}} = 1.023 \times 10^6 \text{ mm}^4 \text{ and}$$
$$A = 2.088 \times 10^3 \text{ mm}^2; c_{yy} = 22.2 \text{ mm}$$

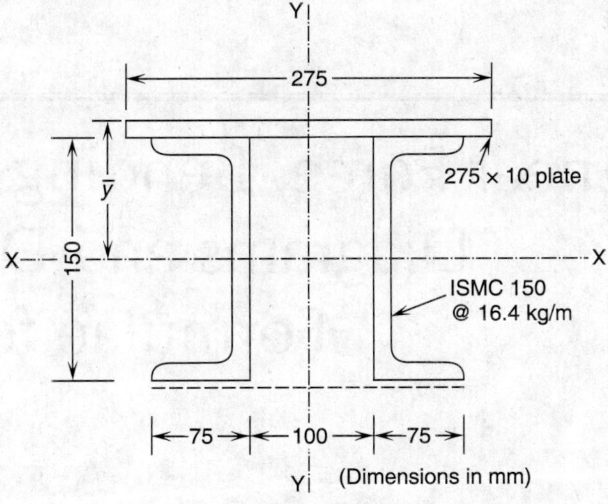

Figure A.22

[**Ans.** $\bar{y} = 53.2$ mm; $I_{\bar{x}\bar{x}} = 2.62 \times 10^7 \text{ mm}^4$ and $I_{\bar{y}\bar{y}} = 4.1146 \times 10^7 \text{ mm}^4$]

Appendix B

Shear Force, Bending Moment Diagrams and Deflection Formulae for Beams

B.1 SIMPLY SUPPORTED BEAM

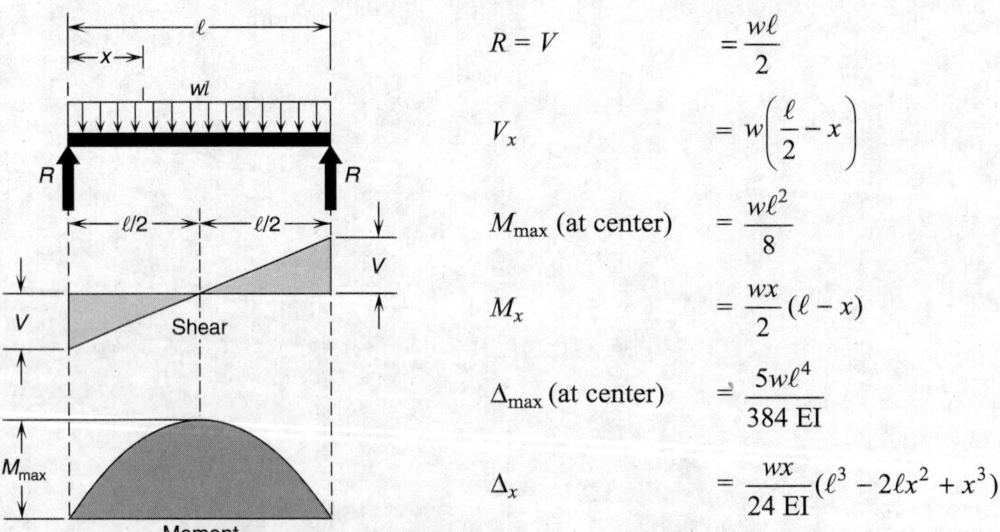

$$R = V = \frac{w\ell}{2}$$

$$V_x = w\left(\frac{\ell}{2} - x\right)$$

$$M_{\max} \text{ (at center)} = \frac{w\ell^2}{8}$$

$$M_x = \frac{wx}{2}(\ell - x)$$

$$\Delta_{\max} \text{ (at center)} = \frac{5w\ell^4}{384\ \text{EI}}$$

$$\Delta_x = \frac{wx}{24\ \text{EI}}(\ell^3 - 2\ell x^2 + x^3)$$

Figure B.1 Simply supported beam: uniformly distributed load.

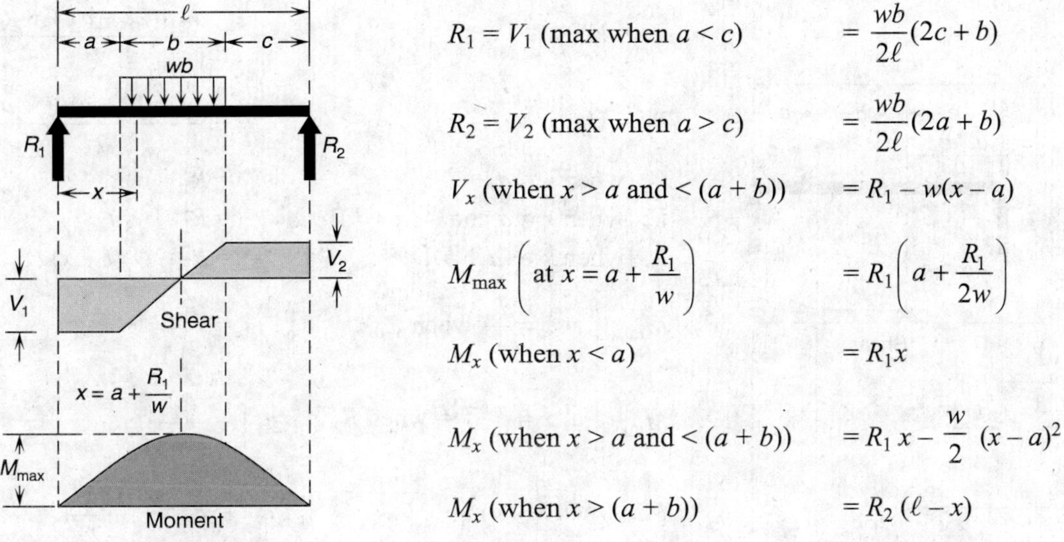

$$R_1 = V_1 \text{ (max when } a < c) = \frac{wb}{2\ell}(2c + b)$$

$$R_2 = V_2 \text{ (max when } a > c) = \frac{wb}{2\ell}(2a + b)$$

$$V_x \text{ (when } x > a \text{ and } < (a + b)) = R_1 - w(x - a)$$

$$M_{\max}\left(\text{at } x = a + \frac{R_1}{w}\right) = R_1\left(a + \frac{R_1}{2w}\right)$$

$$M_x \text{ (when } x < a) = R_1 x$$

$$M_x \text{ (when } x > a \text{ and } < (a + b)) = R_1 x - \frac{w}{2}(x - a)^2$$

$$M_x \text{ (when } x > (a + b)) = R_2 (\ell - x)$$

Figure B.2 Simply supported beam: partially distributed uniform load.

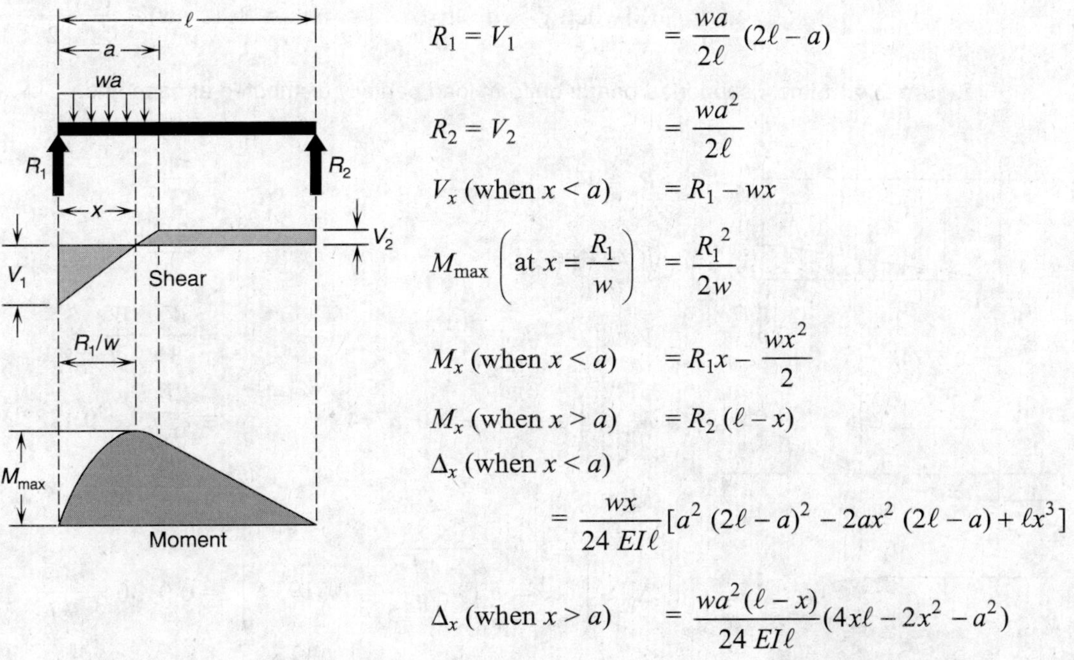

$$R_1 = V_1 = \frac{wa}{2\ell}(2\ell - a)$$

$$R_2 = V_2 = \frac{wa^2}{2\ell}$$

$$V_x \text{ (when } x < a) = R_1 - wx$$

$$M_{\max}\left(\text{at } x = \frac{R_1}{w}\right) = \frac{R_1^{\,2}}{2w}$$

$$M_x \text{ (when } x < a) = R_1 x - \frac{wx^2}{2}$$

$$M_x \text{ (when } x > a) = R_2 (\ell - x)$$

$$\Delta_x \text{ (when } x < a)$$

$$= \frac{wx}{24\, EI\ell}[a^2 (2\ell - a)^2 - 2ax^2 (2\ell - a) + \ell x^3]$$

$$\Delta_x \text{ (when } x > a) = \frac{wa^2(\ell - x)}{24\, EI\ell}(4x\ell - 2x^2 - a^2)$$

Figure B.3 Simply supported beam: uniform load partially distributed at one end.

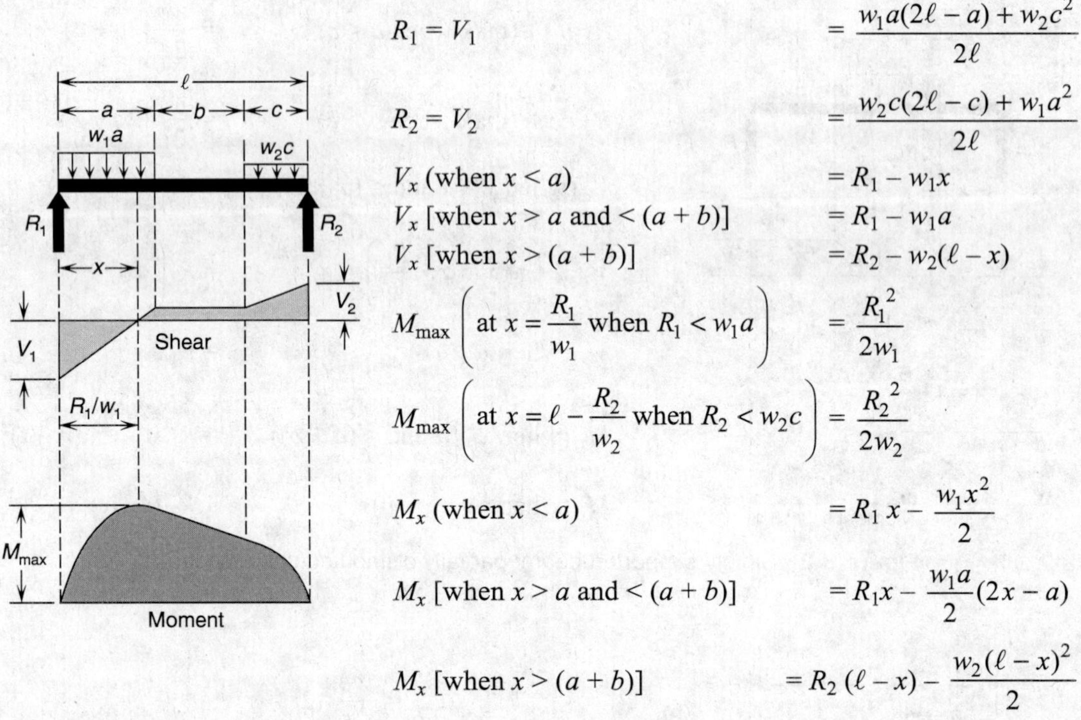

$$R_1 = V_1 = \frac{w_1 a(2\ell - a) + w_2 c^2}{2\ell}$$

$$R_2 = V_2 = \frac{w_2 c(2\ell - c) + w_1 a^2}{2\ell}$$

$$V_x \text{ (when } x < a) = R_1 - w_1 x$$

$$V_x \text{ [when } x > a \text{ and } < (a+b)] = R_1 - w_1 a$$

$$V_x \text{ [when } x > (a+b)] = R_2 - w_2(\ell - x)$$

$$M_{max}\left(\text{at } x = \frac{R_1}{w_1} \text{ when } R_1 < w_1 a\right) = \frac{R_1^2}{2w_1}$$

$$M_{max}\left(\text{at } x = \ell - \frac{R_2}{w_2} \text{ when } R_2 < w_2 c\right) = \frac{R_2^2}{2w_2}$$

$$M_x \text{ (when } x < a) = R_1 x - \frac{w_1 x^2}{2}$$

$$M_x \text{ [when } x > a \text{ and } < (a+b)] = R_1 x - \frac{w_1 a}{2}(2x - a)$$

$$M_x \text{ [when } x > (a+b)] = R_2(\ell - x) - \frac{w_2(\ell - x)^2}{2}$$

Figure B.4 Simply supported beam: uniform load partially distributed at each end.

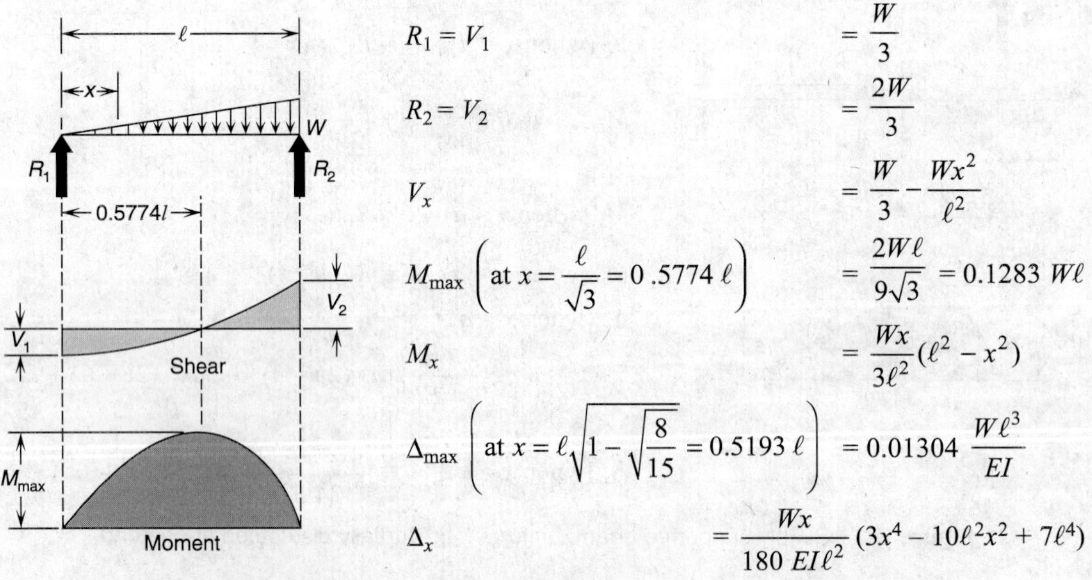

$$R_1 = V_1 = \frac{W}{3}$$

$$R_2 = V_2 = \frac{2W}{3}$$

$$V_x = \frac{W}{3} - \frac{Wx^2}{\ell^2}$$

$$M_{max}\left(\text{at } x = \frac{\ell}{\sqrt{3}} = 0.5774\,\ell\right) = \frac{2W\ell}{9\sqrt{3}} = 0.1283\,W\ell$$

$$M_x = \frac{Wx}{3\ell^2}(\ell^2 - x^2)$$

$$\Delta_{max}\left(\text{at } x = \ell\sqrt{1 - \sqrt{\frac{8}{15}}} = 0.5193\,\ell\right) = 0.01304\,\frac{W\ell^3}{EI}$$

$$\Delta_x = \frac{Wx}{180\,EI\ell^2}(3x^4 - 10\ell^2 x^2 + 7\ell^4)$$

Figure B.5 Simply supported beam: load increasing uniformly to one end.

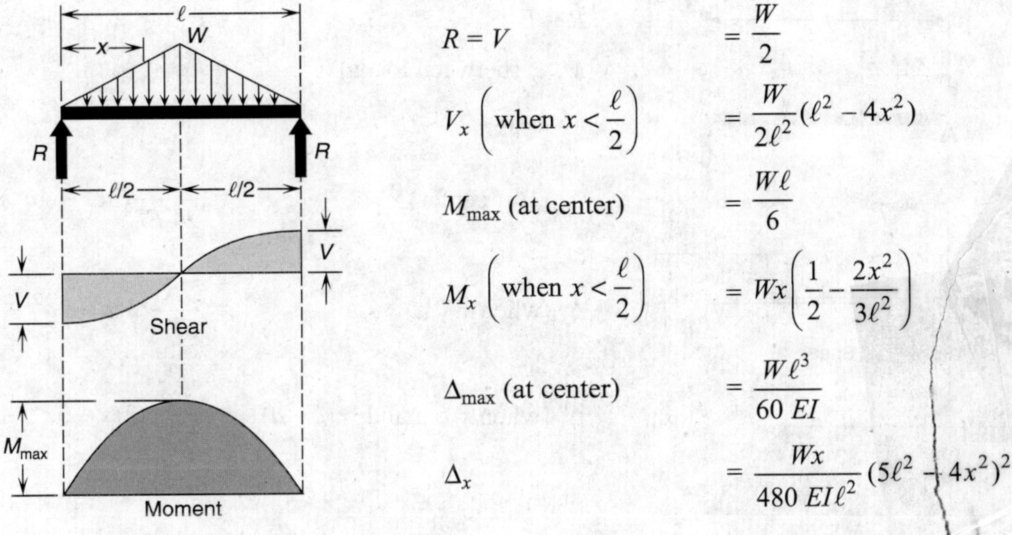

$$R = V = \frac{W}{2}$$

$$V_x \left(\text{when } x < \frac{\ell}{2}\right) = \frac{W}{2\ell^2}(\ell^2 - 4x^2)$$

$$M_{\max} \text{ (at center)} = \frac{W\ell}{6}$$

$$M_x \left(\text{when } x < \frac{\ell}{2}\right) = Wx\left(\frac{1}{2} - \frac{2x^2}{3\ell^2}\right)$$

$$\Delta_{\max} \text{ (at center)} = \frac{W\ell^3}{60\,EI}$$

$$\Delta_x = \frac{Wx}{480\,EI\ell^2}(5\ell^2 - 4x^2)^2$$

Figure B.6 Simply supported beam: load uniformly increasing to centre.

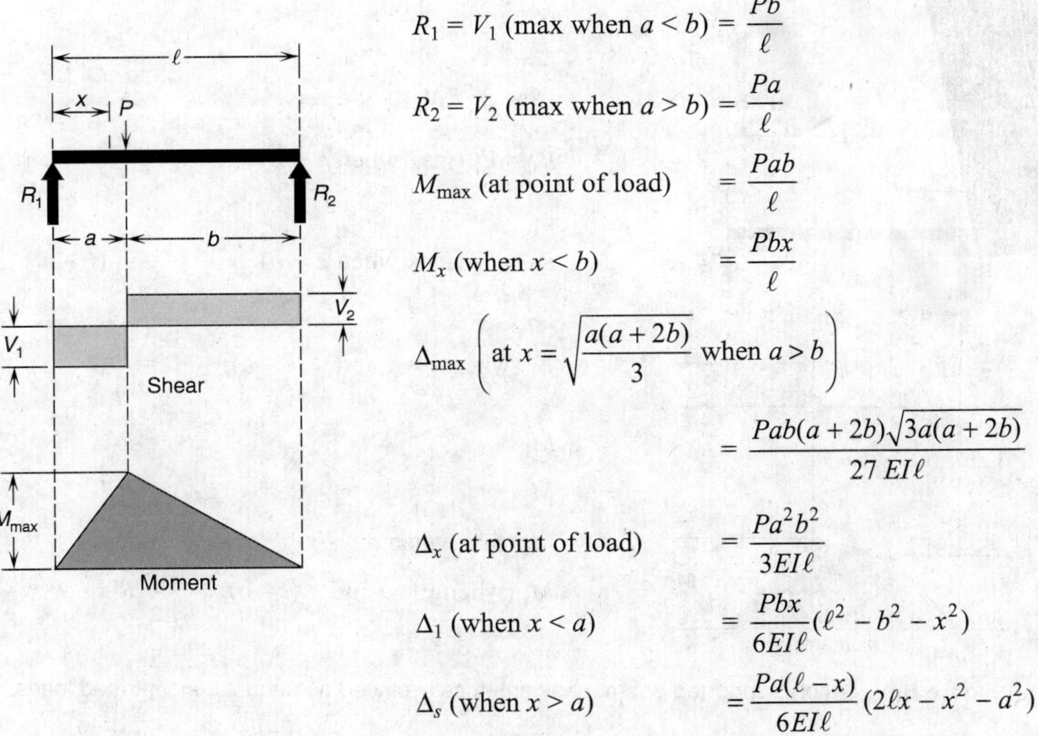

$$R_1 = V_1 \text{ (max when } a < b) = \frac{Pb}{\ell}$$

$$R_2 = V_2 \text{ (max when } a > b) = \frac{Pa}{\ell}$$

$$M_{\max} \text{ (at point of load)} = \frac{Pab}{\ell}$$

$$M_x \text{ (when } x < b) = \frac{Pbx}{\ell}$$

$$\Delta_{\max}\left(\text{at } x = \sqrt{\frac{a(a+2b)}{3}} \text{ when } a > b\right) = \frac{Pab(a+2b)\sqrt{3a(a+2b)}}{27\,EI\ell}$$

$$\Delta_x \text{ (at point of load)} = \frac{Pa^2b^2}{3EI\ell}$$

$$\Delta_1 \text{ (when } x < a) = \frac{Pbx}{6EI\ell}(\ell^2 - b^2 - x^2)$$

$$\Delta_s \text{ (when } x > a) = \frac{Pa(\ell - x)}{6EI\ell}(2\ell x - x^2 - a^2)$$

Figure B.7 Simply supported beam: concentrated load at an arbitrary point.

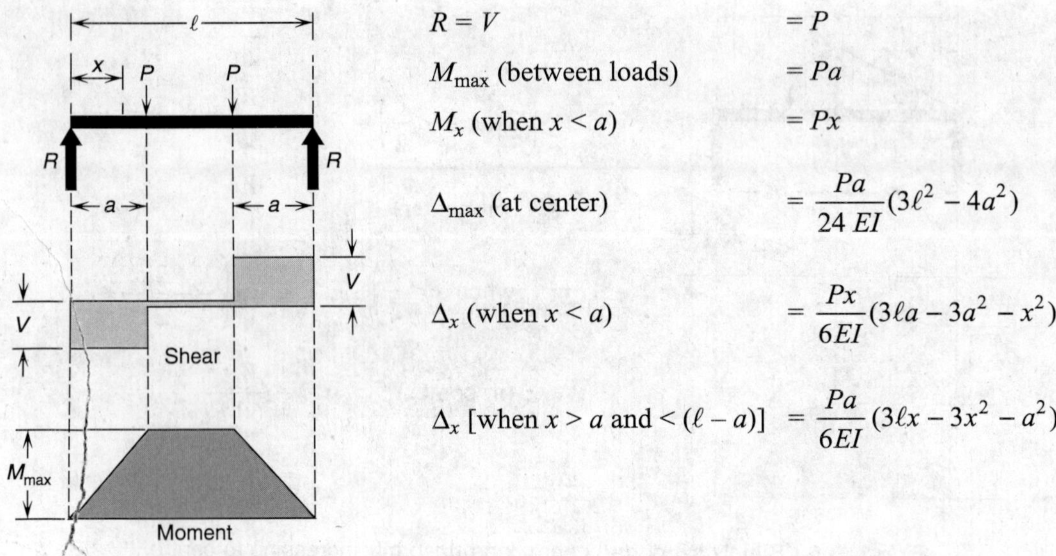

$$R = V = P$$

$$M_{max} \text{ (between loads)} = Pa$$

$$M_x \text{ (when } x < a) = Px$$

$$\Delta_{max} \text{ (at center)} = \frac{Pa}{24\,EI}(3\ell^2 - 4a^2)$$

$$\Delta_x \text{ (when } x < a) = \frac{Px}{6EI}(3\ell a - 3a^2 - x^2)$$

$$\Delta_x \text{ [when } x > a \text{ and} < (\ell - a)] = \frac{Pa}{6EI}(3\ell x - 3x^2 - a^2)$$

Figure B.8 Simply supported beam: symmetrically placed two equal concentrated loads.

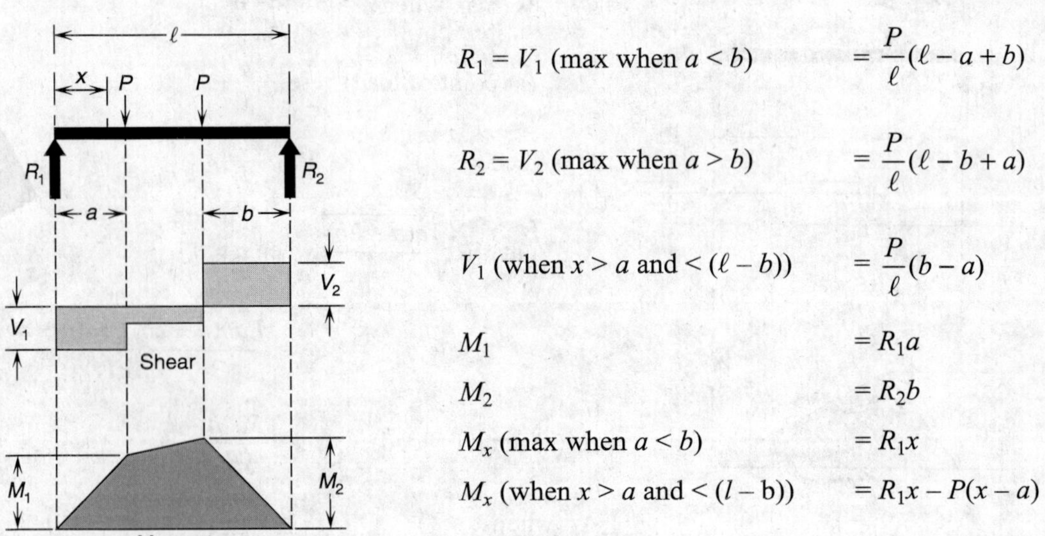

$$R_1 = V_1 \text{ (max when } a < b) = \frac{P}{\ell}(\ell - a + b)$$

$$R_2 = V_2 \text{ (max when } a > b) = \frac{P}{\ell}(\ell - b + a)$$

$$V_1 \text{ (when } x > a \text{ and} < (\ell - b)) = \frac{P}{\ell}(b - a)$$

$$M_1 = R_1 a$$

$$M_2 = R_2 b$$

$$M_x \text{ (max when } a < b) = R_1 x$$

$$M_x \text{ (when } x > a \text{ and} < (l - b)) = R_1 x - P(x - a)$$

Figure B.9 Simply supported beam: unsymmetrically placed two equal concentrated loads.

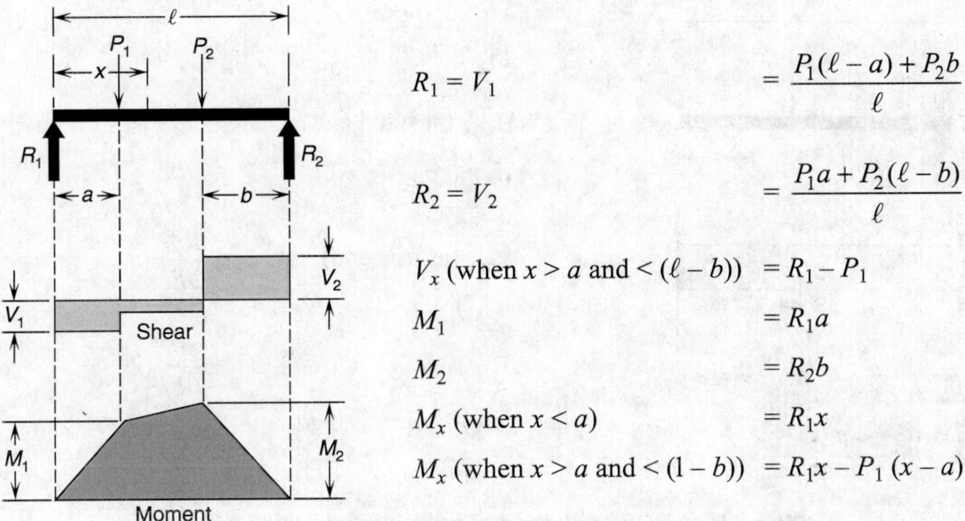

$$R_1 = V_1 \qquad = \frac{P_1(\ell - a) + P_2 b}{\ell}$$

$$R_2 = V_2 \qquad = \frac{P_1 a + P_2(\ell - b)}{\ell}$$

$$V_x \text{ (when } x > a \text{ and } < (\ell - b)) \quad = R_1 - P_1$$

$$M_1 \qquad = R_1 a$$

$$M_2 \qquad = R_2 b$$

$$M_x \text{ (when } x < a) \qquad = R_1 x$$

$$M_x \text{ (when } x > a \text{ and } < (1 - b)) \quad = R_1 x - P_1 (x - a)$$

Figure B.10 Simply supported beam: unsymmetrically placed two unequal concentrated loads.

B.2 CANTILEVER BEAM

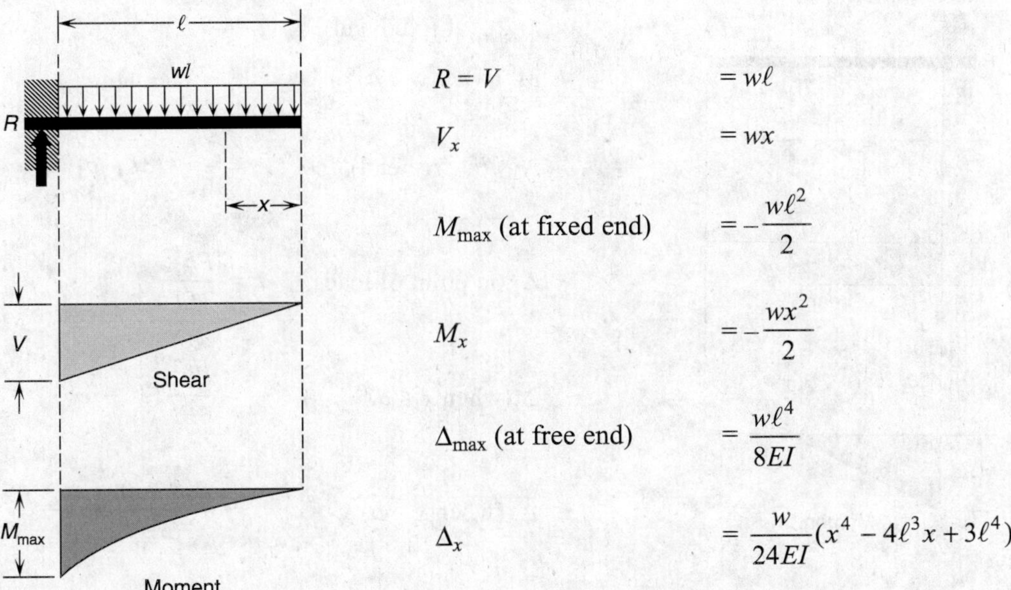

$$R = V \qquad = w\ell$$

$$V_x \qquad = wx$$

$$M_{max} \text{ (at fixed end)} \qquad = -\frac{w\ell^2}{2}$$

$$M_x \qquad = -\frac{wx^2}{2}$$

$$\Delta_{max} \text{ (at free end)} \qquad = \frac{w\ell^4}{8EI}$$

$$\Delta_x \qquad = \frac{w}{24EI}(x^4 - 4\ell^3 x + 3\ell^4)$$

Figure B.11 Cantilever beam: uniformly distributed load.

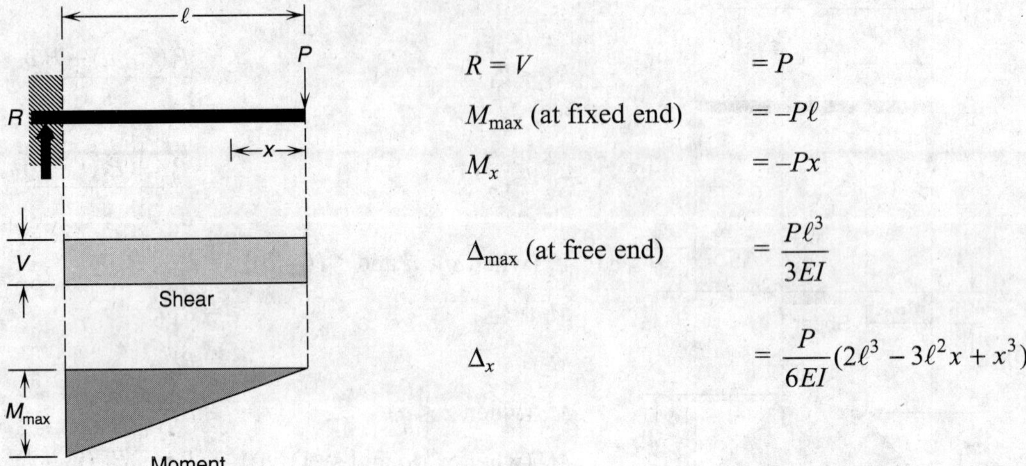

Figure B.12 Cantilever beam: concentrated load at free end.

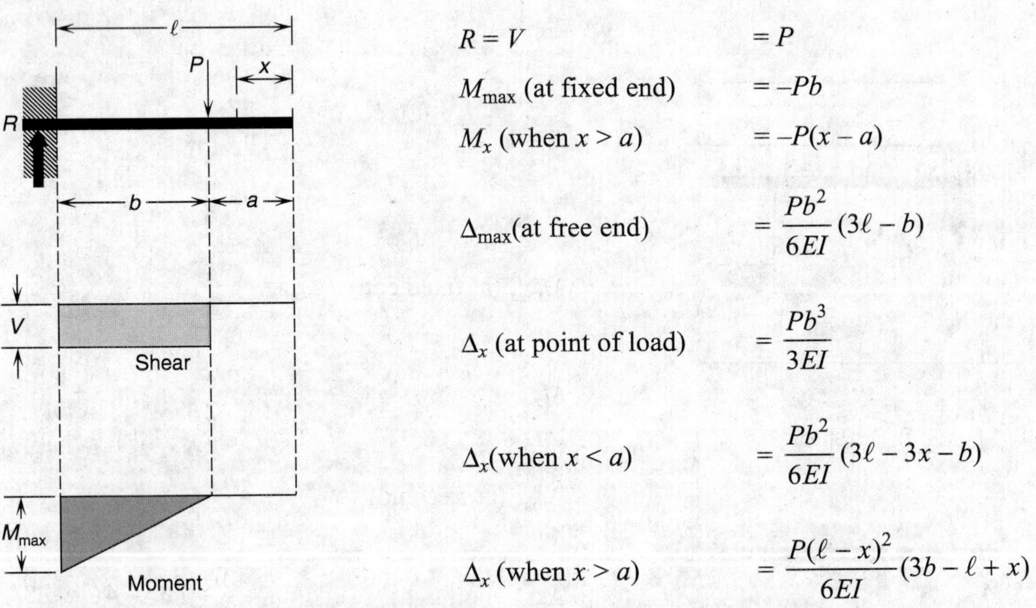

Figure B.13 Cantilever beam: concentrated load at an arbitrary point.

B.3 BEAM FIXED AT ONE END AND SIMPLY SUPPORTED AT THE OTHER

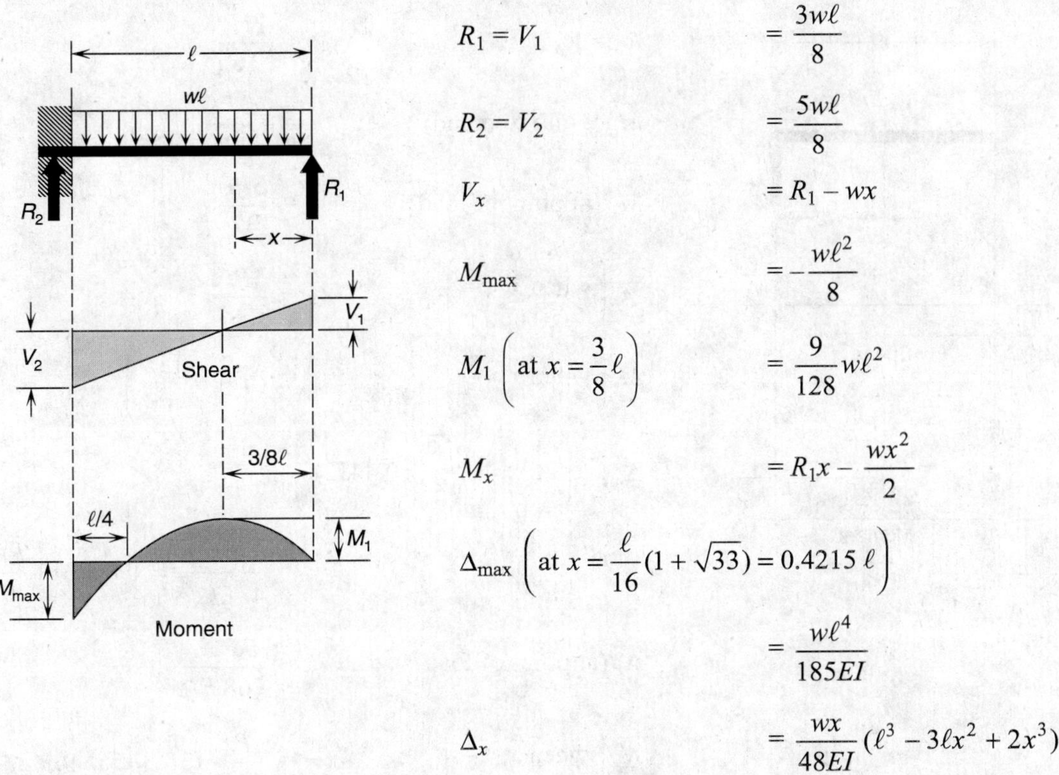

$$R_1 = V_1 = \frac{3w\ell}{8}$$

$$R_2 = V_2 = \frac{5w\ell}{8}$$

$$V_x = R_1 - wx$$

$$M_{max} = -\frac{w\ell^2}{8}$$

$$M_1 \left(\text{at } x = \frac{3}{8}\ell\right) = \frac{9}{128} w\ell^2$$

$$M_x = R_1 x - \frac{wx^2}{2}$$

$$\Delta_{max} \left(\text{at } x = \frac{\ell}{16}(1 + \sqrt{33}) = 0.4215\,\ell\right) = \frac{w\ell^4}{185EI}$$

$$\Delta_x = \frac{wx}{48EI}(\ell^3 - 3\ell x^2 + 2x^3)$$

Figure B.14 Beam fixed at one end, simply supported at other: uniformly distributed load.

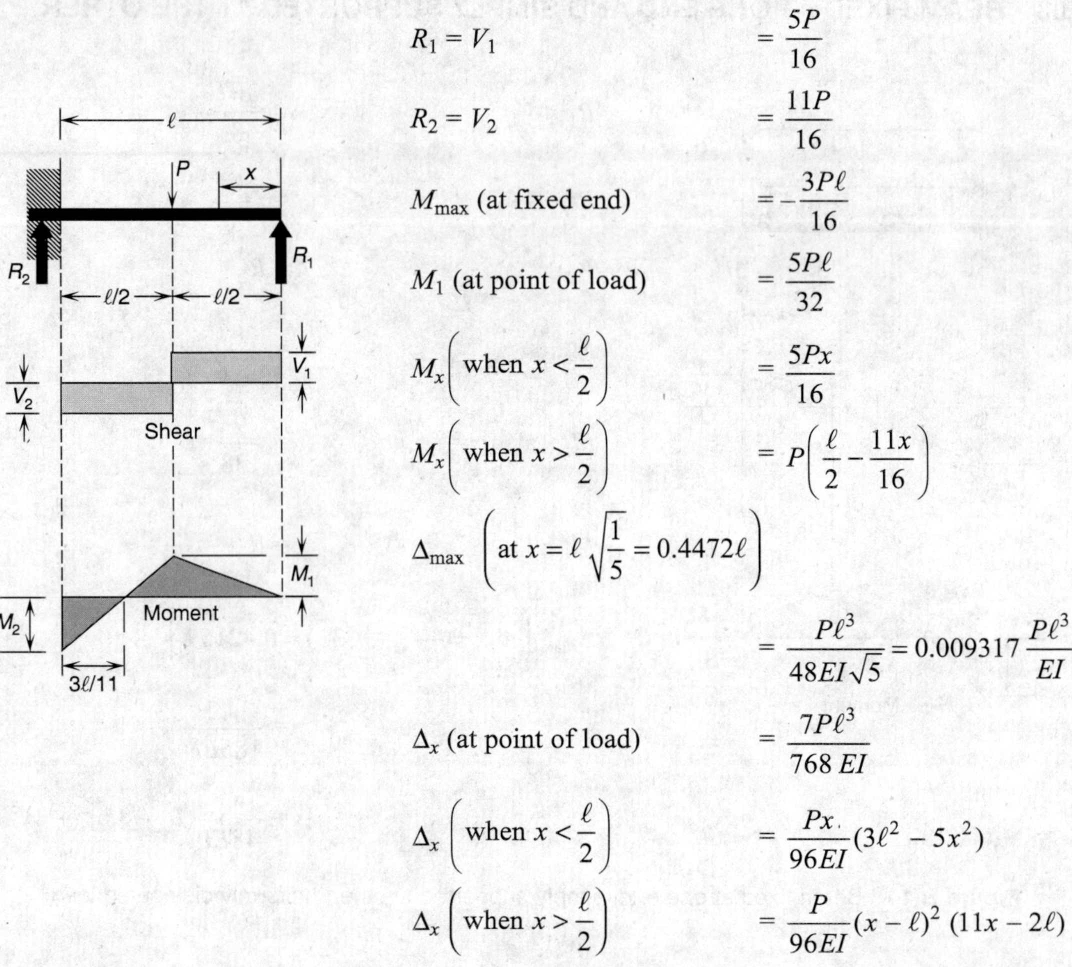

$$R_1 = V_1 = \frac{5P}{16}$$

$$R_2 = V_2 = \frac{11P}{16}$$

$$M_{\max} \text{ (at fixed end)} = -\frac{3P\ell}{16}$$

$$M_1 \text{ (at point of load)} = \frac{5P\ell}{32}$$

$$M_x\left(\text{when } x < \frac{\ell}{2}\right) = \frac{5Px}{16}$$

$$M_x\left(\text{when } x > \frac{\ell}{2}\right) = P\left(\frac{\ell}{2} - \frac{11x}{16}\right)$$

$$\Delta_{\max}\left(\text{at } x = \ell\sqrt{\frac{1}{5}} = 0.4472\ell\right)$$

$$= \frac{P\ell^3}{48EI\sqrt{5}} = 0.009317\frac{P\ell^3}{EI}$$

$$\Delta_x \text{ (at point of load)} = \frac{7P\ell^3}{768\,EI}$$

$$\Delta_x\left(\text{when } x < \frac{\ell}{2}\right) = \frac{Px}{96EI}(3\ell^2 - 5x^2)$$

$$\Delta_x\left(\text{when } x > \frac{\ell}{2}\right) = \frac{P}{96EI}(x - \ell)^2\,(11x - 2\ell)$$

Figure B.15 Beam fixed at one end, simply supported at other: concentrated load at centre.

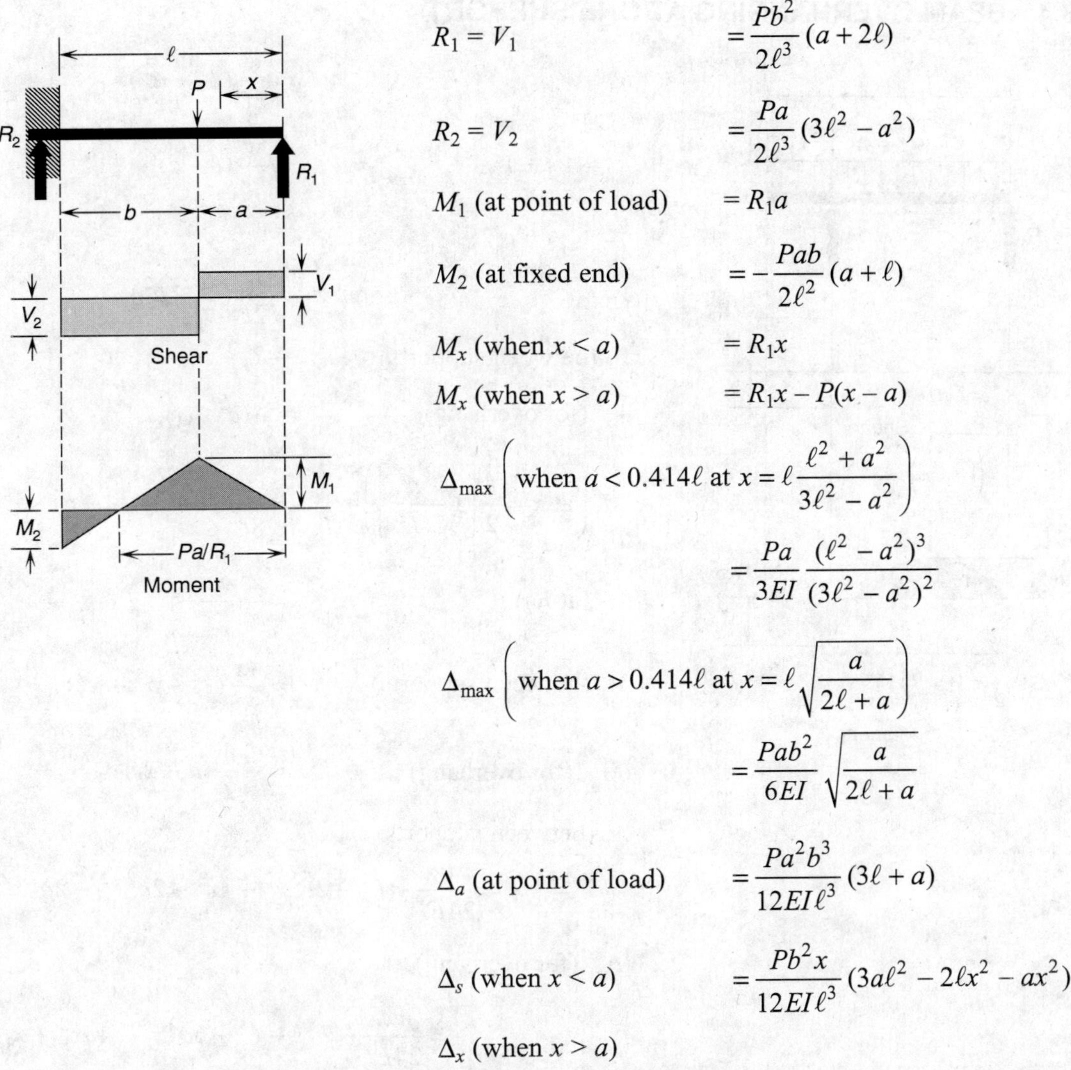

$R_1 = V_1 \quad = \dfrac{Pb^2}{2\ell^3}(a + 2\ell)$

$R_2 = V_2 \quad = \dfrac{Pa}{2\ell^3}(3\ell^2 - a^2)$

M_1 (at point of load) $= R_1 a$

M_2 (at fixed end) $= -\dfrac{Pab}{2\ell^2}(a + \ell)$

M_x (when $x < a$) $= R_1 x$

M_x (when $x > a$) $= R_1 x - P(x - a)$

$\Delta_{max}\left(\text{when } a < 0.414\ell \text{ at } x = \ell \dfrac{\ell^2 + a^2}{3\ell^2 - a^2}\right)$

$= \dfrac{Pa}{3EI}\dfrac{(\ell^2 - a^2)^3}{(3\ell^2 - a^2)^2}$

$\Delta_{max}\left(\text{when } a > 0.414\ell \text{ at } x = \ell\sqrt{\dfrac{a}{2\ell + a}}\right)$

$= \dfrac{Pab^2}{6EI}\sqrt{\dfrac{a}{2\ell + a}}$

Δ_a (at point of load) $= \dfrac{Pa^2 b^3}{12EI\ell^3}(3\ell + a)$

Δ_s (when $x < a$) $= \dfrac{Pb^2 x}{12EI\ell^3}(3a\ell^2 - 2\ell x^2 - ax^2)$

Δ_x (when $x > a$)

$= \dfrac{Pa}{12EI\ell^3}(\ell - x)^2(3\ell^2 x - a^2 x - 2a^2\ell)$

Figure B.16 Beam fixed at one end, simply supported at other: concentrated load at any point.

B.4 BEAM OVERHANGING AT ONE SUPPORT

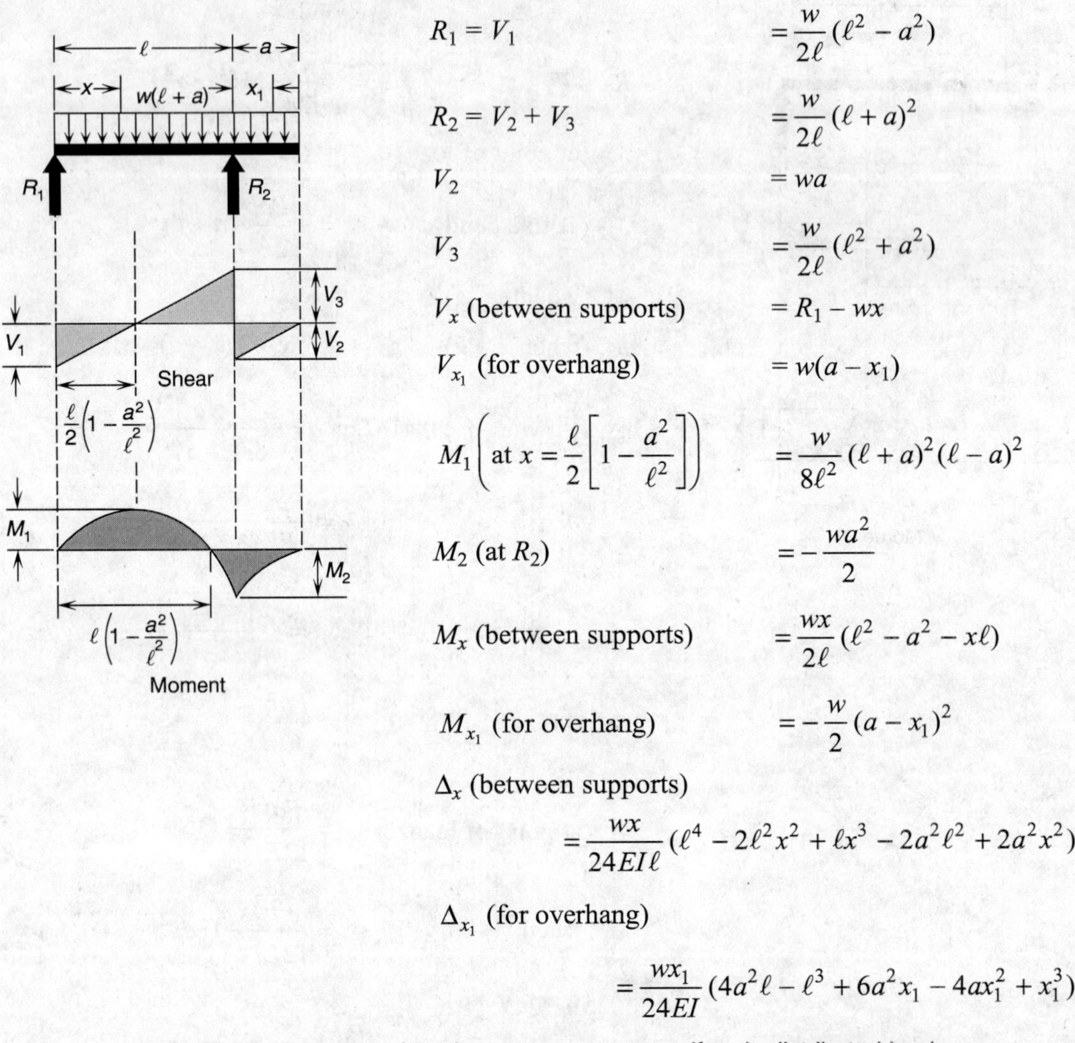

$R_1 = V_1 \qquad = \dfrac{w}{2\ell}(\ell^2 - a^2)$

$R_2 = V_2 + V_3 \qquad = \dfrac{w}{2\ell}(\ell + a)^2$

$V_2 \qquad = wa$

$V_3 \qquad = \dfrac{w}{2\ell}(\ell^2 + a^2)$

V_x (between supports) $\qquad = R_1 - wx$

V_{x_1} (for overhang) $\qquad = w(a - x_1)$

$M_1\left(\text{at } x = \dfrac{\ell}{2}\left[1 - \dfrac{a^2}{\ell^2}\right]\right) \qquad = \dfrac{w}{8\ell^2}(\ell + a)^2(\ell - a)^2$

M_2 (at R_2) $\qquad = -\dfrac{wa^2}{2}$

M_x (between supports) $\qquad = \dfrac{wx}{2\ell}(\ell^2 - a^2 - x\ell)$

M_{x_1} (for overhang) $\qquad = -\dfrac{w}{2}(a - x_1)^2$

Δ_x (between supports)

$$= \frac{wx}{24EI\ell}(\ell^4 - 2\ell^2 x^2 + \ell x^3 - 2a^2\ell^2 + 2a^2x^2)$$

Δ_{x_1} (for overhang)

$$= \frac{wx_1}{24EI}(4a^2\ell - \ell^3 + 6a^2x_1 - 4ax_1^2 + x_1^3)$$

Figure B.17 Beam overhanging at one support: uniformly distributed load.

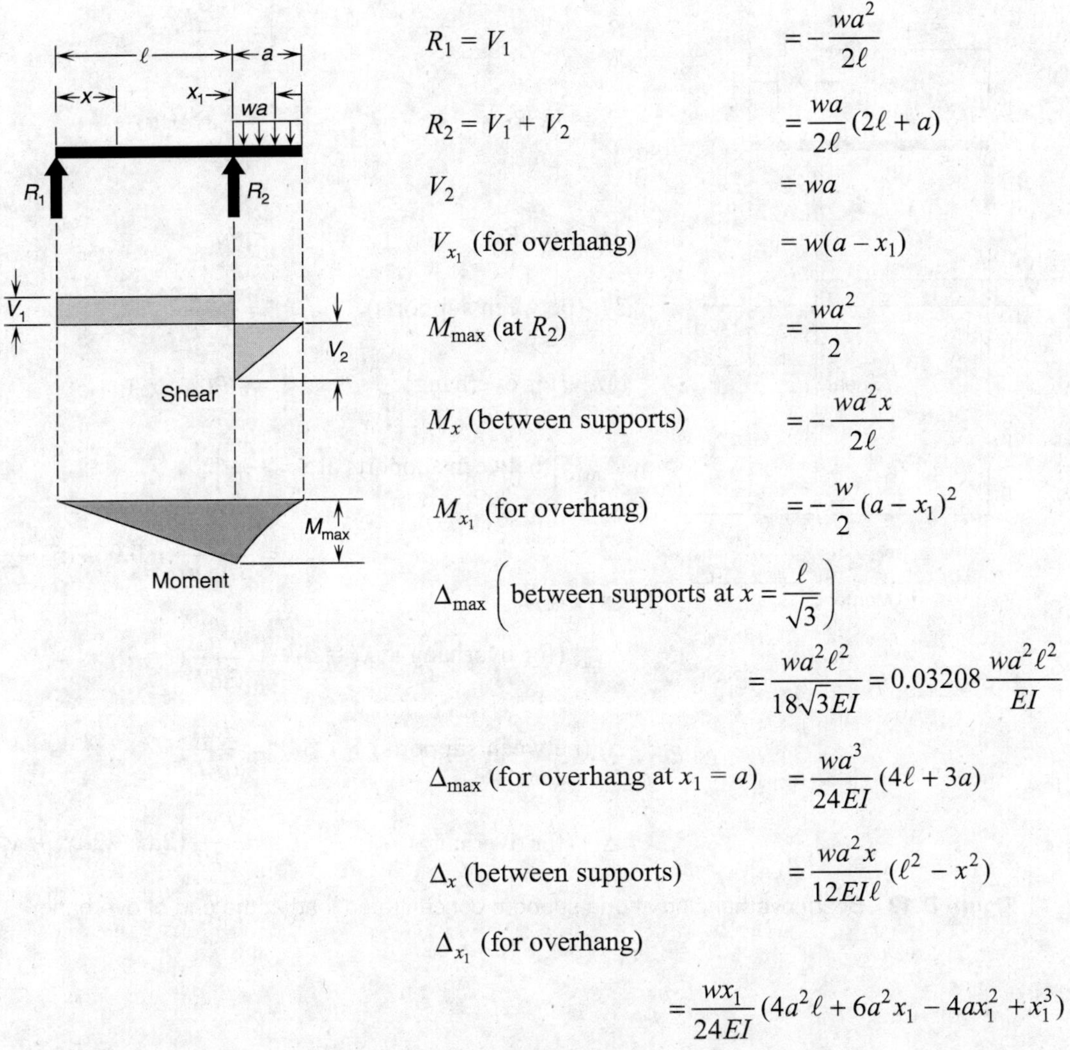

$$R_1 = V_1 \qquad = -\frac{wa^2}{2\ell}$$

$$R_2 = V_1 + V_2 \qquad = \frac{wa}{2\ell}(2\ell + a)$$

$$V_2 \qquad = wa$$

$$V_{x_1} \text{ (for overhang)} \qquad = w(a - x_1)$$

$$M_{max} \text{ (at } R_2\text{)} \qquad = \frac{wa^2}{2}$$

$$M_x \text{ (between supports)} \qquad = -\frac{wa^2 x}{2\ell}$$

$$M_{x_1} \text{ (for overhang)} \qquad = -\frac{w}{2}(a - x_1)^2$$

$$\Delta_{max} \left(\text{between supports at } x = \frac{\ell}{\sqrt{3}} \right)$$

$$= \frac{wa^2\ell^2}{18\sqrt{3}EI} = 0.03208\frac{wa^2\ell^2}{EI}$$

$$\Delta_{max} \text{ (for overhang at } x_1 = a\text{)} \qquad = \frac{wa^3}{24EI}(4\ell + 3a)$$

$$\Delta_x \text{ (between supports)} \qquad = \frac{wa^2 x}{12EI\ell}(\ell^2 - x^2)$$

$$\Delta_{x_1} \text{ (for overhang)}$$

$$= \frac{wx_1}{24EI}(4a^2\ell + 6a^2 x_1 - 4ax_1^2 + x_1^3)$$

Figure B.18 Beam overhanging at one support: uniformly distributed load on the overhang.

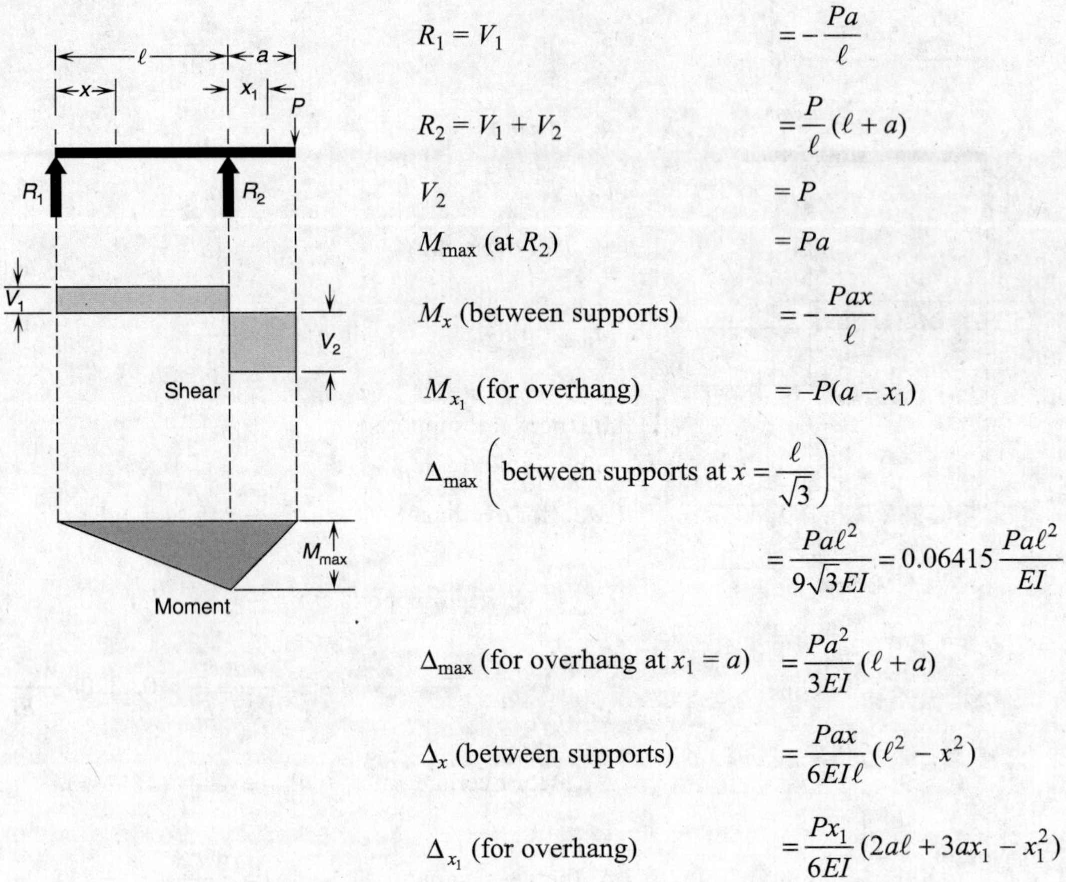

$$R_1 = V_1 \qquad = -\frac{Pa}{\ell}$$

$$R_2 = V_1 + V_2 \qquad = \frac{P}{\ell}(\ell + a)$$

$$V_2 \qquad = P$$

$$M_{max} \text{ (at } R_2) \qquad = Pa$$

$$M_x \text{ (between supports)} \qquad = -\frac{Pax}{\ell}$$

$$M_{x_1} \text{ (for overhang)} \qquad = -P(a - x_1)$$

$$\Delta_{max}\left(\text{between supports at } x = \frac{\ell}{\sqrt{3}}\right)$$

$$= \frac{Pa\ell^2}{9\sqrt{3}EI} = 0.06415\frac{Pa\ell^2}{EI}$$

$$\Delta_{max} \text{ (for overhang at } x_1 = a) \qquad = \frac{Pa^2}{3EI}(\ell + a)$$

$$\Delta_x \text{ (between supports)} \qquad = \frac{Pax}{6EI\ell}(\ell^2 - x^2)$$

$$\Delta_{x_1} \text{ (for overhang)} \qquad = \frac{Px_1}{6EI}(2a\ell + 3ax_1 - x_1^2)$$

Figure B.19 Beam overhanging at one support: concentrated load at the end of overhang.

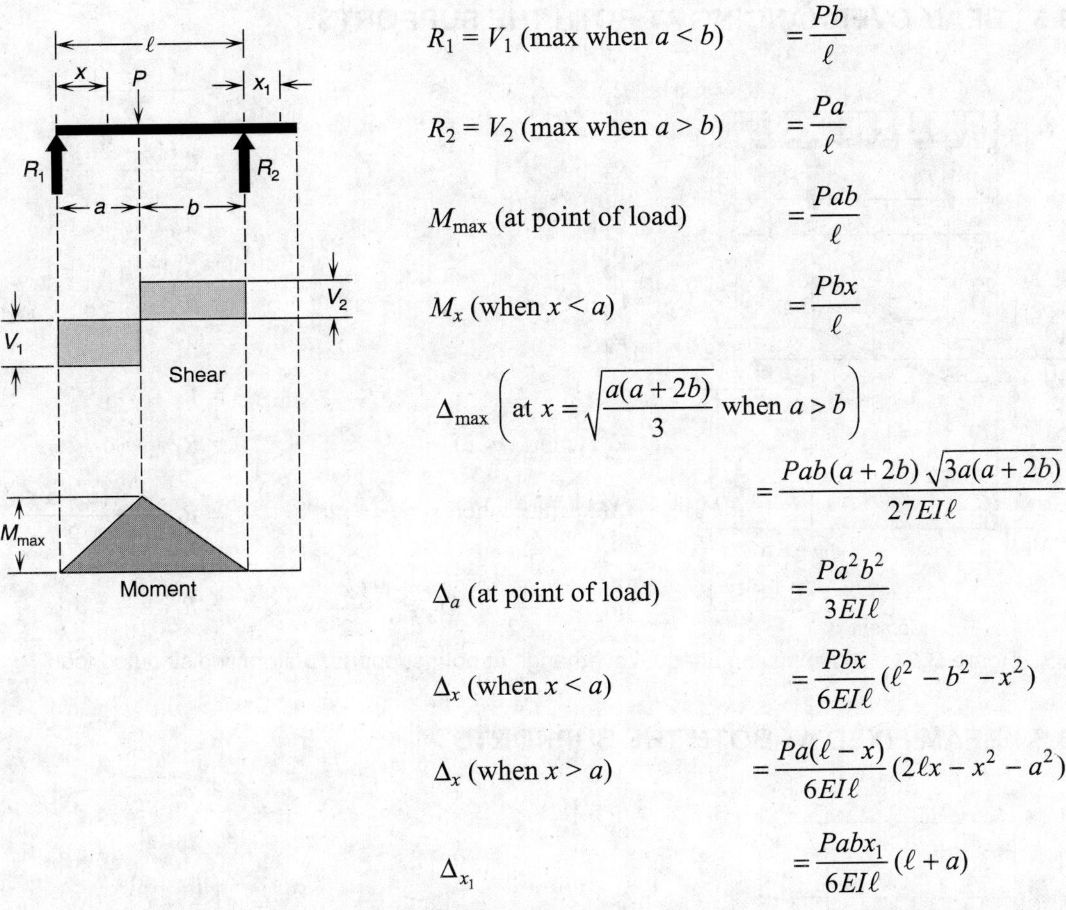

$R_1 = V_1$ (max when $a < b$) $\quad = \dfrac{Pb}{\ell}$

$R_2 = V_2$ (max when $a > b$) $\quad = \dfrac{Pa}{\ell}$

M_{max} (at point of load) $\quad = \dfrac{Pab}{\ell}$

M_x (when $x < a$) $\quad = \dfrac{Pbx}{\ell}$

$$\Delta_{max}\left(\text{at } x = \sqrt{\frac{a(a+2b)}{3}} \text{ when } a > b\right) = \frac{Pab(a+2b)\sqrt{3a(a+2b)}}{27EI\ell}$$

Δ_a (at point of load) $\quad = \dfrac{Pa^2b^2}{3EI\ell}$

Δ_x (when $x < a$) $\quad = \dfrac{Pbx}{6EI\ell}(\ell^2 - b^2 - x^2)$

Δ_x (when $x > a$) $\quad = \dfrac{Pa(\ell - x)}{6EI\ell}(2\ell x - x^2 - a^2)$

$\Delta_{x_1} \quad = \dfrac{Pabx_1}{6EI\ell}(\ell + a)$

Figure B.20 Beam overhanging at one support: concentrated load at any point between supports.

B.5 BEAM OVERHANGING AT BOTH THE SUPPORTS

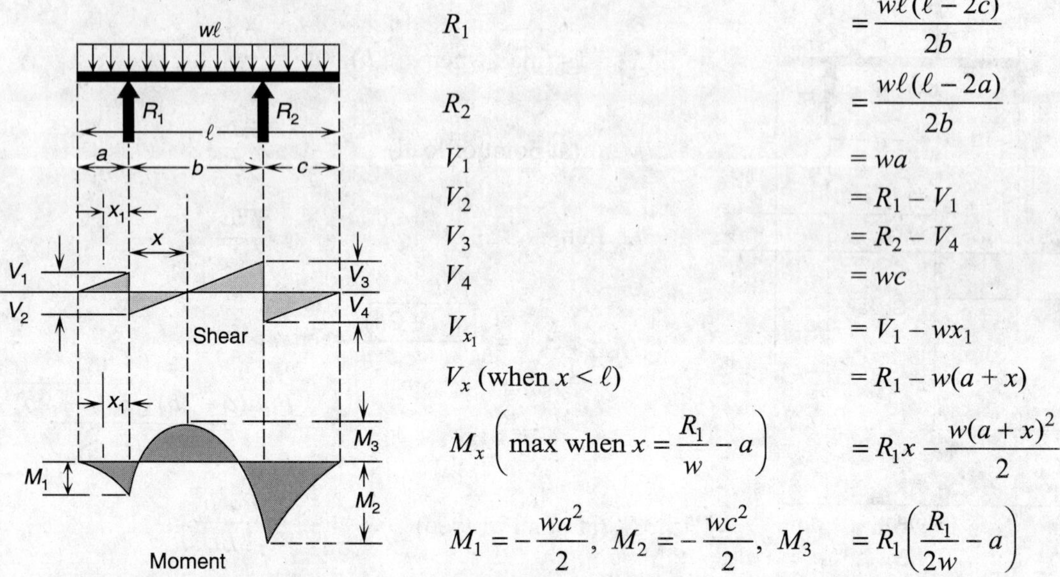

$$R_1 = \frac{w\ell(\ell - 2c)}{2b}$$

$$R_2 = \frac{w\ell(\ell - 2a)}{2b}$$

$$V_1 = wa$$

$$V_2 = R_1 - V_1$$

$$V_3 = R_2 - V_4$$

$$V_4 = wc$$

$$V_{x_1} = V_1 - wx_1$$

$$V_x \text{ (when } x < \ell) = R_1 - w(a + x)$$

$$M_x \left(\text{max when } x = \frac{R_1}{w} - a \right) = R_1 x - \frac{w(a + x)^2}{2}$$

$$M_1 = -\frac{wa^2}{2}, \; M_2 = -\frac{wc^2}{2}, \; M_3 = R_1\left(\frac{R_1}{2w} - a\right)$$

Figure B.21 Beam having unequal overhangs at both supports: uniformly distributed load.

B.6 BEAM FIXED AT BOTH THE SUPPORTS

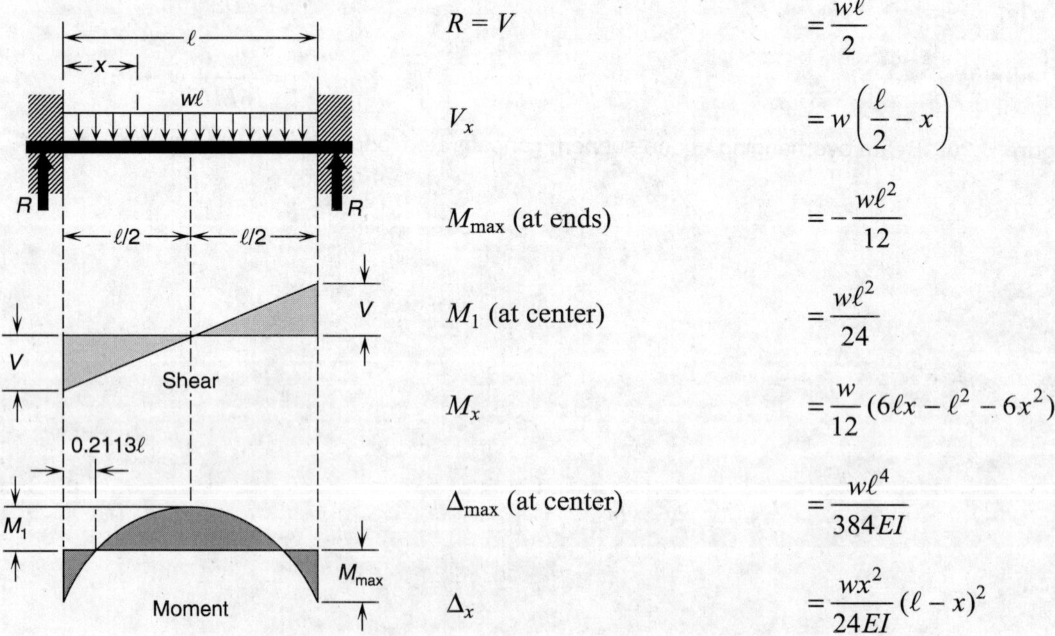

$$R = V = \frac{w\ell}{2}$$

$$V_x = w\left(\frac{\ell}{2} - x\right)$$

$$M_{max} \text{ (at ends)} = -\frac{w\ell^2}{12}$$

$$M_1 \text{ (at center)} = \frac{w\ell^2}{24}$$

$$M_x = \frac{w}{12}(6\ell x - \ell^2 - 6x^2)$$

$$\Delta_{max} \text{ (at center)} = \frac{w\ell^4}{384EI}$$

$$\Delta_x = \frac{wx^2}{24EI}(\ell - x)^2$$

Figure B.22 Beam fixed at both ends: uniformly distributed load.

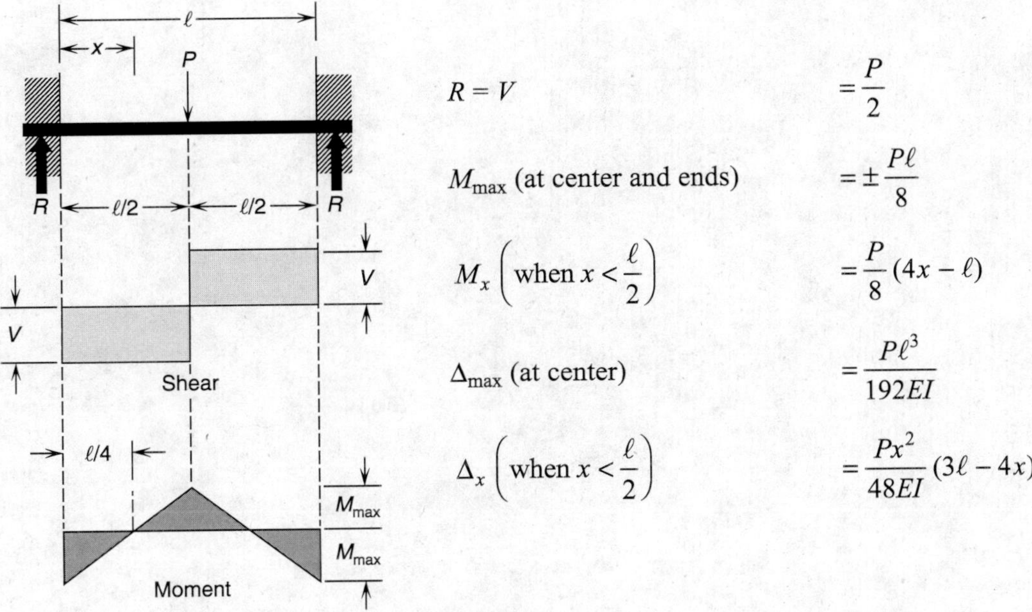

$$R = V \quad = \frac{P}{2}$$

$$M_{max} \text{ (at center and ends)} \quad = \pm\frac{P\ell}{8}$$

$$M_x \left(\text{when } x < \frac{\ell}{2}\right) \quad = \frac{P}{8}(4x - \ell)$$

$$\Delta_{max} \text{ (at center)} \quad = \frac{P\ell^3}{192EI}$$

$$\Delta_x \left(\text{when } x < \frac{\ell}{2}\right) \quad = \frac{Px^2}{48EI}(3\ell - 4x)$$

Figure B.23 Beam fixed at both ends: concentrated load at centre.

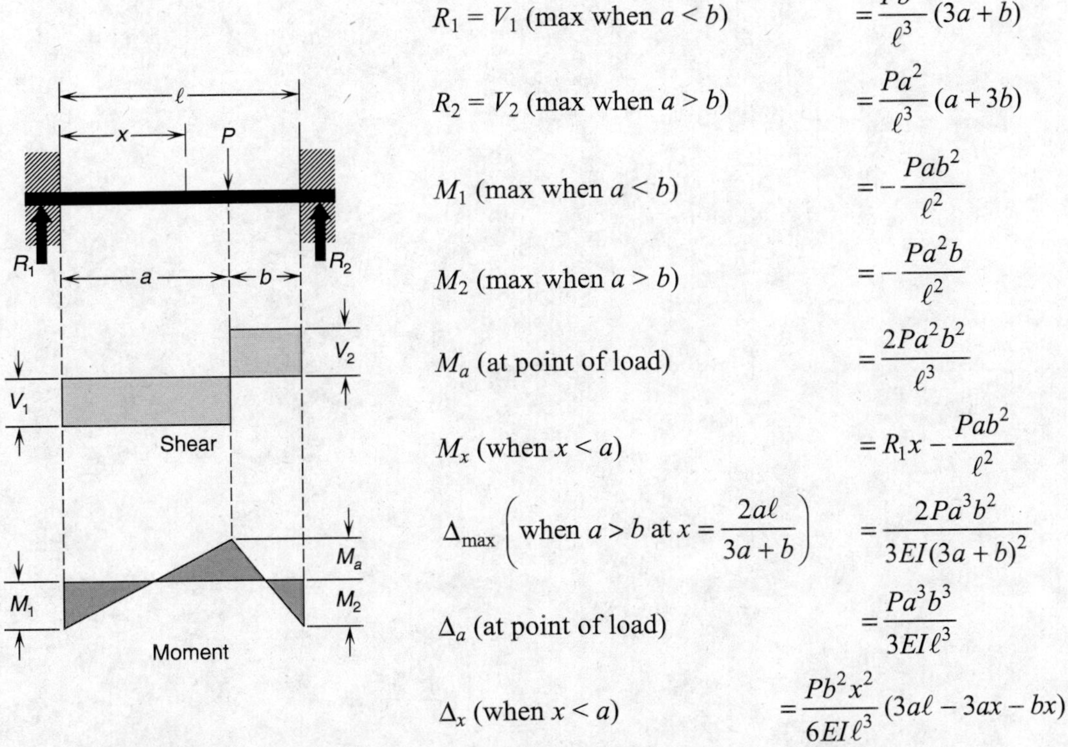

$$R_1 = V_1 \text{ (max when } a < b) \quad = \frac{Pb^2}{\ell^3}(3a + b)$$

$$R_2 = V_2 \text{ (max when } a > b) \quad = \frac{Pa^2}{\ell^3}(a + 3b)$$

$$M_1 \text{ (max when } a < b) \quad = -\frac{Pab^2}{\ell^2}$$

$$M_2 \text{ (max when } a > b) \quad = -\frac{Pa^2b}{\ell^2}$$

$$M_a \text{ (at point of load)} \quad = \frac{2Pa^2b^2}{\ell^3}$$

$$M_x \text{ (when } x < a) \quad = R_1x - \frac{Pab^2}{\ell^2}$$

$$\Delta_{max} \left(\text{when } a > b \text{ at } x = \frac{2a\ell}{3a + b}\right) \quad = \frac{2Pa^3b^2}{3EI(3a + b)^2}$$

$$\Delta_a \text{ (at point of load)} \quad = \frac{Pa^3b^3}{3EI\ell^3}$$

$$\Delta_x \text{ (when } x < a) \quad = \frac{Pb^2x^2}{6EI\ell^3}(3a\ell - 3ax - bx)$$

Figure B.24 Beam fixed at both ends: concentrated load at an arbitrary point.

References

Bores, A.P., Sidebottom, O.N., Seely, F.B. and Smith, J.O., *Advanced Mechanics of Materials*, 3rd ed., Wiley, New York, 1978.

Cook, R.D. and Young, W.C., *Advanced Mechanics of Materials*, Macmillan, New York, 1985.

Crandall, S.H., *Engineering Analysis*, McGraw-Hill, New York, 1956.

Crandall, S.H., Dahl, N.C. and Lardner, T.J., *An introduction to Mechanics of Solids,* 2nd ed., McGraw-Hill, New York, 1972.

Gambhir, M.L., *Stability Analysis and Design of Structures*, Springer-Verlag, Berlin, Heidelberg, 2004.

Hartog, J.P., *Advanced Strength of Materials,* McGraw-Hill, New York, 1952.

Hearn, E.J., *Mechanics of Materials*, 3rd ed., Vol. 1, Elsevier, New York, 1997.

Ketter, R.L., George, C.L. and Sherwood, P.P., *Structural Analysis and Design*, McGraw-Hill, New York, 1979.

Nash, W.A., *Strength of Materials,* 4th ed., McGraw-Hill, New York, 1998.

Popov, Egor P. and Balan, Toader A., *Engineering Mechanics of Solids*, Prentice-Hall of India, 2nd ed., New Delhi, 2006.

Roark, R.J., *Formulas for Stress and Strain*, 4th ed., McGraw-Hill, New York, 1965.

Roark, R.J. and Young, W.C., *Formulas for Stress and Strain*, 5th ed., McGraw-Hill, New York, 1975.

Ryder, G.H., *Strength of Materials,* Macmillan, London, 1965.

Srinath, L.S., Desayi, P., Murthy, N.S. and Ramu, S.A., *Strength of Materials,* Macmillan, New Delhi, 1997.

Timoshenko, S., *Strength of Materials,* 3rd ed., Van Nostrand Reinhold, New York, 1955.

Timoshenko, S. and Goodier, J.N., *Theory of Elasticity*, 3rd ed., McGraw-Hill, New York, 1970.

Timoshenko, S. and Young, D.H., *Elements of Strength of Materials,* 5th ed., D. Van Nostrand Co., New York, 1968.

Index